$115.95 per copy, (In United States).
Price subject to change without prior notice.

0165

MW01166134

RSMeans

Heavy Construction Cost Data

19th Annual Edition

2005

RSMeans
Construction Publishers & Consultants
63 Smiths Lane
Kingston, MA 02364-0800
(781) 585-7880

Copyright©2004 by Reed Construction Data, Inc.
All rights reserved.

Printed in the United States of America
ISSN 0893-5602
ISBN 0-87629-767-X

Senior Editor
Eugene R. Spencer

Contributing Editors
Barbara Balboni
Robert A. Bastoni
John H. Chiang, PE
Robert J. Kuchta
Robert C. McNichols
Robert W. Mewis, CCC
Melville J. Mossman, PE
John J. Moylan
Jeannene D. Murphy
Stephen C. Plotner
Michael J. Regan
Marshall J. Stetson
Phillip R. Waier, PE

**Senior Engineering
Operations Manager**
John H. Ferguson, PE

Editorial Advisory Board
James E. Armstrong, CPE, CEM
Program Manager,
Energy Conservation
NSTAR

William R. Barry, CCC
Cost Consultant

Robert F. Cox, PhD
Assistant Professor
ME Rinker Sr. School of Bldg. Constr.
University of Florida

Roy F. Gilley, AIA
Principal
Gilley-Hinkel Architects

Kenneth K. Humphreys, PhD,
   PE, CCE
Secretary-Treasurer
International Cost
Engineering Council

Patricia L. Jackson, PE
President
Jackson A&E Assoc., Inc.

Martin F. Joyce
Executive Vice President
Bond Brothers, Inc.

**Senior Vice President
& General Manager**
John Ware

Product Manager
Jane Crudden

**Vice President
of Sales**
John M. Shea

Production Manager
Michael Kokernak

Production Coordinator
Marion E. Schofield

Technical Support
Thomas J. Dion
Jonathan Forgit
Mary Lou Geary
Gary L. Hoitt
Alice McSharry
Paula Reale-Camelio
Robin Richardson
Kathryn S. Rodriguez
Sheryl A. Rose

Book & Cover Design
Norman R. Forgit

 This book is recyclable.

 This book is printed on recycled stock.

First Printing

Foreword

RSMeans is a product line of Reed Construction Data, Inc., a leading provider of construction information, products, and services in North America and globally. Reed Construction Data's project information products include more than 100 regional editions, national construction data, sales leads, and local plan rooms in major business centers. Reed Construction Data's PlansDirect provides surveys, plans and specifications. The First Source suite of products consists of *First Source for Products,* SPEC-DATA™, MANU-SPEC™, CADBlocks, Manufacturer Catalogs and First Source Exchange (www.firstsourceexchange.com) for the selection of nationally available building products. Reed Construction Data also publishes ProFile, a database of more than 20,000 U.S. architectural firms. RSMeans provides construction cost data, training, and consulting services in print, CD-ROM and online. Reed Construction Data, headquartered in Atlanta, is owned by Reed Business Information (www.reedconstructiondata.com), a leading provider of critical information and marketing solutions to business professionals in the media, manufacturing, electronics, construction and retail industries. Its market-leading properties include more than 135 business-to-business publications, over 125 Webzines and Web portals, as well as online services, custom publishing, directories, research and direct-marketing lists. Reed Business Information is a member of the Reed Elsevier plc group (NYSE: RUK and ENL)—a world-leading publisher and information provider operating in the science and medical, legal, education and business-to-business industry sectors.

Our Mission

Since 1942, RSMeans has been actively engaged in construction cost publishing and consulting throughout North America.

Today, over 60 years after RSMeans began, our primary objective remains the same: to provide you, the construction and facilities professional, with the most current and comprehensive construction cost data possible.

Whether you are a contractor, an owner, an architect, an engineer, a facilities manager, or anyone else who needs a fast and reliable construction cost estimate, you'll find this publication to be a highly useful and necessary tool.

Today, with the constant flow of new construction methods and materials, it's difficult to find the time to look at and evaluate all the different construction cost possibilities. In addition, because labor and material costs keep changing, last year's cost information is not a reliable basis for today's estimate or budget.

That's why so many construction professionals turn to RSMeans. We keep track of the costs for you, along with a wide range of other key information, from city cost indexes . . . to productivity rates . . . to crew composition . . . to contractor's overhead and profit rates.

RSMeans performs these functions by collecting data from all facets of the industry, and organizing it in a format that is instantly accessible to you. From the preliminary budget to the detailed unit price estimate, you'll find the data in this book useful for all phases of construction cost determination.

The Staff, the Organization, and Our Services

When you purchase one of RSMeans' publications, you are in effect hiring the services of a full-time staff of construction and engineering professionals.

Our thoroughly experienced and highly qualified staff works daily at collecting, analyzing, and disseminating comprehensive cost information for your needs. These staff members have years of practical construction experience and engineering training prior to joining the firm. As a result, you can count on them not only for the cost figures, but also for additional background reference information that will help you create a realistic estimate.

The RSMeans organization is always prepared to help you solve construction problems through its five major divisions: Construction and Cost Data Publishing, Electronic Products and Services, Consulting Services, Insurance Services, and Educational Services.

Besides a full array of construction cost estimating books, Means also publishes a number of other reference works for the construction industry. Subjects include construction estimating and project and business management; special topics such as HVAC, roofing, plumbing, and hazardous waste remediation; and a library of facility management references.

In addition, you can access all of our construction cost data through your computer with RSMeans *CostWorks 2005* CD-ROM, an electronic tool that offers over 50,000 lines of Means detailed construction cost data, along with assembly and whole building cost data. You can also access Means cost information from our Web site at www.rsmeans.com

What's more, you can increase your knowledge and improve your construction estimating and management performance with a Means Construction Seminar or In-House Training Program. These two-day seminar programs offer unparalleled opportunities for everyone in your organization to get updated on a wide variety of construction-related issues.

RSMeans also is a worldwide provider of construction cost management and analysis services for commercial and government owners and of claims and valuation services for insurers.

In short, RSMeans can provide you with the tools and expertise for constructing accurate and dependable construction estimates and budgets in a variety of ways.

Robert Snow Means Established a Tradition of Quality That Continues Today

Robert Snow Means spent years building RSMeans, making certain he always delivered a quality product.

Today, at RSMeans, we do more than talk about the quality of our data and the usefulness of our books. We stand behind all of our data, from historical cost indexes to construction materials and techniques to current costs.

If you have any questions about our products or services, please call us toll-free at 1-800-334-3509. Our customer service representatives will be happy to assist you or visit our Web site at www.rsmeans.com

Table of Contents

Foreword	ii
How the Book Is Built: An Overview	v
How To Use the Book: The Details	vi
Unit Price Section	1
Assemblies Section	315
Reference Section	361
Reference Numbers	362
Crew Listings	409
Historical Cost Indexes	437
City Cost Indexes	438
Location Factors	458
Abbreviations	464
Index	467
Other Means Publications and Reference Works	Yellow Pages
Installing Contractor's Overhead & Profit	Inside Back Cover

UNIT PRICES

GENERAL REQUIREMENTS	1
SITE CONSTRUCTION	2
CONCRETE	3
MASONRY	4
METALS	5
WOOD & PLASTICS	6
THERMAL & MOISTURE PROTECTION	7
DOORS & WINDOWS	8
FINISHES	9
EQUIPMENT	11
SPECIAL CONSTRUCTION	13
CONVEYING SYSTEMS	14
MECHANICAL	15
ELECTRICAL	16

ASSEMBLIES

SUBSTRUCTURE	A
BUILDING SITEWORK	G

REFERENCE INFORMATION

REFERENCE NUMBERS

CREWS

COST INDEXES

INDEX

iii

Related Means Products and Services

Eugene Spencer, Senior Editor of this cost data book, suggests the following Means products and services as **companion** information resources to *Heavy Construction Cost Data*:

Construction Cost Data Books

Building Construction Cost Data 2005
Concrete & Masonry Cost Data 2005

Reference Books

Estimating Handbook, 2nd Ed.
Environmental Remediation Estimating Methods, 2nd Ed.
Estimating for Contractors
Cyberplaces 2nd Ed.: The Internet Guide for Architects, Engineers, Contractors & Facility Managers
Concrete Repair and Maintenance Illustrated

Seminars and In-House Training

Unit Price Estimating
Metric Estimating
Scheduling & Project Management
Facilities Maintenance & Repair Estimating
Plan Reading & Material Takeoff

Means Data on CD-ROM

Get the information found in Means traditional cost books on Means new, easy to use CD-ROM, Means *CostWorks 2005*. Means *CostWorks* users can now enhance their Costlist with the Means *CostWorks Estimator*. For more information see the special full color brochure inserted in this book.

Means Consulting Services

Means Consulting Group provides a number of construction cost-related services including: Cost Estimates, Customized Databases, Benchmark Studies, Estimating Audits, Feasibility Studies, Litigation Support, Staff Assessments and Customized Training.

Means Data for Computer Estimating

Construction costs for software applications.
Over 25 unit price and assemblies cost databases are available through a number of leading estimating and facilities management software providers (listed below). For more information see the "yellow" pages at the back of this publication.

- 3D International
- 4Clicks-Solutions, LLC
- Aepco, Inc.
- Applied Flow Technology
- ArenaSoft Estimating
- ARES Corporation
- Barchan Associates
- BSD – Building Systems Design, Inc.
- CMS – Computerized Micro Solutions
- Corecon Technologies, Inc.
- CorVet Systems
- Discover Software
- Estimating Systems, Inc.
- Entech Engineering
- FAMIS Software, Inc.
- ISES Corporation
- Magellan K-12
- Maximus Asset Solutions
- MC^2 – Management Computer Controls
- Neddam Software Technologies
- Prime Time
- Quest Solutions, Inc.
- RIB Software (Americas), Inc.
- Shaw Beneco Enterprises, Inc.
- Schwab Technology Solutions, Inc.
- Timberline Software Corp.
- TMA Systems, LLC
- US Cost, Inc.
- Vanderweil Facility Advisors
- Vertigraph, Inc.

How the Book Is Built: An Overview

A Powerful Construction Tool

You have in your hands one of the most powerful construction tools available today. A successful project is built on the foundation of an accurate and dependable estimate. This book will enable you to construct just such an estimate.

For the casual user the book is designed to be:

- quickly and easily understood so you can get right to your estimate
- filled with valuable information so you can understand the necessary factors that go into the cost estimate

For the regular user, the book is designed to be:

- a handy desk reference that can be quickly referred to for key costs
- a comprehensive, fully reliable source of current construction costs and productivity rates, so you'll be prepared to estimate any project
- a source book for preliminary project cost, product selections, and alternate materials and methods

To meet all of these requirements we have organized the book into the following clearly defined sections.

How To Use the Book: The Details

This section contains an in-depth explanation of how the book is arranged . . . and how you can use it to determine a reliable construction cost estimate. It includes information about how we develop our cost figures and how to completely prepare your estimate.

Unit Price Section

All cost data has been divided into the 16 divisions according to the MasterFormat system of classification and numbering as developed by the Construction Specifications Institute (CSI) and Construction Specifications Canada (CSC). For a listing of these divisions and an outline of their subdivisions, see the Unit Price Section Table of Contents.

Estimating tips are included at the beginning of each division.

Assemblies Section:

The cost data in this section has been organized in an "Assemblies" format. These assemblies combine components of construction into functional systems and arrange them according to the 7 divisions of the UNIFORMAT II classification system. For a complete explanation of a typical "Assemblies" page see "How To Use the Assemblies Cost Tables."

Reference Section

This section includes information on Reference Numbers, Crew Listings, Historical Cost Indexes, City Cost Indexes, Location Factors, and a listing of Abbreviations. It is visually identified by a vertical gray bar on the edge of pages.

Reference Numbers: At the beginning of selected major classifications in the Unit Price Section are "reference numbers" shown in bold squares. These numbers refer you to related information in the Reference Section.

In this section, you'll find reference tables, explanations, and estimating information that support how we develop the unit price data. Also included are alternate pricing methods, technical data, and estimating procedures, along with information on design and economy in construction. You'll also find helpful tips on what to expect and what to avoid when estimating and constructing your project.

It is recommended that you refer to the Reference Section if a "reference number" appears within the section you are estimating.

Crew Listings: This section lists all the crews referenced in the book. For the purposes of this book, a crew is composed of more than one trade classification and/or the addition of power equipment to any trade classification. Power equipment is included in the cost of the crew. Costs are shown both with bare labor rates and with the installing contractor's overhead and profit added. For each, the total crew cost per eight-hour day and the composite cost per labor-hour are listed.

Historical Cost Indexes: These indexes provide you with data to adjust construction costs over time. If you know costs for a project completed in the past, you can use these indexes to calculate a rough estimate of what it would cost to construct the same project today.

City Cost Indexes: Obviously, costs vary depending on the regional economy. You can adjust the "national average" costs in this book to 316 major cities throughout the U.S. and Canada by using the data in this section. No exchange rate conversion is necessary. How to use information is included.

Location Factors, to quickly adjust the data to over 930 zip code areas, are included.

Abbreviations: A listing of the abbreviations used throughout this book, along with the terms they represent, is included.

Index

A comprehensive listing of all terms and subjects in this book to help you find what you need quickly when you are not sure where it falls in MasterFormat.

The Scope of This Book

This book is designed to be as comprehensive and as easy to use as possible. To that end we have made certain assumptions and limited its scope in three key ways:

1. We have established material prices based on a "national average."
2. We have computed labor costs based on a 30-city "national average" of union wage rates.
3. We have targeted the data for projects of a certain size range.

For a more detailed explanation of how the cost data is developed, see "How To Use the Book: The Details."

Project Size

This book is intended primarily for heavy construction public works, and large site development projects. Generally, the productivities do not apply to small site projects.

How to Use the Book: The Details

What's Behind the Numbers? The Development of Cost Data

The staff at RSMeans continuously monitors developments in the construction industry in order to ensure reliable, thorough and up-to-date cost information.

While *overall* construction costs may vary relative to general economic conditions, price fluctuations within the industry are dependent upon many factors. Individual price variations may, in fact, be opposite to overall economic trends. Therefore, costs are continually monitored and complete updates are published yearly. Also, new items are frequently added in response to changes in materials and methods.

Costs—$ (U.S.)

All costs represent U.S. national averages and are given in U.S. dollars. The Means City Cost Indexes can be used to adjust costs to a particular location. The City Cost Indexes for Canada can be used to adjust U.S. national averages to local costs in Canadian dollars. No exchange rate conversion is necessary.

Material Costs

The RSMeans staff contacts manufacturers, dealers, distributors, and contractors all across the U.S. and Canada to determine national average material costs. If you have access to current material costs for your specific location, you may wish to make adjustments to reflect differences from the national average. Included within material costs are fasteners for a normal installation. RSMeans engineers use manufacturers' recommendations, written specifications and/ or standard construction practice for size and spacing of fasteners. Adjustments to material costs may be required for your specific application or location. Material costs do not include sales tax.

Labor Costs

Labor costs are based on the average of wage rates from 30 major U.S. cities. Rates are determined from labor union agreements or prevailing wages for construction trades for the current year. Rates along with overhead and profit markups are listed on the inside back cover of this book.

- If wage rates in your area vary from those used in this book, or if rate increases are expected within a given year, labor costs should be adjusted accordingly.

Labor costs reflect productivity based on actual working conditions. These figures include time spent during a normal workday on tasks other than actual installation, such as material receiving and handling, mobilization at site, site movement, breaks, and cleanup.

Productivity data is developed over an extended period so as not to be influenced by abnormal variations and reflects a typical average.

Equipment Costs

Equipment costs include not only rental, but also operating costs for equipment under normal use. The operating costs include parts and labor for routine servicing such as repair and replacement of pumps, filters and worn lines. Normal operating expendables such as fuel, lubricants, tires and electricity (where applicable) are also included. Extraordinary operating expendables with highly variable wear patterns such as diamond bits and blades are excluded. These costs are included under materials. Equipment rental rates are obtained from industry sources throughout North America—contractors, suppliers, dealers, manufacturers, and distributors.

Crew Equipment Cost/Day—The power equipment required for each crew is included in the crew cost. The daily cost for crew equipment is based on dividing the weekly bare rental rate by 5 (number of working days per week), and then adding the hourly operating cost times 8 (hours per day). This "Crew Equipment Cost/Day" is listed in Subdivision 01590.

Mobilization/Demobilization—The cost to move construction equipment from an equipment yard or rental company to the job site and back again is not included in equipment costs. Mobilization (to the site) and demobilization (from the site) costs can be found in Section 02305-250. If a piece of equipment is already at the job site, it is not appropriate to utilize mob/demob costs again in an estimate.

General Conditions

Cost data in this book is presented in two ways: Bare Costs and Total Cost including O&P (Overhead and Profit). General Conditions, when applicable, should also be added to the Total Cost including O&P. The costs for General Conditions are listed in Division 1 of the Unit Price Section and the Reference Section of this book. General Conditions for the *Installing Contractor* may range from 0% to 10% of the Total Cost including O&P. For the *General* or *Prime Contractor*, costs for General Conditions may range from 5% to 15% of the Total Cost including O&P, with a figure of 10% as the most typical allowance.

Overhead and Profit

Total Cost including O&P for the *Installing Contractor* is shown in the last column on both the Unit Price and the Assemblies pages of this book. This figure is the sum of the bare material cost plus 10% for profit, the base labor cost plus total overhead and profit, and the bare equipment cost plus 10% for profit. Details for the calculation of Overhead and Profit on labor are shown on the inside back cover and in the Reference Section of this book. (See the "How To Use the Unit Price Pages" for an example of this calculation.)

Factors Affecting Costs

Costs can vary depending upon a number of variables. Here's how we have handled the main factors affecting costs.

Quality—The prices for materials and the workmanship upon which productivity is based represent sound construction work. They are also in line with U.S. government specifications.

Overtime—We have made no allowance for overtime. If you anticipate premium time or work beyond normal working hours, be sure to make an appropriate adjustment to your labor costs.

Productivity—The productivity, daily output, and labor-hour figures for each line item are based on working an eight-hour day in daylight hours in moderate temperatures. For work that extends beyond normal work hours or is performed under adverse conditions, productivity may decrease. (See the section in "How To Use the Unit Price Pages" for more on productivity.)

Size of Project—The size, scope of work, and type of construction project will have a significant impact on cost. Economies of scale can reduce costs for large projects. Unit costs can often run higher for small projects. Costs in this book are intended for the size and type of project as previously described in "How the Book is Built: An Overview." Costs for projects of a significantly different size or type should be adjusted accordingly.

Location—Material prices in this book are for metropolitan areas. However, in dense urban areas, traffic and site storage limitations may increase costs. Beyond a 20-mile radius of large cities, extra trucking or transportation charges may also increase the material costs slightly. On the other hand, lower wage rates may be in effect. Be sure to consider both these factors when preparing an estimate, particularly if the job site is located in a central city or remote rural location.

In addition, highly specialized subcontract items may require travel and per diem expenses for mechanics.

Other factors—
- season of year
- contractor management
- weather conditions
- local union restrictions
- building code requirements
- availability of:
  - adequate energy
  - skilled labor
  - building materials
- owner's special requirements/restrictions
- safety requirements
- environmental considerations
- traffic control

Unpredictable Factors—General business conditions influence "in-place" costs of all items. Substitute materials and construction methods may have to be employed. These may affect the installed cost and/or life cycle costs. Such factors may be difficult to evaluate and cannot necessarily be predicted on the basis of the job's location in a particular section of the country. Thus, where these factors apply, you may find significant, but unavoidable cost variations for which you will have to apply a measure of judgment to your estimate.

Rounding of Costs

In general, all unit prices in excess of $5.00 have been rounded to make them easier to use and still maintain adequate precision of the results. The rounding rules we have chosen are in the following table.

Prices from . . .	Rounded to the nearest . . .
$.01 to $5.00	$.01
$5.01 to $20.00	$.05
$20.01 to $100.00	$.50
$100.01 to $300.00	$1.00
$300.01 to $1,000.00	$5.00
$1,000.01 to $10,000.00	$25.00
$10,000.01 to $50,000.00	$100.00
$50,000.01 and above	$500.00

How Subcontracted Items Affect Costs

A considerable portion of all large construction jobs is usually subcontracted. In fact, the percentage done by subcontractors is constantly increasing and may run over 90%. Since the workmen employed by these companies do nothing else but install their particular product, they soon become expert in that line. The result is, installation by these firms is accomplished so efficiently that the total in-place cost, even adding the general contractor's overhead and profit, is no more and often less than if the principal contractor had handled the installation himself. There is, moreover, the big advantage of having the work done right.

Companies that deal with construction specialties are anxious to have their product perform well and consequently the installation will be the best possible.

Contingencies

The contractor should consider inflationary price trends and possible material shortages during the course of the job. These escalation factors are dependent upon both economic conditions and the anticipated time between the estimate and actual construction. If drawings are not complete or approved or a budget cost is wanted, it is wise to add 5% to 10%. Contingencies then are a matter of judgment. Additional allowances for contingencies are shown in Division 1.

Estimating Precision

When making a construction cost estimate, ignore the cents column. Only use the total per unit cost to the nearest dollar. The cents will average up in a column of figures. A construction cost estimate of $257,323.37 is cumbersome. A figure of $257,325 is certainly more sensible and $257,000 is better and just as likely to be right.

If you follow this simple instruction, the time saved is tremendous with an added important advantage. Using round figures leaves the professional estimator free to exercise judgment and common sense rather than being overcome and befuddled by a mass of computations.

When the estimate is done, make a rough check of the big items for correct location of the decimal point. That is important. A large error can creep in if you write down $300 when it should be $3,000. Also check the list to be sure you have not omitted any large item. A common error is to overlook, for example, compaction or to forget seeding or mulching, or to otherwise commit a gross omission. No amount of accuracy in prices can compensate for such an oversight.

It is important to keep bare costs and costs that already include the subcontractor's overhead and profit separate since different markups will have to be applied to each category. Organize your estimating procedures to minimize confusion and simplify checking to ensure against omissions and/or duplications.

The Quantity Take-Off

Here are a few simplified rules for handling numbers and measurements when "taking off" quantities from a set of plans. A correct quantity take-off is critical to the success of an estimate, since no estimate will be reliable if a mistake is made in the quantity take-off, no matter how accurate the unit price information.

1. Use preprinted forms for orderly sequence of dimensions and locations. (RSMeans provides various forms for this purpose, including the RSMeans *Quantity Sheet* or the *Consolidated Cost Analysis Sheet.*)
2. Be consistent when listing dimensions, for example:
   Length × Width × Height.
3. Use printed dimensions where given.
4. When possible, add up printed dimensions for a single entry.
5. Measure all other dimensions carefully.
6. Use each set of dimensions to calculate multiple quantities where possible.
7. Convert feet and inch measurements to decimal feet when listing. Memorize any decimal equivalents to .01 parts of a foot.
8. Do not "round off" until the final summary of quantities.
9. Mark drawings as quantities are assembled.
10. Keep similar items together, different items separate.

11. Identify sections and drawing numbers to aid in future checking for completeness.
12. Measure or list everything that drawings show.
13. It may be necessary to list items not called for because the quantity surveyor estimator is able to identify additional materials required to make the job complete.
14. Be alert for: (a) notes on plans such as N.T.S. (Not To Scale); (b) changes in the scale used throughout the drawings; (c) drawings reduced to 1/2 or 1/4 the original size; (d) discrepancies between the specifications and the plans.
15. Use only the front side of each piece of paper or form, except for certain preprinted forms.

Working with Plans

When working with plans for an estimate, try to approach each job in the same manner.

Keep a uniform and consistent system. This will greatly reduce the chance of omission.

The following short cuts can be used successfully when making a quantity take-off:

1. Abbreviate when possible.
2. List all gross dimensions that can be either used again for different quantities or used as a rough check of other quantities for approximate verifications.
3. Convert to decimals when working with feet and inches.
4. Multiply the large numbers first to reduce rounding errors.
5. Do not convert units until the final answer is obtained. When estimating concrete work, keep all volumes to the nearest Cubic Foot, then summarize and convert to Cubic Yards.

6. Take advantage of design symmetry or repetition:
   Repetitive designs
   Symmetrical design around a center line
   Similar layouts

When figuring alternates, it is best to total all items that are involved in the basic system, then total all items that are involved in the alternates. Thus you work with positive numbers in all cases. When adds and deducts are used, it is often confusing to know whether to add or subtract a given item, especially on a complicated or involved alternate.

Final Checklist

Estimating can be a straightforward process provided you remember the basics. Here's a checklist of some of the items you should remember to do before completing your estimate.

Did you remember to . . .

- factor in the City Cost Index for your locale
- take into consideration which items have been marked up and by how much
- mark up the entire estimate sufficiently for your purposes
- read the background information on techniques and technical matters that could impact your project time span and cost
- include all components of your project in the final estimate
- double check your figures to be sure of your accuracy
- call RSMeans if you have any questions about your estimate or the data you've found in our publications

Remember, RSMeans stands behind its publications. If you have any questions about your estimate . . . about the costs you've used from our books . . . or even about the technical aspects of the job that may affect your estimate, feel free to call the RSMeans editors at 1-800-334-3509.

Unit Price Section

Table of Contents

Div No.		Page
	General Requirements	**5**
01100	Summary	6
01200	Price & Payment Procedures	6
01300	Administrative Requirements	7
01400	Quality Requirements	9
01500	Temporary Facilities & Controls	11
	Site Construction	**29**
02050	Basic Site Materials & Methods	30
02100	Site Remediation	32
02200	Site Preparation	33
02300	Earthwork	51
02400	Tunneling, Boring & Jacking	68
02450	Foundation & L. B. Elements	69
02500	Utility Services	75
02600	Drainage & Containment	94
02700	Bases, Ballasts, Pavements, & Appurtenances	102
02800	Site Improvements & Amenities	110
02900	Planting	122
02950	Site Restoration & Rehab.	130
	Concrete	**137**
03050	Basic Conc. Mat. & Methods	138
03100	Concrete Forms & Accessories	140
03200	Concrete Reinforcement	151
03300	Cast-In-Place Concrete	157
03400	Precast Concrete	164
03500	Cementitious Decks & Underlay.	166
03600	Grouts	166
03900	Concrete Restor. & Cleaning	166
	Masonry	**167**
04050	Basic Masonry Mat. & Methods	168
04200	Masonry Units	172
04500	Refractories	174
04800	Masonry Assemblies	174
04900	Masonry Restoration & Cleaning	179

Div. No.		Page
	Metals	**183**
05050	Basic Materials & Methods	184
05100	Structural Metal Framing	191
05200	Metal Joists	196
05300	Metal Deck	197
05500	Metal Fabrications	198
05650	Railroad Track & Accessories	203
05800	Expansion Control	204
05900	Metal Restoration & Cleaning	205
	Wood & Plastics	**207**
06050	Basic Wd. & Plastic Mat. & Meth.	208
06100	Rough Carpentry	212
	Thermal & Moisture Protection	**221**
07050	Basic Materials & Methods	222
07100	Dampproofing & Waterproofing	222
07200	Thermal Protection	224
07500	Membrane Roofing	228
07700	Roof Specialties & Accessories	228
07900	Joint Sealers	229
	Doors & Windows	**231**
08050	Basic Materials & Methods	232
08700	Hardware	233
	Finishes	**235**
09050	Basic Materials & Methods	236
09600	Flooring	237
09900	Paints & Coatings	237
	Equipment	**241**
11280	Hydraulic Gates & Valves	242
11300	Fluid Waste Treatment & Disposal Equipment	243
11480	Athletic Recreational & Therapeutic Equipment	244

Div. No.		Page
	Special Construction	**245**
13005	Selective Demolition	246
13010	Air Supported Structures	247
13120	Pre-Engineered Structures	248
13200	Storage Tanks	251
	Conveying Systems	**253**
14500	Material Transport	254
14600	Hoists & Cranes	254
	Mechanical	**255**
15050	Basic Materials & Methods	256
15100	Building Services Piping	257
15200	Process Piping	289
15400	Plumbing Fixtures & Equipment	290
15950	Testing/Adjusting/Balancing	293
	Electrical	**295**
16050	Basic Elect. Mat. & Methods	296
16100	Wiring Methods	297
16200	Electrical Power	300
16300	Transmission & Distribution	302
16400	Low-Voltage Distribution	307
16500	Lighting	308

How to Use the Unit Price Pages

The following is a detailed explanation of a sample entry in the Unit Price Section. Next to each bold number below is the item being described with appropriate component of the sample entry following in parenthesis. Some prices are listed as bare costs, others as costs that include overhead and profit of the installing contractor. In most cases, if the work is to be subcontracted, the general contractor will need to add an additional markup (RSMeans suggests using 10%) to the figures in the column "Total Incl. O&P."

1 Division Number/Title (03300/Cast-In-Place Concrete)

Use the Unit Price Section Table of Contents to locate specific items. The sections are classified according to the CSI MasterFormat (1995 Edition).

2 Line Numbers (03310 240 3900)

Each unit price line item has been assigned a unique 12-digit code based on the CSI MasterFormat classification.

```
            ┌── Level One - CSI-MasterFormat Division
            │ ┌── Level Two - CSI
  03300
  03310-240-3900
            │   │   └── Means 12-digit Line Number
            │   └── Level Four - Means
            └── Level Three - CSI
```

3 Description (Concrete-In-Place, etc.)

Each line item is described in detail. Sub-items and additional sizes are indented beneath the appropriate line items. The first line or two after the main item (in boldface) may contain descriptive information that pertains to all line items beneath this boldface listing.

4 Reference Number Information

R03310 -010

You'll see reference numbers shown in bold rectangles at the beginning of some sections. These refer to related items in the Reference Section, visually identified by a vertical gray bar on the edge of pages.

The relation may be: (1) an estimating procedure that should be read before estimating, (2) an alternate pricing method, or (3) technical information.

The "R" designates the Reference Section. The numbers refer to the MasterFormat classification system.

It is strongly recommended that you review all reference numbers that appear within the section in which you are working.

Note: **Not all reference numbers appear in all Means publications.**

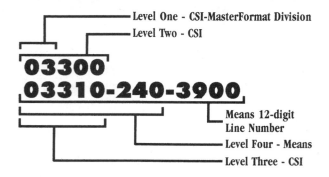

03300 | Cast-In-Place Concrete

03310 | Structural Concrete

			CREW	DAILY OUTPUT	LABOR-HOURS	UNIT	MAT.	LABOR	EQUIP.	TOTAL	TOTAL INCL O&P	
240	0010	**CONCRETE IN PLACE** Including forms (4 uses) reinforcing										240
	0050	steel and finishing unless otherwise indicated										
	0300	Beams, 5 kip per L.F., 10' span	C-14A	15	.04	C.Y.	284		46.50	770.50	1,050	
	0350	25' span	"	18.55	10.782		214	370	39.50	623.50	860	
	3850	Over 5 C.Y.		81.04	1.382		226	45	.29	271.29	320	
	3900	Footings, strip, 18" x 9", unreinforced		40	2.800		101	91.50	.59	193.09	255	
	3920	18" x 9", reinforced		35	3.200		120	105	.67	225.67	297	
	3925	20" x 10", unreinforced		45	2.489		98	81.50	.52	180.02	236	
	3930	20" x 10", reinforced		40	2.800		114	91.50	.59	206.09	269	
	3935	24" x 12", unreinforced		55	2.036		97	66.50	.43	163.93		
	3940	24" x 12", reinforced		48	2.333		113	76	.49	189.49		
	3945	36" x 12", unreinforced		70	1.600		93	52.50	.34	145.84	184	
	3950	36" x 12", reinforced		60	1.867		108	61	.39	169.39		
	4000	Foundation mat, under 10 C.Y.		38.67	2.896		164	94.50	.61	259.11	330	

Crew (C-14C)

The "Crew" column designates the typical trade or crew used to install the item. If an installation can be accomplished by one trade and requires no power equipment, that trade and the number of workers are listed (for example, "2 Carpenters"). If an installation requires a composite crew, a crew code designation is listed (for example, "C-14C"). You'll find full details on all composite crews in the Crew Listings.

* For a complete list of all trades utilized in this book and their abbreviations, see the inside back cover.

Crews

Crew No.	Bare Costs		Incl. Subs O & P		Cost Per Labor-Hour	
Crew C-14C	Hr.	Daily	Hr.	Daily	Bare Costs	Incl. O&P
1 Carpenter Foreman (out)	$36.25	$290.00	$56.45	$451.60	$32.66	$51.14
6 Carpenters	34.25	1644.00	53.35	2560.80		
2 Rodmen (reinf.)	37.95	607.20	62.60	1001.60		
4 Laborers	26.70	854.40	41.55	1329.60		
1 Cement Finisher	32.85	262.80	48.35	386.80		
1 Gas Engine Vibrator		24.00		26.40	.21	.24
112 L.H., Daily Totals		$3682.40		$5756.80	$32.87	$51.38

Productivity: Daily Output (40.0)/Labor-Hours (2.80)

The "Daily Output" represents the typical number of units the designated crew will install in a normal 8-hour day. To find out the number of days the given crew would require to complete the installation, divide your quantity by the daily output. For example:

Quantity	÷	Daily Output	=	Duration
100 C.Y.	÷	40.0/ Crew Day	=	2.50 Crew Days

The "Labor-Hours" figure represents the number of labor-hours required to install one unit of work. To find out the number of labor-hours required for your particular task, multiply the quantity of the item times the number of labor-hours shown. For example:

Quantity	x	Productivity Rate	=	Duration
100 C.Y.	x	2.80 Labor-Hours/ C.Y.	=	280 Labor-Hours

Unit (C.Y.)

The abbreviated designation indicates the unit of measure upon which the price, production, and crew are based (C.Y. = Cubic Yard). For a complete listing of abbreviations refer to the Abbreviations Listing in the Reference Section of this book.

Bare Costs:

Mat. (Bare Material Cost) (101)

The unit material cost is the "bare" material cost with no overhead and profit included. *Costs shown reflect national average material prices for January of the current year and include delivery to the job site. No sales taxes are included.*

Labor (91.50)

The unit labor cost is derived by multiplying bare labor-hour costs for Crew C-14C by labor-hour units. The bare labor-hour cost is found in the Crew Section under C-14C. (If a trade is listed, the hourly labor cost—the wage rate—is found on the inside back cover.)

Labor-Hour Cost Crew C-14C	x	Labor-Hour Units	=	Labor
$32.66	x	2.80	=	$91.00

Equip. (Equipment) (.59)

Equipment costs for each crew are listed in the description of each crew. Tools or equipment whose value justifies purchase or ownership by a contractor are considered overhead as shown on the inside back cover. The unit equipment cost is derived by multiplying the bare equipment hourly cost by the labor-hour units.

Equipment Cost Crew C-14C	x	Labor-Hour Units	=	Equip.
.21	x	2.80	=	$.59

Total (193.09)

The total of the bare costs is the arithmetic total of the three previous columns: mat., labor, and equip.

Material	+	Labor	+	Equip.	=	Total
$101	+	$91.50	+	$.59	=	$193.09

Total Costs Including O&P

This figure is the sum of the bare material cost plus 10% for profit; the bare labor cost plus total overhead and profit (per the inside back cover or, if a crew is listed, from the crew listings); and the bare equipment cost plus 10% for profit.

Material is Bare Material Cost + 10% = 101 + 10.10	=	$111.10
Labor for Crew C-14C = Labor-Hour Cost (51.14) x Labor-Hour Units (2.80)	=	$143.19
Equip. is Bare Equip. Cost + 10% = .59 + .06	=	$ .65
Total (Rounded)	=	$255

Division 1
General Requirements

Estimating Tips

The General Requirements of any contract are very important to both the bidder and the owner. These lay the ground rules under which the contract will be executed and have a significant influence on the cost of operations. Therefore, it is extremely important to thoroughly read and understand the General Requirements both before preparing an estimate and when the estimate is complete, to ascertain that nothing in the contract is overlooked. Caution should be exercised when applying items listed in Division 1 to an estimate. Many of the items are included in the unit prices listed in the other divisions such as mark-ups on labor and company overhead.

01200 Price & Payment Procedures

- When estimating historic preservation projects (depending on the condition of the existing structure and the owner's requirements), a 15-20% contingency or allowance is recommended, regardless of the stage of the drawings.

01300 Administrative Requirements

- Before determining a final cost estimate, it is a good practice to review all the items listed in subdivision 01300 to make final adjustments for items that may need customizing to specific job conditions.
- Historic preservation projects may require specialty labor and methods, as well as extra time to protect existing materials that must be preserved and/or restored. Some additional expenses may be incurred in architectural fees for facility surveys and other special inspections and analyses.
- Requirements for initial and periodic submittals can represent a significant cost to the General Requirements of a job. Thoroughly check the submittal specifications when estimating a project to determine any costs that should be included.

01400 Quality Requirements

- All projects will require some degree of Quality Control. This cost is not included in the unit cost of construction listed in each division. Depending upon the terms of the contract, the various costs of inspection and testing can be the responsibility of either the owner or the contractor. Be sure to include the required costs in your estimate.

01500 Temporary Facilities & Controls

- Barricades, access roads, safety nets, scaffolding, security and many more requirements for the execution of a safe project are elements of direct cost. These costs can easily be overlooked when preparing an estimate. When looking through the major classifications of this subdivision, determine which items apply to each division in your estimate.

01590 Equipment Rental

- This subdivision contains transportation, handling, storage, protection and product options and substitutions. Listed in this cost manual are average equipment rental rates for all types of equipment. This is useful information when estimating the time and materials requirement of any particular operation in order to establish a unit or total cost.
- A good rule of thumb is that weekly rental is 3 times daily rental and that monthly rental is 3 times weekly rental.
- The figures in the column for Crew Equipment Cost represent the rental rate used in determining the daily cost of equipment in a crew. It is calculated by dividing the weekly rate by 5 days and adding the hourly operating cost times 8 hours.

01740 Execution Requirements

- When preparing an estimate, read the specifications to determine the requirements for Contract Closeout thoroughly. Final cleaning, record documentation, operation and maintenance data, warranties and bonds, and spare parts and maintenance materials can all be elements of cost for the completion of a contract. Do not overlook these in your estimate.

01830 Operations & Maintenance

- If maintenance and repair are included in your contract, they require special attention. To estimate the cost to remove and replace any unit usually requires a site visit to determine the accessibility and the specific difficulty at that location. Obstructions, dust control, safety, and often overtime hours must be considered when preparing your estimate.

Reference Numbers

Reference numbers are shown in bold squares at the beginning of some major classifications. These numbers refer to related items in the Reference Section. The reference information may be an estimating procedure, an alternate pricing method or technical information.

Note: Not all subdivisions listed here necessarily appear in this publication.

01100 | Summary

01107	Professional Consultant	CREW	DAILY OUTPUT	LABOR-HOURS	UNIT	2005 BARE COSTS				TOTAL INCL O&P	
						MAT.	LABOR	EQUIP.	TOTAL		
300	0010	**ENGINEERING FEES**										**300**
	0020	Educational planning consultant, minimum				Project					.50%
	0100	Maximum				"					2.50%
	0200	Electrical, minimum				Contrct					4.10%
	0300	Maximum									10.10%
	0400	Elevator & conveying systems, minimum									2.50%
	0500	Maximum									5%
	0600	Food service & kitchen equipment, minimum									8%
	0700	Maximum									12%
	0800	Landscaping & site development, minimum									2.50%
	0900	Maximum									6%
	1000	Mechanical (plumbing & HVAC), minimum									4.10%
	1100	Maximum				↓					10.10%
	1200	Structural, minimum				Project					1%
	1300	Maximum				"					2.50%
700	0010	**SURVEYING** Conventional, topographical, minimum	A-7	3.30	7.273	Acre	16.30	269	18.40	303.70	455	**700**
	0100	Maximum	A-8	.60	53.333		49	1,925	101	2,075	3,125
	0300	Lot location and lines, minimum, for large quantities	A-7	2	12		25.50	445	30.50	501	745
	0320	Average	"	1.25	19.200		46	710	48.50	804.50	1,200
	0400	Maximum, for small quantities	A-8	1	32	↓	73.50	1,150	61	1,284.50	1,925
	0600	Monuments, 3' long	A-7	10	2.400	Ea.	20	88.50	6.05	114.55	166
	0800	Property lines, perimeter, cleared land	"	1,000	.024	L.F.	.03	.89	.06	.98	1.47
	1100	Crew for layout of building, trenching or pipe laying, 2 person crew	A-6	1	16	Day		545	60.50	605.50	900
	1200	3 person crew	A-7	1	24			885	60.50	945.50	1,450
	1400	Crew for roadway layout, 4 person crew	A-8	1	32	↓		1,150	61	1,211	1,850
	1500	Aerial surveying, including ground control, minimum fee, 10 acres				Total					5,700
	1510	100 acres									9,500
	1550	From existing photography, deduct				↓					1,370
	1600	2' contours, 10 acres				Acre					460
	1650	20 acres									315
	1800	50 acres									95
	1850	100 acres				↓					85
	2150	For 1' contours and									
	2160	dense urban areas, add to above				Acre					40%
	3000	Inertial guidance system for									
	3010	locating coordinates, rent per day				Ea.					4,000

01200 | Price & Payment Procedures

01250	Contract Modification Procedures	CREW	DAILY OUTPUT	LABOR-HOURS	UNIT	2005 BARE COSTS				TOTAL INCL O&P	
						MAT.	LABOR	EQUIP.	TOTAL		
500	0010	**JOB CONDITIONS** Modifications to total										**500**
	0020	project cost summaries									
	0100	Economic conditions, favorable, deduct				Project					2%
	0200	Unfavorable, add									5%
	0500	General Contractor management, experienced, deduct									2%
	0600	Inexperienced, add									10%
	0700	Labor availability, surplus, deduct									1%
	0800	Shortage, add				↓					10%

Important: See the Reference Section for critical supporting data - Reference Nos., Crews, & City Cost Indexes

01200 | Price & Payment Procedures

01250 | Contract Modification Procedures

			CREW	DAILY OUTPUT	LABOR-HOURS	UNIT	2005 BARE COSTS				TOTAL INCL O&P	
							MAT.	LABOR	EQUIP.	TOTAL		
500	1100	Subcontractor availability, surplus, deduct				Project					5%	500
	1200	Shortage, add				↓					12%	
600	0010	**OVERTIME** For early completion of projects or where	R01100 -110									600
	0020	labor shortages exist, add to usual labor, up to				Costs		100%				

01255 | Cost Indexes

200	0010	**CONSTRUCTION COST INDEX** (Reference) over 930 zip code locations in										200
	0020	The U.S. and Canada, total bldg cost, min. (Clarksdale, MS)				%					66.80%	
	0050	Average									100%	
	0100	Maximum (New York, NY)				↓					132.40%	
400	0010	**HISTORICAL COST INDEXES** (Reference) Back to 1955										400

01290 | Payment Procedures

800	0010	**TAXES** Sales tax, State, average	R01100 -090			%	4.85%					800
	0050	Maximum					7.25%					
	0200	Social Security, on first $87,900 of wages	R01100 -100					7.65%				
	0300	Unemployment, MA, combined Federal and State, minimum						1.92%				
	0350	Average				↓		6.20%				
	0400	Maximum				↓		11.76%				

01300 | Administrative Requirements

01310 | Project Management/Coordination

			CREW	DAILY OUTPUT	LABOR-HOURS	UNIT	2005 BARE COSTS				TOTAL INCL O&P	
							MAT.	LABOR	EQUIP.	TOTAL		
150	0010	**PERMITS** Rule of thumb, most cities, minimum				Job					.50%	150
	0100	Maximum				"					2%	
200	0010	**PERFORMANCE BOND** For buildings, minimum	R01100 -080			Job					.60%	200
	0200	Maximum									1.50%	
	0300	Highways & Bridges, resurfacing, minimum									.40%	
	0350	Maximum				↓					.94%	
350	0010	**INSURANCE** Builders risk, standard, minimum	R01100 -040			Job					.24%	350
	0050	Maximum									.64%	
	0200	All-risk type, minimum	R01100 -060								.25%	
	0250	Maximum				↓					.62%	
	0400	Contractor's equipment floater, minimum				Value					.50%	
	0450	Maximum				"					1.50%	
	0800	Workers' compensation & employer's liability, average										
	0850	by trade, carpentry, general				Payroll		18.42%				
	0900	Clerical						.63%				
	0950	Concrete						16.07%				
	1000	Electrical						6.46%				
	1050	Excavation						10.43%				
	1250	Masonry						15.06%				
	1300	Painting & decorating						13.12%				
	1350	Pile driving						22.45%				
	1450	Plumbing						7.96%				
	1500	Roofing						33.14%				
	1600	Steel erection, structural						38.60%				
	1700	Waterproofing, brush or hand caulking						7.12%				
	1800	Wrecking				↓		39.48%				

01310		Project Management/Coordination		CREW	DAILY OUTPUT	LABOR-HOURS	UNIT	2005 BARE COSTS				TOTAL INCL O&P	
								MAT.	LABOR	EQUIP.	TOTAL		
350	2000	Range of 35 trades in 50 states, excl. wrecking, min.	R01100 -040				Payroll		2.40%				**350**
	2100	Average							16.20%				
	2200	Maximum	R01100 -060				▼		137.60%				
400	0010	**MAIN OFFICE EXPENSE** Average for General Contractors	R01100 -050										**400**
	0020	As a percentage of their annual volume											
	0030	Annual volume to $50,000, minimum					% Vol.				20%		
	0040	Maximum									30%		
	0060	To $100,000, minimum									17%		
	0070	Maximum									22%		
	0080	To $250,000, minimum									16%		
	0090	Maximum									19%		
	0110	To $500,000, minimum									14%		
	0120	Maximum									16%		
	0125	Annual volume under 1 million dollars									13.60%		
	0130	To $1,000,000, minimum									8%		
	0140	Maximum									10%		
	0150	Up to 4.0 million dollars									6.80%		
	0200	Up to 7.0 million dollars									5.60%		
	0250	Up to 10 million dollars									5.10%		
	0300	Over 10 million dollars		▼			▼				3.90%		
500	0010	**MARK-UP** For General Contractors for change	R01100 -070										**500**
	0100	of scope of job as bid											
	0200	Extra work, by subcontractors, add	R01100 -090				%					10%	
	0250	By General Contractor, add										15%	
	0400	Omitted work, by subcontractors, deduct all but										5%	
	0450	By General Contractor, deduct all but										7.50%	
	0600	Overtime work, by subcontractors, add										15%	
	0650	By General Contractor, add					▼					10%	
	1150	Overhead markup, see division 01310-620		▼									
620	0010	**OVERHEAD & PROFIT** Allowance to add to items in this											**620**
	0020	book that do not include Subs O&P, average					%					25%	
	0100	Allowance to add to items in this book that											
	0110	do include Subs O&P, minimum					%					5%	
	0150	Average										10%	
	0200	Maximum										15%	
	0300	Typical, by size of project, under $100,000									30%		
	0350	$500,000 project									25%		
	0400	$2,000,000 project									20%		
	0450	Over $10,000,000 project					▼				15%		
700	0010	**FIELD PERSONNEL** Clerk average					Week		320		320	500	**700**
	0100	Field engineer, minimum							765		765	1,200	
	0120	Average							995		995	1,550	
	0140	Maximum							1,150		1,150	1,800	
	0160	General purpose laborer, average							1,050		1,050	1,625	
	0180	Project manager, minimum							1,425		1,425	2,225	
	0200	Average							1,625		1,625	2,525	
	0220	Maximum							1,850		1,850	2,875	
	0240	Superintendent, minimum							1,375		1,375	2,150	
	0260	Average							1,500		1,500	2,325	
	0280	Maximum							1,725		1,725	2,675	
	0290	Timekeeper, average					▼		890		890	1,375	

Important: See the Reference Section for critical supporting data - Reference Nos., Crews, & City Cost Indexes

01300 | Administrative Requirements

			CREW	DAILY OUTPUT	LABOR-HOURS	UNIT	MAT.	LABOR	EQUIP.	TOTAL	TOTAL INCL O&P	
01320		**Const. Progress Documentation**						2005 BARE COSTS				
200	0010	**SCHEDULING** Critical path, as % of architectural fee, minimum				%					.50%	**200**
	0100	Maximum				"					1%	
	0300	Computer-update, micro, no plots, minimum				Ea.					450	
	0400	Including plots, maximum				"					1,550	
	0600	Rule of thumb, CPM scheduling, small job ($10 Million)				Job					.05%	
	0650	Large job ($50 Million +)									.03%	
	0700	Including cost control, small job									.08%	
	0750	Large job									.04%	
01321		**Construction Photos**										
500	0010	**PHOTOGRAPHS** 8" x 10", 4 shots, 2 prints ea., std. mounting				Set	292			292	320	**500**
	0100	Hinged linen mounts					315			315	350	
	0200	8" x 10", 4 shots, 2 prints each, in color					340			340	370	
	0300	For I.D. slugs, add to all above					4.09			4.09	4.50	
	0500	Aerial photos, initial fly-over, 6 shots, 1 print ea., 8" x 10"					690			690	760	
	0550	11" x 14" prints					770			770	845	
	0600	16" x 20" prints					990			990	1,100	
	0700	For full color prints, add					40%				40%	
	0750	Add for traffic control area					285			285	315	
	0900	For over 30 miles from airport, add per				Mile	5.15			5.15	5.70	
	1000	Vertical photography, 4 to 6 shots with										
	1010	different scales, 1 print each				Set	1,050			1,050	1,175	
	1500	Time lapse equipment, camera and projector, buy					3,675			3,675	4,050	
	1550	Rent per month					545			545	600	
	1700	Cameraman and film, including processing, B.&W.				Day	1,350			1,350	1,475	
	1720	Color				"	1,350			1,350	1,475	

01400 | Quality Requirements

			CREW	DAILY OUTPUT	LABOR-HOURS	UNIT	MAT.	LABOR	EQUIP.	TOTAL	TOTAL INCL O&P	
01450		**Quality Control**						2005 BARE COSTS				
500	0010	**TESTING** and Inspectional Services										**500**
	0200	Asphalt testing, compressive strength Marshall stability, set of 3				Ea.					165	
	0220	Density, set of 3									95	
	0250	Extraction, individual tests on sample	R02315 -300								150	
	0300	Penetration									45	
	0350	Mix design, 5 specimens									200	
	0360	Additional specimen									40	
	0400	Specific gravity									45	
	0420	Swell test									70	
	0450	Water effect and cohesion, set of 6									200	
	0470	Water effect and plastic flow									70	
	0600	Concrete testing, aggregates, abrasion, ASTM C 131									150	
	0650	Absorption, ASTM C 127									46	
	0800	Petrographic analysis, ASTM C 295									850	
	0900	Specific gravity, ASTM C 127									55	
	1000	Sieve analysis, washed, ASTM C 136									65	
	1050	Unwashed									65	
	1200	Sulfate soundness									125	
	1300	Weight per cubic foot									40	
	1500	Cement, physical tests, ASTM C 150									350	

		01450	Quality Control			DAILY OUTPUT	LABOR-HOURS	UNIT	2005 BARE COSTS				TOTAL INCL O&P	
					CREW				MAT.	LABOR	EQUIP.	TOTAL		
500	1600		Chemical tests, ASTM C 150	R02315 -300				Ea.					270	500
	1800		Compressive test, cylinder, delivered to lab, ASTM C 39										13	
	1900		Picked up by lab, minimum										15	
	1950		Average										20	
	2000		Maximum										30	
	2200		Compressive strength, cores (not incl. drilling), ASTM C 42					↓					40	
	2250		Core drilling, 4" diameter (plus technician)					Inch					25	
	2260		Technician for core drilling					Hr.					50	
	2300		Patching core holes					Ea.					24	
	2400		Drying shrinkage at 28 days										260	
	2500		Flexural test beams, ASTM C 78										65	
	2600		Mix design, one batch mix										285	
	2650		Added trial batches										132	
	2800		Modulus of elasticity, ASTM C 469										180	
	2900		Tensile test, cylinders, ASTM C 496										50	
	3000		Water-Cement ratio curve, 3 batches										155	
	3100		4 batches										205	
	3300		Masonry testing, absorption, per 5 brick, ASTM C 67										50	
	3350		Chemical resistance, per 2 brick										55	
	3400		Compressive strength, per 5 brick, ASTM C 67										75	
	3420		Efflorescence, per 5 brick, ASTM C 67										75	
	3440		Imperviousness, per 5 brick										96	
	3470		Modulus of rupture, per 5 brick										95	
	3500		Moisture, block only										35	
	3550		Mortar, compressive strength, set of 3										25	
	4100		Reinforcing steel, bend test										61	
	4200		Tensile test, up to #8 bar										40	
	4220		#9 to #11 bar										45	
	4240		#14 bar and larger										70	
	4400		Soil testing, Atterberg limits, liquid and plastic limits										65	
	4510		Hydrometer analysis										120	
	4530		Specific gravity, ASTM D 354										48	
	4600		Sieve analysis, washed, ASTM D 422										60	
	4700		Unwashed, ASTM D 422										65	
	4710		Consolidation test (ASTM D2435), minimum										275	
	4715		Maximum										475	
	4720		Density and classification of undisturbed sample										80	
	4735		Soil density, nuclear method, ASTM D2922										38.67	
	4740		Sand cone method ASTM D1556										30.17	
	4750		Moisture content, ASTM D 2216										10	
	4780		Permeability test, double ring infiltrometer										550	
	4800		Permeability, var. or constant head, undist., ASTM D 2434										250	
	4850		Recompacted										275	
	4900		Proctor compaction, 4" standard mold, ASTM D 698										135	
	4950		6" modified mold										75	
	5100		Shear tests, triaxial, minimum										450	
	5150		Maximum										600	
	5300		Direct shear, minimum, ASTM D 3080										350	
	5350		Maximum										450	
	5550		Technician for inspection, per day, earthwork										220	
	5570		Concrete										270	
	5650		Bolting										280	
	5750		Roofing										255	
	5790		Welding						↓					270	
	5820		Non-destructive testing, dye penetrant						Day					330	
	5840		Magnetic particle						↓					330	

Important: See the Reference Section for critical supporting data - Reference Nos., Crews, & City Cost Indexes

01400 | Quality Requirements

		01450	Quality Control	CREW	DAILY OUTPUT	LABOR-HOURS	UNIT	2005 BARE COSTS				TOTAL INCL O&P	
								MAT.	LABOR	EQUIP.	TOTAL		
500	5860		Radiography	R02315 -300			Day					495	500
	5880		Ultrasonic									340	
	5900		Vibration monitoring, seismograph and technician									495	
	5910		Seismograph, rental only				Week					275	
	6000		Welding certification, minimum				Ea.					100	
	6100		Maximum				"					275	
	7000		Underground storage tank										
	7500		Volumetric tightness test ,<=12,000 gal				Ea.					400	
	7510		<=30,000 gal				"					660	
	7600		Vadose zone (soil gas) sampling, 10-40 samples, min.				Day					1,500	
	7610		Maximum				"					2,500	
	7700		Ground water monitoring incl. drilling 3 wells, min.				Total					5,000	
	7710		Maximum				"					7,000	
	8000		X-ray concrete slabs				Ea.					200	

01500 | Temporary Facilities & Controls

		01510	Temporary Utilities	CREW	DAILY OUTPUT	LABOR-HOURS	UNIT	2005 BARE COSTS				TOTAL INCL O&P	
								MAT.	LABOR	EQUIP.	TOTAL		
050	0010	**TEMPORARY POWER EQUIP (PRO-RATED PER JOB)**											050
	0020		Service, overhead feed, 3 use										
	0030		100 Amp	1 Elec	1.25	6.400	Ea.	550	261		811	995	
	0040		200 Amp		1	8		700	325		1,025	1,250	
	0050		400 Amp		.75	10.667		1,350	435		1,785	2,150	
	0060		600 Amp		.50	16		1,700	650		2,350	2,825	
	0100		Underground feed, 3 use										
	0110		100 Amp	1 Elec	2	4	Ea.	525	163		688	825	
	0120		200 Amp		1.15	6.957		720	283		1,003	1,200	
	0130		400 Amp		1	8		1,350	325		1,675	1,950	
	0140		600 Amp		.75	10.667		1,550	435		1,985	2,350	
	0150		800 Amp		.50	16		2,600	650		3,250	3,825	
	0160		1000 Amp		.35	22.857		2,700	930		3,630	4,325	
	0170		1200 Amp		.25	32		2,950	1,300		4,250	5,200	
	0180		2000 Amp		.20	40		3,800	1,625		5,425	6,600	
	0200		Transformers, 3 use										
	0210		30 KVA	1 Elec	1	8	Ea.	675	325		1,000	1,225	
	0220		45 KVA		.75	10.667		775	435		1,210	1,500	
	0230		75 KVA		.50	16		1,175	650		1,825	2,275	
	0240		112.5 KVA		.40	20		1,600	815		2,415	2,975	
	0250		Feeder, PVC, CU wire										
	0260		60 Amp w/trench	1 Elec	96	.083	L.F.	3.78	3.40		7.18	9.20	
	0270		100 Amp w/trench		85	.094		5.05	3.84		8.89	11.30	
	0280		200 Amp w/trench		59	.136		10.65	5.55		16.20	19.90	
	0290		400 Amp w/trench		42	.190		22.50	7.75		30.25	36	
	0300		Feeder, PVC, aluminum wire										
	0310		60 Amp w/trench	1 Elec	96	.083	L.F.	3.53	3.40		6.93	8.95	
	0320		100 Amp w/trench		85	.094		4.08	3.84		7.92	10.20	
	0330		200 Amp w/trench		59	.136		8.60	5.55		14.15	17.65	
	0340		400 Amp w/trench		42	.190		14.75	7.75		22.50	28	
	0350		Feeder, EMT, CU wire										
	0360		60 Amp	1 Elec	90	.089	L.F.	3.63	3.62		7.25	9.40	

1 GENERAL REQUIREMENTS

		01510	Temporary Utilities	CREW	DAILY OUTPUT	LABOR-HOURS	UNIT	MAT.	2005 BARE COSTS LABOR	EQUIP.	TOTAL	TOTAL INCL O&P	
050	0370		100 Amp	1 Elec	80	.100	L.F.	6.80	4.08		10.88	13.55	050
	0380		200 Amp		60	.133		13.10	5.45		18.55	22.50	
	0390		400 Amp	↓	35	.229	↓	17.15	9.30		26.45	33	
	0400		Feeder, EMT, Al wire										
	0410		60 Amp	1 Elec	90	.089	L.F.	2.93	3.62		6.55	8.60	
	0420		100 Amp		80	.100		5.90	4.08		9.98	12.55	
	0430		200 Amp		60	.133		11.05	5.45		16.50	20.50	
	0440		400 Amp	↓	35	.229	↓	15.05	9.30		24.35	30.50	
	0500		Equipment, 3 use										
	0510		Spider box 50 Amp	1 Elec	8	1	Ea.	168	41		209	246	
	0520		Lighting cord 100'		8	1		36.50	41		77.50	101	
	0530		Light stanchion	↓	8	1	↓	74	41		115	142	
	0540		Temporary cords, 100', 3 use										
	0550		Feeder cord, 50 Amp	1 Elec	16	.500	Ea.	156	20.50		176.50	203	
	0560		Feeder cord, 100 Amp		12	.667		450	27		477	535	
	0570		Tap cord, 50 Amp		12	.667		430	27		457	515	
	0580		Tap cord, 100 Amp	↓	6	1.333	↓	615	54.50		669.50	760	
	0590		Temporary cords, 50', 3 use										
	0600		Feeder cord, 50 Amp	1 Elec	16	.500	Ea.	91	20.50		111.50	131	
	0610		Feeder cord, 100 Amp		12	.667		224	27		251	288	
	0620		Tap cord, 50 Amp		12	.667		195	27		222	255	
	0630		Tap cord, 100 Amp	↓	6	1.333	↓	320	54.50		374.50	435	
	0700		Connections										
	0710		Compressor or pump										
	0720		30 Amp	1 Elec	7	1.143	Ea.	18.30	46.50		64.80	89.50	
	0730		60 Amp		5.30	1.509		42	61.50		103.50	138	
	0740		100 Amp	↓	4	2	↓	54	81.50		135.50	181	
	0750		Tower crane										
	0760		60 Amp	1 Elec	4.50	1.778	Ea.	42	72.50		114.50	154	
	0770		100 Amp	"	3	2.667	"	54	109		163	222	
	0780		Manlift										
	0790		Single	1 Elec	3	2.667	Ea.	43	109		152	210	
	0800		Double	"	2	4	"	55.50	163		218.50	305	
	0810		Welder										
	0820		50 Amp w/disconnect	1 Elec	5	1.600	Ea.	270	65		335	395	
	0830		100 Amp w/disconnect		3.80	2.105		415	86		501	585	
	0840		200 Amp w/disconnect		2.50	3.200		660	130		790	920	
	0850		400 Amp w/disconnect	↓	1	8	↓	2,100	325		2,425	2,775	
800	0010	**TEMPORARY UTILITIES**											800
	0100	Heat, incl. fuel and operation, per week, 12 hrs. per day		1 Skwk	100	.080	CSF Flr	6.25	2.79		9.04	11.25	
	0200	24 hrs. per day		"	60	.133		9.40	4.65		14.05	17.60	
	0350	Lighting, incl. service lamps, wiring & outlets, minimum		1 Elec	34	.235		2.32	9.60		11.92	16.80	
	0360	Maximum		"	17	.471		5.05	19.20		24.25	34	
	0400	Power for temp lighting only, per month, min/month 6.6 KWH									.75	1.18	
	0450	Maximum/month 23.6 KWH									2.85	3.14	
	0600	Power for job duration incl. elevator, etc., minimum									47	51.70	
	0650	Maximum					↓				110	121	
	0700	Temporary construction water bill per mo. average					Month	57			57	63	
	1000	Toilet, portable, see division 01590-400											

		01520	Construction Facilities	CREW	DAILY OUTPUT	LABOR-HOURS	UNIT	MAT.	LABOR	EQUIP.	TOTAL	TOTAL INCL O&P	
500	0010	**OFFICE** Trailer, furnished, no hookups, 20' x 8', buy		2 Skwk	1	16	Ea.	6,200	560		6,760	7,700	500
	0250	Rent per month					↓		154		154	169	
	0300	32' x 8', buy		2 Skwk	.70	22.857		9,250	795		10,045	11,500	
	0350	Rent per month					↓		163		163	180	

01520 | Construction Facilities

			DAILY OUTPUT	LABOR-HOURS	UNIT	2005 BARE COSTS				TOTAL INCL O&P	
			CREW			MAT.	LABOR	EQUIP.	TOTAL		
500	0400	50' x 10', buy	2 Skwk	.60	26.667	Ea.	17,000	930		17,930	20,200	500
	0450	Rent per month					254			254	280	
	0500	50' x 12', buy	2 Skwk	.50	32		20,300	1,125		21,425	24,000	
	0550	Rent per month					286			286	315	
	0700	For air conditioning, rent per month, add				▼	39.50			39.50	43.50	
	0800	For delivery, add per mile				Mile	1.53			1.53	1.68	
	1000	Portable buildings, prefab, on skids, economy, 8' x 8'	2 Carp	265	.060	S.F.	81.50	2.07		83.57	93	
	1100	Deluxe, 8' x 12'	"	150	.107	"	88.50	3.65		92.15	103	
	1200	Storage boxes, 20' x 8', buy	2 Skwk	1.80	8.889	Ea.	3,175	310		3,485	3,975	
	1250	Rent per month					74.50			74.50	82	
	1300	40' x 8', buy	2 Skwk	1.40	11.429		4,250	400		4,650	5,300	
	1350	Rent per month				▼	105			105	116	
	5000	Air supported structures, see division 13011-200										
550	0010	**FIELD OFFICE EXPENSE**										550
	0100	Field office expense, office equipment rental average				Month	143			143	157	
	0120	Office supplies, average				"	85			85	93.50	
	0125	Office trailer rental, see division 01520-500										
	0140	Telephone bill; avg. bill/month incl. long dist.				Month	204			204	224	
	0160	Field office lights & HVAC				"	98			98	108	
900	0010	**WEATHER STATION** Remote recording, minimum				Ea.	5,450			5,450	6,000	900
	0100	Maximum				"	25,500			25,500	28,100	

01530 | Temporary Construction

			CREW	DAILY OUTPUT	LABOR-HOURS	UNIT	MAT.	LABOR	EQUIP.	TOTAL	TOTAL INCL O&P	
700	0010	**PROTECTION** Stair tread, 2" x 12" planks, 1 use	1 Carp	75	.107	Tread	4.80	3.65		8.45	11	700
	0100	Exterior plywood, 1/2" thick, 1 use	↓	65	.123		1.86	4.22		6.08	8.60	
	0200	3/4" thick, 1 use	▼	60	.133	▼	2.82	4.57		7.39	10.20	
900	0010	**WINTER PROTECTION** Reinforced plastic on wood										900
	0100	framing to close openings	2 Clab	750	.021	S.F.	.38	.57		.95	1.31	
	0200	Tarpaulins hung over scaffolding, 8 uses, not incl. scaffolding	↓	1,500	.011		.18	.28		.46	.64	
	0300	Prefab fiberglass panels, steel frame, 8 uses	▼	1,200	.013	▼	.74	.36		1.10	1.36	

01540 | Construction Aids

			CREW	DAILY OUTPUT	LABOR-HOURS	UNIT	MAT.	LABOR	EQUIP.	TOTAL	TOTAL INCL O&P	
700	0010	**SAFETY NETS** No supports, stock sizes, nylon, 4" mesh				S.F.	1.03			1.03	1.13	700
	0100	Polypropylene, 6" mesh					1.59			1.59	1.75	
	0200	Small mesh debris nets, 1/4" & 3/4" mesh, stock sizes					.74			.74	.81	
	0220	Combined 4" mesh and 1/4" mesh, stock sizes					2.05			2.05	2.26	
	0300	Monthly rental, 4" mesh, stock sizes, 1st month					.40			.40	.44	
	0320	2nd month rental					.20			.20	.22	
	0340	Maximum rental/year				▼	.90			.90	.99	
750	0010	**SCAFFOLDING**										750
	0910	Steel tubular, heavy duty shoring, buy [R01540-100]										
	0920	Frames 5' high 2' wide				Ea.	82.50			82.50	91	
	0925	5' high 4' wide					93.50			93.50	103	
	0930	6' high 2' wide					94.50			94.50	104	
	0935	6' high 4' wide				▼	111			111	122	
	0940	Accessories										
	0945	Cross braces				Ea.	17.60			17.60	19.35	
	0950	U-head, 8" x 8"					19.25			19.25	21	
	0955	J-head, 4" x 8"					14.10			14.10	15.50	
	0960	Base plate, 8" x 8"					15.60			15.60	17.20	
	0965	Leveling jack				▼	33.50			33.50	37	
	1000	Steel tubular, regular, buy										
	1100	Frames 3' high 5' wide	↓			Ea.	64			64	70	

01540 | Construction Aids

		CREW	DAILY OUTPUT	LABOR-HOURS	UNIT	MAT.	LABOR	EQUIP.	TOTAL	TOTAL INCL O&P	
						2005 BARE COSTS					
750	1150	5' high 5' wide				Ea.	73.50			73.50	81	**750**
	1200	6'-4" high 5' wide					92.50			92.50	102	
	1350	7'-6" high 6' wide					160			160	175	
	1500	Accessories cross braces					16.50			16.50	18.15	
	1550	Guardrail post					16.50			16.50	18.15	
	1600	Guardrail 7' section					8			8	8.80	
	1650	Screw jacks & plates					26.50			26.50	29	
	1700	Sidearm brackets					31			31	34	
	1750	8" casters					36.50			36.50	40	
	1800	Plank 2" x 10" x 16'-0"					47			47	51.50	
	1900	Stairway section					270			270	296	
	1910	Stairway starter bar					32			32	35	
	1920	Stairway inside handrail					58.50			58.50	64	
	1930	Stairway outside handrail					80.50			80.50	88.50	
	1940	Walk-thru frame guardrail				▼	40.50			40.50	44.50	
	2000	Steel tubular, regular, rent/mo.										
	2100	Frames 3' high 5' wide				Ea.	3.75			3.75	4.13	
	2150	5' high 5' wide					3.75			3.75	4.13	
	2200	6'-4" high 5' wide					3.75			3.75	4.13	
	2250	7'-6" high 6' wide					7			7	7.70	
	2500	Accessories, cross braces					.60			.60	.66	
	2550	Guardrail post					1			1	1.10	
	2600	Guardrail 7' section					.75			.75	.83	
	2650	Screw jacks & plates					1.50			1.50	1.65	
	2700	Sidearm brackets					1.50			1.50	1.65	
	2750	8" casters					6			6	6.60	
	2800	Outrigger for rolling tower					3			3	3.30	
	2850	Plank 2" x 10" x 16'-0"					5			5	5.50	
	2900	Stairway section					10			10	11	
	2910	Stairway starter bar					.10			.10	.11	
	2920	Stairway inside handrail					5			5	5.50	
	2930	Stairway outside handrail					5			5	5.50	
	2940	Walk-thru frame guardrail				▼	2			2	2.20	
	3000	Steel tubular, heavy duty shoring, rent/mo.										
	3250	5' high 2' & 4' wide				Ea.	5			5	5.50	
	3300	6' high 2' & 4' wide					5			5	5.50	
	3500	Accessories, cross braces					1			1	1.10	
	3600	U - head, 8" x 8"					1			1	1.10	
	3650	J - head, 4" x 8"					1			1	1.10	
	3700	Base plate, 8" x 8"					1			1	1.10	
	3750	Leveling jack				▼	2			2	2.20	
	4000	Scaffolding, regular, rent/mo, no plank, w/ erect or dismantle										
	4100	Building exterior 2 stories	3 Carp	25	.960	C.S.F.	24.50	33		57.50	77.50	
	4150	4 stories	"	23	1.043		24.50	35.50		60	82	
	4200	6 stories	4 Carp	24	1.333		24.50	45.50		70	97.50	
	4250	8 stories		22	1.455		24.50	50		74.50	104	
	4300	10 stories		20	1.600		24.50	55		79.50	112	
	4350	12 stories		18	1.778		24.50	61		85.50	122	
	4500	1st tier, 3' high x 5' wide frames, 1 use/mo, erect or dismantle	1 Carp	2.38	3.361		37.50	115		152.50	220	
	4550	2 uses/month		2.38	3.361		18.70	115		133.70	200	
	4600	4 uses/month		2.38	3.361		9.35	115		124.35	189	
	4650	8 uses/month		2.38	3.361		4.67	115		119.67	184	
	5000	5' high x 5' wide frames, 1 use/month		3.97	2.015		27.50	69		96.50	138	
	5050	2 uses/month		3.97	2.015		13.65	69		82.65	123	
	5100	4 uses/month		3.97	2.015		6.80	69		75.80	116	
	5150	8 uses/month		3.97	2.015	▼	3.41	69		72.41	112	

R01540 -100

Important: See the Reference Section for critical supporting data - Reference Nos., Crews, & City Cost Indexes

GENERAL REQUIREMENTS 1

01540	Construction Aids		CREW	DAILY OUTPUT	LABOR-HOURS	UNIT	2005 BARE COSTS				TOTAL INCL O&P	
							MAT.	LABOR	EQUIP.	TOTAL		
750	5500	6'-4" high x 5' wide frames, 1 use/month	R01540-100	1 Carp	5.02	1.594	C.S.F.	24.50	54.50		79	112	**750**
	5550	2 uses/month			5.02	1.594		12.15	54.50		66.65	98.50	
	5600	4 uses/month			5.02	1.594		6.10	54.50		60.60	91.50	
	5650	8 uses/month			5.02	1.594		3.04	54.50		57.54	88.50	
	5700	Planks, 2x10x16'-0", labor only, erect/remove to 50' H		3 Carp	144	.167	Ea.		5.70		5.70	8.90	
	5800	Over 50' high		4 Carp	160	.200	"		6.85		6.85	10.65	
	6000	Heavy duty shoring for elevated slab forms to 8'-2" high, floor area											
	6010	incl. erection or dismantle labor											
	6100	1 use/month		4 Carp	36	.889	C.S.F.	29.50	30.50		60	80	
	6150	2 uses/month			36	.889		14.80	30.50		45.30	64	
	6200	3 uses/month			36	.889		9.85	30.50		40.35	58.50	
	6500	To 14'-8" high											
	6600	1 use/month		4 Carp	18	1.778	C.S.F.	43	61		104	143	
	6650	2 uses/month			18	1.778		21.50	61		82.50	119	
	6700	3 uses/month			18	1.778		14.40	61		75.40	111	
755	0010	**SCAFFOLDING SPECIALTIES**											**755**
	1200	Sidewalk bridge, heavy duty steel posts & beams, including											
	1210	parapet protection & waterproofing											
	1230	3 posts		3 Carp	10	2.400	L.F.	53.50	82		135.50	187	
	1500	Sidewalk bridge using tubular steel											
	1510	scaffold frames, including planking		3 Carp	45	.533	L.F.	4.72	18.25		22.97	33.50	
	1600	For 2 uses per month, deduct from all above						50%					
	1700	For 1 use every 2 months, add to all above						100%					
	1900	Catwalks, 20" wide, no guardrails, 7' span, buy					Ea.	138			138	151	
	2000	10' span, buy					"	176			176	194	
	2800	Hand winch-operated masons scaffolding, no plank											
	2810	plank moving not required											
	2900	98' long, 10'-6" high, buy					Ea.	28,900			28,900	31,700	
	3000	Rent per month						1,150			1,150	1,275	
	3100	28'-6" high, buy						35,400			35,400	38,900	
	3200	Rent per month						1,425			1,425	1,550	
	3400	196' long, 28'-6" high, buy						68,500			68,500	75,000	
	3500	Rent per month						2,725			2,725	3,000	
	3600	64'-6" high, buy						93,500			93,500	103,000	
	3700	Rent per month						3,750			3,750	4,125	
	3720	Putlog, standard, 8' span, with hangers, buy						67			67	74	
	3730	Rent per month						10			10	11	
	3750	12' span, buy						101			101	111	
	3755	Rent per month						15			15	16.50	
	3760	Trussed type, 16' span, buy						231			231	254	
	3770	Rent per month						20			20	22	
	3790	22' span, buy						277			277	305	
	3795	Rent per month						30			30	33	
	3800	Rolling ladders with handrails, 30" wide, buy, 2 step						173			173	191	
	4000	7 step						585			585	645	
	4050	10 step						765			765	845	
	4100	Rolling towers, buy, 5' wide, 7' long, 10' high						1,250			1,250	1,375	
	4200	For 5' high added sections, to buy, add						209			209	230	
	4300	Complete incl. wheels, railings, outriggers,											
	4350	21' high, to buy					Ea.	2,125			2,125	2,325	
	4400	Rent/month					"	159			159	175	
780	0010	**SWING STAGING**, 500 lb cap., 2' wide to 24' long, hand operated hoist											**780**
	0020	steel cable type, with 60' cables, buy					Ea.	4,450			4,450	4,900	
	0030	Rent per month					"	445			445	490	
	0600	Lightweight (not for masons) 24' long for 150' height,											

		01540	**Construction Aids**	CREW	DAILY OUTPUT	LABOR-HOURS	UNIT	2005 BARE COSTS				TOTAL INCL O&P	
								MAT.	LABOR	EQUIP.	TOTAL		
780	0610		manual type, buy				Ea.	4,725			4,725	5,175	780
	0620		Rent per month					470			470	515	
	0700		Powered, electric or air, to 150' high, buy					16,700			16,700	18,400	
	0710		Rent per month					1,175			1,175	1,275	
	0780		To 300' high, buy					19,600			19,600	21,600	
	0800		Rent per month					1,375			1,375	1,500	
	1000		Bosun's chair or work basket 3' x 3.5', to 300' high, electric, buy					7,350			7,350	8,075	
	1010		Rent per month					515			515	565	
	2200		Move swing staging (setup and remove)	E-4	2	16	Move		620	40.50	660.50	1,150	
790	0010		**SURVEYOR STAKES** Hardwood, 1" x 1" x 48" long				C	46.50			46.50	51.50	790
	0100		2" x 2" x 18" long					61.50			61.50	68	
	0150		2" x 2" x 24" long					73.50			73.50	81	
	0200		2" x 2" x 30" long					72.50			72.50	79.50	
800	0010		**TARPAULINS** Cotton duck, 10 oz. to 13.13 oz. per S.Y., minimum				S.F.	.49			.49	.54	800
	0050		Maximum					.58			.58	.64	
	0100		Polyvinyl coated nylon, 14 oz. to 18 oz., minimum					.48			.48	.53	
	0150		Maximum					.68			.68	.75	
	0200		Reinforced polyethylene 3 mils thick, white					.11			.11	.12	
	0300		4 mils thick, white, clear or black					.14			.14	.15	
	0400		5.5 mils thick, clear					.19			.19	.21	
	0500		White, fire retardant					.18			.18	.20	
	0600		7.5 mils, oil resistant, fire retardant					.19			.19	.21	
	0700		8.5 mils, black					.24			.24	.26	
	0710		Woven polyethylene, 6 mils thick					.48			.48	.53	
	0730		Polyester reinforced w/ integral fastening system 11 mils thick					1.07			1.07	1.18	
	0740		Mylar polyester, non-reinforced, 7 mils thick					1.17			1.17	1.29	
820	0010		**SMALL TOOLS** As % of contractor's work, minimum				Total					.50%	820
	0100		Maximum				"					2%	

01550 | Vehicular Access & Parking

			CREW	DAILY OUTPUT	LABOR-HOURS	UNIT	MAT.	LABOR	EQUIP.	TOTAL	TOTAL INCL O&P	
700	0010	**ROADS AND SIDEWALKS** Temporary										700
	0050	Roads, gravel fill, no surfacing, 4" gravel depth	B-14	715	.067	S.Y.	3.23	1.89	.30	5.42	6.80	
	0100	8" gravel depth	"	615	.078	"	6.45	2.19	.35	8.99	10.90	
	1000	Ramp, 3/4" plywood on 2" x 6" joists, 16" O.C.	2 Carp	300	.053	S.F.	1.48	1.83		3.31	4.47	
	1100	On 2" x 10" joists, 16" O.C.	"	275	.058		2.11	1.99		4.10	5.40	
	2200	Sidewalks, 2" x 12" planks, 2 uses	1 Carp	350	.023		.80	.78		1.58	2.10	
	2300	Exterior plywood, 2 uses, 1/2" thick		750	.011		.31	.37		.68	.91	
	2400	5/8" thick		650	.012		.39	.42		.81	1.09	
	2500	3/4" thick		600	.013		.47	.46		.93	1.23	

01560 | Barriers & Enclosures

			CREW	DAILY OUTPUT	LABOR-HOURS	UNIT	MAT.	LABOR	EQUIP.	TOTAL	TOTAL INCL O&P	
100	0010	**BARRICADES** 5' high, 3 rail @ 2" x 8", fixed	2 Carp	30	.533	L.F.	11	18.25		29.25	40.50	100
	0150	Movable	"	20	.800	"	11	27.50		38.50	54.50	
	0300	Stock units, 6' high, 8' wide, plain, buy				Ea.	435			435	480	
	0350	With reflective tape, buy				"	525			525	580	
	0400	Break-a-way 3" PVC pipe barricade										
	0410	with 3 ea. 1' x 4' reflectorized panels, buy				Ea.	305			305	335	
	0500	Plywood with steel legs, 32" wide					72			72	79	
	0600	Telescoping Christmas tree, 9' high, 5 flags, buy					122			122	134	
	0800	Traffic cones, PVC, 18" high					6.05			6.05	6.65	
	0850	28" high					18.45			18.45	20.50	
	1000	Guardrail, wooden, 3' high, 1" x 6", on 2" x 4" posts	2 Carp	200	.080	L.F.	1.02	2.74		3.76	5.40	
	1100	2" x 6", on 4" x 4" posts	"	165	.097		1.85	3.32		5.17	7.20	

Important: See the Reference Section for critical supporting data - Reference Nos., Crews, & City Cost Indexes

			CREW	DAILY OUTPUT	LABOR-HOURS	UNIT	2005 BARE COSTS				TOTAL INCL O&P	
	01560	**Barriers & Enclosures**					MAT.	LABOR	EQUIP.	TOTAL		
100	1200	Portable metal with base pads, buy				L.F.	15.50			15.50	17.05	**100**
	1250	Typical installation, assume 10 reuses	2 Carp	600	.027	↓	1.60	.91		2.51	3.18	
	1300	Barricade tape, polyethelyne, 7 mil, 3" wide x 500' long roll				Ea.	25			25	27.50	
	5000	Barricades, see also division 01590-400										
250	0010	**TEMPORARY FENCING** Chain link, 11 ga, 5' high	2 Clab	400	.040	L.F.	4.65	1.07		5.72	6.75	**250**
	0100	6' high		300	.053		4.48	1.42		5.90	7.15	
	0200	Rented chain link, 6' high, to 1000' (up to 12 mo.)		400	.040		2.94	1.07		4.01	4.89	
	0250	Over 1000' (up to 12 mo.)	↓	300	.053		2.13	1.42		3.55	4.56	
	0350	Plywood, painted, 2" x 4" frame, 4' high	A-4	135	.178		5.30	5.85		11.15	14.85	
	0400	4" x 4" frame, 8' high	"	110	.218		9.90	7.20		17.10	22	
	0500	Wire mesh on 4" x 4" posts, 4' high	2 Carp	100	.160		7.90	5.50		13.40	17.25	
	0550	8' high	"	80	.200	↓	12.10	6.85		18.95	24	
400	0010	**TEMPORARY CONSTRUCTION** See also division 01530										**400**
800	0010	**WATCHMAN** Service, monthly basis, uniformed person, minimum				Hr.					9	**800**
	0100	Maximum									16	
	0200	Person and command dog, minimum									12	
	0300	Maximum				↓					17.70	
	0500	Sentry dog, leased, with job patrol (yard dog), 1 dog				Week					230	
	0600	2 dogs				"					300	
	0800	Purchase, trained sentry dog, minimum				Ea.					1,000	
	0900	Maximum				"					2,500	
	01580	**Project Signs**										
700	0010	**SIGNS** Hi Intensity reflectorized, no posts, buy				S.F.	16.40			16.40	18	**700**

GENERAL REQUIREMENTS

1

01590 | Equipment Rental

			UNIT	HOURLY OPER. COST	RENT PER DAY	RENT PER WEEK	RENT PER MONTH	CREW EQUIPMENT COST/DAY	
100	0010	**CONCRETE EQUIPMENT RENTAL**							**100**
	0100	without operators							
	0150	For batch plant, see div. 01590-500							
	0200	Bucket, concrete lightweight, 1/2 C.Y.	Ea.	.55	15.35	46	138	13.60	
	0300	1 C.Y.		.60	18.65	56	168	16	
	0400	1-1/2 C.Y.		.75	25.50	76	228	21.20	
	0500	2 C.Y.		.80	30.50	91	273	24.60	
	0580	8 C.Y.		4.40	200	600	1,800	155.20	
	0600	Cart, concrete, self propelled, operator walking, 10 C.F.		2.10	55	165	495	49.80	
	0700	Operator riding, 18 C.F.		3.35	83.50	250	750	76.80	
	0800	Conveyer for concrete, portable, gas, 16" wide, 26' long		7.35	117	350	1,050	128.80	
	0900	46' long		7.70	142	425	1,275	146.60	
	1000	56' long		7.85	150	450	1,350	152.80	
	1100	Core drill, electric, 2-1/2 H.P., 1" to 8" bit diameter		1.61	62.50	187	560	50.30	
	1150	11 HP, 8" to 18" cores		6.61	83.50	250.80	750	103.05	
	1200	Finisher, concrete floor, gas, riding trowel, 48" diameter		4.95	86.50	260	780	91.60	
	1300	Gas, manual, 3 blade, 36" trowel		1.05	16.65	50	150	18.40	
	1400	4 blade, 48" trowel		1.50	21	63	189	24.60	
	1500	Float, hand-operated (Bull float) 48" wide		.08	13.35	40	120	8.65	
	1570	Curb builder, 14 H.P., gas, single screw		9.50	195	585	1,750	193	
	1590	Double screw		9.90	228	685	2,050	216.20	
	1600	Grinder, concrete and terrazzo, electric, floor		2.10	88.50	266	800	70	
	1700	Wall grinder		1.05	44.50	133	400	35	
	1800	Mixer, powered, mortar and concrete, gas, 6 C.F., 18 H.P.		5.25	102	305	915	103	
	1900	10 C.F., 25 H.P.		6.40	120	360	1,075	123.20	
	2000	16 C.F.		6.70	143	430	1,300	139.60	
	2100	Concrete, stationary, tilt drum, 2 C.Y.		5.25	200	600	1,800	162	
	2120	Pump, concrete, truck mounted 4" line 80' boom		21.65	885	2,655	7,975	704.20	
	2140	5" line, 110' boom		28.05	1,175	3,510	10,500	926.40	
	2160	Mud jack, 50 C.F. per hr.		4.82	115	346	1,050	107.75	
	2180	225 C.F. per hr.		7.56	154	461.60	1,375	152.80	
	2190	Shotcrete pump rig, 12 CY/hr		11.10	225	675	2,025	223.80	
	2600	Saw, concrete, manual, gas, 18 H.P.		3.60	35	105	315	49.80	
	2650	Self-propelled, gas, 30 H.P.		7.05	96.50	290	870	114.40	
	2700	Vibrators, concrete, electric, 60 cycle, 2 H.P.		.37	10.65	32	96	9.35	
	2800	3 H.P.		.56	15.65	47	141	13.90	
	2900	Gas engine, 5 H.P.		.95	18.35	55	165	18.60	
	3000	8 H.P.		1.35	22	66	198	24	
	3050	Vibrating screed, gas engine, 8 H.P.		1.68	49.50	148	445	43.05	
	3100	Concrete transit mixer, hydraulic drive							
	3120	6 x 4, 250 H.P., 8 C.Y., rear discharge		34.95	550	1,650	4,950	609.60	
	3200	Front discharge		40.65	670	2,015	6,050	728.20	
	3300	6 x 6, 285 H.P., 12 C.Y., rear discharge		40.60	670	2,010	6,025	726.80	
	3400	Front discharge		41.55	680	2,040	6,125	740.40	
200	0010	**EARTHWORK EQUIPMENT RENTAL** Without operators	R01590 -100						**200**
	0040	Aggregate spreader, push type 8' to 12' wide	Ea.	1.80	34.50	103	310	35	
	0045	Tailgate type, 8' wide	"	1.75	31.50	95	285	33	
	0050	Augers for vertical drilling	R02315 -450						
	0055	Earth auger, truck-mounted, for fence & sign posts	Ea.	8.30	485	1,455	4,375	357.40	
	0060	For borings and monitoring wells		29.20	620	1,855	5,575	604.60	
	0070	Earth auger, portable, trailer mounted	R02455 -900	1.55	22.50	67	201	25.80	
	0075	Earth auger, truck-mounted, for caissons, water wells, utility poles		153.10	3,250	9,770	29,300	3,179	
	0080	Auger, horizontal boring machine, 12" to 36" diameter, 45 H.P.		15.85	182	545	1,625	235.80	
	0090	12" to 48" diameter, 65 H.P.		22.30	325	970	2,900	372.40	
	0095	Auger, for fence posts, gas engine, hand held		.30	4.33	13	39	5	
	0100	Excavator, diesel hydraulic, crawler mounted, 1/2 C.Y. cap.		15.80	345	1,030	3,100	332.40	
	0120	5/8 C.Y. capacity		19.30	465	1,390	4,175	432.40	
	0140	3/4 C.Y. capacity		22.50	490	1,470	4,400	474	

Important: See the Reference Section for critical supporting data - Reference Nos., Crews, & City Cost Indexes

01590 | Equipment Rental

			UNIT	HOURLY OPER. COST	RENT PER DAY	RENT PER WEEK	RENT PER MONTH	CREW EQUIPMENT COST/DAY	
200	0150	1 C.Y. capacity	Ea.	27.10	570	1,705	5,125	557.80	200
	0200	1-1/2 C.Y. capacity		33.55	755	2,260	6,775	720.40	
	0300	2 C.Y. capacity		42.35	950	2,855	8,575	909.80	
	0320	2-1/2 C.Y. capacity		55.75	1,275	3,845	11,500	1,215	
	0340	3-1/2 C.Y. capacity		94.45	2,100	6,320	19,000	2,020	
	0341	Attachments							
	0342	Bucket thumbs		2.50	208	625	1,875	145	
	0345	Grapples		1	200	600.40	1,800	128.10	
	0350	Gradall type, truck mounted, 3 ton @ 15' radius, 5/8 C.Y.		38.35	885	2,655	7,975	837.80	
	0370	1 C.Y. capacity		44.25	1,025	3,085	9,250	971	
	0400	Backhoe-loader, 40 to 45 H.P., 5/8 C.Y. capacity		8.40	177	530	1,600	173.20	
	0450	45 H.P. to 60 H.P., 3/4 C.Y. capacity		10.60	218	655	1,975	215.80	
	0460	80 H.P., 1-1/4 C.Y. capacity		13.30	250	750	2,250	256.40	
	0470	112 H.P., 1-1/2 C.Y. capacity		18.10	390	1,165	3,500	377.80	
	0480	Attachments							
	0482	Compactor, 20,000 lb		4.25	117	350	1,050	104	
	0485	Hydraulic hammer, 750 ft-lbs		1.95	68.50	205	615	56.60	
	0486	Hydraulic hammer, 1200 ft-lbs		4	133	400	1,200	112	
	0500	Brush chipper, gas engine, 6" cutter head, 35 H.P.		5.90	93.50	280	840	103.20	
	0550	12" cutter head, 130 H.P.		9.35	150	450	1,350	164.80	
	0600	15" cutter head, 165 H.P.		13.40	158	475	1,425	202.20	
	0750	Bucket, clamshell, general purpose, 3/8 C.Y.		1	35	105	315	29	
	0800	1/2 C.Y.		1.10	41.50	125	375	33.80	
	0850	3/4 C.Y.		1.25	51.50	155	465	41	
	0900	1 C.Y.		1.30	55	165	495	43.40	
	0950	1-1/2 C.Y.		2.05	75	225	675	61.40	
	1000	2 C.Y.		2.15	85	255	765	68.20	
	1010	Bucket, dragline, medium duty, 1/2 C.Y.		.60	22.50	67	201	18.20	
	1020	3/4 C.Y.		.60	23.50	70	210	18.80	
	1030	1 C.Y.		.65	25.50	76	228	20.40	
	1040	1-1/2 C.Y.		.95	38.50	115	345	30.60	
	1050	2 C.Y.		1.05	43.50	130	390	34.40	
	1070	3 C.Y.		1.55	58.50	175	525	47.40	
	1200	Compactor, manually guided 2-drum vibratory smooth roller, 7.5 H.P.		4.60	143	430	1,300	122.80	
	1250	Rammer compactor, gas, 1000 lb. blow		1.60	36.50	110	330	34.80	
	1300	Vibratory plate, gas, 18" plate, 3000 lb. blow		1.50	22	66	198	25.20	
	1350	21" plate, 5000 lb. blow		1.80	37.50	112	335	36.80	
	1370	Curb builder/extruder, 14 H.P., gas, single screw		9.50	197	590	1,775	194	
	1390	Double screw		9.90	228	685	2,050	216.20	
	1500	Disc harrow attachment, for tractor		.36	60.50	181	545	39.10	
	1750	Extractor, piling, see lines 2500 to 2750							
	1810	Feller buncher, shearing & accumulating trees, 100 H.P.	Ea.	19.80	455	1,370	4,100	432.40	
	1860	Grader, self-propelled, 25,000 lb.		18.20	400	1,205	3,625	386.60	
	1910	30,000 lb.		20.70	485	1,455	4,375	456.60	
	1920	40,000 lb.		30.50	730	2,190	6,575	682	
	1930	55,000 lb.		40.25	1,025	3,080	9,250	938	
	1950	Hammer, pavement demo., hyd., gas, self-prop., 1000 to 1250 lb.		20.15	390	1,170	3,500	395.20	
	2000	Diesel 1300 to 1500 lb.		27.80	590	1,765	5,300	575.40	
	2050	Pile driving hammer, steam or air, 4150 ft.-lb. @ 225 BPM		6.35	275	825	2,475	215.80	
	2100	8750 ft.-lb. @ 145 BPM		8.25	450	1,350	4,050	336	
	2150	15,000 ft.-lb. @ 60 BPM		8.60	485	1,455	4,375	359.80	
	2200	24,450 ft.-lb. @ 111 BPM		11.35	535	1,605	4,825	411.80	
	2250	Leads, 15,000 ft.-lb. hammers	L.F.	.03	1.69	5.06	15.20	1.25	
	2300	24,450 ft.-lb. hammers and heavier	"	.05	2.67	8	24	2	
	2350	Diesel type hammer, 22,400 ft.-lb.	Ea.	24	620	1,860	5,575	564	
	2400	41,300 ft.-lb.		32.20	675	2,025	6,075	662.60	
	2450	141,000 ft.-lb.		60.90	1,475	4,430	13,300	1,373	
	2500	Vib. elec. hammer/extractor, 200 KW diesel generator, 34 H.P.		26.25	655	1,965	5,900	603	

R01590 -100
R02315 -450
R02455 -900

GENERAL REQUIREMENTS 1

01590 | Equipment Rental

			UNIT	HOURLY OPER. COST	RENT PER DAY	RENT PER WEEK	RENT PER MONTH	CREW EQUIPMENT COST/DAY	
200	2550	80 H.P.	R01590 -100	Ea.	44.80	960	2,885	8,650	935.40	**200**
	2600	150 H.P.			64	1,475	4,455	13,400	1,403	
	2700	Extractor, steam or air, 700 ft.-lb.	R02315 -450		14.50	410	1,235	3,700	363	
	2750	1000 ft.-lb.			16.60	515	1,545	4,625	441.80	
	2800	Log chipper, up to 22" diam, 600 H.P.	R02455 -900		38.53	1,250	3,770	11,300	1,062	
	2850	Logger, for skidding & stacking logs, 150 H.P.			33.80	790	2,375	7,125	745.40	
	2900	Rake, spring tooth, with tractor			7.99	213	638	1,925	191.50	
	3000	Roller, vibratory, tandem, smooth drum, 20 H.P.			5.05	112	335	1,000	107.40	
	3050	35 H.P.			7.15	203	610	1,825	179.20	
	3100	Towed type vibratory compactor, smooth drum, 50 H.P.			31.95	560	1,680	5,050	591.60	
	3150	Sheepsfoot, 50 H.P.			32.65	580	1,745	5,225	610.20	
	3170	Landfill compactor, 220 HP			46.75	1,150	3,470	10,400	1,068	
	3200	Pneumatic tire roller, 80 H.P.			8.35	292	875	2,625	241.80	
	3250	120 H.P.			13.60	505	1,515	4,550	411.80	
	3300	Sheepsfoot vibratory roller, 200 H.P.			35.35	855	2,560	7,675	794.80	
	3320	340 H.P.			48.90	1,225	3,670	11,000	1,125	
	3350	Smooth drum vibratory roller, 75 H.P.			13.60	395	1,180	3,550	344.80	
	3400	125 H.P.			17.15	495	1,480	4,450	433.20	
	3410	Rotary mower, brush, 60", with tractor			11.35	230	690	2,075	228.80	
	3450	Scrapers, towed type, 9 to 12 C.Y. capacity			4.12	196	589	1,775	150.75	
	3500	12 to 17 C.Y. capacity			1.51	262	785.60	2,350	169.20	
	3550	Scrapers, self-propelled, 4 x 4 drive, 2 engine, 14 C.Y. capacity			82.40	1,500	4,465	13,400	1,552	
	3600	2 engine, 24 C.Y. capacity			119.45	2,275	6,805	20,400	2,317	
	3640	32 - 44 C.Y. capacity			142.65	2,650	7,935	23,800	2,728	
	3650	Self-loading, 11 C.Y. capacity			40.55	825	2,475	7,425	819.40	
	3700	22 C.Y. capacity			76.40	1,725	5,185	15,600	1,648	
	3710	Screening plant 110 H.P. w/ 5' x 10' screen			21.75	370	1,105	3,325	395	
	3720	5' x 16' screen			23.80	465	1,390	4,175	468.40	
	3850	Shovels, see Cranes division 01590-600								
	3860	Shovel/backhoe bucket, 1/2 C.Y.		Ea.	1.80	55	165	495	47.40	
	3870	3/4 C.Y.			1.85	61.50	185	555	51.80	
	3880	1 C.Y.			1.95	71.50	215	645	58.60	
	3890	1-1/2 C.Y.			2.10	85	255	765	67.80	
	3910	3 C.Y.			2.40	118	355	1,075	90.20	
	3950	Stump chipper, 18" deep, 30 H.P.			4.56	57	171	515	70.70	
	4110	Tractor, crawler, with bulldozer, torque converter, diesel 80 H.P.			16.10	310	930	2,800	314.80	
	4150	105 H.P.			22.10	460	1,385	4,150	453.80	
	4200	140 H.P.			27.25	590	1,770	5,300	572	
	4260	200 H.P.			40.95	985	2,960	8,875	919.60	
	4310	300 H.P.			53.10	1,275	3,850	11,600	1,195	
	4360	410 H.P.			71.95	1,575	4,740	14,200	1,524	
	4370	500 H.P.			95.25	2,125	6,365	19,100	2,035	
	4380	700 H.P.			142.20	3,325	10,005	30,000	3,139	
	4400	Loader, crawler, torque conv., diesel, 1-1/2 C.Y., 80 H.P.			14.95	320	965	2,900	312.60	
	4450	1-1/2 to 1-3/4 C.Y., 95 H.P.			17.35	385	1,150	3,450	368.80	
	4510	1-3/4 to 2-1/4 C.Y., 130 H.P.			23.60	605	1,815	5,450	551.80	
	4530	2-1/2 to 3-1/4 C.Y., 190 H.P.			35.50	840	2,520	7,550	788	
	4560	3-1/2 to 5 C.Y., 275 H.P.			46.95	1,200	3,590	10,800	1,094	
	4610	Tractor loader, wheel, torque conv., 4 x 4, 1 to 1-1/4 C.Y., 65 H.P.			9.85	192	575	1,725	193.80	
	4620	1-1/2 to 1-3/4 C.Y., 80 H.P.			12.50	235	705	2,125	241	
	4650	1-3/4 to 2 C.Y., 100 H.P.			14.20	278	835	2,500	280.60	
	4710	2-1/2 to 3-1/2 C.Y., 130 H.P.			15.25	305	915	2,750	305	
	4730	3 to 4-1/2 C.Y., 170 H.P.			20.95	465	1,390	4,175	445.60	
	4760	5-1/4 to 5-3/4 C.Y., 270 H.P.			34.55	705	2,120	6,350	700.40	
	4810	7 to 8 C.Y., 375 H.P.			58.95	1,250	3,755	11,300	1,223	
	4870	12-1/2 C.Y., 690 H.P.			82.30	1,950	5,860	17,600	1,830	
	4880	Wheeled, skid steer, 10 C.F., 30 H.P. gas			8.80	138	415	1,250	153.40	
	4890	1 C.Y., 78 H.P., diesel			10.35	192	575	1,725	197.80	

Important: See the Reference Section for critical supporting data - Reference Nos., Crews, & City Cost Indexes

01590 | Equipment Rental

				UNIT	HOURLY OPER. COST	RENT PER DAY	RENT PER WEEK	RENT PER MONTH	CREW EQUIPMENT COST/DAY	
200	4891	Attachments for all skid steer loaders	R01590 -100							200
	4892	Auger		Ea.	.40	66.50	200	600	43.20	
	4893	Backhoe	R02315 -450		.65	109	326	980	70.40	
	4894	Broom			.63	105	314	940	67.85	
	4895	Forks	R02455 -900		.22	37.50	112	335	24.15	
	4896	Grapple			.50	83	249	745	53.80	
	4897	Concrete hammer			.98	163	489	1,475	105.65	
	4898	Tree spade			1	167	500	1,500	108	
	4899	Trencher			.68	114	342	1,025	73.85	
	4900	Trencher, chain, boom type, gas, operator walking, 12 H.P.			2.75	43.50	130	390	48	
	4910	Operator riding, 40 H.P.			8.55	242	725	2,175	213.40	
	5000	Wheel type, diesel, 4' deep, 12" wide			46.90	735	2,205	6,625	816.20	
	5100	Diesel, 6' deep, 20" wide			63.75	1,575	4,750	14,300	1,460	
	5150	Ladder type, diesel, 5' deep, 8" wide			23.10	705	2,115	6,350	607.80	
	5200	Diesel, 8' deep, 16" wide			56.80	1,600	4,830	14,500	1,420	
	5210	Tree spade, self-propelled			9.60	267	800	2,400	236.80	
	5250	Truck, dump, tandem, 12 ton payload			20.45	272	815	2,450	326.60	
	5300	Three axle dump, 16 ton payload			27.95	420	1,265	3,800	476.60	
	5350	Dump trailer only, rear dump, 16-1/2 C.Y.			4.25	115	345	1,025	103	
	5400	20 C.Y.			4.60	132	395	1,175	115.80	
	5450	Flatbed, single axle, 1-1/2 ton rating			11.85	56.50	170	510	128.80	
	5500	3 ton rating			15.15	91.50	275	825	176.20	
	5550	Off highway rear dump, 25 ton capacity			40.40	980	2,945	8,825	912.20	
	5600	35 ton capacity			41.30	1,000	3,015	9,050	933.40	
	5610	50 ton capacity			53.15	1,250	3,780	11,300	1,181	
	5620	65 ton capacity			56.85	1,375	4,105	12,300	1,276	
	5630	100 ton capacity			73	1,775	5,325	16,000	1,649	
	6000	Vibratory plow, 25 H.P., walking			3.90	58.50	175	525	66.20	
400	0010	**GENERAL EQUIPMENT RENTAL** Without operators	R01590 -100							400
	0150	Aerial lift, scissor type, to 15' high, 1000 lb. cap., electric		Ea.	2.30	41.50	125	375	43.40	
	0160	To 25' high, 2000 lb. capacity	R02250 -400		2.70	60	180	540	57.60	
	0170	Telescoping boom to 40' high, 500 lb. capacity, gas			11.45	268	805	2,425	252.60	
	0180	To 45' high, 500 lb. capacity	R02065 -300		12.30	310	925	2,775	283.40	
	0190	To 60' high, 600 lb. capacity			14.25	410	1,230	3,700	360	
	0195	Air compressor, portable, 6.5 CFM, electric			.42	14.65	44	132	12.15	
	0196	Gasoline			.51	22	66	198	17.30	
	0200	Air compressor, portable, gas engine, 60 C.F.M.			4.50	32.50	97	291	55.40	
	0300	160 C.F.M.			7.80	41.50	125	375	87.40	
	0400	Diesel engine, rotary screw, 250 C.F.M.			7.80	83.50	250	750	112.40	
	0500	365 C.F.M.			10.35	107	320	960	146.80	
	0550	450 C.F.M.			12.55	127	380	1,150	176.40	
	0600	600 C.F.M.			21.35	212	635	1,900	297.80	
	0700	750 C.F.M.			23.05	222	665	2,000	317.40	
	0800	For silenced models, small sizes, add			3%	5%	5%	5%		
	0900	Large sizes, add			5%	7%	7%	7%		
	0920	Air tools and accessories								
	0930	Breaker, pavement, 60 lb.		Ea.	.40	11	33	99	9.80	
	0940	80 lb.			.40	12.65	38	114	10.80	
	0950	Drills, hand (jackhammer) 65 lb.			.45	15	45	135	12.60	
	0960	Track or wagon, swing boom, 4" drifter			35.05	640	1,920	5,750	664.40	
	0970	5" drifter			47.70	735	2,205	6,625	822.60	
	0975	Track mounted quarry drill, 6" diameter drill			50.45	795	2,380	7,150	879.60	
	0980	Dust control per drill			.78	12	36	108	13.45	
	0990	Hammer, chipping, 12 lb.			.40	21.50	64	192	16	
	1000	Hose, air with couplings, 50' long, 3/4" diameter			.03	5.65	17	51	3.65	
	1100	1" diameter			.03	5.65	17	51	3.65	
	1200	1-1/2" diameter			.04	7.35	22	66	4.70	
	1300	2" diameter			.10	17.35	52	156	11.20	

01590 | Equipment Rental

			UNIT	HOURLY OPER. COST	RENT PER DAY	RENT PER WEEK	RENT PER MONTH	CREW EQUIPMENT COST/DAY	
400	1400	2-1/2" diameter	Ea.	.12	19.65	59	177	12.75	400
	1410	3" diameter	R01590 -100	.16	27.50	82	246	17.70	
	1450	Drill, steel, 7/8" x 2'	R02250 -400	.05	8	24	72	5.20	
	1460	7/8" x 6'		.06	9.35	28	84	6.10	
	1520	Moil points	R02065 -300	.03	4.33	13	39	2.85	
	1525	Pneumatic nailer w/accessories		.41	27	81	243	19.50	
	1530	Sheeting driver for 60 lb. breaker		.10	6.95	20.80	62.50	4.95	
	1540	For 90 lb. breaker		.15	10	30	90	7.20	
	1550	Spade, 25 lb.		.35	6.35	19	57	6.60	
	1560	Tamper, single, 35 lb.		.58	38.50	115	345	27.65	
	1570	Triple, 140 lb.		.87	57.50	173	520	41.55	
	1580	Wrenches, impact, air powered, up to 3/4" bolt		.25	7.35	22	66	6.40	
	1590	Up to 1-1/4" bolt		.35	15.35	46	138	12	
	1600	Barricades, barrels, reflectorized, 1 to 50 barrels		.02	2.60	7.80	23.50	1.70	
	1610	100 to 200 barrels		.01	1.93	5.80	17.40	1.25	
	1620	Barrels with flashers, 1 to 50 barrels		.02	3.27	9.80	29.50	2.10	
	1630	100 to 200 barrels		.02	2.60	7.80	23.50	1.70	
	1640	Barrels with steady burn type C lights		.03	4.33	13	39	2.85	
	1650	Illuminated board, trailer mounted, with generator		.65	117	350	1,050	75.20	
	1670	Portable barricade, stock, with flashers, 1 to 6 units		.02	3.27	9.80	29.50	2.10	
	1680	25 to 50 units		.02	3.03	9.10	27.50	2	
	1690	Butt fusion machine, electric		22.35	425	1,270	3,800	432.80	
	1695	Electro fusion machine		8.85	168	505	1,525	171.80	
	1700	Carts, brick, hand powered, 1000 lb. capacity		.30	49.50	148	445	32	
	1800	Gas engine, 1500 lb., 7-1/2' lift		2.77	94	282	845	78.55	
	1822	Dehumidifier, medium, 6 lb/hr, 150 CFM		.68	41.50	124	370	30.25	
	1824	Large, 18 lb/hr, 600 CFM		1.36	82.50	248	745	60.50	
	1830	Distributor, asphalt, trailer mtd, 2000 gal., 38 H.P. diesel		7.35	243	730	2,200	204.80	
	1840	3000 gal., 38 H.P. diesel		8.35	278	835	2,500	233.80	
	1850	Drill, rotary hammer, electric, 1-1/2" diameter		.40	24.50	73	219	17.80	
	1860	Carbide bit for above		.03	5.35	16	48	3.45	
	1865	Rotary, crawler, 250 H.P.		87.35	1,600	4,835	14,500	1,666	
	1870	Emulsion sprayer, 65 gal., 5 H.P. gas engine		1.80	72	216	650	57.60	
	1880	200 gal., 5 H.P. engine		4.85	120	360	1,075	110.80	
	1920	Floodlight, mercury vapor, or quartz, on tripod							
	1930	1000 watt	Ea.	.30	12	36	108	9.60	
	1940	2000 watt		.52	22	66	198	17.35	
	1950	Floodlights, trailer mounted with generator, 1 - 300 watt light		2.45	65	195	585	58.60	
	1960	2 - 1000 watt lights		3.30	108	325	975	91.40	
	2000	4 - 300 watt lights		2.75	76.50	230	690	68	
	2020	Forklift, wheeled, for brick, 18', 3000 lb., 2 wheel drive, gas		13.70	177	530	1,600	215.60	
	2040	28', 4000 lb., 4 wheel drive, diesel		11.75	232	695	2,075	233	
	2050	For rough terrain, 8000 lb., 16' lift, 68 H.P.		15.70	360	1,075	3,225	340.60	
	2060	For plant, 4 T. capacity, 80 H.P., 2 wheel drive, gas		7.80	83.50	250	750	112.40	
	2080	10 T. capacity, 120 H.P., 2 wheel drive, diesel		11.95	157	470	1,400	189.60	
	2100	Generator, electric, gas engine, 1.5 KW to 3 KW		1.75	15	45	135	23	
	2200	5 KW		2.40	21.50	65	195	32.20	
	2300	10 KW		3.95	46.50	140	420	59.60	
	2400	25 KW		7.65	71.50	215	645	104.20	
	2500	Diesel engine, 20 KW		5.95	61.50	185	555	84.60	
	2600	50 KW		11.05	68.50	205	615	129.40	
	2700	100 KW		16.20	78.50	235	705	176.60	
	2800	250 KW		45.65	142	425	1,275	450.20	
	2850	Hammer, hydraulic, for mounting on boom; to 500 ft.-lb.		1.75	63.50	190	570	52	
	2860	1000 ft.-lb.		3.10	102	305	915	85.80	
	2900	Heaters, space, oil or electric, 50 MBH		.92	9.65	29	87	13.15	
	3000	100 MBH		1.66	13.65	41	123	21.50	
	3100	300 MBH		5.33	35	105	315	63.65	

Important: See the Reference Section for critical supporting data - Reference Nos., Crews, & City Cost Indexes

GENERAL REQUIREMENTS

		01590 | Equipment Rental		UNIT	HOURLY OPER. COST	RENT PER DAY	RENT PER WEEK	RENT PER MONTH	CREW EQUIPMENT COST/DAY	
400	3150	500 MBH	R01590 -100	Ea.	10.70	50	150	450	115.60	400
	3200	Hose, water, suction with coupling, 20' long, 2" diameter			.02	4	12	36	2.55	
	3210	3" diameter	R02250 -400		.03	6.65	20	60	4.25	
	3220	4" diameter			.03	8	24	72	5.05	
	3230	6" diameter	R02065 -300		.10	20	60	180	12.80	
	3240	8" diameter			.25	41.50	125	375	27	
	3250	Discharge hose with coupling, 50' long, 2" diameter			.01	3	9	27	1.90	
	3260	3" diameter			.01	4	12	36	2.50	
	3270	4" diameter			.02	5.65	17	51	3.55	
	3280	6" diameter			.06	14	42	126	8.90	
	3290	8" diameter			.31	52	156	470	33.70	
	3295	Insulation blower			.11	7	21	63	5.10	
	3300	Ladders, extension type, 16' to 36' long			.17	28.50	86	258	18.55	
	3400	40' to 60' long			.26	36.50	110	330	24.10	
	3405	Lance for cutting concrete			2.78	105	315	945	85.25	
	3407	Lawn mower, rotary, 22", 5HP			.97	27	81	243	23.95	
	3408	48" self propelled			3.06	100	300	900	84.50	
	3410	Level, laser type, for pipe and sewer leveling			1.27	84.50	253	760	60.75	
	3430	Electronic			.76	50.50	151	455	36.30	
	3440	Laser type, rotating beam for grade control			1.04	69	207	620	49.70	
	3460	Builders level with tripod and rod			.07	12.35	37	111	7.95	
	3500	Light towers, towable, with diesel generator, 2000 watt			2.75	76.50	230	690	68	
	3600	4000 watt			3.30	108	325	975	91.40	
	3700	Mixer, powered, plaster and mortar, 6 C.F., 7 H.P.			1.30	26.50	80	240	26.40	
	3800	10 C.F., 9 H.P.			1.55	41.50	125	375	37.40	
	3850	Nailer, pneumatic			.41	27	81	243	19.50	
	3900	Paint sprayers complete, 8 CFM			.69	45.50	137	410	32.90	
	4000	17 CFM			1.10	73.50	220	660	52.80	
	4020	Pavers, bituminous, rubber tires, 8' wide, 50 H.P., diesel			26.80	775	2,320	6,950	678.40	
	4030	10' wide, 150 H.P.			59.45	1,300	3,895	11,700	1,255	
	4050	Crawler, 8' wide, 100 H.P., diesel			55.60	1,450	4,335	13,000	1,312	
	4060	10' wide, 150 H.P.			66.95	1,875	5,590	16,800	1,654	
	4070	Concrete paver, 12' to 24' wide, 250 H.P.			57.90	1,325	3,995	12,000	1,262	
	4080	Placer-spreader-trimmer, 24' wide, 300 H.P.			87.60	2,025	6,045	18,100	1,910	
	4100	Pump, centrifugal gas pump, 1-1/2", 4 MGPH			2.60	36.50	110	330	42.80	
	4200	2", 8 MGPH			3.25	41.50	125	375	51	
	4300	3", 15 MGPH			3.45	43.50	130	390	53.60	
	4400	6", 90 MGPH			15.75	152	455	1,375	217	
	4500	Submersible electric pump, 1-1/4", 55 GPM			.35	17	51	153	13	
	4600	1-1/2", 83 GPM			.41	19.65	59	177	15.10	
	4700	2", 120 GPM			.43	24.50	73	219	18.05	
	4800	3", 300 GPM			.90	33.50	100	300	27.20	
	4900	4", 560 GPM			6.16	135	405	1,225	130.30	
	5000	6", 1590 GPM			9.27	200	600	1,800	194.15	
	5100	Diaphragm pump, gas, single, 1-1/2" diameter			.78	38.50	116	350	29.45	
	5200	2" diameter			2.70	48.50	145	435	50.60	
	5300	3" diameter			2.75	48.50	145	435	51	
	5400	Double, 4" diameter			3.55	76.50	230	690	74.40	
	5500	Trash pump, self-priming, gas, 2" diameter			2.55	26.50	80	240	36.40	
	5600	Diesel, 4" diameter			6.05	53.50	160	480	80.40	
	5650	Diesel, 6" diameter			18.35	112	335	1,000	213.80	
	5655	Grout Pump			8.30	58.50	175	525	101.40	
	5660	Rollers, see division 01590-200								
	5700	Salamanders, L.P. gas fired, 100,000 BTU		Ea.	1.66	10.65	32	96	19.70	
	5705	50,000 BTU			1.25	7.65	23	69	14.60	
	5720	Sandblaster, portable, open top, 3 C.F. capacity			.40	20	60	180	15.20	
	5730	6 C.F. capacity			.65	29	87	261	22.60	
	5740	Accessories for above			.11	18.35	55	165	11.90	

01590 | Equipment Rental

			UNIT	HOURLY OPER. COST	RENT PER DAY	RENT PER WEEK	RENT PER MONTH	CREW EQUIPMENT COST/DAY	
400	5750	Sander, floor	R01590 -100	Ea.	.72	17	51	153	15.95	400
	5760	Edger			.71	25	75	225	20.70	
	5800	Saw, chain, gas engine, 18" long	R02250 -400		1.20	16	48	144	19.20	
	5900	36" long			.55	48.50	145	435	33.40	
	5950	60" long	R02065 -300		.55	50	150	450	34.40	
	6000	Masonry, table mounted, 14" diameter, 5 H.P.			1.26	54.50	163	490	42.70	
	6050	Portable cut-off, 8 H.P.			1.30	28.50	85	255	27.40	
	6100	Circular, hand held, electric, 7-1/4" diameter			.20	7.35	22	66	6	
	6200	12" diameter			.27	11	33	99	8.75	
	6250	Wall saw, w/hydraulic power, 10 H.P			2.04	95	284.80	855	73.30	
	6275	Shot blaster, walk behind, 20" wide			1.42	545	1,635	4,900	338.35	
	6300	Steam cleaner, 100 gallons per hour			2.25	61.50	185	555	55	
	6310	200 gallons per hour			3	75	225	675	69	
	6340	Tar Kettle/Pot, 400 gallon			2.69	51.50	155	465	52.50	
	6350	Torch, cutting, acetylene-oxygen, 150' hose			1.50	10	30	90	18	
	6360	Hourly operating cost includes tips and gas			8.10				64.80	
	6410	Toilet, portable chemical			.11	17.65	53	159	11.50	
	6420	Recycle flush type			.13	21.50	65	195	14.05	
	6430	Toilet, fresh water flush, garden hose,			.15	24.50	73	219	15.80	
	6440	Hoisted, non-flush, for high rise			.13	21.50	64	192	13.85	
	6450	Toilet, trailers, minimum			.22	36.50	110	330	23.75	
	6460	Maximum			.66	110	330	990	71.30	
	6465	Tractor, farm with attachment		▼	10.25	222	665	2,000	215	
	6470	Trailer, office, see division 01520-500								
	6500	Trailers, platform, flush deck, 2 axle, 25 ton capacity		Ea.	4.25	90	270	810	88	
	6600	40 ton capacity			5.55	125	375	1,125	119.40	
	6700	3 axle, 50 ton capacity			6	138	415	1,250	131	
	6800	75 ton capacity			7.45	182	545	1,625	168.60	
	6810	Trailer mounted cable reel for H.V. line work			4.47	213	639	1,925	163.55	
	6820	Trailer mounted cable tensioning rig			8.75	415	1,250	3,750	320	
	6830	Cable pulling rig		▼	55.44	2,375	7,120	21,400	1,868	
	6850	Trailer, storage, see division 01520-500								
	6900	Water tank, engine driven discharge, 5000 gallons		Ea.	5.60	120	360	1,075	116.80	
	6925	10,000 gallons			7.70	168	505	1,525	162.60	
	6950	Water truck, off highway, 6000 gallons			49.25	700	2,095	6,275	813	
	7010	Tram car for H.V. line work, powered, 2 conductor			5.73	116	347	1,050	115.25	
	7020	Transit (builder's level) with tripod			.07	12.35	37	111	7.95	
	7030	Trench box, 3000 lbs. 6'x8'			.67	90.50	272	815	59.75	
	7040	7200 lbs. 6'x20'			.82	136	409	1,225	88.35	
	7050	8000 lbs., 8' x 16'			.88	147	442	1,325	95.45	
	7060	9500 lbs., 8'x20'			1.19	199	597	1,800	128.90	
	7065	11,000 lbs., 8'x24'			1.34	224	671	2,025	144.90	
	7070	12,000 lbs., 10' x 20'			1.77	294	883	2,650	190.75	
	7100	Truck, pickup, 3/4 ton, 2 wheel drive			5.75	53.50	160	480	78	
	7200	4 wheel drive			5.90	63.50	190	570	85.20	
	7250	Crew carrier, 9 passenger			5.44	92	276.40	830	98.80	
	7290	Tool van, 24,000 G.V.W.			8.40	91.50	274.80	825	122.15	
	7300	Tractor, 4 x 2, 30 ton capacity, 195 H.P.			13.65	165	495	1,475	208.20	
	7410	250 H.P.			18.20	243	730	2,200	291.60	
	7500	6 x 2, 40 ton capacity, 240 H.P.			17.45	270	810	2,425	301.60	
	7600	6 x 4, 45 ton capacity, 240 H.P.			22.20	293	880	2,650	353.60	
	7620	Vacuum truck, hazardous material, 2500 gallon			6.51	300	902	2,700	232.50	
	7625	5,000 gallon			11.37	400	1,202.80	3,600	331.50	
	7640	Tractor, with A frame, boom and winch, 225 H.P.			15.75	223	670	2,000	260	
	7650	Vacuum, H.E.P.A., 16 gal., wet/dry			.27	24	72	216	16.55	
	7655	55 gal, wet/dry			.60	36	108	325	26.40	
	7660	Water tank, portable			1	9.35	28	84	13.60	
	7690	Large production vacuum loader, 3150 CFM			15	600	1,800	5,400	480	

Important: See the Reference Section for critical supporting data - Reference Nos., Crews, & City Cost Indexes

01590 | Equipment Rental

			UNIT	HOURLY OPER. COST	RENT PER DAY	RENT PER WEEK	RENT PER MONTH	CREW EQUIPMENT COST/DAY	
400	7700	Welder, electric, 200 amp	Ea.	3.79	60.50	181	545	66.50	400
	7800	300 amp		5.28	64	192	575	80.65	
	7900	Gas engine, 200 amp		5.70	28	84	252	62.40	
	8000	300 amp		7.70	32.50	98	294	81.20	
	8100	Wheelbarrow, any size		.06	10.65	32	96	6.90	
	8200	Wrecking ball, 4000 lb.		1.85	68.50	205	615	55.80	
500	0010	**HIGHWAY EQUIPMENT RENTAL**							500
	0050	Asphalt batch plant, portable drum mixer, 100 ton/hr.	Ea.	55.75	1,275	3,850	11,600	1,216	
	0060	200 ton/hr.		61.85	1,350	4,045	12,100	1,304	
	0070	300 ton/hr.		72.05	1,600	4,785	14,400	1,533	
	0100	Backhoe attachment, long stick, up to 185 HP, 10.5' long		.30	20	60	180	14.40	
	0140	Up to 250 HP, 12' long		.33	21.50	65	195	15.65	
	0180	Over 250 HP, 15' long		.43	28.50	85	255	20.45	
	0200	Special dipper arm, up to 100 HP, 32' long		.88	58.50	175	525	42.05	
	0240	Over 100 HP, 33' long		1.10	73.50	220	660	52.80	
	0300	Concrete batch plant, portable, electric, 200 CY/Hr		10.40	570	1,715	5,150	426.20	
	0500	Grader attachment, ripper/scarifier, rear mounted							
	0520	Up to 135 HP	Ea.	2.75	58.50	175	525	57	
	0540	Up to 180 HP		3.30	75	225	675	71.40	
	0580	Up to 250 HP		3.65	85	255	765	80.20	
	0700	Pvmt. removal bucket, for hyd. excavator, up to 90 HP		1.40	43.50	130	390	37.20	
	0740	Up to 200 HP		1.60	65	195	585	51.80	
	0780	Over 200 HP		1.70	76.50	230	690	59.60	
	0900	Aggregate spreader, self-propelled, 187 HP		37.15	790	2,370	7,100	771.20	
	1000	Chemical spreader, 3 C.Y.		2.20	61.50	185	555	54.60	
	1900	Hammermill, traveling, 250 HP		39.98	1,700	5,120	15,400	1,344	
	2000	Horizontal borer, 3" diam, 13 HP gas driven		3.95	51.50	155	465	62.60	
	2200	Hydromulchers, gas power, 3000 gal., for truck mounting		10.50	187	560	1,675	196	
	2400	Joint & crack cleaner, walk behind, 25 HP		2.05	46.50	140	420	44.40	
	2500	Filler, trailer mounted, 400 gal., 20 HP		6	182	545	1,625	157	
	3000	Paint striper, self propelled, double line, 30 HP		5.15	155	465	1,400	134.20	
	3200	Post drivers, 6" I-Beam frame, for truck mounting		8.15	415	1,250	3,750	315.20	
	3400	Road sweeper, self propelled, 8' wide, 90 HP		23.50	430	1,285	3,850	445	
	4000	Road mixer, self-propelled, 130 HP		29.05	570	1,715	5,150	575.40	
	4100	310 HP		54	1,850	5,570	16,700	1,546	
	4200	Cold mix paver, incl pug mill and bitumen tank,							
	4220	165 HP	Ea.	66.95	1,875	5,600	16,800	1,656	
	4250	Paver, asphalt, wheel or crawler, 130 H.P., diesel		65.65	1,775	5,325	16,000	1,590	
	4300	Paver, road widener, gas 1' to 6', 67 HP		29.65	630	1,890	5,675	615.20	
	4400	Diesel, 2' to 14', 88 HP		39.95	955	2,870	8,600	893.60	
	4600	Slipform pavers, curb and gutter, 2 track, 75 HP		24.85	640	1,920	5,750	582.80	
	4700	4 track, 165 HP		34.75	745	2,235	6,700	725	
	4800	Median barrier, 215 HP		35.35	775	2,325	6,975	747.80	
	4901	Trailer, low bed, 75 ton capacity		7.90	178	535	1,600	170.20	
	5000	Road planer, walk behind, 10" cutting width, 10 HP		2	26	78	234	31.60	
	5100	Self propelled, 12" cutting width, 64 HP		5.30	195	585	1,750	159.40	
	5200	Pavement profiler, 4' to 6' wide, 450 HP		146.05	2,675	8,040	24,100	2,776	
	5300	8' to 10' wide, 750 HP		233.70	4,000	11,980	35,900	4,266	
	5400	Roadway plate, steel, 1"x8'x20'		.06	10	30	90	6.50	
	5600	Stabilizer, self-propelled, 150 HP		27.40	540	1,625	4,875	544.20	
	5700	310 HP		44.55	1,150	3,445	10,300	1,045	
	5800	Striper, thermal, truck mounted 120 gal. paint, 150 H.P.		32.40	485	1,450	4,350	549.20	
	6000	Tar kettle, 330 gal., trailer mounted		2.37	36.50	110	330	40.95	
	7000	Tunnel locomotive, diesel, 8 to 12 ton		20.95	540	1,615	4,850	490.60	
	7005	Electric, 10 ton		20.65	610	1,830	5,500	531.20	
	7010	Muck cars, 1/2 C.Y. capacity		1.50	20	60	180	24	
	7020	1 C.Y. capacity		1.70	28	84	252	30.40	
	7030	2 C.Y. capacity		1.85	32.50	98	294	34.40	

Reference codes: R01590-100, R02250-400, R02065-300, R01590-100

01590 | Equipment Rental

				UNIT	HOURLY OPER. COST	RENT PER DAY	RENT PER WEEK	RENT PER MONTH	CREW EQUIPMENT COST/DAY	
500	7040	Side dump, 2 C.Y. capacity	R01590 -100	Ea.	2	40	120	360	40	**500**
	7050	3 C.Y. capacity			2.70	45	135	405	48.60	
	7060	5 C.Y. capacity			3.85	58.50	175	525	65.80	
	7100	Ventilating blower for tunnel, 7-1/2 H.P.			1.22	44	132	395	36.15	
	7110	10 H.P.			1.41	45.50	137	410	38.70	
	7120	20 H.P.			2.30	56.50	170	510	52.40	
	7140	40 H.P.			4.02	83.50	250	750	82.15	
	7160	60 H.P.			6.09	127	380	1,150	124.70	
	7175	75 H.P.			7.79	168	505	1,525	163.30	
	7180	200 H.P.			17.45	245	735	2,200	286.60	
	7800	Windrow loader, elevating			33.40	875	2,620	7,850	791.20	
600	0010	**LIFTING AND HOISTING EQUIPMENT RENTAL**	R01590 -100							**600**
	0100	without operators								
	0120	Aerial lift truck, 2 person, to 80'		Ea.	18.60	580	1,740	5,225	496.80	
	0140	Boom work platform, 40' snorkel			9.80	203	610	1,825	200.40	
	0150	Crane, flatbed mntd, 3 ton cap.			12.45	183	550	1,650	209.60	
	0200	Crane, climbing, 106' jib, 6000 lb. capacity, 410 FPM			43.35	1,350	4,070	12,200	1,161	
	0300	101' jib, 10,250 lb. capacity, 270 FPM			48.75	1,725	5,150	15,500	1,420	
	0400	Tower, static, 130' high, 106' jib,								
	0500	6200 lb. capacity at 400 FPM		Ea.	46.50	1,575	4,700	14,100	1,312	
	0600	Crawler mounted, lattice boom, 1/2 C.Y., 15 tons at 12' radius			20.76	570	1,710	5,125	508.10	
	0700	3/4 C.Y., 20 tons at 12' radius			27.68	680	2,045	6,125	630.45	
	0800	1 C.Y., 25 tons at 12' radius			36.90	895	2,690	8,075	833.20	
	0900	1-1/2 C.Y., 40 tons at 12' radius			41.15	930	2,795	8,375	888.20	
	1000	2 C.Y., 50 tons at 12' radius			51.35	1,325	3,975	11,900	1,206	
	1100	3 C.Y., 75 tons at 12' radius			49.10	1,300	3,885	11,700	1,170	
	1200	100 ton capacity, 60' boom			62.80	1,700	5,085	15,300	1,519	
	1300	165 ton capacity, 60' boom			92.20	2,175	6,560	19,700	2,050	
	1400	200 ton capacity, 70' boom			98.10	2,350	7,015	21,000	2,188	
	1500	350 ton capacity, 80' boom			141.90	3,325	9,945	29,800	3,124	
	1600	Truck mounted, lattice boom, 6 x 4, 20 tons at 10' radius			27.76	960	2,875	8,625	797.10	
	1700	25 tons at 10' radius			29.66	1,025	3,070	9,200	851.30	
	1800	8 x 4, 30 tons at 10' radius			36.66	1,075	3,260	9,775	945.30	
	1900	40 tons at 12' radius			32.74	1,150	3,450	10,400	951.90	
	2000	8 x 4, 60 tons at 15' radius			39.81	1,250	3,770	11,300	1,072	
	2050	82 tons at 15' radius			46.50	1,500	4,470	13,400	1,266	
	2100	90 tons at 15' radius			44.08	1,625	4,860	14,600	1,325	
	2200	115 tons at 15' radius			50.55	1,800	5,400	16,200	1,484	
	2300	150 tons at 18' radius			43.10	1,825	5,500	16,500	1,445	
	2350	165 tons at 18' radius			77.25	2,150	6,470	19,400	1,912	
	2400	Truck mounted, hydraulic, 12 ton capacity			32.45	570	1,715	5,150	602.60	
	2500	25 ton capacity			32.60	595	1,780	5,350	616.80	
	2550	33 ton capacity			33.45	625	1,870	5,600	641.60	
	2560	40 ton capacity			31.85	595	1,790	5,375	612.80	
	2600	55 ton capacity			47.05	865	2,595	7,775	895.40	
	2700	80 ton capacity			56	910	2,735	8,200	995	
	2720	100 ton capacity			80.75	2,325	6,945	20,800	2,035	
	2740	120 ton capacity			85.55	2,525	7,550	22,700	2,194	
	2760	150 ton capacity			104.65	3,175	9,535	28,600	2,744	
	2800	Self-propelled, 4 x 4, with telescoping boom, 5 ton			15.50	320	965	2,900	317	
	2900	12-1/2 ton capacity			25.10	510	1,525	4,575	505.80	
	3000	15 ton capacity			25.70	585	1,755	5,275	556.60	
	3050	20 ton capacity			26.80	610	1,835	5,500	581.40	
	3100	25 ton capacity			27.25	595	1,785	5,350	575	
	3150	40 ton capacity			45.20	845	2,540	7,625	869.60	
	3200	Derricks, guy, 20 ton capacity, 60' boom, 75' mast			12.72	330	996	3,000	300.95	
	3300	100' boom, 115' mast			20.60	570	1,710	5,125	506.80	

Important: See the Reference Section for critical supporting data - Reference Nos., Crews, & City Cost Indexes

01590 | Equipment Rental

			UNIT	HOURLY OPER. COST	RENT PER DAY	RENT PER WEEK	RENT PER MONTH	CREW EQUIPMENT COST/DAY	
600	3400	Stiffleg, 20 ton capacity, 70' boom, 37' mast	Ea.	14.71	425	1,280	3,850	373.70	600
	3500	100' boom, 47' mast		23.19	695	2,080	6,250	601.50	
	3550	Helicopter, small, lift to 1250 lbs. maximum, w/pilot		67.85	2,675	8,050	24,200	2,153	
	3600	Hoists, chain type, overhead, manual, 3/4 ton		.10	2	6	18	2	
	3900	10 ton		.60	13	39	117	12.60	
	4000	Hoist and tower, 5000 lb. cap., portable electric, 40' high		4.19	191	574	1,725	148.30	
	4100	For each added 10' section, add		.09	15	45	135	9.70	
	4200	Hoist and single tubular tower, 5000 lb. electric, 100' high		5.66	267	801	2,400	205.50	
	4300	For each added 6'-6" section, add		.15	25	75	225	16.20	
	4400	Hoist and double tubular tower, 5000 lb., 100' high		6.06	294	882	2,650	224.90	
	4500	For each added 6'-6" section, add		.17	28.50	85	255	18.35	
	4550	Hoist and tower, mast type, 6000 lb., 100' high		6.56	305	915	2,750	235.50	
	4570	For each added 10' section, add		.11	18.35	55	165	11.90	
	4600	Hoist and tower, personnel, electric, 2000 lb., 100' @ 125 FPM		13.52	815	2,440	7,325	596.15	
	4700	3000 lb., 100' @ 200 FPM		15.40	915	2,750	8,250	673.20	
	4800	3000 lb., 150' @ 300 FPM		17.10	1,025	3,090	9,275	754.80	
	4900	4000 lb., 100' @ 300 FPM		17.73	1,050	3,150	9,450	771.85	
	5000	6000 lb., 100' @ 275 FPM		19.14	1,100	3,300	9,900	813.10	
	5100	For added heights up to 500', add	L.F.	.01	1.67	5	15	1.10	
	5200	Jacks, hydraulic, 20 ton	Ea.	.05	6.35	19	57	4.20	
	5500	100 ton	"	.30	18.65	56	168	13.60	
	6000	Jacks, hydraulic, climbing with 50' jackrods							
	6010	and control consoles, minimum 3 mo. rental							
	6100	30 ton capacity	Ea.	1.65	110	329	985	79	
	6150	For each added 10' jackrod section, add		.05	3.33	10	30	2.40	
	6300	50 ton capacity		2.65	177	530	1,600	127.20	
	6350	For each added 10' jackrod section, add		.06	4	12	36	2.90	
	6500	125 ton capacity		6.95	465	1,390	4,175	333.60	
	6550	For each added 10' jackrod section, add		.48	31.50	95	285	22.85	
	6600	Cable jack, 10 ton capacity with 200' cable		1.38	91.50	275	825	66.05	
	6650	For each added 50' of cable, add		.15	10	30	90	7.20	
700	0010	**WELLPOINT EQUIPMENT RENTAL** See also division 02240							700
	0020	Based on 2 months rental							
	0100	Combination jetting & wellpoint pump, 60 H.P. diesel	Ea.	9.17	272	817	2,450	236.75	
	0200	High pressure gas jet pump, 200 H.P., 300 psi	"	16.39	233	698	2,100	270.70	
	0300	Discharge pipe, 8" diameter	L.F.	.01	.44	1.32	3.96	.35	
	0350	12" diameter		.01	.65	1.96	5.90	.45	
	0400	Header pipe, flows up to 150 G.P.M., 4" diameter		.01	.40	1.20	3.60	.30	
	0500	400 G.P.M., 6" diameter		.01	.47	1.42	4.26	.35	
	0600	800 G.P.M., 8" diameter		.01	.65	1.96	5.90	.45	
	0700	1500 G.P.M., 10" diameter		.01	.69	2.06	6.20	.50	
	0800	2500 G.P.M., 12" diameter		.02	1.30	3.89	11.65	.95	
	0900	4500 G.P.M., 16" diameter		.02	1.66	4.98	14.95	1.15	
	0950	For quick coupling aluminum and plastic pipe, add		.03	1.72	5.15	15.45	1.25	
	1100	Wellpoint, 25' long, with fittings & riser pipe, 1-1/2" or 2" diameter	Ea.	.05	3.43	10.28	31	2.45	
	1200	Wellpoint pump, diesel powered, 4" diameter, 20 H.P.		4.45	157	471	1,425	129.80	
	1300	6" diameter, 30 H.P.		5.81	195	584	1,750	163.30	
	1400	8" suction, 40 H.P.		7.91	267	801	2,400	223.50	
	1500	10" suction, 75 H.P.		10.86	310	936	2,800	274.10	
	1600	12" suction, 100 H.P.		16.25	500	1,500	4,500	430	
	1700	12" suction, 175 H.P.		21.61	550	1,650	4,950	502.90	
800	0010	**MARINE EQUIPMENT RENTAL**							800
	0200	Barge, 400 Ton, 30' wide x 90' long	Ea.	16.15	240	720	2,150	273.20	
	0240	800 Ton, 45' wide x 90' long		26.65	345	1,030	3,100	419.20	
	2000	Tugboat, diesel, 100 HP		16.95	168	505	1,525	236.60	
	2040	250 HP		31.45	315	945	2,825	440.60	
	2080	380 HP		73.10	925	2,780	8,350	1,141	

R01590 -100 (reference at row 3400)
R02240 -900 (reference at WELLPOINT row)

Division Notes

	CREW	DAILY OUTPUT	LABOR-HOURS	UNIT	2005 BARE COSTS				TOTAL INCL O&P
					MAT.	LABOR	EQUIP.	TOTAL	

Division 2
Site Construction

Estimating Tips

02200 Site Preparation

- If possible visit the site and take an inventory of the type, quantity and size of the trees. Certain trees may have a landscape resale value or firewood value. Stump disposal can be very expensive, particularly if they cannot be buried at the site. Consider using a bulldozer in lieu of hand cutting trees.

- Estimators should visit the site to determine the need for haul road, access, storage of materials, and security considerations. When estimating for access roads on unstable soil, consider using a geotextile stabilization fabric. It can greatly reduce the quantity of crushed stone or gravel. Sites of limited size and access can cause cost overruns due to lost productivity. Theft and damage is another consideration if the location is isolated. A temporary fence or security guards may be required. Investigate the site thoroughly.

02210 Subsurface Investigation

In preparing estimates on structures involving earthwork or foundations, all information concerning soil characteristics should be obtained. Look particularly for hazardous waste, evidence of prior dumping of debris, and previous stream beds.

02220 Selective Demolition

The costs shown for selective demolition do not include rubbish handling or disposal. These items should be estimated separately using Means data or other sources.

- Historic preservation often requires that the contractor remove materials from the existing structure, rehab them and replace them. The estimator must be aware of any related measures and precautions that must be taken when doing selective demolition, and cutting and patching. Requirements may include special handling and storage, as well as security.

- In addition to Section 02220, you can find selective demolition items in each division. Example: Roofing demolition is in division 7.

02300 Earthwork

- Estimating the actual cost of performing earthwork requires careful consideration of the variables involved. This includes items such as type of soil, whether or not water will be encountered, dewatering, whether or not banks need bracing, disposal of excavated earth, length of haul to fill or spoil sites, etc. If the project has large quantities of cut or fill, consider raising or lowering the site to reduce costs while paying close attention to the effect on site drainage and utilities if doing this.

- If the project has large quantities of fill, creating a borrow pit on the site can significantly lower the costs.

- It is very important to consider what time of year the project is scheduled for completion. Bad weather can create large cost overruns from dewatering, site repair and lost productivity from cold weather.

02500 Utility Services
02600 Drainage & Containment

- Never assume that the water, sewer and drainage lines will go in at the early stages of the project. Consider the site access needs before dividing the site in half with open trenches, loose pipe, and machinery obstructions. Always inspect the site to establish that the site drawings are complete. Check off all existing utilities on your drawing as you locate them. If you find any discrepancies, mark up the site plan for further research. Differing site conditions can be very costly if discovered later in the project.

- See also Section 02955 for restoration of pipe where removal/replacement may be undesirable. Use of new types of piping materials can reduce the overall project cost. Owners/design engineers should consider the installing construction as a valuable source of current information on piping products that could lead to significant utility cost savings.

02700 Bases, Ballasts, Pavements/Appurtenances

- When estimating paving, keep in mind the project schedule. If an asphaltic paving project is in a colder climate and runs through to the spring, consider placing the base course in the autumn, then topping it in the spring just prior to completion. This could save considerable costs in spring repair. Keep in mind that prices for asphalt and concrete are generally higher in the cold seasons.

- See also Sections 02960/02965.

02900 Planting

- The timing of planting and guarantee specifications often dictate the costs for establishing tree and shrub growth and a stand of grass or ground cover. Establish the work performance schedule to coincide with the local planting season. Maintenance and growth guarantees can add from 20% to 100% to the total landscaping cost. The cost to replace trees and shrubs can be as high as 5% of the total cost depending on the planting zone, soil conditions and time of year.

02960 & 02965 Flexible Pavement Surfacing Recovery

- Recycling of asphalt pavement is becoming very popular and is an alternative to removal and replacement of asphalt pavement. It can be a good value engineering proposal if removed pavement can be recycled either at the site or another site that is reasonably close to the project site.

Reference Numbers

Reference numbers are shown in bold squares at the beginning of some major classifications. These numbers refer to related items in the Reference Section. The reference information may be an estimating procedure, an alternate pricing method or technical information.

Note: Not all subdivisions listed here necessarily appear in this publication.

2 SITE CONSTRUCTION

02065 | Cement & Concrete

		CREW	DAILY OUTPUT	LABOR-HOURS	UNIT	2005 BARE COSTS MAT.	LABOR	EQUIP.	TOTAL	TOTAL INCL O&P	
300	0010	**PLANT MIXED BITUMINOUS CONCRETE**										300
	0020	Asphaltic concrete plant mix (145 LB per c.f.)				Ton	34.50			34.50	38	
	0040	Asphaltic concrete less than 300 tons add trucking costs										
	0050	See 02315-490-0010 for hauling costs										
	0200	All weather patching mix, hot					33.50			33.50	37	
	0250	Cold patch					40			40	44	
	0300	Berm mix					33.50			33.50	37	
	0400	Base mix					34.50			34.50	38	
	0500	Binder mix					34.50			34.50	38	
	0600	Sand or sheet mix					39.50			39.50	43.50	
400	0010	**RECYCLED PLANT MIXED BITUMINOUS CONCRETE**										400
	0200	Reclaimed pavement in stockpile				Ton	13.65			13.65	15	
	0400	Recycled pavement, at plant, ratio old: new, 70:30					19.65			19.65	21.50	
	0600	Ratio old: new, 30:70					27.50			27.50	30.50	

02080 | Utility Materials

		CREW	DAILY OUTPUT	LABOR-HOURS	UNIT	MAT.	LABOR	EQUIP.	TOTAL	INCL O&P	
100	0010	**FIRE HYDRANTS**										100
	0020	Mechanical joints unless otherwise noted										
	1000	Fire hydrants, two way; excavation and backfill not incl.										
	1100	4-1/2" valve size, depth 2'-0"	B-21	10	2.800	Ea.	1,100	86.50	15.85	1,202.35	1,350	
	1120	2'-6"		10	2.800		1,100	86.50	15.85	1,202.35	1,350	
	1140	3'-0"		10	2.800		1,100	86.50	15.85	1,202.35	1,350	
	1160	3'-6"		9	3.111		1,225	96	17.60	1,338.60	1,525	
	1200	4'-6"		9	3.111		1,250	96	17.60	1,363.60	1,525	
	1220	5'-0"		8	3.500		1,250	108	19.80	1,377.80	1,575	
	1240	5'-6"		8	3.500		1,250	108	19.80	1,377.80	1,575	
	1260	6'-0"		7	4		1,200	124	22.50	1,346.50	1,550	
	1280	6'-6"		7	4		1,250	124	22.50	1,396.50	1,600	
	1300	7'-0"		6	4.667		1,275	144	26.50	1,445.50	1,675	
	1340	8'-0"		6	4.667		1,375	144	26.50	1,545.50	1,750	
	1420	10'-0"		5	5.600		1,450	173	31.50	1,654.50	1,900	
	2000	5-1/4" valve size, depth 2'-0"		10	2.800		865	86.50	15.85	967.35	1,100	
	2080	4'-0"		9	3.111		975	96	17.60	1,088.60	1,250	
	2160	6'-0"		7	4		1,100	124	22.50	1,246.50	1,425	
	2240	8'-0"		6	4.667		1,200	144	26.50	1,370.50	1,575	
	2320	10'-0"		5	5.600		1,325	173	31.50	1,529.50	1,750	
	2350	For threeway valves, add					7%					
	2400	Lower barrel extensions with stems, 1'-0"	B-20	14	1.714		345	51.50		396.50	460	
	2440	2'-0"		13	1.846		420	55.50		475.50	545	
	2480	3'-0"		12	2		495	60		555	640	
	2520	4'-0"		10	2.400		570	72		642	735	
	5000	Indicator post										
	5020	Adjustable, valve size 4" to 14", 4' bury	B-21	10	2.800	Ea.	380	86.50	15.85	482.35	565	
	5060	8' bury		7	4		485	124	22.50	631.50	745	
	5080	10' bury		6	4.667		495	144	26.50	665.50	795	
	5100	12' bury		5	5.600		745	173	31.50	949.50	1,125	
	5120	14' bury		4	7		815	216	39.50	1,070.50	1,275	
	5500	Non-adjustable, valve size 4" to 14", 3' bury		10	2.800		380	86.50	15.85	482.35	565	
	5520	3'-6" bury		10	2.800		380	86.50	15.85	482.35	565	
	5540	4' bury		9	3.111		380	96	17.60	493.60	585	
400	0010	**UTILITY BOXES** Precast concrete, 6" thick										400
	0040	4' x 6' x 6' high, I.D.	B-13	2	28	Ea.	1,225	805	310	2,340	2,950	
	0050	5' x 10' x 6' high, I.D.		2	28		1,525	805	310	2,640	3,275	
	0100	6' x 10' x 6' high, I.D.		2	28		1,575	805	310	2,690	3,325	
	0150	5' x 12' x 6' high, I.D.		2	28		1,675	805	310	2,790	3,450	
	0200	6' x 12' x 6' high, I.D.		1.80	31.111		1,875	895	345	3,115	3,825	

02080	Utility Materials	CREW	DAILY OUTPUT	LABOR-HOURS	UNIT	2005 BARE COSTS				TOTAL INCL O&P
						MAT.	LABOR	EQUIP.	TOTAL	
400 0250	6' x 13' x 6' high, I.D.	B-13	1.50	37.333	Ea.	2,475	1,075	410	3,960	4,825 **400**
0300	8' x 14' x 7' high, I.D.	↓	1	56	↓	2,675	1,600	615	4,890	6,075
0350	Hand hole, precast concrete, 1-1/2" thick									
0400	1'-0" x 2'-0" x 1'-9", I.D., light duty	B-1	4	6	Ea.	260	164		424	540
0450	4'-6" x 3'-2" x 2'-0", O.D., heavy duty	B-6	3	8		800	231	72	1,103	1,325
0460	Meter pit, 4' x 4', 4' deep		2	12		680	345	108	1,133	1,400
0470	6' deep		1.60	15		970	430	135	1,535	1,900
0480	8' deep		1.40	17.143		1,275	495	154	1,924	2,325
0490	10' deep		1.20	20		1,625	575	180	2,380	2,875
0500	15' deep		1	24		2,375	690	216	3,281	3,925
0510	6' x 6', 4' deep		1.40	17.143		1,000	495	154	1,649	2,025
0520	6' deep		1.20	20		1,500	575	180	2,255	2,750
0530	8' deep		1	24		2,025	690	216	2,931	3,525
0540	10' deep		.80	30		2,525	865	270	3,660	4,400
0550	15' deep	↓	.60	40	↓	3,825	1,150	360	5,335	6,400
500 0010	**VALVES** Water distribution, see also div. 15110									**500**
3000	Butterfly valves with boxes, cast iron, mech. jt. R15100 -050									
3100	4" diameter	B-20	6	4	Ea.	214	120		334	420
3180	8" diameter	B-21	4	7		298	216	39.50	553.50	705
3340	12" diameter		3	9.333		570	288	53	911	1,125
3400	14" diameter		2	14		980	435	79	1,494	1,825
3460	18" diameter		1.50	18.667		1,650	575	106	2,331	2,800
3480	20" diameter		1	28		2,075	865	158	3,098	3,800
3500	24" diameter	↓	.50	56	↓	2,875	1,725	315	4,915	6,200
3600	With lever operator									
3610	4" diameter	B-20	6	4	Ea.	116	120		236	315
3616	8" diameter	B-21	4	7		235	216	39.50	490.50	640
3620	12" diameter		3	9.333		700	288	53	1,041	1,275
3624	16" diameter		2	14		1,525	435	79	2,039	2,425
3630	24" diameter	↓	.50	56	↓	2,500	1,725	315	4,540	5,775
3700	Check valves, flanged									
3710	4" diameter	B-20	6	4	Ea.	425	120		545	650
3714	6" diameter	"	5	4.800		720	144		864	1,025
3716	8" diameter	B-21	4	7		1,475	216	39.50	1,730.50	1,975
3720	12" diameter		3	9.333		3,575	288	53	3,916	4,450
3724	16" diameter		2	14		7,900	435	79	8,414	9,425
3726	18" diameter		1.50	18.667		11,700	575	106	12,381	13,800
3730	24" diameter	↓	.50	56	↓	19,500	1,725	315	21,540	24,500
3800	Gate valves, C.I., 250 PSI, mechanical joint, w/boxes									
3810	4" diameter	B-21A	8	5	Ea.	166	165	63.50	394.50	505
3814	6" diameter		6.80	5.882		214	194	74.50	482.50	615
3816	8" diameter		5.60	7.143		340	236	90.50	666.50	830
3820	12" diameter	↓	4.08	9.804		730	325	124	1,179	1,425
3824	16" diameter	B-21	1	28		900	865	158	1,923	2,500
3828	20" diameter		.80	35		1,400	1,075	198	2,673	3,450
3830	24" diameter		.50	56		1,725	1,725	315	3,765	4,925
3831	30" diameter		.35	80		2,075	2,475	455	5,005	6,600
3832	36" diameter	↓	.30	93.333		2,475	2,875	530	5,880	7,775
3880	Sleeve, for tapping mains, 8" x 4", add					505			505	555
3884	10" x 6", add					1,025			1,025	1,125
3888	12" x 6", add					1,025			1,025	1,125
3892	12" x 8", add				↓	1,100			1,100	1,200
3900	Globe valves, flanged, iron body, class 125									
3910	4" diameter	B-20	10	2.400	Ea.	1,125	72		1,197	1,350
3914	6" diameter	"	9	2.667	↓	2,175	80		2,255	2,525

02050 | Basic Site Materials & Methods

		CREW	DAILY OUTPUT	LABOR-HOURS	UNIT	2005 BARE COSTS				TOTAL INCL O&P
						MAT.	LABOR	EQUIP.	TOTAL	
500	3916	8" diameter R15100-050	B-21	6	4.667	Ea.	4,050	144	26.50	4,220.50	4,700	**500**
	3920	12" diameter		4	7		8,900	216	39.50	9,155.50	10,200
	3924	16" diameter		2	14		8,025	435	79	8,539	9,575
	3928	20" diameter		.60	46.667		8,700	1,450	264	10,414	12,100
	3930	24" diameter		.50	56		10,300	1,725	315	12,340	14,400
600	0010	**UTILITY ACCESSORIES**										**600**
	0400	Underground tape, detectable, reinforced, alum. foil core, 2"	1 Clab	150	.053	C.L.F.	1.43	1.42		2.85	3.79
	0500	6"	"	140	.057	"	3.58	1.53		5.11	6.30

02100 | Site Remediation

		CREW	DAILY OUTPUT	LABOR-HOURS	UNIT	2005 BARE COSTS				TOTAL INCL O&P
						MAT.	LABOR	EQUIP.	TOTAL	
300	0010	**HAZARDOUS WASTE CLEANUP/PICKUP/DISPOSAL**										**300**
	0100	For contractor equipment, i.e. dozer,									
	0110	Front end loader, dump truck, etc., see div. 01590-200									
	1000	Solid pickup									
	1100	55 gal. drums				Ea.					220
	1120	Bulk material, minimum				Ton					165
	1130	Maximum				"					550
	1200	Transportation to disposal site									
	1220	Truckload = 80 drums or 25 C.Y. or 18 tons									
	1260	Minimum				Mile					2.50
	1270	Maximum				"					4.40
	3000	Liquid pickup, vacuum truck, stainless steel tank									
	3100	Minimum charge, 4 hours									
	3110	1 compartment, 2200 gallon				Hr.					110
	3120	2 compartment, 5000 gallon				"					110
	3400	Transportation in 6900 gallon bulk truck				Mile					4.75
	3410	In teflon lined truck				"					5.50
	5000	Heavy sludge or dry vacuumable material				Hr.					110
	6000	Dumpsite disposal charge, minimum				Ton					110
	6020	Maximum				"					440

		CREW	DAILY OUTPUT	LABOR-HOURS	UNIT	MAT.	LABOR	EQUIP.	TOTAL	TOTAL INCL O&P
200	0010	**REMOVAL OF UNDERGROUND STORAGE TANKS** R02115-200										**200**
	0011	Petroleum storage tanks, non-leaking									
	0100	Excavate & load onto trailer									
	0110	3000 gal. to 5000 gal. tank	B-14	4	12	Ea.		335	54	389	580
	0120	6000 gal to 8000 gal tank	B-3A	3	13.333			375	240	615	845
	0130	9000 gal to 12000 gal tank	"	2	20			565	360	925	1,275
	0190	Known leaking tank add				%				100%	100%
	0200	Remove sludge, water and remaining product from tank bottom									
	0201	of tank with vacuum truck									
	0300	3000 gal to 5000 gal tank	A-13	5	1.600	Ea.		53	96	149	186
	0310	6000 gal to 8000 gal tank		4	2			66	120	186	232
	0320	9000 gal to 12000 gal tank		3	2.667			88	160	248	310
	0390	Dispose of sludge off-site, average				Gal.					4.40
	0400	Insert inert solid CO2 "dry ice" into tank									

Important: See the Reference Section for critical supporting data - Reference Nos., Crews, & City Cost Indexes

02115	Underground Storage Tank Removal		CREW	DAILY OUTPUT	LABOR-HOURS	UNIT	2005 BARE COSTS				TOTAL INCL O&P	
							MAT.	LABOR	EQUIP.	TOTAL		
200	0401	For cleaning/transporting tanks (1.5 lbs./100 gal. cap)	R02115 -200	1 Clab	500	.016	Lb.	1.34	.43		1.77	2.13	200
	1020	Haul tank to certified salvage dump, 100 miles round trip											
	1023	3000 gal. to 5000 gal. tank					Ea.				550	690	
	1026	6000 gal. to 8000 gal. tank									650	825	
	1029	9,000 gal. to 12,000 gal. tank					↓				875	1,100	
	1100	Disposal of contaminated soil to landfill											
	1110	Minimum					C.Y.					110	
	1111	Maximum					"					330	
	1120	Disposal of contaminated soil to											
	1121	bituminous concrete batch plant											
	1130	Minimum					C.Y.					55	
	1131	Maximum					"					110	
	2010	Decontamination of soil on site incl poly tarp on top/bottom											
	2011	Soil containment berm, and chemical treatment											
	2020	Minimum		B-11C	100	.160	C.Y.	5.60	4.91	2.16	12.67	16	
	2021	Maximum		"	100	.160		7.25	4.91	2.16	14.32	17.85	
	2050	Disposal of decontaminated soil, minimum										66	
	2055	Maximum		↓			↓					135	

02210	Subsurface Investigation	CREW	DAILY OUTPUT	LABOR-HOURS	UNIT	2005 BARE COSTS				TOTAL INCL O&P	
						MAT.	LABOR	EQUIP.	TOTAL		
120	0010 **BORING AND EXPLORATORY DRILLING**										120
	0020 Borings, initial field stake out & determination of elevations	A-6	1	16	Day		545	60.50	605.50	900	
	0100 Drawings showing boring details				Total		185		185	270	
	0200 Report and recommendations from P.E.						415		415	595	
	0300 Mobilization and demobilization, minimum	B-55	4	6	↓		160	195	355	465	
	0350 For over 100 miles, per added mile		450	.053	Mile		1.43	1.73	3.16	4.11	
	0600 Auger holes in earth, no samples, 2-1/2" diameter		78.60	.305	L.F.		8.15	9.95	18.10	23.50	
	0650 4" diameter		67.50	.356			9.50	11.55	21.05	27.50	
	0800 Cased borings in earth, with samples, 2-1/2" diameter		55.50	.432		14.20	11.55	14.05	39.80	49	
	0850 4" diameter	↓	32.60	.736		22.50	19.70	24	66.20	81.50	
	1000 Drilling in rock, "BX" core, no sampling	B-56	34.90	.458			13.70	28	41.70	52	
	1050 With casing & sampling		31.70	.505		14.20	15.10	31	60.30	72.50	
	1200 "NX" core, no sampling		25.92	.617			18.45	38	56.45	69.50	
	1250 With casing and sampling	↓	25	.640	↓	17.35	19.10	39	75.45	91	
	1400 Drill rig and crew with truck mounted auger	B-55	1	24	Day		640	780	1,420	1,850	
	1450 With crawler type drill	B-56	1	16	"		480	980	1,460	1,800	
	1500 For inner city borings add, minimum									10%	
	1510 Maximum									20%	
200	0010 **CORE DRILLING** Reinforced concrete slab, up to 6" thick slab										200
	0020 Including bit, layout and set up										
	0100 1" diameter core	B-89A	28	.571	Ea.	2.40	17.60	3.68	23.68	34	
	0150 Each added inch thick, add		300	.053		.43	1.64	.34	2.41	3.40	
	0300 3" diameter core		23	.696		5.35	21.50	4.48	31.33	44.50	
	0350 Each added inch thick, add		186	.086		.96	2.65	.55	4.16	5.80	
	0500 4" diameter core		19	.842		5.35	26	5.40	36.75	52.50	
	0550 Each added inch thick, add		170	.094		1.21	2.90	.61	4.72	6.50	
	0700 6" diameter core		14	1.143		8.80	35	7.35	51.15	72.50	
	0750 Each added inch thick, add	↓	140	.114	↓	1.49	3.52	.74	5.75	7.90	

SITE CONSTRUCTION 2

33

2 SITE CONSTRUCTION

	02210	Subsurface Investigation	CREW	DAILY OUTPUT	LABOR-HOURS	UNIT	2005 BARE COSTS MAT.	LABOR	EQUIP.	TOTAL	TOTAL INCL O&P	
200	0900	8" diameter core	B-89A	11	1.455	Ea.	12	45	9.35	66.35	93	200
	0950	Each added inch thick, add		95	.168		2.02	5.20	1.08	8.30	11.45	
	1100	10" diameter core		10	1.600		16	49.50	10.30	75.80	105	
	1150	Each added inch thick, add		80	.200		2.65	6.15	1.29	10.09	13.95	
	1300	12" diameter core		9	1.778		19.20	54.50	11.45	85.15	119	
	1350	Each added inch thick, add		68	.235		3.18	7.25	1.52	11.95	16.40	
	1500	14" diameter core		7	2.286		23.50	70.50	14.70	108.70	151	
	1550	Each added inch thick, add		55	.291		4.04	8.95	1.87	14.86	20.50	
	1700	18" diameter core		4	4		30	123	26	179	254	
	1750	Each added inch thick, add		28	.571		5.30	17.60	3.68	26.58	37.50	
	1760	For horizontal holes, add to above								30%	30%	
	1770	Prestressed hollow core plank, 6" thick										
	1780	1" diameter core	B-89A	52	.308	Ea.	1.59	9.45	1.98	13.02	18.70	
	1790	Each added inch thick, add		350	.046		.28	1.41	.29	1.98	2.82	
	1800	3" diameter core		50	.320		3.51	9.85	2.06	15.42	21.50	
	1810	Each added inch thick, add		240	.067		.58	2.05	.43	3.06	4.30	
	1820	4" diameter core		48	.333		4.67	10.25	2.15	17.07	23.50	
	1830	Each added inch thick, add		216	.074		.81	2.28	.48	3.57	4.96	
	1840	6" diameter core		44	.364		5.80	11.20	2.34	19.34	26.50	
	1850	Each added inch thick, add		175	.091		.96	2.81	.59	4.36	6.10	
	1860	8" diameter core		32	.500		7.75	15.40	3.22	26.37	36	
	1870	Each added inch thick, add		118	.136		1.34	4.17	.87	6.38	8.95	
	1880	10" diameter core		28	.571		10.45	17.60	3.68	31.73	43	
	1890	Each added inch thick, add		99	.162		1.44	4.97	1.04	7.45	10.45	
	1900	12" diameter core		22	.727		12.75	22.50	4.68	39.93	54	
	1910	Each added inch thick, add		85	.188		2.12	5.80	1.21	9.13	12.65	
	1950	Minimum charge for above, 3" diameter core		7	2.286	Total		70.50	14.70	85.20	125	
	2000	4" diameter core		6.80	2.353			72.50	15.15	87.65	130	
	2050	6" diameter core		6	2.667			82	17.15	99.15	147	
	2100	8" diameter core		5.50	2.909			89.50	18.75	108.25	160	
	2150	10" diameter core		4.75	3.368			104	21.50	125.50	185	
	2200	12" diameter core		3.90	4.103			126	26.50	152.50	225	
	2250	14" diameter core		3.38	4.734			146	30.50	176.50	261	
	2300	18" diameter core		3.15	5.079			156	32.50	188.50	279	
	3010	Bits for core drill, diamond, premium, 1" diameter				Ea.	107			107	118	
	3020	3" diameter					264			264	291	
	3040	4" diameter					294			294	325	
	3050	6" diameter					470			470	520	
	3080	8" diameter					645			645	710	
	3120	12" diameter					1,025			1,025	1,125	
	3180	18" diameter					2,100			2,100	2,300	
	3240	24" diameter					2,800			2,800	3,075	
700	0010	**TEST PITS**										700
	0020	Hand digging, light soil	1 Clab	4.50	1.778	C.Y.		47.50		47.50	74	
	0100	Heavy soil	"	2.50	3.200			85.50		85.50	133	
	0120	Loader-backhoe, light soil	B-11M	28	.571			17.55	9.15	26.70	37	
	0130	Heavy soil	"	20	.800			24.50	12.80	37.30	51.50	
	1000	Subsurface exploration, mobilization				Mile				5.50	6.30	
	1010	Difficult access for rig, add				Hr.				110	126	
	1020	Auger borings, drill rig, incl. samples				L.F.				12.65	14.60	
	1030	Hand auger								18.70	21.50	
	1050	Drill and sample every 5', split spoon								16.50	19	
	1060	Extra samples				Ea.				22	25.50	

Important: See the Reference Section for critical supporting data - Reference Nos., Crews, & City Cost Indexes

02220	Site Demolition		CREW	DAILY OUTPUT	LABOR-HOURS	UNIT	MAT.	2005 BARE COSTS LABOR	EQUIP.	TOTAL	TOTAL INCL O&P	
110	0010	**BUILDING DEMOLITION** Large urban projects, incl. 20 mi. haul	R02220 -510										110
	0011	No foundation or dump fees, C.F. is vol. of building standing										
	0012	Steel		B-8	21,500	.003	C.F.		.09	.11	.20	.25
	0050	Concrete			15,300	.004			.12	.15	.27	.36
	0080	Masonry			20,100	.003			.09	.12	.21	.27
	0100	Mixture of types, average			20,100	.003			.09	.12	.21	.27
	0500	Small bldgs, or single bldgs, no salvage included, steel		B-3	14,800	.003			.09	.12	.21	.27
	0600	Concrete			11,300	.004			.12	.15	.27	.36
	0650	Masonry			14,800	.003			.09	.12	.21	.27
	0700	Wood			14,800	.003			.09	.12	.21	.27
	1000	Single family, one story house, wood, minimum					Ea.				2,525	2,975
	1020	Maximum									4,400	5,275
	1200	Two family, two story house, wood, minimum									3,300	3,950
	1220	Maximum									6,375	7,700
	1300	Three family, three story house, wood, minimum									4,400	5,275
	1320	Maximum									7,700	9,250
	5000	For buildings with no interior walls, deduct									50%	
120	0010	**EXPLOSIVE/IMPLOSIVE DEMOLITION** Large projects,	R02220 -510										120
	0020	no disposal fee based on building volume, steel building		B-5B	16,900	.003	C.F.		.09	.11	.20	.25
	0100	Concrete building			16,900	.003			.09	.11	.20	.25
	0200	Masonry building			16,900	.003			.09	.11	.20	.25
	0400	Disposal of material, minimum		B-3	445	.108	C.Y.		3.09	3.91	7	9.05
	0500	Maximum		"	365	.132	"		3.77	4.77	8.54	11.05
130	0010	**BLDG. FOOTINGS AND FOUNDATIONS DEMOLITION**	R02220 -510										130
	0200	Floors, concrete slab on grade,										
	0240	4" thick, plain concrete		B-9C	500	.080	S.F.		2.17	.28	2.45	3.68
	0280	Reinforced, wire mesh			470	.085			2.31	.30	2.61	3.92
	0300	Rods			400	.100			2.71	.35	3.06	4.61
	0400	6" thick, plain concrete			375	.107			2.89	.38	3.27	4.91
	0420	Reinforced, wire mesh			340	.118			3.19	.42	3.61	5.40
	0440	Rods			300	.133			3.61	.47	4.08	6.10
	1000	Footings, concrete, 1' thick, 2' wide		B-5	300	.187	L.F.		5.45	3.10	8.55	11.80
	1080	1'-6" thick, 2' wide			250	.224			6.55	3.72	10.27	14.20
	1120	3' wide			200	.280			8.20	4.65	12.85	17.70
	1140	2' thick, 3' wide			175	.320			9.35	5.30	14.65	20.50
	1200	Average reinforcing, add									10%	10%
	1220	Heavy reinforcing, add									20%	20%
	2000	Walls, block, 4" thick		1 Clab	180	.044	S.F.		1.19		1.19	1.85
	2040	6" thick			170	.047			1.26		1.26	1.96
	2080	8" thick			150	.053			1.42		1.42	2.22
	2100	12" thick			150	.053			1.42		1.42	2.22
	2200	For horizontal reinforcing, add									10%	10%
	2220	For vertical reinforcing, add									20%	20%
	2400	Concrete, plain concrete, 6" thick		B-9	160	.250			6.80	.89	7.69	11.50
	2420	8" thick			140	.286			7.75	1.01	8.76	13.15
	2440	10" thick			120	.333			9.05	1.18	10.23	15.35
	2500	12" thick			100	.400			10.85	1.42	12.27	18.40
	2600	For average reinforcing, add									10%	10%
	2620	For heavy reinforcing, add									20%	20%
	4000	For congested sites or small quantities, add up to									200%	200%
	4200	Add for disposal, on site		B-11A	232	.069	C.Y.		2.12	3.96	6.08	7.60
	4250	To five miles		B-30	220	.109	"		3.26	7.60	10.86	13.30
220	0010	**FENCING DEMOLITION**	R02220 -510										220
	1600	Fencing, barbed wire, 3 strand		2 Clab	430	.037	L.F.		.99		.99	1.55

02220	Site Demolition		CREW	DAILY OUTPUT	LABOR-HOURS	UNIT	2005 BARE COSTS MAT.	LABOR	EQUIP.	TOTAL	TOTAL INCL O&P	
220	1650	5 strand		2 Clab	280	.057	L.F.		1.53		1.53	2.37	220
	1700	Chain link, posts & fabric, remove only, 8' to 10' high	R02220 -510	B-6	445	.054			1.55	.48	2.03	2.92	
	1750	Remove and reset		"	70	.343			9.90	3.08	12.98	18.60	
	1775	Fencing, wood, all types, 4'to 6' high		2 Clab	432	.037			.99		.99	1.54	
	1790	Remove and store		B-80	235	.136			3.92	2.27	6.19	8.50	
230	0010	**HYDRODEMOLITION**	R02220 -510										230
	0015	Hydrodemolition, concrete pavement, 4000 PSI, 2" depth		B-5	500	.112	S.F.		3.28	1.86	5.14	7.10	
	0120	4" depth			450	.124			3.64	2.07	5.71	7.85	
	0130	6" depth			400	.140			4.10	2.32	6.42	8.85	
	0410	6000 PSI, 2" depth			410	.137			4	2.27	6.27	8.65	
	0420	4" depth			350	.160			4.68	2.66	7.34	10.10	
	0430	6" depth			300	.187			5.45	3.10	8.55	11.80	
	0510	8000 PSI, 2" depth			330	.170			4.97	2.82	7.79	10.75	
	0520	4" depth			280	.200			5.85	3.32	9.17	12.65	
	0530	6" depth			240	.233			6.85	3.87	10.72	14.75	
240	0010	**MINOR SITE DEMOLITION**	R02220 -510										240
	0015	No hauling, abandon catch basin or manhole		B-6	7	3.429	Ea.		99	31	130	186	
	0020	Remove existing catch basin or manhole, masonry			4	6			173	54	227	325	
	0030	Catch basin or manhole frames and covers, stored			13	1.846			53	16.60	69.60	100	
	0040	Remove and reset			7	3.429			99	31	130	186	
	0100	Roadside delineators, remove only		B-80	175	.183			5.25	3.05	8.30	11.45	
	0110	Remove and reset		"	100	.320			9.20	5.35	14.55	20	
	0800	Guiderail, corrugated steel, remove only		B-80A	100	.240	L.F.		6.40	1.76	8.16	11.90	
	0850	Remove and reset		"	40	.600	"		16	4.40	20.40	30	
	0860	Guide posts, remove only		B-80B	120	.267	Ea.		7.55	1.75	9.30	13.55	
	0870	Remove and reset		B-55	50	.480			12.85	15.60	28.45	37	
	0900	Hydrants, fire, remove only		B-21A	5	8			264	101	365	510	
	0950	Remove and reset		"	2	20			660	253	913	1,275	
	1000	Masonry walls, block or tile, solid, remove		B-5	1,800	.031	C.F.		.91	.52	1.43	1.97	
	1100	Cavity wall			2,200	.025			.74	.42	1.16	1.61	
	1200	Brick, solid			900	.062			1.82	1.03	2.85	3.94	
	1300	With block back-up			1,130	.050			1.45	.82	2.27	3.14	
	1400	Stone, with mortar			900	.062			1.82	1.03	2.85	3.94	
	1500	Dry set			1,500	.037			1.09	.62	1.71	2.36	
	1600	Median barrier, precast concrete, remove and store		B-3	430	.112	L.F.		3.20	4.05	7.25	9.35	
	1610	Remove and reset		"	390	.123			3.53	4.47	8	10.30	
	2900	Pipe removal, sewer/water, no excavation, 12" diameter		B-6	175	.137			3.95	1.23	5.18	7.45	
	2930	15"-18" diameter			150	.160			4.61	1.44	6.05	8.70	
	2960	21"-24" diameter			120	.200			5.75	1.80	7.55	10.85	
	3000	27"-36" diameter			90	.267			7.70	2.40	10.10	14.45	
	3200	Steel, welded connections, 4" diameter			160	.150			4.32	1.35	5.67	8.15	
	3300	10" diameter			80	.300			8.65	2.70	11.35	16.25	
	3500	Railroad track removal, ties and track		B-13	330	.170			4.89	1.87	6.76	9.55	
	3600	Ballast		B-14	500	.096	C.Y.		2.70	.43	3.13	4.65	
	3700	Remove and re-install, ties & track using new bolts & spikes			50	.960	L.F.		27	4.32	31.32	46.50	
	3800	Turnouts using new bolts and spikes			1	48	Ea.		1,350	216	1,566	2,325	
	3850	Runways, remove rubber skid marks, 4-6 passes		B-59A	35	.686	M.S.F.	27	18.50	9.30	54.80	68.50	
	3860	6-10 passes		"	35	.686	"	40.50	18.50	9.30	68.30	83	
	4000	Sidewalk removal, bituminous, 2-1/2" thick		B-6	325	.074	S.Y.		2.13	.66	2.79	4	
	4050	Brick, set in mortar			185	.130			3.74	1.17	4.91	7.05	
	4100	Concrete, plain, 4"			160	.150			4.32	1.35	5.67	8.15	
	4200	Mesh reinforced			150	.160			4.61	1.44	6.05	8.70	
	5000	Slab on grade removal, plain		B-5	45	1.244	C.Y.		36.50	20.50	57	78.50	
	5100	Mesh reinforced			33	1.697			49.50	28	77.50	108	
	5200	Rod reinforced			25	2.240			65.50	37	102.50	142	

Important: See the Reference Section for critical supporting data - Reference Nos., Crews, & City Cost Indexes

		02220	Site Demolition		CREW	DAILY OUTPUT	LABOR-HOURS	UNIT	2005 BARE COSTS				TOTAL INCL O&P	
									MAT.	LABOR	EQUIP.	TOTAL		
240	5500	For congested sites or small quantities, add up to		R02220 -510				C.Y.				200%	200%	240
	5550	For disposal on site, add			B-11A	232	.069			2.12	3.96	6.08	7.60	
	5600	To 5 miles, add			B-34D	76	.105			2.90	5.50	8.40	10.45	
250	0010	**DEMOLISH, REMOVE PAVEMENT AND CURB**		R02220 -510										250
	5010	Pavement removal, bituminous roads, 3" thick			B-38	690	.058	S.Y.		1.74	1.20	2.94	3.98	
	5050	4" to 6" thick				420	.095			2.85	1.96	4.81	6.55	
	5100	Bituminous driveways				640	.063			1.87	1.29	3.16	4.29	
	5200	Concrete to 6" thick, hydraulic hammer, mesh reinforced				255	.157			4.70	3.24	7.94	10.75	
	5300	Rod reinforced				200	.200			6	4.13	10.13	13.75	
	5400	Concrete, 7" to 24" thick, plain				33	1.212	C.Y.		36.50	25	61.50	83	
	5500	Reinforced				24	1.667	"		50	34.50	84.50	115	
	5600	With hand held air equipment, bituminous, to 6" thick			B-39	1,900	.025	S.F.		.71	.07	.78	1.18	
	5700	Concrete to 6" thick, no reinforcing				1,600	.030			.84	.09	.93	1.40	
	5800	Mesh reinforced				1,400	.034			.96	.10	1.06	1.60	
	5900	Rod reinforced				765	.063			1.76	.19	1.95	2.93	
	6000	Curbs, concrete, plain			B-6	360	.067	L.F.		1.92	.60	2.52	3.61	
	6100	Reinforced				275	.087			2.52	.78	3.30	4.73	
	6200	Granite				360	.067			1.92	.60	2.52	3.61	
	6300	Bituminous				528	.045			1.31	.41	1.72	2.46	
310	0010	**SELECTIVE DEMOLITION, CUTOUT**		R02220 -510										310
	0020	Concrete, elev. slab, light reinforcement, under 6 CF			B-9C	65	.615	C.F.		16.70	2.18	18.88	28.50	
	0050	Light reinforcing, over 6 C.F.			"	75	.533	"		14.45	1.89	16.34	24.50	
	0200	Slab on grade to 6" thick, not reinforced, under 8 S.F.			B-9	85	.471	S.F.		12.75	1.67	14.42	21.50	
	0250	8 - 16 S.F.			"	175	.229	"		6.20	.81	7.01	10.55	
	0255	For over 16 SF see 02220-130-0400												
	0600	Walls, not reinforced, under 6 C.F.			B-9	60	.667	C.F.		18.05	2.36	20.41	30.50	
	0650	6 - 12 C.F.			"	80	.500	"		13.55	1.77	15.32	23	
	0655	For over 12 CF see 02220-130-2500												
	1000	Concrete, elevated slab, bar reinforced, under 6 C.F.			B-9C	45	.889	C.F.		24	3.15	27.15	41	
	1050	Bar reinforced, over 6 C.F.			"	50	.800	"		21.50	2.83	24.33	36.50	
	1200	Slab on grade to 6" thick, bar reinforced, under 8 S.F.			B-9	75	.533	S.F.		14.45	1.89	16.34	24.50	
	1250	8 - 16 S.F.			"	150	.267	"		7.25	.94	8.19	12.30	
	1255	For over 16 SF see 02220-130-0440												
	1400	Walls, bar reinforced, under 6 C.F.			B-9C	50	.800	C.F.		21.50	2.83	24.33	36.50	
	1450	6 - 12 CF			"	70	.571	"		15.50	2.02	17.52	26	
	1455	For over 12 CF see 02220-130-2500 & 2600												
	2000	Brick, to 4 S.F. opening, not including toothing												
	2040	4" thick			B-9C	30	1.333	Ea.		36	4.72	40.72	61	
	2060	8" thick				18	2.222			60	7.85	67.85	102	
	2080	12" thick				10	4			108	14.15	122.15	185	
	2400	Concrete block, to 4 S.F. opening, 2" thick				35	1.143			31	4.05	35.05	52.50	
	2420	4" thick				30	1.333			36	4.72	40.72	61	
	2440	8" thick				27	1.481			40	5.25	45.25	68.50	
	2460	12" thick				24	1.667			45	5.90	50.90	77	
	2600	Gypsum block, to 4 S.F. opening, 2" thick			B-9	80	.500			13.55	1.77	15.32	23	
	2620	4" thick				70	.571			15.50	2.02	17.52	26	
	2640	8" thick				55	.727			19.70	2.57	22.27	33.50	
	2800	Terra cotta, to 4 S.F. opening, 4" thick				70	.571			15.50	2.02	17.52	26	
	2840	8" thick				65	.615			16.70	2.18	18.88	28.50	
	2880	12" thick				50	.800			21.50	2.83	24.33	36.50	
	3000	Toothing masonry cutouts, brick, soft old mortar			1 Brhe	40	.200	V.L.F.		5.40		5.40	8.20	
	3100	Hard mortar				30	.267			7.15		7.15	10.95	
	3200	Block, soft old mortar				70	.114			3.07		3.07	4.69	
	3400	Hard mortar				50	.160			4.30		4.30	6.55	
	6000	Walls, interior, not including re-framing,												

02220 | Site Demolition

		CREW	DAILY OUTPUT	LABOR-HOURS	UNIT	2005 BARE COSTS				TOTAL INCL O&P	
						MAT.	LABOR	EQUIP.	TOTAL		
310	6010	openings to 5 S.F.									**310**
	6100	Drywall to 5/8″ thick R02220 -510	1 Clab	24	.333	Ea.		8.90		8.90	13.85
	6200	Paneling to 3/4″ thick		20	.400			10.70		10.70	16.60
	6300	Plaster, on gypsum lath		20	.400			10.70		10.70	16.60
	6340	On wire lath	▼	14	.571	▼		15.25		15.25	23.50
	7000	Wood frame, not including re-framing, openings to 5 S.F.									
	7200	Floors, sheathing and flooring to 2″ thick	1 Clab	5	1.600	Ea.		42.50		42.50	66.50
	7310	Roofs, sheathing to 1″ thick, not including roofing		6	1.333			35.50		35.50	55.50
	7410	Walls, sheathing to 1″ thick, not including siding	▼	7	1.143	▼		30.50		30.50	47.50
320	0010	**SELECTIVE DEMOLITION, DISPOSAL ONLY**									**320**
	0015	Urban bldg w/salvage value allowed R02220 -510									
	0020	Including loading and 5 mile haul to dump									
	0200	Steel frame	B-3	430	.112	C.Y.		3.20	4.05	7.25	9.35
	0300	Concrete frame		365	.132			3.77	4.77	8.54	11.05
	0400	Masonry construction		445	.108			3.09	3.91	7	9.05
	0500	Wood frame	▼	247	.194	▼		5.55	7.05	12.60	16.30
330	0010	**SELECTIVE DEMOLITION, DUMP CHARGES**									**330**
	0020	Dump charges, typical urban city, tipping fees only R02220 -510									
	0100	Building construction materials				Ton					70
	0200	Trees, brush, lumber									50
	0300	Rubbish only									60
	0500	Reclamation station, usual charge	▼			▼					85
340	0010	**SELECTIVE DEMOLITION, GUTTING**									**340**
	0020	Building interior, including disposal, dumpster fees not included R02220 -510									
	0500	Residential building									
	0560	Minimum	B-16	400	.080	SF Flr.		2.19	1.19	3.38	4.71
	0580	Maximum	″	360	.089	″		2.44	1.32	3.76	5.25
	0900	Commercial building									
	1000	Minimum	B-16	350	.091	SF Flr.		2.51	1.36	3.87	5.40
	1020	Maximum	″	250	.128	″		3.51	1.91	5.42	7.55
350	0010	**SELECTIVE DEMOLITION, RUBBISH HANDLING**									**350**
	0020	The following are to be added to the demolition prices R02220 -510									
	0400	Chute, circular, prefabricated steel, 18″ diameter	B-1	40	.600	L.F.	28	16.40		44.40	56.50
	0440	30″ diameter	″	30	.800	″	37.50	22		59.50	75.50
	0725	Dumpster, weekly rental, 1 dump/week, 20 C.Y. capacity (8 Tons)				Week					440
	0800	30 C.Y. capacity (10 Tons)									665
	0840	40 C.Y. capacity (13 Tons)				▼					805
	0900	Alternate pricing for dumpsters									
	0910	Delivery, average for all sizes				Ea.					52
	0920	Haul, average for all sizes									130
	0930	Rent per day, average for all sizes									3.10
	0940	Rent per month, average for all sizes				▼					31
	0950	Disposal fee per ton, average for all sizes				Ton					47
	1000	Dust partition, 6 mil polyethylene, 1″ x 3″ frame	2 Carp	2,000	.008	S.F.	.17	.27		.44	.62
	1080	2″ x 4″ frame	″	2,000	.008	″	.28	.27		.55	.74
	2000	Load, haul, and dump, 50′ haul	2 Clab	24	.667	C.Y.		17.80		17.80	27.50
	2040	100′ haul		16.50	.970			26		26	40.50
	2080	Over 100′ haul, add per 100 L.F.		35.50	.451			12.05		12.05	18.75
	2120	In elevators, per 10 floors, add	▼	140	.114			3.05		3.05	4.75
	3000	Loading & trucking, including 2 mile haul, chute loaded	B-16	45	.711			19.50	10.60	30.10	41.50
	3040	Hand loading truck, 50′ haul	″	48	.667			18.25	9.95	28.20	39.50
	3080	Machine loading truck	B-17	120	.267			7.60	4.52	12.12	16.60
	5000	Haul, per mile, up to 8 C.Y. truck	B-34B	1,165	.007	▼		.19	.41	.60	.74
	5100	Over 8 C.Y. truck	″	1,550	.005	▼		.14	.31	.45	.56

Important: See the Reference Section for critical supporting data - Reference Nos., Crews, & City Cost Indexes

02200 | Site Preparation

02220 | Site Demolition

		CREW	DAILY OUTPUT	LABOR-HOURS	UNIT	2005 BARE COSTS MAT.	LABOR	EQUIP.	TOTAL	TOTAL INCL O&P	
360	**0010**	**SELECTIVE DEMOLITION, SAW CUTTING** R02220 -510									**360**
	0015	Asphalt, up to 3" deep	B-89	1,050	.015	L.F.	.26	.46	.29	1.01	1.30	
	0020	Each additional inch of depth		1,800	.009		.06	.27	.17	.50	.66	
	0400	Concrete slabs, mesh reinforcing, up to 3" deep		980	.016		.35	.49	.31	1.15	1.47	
	0420	Each additional inch of depth		1,600	.010		.12	.30	.19	.61	.79	
	0800	Concrete walls, hydraulic saw, plain, per inch of depth	B-89B	250	.064		.32	1.92	1.76	4	5.20	
	0820	Rod reinforcing, per inch of depth		150	.107		.44	3.19	2.93	6.56	8.55	
	1200	Masonry walls, hydraulic saw, brick, per inch of depth		300	.053		.32	1.60	1.47	3.39	4.38	
	1220	Block walls, solid, per inch of depth		250	.064		.33	1.92	1.76	4.01	5.20	
	2000	Brick or masonry w/hand held saw, per inch of depth	A-1	125	.064		.26	1.71	.40	2.37	3.39	
	3020	Blades for saw, diamond, 12" diameter				Ea.	575			575	630	
	3040	18" diameter					930			930	1,025	
	3080	24" diameter					1,400			1,400	1,550	
	3120	30" diameter					1,500			1,500	1,675	
	3160	36" diameter					2,125			2,125	2,350	
	3200	42" diameter					3,075			3,075	3,375	
	5000	Wood sheathing to 1" thick, on walls	1 Carp	200	.040	L.F.		1.37		1.37	2.13	
	5020	On roof	"	250	.032	"		1.10		1.10	1.71	
	9950	See also div. 02210-200 core drilling										
370	**0010**	**SELECTIVE DEMOLITION, TORCH CUTTING** R02220 -510									**370**
	0020	Steel, 1" thick plate	1 Clab	360	.022	L.F.	.18	.59		.77	1.12	
	0040	1" diameter bar	"	210	.038	Ea.		1.02		1.02	1.58	
	1000	Oxygen lance cutting, reinforced concrete walls										
	1040	12" to 16" thick walls	1 Clab	10	.800	L.F.		21.50		21.50	33	
	1080	24" thick walls	"	6	1.333	"		35.50		35.50	55.50	
372	**0010**	**SELECTIVE DEMOLITION, UTILITY MATERIALS** R02220 -510									**372**
	0020	See other utility items in 02220-240-0010										
	0100	Demo Fire Hydrant extensions	B-20	14	1.714	Ea.		51.50		51.50	80.50	
	0200	Demo Precast Utility boxes up to 8'x14'x7'	B-13	2	28			805	310	1,115	1,600	
	0300	Demo handholes and meter pits	B-6	2	12			345	108	453	650	
	0400	Demo utility valves 4"-12"	B-20	4	6			180		180	281	
	0500	Demo utility valves 14"-24"	B-21	2	14			435	79	514	755	
374	**0010**	**SELECTIVE DEMOLITION, RIP-RAP & ROCK LINING** R02220 -510									**374**
	0100	Demo slope protection broken stone	B-13	62	.903	C.Y.		26	9.95	35.95	51	
	0200	Demo slope protection 3/8 to 1/4 CY pieces		60	.933	S.Y.		27	10.30	37.30	53	
	0300	Demo slope protection 18 inch depth		60	.933	"		27	10.30	37.30	53	
	0400	Demo slope protection dumped stone		93	.602	Ton		17.35	6.65	24	34	
	0500	Demo slope protection gabions 6-12 inches deep		60	.933	S.Y.		27	10.30	37.30	53	
	0600	Demo slope protection gabions 18-36 inches deep		30	1.867	"		53.50	20.50	74	106	
376	**0010**	**SELECTIVE DEMOLITION, SHORE PROTECT/MOORING STRUCTURES**									**376**
	0100	Demo breakwaters,bulkheads concrete maximun	B-9	12	3.333	L.F.		90.50	11.80	102.30	154	
	0200	Demo breakwaters,bulkheads concrete 12' min		10	4			108	14.15	122.15	185	
	0300	Demo breakwaters,bulkheads concrete 12' max		9	4.444			120	15.75	135.75	204	
	0400	Demo breakwaters,bulkheads steel shore driven	B-40B	54	.889			26	17.65	43.65	59.50	
	0500	Demo breakwaters,bulkheads steel barge driven	B-76A	30	2.133			61	49.50	110.50	148	
	0600	Demo jetties, docks floating	B-21B	600	.067	S.F.		1.93	1	2.93	4.08	
	0700	Demo jetties, docks pier supported 3"-4" decking		300	.133			3.86	2.01	5.87	8.15	
	0800	Demo jetties, docks floating, prefab small boat min		600	.067			1.93	1	2.93	4.08	
	0900	Demo jetties, docks floating, prefab small boat max		300	.133			3.86	2.01	5.87	8.15	
	1000	Demo jetties, docks floating, prefab per slip min		3	13.333	Ea.		385	201	586	815	
	1010	Demo jetties, docks floating, prefab per slip max		3	13.333	"		385	201	586	815	
378	**0010**	**SELECTIVE DEMOLITION, PILES**									**378**
	0100	Demo cast in place piles 8"-10" corrugated	B-19	600	.107	V.L.F.		3.61	2.55	6.16	8.55	

SITE CONSTRUCTION 2

39

2 SITE CONSTRUCTION

		02220	Site Demolition	CREW	DAILY OUTPUT	LABOR-HOURS	UNIT	2005 BARE COSTS MAT.	LABOR	EQUIP.	TOTAL	TOTAL INCL O&P	
378	0200		Demo cast in place piles 12"-14" corrugated	B-19	500	.128	V.L.F.		4.33	3.06	7.39	10.25	378
	0300		Demo cast in place piles 16" corrugated		400	.160			5.40	3.83	9.23	12.85	
	0400		Demo cast in place piles 12" fluted		600	.107			3.61	2.55	6.16	8.55	
	0500		Demo cast in place piles 14"-18" fluted		500	.128			4.33	3.06	7.39	10.25	
	0600		Demo cast in place piles 12" end bearing		600	.107			3.61	2.55	6.16	8.55	
	0700		Demo cast in place piles 14"-18" end bearing		500	.128			4.33	3.06	7.39	10.25	
	0800		Demo precast prestressed piles 12"-14" dia		700	.091			3.09	2.19	5.28	7.35	
	0900		Demo precast prestressed piles 16"-24" dia		600	.107			3.61	2.55	6.16	8.55	
	1000		Demo precast prestressed piles 36"-66" dia		300	.213			7.20	5.10	12.30	17.10	
	1100		Demo precast prestressed piles 10"-14" thick		600	.107			3.61	2.55	6.16	8.55	
	1200		Demo precast prestressed piles 16"-24" thick		500	.128			4.33	3.06	7.39	10.25	
	1300		Demo pressure grounted pile 5"		150	.427			14.40	10.20	24.60	34	
	1400		Demo steel piles 8"-12" tip		600	.107			3.61	2.55	6.16	8.55	
	1500		Demo steel piles H sections HP8 to HP12	↓	600	.107			3.61	2.55	6.16	8.55	
	1600		Demo steel piles H section HP14	B-19A	600	.107			3.61	3.25	6.86	9.35	
	1700		Demo steel pipe piles 8"-12"	B-19	600	.107			3.61	2.55	6.16	8.55	
	1800		Demo steel pipe piles 14"-18" plain		500	.128			4.33	3.06	7.39	10.25	
	1900		Demo steel pipe piles 14"-18" concrete filled		400	.160			5.40	3.83	9.23	12.85	
	2000		Demo timber piles to 14" dia	↓	600	.107	↓		3.61	2.55	6.16	8.55	
380	0010		**SELECTIVE DEMOLITION, UTILITY VALVES & ACCESSORIES**										380
	0100		Demo utility valves 4"-12" dia	B-20	4	6	Ea.		180		180	281	
	0200		Demo utility valves 14"-24" dia	B-21	2	14			435	79	514	755	
	0300		Demo crosses 4"-12"	B-20	8	3			90		90	140	
	0400		Demo crosses 14"-24"	B-21	4	7			216	39.50	255.50	380	
	0500		Demo utility cut-in valves 4"-12" dia	B-20	20	1.200			36		36	56	
	0600		Demo curb boxes	"	20	1.200	↓		36		36	56	
381	0010		**SELECTIVE DEMOLITION, WATER & SEWER PIPING AND FITTINGS**										381
	0020		See other utility items in 02220-240-0010										
	0090		Demo concrete pipe 4"-10" dia	B-6	250	.096	L.F.		2.77	.86	3.63	5.20	
	0100		Demo concrete pipe 42"-48" dia	B-13B	96	.583			16.80	9.35	26.15	36.50	
	0200		Demo concrete pipe 60"-84" dia	"	80	.700			20	11.20	31.20	43.50	
	0300		Demo concrete pipe 96" dia	B-13C	80	.700			20	19	39	52	
	0400		Demo concrete pipe 108"-144" dia	"	64	.875	↓		25	23.50	48.50	65	
	0450		Demo concrete fittings 12" dia	B-6	24	1	Ea.		29	9	38	54.50	
	0480		Demo concrete end pieces 12" dia		200	.120	L.F.		3.46	1.08	4.54	6.50	
	0490		Demo concrete end pieces 18" dia		150	.160			4.61	1.44	6.05	8.70	
	0500		Demo concrete end pieces 24"-36" dia		100	.240	↓		6.90	2.16	9.06	13	
	0600		Demo concrete fittings 24"-36" dia	↓	12	2	Ea.		57.50	18	75.50	108	
	0700		Demo concrete fittings 48"-84" dia	B-13B	12	4.667			134	74.50	208.50	289	
	0800		Demo concrete fittings 96" dia	"	8	7			202	112	314	435	
	0900		Demo concrete fittings 108"-144" dia	B-13C	4	14	↓		405	380	785	1,050	
	1000		Demo ductile iron pipe 4" dia	B-21B	200	.200	L.F.		5.80	3.01	8.81	12.25	
	1100		Demo ductile iron pipe 6"-12" dia		175	.229			6.60	3.44	10.04	14	
	1200		Demo ductile iron pipe 14"-24" dia		120	.333	↓		9.65	5	14.65	20.50	
	1300		Demo ductile iron fittings 4"-12" dia		24	1.667	Ea.		48	25	73	102	
	1400		Demo ductile iron fittings 14"-16" dia		18	2.222			64.50	33.50	98	137	
	1500		Demo ductile iron fittings 18"-24" dia	↓	12	3.333	↓		96.50	50	146.50	204	
	1600		Demo plastic pipe 3/4"-4" dia	B-20	700	.034	L.F.		1.03		1.03	1.61	
	1700		Demo plastic pipe 6"-8" dia		500	.048			1.44		1.44	2.25	
	1800		Demo plastic pipe 10"-18" dia		300	.080			2.41		2.41	3.75	
	1900		Demo plastic pipe 20"-36" dia		200	.120			3.61		3.61	5.60	
	1910		Demo plastic pipe 42"-48" dia		180	.133			4.01		4.01	6.25	
	1920		Demo plastic pipe 54"-60" dia	↓	160	.150	↓		4.51		4.51	7	
	2000		Demo plastic fittings 4"-8" dia	B-6	75	.320	Ea.		9.20	2.88	12.08	17.35	

Important: See the Reference Section for critical supporting data - Reference Nos., Crews, & City Cost Indexes

02220 | Site Demolition

		CREW	DAILY OUTPUT	LABOR-HOURS	UNIT	2005 BARE COSTS MAT.	LABOR	EQUIP.	TOTAL	TOTAL INCL O&P	
381	2100	Demo plastic fittings 10"-14" dia	B-6	50	.480	Ea.		13.85	4.32	18.17	26.50	**381**
	2200	Demo plastic fittings 16"-24" dia		20	1.200			34.50	10.80	45.30	65	
	2210	Demo plastic fittings 30"-36" dia		15	1.600			46	14.40	60.40	87	
	2220	Demo plastic fittings 42 "-48" dia		12	2			57.50	18	75.50	108	
	2300	Demo copper pipe 3/4"-2" dia	Q-1	500	.032	L.F.		1.18		1.18	1.77	
	2400	Demo copper pipe 2 1/2"-3" dia		300	.053			1.96		1.96	2.95	
	2500	Demo copper pipe 4"-6" dia		200	.080			2.94		2.94	4.42	
	2600	Demo copper fittings 3/4"-2"		15	1.067	Ea.		39		39	59	
	2700	Demo cast iron pipe 4" dia		200	.080	L.F.		2.94		2.94	4.42	
	2800	Demo cast iron pipe 5"-6" dia	Q-2	200	.120			4.58		4.58	6.90	
	2900	Demo cast iron pipe 8"-12" dia	Q-3	200	.160			6.25		6.25	9.35	
	3000	Demo cast iron fittings 4" dia	Q-1	30	.533	Ea.		19.60		19.60	29.50	
	3100	Demo cast iron fittings 5"-6" dia	Q-2	30	.800			30.50		30.50	46	
	3200	Demo cast iron fittings 8"-15" dia	Q-3	30	1.067			41.50		41.50	62.50	
382	0010	**SELECTIVE DEMOLITION, METAL DRAINAGE PIPING**										**382**
	0020	See other utility items in 02220-240-0010, 02220-380-0010,										
	0025	02220-381-0010										
	0100	Demo CMP pipe, aluminium, 6"-10" dia	B-21	800	.035	L.F.		1.08	.20	1.28	1.90	
	0110	Demo CMP pipe, aluminium, 12" dia		600	.047			1.44	.26	1.70	2.52	
	0120	Demo CMP pipe, aluminium, 18" dia		600	.047			1.44	.26	1.70	2.52	
	0140	Demo CMP pipe, steel, 6"-10" dia		800	.035			1.08	.20	1.28	1.90	
	0150	Demo CMP pipe, steel, 12" dia		600	.047			1.44	.26	1.70	2.52	
	0160	Demo CMP pipe, steel, 18" dia		400	.070			2.16	.40	2.56	3.79	
	0170	Demo CMP pipe, steel, 24" dia	B-13	300	.187			5.35	2.06	7.41	10.55	
	0180	Demo CMP pipe, steel, 30" - 36" dia		250	.224			6.45	2.47	8.92	12.65	
	0190	Demo CMP pipe, steel, 48" - 60" dia		200	.280			8.05	3.08	11.13	15.80	
	0200	Demo CMP pipe, steel, 72" dia	B-13B	100	.560			16.10	8.95	25.05	35	
	0210	Demo CMP end sections, steel, 10"-18" dia	B-21	40	.700	Ea.		21.50	3.96	25.46	38	
	0220	Demo CMP end sections, steel, 24"-36" dia	B-13	30	1.867			53.50	20.50	74	106	
	0230	Demo CMP end sections, steel, 48" dia		20	2.800			80.50	31	111.50	158	
	0240	Demo CMP end sections, steel, 60" dia		10	5.600			161	61.50	222.50	315	
	0250	Demo CMP end sections, steel, 72" dia	B-13B	10	5.600			161	89.50	250.50	345	
	0260	Demo CMP fittings, 8"-12" dia	B-21	60	.467			14.40	2.64	17.04	25.50	
	0270	Demo CMP fittings, 18" dia	"	40	.700			21.50	3.96	25.46	38	
	0280	Demo CMP fittings, 24"-48" dia	B-13	30	1.867			53.50	20.50	74	106	
	0290	Demo CMP fittings, 60" dia	"	20	2.800			80.50	31	111.50	158	
	0300	Demo CMP fittings, 72" dia	B-13B	10	5.600			161	89.50	250.50	345	

02230 | Site Clearing

		CREW	DAILY OUTPUT	LABOR-HOURS	UNIT	MAT.	LABOR	EQUIP.	TOTAL	TOTAL INCL O&P	
100	0010	**CLEAR AND GRUB**										**100**
	0020	Cut & chip light trees to 6" diam.	B-7	1	48	Acre		1,350	1,025	2,375	3,225	
	0150	Grub stumps and remove	B-30	2	12			360	835	1,195	1,475	
	0200	Cut & chip medium, trees to 12" diam.	B-7	.70	68.571			1,950	1,450	3,400	4,600	
	0250	Grub stumps and remove	B-30	1	24			720	1,675	2,395	2,950	
	0300	Cut & chip heavy, trees to 24" diam.	B-7	.30	160			4,550	3,400	7,950	10,800	
	0350	Grub stumps and remove	B-30	.50	48			1,425	3,350	4,775	5,850	
	0400	If burning is allowed, reduce cut & chip									40%	
	3000	Chipping stumps, to 18" deep, 12" diam.	B-86	20	.400	Ea.		13.85	3.54	17.39	25	
	3040	18" diameter		16	.500			17.35	4.42	21.77	31	
	3080	24" diameter		14	.571			19.80	5.05	24.85	35.50	
	3100	30" diameter		12	.667			23	5.90	28.90	41.50	
	3120	36" diameter		10	.800			27.50	7.05	34.55	50	
	3160	48" diameter		8	1			34.50	8.85	43.35	61.50	
	5000	Tree thinning, feller buncher, conifer										
	5080	Up to 8" diameter	B-93	240	.033	Ea.		1.15	1.80	2.95	3.72	

02230	Site Clearing	CREW	DAILY OUTPUT	LABOR-HOURS	UNIT	2005 BARE COSTS				TOTAL INCL O&P	
						MAT.	LABOR	EQUIP.	TOTAL		
100 5120	12" diameter	B-93	160	.050	Ea.		1.73	2.70	4.43	5.60	**100**
5240	Hardwood, up to 4" diameter		240	.033			1.15	1.80	2.95	3.72	
5280	8" diameter		180	.044			1.54	2.40	3.94	4.96	
5320	12" diameter		120	.067			2.31	3.60	5.91	7.45	
7000	Tree removal, congested area, aerial lift truck										
7040	8" diameter	B-85	7	5.714	Ea.		163	96	259	355	
7080	12" diameter		6	6.667			190	112	302	415	
7120	18" diameter		5	8			228	134	362	495	
7160	24" diameter		4	10			285	168	453	625	
7240	36" diameter		3	13.333			380	223	603	830	
7280	48" diameter		2	20			570	335	905	1,250	
9000	Site clearing with 335 H.P. dozer, trees to 6" diameter	B-10M	280	.043			1.37	4.27	5.64	6.80	
9010	To 12" diameter		150	.080			2.56	7.95	10.51	12.65	
9020	To 24" diameter		100	.120			3.84	11.95	15.79	19	
9030	To 36" diameter		50	.240			7.70	24	31.70	38	
9100	Grub stumps, trees to 6" diameter		400	.030			.96	2.99	3.95	4.75	
9103	To 36" diameter		195	.062			1.97	6.15	8.12	9.75	
200 0010	**SELECTIVE CLEARING**										**200**
0020	Clearing brush with brush saw	A-1C	.25	32	Acre		855	77	932	1,400	
0100	By hand	1 Clab	.12	66.667			1,775		1,775	2,775	
0300	With dozer, ball and chain, light clearing	B-11A	2	8			245	460	705	880	
0400	Medium clearing		1.50	10.667			325	615	940	1,175	
0500	With dozer and brush rake, light		10	1.600			49	92	141	176	
0550	Medium brush to 4" diameter		8	2			61.50	115	176.50	220	
0600	Heavy brush to 4" diameter		6.40	2.500			76.50	144	220.50	275	
1000	Brush mowing, tractor w/rotary mower, no removal										
1020	Light density	B-84	2	4	Acre		139	114	253	335	
1040	Medium density		1.50	5.333			185	153	338	445	
1080	Heavy density		1	8			277	229	506	670	
300 0010	**SELECTIVE TREE REMOVAL** With tractor, large tract, firm										**300**
0020	level terrain, no boulders, less than 12" diam. trees										
0300	300 HP dozer, up to 400 trees/acre, 0 to 25% hardwoods	B-10M	.75	16	Acre		510	1,600	2,110	2,525	
0340	25% to 50% hardwoods		.60	20			640	2,000	2,640	3,175	
0370	75% to 100% hardwoods		.45	26.667			855	2,650	3,505	4,225	
0400	500 trees/acre, 0% to 25% hardwoods		.60	20			640	2,000	2,640	3,175	
0440	25% to 50% hardwoods		.48	25			800	2,500	3,300	3,975	
0470	75% to 100% hardwoods		.36	33.333			1,075	3,325	4,400	5,275	
0500	More than 600 trees/acre, 0 to 25% hardwoods		.52	23.077			740	2,300	3,040	3,650	
0540	25% to 50% hardwoods		.42	28.571			915	2,850	3,765	4,525	
0570	75% to 100% hardwoods		.31	38.710			1,250	3,850	5,100	6,125	
0900	Large tract clearing per tree										
1500	300 HP dozer, to 12" diameter, softwood	B-10M	320	.037	Ea.		1.20	3.73	4.93	5.95	
1550	Hardwood		100	.120			3.84	11.95	15.79	19	
1600	12" to 24" diameter, softwood		200	.060			1.92	5.95	7.87	9.45	
1650	Hardwood		80	.150			4.80	14.95	19.75	24	
1700	24" to 36" diameter, softwood		100	.120			3.84	11.95	15.79	19	
1750	Hardwood		50	.240			7.70	24	31.70	38	
1800	36" to 48" diameter, softwood		70	.171			5.50	17.05	22.55	27	
1850	Hardwood		35	.343			10.95	34	44.95	54	
2000	Stump removal on site by hydraulic backhoe, 1-1/2 C.Y.										
2040	4" to 6" diameter	B-17	60	.533	Ea.		15.20	9.05	24.25	33.50	
2050	8" to 12" diameter	B-30	33	.727			22	50.50	72.50	89	
2100	14" to 24" diameter		25	.960			28.50	67	95.50	117	
2150	26" to 36" diameter		16	1.500			45	105	150	183	
3000	Remove selective trees, on site using chain saws and chipper,										

Important: See the Reference Section for critical supporting data - Reference Nos., Crews, & City Cost Indexes

02230	Site Clearing	CREW	DAILY OUTPUT	LABOR-HOURS	UNIT	2005 BARE COSTS MAT.	LABOR	EQUIP.	TOTAL	TOTAL INCL O&P	
300	3050	not incl. stumps, up to 6" diameter	B-7	18	2.667	Ea.		75.50	56.50	132	180	300
	3100	8" to 12" diameter		12	4			113	85	198	269	
	3150	14" to 24" diameter		10	4.800			136	102	238	320	
	3200	26" to 36" diameter		8	6			170	127	297	405	
	3300	Machine load, 2 mile haul to dump, 12" diam. tree, add								160	240	
500	0010	**STRIPPING & STOCKPILING OF SOIL**										500
	0020	200 H.P. dozer, ideal conditions	B-10B	2,300	.005	C.Y.		.17	.40	.57	.69	
	0100	Adverse conditions	"	1,150	.010			.33	.80	1.13	1.39	
	0200	300 HP dozer, ideal conditions	B-10M	3,000	.004			.13	.40	.53	.63	
	0300	Adverse conditions	"	1,650	.007			.23	.72	.95	1.15	
	1400	Loam or topsoil, remove and stockpile on site										
	1420	6" deep, 200' haul	B-10B	865	.014	C.Y.		.44	1.06	1.50	1.84	
	1430	300' haul		520	.023			.74	1.77	2.51	3.07	
	1440	500' haul		225	.053			1.71	4.09	5.80	7.10	
	1450	Alternate method: 6" deep, 200' haul		5,090	.002	S.Y.		.08	.18	.26	.31	
	1460	500' haul		1,325	.009	"		.29	.69	.98	1.20	

02240	Dewatering	CREW	DAILY OUTPUT	LABOR-HOURS	UNIT	MAT.	LABOR	EQUIP.	TOTAL	TOTAL INCL O&P	
330	0010	**CUT DRAINAGE DITCH**										330
	0020	Cut drainage ditch, common earth, 30'w x 1'deep	B-11L	6,000	.003	L.F.		.08	.08	.16	.21	
	0200	Clay and till		4,200	.004			.12	.11	.23	.30	
	0250	Clean wet drainage ditch, 30' wide		10,000	.002			.05	.05	.10	.13	
500	0010	**DEWATERING**										500
	0020	Excavate drainage trench, 2' wide, 2' deep	B-11C	90	.178	C.Y.		5.45	2.40	7.85	11	
	0100	2' wide, 3' deep, with backhoe loader	"	135	.119			3.64	1.60	5.24	7.30	
	0200	Excavate sump pits by hand, light soil	1 Clab	7.10	1.127			30		30	47	
	0300	Heavy soil	"	3.50	2.286			61		61	95	
	0500	Pumping 8 hr., attended 2 hrs. per day, including 20 L.F.										
	0550	of suction hose & 100 L.F. discharge hose										
	0600	2" diaphragm pump used for 8 hours	B-10H	4	3	Day		96	14.25	110.25	162	
	0620	Add per additional pump							37.50	37.50	43	
	0650	4" diaphragm pump used for 8 hours	B-10I	4	3			96	21.50	117.50	170	
	0670	Add per additional pump							80	80	91	
	0800	8 hrs. attended, 2" diaphragm pump	B-10H	1	12			385	57	442	650	
	0820	Add per additional pump							37.50	37.50	43	
	0900	3" centrifugal pump	B-10J	1	12			385	63	448	655	
	0920	Add per additional pump							53	53	58.50	
	1000	4" diaphragm pump	B-10I	1	12			385	86.50	471.50	680	
	1020	Add per additional pump							91	91	105	
	1100	6" centrifugal pump	B-10K	1	12			385	248	633	855	
	1120	Add per additional pump							118	118	134	
	1300	CMP, incl. excavation 3' deep, 12" diameter	B-6	115	.209	L.F.	9.65	6	1.88	17.53	22	
	1400	18" diameter		100	.240	"	12	6.90	2.16	21.06	26	
	1600	Sump hole construction, incl. excavation and gravel, pit		1,250	.019	C.F.	.70	.55	.17	1.42	1.81	
	1700	With 12" gravel collar, 12" pipe, corrugated, 16 ga.		70	.343	L.F.	15.10	9.90	3.08	28.08	35	
	1800	15" pipe, corrugated, 16 ga.		55	.436		19.30	12.60	3.92	35.82	44.50	
	1900	18" pipe, corrugated, 16 ga.		50	.480		22.50	13.85	4.32	40.67	51.50	
	2000	24" pipe, corrugated, 14 ga.		40	.600		27	17.30	5.40	49.70	62.50	
	2200	Wood lining, up to 4' x 4', add		300	.080	SFCA	12.80	2.31	.72	15.83	18.40	
	9950	See div. 02240-900 for wellpoints										
	9960	See div. 02240-700 for deep well systems										
	9970	See div. 15440 for pumps										
700	0010	**WELLS** For dewatering 10' to 20' deep, 2' diameter										700
	0020	with steel casing, minimum	B-6	165	.145	V.L.F.	2.35	4.19	1.31	7.85	10.50	

02200 | Site Preparation

02240 | Dewatering

		CREW	DAILY OUTPUT	LABOR-HOURS	UNIT	MAT.	LABOR	EQUIP.	TOTAL	TOTAL INCL O&P	
700	0050	Average	B-6	98	.245	V.L.F.	4.68	7.05	2.20	13.93	18.40	**700**
	0100	Maximum	↓	49	.490	↓	12.50	14.10	4.40	31	40	
	0300	For pumps for dewatering, see division 01590-400										
	0500	For domestic water wells, see division 02520-510										
900	0010	**WELLPOINTS** For wellpoint equipment rental, see div. 01590-700 R02240 -900										**900**
	0100	Installation and removal of single stage system										
	0110	Labor only, .75 labor-hours per L.F., minimum	1 Clab	10.70	.748	LF Hdr		19.95		19.95	31	
	0200	2.0 labor-hours per L.F., maximum	"	4	2	"		53.50		53.50	83	
	0400	Pump operation, 4 @ 6 hr. shifts										
	0410	Per 24 hour day	4 Eqlt	1.27	25.197	Day		835		835	1,250	
	0500	Per 168 hour week, 160 hr. straight, 8 hr. double time	↓	.18	177	Week		5,875		5,875	8,850	
	0550	Per 4.3 week month	↓	.04	800	Month		26,400		26,400	39,800	
	0600	Complete installation, operation, equipment rental, fuel &										
	0610	removal of system with 2" wellpoints 5' O.C.										
	0700	100' long header, 6" diameter, first month	4 Eqlt	3.23	9.907	LF Hdr	125	325		450	630	
	0800	Thereafter, per month		4.13	7.748		99.50	256		355.50	495	
	1000	200' long header, 8" diameter, first month		6	5.333		110	176		286	385	
	1100	Thereafter, per month		8.39	3.814		56	126		182	252	
	1300	500' long header, 8" diameter, first month		10.63	3.010		43.50	99.50		143	198	
	1400	Thereafter, per month		20.91	1.530		31	50.50		81.50	111	
	1600	1,000' long header, 10" diameter, first month		11.62	2.754		37.50	91		128.50	178	
	1700	Thereafter, per month	↓	41.81	.765	↓	18.70	25.50		44.20	58.50	
	1900	Note: above figures include pumping 168 hrs. per week										
	1910	and include the pump operator and one stand-by pump.	↓									

02250 | Shoring & Underpinning

		CREW	DAILY OUTPUT	LABOR-HOURS	UNIT	MAT.	LABOR	EQUIP.	TOTAL	TOTAL INCL O&P	
100	0010	**GROUTING, PRESSURE**										**100**
	0020	Grouting, pressure, cement & sand, 1:1 mix, minimum	B-61	124	.323	Bag	9.10	9.15	2.01	20.26	26.50	
	0100	Maximum		51	.784	"	9.10	22.50	4.89	36.49	50	
	0200	Cement and sand, 1:1 mix, minimum		250	.160	C.F.	18.20	4.54	1	23.74	28	
	0300	Maximum		100	.400		27.50	11.35	2.50	41.35	50.50	
	0400	Epoxy cement grout, minimum		137	.292		126	8.30	1.82	136.12	154	
	0500	Maximum	↓	57	.702	↓	126	19.90	4.38	150.28	175	
	0600	Structural epoxy grout				Gal.	49.50			49.50	54.50	
	0700	Alternate pricing method: (Add for materials)										
	0710	5 person crew and equipment	B-61	1	40	Day		1,125	250	1,375	2,025	
400	0010	**SHEET PILING**										**400**
	0020	Sheet piling steel, not incl. wales, 22 psf, 15' excav., left in place	B-40	10.81	5.920	Ton	865	200	212	1,277	1,500	
	0100	Drive, extract & salvage R02250 -400		6	10.667		385	360	380	1,125	1,425	
	0300	20' deep excavation, 27 psf, left in place		12.95	4.942		865	167	177	1,209	1,400	
	0400	Drive, extract & salvage R02250 -450		6.55	9.771		385	330	350	1,065	1,325	
	0600	25' deep excavation, 38 psf, left in place		19	3.368		865	114	121	1,100	1,275	
	0700	Drive, extract & salvage		10.50	6.095		385	206	218	809	990	
	0900	40' deep excavation, 38 psf, left in place		21.20	3.019		865	102	108	1,075	1,225	
	1000	Drive, extract & salvage		12.25	5.224	↓	385	177	187	749	910	
	1200	15' deep excavation, 22 psf, left in place		983	.065	S.F.	10.05	2.20	2.33	14.58	17.10	
	1300	Drive, extract & salvage		545	.117		4.31	3.97	4.20	12.48	15.70	
	1500	20' deep excavation, 27 psf, left in place		960	.067		12.65	2.25	2.39	17.29	20	
	1600	Drive, extract & salvage		485	.132		5.60	4.46	4.72	14.78	18.45	
	1800	25' deep excavation, 38 psf, left in place		1,000	.064		18.60	2.16	2.29	23.05	26.50	
	1900	Drive, extract & salvage	↓	553	.116	↓	7.70	3.91	4.14	15.75	19.25	
	2100	Rent steel sheet piling and wales, first month				Ton	206			206	226	
	2200	Per added month					20.50			20.50	22.50	
	2300	Rental piling left in place, add to rental					685			685	755	
	2500	Wales, connections & struts, 2/3 salvage					210			210	231	
	2700	High strength piling, 50,000 psi, add				↓	46.50			46.50	51	

2 SITE CONSTRUCTION

Important: See the Reference Section for critical supporting data - Reference Nos., Crews, & City Cost Indexes

		02250	Shoring & Underpinning		CREW	DAILY OUTPUT	LABOR-HOURS	UNIT	2005 BARE COSTS				TOTAL INCL O&P	
									MAT.	LABOR	EQUIP.	TOTAL		
400	2800	55,000 psi, add	R02250 -400					Ton	49.50			49.50	54.50	400
	3000	Tie rod, not upset, 1-1/2" to 4" diameter with turnbuckle							1,525			1,525	1,675	
	3100	No turnbuckle	R02250 -450						1,175			1,175	1,300	
	3300	Upset, 1-3/4" to 4" diameter with turnbuckle							1,725			1,725	1,900	
	3400	No turnbuckle							1,500			1,500	1,650	
	3600	Lightweight, 18" to 28" wide, 7 ga., 9.22 psf, and												
	3610	9 ga., 8.6 psf, minimum						Lb.	.63			.63	.69	
	3700	Average							.68			.68	.75	
	3750	Maximum							.79			.79	.87	
	3900	Wood, solid sheeting, incl. wales, braces and spacers,												
	3910	drive, extract & salvage, 8' deep excavation		B-31	330	.121	S.F.	1.72	3.47	.39	5.58	7.70		
	4000	10' deep, 50 S.F./hr. in & 150 S.F./hr. out			300	.133		1.77	3.81	.43	6.01	8.35		
	4100	12' deep, 45 S.F./hr. in & 135 S.F./hr. out			270	.148		1.82	4.24	.48	6.54	9.15		
	4200	14' deep, 42 S.F./hr. in & 126 S.F./hr. out			250	.160		1.88	4.58	.52	6.98	9.80		
	4300	16' deep, 40 S.F./hr. in & 120 S.F./hr. out			240	.167		1.94	4.77	.54	7.25	10.10		
	4400	18' deep, 38 S.F./hr. in & 114 S.F./hr. out			230	.174		2	4.98	.56	7.54	10.55		
	4500	20' deep, 35 S.F./hr. in & 105 S.F./hr. out			210	.190		2.07	5.45	.62	8.14	11.45		
	4520	Left in place, 8' deep, 55 S.F./hr.			440	.091		3.10	2.60	.29	5.99	7.80		
	4540	10' deep, 50 S.F./hr.			400	.100		3.26	2.86	.32	6.44	8.40		
	4560	12' deep, 45 S.F./hr.			360	.111		3.44	3.18	.36	6.98	9.15		
	4565	14' deep, 42 S.F./hr.			335	.119		3.65	3.42	.39	7.46	9.75		
	4570	16' deep, 40 S.F./hr.			320	.125		3.88	3.58	.40	7.86	10.25		
	4580	18' deep, 38 S.F./hr.			305	.131		4.13	3.75	.42	8.30	10.85		
	4590	20' deep, 35 S.F./hr.			280	.143		4.43	4.09	.46	8.98	11.75		
	4700	Alternate pricing, left in place, 8' deep			1.76	22.727	M.B.F.	695	650	73.50	1,418.50	1,850		
	4800	Drive, extract and salvage, 8' deep			1.32	30.303	"	620	865	98	1,583	2,150		
	5000	For treated lumber add cost of treatment to lumber												
	5010	See division 06070-400												
500	0010	**SHORING**												500
	0020	Shoring, existing building, with timber, no salvage allowance		B-51	2.20	21.818	M.B.F.	800	590	58.50	1,448.50	1,850		
	1000	On cribbing with 35 ton screw jacks, per box and jack		"	3.60	13.333	Jack	44.50	360	35.50	440	650		
	1100	Masonry openings in walls, see div. 02220-310												
600	0010	**SLABJACKING**												600
	0100	4" thick slab		D-4	1,500	.021	S.F.	.23	.65	.09	.97	1.35		
	0150	6" thick slab			1,200	.027		.33	.81	.11	1.25	1.72		
	0200	8" thick slab			1,000	.032		.39	.98	.14	1.51	2.06		
	0250	10" thick slab			900	.036		.42	1.09	.15	1.66	2.28		
	0300	12" thick slab			850	.038		.46	1.15	.16	1.77	2.42		
800	0010	**UNDERPINNING FOUNDATIONS** Including excavation,												800
	0020	forming, reinforcing, concrete and equipment												
	0100	5' to 16' below grade, 100 to 500 C.Y.		B-52	2.30	24.348	C.Y.	225	765	171	1,161	1,600		
	0200	Over 500 C.Y.			2.50	22.400		203	705	158	1,066	1,500		
	0400	16' to 25' below grade, 100 to 500 C.Y.			2	28		248	880	197	1,325	1,850		
	0500	Over 500 C.Y.			2.10	26.667		234	835	188	1,257	1,775		
	0700	26' to 40' below grade, 100 to 500 C.Y.			1.60	35		270	1,100	246	1,616	2,275		
	0800	Over 500 C.Y.			1.80	31.111		248	975	219	1,442	2,025		
	0900	For under 50 C.Y., add						10%	40%					
	1000	For 50 C.Y. to 100 C.Y., add						5%	20%					
900	0010	**VIBROFLOTATION**												900
	0900	Vibroflotation compacted sand cylinder, minimum		B-60	750	.075	V.L.F.		2.28	1.55	3.83	5.20		
	0950	Maximum			325	.172			5.25	3.58	8.83	12		
	1100	Vibro replacement compacted stone cylinder, minimum			500	.112			3.43	2.33	5.76	7.80		
	1150	Maximum			250	.224			6.85	4.65	11.50	15.55		
	1300	Mobilization and demobilization, minimum			.47	119	Total		3,650	2,475	6,125	8,300		
	1400	Maximum			.14	400	"		12,200	8,300	20,500	27,800		

SITE CONSTRUCTION 2

45

02260 | Excavation Support/Protection

		CREW	DAILY OUTPUT	LABOR-HOURS	UNIT	MAT.	LABOR	EQUIP.	TOTAL	TOTAL INCL O&P	
200	0010	**COFFERDAMS**, incl. mobilization and temporary sheeting										**200**
	0020	Shore driven	B-40	960	.067	S.F.	12.35	2.25	2.39	16.99	19.80	
	0060	Barge driven	"	550	.116	"	12.35	3.93	4.17	20.45	24.50	
	0080	Soldier beams & lagging H piles with 3" wood sheeting										
	0090	horizontal between piles, including removal of wales & braces										
	0100	No hydrostatic head, 15' deep, 1 line of braces, minimum	B-50	545	.205	S.F.	8	6.65	3.11	17.76	23	
	0200	Maximum		495	.226		8.90	7.30	3.42	19.62	25.50	
	0400	15' to 22' deep with 2 lines of braces, 10" H, minimum		360	.311		9.40	10.05	4.70	24.15	31.50	
	0500	Maximum		330	.339		10.65	10.95	5.15	26.75	35	
	0700	23' to 35' deep with 3 lines of braces, 12" H, minimum		325	.345		12.30	11.15	5.20	28.65	37	
	0800	Maximum		295	.380		13.35	12.25	5.75	31.35	40.50	
	1000	36' to 45' deep with 4 lines of braces, 14" H, minimum		290	.386		13.80	12.50	5.85	32.15	41.50	
	1100	Maximum		265	.423		14.55	13.65	6.40	34.60	45	
	1300	No hydrostatic head, left in place, 15' dp., 1 line of braces, min.		635	.176		10.65	5.70	2.67	19.02	24	
	1400	Maximum		575	.195		11.45	6.30	2.94	20.69	26	
	1600	15' to 22' deep with 2 lines of braces, minimum		455	.246		16	7.95	3.72	27.67	34.50	
	1700	Maximum		415	.270		17.80	8.70	4.08	30.58	38	
	1900	23' to 35' deep with 3 lines of braces, minimum		420	.267		19.05	8.60	4.03	31.68	39	
	2000	Maximum		380	.295		21	9.50	4.45	34.95	43	
	2200	36' to 45' deep with 4 lines of braces, minimum		385	.291		23	9.40	4.40	36.80	45	
	2300	Maximum	▼	350	.320		26.50	10.35	4.84	41.69	51.50	
	2350	Lagging only, 3" thick wood between piles 8' O.C., minimum	B-46	400	.120		1.78	3.64	.08	5.50	7.90	
	2370	Maximum		250	.192		2.67	5.80	.13	8.60	12.45	
	2400	Open sheeting no bracing, for trenches to 10' deep, min.		1,736	.028		.80	.84	.02	1.66	2.25	
	2450	Maximum	▼	1,510	.032		.89	.96	.02	1.87	2.55	
	2500	Tie-back method, add to open sheeting, add, minimum								20%	20%	
	2550	Maximum				▼				60%	60%	
	2700	Tie-backs only, based on tie-backs total length, minimum	B-46	86.80	.553	L.F.	10.70	16.75	.39	27.84	39	
	2750	Maximum		38.50	1.247	"	18.80	38	.87	57.67	82.50	
	3500	Tie-backs only, typical average, 25' long		2	24	Ea.	470	730	16.80	1,216.80	1,700	
	3600	35' long	▼	1.58	30.380	"	625	920	21.50	1,566.50	2,200	
	4500	Trench box, 7' deep, 16' x 8', see division 01590-400-7050				Day				175	193	
	4600	20' x 10', see division 01590-400-7070				"				295	320	
	5200	Wood sheeting, in trench, jacks at 4' O.C., 8' deep	B-1	800	.030	S.F.	.62	.82		1.44	1.96	
	5250	12' deep		700	.034		.73	.94		1.67	2.26	
	5300	15' deep	▼	600	.040	▼	1	1.09		2.09	2.80	
	6000	See also div. 02250-400										
720	0010	**ROCK BOLTING**										**720**
	2020	Hollow core, prestressable anchor, 1" diameter, 5' long	2 Skwk	32	.500	Ea.	75	17.45		92.45	110	
	2025	10' long		24	.667		150	23		173	201	
	2060	2" diameter, 5' long		32	.500		280	17.45		297.45	335	
	2065	10' long		24	.667		560	23		583	650	
	2100	Super high-tensile, 3/4" diameter, 5' long		32	.500		17	17.45		34.45	45.50	
	2105	10' long		24	.667		34	23		57	73.50	
	2160	2" diameter, 5' long		32	.500		152	17.45		169.45	194	
	2165	10' long	▼	24	.667		305	23		328	370	
	4400	Drill hole for rock bolt, 1-3/4" diam., 5' long (for 3/4" bolt)	B-56	17	.941			28	57.50	85.50	107	
	4405	10' long		9	1.778			53	109	162	201	
	4420	2" diameter, 5' long (for 1" bolt)		13	1.231			37	75.50	112.50	139	
	4425	10' long		7	2.286			68.50	140	208.50	258	
	4460	3-1/2" diameter, 5' long (for 2" bolt)		10	1.600			48	98	146	181	
	4465	10' long	▼	5	3.200	▼		95.50	196	291.50	360	
730	0010	**SOIL NAILING**										**730**
	0020	Soil nailing does not include guniting of surfaces										

Important: See the Reference Section for critical supporting data - Reference Nos., Crews, & City Cost Indexes

			DAILY	LABOR-		2005 BARE COSTS				TOTAL
02260		**Excavation Support/Protection**								
			CREW	OUTPUT	HOURS	UNIT	MAT.	LABOR	EQUIP.	TOTAL	INCL O&P
730	0030	See 03370-300 for guniting of surfaces										730
	0035	Layout and vertical and horizontal control per day	A-6	1	16	Day		545	60.50	605.50	900
	0038	Layout and vertical and horizontal control average holes per day	"	60	.267	Ea.		9.05	1.01	10.06	15
	0050	For Grade 150 soil nail add $ 1.25/ LF to base material cost									
	0060	Material delivery add $ 1.68 to $ 1.85 per truck mile for shipping									
	0090	Average Soil nailing, grade 75, 15 min setup per hole & 80'/hr drilling									
	0100	Soil nailing , drill hole, install # 8 nail, grout 20' depth average	B-47G	16	2	Ea.	202	57.50	68.50	328	390
	0110	Soil nailing , drill hole, install # 8 nail, grout 25' depth average		14.20	2.254		253	65	77.50	395.50	465
	0120	Soil nailing , drill hole, install # 8 nail, grout 30' depth average		12.80	2.500		350	72	86	508	590
	0130	Soil nailing , drill hole, install # 8 nail, grout 35' depth average		11.60	2.759		410	79.50	94.50	584	675
	0140	Soil nailing , drill hole, install # 8 nail, grout 40' depth average		10.70	2.991		465	86	103	654	760
	0150	Soil nailing , drill hole, install # 8 nail, grout 45' depth average		9.90	3.232		525	93	111	729	840
	0160	Soil nailing , drill hole, install # 8 nail, grout 50' depth average		9.10	3.516		615	101	121	837	965
	0170	Soil nailing , drill hole, install # 8 nail, grout 55' depth average		8.50	3.765		675	108	129	912	1,050
	0180	Soil nailing , drill hole, install # 8 nail, grout 60' depth average		8	4		730	115	137	982	1,125
	0190	Soil nailing , drill hole, install # 8 nail, grout 65' depth average		7.50	4.267		790	123	147	1,060	1,225
	0200	Soil nailing , drill hole, install # 8 nail, grout 70' depth average		7.10	4.507		845	130	155	1,130	1,300
	0210	Soil nailing , drill hole, install # 8 nail, grout 75' depth average	▼	6.70	4.776	▼	905	138	164	1,207	1,375
	0290	Average Soil nailing, grade 75, 15 min setup per hole & 90'/hr drilling									
	0300	Soil nailing , drill hole, install # 8 nail, grout 20' depth average	B-47G	16.60	1.928	Ea.	202	55.50	66	323.50	380
	0310	Soil nailing , drill hole, install # 8 nail, grout 25' depth average		15	2.133		253	61.50	73.50	388	455
	0320	Soil nailing , drill hole, install # 8 nail, grout 30' depth average		13.70	2.336		350	67.50	80	497.50	580
	0330	Soil nailing , drill hole, install # 8 nail, grout 35' depth average		12.30	2.602		410	75	89.50	574.50	665
	0340	Soil nailing , drill hole, install # 8 nail, grout 40' depth average		11.40	2.807		465	81	96.50	642.50	745
	0350	Soil nailing , drill hole, install # 8 nail, grout 45' depth average		10.70	2.991		525	86	103	714	820
	0360	Soil nailing , drill hole, install # 8 nail, grout 50' depth average		9.80	3.265		615	94	112	821	945
	0370	Soil nailing , drill hole, install # 8 nail, grout 55' depth average		9.20	3.478		675	100	119	894	1,025
	0380	Soil nailing , drill hole, install # 8 nail, grout 60' depth average		8.70	3.678		730	106	126	962	1,100
	0390	Soil nailing , drill hole, install # 8 nail, grout 65' depth average		8.10	3.951		790	114	136	1,040	1,200
	0400	Soil nailing , drill hole, install # 8 nail, grout 70' depth average		7.70	4.156		845	120	143	1,108	1,275
	0410	Soil nailing , drill hole, install # 8 nail, grout 75' depth average	▼	7.40	4.324	▼	905	125	149	1,179	1,350
	0490	Average Soil nailing, grade 75, 15 min setup per hole & 100'/hr drilling									
	0500	Soil nailing , drill hole, install # 8 nail, grout 20' depth average	B-47G	17.80	1.798	Ea.	202	52	61.50	315.50	370
	0510	Soil nailing , drill hole, install # 8 nail, grout 25' depth average		16	2		253	57.50	68.50	379	445
	0520	Soil nailing , drill hole, install # 8 nail, grout 30' depth average		14.60	2.192		350	63	75.50	488.50	565
	0530	Soil nailing , drill hole, install # 8 nail, grout 35' depth average		13.30	2.406		410	69.50	82.50	562	650
	0540	Soil nailing , drill hole, install # 8 nail, grout 40' depth average		12.30	2.602		465	75	89.50	629.50	730
	0550	Soil nailing , drill hole, install # 8 nail, grout 45' depth average		11.40	2.807		525	81	96.50	702.50	805
	0560	Soil nailing , drill hole, install # 8 nail, grout 50' depth average		10.70	2.991		615	86	103	804	920
	0570	Soil nailing , drill hole, install # 8 nail, grout 55' depth average		10	3.200		675	92	110	877	1,000
	0580	Soil nailing , drill hole, install # 8 nail, grout 60' depth average		9.40	3.404		730	98	117	945	1,075
	0590	Soil nailing , drill hole, install # 8 nail, grout 65' depth average		8.90	3.596		790	104	123	1,017	1,175
	0600	Soil nailing , drill hole, install # 8 nail, grout 70' depth average		8.40	3.810		845	110	131	1,086	1,250
	0610	Soil nailing , drill hole, install # 8 nail, grout 75' depth average	▼	8	4	▼	905	115	137	1,157	1,325
	0690	Average Soil nailing, grade 75, 15 min setup per hole & 110'/hr drilling									
	0700	Soil nailing , drill hole, install # 8 nail, grout 20' depth average	B-47G	18.50	1.730	Ea.	202	50	59.50	311.50	365
	0710	Soil nailing , drill hole, install # 8 nail, grout 25' depth average		16.60	1.928		253	55.50	66	374.50	435
	0720	Soil nailing , drill hole, install # 8 nail, grout 30' depth average		15.50	2.065		350	59.50	71	480.50	555
	0730	Soil nailing , drill hole, install # 8 nail, grout 35' depth average		14.10	2.270		410	65.50	78	553.50	635
	0740	Soil nailing , drill hole, install # 8 nail, grout 40' depth average		13	2.462		465	71	84.50	620.50	715
	0750	Soil nailing , drill hole, install # 8 nail, grout 45' depth average		12	2.667		525	77	91.50	693.50	795
	0760	Soil nailing , drill hole, install # 8 nail, grout 50' depth average		11.40	2.807		615	81	96.50	792.50	905
	0770	Soil nailing , drill hole, install # 8 nail, grout 55' depth average		10.70	2.991		675	86	103	864	985
	0780	Soil nailing , drill hole, install # 8 nail, grout 60' depth average		10	3.200		730	92	110	932	1,075
	0790	Soil nailing , drill hole, install # 8 nail, grout 65' depth average		9.40	3.404		790	98	117	1,005	1,150
	0800	Soil nailing , drill hole, install # 8 nail, grout 70' depth average	▼	9.10	3.516		845	101	121	1,067	1,225

SITE CONSTRUCTION **2**

	02260	Excavation Support/Protection	CREW	DAILY OUTPUT	LABOR-HOURS	UNIT	2005 BARE COSTS				TOTAL INCL O&P	
							MAT.	LABOR	EQUIP.	TOTAL		
730	0810	Soil nailing , drill hole, install # 8 nail, grout 75' depth average	B-47G	8.60	3.721	Ea.	905	107	128	1,140	1,300	**730**
	0890	Average Soil nailing, grade 75, 15 min setup per hole & 120'/hr drilling										
	0900	Soil nailing , drill hole, install # 8 nail, grout 20' depth average	B-47G	19.20	1.667	Ea.	202	48	57	307	360	
	0910	Soil nailing , drill hole, install # 8 nail, grout 25' depth average		17.10	1.871		253	54	64.50	371.50	430	
	0920	Soil nailing , drill hole, install # 8 nail, grout 30' depth average		16	2		350	57.50	68.50	476	550	
	0930	Soil nailing , drill hole, install # 8 nail, grout 35' depth average		14.60	2.192		410	63	75.50	548.50	630	
	0940	Soil nailing , drill hole, install # 8 nail, grout 40' depth average		13.70	2.336		465	67.50	80	612.50	710	
	0950	Soil nailing , drill hole, install # 8 nail, grout 45' depth average		12.60	2.540		525	73	87	685	785	
	0960	Soil nailing , drill hole, install # 8 nail, grout 50' depth average		12	2.667		615	77	91.50	783.50	895	
	0970	Soil nailing , drill hole, install # 8 nail, grout 55' depth average		11.20	2.857		675	82.50	98	855.50	975	
	0980	Soil nailing , drill hole, install # 8 nail, grout 60' depth average		10.70	2.991		730	86	103	919	1,050	
	0990	Soil nailing , drill hole, install # 8 nail, grout 65' depth average		10	3.200		790	92	110	992	1,125	
	1000	Soil nailing , drill hole, install # 8 nail, grout 70' depth average		9.60	3.333		845	96	114	1,055	1,200	
	1010	Soil nailing , drill hole, install # 8 nail, grout 75' depth average	▼	9.10	3.516	▼	905	101	121	1,127	1,275	
	1190	Difficult Soil nailing, grade 75, 20 min setup per hole & 80'/hr drilling			.							
	1200	Soil nailing , drill hole, install # 8 nail, grout 20' depth difficult	B-47G	13.70	2.336	Ea.	202	67.50	80	349.50	415	
	1210	Soil nailing , drill hole, install # 8 nail, grout 25' depth difficult		12.30	2.602		253	75	89.50	417.50	495	
	1220	Soil nailing , drill hole, install # 8 nail, grout 30' depth difficult		11.20	2.857		350	82.50	98	530.50	620	
	1230	Soil nailing , drill hole, install # 8 nail, grout 35' depth difficult		10.40	3.077		410	88.50	106	604.50	705	
	1240	Soil nailing , drill hole, install # 8 nail, grout 40' depth difficult		9.60	3.333		465	96	114	675	790	
	1250	Soil nailing , drill hole, install # 8 nail, grout 45' depth difficult		8.90	3.596		525	104	123	752	870	
	1260	Soil nailing , drill hole, install # 8 nail, grout 50' depth difficult		8.30	3.855		615	111	132	858	990	
	1270	Soil nailing , drill hole, install # 8 nail, grout 55' depth difficult		7.90	4.051		675	117	139	931	1,075	
	1280	Soil nailing , drill hole, install # 8 nail, grout 60' depth difficult		7.40	4.324		730	125	149	1,004	1,150	
	1290	Soil nailing , drill hole, install # 8 nail, grout 65' depth difficult		7	4.571		790	132	157	1,079	1,250	
	1300	Soil nailing , drill hole, install # 8 nail, grout 70' depth difficult		6.60	4.848		845	140	167	1,152	1,325	
	1310	Soil nailing , drill hole, install # 8 nail, grout 75' depth difficult	▼	6.30	5.079	▼	905	146	174	1,225	1,425	
	1390	Difficult soil nailing, grade 75, 20 min setup per hole & 90'/hr drilling										
	1400	Soil nailing , drill hole, install # 8 nail, grout 20' depth difficult	B-47G	14.10	2.270	Ea.	202	65.50	78	345.50	410	
	1410	Soil nailing , drill hole, install # 8 nail, grout 25' depth difficult		13	2.462		253	71	84.50	408.50	480	
	1420	Soil nailing , drill hole, install # 8 nail, grout 30' depth difficult		12	2.667		350	77	91.50	518.50	605	
	1430	Soil nailing , drill hole, install # 8 nail, grout 35' depth difficult		10.90	2.936		410	84.50	101	595.50	690	
	1440	Soil nailing , drill hole, install # 8 nail, grout 40' depth difficult		10.20	3.137		465	90.50	108	663.50	775	
	1450	Soil nailing , drill hole, install # 8 nail, grout 45' depth difficult		9.60	3.333		525	96	114	735	850	
	1460	Soil nailing , drill hole, install # 8 nail, grout 50' depth difficult		8.90	3.596		615	104	123	842	970	
	1470	Soil nailing , drill hole, install # 8 nail, grout 55' depth difficult		8.40	3.810		675	110	131	916	1,050	
	1480	Soil nailing , drill hole, install # 8 nail, grout 60' depth difficult		8	4		730	115	137	982	1,125	
	1490	Soil nailing , drill hole, install # 8 nail, grout 65' depth difficult		7.50	4.267		790	123	147	1,060	1,225	
	1500	Soil nailing , drill hole, install # 8 nail, grout 70' depth difficult		7.20	4.444		845	128	153	1,126	1,300	
	1510	Soil nailing , drill hole, install # 8 nail, grout 75' depth difficult	▼	6.90	4.638	▼	905	134	159	1,198	1,375	
	1590	Difficult soil nailing, grade 75, 20 min setup per hole & 100'/hr drilling										
	1600	Soil nailing , drill hole, install # 8 nail, grout 20' depth difficult	B-47G	15	2.133	Ea.	202	61.50	73.50	337	400	
	1610	Soil nailing , drill hole, install # 8 nail, grout 25' depth difficult		13.70	2.336		253	67.50	80	400.50	470	
	1620	Soil nailing , drill hole, install # 8 nail, grout 30' depth difficult		12.60	2.540		350	73	87	510	595	
	1630	Soil nailing , drill hole, install # 8 nail, grout 35' depth difficult		11.70	2.735		410	78.50	94	582.50	675	
	1640	Soil nailing , drill hole, install # 8 nail, grout 40' depth difficult		10.90	2.936		465	84.50	101	650.50	755	
	1650	Soil nailing , drill hole, install # 8 nail, grout 45' depth difficult		10.20	3.137		525	90.50	108	723.50	835	
	1660	Soil nailing , drill hole, install # 8 nail, grout 50' depth difficult		9.60	3.333		615	96	114	825	950	
	1670	Soil nailing , drill hole, install # 8 nail, grout 55' depth difficult		9.10	3.516		675	101	121	897	1,025	
	1680	Soil nailing , drill hole, install # 8 nail, grout 60' depth difficult		8.60	3.721		730	107	128	965	1,100	
	1690	Soil nailing , drill hole, install # 8 nail, grout 65' depth difficult		8.10	3.951		790	114	136	1,040	1,200	
	1700	Soil nailing , drill hole, install # 8 nail, grout 70' depth difficult		7.70	4.156		845	120	143	1,108	1,275	
	1710	Soil nailing , drill hole, install # 8 nail, grout 75' depth difficult	▼	7.40	4.324	▼	905	125	149	1,179	1,350	
	1790	Difficult soil nailing, grade 75, 20 min setup per hole & 110'/hr drilling										
	1800	Soil nailing , drill hole, install # 8 nail, grout 20' depth difficult	B-47G	15.50	2.065	Ea.	202	59.50	71	332.50	395	
	1810	Soil nailing , drill hole, install # 8 nail, grout 25' depth difficult	▼	14.10	2.270	▼	253	65.50	78	396.50	465	

Important: See the Reference Section for critical supporting data - Reference Nos., Crews, & City Cost Indexes

2 SITE CONSTRUCTION

		CREW	DAILY OUTPUT	LABOR-HOURS	UNIT	2005 BARE COSTS				TOTAL INCL O&P
02260	Excavation Support/Protection					MAT.	LABOR	EQUIP.	TOTAL	
730 1820	Soil nailing , drill hole, install # 8 nail, grout 30' depth difficult	B-47G	13.30	2.406	Ea.	350	69.50	82.50	502	585 **730**
1830	Soil nailing , drill hole, install # 8 nail, grout 35' depth difficult		12.30	2.602		410	75	89.50	574.50	665
1840	Soil nailing , drill hole, install # 8 nail, grout 40' depth difficult		11.40	2.807		465	81	96.50	642.50	745
1850	Soil nailing , drill hole, install # 8 nail, grout 45' depth difficult		10.70	2.991		525	86	103	714	820
1860	Soil nailing , drill hole, install # 8 nail, grout 50' depth difficult		10.20	3.137		615	90.50	108	813.50	935
1870	Soil nailing , drill hole, install # 8 nail, grout 55' depth difficult		9.60	3.333		675	96	114	885	1,025
1880	Soil nailing , drill hole, install # 8 nail, grout 60' depth difficult		9.10	3.516		730	101	121	952	1,100
1890	Soil nailing , drill hole, install # 8 nail, grout 65' depth difficult		8.60	3.721		790	107	128	1,025	1,175
1900	Soil nailing , drill hole, install # 8 nail, grout 70' depth difficult		8.30	3.855		845	111	132	1,088	1,250
1910	Soil nailing , drill hole, install # 8 nail, grout 75' depth difficult		7.90	4.051		905	117	139	1,161	1,325
1990	Difficult soil nailing, grade 75, 20 min setup per hole & 120'/hr drilling									
2000	Soil nailing , drill hole, install # 8 nail, grout 20' depth difficult	B-47G	16	2	Ea.	202	57.50	68.50	328	390
2010	Soil nailing , drill hole, install # 8 nail, grout 25' depth difficult		14.60	2.192		253	63	75.50	391.50	460
2020	Soil nailing , drill hole, install # 8 nail, grout 30' depth difficult		13.70	2.336		350	67.50	80	497.50	580
2030	Soil nailing , drill hole, install # 8 nail, grout 35' depth difficult		12.60	2.540		410	73	87	570	660
2040	Soil nailing , drill hole, install # 8 nail, grout 40' depth difficult		12	2.667		465	77	91.50	633.50	735
2050	Soil nailing , drill hole, install # 8 nail, grout 45' depth difficult		11.20	2.857		525	82.50	98	705.50	810
2060	Soil nailing , drill hole, install # 8 nail, grout 50' depth difficult		10.70	2.991		615	86	103	804	920
2070	Soil nailing , drill hole, install # 8 nail, grout 55' depth difficult		10	3.200		675	92	110	877	1,000
2080	Soil nailing , drill hole, install # 8 nail, grout 60' depth difficult		9.60	3.333		730	96	114	940	1,075
2090	Soil nailing , drill hole, install # 8 nail, grout 65' depth difficult		9.10	3.516		790	101	121	1,012	1,150
2100	Soil nailing , drill hole, install # 8 nail, grout 70' depth difficult		8.70	3.678		845	106	126	1,077	1,225
2110	Soil nailing , drill hole, install # 8 nail, grout 75' depth difficult		8.30	3.855		905	111	132	1,148	1,300
2990	Very difficult soil nailing, grade 75, 25 min setup & 80'/hr drilling									
3000	Soil nailing, drill, install # 8 nail, grout 20' depth very difficult	B-47G	12	2.667	Ea.	202	77	91.50	370.50	440
3010	Soil nailing, drill, install # 8 nail, grout 25' depth very difficult		10.90	2.936		253	84.50	101	438.50	520
3020	Soil nailing, drill, install # 8 nail, grout 30' depth very difficult		10	3.200		350	92	110	552	650
3030	Soil nailing, drill, install # 8 nail, grout 35' depth very difficult		9.40	3.404		410	98	117	625	730
3040	Soil nailing, drill, install # 8 nail, grout 40' depth very difficult		8.70	3.678		465	106	126	697	815
3050	Soil nailing, drill, install # 8 nail, grout 45' depth very difficult		8.10	3.951		525	114	136	775	900
3060	Soil nailing, drill, install # 8 nail, grout 50' depth very difficult		7.60	4.211		615	121	145	881	1,025
3070	Soil nailing, drill, install # 8 nail, grout 55' depth very difficult		7.30	4.384		675	126	151	952	1,100
3080	Soil nailing, drill, install # 8 nail, grout 60' depth very difficult		6.90	4.638		730	134	159	1,023	1,175
3090	Soil nailing, drill, install # 8 nail, grout 65' depth very difficult		6.50	4.923		790	142	169	1,101	1,275
3100	Soil nailing, drill, install # 8 nail, grout 70' depth very difficult		6.20	5.161		845	149	177	1,171	1,350
3110	Soil nailing, drill, install # 8 nail, grout 75' depth very difficult		5.90	5.424		905	156	186	1,247	1,450
3190	Very difficult soil nailing, grade 75, 25 min setup & 90'/hr drilling									
3200	Soil nailing, drill, install # 8 nail, grout 20' depth very difficult	B-47G	12.30	2.602	Ea.	202	75	89.50	366.50	440
3210	Soil nailing, drill, install # 8 nail, grout 25' depth very difficult		11.40	2.807		253	81	96.50	430.50	510
3220	Soil nailing, drill, install # 8 nail, grout 30' depth very difficult		10.70	2.991		350	86	103	539	630
3230	Soil nailing, drill, install # 8 nail, grout 35' depth very difficult		9.80	3.265		410	94	112	616	720
3240	Soil nailing, drill, install # 8 nail, grout 40' depth very difficult		9.20	3.478		465	100	119	684	800
3250	Soil nailing, drill, install # 8 nail, grout 45' depth very difficult		8.70	3.678		525	106	126	757	875
3260	Soil nailing, drill, install # 8 nail, grout 50' depth very difficult		8.10	3.951		615	114	136	865	1,000
3270	Soil nailing, drill, install # 8 nail, grout 55' depth very difficult		7.70	4.156		675	120	143	938	1,075
3280	Soil nailing, drill, install # 8 nail, grout 60' depth very difficult		7.40	4.324		730	125	149	1,004	1,150
3290	Soil nailing, drill, install # 8 nail, grout 65' depth very difficult		7	4.571		790	132	157	1,079	1,250
3300	Soil nailing, drill, install # 8 nail, grout 70' depth very difficult		6.70	4.776		845	138	164	1,147	1,325
3310	Soil nailing, drill, install # 8 nail, grout 75' depth very difficult		6.40	5		905	144	172	1,221	1,400
3390	Very difficult soil nailing, grade 75, 25 min setup & 100'/hr drilling									
3400	Soil nailing, drill, install # 8 nail, grout 20' depth very difficult	B-47G	13	2.462	Ea.	202	71	84.50	357.50	425
3410	Soil nailing, drill, install # 8 nail, grout 25' depth very difficult		12	2.667		253	77	91.50	421.50	495
3420	Soil nailing, drill, install # 8 nail, grout 30' depth very difficult		11.20	2.857		350	82.50	98	530.50	620
3430	Soil nailing, drill, install # 8 nail, grout 35' depth very difficult		10.40	3.077		410	88.50	106	604.50	705
3440	Soil nailing, drill, install # 8 nail, grout 40' depth very difficult		9.80	3.265		465	94	112	671	785
3450	Soil nailing, drill, install # 8 nail, grout 45' depth very difficult		9.20	3.478		525	100	119	744	860

			DAILY	LABOR-		2005 BARE COSTS				TOTAL	
02260	**Excavation Support/Protection**	CREW	OUTPUT	HOURS	UNIT	MAT.	LABOR	EQUIP.	TOTAL	INCL O&P	
730	3460	Soil nailing, drill, install # 8 nail, grout 50' depth very difficult	B-47G	8.70	3.678	Ea.	615	106	126	847	975	**730**
	3470	Soil nailing, drill, install # 8 nail, grout 55' depth very difficult		8.30	3.855		675	111	132	918	1,050	
	3480	Soil nailing, drill, install # 8 nail, grout 60' depth very difficult		7.90	4.051		730	117	139	986	1,150	
	3490	Soil nailing, drill, install # 8 nail, grout 65' depth very difficult		7.50	4.267		790	123	147	1,060	1,225	
	3500	Soil nailing, drill, install # 8 nail, grout 70' depth very difficult		7.20	4.444		845	128	153	1,126	1,300	
	3510	Soil nailing, drill, install # 8 nail, grout 75' depth very difficult		6.90	4.638		905	134	159	1,198	1,375	
	3590	Very difficult soil nailing, grade 75, 25 min setup & 110'/hr drilling										
	3600	Soil nailing, drill, install # 8 nail, grout 20' depth very difficult	B-47G	13.30	2.406	Ea.	202	69.50	82.50	354	420	
	3610	Soil nailing, drill, install # 8 nail, grout 25' depth very difficult		13.70	2.336		253	67.50	80	400.50	470	
	3620	Soil nailing, drill, install # 8 nail, grout 30' depth very difficult		11.70	2.735		350	78.50	94	522.50	610	
	3630	Soil nailing, drill, install # 8 nail, grout 35' depth very difficult		10.90	2.936		410	84.50	101	595.50	690	
	3640	Soil nailing, drill, install # 8 nail, grout 40' depth very difficult		10.20	3.137		465	90.50	108	663.50	775	
	3650	Soil nailing, drill, install # 8 nail, grout 45' depth very difficult		9.60	3.333		525	96	114	735	850	
	3660	Soil nailing, drill, install # 8 nail, grout 50' depth very difficult		9.20	3.478		615	100	119	834	960	
	3670	Soil nailing, drill, install # 8 nail, grout 55' depth very difficult		8.70	3.678		675	106	126	907	1,050	
	3680	Soil nailing, drill, install # 8 nail, grout 60' depth very difficult		8.30	3.855		730	111	132	973	1,125	
	3690	Soil nailing, drill, install # 8 nail, grout 65' depth very difficult		7.90	4.051		790	117	139	1,046	1,200	
	3700	Soil nailing, drill, install # 8 nail, grout 70' depth very difficult		7.60	4.211		845	121	145	1,111	1,275	
	3710	Soil nailing, drill, install # 8 nail, grout 75' depth very difficult		7.30	4.384		905	126	151	1,182	1,350	
	3790	Very difficult soil nailing, grade 75, 25 min setup & 120'/hr drilling										
	3800	Soil nailing, drill, install # 8 nail, grout 20' depth very difficult	B-47G	13.70	2.336	Ea.	202	67.50	80	349.50	415	
	3810	Soil nailing, drill, install # 8 nail, grout 25' depth very difficult		12.60	2.540		253	73	87	413	485	
	3820	Soil nailing, drill, install # 8 nail, grout 30' depth very difficult		12	2.667		350	77	91.50	518.50	605	
	3830	Soil nailing, drill, install # 8 nail, grout 35' depth very difficult		11.20	2.857		410	82.50	98	590.50	685	
	3840	Soil nailing, drill, install # 8 nail, grout 40' depth very difficult		10.70	2.991		465	86	103	654	760	
	3850	Soil nailing, drill, install # 8 nail, grout 45' depth very difficult		10	3.200		525	92	110	727	840	
	3860	Soil nailing, drill, install # 8 nail, grout 50' depth very difficult		9.60	3.333		615	96	114	825	950	
	3870	Soil nailing, drill, install # 8 nail, grout 55' depth very difficult		9.10	3.516		675	101	121	897	1,025	
	3880	Soil nailing, drill, install # 8 nail, grout 60' depth very difficult		8.70	3.678		730	106	126	962	1,100	
	3890	Soil nailing, drill, install # 8 nail, grout 65' depth very difficult		8.30	3.855		790	111	132	1,033	1,175	
	3900	Soil nailing, drill, install # 8 nail, grout 70' depth very difficult		8	4		845	115	137	1,097	1,250	
	3910	Soil nailing, drill, install # 8 nail, grout 75' depth very difficult		7.60	4.211		905	121	145	1,171	1,350	
	3990	Severe soil nailing, grade 75, 30 min setup per hole & 80'/hr drilling										
	4000	Soil nailing , drill hole, install # 8 nail, grout 20' depth severe	B-47G	10.70	2.991	Ea.	202	86	103	391	470	
	4010	Soil nailing , drill hole, install # 8 nail, grout 25' depth severe		9.80	3.265		253	94	112	459	545	
	4020	Soil nailing , drill hole, install # 8 nail, grout 30' depth severe		9.10	3.516		350	101	121	572	675	
	4030	Soil nailing , drill hole, install # 8 nail, grout 35' depth severe		8.60	3.721		410	107	128	645	755	
	4040	Soil nailing , drill hole, install # 8 nail, grout 40' depth severe		8	4		465	115	137	717	845	
	4050	Soil nailing , drill hole, install # 8 nail, grout 45' depth severe		7.50	4.267		525	123	147	795	925	
	4060	Soil nailing , drill hole, install # 8 nail, grout 50' depth severe		7.10	4.507		615	130	155	900	1,050	
	4070	Soil nailing , drill hole, install # 8 nail, grout 55' depth severe		6.80	4.706		675	135	162	972	1,125	
	4080	Soil nailing , drill hole, install # 8 nail, grout 60' depth severe		6.40	5		730	144	172	1,046	1,225	
	4090	Soil nailing , drill hole, install # 8 nail, grout 65' depth severe		6.10	5.246		790	151	180	1,121	1,300	
	4100	Soil nailing , drill hole, install # 8 nail, grout 70' depth severe		5.80	5.517		845	159	189	1,193	1,375	
	4110	Soil nailing , drill hole, install # 8 nail, grout 75' depth severe		5.60	5.714		905	165	196	1,266	1,475	
	4190	Severe soil nailing, grade 75, 30 min setup per hole & 90'/hr drilling										
	4200	Soil nailing, drill hole, install # 8 nail, grout 20' depth severe	B-47G	10.90	2.936	Ea.	202	84.50	101	387.50	465	
	4210	Soil nailing, drill hole, install # 8 nail, grout 25' depth severe		10.20	3.137		253	90.50	108	451.50	535	
	4220	Soil nailing, drill hole, install # 8 nail, grout 30' depth severe		9.60	3.333		350	96	114	560	660	
	4230	Soil nailing, drill hole, install # 8 nail, grout 35' depth severe		8.90	3.596		410	104	123	637	745	
	4240	Soil nailing, drill hole, install # 8 nail, grout 40' depth severe		8.40	3.810		465	110	131	706	830	
	4250	Soil nailing, drill hole, install # 8 nail, grout 45' depth severe		8	4		525	115	137	777	905	
	4260	Soil nailing, drill hole, install # 8 nail, grout 50' depth severe		7.50	4.267		615	123	147	885	1,025	
	4270	Soil nailing, drill hole, install # 8 nail, grout 55' depth severe		7.20	4.444		675	128	153	956	1,100	
	4280	Soil nailing, drill hole, install # 8 nail, grout 60' depth severe		6.90	4.638		730	134	159	1,023	1,175	
	4290	Soil nailing, drill hole, install # 8 nail, grout 65' depth severe		6.50	4.923		790	142	169	1,101	1,275	

Important: See the Reference Section for critical supporting data - Reference Nos., Crews, & City Cost Indexes

02200 | Site Preparation

02260 | Excavation Support/Protection

		CREW	DAILY OUTPUT	LABOR-HOURS	UNIT	MAT.	LABOR	EQUIP.	TOTAL	TOTAL INCL O&P	
730	4300	Soil nailing , drill hole, install # 8 nail, grout 70' depth severe	B-47G	6.20	5.161	Ea.	845	149	177	1,171	1,350	730
	4310	Soil nailing , drill hole, install # 8 nail, grout 75' depth severe	↓	6	5.333	↓	905	154	183	1,242	1,425	
	4390	Severe soil nailing, grade 75, 30 min setup per hole & 100'/hr drilling										
	4400	Soil nailing , drill hole, install # 8 nail, grout 20' depth severe	B-47G	11.40	2.807	Ea.	202	81	96.50	379.50	455	
	4410	Soil nailing , drill hole, install # 8 nail, grout 25' depth severe		10.70	2.991		253	86	103	442	525	
	4420	Soil nailing , drill hole, install # 8 nail, grout 30' depth severe		10	3.200		350	92	110	552	650	
	4430	Soil nailing , drill hole, install # 8 nail, grout 35' depth severe		9.40	3.404		410	98	117	625	730	
	4440	Soil nailing , drill hole, install # 8 nail, grout 40' depth severe		8.90	3.596		465	104	123	692	810	
	4450	Soil nailing , drill hole, install # 8 nail, grout 45' depth severe		8.40	3.810		525	110	131	766	890	
	4460	Soil nailing , drill hole, install # 8 nail, grout 50' depth severe		8	4		615	115	137	867	1,000	
	4470	Soil nailing , drill hole, install # 8 nail, grout 55' depth severe		7.60	4.211		675	121	145	941	1,075	
	4480	Soil nailing , drill hole, install # 8 nail, grout 60' depth severe		7.30	4.384		730	126	151	1,007	1,175	
	4490	Soil nailing , drill hole, install # 8 nail, grout 65' depth severe		7	4.571		790	132	157	1,079	1,250	
	4500	Soil nailing , drill hole, install # 8 nail, grout 70' depth severe		6.70	4.776		845	138	164	1,147	1,325	
	4510	Soil nailing , drill hole, install # 8 nail, grout 75' depth severe	↓	6.40	5	↓	905	144	172	1,221	1,400	
	4590	Severe soil nailing, grade 75, 30 min setup per hole & 110'/hr drilling										
	4600	Soil nailing , drill hole, install # 8 nail, grout 20' depth severe	B-47G	11.70	2.735	Ea.	202	78.50	94	374.50	445	
	4610	Soil nailing , drill hole, install # 8 nail, grout 25' depth severe		10.90	2.936		253	84.50	101	438.50	520	
	4620	Soil nailing , drill hole, install # 8 nail, grout 30' depth severe		10.40	3.077		350	88.50	106	544.50	640	
	4630	Soil nailing , drill hole, install # 8 nail, grout 35' depth severe		9.80	3.265		410	94	112	616	720	
	4640	Soil nailing , drill hole, install # 8 nail, grout 40' depth severe		9.20	3.478		465	100	119	684	800	
	4650	Soil nailing , drill hole, install # 8 nail, grout 45' depth severe		8.70	3.678		525	106	126	757	875	
	4660	Soil nailing , drill hole, install # 8 nail, grout 50' depth severe		8.40	3.810		615	110	131	856	990	
	4670	Soil nailing , drill hole, install # 8 nail, grout 55' depth severe		8	4		675	115	137	927	1,075	
	4680	Soil nailing , drill hole, install # 8 nail, grout 60' depth severe		7.60	4.211		730	121	145	996	1,150	
	4690	Soil nailing , drill hole, install # 8 nail, grout 65' depth severe		7.30	4.384		790	126	151	1,067	1,225	
	4700	Soil nailing , drill hole, install # 8 nail, grout 70' depth severe		7.10	4.507		845	130	155	1,130	1,300	
	4710	Soil nailing , drill hole, install # 8 nail, grout 75' depth severe	↓	6.80	4.706	↓	905	135	162	1,202	1,375	
	4790	Severe soil nailing, grade 75, 30 min setup per hole & 120'/hr drilling										
	4800	Soil nailing , drill hole, install # 8 nail, grout 20' depth severe	B-47G	12	2.667	Ea.	202	77	91.50	370.50	440	
	4810	Soil nailing , drill hole, install # 8 nail, grout 25' depth severe		11.20	2.857		253	82.50	98	433.50	515	
	4820	Soil nailing , drill hole, install # 8 nail, grout 30' depth severe		10.70	2.991		350	86	103	539	630	
	4830	Soil nailing , drill hole, install # 8 nail, grout 35' depth severe		10	3.200		410	92	110	612	715	
	4840	Soil nailing , drill hole, install # 8 nail, grout 40' depth severe		9.60	3.333		465	96	114	675	790	
	4850	Soil nailing , drill hole, install # 8 nail, grout 45' depth severe		9.10	3.516		525	101	121	747	865	
	4860	Soil nailing , drill hole, install # 8 nail, grout 50' depth severe		8.70	3.678		615	106	126	847	975	
	4870	Soil nailing , drill hole, install # 8 nail, grout 55' depth severe		8.30	3.855		675	111	132	918	1,050	
	4880	Soil nailing , drill hole, install # 8 nail, grout 60' depth severe		8	4		730	115	137	982	1,125	
	4890	Soil nailing , drill hole, install # 8 nail, grout 65' depth severe		7.60	4.211		790	121	145	1,056	1,225	
	4900	Soil nailing , drill hole, install # 8 nail, grout 70' depth severe		7.40	4.324		845	125	149	1,119	1,275	
	4910	Soil nailing , drill hole, install # 8 nail, grout 75' depth severe	↓	7.10	4.507	↓	905	130	155	1,190	1,375	

02300 | Earthwork

02305 | Equipment

		CREW	DAILY OUTPUT	LABOR-HOURS	UNIT	MAT.	LABOR	EQUIP.	TOTAL	TOTAL INCL O&P	
250	0010	**MOBILIZATION OR DEMOB.** (One or the other, unless noted) R01590 -100										250
	0015	Up to 25 mi haul dist (50 mi round trip for mob/demob crew)										
	0020	Dozer, loader, backhoe, excav., grader, paver, roller, 70 to 150 H.P.	B-34N	4	2	Ea.		55	112	167	207	
	0100	Above 150 HP	B-34K	3	2.667	↓		73.50	175	248.50	305	

SITE CONSTRUCTION 2

02305 | Equipment

		CREW	DAILY OUTPUT	LABOR-HOURS	UNIT	MAT.	LABOR	EQUIP.	TOTAL	TOTAL INCL O&P
250	**0300** Scraper, towed type (incl. tractor), 6 C.Y. capacity R01590-100	B-34K	3	2.667	Ea.		73.50	175	248.50	305	**250**
	0400 10 C.Y.		2.50	3.200			88	210	298	365
	0600 Self-propelled scraper, 15 C.Y.		2.50	3.200			88	210	298	365
	0700 24 C.Y.		2	4			110	262	372	455
	0900 Shovel or dragline, 3/4 C.Y.		3.60	2.222			61	146	207	254
	1000 1-1/2 C.Y.		3	2.667			73.50	175	248.50	305
	1100 Small equipment, placed in rear of, or towed by pickup truck	A-3A	8	1			27	10.65	37.65	52.50
	1150 Equip up to 70 HP, on flatbed trailer behind pickup truck	A-3D	4	2			53.50	43.50	97	129
	2000 Mob & demob truck-mounted crane up to 75 ton, driver only	1 EQHV	3.60	2.222			80		80	120
	2100 Crane, truck-mounted, over 75 ton	A-3E	2.50	6.400			203	34	237	350
	2200 Crawler-mounted, up to 75 ton	A-3F	2	8			254	278	532	690
	2300 Over 75 ton	A-3G	1.50	10.667			340	405	745	960
	2500 For each additional 5 miles haul distance, add						10%	10%		
	3000 For large pieces of equipment, allow for assembly/knockdown									
	3001 For mob/demob of vibrofloatation equip, see section 02250-900									
	3100 For mob/demob of micro-tunneling equip, see section 02441-400									
	3200 For mob/demob of pile driving equip, see section 02455-650									
	3300 For mob/demob of caisson drilling equip, see 02465-950									

02310 | Grading

		CREW	DAILY OUTPUT	LABOR-HOURS	UNIT	MAT.	LABOR	EQUIP.	TOTAL	TOTAL INCL O&P
100	**0010 FINISH GRADING**										**100**
	0012 Finish grading area to be paved with grader, small area	B-11L	400	.040	S.Y.		1.23	1.14	2.37	3.14
	0100 Large area		2,000	.008			.25	.23	.48	.63
	0200 Grade subgrade for base course, roadways		3,500	.005			.14	.13	.27	.35
	1020 For large parking lots	B-32C	5,000	.010			.30	.31	.61	.80
	1050 For small irregular areas	"	2,000	.024			.74	.78	1.52	2
	1100 Fine grade for slab on grade, machine	B-11L	1,040	.015			.47	.44	.91	1.20
	1150 Hand grading	B-18	700	.034			.94	.05	.99	1.52
	1200 Fine grade granular base for sidewalks and bikeways	B-62	1,200	.020			.58	.13	.71	1.03
	2550 Hand grade select gravel	2 Clab	60	.267	C.S.F.		7.10		7.10	11.10
	3000 Hand grade select gravel, including compaction, 4" deep	B-18	555	.043	S.Y.		1.18	.07	1.25	1.91
	3100 6" deep		400	.060			1.64	.09	1.73	2.66
	3120 8" deep		300	.080			2.19	.12	2.31	3.55
	3300 Finishing grading slopes, gentle	B-11L	8,900	.002			.06	.05	.11	.14
	3310 Steep slopes	"	7,100	.002			.07	.06	.13	.18

02315 | Excavation and Fill

		CREW	DAILY OUTPUT	LABOR-HOURS	UNIT	MAT.	LABOR	EQUIP.	TOTAL	TOTAL INCL O&P
110	**0010 BACKFILL, GENERAL** R02315-300										**110**
	0015 By hand, no compaction, light soil	1 Clab	14	.571	L.C.Y.		15.25		15.25	23.50
	0100 Heavy soil		11	.727	"		19.40		19.40	30
	0300 Compaction in 6" layers, hand tamp, add to above		20.60	.388	E.C.Y.		10.35		10.35	16.15
	0400 Roller compaction operator walking, add	B-10A	100	.120			3.84	1.23	5.07	7.20
	0500 Air tamp, add	B-9D	190	.211			5.70	.93	6.63	9.95
	0600 Vibrating plate, add	A-1D	60	.133			3.56	.42	3.98	6
	0800 Compaction in 12" layers, hand tamp, add to above	1 Clab	34	.235			6.30		6.30	9.80
	0900 Roller compaction operator walking, add	B-10A	150	.080			2.56	.82	3.38	4.79
	1000 Air tamp, add	B-9	285	.140			3.80	.50	4.30	6.45
	1100 Vibrating plate, add	A-1E	90	.089			2.37	.41	2.78	4.14
	3000 For flowable fill, see div. 03310-220									
120	**0010 BACKFILL, STRUCTURAL** Dozer or F.E. loader										**120**
	0020 From existing stockpile, no compaction									
	2000 80 H.P., 50' haul, sand & gravel	B-10L	1,100	.011	L.C.Y.		.35	.29	.64	.84
	2020 Common earth		975	.012			.39	.32	.71	.96
	2040 Clay		850	.014			.45	.37	.82	1.10
	2200 150' haul, sand & gravel		550	.022			.70	.57	1.27	1.69

Important: See the Reference Section for critical supporting data - Reference Nos., Crews, & City Cost Indexes

2 SITE CONSTRUCTION

02315 | Excavation and Fill

		CREW	DAILY OUTPUT	LABOR-HOURS	UNIT	2005 BARE COSTS				TOTAL INCL O&P
						MAT.	LABOR	EQUIP.	TOTAL	
120	2220	Common earth	B-10L	490	.024	L.C.Y.		.78	.64	1.42	1.90
	2240	Clay		425	.028			.90	.74	1.64	2.19
	2400	300' haul, sand & gravel		370	.032			1.04	.85	1.89	2.52
	2420	Common earth		330	.036			1.16	.95	2.11	2.82
	2440	Clay		290	.041			1.32	1.09	2.41	3.20
	3000	105 H.P., 50' haul, sand & gravel	B-10W	1,350	.009			.28	.34	.62	.80
	3020	Common earth		1,225	.010			.31	.37	.68	.89
	3040	Clay		1,100	.011			.35	.41	.76	.98
	3200	150' haul, sand & gravel		670	.018			.57	.68	1.25	1.62
	3220	Common earth		610	.020			.63	.74	1.37	1.78
	3240	Clay		550	.022			.70	.83	1.53	1.97
	3300	300' haul, sand & gravel		465	.026			.83	.98	1.81	2.33
	3320	Common earth		415	.029			.93	1.09	2.02	2.61
	3340	Clay		370	.032			1.04	1.23	2.27	2.93
	4000	200 H.P., 50' haul, sand & gravel	B-10B	2,500	.005			.15	.37	.52	.63
	4020	Common earth		2,200	.005			.17	.42	.59	.73
	4040	Clay		1,950	.006			.20	.47	.67	.82
	4200	150' haul, sand & gravel		1,225	.010			.31	.75	1.06	1.31
	4220	Common earth		1,100	.011			.35	.84	1.19	1.45
	4240	Clay		975	.012			.39	.94	1.33	1.64
	4400	300' haul, sand & gravel		805	.015			.48	1.14	1.62	1.99
	4420	Common earth		735	.016			.52	1.25	1.77	2.17
	4440	Clay		660	.018			.58	1.39	1.97	2.41
	5000	300 H.P., 50' haul, sand & gravel	B-10M	3,170	.004			.12	.38	.50	.60
	5020	Common earth		2,900	.004			.13	.41	.54	.65
	5040	Clay		2,700	.004			.14	.44	.58	.71
	5200	150' haul, sand & gravel		2,200	.005			.17	.54	.71	.87
	5220	Common earth		1,950	.006			.20	.61	.81	.97
	5240	Clay		1,700	.007			.23	.70	.93	1.11
	5400	300' haul, sand & gravel		1,500	.008			.26	.80	1.06	1.27
	5420	Common earth		1,350	.009			.28	.89	1.17	1.40
	5440	Clay		1,225	.010			.31	.98	1.29	1.55
	6000	For compaction, see div. 02315-310									
	6010	For trench backfill, see div. 02315-610 & 02315-620									
210	0010	**BORROW, LOADING AND/OR SPREADING**									
	4000	Common earth, shovel, 1 C.Y. bucket	B-12N	840	.019	B.C.Y.	8.10	.60	1.06	9.76	11
	4010	1-1/2 C.Y. bucket	B-12O	1,135	.014		8.10	.44	.84	9.38	10.50
	4020	3 C.Y. bucket	B-12T	1,800	.009		8.10	.28	.70	9.08	10.10
	4030	Front end loader, wheel mounted									
	4050	3/4 C.Y. bucket	B-10R	550	.022	B.C.Y.	8.10	.70	.35	9.15	10.35
	4060	1-1/2 C.Y. bucket	B-10S	970	.012		8.10	.40	.25	8.75	9.75
	4070	3 C.Y. bucket	B-10T	1,575	.008		8.10	.24	.19	8.53	9.50
	4080	5 C.Y. bucket	B-10U	2,600	.005		8.10	.15	.27	8.52	9.40
	5000	Select granular fill, shovel, 1 C.Y. bucket	B-12N	925	.017		7.90	.54	.96	9.40	10.55
	5010	1-1/2 C.Y. bucket	B-12O	1,250	.013		7.90	.40	.76	9.06	10.10
	5020	3 C.Y. bucket	B-12T	1,980	.008		7.90	.25	.64	8.79	9.75
	5030	Front end loader, wheel mounted									
	5050	3/4 C.Y. bucket	B-10R	800	.015	B.C.Y.	7.90	.48	.24	8.62	9.65
	5060	1-1/2 C.Y. bucket	B-10S	1,065	.011		7.90	.36	.23	8.49	9.45
	5070	3 C.Y. bucket	B-10T	1,735	.007		7.90	.22	.18	8.30	9.20
	5080	5 C.Y. bucket	B-10U	2,850	.004		7.90	.13	.25	8.28	9.10
	6000	Clay, till, or blasted rock, shovel, 1 C.Y. bucket	B-12N	715	.022		6	.70	1.25	7.95	9.05
	6010	1-1/2 C.Y. bucket	B-12O	965	.017		6	.52	.99	7.51	8.50
	6020	3 C.Y. bucket	B-12T	1,530	.010		6	.33	.82	7.15	8
	6030	Front end loader, wheel mounted									
	6035	3/4 C.Y. bucket	B-10R	465	.026	B.C.Y.	6	.83	.42	7.25	8.30

SITE CONSTRUCTION 2

SITE CONSTRUCTION 2

02315	Excavation and Fill	CREW	DAILY OUTPUT	LABOR-HOURS	UNIT	2005 BARE COSTS				TOTAL INCL O&P	
						MAT.	LABOR	EQUIP.	TOTAL		
210											**210**
6040	1-1/2 C.Y. bucket	B-10S	825	.015	B.C.Y.	6	.47	.29	6.76	7.65	
6045	3 C.Y. bucket	B-10T	1,340	.009		6	.29	.23	6.52	7.30	
6050	5 C.Y. bucket	B-10U	2,200	.005	↓	6	.17	.32	6.49	7.20	
6060	Front end loader, track mounted										
6065	1-1/2 C.Y. bucket	B-10N	715	.017	B.C.Y.	6	.54	.44	6.98	7.90	
6070	3 C.Y. bucket	B-10P	1,190	.010		6	.32	.66	6.98	7.80	
6075	5 C.Y. bucket	B-10Q	1,835	.007		6	.21	.60	6.81	7.60	
7000	Topsoil or loam from stockpile, shovel, 1 C.Y. bucket	B-12N	840	.019		21	.60	1.06	22.66	25.50	
7010	1-1/2 C.Y. bucket	B-12O	1,135	.014		21	.44	.84	22.28	25	
7020	3 C.Y. bucket	B-12T	1,800	.009	↓	21	.28	.70	21.98	24.50	
7030	Front end loader, wheel mounted										
7050	3/4 C.Y. bucket	B-10R	550	.022	B.C.Y.	21	.70	.35	22.05	25	
7060	1-1/2 C.Y. bucket	B-10S	970	.012		21	.40	.25	21.65	24.50	
7070	3 C.Y. bucket	B-10T	1,575	.008		21	.24	.19	21.43	24	
7080	5 C.Y. bucket	B-10U	2,600	.005	↓	21	.15	.27	21.42	24	
8900	For larger hauling units, deduct from above								30%		
9000	Hauling only, excavated or borrow material, see div. 02315-490										
9200	For flowable fill, see section 03310-220										
310											**310**
0010	**COMPACTION, GENERAL** R02315-300										
5000	Riding, vibrating roller, 6" lifts, 2 passes	B-10Y	3,000	.004	E.C.Y.		.13	.11	.24	.32	
5020	3 passes		2,300	.005			.17	.15	.32	.42	
5040	4 passes		1,900	.006			.20	.18	.38	.51	
5060	12" lifts, 2 passes		5,200	.002			.07	.07	.14	.18	
5080	3 passes		3,500	.003			.11	.10	.21	.28	
5100	4 passes	↓	2,600	.005			.15	.13	.28	.37	
5600	Sheepsfoot or wobbly wheel roller, 6" lifts, 2 passes	B-10G	2,400	.005			.16	.33	.49	.60	
5620	3 passes		1,735	.007			.22	.46	.68	.84	
5640	4 passes		1,300	.009			.30	.61	.91	1.12	
5680	12" lifts, 2 passes		5,200	.002			.07	.15	.22	.28	
5700	3 passes		3,500	.003			.11	.23	.34	.42	
5720	4 passes	↓	2,600	.005			.15	.31	.46	.56	
6000	Towed sheepsfoot or wobbly wheel roller, 6" lifts, 2 passes	B-10D	10,000	.001			.04	.15	.19	.23	
6020	3 passes		2,000	.006			.19	.76	.95	1.13	
6030	4 passes		1,500	.008			.26	1.02	1.28	1.51	
6050	12" lifts, 2 passes		6,000	.002			.06	.26	.32	.38	
6060	3 passes		4,000	.003			.10	.38	.48	.57	
6070	4 passes	↓	3,000	.004			.13	.51	.64	.75	
6200	Vibrating roller, 6" lifts, 2 passes	B-10C	2,600	.005			.15	.58	.73	.86	
6210	3 passes		1,735	.007			.22	.87	1.09	1.30	
6220	4 passes		1,300	.009			.30	1.16	1.46	1.73	
6250	12" lifts, 2 passes		5,200	.002			.07	.29	.36	.43	
6260	3 passes		3,465	.003			.11	.44	.55	.65	
6270	4 passes	↓	2,600	.005			.15	.58	.73	.86	
7000	Walk behind, vibrating plate 18" wide, 6" lifts, 2 passes	A-1D	200	.040			1.07	.13	1.20	1.80	
7020	3 passes		185	.043			1.15	.14	1.29	1.95	
7040	4 passes	↓	140	.057			1.53	.18	1.71	2.57	
7200	12" lifts, 2 passes	A-1E	560	.014			.38	.07	.45	.66	
7220	3 passes		375	.021			.57	.10	.67	1	
7240	4 passes	↓	280	.029			.76	.13	.89	1.33	
7500	Vibrating roller 24" wide, 6" lifts, 2 passes	B-10A	420	.029			.91	.29	1.20	1.71	
7520	3 passes		280	.043			1.37	.44	1.81	2.57	
7540	4 passes		210	.057			1.83	.58	2.41	3.42	
7600	12" lifts, 2 passes		840	.014			.46	.15	.61	.86	
7620	3 passes		560	.021			.69	.22	.91	1.28	
7640	4 passes	↓	420	.029	↓		.91	.29	1.20	1.71	
8000	Rammer tamper, 6" to 11", 4" lifts, 2 passes	A-1F	130	.062			1.64	.27	1.91	2.85	

02315 | Excavation and Fill

		CREW	DAILY OUTPUT	LABOR-HOURS	UNIT	2005 BARE COSTS				TOTAL INCL O&P
						MAT.	LABOR	EQUIP.	TOTAL	
310 8050	3 passes R02315 -300	A-1F	97	.082	E.C.Y.		2.20	.36	2.56	3.83 **310**
8100	4 passes		65	.123			3.29	.54	3.83	5.70
8200	8" lifts, 2 passes		260	.031			.82	.13	.95	1.43
8250	3 passes		195	.041			1.10	.18	1.28	1.90
8300	4 passes		130	.062			1.64	.27	1.91	2.85
8400	13" to 18", 4" lifts, 2 passes	A-1G	390	.021			.55	.09	.64	.95
8450	3 passes		290	.028			.74	.12	.86	1.28
8500	4 passes		195	.041			1.10	.18	1.28	1.90
8600	8" lifts, 2 passes		780	.010			.27	.04	.31	.48
8650	3 passes		585	.014			.37	.06	.43	.64
8700	4 passes		390	.021			.55	.09	.64	.95
9000	Water, 3000 gal. truck, 3 mile haul	B-45	1,888	.008		.20	.26	.28	.74	.93
9010	6 mile haul		1,444	.011		.20	.34	.36	.90	1.14
9020	12 mile haul		1,000	.016		.20	.50	.53	1.23	1.55
9030	6000 gal. wagon, 3 mile haul	B-59	2,000	.004		.20	.11	.16	.47	.57
9040	6 mile haul	"	1,600	.005		.20	.14	.20	.54	.65
412 0010	**DRILLING ONLY ROCK** R02315 -450									**412**
0015	Drilling only rock, 2" hole for rock bolts	B-47	316	.076	L.F.		2.24	3.16	5.40	6.90
0800	2-1/2" hole for pre-splitting		600	.040			1.18	1.66	2.84	3.64
4600	Quarry operations, 2-1/2" to 3-1/2" diameter		715	.034			.99	1.40	2.39	3.05
4610	6" diameter drill holes	B-47A	1,350	.018			.56	.61	1.17	1.52
416 0010	**DRILLING AND BLASTING ROCK**									**416**
0020	Rock, open face, under 1500 CY	B-47	225	.107	B.C.Y.	1.89	3.14	4.43	9.46	11.80
0100	Over 1500 C.Y.		300	.080		1.89	2.36	3.33	7.58	9.35
0200	Areas where blasting mats are required, under 1500 C.Y.		175	.137		1.89	4.04	5.70	11.63	14.55
0250	Over 1500 C.Y.		250	.096		1.89	2.83	3.99	8.71	10.80
0300	Bulk drilling and blasting, can vary greatly, average									5.50
0500	Pits, average									22
1300	Deep hole method, up to 1500 C.Y.	B-47	50	.480		1.89	14.15	19.95	35.99	46
1400	Over 1500 C.Y.		66	.364		1.89	10.70	15.10	27.69	35
1900	Restricted areas, up to 1500 C.Y.		13	1.846		1.89	54.50	76.50	132.89	170
2000	Over 1500 C.Y.		20	1.200		1.89	35.50	50	87.39	112
2200	Trenches, up to 1500 C.Y.		22	1.091		5.50	32	45.50	83	106
2300	Over 1500 C.Y.		26	.923		5.50	27	38.50	71	90
2500	Pier holes, up to 1500 C.Y.		22	1.091		1.89	32	45.50	79.39	102
2600	Over 1500 C.Y.		31	.774		1.89	23	32	56.89	72.50
2800	Boulders under 1/2 C.Y., loaded on truck, no hauling	B-100	80	.150			4.80	6.90	11.70	14.90
2900	Boulders, drilled, blasted	B-47	100	.240		1.89	7.10	10	18.99	24
3100	Jackhammer operators with foreman compressor, air tools	B-9	1	40	Day		1,075	142	1,217	1,825
3300	Track drill, compressor, operator and foreman	B-47	1	24	"		710	1,000	1,710	2,200
3500	Blasting caps				Ea.	3.42			3.42	3.76
3700	Explosives					.29			.29	.32
3900	Blasting mats, rent, for first day					94			94	104
4000	Per added day					31.50			31.50	34.50
4200	Preblast survey for 6 room house, individual lot, minimum	A-6	2.40	6.667			226	25.50	251.50	375
4300	Maximum	"	1.35	11.852			400	45	445	665
4500	City block within zone of influence, minimum	A-8	25,200	.001	S.F.		.05		.05	.07
4600	Maximum	"	15,100	.002	"		.08		.08	.12
5000	Excavate and load boulders, less than 0.5 C.Y.	B-10T	80	.150	B.C.Y.		4.80	3.81	8.61	11.50
5020	0.5 C.Y. to 1 C.Y.	B-10U	100	.120			3.84	7	10.84	13.55
5200	Excavate and load blasted rock, 3 C.Y. power shovel	B-12T	1,530	.010			.33	.82	1.15	1.41
5400	Haul boulders, 25 Ton off-highway dump, 1 mile round trip	B-34E	330	.024			.67	2.76	3.43	4.06
5420	2 mile round trip		275	.029			.80	3.32	4.12	4.87
5440	3 mile round trip		225	.036			.98	4.05	5.03	5.95
5460	4 mile round trip		200	.040			1.10	4.56	5.66	6.70

SITE CONSTRUCTION 2

2 SITE CONSTRUCTION

02315 | Excavation and Fill

			DAILY	LABOR-		2005 BARE COSTS				TOTAL
		CREW	OUTPUT	HOURS	UNIT	MAT.	LABOR	EQUIP.	TOTAL	INCL O&P
416 5600	Bury boulders on site, less than 0.5 C.Y., 300 H.P. dozer									**416**
5620	150' haul	B-10M	310	.039	B.C.Y.		1.24	3.85	5.09	6.10
5640	300' haul		210	.057			1.83	5.70	7.53	9.05
5800	0.5 to 1 C.Y., 300 H.P. dozer, 150' haul		300	.040			1.28	3.98	5.26	6.35
5820	300' haul		200	.060			1.92	5.95	7.87	9.45
418 0010	**RIPPING** R02315 -450									**418**
0020	Ripping, trap rock, soft, 300 HP dozer, ideal conditions	B-11S	700	.017	B.C.Y.		.55	1.81	2.36	2.82
1500	Adverse conditions		660	.018			.58	1.92	2.50	2.99
1600	Medium hard, 300 HP dozer, ideal conditons		600	.020			.64	2.11	2.75	3.29
1700	Adverse conditions		540	.022			.71	2.34	3.05	3.66
2000	Very hard, 410 HP dozer, ideal conditions	B-11T	350	.034			1.10	4.58	5.68	6.70
2100	Adverse conditions	"	310	.039			1.24	5.15	6.39	7.60
2200	Shale, soft, 300 HP dozer, ideal conditons	B-11S	1,500	.008			.26	.84	1.10	1.32
2300	Adverse conditions	"	1,350	.009			.28	.94	1.22	1.46
2310	Grader rear ripper, 180 H.P. ideal conditions	B-11J	740	.022			.66	.71	1.37	1.79
2320	Adverse conditions	"	630	.025			.78	.84	1.62	2.11
2400	Medium hard, 300 HP dozer, ideal conditons	B-11S	1,200	.010			.32	1.06	1.38	1.65
2500	Adverse conditions	"	1,080	.011			.36	1.17	1.53	1.83
2510	Grader rear ripper, 180 H.P. ideal conditions	B-11J	625	.026			.79	.84	1.63	2.13
2520	Adverse conditions	"	530	.030			.93	1	1.93	2.52
2600	Very hard, 410 HP dozer, ideal conditons	B-11T	800	.015			.48	2.01	2.49	2.94
2700	Adverse conditions	"	720	.017			.53	2.23	2.76	3.26
2800	Till, boulder clay/hardpan, soft, 300 H.P. dozer, ideal conditions	B-11S	7,000	.002			.05	.18	.23	.28
2810	Adverse conditions	"	6,300	.002			.06	.20	.26	.31
2815	Grader rear ripper, 180 H.P. ideal conditions	B-11J	1,500	.011			.33	.35	.68	.89
2816	Adverse conditions	"	1,275	.013			.39	.41	.80	1.05
2820	Medium hard, 300 H.P. dozer, ideal conditions	B-11S	6,000	.002			.06	.21	.27	.33
2830	Adverse conditions	"	5,400	.002			.07	.23	.30	.37
2835	Grader rear ripper, 180 H.P. ideal conditions	B-11J	740	.022			.66	.71	1.37	1.79
2836	Adverse conditions	"	630	.025			.78	.84	1.62	2.11
2840	Very hard, 410 H.P. dozer, ideal conditions	B-11T	5,000	.002			.08	.32	.40	.47
2850	Adverse conditions	"	4,500	.003			.09	.36	.45	.52
3000	Dozing ripped material, 200 HP, 100' haul	B-10B	700	.017			.55	1.31	1.86	2.27
3050	300' haul	"	250	.048			1.54	3.68	5.22	6.40
3200	300 HP, 100' haul	B-10M	1,150	.010			.33	1.04	1.37	1.65
3250	300' haul	"	400	.030			.96	2.99	3.95	4.75
3400	410 HP, 100' haul	B-10X	1,680	.007			.23	.91	1.14	1.35
3450	300' haul	"	600	.020			.64	2.54	3.18	3.76
3635	Ripping, trap rock, soft, 300 HP dozer, ideal conditions	B-11S	700	.017			.55	1.81	2.36	2.82
424 0010	**EXCAVATING, BULK BANK MEASURE** Common earth piled R02315 -400									**424**
0020	For loading onto trucks, add								15%	15%
0050	For mobilization and demobilization, see division 02305-250 R02315 -450									
0100	For hauling, see division 02315-490									
0200	Backhoe, hydraulic, crawler mtd., 1 C.Y. cap. = 75 C.Y./hr.	B-12A	600	.027	B.C.Y.		.83	.93	1.76	2.30
0250	1-1/2 C.Y. cap. = 100 C.Y./hr.	B-12B	800	.020			.63	.90	1.53	1.95
0260	2 C.Y. cap. = 130 C.Y./hr.	B-12C	1,040	.015			.48	.87	1.35	1.70
0300	3 C.Y. cap. = 160 C.Y./hr.	B-12D	1,280	.013			.39	1.58	1.97	2.34
0310	Wheel mounted, 1/2 C.Y. cap. = 30 C.Y./hr.	B-12E	240	.067			2.09	1.39	3.48	4.71
0360	3/4 C.Y. cap. = 45 C.Y./hr.	B-12F	360	.044			1.39	1.32	2.71	3.58
0500	Clamshell, 1/2 C.Y. cap. = 20 C.Y./hr.	B-12G	160	.100			3.13	3.39	6.52	8.50
0550	1 C.Y. cap. = 35 C.Y./hr.	B-12H	280	.057			1.79	3.13	4.92	6.15
0950	Dragline, 1/2 C.Y. cap. = 30 C.Y./hr.	B-12I	240	.067			2.09	2.71	4.80	6.15
1000	3/4 C.Y. cap. = 35 C.Y./hr.	"	280	.057			1.79	2.32	4.11	5.30
1050	1-1/2 C.Y. cap. = 65 C.Y./hr.	B-12P	520	.031			.96	1.77	2.73	3.41
1100	3 C.Y. cap. = 112 C.Y./hr.	B-12V	900	.018			.56	1.35	1.91	2.34

Important: See the Reference Section for critical supporting data - Reference Nos., Crews, & City Cost Indexes

02315 | Excavation and Fill

		CREW	DAILY OUTPUT	LABOR-HOURS	UNIT	2005 BARE COSTS				TOTAL INCL O&P
						MAT.	LABOR	EQUIP.	TOTAL	
424										**424**
1200	Front end loader, track mtd., 1-1/2 C.Y. cap. = 70 C.Y./hr. [R02315 -400]	B-10N	560	.021	B.C.Y.		.69	.56	1.25	1.65
1250	2-1/2 C.Y. cap. = 95 C.Y./hr.	B-10O	760	.016			.51	.73	1.24	1.57
1300	3 C.Y. cap. = 130 C.Y./hr. [R02315 -450]	B-10P	1,040	.012			.37	.76	1.13	1.39
1350	5 C.Y. cap. = 160 C.Y./hr.	B-10Q	1,280	.009			.30	.86	1.16	1.40
1500	Wheel mounted, 3/4 C.Y. cap. = 45 C.Y./hr.	B-10R	360	.033			1.07	.54	1.61	2.21
1550	1-1/2 C.Y. cap. = 80 C.Y./hr.	B-10S	640	.019			.60	.38	.98	1.32
1600	2-1/4 C.Y. cap. = 100 C.Y./hr.	B-10T	800	.015			.48	.38	.86	1.15
1601	3 C.Y. cap. = 140 C.Y./hr.	"	1,120	.011			.34	.27	.61	.82
1650	5 C.Y. cap. = 185 C.Y./hr.	B-10U	1,480	.008			.26	.47	.73	.91
1800	Hydraulic excavator, truck mtd, 1/2 C.Y. = 30 C.Y./hr.	B-12J	240	.067			2.09	3.49	5.58	7.05
1850	48 inch bucket, 1 C.Y. = 45 C.Y./hr.	B-12K	360	.044			1.39	2.70	4.09	5.10
3700	Shovel, 1/2 C.Y. capacity = 55 C.Y./hr.	B-12L	440	.036			1.14	1.26	2.40	3.13
3750	3/4 C.Y. capacity = 85 C.Y./hr.	B-12M	680	.024			.74	1	1.74	2.23
3800	1 C.Y. capacity = 120 C.Y./hr.	B-12N	960	.017			.52	.93	1.45	1.82
3850	1-1/2 C.Y. capacity = 160 C.Y./hr.	B-12O	1,280	.013			.39	.75	1.14	1.42
3900	3 C.Y. cap. = 250 C.Y./hr.	B-12T	2,000	.008			.25	.63	.88	1.07
4000	For soft soil or sand, deduct								15%	15%
4100	For heavy soil or stiff clay, add								60%	60%
4200	For wet excavation with clamshell or dragline, add								100%	100%
4250	All other equipment, add								50%	50%
4400	Clamshell in sheeting or cofferdam, minimum	B-12H	160	.100			3.13	5.50	8.63	10.85
4450	Maximum	"	60	.267			8.35	14.60	22.95	29
8000	For hauling excavated material, see div. 02315-490									
432	**0010**	**EXCAVATING, BULK, DOZER** Open site [R02315 -400]				B.C.Y.					**432**
2000	80 H.P., 50' haul, sand & gravel	B-10L	460	.026	B.C.Y.		.83	.68	1.51	2.02
2020	Common earth		400	.030			.96	.79	1.75	2.33
2040	Clay		250	.048			1.54	1.26	2.80	3.73
2200	150' haul, sand & gravel		230	.052			1.67	1.37	3.04	4.05
2220	Common earth		200	.060			1.92	1.57	3.49	4.65
2240	Clay		125	.096			3.07	2.52	5.59	7.45
2400	300' haul, sand & gravel		120	.100			3.20	2.62	5.82	7.75
2420	Common earth		100	.120			3.84	3.15	6.99	9.30
2440	Clay		65	.185			5.90	4.84	10.74	14.35
3000	105 H.P., 50' haul, sand & gravel	B-10W	700	.017			.55	.65	1.20	1.54
3020	Common earth		610	.020			.63	.74	1.37	1.78
3040	Clay		385	.031			1	1.18	2.18	2.82
3200	150' haul, sand & gravel		310	.039			1.24	1.46	2.70	3.49
3220	Common earth		270	.044			1.42	1.68	3.10	4.01
3240	Clay		170	.071			2.26	2.67	4.93	6.35
3300	300' haul, sand & gravel		140	.086			2.74	3.24	5.98	7.75
3320	Common earth		120	.100			3.20	3.78	6.98	9.05
3340	Clay		100	.120			3.84	4.54	8.38	10.85
4000	200 H.P., 50' haul, sand & gravel	B-10B	1,400	.009			.27	.66	.93	1.14
4020	Common earth		1,230	.010			.31	.75	1.06	1.29
4040	Clay		770	.016			.50	1.19	1.69	2.07
4200	150' haul, sand & gravel		595	.020			.65	1.55	2.20	2.68
4220	Common earth		516	.023			.74	1.78	2.52	3.09
4240	Clay		325	.037			1.18	2.83	4.01	4.91
4400	300' haul, sand & gravel		310	.039			1.24	2.97	4.21	5.15
4420	Common earth		270	.044			1.42	3.41	4.83	5.90
4440	Clay		170	.071			2.26	5.40	7.66	9.40
5000	300 H.P., 50' haul, sand & gravel	B-10M	1,900	.006			.20	.63	.83	1
5020	Common earth		1,650	.007			.23	.72	.95	1.15
5040	Clay		1,025	.012			.37	1.17	1.54	1.85
5200	150' haul, sand & gravel		920	.013			.42	1.30	1.72	2.06

SITE CONSTRUCTION 2

02315	Excavation and Fill	CREW	DAILY OUTPUT	LABOR-HOURS	UNIT	2005 BARE COSTS				TOTAL INCL O&P	
						MAT.	LABOR	EQUIP.	TOTAL		
432	5220	Common earth	B-10M R02315 -400	800	.015	B.C.Y.		.48	1.49	1.97	2.37	**432**
	5240	Clay		500	.024			.77	2.39	3.16	3.80	
	5400	300' haul, sand & gravel		470	.026			.82	2.54	3.36	4.04	
	5420	Common earth		410	.029			.94	2.91	3.85	4.63	
	5440	Clay		250	.048			1.54	4.78	6.32	7.60	
	5500	460 H.P., 50' haul, sand & gravel	B-10X	1,930	.006			.20	.79	.99	1.17	
	5510	Common earth		1,680	.007			.23	.91	1.14	1.35	
	5520	Clay		1,050	.011			.37	1.45	1.82	2.16	
	5530	150' haul, sand & gravel		1,290	.009			.30	1.18	1.48	1.75	
	5540	Common earth		1,120	.011			.34	1.36	1.70	2.02	
	5550	Clay		700	.017			.55	2.18	2.73	3.22	
	5560	300' haul, sand & gravel		660	.018			.58	2.31	2.89	3.42	
	5570	Common earth		575	.021			.67	2.65	3.32	3.94	
	5580	Clay		350	.034			1.10	4.35	5.45	6.45	
	6000	700 H.P., 50' haul, sand & gravel	B-10V	3,500	.003			.11	.90	1.01	1.16	
	6010	Common earth		3,035	.004			.13	1.03	1.16	1.33	
	6020	Clay		1,925	.006			.20	1.63	1.83	2.09	
	6030	150' haul, sand & gravel		2,025	.006			.19	1.55	1.74	2	
	6040	Common earth		1,750	.007			.22	1.79	2.01	2.30	
	6050	Clay		1,100	.011			.35	2.85	3.20	3.67	
	6060	300' haul, sand & gravel		1,030	.012			.37	3.05	3.42	3.92	
	6070	Common earth		900	.013			.43	3.49	3.92	4.49	
	6080	Clay		550	.022			.70	5.70	6.40	7.35	
	6090	For dozer with ripper, see div. 02315-418										
442	0010	**EXCAVATION, BULK, DRAG LINE**										**442**
	0011	Excavate and load on truck, bank measure										
	0012	Bucket drag line, 3/4 C.Y., sand/gravel	B-12I	440	.036	B.C.Y.		1.14	1.48	2.62	3.36	
	0100	Light clay		310	.052			1.62	2.09	3.71	4.77	
	0110	Heavy clay		250	.064			2	2.60	4.60	5.90	
	0120	Unclassified soil		280	.057			1.79	2.32	4.11	5.30	
	0200	1-1/2 C.Y. bucket, sand/gravel	B-12P	575	.028			.87	1.60	2.47	3.09	
	0210	Light clay		440	.036			1.14	2.09	3.23	4.04	
	0220	Heavy clay		352	.045			1.42	2.61	4.03	5.05	
	0230	Unclassified soil		300	.053			1.67	3.06	4.73	5.90	
	0300	3 C.Y., sand/gravel	B-12V	720	.022			.70	1.69	2.39	2.92	
	0310	Light clay		700	.023			.72	1.74	2.46	3	
	0320	Heavy clay		600	.027			.83	2.03	2.86	3.51	
	0330	Unclassified soil		550	.029			.91	2.21	3.12	3.82	
452	0010	**EXCAVATION, BULK, SCRAPERS**	R02315 -450									**452**
	0100	Elev. scraper 11 C.Y., sand & gravel 1500' haul, 1/4 dozer	B-33F	690	.020	B.C.Y.		.66	1.62	2.28	2.78	
	0150	3000' haul		610	.023			.74	1.83	2.57	3.15	
	0200	5000' haul		505	.028			.90	2.21	3.11	3.80	
	0300	Common earth, 1500' haul		600	.023			.76	1.86	2.62	3.20	
	0350	3000' haul		530	.026			.86	2.11	2.97	3.62	
	0400	5000' haul		440	.032			1.03	2.54	3.57	4.36	
	0500	Clay, 1500' haul		375	.037			1.21	2.98	4.19	5.10	
	0550	3000' haul		330	.042			1.37	3.39	4.76	5.80	
	0600	5000' haul		275	.051			1.65	4.07	5.72	6.95	
	1000	Self propelled scraper, 14 C.Y. 1/4 push dozer, sand										
	1050	Sand and gravel, 1500' haul	B-33D	920	.015	B.C.Y.		.49	2.01	2.50	2.96	
	1100	3000' haul		805	.017			.56	2.30	2.86	3.38	
	1200	5000' haul		645	.022			.70	2.87	3.57	4.23	
	1300	Common earth, 1500' haul		800	.017			.57	2.31	2.88	3.40	
	1350	3000' haul		700	.020			.65	2.64	3.29	3.89	
	1400	5000' haul		560	.025			.81	3.31	4.12	4.87	
	1500	Clay, 1500' haul		500	.028			.91	3.70	4.61	5.45	

Important: See the Reference Section for critical supporting data - Reference Nos., Crews, & City Cost Indexes

SITE CONSTRUCTION | 2

02315 | Excavation and Fill

			CREW	DAILY OUTPUT	LABOR-HOURS	UNIT	2005 BARE COSTS				TOTAL INCL O&P	
							MAT.	LABOR	EQUIP.	TOTAL		
452	1550	3000' haul	B-33D	440	.032	B.C.Y.		1.03	4.21	5.24	6.20	452
	1600	5000' haul		350	.040			1.30	5.30	6.60	7.75	
	2000	21 C.Y., 1/4 push dozer, sand & gravel, 1500' haul	B-33E	1,180	.012			.38	2.22	2.60	3.02	
	2100	3000' haul		910	.015			.50	2.87	3.37	3.92	
	2200	5000' haul		750	.019			.60	3.49	4.09	4.76	
	2300	Common earth, 1500' haul		1,030	.014			.44	2.54	2.98	3.46	
	2350	3000' haul		790	.018			.57	3.31	3.88	4.51	
	2400	5000' haul		650	.022			.70	4.02	4.72	5.50	
	2500	Clay, 1500' haul		645	.022			.70	4.06	4.76	5.55	
	2550	3000' haul		495	.028			.92	5.30	6.22	7.20	
	2600	5000' haul		405	.035			1.12	6.45	7.57	8.80	
	2700	Towed, 10 C.Y., 1/4 push dozer, sand & gravel, 1500' haul	B-33B	560	.025			.81	2.97	3.78	4.50	
	2720	3000' haul		450	.031			1.01	3.70	4.71	5.60	
	2730	5000' haul		365	.038			1.24	4.56	5.80	6.90	
	2750	Common earth, 1500' haul		420	.033			1.08	3.96	5.04	6	
	2770	3000' haul		400	.035			1.13	4.16	5.29	6.30	
	2780	5000' haul		310	.045			1.46	5.35	6.81	8.10	
	2800	Clay, 1500' haul		315	.044			1.44	5.30	6.74	8	
	2820	3000' haul		300	.047			1.51	5.55	7.06	8.40	
	2840	5000' haul		225	.062			2.01	7.40	9.41	11.20	
	2900	15 C.Y., 1/4 push dozer, sand & gravel, 1500' haul	B-33C	800	.017			.57	2.08	2.65	3.15	
	2920	3000' haul		640	.022			.71	2.60	3.31	3.94	
	2940	5000' haul		520	.027			.87	3.20	4.07	4.84	
	2960	Common earth, 1500' haul		600	.023			.76	2.77	3.53	4.20	
	2980	3000' haul		560	.025			.81	2.97	3.78	4.50	
	3000	5000' haul		440	.032			1.03	3.78	4.81	5.70	
	3020	Clay, 1500' haul		450	.031			1.01	3.70	4.71	5.60	
	3040	3000' haul		420	.033			1.08	3.96	5.04	6	
	3060	5000' haul		320	.044			1.42	5.20	6.62	7.85	
462	0010	**EXCAVATION, STRUCTURAL**										462
	0015	Hand, pits to 6' deep, sandy soil	1 Clab	8	1	B.C.Y.		26.50		26.50	41.50	
	0100	Heavy soil or clay	"	4	2			53.50		53.50	83	
	0200	Pits to 2' deep, normal soil	B-2	24	1.667			45		45	70.50	
	0300	Pits 6' to 12' deep, sandy soil	1 Clab	5	1.600			42.50		42.50	66.50	
	0500	Heavy soil or clay		3	2.667			71		71	111	
	0700	Pits 12' to 18' deep, sandy soil		4	2			53.50		53.50	83	
	0900	Heavy soil or clay		2	4			107		107	166	
	1000	Hand trimming, bottom of excavation	B-2	2,400	.017	S.F.		.45		.45	.70	
	1010	Slopes and sides		2,400	.017	"		.45		.45	.70	
	1030	Around obstructions		8	5	B.C.Y.		136		136	211	
	1100	Hand loading trucks from stock pile, sandy soil	1 Clab	12	.667			17.80		17.80	27.50	
	1300	Heavy soil or clay	"	8	1			26.50		26.50	41.50	
	1500	For wet or muck hand excavation, add to above				%				50%	50%	
	1550	Excavation rock by hand/air tool	B-9	3.40	11.765	B.C.Y.		320	41.50	361.50	540	
	6000	Machine excavation, for spread and mat footings, elevator pits,										
	6001	and small building foundations										
	6030	Common earth, hydraulic backhoe, 1/2 C.Y. bucket	B-12E	55	.291	B.C.Y.		9.10	6.05	15.15	20.50	
	6035	3/4 C.Y. bucket	B-12F	90	.178			5.55	5.25	10.80	14.30	
	6040	1 C.Y. bucket	B-12A	108	.148			4.64	5.15	9.79	12.80	
	6050	1-1/2 C.Y. bucket	B-12B	144	.111			3.48	5	8.48	10.80	
	6060	2 C.Y. bucket	B-12C	200	.080			2.50	4.55	7.05	8.85	
	6070	Sand and gravel, 3/4 C.Y. bucket	B-12F	100	.160			5	4.74	9.74	12.85	
	6080	1 C.Y. bucket	B-12A	120	.133			4.17	4.65	8.82	11.50	
	6090	1-1/2 C.Y. bucket	B-12B	160	.100			3.13	4.50	7.63	9.75	
	6100	2 C.Y. bucket	B-12C	220	.073			2.28	4.14	6.42	8.05	

R02315-450

2 SITE CONSTRUCTION

02315	Excavation and Fill	CREW	DAILY OUTPUT	LABOR-HOURS	UNIT	2005 BARE COSTS				TOTAL INCL O&P
						MAT.	LABOR	EQUIP.	TOTAL	
462	6110	Clay, till, or blasted rock, 3/4 C.Y. bucket R02315 -450	B-12F	80	.200	B.C.Y.		6.25	5.95	12.20	16.05	**462**
	6120	1 C.Y. bucket	B-12A	95	.168			5.25	5.85	11.10	14.50	
	6130	1-1/2 C.Y. bucket	B-12B	130	.123			3.85	5.55	9.40	12	
	6140	2 C.Y. bucket	B-12C	175	.091			2.86	5.20	8.06	10.05	
	9010	For mobilization or demobilization, see div. 02305-250									
	9020	For dewatering, see div. 02240-500									
	9022	For larger structures, see Bulk Excavation, div. 02315-424									
	9024	For loading onto trucks, add								15%	
	9026	For hauling, see div. 02315-490									
	9030	For sheeting or soldier beams/lagging, see div. 02250 & 02260									
	9040	For trench excavation of strip footings, see div. 02315-610									
490	0010	**HAULING**, excavated or borrow, loose cubic yards									**490**
	0012	no loading included, highway haulers									
	0020	6 C.Y. dump truck, 1/4 mile round trip, 5.0 loads/hr.	B-34A	195	.041	L.C.Y.		1.13	1.68	2.81	3.56	
	0030	1/2 mile round trip, 4.1 loads/hr.		160	.050			1.38	2.04	3.42	4.35	
	0040	1 mile round trip, 3.3 loads/hr.		130	.062			1.70	2.51	4.21	5.35	
	0100	2 mile round trip, 2.6 loads/hr.		100	.080			2.20	3.27	5.47	6.95	
	0150	3 mile round trip, 2.1 loads/hr.		80	.100			2.76	4.08	6.84	8.70	
	0200	4 mile round trip, 1.8 loads/hr.		70	.114			3.15	4.67	7.82	9.95	
	0310	12 C.Y. dump truck, 1/4 mile round trip 3.7 loads/hr.	B-34B	288	.028			.77	1.66	2.43	2.99	
	0320	1/2 mile round trip, 3.2 loads/hr.		250	.032			.88	1.91	2.79	3.44	
	0330	1 mile round trip 2.7, loads/hr.		210	.038			1.05	2.27	3.32	4.10	
	0400	2 mile round trip, 2.2 loads/hr.		180	.044			1.22	2.65	3.87	4.78	
	0450	3 mile round trip, 1.9 loads/hr.		170	.047			1.30	2.80	4.10	5.05	
	0500	4 mile round trip, 1.6 loads/hr.		125	.064			1.76	3.81	5.57	6.90	
	0540	5 mile round trip, 1 load/hr.		78	.103			2.83	6.10	8.93	11	
	0550	10 mile round trip, 0.60 load/hr.		58	.138			3.80	8.20	12	14.85	
	0560	20 mile round trip, 0.4 load/hr.		39	.205			5.65	12.20	17.85	22	
	0600	16.5 C.Y. dump trailer, 1 mile round trip, 2.6 loads/hr.	B-34C	280	.029			.79	1.45	2.24	2.79	
	0700	2 mile round trip, 2.1 loads/hr.		225	.036			.98	1.80	2.78	3.47	
	1000	3 mile round trip, 1.8 loads/hr.		193	.041			1.14	2.10	3.24	4.05	
	1100	4 mile round trip, 1.6 loads/hr.		172	.047			1.28	2.35	3.63	4.54	
	1110	5 mile round trip, 1 load/hr.		108	.074			2.04	3.75	5.79	7.25	
	1120	10 mile round trip, .60 load/hr.		80	.100			2.76	5.05	7.81	9.75	
	1130	20 mile round trip, .4 load/hr.		54	.148			4.08	7.50	11.58	14.45	
	1150	20 C.Y. dump trailer, 1 mile round trip, 2.5 loads/hr.	B-34D	325	.025			.68	1.28	1.96	2.44	
	1200	2 mile round trip, 2 loads/hr.		260	.031			.85	1.61	2.46	3.06	
	1220	3 mile round trip, 1.7 loads/hr.		221	.036			1	1.89	2.89	3.60	
	1240	4 mile round trip, 1.5 loads/hr.		195	.041			1.13	2.14	3.27	4.07	
	1245	5 mile round trip, 1.1 load/hr.		143	.056			1.54	2.92	4.46	5.55	
	1250	10 mile round trip, .75 load/hr.		110	.073			2	3.80	5.80	7.20	
	1255	20 mile round trip, .5 load/hr.		78	.103			2.83	5.35	8.18	10.20	
	1300	Hauling in medium traffic, add								20%	20%	
	1400	Heavy traffic, add								30%	30%	
	1600	Grading at dump, or embankment if required, by dozer	B-10B	1,000	.012			.38	.92	1.30	1.59	
	1800	Spotter at fill or cut, if required	1 Clab	8	1	Hr.		26.50		26.50	41.50	
	2000	Off highway haulers									
	2010	22 C.Y. rear/bottom dump, 1000' rnd trip, 4.5 loads/hr.	B-34F	645	.012	L.C.Y.		.34	1.45	1.79	2.11	
	2020	1/2 mile round trip, 4.2 loads/hr.		600	.013			.37	1.56	1.93	2.27	
	2030	1 mile round trip, 3.9 loads/hr.		555	.014			.40	1.68	2.08	2.46	
	2040	2 mile round trip, 3.3 loads/hr.		470	.017			.47	1.99	2.46	2.89	
	2050	34 C.Y. rear or bottom dump, 1000' round trip, 4 loads/hr.	B-34G	885	.009			.25	1.33	1.58	1.85	
	2060	1/2 mile round trip, 3.8 loads/hr.		840	.010			.26	1.41	1.67	1.95	
	2070	1 mile round trip, 3.5 loads/hr.		775	.010			.28	1.52	1.80	2.11	
	2080	2 mile round trip, 3.0 loads/hr.		665	.012			.33	1.78	2.11	2.46	

Important: See the Reference Section for critical supporting data - Reference Nos., Crews, & City Cost Indexes

02315 | Excavation and Fill

			CREW	DAILY OUTPUT	LABOR-HOURS	UNIT	2005 BARE COSTS				TOTAL INCL O&P	
							MAT.	LABOR	EQUIP.	TOTAL		
490	2090	42 C.Y. rear or bottom bump, 1000' round trip, 3.8 loads/hr.	B-34H	1,040	.008	L.C.Y.		.21	1.23	1.44	1.67	**490**
	2100	1/2 mile round trip, 3.6 loads/hr.		980	.008			.22	1.30	1.52	1.77	
	2110	1 mile round trip, 3.3 loads/hr.		900	.009			.24	1.42	1.66	1.93	
	2120	2 mile round trip, 2.8 loads/hr.		765	.010			.29	1.67	1.96	2.28	
	2130	60 C.Y. rear or bottom dump, 1000' round trip, 3.6 loads/hr.	B-34J	1,400	.006			.16	1.18	1.34	1.53	
	2140	1/2 mile round trip, 3.4 loads/hr.		1,325	.006			.17	1.25	1.42	1.62	
	2150	1 mile round trip, 3.1 loads/hr.		1,200	.007			.18	1.37	1.55	1.79	
	2160	2 mile round trip, 2.6 loads/hr.		1,015	.008			.22	1.62	1.84	2.12	
	3000	Rough terrain or steep grades, add to above								100%		
	4500	Dust control, light	B-59	1	8	Day		220	325	545	695	
	4501	Heavy	"	.50	16			440	650	1,090	1,375	
	4600	Haul road maintenance	B-86A	1	8			277	455	732	920	
	4700	Highway hauling beyond 20 miles, per loaded mile, minimum				Mile				1.05	1.15	
	4750	Maximum				"				2.10	2.31	
520	0010	**FILL**, spread dumped material, no compaction										**520**
	0020	By dozer, no compaction	B-10B	1,000	.012	L.C.Y.		.38	.92	1.30	1.59	
	0100	By hand	1 Clab	12	.667	"		17.80		17.80	27.50	
	0150	Spread fill, from stockpile with 2-1/2 C.Y. F.E. loader										
	0170	130 H.P., 300' haul	B-10P	600	.020	L.C.Y.		.64	1.31	1.95	2.41	
	0190	With dozer 300 H.P., 300' haul	B-10M	600	.020	"		.64	1.99	2.63	3.16	
	0400	For compaction of embankment, see div. 02315-310										
	0500	Gravel fill, compacted, under floor slabs, 4" deep	B-37	10,000	.005	S.F.	.16	.13	.01	.30	.39	
	0600	6" deep		8,600	.006		.24	.16	.01	.41	.51	
	0700	9" deep		7,200	.007		.39	.19	.01	.59	.74	
	0800	12" deep		6,000	.008		.55	.22	.02	.79	.98	
	1000	Alternate pricing method, 4" deep		120	.400	E.C.Y.	11.80	11.25	.90	23.95	31.50	
	1100	6" deep		160	.300		11.80	8.45	.67	20.92	27	
	1200	9" deep		200	.240		11.80	6.75	.54	19.09	24	
	1300	12" deep		220	.218		11.80	6.15	.49	18.44	23	
	1500	For fill under exterior paving, see division 02720-200										
	1600	For flowable fill, see division 03310-220										
610	0010	**EXCAVATING, TRENCH** or continuous footing, common earth										**610**
	0020	No sheeting or dewatering included										
	0050	1' to 4' deep, 3/8 C.Y. tractor loader/backhoe	B-11C	150	.107	B.C.Y.		3.27	1.44	4.71	6.60	
	0060	1/2 C.Y. tractor loader/backhoe	B-11M	200	.080			2.45	1.28	3.73	5.15	
	0062	3/4 C.Y. hydraulic backhoe	B-12F	270	.059			1.85	1.76	3.61	4.76	
	0090	4' to 6' deep, 1/2 C.Y. tractor loader/backhoe	B-11M	200	.080			2.45	1.28	3.73	5.15	
	0100	5/8 C.Y. hydraulic backhoe	B-12Q	250	.064			2	1.73	3.73	4.96	
	0110	3/4 C.Y. hydraulic backhoe	B-12F	300	.053			1.67	1.58	3.25	4.29	
	0120	1 C.Y. hydraulic backhoe	B-12A	400	.040			1.25	1.39	2.64	3.44	
	0130	1-1/2 C.Y. hydraulic backhoe	B-12B	540	.030			.93	1.33	2.26	2.89	
	0300	1/2 C.Y. hydraulic excavator, truck mounted	B-12J	200	.080			2.50	4.19	6.69	8.45	
	0500	6' to 10' deep, 3/4 C.Y. hydraulic backhoe, 6' to 10' deep	B-12F	225	.071			2.23	2.11	4.34	5.70	
	0510	1 C.Y. hydraulic backhoe	B-12A	400	.040			1.25	1.39	2.64	3.44	
	0600	1 C.Y. hydraulic excavator, truck mounted	B-12K	400	.040			1.25	2.43	3.68	4.58	
	0610	1-1/2 C.Y. hydraulic backhoe	B-12B	600	.027			.83	1.20	2.03	2.60	
	0620	2-1/2 C.Y. hydraulic backhoe	B-12S	1,000	.016			.50	1.22	1.72	2.11	
	0900	10' to 14' deep, 3/4 C.Y. hydraulic backhoe	B-12F	200	.080			2.50	2.37	4.87	6.45	
	0910	1 C.Y. hydraulic backhoe	B-12A	360	.044			1.39	1.55	2.94	3.83	
	1000	1-1/2 C.Y. hydraulic backhoe	B-12B	540	.030			.93	1.33	2.26	2.89	
	1020	2-1/2 C.Y. hydraulic backhoe	B-12S	1,000	.016			.50	1.22	1.72	2.11	
	1030	3 C.Y. hydraulic backhoe	B-12D	1,400	.011			.36	1.44	1.80	2.14	
	1300	14' to 20' deep, 1 C.Y. hydraulic backhoe	B-12A	320	.050			1.57	1.74	3.31	4.31	
	1310	1-1/2 C.Y. hydraulic backhoe	B-12B	480	.033			1.04	1.50	2.54	3.24	
	1320	2-1/2 C.Y. hydraulic backhoe	B-12S	850	.019			.59	1.43	2.02	2.47	

R02315-400

02315	Excavation and Fill	CREW	DAILY OUTPUT	LABOR-HOURS	UNIT	2005 BARE COSTS				TOTAL INCL O&P	
						MAT.	LABOR	EQUIP.	TOTAL		
610	1330	3 C.Y. hydraulic backhoe	B-12D	1,000	.016	B.C.Y.		.50	2.02	2.52	2.99	**610**
	1400	By hand with pick and shovel 2' to 6' deep, light soil	1 Clab	8	1			26.50		26.50	41.50	
	1500	Heavy soil	"	4	2	↓		53.50		53.50	83	
	1700	For tamping backfilled trenches, air tamp, add	A-1G	100	.080	E.C.Y.		2.14	.35	2.49	3.70	
	1900	Vibrating plate, add	B-18	180	.133	"		3.65	.20	3.85	5.95	
	2100	Trim sides and bottom for concrete pours, common earth		1,500	.016	S.F.		.44	.02	.46	.71	
	2300	Hardpan	↓	600	.040	"		1.09	.06	1.15	1.77	
	2400	Pier and spread footing excavation, add to above				B.C.Y.				30%	30%	
	3000	Backfill trench, F.E. loader, wheel mtd., 1 C.Y. bucket										
	3020	Minimal haul	B-10R	400	.030	L.C.Y.		.96	.48	1.44	1.99	
	3040	100' haul		200	.060			1.92	.97	2.89	3.99	
	3060	200' haul	↓	100	.120			3.84	1.94	5.78	8	
	3080	2-1/4 C.Y. bucket, minimum haul	B-10T	600	.020			.64	.51	1.15	1.53	
	3090	100' haul		300	.040			1.28	1.02	2.30	3.07	
	3100	200' haul	↓	150	.080	↓		2.56	2.03	4.59	6.15	
	4000	For backfill with dozer, see div. 02315-120										
	4010	For compaction of backfill, see div. 02315-310	↓									
620	0010	**EXCAVATING, UTILITY TRENCH** Common earth										**620**
	0050	Trenching with chain trencher, 12 H.P., operator walking										
	0100	4" wide trench, 12" deep	B-53	800	.010	L.F.		.33	.06	.39	.57	
	0150	18" deep		750	.011			.35	.06	.41	.60	
	0200	24" deep		700	.011			.38	.07	.45	.65	
	0300	6" wide trench, 12" deep		650	.012			.41	.07	.48	.69	
	0350	18" deep		600	.013			.44	.08	.52	.75	
	0400	24" deep		550	.015			.48	.09	.57	.82	
	0450	36" deep		450	.018			.59	.11	.70	1.01	
	0600	8" wide trench, 12" deep		475	.017			.56	.10	.66	.95	
	0650	18" deep		400	.020			.66	.12	.78	1.13	
	0700	24" deep		350	.023			.76	.14	.90	1.29	
	0750	36" deep	↓	300	.027	↓		.88	.16	1.04	1.51	
	0830	Fly wheel trencher, 18" wide trench, 6' deep, light soil	B-54A	1,992	.005	B.C.Y.		.16	.41	.57	.69	
	0840	Medium soil		1,594	.006			.20	.51	.71	.86	
	0850	Heavy soil	↓	1,295	.007			.24	.63	.87	1.06	
	0860	24" wide trench, 9' deep, light soil	B-54B	4,981	.002			.07	.29	.36	.42	
	0870	Medium soil		4,000	.002			.08	.37	.45	.53	
	0880	Heavy soil	↓	3,237	.003	↓		.10	.45	.55	.66	
	1000	Backfill by hand including compaction, add										
	1050	4" wide trench, 12" deep	A-1G	800	.010	L.F.		.27	.04	.31	.47	
	1100	18" deep		530	.015			.40	.07	.47	.70	
	1150	24" deep		400	.020			.53	.09	.62	.93	
	1300	6" wide trench, 12" deep		540	.015			.40	.06	.46	.69	
	1350	18" deep		405	.020			.53	.09	.62	.91	
	1400	24" deep		270	.030			.79	.13	.92	1.37	
	1450	36" deep		180	.044			1.19	.19	1.38	2.06	
	1600	8" wide trench, 12" deep		400	.020			.53	.09	.62	.93	
	1650	18" deep		265	.030			.81	.13	.94	1.39	
	1700	24" deep		200	.040			1.07	.17	1.24	1.85	
	1750	36" deep	↓	135	.059	↓		1.58	.26	1.84	2.74	
	2000	Chain trencher, 40 H.P. operator riding										
	2050	6" wide trench and backfill, 12" deep	B-54	1,200	.007	L.F.		.22	.18	.40	.53	
	2100	18" deep		1,000	.008			.26	.21	.47	.63	
	2150	24" deep		975	.008			.27	.22	.49	.65	
	2200	36" deep		900	.009			.29	.24	.53	.70	
	2250	48" deep		750	.011			.35	.28	.63	.84	
	2300	60" deep	↓	650	.012	↓		.41	.33	.74	.97	

R02315 -400

02300 | Earthwork

02315 | Excavation and Fill

		CREW	DAILY OUTPUT	LABOR-HOURS	UNIT	MAT.	LABOR	EQUIP.	TOTAL	TOTAL INCL O&P	
620	2400	8" wide trench and backfill, 12" deep	B-54	1,000	.008	L.F.		.26	.21	.47	.63	**620**
	2450	18" deep		950	.008			.28	.22	.50	.67	
	2500	24" deep		900	.009			.29	.24	.53	.70	
	2550	36" deep		800	.010			.33	.27	.60	.79	
	2600	48" deep		650	.012			.41	.33	.74	.97	
	2700	12" wide trench and backfill, 12" deep		975	.008			.27	.22	.49	.65	
	2750	18" deep		860	.009			.31	.25	.56	.73	
	2800	24" deep		800	.010			.33	.27	.60	.79	
	2850	36" deep		725	.011			.36	.29	.65	.87	
	3000	16" wide trench and backfill, 12" deep		835	.010			.32	.26	.58	.76	
	3050	18" deep		750	.011			.35	.28	.63	.84	
	3100	24" deep		700	.011			.38	.31	.69	.91	
	3200	Compaction with vibratory plate, add								50%	50%	
	5100	Hand excavate and trim for pipe bells after trench excavation										
	5200	8" pipe	1 Clab	155	.052	L.F.		1.38		1.38	2.14	
	5300	18" pipe	"	130	.062	"		1.64		1.64	2.56	
	9100	For clay or till, add up to								150%	150%	
630	0010	**EXCAVATING, UTILITY TRENCH, PLOW**										**630**
	0100	Single cable, plowed into fine material	B-11Q	3,800	.003	L.F.		.10	.15	.25	.32	
	0200	Two cable		3,200	.004			.12	.18	.30	.38	
	0300	Single cable, plowed into coarse material		2,000	.006			.19	.29	.48	.60	
640	0010	**UTILITY BEDDING** For pipe and conduit, not incl. compaction										**640**
	0050	Crushed or screened bank run gravel	B-6	150	.160	L.C.Y.	20	4.61	1.44	26.05	30.50	
	0100	Crushed stone 3/4" to 1/2"		150	.160		27.50	4.61	1.44	33.55	38.50	
	0200	Sand, dead or bank		150	.160		4.13	4.61	1.44	10.18	13.20	
	0500	Compacting bedding in trench	A-1D	90	.089	E.C.Y.		2.37	.28	2.65	4	
	0600	If material source exceeds 2 miles, add for extra mileage.										
	0610	See 02315-490-0010 for hauling mileage add.										

02325 | Dredging

		CREW	DAILY OUTPUT	LABOR-HOURS	UNIT	MAT.	LABOR	EQUIP.	TOTAL	TOTAL INCL O&P	
250	0010	**DREDGING**										**250**
	0020	Dredging mobilization and demobilization., add to below, minimum	B-8	.53	120	Total		3,575	4,450	8,025	10,400	
	0100	Maximum	"	.10	640	"		18,900	23,600	42,500	55,000	
	0300	Barge mounted clamshell excavation into scows,										
	0310	Dumped 20 miles at sea, minimum	B-57	310	.155	B.C.Y.		4.67	4.14	8.81	11.70	
	0400	Maximum	"	213	.225	"		6.80	6.05	12.85	17.05	
	0500	Barge mounted dragline or clamshell, hopper dumped,										
	0510	pumped 1000' to shore dump, minimum	B-57	340	.141	B.C.Y.		4.26	3.78	8.04	10.65	
	0525	All pumping uses 2000 gallons of water per cubic yard										
	0600	Maximum	B-57	243	.198	B.C.Y.		5.95	5.30	11.25	14.90	
	1000	Hydraulic method, pumped 1000' to shore dump, minimum		460	.104			3.15	2.79	5.94	7.90	
	1100	Maximum		310	.155			4.67	4.14	8.81	11.70	
	1400	Into scows dumped 20 miles, minimum		425	.113			3.41	3.02	6.43	8.50	
	1500	Maximum		243	.198			5.95	5.30	11.25	14.90	
	1600	For inland rivers and canals in South, deduct								30%	30%	

02340 | Soil Stabilization

		CREW	DAILY OUTPUT	LABOR-HOURS	UNIT	MAT.	LABOR	EQUIP.	TOTAL	TOTAL INCL O&P	
100	0010	**ASPHALT SOIL STABILIZATION** Including scarifying and compaction										**100**
	0020	Asphalt, 1-1/2" deep, 1/2 gal/S.Y.	B-75	4,000	.014	S.Y.	.96	.44	.82	2.22	2.63	
	0040	3/4 gal/S.Y.		4,000	.014		1.44	.44	.82	2.70	3.15	
	0100	3" deep, 1 gal/S.Y.		3,500	.016		1.92	.51	.94	3.37	3.91	
	0140	1-1/2 gal/S.Y.		3,500	.016		2.88	.51	.94	4.33	4.97	
	0200	6" deep, 2 gal/S.Y.		3,000	.019		3.84	.59	1.10	5.53	6.30	

2

02340 | Soil Stabilization

		CREW	DAILY OUTPUT	LABOR-HOURS	UNIT	MAT.	LABOR	EQUIP.	TOTAL	TOTAL INCL O&P	
						2005 BARE COSTS					
100	0240	3 gal/S.Y.	B-75	3,000	.019	S.Y.	5.75	.59	1.10	7.44	8.45	**100**
	0300	8" deep, 2-2/3 gal/S.Y.		2,800	.020		5.15	.63	1.17	6.95	7.90	
	0340	4 gal/S.Y.		2,800	.020		7.70	.63	1.17	9.50	10.70	
	0540	6 gal/S.Y.		2,600	.022		11.50	.68	1.26	13.44	15.10	
200	0010	**CEMENT SOIL STABILIZATION** Including scarifying and compaction										**200**
	1020	Cement, 4% mix, by volume, 6" deep	B-74	1,100	.058	S.Y.	.98	1.81	3.28	6.07	7.45	
	1030	8" deep		1,050	.061		1.28	1.90	3.44	6.62	8.05	
	1060	12" deep		960	.067		1.91	2.08	3.76	7.75	9.40	
	1100	6% mix, 6" deep		1,100	.058		1.40	1.81	3.28	6.49	7.90	
	1120	8" deep		1,050	.061		1.83	1.90	3.44	7.17	8.70	
	1160	12" deep		960	.067		2.76	2.08	3.76	8.60	10.35	
	1200	9% mix, 6" deep		1,100	.058		2.13	1.81	3.28	7.22	8.70	
	1220	8" deep		1,050	.061		2.76	1.90	3.44	8.10	9.70	
	1260	12" deep		960	.067		4.17	2.08	3.76	10.01	11.85	
	1300	12% mix, 6" deep		1,100	.058		2.76	1.81	3.28	7.85	9.40	
	1320	8" deep		1,050	.061		3.70	1.90	3.44	9.04	10.75	
	1360	12" deep		960	.067		5.55	2.08	3.76	11.39	13.40	
300	0010	**GEOTEXTILE SOIL STABILIZATION**										**300**
	1500	Geotextile fabric, woven, 200 lb. tensile strength	2 Clab	2,500	.006	S.Y.	1.92	.17		2.09	2.38	
	1510	Heavy Duty, 600 lb. tensile strength		2,400	.007		1.71	.18		1.89	2.16	
	1550	Non-woven, 120 lb. tensile strength		2,500	.006		.97	.17		1.14	1.34	
500	0010	**LIME SOIL STABILIZATION** Including scarifying and compaction										**500**
	2020	Hydrated lime, for base, 2% mix by weight, 6" deep	B-74	1,800	.036	S.Y.	2.40	1.11	2	5.51	6.50	
	2030	8" deep		1,700	.038		3.20	1.17	2.12	6.49	7.65	
	2060	12" deep		1,550	.041		2.54	1.29	2.33	6.16	7.30	
	2100	4% mix, 6" deep		1,800	.036		2.58	1.11	2	5.69	6.70	
	2120	8" deep		1,700	.038		3.48	1.17	2.12	6.77	7.95	
	2160	12" deep		1,550	.041		5.20	1.29	2.33	8.82	10.25	
	2200	6% mix, 6" deep		1,800	.036		3.91	1.11	2	7.02	8.20	
	2220	8" deep		1,700	.038		5.20	1.17	2.12	8.49	9.85	
	2260	12" deep		1,550	.041		7.75	1.29	2.33	11.37	13.05	
700	0010	**CALCIUM CHLORIDE**										**700**
	0020	Calcium chloride delivered, 100 LB bags, truckload lots				Ton	460			460	505	
	0030	Solution, 4 lb. flake per gallon, tank truck delivery				Gal.	.98			.98	1.08	

02360 | Soil Treatment

		CREW	DAILY OUTPUT	LABOR-HOURS	UNIT	MAT.	LABOR	EQUIP.	TOTAL	TOTAL INCL O&P	
200	0010	**TERMITE PRETREATMENT**										**200**
	0020	Slab and walls, residential	1 Skwk	1,200	.007	SF Flr.	.28	.23		.51	.67	
	0100	Commercial, minimum		2,496	.003		.30	.11		.41	.50	
	0200	Maximum		1,645	.005		.45	.17		.62	.76	
	0400	Insecticides for termite control, minimum		14.20	.563	Gal.	11.60	19.65		31.25	43.50	
	0500	Maximum		11	.727	"	19.85	25.50		45.35	61.50	
	3000	Soil poisoning (sterilization)	1 Clab	4,496	.002	S.F.	.26	.05		.31	.36	
	3100	Herbicide application from truck	B-59	19,000	.001	S.Y.		.01	.02	.03	.04	

02370 | Erosion & Sedimentation Control

		CREW	DAILY OUTPUT	LABOR-HOURS	UNIT	MAT.	LABOR	EQUIP.	TOTAL	TOTAL INCL O&P	
450	0010	**RIP-RAP & ROCK LINING**, Random, broken stone										**450**
	0100	Machine placed for slope protection	B-12G	62	.258	L.C.Y.	25	8.10	8.75	41.85	49.50	
	0110	3/8 to 1/4 C.Y. pieces, grouted	B-13	80	.700	S.Y.	36.50	20	7.70	64.20	79.50	
	0200	18" minimum thickness, not grouted	"	53	1.057	"	15.55	30.50	11.65	57.70	77	
	0300	Dumped, 50 lb. average	B-11A	800	.020	Ton	17.90	.61	1.15	19.66	22	
	0350	100 lb. average		700	.023		25.50	.70	1.31	27.51	30.50	
	0370	300 lb. average		600	.027		30	.82	1.53	32.35	36	
	0400	Gabions, galvanized steel mesh mats or boxes, stone filled, 6" deep	B-13	200	.280	S.Y.	16.90	8.05	3.08	28.03	34.50	

		02370	Erosion & Sedimentation Control	CREW	DAILY OUTPUT	LABOR-HOURS	UNIT	2005 BARE COSTS MAT.	LABOR	EQUIP.	TOTAL	TOTAL INCL O&P	
450	0500		9" deep	B-13	163	.344	S.Y.	26	9.90	3.78	39.68	48	450
	0600		12" deep		153	.366		27.50	10.55	4.03	42.08	50.50	
	0700		18" deep		102	.549		35	15.80	6.05	56.85	69.50	
	0800		36" deep		60	.933		59.50	27	10.30	96.80	118	
700	0010		**SYNTHETIC EROSION CONTROL**										700
	0020		Jute mesh, 100 SY per roll, 4' wide, stapled	B-80A	2,400	.010	S.Y.	.69	.27	.07	1.03	1.26	
	0060		Nylon, 3 dimensional geomatrix, 9 mil thick		700	.034		4.04	.92	.25	5.21	6.15	
	0062		12 mil thick		515	.047		5.50	1.24	.34	7.08	8.35	
	0064		18 mil thick		460	.052		5.80	1.39	.38	7.57	8.95	
	0070		Paper biodegradable mesh	B-1	2,500	.010		.07	.26		.33	.49	
	0080		Paper mulch	B-64	20,000	.001		.05	.02	.01	.08	.10	
	0100		Plastic netting, stapled, 2" x 1" mesh, 20 mil	B-1	2,500	.010		.63	.26		.89	1.10	
	0120		Revegetation mat, webbed	2 Clab	1,000	.016		5.05	.43		5.48	6.20	
	0200		Polypropylene mesh, stapled, 6.5 oz./S.Y.	B-1	2,500	.010		1.30	.26		1.56	1.84	
	0300		Tobacco netting, or jute mesh #2, stapled	"	2,500	.010		.07	.26		.33	.49	
	0400		Soil sealant, liquid sprayed from truck	B-81	5,000	.005		.43	.14	.08	.65	.78	
	1000		Silt fence, polypropylene, 3' high, ideal conditions	2 Clab	1,600	.010	L.F.	.32	.27		.59	.77	
	1100		Adverse conditions	"	950	.017	"	.32	.45		.77	1.05	
	1130		Cellular confinement, poly, 3-dimen, 8' x 20' panels, 4" deep cell	B-6	1,600	.015	S.F.	1.69	.43	.13	2.25	2.67	
	1140		8" deep cells	"	1,200	.020	"	2.32	.58	.18	3.08	3.64	
	1200		Place and remove hay bales	A-2	3	8	Ton	52	214	43	309	435	
	1250		Hay bales, staked	"	2,500	.010	L.F.	2.08	.26	.05	2.39	2.75	

		02390	Shore Protect/Mooring Structures	CREW	DAILY OUTPUT	LABOR-HOURS	UNIT	MAT.	LABOR	EQUIP.	TOTAL	TOTAL INCL O&P	
110	0010		**BREAKWATERS, BULKHEADS** Reinforced concrete,										110
	0015		include footing and tie-backs										
	0020		Up to 6' high, minimum	C-17C	28	2.964	L.F.	40.50	105	13.35	158.85	221	
	0060		Maximum		24.25	3.423		65	121	15.40	201.40	276	
	0100		12' high, minimum		20	4.150		105	146	18.70	269.70	365	
	0160		Maximum		18.50	4.486		122	158	20	300	400	
	0180		Precast bulkhead, complete, including										
	0190		vertical and battered piles, face panels, and cap										
	0195		Using 16' vertical piles				L.F.				240	280	
	0196		Using 20' vertical piles				"				255	295	
	0200		Steel sheeting, with 4' x 4' x 8" concrete deadmen, @ 10' O.C.										
	0210		12' high, shore driven	B-40	27	2.370	L.F.	57.50	80	85	222.50	285	
	0260		Barge driven	B-76	15	4.800	"	86.50	162	150	398.50	520	
	6000		Crushed stone placed behind bulkhead by clam bucket	B-12H	120	.133	L.C.Y.	18.80	4.17	7.30	30.27	35	
120	0010		**BREAKWATERS, BULKHEADS, RESIDENTIAL CANAL**										120
	0020		Aluminum panel sheeting, incl. concrete cap and anchor										
	0030		Coarse compact sand, 4'-0" high, 2'-0" embedment	B-40	200	.320	L.F.	42	10.80	11.45	64.25	76	
	0040		3'-6" embedment		140	.457		49.50	15.45	16.35	81.30	97	
	0060		6'-0" embedment		90	.711		63	24	25.50	112.50	136	
	0120		6'-0" high, 2'-6" embedment		170	.376		55	12.70	13.50	81.20	96	
	0140		4'-0" embedment		125	.512		65	17.30	18.35	100.65	119	
	0160		5'-6" embedment		95	.674		85.50	23	24	132.50	158	
	0220		8'-0" high, 3'-6" embedment		140	.457		71	15.45	16.35	102.80	121	
	0240		5'-0" embedment		100	.640		71	21.50	23	115.50	138	
	0420		Medium compact sand, 3'-0" high, 2'-0" embedment		235	.272		62.50	9.20	9.75	81.45	94.50	
	0440		4'-0" embedment		150	.427		77	14.40	15.25	106.65	124	
	0460		5'-6" embedment		115	.557		93	18.80	19.90	131.70	154	
	0520		5'-0" high, 3'-6" embedment		165	.388		89.50	13.10	13.90	116.50	134	
	0540		5'-0" embedment		120	.533		103	18.05	19.10	140.15	163	
	0560		6'-6" embedment		105	.610		134	20.50	22	176.50	204	
	0620		7'-0" high, 4'-6" embedment		135	.474		134	16	16.95	166.95	191	
	0640		6'-0" embedment		110	.582		148	19.65	21	188.65	217	

2 SITE CONSTRUCTION

		02390	Shore Protect/Mooring Structures	CREW	DAILY OUTPUT	LABOR-HOURS	UNIT	MAT.	2005 BARE COSTS LABOR	EQUIP.	TOTAL	TOTAL INCL O&P	
120	0720		Loose silty sand, 3'-0" high, 3'-0" embedment	B-40	205	.312	L.F.	91	10.55	11.20	112.75	129	120
	0740		4'-6" embedment		155	.413		110	13.95	14.80	138.75	160	
	0760		6'-0" embedment		125	.512		128	17.30	18.35	163.65	189	
	0820		4'-6" high, 4'-6" embedment		155	.413		123	13.95	14.80	151.75	174	
	0840		6'-0" embedment		125	.512		143	17.30	18.35	178.65	205	
	0860		7'-0" embedment		115	.557		173	18.80	19.90	211.70	242	
	0920		6'-0" high, 5'-6" embedment		130	.492		161	16.65	17.60	195.25	224	
	0940		7'-0" embedment		115	.557		178	18.80	19.90	216.70	248	
310	0010	**JETTIES, DOCK ACCESSORIES**											310
	0100		Cleats, aluminum, "S" type, 12" long	1 Clab	8	1	Ea.	14.25	26.50		40.75	57	
	0140		10" long		6.70	1.194		34.50	32		66.50	87.50	
	0180		15" long		6	1.333		45.50	35.50		81	106	
	0400		Dock wheel for corners and piles, vinyl, 12" diameter		4	2		63	53.50		116.50	153	
	1000		Electrical receptacle with circuit breaker,										
	1020		Pile mounted, double 30 amp, 125 volt	1 Elec	2	4	Unit	410	163		573	695	
	1060		Double 50 amp, 125/240 volt		1.60	5		555	204		759	915	
	1120		Free standing, add		4	2		142	81.50		223.50	278	
	1140		Double free standing, add		2.70	2.963		570	121		691	805	
	1160		Light, 2 louvered, with photo electric switch, add		8	1		94	41		135	164	
	1180		Telephone jack on stanchion		8	1		68.50	41		109.50	136	
	1300		Fender, Vinyl, 4" high	1 Clab	160	.050	L.F.	2.81	1.34		4.15	5.15	
	1380		Corner piece		80	.100	Ea.	12.60	2.67		15.27	18.05	
	1400		Hose holder, cast aluminum		16	.500	"	27	13.35		40.35	51	
	1500		Ladder, aluminum, heavy duty										
	1520		Crown top, 5 to 7 step, minimum	1 Clab	5.30	1.509	Ea.	173	40.50		213.50	254	
	1560		Maximum		2	4		220	107		327	405	
	1580		Bracket for portable clamp mounting		8	1		90.50	26.50		117	141	
	1800		Line holder, treated wood, small		16	.500		10.70	13.35		24.05	33	
	1840		Large		13.30	.601		18.20	16.05		34.25	45	
	2000		Mooring whip, fiberglass bolted to dock,										
	2020		1200 lb. boat	1 Clab	8.80	.909	Pr.	261	24.50		285.50	325	
	2040		10,000 lb. boat		6.70	1.194		530	32		562	630	
	2080		60,000 lb. boat		4	2		620	53.50		673.50	765	
	2400		Shock absorbing tubing, vertical bumpers										
	2420		3" diam., vinyl, white	1 Clab	80	.100	L.F.	13	2.67		15.67	18.45	
	2440		Polybutyl, clear		80	.100	"	14.45	2.67		17.12	20	
	2480		Mounts, polybutyl		20	.400	Ea.	15.25	10.70		25.95	33.50	
	2490		Deluxe		20	.400	"	34	10.70		44.70	54	
320	0010	**JETTIES, FLOATING DOCK ACCESSORIES**											320
	0200		Dock connectors, stressed cables with rubber spacers										
	0220		25" long, 3' wide dock	1 Clab	2	4	Joint	525	107		632	740	
	0240		5' wide dock		2	4		610	107		717	835	
	0400		38" long, 4' wide dock		1.75	4.571		570	122		692	820	
	0440		6' wide dock		1.45	5.517		650	147		797	945	
	1000		Gangway, aluminum, one end rolling, no hand rails										
	1020		3' wide, minimum	1 Clab	67	.119	L.F.	146	3.19		149.19	165	
	1040		Maximum		32	.250		168	6.70		174.70	195	
	1100		4' wide, minimum		40	.200		155	5.35		160.35	178	
	1140		Maximum		24	.333		178	8.90		186.90	210	
	1180		For handrails, add					34			34	37	
	2000		Pile guides, beads on stainless cable	1 Clab	4	2	Ea.	345	53.50		398.50	460	
	2020		Rod type, 8" diameter piles, minimum		4	2		34.50	53.50		88	121	
	2040		Maximum		2	4		50	107		157	221	
	2100		10" to 14" diameter piles minimum		3.20	2.500		82.50	67		149.50	195	
	2140		Maximum		1.75	4.571		144	122		266	350	
	2200		Roller type, 4 rollers, minimum		4	2		310	53.50		363.50	425	

Important: See the Reference Section for critical supporting data - Reference Nos., Crews, & City Cost Indexes

		02390 | **Shore Protect/Mooring Structures**	CREW	DAILY OUTPUT	LABOR-HOURS	UNIT	MAT.	LABOR	EQUIP.	TOTAL	TOTAL INCL O&P	
								2005 BARE COSTS				
320	2240	Maximum	1 Clab	1.75	4.571	Ea.	440	122		562	675	**320**
330	0010	**JETTIES, DOCKS, FIXED**										**330**
	0020	Pile supported, treated wood,										
	0030	5' x 20' platform, minimum	F-3	115	.348	S.F.	10.95	12.05	5.25	28.25	36.50	
	0060	Maximum		75	.533		23	18.45	8.05	49.50	63	
	0100	6' x 20' platform, minimum		130	.308		9.75	10.65	4.64	25.04	32.50	
	0160	Maximum		85	.471		23	16.25	7.10	46.35	58	
	0200	8' x 20' platform, minimum		150	.267		9.10	9.20	4.02	22.32	28.50	
	0260	Maximum		100	.400		21	13.85	6.05	40.90	51	
	0420	5' x 30' platform, minimum		155	.258		10.25	8.90	3.89	23.04	29.50	
	0460	Maximum	↓	100	.400	↓	24	13.85	6.05	43.90	54	
	0500	For Greenhart lumber, add									40%	
	0550	Diagonal planking, add									25%	
	1000	Pipe supported dock, 1" aluminum pipe										
	1020	Aluminum planks, galv. stl. framing, 2' 8" wide, minimum	F-3	320	.125	S.F.	20.50	4.32	1.88	26.70	31.50	
	1060	Maximum		120	.333		25.50	11.55	5	42.05	51.50	
	1100	5' wide, minimum		250	.160		11.40	5.55	2.41	19.36	23.50	
	1160	Maximum	↓	160	.250	↓	12.50	8.65	3.77	24.92	31.50	
	1200	Wood deck and galv. steel framing										
	1220	3' wide, minimum	F-3	320	.125	S.F.	20.50	4.32	1.88	26.70	31.50	
	1260	Maximum		120	.333		24.50	11.55	5	41.05	50.50	
	1300	6' wide, minimum		260	.154		18.65	5.30	2.32	26.27	31.50	
	1360	Maximum		100	.400		23	13.85	6.05	42.90	53.50	
	1400	10' wide, minimum		200	.200		20.50	6.90	3.01	30.41	36.50	
	1460	Maximum	↓	80	.500	↓	25	17.30	7.55	49.85	63	
	1800	1-1/2" galv. steel pipe, treated wood dock and framing										
	1820	4' wide, minimum	F-3	275	.145	S.F.	13.80	5.05	2.19	21.04	25.50	
	1860	Maximum		170	.235		22.50	8.15	3.55	34.20	41.50	
	1900	5' wide, minimum		250	.160		11.95	5.55	2.41	19.91	24.50	
	1960	Maximum		160	.250		19.55	8.65	3.77	31.97	39	
	2000	6' wide, minimum		225	.178		11.40	6.15	2.68	20.23	25	
	2060	Maximum	↓	150	.267	↓	16.15	9.20	4.02	29.37	36.50	
350	0010	**JETTIES, DOCKS,** floating including anchors										**350**
	1000	Polystyrene flotation, minimum	F-3	200	.200	S.F.	12.55	6.90	3.01	22.46	28	
	1040	Maximum		135	.296	"	16.30	10.25	4.47	31.02	38.50	
	1100	Alternate method of figuring, minimum		1.13	35.398	Slip	3,125	1,225	535	4,885	5,925	
	1140	Maximum		.70	57.143	"	4,500	1,975	860	7,335	8,925	
	1200	Galv. steel frame and wood deck, 3' wide, minimum		320	.125	S.F.	10.65	4.32	1.88	16.85	20.50	
	1240	Maximum		200	.200		17.15	6.90	3.01	27.06	33	
	1300	4' wide, minimum		320	.125		10.05	4.32	1.88	16.25	19.80	
	1340	Maximum		200	.200		16.05	6.90	3.01	25.96	31.50	
	1500	8' wide, minimum		250	.160		8.45	5.55	2.41	16.41	20.50	
	1540	Maximum		160	.250		12.80	8.65	3.77	25.22	31.50	
	1700	Treated wood frames and deck, 3' wide, minimum		250	.160		11.05	5.55	2.41	19.01	23.50	
	1740	Maximum	↓	125	.320	↓	38	11.05	4.82	53.87	64	
	2000	Polyethylene drums, treated wood frame and deck										
	2100	6' wide, minimum	F-3	250	.160	S.F.	12.55	5.55	2.41	20.51	25	
	2140	Maximum		125	.320		16.65	11.05	4.82	32.52	41	
	2200	8' wide, minimum		233	.172		11.25	5.95	2.59	19.79	24.50	
	2240	Maximum		120	.333		15.45	11.55	5	32	40.50	
	2300	10' wide, minimum		200	.200		9.75	6.90	3.01	19.66	25	
	2340	Maximum	↓	110	.364	↓	13.50	12.55	5.50	31.55	40.50	
360	0010	**JETTIES, PIERS,** Municipal with 3" x 12" framing and 3" decking,										**360**
	0020	wood piles and cross bracing, alternate bents battered	B-76	60	1.200	S.F.	66	40.50	37.50	144	179	

02390	Shore Protect/Mooring Structures	CREW	DAILY OUTPUT	LABOR-HOURS	UNIT	2005 BARE COSTS				TOTAL INCL O&P	
						MAT.	LABOR	EQUIP.	TOTAL		
360	**0200** Treated piles, not including mobilization										**360**
0210	50' long, 20 lb. creosote, shore driven	B-19	540	.119	V.L.F.	9.75	4.01	2.83	16.59	20	
0220	Barge driven	B-76	320	.225		9.75	7.60	7.05	24.40	30.50	
0230	2.5 lb. CCA, shore driven	B-19	540	.119		11.95	4.01	2.83	18.79	22.50	
0240	Barge driven	B-76	320	.225		11.95	7.60	7.05	26.60	33	
0250	30' long, 20 lb. creosote, shore driven	B-19	540	.119		9.10	4.01	2.83	15.94	19.50	
0260	Barge driven	B-76	320	.225		9.10	7.60	7.05	23.75	30	
0270	2.5 lb. CCA, shore driven	B-19	540	.119		6.80	4.01	2.83	13.64	16.95	
0280	Barge driven	B-76	320	.225		6.80	7.60	7.05	21.45	27.50	
0300	Mobilization, barge, by tug boat	B-83	25	.640	Mile		19.65	17.65	37.30	49.50	
0350	Standby time for shore pile driving crew				Hr.				430	535	
0360	Standby time for barge driving rig				"				590	695	

02410	Tunnel Excavation	CREW	DAILY OUTPUT	LABOR-HOURS	UNIT	2005 BARE COSTS				TOTAL INCL O&P	
						MAT.	LABOR	EQUIP.	TOTAL		
410	**0010 ROCK EXCAVATION, TUNNEL BORING GENERAL** `R02400 -110`										**410**
0020	Bored Tunnels, incl. mucking 20' diameter				L.F.				385	440	
0100	Rock excavation, minimum								385	440	
0110	Average								825	945	
0120	Maximum								2,200	2,525	
0200	Mixed rock and earth, minimum								660	725	
0210	Average								1,650	1,900	
0220	Maximum								2,425	2,650	
500	**0010 SOFT GROUND SHIELD DRIVEN BORING**										**500**
0300	Earth excavation, minimum				L.F.				330	365	
0310	Average								770	845	
0320	Maximum								1,975	2,175	

02425	Tunnel Linings										
100	**0010 CAST IN PLACE CONCRETE TUNNEL LININGS**										**100**
0500	Tunnel liner invert, incl. reinf., for bored tunnels, 20' diam., rock				L.F.				385	425	
0510	Earth								248	275	
0520	Arch, incl. reinf., for bored tunnels, rock								360	395	
0530	Earth								350	385	

02430	Tunnel Grouting										
100	**0010 TUNNEL LINER GROUTING**										**100**
0800	Contact grouting incl. drilling and connecting, minimum				C.F.				6.90	7.60	
0820	Maximum				"				44	48.50	

02441	Microtunneling										
400	**0010 MICROTUNNELING** Not including excavation, backfill, shoring,										**400**
0020	or dewatering, average 50'/day, slurry method										
0100	24" to 48" outside diameter, minimum				L.F.					640	
0110	Adverse conditions, add				%					50%	
1000	Rent microtunneling machine, average monthly lease				Month					85,500	
1010	Operating technician				Day					640	

Important: See the Reference Section for critical supporting data - Reference Nos., Crews, & City Cost Indexes

02400 | Tunneling, Boring & Jacking

02441 | Microtunneling

		CREW	DAILY OUTPUT	LABOR-HOURS	UNIT	2005 BARE COSTS				TOTAL INCL O&P	
						MAT.	LABOR	EQUIP.	TOTAL		
400	1100	Mobilization and demobilization, minimum				Job					42,800	400
	1110	Maximum				"					430,000	

02442 | Cut and Cover Tunnels

		CREW	DAILY OUTPUT	LABOR-HOURS	UNIT	MAT.	LABOR	EQUIP.	TOTAL	INCL O&P	
150	0010	**CUT AND COVER TUNNELS**										150
	2000	Cut and cover, excavation, not incl. hauling or backfill	B-57	80	.600	C.Y.		18.10	16.05	34.15	45	
	2100	Decking on steel frames, concrete				S.F.				21	23	
	2110	Steel plates								27.50	30.50	
	2120	Wood				↓				14.85	16.50	
	2500	Shotcrete applied in tunnels				C.Y.				220	253	

02444 | Shaft Construction

		CREW	DAILY OUTPUT	LABOR-HOURS	UNIT	MAT.	LABOR	EQUIP.	TOTAL	INCL O&P	
150	0010	**SHAFT CONSTRUCTION**										150
	0400	Shaft excavation, rock, minimum				C.Y.				55	60.50	
	0410	Average								77	84.50	
	0420	Maximum								154	171	
	0430	Earth, minimum								33	36.50	
	0440	Average								49.50	55	
	0450	Maximum				↓				138	154	
	3000	Ventilation for tunnel construction										
	3100	Duct, 48" diameter, 20 ga., spun on site				L.F.				14.85	16.50	
	3200	Fan, 48" diameter, 125 H.P., incl. starter				Ea.				19,000	22,000	

02445 | Boring or Jacking Conduits

		CREW	DAILY OUTPUT	LABOR-HOURS	UNIT	MAT.	LABOR	EQUIP.	TOTAL	INCL O&P	
300	0010	**HORIZONTAL BORING** Casing only, 100' minimum,										300
	0020	not incl. jacking pits or dewatering										
	0100	Roadwork, 1/2" thick wall, 24" diameter casing	B-42	20	3.200	L.F.	67	96	53.50	216.50	284	
	0200	36" diameter		16	4		107	120	67	294	380	
	0300	48" diameter		15	4.267		157	128	71.50	356.50	455	
	0500	Railroad work, 24" diameter		15	4.267		67	128	71.50	266.50	355	
	0600	36" diameter		14	4.571		107	137	76.50	320.50	415	
	0700	48" diameter	↓	12	5.333		157	160	89	406	520	
	0900	For ledge, add								155	190	
	1000	Small diameter boring, 3", sandy soil	B-82	900	.018		18.45	.53	.07	19.05	21.50	
	1040	Rocky soil	"	500	.032	↓	18.45	.96	.13	19.54	22	
	1100	Prepare jacking pits, incl. mobilization & demobilization, minimum				Ea.				2,850	3,375	
	1101	Maximum				"				16,100	19,300	

02450 | Foundation & Load Bearing Elements

02455 | Driven Piles

		CREW	DAILY OUTPUT	LABOR-HOURS	UNIT	2005 BARE COSTS				TOTAL INCL O&P	
						MAT.	LABOR	EQUIP.	TOTAL		
100	0010	**CAST IN PLACE CONCRETE PILES**, 200 piles, 60' long R02450 -820										100
	0020	unless specified otherwise, not incl. pile caps or mobilization										
	0050	Cast in place augered piles, no casing or reinforcing										
	0060	8" diameter	B-43	540	.089	V.L.F.	3.02	2.59	5.90	11.51	13.80	
	0065	10" diameter		480	.100		4.80	2.91	6.60	14.31	17.10	
	0070	12" diameter		420	.114		6.75	3.33	7.55	17.63	21	
	0075	14" diameter		360	.133		9.10	3.88	8.85	21.83	25.50	
	0080	16" diameter	↓	300	.160	↓	12.30	4.66	10.60	27.56	32.50	

2 SITE CONSTRUCTION

		02455	Driven Piles	CREW	DAILY OUTPUT	LABOR-HOURS	UNIT	2005 BARE COSTS MAT.	LABOR	EQUIP.	TOTAL	TOTAL INCL O&P	
100	0085		18" diameter	B-43	240	.200	V.L.F.	15.20	5.85	13.25	34.30	40.50	100
	0100		Cast in place, thin wall shell pile, straight sided,										
	0110		not incl. reinforcing, 8" diam., 16 ga., 5.8 lb./L.F.	B-19	700	.091	V.L.F.	5.65	3.09	2.19	10.93	13.60	
	0200		10" diameter, 16 ga. corrugated, 7.3 lb./L.F.		650	.098		7.40	3.33	2.35	13.08	16.05	
	0300		12" diameter, 16 ga. corrugated, 8.7 lb./L.F.		600	.107		9.60	3.61	2.55	15.76	19.15	
	0400		14" diameter, 16 ga. corrugated, 10.0 lb./L.F.		550	.116		11.30	3.93	2.78	18.01	22	
	0500		16" diameter, 16 ga. corrugated, 11.6 lb./L.F.		500	.128		13.85	4.33	3.06	21.24	25.50	
	0800		Cast in place friction pile, 50' long, fluted,										
	0810		tapered steel, 4000 psi concrete, no reinforcing										
	0900		12" diameter, 7 ga.	B-19	600	.107	V.L.F.	17	3.61	2.55	23.16	27.50	
	1000		14" diameter, 7 ga.		560	.114		18.50	3.86	2.73	25.09	29.50	
	1100		16" diameter, 7 ga.		520	.123		21.50	4.16	2.94	28.60	34	
	1200		18" diameter, 7 ga.		480	.133		25.50	4.51	3.19	33.20	38.50	
	1300		End bearing, fluted, constant diameter,										
	1320		4000 psi concrete, no reinforcing										
	1340		12" diameter, 7 ga.	B-19	600	.107	V.L.F.	17.75	3.61	2.55	23.91	28	
	1360		14" diameter, 7 ga.		560	.114		22	3.86	2.73	28.59	33.50	
	1380		16" diameter, 7 ga.		520	.123		25.50	4.16	2.94	32.60	38.50	
	1400		18" diameter, 7 ga.		480	.133		28.50	4.51	3.19	36.20	41.50	
	1500		For reinforcing steel, add				Lb.	.80			.80	.88	
	1700		For ball or pedestal end, add	B-19	11	5.818	C.Y.	91	197	139	427	570	
	1900		For lengths above 60', concrete, add	"	11	5.818	"	95	197	139	431	570	
	2000		For steel thin shell, pipe only				Lb.	.69			.69	.76	
450	0010		**PRESTRESSED CONCRETE PILES**, 200 piles										450
	0020		unless specified otherwise, not incl. pile caps or mobilization										
	2200		Precast, prestressed, 50' long, 12" diam., 2-3/8" wall	B-19	720	.089	V.L.F.	11.95	3	2.13	17.08	20.50	
	2300		14" diameter, 2-1/2" wall		680	.094		15.75	3.18	2.25	21.18	25	
	2500		16" diameter, 3" wall		640	.100		22	3.38	2.39	27.77	32	
	2800		20" diameter, 3-1/2" wall	B-19A	560	.114		31.50	3.86	3.49	38.85	45	
	2900		24" diameter, 3-1/2" wall	"	520	.123		38.50	4.16	3.76	46.42	53.50	
	2920		36" diameter, 4-1/2" wall	B-19	400	.160		54.50	5.40	3.83	63.73	73	
	2940		54" diameter, 5" wall		340	.188		73	6.35	4.50	83.85	95.50	
	2960		66" diameter, 6" wall		220	.291		110	9.85	6.95	126.80	144	
	3100		Precast, prestressed, 40' long, 10" thick, square		700	.091		8.65	3.09	2.19	13.93	16.90	
	3200		12" thick, square		680	.094		10.90	3.18	2.25	16.33	19.55	
	3275		Shipping for 60 foot long concrete piles, add to material cost					.43			.43	.47	
	3400		14" thick, square	B-19	600	.107		12.90	3.61	2.55	19.06	22.50	
	3500		Octagonal		640	.100		17.85	3.38	2.39	23.62	27.50	
	3700		16" thick, square		560	.114		20.50	3.86	2.73	27.09	31.50	
	3800		Octagonal		600	.107		21.50	3.61	2.55	27.66	32	
	4000		18" thick, square	B-19A	520	.123		24.50	4.16	3.76	32.42	38	
	4100		Octagonal	B-19	560	.114		25.50	3.86	2.73	32.09	37	
	4300		20" thick, square	B-19A	480	.133		30	4.51	4.07	38.58	44.50	
	4400		Octagonal	B-19	520	.123		28	4.16	2.94	35.10	41	
	4600		24" thick, square	B-19A	440	.145		43	4.92	4.44	52.36	60	
	4700		Octagonal	B-19	480	.133		40.50	4.51	3.19	48.20	55	
	4750		Mobilization for 10,000 L.F. pile job, add		3,300	.019			.66	.46	1.12	1.56	
	4800		25,000 L.F. pile job, add		8,500	.008			.25	.18	.43	.61	
	4850		Mobilization by water for barge driving rig, add								100%		
	5000		Pressure grouted pin pile, 5" diam., cased, up to 50 ton,										
	5040		End bearing, less than 20'	B-48	90	.622	V.L.F.	27	18.45	38	83.45	100	
	5080		More than 40'		135	.415		15.75	12.30	25.50	53.55	64.50	
	5120		Friction, loose sand and gravel		107	.523		27	15.55	32	74.55	88.50	
	5160		Dense sand and gravel		135	.415		15.75	12.30	25.50	53.55	64.50	
	5200		Uncased, up to 10 ton capacity, 20'		135	.415		15.95	12.30	25.50	53.75	64.50	

Important: See the Reference Section for critical supporting data - Reference Nos., Crews, & City Cost Indexes

02455	Driven Piles	CREW	DAILY OUTPUT	LABOR-HOURS	UNIT	MAT.	LABOR	EQUIP.	TOTAL	TOTAL INCL O&P
0010	**STEEL PILES**, Not including mobilization or demobilization R02450 -820									
0100	Step tapered, round, concrete filled									
0110	8" tip, 60 ton capacity, 30' depth	B-19	760	.084	V.L.F.	6.90	2.85	2.01	11.76	14.35
0120	60' depth		740	.086		7.80	2.92	2.07	12.79	15.50
0130	80' depth		700	.091		8.05	3.09	2.19	13.33	16.20
0150	10" tip, 90 ton capacity, 30' depth		700	.091		8.50	3.09	2.19	13.78	16.70
0160	60' depth		690	.093		8.75	3.14	2.22	14.11	17.05
0170	80' depth		670	.096		9.40	3.23	2.28	14.91	18
0190	12" tip, 120 ton capacity, 30' depth		660	.097		11.65	3.28	2.32	17.25	20.50
0200	60' depth, 12" diameter		630	.102		11.70	3.43	2.43	17.56	21
0210	80' depth		590	.108		10.35	3.67	2.59	16.61	20
0250	"H" Sections, 50' long, HP8 x 36		640	.100		11.85	3.38	2.39	17.62	21
0400	HP10 X 42		610	.105		13.85	3.55	2.51	19.91	23.50
0500	HP10 X 57		610	.105		18.75	3.55	2.51	24.81	29
0700	HP12 X 53		590	.108		17.70	3.67	2.59	23.96	28
0800	HP12 X 74	B-19A	590	.108		24.50	3.67	3.31	31.48	36.50
1000	HP14 X 73		540	.119		24.50	4.01	3.62	32.13	37.50
1100	HP14 X 89		540	.119		30	4.01	3.62	37.63	43.50
1300	HP14 X 102		510	.125		34	4.24	3.83	42.07	48.50
1400	HP14 X 117		510	.125		39	4.24	3.83	47.07	54
1600	Splice on standard points, not in leads, 8" or 10"	1 Sswl	5	1.600	Ea.	77.50	61		138.50	195
2600	Pipe piles, 50' lg. 8" diam., 29 lb. per L.F., no concrete	B-19	500	.128	V.L.F.	12.80	4.33	3.06	20.19	24.50
2700	Concrete filled		460	.139		13.60	4.70	3.33	21.63	26
2900	10" diameter, 34 lb. per L.F., no concrete		500	.128		15.95	4.33	3.06	23.34	28
3000	Concrete filled		450	.142		17.60	4.81	3.40	25.81	30.50
3200	12" diameter, 44 lb. per L.F., no concrete		475	.135		19.75	4.55	3.22	27.52	32.50
3300	Concrete filled		415	.154		20.50	5.20	3.69	29.39	35
3500	14" diameter, 46 lb. per L.F., no concrete		430	.149		21	5.05	3.56	29.61	35.50
3600	Concrete filled		355	.180		22.50	6.10	·4.31	32.91	39.50
3800	16" diameter, 52 lb. per L.F., no concrete		385	.166		23.50	5.60	3.97	33.07	39.50
3900	Concrete filled		335	.191		26	6.45	4.57	37.02	44
4100	18" diameter, 59 lb. per L.F., no concrete		355	.180		31	6.10	4.31	41.41	48.50
4200	Concrete filled		310	.206		30.50	7	4.94	42.44	50.50
4400	Splices for pipe piles, not in leads, 8" diameter	1 Sswl	4.67	1.713	Ea.	57.50	65.50		123	180
4500	14" diameter		3.79	2.111		75.50	80.50		156	227
4600	16" diameter		3.03	2.640		93	101		194	282
4650	18" diameter		4.50	1.778		123	68		191	256
4800	Points, standard, 8" diameter		4.61	1.735		64.50	66		130.50	189
4900	14" diameter		4.05	1.975		89.50	75.50		165	234
5000	16" diameter		3.37	2.374		109	90.50		199.50	282
5050	18" diameter		5	1.600		170	61		231	296
5200	Points, heavy duty, 10" diameter		2.89	2.768		44.50	106		150.50	238
5300	14" or 16" diameter		2.02	3.960		71	151		222	350
5500	For reinforcing steel, add		1,150	.007	Lb.	.53	.27		.80	1.06
5700	For thick wall sections, add				"	.60			.60	.66
6020	Steel pipe pile end plates, 8" diameter	1 Sswl	14	.571	Ea.	49	22		71	93
6050	10" diameter		14	.571		51.50	22		73.50	95.50
6100	12" diameter		12	.667		54.50	25.50		80	106
6150	14" diameter		10	.800		56.50	30.50		87	117
6200	16" diameter		9	.889		60	34		94	127
6250	18" diameter		8	1		67.50	38		105.50	143
6300	Steel pipe pile shoes, 8" diameter		12	.667		104	25.50		129.50	161
6350	10" diameter		12	.667		108	25.50		133.50	165
6400	12" diameter		10	.800		115	30.50		145.50	182
6450	14" diameter		9	.889		133	34		167	207
6500	16" diameter		8	1		148	38		186	232

SITE CONSTRUCTION 2

		02455	Driven Piles		CREW	DAILY OUTPUT	LABOR-HOURS	UNIT	2005 BARE COSTS				TOTAL INCL O&P	
									MAT.	LABOR	EQUIP.	TOTAL		
600	6550		18" diameter	R02450 -820	1 Sswl	6	1.333	Ea.	166	51		217	273	600
650	0010	**TIMBER PILES,** Friction or end bearing, not including		R02450 -820										650
	0050	mobilization or demobilization												
	0100	Untreated piles, up to 30' long, 12" butts, 8" points			B-19	625	.102	V.L.F.	6.30	3.46	2.45	12.21	15.15	
	0200	30' to 39' long, 12" butts, 8" points				700	.091		6.30	3.09	2.19	11.58	14.30	
	0300	40' to 49' long, 12" butts, 7" points				720	.089		6.30	3	2.13	11.43	14.10	
	0400	50' to 59' long, 13"butts, 7" points				800	.080		6.40	2.70	1.91	11.01	13.45	
	0500	60' to 69' long, 13" butts, 7" points				840	.076		7.20	2.58	1.82	11.60	14	
	0600	70' to 80' long, 13" butts, 6" points				840	.076		8	2.58	1.82	12.40	14.90	
	0700	70' to 80' long, 16" butts, 6" points				800	.080	Ea.	11.45	2.70	1.91	16.06	19	
	0800	Treated piles, 12 lb. per C.F.,												
	0810	friction or end bearing, ASTM class B												
	1000	Up to 30' long, 12" butts, 8" points			B-19	625	.102	V.L.F.	9.30	3.46	2.45	15.21	18.45	
	1100	30' to 39' long, 12" butts, 8" points				700	.091		9.10	3.09	2.19	14.38	17.35	
	1200	40' to 49' long, 12" butts, 7" points				720	.089		9.75	3	2.13	14.88	17.85	
	1300	50' to 59' long, 13" butts, 7" points				800	.080		10.75	2.70	1.91	15.36	18.25	
	1400	60' to 69' long, 13" butts, 6" points			B-19A	840	.076		14.70	2.58	2.32	19.60	23	
	1500	70' to 80' long, 13" butts, 6" points			"	840	.076		18.45	2.58	2.32	23.35	27	
	1600	Treated piles, C.C.A., 2.5# per C.F.												
	1610	8" butts, 10' long			B-19	400	.160	V.L.F.	6.80	5.40	3.83	16.03	20.50	
	1620	11' to 16' long				500	.128		6.80	4.33	3.06	14.19	17.70	
	1630	17' to 20' long				575	.111		6.80	3.76	2.66	13.22	16.40	
	1640	10" butts, 10' to 16' long				500	.128		7.75	4.33	3.06	15.14	18.75	
	1650	17' to 20' long				575	.111		7.75	3.76	2.66	14.17	17.45	
	1660	21' to 40' long				700	.091		7.75	3.09	2.19	13.03	15.85	
	1670	12" butts, 10' to 20' long				575	.111		8.35	3.76	2.66	14.77	18.15	
	1680	21' to 35' long				650	.098		8.35	3.33	2.35	14.03	17.10	
	1690	36' to 40' long				700	.091		8.35	3.09	2.19	13.63	16.55	
	1695	14" butts. to 40' long				700	.091		11.95	3.09	2.19	17.23	20.50	
	1700	Boot for pile tip, minimum			1 Pile	27	.296	Ea.	19.10	9.85		28.95	37.50	
	1800	Maximum				21	.381		57	12.70		69.70	84	
	2000	Point for pile tip, minimum				20	.400		19.05	13.30		32.35	43	
	2100	Maximum				15	.533		68.50	17.75		86.25	105	
	2300	Splice for piles over 50' long, minimum			B-46	35	1.371		47.50	41.50	.96	89.96	121	
	2400	Maximum				20	2.400		57	73	1.68	131.68	182	
	2600	Concrete encasement with wire mesh and tube				331	.145	V.L.F.	8.95	4.40	.10	13.45	17	
	2700	Mobilization for 10,000 L.F. pile job, add			B-19	3,300	.019			.66	.46	1.12	1.56	
	2800	25,000 L.F. pile job, add			"	8,500	.008			.25	.18	.43	.61	
	2900	Mobilization by water for barge driving rig, add										100%		
800	0010	**PILING SPECIAL COSTS**		R02450 -820										800
	0011	Pile caps, see Division 03310-240												
	0500	Cutoffs, concrete piles, plain			1 Pile	5.50	1.455	Ea.		48.50		48.50	80	
	0600	With steel thin shell, add				38	.211			7		7	11.55	
	0700	Steel pile or "H" piles				19	.421			14		14	23	
	0800	Wood piles				38	.211			7		7	11.55	
	0900	Pre-augering up to 30' deep, average soil, 24" diameter			B-43	180	.267	L.F.		7.75	17.65	25.40	31.50	
	0920	36" diameter				115	.417			12.15	27.50	39.65	49	
	0960	48" diameter				70	.686			19.95	45.50	65.45	80.50	
	0980	60" diameter				50	.960			28	63.50	91.50	113	
	1000	Testing, any type piles, test load is twice the design load												
	1050	50 ton design load, 100 ton test						Ea.				15,000	16,050	
	1100	100 ton design load, 200 ton test										19,250	20,300	
	1150	150 ton design load, 300 ton test										24,100	25,700	
	1200	200 ton design load, 400 ton test										26,200	28,800	
	1250	400 ton design load, 800 ton test										30,300	33,700	

Important: See the Reference Section for critical supporting data - Reference Nos., Crews, & City Cost Indexes

02455 | Driven Piles

		CREW	DAILY OUTPUT	LABOR-HOURS	UNIT	2005 BARE COSTS MAT.	LABOR	EQUIP.	TOTAL	TOTAL INCL O&P	
800	1500	Wet conditions, soft damp ground R02450 -820										800
	1600	Requiring mats for crane, add								40%	40%	
	1700	Barge mounted driving rig, add								30%	30%	
900	0010	**MOBILIZATION** R02450 -820										900
	0020	Set up & remove, air compressor, 600 C.F.M.	A-5	3.30	5.455	Ea.		146	9.75	155.75	237	
	0100	1200 C.F.M.	"	2.20	8.182			219	14.65	233.65	355	
	0200	Crane, with pile leads and pile hammer, 75 ton	B-19	.60	106			3,600	2,550	6,150	8,550	
	0300	150 ton	"	.36	177			6,000	4,250	10,250	14,300	
	0500	Drill rig, for caissons, to 36", minimum	B-43	2	24			700	1,600	2,300	2,825	
	0520	Maximum		.50	96			2,800	6,350	9,150	11,300	
	0600	Up to 84"		1	48			1,400	3,175	4,575	5,650	
	0800	Auxiliary boiler, for steam small	A-5	1.66	10.843			290	19.40	309.40	470	
	0900	Large	"	.83	21.687			580	39	619	945	
	1100	Rule of thumb: complete pile driving set up, small	B-19	.45	142			4,800	3,400	8,200	11,400	
	1200	Large	"	.27	237			8,000	5,675	13,675	19,000	
	1300	Mobilization by water for barge driving rig										
	1310	Minimum				Ea.				6,250	6,900	
	1320	Maximum				"				40,700	44,900	

02465 | Bored Piles

		CREW	DAILY OUTPUT	LABOR-HOURS	UNIT	2005 BARE COSTS MAT.	LABOR	EQUIP.	TOTAL	TOTAL INCL O&P	
800	0010	**DRILLED CAISSONS** Incl. excav., concrete, 50 lbs. reinf. R02450 -820										800
	0020	per C.Y., not incl. mobilization, boulder removal, disposal										
	0100	Open style, machine drilled, to 50' deep, in stable ground, no										
	0110	casings or ground water, 18" diam., 0.065 C.Y./L.F.	B-43	200	.240	V.L.F.	6.55	7	15.90	29.45	35.50	
	0200	24" diameter, 0.116 C.Y./L.F.		190	.253		11.70	7.35	16.75	35.80	42.50	
	0300	30" diameter, 0.182 C.Y./L.F.		150	.320		18.35	9.30	21	48.65	58	
	0400	36" diameter, 0.262 C.Y./L.F.		125	.384		26.50	11.20	25.50	63.20	74	
	0500	48" diameter, 0.465 C.Y./L.F.		100	.480		47	14	32	93	108	
	0600	60" diameter, 0.727 C.Y./L.F.		90	.533		73.50	15.55	35.50	124.55	144	
	0700	72" diameter, 1.05 C.Y./L.F.		80	.600		106	17.50	39.50	163	188	
	0800	84" diameter, 1.43 C.Y./L.F.		75	.640		144	18.65	42.50	205.15	234	
	1000	For bell excavation and concrete, add										
	1020	4' bell diameter, 24" shaft, 0.444 C.Y.	B-43	20	2.400	Ea.	36	70	159	265	325	
	1040	6' bell diameter, 30" shaft, 1.57 C.Y.		5.70	8.421		127	245	560	932	1,125	
	1060	8' bell diameter, 36" shaft, 3.72 C.Y.		2.40	20		300	585	1,325	2,210	2,675	
	1080	9' bell diameter, 48" shaft, 4.48 C.Y.		2	24		365	700	1,600	2,665	3,225	
	1100	10' bell diameter, 60" shaft, 5.24 C.Y.		1.70	28.235		425	820	1,875	3,120	3,800	
	1120	12' bell diameter, 72" shaft, 8.74 C.Y.		1	48		710	1,400	3,175	5,285	6,425	
	1140	14' bell diameter, 84" shaft, 13.6 C.Y.		.70	68.571		1,100	2,000	4,550	7,650	9,275	
	1200	Open style, machine drilled, to 50' deep, in wet ground, pulled										
	1300	casing and pumping, 18" diameter, 0.065 C.Y./L.F.	B-48	160	.350	V.L.F.	6.55	10.40	21.50	38.45	46.50	
	1400	24" diameter, 0.116 C.Y./L.F.		125	.448		11.70	13.30	27.50	52.50	63.50	
	1500	30" diameter, 0.182 C.Y./L.F.		85	.659		18.35	19.55	40	77.90	94	
	1600	36" diameter, 0.262 C.Y./L.F.		60	.933		26.50	27.50	57	111	134	
	1700	48" diameter, 0.465 C.Y./L.F.	B-49	55	1.600		47	49.50	73.50	170	209	
	1800	60" diameter, 0.727 C.Y./L.F.		35	2.514		73.50	78	115	266.50	330	
	1900	72" diameter, 1.05 C.Y./L.F.		30	2.933		106	91	134	331	405	
	2000	84" diameter, 1.43 C.Y./L.F.		25	3.520		144	109	161	414	505	
	2100	For bell excavation and concrete, add										
	2120	4' bell diameter, 24" shaft, 0.444 C.Y.	B-48	19.80	2.828	Ea.	36	84	173	293	360	
	2140	6' bell diameter, 30" shaft, 1.57 C.Y.		5.70	9.825		127	292	600	1,019	1,250	
	2160	8' bell diameter, 36" shaft, 3.72 C.Y.		2.40	23.333		300	695	1,425	2,420	2,950	
	2180	9' bell diameter, 48" shaft, 4.48 C.Y.	B-49	3.30	26.667		365	825	1,225	2,415	3,025	
	2200	10' bell diameter, 60" shaft, 5.24 C.Y.		2.80	31.429		425	975	1,450	2,850	3,550	

SITE CONSTRUCTION 2

		02465	Bored Piles	CREW	DAILY OUTPUT	LABOR-HOURS	UNIT	MAT.	LABOR	EQUIP.	TOTAL	TOTAL INCL O&P	
									2005 BARE COSTS				
800	2220		12' bell diameter, 72" shaft, 8.74 C.Y.	B-49	1.60	55	Ea.	710	1,700	2,525	4,935	6,200	800
	2240		14' bell diameter, 84" shaft, 13.6 C.Y.	↓	1	88	↓	1,100	2,725	4,025	7,850	9,875	
	2300		Open style, machine drilled, to 50' deep, in soft rocks and										
	2400		medium hard shales, 18" diameter, 0.065 C.Y./L.F.	B-49	50	1.760	V.L.F.	6.55	54.50	80.50	141.55	181	
	2500		24" diameter, 0.116 C.Y./L.F.		30	2.933		11.70	91	134	236.70	300	
	2600		30" diameter, 0.182 C.Y./L.F.		20	4.400		18.35	136	202	356.35	455	
	2700		36" diameter, 0.262 C.Y./L.F.		15	5.867		26.50	182	269	477.50	605	
	2800		48" diameter, 0.465 C.Y./L.F.		10	8.800		47	272	405	724	915	
	2900		60" diameter, 0.727 C.Y./L.F.		7	12.571		73.50	390	575	1,038.50	1,325	
	3000		72" diameter, 1.05 C.Y./L.F.		6	14.667		106	455	670	1,231	1,550	
	3100		84" diameter, 1.43 C.Y./L.F.	↓	5	17.600	↓	144	545	805	1,494	1,900	
	3200		For bell excavation and concrete, add										
	3220		4' bell diameter, 24" shaft, 0.444 C.Y.	B-49	10.90	8.073	Ea.	36	250	370	656	830	
	3240		6' bell diameter, 30" shaft, 1.57 C.Y.		3.10	28.387		127	880	1,300	2,307	2,925	
	3260		8' bell diameter, 36" shaft, 3.72 C.Y.		1.30	67.692		300	2,100	3,100	5,500	7,000	
	3280		9' bell diameter, 48" shaft, 4.48 C.Y.		1.10	80		365	2,475	3,675	6,515	8,275	
	3300		10' bell diameter, 60" shaft, 5.24 C.Y.		.90	97.778		425	3,025	4,475	7,925	10,100	
	3320		12' bell diameter, 72" shaft, 8.74 C.Y.		.60	146		710	4,550	6,725	11,985	15,200	
	3340		14' bell diameter, 84" shaft, 13.6 C.Y.		.40	220	↓	1,100	6,800	10,100	18,000	22,900	
	3600		For rock excavation, sockets, add, minimum		120	.733	C.F.		22.50	33.50	56	72	
	3650		Average		95	.926			28.50	42.50	71	91	
	3700		Maximum	↓	48	1.833	↓		56.50	84	140.50	181	
	3900		For 50' to 100' deep, add				V.L.F.				7%	7%	
	4000		For 100' to 150' deep, add								25%	25%	
	4100		For 150' to 200' deep, add				↓				30%	30%	
	4200		For casings left in place, add				Lb.	.67			.67	.74	
	4300		For other than 50 lb. reinf. per C.Y., add or deduct				"	.71			.71	.78	
	4400		For steel "I" beam cores, add	B-49	8.30	10.602	Ton	1,400	330	485	2,215	2,600	
	4500		Load and haul excess excavation, 2 miles	B-34B	178	.045	L.C.Y.		1.24	2.68	3.92	4.83	
	4600		For mobilization, 50 mile radius, rig to 36"	B-43	2	24	Ea.		700	1,600	2,300	2,825	
	4650		Rig to 84"	B-48	1.75	32			950	1,950	2,900	3,600	
	4700		For low headroom, add								50%		
	4750		For difficult access, add								25%		
	5000		Bottom inspection	1 Skwk	1.20	6.667	↓		232		232	360	
950	0010		**DRILLED CONCRETE PIERS AND SHAFTS** or Displacement										950
	0100		Caissons, incl. mobilization and demobilization, up to 50 miles										
	0200		Uncased shafts, 30 to 80 tons cap., 17" diam., 10' depth	B-44	88	.727	V.L.F.	16.20	24.50	10.75	51.45	68.50	
	0300		25' depth		165	.388		11.55	12.95	5.75	30.25	39.50	
	0400		80-150 ton capacity, 22" diameter, 10' depth		80	.800		20.50	26.50	11.85	58.85	78.50	
	0500		20' depth		130	.492		16.20	16.45	7.30	39.95	52.50	
	0700		Cased shafts, 10 to 30 ton capacity, 10-5/8" diam., 20' depth		175	.366		11.55	12.20	5.40	29.15	38.50	
	0800		30' depth		240	.267		10.80	8.90	3.94	23.64	30.50	
	0850		30 to 60 ton capacity, 12" diameter, 20' depth		160	.400		16.20	13.35	5.90	35.45	46	
	0900		40' depth		230	.278		12.45	9.30	4.12	25.87	33	
	1000		80 to 100 ton capacity, 16" diameter, 20' depth		160	.400		23	13.35	5.90	42.25	53.50	
	1100		40' depth		230	.278		21.50	9.30	4.12	34.92	43.50	
	1200		110 to 140 ton capacity, 17-5/8" diameter, 20' depth		160	.400		25	13.35	5.90	44.25	55.50	
	1300		40' depth		230	.278		23	9.30	4.12	36.42	45	
	1400		140 to 175 ton capacity, 19" diameter, 20' depth		130	.492		27	16.45	7.30	50.75	64	
	1500		40' depth	↓	210	.305	↓	25	10.15	4.51	39.66	49	
	1700		Over 30' long, L.F. cost tends to be lower										
	1900		Maximum depth is about 90'	↓									

R02450-820 (row 2220)

R02465-800 (row 0010)

02510	Water Distribution	CREW	DAILY OUTPUT	LABOR-HOURS	UNIT	2005 BARE COSTS				TOTAL INCL O&P
						MAT.	LABOR	EQUIP.	TOTAL	
600	**0010**	**VALVES**									**600**
9000	Valves, gate valve, N.R.S. post type, 4" diameter	B-21	32	.875	Ea.	365	27	4.95	396.95	450
9040	8" diameter		16	1.750		705	54	9.90	768.90	870
9080	12" diameter		13	2.154		1,425	66.50	12.20	1,503.70	1,675
9120	O.S.&Y., 4" diameter		32	.875		291	27	4.95	322.95	365
9160	8" diameter		16	1.750		695	54	9.90	758.90	860
9200	12" diameter		13	2.154		1,475	66.50	12.20	1,553.70	1,750
9220	14" diameter	▼	11	2.545		4,150	78.50	14.40	4,242.90	4,725
9400	Check valves, rubber disc, 2-1/2" diameter	B-20	44	.545		350	16.40		366.40	410
9440	4" diameter	B-21	32	.875		430	27	4.95	461.95	520
9500	8" diameter		16	1.750		1,050	54	9.90	1,113.90	1,250
9540	12" diameter		13	2.154		2,675	66.50	12.20	2,753.70	3,075
9700	Detector check valves, reducing, 4" diameter		32	.875		705	27	4.95	736.95	820
9740	8" diameter		16	1.750		1,775	54	9.90	1,838.90	2,050
9800	Galvanized, 4" diameter		32	.875		920	27	4.95	951.95	1,075
9840	8" diameter	▼	16	1.750	▼	2,150	54	9.90	2,213.90	2,450
710	**0010**	**TAPPING, CROSSES AND SLEEVES**									**710**
4000	Drill and tap pressurized main (labor only)									
4100	6" main, 1" to 2" service	Q-1	3	5.333	Ea.		196		196	295
4150	8" main, 1" to 2" service	"	2.75	5.818	"		214		214	320
4500	Tap and insert gate valve									
4600	8" main, 4" branch	B-21	3.20	8.750	Ea.		270	49.50	319.50	475
4650	6" branch		2.70	10.370			320	58.50	378.50	560
4651	Piping, drill, tap & insert gate valve, 8" main, 6" branch		2.70	10.370			320	58.50	378.50	560
4700	10" Main, 4" branch		2.70	10.370			320	58.50	378.50	560
4750	6" branch		2.35	11.915			370	67.50	437.50	645
4800	12" main, 6" branch		2.35	11.915			370	67.50	437.50	645
4850	8" branch	▼	2.35	11.915	▼		370	67.50	437.50	645
7000	Tapping crosses, sleeves, valves; with rubber gaskets									
7020	Crosses, 4" x 4"	B-21	37	.757	Ea.	455	23.50	4.28	482.78	540
7060	8" x 6"		21	1.333		660	41	7.55	708.55	795
7080	8" x 8"		21	1.333		715	41	7.55	763.55	855
7100	10" x 6"		21	1.333		1,300	41	7.55	1,348.55	1,525
7160	12" x 12"		18	1.556		1,650	48	8.80	1,706.80	1,900
7180	14" x 6"		16	1.750		3,275	54	9.90	3,338.90	3,700
7240	16" x 10"		14	2		3,550	62	11.30	3,623.30	4,000
7280	18" x 6"		10	2.800		5,225	86.50	15.85	5,327.35	5,900
7320	18" x 18"		10	2.800		4,825	86.50	15.85	4,927.35	5,450
7340	20" x 6"		8	3.500		4,225	108	19.80	4,352.80	4,850
7360	20" x 12"		8	3.500		4,525	108	19.80	4,652.80	5,200
7420	24" x 12"		6	4.667		5,550	144	26.50	5,720.50	6,350
7440	24" x 18"		6	4.667		8,275	144	26.50	8,445.50	9,350
7600	Cut-in sleeves with rubber gaskets, 4"		18	1.556		150	48	8.80	206.80	249
7640	8"		10	2.800		260	86.50	15.85	362.35	435
7680	12"		9	3.111		430	96	17.60	543.60	640
7800	Cut-in valves with rubber gaskets, 4"		18	1.556		380	48	8.80	436.80	505
7840	8"		10	2.800		795	86.50	15.85	897.35	1,025
7880	12"		9	3.111		1,200	96	17.60	1,313.60	1,500
7900	Tapping Valve 4 inch, MJ, ductile iron		18	1.556		380	48	8.80	436.80	505
7920	Tapping Valve 6 inch, MJ, ductile iron		12	2.333		455	72	13.20	540.20	625
8000	Sleeves with rubber gaskets, 4" x 4"		37	.757		775	23.50	4.28	802.78	895
8030	8" x 4"		21	1.333		470	41	7.55	518.55	590
8040	8" x 6"		21	1.333		520	41	7.55	568.55	645
8060	8" x 8"		21	1.333		610	41	7.55	658.55	740
8070	10" x 4"		21	1.333		475	41	7.55	523.55	595
8080	10" x 6"	▼	21	1.333	▼	530	41	7.55	578.55	655

SITE CONSTRUCTION 2

02510	Water Distribution	CREW	DAILY OUTPUT	LABOR-HOURS	UNIT	2005 BARE COSTS				TOTAL INCL O&P	
						MAT.	LABOR	EQUIP.	TOTAL		
710	8090	10" x 8"	B-21	21	1.333	Ea.	890	41	7.55	938.55	1,050	**710**
	8140	12" x 12"		18	1.556		1,200	48	8.80	1,256.80	1,375	
	8160	14" x 6"		16	1.750		2,600	54	9.90	2,663.90	2,975	
	8220	16" x 10"		14	2		2,850	62	11.30	2,923.30	3,225	
	8260	18" x 6"		10	2.800		4,175	86.50	15.85	4,277.35	4,750	
	8300	18" x 18"		10	2.800		4,325	86.50	15.85	4,427.35	4,925	
	8320	20" x 6"		8	3.500		3,375	108	19.80	3,502.80	3,925	
	8340	20" x 12"		8	3.500		3,625	108	19.80	3,752.80	4,200	
	8400	24" x 12"		6	4.667		4,450	144	26.50	4,620.50	5,125	
	8420	24" x 18"	▼	6	4.667		6,625	144	26.50	6,795.50	7,525	
	8800	Curb box, 6' long	B-20	20	1.200		69.50	36		105.50	133	
	8820	8' long	"	18	1.333	▼	81	40		121	152	
720	0010	**WATER SUPPLY, CONCRETE PIPE**										**720**
	0020	Not including excavation or backfill, without gaskets										
	3000	Conc. cylinder pipe (CCP), 150 PSI, 40' L, 12 diam.	B-13	192	.292	L.F.	51	8.40	3.21	62.61	73	
	3010	24" diameter	"	128	.438		63.50	12.60	4.82	80.92	94.50	
	3040	36" diameter	B-13B	96	.583		78.50	16.80	9.35	104.65	123	
	3050	48" diameter		64	.875		108	25	14	147	172	
	3060	Prestressed (PCCP), 150 PSI, 24' L, 60" diameter		60	.933		176	27	14.90	217.90	252	
	3070	72" diameter		60	.933		221	27	14.90	262.90	300	
	3080	84" diameter	▼	40	1.400		288	40.50	22.50	351	400	
	3090	96" diameter	B-13C	40	1.400		425	40.50	38	503.50	575	
	3100	108" diameter		32	1.750		585	50.50	47.50	683	775	
	3102	120" diameter		16	3.500		865	101	95	1,061	1,200	
	3104	144" diameter	▼	16	3.500	▼	1,075	101	95	1,271	1,425	
	3110	Conc. cylinder pipe (CCP), 150 PSI, elbow, 90°, 12" diameter	B-13	32	1.750	Ea.	177	50.50	19.25	246.75	294	
	3140	24" diameter	"	6	9.333		320	269	103	692	880	
	3150	36" diameter	B-13B	15	3.733		565	107	59.50	731.50	855	
	3160	48" diameter		12	4.667		625	134	74.50	833.50	980	
	3170	Prestressed (PCCP), 150 PSI, elbow, 90°, 60" diameter		10	5.600		975	161	89.50	1,225.50	1,425	
	3180	72" diameter		4	14		9,950	405	224	10,579	11,800	
	3190	84" diameter		3.20	17.500		14,000	505	280	14,785	16,500	
	3200	96" diameter	▼	2	28		27,400	805	450	28,655	31,900	
	3210	108" diameter	B-13C	1.20	46.667		32,400	1,350	1,275	35,025	39,100	
	3220	120" diameter		.40	140		33,600	4,025	3,800	41,425	47,400	
	3225	144" diameter	▼	.30	184		45,800	5,300	5,000	56,100	64,000	
	3230	Concrete cylinder pipe (CCP), 150 PSI, elbow, 45°, 12" diameter	B-13	24	2.333		149	67	25.50	241.50	296	
	3250	24" diameter	"	6	9.333		320	269	103	692	880	
	3260	36" diameter	B-13B	4	14		1,725	405	224	2,354	2,775	
	3270	48" diameter		3	18.667		3,100	535	298	3,933	4,575	
	3280	Prestressed, (PCCP), 150 PSI, elbow, 45°, 60" diameter		2	28		4,325	805	450	5,580	6,500	
	3290	72" diameter		1.60	35		7,675	1,000	560	9,235	10,600	
	3300	84" diameter		1.30	42.945		9,600	1,225	685	11,510	13,300	
	3310	96" diameter	▼	1	56		11,500	1,600	895	13,995	16,100	
	3320	108" diameter	B-13C	.66	84.337		14,500	2,425	2,300	19,225	22,200	
	3330	120" diameter		.40	140		16,300	4,025	3,800	24,125	28,300	
	3340	144" diameter	▼	.30	184	▼	16,400	5,300	5,000	26,700	31,700	
730	0010	**WATER SUPPLY, DUCTILE IRON PIPE** cement lined										**730**
	0020	Not including excavation or backfill										
	2000	Pipe, class 50 water piping, 18' lengths										
	2020	Mechanical joint, 4" diameter	B-21A	200	.200	L.F.	12.50	6.60	2.53	21.63	26.50	
	2040	6" diameter		160	.250		14.50	8.25	3.16	25.91	32	
	2060	8" diameter		133.33	.300		16.05	9.90	3.80	29.75	37	
	2080	10" diameter		114.29	.350		21.50	11.55	4.43	37.48	46.50	
	2100	12" diameter	▼	105.26	.380	▼	26.50	12.55	4.81	43.86	54	

02510	Water Distribution	CREW	DAILY OUTPUT	LABOR-HOURS	UNIT	2005 BARE COSTS				TOTAL INCL O&P
						MAT.	LABOR	EQUIP.	TOTAL	
730 2120	14" diameter	B-21A	100	.400	L.F.	34	13.20	5.05	52.25	63
2140	16" diameter		72.73	.550		37	18.15	6.95	62.10	76
2160	18" diameter		68.97	.580		46.50	19.10	7.35	72.95	88.50
2170	20" diameter		57.14	.700		54.50	23	8.85	86.35	105
2180	24" diameter		47.06	.850		70	28	10.75	108.75	131
3000	Tyton, Push-on joint, 4" diameter		400	.100		7.35	3.30	1.27	11.92	14.50
3020	6" diameter		333.33	.120		8.40	3.96	1.52	13.88	16.85
3040	8" diameter		200	.200		11.50	6.60	2.53	20.63	25.50
3060	10" diameter		181.82	.220		18.15	7.25	2.78	28.18	34
3080	12" diameter		160	.250		19.15	8.25	3.16	30.56	37
3100	14" diameter		133.33	.300		21	9.90	3.80	34.70	42
3120	16" diameter		114.29	.350		29.50	11.55	4.43	45.48	55
3140	18" diameter		100	.400		32.50	13.20	5.05	50.75	61.50
3160	20" diameter		88.89	.450		36	14.85	5.70	56.55	68.50
3180	24" diameter	↓	76.92	.520	↓	47	17.15	6.60	70.75	85
8000	Fittings, mechanical joint									
8006	90° bend, 4" diameter	B-20A	16	2	Ea.	126	64.50		190.50	237
8020	6" diameter		12.80	2.500		169	80.50		249.50	310
8040	8" diameter	↓	10.67	2.999		245	96.50		341.50	420
8060	10" diameter	B-21A	11.43	3.500		480	115	44.50	639.50	755
8080	12" diameter		10.53	3.799		495	125	48	668	790
8100	14" diameter		10	4		685	132	50.50	867.50	1,000
8120	16" diameter		7.27	5.502		735	181	69.50	985.50	1,175
8140	18" diameter		6.90	5.797		1,000	191	73.50	1,264.50	1,475
8160	20" diameter		5.71	7.005		2,550	231	88.50	2,869.50	3,250
8180	24" diameter	↓	4.70	8.511		3,150	281	108	3,539	4,000
8200	Wye or tee, 4" diameter	B-20A	10.67	2.999		153	96.50		249.50	315
8220	6" diameter		8.53	3.751		192	121		313	395
8240	8" diameter	↓	7.11	4.501		284	145		429	535
8260	10" diameter	B-21A	7.62	5.249		490	173	66.50	729.50	870
8280	12" diameter		7.02	5.698		610	188	72	870	1,050
8300	14" diameter		6.67	5.997		1,250	198	76	1,524	1,750
8320	16" diameter		4.85	8.247		1,300	272	104	1,676	1,950
8340	18" diameter		4.60	8.696		2,650	287	110	3,047	3,450
8360	20" diameter		3.81	10.499		3,375	345	133	3,853	4,375
8380	24" diameter	↓	3.14	12.739		3,850	420	161	4,431	5,075
8398	45° bends, 4" diameter	B-20A	16	2		133	64.50		197.50	245
8400	6" diameter	"	12.80	2.500		146	80.50		226.50	284
8410	12" diameter	B-21A	10.53	3.799		355	125	48	528	635
8420	16" diameter		7.27	5.502		460	181	69.50	710.50	860
8430	20" diameter		5.71	7.005		2,125	231	88.50	2,444.50	2,800
8440	24" diameter	↓	4.70	8.511		2,575	281	108	2,964	3,375
8450	Decreaser, 6" x 4" diameter	B-20A	14.22	2.250		104	72.50		176.50	226
8460	8" x 6" diameter	"	11.64	2.749		213	88.50		301.50	370
8470	10" x 6 " diameter	B-21A	13.33	3.001		217	99	38	354	430
8480	12" x 6" diameter		12.70	3.150		245	104	40	389	470
8490	16" x 6" diameter		10	4		885	132	50.50	1,067.50	1,225
8500	20" x 6" diameter	↓	8.42	4.751	↓	1,400	157	60	1,617	1,850
8550	Butterfly valves with boxes, cast iron									
8560	4" diameter	B-20	6	4	Ea.	760	120		880	1,025
8580	8" diameter	B-21	4	7		760	216	39.50	1,015.50	1,225
8590	10" diameter		3.50	8		1,050	247	45.50	1,342.50	1,575
8600	12" diameter		3	9.333		1,350	288	53	1,691	1,975
8610	14" diameter		2	14		1,875	435	79	2,389	2,800
8620	16" diameter	↓	2	14		2,400	435	79	2,914	3,400
9600	Steel sleeve and tap, 4" diameter	B-20	3	8	↓	430	241		671	845

02510	Water Distribution	CREW	DAILY OUTPUT	LABOR-HOURS	UNIT	2005 BARE COSTS				TOTAL INCL O&P	
						MAT.	LABOR	EQUIP.	TOTAL		
730	9620	6" diameter	B-20	2	12	Ea.	505	360		865	1,125	**730**
	9630	8" diameter	↓	2	12	↓	680	360		1,040	1,300	
740	0010	**WATER SUPPLY, POLYETHYLENE PIPE, C901**										**740**
	0020	Not including excavation or backfill										
	1000	Piping, 160 P.S.I., 3/4" diameter	B-20	525	.046	L.F.	.31	1.38		1.69	2.48	
	1120	1" diameter		485	.049		.39	1.49		1.88	2.75	
	1140	1-1/2" diameter		450	.053		.71	1.60		2.31	3.28	
	1160	2" diameter	↓	365	.066	↓	1.09	1.98		3.07	4.28	
	2000	Fittings, insert type, nylon, 160 & 250 psi, cold water										
	2220	Clamp ring, stainless steel, 3/4" diameter	B-20	345	.070	Ea.	.99	2.09		3.08	4.35	
	2240	1" diameter		321	.075		1.04	2.25		3.29	4.64	
	2260	1-1/2" diameter		285	.084		1.29	2.53		3.82	5.35	
	2280	2" diameter		255	.094		1.45	2.83		4.28	6	
	2300	Coupling, 3/4" diameter		66	.364		.72	10.95		11.67	17.85	
	2320	1" diameter		57	.421		.96	12.65		13.61	21	
	2340	1-1/2" diameter		51	.471		2.24	14.15		16.39	24.50	
	2360	2" diameter		48	.500		2.69	15.05		17.74	26.50	
	2400	Elbow, 90°, 3/4" diameter		66	.364		1.28	10.95		12.23	18.45	
	2420	1" diameter		57	.421		1.37	12.65		14.02	21	
	2440	1-1/2" diameter		51	.471		3.41	14.15		17.56	26	
	2460	2" diameter		48	.500		5.70	15.05		20.75	30	
	2500	Tee, 3/4" diameter		42	.571		1.37	17.20		18.57	28.50	
	2520	1" diameter		39	.615		2.25	18.50		20.75	31.50	
	2540	1-1/2" diameter		33	.727		5.05	22		27.05	39.50	
	2560	2" diameter	↓	30	.800	↓	6.85	24		30.85	45	
750	0010	**WATER SUPPLY, POLYVINYL CHLORIDE PIPE**										**750**
	0020	Not including excavation or backfill, unless specified										
	2100	AWWA Class 160, S.D.R. 26, 1-1/2" diameter	B-20	750	.032	L.F.	.45	.96		1.41	2	
	2120	2" diameter		686	.035		1.09	1.05		2.14	2.84	
	2140	2-1/2" diameter		500	.048		1.62	1.44		3.06	4.03	
	2160	3" diameter		430	.056		2.32	1.68		4	5.15	
	2180	4" diameter		375	.064		3.78	1.93		5.71	7.15	
	2200	6" diameter		316	.076		8.15	2.28		10.43	12.50	
	2210	8" diameter	↓	260	.092	↓	13.80	2.78		16.58	19.50	
	3010	AWWA C905, PR 100, DR 41										
	3030	14" diameter	B-20A	213	.150	L.F.	12.60	4.84		17.44	21.50	
	3040	16" diameter		200	.160		16.70	5.15		21.85	26	
	3050	18" diameter		160	.200		21	6.45		27.45	33	
	3060	20" diameter		133	.241		26	7.75		33.75	41	
	3070	24" diameter		107	.299		36.50	9.65		46.15	55	
	3080	30" diameter		80	.400		56.50	12.90		69.40	82	
	3090	36" diameter		80	.400		80	12.90		92.90	108	
	3100	42" diameter		60	.533		109	17.20		126.20	146	
	3200	48" diameter		60	.533		143	17.20		160.20	183	
	4520	Class 150, SDR 18, AWWA C900, 4"		380	.084		2.38	2.71		5.09	6.75	
	4530	6"		316	.101		4.70	3.26		7.96	10.15	
	4540	8"		264	.121		8.15	3.91		12.06	14.90	
	4550	10"		220	.145		12.30	4.69		16.99	20.50	
	4560	12"	↓	186	.172	↓	17.40	5.55		22.95	27.50	
	8000	Fittings with rubber gasket										
	8003	Class 150, D.R. 18										
	8006	90° Bend , 4" diameter	B-20	100	.240	Ea.	39.50	7.20		46.70	55	
	8020	6" diameter		90	.267		70	8		78	89.50	
	8040	8" diameter		80	.300		136	9		145	163	
	8060	10" diameter	↓	50	.480	↓	235	14.45		249.45	282	

Important: See the Reference Section for critical supporting data - Reference Nos., Crews, & City Cost Indexes

02510 | Water Distribution

		CREW	DAILY OUTPUT	LABOR-HOURS	UNIT	2005 BARE COSTS				TOTAL INCL O&P
						MAT.	LABOR	EQUIP.	TOTAL	
750 8080	12" diameter	B-20	30	.800	Ea.	300	24		324	370
8100	Tee, 4" diameter		90	.267		54.50	8		62.50	72
8120	6" diameter		80	.300		121	9		130	147
8140	8" diameter		70	.343		173	10.30		183.30	206
8160	10" diameter		40	.600		196	18.05		214.05	243
8180	12" diameter		20	1.200		420	36		456	515
8200	45° Bend, 4" diameter		100	.240		38.50	7.20		45.70	54
8220	6" diameter		90	.267		68.50	8		76.50	87.50
8240	8" diameter		50	.480		130	14.45		144.45	166
8260	10" diameter		50	.480		199	14.45		213.45	241
8280	12" diameter		30	.800		258	24		282	320
8300	Reducing tee 6"x4"		100	.240		118	7.20		125.20	141
8320	8" x 6"		90	.267		210	8		218	244
8330	10" x 6"		90	.267		270	8		278	310
8340	10"x8"		90	.267		281	8		289	325
8350	12" x 6"		90	.267		320	8		328	365
8360	12" x 8"		90	.267		335	8		343	385
8400	Tapped service tee (threaded type) 6" x 6" x 3/4		100	.240		42	7.20		49.20	57.50
8420	6" x 6" x 3/4"		90	.267		42	8		50	58.50
8430	6" x 6" x 1"		90	.267		42	8		50	58.50
8440	6" x 6" x 1 1/2"		90	.267		42	8		50	58.50
8450	6" x 6" x 2"		90	.267		42	8		50	58.50
8460	8" x 8" x 3/4"		90	.267		177	8		185	208
8470	8" x 8" x 1"		90	.267		177	8		185	208
8480	8" x 8" x 1 1/2"		90	.267		177	8		185	208
8490	8" x 8" x 2"		90	.267		177	8		185	208
8500	Repair coupling 4"		100	.240		25	7.20		32.20	39
8520	6" diameter		90	.267		38.50	8		46.50	54.50
8540	8" diameter		50	.480		84.50	14.45		98.95	116
8560	10" diameter		50	.480		143	14.45		157.45	180
8580	12" diameter		50	.480		209	14.45		223.45	253
8600	Plug end 4"		100	.240		21	7.20		28.20	35
8620	6" diameter		90	.267		38	8		46	54.50
8640	8" diameter		50	.480		65	14.45		79.45	94
8660	10" diameter		50	.480		68.50	14.45		82.95	97.50
8680	12" diameter		50	.480		84.50	14.45		98.95	116
760 0010	**WATER SUPPLY, HDPE**, butt fusion joints, SDR 21, 40' lengths									
0100	4" diameter	B-22A	400	.095	L.F.	1.84	2.88	1.76	6.48	8.40
0200	6" diameter		380	.100		2.65	3.03	1.85	7.53	9.65
0300	8" diameter		320	.119		4.48	3.60	2.20	10.28	12.90
0400	10" diameter		300	.127		6.95	3.84	2.34	13.13	16.20
0500	12" diameter		260	.146		9.80	4.43	2.70	16.93	20.50
0600	14" diameter		220	.173		11.80	5.25	3.19	20.24	24.50
0700	16" diameter		180	.211		15.45	6.40	3.90	25.75	31
0800	18" diameter		140	.271		19.50	8.20	5	32.70	39.50
0900	24" diameter		100	.380		35	11.50	7.05	53.55	63.50
1000	Fittings									
1100	Elbows, 90 degrees									
1200	4" diameter	B-22B	32	.500	Ea.	25.50	15.40	5.35	46.25	58
1300	6" diameter		28	.571		67	17.60	6.15	90.75	108
1400	8" diameter		24	.667		168	20.50	7.15	195.65	225
1500	10" diameter		18	.889		290	27.50	9.55	327.05	375
1600	12" diameter		12	1.333		490	41	14.30	545.30	620
1700	14" diameter		9	1.778		490	54.50	19.10	563.60	640
1800	16" diameter		6	2.667		590	82	28.50	700.50	805
1900	18" diameter		4	4		795	123	43	961	1,125

SITE CONSTRUCTION 2

02510 | Water Distribution

		CREW	DAILY OUTPUT	LABOR-HOURS	UNIT	2005 BARE COSTS				TOTAL INCL O&P	
						MAT.	LABOR	EQUIP.	TOTAL		
760	2000	24" diameter	B-22B	3	5.333	Ea.	1,625	164	57.50	1,846.50	2,100	760
	2100	Tees									
	2200	4" diameter	B-22B	30	.533	Ea.	35.50	16.40	5.75	57.65	71
	2300	6" diameter		26	.615		93	18.95	6.60	118.55	140
	2400	8" diameter		22	.727		236	22.50	7.80	266.30	305
	2500	10" diameter		15	1.067		310	33	11.45	354.45	410
	2600	12" diameter		10	1.600		425	49.50	17.20	491.70	565
	2700	14" diameter		8	2		505	61.50	21.50	588	675
	2800	16" diameter		6	2.667		590	82	28.50	700.50	810
	2900	18" diameter		4	4		800	123	43	966	1,125
	3000	24" diameter	↓	2	8	↓	1,650	246	86	1,982	2,300
770	0010	**WATER SUPPLY, BLACK STEEL PIPE**										770
	0011	Not including excavation or backfill									
	1000	Pipe, black steel, plain end, welded, 1/4" wall thk, 8" diam.	B-35A	108	.519	L.F.	8.30	16.60	11.60	36.50	47.50
	1010	10" diameter		104	.538		10.35	17.20	12.05	39.60	51
	1020	12" diameter		195	.287		12.40	9.20	6.40	28	35
	1030	18" diameter		175	.320		19	10.25	7.15	36.40	44.50
	1040	5/16" wall thickness, 12" diameter		195	.287		16	9.20	6.40	31.60	38.50
	1050	18" diameter		59.20	.946		28	30	21	79	101
	1060	36" diameter		28.96	1.934		56	62	43	161	204
	1070	3/8" wall thickness, 18" diameter		43.20	1.296		28	41.50	29	98.50	127
	1080	24" diameter		36	1.556		37.50	49.50	35	122	156
	1090	30" diameter		30.40	1.842		49.50	59	41	149.50	190
	1100	1/2" wall thickness, 36" diameter		26.08	2.147		79.50	68.50	48	196	246
	1110	48" diameter		21.68	2.583		118	82.50	57.50	258	320
	1135	7/16" wall thickness, 48" diameter		20.80	2.692		104	86	60	250	310
	1140	5/8" wall thickness, 48" diameter	↓	21.68	2.583	↓	147	82.50	57.50	287	355
780	0010	**WATER SUPPLY, COPPER PIPE**	R15100 -050									780
	0020	Not including excavation or backfill									
	2000	Tubing, type K, 20' joints, 3/4" diameter	Q-1	400	.040	L.F.	2.70	1.47		4.17	5.20
	2200	1" diameter		320	.050		3.57	1.84		5.41	6.70
	3000	1-1/2" diameter		265	.060		5.85	2.22		8.07	9.80
	3020	2" diameter		230	.070		9.10	2.56		11.66	13.85
	3040	2-1/2" diameter		146	.110		13.25	4.03		17.28	20.50
	3060	3" diameter		134	.119		18.50	4.39		22.89	27
	4012	4" diameter	↓	95	.168		30.50	6.20		36.70	43.50
	4016	6" diameter	Q-2	95	.253	↓	93	9.65		102.65	117
	5000	Tubing, type L									
	5108	2" diameter	Q-1	105	.152	L.F.	7.55	5.60		13.15	16.70
	6010	3" diameter		140	.114		15	4.20		19.20	23
	6012	4" diameter	↓	95	.168		25	6.20		31.20	37
	6016	6" diameter	Q-2	100	.240	↓	86	9.15		95.15	108
	7020	Fittings, brass, corporation stops, 3/4" diameter	1 Plum	19	.421	Ea.	16.65	17.20		33.85	44.50
	7040	1" diameter		16	.500		25.50	20.50		46	58.50
	7060	1-1/2" diameter		13	.615		76	25		101	122
	7080	2" diameter		11	.727		126	29.50		155.50	184
	7100	Curb stops, 3/4" diameter		19	.421		36	17.20		53.20	65.50
	7120	1" diameter		16	.500		49.50	20.50		70	85
	7140	1-1/2" diameter		13	.615		110	25		135	159
	7160	2" diameter		11	.727		180	29.50		209.50	243
	7180	Curb box, cast iron, 1/2" to 1" curb stops		12	.667		50	27		77	96
	7200	1-1/4" to 2" curb stops	↓	8	1		94.50	41		135.50	166
	7220	Saddles, 3/4" diameter, add					27			27	30
	7240	2" diameter, add				↓	32			32	35.50
	7250	For copper fittings, see Div. 15107-460									

Important: See the Reference Section for critical supporting data - Reference Nos., Crews, & City Cost Indexes

2 SITE CONSTRUCTION

02520	Wells	CREW	DAILY OUTPUT	LABOR-HOURS	UNIT	2005 BARE COSTS				TOTAL INCL O&P
						MAT.	LABOR	EQUIP.	TOTAL	
510 0010	**WELLS & ACCESSORIES**, domestic									**510**
0100	Drilled, 4" to 6" diameter	B-23	120	.333	L.F.		9.05	28	37.05	45
0200	8" diameter	"	95.20	.420	"		11.40	35	46.40	56.50
0400	Gravel pack well, 40' deep, incl. gravel & casing, complete									
0500	24" diameter casing x 18" diameter screen	B-23	.13	307	Total	22,900	8,350	25,800	57,050	66,500
0600	36" diameter casing x 18" diameter screen		.12	333	"	24,700	9,025	28,000	61,725	72,000
0800	Observation wells, 1-1/4" riser pipe		163	.245	V.L.F.	12.60	6.65	20.50	39.75	47
0900	For flush Buffalo roadway box, add	1 Skwk	16.60	.482	Ea.	34.50	16.80		51.30	64
1200	Test well, 2-1/2" diameter, up to 50' deep (15 to 50 GPM)	B-23	1.51	26.490	"	515	720	2,225	3,460	4,150
1300	Over 50' deep, add	"	121.80	.328	L.F.	13.75	8.90	27.50	50.15	59.50
1500	Pumps, installed in wells to 100' deep, 4" submersible									
1510	1/2 H.P.	Q-1	3.22	4.969	Ea.	330	183		513	640
1520	3/4 H.P.		2.66	6.015		390	221		611	765
1600	1 H.P.		2.29	6.987		415	257		672	840
1700	1-1/2 H.P.	Q-22	1.60	10		1,075	370	375	1,820	2,150
1800	2 H.P.		1.33	12.030		1,250	440	455	2,145	2,550
1900	3 H.P.		1.14	14.035		1,425	515	530	2,470	2,925
2000	5 H.P.		1.14	14.035		1,950	515	530	2,995	3,500
2050	Remove and install motor only, 4 H.P.		1.14	14.035		700	515	530	1,745	2,125
3000	Pump, 6" submersible, 25' to 150' deep, 25 H.P., 249 to 297 GPM		.89	17.978		4,250	660	675	5,585	6,425
3100	25' to 500' deep, 30 H.P., 100 to 300 GPM		.73	21.918		4,850	805	825	6,480	7,425
8000	Steel well casing	B-23A	3,020	.008	Lb.	.59	.24	1.08	1.91	2.21
8110	Well screen assembly, stainless steel, 2" diameter		273	.088	L.F.	47.50	2.64	11.95	62.09	69
8120	3" diameter		253	.095		65.50	2.85	12.85	81.20	90.50
8130	4" diameter		200	.120		75	3.60	16.30	94.90	106
8140	5" diameter		168	.143		87.50	4.29	19.40	111.19	125
8150	6" diameter		126	.190		105	5.70	26	136.70	152
8160	8" diameter		98.50	.244		137	7.30	33	177.30	198
8170	10" diameter		73	.329		172	9.85	44.50	226.35	253
8180	12" diameter		62.50	.384		202	11.55	52	265.55	297
8190	14" diameter		54.30	.442		229	13.25	60	302.25	340
8200	16" diameter		48.30	.497		253	14.90	67.50	335.40	375
8210	18" diameter		39.20	.612		315	18.40	83	416.40	465
8220	20" diameter		31.20	.769		360	23	104	487	545
8230	24" diameter		23.80	1.008		445	30.50	137	612.50	690
8240	26" diameter		21	1.143		500	34.50	155	689.50	775
8300	Slotted PVC, 1-1/4" diameter		521	.046		1.69	1.38	6.25	9.32	10.90
8310	1-1/2" diameter		488	.049		2.46	1.48	6.65	10.59	12.35
8320	2" diameter		273	.088		3.41	2.64	11.95	18	21
8330	3" diameter		253	.095		3.62	2.85	12.85	19.32	22.50
8340	4" diameter		200	.120		4.17	3.60	16.30	24.07	28
8350	5" diameter		168	.143		4.38	4.29	19.40	28.07	33
8360	6" diameter		126	.190		6.10	5.70	26	37.80	44
8370	8" diameter		98.50	.244		9.20	7.30	33	49.50	58
8400	Artificial gravel pack, 2" screen, 6" casing	B-23B	174	.138		2.77	4.14	19.95	26.86	31.50
8405	8" casing		111	.216		3.78	6.50	31.50	41.78	48.50
8410	10" casing		74.50	.322		4.75	9.65	46.50	60.90	71.50
8415	12" casing		60	.400		6.40	12	58	76.40	89
8420	14" casing		50.20	.478		8.30	14.35	69	91.65	107
8425	16" casing		40.70	.590		11.45	17.70	85.50	114.65	134
8430	18" casing		36	.667		13.30	20	96.50	129.80	152
8435	20" casing		29.50	.814		15.30	24.50	118	157.80	184
8440	24" casing		25.70	.934		17	28	135	180	211
8445	26" casing		24.60	.976		19	29.50	141	189.50	221
8450	30" casing		20	1.200		22	36	174	232	271
8455	36" casing		16.40	1.463		23.50	44	212	279.50	325

SITE CONSTRUCTION **2**

02520 | Wells

		CREW	DAILY OUTPUT	LABOR-HOURS	UNIT	2005 BARE COSTS MAT.	LABOR	EQUIP.	TOTAL	TOTAL INCL O&P	
510	8500	Develop well	B-23B	8	3	Hr.	215	90	435	740	855	510
	8550	Pump test well	↓	8	3		57	90	435	582	680	
	8560	Standby well	B-23A	8	3		55.50	90	405	550.50	650	
	8570	Standby, drill rig		8	3	↓		90	405	495	590	
	8580	Surface seal well, concrete filled	↓	1	24	Ea.	540	720	3,250	4,510	5,275	
	8590	Well test pump, install & remove	B-23	1	40			1,075	3,350	4,425	5,375	
	8600	Well sterilization, chlorine	2 Clab	1	16	↓	400	425		825	1,100	
	9950	See div. 02240-900 for wellpoints										
	9960	See div. 02240-700 for drainage wells										
520	0010	**WATER SUPPLY WELLS, PUMPS** with pressure control										520
	1000	Deep well, jet, 42 gal. galvanized tank										
	1040	3/4 HP	1 Plum	.80	10	Ea.	585	410		995	1,250	
	3000	Shallow well, jet, 30 gal. galvanized tank										
	3040	1/2 HP	1 Plum	2	4	Ea.	370	163		533	655	

02530 | Sanitary Sewerage

		CREW	DAILY OUTPUT	LABOR-HOURS	UNIT	MAT.	LABOR	EQUIP.	TOTAL	TOTAL INCL O&P	
300	0010	**PACKAGED LIFT STATIONS**										300
	2500	Sewage lift station, 200,000 GPD	E-8	.20	520	Ea.	119,000	19,400	8,250	146,650	173,500	
	2510	500,000 GPD	↓	.15	684		133,500	25,500	10,900	169,900	203,000	
	2520	800,000 GPD	↓	.13	812	↓	160,500	30,300	12,900	203,700	243,000	
720	0010	**SEWAGE COLLECTION, VENT CAST IRON PIPE**										720
	0020	Not including excavation or backfill										
	2022	Sewage vent cast iron, B & S, 4" diameter	Q-1	66	.242	L.F.	7.85	8.90		16.75	22	
	2024	5" diameter	Q-2	88	.273		8.10	10.40		18.50	24.50	
	2026	6" diameter	"	84	.286		13.40	10.90		24.30	31	
	2028	8" diameter	Q-3	70	.457		21.50	17.80		39.30	50.50	
	2030	10" diameter		66	.485		35.50	18.90		54.40	67.50	
	2032	12" diameter		57	.561		51	22		73	89	
	2034	15" diameter	↓	49	.653	↓	73.50	25.50		99	119	
	8001	Fittings, bends and elbows										
	8110	4" diameter	Q-1	13	1.231	Ea.	25.50	45.50		71	96	
	8112	5" diameter	Q-2	18	1.333		36.50	51		87.50	117	
	8114	6" diameter	"	17	1.412		43.50	54		97.50	129	
	8116	8" diameter	Q-3	11	2.909		125	113		238	305	
	8118	10" diameter		10	3.200		183	125		308	390	
	8120	12" diameter		9	3.556		246	138		384	480	
	8122	15" diameter	↓	7	4.571	↓	720	178		898	1,050	
	8500	Wyes and tees										
	8510	4" diameter	Q-1	8	2	Ea.	39.50	73.50		113	155	
	8512	5" diameter	Q-2	12	2		65	76.50		141.50	187	
	8514	6" diameter	"	11	2.182		84	83		167	217	
	8516	8" diameter	Q-3	7	4.571		201	178		379	490	
	8518	10" diameter		6	5.333		320	208		528	665	
	8520	12" diameter		4	8		530	310		840	1,050	
	8522	15" diameter	↓	3	10.667	↓	1,050	415		1,465	1,775	
730	0010	**SEWAGE COLLECTION, CONCRETE PIPE**										730
	0020	See 02630-530 for sewage/drainage collection, concrete pipe										
770	0010	**SEWAGE COLLECTION, PLASTIC PIPE**										770
	0020	Not including excavation & backfill										
	1100	Piping, DWV Sch 40 ABS, 4" diameter	B-20	375	.064	L.F.	1.60	1.93		3.53	4.75	
	1110	6" diameter		350	.069	"	10.95	2.06		13.01	15.25	
	1120	Fitting, 1/4 bend, 4"		19	1.263	Ea.	16.10	38		54.10	76.50	
	1130	6"	↓	15	1.600	↓	10.95	48		58.95	87	

Important: See the Reference Section for critical supporting data - Reference Nos., Crews, & City Cost Indexes

02500 | Utility Services

02530 | Sanitary Sewerage

			DAILY	LABOR-		2005 BARE COSTS				TOTAL	
		CREW	OUTPUT	HOURS	UNIT	MAT.	LABOR	EQUIP.	TOTAL	INCL O&P	
770	1140	Tee, 4"	B-20	12	2	Ea.	16.10	60		76.10	111	770
	3000	Piping, HDPE Corrugated Type S with watertight gaskets, 4" diameter		425	.056	L.F.	.83	1.70		2.53	3.55	
	3020	6" diameter		400	.060		1.91	1.80		3.71	4.91	
	3040	8" diameter		380	.063		3.66	1.90		5.56	7	
	3060	10" diameter		370	.065		5.05	1.95		7	8.60	
	3080	12" diameter		340	.071		5.65	2.12		7.77	9.50	
	3100	15" diameter	↓	300	.080		7.65	2.41		10.06	12.15	
	3120	18" diameter	B-21	275	.102		10.90	3.15	.58	14.63	17.50	
	3140	24" diameter		250	.112		16.90	3.46	.63	20.99	24.50	
	3160	30" diameter		200	.140		27	4.33	.79	32.12	37	
	3180	36" diameter		180	.156		34	4.81	.88	39.69	46	
	3200	42" diameter		175	.160		47.50	4.95	.91	53.36	61	
	3220	48" diameter		170	.165		61.50	5.10	.93	67.53	77	
	3240	54" diameter		160	.175		95	5.40	.99	101.39	114	
	3260	60" diameter	↓	150	.187	↓	111	5.75	1.06	117.81	132	
	3300	Watertight elbows 12" diam	B-20	11	2.182	Ea.	62.50	65.50		128	171	
	3320	15" diam	"	9	2.667		96	80		176	231	
	3340	18" diam	B-21	9	3.111		159	96	17.60	272.60	340	
	3360	24" diam		9	3.111		335	96	17.60	448.60	540	
	3380	30" diam		8	3.500		540	108	19.80	667.80	780	
	3400	36" diam		8	3.500		690	108	19.80	817.80	950	
	3420	42" diam		6	4.667		875	144	26.50	1,045.50	1,200	
	3440	48" diam	↓	6	4.667		1,325	144	26.50	1,495.50	1,725	
	3460	Watertight tee 12" diam	B-20	7	3.429		140	103		243	315	
	3480	15" diam	"	6	4		167	120		287	370	
	3500	18" diam	B-21	6	4.667		244	144	26.50	414.50	520	
	3520	24" diam		5	5.600		320	173	31.50	524.50	660	
	3540	30" diam		5	5.600		605	173	31.50	809.50	970	
	3560	36" diam		4	7		785	216	39.50	1,040.50	1,250	
	3580	42" diam		4	7		1,350	216	39.50	1,605.50	1,875	
	3600	48" diam	↓	4	7	↓	2,225	216	39.50	2,480.50	2,825	
780	0010	**SEWAGE COLLECTION, POLYVINYL CHLORIDE PIPE**										780
	0020	Not including excavation or backfill										
	2000	10' lengths, S.D.R. 35, B&S, 4" diameter	B-20	375	.064	L.F.	1.92	1.93		3.85	5.10	
	2040	6" diameter		350	.069		3.44	2.06		5.50	7	
	2080	8" diameter	↓	335	.072		5.80	2.15		7.95	9.75	
	2120	10" diameter	B-21	330	.085		8.75	2.62	.48	11.85	14.25	
	2160	12" diameter		320	.087		9.75	2.70	.50	12.95	15.50	
	2200	15" diameter	↓	190	.147	↓	14.75	4.56	.83	20.14	24	
	3040	Fittings, bends or elbows, 4" diameter	B-20	19	1.263	Ea.	3.69	38		41.69	63	
	3080	6" diameter		15	1.600		16.10	48		64.10	92.50	
	3120	Tees, 4" diameter	↓	12	2		6.40	60		66.40	101	
	3160	6" diameter	2 Skwk	10	1.600		20	56		76	109	
	3200	Wyes, 4" diameter	"	12	1.333		7.25	46.50		53.75	80.50	
	3240	6" diameter	B-20	10	2.400	↓	21	72		93	136	
	4000	Piping, DWV PVC, no exc/bkfill, 10' L, Sch 40, 4" dia		375	.064	L.F.	1.88	1.93		3.81	5.05	
	4010	6" dia		350	.069		4.06	2.06		6.12	7.70	
	4020	8" dia	↓	335	.072	↓	12.50	2.15		14.65	17.10	

02540 | Septic Tank Systems

400	0010	**SEPTIC TANKS**										400
	0015	Septic tanks, not incl exc or piping, precast, 1,000 gal	B-21	8	3.500	Ea.	560	108	19.80	687.80	805	
	0020	1,250 gallon		8	3.500		780	108	19.80	907.80	1,050	
	0060	1,500 gallon	↓	7	4	↓	920	124	22.50	1,066.50	1,225	

83

2 SITE CONSTRUCTION

02540 | Septic Tank Systems

		CREW	DAILY OUTPUT	LABOR-HOURS	UNIT	MAT.	LABOR	EQUIP.	TOTAL	TOTAL INCL O&P
400										**400**
0100	2,000 gallon	B-21	5	5.600	Ea.	1,100	173	31.50	1,304.50	1,525
0140	2,500 gallon		5	5.600		1,475	173	31.50	1,679.50	1,925
0180	4,000 gallon	↓	4	7		4,300	216	39.50	4,555.50	5,100
0200	5,000 gallon	B-13	3.50	16		5,625	460	176	6,261	7,100
0220	5,000 gal., 4 piece	"	3	18.667		6,900	535	206	7,641	8,625
0300	15,000 gallon, 4 piece	B-13B	1.70	32.941		15,600	950	525	17,075	19,100
0400	25,000 gallon, 4 piece		1.10	50.909		33,000	1,475	815	35,290	39,400
0500	40,000 gallon, 4 piece	↓	.80	70		38,800	2,025	1,125	41,950	47,000
0520	50,000 gallon, 5 piece	B-13C	.60	93.333		44,600	2,675	2,525	49,800	56,000
0540	75,000 gallon, cast in place	C-14C	.25	448		54,500	14,600	94	69,194	83,000
0560	100,000 gallon	"	.15	746		67,000	24,400	157	91,557	112,500
0600	High density polyethylene, 1,000 gallon	B-21	6	4.667		865	144	26.50	1,035.50	1,200
0700	1,500 gallon		4	7		1,125	216	39.50	1,380.50	1,600
0900	Galley, 4' x 4' x 4'	↓	16	1.750		205	54	9.90	268.90	320
1000	Distribution boxes, concrete, 7 outlets	2 Clab	16	1		104	26.50		130.50	156
1100	9 outlets	"	8	2		281	53.50		334.50	395
1150	Leaching field chambers, 13' x 3'-7" x 1'-4", standard	B-13	16	3.500		565	101	38.50	704.50	820
1200	Heavy duty, 8' x 4' x 1'-6"		14	4		390	115	44	549	655
1300	13' x 3'-9" x 1'-6"		12	4.667		1,375	134	51.50	1,560.50	1,775
1350	20' x 4' x 1'-6"	↓	5	11.200		910	320	123	1,353	1,625
1400	Leaching pit, precast concrete, 3' diameter, 3' deep	B-21	8	3.500		370	108	19.80	497.80	595
1500	6' diameter, 3' section		4.70	5.957		645	184	33.50	862.50	1,025
1600	Leaching pit, 6'-6" diameter, 6' deep		5	5.600		625	173	31.50	829.50	990
1620	8' deep		4	7		735	216	39.50	990.50	1,200
1700	8' diameter, H-20 load, 6' deep		4	7		1,000	216	39.50	1,255.50	1,500
1720	8' deep		3	9.333		1,250	288	53	1,591	1,875
2000	Velocity reducing pit, precast conc., 6' diameter, 3' deep	↓	4.70	5.957	↓	310	184	33.50	527.50	665
2200	Excavation for septic tank, 3/4 C.Y. backhoe	B-12F	145	.110	C.Y.		3.45	3.27	6.72	8.90
2400	4' trench for disposal field, 3/4 C.Y. backhoe	"	335	.048	L.F.		1.49	1.42	2.91	3.84
2600	Gravel fill, run of bank	B-6	150	.160	C.Y.	16.15	4.61	1.44	22.20	26.50
2800	Crushed stone, 3/4"	"	150	.160	"	24	4.61	1.44	30.05	34.50

02550 | Piped Energy Distribution

		CREW	DAILY OUTPUT	LABOR-HOURS	UNIT	MAT.	LABOR	EQUIP.	TOTAL	TOTAL INCL O&P
100										**100**
0010	**CHILLED/HVAC HOT WATER DISTRIBUTION**									
1005	Pipe, black steel w/2" polyurethane insul, 20' lengths									
1010	Align & tackweld on sleepers (NIC), 1-1/4"	B-35	864	.056	L.F.	12.35	1.83	.64	14.82	17.10
1020	1-1/2"		824	.058		13.75	1.91	.67	16.33	18.80
1030	2"		680	.071		14.50	2.32	.82	17.64	20.50
1040	2-1/2"		560	.086		15.60	2.82	.99	19.41	22.50
1050	3"		528	.091		15.95	2.99	1.05	19.99	23.50
1060	4"		384	.125		17.50	4.11	1.45	23.06	27
1070	5"		360	.133		26	4.38	1.54	31.92	37
1080	6"		296	.162		28	5.35	1.87	35.22	41
1090	8"		264	.182		37.50	5.95	2.10	45.55	53
1100	12"		216	.222		65.50	7.30	2.57	75.37	86
1110	On trench bottom, 18"	↓	176.13	.273		101	8.95	3.15	113.10	128
1120	24"	B-35A	145	.386		151	12.35	8.65	172	194
1130	30"		121	.463		217	14.80	10.35	242.15	273
1140	36"		100	.560	↓	291	17.90	12.50	321.40	360
1150	Fittings, elbows, on sleepers, 1-1/2"	Q-17	21	.762	Ea.	320	28	3.84	351.84	400
1160	3"		9.36	1.709		420	63	8.60	491.60	565
1170	4"		8	2		515	73.50	10.10	598.60	685
1180	6"		6	2.667		665	98.50	13.45	776.95	895
1190	8"		4.64	3.448		805	127	17.40	949.40	1,100
1200	Fittings, tees, 1-1/2"	↓	17	.941	↓	640	34.50	4.74	679.24	760

Important: See the Reference Section for critical supporting data - Reference Nos., Crews, & City Cost Indexes

02550 | Piped Energy Distribution

		CREW	DAILY OUTPUT	LABOR-HOURS	UNIT	2005 BARE COSTS				TOTAL INCL O&P
						MAT.	LABOR	EQUIP.	TOTAL	
100 1210	3"	Q-17	8.50	1.882	Ea.	795	69.50	9.50	874	985
1220	4"	↓	6	2.667		855	98.50	13.45	966.95	1,100
1230	6"	Q-17A	6.72	3.571		1,075	130	102	1,307	1,475
1240	8"	"	6.40	3.750		1,250	137	107	1,494	1,700
1250	Fittings, reducer, 3"	Q-17	16	1		169	37	5.05	211.05	247
1260	4"		12	1.333		201	49	6.70	256.70	300
1270	6"		12	1.333		287	49	6.70	342.70	395
1280	8"	↓	10	1.600		370	59	8.05	437.05	500
1290	Fittings, anchor, 4"	Q-17A	12	2		490	73	57	620	715
1300	6"		10.50	2.286		565	83.50	65	713.50	820
1310	8"	↓	10	2.400		630	87.50	68.50	786	900
1320	Fittings, cap, 1-1/2"	Q-17	42	.381		55	14.05	1.92	70.97	83.50
1330	3"		14.64	1.093		69.50	40.50	5.50	115.50	143
1340	4"		16	1		79.50	37	5.05	121.55	149
1350	6"		16	1		137	37	5.05	179.05	212
1360	8"		13.50	1.185		175	43.50	5.95	224.45	264
1365	12"	↓	11	1.455		267	53.50	7.35	327.85	380
1370	Elbow, on trench bottom, 12"	Q-17A	12	2		1,275	73	57	1,405	1,600
1380	18"		8	3		2,300	110	85.50	2,495.50	2,775
1390	24"		6	4		3,275	146	114	3,535	3,950
1400	30"		5.36	4.478		4,475	164	127	4,766	5,300
1410	36"		4	6		6,075	219	171	6,465	7,200
1420	Fittings, tee, 12"		6.72	3.571		1,700	130	102	1,932	2,150
1430	18"		6	4		2,825	146	114	3,085	3,450
1440	24"		4.64	5.172		4,400	189	147	4,736	5,300
1450	30"		4.16	5.769		6,750	211	164	7,125	7,925
1460	36"		3.36	7.143		9,925	261	203	10,389	11,500
1470	Fittings, reducer, 12"		10.64	2.256		540	82.50	64	686.50	790
1480	18"		9.68	2.479		855	90.50	70.50	1,016	1,150
1490	24"		8	3		1,325	110	85.50	1,520.50	1,700
1500	30"		7.04	3.409		2,025	125	97	2,247	2,550
1510	36"		5.36	4.478		2,900	164	127	3,191	3,575
1520	Fittings, anchor, 12"		11	2.182		680	79.50	62	821.50	940
1530	18"		6.72	3.571		730	130	102	962	1,100
1540	24"		6.32	3.797		1,025	139	108	1,272	1,450
1550	30"		4.64	5.172		1,225	189	147	1,561	1,800
1560	36"	↓	4	6	↓	1,500	219	171	1,890	2,175
1565	Weld in place and install shrink collar									
1570	On sleepers, 1-1/2"	Q-17A	18.50	1.297	Ea.	14.75	47.50	37	99.25	128
1580	3"		6.72	3.571		25	130	102	257	335
1590	4"		5.36	4.478		25	164	127	316	415
1600	6"		4	6		29.50	219	171	419.50	550
1610	8"		3.36	7.143		45.50	261	203	509.50	670
1620	12"		2.64	9.091		59	330	259	648	850
1630	On trench bottom, 18"		2	12		88.50	440	340	868.50	1,125
1640	24"		1.36	17.647		118	645	500	1,263	1,650
1650	30"		1.04	23.077		148	845	655	1,648	2,175
1660	36"	↓	1	24	↓	177	875	685	1,737	2,275
200 0010	**PIPE CONDUIT, PREFABRICATED / PREINSULATED**									
0020	Does not include trenching, fittings or crane.									
0300	For cathodic protection, add 12 to 14%									
0310	of total built-up price (casing plus service pipe)									
0580	Polyurethane insulated system, 250°F. max. temp.									
0620	Black steel service pipe, standard wt., 1/2" insulation									
0660	3/4" diam. pipe size	Q-17	54	.296	L.F.	24.50	10.90	1.49	36.89	45
0670	1" diam. pipe size	↓	50	.320	↓	27	11.80	1.61	40.41	49

02550	Piped Energy Distribution	CREW	DAILY OUTPUT	LABOR-HOURS	UNIT	2005 BARE COSTS				TOTAL INCL O&P	
						MAT.	LABOR	EQUIP.	TOTAL		
200 0680	1-1/4" diam. pipe size	Q-17	47	.340	L.F.	30	12.55	1.72	44.27	53.50	**200**
0690	1-1/2" diam. pipe size		45	.356		32.50	13.10	1.79	47.39	57.50	
0700	2" diam. pipe size		42	.381		34	14.05	1.92	49.97	60	
0710	2-1/2" diam. pipe size		34	.471		34.50	17.35	2.37	54.22	66	
0720	3" diam. pipe size		28	.571		39.50	21	2.88	63.38	78	
0730	4" diam. pipe size		22	.727		50	27	3.67	80.67	99.50	
0740	5" diam. pipe size	↓	18	.889		64	33	4.48	101.48	124	
0750	6" diam. pipe size	Q-18	23	1.043		74	40	3.51	117.51	145	
0760	8" diam. pipe size		19	1.263		109	48.50	4.24	161.74	197	
0770	10" diam. pipe size		16	1.500		143	57.50	5.05	205.55	250	
0780	12" diam. pipe size		13	1.846		178	70.50	6.20	254.70	310	
0790	14" diam. pipe size		11	2.182		199	83.50	7.35	289.85	350	
0800	16" diam. pipe size		10	2.400		228	91.50	8.05	327.55	400	
0810	18" diam. pipe size		8	3		262	115	10.10	387.10	470	
0820	20" diam. pipe size		7	3.429		292	131	11.50	434.50	530	
0830	24" diam. pipe size	↓	6	4		355	153	13.45	521.45	640	
0900	For 1" thick insulation, add					10%					
0940	For 1-1/2" thick insulation, add					13%					
0980	For 2" thick insulation, add				↓	20%					
1500	Gland seal for system, 3/4" diam. pipe size	Q-17	32	.500	Ea.	395	18.45	2.52	415.97	465	
1510	1" diam. pipe size		32	.500		395	18.45	2.52	415.97	465	
1540	1-1/4" diam. pipe size		30	.533		425	19.65	2.69	447.34	495	
1550	1-1/2" diam. pipe size		30	.533		425	19.65	2.69	447.34	495	
1560	2" diam. pipe size		28	.571		505	21	2.88	528.88	590	
1570	2-1/2" diam. pipe size		26	.615		545	22.50	3.10	570.60	635	
1580	3" diam. pipe size		24	.667		585	24.50	3.36	612.86	680	
1590	4" diam. pipe size		22	.727		690	27	3.67	720.67	805	
1600	5" diam. pipe size	↓	19	.842		850	31	4.24	885.24	985	
1610	6" diam. pipe size	Q-18	26	.923		905	35.50	3.10	943.60	1,050	
1620	8" diam. pipe size		25	.960		1,050	36.50	3.23	1,089.73	1,200	
1630	10" diam. pipe size		23	1.043		1,275	40	3.51	1,318.51	1,475	
1640	12" diam. pipe size		21	1.143		1,400	43.50	3.84	1,447.34	1,625	
1650	14" diam. pipe size		19	1.263		1,575	48.50	4.24	1,627.74	1,825	
1660	16" diam. pipe size		18	1.333		1,850	51	4.48	1,905.48	2,100	
1670	18" diam. pipe size		16	1.500		1,975	57.50	5.05	2,037.55	2,275	
1680	20" diam. pipe size		14	1.714		2,250	65.50	5.75	2,321.25	2,575	
1690	24" diam. pipe size	↓	12	2	↓	2,500	76.50	6.70	2,583.20	2,875	
2000	Elbow, 45° for system										
2020	3/4" diam. pipe size	Q-17	14	1.143	Ea.	250	42	5.75	297.75	345	
2040	1" diam. pipe size		13	1.231		257	45.50	6.20	308.70	360	
2050	1-1/4" diam. pipe size		11	1.455		292	53.50	7.35	352.85	410	
2060	1-1/2" diam. pipe size		9	1.778		305	65.50	8.95	379.45	445	
2070	2" diam. pipe size		6	2.667		320	98.50	13.45	431.95	515	
2080	2-1/2" diam. pipe size		4	4		345	147	20	512	625	
2090	3" diam. pipe size		3.50	4.571		400	168	23	591	720	
2100	4" diam. pipe size		3	5.333		465	197	27	689	835	
2110	5" diam. pipe size	↓	2.80	5.714		595	211	29	835	1,000	
2120	6" diam. pipe size	Q-18	4	6		675	229	20	924	1,100	
2130	8" diam. pipe size		3	8		980	305	27	1,312	1,575	
2140	10" diam. pipe size		2.40	10		1,250	380	33.50	1,663.50	1,975	
2150	12" diam. pipe size		2	12		1,650	460	40.50	2,150.50	2,550	
2160	14" diam. pipe size		1.80	13.333		2,050	510	45	2,605	3,075	
2170	16" diam. pipe size		1.60	15		2,425	575	50.50	3,050.50	3,600	
2180	18" diam. pipe size		1.30	18.462		3,050	705	62	3,817	4,475	
2190	20" diam. pipe size		1	24		3,850	915	80.50	4,845.50	5,700	
2200	24" diam. pipe size	↓	.70	34.286	↓	4,825	1,300	115	6,240	7,425	

Important: See the Reference Section for critical supporting data - Reference Nos., Crews, & City Cost Indexes

02500 | Utility Services

02550 | Piped Energy Distribution

		CREW	DAILY OUTPUT	LABOR-HOURS	UNIT	2005 BARE COSTS				TOTAL INCL O&P
						MAT.	LABOR	EQUIP.	TOTAL	
200	2260	For elbow, 90°, add				Ea.	25%					200
	2300	For tee, straight, add					85%	30%			
	2340	For tee, reducing, add					170%	30%			
	2380	For weldolet, straight, add					50%				
	2800	Calcium silicate insulated system, high temp. (1200°F)									
	2840	Steel casing with protective exterior coating									
	2850	6-5/8" diameter	Q-18	52	.462	L.F.	45.50	17.65	1.55	64.70	78.50
	2860	8-5/8" diameter		50	.480		50	18.35	1.61	69.96	84.50
	2870	10-3/4" diameter		47	.511		58.50	19.50	1.72	79.72	96
	2880	12-3/4" diameter		44	.545		63.50	21	1.83	86.33	104
	2890	14" diameter		41	.585		71.50	22.50	1.97	95.97	115
	2900	16" diameter		39	.615		77	23.50	2.07	102.57	122
	2910	18" diameter		36	.667		85	25.50	2.24	112.74	134
	2920	20" diameter		34	.706		95.50	27	2.37	124.87	148
	2930	22" diameter		32	.750		133	28.50	2.52	164.02	192
	2940	24" diameter		29	.828		151	31.50	2.78	185.28	217
	2950	26" diameter		26	.923		173	35.50	3.10	211.60	246
	2960	28" diameter		23	1.043		215	40	3.51	258.51	300
	2970	30" diameter		21	1.143		228	43.50	3.84	275.34	320
	2980	32" diameter		19	1.263		257	48.50	4.24	309.74	360
	2990	34" diameter		18	1.333		260	51	4.48	315.48	365
	3000	36" diameter		16	1.500		279	57.50	5.05	341.55	395
	3040	For multi-pipe casings, add					10%				
	3060	For oversize casings, add					2%				
	3400	Steel casing gland seal, single pipe									
	3420	6-5/8" diameter	Q-18	25	.960	Ea.	635	36.50	3.23	674.73	760
	3440	8-5/8" diameter		23	1.043		745	40	3.51	788.51	885
	3450	10-3/4" diameter		21	1.143		835	43.50	3.84	882.34	990
	3460	12-3/4" diameter		19	1.263		995	48.50	4.24	1,047.74	1,175
	3470	14" diameter		17	1.412		1,100	54	4.74	1,158.74	1,275
	3480	16" diameter		16	1.500		1,275	57.50	5.05	1,337.55	1,500
	3490	18" diameter		15	1.600		1,425	61	5.40	1,491.40	1,650
	3500	20" diameter		13	1.846		1,575	70.50	6.20	1,651.70	1,850
	3510	22" diameter		12	2		1,775	76.50	6.70	1,858.20	2,075
	3520	24" diameter		11	2.182		1,975	83.50	7.35	2,065.85	2,300
	3530	26" diameter		10	2.400		2,250	91.50	8.05	2,349.55	2,625
	3540	28" diameter		9.50	2.526		2,550	96.50	8.50	2,655	2,950
	3550	30" diameter		9	2.667		2,600	102	8.95	2,710.95	3,025
	3560	32" diameter		8.50	2.824		2,925	108	9.50	3,042.50	3,375
	3570	34" diameter		8	3		3,175	115	10.10	3,300.10	3,675
	3580	36" diameter		7	3.429		3,375	131	11.50	3,517.50	3,925
	3620	For multi-pipe casings, add					5%				
	4000	Steel casing anchors, single pipe									
	4020	6-5/8" diameter	Q-18	8	3	Ea.	570	115	10.10	695.10	815
	4040	8-5/8" diameter		7.50	3.200		595	122	10.75	727.75	850
	4050	10-3/4" diameter		7	3.429		785	131	11.50	927.50	1,075
	4060	12-3/4" diameter		6.50	3.692		835	141	12.40	988.40	1,150
	4070	14" diameter		6	4		980	153	13.45	1,146.45	1,325
	4080	16" diameter		5.50	4.364		1,150	167	14.65	1,331.65	1,525
	4090	18" diameter		5	4.800		1,275	183	16.15	1,474.15	1,725
	4100	20" diameter		4.50	5.333		1,425	204	17.90	1,646.90	1,875
	4110	22" diameter		4	6		1,575	229	20	1,824	2,100
	4120	24" diameter		3.50	6.857		1,725	262	23	2,010	2,325
	4130	26" diameter		3	8		1,950	305	27	2,282	2,650
	4140	28" diameter		2.50	9.600		2,125	365	32.50	2,522.50	2,925
	4150	30" diameter		2	12		2,275	460	40.50	2,775.50	3,225

SITE CONSTRUCTION 2

		02550	**Piped Energy Distribution**	CREW	DAILY OUTPUT	LABOR-HOURS	UNIT	2005 BARE COSTS				TOTAL INCL O&P	
								MAT.	LABOR	EQUIP.	TOTAL		
200	4160		32" diameter	Q-18	1.50	16	Ea.	2,725	610	54	3,389	3,975	200
	4170		34" diameter		1	24		3,050	915	80.50	4,045.50	4,825	
	4180		36" diameter		1	24		3,325	915	80.50	4,320.50	5,125	
	4220		For multi-pipe, add					5%	20%				
	4800		Steel casing elbow										
	4820		6-5/8" diameter	Q-18	15	1.600	Ea.	785	61	5.40	851.40	960	
	4830		8-5/8" diameter		15	1.600		835	61	5.40	901.40	1,025	
	4850		10-3/4" diameter		14	1.714		1,000	65.50	5.75	1,071.25	1,200	
	4860		12-3/4" diameter		13	1.846		1,200	70.50	6.20	1,276.70	1,450	
	4870		14" diameter		12	2		1,275	76.50	6.70	1,358.20	1,525	
	4880		16" diameter		11	2.182		1,400	83.50	7.35	1,490.85	1,650	
	4890		18" diameter		10	2.400		1,600	91.50	8.05	1,699.55	1,900	
	4900		20" diameter		9	2.667		1,700	102	8.95	1,810.95	2,050	
	4910		22" diameter		8	3		1,825	115	10.10	1,950.10	2,200	
	4920		24" diameter		7	3.429		2,025	131	11.50	2,167.50	2,425	
	4930		26" diameter		6	4		2,200	153	13.45	2,366.45	2,675	
	4940		28" diameter		5	4.800		2,375	183	16.15	2,574.15	2,925	
	4950		30" diameter		4	6		2,400	229	20	2,649	3,000	
	4960		32" diameter		3	8		2,725	305	27	3,057	3,500	
	4970		34" diameter		2	12		2,975	460	40.50	3,475.50	4,000	
	4980		36" diameter		2	12		3,175	460	40.50	3,675.50	4,225	
	5500		Black steel service pipe, std. wt., 1" thick insulation										
	5510		3/4" diameter pipe size	Q-17	54	.296	L.F.	19.90	10.90	1.49	32.29	40	
	5540		1" diameter pipe size		50	.320		20.50	11.80	1.61	33.91	42.50	
	5550		1-1/4" diameter pipe size		47	.340		23	12.55	1.72	37.27	46	
	5560		1-1/2" diameter pipe size		45	.356		25.50	13.10	1.79	40.39	49.50	
	5570		2" diameter pipe size		42	.381		28	14.05	1.92	43.97	54	
	5580		2-1/2" diameter pipe size		34	.471		29.50	17.35	2.37	49.22	61	
	5590		3" diameter pipe size		28	.571		33.50	21	2.88	57.38	71.50	
	5600		4" diameter pipe size		22	.727		43	27	3.67	73.67	92	
	5610		5" diameter pipe size		18	.889		58.50	33	4.48	95.98	118	
	5620		6" diameter pipe size	Q-18	23	1.043		63.50	40	3.51	107.01	134	
	6000		Black steel service pipe, std. wt., 1-1/2" thick insul.										
	6010		3/4" diameter pipe size	Q-17	54	.296	L.F.	20	10.90	1.49	32.39	40	
	6040		1" diameter pipe size		50	.320		22.50	11.80	1.61	35.91	44.50	
	6050		1-1/4" diameter pipe size		47	.340		25.50	12.55	1.72	39.77	48.50	
	6060		1-1/2" diameter pipe size		45	.356		27.50	13.10	1.79	42.39	52	
	6070		2" diameter pipe size		42	.381		30	14.05	1.92	45.97	56	
	6080		2-1/2" diameter pipe size		34	.471		32	17.35	2.37	51.72	63.50	
	6090		3" diameter pipe size		28	.571		36	21	2.88	59.88	74	
	6100		4" diameter pipe size		22	.727		46	27	3.67	76.67	95	
	6110		5" diameter pipe size		18	.889		58.50	33	4.48	95.98	118	
	6120		6" diameter pipe size	Q-18	23	1.043		66.50	40	3.51	110.01	137	
	6130		8" diameter pipe size		19	1.263		95.50	48.50	4.24	148.24	182	
	6140		10" diameter pipe size		16	1.500		125	57.50	5.05	187.55	229	
	6150		12" diameter pipe size		13	1.846		149	70.50	6.20	225.70	277	
	6190		For 2" thick insulation, add					15%					
	6220		For 2-1/2" thick insulation, add					25%					
	6260		For 3" thick insulation, add					30%					
	6800		Black steel service pipe, ex. hvy. wt., 1" thick insul.										
	6820		3/4" diameter pipe size	Q-17	50	.320	L.F.	21	11.80	1.61	34.41	42.50	
	6840		1" diameter pipe size		47	.340		22.50	12.55	1.72	36.77	45	
	6850		1-1/4" diameter pipe size		44	.364		25.50	13.40	1.83	40.73	50.50	
	6860		1-1/2" diameter pipe size		42	.381		27.50	14.05	1.92	43.47	53	
	6870		2" diameter pipe size		40	.400		29	14.75	2.02	45.77	56	
	6880		2-1/2" diameter pipe size		31	.516	Ea.	36	19	2.60	57.60	71	

Important: See the Reference Section for critical supporting data - Reference Nos., Crews, & City Cost Indexes

			DAILY	**LABOR-**				**2005 BARE COSTS**			**TOTAL**	
	02550	**Piped Energy Distribution**	**CREW**	**OUTPUT**	**HOURS**	**UNIT**	**MAT.**	**LABOR**	**EQUIP.**	**TOTAL**	**INCL O&P**	
200	6890	3″ diameter pipe size	Q-17	27	.593	L.F.	40.50	22	2.99	65.49	81	**200**
	6900	4″ diameter pipe size		21	.762		53	28	3.84	84.84	105	
	6910	5″ diameter pipe size		17	.941		74.50	34.50	4.74	113.74	139	
	6920	6″ diameter pipe size	Q-18	22	1.091		82.50	41.50	3.67	127.67	157	
	7400	Black steel service pipe, ex. hvy. wt., 1-1/2″ thick insul.										
	7420	3/4″ diameter pipe size	Q-17	50	.320	L.F.	21	11.80	1.61	34.41	42.50	
	7440	1″ diameter pipe size		47	.340		24	12.55	1.72	38.27	47	
	7450	1-1/4″ diameter pipe size		44	.364		27.50	13.40	1.83	42.73	52.50	
	7460	1-1/2″ diameter pipe size		42	.381		30.50	14.05	1.92	46.47	56.50	
	7470	2″ diameter pipe size		40	.400		30.50	14.75	2.02	47.27	57.50	
	7480	2-1/2″ diameter pipe size		31	.516		36.50	19	2.60	58.10	71.50	
	7490	3″ diameter pipe size		27	.593		43	22	2.99	67.99	83.50	
	7500	4″ diameter pipe size		21	.762		56	28	3.84	87.84	108	
	7510	5″ diameter pipe size		17	.941		77	34.50	4.74	116.24	142	
	7520	6″ diameter pipe size	Q-18	22	1.091		85	41.50	3.67	130.17	160	
	7530	8″ diameter pipe size		18	1.333		127	51	4.48	182.48	221	
	7540	10″ diameter pipe size		15	1.600		151	61	5.40	217.40	264	
	7550	12″ diameter pipe size		13	1.846		186	70.50	6.20	262.70	315	
	7590	For 2″ thick insulation, add					13%					
	7640	For 2-1/2″ thick insulation, add					18%					
	7680	For 3″ thick insulation, add					24%					
464	0010	**PIPING, GAS SERVICE & DISTRIBUTION, POLYETHYLENE**										**464**
	0020	not including excavation or backfill										
	1000	60 psi coils, comp cplg @ 100′, 1/2″ diameter, SDR 9.3	B-20A	608	.053	L.F.	.47	1.70		2.17	3.11	
	1040	1-1/4″ diameter, SDR 11		544	.059		.87	1.90		2.77	3.85	
	1100	2″ diameter, SDR 11		488	.066		1.08	2.11		3.19	4.42	
	1160	3″ diameter, SDR 11		408	.078		2.25	2.53		4.78	6.35	
	1500	60 PSI 40′ joints with coupling, 3″ diameter, SDR 11	B-21A	408	.098		2.25	3.23	1.24	6.72	8.75	
	1540	4″ diameter, SDR 11		352	.114		5.15	3.75	1.44	10.34	13	
	1600	6″ diameter, SDR 11		328	.122		16.10	4.02	1.54	21.66	25.50	
	1640	8″ diameter, SDR 11		272	.147		22	4.85	1.86	28.71	33.50	
466	0010	**PIPING, GAS SERVICE & DISTRIBUTION, STEEL**										**466**
	0020	not including excavation or backfill, tar coated and wrapped										
	4000	Schedule 40, plain end										
	4040	1″ diameter	Q-4	300	.107	L.F.	2.64	4.15	.27	7.06	9.45	
	4080	2″ diameter		280	.114		4.14	4.45	.29	8.88	11.60	
	4120	3″ diameter		260	.123		6.85	4.79	.31	11.95	15.10	
	4160	4″ diameter	B-35	255	.188		8.90	6.20	2.18	17.28	21.50	
	4200	5″ diameter		220	.218		12.95	7.15	2.52	22.62	28	
	4240	6″ diameter		180	.267		15.85	8.75	3.08	27.68	34	
	4280	8″ diameter		140	.343		25	11.25	3.96	40.21	49	
	4320	10″ diameter		100	.480		71	15.75	5.55	92.30	108	
	4360	12″ diameter		80	.600		100	19.70	6.95	126.65	148	
	4400	14″ diameter		75	.640		107	21	7.40	135.40	158	
	4440	16″ diameter		70	.686		117	22.50	7.95	147.45	172	
	4480	18″ diameter		65	.738		151	24.50	8.55	184.05	212	
	4520	20″ diameter		60	.800		234	26.50	9.25	269.75	310	
	4560	24″ diameter		50	.960		268	31.50	11.10	310.60	355	
	5000	Threaded and coupled										
	5002	4″ diameter	B-20	144	.167	L.F.	17.80	5		22.80	27.50	
	5004	5″ diameter	″	140	.171		24	5.15		29.15	34	
	5006	6″ diameter	B-21	126	.222		38.50	6.85	1.26	46.61	54.50	
	5008	8″ diameter		108	.259		59	8	1.47	68.47	78.50	
	5012	12″ diameter		72	.389		96	12	2.20	110.20	126	
	6000	Schedule 80, plain end										

2 SITE CONSTRUCTION

02550 | Piped Energy Distribution

		CREW	DAILY OUTPUT	LABOR-HOURS	UNIT	2005 BARE COSTS				TOTAL INCL O&P
						MAT.	LABOR	EQUIP.	TOTAL	
466 6002	4" diameter	B-35	144	.333	L.F.	33	10.95	3.85	47.80	57.50 **466**
6006	6" diameter		126	.381		62	12.50	4.40	78.90	92
6008	8" diameter		108	.444		82.50	14.60	5.15	102.25	119
6012	12" diameter		72	.667		165	22	7.70	194.70	224
6016	16" diameter, 1/2" wall		48	1		79.50	33	11.55	124.05	150
6020	20" diameter, 1/2" wall		38	1.263		140	41.50	14.60	196.10	233
6022	24" diameter, 1/2" wall	▼	36	1.333	▼	159	44	15.40	218.40	259
8008	Elbow, weld joint, standard weight									
8020	4" diameter	Q-16	6.80	3.529	Ea.	50.50	135	11.85	197.35	271
8026	8" diameter		3.40	7.059		223	269	23.50	515.50	675
8030	12" diameter		2.30	10.435		450	400	35	885	1,125
8034	16" diameter		1.50	16		815	610	54	1,479	1,875
8038	20" diameter		1.20	20		1,550	765	67	2,382	2,925
8040	24" diameter	▼	1.02	23.529	▼	2,050	895	79	3,024	3,675
8100	Extra heavy									
8102	4" diameter	Q-16	5.30	4.528	Ea.	101	173	15.20	289.20	390
8108	8" diameter		2.60	9.231		335	350	31	716	930
8112	12" diameter		1.80	13.333		600	510	45	1,155	1,475
8116	16" diameter		1.20	20		1,075	765	67	1,907	2,425
8120	20" diameter		.94	25.532		2,050	975	86	3,111	3,825
8122	24" diameter	▼	.80	30	▼	2,725	1,150	101	3,976	4,825
8200	Malleable, standard weight									
8202	4" diameter	B-20	12	2	Ea.	112	60		172	217
8208	8" diameter		6	4		340	120		460	560
8212	12" diameter	▼	4	6	▼	410	180		590	730
8300	Extra heavy									
8302	4" diameter	B-20	12	2	Ea.	224	60		284	340
8308	8" diameter	B-21	6	4.667		425	144	26.50	595.50	720
8312	12" diameter	"	4	7	▼	545	216	39.50	800.50	980
8500	Tee weld, standard weight									
8510	4" diameter	Q-16	4.50	5.333	Ea.	83.50	203	17.90	304.40	415
8514	6" diameter		3	8		142	305	27	474	645
8516	8" diameter		2.30	10.435		248	400	35	683	910
8520	12" diameter		1.50	16		625	610	54	1,289	1,650
8524	16" diameter		1	24		1,225	915	80.50	2,220.50	2,825
8528	20" diameter		.80	30		3,000	1,150	101	4,251	5,125
8530	24" diameter	▼	.70	34.286	▼	3,875	1,300	115	5,290	6,350
8810	Malleable, standard weight									
8812	4" diameter	B-20	8	3	Ea.	173	90		263	330
8818	8" diameter	B-21	4	7	"	340	216	39.50	595.50	750
8900	Extra heavy									
8902	4" diameter	B-20	8	3	Ea.	259	90		349	425
8908	8" diameter	B-21	4	7		425	216	39.50	680.50	845
8912	12" diameter	"	2.70	10.370	▼	700	320	58.50	1,078.50	1,325
468 0010	**PIPING, VALVES & METERS, GAS DISTRIBUTION**									**468**
0020	not including excavation or backfill									
0100	Gas stops, with or without checks									
0140	1-1/4" size	1 Plum	12	.667	Ea.	45.50	27		72.50	91
0180	1-1/2" size		10	.800		61.50	32.50		94	117
0200	2" size	▼	8	1	▼	92	41		133	163
0600	Pressure regulator valves, iron and bronze									
0680	2" diameter	1 Plum	11	.727	Ea.	224	29.50		253.50	291
0700	3" diameter	Q-1	13	1.231		455	45.50		500.50	570
0740	4" diameter	"	8	2	▼	1,300	73.50		1,373.50	1,525
2000	Lubricated semi-steel plug valve									
2040	3/4" diameter	1 Plum	16	.500	Ea.	78.50	20.50		99	117

Important: See the Reference Section for critical supporting data - Reference Nos., Crews, & City Cost Indexes

2 SITE CONSTRUCTION

02550	Piped Energy Distribution	CREW	DAILY OUTPUT	LABOR-HOURS	UNIT	MAT.	2005 BARE COSTS LABOR	EQUIP.	TOTAL	TOTAL INCL O&P	
468	2080	1" diameter	1 Plum	14	.571	Ea.	76.50	23.50		100	119	468
	2100	1-1/4" diameter		12	.667		79	27		106	128	
	2140	1-1/2" diameter		11	.727		79	29.50		108.50	131	
	2180	2" diameter		8	1		73.50	41		114.50	143	
	2300	2-1/2" diameter	Q-1	5	3.200		118	118		236	305	
	2340	3" diameter	"	4.50	3.556		182	131		313	395	
550	0010	**GAS STATION PRODUCT LINE**										550
	0020	Primary containment pipe, fiberglass-reinforced										
	0030	Plastic pipe 15' & 30' lengths										
	0040	2" diameter	Q-6	425	.056	L.F.	3.23	2.16		5.39	6.80	
	0050	3" diameter		400	.060		4.23	2.29		6.52	8.10	
	0060	4" diameter		375	.064		5.45	2.45		7.90	9.70	
	0100	Fittings										
	0110	Elbows, 90° & 45°, bell-ends, 2"	Q-6	24	1	Ea.	32.50	38		70.50	93.50	
	0120	3" diameter		22	1.091		34.50	41.50		76	100	
	0130	4" diameter		20	1.200		45.50	46		91.50	119	
	0200	Tees, bell ends, 2"		21	1.143		39.50	43.50		83	109	
	0210	3" diameter		18	1.333		40	51		91	121	
	0220	4" diameter		15	1.600		55	61		116	153	
	0230	Flanges bell ends, 2"		24	1		13.10	38		51.10	72	
	0240	3" diameter		22	1.091		16.50	41.50		58	80.50	
	0250	4" diameter		20	1.200		22.50	46		68.50	93.50	
	0260	Sleeve couplings, 2"		21	1.143		8.35	43.50		51.85	74.50	
	0270	3" diameter		18	1.333		11.85	51		62.85	89.50	
	0280	4" diameter		15	1.600		16.35	61		77.35	110	
	0290	Threaded adapters 2"		21	1.143		10.95	43.50		54.45	77.50	
	0300	3" diameter		18	1.333		19.25	51		70.25	97.50	
	0310	4" diameter		15	1.600		26	61		87	121	
	0320	Reducers, 2"		27	.889		14.60	34		48.60	67	
	0330	3" diameter		22	1.091		16.90	41.50		58.40	81	
	0340	4" diameter		20	1.200		22	46		68	93	
	1010	Gas station product line for secondary containment (double wall)										
	1100	Fiberglass reinforced plastic pipe 25' lengths										
	1120	Pipe, plain end, 3"	Q-6	375	.064	L.F.	5.55	2.45		8	9.80	
	1130	4" diameter		350	.069		9.10	2.62		11.72	13.95	
	1140	5" diameter		325	.074		11.85	2.82		14.67	17.25	
	1150	6" diameter		300	.080		12.20	3.06		15.26	18	
	1200	Fittings										
	1230	Elbows, 90° & 45°, 3"	Q-6	18	1.333	Ea.	40	51		91	121	
	1240	4" diameter		16	1.500		69.50	57.50		127	163	
	1250	5" diameter		14	1.714		162	65.50		227.50	277	
	1260	6" diameter		12	2		164	76.50		240.50	295	
	1270	Tees, 3"		15	1.600		59	61		120	157	
	1280	4" diameter		12	2		87	76.50		163.50	211	
	1290	5" diameter		9	2.667		176	102		278	345	
	1300	6" diameter		6	4		184	153		337	430	
	1310	Couplings, 3"		18	1.333		28	51		79	108	
	1320	4" diameter		16	1.500		72.50	57.50		130	166	
	1330	5" diameter		14	1.714		150	65.50		215.50	265	
	1340	6" diameter		12	2		156	76.50		232.50	287	
	1350	Cross-over nipples, 3"		18	1.333		6.40	51		57.40	83.50	
	1360	4" diameter		16	1.500		7.50	57.50		65	94.50	
	1370	5" diameter		14	1.714		11.15	65.50		76.65	111	
	1380	6" diameter		12	2		11.70	76.50		88.20	128	
	1400	Telescoping, reducers, concentric 4" x 3"		18	1.333		21.50	51		72.50	100	
	1410	5" x 4"		17	1.412		55.50	54		109.50	143	

02550	Piped Energy Distribution	CREW	DAILY OUTPUT	LABOR-HOURS	UNIT	2005 BARE COSTS				TOTAL INCL O&P	
						MAT.	LABOR	EQUIP.	TOTAL		
550 1420	6" x 5"	Q-6	16	1.500	Ea.	134	57.50		191.50	233	**550**

02580	Elec/Communication Structures										
100 0010	**RADIO TOWERS**										**100**
0020	Guyed, 50'h, 40 lb. sec., 70 MPH basic wind spd.	2 Sswk	1	16	Ea.	2,225	610		2,835	3,550	
0100	Wind load 90 MPH basic wind speed	"	1	16		2,225	610		2,835	3,550	
0300	190' high, 40 lb. section, wind load 70 MPH basic wind speed	K-2	.33	72.727		6,000	2,550	535	9,085	11,600	
0400	200' high, 70 lb. section, wind load 90 MPH basic wind speed		.33	72.727		12,100	2,550	535	15,185	18,300	
0600	300' high, 70 lb. section, wind load 70 MPH basic wind speed		.20	120		17,200	4,200	880	22,280	27,100	
0700	270' high, 90 lb. section, wind load 90 MPH basic wind speed		.20	120		19,800	4,200	880	24,880	30,000	
0800	400' high, 100 lb. section, wind load 70 MPH basic wind speed		.14	171		28,900	6,000	1,250	36,150	43,400	
0900	Self-supporting, 60' high, wind load 70 MPH basic wind speed		.80	30		4,650	1,050	220	5,920	7,150	
0910	60' high, wind load 90 MPH basic wind speed		.45	53.333		8,350	1,875	390	10,615	12,800	
1000	120' high, wind load 70 MPH basic wind speed		.40	60		11,400	2,100	440	13,940	16,600	
1200	190' high, wind load 90 MPH basic wind speed	▼	.20	120		27,600	4,200	880	32,680	38,600	
2000	For states west of Rocky Mountains, add for shipping				▼	10%					
410 0010	**UNDERGROUND DUCTS AND MANHOLES**, In slab or duct bank										**410**
0011	Not including excavation, backfill and cast in place concrete										
1000	Direct burial										
1010	PVC, schedule 40, w/coupling, 1/2" diameter	1 Elec	340	.024	L.F.	.32	.96		1.28	1.78	
1020	3/4" diameter		290	.028		.42	1.12		1.54	2.13	
1030	1" diameter		260	.031		.63	1.25		1.88	2.56	
1040	1-1/2" diameter		210	.038		1.04	1.55		2.59	3.45	
1050	2" diameter	▼	180	.044		1.34	1.81		3.15	4.17	
1060	3" diameter	2 Elec	240	.067		2.65	2.72		5.37	6.95	
1070	4" diameter		160	.100		3.63	4.08		7.71	10.05	
1080	5" diameter		120	.133		5.35	5.45		10.80	13.95	
1090	6" diameter	▼	90	.178	▼	7.15	7.25		14.40	18.65	
1110	Elbows, 1/2" diameter	1 Elec	48	.167	Ea.	1.58	6.80		8.38	11.85	
1120	3/4" diameter		38	.211		1.59	8.60		10.19	14.50	
1130	1" diameter		32	.250		2.69	10.20		12.89	18.10	
1140	1-1/2" diameter		21	.381		5.20	15.50		20.70	28.50	
1150	2" diameter		16	.500		7.55	20.50		28.05	39	
1160	3" diameter		12	.667		23	27		50	66	
1170	4" diameter		9	.889		40	36		76	98	
1180	5" diameter		8	1		70	41		111	138	
1190	6" diameter		5	1.600		119	65		184	228	
1210	Adapters, 1/2" diameter		52	.154		.51	6.25		6.76	9.90	
1220	3/4" diameter		43	.186		.94	7.60		8.54	12.35	
1230	1" diameter		39	.205		1.19	8.35		9.54	13.75	
1240	1-1/2" diameter		35	.229		1.85	9.30		11.15	15.90	
1250	2" diameter		26	.308		2.66	12.55		15.21	21.50	
1260	3" diameter		20	.400		6.75	16.30		23.05	32	
1270	4" diameter		14	.571		11.80	23.50		35.30	47.50	
1280	5" diameter		12	.667		23	27		50	66	
1290	6" diameter		9	.889		28	36		64	85	
1340	Bell end & cap, 1-1/2" diameter		35	.229		9	9.30		18.30	24	
1350	Bell end & plug, 2" diameter		26	.308		9.50	12.55		22.05	29	
1360	3" diameter		20	.400		11.90	16.30		28.20	37.50	
1370	4" diameter		14	.571		13.90	23.50		37.40	50	
1380	5" diameter		12	.667		21	27		48	63.50	
1390	6" diameter		9	.889		24	36		60	80	
1450	Base spacer, 2" diameter		56	.143		1.61	5.80		7.41	10.40	
1460	3" diameter	▼	46	.174	▼	1.78	7.10		8.88	12.50	

Important: See the Reference Section for critical supporting data - Reference Nos., Crews, & City Cost Indexes

			DAILY	LABOR-		2005 BARE COSTS				TOTAL	
02580	**Elec/Communication Structures**	CREW	OUTPUT	HOURS	UNIT	MAT.	LABOR	EQUIP.	TOTAL	INCL O&P	
410	1470	4" diameter	1 Elec	41	.195	Ea.	1.88	7.95		9.83	13.90	**410**
	1480	5" diameter		37	.216		2.31	8.80		11.11	15.65	
	1490	6" diameter		34	.235		3.24	9.60		12.84	17.80	
	1550	Intermediate spacer, 2" diameter		60	.133		1.40	5.45		6.85	9.65	
	1560	3" diameter		46	.174		1.78	7.10		8.88	12.50	
	1570	4" diameter		41	.195		1.88	7.95		9.83	13.90	
	1580	5" diameter		37	.216		2.31	8.80		11.11	15.65	
	1590	6" diameter		34	.235		3.24	9.60		12.84	17.80	
420	0010	**ELECTRIC & TELEPHONE UNDERGROUND,** Not including excavation										**420**
	0200	backfill and cast in place concrete										
	0400	Hand holes, precast concrete, with concrete cover										
	0600	2' x 2' x 3' deep	R-3	2.40	8.333	Ea.	263	335	66	664	855	
	0800	3' x 3' x 3' deep		1.90	10.526		340	420	83.50	843.50	1,100	
	1000	4' x 4' x 4' deep		1.40	14.286		680	570	113	1,363	1,725	
	1200	Manholes, precast with iron racks & pulling irons, C.I. frame										
	1400	and cover, 4' x 6' x 7' deep	B-13	2	28	Ea.	1,300	805	310	2,415	3,050	
	1600	6' x 8' x 7' deep		1.90	29.474		1,600	850	325	2,775	3,425	
	1800	6' x 10' x 7' deep		1.80	31.111		1,800	895	345	3,040	3,725	
	4200	Underground duct, banks ready for concrete fill, min. of 7.5"										
	4400	between conduits, ctr. to ctr.(for wire & cable see div. 16120)										
	4580	PVC, type EB, 1 @ 2" diameter	2 Elec	480	.033	L.F.	.73	1.36		2.09	2.82	
	4600	2 @ 2" diameter		240	.067		1.46	2.72		4.18	5.65	
	4800	4 @ 2" diameter		120	.133		2.92	5.45		8.37	11.30	
	4900	1 @ 3" diameter		400	.040		1	1.63		2.63	3.53	
	5000	2 @ 3" diameter		200	.080		2	3.26		5.26	7.05	
	5200	4 @ 3" diameter		100	.160		3.99	6.50		10.49	14.10	
	5300	1 @ 4" diameter		320	.050		1.53	2.04		3.57	4.71	
	5400	2 @ 4" diameter		160	.100		3.06	4.08		7.14	9.40	
	5600	4 @ 4" diameter		80	.200		6.10	8.15		14.25	18.85	
	5800	6 @ 4" diameter		54	.296		9.15	12.05		21.20	28	
	5810	1 @ 5" diameter		260	.062		2.27	2.51		4.78	6.25	
	5820	2 @ 5" diameter		130	.123		4.54	5		9.54	12.45	
	5840	4 @ 5" diameter		70	.229		9.10	9.30		18.40	24	
	5860	6 @ 5" diameter		50	.320		13.65	13.05		26.70	34.50	
	5870	1 @ 6" diameter		200	.080		3.24	3.26		6.50	8.40	
	5880	2 @ 6" diameter		100	.160		6.50	6.50		13	16.85	
	5900	4 @ 6" diameter		50	.320		13	13.05		26.05	33.50	
	5920	6 @ 6" diameter		30	.533		19.45	21.50		40.95	54	
	6200	Rigid galvanized steel, 2 @ 2" diameter		180	.089		16.10	3.62		19.72	23	
	6400	4 @ 2" diameter		90	.178		32	7.25		39.25	46.50	
	6800	2 @ 3" diameter		100	.160		37.50	6.50		44	50.50	
	7000	4 @ 3" diameter		50	.320		75	13.05		88.05	102	
	7200	2 @ 4" diameter		70	.229		51.50	9.30		60.80	70.50	
	7400	4 @ 4" diameter		34	.471		103	19.20		122.20	142	
	7600	6 @ 4" diameter		22	.727		155	29.50		184.50	214	
	7620	2 @ 5" diameter		60	.267		109	10.85		119.85	136	
	7640	4 @ 5" diameter		30	.533		218	21.50		239.50	273	
	7660	6 @ 5" diameter		18	.889		325	36		361	415	
	7680	2 @ 6" diameter		40	.400		158	16.30		174.30	199	
	7700	4 @ 6" diameter		20	.800		315	32.50		347.50	400	
	7720	6 @ 6" diameter		14	1.143		475	46.50		521.50	595	
	8000	Fittings, PVC type EB, elbow, 2" diameter	1 Elec	16	.500	Ea.	10.55	20.50		31.05	42	
	8200	3" diameter		14	.571		11.15	23.50		34.65	47	
	8400	4" diameter		12	.667		15.80	27		42.80	58	
	8420	5" diameter		10	.800		37.50	32.50		70	89.50	
	8440	6" diameter		9	.889		71	36		107	132	

SITE CONSTRUCTION **2**

02580 | Elec/Communication Structures

		CREW	DAILY OUTPUT	LABOR-HOURS	UNIT	MAT.	LABOR	EQUIP.	TOTAL	TOTAL INCL O&P	
420											**420**
8500	Coupling, 2" diameter				Ea.	.76			.76	.84	
8600	3" diameter					2.64			2.64	2.90	
8700	4" diameter					4.14			4.14	4.55	
8720	5" diameter					7.45			7.45	8.20	
8740	6" diameter					21.50			21.50	24	
8800	Adapter, 2" diameter	1 Elec	26	.308		1.35	12.55		13.90	20	
9000	3" diameter		20	.400		3.75	16.30		20.05	28.50	
9200	4" diameter		16	.500		5	20.50		25.50	36	
9220	5" diameter		13	.615		12.50	25		37.50	51.50	
9240	6" diameter		10	.800		16.50	32.50		49	66.50	
9400	End bell, 2" diameter		16	.500		6.55	20.50		27.05	37.50	
9600	3" diameter		14	.571		7.85	23.50		31.35	43	
9800	4" diameter		12	.667		9.25	27		36.25	50.50	
9810	5" diameter		10	.800		13.95	32.50		46.45	64	
9820	6" diameter		8	1		26.50	41		67.50	89.50	
9830	5° angle coupling, 2" diameter		26	.308		11	12.55		23.55	31	
9840	3" diameter		20	.400		13.95	16.30		30.25	40	
9850	4" diameter		16	.500		16.50	20.50		37	48.50	
9860	5" diameter		13	.615		18.05	25		43.05	57.50	
9870	6" diameter		10	.800		18.50	32.50		51	69	
9880	Expansion joint, 2" diameter		16	.500		32.50	20.50		53	66	
9890	3" diameter		18	.444		57	18.10		75.10	89.50	
9900	4" diameter		12	.667		82	27		109	131	
9910	5" diameter		10	.800		128	32.50		160.50	190	
9920	6" diameter	▼	8	1		172	41		213	250	
9930	Heat bender, 2" diameter					425			425	470	
9940	6" diameter					1,175			1,175	1,300	
9950	Cement, quart				▼	14.70			14.70	16.15	
500	**UTILITY POLES**										**500**
0010											
6200	Poles, wood, preservative treatment, see also div. 16520, 20' high	R-3	3.10	6.452	Ea.	241	258	51	550	705	
6400	25' high		2.90	6.897		255	276	54.50	585.50	750	
6600	30' high		2.60	7.692		281	310	61	652	835	
6800	35' high		2.40	8.333		365	335	66	766	970	
7000	40' high		2.30	8.696		450	350	69	869	1,100	
7200	45' high	▼	1.70	11.765	▼	550	470	93.50	1,113.50	1,400	
7400	Cross arms with hardware & insulators										
7600	4' long	1 Elec	2.50	3.200	Ea.	114	130		244	320	
7800	5' long		2.40	3.333		133	136		269	350	
8000	6' long	▼	2.20	3.636	▼	153	148		301	390	

02620 | Subdrainage

		CREW	DAILY OUTPUT	LABOR-HOURS	UNIT	MAT.	LABOR	EQUIP.	TOTAL	TOTAL INCL O&P	
300	**GEOTEXTILES FOR SUBSURFACE DRAINAGE**										**300**
0010											
0100	Fabric, laid in trench, polypropylene, ideal conditions	2 Clab	2,400	.007	S.Y.	1.28	.18		1.46	1.69	
0110	Adverse conditions		1,600	.010	"	1.28	.27		1.55	1.83	
0170	Fabric ply bonded to 3 dimen. nylon mat, .4" thk, ideal conditions		2,000	.008	S.F.	.89	.21		1.10	1.31	
0180	Adverse conditions		1,200	.013	"	1.11	.36		1.47	1.77	
0185	Soil drainage mat on vertical wall, 0.44" thick	▼	265	.060	S.Y.	15.60	1.61		17.21	19.65	

2 SITE CONSTRUCTION

		02620	Subdrainage	CREW	DAILY OUTPUT	LABOR-HOURS	UNIT	2005 BARE COSTS MAT.	LABOR	EQUIP.	TOTAL	TOTAL INCL O&P	
300	0188		0.25" thick	2 Clab	300	.053	S.Y.	9.50	1.42		10.92	12.65	300
	0190		0.8" thick, ideal conditions	↓	2,400	.007	S.F.	1.40	.18		1.58	1.82	
	0200		Adverse conditions	↓	1,600	.010	"	1.84	.27		2.11	2.44	
	0300		Drainage material, 3/4" gravel fill in trench	B-6	260	.092	C.Y.	53.50	2.66	.83	56.99	63.50	
	0400		Pea stone	"	260	.092	"	18.80	2.66	.83	22.29	25.50	
610	0010	**PIPING, SUBDRAINAGE, CONCRETE**	R02510-810										610
	0021		Not including excavation and backfill										
	3000		Porous wall concrete underdrain, std. strength, 4" diameter	B-20	335	.072	L.F.	1.95	2.15		4.10	5.50	
	3020		6" diameter	"	315	.076		2.54	2.29		4.83	6.35	
	3040		8" diameter	B-21	310	.090		3.13	2.79	.51	6.43	8.30	
	3060		12" diameter		285	.098		6.65	3.04	.56	10.25	12.60	
	3080		15" diameter		230	.122		7.60	3.76	.69	12.05	14.95	
	3100		18" diameter	↓	165	.170		10.05	5.25	.96	16.26	20	
	4000		Extra strength, 6" diameter	B-20	315	.076		2.57	2.29		4.86	6.40	
	4020		8" diameter	B-21	310	.090		3.86	2.79	.51	7.16	9.10	
	4040		10" diameter		285	.098		7.70	3.04	.56	11.30	13.80	
	4060		12" diameter		230	.122		8.35	3.76	.69	12.80	15.80	
	4080		15" diameter		200	.140		9.25	4.33	.79	14.37	17.75	
	4100		18" diameter	↓	165	.170	↓	13.50	5.25	.96	19.71	24	
620	0010	**PIPING, SUBDRAINAGE, CORRUGATED METAL**											620
	0021		Not including excavation and backfill										
	2010		Aluminum, perforated										
	2020		6" diameter, 18 ga.	B-14	380	.126	L.F.	2.76	3.55	.57	6.88	9.15	
	2200		8" diameter, 16 ga.		370	.130		3.94	3.64	.58	8.16	10.60	
	2220		10" diameter, 16 ga.		360	.133		4.93	3.75	.60	9.28	11.85	
	2240		12" diameter, 16 ga.	↓	285	.168	↓	5.50	4.73	.76	10.99	14.20	
	3000		Uncoated galvanized, perforated										
	3020		6" diameter, 18 ga.	B-20	380	.063	L.F.	4.37	1.90		6.27	7.75	
	3200		8" diameter, 16 ga.	"	370	.065		6	1.95		7.95	9.65	
	3220		10" diameter, 16 ga.	B-21	360	.078		9	2.40	.44	11.84	14.10	
	3240		12" diameter, 16 ga.		285	.098		9.45	3.04	.56	13.05	15.70	
	3260		18" diameter, 16 ga.	↓	205	.137	↓	14.40	4.22	.77	19.39	23.50	
	4000		Steel, perforated, asphalt coated										
	4020		6" diameter 18 ga.	B-20	380	.063	L.F.	3.49	1.90		5.39	6.80	
	4030		8" diameter 18 ga	"	370	.065		5.45	1.95		7.40	9.05	
	4040		10" diameter 16 ga	B-21	360	.078		6.30	2.40	.44	9.14	11.10	
	4050		12" diameter 16 ga		285	.098		7.20	3.04	.56	10.80	13.25	
	4060		18" diameter 16 ga	↓	205	.137	↓	9.85	4.22	.77	14.84	18.20	
630	0010	**PIPING, SUBDRAINAGE, PLASTIC**											630
	0020		Not including excavation and backfill										
	1110		10"				Ea.	3.01			3.01	3.31	
	2100		Perforated PVC, 4" diameter	B-14	314	.153	L.F.	.71	4.29	.69	5.69	8.20	
	2110		6" diameter		300	.160		1.31	4.49	.72	6.52	9.20	
	2120		8" diameter		290	.166		1.49	4.65	.74	6.88	9.65	
	2130		10" diameter		280	.171		2.31	4.82	.77	7.90	10.85	
	2140		12" diameter	↓	270	.178	↓	3.20	4.99	.80	8.99	12.10	
660	0010	**PIPING, SUBDRAINAGE, CORR. PLASTIC TUBING, PERF. OR PLAIN**											660
	0020		In rolls, not including excavation and backfill										
	0030		3" diameter	2 Clab	1,200	.013	L.F.	.46	.36		.82	1.06	
	0040		4" diameter		1,200	.013		.58	.36		.94	1.18	
	0041		With silt sock		1,200	.013		.68	.36		1.04	1.30	
	0060		6" diameter		900	.018		1.84	.47		2.31	2.76	
	0080		8" diameter	↓	700	.023	↓	2.61	.61		3.22	3.82	
	0200		Fittings										

02620 | Subdrainage

		CREW	DAILY OUTPUT	LABOR-HOURS	UNIT	MAT.	LABOR	EQUIP.	TOTAL	TOTAL INCL O&P	
660	**0230** Elbows, 3" diameter	1 Clab	32	.250	Ea.	4.53	6.70		11.23	15.40	**660**
	0240 4" diameter		32	.250		4.98	6.70		11.68	15.90	
	0250 5" diameter		32	.250		5.90	6.70		12.60	16.90	
	0260 6" diameter		32	.250		7.95	6.70		14.65	19.10	
	0280 8" diameter		32	.250		9.75	6.70		16.45	21	
	0330 Tees, 3" diameter		27	.296		4.42	7.90		12.32	17.15	
	0340 4" diameter		27	.296		4.64	7.90		12.54	17.40	
	0350 5" diameter		27	.296		5.75	7.90		13.65	18.60	
	0360 6" diameter		27	.296		7.75	7.90		15.65	21	
	0370 6" x 6" x 4"		27	.296		7.95	7.90		15.85	21	
	0380 8" diameter		27	.296		9.30	7.90		17.20	22.50	
	0390 8" x 8" x 6"		27	.296		16.60	7.90		24.50	30.50	
	0430 End cap, 3" diameter		32	.250		1.71	6.70		8.41	12.30	
	0440 4" diameter		32	.250		1.88	6.70		8.58	12.45	
	0460 6" diameter		32	.250		2.82	6.70		9.52	13.50	
	0480 8" diameter		32	.250		5.90	6.70		12.60	16.90	
	0530 Coupler, 3" diameter		32	.250		1.41	6.70		8.11	11.95	
	0540 4" diameter		32	.250		1.48	6.70		8.18	12.05	
	0550 5" diameter		32	.250		1.83	6.70		8.53	12.40	
	0560 6" diameter		32	.250		2.47	6.70		9.17	13.10	
	0580 8" diameter		32	.250		4.58	6.70		11.28	15.45	
	0590 Heavy duty highway type, add					10%					
	0660 Reducer, 6" to 4"	1 Clab	32	.250		3.17	6.70		9.87	13.90	
	0680 8" to 6"		32	.250		5.05	6.70		11.75	16	
	0730 "Y" fitting, 3" diameter		27	.296		5.55	7.90		13.45	18.40	
	0740 4" diameter		27	.296		6.40	7.90		14.30	19.35	
	0750 5" diameter		27	.296		8.05	7.90		15.95	21	
	0760 6" diameter		27	.296		9.75	7.90		17.65	23	
	0780 8" diameter		27	.296		11.95	7.90		19.85	25.50	
	0860 Silt sock only for above tubing, 6" dia.				L.F.	.63			.63	.69	
	0880 8" diameter				"	1.04			1.04	1.14	

02630 | Storm Drainage

		CREW	DAILY OUTPUT	LABOR-HOURS	UNIT	MAT.	LABOR	EQUIP.	TOTAL	TOTAL INCL O&P	
110	**0010 CATCH BASIN GRATES AND FRAMES** not including footing, excavation										**110**
	1580 Curb inlet frame, grate, and curb box										
	1582 Large 24" x 36" heavy duty	B-24	2	12	Ea.	395	375		770	1,000	
	1590 Small 10" x 21" medium duty	"	2	12		305	375		680	905	
	1600 Frames & covers, C.I., 24" square, 500 lb.	B-6	7.80	3.077		239	88.50	27.50	355	430	
	1700 26" D shape, 600 lb.		7	3.429		380	99	31	510	605	
	1800 Light traffic, 18" diameter, 100 lb.		10	2.400		128	69	21.50	218.50	271	
	1900 24" diameter, 300 lb.		8.70	2.759		196	79.50	25	300.50	365	
	2000 36" diameter, 900 lb.		5.80	4.138		425	119	37	581	690	
	2100 Heavy traffic, 24" diameter, 400 lb.		7.80	3.077		193	88.50	27.50	309	380	
	2200 36" diameter, 1150 lb.		3	8		610	231	72	913	1,100	
	2300 Mass. State standard, 26" diameter, 475 lb.		7	3.429		465	99	31	595	695	
	2400 30" diameter, 620 lb.		7	3.429		325	99	31	455	540	
	2500 Watertight, 24" diameter, 350 lb.		7.80	3.077		325	88.50	27.50	441	520	
	2600 26" diameter, 500 lb.		7	3.429		310	99	31	440	525	
	2700 32" diameter, 575 lb.		6	4		665	115	36	816	945	
	2800 3 piece cover & frame, 10" deep,										
	2900 1200 lbs., for heavy equipment	B-6	3	8	Ea.	990	231	72	1,293	1,525	
	3000 Raised for paving 1-1/4" to 2" high,										
	3100 4 piece expansion ring										
	3200 20" to 26" diameter	1 Clab	3	2.667	Ea.	114	71		185	237	
	3300 30" to 36" diameter	"	3	2.667	"	159	71		230	286	

2 SITE CONSTRUCTION

Important: See the Reference Section for critical supporting data - Reference Nos., Crews, & City Cost Indexes

02630 | Storm Drainage

			CREW	DAILY OUTPUT	LABOR-HOURS	UNIT	MAT.	LABOR	EQUIP.	TOTAL	TOTAL INCL O&P	
110	3320	Frames and covers, existing, raised for paving, 2", including										110
	3340	row of brick, concrete collar, up to 12" wide frame	B-6	18	1.333	Ea.	35.50	38.50	12	86	112	
	3360	20" to 26" wide frame		11	2.182		56.50	63	19.60	139.10	180	
	3380	30" to 36" wide frame	↓	9	2.667		70	77	24	171	222	
	3400	Inverts, single channel brick	D-1	3	5.333		78.50	166		244.50	340	
	3500	Concrete		5	3.200		61	99.50		160.50	220	
	3600	Triple channel, brick		2	8		119	249		368	510	
	3700	Concrete	↓	3	5.333	↓	105	166		271	370	
400	0010	**STORM DRAINAGE MANHOLES, FRAMES & COVERS** not including										400
	0020	footing, excavation, backfill (See line items for frame & cover)										
	0050	Brick, 4' inside diameter, 4' deep	D-1	1	16	Ea.	320	495		815	1,100	
	0100	6' deep		.70	22.857		445	710		1,155	1,575	
	0150	8' deep		.50	32	↓	565	995		1,560	2,150	
	0200	For depths over 8', add		4	4	V.L.F.	174	124		298	380	
	0400	Concrete blocks (radial), 4' I.D., 4' deep		1.50	10.667	Ea.	255	330		585	785	
	0500	6' deep		1	16		335	495		830	1,125	
	0600	8' deep		.70	22.857	↓	415	710		1,125	1,525	
	0700	For depths over 8', add	↓	5.50	2.909	V.L.F.	41.50	90.50		132	184	
	0800	Concrete, cast in place, 4' x 4', 8" thick, 4' deep	C-14H	2	24	Ea.	445	810	12	1,267	1,750	
	0900	6' deep		1.50	32		650	1,075	16	1,741	2,400	
	1000	8' deep		1	48	↓	920	1,625	24	2,569	3,575	
	1100	For depths over 8', add		8	6	V.L.F.	107	202	3	312	435	
	1110	Precast, 4' I.D., 4' deep	B-22	4.10	7.317	Ea.	675	229	58	962	1,175	
	1120	6' deep		3	10		870	315	79.50	1,264.50	1,525	
	1130	8' deep		2	15	↓	1,025	470	119	1,614	1,975	
	1140	For depths over 8', add	↓	16	1.875	V.L.F.	143	58.50	14.85	216.35	264	
	1150	5' I.D., 4' deep	B-6	3	8	Ea.	720	231	72	1,023	1,225	
	1160	6' deep		2	12		970	345	108	1,423	1,725	
	1170	8' deep		1.50	16	↓	1,225	460	144	1,829	2,225	
	1180	For depths over 8', add		12	2	V.L.F.	159	57.50	18	234.50	283	
	1190	6' I.D., 4' deep		2	12	Ea.	1,175	345	108	1,628	1,950	
	1200	6' deep		1.50	16		1,525	460	144	2,129	2,550	
	1210	8' deep		1	24	↓	1,875	690	216	2,781	3,375	
	1220	For depths over 8', add	↓	8	3	V.L.F.	246	86.50	27	359.50	435	
	1250	Slab tops, precast, 8" thick										
	1300	4' diameter manhole	B-6	8	3	Ea.	165	86.50	27	278.50	345	
	1400	5' diameter manhole		7.50	3.200		325	92	29	446	535	
	1500	6' diameter manhole	↓	7	3.429		405	99	31	535	630	
	3800	Steps, heavyweight cast iron, 7" x 9"	1 Bric	40	.200		12.20	7.05		19.25	24	
	3900	8" x 9"		40	.200		18.35	7.05		25.40	31	
	3928	12" x 10-1/2"		40	.200		16.40	7.05		23.45	29	
	4000	Standard sizes, galvanized steel		40	.200		15.30	7.05		22.35	27.50	
	4100	Aluminum	↓	40	.200	↓	18.90	7.05		25.95	32	
510	0010	**PIPING, STORM DRAINAGE, CORRUGATED METAL**										510
	0020	Not including excavation or backfill										
	2000	Corrugated metal pipe, galvanized and coated										
	2020	Bituminous coated with paved invert, 20' lengths										
	2040	8" diameter, 16 ga.	B-14	330	.145	L.F.	8.95	4.09	.65	13.69	16.85	
	2060	10" diameter, 16 ga.		260	.185		10.75	5.20	.83	16.78	20.50	
	2080	12" diameter, 16 ga.		210	.229		12.85	6.40	1.03	20.28	25	
	2100	15" diameter, 16 ga.		200	.240		15.65	6.75	1.08	23.48	29	
	2120	18" diameter, 16 ga.		190	.253		20	7.10	1.14	28.24	34.50	
	2140	24" diameter, 14 ga.	↓	160	.300		24.50	8.45	1.35	34.30	41.50	
	2160	30" diameter, 14 ga.	B-13	120	.467		32.50	13.45	5.15	51.10	61.50	
	2180	36" diameter, 12 ga.	↓	120	.467	↓	47.50	13.45	5.15	66.10	78.50	

2 SITE CONSTRUCTION

02630 | Storm Drainage

		CREW	DAILY OUTPUT	LABOR-HOURS	UNIT	2005 BARE COSTS				TOTAL INCL O&P
						MAT.	LABOR	EQUIP.	TOTAL	
510	2200	48" diameter, 12 ga.	B-13	100	.560	L.F.	72.50	16.10	6.15	94.75	111	510
	2220	60" diameter, 10 ga.	B-13B	75	.747		93.50	21.50	11.95	126.95	149
	2240	72" diameter, 8 ga.	"	45	1.244	↓	140	36	19.90	195.90	231
	2250	End sections, 8" diameter, 16 ga.	B-14	20	2.400	Ea.	52.50	67.50	10.80	130.80	173
	2255	10" diameter, 16 ga.		20	2.400		59.50	67.50	10.80	137.80	181
	2260	12" diameter, 16 ga.		18	2.667		73.50	75	12	160.50	210
	2265	15" diameter, 16 ga.		18	2.667		91	75	12	178	229
	2270	18" diameter, 16 ga.	↓	16	3		115	84.50	13.50	213	271
	2275	24" diameter, 16 ga.	B-13	16	3.500		164	101	38.50	303.50	380
	2280	30" diameter, 16 ga.		14	4		310	115	44	469	570
	2285	36" diameter, 14 ga.		14	4		410	115	44	569	675
	2290	48" diameter, 14 ga.		10	5.600		850	161	61.50	1,072.50	1,250
	2292	60" diameter, 14 ga.	↓	6	9.333		1,175	269	103	1,547	1,800
	2294	72" diameter, 14 ga.	B-13B	5	11.200		2,025	320	179	2,524	2,925
	2300	Bends or elbows, 8" diameter	B-14	28	1.714		79.50	48	7.70	135.20	171
	2320	10" diameter		25	1.920		98.50	54	8.65	161.15	201
	2340	12" diameter, 16 ga.		30	1.600		114	45	7.20	166.20	202
	2342	18" diameter, 16 ga.		20	2.400		174	67.50	10.80	252.30	310
	2344	24" diameter, 14 ga.		16	3		243	84.50	13.50	341	410
	2346	30" diameter, 14 ga.	↓	15	3.200		300	90	14.40	404.40	485
	2348	36" diameter, 14 ga.	B-13	15	3.733		405	107	41	553	655
	2350	48" diameter, 12 ga.	"	12	4.667		580	134	51.50	765.50	905
	2352	60" diameter, 10 ga.	B-13B	10	5.600		780	161	89.50	1,030.50	1,200
	2354	72" diameter, 10 ga.	"	6	9.333		965	269	149	1,383	1,650
	2360	Wyes or tees, 8" diameter	B-14	25	1.920		111	54	8.65	173.65	215
	2380	10" diameter		21	2.286		138	64	10.30	212.30	263
	2400	12" diameter, 16 ga.		22.48	2.135		115	60	9.60	184.60	231
	2410	18" diameter, 16 ga.		16	3		232	84.50	13.50	330	400
	2412	24" diameter, 14 ga.	↓	16	3		350	84.50	13.50	448	530
	2414	30" diameter, 14 ga.	B-13	12	4.667		455	134	51.50	640.50	765
	2416	36" diameter, 14 ga.		11	5.091		560	147	56	763	905
	2418	48" diameter, 12 ga.	↓	10	5.600		845	161	61.50	1,067.50	1,250
	2420	60" diameter, 10 ga.	B-13B	8	7		1,225	202	112	1,539	1,775
	2422	72" diameter, 10 ga.	"	5	11.200	↓	1,475	320	179	1,974	2,325
	2500	Galvanized, uncoated, 20' lengths									
	2520	8" diameter, 16 ga.	B-14	355	.135	L.F.	6.90	3.80	.61	11.31	14.10
	2540	10" diameter, 16 ga.		280	.171		7.60	4.82	.77	13.19	16.65
	2560	12" diameter, 16 ga.		220	.218		8.65	6.15	.98	15.78	20
	2580	15" diameter, 16 ga.		220	.218		11	6.15	.98	18.13	22.50
	2600	18" diameter, 16 ga.		205	.234		14.55	6.60	1.05	22.20	27.50
	2620	24" diameter, 14 ga.	↓	175	.274		21	7.70	1.23	29.93	36.50
	2640	30" diameter, 14 ga.	B-13	130	.431		27	12.40	4.74	44.14	54
	2660	36" diameter, 12 ga.		130	.431		44	12.40	4.74	61.14	73
	2680	48" diameter, 12 ga.	↓	110	.509		59	14.65	5.60	79.25	93.50
	2690	60" diameter, 10 ga.	B-13B	78	.718		93	20.50	11.50	125	147
	2695	72" diameter, 10 ga.	"	60	.933	↓	117	27	14.90	158.90	187
	2711	Bends or elbows, 12" diameter, 16 ga.	B-14	30	1.600	Ea.	104	45	7.20	156.20	192
	2712	15" diameter, 16 ga.		25.04	1.917		128	54	8.65	190.65	234
	2714	18" diameter, 16 ga.		20	2.400		144	67.50	10.80	222.30	275
	2716	24" diameter, 14 ga.		16	3		210	84.50	13.50	308	375
	2718	30" diameter, 14 ga.	↓	15	3.200		263	90	14.40	367.40	445
	2720	36" diameter, 14 ga.	B-13	15	3.733		360	107	41	508	605
	2722	48" diameter, 12 ga.		12	4.667		495	134	51.50	680.50	810
	2724	60" diameter, 10 ga.		10	5.600		760	161	61.50	982.50	1,150
	2726	72" diameter, 10 ga.	↓	6	9.333		940	269	103	1,312	1,550
	2728	Wyes or tees, 12" diameter, 16 ga.	B-14	22.48	2.135	↓	138	60	9.60	207.60	256

2

SITE CONSTRUCTION

02630 | Storm Drainage

		CREW	DAILY OUTPUT	LABOR-HOURS	UNIT	MAT.	LABOR	EQUIP.	TOTAL	TOTAL INCL O&P	
510	2730	18" diameter, 16 ga.	B-14	15	3.200	Ea.	201	90	14.40	305.40	375	510
	2732	24" diameter, 14 ga.		15	3.200		310	90	14.40	414.40	500	
	2734	30" diameter, 14 ga.		14	3.429		410	96.50	15.45	521.95	615	
	2736	36" diameter, 14 ga.	B-13	14	4		530	115	44	689	810	
	2738	48" diameter, 12 ga.		12	4.667		780	134	51.50	965.50	1,125	
	2740	60" diameter, 10 ga.		10	5.600		1,150	161	61.50	1,372.50	1,575	
	2742	72" diameter, 10 ga.		6	9.333		1,350	269	103	1,722	2,000	
	2780	End sections, 8" diameter	B-14	24	2		71.50	56	9	136.50	175	
	2785	10" diameter		22	2.182		73	61.50	9.80	144.30	186	
	2790	12" diameter		35	1.371		82.50	38.50	6.15	127.15	157	
	2800	18" diameter		30	1.600		98	45	7.20	150.20	185	
	2810	24" diameter	B-13	25	2.240		143	64.50	24.50	232	284	
	2820	30" diameter		25	2.240		272	64.50	24.50	361	425	
	2825	36" diameter		20	2.800		360	80.50	31	471.50	555	
	2830	48" diameter		10	5.600		745	161	61.50	967.50	1,125	
	2835	60" diameter	B-13B	5	11.200		970	320	179	1,469	1,775	
	2840	72" diameter	"	4	14		1,750	405	224	2,379	2,800	
	2850	Couplings, 12" diameter					16			16	17.60	
	2855	18" diameter					23.50			23.50	25.50	
	2860	24" diameter					31.50			31.50	34.50	
	2865	30" diameter					38.50			38.50	42.50	
	2870	36" diameter					54			54	59.50	
	2875	48" diameter					73.50			73.50	81	
	2880	60" diameter					116			116	127	
	2885	72" diameter					160			160	176	
520	0010	**PIPING, DRAINAGE & SEWAGE, CORRUGATED HDPE TYPE S**										520
	0020	Not including excavation & backfill, bell & spigot										
	1000	With gaskets, 4" diameter	B-20	425	.056	L.F.	.72	1.70		2.42	3.43	
	1010	6" diameter		400	.060		1.66	1.80		3.46	4.64	
	1020	8" diameter		380	.063		3.18	1.90		5.08	6.45	
	1030	10" diameter		370	.065		4.39	1.95		6.34	7.85	
	1040	12" diameter		340	.071		4.90	2.12		7.02	8.70	
	1050	15" diameter		300	.080		6.65	2.41		9.06	11.10	
	1060	18" diameter	B-21	275	.102		9.50	3.15	.58	13.23	15.95	
	1070	24" diameter		250	.112		14.70	3.46	.63	18.79	22.50	
	1080	30" diameter		200	.140		23.50	4.33	.79	28.62	33	
	1090	36" diameter		180	.156		29.50	4.81	.88	35.19	41	
	1100	42" diameter		175	.160		41.50	4.95	.91	47.36	54	
	1110	48" diameter		170	.165		53.50	5.10	.93	59.53	68	
	1120	54" diameter		160	.175		82.50	5.40	.99	88.89	100	
	1130	60" diameter		150	.187		96.50	5.75	1.06	103.31	116	
	1135	Add 15% to material pipe cost for water tight connection bell & spigot										
	1140	HDPE type s, elbows 12" diam	B-20	11	2.182	Ea.	54	65.50		119.50	162	
	1150	HDPE type s, elbows 15" diam	"	9	2.667		83.50	80		163.50	217	
	1160	HDPE type s, elbows 18" diam	B-21	9	3.111		138	96	17.60	251.60	320	
	1170	HDPE type s, elbows 24" diam		9	3.111		293	96	17.60	406.60	490	
	1180	HDPE type s, elbows 30" diam		8	3.500		470	108	19.80	597.80	705	
	1190	HDPE type s, elbows 36" diam		8	3.500		600	108	19.80	727.80	850	
	1240	HDPE type s, Tee 12" diam	B-20	7	3.429		122	103		225	295	
	1260	HDPE type s, Tee, 15" diam	"	6	4		145	120		265	345	
	1280	HDPE type s, Tee, 18" diam	B-21	6	4.667		212	144	26.50	382.50	485	
	1300	HDPE type s, Tee, 24" diam		5	5.600		279	173	31.50	483.50	610	
	1320	HDPE type s, Tee, 30" diam		5	5.600		525	173	31.50	729.50	885	
	1400	Add to basic installation cost for each split coupling joint										
	1402	HDPE type s, split coupling, 12" diam	B-20	17	1.412	Ea.	3.95	42.50		46.45	70.50	

SITE CONSTRUCTION 2

SITE CONSTRUCTION 2

		02630	Storm Drainage	CREW	DAILY OUTPUT	LABOR-HOURS	UNIT	2005 BARE COSTS				TOTAL INCL O&P	
								MAT.	LABOR	EQUIP.	TOTAL		
520	1420		HDPE type s, split coupling, 15" diam	B-20	15	1.600	Ea.	5.25	48		53.25	81	**520**
530	0010	**SEWAGE/DRAINAGE COLLECTION, CONCRETE PIPE**	R02510 -810										**530**
	0020	Not including excavation or backfill											
	0050	Box culvert, cast in place, 6' x 6'		C-15	16	4.500	L.F.	103	144		247	335	
	0060	8' x 8'			14	5.143		151	165		316	420	
	0070	12' x 12'			10	7.200		296	231		527	680	
	0100	Box culvert, precast, base price, 8' long, 6' x 3'		B-69	140	.343		160	10	7.10	177.10	199	
	0150	6' x 7'			125	.384		242	11.20	7.95	261.15	293	
	0200	8' x 3'			133	.361		221	10.50	7.50	239	267	
	0250	8' x 8'			100	.480		298	14	9.95	321.95	360	
	0300	10' x 3'			110	.436		325	12.70	9.05	346.75	390	
	0350	10' x 8'			80	.600		370	17.50	12.45	399.95	450	
	0400	12' x 3'			100	.480		320	14	9.95	343.95	380	
	0450	12' x 8'			67	.716		545	21	14.85	580.85	650	
	0500	Set up charge at plant, add to base price					Job	3,050			3,050	3,350	
	0510	Inserts and keyway, add					Ea.	268			268	294	
	0520	Sloped or skewed end, add					"	425			425	470	
	1000	Non-reinforced pipe, extra strength, B&S or T&G joints											
	1010	6" diameter		B-14	265.04	.181	L.F.	4.13	5.10	.82	10.05	13.30	
	1020	8" diameter			224	.214		4.54	6	.96	11.50	15.35	
	1040	12" diameter			200	.240		6.20	6.75	1.08	14.03	18.45	
	1050	15" diameter			180	.267		7.25	7.50	1.20	15.95	21	
	1060	18" diameter			144	.333		8.90	9.35	1.50	19.75	26	
	1070	21" diameter			112	.429		10.95	12.05	1.93	24.93	33	
	1080	24" diameter			100	.480		13.40	13.50	2.16	29.06	38	
	1560	Reinforced culvert, class 2, no gaskets											
	1590	27" diameter		B-21	88	.318	L.F.	27	9.85	1.80	38.65	46.50	
	1592	30" diameter		B-13	80	.700		29	20	7.70	56.70	71.50	
	1594	36" diameter		"	72	.778		34	22.50	8.55	65.05	81.50	
	2000	Reinforced culvert, class 3, no gaskets											
	2010	12" diameter		B-14	150	.320	L.F.	11.20	9	1.44	21.64	28	
	2020	15" diameter			150	.320		14.35	9	1.44	24.79	31.50	
	2030	18" diameter			132	.364		15.10	10.20	1.64	26.94	34	
	2035	21" diameter			120	.400		19.60	11.25	1.80	32.65	41	
	2040	24" diameter			100	.480		22.50	13.50	2.16	38.16	48	
	2045	27" diameter		B-13	92	.609		28	17.50	6.70	52.20	65.50	
	2050	30" diameter			88	.636		30.50	18.30	7	55.80	69	
	2060	36" diameter			72	.778		43.50	22.50	8.55	74.55	92	
	2070	42" diameter		B-13B	72	.778		59.50	22.50	12.45	94.45	114	
	2080	48" diameter			64	.875		73.50	25	14	112.50	135	
	2090	60" diameter			48	1.167		118	33.50	18.65	170.15	201	
	2100	72" diameter			40	1.400		166	40.50	22.50	229	270	
	2120	84" diameter			32	1.750		281	50.50	28	359.50	420	
	2140	96" diameter			24	2.333		335	67	37.50	439.50	515	
	2200	With gaskets, class 3, 12" diameter		B-21	168	.167		12.25	5.15	.94	18.34	22.50	
	2220	15" diameter			160	.175		14.70	5.40	.99	21.09	25.50	
	2230	18" diameter			152	.184		18.40	5.70	1.04	25.14	30.50	
	2240	24" diameter			136	.206		27.50	6.35	1.17	35.02	41.50	
	2260	30" diameter		B-13	88	.636		37	18.30	7	62.30	76	
	2270	36" diameter		"	72	.778		55	22.50	8.55	86.05	104	
	2290	48" diameter		B-13B	64	.875		90	25	14	129	153	
	2310	72" diameter		"	40	1.400		233	40.50	22.50	296	345	
	2330	Flared ends, 6'-1" long, 12" diameter		B-21	190	.147		34	4.56	.83	39.39	45	
	2340	15" diameter			155	.181		38.50	5.60	1.02	45.12	52.50	
	2400	6'-2" long, 18" diameter			122	.230		40	7.10	1.30	48.40	56.50	

Important: See the Reference Section for critical supporting data - Reference Nos., Crews, & City Cost Indexes

02630	Storm Drainage		CREW	DAILY OUTPUT	LABOR-HOURS	UNIT	2005 BARE COSTS				TOTAL INCL O&P	
							MAT.	LABOR	EQUIP.	TOTAL		
530	2420	24" diameter		B-21	88	.318	L.F.	46.50	9.85	1.80	58.15	68	**530**
	2440	36" diameter	R02510 -810	B-13	60	.933	↓	84.50	27	10.30	121.80	145	
	2500	Class 4											
	2510	12" diameter		B-21	168	.167	L.F.	13.20	5.15	.94	19.29	23.50	
	2512	15" diameter			160	.175		15.60	5.40	.99	21.99	26.50	
	2514	18" diameter			152	.184		18.45	5.70	1.04	25.19	30.50	
	2516	21" diameter			144	.194		23	6	1.10	30.10	35.50	
	2518	24" diameter			136	.206		27	6.35	1.17	34.52	40.50	
	2520	27" diameter		↓	120	.233		33.50	7.20	1.32	42.02	49.50	
	2522	30" diameter		B-13	88	.636		37	18.30	7	62.30	76	
	2524	36" diameter		"	72	.778	↓	55	22.50	8.55	86.05	104	
	2600	Class 5											
	2610	12" diameter		B-21	168	.167	L.F.	13.45	5.15	.94	19.54	24	
	2612	15" diameter			160	.175		15.85	5.40	.99	22.24	27	
	2614	18" diameter			152	.184		18.90	5.70	1.04	25.64	31	
	2616	21" diameter			144	.194		23.50	6	1.10	30.60	36.50	
	2618	24" diameter			136	.206		27.50	6.35	1.17	35.02	41.50	
	2620	27" diameter		↓	120	.233		35.50	7.20	1.32	44.02	51.50	
	2622	30" diameter		B-13	88	.636		39	18.30	7	64.30	78.50	
	2624	36" diameter		"	72	.778		54.50	22.50	8.55	85.55	104	
	2800	Add for rubber joints,						12%					
	3080	Radius pipe, add to pipe prices, 12" to 60" diameter						50%					
	3090	Over 60" diameter, add					↓	20%					
	3500	Reinforced elliptical, 8' lengths, C507 class 3											
	3520	14" x 23" inside, round equivalent 18" diameter		B-21	82	.341	L.F.	26.50	10.55	1.93	38.98	47.50	
	3530	24" x 38" inside, round equivalent 30" diameter		B-13	58	.966		47	28	10.65	85.65	106	
	3540	29" x 45" inside, round equivalent 36" diameter			52	1.077		60.50	31	11.85	103.35	128	
	3550	38" x 60" inside, round equivalent 48" diameter			38	1.474		79	42.50	16.25	137.75	170	
	3560	48" x 76" inside, round equivalent 60" diameter			26	2.154		120	62	23.50	205.50	254	
	3570	58" x 91" inside, round equivalent 72" diameter		↓	22	2.545	↓	171	73.50	28	272.50	335	
	3780	Concrete slotted pipe, class 4 mortar joint											
	3800	12" diameter		B-21	168	.167	L.F.	13.15	5.15	.94	19.24	23.50	
	3840	18" diameter		"	152	.184	"	20.50	5.70	1.04	27.24	32.50	
	3900	Class 4 O-ring											
	3940	12" diameter		B-21	168	.167	L.F.	13.75	5.15	.94	19.84	24	
	3960	18" diameter		"	152	.184	"	18.45	5.70	1.04	25.19	30.50	
	6200	Gasket, conc. pipe joint, 12"					Ea.	3.16			3.16	3.48	
	6220	24"						5.80			5.80	6.35	
	6240	36"						8.40			8.40	9.25	
	6260	48"						13.25			13.25	14.60	
	6270	60"						18.90			18.90	21	
	6280	72"		↓			↓	21			21	23	

02640	Culverts/Manuf Const											
810	0010	**OVAL ARCH CULVERTS**											**810**
	3000	Corrugated galvanized or aluminum, coated & paved											
	3020	17" x 13", 16 ga., 15" equivalent		B-14	200	.240	L.F.	26.50	6.75	1.08	34.33	40.50	
	3040	21" x 15", 16 ga., 18" equivalent			150	.320		34	9	1.44	44.44	53	
	3060	28" x 20", 14 ga., 24" equivalent			125	.384		49	10.80	1.73	61.53	72	
	3080	35" x 24", 14 ga., 30" equivalent		↓	100	.480		59.50	13.50	2.16	75.16	89	
	3100	42" x 29", 12 ga., 36" equivalent		B-13	100	.560		88.50	16.10	6.15	110.75	129	
	3120	49" x 33", 12 ga., 42" equivalent			90	.622		107	17.90	6.85	131.75	152	
	3140	57" x 38", 12 ga., 48" equivalent		↓	75	.747	↓	119	21.50	8.20	148.70	173	
	3160	Steel, plain oval arch culverts, plain											
	3180	17" x 13", 16 ga., 15" equivalent		B-14	225	.213	L.F.	14.30	6	.96	21.26	26	
	3200	21" x 15", 16 ga., 18" equivalent		↓	175	.274	↓	16.95	7.70	1.23	25.88	32	

SITE CONSTRUCTION 2

101

02640 | Culverts/Manuf Const

		CREW	DAILY OUTPUT	LABOR-HOURS	UNIT	MAT.	2005 BARE COSTS LABOR	EQUIP.	TOTAL	TOTAL INCL O&P	
810	3220	28" x 20", 14 ga., 24" equivalent	B-14	150	.320	L.F.	27	9	1.44	37.44	45.50	810
	3240	35" x 24", 14 ga., 30" equivalent	B-13	108	.519		34	14.95	5.70	54.65	67	
	3260	42" x 29", 12 ga., 36" equivalent		108	.519		56.50	14.95	5.70	77.15	91.50	
	3280	49" x 33", 12 ga., 42" equivalent		92	.609		66.50	17.50	6.70	90.70	108	
	3300	57" x 38", 12 ga., 48" equivalent		75	.747	▼	58.50	21.50	8.20	88.20	107	
	3320	End sections, 17" x 13"		22	2.545	Ea.	78.50	73.50	28	180	230	
	3340	42" x 29"	▼	17	3.294	"	320	95	36.50	451.50	540	
	3360	Multi-plate arch, steel	B-20	1,690	.014	Lb.	.90	.43		1.33	1.65	

02660 | Ponds & Reservoirs

		CREW	DAILY OUTPUT	LABOR-HOURS	UNIT	MAT.	2005 BARE COSTS LABOR	EQUIP.	TOTAL	TOTAL INCL O&P	
600	0010	**SLURRY TRENCH** Excavated slurry trench in wet soils										600
	0020	backfilled with 3000 PSI concrete, no reinforcing steel										
	0050	Minimum	C-7	333	.216	C.F.	6	6.25	2.88	15.13	19.30	
	0100	Maximum		200	.360	"	10.05	10.40	4.80	25.25	32.50	
	0200	Alternate pricing method, minimum		150	.480	S.F.	12	13.85	6.40	32.25	41.50	
	0300	Maximum	▼	120	.600		18	17.30	8	43.30	55	
	0500	Reinforced slurry trench, minimum	B-48	177	.316		9	9.40	19.30	37.70	45.50	
	0600	Maximum	"	69	.812	▼	30	24	49.50	103.50	125	
	0800	Haul for disposal, 2 mile haul, excavated material, add	B-34B	99	.081	C.Y.		2.23	4.81	7.04	8.70	
	0900	Haul bentonite castings for disposal, add	"	40	.200	"		5.50	11.90	17.40	21.50	
610	0010	**POND AND RESERVOIR LINERS, HDPE,** 100,000 S.F. or more										610
	1100	30 mil thick	3 Skwk	1,850	.013	S.F.	.27	.45		.72	1	
	1200	60 mil thick		1,600	.015		.55	.52		1.07	1.42	
	1300	120 mil thick	▼	1,440	.017	▼	1.24	.58		1.82	2.26	

02700 | Bases, Ballasts, Pavements & Appurtenances

02710 | Bound Base Courses

		CREW	DAILY OUTPUT	LABOR-HOURS	UNIT	MAT.	2005 BARE COSTS LABOR	EQUIP.	TOTAL	TOTAL INCL O&P	
100	0010	**AGGREGATE-BITUMINOUS BASE COURSE** for roadways R02065-300										100
	0020	and large paved areas										
	0500	Bituminous concrete, 4" thick	B-25	4,545	.019	S.Y.	7.75	.56	.44	8.75	9.90	
	0550	6" thick		3,700	.024		11.35	.69	.54	12.58	14.15	
	0560	8" thick		3,000	.029		15.15	.85	.67	16.67	18.70	
	0570	10" thick	▼	2,545	.035	▼	18.80	1	.79	20.59	23	
	8900	For small and irregular areas, add	▼					50%	50%			
200	0010	**ASPHALT-TREATED PERMEABLE BASE COURSE** for roadways R02065-300										200
	0020	and large paved areas										
	0700	Liquid application to gravel base, asphalt emulsion	B-45	6,000	.003	Gal.	2.92	.08	.09	3.09	3.44	
	0800	Prime and seal, cut back asphalt		6,000	.003	"	3.45	.08	.09	3.62	4.02	
	1000	Macadam penetration crushed stone, 2 gal. per S.Y., 4" thick		6,000	.003	S.Y.	5.85	.08	.09	6.02	6.65	
	1100	6" thick, 3 gal. per S.Y.		4,000	.004		8.75	.12	.13	9	10	
	1200	8" thick, 4 gal. per S.Y.	▼	3,000	.005	▼	11.70	.17	.18	12.05	13.30	
	8900	For small and irregular areas, add	▼					50%	50%			

02720 | Unbound Base Courses & Ballasts

		CREW	DAILY OUTPUT	LABOR-HOURS	UNIT	MAT.	2005 BARE COSTS LABOR	EQUIP.	TOTAL	TOTAL INCL O&P	
200	0010	**AGGREGATE BASE COURSE** For roadways and large areas R02065-300										200
	0050	Crushed 3/4" stone base, compacted, 3" deep	B-36C	5,200	.008	S.Y.	2.65	.25	.49	3.39	3.83	
	0100	6" deep		5,000	.008		5.30	.26	.51	6.07	6.80	
	0200	9" deep	▼	4,600	.009	▼	7.95	.28	.56	8.79	9.80	

Important: See the Reference Section for critical supporting data - Reference Nos., Crews, & City Cost Indexes

02720 | Unbound Base Courses & Ballasts

			CREW	DAILY OUTPUT	LABOR-HOURS	UNIT	2005 BARE COSTS				TOTAL INCL O&P	
							MAT.	LABOR	EQUIP.	TOTAL		
200	0300	12" deep	B-36C	4,200	.010	S.Y.	10.65	.31	.61	11.57	12.85	200
	0301	Crushed 1-1/2" stone base, compacted to 4" deep	B-36B	6,000	.011		4.17	.33	.49	4.99	5.65	
	0302	6" deep		5,400	.012		6.25	.37	.54	7.16	8.05	
	0303	8" deep		4,500	.014		8.35	.44	.65	9.44	10.60	
	0304	12" deep		3,800	.017		12.50	.52	.77	13.79	15.40	
	0350	Bank run gravel, spread and compacted										
	0370	6" deep	B-32	6,000	.005	S.Y.	3.47	.17	.26	3.90	4.36	
	0390	9" deep		4,900	.007		5.20	.21	.32	5.73	6.35	
	0400	12" deep		4,200	.008		6.95	.25	.37	7.57	8.40	
	0600	Cold laid asphalt pavement, see div. 02740-400										
	1500	Alternate method to figure base course										
	1510	Crushed stone, 3/4", compacted, 3" deep	B-36C	435	.092	E.C.Y.	27.50	2.95	5.85	36.30	41	
	1511	6" deep	B-36B	835	.077		27.50	2.38	3.50	33.38	37.50	
	1512	9" deep		1,150	.056		27.50	1.73	2.54	31.77	35.50	
	1513	12" deep		1,400	.046		27.50	1.42	2.09	31.01	34.50	
	1520	Crushed stone, 1-1/2", compacted 4" deep		665	.096		27.50	2.99	4.40	34.89	39.50	
	1521	6" deep		900	.071		27.50	2.21	3.25	32.96	37	
	1522	8" deep		1,000	.064		27.50	1.99	2.92	32.41	36.50	
	1523	12" deep		1,265	.051		27.50	1.57	2.31	31.38	35	
	1530	Gravel, bank run, compacted, 6" deep	B-36C	835	.048		17.80	1.53	3.06	22.39	25.50	
	1531	9" deep		1,150	.035		17.80	1.11	2.22	21.13	23.50	
	1532	12" deep		1,400	.029		17.80	.92	1.83	20.55	23	
	2000	Alternate method to figure base course										
	2005	Bituminous concrete, 4" thick	B-25	1,000	.088	Ton	34.50	2.56	2.01	39.07	44	
	2006	6" thick		1,220	.072		34.50	2.10	1.65	38.25	43	
	2007	8" thick		1,320	.067		34.50	1.94	1.52	37.96	42.50	
	2008	10" thick		1,400	.063		34.50	1.83	1.44	37.77	42.50	
	2010	Crushed stone, 3/4" maximum size, 3" deep	B-36	540	.074		18.10	2.24	2.10	22.44	25.50	
	2011	6" deep		1,625	.025		16.45	.75	.70	17.90	20	
	2012	9" deep		1,785	.022		16.45	.68	.64	17.77	19.85	
	2013	12" deep		1,950	.021		16.45	.62	.58	17.65	19.70	
	2020	Crushed stone, 1-1/2" maximum size, 4" deep		720	.056		24	1.68	1.58	27.26	30.50	
	2021	6" deep		815	.049		24	1.49	1.39	26.88	30	
	2022	8" deep		835	.048		24	1.45	1.36	26.81	29.50	
	2023	12" deep		975	.041		24	1.24	1.16	26.40	29	
	2030	Bank run gravel, 6" deep	B-32A	875	.027		11.95	.88	1.02	13.85	15.60	
	2031	9" deep		970	.025		11.95	.79	.92	13.66	15.35	
	2032	12" deep		1,060	.023		11.95	.72	.84	13.51	15.15	
	6000	Stabilization fabric, polypropylene, 6 oz./S.Y.	B-6	10,000	.002	S.Y.	.91	.07	.02	1	1.13	
	6900	For small and irregular areas, add						50%	50%			
	7000	Prepare and roll sub-base, small areas to 2500 S.Y.	B-32A	1,500	.016	S.Y.		.51	.59	1.10	1.43	
	8000	Large areas over 2500 S.Y.	B-32	3,700	.009	"		.28	.42	.70	.89	

Codes in table: R02065-300

02740 | Flexible Pavement

			CREW	DAILY OUTPUT	LABOR-HOURS	UNIT	2005 BARE COSTS				TOTAL INCL O&P	
							MAT.	LABOR	EQUIP.	TOTAL		
310	0010	**ASPHALTIC CONCRETE PAVEMENT, HIGHWAYS**										310
	0020	and large paved areas										
	0080	Binder course, 1-1/2" thick	B-25	7,725	.011	S.Y.	2.63	.33	.26	3.22	3.69	
	0120	2" thick		6,345	.014		3.50	.40	.32	4.22	4.82	
	0160	3" thick		4,905	.018		5.20	.52	.41	6.13	6.95	
	0200	4" thick		4,140	.021		6.95	.62	.49	8.06	9.15	
	0300	Wearing course, 1" thick	B-25B	10,575	.009		1.86	.27	.21	2.34	2.68	
	0340	1-1/2" thick		7,725	.012		2.83	.37	.28	3.48	3.98	
	0380	2" thick		6,345	.015		3.80	.45	.35	4.60	5.25	
	0420	2-1/2" thick		5,480	.018		4.69	.52	.40	5.61	6.40	
	0460	3" thick		4,900	.020		5.60	.58	.45	6.63	7.55	
	0500	Open graded friction course	B-25C	5,000	.010		1.66	.28	.35	2.29	2.66	

Codes in table: R02065-300, R02700-120

SITE CONSTRUCTION 2

02740 | Flexible Pavement

			CREW	DAILY OUTPUT	LABOR-HOURS	UNIT	MAT.	LABOR	EQUIP.	TOTAL	TOTAL INCL O&P	
							2005 BARE COSTS					
310	0800	Alternate method of figuring paving costs										310
	0810	Binder course, 1-1/2" thick	B-25	630	.140	Ton	34.50	4.06	3.19	41.75	48	
	0811	2" thick		690	.128		34.50	3.71	2.91	41.12	47	
	0812	3" thick		800	.110		34.50	3.20	2.51	40.21	45.50	
	0813	4" thick		850	.104		34.50	3.01	2.37	39.88	45	
	0850	Wearing course, 1" thick	B-25B	575	.167		34.50	4.93	3.81	43.24	49.50	
	0851	1-1/2" thick		630	.152		34.50	4.50	3.48	42.48	48.50	
	0852	2" thick		690	.139		34.50	4.11	3.17	41.78	48	
	0853	2-1/2" thick		745	.129		34.50	3.80	2.94	41.24	47	
	0854	3" thick		800	.120		34.50	3.54	2.74	40.78	46.50	
	1000	Pavement replacement over trench, 2" thick	B-37	90	.533	S.Y.	3.87	15	1.19	20.06	28.50	
	1050	4" thick		70	.686		7.65	19.25	1.54	28.44	40	
	1080	6" thick		55	.873		12.20	24.50	1.95	38.65	53.50	
	1200	For paving projects 300 tons or less add for trucking										
	1300	See 02315-490-0010 for hauling costs										
315	0011	**ASPHALTIC CONCRETE PAVEMENT, LOTS & DRIVEWAYS**										315
	0020	6" stone base, 2" binder course, 1" topping	B-25C	9,000	.005	S.F.	1.38	.16	.20	1.74	1.98	
	0300	Binder course, 1-1/2" thick		35,000	.001		.31	.04	.05	.40	.46	
	0400	2" thick		25,000	.002		.41	.06	.07	.54	.62	
	0500	3" thick		15,000	.003		.63	.10	.12	.85	.97	
	0600	4" thick		10,800	.004		.82	.13	.16	1.11	1.28	
	0800	Sand finish course, 3/4" thick		41,000	.001		.19	.03	.04	.26	.31	
	0900	1" thick		34,000	.001		.23	.04	.05	.32	.38	
	1000	Fill pot holes, hot mix, 2" thick	B-16	4,200	.008		.42	.21	.11	.74	.90	
	1100	4" thick		3,500	.009		.61	.25	.14	1	1.21	
	1120	6" thick		3,100	.010		.82	.28	.15	1.25	1.51	
	1140	Cold patch, 2" thick	B-51	3,000	.016		.50	.43	.04	.97	1.27	
	1160	4" thick		2,700	.018		.95	.48	.05	1.48	1.85	
	1180	6" thick		1,900	.025		1.48	.68	.07	2.23	2.76	
400	0010	**COLD MIXED BITUMINOUS PAVEMENT** 0.5 gal. asphalt/S.Y. per in. depth										400
	0020	Well graded granular aggregate										
	0100	Blade mixed in windrows, spread & compacted 4" course	B-90A	1,600	.035	S.Y.	4.57	1.10	.83	6.50	7.65	
	0200	Traveling plant mixed in windrows, compacted 4" course	B-90B	3,000	.016		4.57	.50	.66	5.73	6.55	
	0300	Rotary plant mixed in place, compacted 4" course	"	3,500	.014		4.57	.43	.56	5.56	6.30	
	0400	Central stationary plant, mixed, compacted 4" course	B-36	7,200	.006		9.15	.17	.16	9.48	10.50	

02750 | Rigid Pavement

			CREW	DAILY OUTPUT	LABOR-HOURS	UNIT	MAT.	LABOR	EQUIP.	TOTAL	TOTAL INCL O&P	
300	0010	**PLAIN CEMENT CONCRETE PAVEMENT**										300
	0015	Including joints, finishing and curing										
	0020	Fixed form, 12' pass, unreinforced, 6" thick	B-26	3,000	.029	S.Y.	19.80	.88	.79	21.47	24	
	0030	7" thick		2,850	.031		25	.92	.83	26.75	30	
	0100	8" thick		2,750	.032		27.50	.96	.86	29.32	33	
	0110	8" thick, small area		1,375	.064		27.50	1.91	1.72	31.13	35.50	
	0200	9" thick		2,500	.035		32	1.05	.95	34	37.50	
	0300	10" thick		2,100	.042		34.50	1.25	1.13	36.88	41	
	0310	10" thick, small area		1,050	.084		34.50	2.51	2.25	39.26	44.50	
	0400	12" thick		1,800	.049		37	1.46	1.31	39.77	44	
	0500	15" thick		1,500	.059		41.50	1.75	1.58	44.83	50	
	0600	For continuous welded steel reinforcement over 10' wide, add									3.85	4.55	
	0610	For under 10' pass, add									7.30	8.55	
	0650	For 24' pass, deduct						20%	20%	20%		
	0655	For slipforming, deduct						10%	10%	10%		
	0700	Finishing, broom finish small areas	2 Cefi	120	.133	S.Y.		4.38		4.38	6.45	
	0710	Transverse joint support dowels	C-1	350	.091	Ea.	3	2.96		5.96	7.90	
	0720	Transverse contraction joints, saw cut & grind	A-1B	120	.067	L.F.		1.78	.95	2.73	3.82	

Reference boxes: R02065-300, R02700-120

Important: See the Reference Section for critical supporting data - Reference Nos., Crews, & City Cost Indexes

02700 | Bases, Ballasts, Pavements & Appurtenances

02750 | Rigid Pavement

		CREW	DAILY OUTPUT	LABOR-HOURS	UNIT	MAT.	LABOR	EQUIP.	TOTAL	TOTAL INCL O&P	
300	0730 Transverse expansion joints, incl. premolded bit. jt. filler	C-1	150	.213	L.F.	1.45	6.90		8.35	12.35	300
	0740 Transverse construction joint using bulkhead	"	73	.438	↓	2.10	14.20		16.30	24.50	
	0750 Longitudinal joint tie bars, grouted	B-23	70	.571	Ea.	3.26	15.50	48	66.76	80	
	1000 Curing, with sprayed membrane by hand	2 Clab	1,500	.011	S.Y.	.47	.28		.75	.96	
	1650 For integral coloring, see div. 03310-220										
	3200 Concrete grooving, continuous for roadways	B-71	700	.080	S.Y.		2.43	7.05	9.48	11.50	

02760 | Paving Specialties

		CREW	DAILY OUTPUT	LABOR-HOURS	UNIT	MAT.	LABOR	EQUIP.	TOTAL	TOTAL INCL O&P	
300	0010 **PAINTED TRAFFIC LINES AND MARKINGS**										300
	0020 Acrylic waterborne, white or yellow, 4" wide	B-78	20,000	.002	L.F.	.15	.06	.02	.23	.29	
	0200 6" wide		11,000	.004		.14	.12	.04	.30	.37	
	0500 8" wide		10,000	.005		.20	.13	.04	.37	.45	
	0600 12" wide		4,000	.012	↓	.36	.32	.10	.78	1.01	
	0620 Arrows or gore lines		2,300	.021	S.F.	.59	.56	.17	1.32	1.72	
	0640 Temporary paint, white or yellow	↓	15,000	.003	L.F.	.18	.09	.03	.30	.36	
	0660 Removal	1 Clab	300	.027			.71		.71	1.11	
	0710 Thermoplastic, white or yellow, 4" wide	B-79	15,000	.003		.67	.07	.06	.80	.91	
	0730 6" wide		14,000	.003		.97	.08	.06	1.11	1.26	
	0740 8" wide		12,000	.003		1.31	.09	.07	1.47	1.66	
	0750 12" wide		6,000	.007	↓	1.94	.18	.15	2.27	2.58	
	0760 Arrows		660	.061	S.F.	1.80	1.64	1.34	4.78	6	
	0770 Gore lines		2,500	.016		1.19	.43	.35	1.97	2.37	
	0780 Letters	↓	660	.061		1.49	1.64	1.34	4.47	5.65	
	0782 Thermoplastic material				Ton	955			955	1,050	
	0784 Glass beads, add				M.L.F.	8.95			8.95	9.85	
500	0010 **PAVEMENT MARKINGS**										500
	0790 Layout of pavement marking	A-2	25,000	.001	L.F.		.03	.01	.04	.05	
	0800 Parking stall, paint, white	B-78	440	.109	Stall	3.18	2.95	.88	7.01	9.05	
	1000 Street letters and numbers	"	1,600	.030	S.F.	.61	.81	.24	1.66	2.20	
	1100 Pavement marking letter, 6"	2 Pord	400	.040	Ea.	1.22	1.22		2.44	3.18	
	1110 12" letter		272	.059		1.78	1.80		3.58	4.67	
	1120 24" letter		160	.100		2.98	3.06		6.04	7.90	
	1130 36" letter		84	.190		6.05	5.85		11.90	15.40	
	1140 42" letter		84	.190		14.60	5.85		20.45	25	
	1150 72" letter		40	.400		41	12.25		53.25	63.50	
	1200 Handicap symbol	↓	40	.400	↓	25.50	12.25		37.75	46.50	

02770 | Curbs and Gutters

		CREW	DAILY OUTPUT	LABOR-HOURS	UNIT	MAT.	LABOR	EQUIP.	TOTAL	TOTAL INCL O&P	
100	0010 **BITUMINOUS CONCRETE CURBS**										100
	0012 Curbs, asphaltic, machine formed, 8" wide, 6" high, 40 L.F./ton	B-27	1,000	.032	L.F.	.62	.87	.22	1.71	2.27	
	0100 8" wide, 8" high, 30 L.F. per ton		900	.036		.71	.97	.24	1.92	2.55	
	0150 Asphaltic berm, 12" W, 3"-6" H, 35 L.F./ton, before pavement	↓	700	.046		.95	1.24	.31	2.50	3.33	
	0200 12" W, 1-1/2" to 4" H, 60 L.F. per ton, laid with pavement	B-2	1,050	.038	↓	.58	1.03		1.61	2.25	
300	0010 **CEMENT CONCRETE CURBS**										300
	0300 Concrete, wood forms, 6" x 18", straight	C-2A	500	.096	L.F.	2.66	3.18		5.84	7.85	
	0400 6" x 18", radius		200	.240		2.77	7.95		10.72	15.30	
	0410 Steel forms, 6" x 18", straight		700	.069		3.55	2.27		5.82	7.40	
	0411 6" x 18", radius	↓	400	.120		4.13	3.97		8.10	10.65	
	0415 Machine formed, 6" x 18", straight	B-69A	2,000	.024		3.62	.71	.29	4.62	5.40	
	0416 6" x 18", radius	"	900	.053	↓	3.77	1.57	.65	5.99	7.25	
	0421 Curb and gutter, straight										
	0422 with 6" high curb and 6" thick gutter, wood forms										
	0430 24" wide, .055 C.Y. per L.F.	C-2A	375	.128	L.F.	13.50	4.24		17.74	21.50	

SITE CONSTRUCTION 2

105

2 SITE CONSTRUCTION

		02770	Curbs and Gutters	CREW	DAILY OUTPUT	LABOR-HOURS	UNIT	2005 BARE COSTS MAT.	LABOR	EQUIP.	TOTAL	TOTAL INCL O&P	
300	0435		30" wide, .066 C.Y. per L.F.	C-2A	340	.141	L.F.	14.70	4.67		19.37	23.50	300
	0440		Steel forms, 24" wide, straight		700	.069		5.90	2.27		8.17	10	
	0441		Radius		300	.160		5.90	5.30		11.20	14.65	
	0442		30" wide, straight		700	.069		7.05	2.27		9.32	11.30	
	0443		Radius		300	.160		7.05	5.30		12.35	15.95	
	0445		Machine formed, 24" wide, straight	B-69A	2,000	.024		5.90	.71	.29	6.90	7.90	
	0446		Radius		900	.053		5.90	1.57	.65	8.12	9.60	
	0447		30" wide, straight		2,000	.024		7.05	.71	.29	8.05	9.20	
	0448		Radius		900	.053		7.05	1.57	.65	9.27	10.90	
	0451		Median mall, 2' x 9" high, straight		2,200	.022		6.15	.64	.26	7.05	8.05	
	0452		Radius	B-69B	900	.053		6.15	1.57	.81	8.53	10.10	
	0453		4' x 9" high, straight		2,000	.024		12.40	.71	.36	13.47	15.10	
	0454		Radius		800	.060		12.40	1.76	.91	15.07	17.30	
	0550		Precast, 6" x 18", straight	B-29	700	.080		7.35	2.30	1.20	10.85	12.95	
	0600		6" x 18", radius	"	325	.172		8.40	4.96	2.58	15.94	19.75	
500	0010	**STONE CURBS**											500
	1000		Granite, split face, straight, 5" x 16"	D-13	500	.096	L.F.	9.45	3.14	1.01	13.60	16.30	
	1100		6" x 18"	"	450	.107		12.40	3.49	1.12	17.01	20	
	1300		Radius curbing, 6" x 18", over 10' radius	B-29	260	.215		15.20	6.20	3.22	24.62	30	
	1400		Corners, 2' radius		80	.700	Ea.	51	20	10.45	81.45	98.50	
	1600		Edging, 4-1/2" x 12", straight		300	.187	L.F.	4.72	5.35	2.79	12.86	16.55	
	1800		Curb inlets, (guttermouth) straight		41	1.366	Ea.	113	39.50	20.50	173	208	
	2000		Indian granite (belgian block)										
	2100		Jumbo, 10-1/2" x 7-1/2" x 4", grey	D-1	150	.107	L.F.	1.72	3.32		5.04	6.95	
	2150		Pink		150	.107		2.24	3.32		5.56	7.50	
	2200		Regular, 9" x 4-1/2" x 4-1/2", grey		160	.100		1.56	3.11		4.67	6.45	
	2250		Pink		160	.100		2.14	3.11		5.25	7.10	
	2300		Cubes, 4" x 4" x 4", grey		175	.091		1.49	2.84		4.33	5.95	
	2350		Pink		175	.091		1.56	2.84		4.40	6.05	
	2400		6" x 6" x 6", pink		155	.103		3.86	3.21		7.07	9.15	
	2500		Alternate pricing method for indian granite										
	2550		Jumbo, 10-1/2" x 7-1/2" x 4" (30lb), grey				Ton	98			98	108	
	2600		Pink					130			130	143	
	2650		Regular, 9" x 4-1/2" x 4-1/2" (20lb), grey					110			110	121	
	2700		Pink					150			150	165	
	2750		Cubes, 4" x 4" x 4" (5lb), grey					180			180	198	
	2800		Pink					200			200	220	
	2850		6" x 6" x 6" (25lb), pink					150			150	165	
	2900		For pallets, add					16.50			16.50	18.15	
		02775	**Sidewalks**										
275	0010	**SIDEWALKS, DRIVEWAYS, & PATIOS** No base											275
	0020		Asphaltic concrete, 2" thick	[R02065 -300] B-37	720	.067	S.Y.	3.65	1.87	.15	5.67	7.05	
	0100		2-1/2" thick	"	660	.073	"	4.63	2.04	.16	6.83	8.45	
	0110		Bedding for brick or stone, mortar, 1" thick	D-1	300	.053	S.F.	.35	1.66		2.01	2.91	
	0120		2" thick	"	200	.080		.87	2.49		3.36	4.75	
	0130		Sand, 2" thick	B-18	8,000	.003		.16	.08	.01	.25	.34	
	0140		4" thick	"	4,000	.006		.32	.16	.01	.49	.63	
	0300		Concrete, 3000 psi, CIP, 6 x 6 - W1.4 x W1.4 mesh,										
	0310		broomed finish, no base, 4" thick	B-24	600	.040	S.F.	1.35	1.25		2.60	3.40	
	0350		5" thick		545	.044		1.80	1.38		3.18	4.08	
	0400		6" thick		510	.047		2.10	1.47		3.57	4.55	
	0450		For bank run gravel base, 4" thick, add	B-18	2,500	.010		.41	.26	.01	.68	.89	
	0520		8" thick, add	"	1,600	.015		.84	.41	.02	1.27	1.59	
	0550		Exposed aggregate finish, add to above, minimum	B-24	1,875	.013		.10	.40		.50	.72	

SITE CONSTRUCTION 2

02775 | Sidewalks

		CREW	DAILY OUTPUT	LABOR-HOURS	UNIT	MAT.	LABOR	EQUIP.	TOTAL	TOTAL INCL O&P	
275	0600	Maximum	B-24	455	.053	S.F.	.35	1.65		2	2.89	275
	0700	Patterned surface, add to above min. [R02065-300]	↓	1,200	.020	↓		.63		.63	.95	
	0800	For integral colors, see Div. 03310-220										
	0950	Concrete tree grate, 5' square	B-6	25	.960	Ea.	320	27.50	8.65	356.15	400	
	0960	Cast iron tree grate with frame, 2 piece, round, 5' diameter		25	.960		905	27.50	8.65	941.15	1,050	
	0980	Square, 5' side	↓	25	.960	↓	935	27.50	8.65	971.15	1,075	
	1000	Crushed stone, 1" thick, white marble	2 Clab	1,700	.009	S.F.	.19	.25		.44	.60	
	1050	Bluestone		1,700	.009		.21	.25		.46	.62	
	1070	Granite chips	↓	1,700	.009		.20	.25		.45	.61	
	1200	For 2" asphaltic conc base and tack coat, add to above	B-37	7,200	.007		.46	.19	.01	.66	.82	
	1660	Limestone pavers, 3" thick	D-1	72	.222		7.90	6.90		14.80	19.15	
	1670	4" thick		70	.229		10.45	7.10		17.55	22.50	
	1680	5" thick	↓	68	.235		13.05	7.30		20.35	25.50	
	1700	Redwood, prefabricated, 4' x 4' sections	2 Carp	316	.051		8.10	1.73		9.83	11.60	
	1750	Redwood planks, 1" thick, on sleepers	"	240	.067		5.65	2.28		7.93	9.80	
	1830	1-1/2" thick	B-28	167	.144		3.74	4.56		8.30	11.20	
	1840	2" thick		167	.144		4.82	4.56		9.38	12.40	
	1850	3" thick		150	.160		7.10	5.10		12.20	15.70	
	1860	4" thick		150	.160		9.35	5.10		14.45	18.15	
	1870	5" thick	↓	150	.160	↓	11.85	5.10		16.95	21	
	2100	River or beach stone, stock	B-1	18	1.333	Ton	26.50	36.50		63	86	
	2150	Quarried	"	18	1.333	"	46	36.50		82.50	108	
	2160	Load, dump, and spread stone with skid steer, 100' haul	B-62	24	1	C.Y.		29	6.40	35.40	51.50	
	2165	200' haul		18	1.333			38.50	8.50	47	68.50	
	2168	300' haul	↓	12	2	↓		57.50	12.80	70.30	103	
	2170	Shale paver, 2-1/4" thick	D-1	200	.080	S.F.	2.71	2.49		5.20	6.75	
	2200	Coarse washed sand bed, 1"	B-62	1,350	.018	S.Y.	1.19	.51	.11	1.81	2.23	
	2250	Stone dust, 4" thick	"	900	.027	"	2.88	.77	.17	3.82	4.54	
	2300	Tile thinset pavers, 3/8" thick	D-1	300	.053	S.F.	2.97	1.66		4.63	5.80	
	2350	3/4" thick	"	280	.057	"	4.74	1.78		6.52	7.90	
	2400	Wood rounds, cypress	B-1	175	.137	Ea.	8.65	3.75		12.40	15.35	
	8000	For temporary barricades, see div. 01560-100										

02780 | Unit Pavers

		CREW	DAILY OUTPUT	LABOR-HOURS	UNIT	MAT.	LABOR	EQUIP.	TOTAL	TOTAL INCL O&P	
100	0010	**ASPHALT BLOCKS**										100
	0020	Rectangular, 6" x 12" x 1-1/4", w/bed & neopr. adhesive	D-1	135	.119	S.F.	4.22	3.68		7.90	10.25	
	0100	3" thick		130	.123		5.90	3.83		9.73	12.35	
	0300	Hexagonal tile, 8" wide, 1-1/4" thick		135	.119		4.22	3.68		7.90	10.25	
	0400	2" thick		130	.123		5.90	3.83		9.73	12.35	
	0500	Square, 8" x 8", 1-1/4" thick		135	.119		4.22	3.68		7.90	10.25	
	0600	2" thick	↓	130	.123		5.90	3.83		9.73	12.35	
	0900	For exposed aggregate (ground finish) add					.29			.29	.32	
	0910	For colors, add				↓	.22			.22	.24	
200	0010	**BRICK PAVING** 4" x 8" x 1-1/2", without joints (4.5 brick/S.F.)	D-1	110	.145	S.F.	2.43	4.52		6.95	9.60	200
	0100	Grouted, 3/8" joint (3.9 brick/S.F.)		90	.178		2.78	5.55		8.33	11.45	
	0200	4" x 8" x 2-1/4", without joints (4.5 bricks/S.F.)		110	.145		3.08	4.52		7.60	10.30	
	0300	Grouted, 3/8" joint (3.9 brick/S.F.)	↓	90	.178		2.84	5.55		8.39	11.50	
	0500	Bedding, asphalt, 3/4" thick	B-25	5,130	.017		.34	.50	.39	1.23	1.58	
	0540	Course washed sand bed, 1" thick	B-18	5,000	.005		.18	.13	.01	.32	.41	
	0580	Mortar, 1" thick	D-1	300	.053		.29	1.66		1.95	2.85	
	0620	2" thick		200	.080		.58	2.49		3.07	4.43	
	1500	Brick on 1" thick sand bed laid flat, 4.5 per S.F.		100	.160		2.49	4.97		7.46	10.35	
	2000	Brick pavers, laid on edge, 7.2 per S.F.		70	.229		2.36	7.10		9.46	13.40	
	2500	For 4" thick concrete bed and joints, add	↓	595	.027		.89	.84		1.73	2.25	
	2800	For steam cleaning, add	A-1H	950	.008	↓	.05	.22	.06	.33	.47	

02700 | Bases, Ballasts, Pavements & Appurtenances

02780 | Unit Pavers

		CREW	DAILY OUTPUT	LABOR-HOURS	UNIT	MAT.	LABOR	EQUIP.	TOTAL	TOTAL INCL O&P	
400	**0010 PRECAST CONCRETE PAVING SLABS**										**400**
	0710 Precast concrete patio blocks, 2-3/8" thick, colors, 8" x 16"	D-1	265	.060	S.F.	1.18	1.88		3.06	4.16	
	0715 12" x 12"		300	.053		1.53	1.66		3.19	4.22	
	0720 16" x 16"		335	.048		1.69	1.48		3.17	4.12	
	0730 24" x 24"		510	.031		2.10	.98		3.08	3.80	
	0740 Green, 8" x 16"	▼	265	.060		1.52	1.88		3.40	4.53	
	0750 Exposed local aggregate, natural	2 Bric	250	.064		5.05	2.26		7.31	9	
	0800 Colors		250	.064		5.50	2.26		7.76	9.50	
	0850 Exposed granite or limestone aggregate		250	.064		6.10	2.26		8.36	10.20	
	0900 Exposed white tumblestone aggregate	▼	250	.064	▼	3.62	2.26		5.88	7.40	
600	**0010 STONE PAVERS**										**600**
	1100 Flagging, bluestone, irregular, 1" thick,	D-1	81	.198	S.F.	4.63	6.15		10.78	14.45	
	1110 1-1/2" thick		90	.178		5.45	5.55		11	14.40	
	1120 Pavers, 1/2" thick		110	.145		7.70	4.52		12.22	15.40	
	1130 3/4" thick		95	.168		9.80	5.25		15.05	18.75	
	1140 1" thick		81	.198		10.50	6.15		16.65	21	
	1150 Snapped random rectangular, 1" thick		92	.174		7	5.40		12.40	15.95	
	1200 1-1/2" thick		85	.188		8.45	5.85		14.30	18.15	
	1250 2" thick		83	.193		9.80	6		15.80	19.95	
	1300 Slate, natural cleft, irregular, 3/4" thick		92	.174		5.15	5.40		10.55	13.95	
	1310 1" thick		85	.188		6	5.85		11.85	15.50	
	1350 Random rectangular, gauged, 1/2" thick		105	.152		11.15	4.74		15.89	19.50	
	1400 Random rectangular, butt joint, gauged, 1/4" thick	▼	150	.107		12	3.32		15.32	18.30	
	1450 For sand rubbed finish, add					5.60			5.60	6.15	
	1550 Granite blocks, 3-1/2" x 3-1/2" x 3-1/2"	D-1	92	.174		6.25	5.40		11.65	15.15	
	1560 4" x 4" x 4"		95	.168		6.60	5.25		11.85	15.20	
	1600 4" to 12" long, 3" to 5" wide, 3" to 5" thick		98	.163		5.20	5.05		10.25	13.50	
	1650 6" to 15" long, 3" to 6" wide, 3" to 5" thick	▼	105	.152	▼	2.78	4.74		7.52	10.25	
700	**0010 STEPS** Incl. excav., borrow & concrete base, where applicable										**700**
	0100 Brick steps	B-24	35	.686	LF Riser	9.10	21.50		30.60	42.50	
	0200 Railroad ties	2 Clab	25	.640		2.89	17.10		19.99	29.50	
	0300 Bluestone treads, 12" x 2" or 12" x 1-1/2"	B-24	30	.800	▼	22	25		47	62	
	0500 Concrete, cast in place, see division 03310-240										
	0600 Precast concrete, see division 03480-800										

02785 | Flexible Pavement Coating

		CREW	DAILY OUTPUT	LABOR-HOURS	UNIT	MAT.	LABOR	EQUIP.	TOTAL	TOTAL INCL O&P	
250	**0010 FOG SEAL**										**250**
	0012 Sealcoating, 2 coat coal tar pitch emulsion over 10,000 SY	B-45	5,000	.003	S.Y.	.46	.10	.11	.67	.78	
	0030 1000 to 10,000 S.Y.	"	3,000	.005		.46	.17	.18	.81	.95	
	0100 Under 1000 S.Y.	B-1	1,050	.023		.46	.63		1.09	1.48	
	0300 Petroleum resistant, over 10,000 S.Y.	B-45	5,000	.003		.57	.10	.11	.78	.90	
	0320 1000 to 10,000 S.Y.	"	3,000	.005		.57	.17	.18	.92	1.07	
	0400 Under 1000 S.Y.	B-1	1,050	.023		.57	.63		1.20	1.60	
	0600 Non-skid pavement renewal, over 10,000 S.Y.	B-45	5,000	.003		.67	.10	.11	.88	1.01	
	0620 1000 to 10,000 S.Y.	"	3,000	.005		.67	.17	.18	1.02	1.18	
	0700 Under 1000 S.Y.	B-1	1,050	.023		.67	.63		1.30	1.71	
	0800 Prepare and clean surface for above	A-2	8,545	.003	▼		.08	.02	.10	.14	
	1000 Hand seal asphalt curbing	B-1	4,420	.005	L.F.	.34	.15		.49	.60	
	1900 Asphalt surface treatment, single course, small area										
	1901 0.30 gal/S.Y. asphalt material, 20#/S.Y. aggregate	B-91	5,000	.013	S.Y.	.79	.40	.29	1.48	1.79	
	1910 Roadway or large area		10,000	.006		.73	.20	.14	1.07	1.26	
	1950 Asphalt surface treatment, dbl. course for small area		3,000	.021		1.46	.66	.48	2.60	3.14	
	1960 Roadway or large area		6,000	.011		1.31	.33	.24	1.88	2.22	
	1980 Asphalt surface treatment, single course, for shoulders	▼	7,500	.009	▼	.84	.26	.19	1.29	1.53	
300	**0010 LATEX MODIFIED EMULSION**										**300**
	3600 Waterproofing, membrane, tar and fabric, small area	B-63	233	.172	S.Y.	5.40	4.80	.66	10.86	14.05	

108 **Important: See the Reference Section for critical supporting data - Reference Nos., Crews, & City Cost Indexes**

02785 | Flexible Pavement Coating

		CREW	DAILY OUTPUT	LABOR-HOURS	UNIT	2005 BARE COSTS				TOTAL INCL O&P	
						MAT.	LABOR	EQUIP.	TOTAL		
300	3640	Large area	B-63	1,435	.028	S.Y.	4.97	.78	.11	5.86	6.75	300
	3680	Preformed rubberized asphalt, small area		100	.400		7.30	11.20	1.54	20.04	27	
	3720	Large area	↓	367	.109		6.65	3.05	.42	10.12	12.45	
	3780	Rubberized asphalt (latex) seal	B-45	5,000	.003	↓	1.05	.10	.11	1.26	1.43	
400	0010	**SAND SEAL**										400
	2080	Sand sealing, sharp sand, asphalt emulsion, small area	B-91	10,000	.006	S.Y.	.57	.20	.14	.91	1.09	
	2120	Roadway or large area	"	18,000	.004	"	.49	.11	.08	.68	.80	
	3000	Sealing random cracks, min 1/2" wide, to 1-1/2", 1,000 L.F.	B-77	2,800	.014	L.F.	.45	.39	.13	.97	1.25	
	3040	10,000 L.F.		4,000	.010	"	.35	.27	.09	.71	.90	
	3080	Alternate method, 1,000 L.F.		200	.200	Gal.	6.85	5.40	1.89	14.14	18	
	3120	10,000 L.F.	↓	325	.123	"	5.50	3.34	1.16	10	12.50	
	3200	Multi-cracks (flooding), 1 coat, small area	B-92	460	.070	S.Y.	2.05	1.89	.69	4.63	5.95	
	3240	Large area		2,850	.011		1.82	.31	.11	2.24	2.60	
	3280	2 coat, small area		230	.139		6.75	3.78	1.38	11.91	14.85	
	3320	Large area		1,425	.022	↓	6.15	.61	.22	6.98	7.95	
	3360	Alternate method, small area		115	.278	Gal.	7.05	7.55	2.75	17.35	22.50	
	3400	Large area	↓	715	.045	"	6.60	1.22	.44	8.26	9.70	
500	0010	**SLURRY SEAL**										500
	0100	Slurry seal, type I, 8 lbs agg./S.Y., 1 coat, small or irregular area	B-90	2,800	.023	S.Y.	.78	.66	.63	2.07	2.56	
	0150	Roadway or large area		10,000	.006		.78	.18	.18	1.14	1.33	
	0200	Type II, 12 lbs aggregate/S.Y., 2 coats, small or irregular area		2,000	.032		1.56	.92	.88	3.36	4.09	
	0250	Roadway or large area		8,000	.008		1.56	.23	.22	2.01	2.31	
	0300	Type III, 20 lbs aggregate/S.Y., 2 coats, small or irregular area		1,800	.036		1.84	1.02	.97	3.83	4.66	
	0350	Roadway or large area		6,000	.011		1.84	.31	.29	2.44	2.81	
	0400	Slurry seal, thermoplastic coal-tar, type I, small or irregular area		2,400	.027		1.63	.77	.73	3.13	3.77	
	0450	Roadway or large area		8,000	.008		1.63	.23	.22	2.08	2.38	
	0500	Type II, small or irregular area		2,400	.027		2.08	.77	.73	3.58	4.27	
	0550	Roadway or large area	↓	7,800	.008	↓	2.08	.24	.22	2.54	2.90	
	0600	Average mobilization cost				Ea.				3,750	3,750	
600	0010	**SURFACE TREATMENT**										600
	3000	Pavement overlay, polypropylene										
	3040	6 oz. per S.Y., ideal conditions	B-63	10,000	.004	S.Y.	1.12	.11	.02	1.25	1.42	
	3080	Adverse conditions		1,000	.040		1.50	1.12	.15	2.77	3.55	
	3120	4 oz. per S.Y., ideal conditions		10,000	.004		.74	.11	.02	.87	1	
	3160	Adverse conditions	↓	1,000	.040		.95	1.12	.15	2.22	2.95	
	3200	Tack coat, emulsion, .05 gal per S.Y., 1000 S.Y	B-45	2,500	.006		.21	.20	.21	.62	.76	
	3240	10,000 S.Y.		10,000	.002		.17	.05	.05	.27	.32	
	3270	.10 gal per S.Y., 1000 S.Y.		2,500	.006		.39	.20	.21	.80	.96	
	3275	.10 gal per S.Y., 10,000 S.Y.		10,000	.002		.39	.05	.05	.49	.57	
	3280	.15 gal per S.Y., 1000 S.Y.		2,500	.006		.58	.20	.21	.99	1.17	
	3320	10,000 S.Y.	↓	10,000	.002	↓	.46	.05	.05	.56	.65	

02790 | Athletic/Recreational Surfaces

		CREW	DAILY OUTPUT	LABOR-HOURS	UNIT	MAT.	LABOR	EQUIP.	TOTAL	INCL O&P	
500	0010	**SYNTHETIC RUNNING TRACK SURFACING**									500
	0020	Running track, asphalt, incl base, 3" thick	B-37	300	.160	S.Y.	12.35	4.49	.36	17.20	21	
	0100	Surface, latex rubber system, 3/8" thick, black	B-20	125	.192		5.55	5.80		11.35	15.10	
	0150	Colors		125	.192		9.85	5.80		15.65	19.85	
	0300	Urethane rubber system, 3/8" thick, black		120	.200		14.80	6		20.80	25.50	
	0400	Color coating	↓	115	.209	↓	17.55	6.30		23.85	29	

02810	Irrigation System	CREW	DAILY OUTPUT	LABOR-HOURS	UNIT	2005 BARE COSTS				TOTAL INCL O&P
						MAT.	LABOR	EQUIP.	TOTAL	
0010	**SPRINKLER IRRIGATION SYSTEM** For lawns									300
0100	Golf course with fully automatic system	C-17	.05	1,600	9 holes	82,500	56,500		139,000	178,500
0200	24' diam. head at 15' O.C incl. piping, auto oper., minimum	B-20	70	.343	Head	18.20	10.30		28.50	36
0300	Maximum		40	.600		42	.18.05		60.05	74
0500	60' diam. head at 40' O.C. incl. piping, auto oper., minimum		28	.857		55	26		81	101
0600	Maximum		23	1.043		154	31.50		185.50	219
0800	Residential system, custom, 1" supply		2,000	.012	S.F.	.27	.36		.63	.86
0900	1-1/2" supply		1,800	.013	"	.32	.40		.72	.97
1020	Pop up spray head w/risers, hi-pop, full circle pattern, 4"	2 Skwk	76	.211	Ea.	3.36	7.35		10.71	15.10
1030	1/2 circle pattern		76	.211		3.43	7.35		10.78	15.15
1040	6", full circle pattern		76	.211		8.60	7.35		15.95	21
1050	1/2 circle pattern		76	.211		8.60	7.35		15.95	21
1060	12", full circle pattern		76	.211		12.25	7.35		19.60	25
1070	1/2 circle pattern		76	.211		12.30	7.35		19.65	25
1080	Pop up bubbler head w/risers, hi-pop bubbler head, 4"		76	.211		3.50	7.35		10.85	15.25
1090	6"		76	.211		8.60	7.35		15.95	21
1100	12"		76	.211		12.25	7.35		19.60	25
1110	Impact full/part circle sprinklers, 28'-54' 25-60 PSI		37	.432		11.45	15.05		26.50	36
1120	Spaced 37'-49' @ 25-50 PSI		37	.432		25	15.05		40.05	51
1130	Spaced 43'-61' @ 30-60 PSI		37	.432		45.50	15.05		60.55	73.50
1140	Spaced 54'-78' @ 40-80 PSI		37	.432		105	15.05		120.05	139
1145	Impact rotor pop-up full/part commercial circle sprinklers									
1150	Spaced 42'-65' 35-80 PSI	2 Skwk	25	.640	Ea.	23	22.50		45.50	60
1160	Spaced 48'-76' 45-85 PSI	"	25	.640	"	8.85	22.50		31.35	44
1165	Impact rotor pop-up part. circle comm., 53'-75', 55-100 PSI, w/ acc									
1170	Plastic case, metal cover	2 Skwk	25	.640	Ea.	148	22.50		170.50	198
1180	Rubber cover		25	.640		112	22.50		134.50	158
1190	Iron case, metal cover		22	.727		136	25.50		161.50	189
1200	Rubber cover		22	.727		147	25.50		172.50	201
1250	Plastic case, 2 nozzle, metal cover		25	.640		108	22.50		130.50	154
1260	Rubber cover		25	.640		115	22.50		137.50	161
1270	Iron case, 2 nozzle, metal cover		22	.727		157	25.50		182.50	213
1280	Rubber cover		22	.727		164	25.50		189.50	220
1282	Impact rotor pop-up full circle comm., 39'-99', 30-100 PSI									
1284	Plastic case, metal cover	2 Skwk	25	.640	Ea.	113	22.50		135.50	159
1286	Rubber cover		25	.640		125	22.50		147.50	173
1288	Iron case, metal cover		22	.727		162	25.50		187.50	219
1290	Rubber cover		22	.727		171	25.50		196.50	228
1292	Plastic case, 2 nozzle, metal cover		22	.727		122	25.50		147.50	174
1294	Rubber cover		22	.727		126	25.50		151.50	179
1296	Iron case, 2 nozzle, metal cover		20	.800		157	28		185	216
1298	Rubber cover		20	.800		167	28		195	228
1305	Electric remote control valve, plastic, 3/4"		18	.889		17.05	31		48.05	67
1310	1"		18	.889		30	31		61	81
1320	1-1/2"		18	.889		58	31		89	112
1330	2"		18	.889		85.50	31		116.50	142
1335	Quick coupling valves, brass, locking cover									
1340	Inlet coupling valve, 3/4"	2 Skwk	18.75	.853	Ea.	42	29.50		71.50	92.50
1350	1"		18.75	.853		51	29.50		80.50	103
1360	Controller valve boxes, 6" round boxes		18.75	.853		3.73	29.50		33.23	50.50
1370	10" round boxes		14.25	1.123		8.15	39		47.15	70
1380	12" square box		9.75	1.641		16.40	57		73.40	107
1388	Electromech. control, 14 day 3-60 min, auto start to 23/day									
1390	4 station	2 Skwk	1.04	15.385	Ea.	177	535		712	1,025
1400	7 station		.64	25		229	870		1,099	1,600
1410	12 station		.40	40		284	1,400		1,684	2,475

Important: See the Reference Section for critical supporting data - Reference Nos., Crews, & City Cost Indexes

02800 | Site Improvements and Amenities

02810 | Irrigation System

			CREW	DAILY OUTPUT	LABOR-HOURS	UNIT	2005 BARE COSTS				TOTAL INCL O&P	
							MAT.	LABOR	EQUIP.	TOTAL		
300	1420	Dual programs, 18 station	2 Skwk	.24	66.667	Ea.	1,400	2,325		3,725	5,175	**300**
	1430	23 station	↓	.16	100	↓	1,825	3,475		5,300	7,450	
	1435	Backflow preventer, bronze, 0-175 PSI, w/valves, test cocks										
	1440	3/4"	2 Skwk	2	8	Ea.	115	279		394	560	
	1450	1"		2	8		120	279		399	565	
	1460	1-1/2"		2	8		239	279		518	695	
	1470	2"	↓	2	8	↓	249	279		528	710	
	1475	Pressure vacuum breaker, brass, 15-150 PSI										
	1480	3/4"	2 Skwk	2	8	Ea.	69.50	279		348.50	510	
	1490	1"		2	8		91	279		370	535	
	1500	1-1/2"		2	8		180	279		459	635	
	1510	2"	↓	2	8	↓	167	279		446	620	

02820 | Fences & Gates

			CREW	DAILY OUTPUT	LABOR-HOURS	UNIT	MAT.	LABOR	EQUIP.	TOTAL	TOTAL INCL O&P	
130	0010	**FENCE, CHAIN LINK INDUSTRIAL**, schedule 40										**130**
	0020	3 strands barb wire, 2" post @ 10' O.C., set in concrete, 6' H										
	0200	9 ga. wire, galv. steel	B-80C	240	.100	L.F.	11.80	2.67	.56	15.03	17.75	
	0300	Aluminized steel		240	.100		15.15	2.67	.56	18.38	21.50	
	0500	6 ga. wire, galv. steel		240	.100		18.50	2.67	.56	21.73	25	
	0600	Aluminized steel		240	.100		21	2.67	.56	24.23	27.50	
	0800	6 ga. wire, 6' high but omit barbed wire, galv. steel		250	.096		17.95	2.57	.54	21.06	24.50	
	0900	Aluminized steel		250	.096		24.50	2.57	.54	27.61	31.50	
	0920	8' H, 6 ga. wire, 2-1/2" line post, galv. steel		180	.133		28.50	3.56	.74	32.80	38	
	0940	Aluminized steel		180	.133	↓	35	3.56	.74	39.30	45	
	1400	Gate for 6' high fence, 1-5/8" frame, 3' wide, galv. steel		10	2.400	Ea.	131	64	13.40	208.40	258	
	1500	Aluminized steel	↓	10	2.400	"	161	64	13.40	238.40	291	
	2000	5'-0" high fence, 9 ga., no barbed wire, 2" line post,										
	2010	10' O.C., 1-5/8" top rail										
	2100	Galvanized steel	B-80C	300	.080	L.F.	10.25	2.14	.45	12.84	15.05	
	2200	Aluminized steel		300	.080	"	11.85	2.14	.45	14.44	16.85	
	2400	Gate, 4' wide, 5' high, 2" frame, galv. steel		10	2.400	Ea.	149	64	13.40	226.40	278	
	2500	Aluminized steel		10	2.400	"	164	64	13.40	241.40	295	
	3100	Overhead slide gate, chain link, 6' high, to 18' wide	↓	38	.632	L.F.	131	16.90	3.52	151.42	175	
	3105	8' high	B-80	30	1.067		132	30.50	17.80	180.30	212	
	3108	10' high		24	1.333		133	38.50	22	193.50	231	
	3110	Cantilever type		48	.667		56.50	19.20	11.10	86.80	104	
	3120	8' high		24	1.333		82	38.50	22	142.50	174	
	3130	10' high	↓	18	1.778	↓	97	51	29.50	177.50	218	
	5000	Double swing gates, incl. posts & hardware										
	5010	5' high, 12' opening	B-80C	3.40	7.059	Opng.	400	189	39.50	628.50	775	
	5020	20' opening		2.80	8.571		545	229	48	822	1,000	
	5060	6' high, 12' opening		3.20	7.500		675	200	42	917	1,100	
	5070	20' opening	↓	2.60	9.231		930	247	51.50	1,228.50	1,450	
	5080	8' high, 12' opening	B-80	2.13	15.002		1,050	430	250	1,730	2,100	
	5090	20' opening		1.45	22.069		1,375	635	370	2,380	2,900	
	5100	10' high, 12' opening		1.31	24.427		1,200	705	405	2,310	2,850	
	5110	20' opening		1.03	31.068		1,800	895	520	3,215	3,925	
	5120	12' high, 12' opening		1.05	30.476		1,750	880	510	3,140	3,825	
	5130	20' opening	↓	.85	37.647	↓	2,250	1,075	630	3,955	4,825	
	5190	For aluminized steel add					20%					
	7001	Snow fence on steel posts 10' O.C., 4' high	B-1	500	.048	L.F.	2.20	1.31		3.51	4.46	
	7055	Braces, galv. steel	B-80A	960	.025		1.66	.67	.18	2.51	3.06	
	7056	Aluminized steel	"	960	.025	↓	1.99	.67	.18	2.84	3.43	
140	0010	**FENCE, CHAIN LINK RESIDENTIAL**, sch. 20, 11 ga. wire, 1-5/8" post										**140**
	0020	10' O.C., 1-3/8" top rail, 2" corner post, galv. stl. 3' high	B-80C	500	.048	L.F.	3.58	1.28	.27	5.13	6.20	

2

SITE CONSTRUCTION

			DAILY	LABOR-			2005 BARE COSTS			TOTAL	
02820	**Fences & Gates**	CREW	OUTPUT	HOURS	UNIT	MAT.	LABOR	EQUIP.	TOTAL	INCL O&P	
140	0050	4' high	B-80C	400	.060	L.F.	4.06	1.60	.33	5.99	7.30	**140**
	0100	6' high		200	.120	↓	4.92	3.21	.67	8.80	11.10	
	0150	Add for gate 3' wide, 1-3/8" frame, 3' high		12	2	Ea.	53.50	53.50	11.15	118.15	154	
	0170	4' high		10	2.400		61.50	64	13.40	138.90	181	
	0190	6' high		10	2.400		84.50	64	13.40	161.90	206	
	0200	Add for gate 4' wide, 1-3/8" frame, 3' high		9	2.667		60	71.50	14.90	146.40	192	
	0220	4' high		9	2.667		68	71.50	14.90	154.40	201	
	0240	6' high		8	3	↓	80	80	16.75	176.75	230	
	0350	Aluminized steel, 11 ga. wire, 3' high		500	.048	L.F.	4.64	1.28	.27	6.19	7.35	
	0380	4' high		400	.060		5.95	1.60	.33	7.88	9.40	
	0400	6' high		200	.120	↓	8.40	3.21	.67	12.28	14.90	
	0450	Add for gate 3' wide, 1-3/8" frame, 3' high		12	2	Ea.	64	53.50	11.15	128.65	165	
	0470	4' high		10	2.400		105	64	13.40	182.40	230	
	0490	6' high		10	2.400		132	64	13.40	209.40	260	
	0500	Add for gate 4' wide, 1-3/8" frame, 3' high		10	2.400		72.50	64	13.40	149.90	194	
	0520	4' high		9	2.667		80.50	71.50	14.90	166.90	215	
	0540	6' high		8	3	↓	93	80	16.75	189.75	244	
	0620	Vinyl covered, 9 ga. wire, 3' high		500	.048	L.F.	3.86	1.28	.27	5.41	6.50	
	0640	4' high		400	.060		4.51	1.60	.33	6.44	7.80	
	0660	6' high		200	.120		5.70	3.21	.67	9.58	12	
	0720	Add for gate 3' wide, 1-3/8" frame, 3' high		12	2	Ea.	74.50	53.50	11.15	139.15	177	
	0740	4' high		10	2.400		84	64	13.40	161.40	206	
	0760	6' high		10	2.400		131	64	13.40	208.40	258	
	0780	Add for gate 4' wide, 1-3/8" frame, 3' high		10	2.400		85.50	64	13.40	162.90	208	
	0800	4' high		9	2.667		93	71.50	14.90	179.40	228	
	0820	6' high	↓	8	3	↓	105	80	16.75	201.75	258	
150	0010	**FENCE, CHAIN LINK, GATES & POSTS** (1/3 post length in ground)										**150**
	6580	Line posts, galvanized, 2-1/2" OD, set in conc., 4'	B-80	80	.400	Ea.	20.50	11.50	6.65	38.65	47.50	
	6585	5'		76	.421		24.50	12.15	7	43.65	53.50	
	6590	6'		74	.432		25	12.45	7.20	44.65	54.50	
	6595	7'		72	.444		27.50	12.80	7.40	47.70	58	
	6600	8'		69	.464		30	13.35	7.75	51.10	62	
	6610	H-beam, 1-7/8", 4'		83	.386		23	11.10	6.45	40.55	49.50	
	6615	5'		81	.395		26.50	11.40	6.60	44.50	53.50	
	6620	6'		78	.410		31	11.80	6.85	49.65	59.50	
	6625	7'		75	.427		34	12.30	7.10	53.40	64	
	6630	8'		73	.438		37	12.65	7.30	56.95	68	
	6635	Vinyl coated, 2-1/2" OD, set in conc., 4'		79	.405		40	11.65	6.75	58.40	69.50	
	6640	5'		77	.416		45.50	11.95	6.95	64.40	76	
	6645	6'		74	.432		57.50	12.45	7.20	77.15	90.50	
	6650	7'		72	.444		62	12.80	7.40	82.20	96	
	6655	8'		69	.464		67	13.35	7.75	88.10	103	
	6660	End gate post, steel, 3" OD, set in conc.,4'		68	.471		47	13.55	7.85	68.40	81	
	6665	5'		65	.492		59.50	14.20	8.20	81.90	96	
	6670	6'		63	.508		60	14.65	8.45	83.10	98	
	6675	7'		61	.525		80.50	15.10	8.75	104.35	121	
	6685	Vinyl, 4'		68	.471		78.50	13.55	7.85	99.90	116	
	6690	5'		65	.492		86	14.20	8.20	108.40	126	
	6695	6'		63	.508		107	14.65	8.45	130.10	149	
	6705	8'		59	.542		127	15.65	9.05	151.70	174	
	6710	Corner post, galv. steel, 4" OD, set in conc., 4'		65	.492		78.50	14.20	8.20	100.90	118	
	6715	6'		63	.508		111	14.65	8.45	134.10	155	
	6720	7'		61	.525		125	15.10	8.75	148.85	170	
	6725	8'		65	.492		144	14.20	8.20	166.40	189	
	6730	Vinyl, 5'		65	.492		137	14.20	8.20	159.40	182	
	6735	6'	↓	63	.508	↓	182	14.65	8.45	205.10	233	

Important: See the Reference Section for critical supporting data - Reference Nos., Crews, & City Cost Indexes

			CREW	DAILY OUTPUT	LABOR-HOURS	UNIT	2005 BARE COSTS				TOTAL INCL O&P	
		**02820	Fences & Gates**					MAT.	LABOR	EQUIP.	TOTAL		
150	6740	7'	B-80	61	.525	Ea.	166	15.10	8.75	189.85	216	**150**
	6745	8'	↓	59	.542		217	15.65	9.05	241.70	273	
	7031	For corner, end, & pull post bracing, add					20	15				
	7795	Cantilever, manual, exp. roller, (pr), 40' wide x 8' high	B-22	1	30		3,325	940	238	4,503	5,350	
	7800	30' wide x 8' high		1	30		2,775	940	238	3,953	4,750	
	7805	24' wide x 8' high	↓	1	30		2,075	940	238	3,253	3,975	
	7810	Motor operators for gates, (no elec wiring), 3' wide swing	2 Skwk	.50	32		1,125	1,125		2,250	2,975	
	7815	Up to 20' wide swing		.50	32		3,825	1,125		4,950	5,925	
	7820	Up to 45' sliding		.50	32	↓	3,875	1,125		5,000	6,000	
	7825	Overhead gate, 6' to 18' wide, sliding/cantilever		45	.356	L.F.	136	12.40		148.40	168	
	7830	Gate operators, digital receiver		7	2.286	Ea.	410	79.50		489.50	575	
	7835	Two button transmitter		24	.667		52	23		75	93.50	
	7840	3 button station		14	1.143		65	40		105	134	
	7845	Master slave system	↓	4	4		230	139		369	470	
	7900	Auger fence post hole, 3' deep, medium soil, by hand	1 Clab	30	.267			7.10		7.10	11.10	
	7925	By machine	B-80	175	.183			5.25	3.05	8.30	11.45	
	7950	Rock, with jackhammer	B-9C	32	1.250			34	4.43	38.43	57.50	
	7975	With rock drill	B-47C	65	.246	↓		7.35	15.65	23	28.50	
170	0010	**FENCE, FABRIC & ACCESSORIES**										**170**
	1000	Fabric, 9 ga., galv., 1.2 oz. coat, 2" chain link, 4'	B-80A	304	.079	L.F.	3.19	2.11	.58	5.88	7.45	
	1150	5'		285	.084		3.87	2.25	.62	6.74	8.45	
	1200	6'		266	.090		6.10	2.41	.66	9.17	11.20	
	1250	7'		247	.097		9	2.59	.71	12.30	14.75	
	1300	8'		228	.105		11.20	2.81	.77	14.78	17.50	
	1400	9 ga., fused, 4'		304	.079		5.25	2.11	.58	7.94	9.65	
	1450	5'		285	.084		6.55	2.25	.62	9.42	11.40	
	1500	6'		266	.090		7.85	2.41	.66	10.92	13.15	
	1550	7'		247	.097		9.15	2.59	.71	12.45	14.95	
	1600	8'		228	.105		13.10	2.81	.77	16.68	19.60	
	1650	Barbed wire, galv., cost per strand		2,280	.011		.10	.28	.08	.46	.64	
	1700	Vinyl coated		2,280	.011	↓	.24	.28	.08	.60	.79	
	1750	Extension arms, 3 strands		143	.168	Ea.	7.30	4.48	1.23	13.01	16.30	
	1800	6 strands, 2-3/8"		119	.202		9.20	5.40	1.48	16.08	20	
	1850	Eye tops, 2-3/8"		143	.168	↓	3.02	4.48	1.23	8.73	11.65	
	1900	Top rail, incl. tie wires, 1-5/8", galv.		912	.026	L.F.	2.87	.70	.19	3.76	4.46	
	1950	Vinyl coated		912	.026		4.52	.70	.19	5.41	6.25	
	2100	Rail, middle/bottom, w/tie wire, 1-5/8", galv.		912	.026		2.42	.70	.19	3.31	3.96	
	2150	Vinyl coated		912	.026		3.27	.70	.19	4.16	4.90	
	2200	Reinforcing wire, coiled spring, 7 ga. galv.		2,279	.011		.15	.28	.08	.51	.70	
	2250	9 ga., vinyl coated	↓	2,282	.011	↓	.22	.28	.08	.58	.77	
300	0010	**FENCE, VINYL**, white, steel reinforced, stainless steel fasteners										**300**
	0020	Picket, 4" x 4" posts @ 6' - 0" OC, 3' high	B-1	140	.171	L.F.	15.90	4.69		20.59	25	
	0030	4' high		130	.185		18.30	5.05		23.35	28	
	0040	5' high		120	.200		21	5.45		26.45	31.50	
	0100	Board (semi-privacy), 5" x 5" posts @ 7' - 6" OC, 5' high		130	.185		22	5.05		27.05	32	
	0120	6' high		125	.192		25	5.25		30.25	35.50	
	0200	Basketweave, 5" x 5" posts @ 7' - 6" OC, 5' high		160	.150		19.30	4.11		23.41	27.50	
	0220	6' high		150	.160		23	4.38		27.38	32	
	0300	Privacy, 5" x 5" posts @ 7' - 6" OC, 5' high		130	.185	↓	23.50	5.05		28.55	34	
	0320	6' high		150	.160	↓	27	4.38		31.38	36.50	
	0350	Gate, 5' high		9	2.667	Ea.	330	73		403	480	
	0360	6' high		9	2.667		340	73		413	490	
	0400	For posts set in concrete, add	↓	25	.960	↓	6.30	26.50		32.80	48	
410	0010	**FENCES, MISC. METAL**										**410**
	0012	Chicken wire, posts @ 4', 1" mesh, 4' high	B-80C	410	.059	L.F.	1.63	1.56	.33	3.52	4.57	

02820 | Fences & Gates

		CREW	DAILY OUTPUT	LABOR-HOURS	UNIT	2005 BARE COSTS				TOTAL INCL O&P
						MAT.	LABOR	EQUIP.	TOTAL	
410	0100	2" mesh, 6' high	B-80C	350	.069	L.F.	1.47	1.83	.38	3.68	4.87	**410**
	0200	Galv. steel, 12 ga., 2" x 4" mesh, posts 5' O.C., 3' high		300	.080		2.21	2.14	.45	4.80	6.25
	0300	5' high		300	.080		2.95	2.14	.45	5.54	7.05
	0400	14 ga., 1" x 2" mesh, 3' high		300	.080		2.35	2.14	.45	4.94	6.40
	0500	5' high	↓	300	.080	↓	3.25	2.14	.45	5.84	7.40
	1000	Kennel fencing, 1-1/2" mesh, 6' long, 3'-6" wide, 6'-2" high	2 Clab	4	4	Ea.	370	107		477	570
	1050	12' long		4	4		445	107		552	650
	1200	Top covers, 1-1/2" mesh, 6' long		15	1.067		75	28.50		103.50	127
	1250	12' long	↓	12	1.333	↓	120	35.50		155.50	188
	4500	Security fence, prison grade, set in concrete, 12' high	B-80	25	1.280	L.F.	30.50	37	21.50	89	114
	4600	16' high	"	20	1.600	"	36.50	46	26.50	109	141
420	0010	**WIRE FENCING, GENERAL**										**420**
	0015	Barbed wire, galvanized, domestic steel, hi-tensile 15-1/2 ga.				M.L.F.	27			27	29.50
	0020	Standard, 12-3/4 ga.					36			36	39.50
	0210	Barbless wire, 2-strand galvanized, 12-1/2 ga.				↓	36			36	39.50
	0500	Helical razor ribbon, stainless steel, 18" dia x 18" spacing				C.L.F.	105			105	115
	0600	Hardware cloth galv., 1/4" mesh, 23 ga., 2' wide				C.S.F.	47			47	51.50
	0700	3' wide					45.50			45.50	50
	0900	1/2" mesh, 19 ga., 2' wide					41.50			41.50	45.50
	1000	4' wide					40.50			40.50	44.50
	1200	Chain link fabric, steel, 2" mesh, 6 ga, galvanized					155			155	170
	1300	9 ga, galvanized					76.50			76.50	84
	1350	Vinyl coated					64			64	70.50
	1360	Aluminized					99			99	109
	1400	2-1/4" mesh, 11.5 ga, galvanized					51.50			51.50	57
	1600	1-3/4" mesh (tennis courts), 11.5 ga (core), vinyl coated					73			73	80.50
	1700	9 ga, galvanized					66			66	72.50
	2100	Welded wire fabric, galvanized, 1" x 2", 14 ga.					44			44	48
	2200	2" x 4", 12-1/2 ga.				↓	14.95			14.95	16.45
510	0010	**FENCE, WOOD** Basket weave, 3/8" x 4" boards, 2" x 4"										**510**
	0020	stringers on spreaders, 4" x 4" posts									
	0050	No. 1 cedar, 6' high	B-80C	160	.150	L.F.	8.05	4.01	.84	12.90	15.95
	0070	Treated pine, 6' high	"	150	.160	"	9.75	4.28	.89	14.92	18.35
	0200	Board fence, 1" x 4" boards, 2" x 4" rails, 4" x 4" post									
	0220	Preservative treated, 2 rail, 3' high	B-80C	145	.166	L.F.	5.95	4.42	.92	11.29	14.40
	0240	4' high		135	.178		6.55	4.75	.99	12.29	15.65
	0260	3 rail, 5' high		130	.185		7.40	4.93	1.03	13.36	16.95
	0300	6' high		125	.192		8.45	5.15	1.07	14.67	18.45
	0320	No. 2 grade western cedar, 2 rail, 3' high		145	.166		6.50	4.42	.92	11.84	15
	0340	4' high		135	.178		7.70	4.75	.99	13.44	16.95
	0360	3 rail, 5' high		130	.185		8.90	4.93	1.03	14.86	18.60
	0400	6' high		125	.192		9.75	5.15	1.07	15.97	19.90
	0420	No. 1 grade cedar, 2 rail, 3' high		145	.166		9.80	4.42	.92	15.14	18.60
	0440	4' high		135	.178		11.10	4.75	.99	16.84	20.50
	0460	3 rail, 5' high		130	.185		12.85	4.93	1.03	18.81	23
	0500	6' high		125	.192		14.35	5.15	1.07	20.57	25
	0860	Open rail fence, split rails, 2 rail 3' high, no. 1 cedar		160	.150		5.40	4.01	.84	10.25	13.05
	0870	No. 2 cedar		160	.150		4.21	4.01	.84	9.06	11.75
	0880	3 rail, 4' high, no. 1 cedar		150	.160		7.30	4.28	.89	12.47	15.60
	0890	No. 2 cedar		150	.160		4.80	4.28	.89	9.97	12.90
	0920	Rustic rails, 2 rail 3' high, no. 1 cedar		160	.150		3.37	4.01	.84	8.22	10.85
	0930	No. 2 cedar		160	.150		3.23	4.01	.84	8.08	10.65
	0940	3 rail, 4' high		150	.160		4.53	4.28	.89	9.70	12.55
	0950	No. 2 cedar		150	.160		3.42	4.28	.89	8.59	11.35
	1240	Stockade fence, no. 1 cedar, 3-1/4" rails, 6' high	↓	160	.150	↓	9.80	4.01	.84	14.65	17.90

Important: See the Reference Section for critical supporting data - Reference Nos., Crews, & City Cost Indexes

02800 | Site Improvements and Amenities

02820 | Fences & Gates

			CREW	DAILY OUTPUT	LABOR-HOURS	UNIT	MAT.	LABOR	EQUIP.	TOTAL	TOTAL INCL O&P	
510	1260	8' high	B-80C	155	.155	L.F.	12.70	4.14	.86	17.70	21.50	510
	1270	Gate, 3'-6" wide		9	2.667	Ea.	166	71.50	14.90	252.40	310	
	1300	No. 2 cedar, treated wood rails, 6' high		160	.150	L.F.	9.80	4.01	.84	14.65	17.90	
	1320	Gate, 3'-6" wide		8	3	Ea.	57.50	80	16.75	154.25	206	
	1360	Treated pine, treated rails, 6' high		160	.150	L.F.	9.60	4.01	.84	14.45	17.65	
	1400	8' high		150	.160	"	14.40	4.28	.89	19.57	23.50	
	1420	Gate, 3'-6" wide		9	2.667	Ea.	63.50	71.50	14.90	149.90	196	
520	0010	FENCE, WOOD RAIL Picket, No. 2 cedar, Gothic, 2 rail, 3' high	B-1	160	.150	L.F.	5.15	4.11		9.26	12.05	520
	0050	Gate, 3'-6" wide	B-80C	9	2.667	Ea.	44.50	71.50	14.90	130.90	175	
	0400	3 rail, 4' high		150	.160	L.F.	5.90	4.28	.89	11.07	14.10	
	0500	Gate, 3'-6" wide		9	2.667	Ea.	53.50	71.50	14.90	139.90	185	
	5000	Fence rail, redwood, 2" x 4", merch grade 8'	B-1	2,400	.010	L.F.	1.01	.27		1.28	1.54	
	6000	Fence post, select redwood, earthpacked & treated, 4" x 4" x 6'		96	.250	Ea.	10	6.85		16.85	21.50	
	6010	4" x 4" x 8'		96	.250		11.85	6.85		18.70	23.50	
	6020	Set in concrete, 4" x 4" x 6'		50	.480		12.85	13.15		26	34.50	
	6030	4" x 4" x 8'		50	.480		15.35	13.15		28.50	37.50	
	6040	Wood post, 4' high, set in concrete, incl. concrete		50	.480		7.50	13.15		20.65	29	
	6050	Earth packed		96	.250		5.10	6.85		11.95	16.30	
	6060	6' high, set in concrete, incl. concrete		50	.480		9.60	13.15		22.75	31	
	6070	Earth packed		96	.250		7	6.85		13.85	18.35	

02830 | Retaining Walls

			CREW	DAILY OUTPUT	LABOR-HOURS	UNIT	MAT.	LABOR	EQUIP.	TOTAL	TOTAL INCL O&P	
100	0010	CAST IN PLACE RETAINING WALLS										100
	1800	Concrete gravity wall with vertical face including excavation & backfill										
	1850	No reinforcing										
	1900	6' high, level embankment	C-17C	36	2.306	L.F.	59.50	81.50	10.40	151.40	203	
	2000	33° slope embankment		32	2.594		54	91.50	11.65	157.15	214	
	2200	8' high, no surcharge		27	3.074		73.50	108	13.85	195.35	264	
	2300	33° slope embankment		24	3.458		89	122	15.55	226.55	305	
	2500	10' high, level embankment		19	4.368		105	154	19.65	278.65	375	
	2600	33° slope embankment		18	4.611		146	163	21	330	435	
	2800	Reinforced concrete cantilever, incl. excavation, backfill & reinf.										
	2900	6' high, 33° slope embankment	C-17C	35	2.371	L.F.	54	83.50	10.65	148.15	201	
	3000	8' high, 33° slope embankment		29	2.862		62.50	101	12.90	176.40	240	
	3100	10' high, 33° slope embankment		20	4.150		81	146	18.70	245.70	335	
	3200	20' high, 500 lb. per L.F. surcharge		7.50	11.067		243	390	50	683	925	
	3500	Concrete cribbing, incl. excavation and backfill										
	3700	12' high, open face	B-13	210	.267	S.F.	27	7.70	2.94	37.64	44.50	
	3900	Closed face	"	210	.267	"	25.50	7.70	2.94	36.14	43	
	4100	Concrete filled slurry trench, see Div. 02660-600										
200	0010	INTERLOCKING SEGMENTAL RETAINING WALLS										200
	7100	Segmental Retaining Wall system, incl pins, and void fill										
	7120	base not included										
	7140	Large unit, 8" high x 18" wide x 20" deep, 3 plane split	B-62	300	.080	S.F.	8.55	2.31	.51	11.37	13.55	
	7150	straight split		300	.080		8.55	2.31	.51	11.37	13.55	
	7160	Medium, ltwt, 8" high x 18" wide x 12" deep, 3 plane split		400	.060		7.55	1.73	.38	9.66	11.40	
	7170	straight split		400	.060		7.55	1.73	.38	9.66	11.40	
	7180	Small unit, 4" x 18" x 10" deep, 3 plane split		400	.060		9.45	1.73	.38	11.56	13.50	
	7190	straight split		400	.060		9.45	1.73	.38	11.56	13.50	
	7200	Cap unit, 3 plane split		300	.080		10.60	2.31	.51	13.42	15.80	
	7210	Cap unit, st split		300	.080		10.60	2.31	.51	13.42	15.80	
	7260	For reinforcing, add								2.20	2.75	
	8000	For higher walls, add components as necessary										

SITE CONSTRUCTION 2

115

SITE CONSTRUCTION 2

02830	Retaining Walls	CREW	DAILY OUTPUT	LABOR-HOURS	UNIT	2005 BARE COSTS				TOTAL INCL O&P
						MAT.	LABOR	EQUIP.	TOTAL	
600										**600**
0010	**METAL BIN RETAINING WALLS**, Aluminized steel bin, excavation									
0020	and backfill not included, 10' wide									
0100	4' high, 5.5' deep	B-13	650	.086	S.F.	16.30	2.48	.95	19.73	23
0200	8' high, 5.5' deep		615	.091		18.75	2.62	1	22.37	25.50
0300	10' high, 7.7' deep		580	.097		19.75	2.78	1.06	23.59	27
0400	12' high, 7.7' deep		530	.106		21.50	3.04	1.16	25.70	29.50
0500	16' high, 7.7' deep	↓	515	.109	↓	22.50	3.13	1.20	26.83	31
700										**700**
0010	**STONE GABION RETAINING WALLS**									
4300	Stone filled gabions, not incl. excavation,									
4310	Stone, delivered, 3' wide									
4340	Galvanized, 6' long, 1' high	B-13	113	.496	Ea.	61.50	14.25	5.45	81.20	95.50
4400	1'-6" high		50	1.120		76.50	32	12.35	120.85	148
4490	3'-0" high		13	4.308		128	124	47.50	299.50	385
4590	9' long, 1' high		50	1.120		86	32	12.35	130.35	158
4650	1'-6" high		22	2.545		113	73.50	28	214.50	268
4690	3'-0" high		6	9.333		194	269	103	566	740
4890	12' long, 1' high		28	2		115	57.50	22	194.50	239
4950	1'-6" high		13	4.308		148	124	47.50	319.50	405
4990	3'-0" high		3	18.667		255	535	206	996	1,325
5200	PVC coated, 6' long, 1' high		113	.496		68.50	14.25	5.45	88.20	103
5250	1'-6" high		50	1.120		84	32	12.35	128.35	156
5300	3' high		13	4.308		141	124	47.50	312.50	400
5500	9' long, 1' high		50	1.120		90.50	32	12.35	134.85	163
5550	1'-6" high		22	2.545		123	73.50	28	224.50	279
5600	3' high		6	9.333		208	269	103	580	755
5800	12' long, 1' high		28	2		125	57.50	22	204.50	251
5850	1'-6" high		13	4.308		161	124	47.50	332.50	420
5900	3' high		3	18.667	↓	273	535	206	1,014	1,350
6000	Galvanized, 6' long, 1' high		75	.747	C.Y.	61	21.50	8.20	90.70	109
6010	1'-6" high		50	1.120		76.50	32	12.35	120.85	148
6020	3'-0" high		25	2.240		64	64.50	24.50	153	197
6030	9' long, 1' high		50	1.120		86	32	12.35	130.35	158
6040	1'-6" high		33.30	1.682		75.50	48.50	18.50	142.50	178
6050	3'-0" high		16.70	3.353		64.50	96.50	37	198	261
6060	12' long, 1' high		37.50	1.493		86	43	16.45	145.45	179
6070	1'-6" high		25	2.240		74	64.50	24.50	163	208
6080	3'-0" high		12.50	4.480		64	129	49.50	242.50	325
6100	PVC coated, 6' long, 1' high		75	.747		102	21.50	8.20	131.70	155
6110	1'-6" high		50	1.120		84	32	12.35	128.35	156
6120	3' high		25	2.240		70.50	64.50	24.50	159.50	204
6130	9' long, 1' high		50	1.120		90.50	32	12.35	134.85	163
6140	1'-6" high		33.30	1.682		82	48.50	18.50	149	185
6150	3' high		16.67	3.359		69.50	96.50	37	203	266
6160	12' long, 1' high		37.50	1.493		94	43	16.45	153.45	187
6170	1'-6" high		25	2.240		73.50	64.50	24.50	162.50	208
6180	3' high	↓	12.50	4.480	↓	68.50	129	49.50	247	330
800										**800**
0010	**STONE RETAINING WALLS**									
0015	Including excavation, concrete footing and									
0020	stone 3' below grade. Price is exposed face area.									
0200	Decorative random stone, to 6' high, 1'-6" thick, dry set	D-1	35	.457	S.F.	35	14.20		49.20	60
0300	Mortar set		40	.400		35	12.45		47.45	57.50
0500	Cut stone, to 6' high, 1'-6" thick, dry set		35	.457		35	14.20		49.20	60
0600	Mortar set		40	.400		35	12.45		47.45	57.50
0800	Random stone, 6' to 10' high, 2' thick, dry set	↓	45	.356	↓	35	11.05		46.05	55.50

Important: See the Reference Section for critical supporting data - Reference Nos., Crews, & City Cost Indexes

SITE CONSTRUCTION 2

02830	Retaining Walls	CREW	DAILY OUTPUT	LABOR-HOURS	UNIT	MAT.	LABOR	EQUIP.	TOTAL	TOTAL INCL O&P	
800											**800**
0900	Mortar set	D-1	50	.320	S.F.	35	9.95		44.95	53.50	
1100	Cut stone, 6' to 10' high, 2' thick, dry set		45	.356		35	11.05		46.05	55.50	
1200	Mortar set		50	.320	↓	35	9.95		44.95	53.50	
5100	Setting stone, dry		100	.160	C.F.		4.97		4.97	7.60	
5600	With mortar	↓	120	.133	"		4.14		4.14	6.30	

02840	Walk/Road/Parking Appurtenances	CREW	DAILY OUTPUT	LABOR-HOURS	UNIT	MAT.	LABOR	EQUIP.	TOTAL	TOTAL INCL O&P	
150	7031 For corner, end, & pull post bracing, add					20%	15%				**150**
200	0010 **GUIDE/GUARD RAIL** Corrugated stl, galv. stl posts, 6'-3" O.C. [R02700-120]	B-80	850	.038	L.F.	15.20	1.08	.63	16.91	19.10	**200**
	0100 Double face		570	.056	"	22	1.62	.94	24.56	28	
	0200 End sections, galvanized, flared		50	.640	Ea.	63	18.45	10.70	92.15	110	
	0300 Wrap around end		50	.640		94.50	18.45	10.70	123.65	144	
	0350 Anchorage units	↓	15	2.133		760	61.50	35.50	857	970	
	0365 End section, flared	B-80A	78	.308		27.50	8.20	2.26	37.96	45.50	
	0370 End section, wrap-around, corrugated steel	"	78	.308	↓	34	8.20	2.26	44.46	53	
	0400 Timber guide rail, 4" x 8" with 6" x 8" wood posts, treated	B-80	960	.033	L.F.	20.50	.96	.56	22.02	24.50	
	0600 Cable guide rail, 3 at 3/4" cables, steel posts, single face		900	.036		7.25	1.02	.59	8.86	10.20	
	0650 Double face		635	.050		15.85	1.45	.84	18.14	20.50	
	0700 Wood posts		950	.034		9.50	.97	.56	11.03	12.55	
	0750 Double face	↓	650	.049		16.70	1.42	.82	18.94	21.50	
	0760 Breakaway wood posts		195	.164	Ea.	268	4.73	2.74	275.47	305	
	0800 Anchorage units, breakaway		15	2.133	"	690	61.50	35.50	787	895	
	0900 Guide rail, steel box beam, 6" x 6"	↓	120	.267	L.F.	23	7.70	4.45	35.15	42	
	0950 End assembly	B-80A	48	.500	Ea.	228	13.35	3.67	245.02	276	
	1100 Median barrier, steel box beam, 6" x 8"	B-80	215	.149	L.F.	28.50	4.29	2.48	35.27	41	
	1120 Shop curved	B-80A	92	.261	"	34	6.95	1.91	42.86	50.50	
	1140 End assembly		48	.500	Ea.	305	13.35	3.67	322.02	360	
	1150 Corrugated beam	↓	400	.060	L.F.	18.15	1.60	.44	20.19	23	
	1400 Resilient guide fence and light shield, 6' high	B-2	130	.308	"	22	8.35		30.35	37.50	
	1500 Concrete posts, individual, 6'-5", triangular	B-80	110	.291	Ea.	48.50	8.40	4.85	61.75	71.50	
	1550 Square		110	.291		52	8.40	4.85	65.25	75	
	1600 Wood guide posts	↓	150	.213	↓	33.50	6.15	3.56	43.21	50.50	
	2000 Median, precast concrete, 3'-6" high, 2' wide, single face	B-29	380	.147	L.F.	38	4.24	2.20	44.44	50.50	
	2200 Double face	"	340	.165		43	4.74	2.46	50.20	57.50	
	2300 Cast in place, steel forms	C-2	170	.282		82.50	9.40		91.90	105	
	2320 Slipformed	C-7	352	.205	↓	34.50	5.90	2.73	43.13	50	
	2400 Speed bumps, thermoplastic, 10-1/2" x 2-1/4" x 48" long	B-2	120	.333	Ea.	89.50	9.05		98.55	113	
	3000 Energy absorbing terminal, 10 bay, 3' wide	B-80B	.10	320		30,800	9,050	2,100	41,950	50,000	
	3010 7 bay, 2' - 6" wide		.20	160		22,600	4,525	1,050	28,175	32,900	
	3020 5 bay, 2' wide	↓	.20	160		17,500	4,525	1,050	23,075	27,400	
	3030 Impact barrier, UTMCD, barrel type	B-16	30	1.067		305	29	15.90	349.90	405	
	3100 Wide hazard protection, foam sandwich, 7 bay, 7' - 6" wide	B-80B	.14	228		22,600	6,475	1,500	30,575	36,500	
	3110 5' wide		.15	213		19,900	6,025	1,400	27,325	32,800	
	3120 3' wide	↓	.18	177	↓	20,500	5,025	1,175	26,700	31,600	
	5000 For bridge railing, see div. 02850-205										
400	0010 **FENDERS**										**400**
	0015 Bumper rails for garages, 12 Ga. rail, 6" wide, with steel										
	0020 posts 12'-6" O.C., minimum	E-4	190	.168	L.F.	12.15	6.50	.43	19.08	25.50	
	0030 Average		165	.194		15.20	7.50	.49	23.19	30.50	
	0100 Maximum		140	.229		18.25	8.85	.58	27.68	36.50	
	0300 12" channel rail, minimum		160	.200		15.20	7.75	.51	23.46	31	
	0400 Maximum	↓	120	.267	↓	23	10.30	.68	33.98	44	

2 SITE CONSTRUCTION

		02840	Walk/Road/Parking Appurtenances	CREW	DAILY OUTPUT	LABOR-HOURS	UNIT	MAT.	LABOR	EQUIP.	TOTAL	TOTAL INCL O&P	
									2005 BARE COSTS				
800	0010		**PARKING BUMPERS**										800
	0020		Parking barriers, timber w/saddles, treated type										
	0100		4" x 4" for cars	B-2	520	.077	L.F.	2.47	2.08		4.55	5.95	
	0200		6" x 6" for trucks		520	.077	"	5.15	2.08		7.23	8.95	
	0600		Flexible fixed stanchion, 2' high, 3" diameter		100	.400	Ea.	23.50	10.85		34.35	43	
	1000		Wheel stops, precast concrete incl. dowels, 6" x 10" x 6'-0"		120	.333		38	9.05		47.05	56	
	1100		8" x 13" x 6'-0"		120	.333		44	9.05		53.05	62.50	
	1200		Thermoplastic, 6" x 10" x 6'-0"		120	.333		69	9.05		78.05	90	
	1300		Pipe bollards, conc filled/paint, 8' L x 4' D hole, 6" diam.	B-6	20	1.200		279	34.50	10.80	324.30	370	
	1400		8" diam.		15	1.600		420	46	14.40	480.40	550	
	1500		12" diam.		12	2		550	57.50	18	625.50	715	
	2030		Folding with individual padlocks	B-2	50	.800		675	21.50		696.50	780	
910	0010		**TRAFFIC BARRIERS, TRAFFIC CONTROL DEVICES**										910
	0020		Crash barriers										
	0100		Traffic channelizing pavement markers, layout only	A-7	2,000	.012	Ea.		.44	.03	.47	.71	
	0110		13" x 7-1/2" x 2-1/2" high, non-plowable install	2 Clab	96	.167		18.40	4.45		22.85	27	
	0200		8" x 8" x 3-1/4" high, non-plowable, install		96	.167		16.70	4.45		21.15	25.50	
	0230		4" x 4" x 3/4" high, non-plowable, install		120	.133		2.23	3.56		5.79	8	
	0240		9-1/4" x 5-7/8" x 1/4" high, plowable, concrete pav't	A-2A	70	.343		13.35	9.15	3.47	25.97	32.50	
	0250		9-1/4" x 5-7/8" x 1/4" high, plowable, asphalt pav't	"	120	.200		2.56	5.35	2.03	9.94	13.30	
	0300		Barrier and curb delineators, reflectorized, 2" x 4"	2 Clab	150	.107		1.45	2.85		4.30	6.05	
	0310		3" x 5"	"	150	.107		2.95	2.85		5.80	7.70	
	0500		Rumble strip, polycarbonate										
	0510		24" x 3-1/2" x 1/2" high	2 Clab	50	.320	Ea.	5.55	8.55		14.10	19.45	
920	0010		**TRAFFIC BARRIERS, HIGHWAY SOUND BARRIERS**										920
	0020		Highway sound barriers, not including footing										
	0100		Precast concrete, concrete columns @ 30' OC, 8" T, 8' H	C-12	400	.120	L.F.	84.50	4.03	1.51	90.04	101	
	0110		12' H		265	.181		127	6.10	2.27	135.37	152	
	0120		16' H		200	.240		169	8.05	3.01	180.06	202	
	0130		20' H		160	.300		211	10.10	3.77	224.87	253	
	0400		Lt. Wt. composite panel, cementitious face, St. posts @ 12' OC, 8' H	B-80B	190	.168		95	4.76	1.10	100.86	114	
	0410		12' H		125	.256		143	7.25	1.68	151.93	170	
	0420		16' H		95	.337		190	9.55	2.21	201.76	226	
	0430		20' H		75	.427		238	12.05	2.79	252.84	283	

02850 | Prefabricated Bridges

				CREW	DAILY OUTPUT	LABOR-HOURS	UNIT	MAT.	LABOR	EQUIP.	TOTAL	TOTAL INCL O&P	
205	0010		**BRIDGES, HIGHWAY**										205
	0020		Structural steel, rolled beams	E-5	8.50	9.412	Ton	1,750	355	165	2,270	2,725	
	0500		Built up, plate girders	E-6	10.50	12.190	"	2,100	460	145	2,705	3,250	
	1000		Concrete in place, no reinforcing, abutment footings	C-17B	30	2.733	C.Y.	126	96.50	8.50	231	298	
	1050		Abutment		23	3.565		126	126	11.10	263.10	345	
	1100		Walls, stems and wing walls		20	4.100		170	145	12.75	327.75	425	
	1150		Parapets		12	6.833		170	241	21.50	432.50	585	
	1170		Pier footings		40	2.050		126	72.50	6.40	204.90	258	
	1180		Piers and columns, see div. 03310-240										
	1190		Pier caps including shoring	C-17B	10	8.200	C.Y.	208	289	25.50	522.50	705	
	1210		Beams, including shoring	C-14	15	9.600		256	320	66.50	642.50	855	
	1220		Haunches		10	14.400		296	480	99.50	875.50	1,175	
	1230		Bridge sidewalks		750	.192	S.F.	5.55	6.40	1.33	13.28	17.55	
	1250		Decks, including finish and cure, 8" thick										
	1260		Shored forms	C-14	3,000	.048	S.F.	4.48	1.59	.33	6.40	7.80	
	1270		Beam supported forms		3,500	.041		3.28	1.37	.28	4.93	6.05	
	1280		Stay in place metal slab forms, 20 gauge		5,000	.029		4.48	.96	.20	5.64	6.65	
	1500		Precast, prestressed concrete										
	1510		Box girders, 35' to 50' span				Ea.				11,200	12,900	
	1520		50' to 65' span								15,000	17,200	

02800 | Site Improvements and Amenities

02850 | Prefabricated Bridges

		CREW	DAILY OUTPUT	LABOR-HOURS	UNIT	MAT.	LABOR	EQUIP.	TOTAL	TOTAL INCL O&P	
205	1530	65' to 85' span				Ea.				19,300	22,100	205
	1540	Deck beams, 12" deep				S.F.				43	49	
	1541	15" deep								48	55.50	
	1542	18" deep								53.50	62	
	1543	20" deep								59	67.50	
	1550	Double T, 40' to 45' span				Ea.				8,125	9,350	
	1555	45' to 55' span								8,550	9,850	
	1560	55' to 65' span								8,875	9,050	
	1580	Price per S.F.				S.F.				18.50	21.30	
	1600	I beams, 60' to 80' span				Ea.				8,775	10,100	
	1610	80' to 100' span								10,400	11,900	
	1620	100' to 120' span								13,900	16,000	
	1630	120' to 140' span								16,600	19,100	
	2000	Reinforcing, in place	4 Rodm	3	10.667	Ton	1,400	405		1,805	2,225	
	2050	Galvanized coated		3	10.667		1,775	405		2,180	2,625	
	2100	Epoxy coated		3	10.667		2,600	405		3,005	3,550	
	2110	See also Div. 03200										
	3000	Expansion dams, steel, double upset 4" x 8" angles welded to										
	3010	double 8" x 8" angles, 1-3/4" compression seal	C-22	30	1.400	L.F.	390	53.50	2.58	446.08	515	
	3040	Double 8" x 8" angles only, 1-3/4" compression seal		35	1.200		291	45.50	2.21	338.71	395	
	3050	Galvanized		35	1.200		360	45.50	2.21	407.71	470	
	3060	Double 8" x 6" angles only, 1-3/4" compression seal		35	1.200		223	45.50	2.21	270.71	320	
	3100	Double 10" channels, 1-3/4" compression seal		35	1.200		204	45.50	2.21	251.71	300	
	3420	For 3" compression seal, add					42			42	46	
	3440	For double slotted extrusions with seal strip, add					194			194	213	
	3490	For galvanizing, add				Lb.	.71			.71	.78	
	4000	Approach railings, steel, galv. pipe, 2 line	C-22	140	.300	L.F.	90.50	11.45	.55	102.50	119	
	4200	Bridge railings, steel, galv. pipe, 3 line w/screen		85	.494		284	18.85	.91	303.76	345	
	4220	4 line w/screen		75	.560		245	21.50	1.03	267.53	305	
	4300	Aluminum, pipe, 3 line w/screen		95	.442		97	16.85	.81	114.66	135	
	8000	For structural excavation, see div. 02315-462										
	8010	For dewatering, see div. 02240										
210	0010	**BRIDGES, PEDESTRIAN** spans over streams, roadways, etc.										210
	0020	including erection, not including foundations										
	0050	Precast concrete, complete in place, 8' wide, 60' span	E-2	215	.260	S.F.	49.50	9.65	6.15	65.30	78	
	0100	100' span		185	.303		54	11.20	7.15	72.35	86.50	
	0150	120' span		160	.350		58.50	12.95	8.30	79.75	95.50	
	0200	150' span		145	.386		61	14.25	9.15	84.40	102	
	0300	Steel, trussed or arch spans, compl. in place, 8' wide, 40' span		320	.175		62.50	6.45	4.14	73.09	84	
	0400	50' span		395	.142		56	5.25	3.35	64.60	74	
	0500	60' span		465	.120		56	4.45	2.85	63.30	72.50	
	0600	80' span		570	.098		66.50	3.63	2.32	72.45	82.50	
	0700	100' span		465	.120		93.50	4.45	2.85	100.80	114	
	0800	120' span		365	.153		118	5.65	3.63	127.28	144	
	0900	150' span		310	.181		126	6.70	4.27	136.97	154	
	1000	160' span		255	.220		126	8.10	5.20	139.30	158	
	1100	10' wide, 80' span		640	.087		69.50	3.23	2.07	74.80	84	
	1200	120' span		415	.135		90	4.99	3.19	98.18	111	
	1300	150' span		445	.126		101	4.65	2.98	108.63	122	
	1400	200' span		205	.273		108	10.10	6.45	124.55	142	
	1600	Wood, laminated type, complete in place, 80' span	C-12	203	.236		42.50	7.95	2.97	53.42	62	
	1700	130' span	"	153	.314		44	10.55	3.94	58.49	69	

02870 | Site Furnishings

		CREW	DAILY OUTPUT	LABOR-HOURS	UNIT	MAT.	LABOR	EQUIP.	TOTAL	TOTAL INCL O&P	
310	0010	**BENCHES**										310
	0012	Seating, benches, park, precast conc, w/backs, wood rails, 4' long	2 Clab	5	3.200	Ea.	320	85.50		405.50	485	

			CREW	DAILY OUTPUT	LABOR-HOURS	UNIT	2005 BARE COSTS				TOTAL INCL O&P	
		02870 | Site Furnishings					MAT.	LABOR	EQUIP.	TOTAL		
310	0100	8' long	2 Clab	4	4	Ea.	670	107		777	900	**310**
	0300	Fiberglass, without back, one piece, 4' long		10	1.600		405	42.50		447.50	515	
	0400	8' long		7	2.286		840	61		901	1,025	
	0500	Steel barstock pedestals w/backs, 2" x 3" wood rails, 4' long		10	1.600		770	42.50		812.50	915	
	0510	8' long		7	2.286		910	61		971	1,100	
	0520	3" x 8" wood plank, 4' long		10	1.600		775	42.50		817.50	920	
	0530	8' long		7	2.286		810	61		871	985	
	0540	Backless, 4" x 4" wood plank, 4' square		10	1.600		755	42.50		797.50	895	
	0550	8' long		7	2.286		715	61		776	885	
	0600	Aluminum pedestals, with backs, aluminum slats, 8' long		8	2		370	53.50		423.50	490	
	0610	15' long		5	3.200		360	85.50		445.50	535	
	0620	Portable, aluminum slats, 8' long		8	2		320	53.50		373.50	440	
	0630	15' long		5	3.200		470	85.50		555.50	655	
	0800	Cast iron pedestals, back & arms, wood slats, 4' long		8	2		290	53.50		343.50	405	
	0820	8' long		5	3.200		845	85.50		930.50	1,075	
	0840	Backless, wood slats, 4' long		8	2		505	53.50		558.50	640	
	0860	8' long		5	3.200		500	85.50		585.50	685	
	1700	Steel frame, fir seat, 10' long	▼	10	1.600	▼	176	42.50		218.50	261	
		02880 | Playfield Equipment										
100	0010	**ATHLETIC OR RECREATIONAL SCREENING**										**100**
	0015	Backstops, baseball, prefabricated, 30' wide, 12' high & 1 overhang	B-1	1	24	Ea.	2,475	655		3,130	3,750	
	0100	40' wide, 12' high & 2 overhangs	"	.75	32		3,100	875		3,975	4,800	
	0300	Basketball, steel, single goal	B-13	3.04	18.421		975	530	203	1,708	2,125	
	0400	Double goal	"	1.92	29.167	▼	590	840	320	1,750	2,300	
	0600	Tennis, wire mesh with pair of ends	B-1	2.48	9.677	Set	1,525	265		1,790	2,075	
	0700	Enclosed court	"	1.30	18.462	Ea.	4,325	505		4,830	5,550	
	0900	Handball or squash court, outdoor, wood	2 Carp	.50	32		3,575	1,100		4,675	5,625	
	1000	Masonry handball/squash court	D-1	.30	53.333	▼	21,700	1,650		23,350	26,400	
110	0010	**TENNIS COURT FENCES**										**110**
	0860	Tennis courts, 11 ga. wire, 2-1/2" post set										
	0870	in concrete, 10' O.C., 1-5/8" top rail										
	0900	10' high	B-80	190	.168	L.F.	11.65	4.85	2.81	19.31	23.50	
	0920	12' high		170	.188	"	13.65	5.40	3.14	22.19	27	
	1000	Add for gate 4' wide, 1-5/8" frame 7' high		10	3.200	Ea.	109	92	53.50	254.50	320	
	1040	Aluminized steel, 11 ga. wire 10' high		190	.168	L.F.	13.85	4.85	2.81	21.51	26	
	1100	12' high		170	.188	"	17.05	5.40	3.14	25.59	30.50	
	1140	Add for gate 4' wide, 1-5/8" frame, 7' high		10	3.200	Ea.	126	92	53.50	271.50	340	
	1250	Vinyl covered, 9 ga. wire, 10' high		190	.168	L.F.	13.65	4.85	2.81	21.31	25.50	
	1300	12' high	▼	170	.188	"	22.50	5.40	3.14	31.04	37	
	1310	Fence, CL, tennis ct, transom gate, single, galv, 4' x 7' x 3'	B-80A	8.72	2.752	Ea.	237	73.50	20	330.50	395	
	1400	Add for gate 4' wide, 1-5/8" frame, 7' high	B-80	10	3.200	"	127	92	53.50	272.50	340	
330	0010	**GOAL POSTS**										**330**
	0020	Goal posts, steel, football, double post	B-1	1.50	16	Pr.	1,525	440		1,965	2,350	
	0100	Deluxe, single post		1.50	16		2,475	440		2,915	3,400	
	0300	Football, convertible to soccer		1.50	16		2,450	440		2,890	3,375	
	0500	Soccer, regulation	▼	2	12	▼	1,950	330		2,280	2,625	
		02890 | Traffic Signs & Signals										
100	0010	**SIGNS** Stock, 24" x 24", no posts, .080" alum. reflectorized	B-80	70	.457	Ea.	55	13.15	7.65	75.80	89	**100**
	0100	High intensity		70	.457		55	13.15	7.65	75.80	89	
	0300	30" x 30", reflectorized		70	.457		112	13.15	7.65	132.80	151	
	0400	High intensity	▼	70	.457		112	13.15	7.65	132.80	151	

2 SITE CONSTRUCTION

02890	Traffic Signs & Signals	CREW	DAILY OUTPUT	LABOR-HOURS	UNIT	2005 BARE COSTS				TOTAL INCL O&P	
						MAT.	LABOR	EQUIP.	TOTAL		
100 0600	Guide and directional signs, 12" x 18", reflectorized	B-80	70	.457	Ea.	35.50	13.15	7.65	56.30	68	**100**
0700	High intensity		70	.457		34	13.15	7.65	54.80	66	
0900	18" x 24", stock signs, reflectorized		70	.457		40	13.15	7.65	60.80	72.50	
1000	High intensity		70	.457		40	13.15	7.65	60.80	72.50	
1200	24" x 24", stock signs, reflectorized		70	.457		49.50	13.15	7.65	70.30	83	
1300	High intensity		70	.457		49.50	13.15	7.65	70.30	83	
1500	Add to above for steel posts, galvanized, 10'-0" upright, bolted		200	.160		17.90	4.61	2.67	25.18	29.50	
1600	12'-0" upright, bolted		140	.229	↓	24	6.60	3.81	34.41	40.50	
1800	Highway road signs, aluminum, over 20 S. F., reflectorized		350	.091	S.F.	23	2.63	1.53	27.16	30.50	
2000	High intensity		350	.091		23	2.63	1.53	27.16	30.50	
2200	Highway, suspended over road, 80 S.F. min., reflectorized		165	.194		23	5.60	3.23	31.83	37	
2300	High intensity		165	.194	↓	23	5.60	3.23	31.83	37	
2350	Roadway delineators and reference markers		500	.064	Ea.	23	1.84	1.07	25.91	29	
2360	Delineator post only, 6'	↓	500	.064	↓	7.25	1.84	1.07	10.16	11.95	
2400	Highway sign bridge structure, 45' to 80'								24,600	26,800	
2410	Cantilever structure, add				↓				20%		
5000	Removal of signs, including supports										
5020	To 10 S.F.	B-80B	16	2	Ea.		56.50	13.10	69.60	101	
5030	11 S.F. to 20 S.F.	"	5	6.400			181	42	223	325	
5040	21 S.F. to 40 S.F.	B-14	1.80	26.667			750	120	870	1,275	
5050	41 S.F. to 100 S.F.	B-13	1.30	43.077	↓		1,250	475	1,725	2,425	
5200	Remove and relocate signs, including supports										
5210	Remove and relocate signs, to 10 S.F.	B-80B	5	6.400	Ea.	248	181	42	471	595	
5220	11 S.F. to 20 S.F.	"	1.70	18.824		555	535	123	1,213	1,575	
5230	21 S.F. to 40 S.F.	B-14	.56	85.714		580	2,400	385	3,365	4,800	
5240	41 S.F. to 100 S.F.	B-13	.32	175	↓	955	5,050	1,925	7,930	10,900	
8000	For temporary barricades and lights, see div. 01560-100										
300 0010	**TRAFFIC SIGNALS** Mid block pedestrian crosswalk,										**300**
0020	with pushbutton and mast arms	R-2	.30	186	Total	54,000	6,675	1,050	61,725	71,000	
0100	Intersection, 8 signals w/three sect. (2 each direction), programmed	"	.15	373		121,000	13,400	2,125	136,525	156,000	
0120	For each additional traffic phase controller, add	L-9	1.20	30		6,525	930		7,455	8,675	
0200	Semi-actuated, detectors in side street only, add		.81	44.444		9,775	1,375		11,150	13,000	
0300	Fully-actuated, detectors in all streets, add		.49	73.469		33,200	2,275		35,475	40,200	
0400	For pedestrian pushbutton, add		.70	51.429	↓	18,900	1,600		20,500	23,400	
0500	Optically programmed signal only, add per head		1.64	21.951		10,900	680		11,580	13,100	
0600	School flashing system, programmed	↓	.41	87.805	Signal	25,500	2,725		28,225	32,500	
1000	Intersection traffic signals, LED, mast, programmable, no lane control										
1010	Includes all labor, material and equip. for complete installation				Ea.	150,000			150,000	165,000	
1100	Intersection traffic signals, LED, mast, programmable, R/L lane control										
1110	Includes all labor, material and equip. for complete installation				Ea.	200,000			200,000	220,000	
1200	Add protective/permissive left turns to existing traffic light										
1210	Includes all labor, material and equip. for complete installation				Ea.	60,000			60,000	66,000	
1300	Replace existing light heads with LED Heads wire hung										
1310	Includes all labor, material and equip. for complete installation				Ea.	40,000			40,000	44,000	
1400	Replace existing light heads with LED Heads mast arm hung										
1410	Includes all labor, material and equip. for complete installation				Ea.	80,000			80,000	88,000	

SITE CONSTRUCTION 2

2

SITE CONSTRUCTION

02905	Plants, Planting, Transplanting	CREW	DAILY OUTPUT	LABOR-HOURS	UNIT	2005 BARE COSTS				TOTAL INCL O&P	
						MAT.	LABOR	EQUIP.	TOTAL		
725	**0010**	**PLANTING**									**725**
	0012	Moving shrubs on site, 12" ball	B-62	28	.857	Ea.		24.50	5.50	30	44
	0100	24" ball	"	22	1.091			31.50	6.95	38.45	56
	0300	Moving trees on site, 36" ball	B-6	3.75	6.400			184	57.50	241.50	350
	0400	60" ball	"	1	24	▼		690	216	906	1,300
925	**0010**	**TREE REMOVAL**				Ea.					**925**
	0100	Dig & lace, shrubs, broadleaf evergreen, 18"-24"	B-1	55	.436			11.95		11.95	18.60
	0200	2'-3'	"	35	.686			18.75		18.75	29
	0300	3'-4'	B-6	30	.800			23	7.20	30.20	43.50
	0400	4'-5'	"	20	1.200			34.50	10.80	45.30	65
	1000	Deciduous, 12"-15"	B-1	110	.218			5.95		5.95	9.30
	1100	18"-24"	⬇	65	.369			10.10		10.10	15.75
	1200	2'-3'	⬇	55	.436			11.95		11.95	18.60
	1300	3'-4'	B-6	50	.480			13.85	4.32	18.17	26.50
	2000	Evergreeen, 18"-24"	B-1	55	.436			11.95		11.95	18.60
	2100	2'-0" to 2'-6"		50	.480			13.15		13.15	20.50
	2200	2'-6" to 3'-0"		35	.686			18.75		18.75	29
	2300	3'-0" to 3'-6"		20	1.200			33		33	51
	3000	Trees, deciduous, small, 2'-3'	⬇	55	.436			11.95		11.95	18.60
	3100	3'-4'	B-6	50	.480			13.85	4.32	18.17	26.50
	3200	4'-5'		35	.686			19.75	6.15	25.90	37.50
	3300	5'-6'		30	.800			23	7.20	30.20	43.50
	4000	Shade, 5'-6'		50	.480			13.85	4.32	18.17	26.50
	4100	6'-8'		35	.686			19.75	6.15	25.90	37.50
	4200	8'-10'		25	.960			27.50	8.65	36.15	52
	4300	2" caliper		12	2			57.50	18	75.50	108
	5000	Evergreen, 4'-5'		35	.686			19.75	6.15	25.90	37.50
	5100	5'-6'		25	.960			27.50	8.65	36.15	52
	5200	6'-7'		19	1.263			36.50	11.35	47.85	68.50
	5300	7'-8'		15	1.600			46	14.40	60.40	87
	5400	8'-10'	⬇	11	2.182	▼		63	19.60	82.60	118

**02910	Plant Preparation**										
500	**0010**	**MULCHING**									**500**
	0200	Hay, 1" deep, hand spread	1 Clab	475	.017	S.Y.	.56	.45		1.01	1.32
	0250	Power mulcher, small	B-64	180	.089	M.S.F.	62	2.38	1.29	65.67	73.50
	0350	Large	B-65	530	.030	"	62	.81	.59	63.40	70.50
	0400	Humus peat, 1" deep, hand spread	1 Clab	700	.011	S.Y.	2.13	.31		2.44	2.81
	0450	Push spreader	"	2,500	.003	"	2.13	.09		2.22	2.47
	0550	Tractor spreader	B-66	700	.011	M.S.F.	237	.38	.25	237.63	261
	0600	Oat straw, 1" deep, hand spread	1 Clab	475	.017	S.Y.	.32	.45		.77	1.05
	0650	Power mulcher, small	B-64	180	.089	M.S.F.	35.50	2.38	1.29	39.17	44
	0700	Large	B-65	530	.030	"	35.50	.81	.59	36.90	41
	0750	Add for asphaltic emulsion	B-45	1,770	.009	Gal.	1.74	.28	.30	2.32	2.67
	0800	Peat moss, 1" deep, hand spread	1 Clab	900	.009	S.Y.	1.69	.24		1.93	2.23
	0850	Push spreader	"	2,500	.003	"	1.69	.09		1.78	1.99
	0950	Tractor spreader	B-66	700	.011	M.S.F.	188	.38	.25	188.63	208
	1000	Polyethylene film, 6 mil.	2 Clab	2,000	.008	S.Y.	.16	.21		.37	.51
	1010	4 mil		2,300	.007		.13	.19		.32	.43
	1020	1-1/2 mil		2,500	.006		.09	.17		.26	.37
	1050	Filter fabric weed barrier	⬇	2,000	.008		.75	.21		.96	1.16
	1100	Redwood nuggets, 3" deep, hand spread	1 Clab	150	.053	▼	4.17	1.42		5.59	6.80
	1150	Skid steer loader	B-63	13.50	2.963	M.S.F.	465	83	11.40	559.40	650
	1200	Stone mulch, hand spread, ceramic chips, economy	1 Clab	125	.064	S.Y.	6.10	1.71		7.81	9.40
	1250	Deluxe	"	95	.084	"	9.40	2.25		11.65	13.85

Important: See the Reference Section for critical supporting data - Reference Nos., Crews, & City Cost Indexes

		02910	Plant Preparation	CREW	DAILY OUTPUT	LABOR-HOURS	UNIT	2005 BARE COSTS MAT.	LABOR	EQUIP.	TOTAL	TOTAL INCL O&P	
500	1300		Granite chips	B-1	10	2.400	C.Y.	30.50	65.50		96	136	**500**
	1400		Marble chips		10	2.400		115	65.50		180.50	228	
	1600		Pea gravel		28	.857		56	23.50		79.50	98	
	1700		Quartz		10	2.400		148	65.50		213.50	264	
	1800		Tar paper, 15 Lb. felt	1 Clab	800	.010	S.Y.	.31	.27		.58	.76	
	1900		Wood chips, 2" deep, hand spread	"	220	.036	"	1.74	.97		2.71	3.42	
	1950		Skid steer loader	B-63	20.30	1.970	M.S.F.	193	55	7.55	255.55	305	
710	0010	**LAWN BED PREPARATION**											**710**
	0100		Rake topsoil, site material, harley rock rake, ideal	B-6	33	.727	M.S.F.		21	6.55	27.55	39	
	0200		Adverse	"	7	3.429			99	31	130	186	
	0300		Screened loam, york rake and finish, ideal	B-62	24	1			29	6.40	35.40	51.50	
	0400		Adverse	"	20	1.200			34.50	7.65	42.15	61.50	
	1000		Remove topsoil & stock pile on site, 75 HP dozer, 6" deep, 50' haul	B-10L	30	.400			12.80	10.50	23.30	31	
	1050		300' haul		6.10	1.967			63	51.50	114.50	153	
	1100		12" deep, 50' haul		15.50	.774			25	20.50	45.50	60	
	1150		300' haul		3.10	3.871			124	102	226	300	
	1200		200 HP dozer, 6" deep, 50' haul	B-10B	125	.096			3.07	7.35	10.42	12.75	
	1250		300' haul		30.70	.391			12.50	30	42.50	52	
	1300		12" deep, 50' haul		62	.194			6.20	14.85	21.05	25.50	
	1350		300' haul		15.40	.779			25	59.50	84.50	104	
	1400		Alternate method, 75 HP dozer, 50' haul	B-10L	860	.014	C.Y.		.45	.37	.82	1.08	
	1450		300' haul	"	114	.105			3.37	2.76	6.13	8.15	
	1500		200 HP dozer, 50' haul	B-10B	2,660	.005			.14	.35	.49	.60	
	1600		300' haul	"	570	.021			.67	1.61	2.28	2.79	
	1800		Rolling topsoil, hand push roller	1 Clab	3,200	.002	S.F.		.07		.07	.10	
	1850		Tractor drawn roller	B-66	10,666	.001	"		.02	.02	.04	.06	
	1900		Remove rocks & debris from grade, by hand	B-62	80	.300	M.S.F.		8.65	1.92	10.57	15.40	
	1920		With rock picker	B-10S	140	.086			2.74	1.72	4.46	6.05	
	2000		Root raking and loading, residential, no boulders	B-6	53.30	.450			13	4.05	17.05	24.50	
	2100		With boulders		32	.750			21.50	6.75	28.25	40.50	
	2200		Municipal, no boulders		200	.120			3.46	1.08	4.54	6.50	
	2300		With boulders		120	.200			5.75	1.80	7.55	10.85	
	2400		Large commercial, no boulders	B-10B	400	.030			.96	2.30	3.26	3.99	
	2500		With boulders	"	240	.050			1.60	3.83	5.43	6.65	
	2600		Rough grade & scarify subsoil to receive topsoil, common earth										
	2610		200 H.P. dozer with scarifier	B-11A	80	.200	M.S.F.		6.15	11.50	17.65	22	
	2620		180 H.P. grader with scarifier	B-11L	110	.145			4.46	4.15	8.61	11.35	
	2700		Clay and till, 200 H.P. dozer with scarifier	B-11A	50	.320			9.80	18.40	28.20	35	
	2710		180 H.P. grader with scarifier	B-11L	40	.400			12.25	11.40	23.65	31.50	
	3000		Scarify subsoil, residential, skid steer loader w/scarifiers, 50 HP	B-66	32	.250			8.25	5.40	13.65	18.40	
	3050		Municipal, skid steer loader w/scarifiers, 50 HP	"	120	.067			2.20	1.44	3.64	4.91	
	3100		Large commercial, 75 HP, dozer w/scarifier	B-10L	240	.050			1.60	1.31	2.91	3.87	
	3200		Grader with scarifier, 135 H.P.	B-11L	280	.057			1.75	1.63	3.38	4.47	
	3500		Screen topsoil from stockpile, vibrating screen, wet material (organic)	B-10P	200	.060	C.Y.		1.92	3.94	5.86	7.25	
	3550		Dry material	"	300	.040			1.28	2.63	3.91	4.84	
	3600		Mixing with conditioners, manure and peat	B-10R	550	.022			.70	.35	1.05	1.45	
	3650		Mobilization add for 2 days or less operation	B-34K	3	2.667	Job		73.50	175	248.50	305	
	3800		Spread conditioned topsoil, 6" deep, by hand	B-1	360	.067	S.Y.	4.71	1.82		6.53	8.05	
	3850		300 HP dozer	B-10M	27	.444	M.S.F.	510	14.20	44.50	568.70	630	
	3900		4" deep, by hand	B-1	470	.051	S.Y.	4.24	1.40		5.64	6.85	
	3920		300 H.P. dozer	B-10M	34	.353	M.S.F.	380	11.30	35	426.30	475	
	3940		180 H.P. grader	B-11L	37	.432	"	380	13.25	12.35	405.60	455	
	4000		Spread soil conditioners, alum. sulfate, 1#/S.Y., hand push spreader	1 Clab	17,500	.001	S.Y.	13.90	.01		13.91	15.30	
	4050		Tractor spreader	B-66	700	.011	M.S.F.	1,550	.38	.25	1,550.63	1,700	
	4100		Fertilizer, 0.2#/S.Y., push spreader	1 Clab	17,500	.001	S.Y.	.07	.01		.08	.10	

02910 | Plant Preparation

			CREW	DAILY OUTPUT	LABOR-HOURS	UNIT	MAT.	LABOR	EQUIP.	TOTAL	TOTAL INCL O&P	
710	4150	Tractor spreader	B-66	700	.011	M.S.F.	7.80	.38	.25	8.43	9.40	710
	4200	Ground limestone, 1#/S.Y., push spreader	1 Clab	17,500	.001	S.Y.	.08	.01		.09	.11	
	4250	Tractor spreader	B-66	700	.011	M.S.F.	8.90	.38	.25	9.53	10.65	
	4300	Lusoil, 3#/S.Y., push spreader	1 Clab	17,500	.001	S.Y.	.43	.01		.44	.49	
	4350	Tractor spreader	B-66	700	.011	M.S.F.	48	.38	.25	48.63	53.50	
	4400	Manure, 18#/S.Y., push spreader	1 Clab	2,500	.003	S.Y.	2.54	.09		2.63	2.92	
	4450	Tractor spreader	B-66	280	.029	M.S.F.	282	.94	.62	283.56	310	
	4500	Perlite, 1" deep, push spreader	1 Clab	17,500	.001	S.Y.	7.85	.01		7.86	8.60	
	4550	Tractor spreader	B-66	700	.011	M.S.F.	870	.38	.25	870.63	960	
	4600	Vermiculite, push spreader	1 Clab	17,500	.001	S.Y.	2.39	.01		2.40	2.65	
	4650	Tractor spreader	B-66	700	.011	M.S.F.	266	.38	.25	266.63	293	
	5000	Spread topsoil, skid steer loader and hand dress	B-62	270	.089	C.Y.	21	2.56	.57	24.13	28	
	5100	Articulated loader and hand dress	B-100	320	.037		21	1.20	1.72	23.92	27	
	5200	Articulated loader and 75HP dozer	B-10M	500	.024		21	.77	2.39	24.16	27.50	
	5300	Road grader and hand dress	B-11L	1,000	.016	▼	21	.49	.46	21.95	25	
	6000	Tilling topsoil, 20 HP tractor, disk harrow, 2" deep	B-66	450	.018	M.S.F.		.59	.38	.97	1.31	
	6050	4" deep		360	.022			.73	.48	1.21	1.64	
	6100	6" deep	▼	270	.030	▼		.98	.64	1.62	2.19	
	6150	26" rototiller, 2" deep	A-1J	1,250	.006	S.Y.		.17	.07	.24	.34	
	6200	4" deep		1,000	.008			.21	.08	.29	.42	
	6250	6" deep	▼	750	.011	▼		.28	.11	.39	.56	
720	0010	**PLANT BED PREPARATION, SHRUB & TREE**										720
	0100	Backfill planting pit, by hand, on site topsoil	2 Clab	18	.889	C.Y.		23.50		23.50	37	
	0200	Prepared planting mix	"	24	.667			17.80		17.80	27.50	
	0300	Skid steer loader, on site topsoil	B-62	340	.071			2.03	.45	2.48	3.63	
	0400	Prepared planting mix	"	410	.059			1.69	.37	2.06	3	
	1000	Excavate planting pit, by hand, sandy soil	2 Clab	16	1			26.50		26.50	41.50	
	1100	Heavy soil or clay	"	8	2			53.50		53.50	83	
	1200	1/2 C.Y. backhoe, sandy soil	B-11C	150	.107			3.27	1.44	4.71	6.60	
	1300	Heavy soil or clay	"	115	.139			4.27	1.88	6.15	8.55	
	2000	Mix planting soil, incl. loam, manure, peat, by hand	2 Clab	60	.267		37	7.10		44.10	52	
	2100	Skid steer loader	B-62	150	.160	▼	37	4.61	1.02	42.63	49	
	3000	Pile sod, skid steer loader	"	2,800	.009	S.Y.		.25	.05	.30	.44	
	3100	By hand	2 Clab	400	.040			1.07		1.07	1.66	
	4000	Remove sod, F.E. loader	B-10S	2,000	.006			.19	.12	.31	.42	
	4100	Sod cutter	B-12K	3,200	.005			.16	.30	.46	.57	
	4200	By hand	2 Clab	240	.067	▼		1.78		1.78	2.77	
810	0010	**LOAM & TOPSOIL**										810
	0300	Fine grade, base course for paving, see div. 02720-200										
	0400	Spread from pile to rough finish grade, F.E. loader, 1.5 CY	B-10S	200	.060	E.C.Y.		1.92	1.20	3.12	4.25	
	0500	Up to 200' radius, by hand	1 Clab	14	.571			15.25		15.25	23.50	
	0600	Top dress by hand, 1 C.Y. for 600 S.F.	"	11.50	.696	▼	21	18.55		39.55	52.50	
	0700	Furnish and place, truck dumped, screened, 4" deep	B-10S	1,300	.009	S.Y.	2.65	.30	.19	3.14	3.57	
	0800	6" deep	"	820	.015	"	3.39	.47	.29	4.15	4.76	

02915 | Shrub and Tree Transplanting

			CREW	DAILY OUTPUT	LABOR-HOURS	UNIT	MAT.	LABOR	EQUIP.	TOTAL	TOTAL INCL O&P	
400	0010	**PLANTING** Trees, shrubs and ground cover										400
	0100	Light soil										
	0110	Bare root seedlings, 3" to 5"	1 Clab	960	.008	Ea.		.22		.22	.35	
	0120	6" to 10"		520	.015			.41		.41	.64	
	0130	11" to 16"		370	.022			.58		.58	.90	
	0140	17" to 24"		210	.038			1.02		1.02	1.58	
	0200	Potted, 2-1/4" diameter		840	.010			.25		.25	.40	
	0210	3" diameter	▼	700	.011	▼		.31		.31	.47	

Important: See the Reference Section for critical supporting data - Reference Nos., Crews, & City Cost Indexes

		CREW	DAILY OUTPUT	LABOR-HOURS	UNIT	2005 BARE COSTS				TOTAL INCL O&P
02915	**Shrub and Tree Transplanting**					MAT.	LABOR	EQUIP.	TOTAL	
0220	4" diameter	1 Clab	620	.013	Ea.		.34		.34	.54
0300	Container, 1 gallon	2 Clab	84	.190			5.10		5.10	7.90
0310	2 gallon		52	.308			8.20		8.20	12.80
0320	3 gallon		40	.400			10.70		10.70	16.60
0330	5 gallon		29	.552			14.75		14.75	23
0400	Bagged and burlapped, 12" diameter ball, by hand		19	.842			22.50		22.50	35
0410	Backhoe/loader, 48 H.P.	B-6	40	.600			17.30	5.40	22.70	32.50
0415	15" diameter, by hand	2 Clab	16	1			26.50		26.50	41.50
0416	Backhoe/loader, 48 H.P.	B-6	30	.800			23	7.20	30.20	43.50
0420	18" diameter by hand	2 Clab	12	1.333			35.50		35.50	55.50
0430	Backhoe/loader, 48 H.P.	B-6	27	.889			25.50	8	33.50	48.50
0440	24" diameter by hand	2 Clab	9	1.778			47.50		47.50	74
0450	Backhoe/loader 48 H.P.	B-6	21	1.143			33	10.25	43.25	62
0470	36" diameter, backhoe/loader, 48 H.P.	"	17	1.412			40.50	12.70	53.20	76.50
0550	Medium soil									
0560	Bare root seedlings, 3" to 5"	1 Clab	672	.012	Ea.		.32		.32	.49
0561	6" to 10"		364	.022			.59		.59	.91
0562	11" to 16"		260	.031			.82		.82	1.28
0563	17" to 24"		145	.055			1.47		1.47	2.29
0570	Potted, 2-1/4" diameter		590	.014			.36		.36	.56
0572	3" diameter		490	.016			.44		.44	.68
0574	4" diameter		435	.018			.49		.49	.76
0590	Container, 1 gallon	2 Clab	59	.271			7.25		7.25	11.25
0592	2 gallon		36	.444			11.85		11.85	18.45
0594	3 gallon		28	.571			15.25		15.25	23.50
0595	5 gallon		20	.800			21.50		21.50	33
0600	Bagged and burlapped, 12" diameter ball, by hand		13	1.231			33		33	51
0605	Backhoe/loader, 48 H.P.	B-6	28	.857			24.50	7.70	32.20	46.50
0607	15" diameter, by hand	2 Clab	11.20	1.429			38		38	59.50
0608	Backhoe/loader, 48 H.P.	B-6	21	1.143			33	10.25	43.25	62
0610	18" diameter, by hand	2 Clab	8.50	1.882			50.50		50.50	78
0615	Backhoe/loader, 48 H.P.	B-6	19	1.263			36.50	11.35	47.85	68.50
0620	24" diameter, by hand	2 Clab	6.30	2.540			68		68	106
0625	Backhoe/loader, 48 H.P.	B-6	14.70	1.633			47	14.70	61.70	88.50
0630	36" diameter, backhoe/loader, 48 H.P.	"	12	2			57.50	18	75.50	108
0700	Heavy or stoney soil									
0710	Bare root seedlings, 3" to 5"	1 Clab	470	.017	Ea.		.45		.45	.71
0711	6" to 10"		255	.031			.84		.84	1.30
0712	11" to 16"		182	.044			1.17		1.17	1.83
0713	17" to 24"		101	.079			2.11		2.11	3.29
0720	Potted, 2-1/4" diameter		360	.022			.59		.59	.92
0722	3" diameter		343	.023			.62		.62	.97
0724	4" diameter		305	.026			.70		.70	1.09
0730	Container, 1 gallon	2 Clab	41.30	.387			10.35		10.35	16.10
0732	2 gallon		25.20	.635			16.95		16.95	26.50
0734	3 gallon		19.60	.816			22		22	34
0735	5 gallon		14	1.143			30.50		30.50	47.50
0750	Bagged and burlapped, 12" diameter ball, by hand		9.10	1.758			47		47	73
0751	Backhoe/loader	B-6	19.60	1.224			35.50	11	46.50	66
0752	15" diameter, by hand	2 Clab	7.80	2.051			55		55	85
0753	Backhoe/loader, 48 H.P.	B-6	14.70	1.633			47	14.70	61.70	88.50
0754	18" diameter, by hand	2 Clab	5.60	2.857			76.50		76.50	119
0755	Backhoe/loader, 48 H.P.	B-6	13.30	1.805			52	16.20	68.20	98
0756	24" diameter, by hand	2 Clab	4.40	3.636			97		97	151
0757	Backhoe/loader, 48 H.P.	B-6	10.30	2.330			67	21	88	126
0758	36" diameter backhoe/loader, 48 H.P.	"	8.40	2.857			82.50	25.50	108	156

400

SITE CONSTRUCTION 2

02915	Shrub and Tree Transplanting	CREW	DAILY OUTPUT	LABOR-HOURS	UNIT	MAT.	LABOR	EQUIP.	TOTAL	TOTAL INCL O&P	
						2005 BARE COSTS					
400 2000	Stake out tree and shrub locations	2 Clab	220	.073	Ea.		1.94		1.94	3.02	**400**

02920 | Lawns & Grasses

		CREW	DAILY OUTPUT	LABOR-HOURS	UNIT	MAT.	LABOR	EQUIP.	TOTAL	TOTAL INCL O&P	
320 0010	**SEEDING, ATHLETIC FIELDS** R02920-500										**320**
0020	Seeding, athletic fields, athletic field mix, 8#/MSF push spreader	1 Clab	8	1	M.S.F.	21.50	26.50		48	65.50	
0100	Tractor spreader	B-66	52	.154		21.50	5.10	3.33	29.93	35.50	
0200	Hydro or air seeding, with mulch & fertil.	B-81	80	.300		24	8.90	5.05	37.95	45	
0400	Birdsfoot trefoil, .45#/M.S.F., push spreader	1 Clab	8	1		7	26.50		33.50	49	
0500	Tractor spreader	B-66	52	.154		7	5.10	3.33	15.43	19	
0600	Hydro or air seeding, with mulch & fertil.	B-81	80	.300		13.45	8.90	5.05	27.40	34	
0800	Bluegrass, 4#/M.S.F., common, push spreader	1 Clab	8	1		14.70	26.50		41.20	57.50	
0900	Tractor spreader	B-66	52	.154		14.70	5.10	3.33	23.13	27.50	
1000	Hydro or air seeding, with mulch & fertil.	B-81	80	.300		24.50	8.90	5.05	38.45	45.50	
1100	Baron, push spreader	1 Clab	8	1		19.45	26.50		45.95	63	
1200	Tractor spreader	B-66	52	.154		19.45	5.10	3.33	27.88	33	
1300	Hydro or air seeding, with mulch & fertil.	B-81	80	.300		26.50	8.90	5.05	40.45	48.50	
1500	Clover, 0.67#/M.S.F., white, push spreader	1 Clab	8	1		1.28	26.50		27.78	43	
1600	Tractor spreader	B-66	52	.154		1.28	5.10	3.33	9.71	12.70	
1700	Hydro or air seeding, with mulch and fertil.	B-81	80	.300		7.05	8.90	5.05	21	27	
1800	Ladino, push spreader	1 Clab	8	1		5.05	26.50		31.55	47	
1900	Tractor spreader	B-66	52	.154		5.05	5.10	3.33	13.48	16.85	
2000	Hydro or air seeding, with mulch and fertil.	B-81	80	.300		22	8.90	5.05	35.95	43.50	
2200	Fescue 5.5#/M.S.F., tall, push spreader	1 Clab	8	1		10	26.50		36.50	52.50	
2300	Tractor spreader	B-66	52	.154		10	5.10	3.33	18.43	22.50	
2400	Hydro or air seeding, with mulch and fertilizer	B-81	80	.300		33	8.90	5.05	46.95	55.50	
2500	Chewing, push spreader	1 Clab	8	1		10	26.50		36.50	52.50	
2600	Tractor spreader	B-66	52	.154		10	5.10	3.33	18.43	22.50	
2700	Hydro or air seeding, with mulch and fertil.	B-81	80	.300		33	8.90	5.05	46.95	55.50	
2900	Crown vetch, 4#/M.S.F., push spreader	1 Clab	8	1		37	26.50		63.50	82	
3000	Tractor spreader	B-66	52	.154		37	5.10	3.33	45.43	52	
3100	Hydro or air seeding, with mulch and fertilizer	B-81	80	.300		50.50	8.90	5.05	64.45	74.50	
3300	Rye, 10#/M.S.F., annual, push spreader	1 Clab	8	1		6.75	26.50		33.25	49	
3400	Tractor spreader	B-66	52	.154		6.75	5.10	3.33	15.18	18.70	
3500	Hydro or air seeding, with mulch and fertilizer	B-81	80	.300		14.80	8.90	5.05	28.75	35.50	
3600	Fine textured, push spreader	1 Clab	8	1		6.75	26.50		33.25	49	
3700	Tractor spreader	B-66	52	.154		6.75	5.10	3.33	15.18	18.70	
3800	Hydro or air seeding, with mulch and fertilizer	B-81	80	.300		14.80	8.90	5.05	28.75	35.50	
4000	Shade mix, 6#/M.S.F., push spreader	1 Clab	8	1		9.60	26.50		36.10	52	
4100	Tractor spreader	B-66	52	.154		9.60	5.10	3.33	18.03	22	
4200	Hydro or air seeding, with mulch and fertilizer	B-81	80	.300		21	8.90	5.05	34.95	42.50	
4400	Slope mix, 6#/M.S.F., push spreader	1 Clab	8	1		9.60	26.50		36.10	52	
4500	Tractor spreader	B-66	52	.154		9.60	5.10	3.33	18.03	22	
4600	Hydro or air seeding, with mulch and fertilizer	B-81	80	.300		24	8.90	5.05	37.95	45.50	
4800	Turf mix, 4#/M.S.F., push spreader	1 Clab	8	1		6.40	26.50		32.90	48.50	
4900	Tractor spreader	B-66	52	.154		6.40	5.10	3.33	14.83	18.35	
5000	Hydro or air seeding, with mulch and fertilizer	B-81	80	.300		16.05	8.90	5.05	30	37	
5200	Utility mix, 7#/M.S.F., push spreader	1 Clab	8	1		11.20	26.50		37.70	54	
5300	Tractor spreader	B-66	52	.154		11.20	5.10	3.33	19.63	23.50	
5400	Hydro or air seeiding, with mulch and fertilizer	B-81	80	.300		42	8.90	5.05	55.95	65	
5600	Wildflower, .10#/M.S.F., push spreader	1 Clab	8	1		3.93	26.50		30.43	46	
5700	Tractor spreader	B-66	52	.154		3.93	5.10	3.33	12.36	15.65	
5800	Hydro or air seeding, with mulch and fertilizer	B-81	80	.300	▼	21.50	8.90	5.05	35.45	43	
7000	Apply fertilizer, 800 lb./acre	B-66	4	2	Ton	345	66	43.50	454.50	525	
7025	Limestone, mechanical spread	1 Clab	1.75	4.571	Acre	3.38	122		125.38	194	
7100	Apply mulch, see div. 02910-500	▼									

Important: See the Reference Section for critical supporting data - Reference Nos., Crews, & City Cost Indexes

2 SITE CONSTRUCTION

02900 | Planting

02920 | Lawns & Grasses

		CREW	DAILY OUTPUT	LABOR-HOURS	UNIT	2005 BARE COSTS				TOTAL INCL O&P
						MAT.	LABOR	EQUIP.	TOTAL	
400	**0010**	**SODDING**									**400**
	0020	Sodding, 1" deep, bluegrass sod, on level ground, over 8 MSF	B-63	22	1.818	M.S.F.	217	51	7	275	325
	0200	4 M.S.F.		17	2.353		243	66	9.05	318.05	380
	0300	1000 S.F.		13.50	2.963		265	83	11.40	359.40	430
	0500	Sloped ground, over 8 M.S.F.		6	6.667		217	186	25.50	428.50	555
	0600	4 M.S.F.		5	8		243	224	30.50	497.50	645
	0700	1000 S.F.		4	10		265	280	38.50	583.50	765
	1000	Bent grass sod, on level ground, over 6 M.S.F.		20	2		485	56	7.70	548.70	630
	1100	3 M.S.F.		18	2.222		540	62	8.55	610.55	700
	1200	Sodding 1000 S.F. or less		14	2.857		615	80	10.95	705.95	815
	1500	Sloped ground, over 6 M.S.F.		15	2.667		485	74.50	10.25	569.75	660
	1600	3 M.S.F.		13.50	2.963		540	83	11.40	634.40	735
	1700	1000 S.F.	▼	12	3.333	▼	615	93	12.80	720.80	840
500	**0010**	**STOLENS, SPRIGGING**									**500**
	0100	6" O.C., by hand	1 Clab	4	2	M.S.F.	13.65	53.50		67.15	98
	0110	Walk behind sprig planter	"	80	.100		13.65	2.67		16.32	19.15
	0120	Towed sprig planter	B-66	350	.023		13.65	.76	.49	14.90	16.70
	0130	9" O.C., by hand	1 Clab	5.20	1.538		10.05	41		51.05	75
	0140	Walk behind sprig planter	"	92	.087		10.05	2.32		12.37	14.65
	0150	Towed sprig planter	B-66	420	.019		10.05	.63	.41	11.09	12.45
	0160	12" O.C., by hand	1 Clab	6	1.333		6.40	35.50		41.90	62.50
	0170	Walk behind sprig planter	"	110	.073		6.40	1.94		8.34	10.05
	0180	Towed sprig planter	B-66	500	.016		6.40	.53	.35	7.28	8.25
	0200	Broadcast, by hand, 2 Bu per M.S.F.	1 Clab	15	.533		5.55	14.25		19.80	28
	0210	4 Bu. per M.S.F.		10	.800		11.15	21.50		32.65	45.50
	0220	6 Bu. per M.S.F.	▼	6.50	1.231		16.70	33		49.70	69.50
	0300	Hydro planter, 6 Bu. per M.S.F.	B-64	100	.160		16.70	4.28	2.32	23.30	27.50
	0320	Manure spreader planting 6 Bu. per M.S.F.	B-66	200	.040	▼	16.70	1.32	.87	18.89	21.50

02930 | Exterior Plants

		CREW	DAILY OUTPUT	LABOR-HOURS	UNIT	2005 BARE COSTS				TOTAL INCL O&P
						MAT.	LABOR	EQUIP.	TOTAL	
430	**0010**	**TREES, DECIDUOUS** zones 2 - 6									**430**
	0100	Acer campestre, (Hedge Maple), Z4, B&B									
	0110	4' to 5'				Ea.	50			50	55
	0120	5' to 6'					60			60	66
	0130	1-1/2" to 2" Cal.					135			135	149
	0140	2" to 2-1/2" Cal.					151			151	166
	0150	2-1/2" to 3" Cal.				▼	179			179	196
	0200	Acer ginnala, (Amur Maple), Z2, cont/BB									
	0210	8' to 10'				Ea.	139			139	153
	0220	10' to 12'					151			151	166
	0230	12' to 14'				▼	184			184	202
	0600	Acer platanoides, (Norway Maple), Z4, B&B									
	0610	8' to 10'				Ea.	162			162	178
	0620	1-1/2" to 2" Cal.					120			120	132
	0630	2" to 2-1/2" Cal.					145			145	159
	0640	2-1/2" to 3" Cal.					178			178	196
	0650	3" to 3-1/2" Cal.					232			232	256
	0660	Bare root, 8' to 10'					202			202	222
	0670	10' to 12'					237			237	261
	0680	12' to 14'				▼	169			169	186
	0700	Acer platanoides columnare, (Column maple), Z4, B&B									
	0710	2" to 2-1/2" Cal.				Ea.	179			179	197
	0720	2-1/2" to 3" Cal.					222			222	244
	0730	3" to 3-1/2" Cal.				▼	222			222	244

		02930	Exterior Plants	CREW	DAILY OUTPUT	LABOR-HOURS	UNIT	2005 BARE COSTS				TOTAL INCL O&P	
								MAT.	LABOR	EQUIP.	TOTAL		
430	0740		3-1/2" to 4" Cal.				Ea.	275			275	305	430
	0750		4" to 4-1/2" Cal.					395			395	435	
	0760		4-1/2" to 5" Cal.					555			555	610	
	0770		5" to 5-1/2" Cal.				↓	745			745	820	
	0800		Acer rubrum, (Red Maple), Z4, B&B										
	0810		1-1/2" to 2" Cal.				Ea.	117			117	129	
	0820		2" to 2-1/2" Cal.					154			154	170	
	0830		2-1/2" to 3" Cal.					185			185	204	
	0840		Bare Root, 8' to 10'					149			149	164	
	0850		10' to 12'				↓	149			149	164	
	0900		Acer saccharum, (Sugar Maple), Z3, B&B										
	0910		1-1/2" to 2" Cal.				Ea.	123			123	135	
	0920		2" to 2-1/2" Cal.					163			163	180	
	0930		2-1/2" to 3" Cal.					171			171	188	
	0940		3" to 3-1/2" Cal.					220			220	242	
	0950		3-1/2" to 4" Cal.					271			271	298	
	0960		Bare Root, 8' to 10'					136			136	150	
	0970		10' to 12'					204			204	224	
	0980		12' to 14'				↓	275			275	305	

		02935	Plant Maintenance										
100	0010	**FERTILIZE**											100
	0100		Dry granular, 4#/M.S.F., hand spread	1 Clab	24	.333	M.S.F.	1.96	8.90		10.86	16	
	0110		Push rotary		140	.057	"	1.96	1.53		3.49	4.53	
	0112		Push rotary, per 1076 feet squared	↓	130	.062	Ea.	1.96	1.64		3.60	4.72	
	0120		Tractor towed spreader, 8'	B-66	500	.016	M.S.F.	1.96	.53	.35	2.84	3.34	
	0130		12' spread		800	.010		1.96	.33	.22	2.51	2.90	
	0140		Truck whirlwind spreader	↓	1,200	.007		1.96	.22	.14	2.32	2.65	
	0180		Water soluable, hydro spread, 1.5 # /MSF	B-64	600	.027		2.06	.71	.39	3.16	3.80	
	0190		Add for weed control				↓	.30			.30	.33	
300	0010	**MOWING**											300
	1650		Mowing brush, tractor with rotary mower										
	1660		Light density	B-84	22	.364	M.S.F.		12.60	10.40	23	30.50	
	1670		Medium density		13	.615			21.50	17.60	39.10	51.50	
	1680		Heavy density	↓	9	.889			31	25.50	56.50	74.50	
	4000		Lawn mowing, improved areas, 16" hand push	1 Clab	48	.167			4.45		4.45	6.95	
	4050		Power mower, 18" - 22"		65	.123			3.29		3.29	5.10	
	4100		22" - 30"		110	.073			1.94		1.94	3.02	
	4150		30" - 32"	↓	140	.057			1.53		1.53	2.37	
	4160		Riding mower, 36" - 44"	B-66	300	.027			.88	.58	1.46	1.97	
	4170		48" - 58"	"	480	.017	↓		.55	.36	.91	1.23	
	4175		Mowing with tractor & attachments										
	4180		3 gang reel, 7'	B-66	930	.009	M.S.F.		.28	.19	.47	.63	
	4190		5 gang reel, 12'		1,200	.007			.22	.14	.36	.49	
	4200		Cutter or sickle-bar, 5', rough terrain		210	.038			1.26	.82	2.08	2.81	
	4210		Cutter or sickle-bar, 5', smooth terrain		340	.024	↓		.78	.51	1.29	1.73	
	4220		Drainage channel, 5' sickle bar	↓	5	1.600	Mile		53	34.50	87.50	118	
	4250		Lawnmower, rotary type, sharpen (all sizes)	1 Clab	10	.800	Ea.		21.50		21.50	33	
	4260		Repair or replace part		7	1.143	"		30.50		30.50	47.50	
	5000		Edge trimming with weed whacker	↓	5,760	.001	L.F.		.04		.04	.06	
410	0010	**TREE PRUNING**											410
	0020		1-1/2" caliper	1 Clab	84	.095	Ea.		2.54		2.54	3.96	
	0030		2" caliper		70	.114			3.05		3.05	4.75	
	0040		2-1/2" caliper	↓	50	.160	↓		4.27		4.27	6.65	

Important: See the Reference Section for critical supporting data - Reference Nos., Crews, & City Cost Indexes

2 SITE CONSTRUCTION

02935 | Plant Maintenance

		CREW	DAILY OUTPUT	LABOR-HOURS	UNIT	MAT.	LABOR	EQUIP.	TOTAL	TOTAL INCL O&P	
410	0050	3" caliper	1 Clab	30	.267	Ea.		7.10		7.10	11.10	**410**
	0060	4" caliper, by hand	2 Clab	21	.762			20.50		20.50	31.50	
	0070	Aerial lift equipment	B-85	38	1.053			30	17.65	47.65	65.50	
	0100	6" caliper, by hand	2 Clab	12	1.333			35.50		35.50	55.50	
	0110	Aerial lift equipment	B-85	20	2			57	33.50	90.50	125	
	0200	9" caliper, by hand	2 Clab	7.50	2.133			57		57	88.50	
	0210	Aerial lift equipment	B-85	12.50	3.200			91	53.50	144.50	199	
	0300	12" caliper, by hand	2 Clab	6.50	2.462			65.50		65.50	102	
	0310	Aerial lift equipment	B-85	10.80	3.704			105	62	167	231	
	0400	18" caliper by hand	2 Clab	5.60	2.857			76.50		76.50	119	
	0410	Aerial lift equipment	B-85	9.30	4.301			122	72	194	268	
	0500	24" caliper, by hand	2 Clab	4.60	3.478			93		93	145	
	0510	Aerial lift equipment	B-85	7.70	5.195			148	87	235	325	
	0600	30" caliper, by hand	2 Clab	3.70	4.324			115		115	180	
	0610	Aerial lift equipment	B-85	6.20	6.452			184	108	292	400	
	0700	36" caliper, by hand	2 Clab	2.70	5.926			158		158	246	
	0710	Aerial lift equipment	B-85	4.50	8.889			253	149	402	555	
	0800	48" caliper, by hand	2 Clab	1.70	9.412	▼		251		251	390	
	0810	Aerial lift equipment	B-85	2.80	14.286			405	239	644	890	
420	0010	**SHRUB PRUNING**										**420**
	6700	Prune, shrub bed	1 Clab	7	1.143	M.S.F.		30.50		30.50	47.50	
	6710	Shrub under 3' height		190	.042	Ea.		1.12		1.12	1.75	
	6720	4' height		90	.089			2.37		2.37	3.69	
	6730	Over 6'		50	.160			4.27		4.27	6.65	
	7350	Prune trees from ground		20	.400			10.70		10.70	16.60	
	7360	High work	▼	8	1	▼		26.50		26.50	41.50	
500	0010	**WATERING**										**500**
	4900	Water lawn or planting bed with hose, 1" of water	1 Clab	16	.500	M.S.F.		13.35		13.35	21	
	4910	50' soaker hoses, in place		82	.098			2.60		2.60	4.05	
	4920	60' soaker hoses, in place		89	.090	▼		2.40		2.40	3.73	
	7500	Water trees or shrubs, under 1" caliper		32	.250	Ea.		6.70		6.70	10.40	
	7550	1" - 3" caliper		17	.471			12.55		12.55	19.55	
	7600	3" - 4" caliper		12	.667			17.80		17.80	27.50	
	7650	Over 4" caliper	▼	10	.800	▼		21.50		21.50	33	
	9000	For sprinkler irrigation systems, see Div. 02810										
610	0010	**WEEDING**										**610**
	0100	Weed planting bed	1 Clab	800	.010	S.Y.		.27		.27	.42	

02945 | Planting Accessories

		CREW	DAILY OUTPUT	LABOR-HOURS	UNIT	MAT.	LABOR	EQUIP.	TOTAL	TOTAL INCL O&P	
510	0010	**TREE GUYING**										**510**
	0015	Tree guying Including stakes, guy wire and wrap										
	0100	Less than 3" caliper, 2 stakes	2 Clab	35	.457	Ea.	15.10	12.20		27.30	35.50	
	0200	3" to 4" caliper, 3 stakes	"	21	.762	"	17.85	20.50		38.35	51	
	1000	Including arrowhead anchor, cable, turnbuckles and wrap										
	1100	Less than 3" caliper, 3 anchors	2 Clab	20	.800	Ea.	47.50	21.50		69	85	
	1200	3" to 6" caliper, 4 anchors		15	1.067		68	28.50		96.50	120	
	1300	6" caliper, 6 anchors		12	1.333		84	35.50		119.50	148	
	1400	8" caliper, 8 anchors	▼	9	1.778	▼	96.50	47.50		144	180	

SITE CONSTRUCTION **2**

02955	Restoration of Underground Piping	CREW	DAILY OUTPUT	LABOR-HOURS	UNIT	MAT.	LABOR	EQUIP.	TOTAL	TOTAL INCL O&P	
210	**0010** **PIPE INTERNAL CLEANING & INSPECTION**										210
	0100 Cleaning, pressure pipe systems										
	0120 Pig method, lengths 1000' to 10,000'										
	0140 4" diameter thru 24" diameter, minimum				L.F.				2.14	2.46	
	0160 Maximum				"				5.35	6.40	
	6000 Sewage/sanitary systems										
	6100 Power rodder with header & cutters										
	6110 Mobilization charge, minimum				Total				320	375	
	6120 Mobilization charge, maximum				"				750	850	
	6140 4" diameter				L.F.				1.07	1.23	
	6150 6" diameter								1.34	1.55	
	6160 8" diameter								1.60	1.87	
	6170 10" diameter								1.87	2.14	
	6180 12" diameter								2.03	2.35	
	6190 14" diameter								2.14	2.46	
	6200 16" diameter								2.41	2.78	
	6210 18" diameter								2.68	3.05	
	6220 20" diameter								2.78	3.21	
	6230 24" diameter								3.10	3.58	
	6240 30" diameter								3.75	4.28	
	6250 36" diameter								4.28	4.92	
	6260 48" diameter								4.60	5.30	
	6270 60" diameter								6	6.90	
	6280 72" diameter				↓				6.65	7.65	
	9000 Inspection, television camera with film										
	9060 500 linear feet				Total				1,075	1,230	
220	**0010** **PIPEBURSTING**, 300' runs, replace with HDPE pipe										220
	0020 Not including excavation, backfill, shoring, or dewatering										
	0100 6" to 15" diameter, minimum				L.F.					80	
	0200 Maximum									160	
	0300 18" to 36" diameter, minimum									187	
	0400 Maximum				↓					320	
	0500 Mobilize and demobilize, minimum				Job					2,675	
	0600 Maximum				"					26,800	
230	**0010** **LINING PIPE** with cement, incl. bypass and cleaning										230
	0020 Less than 10,000 L.F., urban, 6" to 10"	C-17E	130	.615	L.F.	6.90	21.50	.61	29.01	41.50	
	0050 10" to 12"		125	.640		8.45	22.50	.63	31.58	45	
	0070 12" to 16"		115	.696		8.70	24.50	.69	33.89	48.50	
	0100 16" to 20"		95	.842		10.15	29.50	.83	40.48	58	
	0200 24" to 36"		90	.889		10.95	31.50	.88	43.33	61.50	
	0300 48" to 72"		80	1		17.45	35.50	.99	53.94	75.50	
	0500 Rural, 6" to 10"		180	.444		6.90	15.65	.44	22.99	32.50	
	0550 10" to 12"		175	.457		8.45	16.10	.45	25	35	
	0570 12" to 16"		160	.500		8.80	17.65	.50	26.95	38	
	0600 16" to 20"		135	.593		9.30	21	.59	30.89	43.50	
	0700 24" to 36"		125	.640		11.15	22.50	.63	34.28	48	
	0800 48" to 72"		100	.800		17.45	28	.79	46.24	64	
	1000 Greater than 10,000 L.F., urban, 6" to 10"		160	.500		6.90	17.65	.50	25.05	35.50	
	1050 10" to 12"		155	.516		8.35	18.20	.51	27.06	38	
	1070 12" to 16"		140	.571		8.70	20	.57	29.27	41.50	
	1100 16" to 20"		120	.667		9.30	23.50	.66	33.46	47.50	
	1200 24" to 36"		115	.696		11.15	24.50	.69	36.34	51	
	1300 48" to 72"		95	.842		17.45	29.50	.83	47.78	66	
	1500 Rural, 6" to 10"		215	.372		6.90	13.10	.37	20.37	28.50	
	1550 10" to 12"		210	.381		8.45	13.45	.38	22.28	30.50	
	1570 12" to 16"	↓	185	.432	↓	8.70	15.25	.43	24.38	33.50	

Important: See the Reference Section for critical supporting data - Reference Nos., Crews, & City Cost Indexes

02955	Restoration of Underground Piping	CREW	DAILY OUTPUT	LABOR-HOURS	UNIT	2005 BARE COSTS				TOTAL INCL O&P
						MAT.	LABOR	EQUIP.	TOTAL	
230 1600	16" to 20"	C-17E	150	.533	L.F.	9.30	18.80	.53	28.63	40	230
1700	24" to 36"	↓	140	.571	↓	11.15	20	.57	31.72	44.50
1800	48" to 72"	↓	120	.667	↓	17.60	23.50	.66	41.76	56.50
2000	Cured in place pipe, non-pressure, flexible felt resin, 400' runs									
2100	6" diameter				L.F.					32
2200	8" diameter									43
2300	10" diameter									59
2400	12" diameter									75
2500	15" diameter									102
2600	18" diameter									123
2700	21" diameter									144
2800	24" diameter									161
2900	30" diameter									203
3000	36" diameter									268
3100	48" diameter									360
3200	60" diameter									450
3300	72" diameter				↓					535
240 0010	**HDPE PIPE LINING**, excludes cleaning and video inspection										240
0020	Pipe relined with one pipe size smaller than original (4" for 6 ")									
0100	6" diameter, original size	B-6B	600	.080	L.F.	1.84	2.19	1.46	5.49	7.05
0150	8" diameter, original size		600	.080		2.65	2.19	1.46	6.30	7.95
0200	10" diameter, original size		600	.080		4.48	2.19	1.46	8.13	9.95
0250	12" diameter, original size		400	.120		6.95	3.28	2.20	12.43	15.15
0300	14" diameter, original size	↓	400	.120	↓	9.80	3.28	2.20	15.28	18.30
310 0010	**CORROSION RESISTANCE** Wrap & coat, add to pipe, 4" dia.				L.F.	1.46			1.46	1.61	310
0020	5" diameter					1.53			1.53	1.68
0040	6" diameter					1.53			1.53	1.68
0060	8" diameter					2.37			2.37	2.61
0080	10" diameter					3.54			3.54	3.89
0100	12" diameter					3.54			3.54	3.89
0120	14" diameter					5.20			5.20	5.70
0140	16" diameter					5.20			5.20	5.70
0160	18" diameter					5.25			5.25	5.75
0180	20" diameter					5.50			5.50	6.05
0200	24" diameter					6.70			6.70	7.35
0220	Small diameter pipe, 1" diameter, add					1.10			1.10	1.21
0240	2" diameter					1.21			1.21	1.33
0260	2-1/2" diameter					1.33			1.33	1.46
0280	3" diameter				↓	1.33			1.33	1.46
0300	Fittings, field covered, add				S.F.	7			7	7.70
0500	Coating, bituminous, per diameter inch, 1 coat, add				L.F.	.27			.27	.30
0540	3 coat					.42			.42	.46
0560	Coal tar epoxy, per diameter inch, 1 coat, add					.16			.16	.18
0600	3 coat					.34			.34	.37
1000	Polyethylene H.D. extruded, .025" thk., 1/2" diameter add					.24			.24	.26
1020	3/4" diameter					.25			.25	.28
1040	1" diameter					.28			.28	.31
1060	1-1/4" diameter					.34			.34	.37
1080	1-1/2" diameter					.38			.38	.42
1100	.030" thk., 2" diameter					.41			.41	.45
1120	2-1/2" diameter					.48			.48	.53
1140	.035" thk., 3" diameter					.60			.60	.66
1160	3-1/2" diameter					.68			.68	.75
1180	4" diameter				↓	.76			.76	.84

02955 | Restoration of Underground Piping

		CREW	DAILY OUTPUT	LABOR-HOURS	UNIT	2005 BARE COSTS MAT.	LABOR	EQUIP.	TOTAL	TOTAL INCL O&P	
310	**1200**	.040" thk, 5" diameter				L.F.	1.05			1.05	1.16	**310**
	1220	6" diameter					1.09			1.09	1.20	
	1240	8" diameter					1.43			1.43	1.57	
	1260	10" diameter					1.74			1.74	1.91	
	1280	12" diameter					2.08			2.08	2.29	
	1300	.060" thk., 14" diameter					2.87			2.87	3.16	
	1320	16" diameter					3.36			3.36	3.70	
	1340	18" diameter					3.97			3.97	4.37	
	1360	20" diameter					4.46			4.46	4.91	
	1380	Fittings, field wrapped, add				S.F.	4.61			4.61	5.05	
400	**0010**	**PIPE REPAIR**										**400**
	0020	Not including excavation or backfill										
	0100	Clamp, stainless steel, lightweight, for steel pipe										
	0110	3" long, 1/2" diameter pipe	1 Plum	34	.235	Ea.	10.45	9.60		20.05	26	
	0120	3/4" diameter pipe		32	.250		10.80	10.20		21	27	
	0130	1" diameter pipe		30	.267		11.60	10.90		22.50	29	
	0140	1-1/4" diameter pipe		28	.286		12.30	11.65		23.95	31	
	0150	1-1/2" diameter pipe		26	.308		12.70	12.55		25.25	33	
	0160	2" diameter pipe		24	.333		13.85	13.60		27.45	36	
	0170	2-1/2" diameter pipe		23	.348		15.10	14.20		29.30	38	
	0180	3" diameter pipe		22	.364		16.65	14.85		31.50	41	
	0190	3-1/2" diameter pipe		21	.381		17.85	15.55		33.40	43	
	0200	4" diameter pipe	B-20	56	.429		18.65	12.90		31.55	40.50	
	0210	5" diameter pipe		53	.453		21.50	13.60		35.10	44.50	
	0220	6" diameter pipe		48	.500		24	15.05		39.05	49.50	
	0230	8" diameter pipe		30	.800		28	24		52	68	
	0240	10" diameter pipe		28	.857		71.50	26		97.50	119	
	0250	12" diameter pipe		24	1		76	30		106	131	
	0260	14" diameter pipe		22	1.091		79.50	33		112.50	138	
	0270	16" diameter pipe		20	1.200		85.50	36		121.50	150	
	0280	18" diameter pipe		18	1.333		90.50	40		130.50	162	
	0290	20" diameter pipe		16	1.500		100	45		145	180	
	0300	24" diameter pipe		14	1.714		111	51.50		162.50	203	
	0360	For 6" long, add					100%	40%				
	0370	For 9" long, add					200%	100%				
	0380	For 12" long, add					300%	150%				
	0390	For 18" long, add					500%	200%				
	0400	Pipe Freezing for live repairs of systems 3/8 inch to 6 inch										
	0410	Note: Pipe Freezing can also used to install a valve into a live system										
	0420	Pipe Freezing each side 3/8 inch	2 Skwk	8	2	Ea.	500	69.50		569.50	660	
	0425	Pipe Freezing each side 3/8 inch , second location same kit		8	2		19	69.50		88.50	129	
	0430	Pipe Freezing each side 3/4 inch		8	2		430	69.50		499.50	585	
	0435	Pipe Freezing each side 3/4 inch , second location same kit		8	2		19	69.50		88.50	129	
	0440	Pipe Freezing each side 1 1/2 inch		6	2.667		430	93		523	620	
	0445	Pipe Freezing each side 1 1/2 inch , second location same kit		6	2.667		19	93		112	166	
	0450	Pipe Freezing each side 2 inch		6	2.667		635	93		728	845	
	0455	Pipe Freezing each side 2 inch , second location same kit		6	2.667		19	93		112	166	
	0460	Pipe Freezing each side 2 1/2 inch-3 inch		6	2.667		890	93		983	1,125	
	0465	Pipe Frz each side 2 1/2-3 inch , second location same kit		6	2.667		38	93		131	187	
	0470	Pipe Freezing each side 4 inch		4	4		1,450	139		1,589	1,800	
	0475	Pipe Freezing each side 4 inch , second location same kit		4	4		60	139		199	283	
	0480	Pipe Freezing each side 5-6 inch		4	4		3,475	139		3,614	4,050	
	0485	Pipe Freezing each side 5-6 inch , second location same kit		4	4		180	139		319	415	
	0490	Pipe Frz extra 20 lb CO2 cylinders (3/8 to 2 " - 1 ea, 3 " -2 ea)					189			189	208	
	0500	Pipe Freezing extra 50 lb CO2 cylinders (4 " - 2 ea, 5-6 " -6 ea)					425			425	465	
	1000	Clamp, stainless steel, with threaded service tap										

Important: See the Reference Section for critical supporting data - Reference Nos., Crews, & City Cost Indexes

SITE CONSTRUCTION 2

			DAILY	LABOR-			2005 BARE COSTS				TOTAL	
	02955	**Restoration of Underground Piping**	CREW	OUTPUT	HOURS	UNIT	MAT.	LABOR	EQUIP.	TOTAL	INCL O&P	
400	1040	Full seal for iron, steel, PVC pipe										**400**
	1100	6" long, 2" diameter pipe	1 Plum	17	.471	Ea.	135	19.20		154.20	177	
	1110	2-1/2" diameter pipe		16	.500		143	20.50		163.50	188	
	1120	3" diameter pipe		15.60	.513		152	21		173	199	
	1130	3-1/2" diameter pipe	↓	15	.533		159	22		181	207	
	1140	4" diameter pipe	B-20	40	.600		174	18.05		192.05	220	
	1150	6" diameter pipe		34	.706		201	21		222	255	
	1160	8" diameter pipe		21	1.143		251	34.50		285.50	330	
	1170	10" diameter pipe		20	1.200		278	36		314	360	
	1180	12" diameter pipe	↓	17	1.412		315	42.50		357.50	415	
	1205	For 9" long, add					20%	45%				
	1210	For 12" long, add					40%	80%				
	1220	For 18" long, add				↓	70%	110%				
	1600	Clamp, stainless steel, single section										
	1640	Full seal for iron, steel, PVC pipe										
	1700	6" long, 2" diameter pipe	1 Plum	17	.471	Ea.	70	19.20		89.20	106	
	1710	2-1/2" diameter pipe		16	.500		76	20.50		96.50	115	
	1720	3" diameter pipe		15.60	.513		79.50	21		100.50	119	
	1730	3-1/2" diameter pipe	↓	15	.533		87	22		109	129	
	1740	4" diameter pipe	B-20	40	.600		95	18.05		113.05	133	
	1750	6" diameter pipe		34	.706		119	21		140	164	
	1760	8" diameter pipe		21	1.143		159	34.50		193.50	228	
	1770	10" diameter pipe		20	1.200		182	36		218	257	
	1780	12" diameter pipe	↓	17	1.412		198	42.50		240.50	284	
	1800	For 9" long, add					40%	45%				
	1810	For 12" long, add					60%	80%				
	1820	For 18" long, add				↓	120%	110%				
	2000	Clamp, stainless steel, two section										
	2040	Full seal, for iron, steel, PVC pipe										
	2100	6" long, 4" diameter pipe	B-20	24	1	Ea.	135	30		165	195	
	2110	6" diameter pipe		20	1.200		159	36		195	230	
	2120	8" diameter pipe		13	1.846		174	55.50		229.50	279	
	2130	10" diameter pipe		12	2		254	60		314	375	
	2140	12" diameter pipe		10	2.400		278	72		350	415	
	2200	9" long, 4" diameter pipe		16	1.500		174	45		219	262	
	2210	6" diameter pipe		13	1.846		198	55.50		253.50	305	
	2220	8" diameter pipe		9	2.667		206	80		286	350	
	2230	10" diameter pipe		8	3		285	90		375	455	
	2240	12" diameter pipe		7	3.429		315	103		418	510	
	2250	14" diameter pipe		6.40	3.750		475	113		588	700	
	2260	16" diameter pipe		6	4		500	120		620	735	
	2270	18" diameter pipe		5	4.800		515	144		659	790	
	2280	20" diameter pipe		4.60	5.217		555	157		712	855	
	2290	24" diameter pipe	↓	4	6		950	180		1,130	1,325	
	2320	For 12" long, add to 9"					15%	25%				
	2330	For 18" long, add to 9"				↓	70%	55%				
	8000	For internal cleaning and inspection, see Div. 02955-210										
	8100	For pipe testing, see Div. 15955-700										

	02960	**Flex. Pavement Surfacing Recovery**										
100	0010	**PAVEMENT MILLING AND PAVEMENT COLD PLANING**										**100**
	5200	Cold planing & cleaning, 1" to 3" asphalt pavmt., over 25,000 S.Y.	B-71	6,000	.009	S.Y.		.28	.83	1.11	1.34	
	5280	5,000 S.Y. to 10,000 S.Y.	"	4,000	.014	"		.43	1.24	1.67	2.01	
	5300	Asphalt pavement removal from conc. base, no haul										
	5320	Rip, load & sweep 1" to 3"	B-70	8,000	.007	S.Y.		.21	.15	.36	.50	
	5330	3" to 6" deep	"	5,000	.011	↓		.34	.24	.58	.79	

02960 | Flex. Pavement Surfacing Recovery

		CREW	DAILY OUTPUT	LABOR-HOURS	UNIT	2005 BARE COSTS				TOTAL INCL O&P	
						MAT.	LABOR	EQUIP.	TOTAL		
100	**5340**	Profile grooving, asphalt pavement load & sweep, 1″ deep	B-71	12,500	.004	S.Y.		.14	.40	.54	.65	**100**
	5350	3″ deep		9,000	.006			.19	.55	.74	.90	
	5360	6″ deep	↓	5,000	.011	↓		.34	.99	1.33	1.61	
	5400	Mixing material in windrow, 180 H.P. grader	B-11L	9,400	.002	C.Y.		.05	.05	.10	.13	
	5450	For cold laid asphalt pavement, see Div. 02740-400										

02965 | Flex./Bit. Pavement Recycling

		CREW	DAILY OUTPUT	LABOR-HOURS	UNIT	MAT.	LABOR	EQUIP.	TOTAL	TOTAL INCL O&P	
200	**0010**	**COLD IN-PLACE RECYCLED BITUMINOUS PAVEMENT COURSES**									**200**
	5000	Reclamation, pulverizing and blending with existing base									
	5040	Aggregate base, 4″ thick pavement, over 15,000 S.Y.	B-73	2,400	.027	S.Y.		.85	1.59	2.44	3.05	
	5080	5,000 S.Y. to 15,000 S.Y.		2,200	.029			.93	1.73	2.66	3.31	
	5120	8″ thick pavement, over 15,000 S.Y.		2,200	.029			.93	1.73	2.66	3.31	
	5160	5,000 S.Y. to 15,000 S.Y.	↓	2,000	.032	↓		1.02	1.90	2.92	3.65	
	5180	Add for mobilization and demobilization for crew B-73				Ea.				1,700	2,150	
400	**0010**	**HOT IN-PLACE RECYCLED BITUMINOUS PAVEMENT COURSES**									**400**
	5500	Recycle asphalt pavement at site									
	5520	Remove, rejuvenate and spread 4″ deep	B-72	2,500	.026	S.Y.	2.33	.79	3.32	6.44	7.40	
	5521	6″ deep	″	2,000	.032	″	3.41	.99	4.15	8.55	9.80	

02985 | Site Maintenance

		CREW	DAILY OUTPUT	LABOR-HOURS	UNIT	MAT.	LABOR	EQUIP.	TOTAL	TOTAL INCL O&P	
700	**0010**	**SITE MAINTENANCE**									**700**
	1550	General site work maintenance									
	1560	Clearing brush with brush saw & rake	1 Clab	565	.014	S.Y.		.38		.38	.59	
	1570	By hand	″	280	.029			.76		.76	1.19	
	1580	With dozer, ball and chain, light clearing	B-11A	3,675	.004			.13	.25	.38	.48	
	1590	Medium clearing	″	3,110	.005	↓		.16	.30	.46	.57	
	3000	Lawn maintenance									
	3010	Aerate lawn, 18″ cultivating width, walk behind	A-1K	95	.084	M.S.F.		2.25	.25	2.50	3.78	
	3040	48″ cultivating width	B-66	750	.011			.35	.23	.58	.78	
	3060	72″ cultivating width	″	1,100	.007	↓		.24	.16	.40	.53	
	3100	Edge lawn, by hand at walks	1 Clab	16	.500	C.L.F.		13.35		13.35	21	
	3150	At planting beds		7	1.143			30.50		30.50	47.50	
	3200	Using gas powered edger at walks		88	.091	↓		2.43		2.43	3.78	
	3250	At planting beds	↓	24	.333			8.90		8.90	13.85	
	3260	Vacuum, 30″ gas, outdoors with hose	A-1L	96	.083	M.L.F.		2.22	.25	2.47	3.73	
	3400	Weed lawn, by hand	1 Clab	3	2.667	M.S.F.		71		71	111	
	4500	Rake leaves or lawn, by hand, (cavex)	″	7.50	1.067			28.50		28.50	44.50	
	4510	Power rake	A-1L	45	.178	↓		4.75	.53	5.28	8	
	4700	Seeding lawn, see Division 02920-320									
	4750	Sodding, see Division 02920-400									
	5900	Road & walk maintenance									
	5915	De-icing roads and walks									
	5920	Calcium Chloride in truckload lots see Division 02340-700									
	6000	Ice melting comp., 90% Calc. Chlor., effec. to -30°F									
	6010	50-80 lb. poly bags, med. applic. 19 lbs./M.S.F., by hand	1 Clab	60	.133	M.S.F.	15.60	3.56		19.16	22.50	
	6050	With hand operated rotary spreader		110	.073		15.60	1.94		17.54	20	
	6100	Rock salt, med. applic. on road & walkway, by hand		60	.133		3.91	3.56		7.47	9.85	
	6110	With hand operated rotary spreader	↓	110	.073	↓	3.91	1.94		5.85	7.30	
	6600	Shrub maintenance									
	6630	Mulch-see Div. 02910-500									
	6640	Shrub bed fertilize dry granular 3 lbs./M.S.F.	1 Clab	85	.094	M.S.F.	.93	2.51		3.44	4.93	
	6800	Weed, by handhoe		8	1			26.50		26.50	41.50	
	6810	Spray out	↓	32	.250	↓		6.70		6.70	10.40	

Important: See the Reference Section for critical supporting data - Reference Nos., Crews, & City Cost Indexes

		02985	Site Maintenance	CREW	DAILY OUTPUT	LABOR-HOURS	UNIT	MAT.	2005 BARE COSTS LABOR	EQUIP.	TOTAL	TOTAL INCL O&P	
700	6820		Spray after mulch	1 Clab	48	.167	M.S.F.		4.45		4.45	6.95	700
	6840		Shrub purchase and installation see Div. 02915, 02930										
	7000		Highway R.O.W. mowing per month				Mile					31%	
	7050		Highway R.O.W. mowing plus herbicides program, per month				"					17.50%	
	7100		Tree maintenance										
	7140		Clear and grub trees, see Div. 02230-100										
	7160		Cutting and piling trees, see Div. 02230-300										
	7200		Fertilize, tablets, slow release, 30 gram/tree	1 Clab	100	.080	Ea.	.33	2.14		2.47	3.68	
	7280		Guying, including stakes, guy wire & wrap, see Div. 02945-510										
	7300		Planting, trees, Deciduous, in prep. beds, see Div. 02915-400										
	7400		Removal, trees see Div. 02905-925										
	7420		Pest control, spray	1 Clab	24	.333	Ea.	15.15	8.90		24.05	30.50	
	7430		Systemic	"	48	.167	"	14.05	4.45		18.50	22.50	

		02990	Structure Moving										
300	0010		**MOVING BUILDINGS** One day move, up to 24' wide										300
	0020		Reset on new foundation, patch & hook-up, average move				Total					9,300	
	0040		Wood or steel frame bldg., based on ground floor area	B-4	185	.259	S.F.		7.05	1.77	8.82	12.90	
	0060		Masonry bldg., based on ground floor area	"	137	.350			9.50	2.39	11.89	17.40	
	0200		For 24' to 42' wide, add				↓					15%	
	0220		For each additional day on road, add	B-4	1	48	Day		1,300	330	1,630	2,375	
	0240		Construct new basement, move building, 1 day										
	0300		move, patch & hook-up, based on ground floor area	B-3	155	.310	S.F.	6	8.85	11.25	26.10	32.50	

For information about Means Estimating Seminars, see yellow pages 12 and 13 in back of book

SITE CONSTRUCTION **2**

135

		CREW	DAILY OUTPUT	LABOR-HOURS	UNIT	MAT.	LABOR	EQUIP.	TOTAL	TOTAL INCL O&P

(Header spanning MAT., LABOR, EQUIP., TOTAL: **2005 BARE COSTS**)

Division 3
Concrete

Estimating Tips

General

- Carefully check all the plans and specifications. Concrete often appears on drawings other than structural drawings, including mechanical and electrical drawings for equipment pads. The cost of cutting and patching is often difficult to estimate. See Subdivisions 02220 and 03055 for demolition costs.
- Always obtain concrete prices from suppliers near the job site. A volume discount can often be negotiated depending upon competition in the area. Remember to add for waste, particularly for slabs and footings on grade.

03100 Concrete Forms & Accessories

- A primary cost for concrete construction is forming. Most jobs today are constructed with prefabricated forms. The selection of the forms best suited for the job and the total square feet of forms required for efficient concrete forming and placing are key elements in estimating concrete construction. Enough forms must be available for erection to make efficient use of the concrete placing equipment and crew.
- Concrete accessories for forming and placing depend upon the systems used. Study the plans and specifications to assure that all special accessory requirements have been included in the cost estimate such as anchor bolts, inserts and hangers.

03200 Concrete Reinforcement

- Ascertain that the reinforcing steel supplier has included all accessories, cutting, bending and an allowance for lapping, splicing and waste. A good rule of thumb is 10% for lapping, splicing and waste. Also, 10% waste should be allowed for welded wire fabric.

03300 Cast-in-Place Concrete

- When estimating structural concrete, pay particular attention to requirements for concrete additives, curing methods and surface treatments. Special consideration for climate, hot or cold, must be included in your estimate. Be sure to include requirements for concrete placing equipment and concrete finishing.

03400 Precast Concrete
03500 Cementitious Decks & Toppings

- The cost of hauling precast concrete structural members is often an important factor. For this reason, it is important to get a quote from the nearest supplier. It may become economically feasible to set up precasting beds on the site if the hauling costs are prohibitive.

Reference Numbers

Reference numbers are shown in bold squares at the beginning of some major classifications. These numbers refer to related items in the Reference Section. The reference information may be an estimating procedure, an alternate pricing method or technical information.

Note: Not all subdivisions listed here necessarily appear in this publication.

		03055	Selective Demolition		CREW	DAILY OUTPUT	LABOR-HOURS	UNIT	2005 BARE COSTS				TOTAL INCL O&P	
									MAT.	LABOR	EQUIP.	TOTAL		
110	0010		**SELECTIVE CONC. DEMO,** excl saw/torch cutting, load or haul	R02220 -510										110
	0050		Break up into small pieces, minimum reinforcing		B-9	24	1.667	C.Y.		45	5.90	50.90	77	
	0060		Average reinforcing			16	2.500			68	8.85	76.85	115	
	0070		Maximum reinforcing			8	5			136	17.70	153.70	230	
	0150		Remove whole pieces, up to 2 tons per piece		E-18	36	1.111	Ea.		42	22	64	97.50	
	0160		2 - 5 tons per piece			30	1.333			50.50	26.50	77	117	
	0170		5 - 10 tons per piece			24	1.667			63	33	96	147	
	0180		10 - 15 tons per piece			18	2.222			84	44.50	128.50	195	
	0250		Precast unit embedded in masonry, up to 1 CF		D-1	16	1			31		31	47.50	
	0260		1 - 2 CF			12	1.333			41.50		41.50	63	
	0270		2 - 5 CF			10	1.600			49.50		49.50	76	
	0280		5 - 10 CF			8	2			62		62	94.50	

		03060	Basic Concrete Materials											
100	0010		**CONCRETE ADMIXTURES & SURFACE TREATMENTS**											100
	0040		Abrasives, aluminum oxide, over 20 tons					Lb.	1.02			1.02	1.12	
	0070		Under 1 ton						1.11			1.11	1.22	
	0100		Silicon carbide, black, over 20 tons						1.43			1.43	1.57	
	0120		Under 1 ton						1.54			1.54	1.69	
	0200		Air entraining agent, .7 to 1.5 oz. per bag, 55 gallon lots					Gal.	9.25			9.25	10.20	
	0220		5 gallon lots						9.65			9.65	10.60	
	0300		Bonding agent, acrylic latex, 250 S.F. per gallon						21			21	23	
	0320		Epoxy resin, 80 S.F. per gallon						46			46	51	
	0400		Calcium chloride, 50 lb. bags					Ton	360			360	395	
	0420		Less than truckload lots					Bag	27.50			27.50	30	
	0500		Carbon black, liquid, 2 to 8 lbs. per bag of cement					Lb.	3.71			3.71	4.08	
	0600		Colors, integral, 2 to 10 lb. per bag of cement, minimum						2.32			2.32	2.55	
	0610		Average						2.63			2.63	2.89	
	0620		Maximum						3.02			3.02	3.32	
	0700		Curing compound, solvent based, 400 S.F./gal, 55 gal. lots					Gal.	8.65			8.65	9.55	
	0720		5 gallon lots						10.25			10.25	11.25	
	0800		Water based, 250 S.F./gal, 55 gallon lots						4.20			4.20	4.62	
	0820		5 gallon lots						5.25			5.25	5.80	
	0900		Dustproofing compound, (200-600 S.F./gal.), 55 gallon lots						3.99			3.99	4.39	
	0920		5 gallon lots						4.62			4.62	5.10	
	1000		Epoxy dustproof coating, colors, (300-400 S.F. per coat),											
	1010		or transparent, (400-600 S.F. per coat)					Gal.	25			25	27.50	
	1100		Hardeners, metallic, 55 lb. bags, natural (grey)					Lb.	.60			.60	.66	
	1200		Colors, average						1.05			1.05	1.16	
	1300		Non-metallic, 55 lb. bags, natural (grey), minimum						.37			.37	.41	
	1310		Maximum						.42			.42	.46	
	1320		Non-metallic, colors, mininum						.48			.48	.53	
	1340		Maximum						.77			.77	.85	
	1400		Non-metallic, non-slip, 100 lb. bags, minimum						.48			.48	.53	
	1420		Maximum						.96			.96	1.06	
	1500		Solution type, (300 to 400 S.F. per gallon)					Gal.	6.30			6.30	6.95	
	1550		Release agent, for tilt slabs						7.10			7.10	7.80	
	1570		For forms, average						4.42			4.42	4.86	
	1600		Sealer, hardener and dustproofer, epoxy, 150 S.F., minimum						7.10			7.10	7.85	
	1620		Maximum						15.75			15.75	17.35	
	1630		Solvent based, 200 S.F., minimum						8.70			8.70	9.60	
	1640		Maximum						16.20			16.20	17.80	
	1650		Water based, 250 S.F., minimum						10.25			10.25	11.25	
	1660		Maximum						11.80			11.80	13	
	1700		Colors (300-400 S.F. per gallon)						42.50			42.50	46.50	
	1800		Set accelerator for below freezing, 1 to 1-1/2 gal. per C.Y.						5.65			5.65	6.25	

3 CONCRETE

Important: See the Reference Section for critical supporting data - Reference Nos., Crews, & City Cost Indexes

		03060	Basic Concrete Materials	CREW	DAILY OUTPUT	LABOR-HOURS	UNIT	2005 BARE COSTS				TOTAL INCL O&P	
								MAT.	LABOR	EQUIP.	TOTAL		
100	1900		Set retarder, 2 to 4 fl. oz. per bag of cement				Gal.	11.55			11.55	12.75	100
	2000		Waterproofing, integral 1 lb. per bag of cement				Lb.	.85			.85	.94	
	2100		Powdered metallic, 40 lbs. per 100 S.F., minimum					1.06			1.06	1.17	
	2120		Maximum				↓	2.12			2.12	2.33	
110	0010	**AGGREGATE** Expanded shale, C.L. lots, 52 lb. per C.F., minimum					Ton	36			36	39.50	110
	0050		Maximum				"	48			48	53	
	0100		Lightweight vermiculite or perlite, 4 C.F. bag, C.L. lots				Bag	6.70			6.70	7.35	
	0150		L.C.L. lots				"	7.35			7.35	8.10	
	0250		Sand & stone, loaded at pit, crushed bank gravel				Ton	18.60			18.60	20.50	
	0350		Sand, washed, for concrete					25			25	27.50	
	0400		For plaster or brick					13.15			13.15	14.45	
	0450		Stone, 3/4" to 1-1/2"					18.15			18.15	20	
	0500		3/8" roofing stone & 1/2" pea stone					15.10			15.10	16.60	
	0550		For trucking 10 miles, add to the above					5.45			5.45	6	
	0600		30 miles, add to the above				↓	13.60			13.60	14.95	
	0850		Sand & stone, loaded at pit, crushed bank gravel				C.Y.	22			22	24.50	
	0950		Sand, washed, for concrete					13.30			13.30	14.65	
	1000		For plaster or brick					36.50			36.50	40	
	1050		Stone, 3/4" to 1-1/2"					24			24	26.50	
	1100		3/8" roofing stone & 1/2" pea stone					32			32	35	
	1150		For trucking 10 miles, add to the above					6			6	6.60	
	1200		30 miles, add to the above				↓	14.95			14.95	16.45	
	1310		Quartz chips, 50 lb. bags				Cwt.	16.65			16.65	18.35	
	1330		Silica chips, 50 lb. bags					9.15			9.15	10.05	
	1410		White marble, 3/8" to 1/2", 50 lb. bags				↓	13.75			13.75	15.15	
	1430		3/4"				Ton	69.50			69.50	76.50	
200	0010	**CEMENT** Material only	R03310 -060										200
	0240		Portland, type I, plain/air entrained, TL lots, 94 lb bags				Bag	8.25			8.25	9.05	
	0300		Trucked in bulk, per cwt				Cwt.	4.25			4.25	4.68	
	0400		Type III, high early strength, TL lots, 94 lb bags				Bag	9.10			9.10	10	
	0420		L.T.L. or L.C.L. lots					10			10	11	
	0500		White, type III, high early strength, T.L. or C.L. lots, bags					19.85			19.85	22	
	0520		L.T.L. or L.C.L. lots					22.50			22.50	25	
	0600		White, type I, T.L. or C.L. lots, bags					19.45			19.45	21.50	
	0620		L.T.L. or L.C.L. lots				↓	18.40			18.40	20.50	
210	0010	**CRIBBING** See under Retaining Walls, division 02830-100											210
250	0010	**DAMPPROOFING** See division 07100											250
300	0010	**EQUIPMENT** For placing conc. see div. 01590-100											300
700	0010	**SAWING CONCRETE** See division 02220-360											700
850	0010	**WATERPROOFING AND DAMPPROOFING** See division 07100											850
	0050		Integral waterproofing, add to cost of regular concrete				C.Y.	5.10			5.10	5.60	
870	0010	**WINTER PROTECTION** For heated ready mix, add, minimum					4.50			4.50	4.95	870
	0050		Maximum				↓	5.65			5.65	6.20	
	0100		Temporary heat to protect concrete, 24 hours, minimum	2 Clab	50	.320	M.S.F.	94	8.55		102.55	116	
	0150		Maximum	"	25	.640	"	122	17.10		139.10	161	
	0200		Temporary shelter for slab on grade, wood frame/polyethylene sheeting										
	0201		Build or remove, minimum	2 Carp	10	1.600	M.S.F.	257	55		312	370	
	0210		Maximum	"	3	5.333	"	310	183		493	625	
	0300		See also Division 03390-200										

CONCRETE 3

		03110	Structural C.I.P. Forms	CREW	DAILY OUTPUT	LABOR-HOURS	UNIT	2005 BARE COSTS				TOTAL INCL O&P	
								MAT.	LABOR	EQUIP.	TOTAL		
300	0010	**EXPANSION JOINT** See division 03150-250											300
405	0010	**FORMS IN PLACE, BEAMS AND GIRDERS**											405
	0020	See also Elevated Slabs, division 03310-240											
	0500	Exterior spandrel, job-built plywood, 12" wide, 1 use	C-2	225	.213	SFCA	3.45	7.10		10.55	14.85	
	0550	2 use		275	.175		1.83	5.80		7.63	11.05	
	0650	4 use		310	.155		1.12	5.15		6.27	9.30	
	1050	2 use		275	.175		1.65	5.80		7.45	10.85	
	1100	3 use		305	.157		1.20	5.25		6.45	9.45	
	1150	4 use		315	.152		.98	5.10		6.08	9	
	1500	24" wide, 1 use		265	.181		2.76	6.05		8.81	12.45	
	1550	2 use		290	.166		1.56	5.50		7.06	10.30	
	1600	3 use		315	.152		1.10	5.10		6.20	9.10	
	1650	4 use		325	.148		.90	4.92		5.82	8.65	
	2000	Interior beam, job-built plywood, 12" wide, 1 use		300	.160		3.82	5.35		9.17	12.50	
	2050	2 use		340	.141		1.87	4.71		6.58	9.40	
	2100	3 use		364	.132		1.53	4.40		5.93	8.55	
	2150	4 use		377	.127		1.24	4.24		5.48	7.95	
	2500	24" wide, 1 use		320	.150		2.81	5		7.81	10.90	
	2550	2 use		365	.132		1.58	4.38		5.96	8.60	
	2600	3 use		385	.125		1.12	4.16		5.28	7.70	
	2650	4 use		395	.122		.91	4.05		4.96	7.30	
	3000	Encasing steel beam, hung, job-built plywood, 1 use		325	.148		2.73	4.92		7.65	10.65	
	3050	2 use		390	.123		1.50	4.10		5.60	8.05	
	3100	3 use		415	.116		1.09	3.85		4.94	7.20	
	3150	4 use		430	.112		.89	3.72		4.61	6.75	
	3500	Bottoms only, to 30" wide, job-built plywood, 1 use		230	.209		4.15	6.95		11.10	15.40	
	3550	2 use		265	.181		2.33	6.05		8.38	11.95	
	3600	3 use		280	.171		1.66	5.70		7.36	10.75	
	3650	4 use		290	.166		1.35	5.50		6.85	10.10	
	4000	Sides only, vertical, 36" high, job-built plywood, 1 use		335	.143		4.28	4.78		9.06	12.15	
	4050	2 use		405	.119		2.35	3.95		6.30	8.75	
	4100	3 use		430	.112		1.71	3.72		5.43	7.70	
	4150	4 use		445	.108		1.39	3.60		4.99	7.15	
	4500	Sloped sides, 36" high, 1 use		305	.157		4.22	5.25		9.47	12.80	
	4550	2 use		370	.130		2.36	4.32		6.68	9.35	
	4600	3 use		405	.119		1.69	3.95		5.64	8	
	4650	4 use		425	.113		1.37	3.76		5.13	7.35	
	5000	Upstanding beams, 36" high, 1 use		225	.213		5.05	7.10		12.15	16.60	
	5050	2 use		255	.188		2.83	6.25		9.08	12.85	
	5100	3 use		275	.175		2.04	5.80		7.84	11.30	
	5150	4 use		280	.171		1.66	5.70		7.36	10.70	
410	0010	**FORMS IN PLACE, COLUMNS**											410
	0500	Round fiberglass, 4 use per mo., rent, 12" diameter	C-1	160	.200	L.F.	6.40	6.45		12.85	17.10	
	0550	16" diameter		150	.213		7.65	6.90		14.55	19.15	
	0600	18" diameter		140	.229		8.50	7.40		15.90	21	
	0650	24" diameter		135	.237		10.60	7.65		18.25	23.50	
	0700	28" diameter		130	.246		11.80	7.95		19.75	25.50	
	0800	30" diameter		125	.256		12.35	8.30		20.65	26.50	
	0850	36" diameter		120	.267		16.45	8.65		25.10	31.50	
	1500	Round fiber tube, 1 use, 8" diameter		155	.206		1.51	6.70		8.21	12.05	
	1550	10" diameter		155	.206		1.88	6.70		8.58	12.45	
	1600	12" diameter		150	.213		2.17	6.90		9.07	13.15	
	1650	14" diameter		145	.221		3.04	7.15		10.19	14.45	
	1700	16" diameter		140	.229		3.58	7.40		10.98	15.45	
	1750	20" diameter		135	.237		5.50	7.65		13.15	18	

Important: See the Reference Section for critical supporting data - Reference Nos., Crews, & City Cost Indexes

			DAILY	LABOR-		\multicolumn{4}{c}{2005 BARE COSTS}		TOTAL

03110 | Structural C.I.P. Forms

		CREW	DAILY OUTPUT	LABOR-HOURS	UNIT	MAT.	LABOR	EQUIP.	TOTAL	TOTAL INCL O&P
410										**410**
1800	24" diameter	C-1	130	.246	L.F.	7	7.95		14.95	20
1850	30" diameter		125	.256		10.20	8.30		18.50	24
1900	36" diameter		115	.278		12.80	9		21.80	28
1950	42" diameter		100	.320		30.50	10.35		40.85	50
2000	48" diameter		85	.376		39	12.20		51.20	62
2200	For seamless type, add					15%				
3000	Round, steel, 4 use per mo., rent, regular duty, 12" diameter	C-1	145	.221	L.F.	10.45	7.15		17.60	22.50
3050	16" diameter		125	.256		11.75	8.30		20.05	26
3100	Heavy duty, 20" diameter		105	.305		12.90	9.85		22.75	29.50
3150	24" diameter		85	.376		14.15	12.20		26.35	34.50
3200	30" diameter		70	.457		16.25	14.80		31.05	41
3250	36" diameter		60	.533		17.40	17.25		34.65	46
3300	48" diameter		50	.640		24	20.50		44.50	59
3350	60" diameter		45	.711		32	23		55	71
4000	Column capitals, steel, 4 uses/mo., 24" col, 4' cap diameter		12	2.667	Ea.	191	86.50		277.50	345
4050	5' cap diameter		11	2.909		217	94		311	385
4100	6' cap diameter		10	3.200		296	104		400	485
4150	7' cap diameter		9	3.556		360	115		475	580
4500	For second and succeeding months, deduct					50%				
5000	Job-built plywood, 8" x 8" columns, 1 use	C-1	165	.194	SFCA	2.11	6.30		8.41	12.05
5050	2 use		195	.164		1.21	5.30		6.51	9.60
5100	3 use		210	.152		.84	4.93		5.77	8.65
5150	4 use		215	.149		.69	4.82		5.51	8.25
5500	12" x 12" columns, 1 use		180	.178		2.12	5.75		7.87	11.30
5550	2 use		210	.152		1.16	4.93		6.09	9
5600	3 use		220	.145		.85	4.71		5.56	8.30
5650	4 use		225	.142		.69	4.60		5.29	7.90
6000	16" x 16" columns, 1 use		185	.173		2.16	5.60		7.76	11.05
6050	2 use		215	.149		1.13	4.82		5.95	8.75
6100	3 use		230	.139		.86	4.50		5.36	7.95
6150	4 use		235	.136		.70	4.41		5.11	7.60
6500	24" x 24" columns, 1 use		190	.168		2.45	5.45		7.90	11.20
6550	2 use		216	.148		1.35	4.79		6.14	8.95
6600	3 use		230	.139		.98	4.50		5.48	8.10
6650	4 use		238	.134		.80	4.35		5.15	7.70
7000	36" x 36" columns, 1 use		200	.160		2.20	5.20		7.40	10.45
7050	2 use		230	.139		1.23	4.50		5.73	8.35
7100	3 use		245	.131		.88	4.23		5.11	7.55
7150	4 use		250	.128		.72	4.14		4.86	7.25
7500	Steel framed plywood, 4 use per mo., rent, 8" x 8"		340	.094		3.01	3.05		6.06	8.05
7550	10" x 10"		350	.091		2.40	2.96		5.36	7.25
7600	12" x 12"		370	.086		2.81	2.80		5.61	7.45
7650	16" x 16"		400	.080		3.02	2.59		5.61	7.35
7700	20" x 20"		420	.076		1.59	2.47		4.06	5.60
7750	24" x 24"		440	.073		1.49	2.35		3.84	5.30
7755	30" x 30"		440	.073		1.35	2.35		3.70	5.15
415	**FORMS IN PLACE, CULVERT** 5' to 8' square or rectangular, 1 use	C-1	170	.188	SFCA	4.80	6.10		10.90	14.80 **415**
0050	2 use		180	.178		1.71	5.75		7.46	10.85
0100	3 use		190	.168		1.34	5.45		6.79	9.95
0150	4 use		200	.160		1.15	5.20		6.35	9.30
420	**FORMS IN PLACE, ELEVATED SLABS**									**420**
1000	Flat plate, job-built plywood, to 15' high, 1 use	C-2	470	.102	S.F.	4	3.40		7.40	9.70
1050	2 use		520	.092		2.20	3.08		5.28	7.20
1100	3 use		545	.088		1.60	2.94		4.54	6.35
1150	4 use		560	.086		1.30	2.86		4.16	5.90
1500	15' to 20' high ceilings, 4 use		495	.097		1.53	3.23		4.76	6.75

CONCRETE **3**

			DAILY	LABOR-		\multicolumn{4}{c}{2005 BARE COSTS}	TOTAL	
03110	Structural C.I.P. Forms	CREW	OUTPUT	HOURS	UNIT	MAT.	LABOR	EQUIP.	TOTAL	INCL O&P
420 1600	21' to 35' high ceilings, 4 use	C-2	450	.107	S.F.	2.12	3.56		5.68	7.90
2000	Flat slab, drop panels, job-built plywood, to 15' high, 1 use		449	.107		4.44	3.56		8	10.45
2050	2 use		509	.094		2.44	3.14		5.58	7.55
2100	3 use		532	.090		1.77	3.01		4.78	6.65
2150	4 use		544	.088		1.44	2.94		4.38	6.15
2250	15' to 20' high ceilings, 4 use		480	.100		3.73	3.33		7.06	9.30
2350	20' to 35' high ceilings, 4 use		435	.110		4.31	3.68		7.99	10.50
3000	Floor slab hung from steel beams, 1 use		485	.099		2.17	3.30		5.47	7.55
3050	2 use		535	.090		1.52	2.99		4.51	6.35
3100	3 use		550	.087		1.30	2.91		4.21	5.95
3150	4 use		565	.085		1.19	2.83		4.02	5.70
3500	Floor slab, with 20" metal pans, 1 use		415	.116		5.70	3.85		9.55	12.30
3550	2 use		445	.108		3.71	3.60		7.31	9.70
3600	3 use		475	.101		3.04	3.37		6.41	8.60
3650	4 use		500	.096		2.71	3.20		5.91	7.95
3700	Floor slab with 30" pans, 1 use		418	.115		5.65	3.83		9.48	12.20
3720	2 use		455	.105		3.67	3.52		7.19	9.50
3740	3 use		470	.102		3	3.40		6.40	8.60
3760	4 use	▼	480	.100		2.67	3.33		6	8.15
3801	Rent 30" metal pans, 1 use					4.88			4.88	5.35
3821	2 use					4.88			4.88	5.35
3841	3 use					4.88			4.88	5.35
3861	4 use					4.88			4.88	5.35
4000	Floor slab with 19" metal domes, 1 use	C-2	405	.119		6	3.95		9.95	12.75
4050	2 use		435	.110		3.98	3.68		7.66	10.10
4100	3 use		465	.103		3.31	3.44		6.75	9
4150	4 use		495	.097		2.98	3.23		6.21	8.30
4500	With 30" fiberglass domes, 1 use		405	.119		5.85	3.95		9.80	12.55
4520	2 use		450	.107		3.82	3.56		7.38	9.75
4530	3 use		460	.104		3.16	3.48		6.64	8.85
4550	4 use	▼	470	.102		2.82	3.40		6.22	8.40
4701	Rent 30" fiberglass domes, 1 use					3.05			3.05	3.36
4721	2 use					3.05			3.05	3.36
4741	3 use					3.05			3.05	3.36
4761	4 use				▼	3.05			3.05	3.36
5000	Box out for slab openings, over 16" deep, 1 use	C-2	190	.253	SFCA	3.49	8.40		11.89	16.95
5050	2 use		240	.200	"	1.92	6.65		8.57	12.50
5500	Shallow slab box outs, to 10 S.F.		42	1.143	Ea.	9.85	38		47.85	70.50
5550	Over 10 S.F. (use perimeter)		600	.080	L.F.	1.31	2.67		3.98	5.60
6000	Bulkhead forms for slab, with keyway, 1 use, 2 piece		500	.096		1.59	3.20		4.79	6.75
6100	3 piece (see also edge forms)	▼	460	.104		1.73	3.48		5.21	7.30
6200	Bulkhead form for slab, 4-1/2" high, exp metal, incl keyway & stakes	C-1	1,200	.027		.70	.86		1.56	2.11
6210	5-1/2" high		1,100	.029		.78	.94		1.72	2.33
6215	7-1/2" high		960	.033		.94	1.08		2.02	2.71
6220	9-1/2" high		840	.038	▼	1.62	1.23		2.85	3.70
6500	Curb forms, wood, 6" to 12" high, on elevated slabs, 1 use		180	.178	SFCA	1.48	5.75		7.23	10.55
6550	2 use		205	.156		.81	5.05		5.86	8.75
6600	3 use		220	.145		.59	4.71		5.30	8
6650	4 use		225	.142	▼	.48	4.60		5.08	7.70
7000	Edge forms to 6" high, on elevated slab, 4 use		500	.064	L.F.	.17	2.07		2.24	3.42
7070	7" to 12" high, 1 use		162	.198	SFCA	1.14	6.40		7.54	11.20
7080	2 use		198	.162		.63	5.25		5.88	8.85
7090	3 use		222	.144		.46	4.66		5.12	7.75
7101	4 use		350	.091	▼	.37	2.96		3.33	5
7500	Depressed area forms to 12" high, 4 use		300	.107	L.F.	.78	3.45		4.23	6.25
7550	12" to 24" high, 4 use	▼	175	.183	▼	1.06	5.90		6.96	10.35

Important: See the Reference Section for critical supporting data - Reference Nos., Crews, & City Cost Indexes

03100 | Concrete Forms & Accessories

03110 | Structural C.I.P. Forms

			CREW	DAILY OUTPUT	LABOR-HOURS	UNIT	MAT.	LABOR	EQUIP.	TOTAL	TOTAL INCL O&P	
420	8000	Perimeter deck and rail for elevated slabs, straight	C-1	90	.356	L.F.	13.60	11.50		25.10	33	**420**
	8050	Curved		65	.492		18.70	15.95		34.65	45.50	
	8500	Void forms, round fiber, 3" diameter		450	.071		.62	2.30		2.92	4.26	
	8550	4" diameter		425	.075		.72	2.44		3.16	4.58	
	8600	6" diameter		400	.080		1.09	2.59		3.68	5.25	
	8650	8" diameter		375	.085		1.76	2.76		4.52	6.25	
	8700	10" diameter		350	.091		2.28	2.96		5.24	7.10	
	8750	12" diameter		300	.107		2.95	3.45		6.40	8.65	
425	0010	**FORMS IN PLACE, EQUIPMENT FOUNDATIONS** job built										**425**
	0020	1 use	C-2	160	.300	SFCA	2.67	10		12.67	18.50	
	0050	2 use		190	.253		1.47	8.40		9.87	14.70	
	0100	3 use		200	.240		1.07	8		9.07	13.60	
	0150	4 use		205	.234		.87	7.80		8.67	13.10	
430	0010	**FORMS IN PLACE, FOOTINGS** Continuous wall, plywood, 1 use	C-1	375	.085	SFCA	2.31	2.76		5.07	6.85	**430**
	0050	2 use		440	.073		1.27	2.35		3.62	5.05	
	0100	3 use		470	.068		.92	2.20		3.12	4.45	
	0150	4 use		485	.066		.75	2.14		2.89	4.16	
	0500	Dowel supports for footings or beams, 1 use		500	.064	L.F.	.76	2.07		2.83	4.07	
	1000	Integral starter wall, to 4" high, 1 use		400	.080		.81	2.59		3.40	4.92	
	1500	Keyway, 4 use, tapered wood, 2" x 4"	1 Carp	530	.015		.18	.52		.70	1.01	
	1550	2" x 6"		500	.016		.27	.55		.82	1.15	
	2000	Tapered plastic, 2" x 3"		530	.015		.56	.52		1.08	1.43	
	2050	2" x 4"		500	.016		.71	.55		1.26	1.63	
	2250	For keyway hung from supports, add		150	.053		.76	1.83		2.59	3.69	
	3000	Pile cap, square or rectangular, job-built plywood, 1 use	C-1	290	.110	SFCA	2.36	3.57		5.93	8.15	
	3050	2 use		346	.092		1.30	2.99		4.29	6.10	
	3100	3 use		371	.086		.94	2.79		3.73	5.40	
	3150	4 use		383	.084		.77	2.70		3.47	5.05	
	4000	Triangular or hexagonal, 1 use		225	.142		2.78	4.60		7.38	10.20	
	4050	2 use		280	.114		1.53	3.70		5.23	7.45	
	4100	3 use		305	.105		1.11	3.40		4.51	6.50	
	4150	4 use		315	.102		.90	3.29		4.19	6.10	
	5000	Spread footings, job-built lumber, 1 use		305	.105		1.66	3.40		5.06	7.15	
	5050	2 use		371	.086		.92	2.79		3.71	5.35	
	5100	3 use		401	.080		.67	2.58		3.25	4.75	
	5150	4 use		414	.077		.54	2.50		3.04	4.49	
	6000	Supports for dowels, plinths or templates, 2' x 2' footing		25	1.280	Ea.	5.45	41.50		46.95	70.50	
	6050	4' x 4' footing		22	1.455		10.90	47		57.90	85.50	
	6100	8' x 8' footing		20	1.600		22	52		74	105	
	6150	12' x 12' footing		17	1.882		35	61		96	134	
	7000	Plinths, job-built plywood, 1 use		250	.128	SFCA	3.63	4.14		7.77	10.45	
	7100	4 use		270	.119	"	1.19	3.84		5.03	7.25	
435	0010	**FORMS IN PLACE, GRADE BEAM** Job-built plywood, 1 use	C-2	530	.091	SFCA	1.56	3.02		4.58	6.40	**435**
	0050	2 use		580	.083		.86	2.76		3.62	5.25	
	0100	3 use		600	.080		.63	2.67		3.30	4.84	
	0150	4 use		605	.079		.51	2.64		3.15	4.68	
440	0010	**FORMS IN PLACE, MAT FOUNDATION** Job-built plywood, 1 use	C-2	290	.166	SFCA	2.44	5.50		7.94	11.30	**440**
	0050	2 use		310	.155		.92	5.15		6.07	9.05	
	0100	3 use		330	.145		.65	4.85		5.50	8.25	
	0120	4 use		350	.137		.54	4.57		5.11	7.70	
445	0010	**FORMS IN PLACE, SLAB ON GRADE**										**445**
	1000	Bulkhead forms w/keyway, wood, 6" high, 1 use	C-1	510	.063	L.F.	.80	2.03		2.83	4.04	
	1050	2 uses		400	.080		.44	2.59		3.03	4.51	
	1100	4 uses		350	.091		.26	2.96		3.22	4.90	

		03110 | **Structural C.I.P. Forms**	CREW	DAILY OUTPUT	LABOR-HOURS	UNIT	**2005 BARE COSTS**				TOTAL INCL O&P	
							MAT.	LABOR	EQUIP.	TOTAL		
445	1400	Bulkhead form for slab, 4-1/2" high, exp metal, incl keyway & stakes	C-1	1,200	.027	L.F.	.70	.86		1.56	2.11	**445**
	1410	5-1/2" high		1,100	.029		.78	.94		1.72	2.33	
	1420	7-1/2" high		960	.033	↓	.94	1.08		2.02	2.71	
	2000	Curb forms, wood, 6" to 12" high, on grade, 1 use		215	.149	SFCA	2.19	4.82		7.01	9.90	
	2050	2 use		250	.128		1.21	4.14		5.35	7.80	
	2100	3 use		265	.121		.87	3.91		4.78	7.05	
	2150	4 use		275	.116	↓	.71	3.77		4.48	6.65	
	3000	Edge forms, wood, 4 use, on grade, to 6" high		600	.053	L.F.	.27	1.73		2	2.98	
	3050	7" to 12" high		435	.074	SFCA	.74	2.38		3.12	4.52	
	3060	Over 12"		350	.091	"	.80	2.96		3.76	5.50	
	3500	For depressed slabs, 4 use, to 12" high		300	.107	L.F.	.66	3.45		4.11	6.10	
	3550	To 24" high		175	.183		.88	5.90		6.78	10.15	
	4000	For slab blockouts, to 12" high, 1 use		200	.160		.73	5.20		5.93	8.85	
	4050	To 24" high, 1 use		120	.267		.93	8.65		9.58	14.45	
	4100	Plastic (extruded), to 6" high, multiple use, on grade		800	.040		.37	1.29		1.66	2.43	
	5020	Wood, incl. wood stakes, 1" x 3"		900	.036		.57	1.15		1.72	2.41	
	5050	2" x 4"		900	.036	↓	.62	1.15		1.77	2.47	
	6000	Trench forms in floor, wood, 1 use		160	.200	SFCA	2.20	6.45		8.65	12.50	
	6050	2 use		175	.183		1.22	5.90		7.12	10.55	
	6100	3 use		180	.178		.89	5.75		6.64	9.95	
	6150	4 use		185	.173	↓	.72	5.60		6.32	9.50	
	8760	Void form, corrugated fiberboard, 6" x 12", 10' long	↓	240	.133	S.F.	9.20	4.31		13.51	16.80	
450	0010	**FORMS IN PLACE, STAIRS** (Slant length x width), 1 use	C-2	165	.291	S.F.	3.66	9.70		13.36	19.15	**450**
	0050	2 use		170	.282		2.01	9.40		11.41	16.85	
	0100	3 use		180	.267		1.46	8.90		10.36	15.45	
	0150	4 use		190	.253	↓	1.19	8.40		9.59	14.40	
	1000	Alternate pricing method (0.7 L.F./S.F.), 1 use		100	.480	LF Rsr	5.25	16		21.25	31	
	1050	2 use		105	.457		2.89	15.25		18.14	26.50	
	1100	3 use		110	.436		2.10	14.55		16.65	25	
	1150	4 use		115	.417	↓	1.73	13.90		15.63	23.50	
	2000	Stairs, cast on sloping ground (length x width), 1 use		220	.218	S.F.	1.36	7.25		8.61	12.80	
	2100	4 use	↓	240	.200	"	.44	6.65		7.09	10.90	
455	0010	**FORMS IN PLACE, WALLS**										**455**
	0100	Box out for wall openings, to 16" thick, to 10 S.F.	C-2	24	2	Ea.	24.50	66.50		91	131	
	0150	Over 10 S.F. (use perimeter)	"	280	.171	L.F.	2.07	5.70		7.77	11.20	
	0250	Brick shelf, 4" w, add to wall forms, use wall area abv shelf										
	0260	1 use	C-2	240	.200	SFCA	2.18	6.65		8.83	12.80	
	0300	2 use		275	.175		1.20	5.80		7	10.35	
	0350	4 use		300	.160	↓	.87	5.35		6.22	9.25	
	0500	Bulkhead, with keyway, 1 use, 2 piece		265	.181	L.F.	2.76	6.05		8.81	12.45	
	0550	3 piece	↓	175	.274		3.48	9.15		12.63	18.10	
	0600	Bulkhead forms with keyway, 1 piece expanded metal, 8" wall	C-1	1,000	.032		.94	1.04		1.98	2.64	
	0610	10" wall		800	.040		1.62	1.29		2.91	3.80	
	0620	12" wall	↓	525	.061	↓	2.05	1.97		4.02	5.35	
	0700	Buttress, to 8' high, 1 use	C-2	350	.137	SFCA	3.68	4.57		8.25	11.15	
	0750	2 use		430	.112		2.02	3.72		5.74	8	
	0800	3 use		460	.104		1.47	3.48		4.95	7	
	0850	4 use		480	.100	↓	1.21	3.33		4.54	6.55	
	1000	Corbel or haunch, to 12" wide, add to wall forms, 1 use		150	.320	L.F.	2.06	10.65		12.71	18.85	
	1050	2 use		170	.282		1.13	9.40		10.53	15.90	
	1100	3 use		175	.274		.83	9.15		9.98	15.15	
	1150	4 use		180	.267	↓	.67	8.90		9.57	14.60	
	2000	Wall, below grade, job-built plywood, to 8' high, 1 use		300	.160	SFCA	2.37	5.35		7.72	10.90	
	2050	2 use		365	.132		1.50	4.38		5.88	8.50	
	2150	4 use		435	.110		.88	3.68		4.56	6.70	
	2400	Over 8' to 16' high, 1 use		280	.171	↓	4.51	5.70		10.21	13.85	

03110 | Structural C.I.P. Forms

			DAILY OUTPUT	LABOR-HOURS	UNIT	2005 BARE COSTS				TOTAL INCL O&P
		CREW				MAT.	LABOR	EQUIP.	TOTAL	
455	2420	2 use	C-2	345	.139	SFCA	1.89	4.64		6.53	9.30	**455**
	2430	3 use		375	.128		1.57	4.27		5.84	8.40
	2440	4 use		395	.122		1.40	4.05		5.45	7.85
	2445	Exterior wall, 8' to 16' high, 1 use		280	.171		2.18	5.70		7.88	11.30
	2450	2 use		345	.139		1.20	4.64		5.84	8.50
	2500	3 use		375	.128		.85	4.27		5.12	7.60
	2550	4 use		395	.122		.70	4.05		4.75	7.05
	2700	Over 16' high, 1 use		235	.204		2.40	6.80		9.20	13.25
	2750	2 use		290	.166		1.32	5.50		6.82	10.05
	2800	3 use		315	.152		.96	5.10		6.06	8.95
	2850	4 use		330	.145		.78	4.85		5.63	8.40
	3000	For architectural finish, add		1,820	.026		5.40	.88		6.28	7.25
	4000	Radial, smooth curved, job-built plywood, 1 use		245	.196		2.48	6.55		9.03	12.90
	4050	2 use		300	.160		1.36	5.35		6.71	9.80
	4100	3 use		325	.148		.99	4.92		5.91	8.75
	4150	4 use		335	.143		.81	4.78		5.59	8.35
	4200	Below grade, job-built plywood, 1 use		225	.213		2.99	7.10		10.09	14.35
	4210	2 use		225	.213		1.65	7.10		8.75	12.85
	4220	3 use		225	.213		1.33	7.10		8.43	12.50
	4230	4 use		225	.213		.97	7.10		8.07	12.10
	4300	Curved, 2' chords, job-built plywood, 1 use		290	.166		1.95	5.50		7.45	10.75
	4350	2 use		355	.135		1.07	4.51		5.58	8.20
	4400	3 use		385	.125		.78	4.16		4.94	7.30
	4450	4 use		400	.120		.63	4		4.63	6.95
	4500	Over 8' high, 1 use		290	.166		.94	5.50		6.44	9.65
	4525	2 use		355	.135		.52	4.51		5.03	7.55
	4550	3 use		385	.125		.38	4.16		4.54	6.85
	4575	4 use		400	.120		.31	4		4.31	6.60
	4600	Retaining wall, battered, job-built plyw'd, to 8' high, 1 use		300	.160		1.85	5.35		7.20	10.35
	4650	2 use		355	.135		1.02	4.51		5.53	8.10
	4700	3 use		375	.128		.74	4.27		5.01	7.45
	4750	4 use		390	.123		.56	4.10		4.66	7
	4900	Over 8' to 16' high, 1 use		240	.200		2	6.65		8.65	12.60
	4950	2 use		295	.163		1.10	5.40		6.50	9.65
	5000	3 use		305	.157		.80	5.25		6.05	9.05
	5050	4 use		320	.150		.65	5		5.65	8.50
	5100	Retaining wall form, plywood, smooth curve, 1 use		200	.240		3.08	8		11.08	15.85
	5120	2 use		235	.204		1.70	6.80		8.50	12.45
	5130	3 use		250	.192		1.23	6.40		7.63	11.30
	5140	4 use	↓	260	.185		1.01	6.15		7.16	10.70
	5500	For gang wall forming, 192 S.F. sections, deduct					10%	10%			
	5550	384 S.F. sections, deduct				↓	20%	20%			
	5750	Liners for forms (add to wall forms), A.B.S. plastic									
	5800	Aged wood, 4" wide, 1 use	1 Carp	250	.032	SFCA	5.85	1.10		6.95	8.15
	5820	2 use		400	.020		3.22	.69		3.91	4.61
	5840	4 use		750	.011		1.90	.37		2.27	2.66
	5900	Fractured rope rib, 1 use		250	.032		3.69	1.10		4.79	5.75
	6000	4 use		750	.011		1.20	.37		1.57	1.89
	6100	Ribbed look, 1/2" & 3/4" deep, 1 use		300	.027		5.40	.91		6.31	7.30
	6200	4 use		800	.010		1.75	.34		2.09	2.45
	6300	Rustic brick pattern, 1 use		250	.032		3.69	1.10		4.79	5.75
	6400	4 use		750	.011		1.20	.37		1.57	1.89
	6500	Striated, random, 3/8" x 3/8" deep, 1 use		300	.027		3.69	.91		4.60	5.50
	6600	4 use	↓	800	.010	↓	1.20	.34		1.54	1.85
	6800	Rustication strips, A.B.S. plastic, 2 piece snap-on									
	6850	1" deep x 1-3/8" wide, 1 use	C-2	400	.120	L.F.	4.18	4		8.18	10.85

CONCRETE **3**

3 CONCRETE

		**03110	Structural C.I.P. Forms**	CREW	DAILY OUTPUT	LABOR-HOURS	UNIT	2005 BARE COSTS				TOTAL INCL O&P	
							MAT.	LABOR	EQUIP.	TOTAL		
455	6900	2 use	C-2	600	.080	L.F.	2.30	2.67		4.97	6.70	**455**
	6950	4 use		800	.060		1.36	2		3.36	4.60	
	7050	Wood, beveled edge, 3/4" deep, 1 use		600	.080		.10	2.67		2.77	4.26	
	7100	1" deep, 1 use		450	.107		.14	3.56		3.70	5.70	
	7200	For solid board finish, uniform, 1 use, add to wall forms		300	.160	SFCA	.79	5.35		6.14	9.15	
	7300	Non-uniform finish		250	.192		.74	6.40		7.14	10.75	
	7500	Lintel or sill forms, 1 use	1 Carp	30	.267		2.80	9.15		11.95	17.35	
	7520	2 use		34	.235		1.54	8.05		9.59	14.25	
	7540	3 use		36	.222		1.12	7.60		8.72	13.10	
	7560	4 use		37	.216		.91	7.40		8.31	12.55	
	7820	2 use	C-2	1,200	.040		.92	1.33		2.25	3.10	
	7840	3 use		1,240	.039		.67	1.29		1.96	2.75	
	7860	4 use		1,260	.038		.55	1.27		1.82	2.59	
	8000	To 16' high, 1 use		715	.067		2.19	2.24		4.43	5.90	
	8020	2 use		740	.065		1.20	2.16		3.36	4.69	
	8040	3 use		770	.062		.88	2.08		2.96	4.20	
	8060	4 use		790	.061		.72	2.03		2.75	3.94	
	8100	Over 16' high, 1 use		715	.067		2.63	2.24		4.87	6.35	
	8120	2 use		740	.065		1.45	2.16		3.61	4.96	
	8140	3 use		770	.062		1.05	2.08		3.13	4.40	
	8160	4 use		790	.061		.88	2.03		2.91	4.11	
	8600	Pilasters, 1 use		270	.178		2.73	5.95		8.68	12.25	
	8620	2 use		330	.145		1.50	4.85		6.35	9.20	
	8640	3 use		370	.130		1.09	4.32		5.41	7.95	
	8660	4 use		385	.125		.89	4.16		5.05	7.45	
	9010	Steel framed plywood, based on 100 uses of purchased										
	9020	forms, and 4 uses of bracing lumber										
	9060	To 8' high	C-2	600	.080	SFCA	.37	2.67		3.04	4.56	
	9260	Over 8' to 16' high		450	.107		.37	3.56		3.93	5.95	
	9460	Over 16' to 20' high		400	.120		.37	4		4.37	6.65	
	9475	For elevated walls, add							10%				
460	0010	**FORMS IN PLACE, INSULATING CONCRETE**										**460**
	0020	Forms left in place, S.F. is for both sides										
	1000	Panel system, flat cavity, minimum	2 Carp	960	.017	S.F.	1.84	.57		2.41	2.91	
	1010	Maximum		960	.017		2.76	.57		3.33	3.93	
	1020	Grid cavity, minimum		960	.017		2	.57		2.57	3.09	
	1030	Maximum		960	.017		3	.57		3.57	4.19	
	1040	Post and beam cavity, minimum		960	.017		2.20	.57		2.77	3.31	
	1050	Maximum		960	.017		3.30	.57		3.87	4.52	
	1060	Plank system, flat cavity, minimum		1,920	.008		2.10	.29		2.39	2.75	
	1070	Maximum		1,920	.008		3.14	.29		3.43	3.90	
	1120	Block system, flat cavity, minimum		480	.033		1.96	1.14		3.10	3.94	
	1130	Maximum		480	.033		2.94	1.14		4.08	5	
	1140	Grid cavity, minimum		480	.033		1.96	1.14		3.10	3.94	
	1150	Maximum		480	.033		2.94	1.14		4.08	5	
	1160	Post and beam cavity, minimum		480	.033		1.96	1.14		3.10	3.94	
	1170	Maximum		480	.033		2.94	1.14		4.08	5	
820	0010	**SLIPFORMS** Silos, minimum	C-17E	3,885	.021	SFCA	3	.73	.02	3.75	4.45	**820**
	0050	Maximum		1,095	.073		4.30	2.58	.07	6.95	8.80	
	1000	Buildings, minimum		3,660	.022		2	.77	.02	2.79	3.42	
	1050	Maximum		875	.091		3.70	3.22	.09	7.01	9.15	

		**03150	Concrete Accessories**										
080	0010	**ACCESSORIES, ANCHOR BOLTS** J-type, incl. nut and washer										**080**
	0020	1/2" diameter, 6" long	1 Carp	90	.089	Ea.	.94	3.04		3.98	5.75	

Important: See the Reference Section for critical supporting data - Reference Nos., Crews, & City Cost Indexes

03100 | Concrete Forms & Accessories

03150	Concrete Accessories	CREW	DAILY OUTPUT	LABOR-HOURS	UNIT	MAT.	LABOR	EQUIP.	TOTAL	TOTAL INCL O&P	
080 0050	10" long	1 Carp	85	.094	Ea.	1.06	3.22		4.28	6.15	**080**
0100	12" long		85	.094		1.17	3.22		4.39	6.30	
0200	5/8" diameter, 12" long		80	.100		1.18	3.43		4.61	6.65	
0250	18" long		70	.114		1.39	3.91		5.30	7.65	
0300	24" long		60	.133		1.60	4.57		6.17	8.85	
0350	3/4" diameter, 8" long		80	.100		1.39	3.43		4.82	6.90	
0400	12" long		70	.114		1.74	3.91		5.65	8	
0450	18" long		60	.133		2.26	4.57		6.83	9.60	
0500	24" long		50	.160		2.96	5.50		8.46	11.80	
0600	7/8" diameter, 12" long		60	.133		2.29	4.57		6.86	9.60	
0650	18" long		50	.160		3.07	5.50		8.57	11.95	
0700	24" long		40	.200		3.17	6.85		10.02	14.15	
0800	1" diameter, 12" long		55	.145		3.32	4.98		8.30	11.40	
0850	18" long		45	.178		4	6.10		10.10	13.90	
0900	24" long		35	.229		4.88	7.85		12.73	17.55	
0950	36" long		25	.320		6.70	10.95		17.65	24.50	
1200	1-1/2" diameter, 18" long		22	.364		11.85	12.45		24.30	32.50	
1250	24" long		18	.444		14.10	15.20		29.30	39	
1300	36" long	▼	12	.667	▼	17.60	23		40.60	55	
1350	For larger sizes see Division 05090-080										
2000	Galvanized, incl. nut and washer										
2100	1/2" diameter, 6" long	1 Carp	90	.089	Ea.	1.64	3.04		4.68	6.55	
2150	10" long		85	.094		1.86	3.22		5.08	7.05	
2200	12" long		85	.094		2.05	3.22		5.27	7.25	
2500	5/8" diameter, 12" long		80	.100		2.07	3.43		5.50	7.65	
2550	18" long		70	.114		2.44	3.91		6.35	8.80	
2600	24" long		60	.133		2.80	4.57		7.37	10.20	
2700	3/4" diameter, 8" long		80	.100		2.44	3.43		5.87	8.05	
2750	12" long		70	.114		3.05	3.91		6.96	9.45	
2800	18" long		60	.133		3.96	4.57		8.53	11.45	
2850	24" long		50	.160		5.20	5.50		10.70	14.25	
3000	7/8" diameter, 12" long		60	.133		4.01	4.57		8.58	11.50	
3050	18" long		50	.160		5.40	5.50		10.90	14.45	
3100	24" long		40	.200		5.55	6.85		12.40	16.75	
3200	1" diameter, 12" long		55	.145		5.80	4.98		10.78	14.15	
3250	18" long		45	.178		7	6.10		13.10	17.20	
3300	24" long		35	.229		8.55	7.85		16.40	21.50	
3350	36" long		25	.320		11.70	10.95		22.65	30	
3500	1-1/2" diameter, 18" long		22	.364		20.50	12.45		32.95	42.50	
3550	24" long		18	.444		24.50	15.20		39.70	50.50	
3600	36" long	▼	12	.667	▼	36	23		59	75	
3700	For larger sizes, galvanized, see Division 05090-080										
8000	Sleeves, see Division 03150-620										
8800	Templates, steel, 8" bolt spacing	2 Carp	16	1	Ea.	7.15	34.50		41.65	61.50	
8850	12" bolt spacing		15	1.067		7.45	36.50		43.95	65	
8900	16" bolt spacing		14	1.143		8.90	39		47.90	71	
8950	24" bolt spacing		12	1.333		11.90	45.50		57.40	84	
9100	Wood, 8" bolt spacing		16	1		.71	34.50		35.21	54.50	
9150	12" bolt spacing		15	1.067		.94	36.50		37.44	58	
9200	16" bolt spacing		14	1.143		1.18	39		40.18	62.50	
9250	24" bolt spacing	▼	16	1	▼	1.65	34.50		36.15	55.50	
085 0013	**ANCHOR BOLTS** See divisions 04080-070 and 05090-080										**085**
160 0010	**ACCESSORIES, CHAMFER STRIPS**										**160**
2000	Polyvinyl chloride, 1/2" wide with leg	1 Carp	535	.015	L.F.	.22	.51		.73	1.04	

03150		Concrete Accessories	CREW	DAILY OUTPUT	LABOR-HOURS	UNIT	2005 BARE COSTS				TOTAL INCL O&P	
							MAT.	LABOR	EQUIP.	TOTAL		
160	2200	3/4" wide with leg	1 Carp	525	.015	L.F.	.30	.52		.82	1.14	**160**
	2400	1" radius with leg		515	.016		.60	.53		1.13	1.49	
	2800	1-1/2" radius with leg		500	.016		2.01	.55		2.56	3.06	
	5000	Wood, 1/2" wide		535	.015		.09	.51		.60	.90	
	5200	3/4" wide		525	.015		.10	.52		.62	.92	
	5400	1" wide		515	.016		.14	.53		.67	.98	
200	0010	**ACCESSORIES, DOVETAIL ANCHOR SYSTEM**										**200**
	0500	Anchor slot, galv., filled, 24 ga.	1 Carp	425	.019	L.F.	.77	.64		1.41	1.85	
	0600	20 ga.		400	.020		.94	.69		1.63	2.10	
	0800	16 oz. copper, foam filled		375	.021		1.97	.73		2.70	3.31	
	0900	26 ga. stainless steel, foam filled		375	.021		1.23	.73		1.96	2.50	
	1200	Brick anchor, corr., galv., 3-1/2" long, 16 ga.	1 Bric	10.50	.762	C	22.50	27		49.50	66	
	1300	12 ga.		10.50	.762		31.50	27		58.50	75.50	
	1500	Flat, galv., 3-1/2" long, 16 ga.		10.50	.762		24.50	27		51.50	68	
	1600	12 ga.		10.50	.762		46	27		73	91.50	
	6000	Stone anchors, 3-1/2" long, galv., 1/8" x 1" wide		10.50	.762		96.50	27		123.50	147	
	6100	1/4" x 1" wide		10.50	.762		149	27		176	205	
250	0010	**EXPANSION JOINT** Keyed, cold, 24 ga, incl. stakes, 3-1/2" high	1 Carp	200	.040	L.F.	.58	1.37		1.95	2.77	**250**
	0050	4-1/2" high		200	.040		.70	1.37		2.07	2.90	
	0100	5-1/2" high		195	.041		.78	1.41		2.19	3.05	
	0150	7-1/2" high		190	.042		.94	1.44		2.38	3.28	
	0200	9-1/2" high		185	.043		2.05	1.48		3.53	4.57	
	0300	Poured asphalt, plain, 1/2" x 1"	1 Clab	450	.018		.36	.47		.83	1.14	
	0350	1" x 2"		400	.020		1.26	.53		1.79	2.22	
	0500	Neoprene, liquid, cold applied, 1/2" x 1"		450	.018		1.63	.47		2.10	2.53	
	0550	1" x 2"		400	.020		6.25	.53		6.78	7.75	
	0700	Polyurethane, poured, 2 part, 1/2" x 1"		400	.020		2	.53		2.53	3.03	
	0750	1" x 2"		350	.023		7.40	.61		8.01	9.10	
	0900	Rubberized asphalt, hot or cold applied, 1/2" x 1"		450	.018		.44	.47		.91	1.22	
	0950	1" x 2"		400	.020		1.23	.53		1.76	2.18	
	1100	Hot applied, fuel resistant, 1/2" x 1"		450	.018		.76	.47		1.23	1.58	
	1150	1" x 2"		400	.020		1.55	.53		2.08	2.54	
	2000	Premolded, bituminous fiber, 1/2" x 6"	1 Carp	375	.021		.39	.73		1.12	1.57	
	2050	1" x 12"		300	.027		1.37	.91		2.28	2.93	
	2250	Cork with resin binder, 1/2" x 6"		375	.021		1.52	.73		2.25	2.81	
	2300	1" x 12"		300	.027		5.45	.91		6.36	7.40	
	2500	Neoprene sponge, closed cell, 1/2" x 6"		375	.021		1.32	.73		2.05	2.59	
	2550	1" x 12"		300	.027		6.05	.91		6.96	8.10	
	2750	Polyethylene foam, 1/2" x 6"		375	.021		.49	.73		1.22	1.68	
	2800	1" x 12"		300	.027		1.39	.91		2.30	2.95	
	3000	Polyethylene backer rod, 3/8" diameter		460	.017		.04	.60		.64	.97	
	3050	3/4" diameter		460	.017		.06	.60		.66	1	
	3100	1" diameter		460	.017		.07	.60		.67	1.01	
	3500	Polyurethane foam, with polybutylene, 1/2" x 1/2"		475	.017		.60	.58		1.18	1.56	
	3550	1" x 1"		450	.018		1.67	.61		2.28	2.79	
	3750	Polyurethane foam, regular, closed cell, 1/2" x 6"		375	.021		1.19	.73		1.92	2.45	
	3800	1" x 12"		300	.027		1.39	.91		2.30	2.95	
	4000	Polyvinyl chloride foam, closed cell, 1/2" x 6"		375	.021		1.97	.73		2.70	3.31	
	4050	1" x 12"		300	.027		5.50	.91		6.41	7.45	
	4250	Rubber, gray sponge, 1/2" x 6"		375	.021		2.46	.73		3.19	3.85	
	4300	1" x 12"		300	.027		9.65	.91		10.56	12	
	4500	Lead wool for joints, 1 ton lots				Lb.	1.90			1.90	2.09	
	5000	For installation in walls, add						75%				
	5250	For installation in boxouts, add						25%				

Important: See the Reference Section for critical supporting data - Reference Nos., Crews, & City Cost Indexes

03150	Concrete Accessories	CREW	DAILY OUTPUT	LABOR-HOURS	UNIT	2005 BARE COSTS				TOTAL INCL O&P	
						MAT.	LABOR	EQUIP.	TOTAL		
350	**0010**	**ACCESSORIES, HANGERS**									**350**
	8500	Wire, black annealed, 9 ga				Cwt.	105			105	116
	8600	16 ga				"	110			110	121
600	**0010**	**SHORES** Erect and strip, by hand, horizontal members									**600**
	0500	Aluminum joists and stringers	2 Carp	60	.267	Ea.		9.15		9.15	14.25
	0600	Steel, adjustable beams		45	.356			12.20		12.20	18.95
	0700	Wood joists		50	.320			10.95		10.95	17.05
	0800	Wood stringers		30	.533			18.25		18.25	28.50
	1000	Vertical members to 10' high		55	.291			9.95		9.95	15.50
	1050	To 13' high		50	.320			10.95		10.95	17.05
	1100	To 16' high		45	.356			12.20		12.20	18.95
	1500	Reshoring		1,400	.011	S.F.	.38	.39		.77	1.02
	1600	Flying truss system	C-17D	9,600	.009	SFCA		.31	.05	.36	.54
	1760	Horizontal, aluminum joists, 6-1/4" high x 5' to 21' span, buy				L.F.	29.50			29.50	32
	1770	Beams, 7-1/4" high x 4' to 30' span				"	40.50			40.50	44.50
	1810	Horizontal, steel beam, adjustable, 4' to 7' span				Ea.	465			465	515
	1830	6' to 10' span					600			600	660
	1920	9' to 15' span					760			760	840
	1940	12' to 20' span					775			775	855
	1970	Steel stringer, W8x10, 4' to 16' span, buy				L.F.	16.75			16.75	18.40
	3000	Rent for job duration, aluminum joist @ 2' O.C., per mo				SF Flr.	.73			.73	.80
	3050	Steel W8x10					.42			.42	.46
	3060	Steel adjustable					1.50			1.50	1.65
	3500	#1 post shore, steel, 5'-7" to 9'-6" high, 10000# cap., buy				Ea.	310			310	340
	3550	#2 post shore, 7'-3" to 12'-10" high, 7800# capacity					330			330	365
	3600	#3 post shore, 8'-10" to 16'-1" high, 3800# capacity					360			360	395
	5010	Frame shoring systems, steel, 12000#/leg, buy									
	5040	Frame, 2' wide x 6' high				Ea.	287			287	315
	5250	X-brace					45.50			45.50	50
	5550	Base plate					36			36	39.50
	5600	Screw jack					115			115	126
	5650	U-head, 8" x 8"					18.15			18.15	20
620	**0010**	**ACCESSORIES, SLEEVES AND CHASES**									**620**
	0100	Plastic, 1 use, 9" long, 2" diameter	1 Carp	100	.080	Ea.	.53	2.74		3.27	4.85
	0150	4" diameter		90	.089		1.56	3.04		4.60	6.45
	0200	6" diameter		75	.107		2.75	3.65		6.40	8.75
	0250	12" diameter		60	.133		18.05	4.57		22.62	27
	5000	Sheet metal, 2" diameter		100	.080		.68	2.74		3.42	5
	5100	4" diameter		90	.089		.85	3.04		3.89	5.70
	5150	6" diameter		75	.107		1.23	3.65		4.88	7.05
	5200	12" diameter		60	.133		2.46	4.57		7.03	9.80
	6000	Steel pipe, 2" diameter		100	.080		3.49	2.74		6.23	8.10
	6100	4" diameter		90	.089		11.90	3.04		14.94	17.80
	6150	6" diameter		75	.107		25.50	3.65		29.15	34
	6200	12" diameter		60	.133		64	4.57		68.57	77.50
640	**0010**	**ACCESSORIES, SNAP TIES, FLAT WASHER,** 4-3/4" L&W									**640**
	0100	3000 lb., to 8"				C	96			96	106
	0150	9" & 10"					110			110	121
	0200	11" & 12"					115			115	127
	0250	16"					122			122	135
	0300	18"					121			121	133
	0500	With plastic cone, to 8"					89			89	97.50
	0550	9" & 10"					98.50			98.50	108

CONCRETE 3

03150	Concrete Accessories	CREW	DAILY OUTPUT	LABOR-HOURS	UNIT	2005 BARE COSTS				TOTAL INCL O&P	
						MAT.	LABOR	EQUIP.	TOTAL		
640	0600	11" & 12"				C	105			105	116	**640**
	0650	16"					111			111	122	
	0700	18"					116			116	128	
	1000	5000 lb., to 8"					121			121	133	
	1100	9" & 10"					133			133	146	
	1150	11" & 12"					141			141	156	
	1200	16"					155			155	171	
	1250	18"					152			152	167	
	1500	With plastic cone, to 8"					152			152	167	
	1550	9" & 10"					166			166	183	
	1600	11" & 12"					181			181	199	
	1650	16"					200			200	220	
	1700	18"					207			207	228	
660	0010	**STAIR TREAD INSERTS** Cast iron, abrasive, 3" wide	1 Carp	90	.089	L.F.	7.65	3.04		10.69	13.15	**660**
	0020	4" wide		80	.100		10.20	3.43		13.63	16.55	
	0040	6" wide		75	.107		15.30	3.65		18.95	22.50	
	0050	9" wide		70	.114		23	3.91		26.91	31.50	
	0100	12" wide		65	.123		30.50	4.22		34.72	40	
	0300	Cast aluminum, compared to cast iron, deduct					10%					
	0500	Extruded aluminum safety tread, 3" wide	1 Carp	75	.107		8.70	3.65		12.35	15.30	
	0550	4" wide		75	.107		11.65	3.65		15.30	18.50	
	0600	6" wide		75	.107		17.45	3.65		21.10	25	
	0650	9" wide to resurface stairs		70	.114		26	3.91		29.91	35	
	1700	Cement fill for pan-type metal treads, plain	1 Cefi	115	.070	S.F.	1.50	2.29		3.79	5	
	1750	Non-slip	"	100	.080	"	1.65	2.63		4.28	5.70	
850	0010	**ACCESSORIES, WALL AND FOUNDATION**										**850**
	3000	Form oil, coverage varies greatly, minimum				Gal.	4.35			4.35	4.79	
	3050	Maximum				"	6.60			6.60	7.25	
	3500	Form patches, 1-3/4" diameter				C	65.50			65.50	72	
	3550	2-3/4" diameter				"	83			83	91.50	
	4000	Nail stakes, 3/4" diameter, 18" long				Ea.	2.23			2.23	2.45	
	4050	24" long					3.46			3.46	3.81	
	4200	30" long					4.20			4.20	4.62	
	4250	36" long					4.78			4.78	5.25	
860	0010	**WATERSTOP** PVC, ribbed 3/16" thick, 4" wide	1 Carp	155	.052	L.F.	.78	1.77		2.55	3.61	**860**
	0050	6" wide		145	.055		1.34	1.89		3.23	4.41	
	0500	Ribbed, PVC, with center bulb, 9" wide, 3/16" thick		135	.059		2	2.03		4.03	5.35	
	0550	3/8" thick		130	.062		3.03	2.11		5.14	6.60	
	0800	Dumbbell type, PVC, 6" wide, 3/16" thick		150	.053		1.35	1.83		3.18	4.34	
	0850	3/8" thick		145	.055		2.23	1.89		4.12	5.40	
	1000	9" wide, 3/8" thick, PVC, plain		130	.062		2.60	2.11		4.71	6.15	
	1050	Center bulb		130	.062		5.95	2.11		8.06	9.85	
	1250	Split PVC, 3/8" thick, 6" wide		145	.055		4.03	1.89		5.92	7.35	
	1300	9" wide		130	.062		6	2.11		8.11	9.90	
	2000	Rubber, flat dumbbell, 3/8" thick, 6" wide		145	.055		5.65	1.89		7.54	9.15	
	2050	9" wide		135	.059		8.45	2.03		10.48	12.45	
	2500	Flat dumbbell split, 3/8" thick, 6" wide		145	.055		4.03	1.89		5.92	7.35	
	2550	9" wide		135	.059		6	2.03		8.03	9.75	
	3000	Center bulb, 1/4" thick, 6" wide		145	.055		4.90	1.89		6.79	8.35	
	3050	9" wide		135	.059		9.65	2.03		11.68	13.80	
	3500	Center bulb split, 3/8" thick, 6" wide		145	.055		6	1.89		7.89	9.55	
	3550	9" wide		135	.059		10.35	2.03		12.38	14.55	
	5000	Waterstop fittings, rubber, flat										
	5010	Dumbbell or center bulb, 3/8" thick,										

3 CONCRETE

Important: See the Reference Section for critical supporting data - Reference Nos., Crews, & City Cost Indexes

03150	Concrete Accessories	CREW	DAILY OUTPUT	LABOR-HOURS	UNIT	2005 BARE COSTS				TOTAL INCL O&P
						MAT.	LABOR	EQUIP.	TOTAL	
860	5200	Field union, 6" wide	1 Carp	50	.160	Ea.	19.55	5.50		25.05	30	**860**
	5250	9" wide		50	.160		23	5.50		28.50	34
	5500	Flat cross, 6" wide		30	.267		39.50	9.15		48.65	58
	5550	9" wide		30	.267		56	9.15		65.15	76.50
	6000	Flat tee, 6" wide		30	.267		36.50	9.15		45.65	54.50
	6050	9" wide		30	.267		48	9.15		57.15	67
	6500	Flat ell, 6" wide		40	.200		33	6.85		39.85	47
	6550	9" wide		40	.200		44	6.85		50.85	59
	7000	Vertical tee, 6" wide		25	.320		32	10.95		42.95	52
	7050	9" wide		25	.320		40.50	10.95		51.45	61.50
	7500	Vertical ell, 6" wide		35	.229		28.50	7.85		36.35	43
	7550	9" wide	↓	35	.229	↓	38.50	7.85		46.35	54.50

03200 | Concrete Reinforcement

03210	Reinforcing Steel	CREW	DAILY OUTPUT	LABOR-HOURS	UNIT	2005 BARE COSTS				TOTAL INCL O&P
						MAT.	LABOR	EQUIP.	TOTAL	
100	0010	**ACCESSORIES** Materials only										**100**
	0020	See also Form Accessories, division 03150 [R03210 -080]									
	0100	Beam bolsters, (BB) standard, lower, up to 1-1/2" high, plain				C.L.F.	52			52	57.50
	0102	Galvanized					80.50			80.50	88.50
	0104	Stainless					151			151	166
	0106	Plastic					91			91	100
	0108	Epoxy					125			125	137
	0110	2-1/2" to 3" high, plain					68.50			68.50	75
	0120	Galvanized					91			91	100
	0140	Stainless					208			208	228
	0160	Plastic					95			95	105
	0162	Epoxy					133			133	146
	0200	Upper, standard (BBU) to 1-1/2" high, plain					150			150	165
	0210	2-1/2" to 3" high					272			272	299
	0300	Beam bolster with plate (BBP) to 1-1/2" high, plain					173			173	190
	0310	2-1/2" to 3" high					335			335	370
	0500	Slab bolsters, continuous, plain (SB) 3/4" to 1" high, plain					43			43	47
	0502	Galvanized					49.50			49.50	54.50
	0504	Stainless					69.50			69.50	76.50
	0506	Plastic					59			59	65
	0510	1" to 2" high, plain					53.50			53.50	59
	0515	Galvanized					63			63	69
	0520	Stainless					119			119	131
	0525	Plastic					71			71	78
	0530	For bolsters with wire runners (SBR), add					60.50			60.50	66.50
	0540	For bolsters with plates (SBP), add				↓	146			146	161
	0700	Clip or bar ties, 16 ga., plain, 3" long				C	11.90			11.90	13.10
	0710	4" long					12.60			12.60	13.85
	0720	6" long					14.75			14.75	16.20
	0730	8" long				↓	15.40			15.40	16.95
	0900	Flange clips, expandable flanges, 10 ga., 12" O.C., continuous,									
	0910	galvanized, over 500 L.F., 4" to 8"				C.L.F.	44			44	48.50
	0920	9" to 12"					60.50			60.50	66.50
	0930	17" to 24"				↓	61.50			61.50	68

03210	Reinforcing Steel		CREW	DAILY OUTPUT	LABOR-HOURS	UNIT	2005 BARE COSTS				TOTAL INCL O&P	
							MAT.	LABOR	EQUIP.	TOTAL		
100	1200	High chairs, individual, no plates (1 HC), to 3" high, plain	R03210 -080				C	64.50			64.50	71	100
	1202	Galvanized						90			90	99	
	1204	Stainless						184			184	202	
	1206	Plastic						85.50			85.50	94.50	
	1210	5" high, plain						99			99	109	
	1212	Galvanized						129			129	142	
	1214	Stainless						281			281	310	
	1216	Plastic						120			120	132	
	1220	8" high, plain						214			214	236	
	1222	Galvanized						286			286	315	
	1224	Stainless						470			470	520	
	1226	Plastic						269			269	296	
	1230	12" high, plain						445			445	490	
	1232	Galvanized						570			570	625	
	1234	Stainless						855			855	940	
	1236	Plastic						500			500	550	
	1240	15" high, plain						830			830	915	
	1242	Galvanized						970			970	1,075	
	1244	Stainless						1,425			1,425	1,575	
	1246	Plastic						900			900	990	
	1250	For each added inch up to 24" high, plain, add						62			62	68	
	1252	Galvanized, add						71.50			71.50	79	
	1254	Stainless, add						80			80	88	
	1256	Plastic, add						71.50			71.50	79	
	1400	Individual high chairs, with plates, (HCP), to 5" high, add						284			284	310	
	1410	Over 5" high, add						325			325	360	
	1500	Bar chair (BC) for up to 1-3/4" high, plain						36.50			36.50	40	
	1520	Galvanized						42			42	46.50	
	1530	Stainless						112			112	124	
	1540	Plastic						74.50			74.50	82	
	1550	Joist chair (JC), joists up to 6", plain						44			44	48.50	
	1580	Galvanized						56			56	62	
	1600	Stainless						80			80	88	
	1620	Plastic						76.50			76.50	84	
	1630	Epoxy					▼	185			185	204	
	1700	Continuous high chairs, legs 8" O.C. (CHC) to 4" high, plain					C.L.F.	85.50			85.50	94.50	
	1705	Galvanized						105			105	116	
	1710	Stainless						225			225	247	
	1715	Plastic						106			106	116	
	1718	Epoxy						160			160	176	
	1720	6" high, plain						123			123	136	
	1725	Galvanized						177			177	195	
	1730	Stainless						261			261	287	
	1735	Plastic						169			169	186	
	1738	Epoxy						215			215	236	
	1740	8" high, plain						179			179	197	
	1745	Galvanized						249			249	273	
	1750	Stainless						310			310	340	
	1755	Plastic						234			234	258	
	1758	Epoxy						325			325	355	
	1760	12" high, plain						440			440	485	
	1765	Galvanized						530			530	580	
	1770	Stainless						740			740	815	
	1775	Plastic						500			500	550	
	1778	Epoxy						770			770	845	
	1780	15" high, plain				▼	▼	495			495	545	

Important: See the Reference Section for critical supporting data - Reference Nos., Crews, & City Cost Indexes

				DAILY	LABOR-		2005 BARE COSTS				TOTAL	
	03210	**Reinforcing Steel**	CREW	OUTPUT	HOURS	UNIT	MAT.	LABOR	EQUIP.	TOTAL	INCL O&P	
100	1785	Galvanized				C.L.F.	595			595	655	100
	1790	Stainless					590			590	650	
	1795	Plastic					565			565	620	
	1798	Epoxy					1,000			1,000	1,125	
	1800	For each added 1" up to 24" high, plain, add					42			42	46.50	
	1820	Galvanized, add					52			52	57	
	1840	Stainless, add					46.50			46.50	51	
	1860	Plastic, add					46.50			46.50	51	
	1900	For continuous bottom plate, (CHCP), add					202			202	222	
	1940	For upper continuous high chairs, (CHCU), add					202			202	222	
	1960	For galvanized wire runners, add					190			190	208	
	2100	Paper tubing, 4' lengths, for #2 & #3 bar					76			76	83.50	
	2120	For #6 bar				▼	94.50			94.50	104	
	2200	Screed base, 1/2" diameter, 2-1/2" high, plain				C	196			196	216	
	2210	Galvanized					203			203	224	
	2220	5-1/2" high, plain					236			236	260	
	2250	Galvanized					251			251	276	
	2300	3/4" diameter, 2-1/2" high, plain					243			243	268	
	2310	Galvanized					258			258	284	
	2320	5-1/2" high, plain					300			300	330	
	2350	Galvanized					320			320	355	
	2400	Screed holder, 1/2" diam. for 1" I.D. pipe, plain, 6" long					203			203	224	
	2420	12" long					335			335	370	
	2500	3/4" diameter, for 1-1/2" I.D. pipe, 6" long					365			365	400	
	2520	12" long					600			600	660	
	2700	Screw anchor for bolts, plain, 1/2" diameter					133			133	147	
	2720	1" diameter					400			400	440	
	2740	1-1/2" diameter					660			660	725	
	2800	Screw eye bolts, 1/2" x 5" long					152			152	167	
	2820	1" x 9" long					600			600	660	
	2840	1-1/2" x 14" long					1,450			1,450	1,600	
	2900	Screw anchor bolts, 1/2" x up to 7" long					610			610	670	
	2920	1" x up to 12" long					1,975			1,975	2,175	
	3000	Slab lifting inserts, single, 3/4" dia., galv., 4" high					410			410	450	
	3010	6" high					500			500	550	
	3030	7" high					575			575	630	
	3100	1" diameter, 5" high					645			645	710	
	3120	7" high					680			680	750	
	3200	Double lifting inserts, 1" diameter, 5" high					1,275			1,275	1,400	
	3220	7" high					1,350			1,350	1,475	
	3330	1-1/4" diameter, 5" high				▼	1,400			1,400	1,525	
	3500	Sleeper clips for wood sleepers, 20 ga., galv., 2" wide				M	485			485	535	
	3520	4" wide					600			600	660	
	3600	Spacers, plastic for 1" bar clearance, average					74.50			74.50	82	
	3620	For 2" bar clearance, average				▼	90			90	99	
	3800	Subgrade chairs, 1/2" diameter, 3-1/2" high				C	400			400	440	
	3850	12" high					1,125			1,125	1,225	
	3900	3/4" diameter, 3-1/2" high					515			515	565	
	3950	12" high					1,225			1,225	1,350	
	4200	Subgrade stakes, 3/4" diameter, 12" long					415			415	455	
	4250	24" long					560			560	615	
	4300	1" diameter, 12" long					625			625	685	
	4350	24" long				▼	915			915	1,000	
	4500	Tie wire, 16 ga. annealed steel, under 500 lbs.				Cwt.	120			120	132	
	4520	2,000 to 4,000 lbs.				"	113			113	124	
	4550	Tie wire holder, plastic case				Ea.	47.50			47.50	52	

R03210 -080

CONCRETE **3**

3 CONCRETE

		03210 | **Reinforcing Steel**		CREW	DAILY OUTPUT	LABOR-HOURS	UNIT	2005 BARE COSTS				TOTAL INCL O&P	
								MAT.	LABOR	EQUIP.	TOTAL		
100	4600	Aluminum case	R03210 -080				Ea.	57.50			57.50	63	100
200	0010	**COATED REINFORCING** Add to fabricated & delivered price											200
	0100	Epoxy coated, A775					Ton	360			360	395	
	0150	Galvanized, #3						700			700	770	
	0200	#4						700			700	770	
	0250	#5						685			685	755	
	0300	#6 or over						685			685	755	
	1000	For over 20 tons, #6 or larger, minimum						635			635	700	
	1500	Maximum						765			765	840	
500	0010	**REINFORCING STEEL, A615**											500
	0150	Reinforcing, A615 grade 40, mill base					Ton	590			590	645	
	0200	Detailed, cut, bent, and delivered, average						800			800	880	
	0650	Reinforcing steel, A615 grade 60, mill base						590			590	645	
	0700	Detailed, cut, bent, and delivered, average						800			800	880	
	1000	Reinforcing steel, extras, add to mill base											
	1005	Mill extra, add for delivery to shop, average					Ton	27			27	29.50	
	1010	Shop extra, add for for handling & storage, average						22			22	24	
	1020	Shop extra, add for bending, limited percent of bars						22			22	24	
	1030	Average percent of bars						44			44	48.50	
	1050	Large percent of bars						88			88	97	
	1200	Shop extra, add for detailing, under 50 tons						40			40	44	
	1250	50 to 150 tons						30			30	33	
	1300	150 to 500 tons						28.50			28.50	31.50	
	1350	Over 500 tons						27			27	29.50	
	1700	Shop extra, add for listing, average						5			5	5.50	
	2000	Mill extra, add for quantity, under 20 tons						5			5	5.50	
	2100	Shop extra, add for quantity, under 20 tons						40			40	44	
	2200	20 to 49 tons						30			30	33	
	2250	50 to 99 tons						20			20	22	
	2300	100 to 300 tons						12			12	13.20	
	2500	Shop extra, add for size, #3						136			136	150	
	2550	#4						68			68	75	
	2600	#5						34			34	37.50	
	2650	#6						30.50			30.50	33.50	
	2700	#7 to #11						41			41	45	
	2750	#14						51			51	56	
	2800	#18						58			58	63.50	
	2900	Shop extra, add for delivery to job, average						22			22	24	
600	0010	**REINFORCING IN PLACE** A615 Grade 60, incl. access. labor	R03210 -020										600
	0100	Beams & Girders, #3 to #7		4 Rodm	1.60	20	Ton	800	760		1,560	2,125	
	0150	#8 to #18	R03210 -025		2.70	11.852		800	450		1,250	1,625	
	0200	Columns, #3 to #7			1.50	21.333		800	810		1,610	2,200	
	0250	#8 to #18	R03210 -080		2.30	13.913		800	530		1,330	1,750	
	0300	Spirals, hot rolled, 8" to 15" diameter			2.20	14.545		1,125	550		1,675	2,125	
	0320	15" to 24" diameter			2.20	14.545		1,075	550		1,625	2,075	
	0330	24" to 36" diameter			2.30	13.913		1,025	530		1,555	2,000	
	0340	36" to 48" diameter			2.40	13.333		970	505		1,475	1,900	
	0360	48" to 64" diameter			2.50	12.800		1,075	485		1,560	1,975	
	0380	64" to 84" diameter			2.60	12.308		1,125	465		1,590	2,000	
	0390	84" to 96" diameter			2.70	11.852		1,175	450		1,625	2,050	
	0400	Elevated slabs, #4 to #7			2.90	11.034		850	420		1,270	1,625	
	0500	Footings, #4 to #7			2.10	15.238		760	580		1,340	1,800	
	0550	#8 to #18			3.60	8.889		720	335		1,055	1,350	
	0600	Slab on grade, #3 to #7			2.30	13.913		760	530		1,290	1,700	

Important: See the Reference Section for critical supporting data - Reference Nos., Crews, & City Cost Indexes

				DAILY	LABOR-		2005 BARE COSTS				TOTAL	
03210		**Reinforcing Steel**	CREW	OUTPUT	HOURS	UNIT	MAT.	LABOR	EQUIP.	TOTAL	INCL O&P	
600	0700	Walls, #3 to #7	4 Rodm	3	10.667	Ton	760	405		1,165	1,500	**600**
	0750	#8 to #18	↓	4	8	↓	760	305		1,065	1,325	
	0900	Use the following for a rough estimate guide										
	1000	Typical in place, average, under 10 ton job, #3 to #7	4 Rodm	1.80	17.778	Ton	825	675		1,500	2,025	
	1010	#8 to #18		2.70	11.852		845	450		1,295	1,675	
	1050	10 - 50 ton job, #3 to #7		2.10	15.238		810	580		1,390	1,850	
	1060	#8 to #18		3	10.667		830	405		1,235	1,575	
	1100	50 - 100 ton job, #3 - #7		2.20	14.545		790	550		1,340	1,775	
	1110	#8 to #18		3.10	10.323		810	390		1,200	1,525	
	1150	Over 100 ton job, #3 - #7		2.30	13.913		785	530		1,315	1,725	
	1160	#8 - #18	↓	3.20	10		800	380		1,180	1,500	
	1200	High strength steel, Grade 75, #14 bars only, add					60			60	66	
	2000	Unloading & sorting, add to above	C-5	100	.560			20.50	6.15	26.65	40.50	
	2200	Crane cost for handling, add to above, minimum		135	.415			15.25	4.57	19.82	29.50	
	2210	Average		92	.609			22.50	6.70	29.20	43.50	
	2220	Maximum	↓	35	1.600	↓		59	17.60	76.60	114	
	2300	For epoxy coated rebar, add					35%					
	2350	For stainless steel rebar, add					300%					
	2400	Dowels, 2 feet long, deformed, #3	2 Rodm	520	.031	Ea.	.33	1.17		1.50	2.29	
	2410	#4		480	.033		.59	1.26		1.85	2.74	
	2420	#5		435	.037		.92	1.40		2.32	3.31	
	2430	#6		360	.044	↓	1.32	1.69		3.01	4.23	
	2450	Longer and heavier dowels, add		725	.022	Lb.	.44	.84		1.28	1.86	
	2500	Smooth dowels, 12" long, 1/4" or 3/8" diameter		140	.114	Ea.	.69	4.34		5.03	7.90	
	2520	5/8" diameter		125	.128		1.21	4.86		6.07	9.35	
	2530	3/4" diameter	↓	110	.145	↓	1.50	5.50		7	10.75	
	2600	Dowel sleeves for CIP concrete, 2-part system										
	2610	Sleeve base, plastic, for #5 bar, fasten to edge form	1 Rodm	200	.040	Ea.	.33	1.52		1.85	2.86	
	2615	Sleeve, plastic, for #5 bar x 9" long, snap onto base		400	.020		.96	.76		1.72	2.31	
	2620	Sleeve base, for #6 bar		175	.046		.33	1.73		2.06	3.22	
	2625	Sleeve, for #6 bar		350	.023		1.02	.87		1.89	2.55	
	2630	Sleeve base, for #8 bar		150	.053		.40	2.02		2.42	3.78	
	2635	Sleeve, for #8 bar	↓	300	.027		1.11	1.01		2.12	2.89	
	2700	Dowel caps, visual warning only, plastic, #3 to #8	2 Rodm	800	.020		.24	.76		1	1.51	
	2720	#7 to #14		750	.021		.56	.81		1.37	1.96	
	2750	Impalement protective, plastic, #3 to #7		800	.020		1.74	.76		2.50	3.16	
	2760	#7 to #11		775	.021		2.12	.78		2.90	3.62	
	2770	#11 to #16	↓	750	.021	↓	2.35	.81		3.16	3.93	
	3000	For epoxy dowel anchoring, see Div 05090-300										
650	0010	**GLASS FIBER REINFORCED POLYMER** reinforcing bars										**650**
	0050	#2 bar, .043 lbs/ ft.				L.F.	.35			.35	.39	
	0100	#3 bar, .092 lbs/ ft.					.48			.48	.53	
	0150	#4 bar, .160 lbs/ ft.					.69			.69	.76	
	0200	#5 bar, .258 lbs/ ft.					.88			.88	.97	
	0250	#6 bar, .372 lbs/ ft.					1.25			1.25	1.38	
	0300	#7 bar, .497 lbs/ ft.					1.60			1.60	1.76	
	0350	#8 bar, .620 lbs/ ft.					1.88			1.88	2.07	
	0400	#9 bar, .800 lbs/ ft.					2.30			2.30	2.53	
	0450	#10 bar, 1.08 lbs/ ft.				↓	3			3	3.30	
700	0010	**SPLICING REINFORCING BARS** Incl. holding bars in										**700**
	0020	place while splicing										
	0100	Butt weld columns #4 bars	C-5	190	.295	Ea.	3.56	10.85	3.25	17.66	25	
	0110	#6 bars		150	.373		5	13.75	4.11	22.86	32	
	0130	#10 bars		95	.589		12.30	21.50	6.50	40.30	55.50	
	0150	#14 bars	↓	65	.862	↓	19.50	31.50	9.50	60.50	83	

R03210-020
R03210-025
R03210-080

CONCRETE **3**

	03210	Reinforcing Steel	CREW	DAILY OUTPUT	LABOR-HOURS	UNIT	2005 BARE COSTS				TOTAL INCL O&P	
							MAT.	LABOR	EQUIP.	TOTAL		
700	0280	Column splices, bar to bar end bearing										**700**
	0300	#7 or #8 bars	C-5	190	.295	Ea.	32	10.85	3.25	46.10	56.50	
	0310	#9 or #10 bars		170	.329		35	12.15	3.63	50.78	62	
	0320	#11 bars		160	.350		42.50	12.90	3.85	59.25	71.50	
	0330	#14 bars		150	.373		53.50	13.75	4.11	71.36	85	
	0340	#18 bars		140	.400		77	14.75	4.40	96.15	114	
	0500	Transition bar to bar end bearing, #14 to #18 bar					76			76	83.50	
	0520	#14 to #11 bar					58.50			58.50	64	
	0540	#11 to #10 bar					44			44	48.50	
	0550	#10 to #9 bar					36.50			36.50	40.50	
	0560	#9 to #8 bar					34			34	37.50	
	0580	#8 to #7 bar					33			33	36.50	
	0600	For bolted speed sleeve type, deduct						15%				
	0800	Mechanical butt splice, sleeve type w/ filler metal, compression										
	0810	only, all grades, columns only #11 bars	C-5	68	.824	Ea.	37.50	30.50	9.05	77.05	100	
	0900	#14 bars		62	.903		61	33.50	9.95	104.45	131	
	0920	#18 bars		62	.903		111	33.50	9.95	154.45	186	
	1000	125% yield point, grade 60, columns only, #6 bars		68	.824		30	30.50	9.05	69.55	92	
	1020	#7 or #8 bars		68	.824		32	30.50	9.05	71.55	94.50	
	1030	#9 bars		68	.824		33	30.50	9.05	72.55	95.50	
	1040	#10 bars		68	.824		35	30.50	9.05	74.55	97.50	
	1050	#11 bars		68	.824		42.50	30.50	9.05	82.05	106	
	1060	#14 bars		62	.903		53.50	33.50	9.95	96.95	123	
	1070	#18 bars		62	.903		77	33.50	9.95	120.45	149	
	1200	Full tension, grade 60 steel, columns,										
	1220	slabs or beams, #6, #7, #8 bars	C-5	68	.824	Ea.	10.10	30.50	9.05	49.65	70	
	1230	#9 bars		68	.824		11.05	30.50	9.05	50.60	71	
	1240	#10 bars		68	.824		12.30	30.50	9.05	51.85	72.50	
	1250	#11 bars		68	.824		13.05	30.50	9.05	52.60	73.50	
	1260	#14 bars		62	.903		19.50	33.50	9.95	62.95	86	
	1270	#18 bars		62	.903		29.50	33.50	9.95	72.95	97	
	1400	If equipment handling not required, deduct						50%				
	1600	Mechanical threaded type, bar threading not included,										
	1700	Straight bars, #10 & #11	C-5	140	.400	Ea.	13.05	14.75	4.40	32.20	43	
	1750	#14 bars		130	.431		19.50	15.85	4.74	40.09	52	
	1800	#18 bars		75	.747		29.50	27.50	8.20	65.20	86	
	2100	#11 to #18 & #14 to #18 transition		75	.747		32	27.50	8.20	67.70	88.50	
	2400	Bent bars, #10 & #11		105	.533		13.05	19.65	5.85	38.55	52.50	
	2500	#14		90	.622		22	23	6.85	51.85	69	
	2600	#18		70	.800		36.50	29.50	8.80	74.80	97.50	
	2800	#11 to #14 transition		75	.747		21	27.50	8.20	56.70	76.50	
	2900	#11 to #18 & #14 to #18 transition		70	.800		35	29.50	8.80	73.30	95.50	

	03220	**Welded Wire Fabric**										
200	0010	**WELDED WIRE FABRIC** ASTM A185										**200**
	0100	6 x 6 - W1.4 x W1.4 (10 x 10) 21 lb. per C.S.F.	2 Rodm	35	.457	C.S.F.	19.35	17.35		36.70	50	
	0200	6 x 6 - W2.1 x W2.1 (8 x 8) 30 lb. per C.S.F.		31	.516		25.50	19.60		45.10	60.50	
	0300	6 x 6 - W2.9 x W2.9 (6 x 6) 42 lb. per C.S.F.		29	.552		33	21		54	70.50	
	0400	6 x 6 - W4 x W4 (4 x 4) 58 lb. per C.S.F.		27	.593		44	22.50		66.50	85	
	0500	4 x 4 - W1.4 x W1.4 (10 x 10) 31 lb. per C.S.F.		31	.516		17.70	19.60		37.30	52	
	0600	4 x 4 - W2.1 x W2.1 (8 x 8) 44 lb. per C.S.F.		29	.552		21	21		42	57.50	
	0650	4 x 4 - W2.9 x W2.9 (6 x 6) 61 lb. per C.S.F.		27	.593		30	22.50		52.50	70	
	0700	4 x 4 - W4 x W4 (4 x 4) 85 lb. per C.S.F.		25	.640		54	24.50		78.50	99	
	0800	2 x 2 - #14 galv., 21 lb/C.S.F., beam & column wrap		6.50	2.462		28	93.50		121.50	185	
	0900	2 x 2 - #12 galv. for gunite reinforcing		6.50	2.462		31.50	93.50		125	189	

Important: See the Reference Section for critical supporting data - Reference Nos., Crews, & City Cost Indexes

03200 | Concrete Reinforcement

03230 | Stressing Tendons

		CREW	DAILY OUTPUT	LABOR-HOURS	UNIT	2005 BARE COSTS MAT.	LABOR	EQUIP.	TOTAL	TOTAL INCL O&P	
600	0010	**PRESTRESSING STEEL** Post-tensioned in field										600
	0100	Grouted strand, 50' span, 100 kip	C-3	1,200	.053	Lb.	1.13	1.85	.10	3.08	4.34	
	0150	300 kip		2,700	.024		.63	.82	.04	1.49	2.07	
	0300	100' span, 100 kip		1,700	.038		1.13	1.31	.07	2.51	3.43	
	0350	300 kip		3,200	.020		1.01	.70	.04	1.75	2.27	
	0500	200' span, 100 kip		2,700	.024		1.13	.82	.04	1.99	2.62	
	0550	300 kip		3,500	.018		1.01	.64	.03	1.68	2.18	
	0800	Grouted bars, 50' span, 42 kip		2,600	.025		.59	.86	.05	1.50	2.08	
	0850	143 kip		3,200	.020		.57	.70	.04	1.31	1.79	
	1000	75' span, 42 kip		3,200	.020		.59	.70	.04	1.33	1.81	
	1050	143 kip	↓	4,200	.015		.53	.53	.03	1.09	1.48	
	1200	Ungrouted strand, 50' span, 100 kip	C-4	1,275	.025		.46	.97	.03	1.46	2.14	
	1250	300 kip		1,475	.022		.46	.83	.03	1.32	1.92	
	1400	100' span, 100 kip		1,500	.021		.46	.82	.03	1.31	1.89	
	1450	300 kip		1,650	.019		.46	.75	.02	1.23	1.77	
	1600	200' span, 100 kip		1,500	.021		.46	.82	.03	1.31	1.89	
	1650	300 kip		1,700	.019		.46	.72	.02	1.20	1.73	
	1800	Ungrouted bars, 50' span, 42 kip		1,400	.023		.40	.88	.03	1.31	1.92	
	1850	143 kip		1,700	.019		.40	.72	.02	1.14	1.66	
	2000	75' span, 42 kip		1,800	.018		.40	.68	.02	1.10	1.59	
	2050	143 kip		2,200	.015		.40	.56	.02	.98	1.38	
	2220	Ungrouted single strand, 100' slab, 25 kip		1,200	.027		.46	1.03	.03	1.52	2.24	
	2250	35 kip	↓	1,475	.022	↓	.46	.83	.03	1.32	1.92	

03240 | Fibrous Reinforcing

		CREW	DAILY OUTPUT	LABOR-HOURS	UNIT	MAT.	LABOR	EQUIP.	TOTAL	TOTAL INCL O&P	
300	0010	**FIBROUS REINFORCING**										300
	0100	Synthetic fibers, add to concrete				Lb.	3.87			3.87	4.26	
	0110	1-1/2 lb. per C.Y.				C.Y.	6			6	6.60	
	0150	Steel fibers, add to concrete				Lb.	.44			.44	.48	
	0155	25 lb. per C.Y.				C.Y.	11			11	12.10	
	0160	50 lb. per C.Y.					22			22	24	
	0170	75 lb. per C.Y.					34			34	37.50	
	0180	100 lb. per C.Y.				↓	44			44	48.50	

03300 | Cast-In-Place Concrete

03310 | Structural Concrete

		CREW	DAILY OUTPUT	LABOR-HOURS	UNIT	2005 BARE COSTS MAT.	LABOR	EQUIP.	TOTAL	TOTAL INCL O&P	
200	0010	**CONCRETE, FIELD MIX** FOB forms 2250 psi				C.Y.	72.50			72.50	79.50	200
	0020	3000 psi				"	75.50			75.50	83	
220	0010	**CONCRETE, READY MIX** Normal weight [R03310 -060]										220
	0020	2000 psi				C.Y.	77.50			77.50	85.50	
	0100	2500 psi					79.50			79.50	87.50	
	0150	3000 psi					81			81	89	
	0200	3500 psi					82			82	90	
	0300	4000 psi					84			84	92.50	
	0350	4500 psi					86			86	94.50	
	0400	5000 psi					90			90	99	
	0411	6000 psi					103			103	113	
	0412	8000 psi				↓	167			167	184	

CONCRETE 3

		03310	Structural Concrete	CREW	DAILY OUTPUT	LABOR- HOURS	UNIT	2005 BARE COSTS MAT.	LABOR	EQUIP.	TOTAL	TOTAL INCL O&P	
220	0413		10,000 psi				C.Y.	238			238	261	220
	0414		12,000 psi	R03310 -060				287			287	315	
	0900		Roller compacted concrete, 1.5″ - 2″ agg., 200 lb. cement/C.Y.					60.50			60.50	66.50	
	1000		For high early strength cement, add					10%					
	1010		For structural lightweight with regular sand, add					25%					
	1300		For winter concrete, add					4.50			4.50	4.95	
	1400		For hot weather concrete, add					8.10			8.10	8.90	
	1500		For Saturday delivery, add					7			7	7.70	
	2000		For all lightweight aggregate, add				↓	45%					
	3000		For integral colors, 2500 psi, 5 bag mix										
	3100		Red, yellow or brown, 1.8 lb. per bag, add				C.Y.	21			21	23	
	3200		9.4 lb. per bag, add					109			109	120	
	3400		Black, 1.8 lb. per bag, add					23.50			23.50	26	
	3500		7.5 lb. per bag, add					98.50			98.50	108	
	3700		Green, 1.8 lb. per bag, add					27			27	30	
	3800		7.5 lb. per bag, add				↓	113			113	125	
	4000		Flowable fill, ash cement aggregate water										
	4100		40 - 80 psi				C.Y.	62.50			62.50	69	
	4150		Structural: ash cement aggregate water & sand										
	4200		50 psi				C.Y.	66.50			66.50	73.50	
	4250		140 psi					69.50			69.50	76.50	
	4300		500 psi					72.50			72.50	79.50	
	4350		1000 psi					76.50			76.50	84	
	5000		For hot water, add	↓			↓	2			2	2.40	
240	0010		**CONCRETE IN PLACE** Including forms (4 uses), reinforcing										240
	0050		steel and finishing unless otherwise indicated										
	0300		Beams, 5 kip per L.F., 10′ span	C-14A	15.62	12.804	C.Y.	284	440	46.50	770.50	1,050	
	0350		25′ span		18.55	10.782		214	370	39.50	623.50	860	
	0700		Columns, square, 12″ x 12″, minimum reinforcing		11.96	16.722		287	575	61	923	1,275	
	0720		Average reinforcing		10.13	19.743		455	675	72	1,202	1,625	
	0740		Maximum reinforcing		9.03	22.148		680	760	80.50	1,520.50	2,050	
	0800		16″ x 16″, minimum reinforcing		16.22	12.330		228	425	45	698	965	
	0820		Average reinforcing		12.57	15.911		385	545	58	988	1,350	
	0840		Maximum reinforcing		10.25	19.512		600	670	71	1,341	1,800	
	0900		24″ x 24″, minimum reinforcing		23.66	8.453		193	290	31	514	700	
	0920		Average reinforcing		17.71	11.293		345	385	41	771	1,025	
	0940		Maximum reinforcing		14.15	14.134		550	485	51.50	1,086.50	1,425	
	1000		36″ x 36″, minimum reinforcing		33.69	5.936		170	204	21.50	395.50	530	
	1020		Average reinforcing		23.32	8.576		300	294	31	625	825	
	1040		Maximum reinforcing		17.82	11.223		510	385	41	936	1,200	
	1200		16″ diameter, minimum reinforcing		31.49	6.351		225	218	23	466	615	
	1220		Average reinforcing		19.12	10.460		390	360	38	788	1,025	
	1240		Maximum reinforcing		13.77	14.524		590	500	53	1,143	1,500	
	1300		20″ diameter, minimum reinforcing		41.04	4.873		224	167	17.75	408.75	530	
	1320		Average reinforcing		24.05	8.316		375	285	30.50	690.50	895	
	1340		Maximum reinforcing		17.01	11.758		590	405	43	1,038	1,325	
	1400		24″ diameter, minimum reinforcing		51.85	3.857		209	132	14.05	355.05	450	
	1420		Average reinforcing		27.06	7.391		375	253	27	655	840	
	1440		Maximum reinforcing		18.29	10.935		580	375	40	995	1,275	
	1500		36″ diameter, minimum reinforcing		75.04	2.665		210	91.50	9.70	311.20	385	
	1520		Average reinforcing		37.49	5.335		360	183	19.40	562.40	705	
	1540		Maximum reinforcing	↓	22.84	8.757		565	300	32	897	1,125	
	1900		Elevated slabs, flat slab with drops, 125 psf Sup. Load, 20′ span	C-14B	38.45	5.410		224	185	18.95	427.95	560	
	1950		30′ span		50.99	4.079		232	140	14.30	386.30	490	
	2100		Flat plate, 125 psf Sup. Load, 15′ span		30.24	6.878		204	235	24	463	620	
	2150		25′ span	↓	49.60	4.194	↓	209	144	14.70	367.70	470	

03310	Structural Concrete	CREW	DAILY OUTPUT	LABOR-HOURS	UNIT	2005 BARE COSTS				TOTAL INCL O&P
						MAT.	LABOR	EQUIP.	TOTAL	
240										**240**
2300	Waffle const., 30" domes, 125 psf Sup. Load, 20' span	C-14B	37.07	5.611	C.Y.	315	192	19.65	526.65	670
2350	30' span		44.07	4.720		281	162	16.50	459.50	580
2500	One way joists, 30" pans, 125 psf Sup. Load, 15' span		27.38	7.597		395	260	26.50	681.50	870
2550	25' span		31.15	6.677		360	229	23.50	612.50	780
2700	One way beam & slab, 125 psf Sup. Load, 15' span		20.59	10.102		227	345	35.50	607.50	830
2750	25' span		28.36	7.334		209	251	25.50	485.50	650
2900	Two way beam & slab, 125 psf Sup. Load, 15' span		24.04	8.652		215	296	30.50	541.50	735
2950	25' span		35.87	5.799		181	198	20.50	399.50	530
3100	Elevated slabs including finish, not									
3110	including forms or reinforcing									
3150	Regular concrete, 4" slab	C-8	2,613	.021	S.F.	1.09	.64	.27	2	2.47
3200	6" slab		2,585	.022		1.61	.65	.27	2.53	3.05
3250	2-1/2" thick floor fill		2,685	.021		.70	.62	.26	1.58	2.01
3300	Lightweight, 110# per C.F., 2-1/2" thick floor fill		2,585	.022		.90	.65	.27	1.82	2.27
3400	Cellular concrete, 1-5/8" fill, under 5000 S.F.		2,000	.028		.60	.84	.35	1.79	2.32
3450	Over 10,000 S.F.		2,200	.025		.57	.76	.32	1.65	2.13
3500	Add per floor for 3 to 6 stories high		31,800	.002			.05	.02	.07	.10
3520	For 7 to 20 stories high		21,200	.003			.08	.03	.11	.16
3800	Footings, spread under 1 C.Y.	C-14C	38.07	2.942	C.Y.	162	96	.62	258.62	330
3850	Over 5 C.Y.		81.04	1.382		226	45	.29	271.29	320
3900	Footings, strip, 18" x 9", unreinforced		40	2.800		101	91.50	.59	193.09	255
3920	18" x 9", reinforced		35	3.200		120	105	.67	225.67	297
3925	20" x 10", unreinforced		45	2.489		98	81.50	.52	180.02	236
3930	20" x 10", reinforced		40	2.800		114	91.50	.59	206.09	269
3935	24" x 12", unreinforced		55	2.036		97	66.50	.43	163.93	211
3940	24" x 12", reinforced		48	2.333		113	76	.49	189.49	244
3945	36" x 12", unreinforced		70	1.600		93	52.50	.34	145.84	184
3950	36" x 12", reinforced		60	1.867		108	61	.39	169.39	214
4000	Foundation mat, under 10 C.Y.		38.67	2.896		164	94.50	.61	259.11	330
4050	Over 20 C.Y.		56.40	1.986		144	65	.42	209.42	260
4200	Grade walls, 8" thick, 8' high	C-14D	45.83	4.364		143	148	15.90	306.90	405
4250	14' high		27.26	7.337		186	249	26.50	461.50	625
4260	12" thick, 8' high		64.32	3.109		129	106	11.30	246.30	320
4270	14' high		40.01	4.999		146	170	18.20	334.20	445
4300	15" thick, 8' high		80.02	2.499		121	85	9.10	215.10	275
4350	12' high		51.26	3.902		128	133	14.20	275.20	365
4500	18' high		48.85	4.094		143	139	14.90	296.90	390
4520	Handicap access ramp, railing both sides, 3' wide	C-14H	14.58	3.292	L.F.	167	111	1.65	279.65	360
4525	5' wide		12.22	3.928		175	132	1.96	308.96	400
4530	With 6" curb and rails both sides, 3' wide		8.55	5.614		174	189	2.81	365.81	490
4535	5' wide		7.31	6.566		178	221	3.28	402.28	545
4650	Slab on grade, not including finish, 4" thick	C-14E	60.75	1.449	C.Y.	104	48.50	.39	152.89	192
4700	6" thick	"	92	.957	"	98	32	.26	130.26	159
4751	Slab on grade, incl. troweled finish, not incl. forms									
4760	or reinforcing, over 10,000 S.F., 4" thick	C-14F	3,425	.021	S.F.	1.07	.65	.01	1.73	2.15
4820	6" thick		3,350	.021		1.56	.67	.01	2.24	2.71
4840	8" thick		3,184	.023		2.13	.70	.01	2.84	3.39
4900	12" thick		2,734	.026		3.20	.82	.01	4.03	4.74
4950	15" thick		2,505	.029		4.01	.89	.01	4.91	5.75
5000	Slab on grade, incl. textured finish, not incl. forms									
5001	or reinforcing, 4" thick	C-14G	2,873	.019	S.F.	1.05	.59	.01	1.65	2.06
5010	6" thick		2,590	.022		1.65	.66	.01	2.32	2.81
5020	8" thick		2,320	.024		2.15	.74	.01	2.90	3.47
5200	Lift slab in place above the foundation, incl. forms,									
5210	reinforcing, concrete and columns, minimum	C-14B	2,113	.098	S.F.	4.92	3.37	.34	8.63	11.05
5250	Average		1,650	.126		5.35	4.31	.44	10.10	13.10

CONCRETE **3**

03310 | Structural Concrete

		CREW	DAILY OUTPUT	LABOR-HOURS	UNIT	2005 BARE COSTS				TOTAL INCL O&P	
						MAT.	LABOR	EQUIP.	TOTAL		
240	5300	Maximum	C-14B	1,500	.139	S.F.	5.75	4.75	.49	10.99	14.25	240
	5500	Lightweight, ready mix, including screed finish only,										
	5510	not including forms or reinforcing										
	5550	1:4 for structural roof decks	C-14B	260	.800	C.Y.	110	27.50	2.80	140.30	167	
	5600	1:6 for ground slab with radiant heat	C-14F	92	.783		111	24.50	.26	135.76	158	
	5650	1:3:2 with sand aggregate, roof deck	C-14B	260	.800		105	27.50	2.80	135.30	162	
	5700	Ground slab	C-14F	107	.673		105	21	.22	126.22	147	
	5900	Pile caps, incl. forms and reinf., sq. or rect., under 5 C.Y.	C-14C	54.14	2.069		104	67.50	.43	171.93	222	
	5950	Over 10 C.Y.		75	1.493		108	49	.31	157.31	196	
	6000	Triangular or hexagonal, under 5 C.Y.		53	2.113		99.50	69	.44	168.94	219	
	6050	Over 10 C.Y.	↓	85	1.318		112	43	.28	155.28	192	
	6200	Retaining walls, gravity, 4' high see division 02830-100	C-14D	66.20	3.021		103	103	11	217	285	
	6250	10' high		125	1.600		93	54.50	5.80	153.30	194	
	6300	Cantilever, level backfill loading, 8' high		70	2.857		125	97	10.40	232.40	300	
	6350	16' high	↓	91	2.198	↓	120	74.50	8	202.50	258	
	6800	Stairs, not including safety treads, free standing, 3'-6" wide	C-14H	83	.578	LF Nose	14.80	19.50	.29	34.59	47	
	6850	Cast on ground		125	.384	"	8.70	12.95	.19	21.84	30	
	7000	Stair landings, free standing		200	.240	S.F.	3.13	8.10	.12	11.35	16.15	
	7050	Cast on ground	↓	475	.101	"	1.84	3.41	.05	5.30	7.40	
450	0010	**INSULATING CONCRETE** See division 03520-250										450
700	0010	**PLACING CONCRETE** and vibrating, including labor & equipment										700
	0050	Beams, elevated, small beams, pumped	C-20	60	1.067	C.Y.		30.50	12.55	43.05	61	
	0100	With crane and bucket	C-7	45	1.600			46	21.50	67.50	94.50	
	0200	Large beams, pumped	C-20	90	.711			20.50	8.35	28.85	40.50	
	0250	With crane and bucket	C-7	65	1.108			32	14.75	46.75	65.50	
	0400	Columns, square or round, 12" thick, pumped	C-20	60	1.067			30.50	12.55	43.05	61	
	0450	With crane and bucket	C-7	40	1.800			52	24	76	106	
	0600	18" thick, pumped	C-20	90	.711			20.50	8.35	28.85	40.50	
	0650	With crane and bucket	C-7	55	1.309			38	17.45	55.45	77	
	0800	24" thick, pumped	C-20	92	.696			19.95	8.15	28.10	39.50	
	0850	With crane and bucket	C-7	70	1.029			29.50	13.70	43.20	60.50	
	1000	36" thick, pumped	C-20	140	.457			13.10	5.35	18.45	26	
	1050	With crane and bucket	C-7	100	.720			21	9.60	30.60	42.50	
	1400	Elevated slabs, less than 6" thick, pumped	C-20	140	.457			13.10	5.35	18.45	26	
	1450	With crane and bucket	C-7	95	.758			22	10.10	32.10	44.50	
	1500	6" to 10" thick, pumped	C-20	160	.400			11.50	4.70	16.20	23	
	1550	With crane and bucket	C-7	110	.655			18.90	8.75	27.65	38.50	
	1600	Slabs over 10" thick, pumped	C-20	180	.356			10.20	4.18	14.38	20.50	
	1650	With crane and bucket	C-7	130	.554			16	7.40	23.40	32.50	
	1900	Footings, continuous, shallow, direct chute	C-6	120	.400			11.20	.40	11.60	17.70	
	1950	Pumped	C-20	150	.427			12.25	5	17.25	24.50	
	2000	With crane and bucket	C-7	90	.800			23	10.65	33.65	47.50	
	2100	Footings, continuous, deep, direct chute	C-6	140	.343			9.60	.34	9.94	15.20	
	2150	Pumped	C-20	160	.400			11.50	4.70	16.20	23	
	2200	With crane and bucket	C-7	110	.655			18.90	8.75	27.65	38.50	
	2400	Footings, spread, under 1 C.Y., direct chute	C-6	55	.873			24.50	.87	25.37	38.50	
	2450	Pumped	C-20	65	.985			28.50	11.55	40.05	56.50	
	2500	With crane and bucket	C-7	45	1.600			46	21.50	67.50	94.50	
	2600	Over 5 C.Y., direct chute	C-6	120	.400			11.20	.40	11.60	17.70	
	2650	Pumped	C-20	150	.427			12.25	5	17.25	24.50	
	2700	With crane and bucket	C-7	100	.720			21	9.60	30.60	42.50	
	2900	Foundation mats, over 20 C.Y., direct chute	C-6	350	.137			3.85	.14	3.99	6.05	
	2950	Pumped	C-20	400	.160			4.59	1.88	6.47	9.10	
	3000	With crane and bucket	C-7	300	.240	↓		6.95	3.20	10.15	14.10	

Important: See the Reference Section for critical supporting data - Reference Nos., Crews, & City Cost Indexes

03310 | Structural Concrete

		CREW	DAILY OUTPUT	LABOR-HOURS	UNIT	MAT.	LABOR	EQUIP.	TOTAL	TOTAL INCL O&P	
700	3200	Grade beams, direct chute	C-6	150	.320	C.Y.		9	.32	9.32	14.15	**700**
	3250	Pumped	C-20	180	.356			10.20	4.18	14.38	20.50	
	3300	With crane and bucket	C-7	120	.600			17.30	8	25.30	35.50	
	3500	High rise, for more than 5 stories, pumped, add per story	C-20	2,100	.030			.88	.36	1.24	1.73	
	3510	With crane and bucket, add per story	C-7	2,100	.034			.99	.46	1.45	2.02	
	3700	Pile caps, under 5 C.Y., direct chute	C-6	90	.533			14.95	.53	15.48	23.50	
	3750	Pumped	C-20	110	.582			16.70	6.85	23.55	33	
	3800	With crane and bucket	C-7	80	.900			26	12	38	53	
	3850	Pile cap, 5 C.Y. to 10 C.Y., direct chute	C-6	175	.274			7.70	.27	7.97	12.15	
	3900	Pumped	C-20	200	.320			9.20	3.76	12.96	18.25	
	3950	With crane and bucket	C-7	150	.480			13.85	6.40	20.25	28	
	4000	Over 10 C.Y., direct chute	C-6	215	.223			6.25	.22	6.47	9.90	
	4050	Pumped	C-20	240	.267			7.65	3.13	10.78	15.20	
	4100	With crane and bucket	C-7	185	.389			11.25	5.20	16.45	23	
	4300	Slab on grade, 4" thick, direct chute	C-6	110	.436			12.25	.44	12.69	19.35	
	4350	Pumped	C-20	130	.492			14.15	5.80	19.95	28	
	4400	With crane and bucket	C-7	110	.655			18.90	8.75	27.65	38.50	
	4600	Over 6" thick, direct chute	C-6	165	.291			8.15	.29	8.44	12.85	
	4650	Pumped	C-20	185	.346			9.95	4.06	14.01	19.70	
	4700	With crane and bucket	C-7	145	.497			14.35	6.60	20.95	29.50	
	4900	Walls, 8" thick, direct chute	C-6	90	.533			14.95	.53	15.48	23.50	
	4950	Pumped	C-20	100	.640			18.35	7.50	25.85	36.50	
	5000	With crane and bucket	C-7	80	.900			26	12	38	53	
	5050	12" thick, direct chute	C-6	100	.480			13.45	.48	13.93	21	
	5100	Pumped	C-20	110	.582			16.70	6.85	23.55	33	
	5200	With crane and bucket	C-7	90	.800			23	10.65	33.65	47.50	
	5300	15" thick, direct chute	C-6	105	.457			12.85	.46	13.31	20.50	
	5350	Pumped	C-20	120	.533			15.30	6.25	21.55	30.50	
	5400	With crane and bucket	C-7	95	.758	▼		22	10.10	32.10	44.50	
	5600	Wheeled concrete dumping, add to placing costs above										
	5610	Walking cart, 50' haul, add	C-18	32	.281	C.Y.		7.55	1.56	9.11	13.50	
	5620	150' haul, add		24	.375			10.10	2.07	12.17	18	
	5700	250' haul, add	▼	18	.500			13.45	2.77	16.22	24	
	5800	Riding cart, 50' haul, add	C-19	80	.112			3.03	.96	3.99	5.75	
	5810	150' haul, add		60	.150			4.04	1.28	5.32	7.70	
	5900	250' haul, add	▼	45	.200	▼		5.40	1.71	7.11	10.30	

03350 | Concrete Finishing

		CREW	DAILY OUTPUT	LABOR-HOURS	UNIT	MAT.	LABOR	EQUIP.	TOTAL	TOTAL INCL O&P	
300	0010	**FINISHING FLOORS** Monolithic, screed finish	1 Cefi	900	.009	S.F.		.29		.29	.43	**300**
	0100	Screed and bull float (darby) finish		725	.011			.36		.36	.53	
	0150	Screed, float, and broom finish		630	.013			.42		.42	.61	
	0200	Screed, float, and hand trowel		600	.013			.44		.44	.64	
	0250	Machine trowel		550	.015			.48		.48	.70	
	1600	Exposed local aggregate finish, minimum		625	.013		4.08	.42		4.50	5.10	
	1650	Maximum		465	.017		12.90	.57		13.47	15.05	
	1800	Floor abrasives, .25 psf, aluminum oxide		850	.009		.26	.31		.57	.74	
	1850	Silicon carbide		850	.009		.36	.31		.67	.85	
	2000	Floor hardeners, metallic, light service, .50 psf, add		850	.009		.30	.31		.61	.79	
	2050	Medium service, .75 psf		750	.011		.45	.35		.80	1.02	
	2100	Heavy service, 1.0 psf		650	.012		.60	.40		1	1.26	
	2150	Extra heavy, 1.5 psf		575	.014		.90	.46		1.36	1.66	
	2300	Non-metallic, light service, .50 psf		850	.009		.19	.31		.50	.66	
	2350	Medium service, .75 psf		750	.011		.28	.35		.63	.83	
	2400	Heavy service, 1.00 psf		650	.012		.37	.40		.77	1.01	
	2450	Extra heavy, 1.50 psf	▼	575	.014	▼	.56	.46		1.02	1.28	
	2600	Add for colored hardeners, metallic					50%					

03350 | Concrete Finishing

		CREW	DAILY OUTPUT	LABOR-HOURS	UNIT	MAT.	LABOR	EQUIP.	TOTAL	TOTAL INCL O&P	
300	2650	Non-metallic					25%					**300**
	2800	Trap rock wearing surface for monolithic floors										
	2810	2.0 psf	C-10B	1,250	.032	S.F.	.85	.93	.14	1.92	2.50	
	3000	Floor coloring, dusted on, 0.5 psf, add to above, minimum	1 Cefi	1,300	.006		.63	.20		.83	.99	
	3050	Maximum	"	625	.013	↓	2.21	.42		2.63	3.05	
	3100	Colors only, minimum				Lb.	1.05			1.05	1.16	
	3120	Maximum				"	3.68			3.68	4.04	
	3200	Integral colors, see division 03310-220										
	3600	1/2" topping using 5 lb. per bag, regular colors	C-10B	590	.068	S.F.	.21	1.98	.29	2.48	3.54	
	3650	Blue or green	"	590	.068		.21	1.98	.29	2.48	3.54	
	3800	Dustproofing, silicate liquids, 1 coat	1 Cefi	1,900	.004		.08	.14		.22	.29	
	3850	2 coats		1,300	.006		.14	.20		.34	.46	
	4000	Epoxy coating, 1 coat, clear		1,500	.005		.42	.18		.60	.72	
	4050	Colors		1,500	.005		.42	.18		.60	.72	
	4400	Stair finish, float		275	.029			.96		.96	1.41	
	4500	Steel trowel finish		200	.040			1.31		1.31	1.93	
	4600	Silicon carbide finish, .25 psf	↓	150	.053	↓	.57	1.75		2.32	3.21	
325	0010	**CONTROL JOINT**, concrete floor slab										**325**
	0100	Sawcut in green concrete [R03210-080]										
	0120	1" depth	C-27	2,000	.008	L.F.		.26	.06	.32	.44	
	0140	1-1/2" depth		1,800	.009			.29	.06	.35	.50	
	0160	2" depth	↓	1,600	.010			.33	.07	.40	.56	
	0200	Clean out control joint of debris	C-28	6,000	.001	↓		.04		.04	.06	
	0300	Joint sealant										
	0320	Backer rod, polyethylene, 1/4" diameter	1 Cefi	460	.017	L.F.	.02	.57		.59	.86	
	0340	Sealant, polyurethane										
	0360	1/4" x 1/4" (308 LF/Gal)	1 Cefi	270	.030	L.F.	.15	.97		1.12	1.60	
	0380	1/4" x 1/2" (154 LF/Gal)	"	255	.031	"	.31	1.03		1.34	1.86	
350	0010	**FINISHING WALLS** Break ties and patch voids	1 Cefi	540	.015	S.F.	.03	.49		.52	.75	**350**
	0050	Burlap rub with grout		450	.018		.03	.58		.61	.89	
	0100	Carborundum rub, dry		270	.030		.03	.97		1	1.46	
	0150	Wet rub	↓	175	.046		.03	1.50		1.53	2.24	
	0300	Bush hammer, green concrete	B-39	1,000	.048		.03	1.35	.14	1.52	2.28	
	0350	Cured concrete	"	650	.074		.03	2.07	.22	2.32	3.48	
	0600	Float finish, 1/16" thick	1 Cefi	300	.027		.01	.88		.89	1.31	
	0700	Sandblast, light penetration	C-10	1,100	.022		.23	.67		.90	1.25	
	0750	Heavy penetration	"	375	.064	↓	.46	1.97		2.43	3.44	
	0800	Board finish, see division 03110-455										
	0850	Rustication strips, see division 03110-455										
600	0010	**SLAB TEXTURE STAMPING,** buy										**600**
	0020	Approx. 3 S.F.- 5 S.F. each, minimum				Ea.	45			45	49.50	
	0030	Average				"	49.50			49.50	54.50	
	0120	Per S.F. of tool, average				S.F.	54			54	59.50	
	0200	Commonly used chemicals for texture systems										
	0210	Hardeners w/colors average				S.F.	.45			.45	.50	
	0220	Release agents w/colors, average					.18			.18	.20	
	0225	Clear, average					.14			.14	.15	
	0230	Sealers, clear, average					.12			.12	.13	
	0240	Colors, average				↓	.15			.15	.17	

03370 | Specially Placed Concrete

		CREW	DAILY OUTPUT	LABOR-HOURS	UNIT	MAT.	LABOR	EQUIP.	TOTAL	TOTAL INCL O&P	
300	0010	**GUNITE,** dry mix										**300**
	0020	Applied in 1" layers, no mesh included	C-8	2,000	.028	S.F.	.25	.84	.35	1.44	1.94	

Important: See the Reference Section for critical supporting data - Reference Nos., Crews, & City Cost Indexes

			DAILY	LABOR-		2005 BARE COSTS				TOTAL	
	03370	**Specially Placed Concrete**	CREW	OUTPUT	HOURS	UNIT	MAT.	LABOR	EQUIP.	TOTAL	INCL O&P
300	0100	Mesh for gunite 2 x 2, #12, to 3" thick	2 Rodm	800	.020	S.F.	.32	.76		1.08	1.60	**300**
	0150	Over 3" thick	"	500	.032		.32	1.21		1.53	2.35
	0300	Typical in place, including mesh, 2" thick, minimum	C-16	1,000	.072		.82	2.28	.70	3.80	5.20
	0350	Maximum		500	.144		.82	4.56	1.41	6.79	9.55
	0500	4" thick, minimum		750	.096		1.32	3.04	.94	5.30	7.20
	0550	Maximum		350	.206		1.32	6.50	2.01	9.83	13.75
	0900	Prepare old walls, no scaffolding, minimum	C-10	1,000	.024			.74		.74	1.10
	0950	Maximum	"	275	.087			2.69		2.69	4
	1100	For high finish requirement or close tolerance, add, minimum						50%			
	1150	Maximum						110%			
700	0010	**ROLLER COMPACTED CONCRETE** No material, spread and compact										**700**
	0100	Mass placement, 1' lift, 1' layer	B-10C	1,280	.009	C.Y.		.30	1.18	1.48	1.76
	0200	2' lift, 6" layer	"	1,600	.007			.24	.94	1.18	1.40
	0210	Vertical face, formed, 1' lift	B-11V	400	.060			1.60	.31	1.91	2.83
	0220	6" lift	"	200	.120			3.20	.61	3.81	5.65
	0300	Sloped face, nonformed, 1' lift	B-11L	384	.042			1.28	1.19	2.47	3.26
	0360	6" lift	"	192	.083			2.56	2.38	4.94	6.55
	0400	Surface preparation, vacuum truck	B-6A	3,280	.006	S.Y.		.18	.10	.28	.39
	0450	Water clean	B-9A	3,000	.008			.22	.11	.33	.45
	0460	Water blast	B-9B	800	.030			.81	.47	1.28	1.77
	0500	Joint bedding placement, 1" thick	B-11C	975	.016			.50	.22	.72	1.01
	0510	Conveyance of materials, 18 C.Y. truck, 5 min. cycle	B-34F	2,048	.004	C.Y.		.11	.46	.57	.66
	0520	10 min. cycle		1,024	.008			.22	.91	1.13	1.33
	0540	15 min. cycle		680	.012			.32	1.37	1.69	2
	0550	With crane and bucket	C-23A	1,600	.025			.74	1.24	1.98	2.50
	0560	With 4 C.Y. loader, 4 min. cycle	B-10U	480	.025			.80	1.46	2.26	2.83
	0570	8 min. cycle		240	.050			1.60	2.92	4.52	5.65
	0580	12 min. cycle		160	.075			2.40	4.38	6.78	8.45
	0590	With belt conveyor	C-7D	600	.093			2.62	.25	2.87	4.34
	0600	With 17 C.Y. scraper, 5 min. cycle	B-33J	1,440	.006			.19	1.08	1.27	1.48
	0610	10 min. cycle		720	.011			.39	2.16	2.55	2.95
	0620	15 min. cycle		480	.017			.58	3.23	3.81	4.43
	0630	20 min. cycle		360	.022			.77	4.31	5.08	5.90
	0640	Water cure, small job, < 500 CY	B-94C	8	1	Hr.		26.50	7.85	34.35	50
	0650	Large job, over 500 CY	B-59A	8	3	"		81	40.50	121.50	170
	0660	RCC paving, with asphalt paver including material	B-25C	1,000	.048	C.Y.	51.50	1.42	1.77	54.69	60.50
	0670	8" thick layers		4,200	.011	S.Y.	13.45	.34	.42	14.21	15.75
	0680	12" thick layers		2,800	.017	"	19.95	.51	.63	21.09	23.50
	0700	Roller compacted concrete, 1.5" - 2" agg., 200 lb. cement/C.Y.				C.Y.	60.50			60.50	66.50
800	0010	**SHOTCRETE**, Wet mix, placed @ 10 C.Y. per hour, 3000 psi	C-8C	80	.600	C.Y.	81.50	17.65	2.80	101.95	120	**800**
	0100	2nd pour	C-6	15	3.200	"	73	90	3.20	166.20	222
	1010	Fiber reinforced, 1" thick	C-8E	1,160	.028	S.F.	.61	.84	.71	2.16	2.72
	1020	2" thick		600	.053		1.21	1.62	1.37	4.20	5.30
	1030	3" thick		550	.058		1.82	1.76	1.50	5.08	6.35
	1040	4" thick		500	.064		2.42	1.94	1.65	6.01	7.40
	03390	**Concrete Curing**									
200	0010	**CURING** Burlap, 4 uses assumed, 7.5 oz.	2 Clab	55	.291	C.S.F.	7	7.75		14.75	19.80	**200**
	0100	10 oz.		55	.291		11.30	7.75		19.05	24.50
	0200	Waterproof curing paper, 2 ply, reinforced		70	.229		6	6.10		12.10	16.10
	0300	Sprayed membrane curing compound		95	.168		5.25	4.50		9.75	12.80
	0400	Curing blankets, 1" to 2" thick, buy, minimum				S.F.	.35			.35	.39
	0450	Maximum					.53			.53	.58
	0500	Electrically heated pads, 110 volts, 15 watts per S.F., buy					4.68			4.68	5.15
	0600	20 watts per S.F., buy					6.20			6.20	6.85

CONCRETE 3

03300 | Cast-In-Place Concrete

03390 | Concrete Curing

		CREW	DAILY OUTPUT	LABOR-HOURS	UNIT	2005 BARE COSTS MAT.	LABOR	EQUIP.	TOTAL	TOTAL INCL O&P	
200	0710	Electrically, heated pads, 15 watts/S.F., 20 uses, minimum				S.F.	.17			.17	.18	**200**
	0800	Maximum				↓	.28			.28	.31	

03400 | Precast Concrete

03410 | Plant-Precast Structural Concrete

			CREW	DAILY OUTPUT	LABOR-HOURS	UNIT	2005 BARE COSTS MAT.	LABOR	EQUIP.	TOTAL	TOTAL INCL O&P	
100	0011	**BEAMS**, "L" shaped, 20' span, 12" x 20"	C-11	32	2.250	Ea.	985	84	45	1,114	1,275	**100**
	1000	Inverted tee beams, add to above, small beams					15%					
	1050	Large beams					20%					
	1200	Rectangular, 20' span, 12" x 20"	C-11	32	2.250		730	84	45	859	995	
	1250	18" x 36"		24	3		825	112	60	997	1,175	
	1300	24" x 44"		22	3.273		1,025	122	65.50	1,212.50	1,400	
	1400	30' span, 12" x 36"		24	3		1,100	112	60	1,272	1,475	
	1450	18" x 44"		20	3.600		1,400	134	72.50	1,606.50	1,825	
	1500	24" x 52"		16	4.500		1,800	168	90.50	2,058.50	2,375	
	1600	40' span, 12" x 52"		20	3.600		2,050	134	72.50	2,256.50	2,575	
	1650	18" x 52"		16	4.500		2,450	168	90.50	2,708.50	3,075	
	1700	24" x 52"		12	6		2,650	223	120	2,993	3,450	
	2000	"T" shaped, 20' span, 12" x 20"		32	2.250		1,250	84	45	1,379	1,575	
	2050	18" x 36"		24	3		1,400	112	60	1,572	1,800	
	2100	24" x 44"		22	3.273		1,750	122	65.50	1,937.50	2,200	
	2200	30' span, 12" x 36"		24	3		1,875	112	60	2,047	2,325	
	2250	18" x 44"		20	3.600		2,375	134	72.50	2,581.50	2,900	
	2300	24" x 52"		16	4.500		3,050	168	90.50	3,308.50	3,750	
	2500	40' span, 12" x 52"		20	3.600		3,500	134	72.50	3,706.50	4,150	
	2550	18" x 52"		16	4.500		4,150	168	90.50	4,408.50	4,975	
	2600	24" x 52"	↓	12	6	↓	4,500	223	120	4,843	5,475	
210	0010	**COLUMNS** Rectangular to 12' high, small columns	C-11	120	.600	L.F.	41	22.50	12.05	75.55	97.50	**210**
	0050	Large columns		96	.750		72	28	15.05	115.05	144	
	0300	24' high, small columns		192	.375		41	13.95	7.55	62.50	78	
	0350	Large columns		144	.500	↓	72	18.60	10.05	100.65	123	
	0700	24' high, 1 haunch, 12" x 12"		32	2.250	Ea.	1,000	84	45	1,129	1,300	
	0800	20" x 20"	↓	28	2.571	"	1,750	95.50	51.50	1,897	2,150	
400	0010	**JOISTS** 40 psf L.L., 6" deep for 12' spans	C-12	600	.080	L.F.	6.90	2.69	1	10.59	12.85	**400**
	0050	8" deep for 16' spans		575	.083		11.45	2.80	1.05	15.30	18.10	
	0100	10" deep for 20' spans		550	.087		20	2.93	1.10	24.03	28	
	0150	12" deep for 24' spans	↓	525	.091	↓	27.50	3.07	1.15	31.72	36.50	
650	0010	**PRESTRESSED CONCRETE** pretensioned, see division 03400										**650**
	0020	See also division 03230-600										
	0100	Post-tensioned in place, small job	C-17B	8.50	9.647	C.Y.	560	340	30	930	1,175	
	0200	Large job	"	10	8.200	"	420	289	25.50	734.50	940	
660	0010	**PRESTRESSED** Roof and floor members, see division 03400										**660**
750	0010	**TEES** Prestressed										**750**
	0020	Quad tee, short spans, roof	C-11	7,200	.010	S.F.	4.95	.37	.20	5.52	6.30	
	0050	Floor		7,200	.010		4.95	.37	.20	5.52	6.30	
	0200	Double tee, floor members, 60' span		8,400	.009		6.80	.32	.17	7.29	8.25	
	0250	80' span		8,000	.009		8	.34	.18	8.52	9.60	
	0300	Roof members, 30' span	↓	4,800	.015	↓	5.80	.56	.30	6.66	7.65	

3
CONCRETE

			DAILY	LABOR-		2005 BARE COSTS				TOTAL	
	03410	**Plant-Precast Structural Concrete**	CREW	OUTPUT	HOURS	UNIT	MAT.	LABOR	EQUIP.	TOTAL	INCL O&P
750	0350	50' span	C-11	6,400	.011	S.F.	6.60	.42	.23	7.25	8.25	**750**
	0400	Wall members, up to 55' high		3,600	.020		8.25	.74	.40	9.39	10.85
	0500	Single tee roof members, 40' span		3,200	.023		6.50	.84	.45	7.79	9.10
	0550	80' span		5,120	.014		8.05	.52	.28	8.85	10.05
	0600	100' span		6,000	.012		12.10	.45	.24	12.79	14.35
	0650	120' span		6,000	.012		13.20	.45	.24	13.89	15.55
	1000	Double tees, floor members									
	1100	Lightweight, 20" x 8' wide, 45' span	C-11	20	3.600	Ea.	2,125	134	72.50	2,331.50	2,625
	1150	24" x 8' wide, 50' span		18	4		2,300	149	80.50	2,529.50	2,875
	1200	32" x 10' wide, 60' span		16	4.500		3,875	168	90.50	4,133.50	4,675
	1250	Standard weight, 12" x 8' wide, 20' span		22	3.273		790	122	65.50	977.50	1,150
	1300	16" x 8' wide, 25' span		20	3.600		1,050	134	72.50	1,256.50	1,450
	1350	18" x 8' wide, 30' span		20	3.600		1,350	134	72.50	1,556.50	1,775
	1400	20" x 8' wide, 45' span		18	4		1,500	149	80.50	1,729.50	2,000
	1450	24" x 8' wide, 50' span		16	4.500		1,950	168	90.50	2,208.50	2,550
	1500	32" x 10' wide, 60' span		14	5.143		3,575	191	103	3,869	4,375
	2000	Roof members									
	2050	Lightweight, 20" x 8' wide, 40' span	C-11	20	3.600	Ea.	1,775	134	72.50	1,981.50	2,250
	2100	24" x 8' wide, 50' span		18	4		2,375	149	80.50	2,604.50	2,975
	2150	32" x 10' wide, 60' span		16	4.500		3,925	168	90.50	4,183.50	4,700
	2200	Standard weight, 12" x 8' wide, 30' span		22	3.273		1,200	122	65.50	1,387.50	1,575
	2250	16" x 8' wide, 30' span		20	3.600		1,250	134	72.50	1,456.50	1,675
	2300	18" x 8' wide, 30' span		20	3.600		1,400	134	72.50	1,606.50	1,825
	2350	20" x 8' wide, 40' span		18	4		1,425	149	80.50	1,654.50	1,925
	2400	24" x 8' wide, 50' span		16	4.500		1,900	168	90.50	2,158.50	2,500
	2450	32" x 10' wide, 60' span		14	5.143		3,325	191	103	3,619	4,100
855	0010	**PRECAST WINDOW SILLS**										**855**
	0600	Precast concrete, 4" tapers to 3", 9" wide	D-1	70	.229	L.F.	9.45	7.10		16.55	21
	0650	11" wide		60	.267		12.50	8.30		20.80	26.50
	0700	13" wide, 3 1/2" tapers to 2 1/2", 12" wall		50	.320		13.50	9.95		23.45	30

	03470	**Tilt-Up, Precast Concrete**									
600	0010	**TILT-UP** Wall panel construction, walls only, 5-1/2" thick	C-14	1,600	.090	S.F.	3.57	2.99	.62	7.18	9.30	**600**
	0100	7-1/2" thick		1,550	.093		4	3.09	.64	7.73	9.95
	0500	Walls and columns, 5-1/2" thick walls, 12" x 12" columns		1,565	.092		4.35	3.06	.64	8.05	10.25
	0550	7-1/2" thick wall, 12" x 12" columns		1,370	.105		5.80	3.49	.73	10.02	12.65
	0800	Columns only, site precast, minimum		200	.720	L.F.	26	24	4.98	54.98	71.50
	0850	Maximum		105	1.371	"	47.50	45.50	9.50	102.50	134

	03480	**Precast Concrete Specialties**									
200	0010	**CURBS** Roadway type, see division 02770-300										**200**

800	0010	**STAIRS**, Precast concrete treads on steel stringers, 3' wide	C-12	75	.640	Riser	64.50	21.50	8.05	94.05	113	**800**
	0300	Front entrance, 5' wide with 48" platform, 2 risers		16	3	Flight	340	101	37.50	478.50	575
	0350	5 risers		12	4		530	134	50	714	850
	0500	6' wide, 2 risers		15	3.200		360	108	40	508	605
	0550	5 risers		11	4.364		560	147	55	762	905
	0700	7' wide, 2 risers		14	3.429		375	115	43	533	640
	0750	5 risers		10	4.800		590	161	60	811	965
	1200	Basement entrance stairs, steel bulkhead doors, minimum	B-51	22	2.182		700	59	5.85	764.85	870
	1250	Maximum	"	11	4.364		1,000	118	11.70	1,129.70	1,300

CONCRETE 3

03500 | Cementitious Decks & Underlayments

03510 | Cementitious Roof Deck

			CREW	DAILY OUTPUT	LABOR-HOURS	UNIT	2005 BARE COSTS				TOTAL INCL O&P	
							MAT.	LABOR	EQUIP.	TOTAL		
250	0010	**CONCRETE CHANNEL SLABS** 2-3/4" or 3-1/2" thick, straight	C-12	1,575	.030	S.F.	4.48	1.02	.38	5.88	6.95	250
	0050	Chopped up		785	.061		4.90	2.05	.77	7.72	9.40	
	0200	6" thick, span to 20'		1,300	.037		5.15	1.24	.46	6.85	8.15	
	0300	8" thick, span to 24'	↓	1,100	.044	↓	5.95	1.47	.55	7.97	9.40	

03520 | Lightweight Concrete Roof Insulation

			CREW	DAILY OUTPUT	LABOR-HOURS	UNIT	2005 BARE COSTS				TOTAL INCL O&P	
							MAT.	LABOR	EQUIP.	TOTAL		
250	0010	**INSULATING ROOF FILL**, lightweight cellular concrete										250
	0020	Portland cement and foaming agent	C-8	50	1.120	C.Y.	80	33.50	14.10	127.60	155	
	0100	Poured vermiculite or perlite, field mix,										
	0110	1:6 field mix	C-8	50	1.120	C.Y.	91.50	33.50	14.10	139.10	167	
	0200	Ready mix, 1:6 mix, roof fill, 2" thick		10,000	.006	S.F.	.51	.17	.07	.75	.89	
	0250	3" thick	↓	7,700	.007		.76	.22	.09	1.07	1.27	
	0400	Expanded volcanic glass rock, with binder, minimum	2 Carp	1,500	.011		.28	.37		.65	.88	
	0450	Maximum	"	1,200	.013	↓	.83	.46		1.29	1.63	

03600 | Grouts

03610 | Construction Grout

			CREW	DAILY OUTPUT	LABOR-HOURS	UNIT	2005 BARE COSTS				TOTAL INCL O&P	
							MAT.	LABOR	EQUIP.	TOTAL		
400	0010	**GROUT** Column & machine bases, non-shrink, metallic, 1" deep	1 Cefi	35	.229	S.F.	10.50	7.50		18	22.50	400
	0050	2" deep		25	.320		21	10.50		31.50	38.50	
	0300	Non-shrink, non-metallic, 1" deep		35	.229		9.35	7.50		16.85	21.50	
	0350	2" deep	↓	25	.320	↓	18.75	10.50		29.25	36	

03900 | Concrete Restoration & Cleaning

03930 | Concrete Rehabilitation

			CREW	DAILY OUTPUT	LABOR-HOURS	UNIT	2005 BARE COSTS				TOTAL INCL O&P	
							MAT.	LABOR	EQUIP.	TOTAL		
200	0010	**BRIDGE DECK**, apply waterproof membrane				S.Y.					5.60	200
	0020	Apply waterproof sealer				"					3.35	
	0100	Bridge joint, saw and seal				L.F.					10.10	
300	0010	**CRACK REPAIR**, including chipping, sand blasting and cleaning										300
	0100	Epoxy injection, 1/8" wide, 12" deep	B-9	80	.500	L.F.	4.75	13.55	1.77	20.07	28	
	0110	1/4" wide, 12" deep		60	.667		9.50	18.05	2.36	29.91	41	
	0200	Latex injection, 1/8" wide, 12" deep		100	.400		1.31	10.85	1.42	13.58	19.85	
	0210	1/4" wide, 12" deep	↓	75	.533	↓	2.62	14.45	1.89	18.96	27.50	

For information about Means Estimating Seminars, see yellow pages 12 and 13 in back of book

Important: See the Reference Section for critical supporting data - Reference Nos., Crews, & City Cost Indexes

Division 4
Masonry

Estimating Tips

04050 Basic Masonry Materials & Methods

- The terms *mortar* and *grout* are often used interchangeably, and incorrectly. Mortar is used to bed masonry units, seal the entry of air and moisture, provide architectural appearance, and allow for size variations in the units. Grout is used primarily in reinforced masonry construction and is used to bond the masonry to the reinforcing steel. Common mortar types are M(2500 psi), S(1800 psi), N(750 psi), and O(350 psi), and conform to ASTM C270. Grout is either fine or coarse, conforms to ASTM C476, and in-place strengths generally exceed 2500 psi. Mortar and grout are different components of masonry construction and are placed by entirely different methods. An estimator should be aware of their unique uses and costs.

- Waste, specifically the loss/droppings of mortar and the breakage of brick and block, is included in all masonry assemblies in this division. A factor of 25% is added for mortar and 3% for brick and concrete masonry units.

- Scaffolding or staging is not included in any of the Division 4 costs. Refer to section 01540 for scaffolding and staging costs.

04800 Masonry Assemblies

- The most common types of unit masonry are brick and concrete masonry. The major classifications of brick are building brick (ASTM C62), facing brick (ASTM C216) and glazed brick, fire brick and pavers. Many varieties of texture and appearance can exist within these classifications, and the estimator would be wise to check local custom and availability within the project area. On repair and remodeling jobs, matching the existing brick may be the most important criteria.

- Brick and concrete block are priced by the piece and then converted into a price per square foot of wall. Openings less than two square feet are generally ignored by the estimator because any savings in units used is offset by the cutting and trimming required.

- It is often difficult and expensive to find and purchase small lots of historic brick. Costs can vary widely. Many design issues affect costs, selection of mortar mix, and repairs or replacement of masonry materials. Cleaning techniques must be reflected in the estimate.

- All masonry walls, whether interior or exterior, require bracing. The cost of bracing walls during construction should be included by the estimator and this bracing must remain in place until permanent bracing is complete. Permanent bracing of masonry walls is accomplished by masonry itself, in the form of pilasters or abutting wall corners, or by anchoring the walls to the structural frame. Accessories in the form of anchors, anchor slots and ties are used, but their supply and installation can be by different trades. For instance, anchor slots on spandrel beams and columns are supplied and welded in place by the steel fabricator, but the ties from the slots into the masonry are installed by the bricklayer. Regardless of the installation method the estimator must be certain that these accessories are accounted for in pricing.

Reference Numbers

Reference numbers are shown in bold squares at the beginning of some major classifications. These numbers refer to related items in the Reference Section. The reference information may be an estimating procedure, an alternate pricing method or technical information.

Note: Not all subdivisions listed here necessarily appear in this publication.

4 MASONRY

04055 | Selective Demolition

			CREW	DAILY OUTPUT	LABOR-HOURS	UNIT	MAT.	LABOR	EQUIP.	TOTAL	TOTAL INCL O&P	
								2005 BARE COSTS				
110	0010	**SELECTIVE DEMOLITION, MASONRY** R02220 -510										110
	1000	Chimney, 16" x 16", soft old mortar	1 Clab	55	.145	C.F.		3.88		3.88	6.05	
	1020	Hard mortar		40	.200			5.35		5.35	8.30	
	1030	16" x 20", soft old mortar		55	.145			3.88		3.88	6.05	
	1040	Hard mortar		40	.200			5.35		5.35	8.30	
	1050	16" x 24", soft old mortar		55	.145			3.88		3.88	6.05	
	1060	Hard mortar		40	.200			5.35		5.35	8.30	
	1080	20" x 20", soft old mortar		55	.145			3.88		3.88	6.05	
	1110	20" x 24", soft old mortar		55	.145			3.88		3.88	6.05	
	1140	20" x 32", soft old mortar		55	.145			3.88		3.88	6.05	
	1160	Hard mortar		40	.200			5.35		5.35	8.30	
	1200	48" x 48", soft old mortar		55	.145			3.88		3.88	6.05	
	1220	Hard mortar	▼	40	.200	▼		5.35		5.35	8.30	
	1250	Metal, high temp steel jacket, 24" diameter	E-2	130	.431	V.L.F.		15.90	10.20	26.10	38.50	
	1260	60" diameter	"	60	.933			34.50	22	56.50	83.50	
	2000	Columns, 8" x 8", soft old mortar	1 Clab	48	.167			4.45		4.45	6.95	
	2020	Hard mortar		40	.200			5.35		5.35	8.30	
	2060	16" x 16", soft old mortar		16	.500			13.35		13.35	21	
	2100	Hard mortar		14	.571			15.25		15.25	23.50	
	2140	24" x 24", soft old mortar		8	1			26.50		26.50	41.50	
	2160	Hard mortar		6	1.333			35.50		35.50	55.50	
	2200	36" x 36", soft old mortar		4	2			53.50		53.50	83	
	2220	Hard mortar		3	2.667	▼		71		71	111	
	2230	Alternate pricing method, soft old mortar		30	.267	C.F.		7.10		7.10	11.10	
	2240	Hard mortar	▼	23	.348	"		9.30		9.30	14.45	
	3000	Copings, precast or masonry, to 8" wide										
	3020	Soft old mortar	1 Clab	180	.044	L.F.		1.19		1.19	1.85	
	3040	Hard mortar	"	160	.050	"		1.34		1.34	2.08	
	3100	To 12" wide										
	3120	Soft old mortar	1 Clab	160	.050	L.F.		1.34		1.34	2.08	
	3140	Hard mortar	"	140	.057	"		1.53		1.53	2.37	
	4000	Fireplace, brick, 30" x 24" opening										
	4020	Soft old mortar	1 Clab	2	4	Ea.		107		107	166	
	4040	Hard mortar		1.25	6.400			171		171	266	
	4100	Stone, soft old mortar		1.50	5.333			142		142	222	
	4120	Hard mortar		1	8	▼		214		214	330	
	5000	Veneers, brick, soft old mortar		140	.057	S.F.		1.53		1.53	2.37	
	5020	Hard mortar		125	.064			1.71		1.71	2.66	
	5100	Granite and marble, 2" thick		180	.044			1.19		1.19	1.85	
	5120	4" thick		170	.047			1.26		1.26	1.96	
	5140	Stone, 4" thick		180	.044			1.19		1.19	1.85	
	5160	8" thick		175	.046	▼		1.22		1.22	1.90	
	5400	Alternate pricing method, stone, 4" thick		60	.133	C.F.		3.56		3.56	5.55	
	5420	8" thick	▼ ▼	85	.094	"		2.51		2.51	3.91	

04060 | Masonry Mortar

			CREW	DAILY OUTPUT	LABOR-HOURS	UNIT	MAT.	LABOR	EQUIP.	TOTAL	TOTAL INCL O&P	
200	0010	**CEMENT** Gypsum 80 lb. bag, T.L. lots				Bag	23			23	25.50	200
	0050	L.T.L. lots					23			23	25.50	
	0100	Masonry, 70 lb. bag, T.L. lots					6.10			6.10	6.70	
	0150	L.T.L. lots					6.45			6.45	7.10	
	0200	White, 70 lb. bag, T.L. lots					16.35			16.35	18	
	0250	L.T.L. lots				▼	17.45			17.45	19.20	
400	0010	**LIME** Masons, hydrated, 50 lb. bag, T.L. lots				Bag	5.80			5.80	6.35	400
	0050	L.T.L. lots					6			6	6.60	
	0200	Finish, double hydrated, 50 lb. bag, T.L. lots					7			7	7.70	
	0250	L.T.L. lots				▼	7.75			7.75	8.55	

Important: See the Reference Section for critical supporting data - Reference Nos., Crews, & City Cost Indexes

04060	Masonry Mortar	CREW	DAILY OUTPUT	LABOR-HOURS	UNIT	2005 BARE COSTS				TOTAL INCL O&P	
						MAT.	LABOR	EQUIP.	TOTAL		
500	**0010**	**MORTAR**										**500**
	0100	Type M, 1:1:6 mix	1 Brhe	143	.056	C.F.	3.56	1.50		5.06	6.20
	0200	Type N, 1:3 mix		143	.056		3.19	1.50		4.69	5.80
	0300	Type O, 1:3 mix		143	.056		3.28	1.50		4.78	5.90
	0400	Type PM, 1:1:6 mix, 2500 psi		143	.056		3.48	1.50		4.98	6.10
	0500	Type S, 1/2:1:4 mix	↓	143	.056	↓	3.69	1.50		5.19	6.35
	2000	With portland cement and lime									
	2100	Type M, 1:1/4:3 mix	1 Brhe	143	.056	C.F.	4.16	1.50		5.66	6.85
	2200	Type N, 1:1:6 mix, 750 psi		143	.056		3.41	1.50		4.91	6.05
	2300	Type O, 1:2:9 mix (Pointing Mortar)		143	.056		3.31	1.50		4.81	5.95
	2400	Type PL, 1:1/2:4 mix, 2500 psi		143	.056		3.78	1.50		5.28	6.45
	2500	Type K, 1:3:12 mix, 75 psi		143	.056		3.27	1.50		4.77	5.90
	2600	Type S, 1:1/2:4 mix, 1800 psi	↓	143	.056		3.78	1.50		5.28	6.45
	2650	Pre-mixed, type S or N					4.07			4.07	4.48
	2700	Mortar for glass block	1 Brhe	143	.056		7.20	1.50		8.70	10.25
	2800	Gypsum cement mortar				↓	10.80			10.80	11.90
	2900	Mortar for Fire Brick, 80 lb. bag, T.L. Lots				Bag	13.50			13.50	14.85
520	**0010**	**MORTAR FOR MASONRY RESTORATION**, See also Division 04060-500				Lb.	1			1	1.10	**520**
	0050	White				"	1.15			1.15	1.27
540	**0010**	**MORTAR PIGMENTS**, 50 lb. bags (2 bags per M bricks),										**540**
	0020	range 2 to 10 lb. per bag of cement, minimum				Lb.	5.80			5.80	6.35
	0050	Average					7.60			7.60	8.40
	0100	Maximum				↓	15.10			15.10	16.60
750	**0010**	**SAND** For mortar, screened and washed, at the pit				Ton	12.80			12.80	14.05	**750**
	0050	With 10 mile haul					13.85			13.85	15.25
	0100	With 30 mile haul				↓	15.10			15.10	16.65
	0200	Screened and washed, at the pit				C.Y.	17.75			17.75	19.55
	0250	With 10 mile haul					19.25			19.25	21
	0300	With 30 mile haul				↓	21			21	23
900	**0010**	**WATERPROOFING** Admixture, 1 qt. to 2 bags of masonry cement				Qt.	7			7	7.70	**900**

04070 | Masonry Grout

			CREW	DAILY OUTPUT	LABOR-HOURS	UNIT	MAT.	LABOR	EQUIP.	TOTAL	TOTAL INCL O&P	
420	**0010**	**GROUTING** Bond bms. & lintels, 8" dp., pumped, not incl. block										**420**
	0020	8" thick, 0.2 C.F. per L.F.	D-4	1,400	.023	L.F.	.73	.70	.10	1.53	1.98
	0050	10" thick, 0.25 C.F. per L.F.		1,200	.027		.83	.81	.11	1.75	2.27
	0060	12" thick, 0.3 C.F. per L.F.	↓	1,040	.031	↓	.99	.94	.13	2.06	2.66
	0200	Concrete block cores, solid, 4" thk., by hand, 0.067 C.F./S.F. of wall	D-8	1,100	.036	S.F.	.22	1.16		1.38	2.01
	0210	6" thick, pumped, 0.175 C.F. per S.F.	D-4	720	.044		.58	1.36	.19	2.13	2.91
	0250	8" thick, pumped, 0.258 C.F. per S.F.		680	.047		.85	1.44	.20	2.49	3.34
	0300	10" thick, pumped, 0.340 C.F. per S.F.		660	.048		1.12	1.48	.20	2.80	3.71
	0350	12" thick, pumped, 0.422 C.F. per S.F.		640	.050		1.39	1.53	.21	3.13	4.08
	0500	Cavity walls, 2" space, pumped, 0.167 C.F./S.F. of wall		1,700	.019		.55	.57	.08	1.20	1.57
	0550	3" space, 0.250 C.F./S.F.		1,200	.027		.83	.81	.11	1.75	2.27
	0600	4" space, 0.333 C.F. per S.F.		1,150	.028		1.10	.85	.12	2.07	2.63
	0700	6" space, 0.500 C.F. per S.F.		800	.040	↓	1.65	1.22	.17	3.04	3.87
	0800	Door frames, 3' x 7' opening, 2.5 C.F. per opening		60	.533	Opng.	8.25	16.30	2.25	26.80	36
	0850	6' x 7' opening, 3.5 C.F. per opening		45	.711	"	11.55	21.50	3	36.05	49
	2000	Grout, C476, for bond beams, lintels and CMU cores	↓	350	.091	C.F.	3.31	2.79	.39	6.49	8.30

04080 | Anchorage & Reinforcement

			CREW	DAILY OUTPUT	LABOR-HOURS	UNIT	MAT.	LABOR	EQUIP.	TOTAL	TOTAL INCL O&P	
070	**0010**	**ANCHOR BOLTS** Hooked type with nut and washer, 1/2" diam., 8" long	1 Bric	200	.040	Ea.	.55	1.41		1.96	2.76	**070**
	0030	12" long		190	.042		1.17	1.48		2.65	3.55
	0040	5/8" diameter, 8" long		180	.044		.82	1.57		2.39	3.29
	0050	12" long	↓	170	.047	↓	.90	1.66		2.56	3.52

04080	Anchorage & Reinforcement	CREW	DAILY OUTPUT	LABOR-HOURS	UNIT	2005 BARE COSTS				TOTAL INCL O&P	
						MAT.	LABOR	EQUIP.	TOTAL		
070 0060	3/4" diameter, 8" long	1 Bric	160	.050	Ea.	1.39	1.76		3.15	4.22	**070**
0070	12" long	↓	150	.053	↓	1.74	1.88		3.62	4.77	
200 0010	**REINFORCING** Steel bars A615, placed horiz., #3 & #4 bars	1 Bric	450	.018	Lb.	.40	.63		1.03	1.39	**200**
0020	#5 & #6 bars		800	.010		.40	.35		.75	.98	
0050	Placed vertical, #3 & #4 bars		350	.023		.40	.81		1.21	1.67	
0060	#5 & #6 bars	↓	650	.012	↓	.40	.43		.83	1.10	
0500	Joint reinforcing, ladder type, mill std galvanized										
0600	9 ga. sides, 9 ga. ties, 4" wall	1 Bric	30	.267	C.L.F.	8.95	9.40		18.35	24	
0650	6" wall		30	.267		10.15	9.40		19.55	25.50	
0700	8" wall		25	.320		10.55	11.30		21.85	29	
0750	10" wall		20	.400		11.20	14.10		25.30	34	
0800	12" wall	↓	20	.400	↓	12.50	14.10		26.60	35.50	
1000	Truss type										
1100	9 ga. sides, 9 ga. ties, 4" wall	1 Bric	30	.267	C.L.F.	13.55	9.40		22.95	29	
1150	6" wall		30	.267		13.70	9.40		23.10	29.50	
1200	8" wall		25	.320		14.25	11.30		25.55	33	
1250	10" wall		20	.400		14.90	14.10		29	38	
1300	12" wall		20	.400		15.75	14.10		29.85	39	
1500	3/16" sides, 9 ga. ties, 4" wall		30	.267		18.80	9.40		28.20	35	
1550	6" wall		30	.267		19.05	9.40		28.45	35.50	
1600	8" wall		25	.320		19.60	11.30		30.90	38.50	
1650	10" wall		20	.400		19.60	14.10		33.70	43	
1700	12" wall		20	.400		21	14.10		35.10	44.50	
2000	3/16" sides, 3/16" ties, 4" wall		30	.267		24	9.40		33.40	40.50	
2050	6" wall		30	.267		24	9.40		33.40	41	
2100	8" wall		25	.320		25	11.30		36.30	44.50	
2150	10" wall		20	.400		26.50	14.10		40.60	50.50	
2200	12" wall	↓	20	.400	↓	27	14.10		41.10	51.50	
2500	Cavity truss type, galvanized										
2600	9 ga. sides, 9 ga. ties, 4" wall	1 Bric	25	.320	C.L.F.	22.50	11.30		33.80	42	
2650	6" wall		25	.320		23	11.30		34.30	42.50	
2700	8" wall		20	.400		25.50	14.10		39.60	49.50	
2750	10" wall		15	.533		28.50	18.80		47.30	60	
2800	12" wall		15	.533		36.50	18.80		55.30	69	
3000	3/16" sides, 9 ga. ties, 4" wall		25	.320		30.50	11.30		41.80	50.50	
3050	6" wall		25	.320		31	11.30		42.30	51	
3100	8" wall		20	.400		34	14.10		48.10	59	
3150	10" wall		15	.533		37	18.80		55.80	69	
3200	12" wall	↓	15	.533	↓	45	18.80		63.80	78	
3500	For hot dip galvanizing, add					80%					
650 0010	**WALL TIES** To brick veneer, galv., corrugated, 7/8" x 7", 22 Ga.	1 Bric	10.50	.762	C	5.75	27		32.75	47.50	**650**
0100	24 Ga.		10.50	.762		5.25	27		32.25	47	
0150	16 Ga.		10.50	.762		17.65	27		44.65	60.50	
0200	Buck anchors, galv., corrugated, 16 gauge, 2" bend, 8" x 2"		10.50	.762		115	27		142	168	
0250	8" x 3"	↓	10.50	.762	↓	107	27		134	159	
0300	Adjustable, rectangular, 4-1/8" wide										
0350	Anchor and tie, 3/16" wire, mill galv.										
0400	2-3/4" eye, 3-1/4" tie	1 Bric	1.05	7.619	M	420	269		689	870	
0500	4-3/4" tie		1.05	7.619		460	269		729	915	
0520	5-1/2" tie		1.05	7.619		475	269		744	935	
0550	4-3/4" eye, 3-1/4" tie		1.05	7.619		465	269		734	925	
0570	4-3/4" tie		1.05	7.619		510	269		779	970	
0580	5-1/2" tie		1.05	7.619	↓	525	269		794	990	
0670	3/16" diameter		10.50	.762	C	14.60	27		41.60	57	
0680	Cavity wall, Z type, galvanized, 6" long, 1/4" diameter		10.50	.762		28.50	27		55.50	72.50	
0850	3/16" diameter	↓	10.50	.762	↓	16.80	27		43.80	59.50	

Important: See the Reference Section for critical supporting data - Reference Nos., Crews, & City Cost Indexes

04080	Anchorage & Reinforcement	CREW	DAILY OUTPUT	LABOR-HOURS	UNIT	2005 BARE COSTS				TOTAL INCL O&P
						MAT.	LABOR	EQUIP.	TOTAL	
650 0855	8" long, 1/4" diameter	1 Bric	10.50	.762	C	31	27		58	75 **650**
1000	Rectangular type, galvanized, 1/4" diameter, 2" x 6"		10.50	.762		30.50	27		57.50	75
1050	4" x 6"		10.50	.762		34.50	27		61.50	79
1100	3/16" diameter, 2" x 6"		10.50	.762		21.50	27		48.50	64.50
1150	4" x 6"		10.50	.762		23.50	27		50.50	67
1200	Mesh wall tie, 1/2" mesh, hot dip galvanized									
1400	16 gauge, 12" long, 3" wide	1 Bric	9	.889	C	92.50	31.50		124	150
1420	6" wide		9	.889		131	31.50		162.50	192
1440	12" wide		8.50	.941		211	33		244	283
1500	Rigid partition anchors, plain, 8" long, 1" x 1/8"		10.50	.762		67	27		94	115
1550	1" x 1/4"		10.50	.762		132	27		159	186
1580	1-1/2" x 1/8"		10.50	.762		93	27		120	143
1600	1-1/2" x 1/4"		10.50	.762		221	27		248	284
1650	2" x 1/8"		10.50	.762		116	27		143	169
1700	2" x 1/4"		10.50	.762		315	27		342	385
2000	Column flange ties, wire, galvanized									
2300	3/16" diameter, up to 3" wide	1 Bric	10.50	.762	C	68.50	27		95.50	116
2350	To 5" wide		10.50	.762		73	27		100	122
2400	To 7" wide		10.50	.762		76.50	27		103.50	125
2600	To 9" wide		10.50	.762		81	27		108	130
2650	1/4" diameter, up to 3" wide		10.50	.762		94.50	27		121.50	145
2700	To 5" wide		10.50	.762		103	27		130	154
2800	To 7" wide		10.50	.762		108	27		135	160
2850	To 9" wide		10.50	.762		116	27		143	169
2900	For hot dip galvanized, add					35%				
4000	Channel slots, 1-3/8" x 1/2" x 8"									
4100	12 gauge, plain	1 Bric	10.50	.762	C	160	27		187	217
4150	16 gauge, galvanized	"	10.50	.762	"	99.50	27		126.50	150
4200	Channel slot anchors									
4300	16 gauge, galvanized, 1-1/4" x 3-1/2"				C	34.50			34.50	38
4350	1-1/4" x 5-1/2"					41			41	45
4400	1-1/4" x 7-1/2"					47.50			47.50	52.50
4500	1/8" plain, 1-1/4" x 3-1/2"					66			66	72.50
4550	1-1/4" x 5-1/2"					80.50			80.50	88.50
4600	1-1/4" x 7-1/2"					95.50			95.50	105
4700	For corrugation, add					45			45	49.50
4750	For hot dip galvanized, add					35%				
5000	Dowels									
5100	Plain, 1/4" diameter, 3" long				C	24			24	26
5150	4" long					27			27	29.50
5200	6" long					33			33	36.50
5300	3/8" diameter, 3" long					32			32	35.50
5350	4" long					40			40	44
5400	6" long					55.50			55.50	61
5500	1/2" diameter, 3" long					47.50			47.50	52.50
5550	4" long					59			59	64.50
5600	6" long					78.50			78.50	86
5700	5/8" diameter, 3" long					66			66	72.50
5750	4" long					82.50			82.50	91
5800	6" long					116			116	128
6000	3/4" diameter, 3" long					84			84	92.50
6100	4" long					109			109	120
6150	6" long					160			160	176
6300	For hot dip galvanized, add					35%				

04090	Masonry Accessories	CREW	DAILY OUTPUT	LABOR-HOURS	UNIT	2005 BARE COSTS				TOTAL INCL O&P	
						MAT.	LABOR	EQUIP.	TOTAL		
170	**0010** CONTROL JOINT Rubber, 4" and wider wall	1 Bric	400	.020	L.F.	1.70	.71		2.41	2.94	**170**
0050	PVC, 4" wall		400	.020		1.03	.71		1.74	2.20	
0100	Rubber, 6" wall		320	.025		2.17	.88		3.05	3.73	
0120	PVC, 6" wall		320	.025		1.28	.88		2.16	2.75	
0140	Rubber, 8" and wider wall		280	.029		2.79	1.01		3.80	4.60	
0160	PVC, 8" wall		280	.029		1.50	1.01		2.51	3.18	
0180	12" wall	↓	240	.033	↓	3.17	1.17		4.34	5.30	
900	**0010** WALL PLUGS For nailing to brickwork, 26 ga., galvanized, plain	1 Bric	10.50	.762	C	25	27		52	68.50	**900**
0050	Wood filled	"	10.50	.762	"	84	27		111	134	

04210	Clay Masonry Units	CREW	DAILY OUTPUT	LABOR-HOURS	UNIT	2005 BARE COSTS				TOTAL INCL O&P	
						MAT.	LABOR	EQUIP.	TOTAL		
300	**0010** FACE BRICK C216, TL lots, material only										**300**
0300	Standard modular, 4" x 2-2/3" x 8", minimum				M	375			375	410	
0350	Maximum					465			465	510	
0450	Economy, 4" x 4" x 8", minimum					720			720	790	
0500	Maximum					550			550	605	
0510	Economy, 4" x 4" x 12", minimum					720			720	790	
0520	Maximum					550			550	605	
0550	Jumbo, 6" x 4" x 12", minimum					1,125			1,125	1,225	
0600	Maximum					1,425			1,425	1,575	
0610	Jumbo, 8" x 4" x 12", minimum					1,125			1,125	1,225	
0620	Maximum					1,425			1,425	1,575	
0650	Norwegian, 4" x 3-1/5" x 12", minimum					720			720	790	
0700	Maximum					715			715	785	
0710	Norwegian, 6" x 3-1/5" x 12", minimum					720			720	790	
0720	Maximum					715			715	785	
0850	Standard glazed, plain colors, 4" x 2-2/3" x 8", minimum					1,150			1,150	1,275	
0900	Maximum					1,125			1,125	1,225	
1000	Deep trim shades, 4" x 2-2/3" x 8", minimum					1,150			1,150	1,250	
1050	Maximum					1,275			1,275	1,400	
1080	Jumbo utility, 4" x 4" x 12"					1,050			1,050	1,150	
1120	4" x 8" x 8"					1,075			1,075	1,175	
1140	4" x 8" x 16"					2,250			2,250	2,475	
1260	Engineer, 4" x 3-1/5" x 8", minimum					470			470	515	
1270	Maximum					485			485	535	
1350	King, 4" x 2-3/4" x 10", minimum					430			430	470	
1360	Maximum					460			460	505	
1400	Norman, 4" x 2-3/4" x 12"					430			430	470	
1450	Roman, 4" x 2" x 12"					430			430	470	
1500	SCR, 6" x 2-2/3" x 12"					430			430	470	
1550	Double, 4" x 5-1/3" x 8"					430			430	470	
1600	Triple, 4" x 5-1/3" x 12"					430			430	470	
1770	Standard modular, double glazed, 4" x 2-2/3" x 8"					970			970	1,075	
1850	Jumbo, colored glazed ceramic, 6" x 4" x 12"					1,425			1,425	1,575	
2050	Jumbo utility, glazed, 4" x 4" x 12"					1,075			1,075	1,175	
2100	4" x 8" x 8"					1,475			1,475	1,625	
2150	4" x 16" x 8"				↓	2,550			2,550	2,800	

Important: See the Reference Section for critical supporting data - Reference Nos., Crews, & City Cost Indexes

		04210	Clay Masonry Units	CREW	DAILY OUTPUT	LABOR-HOURS	UNIT	2005 BARE COSTS				TOTAL INCL O&P	
								MAT.	LABOR	EQUIP.	TOTAL		
300	2170		For less than truck load lots, add				M	10			10	11	**300**
	2180		For buff or gray brick, add					15			15	16.50	
	3050		Used brick, minimum					325			325	360	
	3100		Maximum					430			430	470	
	3150		Add for brick to match existing work, minimum					5%					
	3200		Maximum					50%					

04220 | Concrete Masonry Units

		04220	Concrete Masonry Units	CREW	DAILY OUTPUT	LABOR-HOURS	UNIT	MAT.	LABOR	EQUIP.	TOTAL	TOTAL INCL O&P	
210	0010		**CONCRETE BLOCK** material only										**210**
	0020		2" x 8" x 16" solid, normal-weight, 2,000 psi				Ea.	.61			.61	.67	
	0050		3,500 psi					.67			.67	.74	
	0100		5,000 psi					.76			.76	.84	
	0150		Lightweight, std.					.73			.73	.80	
	0300		3" x 8" x 16" solid, normal-weight, 2000 psi					.70			.70	.77	
	0350		3,500 psi					.87			.87	.96	
	0400		5,000 psi					1.18			1.18	1.30	
	0450		Lightweight, std.					.69			.69	.76	
	0600		4" x 8" x 16" hollow, normal-weight, 2000 psi					.70			.70	.77	
	0650		3,500 psi					.75			.75	.83	
	0700		5000 psi					.85			.85	.94	
	0750		Lightweight, std.					.86			.86	.95	
	1300		Solid, normal-weight, 2,000 psi					1.02			1.02	1.12	
	1350		3,500 psi					.92			.92	1.01	
	1400		5,000 psi					1.05			1.05	1.16	
	1450		Lightweight, std.					.95			.95	1.05	
	1600		6" x 8" x 16" hollow, normal-weight, 2,000 psi					1.06			1.06	1.17	
	1650		3,500 psi					1.13			1.13	1.24	
	1700		5,000 psi					1.11			1.11	1.22	
	1750		Lightweight, std.					1.18			1.18	1.30	
	2300		Solid, normal-weight, 2,000 psi					1.18			1.18	1.30	
	2350		3,500 psi					1.23			1.23	1.35	
	2400		5,000 psi					1.56			1.56	1.72	
	2450		Lightweight, std.					1.48			1.48	1.63	
	2600		8" x 8" x 16" hollow, normal-weight, 2000 psi					1.11			1.11	1.22	
	2650		3500 psi					1.28			1.28	1.41	
	2700		5,000 psi					1.32			1.32	1.45	
	2750		Lightweight, std.					1.44			1.44	1.58	
	3200		Solid, normal-weight, 2,000 psi					1.81			1.81	1.99	
	3250		3,500 psi					1.89			1.89	2.08	
	3300		5,000 psi					2.15			2.15	2.37	
	3350		Lightweight, std.					2.04			2.04	2.24	
	3400		10" x 8" x 16" hollow, normal-weight, 2000 psi					1.11			1.11	1.22	
	3410		3500 psi					1.28			1.28	1.41	
	3420		5,000 psi					1.32			1.32	1.45	
	3430		Lightweight, std.					1.44			1.44	1.58	
	3480		Solid, normal-weight, 2,000 psi					1.81			1.81	1.99	
	3490		3,500 psi					1.89			1.89	2.08	
	3500		5,000 psi					2.15			2.15	2.37	
	3510		Lightweight, std.					2.04			2.04	2.24	
	3600		12" x 8" x 16" hollow, normal-weight, 2000 psi					1.63			1.63	1.79	
	3650		3,500 psi					1.68			1.68	1.85	
	3700		5,000 psi					1.84			1.84	2.02	
	3750		Lightweight, std.					2.25			2.25	2.48	
	4300		Solid, normal-weight, 2,000 psi					2.74			2.74	3.01	
	4350		3,500 psi					2.25			2.25	2.48	
	4400		5,000 psi					3.20			3.20	3.52	

MASONRY 4

04200 | Masonry Units

		04220	Concrete Masonry Units	CREW	DAILY OUTPUT	LABOR-HOURS	UNIT	MAT.	LABOR	EQUIP.	TOTAL	TOTAL INCL O&P	
210	4500		Lightweight, std.				Ea.	3.40			3.40	3.74	210

(Header spanning: 2005 BARE COSTS over MAT., LABOR, EQUIP., TOTAL)

04500 | Refractories

04550 | Flue Liners

				CREW	DAILY OUTPUT	LABOR-HOURS	UNIT	MAT.	LABOR	EQUIP.	TOTAL	TOTAL INCL O&P	
250	0010	**FLUE LINING** Including mortar joints, 8" x 8"		D-1	125	.128	V.L.F.	3.87	3.98		7.85	10.30	250
	0100	8" x 12"			103	.155		5.45	4.83		10.28	13.35	
	0200	12" x 12"			93	.172		6.60	5.35		11.95	15.40	
	0300	12" x 18"			84	.190		11.45	5.90		17.35	21.50	
	0400	18" x 18"			75	.213		16.15	6.65		22.80	28	
	0500	20" x 20"			66	.242		34	7.55		41.55	49	
	0600	24" x 24"			56	.286		43.50	8.90		52.40	61	
	1000	Round, 18" diameter			66	.242		25	7.55		32.55	39	
	1100	24" diameter		▼	47	.340	▼	39	10.60		49.60	59	

04580 | Refractory Brick

				CREW	DAILY OUTPUT	LABOR-HOURS	UNIT	MAT.	LABOR	EQUIP.	TOTAL	TOTAL INCL O&P	
250	0010	**FIRE BRICK** 9" x 2-1/2" x 4-1/2", low duty, 2000° F		D-1	.60	26.667	M	820	830		1,650	2,175	250
	0050	High duty, 3000° F		"	.60	26.667	"	1,500	830		2,330	2,925	
260	0010	**FIRE CLAY** Gray, high duty, 100 lb. bag					Bag	42			42	46	260
	0050	100 lb. drum, premixed (400 brick per drum)					Drum	52			52	57	

04800 | Masonry Assemblies

04810 | Unit Masonry Assemblies

				CREW	DAILY OUTPUT	LABOR-HOURS	UNIT	MAT.	LABOR	EQUIP.	TOTAL	TOTAL INCL O&P	
040	0010	**ADOBE BRICK** Semi-stabilized, with cement mortar											040
	0060	Brick, 10" x 4" x 14", 2.6/S.F.		D-8	560	.071	S.F.	2.13	2.28		4.41	5.80	
	0080	12" x 4" x 16", 2.3/S.F.			580	.069		3.49	2.20		5.69	7.20	
	0100	10" x 4" x 16", 2.3/S.F.			590	.068		3.17	2.16		5.33	6.80	
	0120	8" x 4" x 16", 2.3/S.F.			560	.071		2.43	2.28		4.71	6.15	
	0140	4" x 4" x 16", 2.3/S.F.			540	.074		2.55	2.36		4.91	6.40	
	0160	6" x 4" x 16", 2.3/S.F.			540	.074		1.94	2.36		4.30	5.75	
	0180	4" x 4" x 12", 3.0/S.F.			520	.077		2.42	2.45		4.87	6.40	
	0200	8" x 4" x 12", 3.0/S.F.		▼	520	.077	▼	2.63	2.45		5.08	6.65	
050	0010	**AUTOCLAVED AERATED CONCRETE BLOCK** Scaffolding not incl											050
	0050	Solid, 4" x 12" x 24", incl mortar		D-8	600	.067	S.F.	1.84	2.13		3.97	5.25	
	0060	6" x 12" x 24"			600	.067		2.47	2.13		4.60	5.95	
	0070	8" x 8" x 24"			575	.070		3.10	2.22		5.32	6.80	
	0080	10" x 12" x 24"			575	.070		4.04	2.22		6.26	7.85	
	0090	12" x 12" x 24"		▼	550	.073	▼	4.71	2.32		7.03	8.75	
100	0010	**BRICK VENEER** Scaffolding not included, truck load lots											100
	0601	Buff or gray face, running bond, (6.75/S.F.)		D-8	1.50	26.667	M	425	850		1,275	1,775	

4 MASONRY

04810	Unit Masonry Assemblies	CREW	DAILY OUTPUT	LABOR-HOURS	UNIT	2005 BARE COSTS				TOTAL INCL O&P	
						MAT.	LABOR	EQUIP.	TOTAL		
100											**100**
0700	Glazed face, 4" x 2-2/3" x 8", running bond	D-8	1.40	28.571	M	1,300	910		2,210	2,825	
0750	Full header every 6th course (7.88/S.F.)	↓	1.35	29.630	↓	1,225	945		2,170	2,800	
1999	Alternate method of figuring by square foot										
2000	Standard, sel. common, 4" x 2-2/3" x 8", (6.75/S.F.)	D-8	230	.174	S.F.	2.88	5.55		8.43	11.60	
2020	Standard, red, 4" x 2-2/3" x 8", running bond (6.75/SF)		220	.182		2.88	5.80		8.68	12	
2600	Buff or gray face, running bond, (6.75/S.F.)		220	.182		3.06	5.80		8.86	12.20	
2700	Glazed face brick, running bond		210	.190		8.30	6.10		14.40	18.40	
2750	Full header every 6th course (7.88/S.F.)	↓	170	.235	↓	9.70	7.50		17.20	22	
3550	For curved walls, add						30%				
110	0010	**OVERSIZED BRICK**, scaffolding not included										**110**
160	0010	**CHIMNEY** See Div. 03310-240 for foundation, add to prices below										**160**
0100	Brick, 16" x 16", 8" flue, scaff. not incl.	D-1	18.20	.879	V.L.F.	15.90	27.50		43.40	59	
0150	16" x 20" with one 8" x 12" flue		16	1		25	31		56	75	
0200	16" x 24" with two 8" x 8" flues		14	1.143		36.50	35.50		72	94.50	
0250	20" x 20" with one 12" x 12" flue		13.70	1.168		28	36.50		64.50	86	
0300	20" x 24" with two 8" x 12" flues		12	1.333		41	41.50		82.50	108	
0350	20" x 32" with two 12" x 12" flues	↓	10	1.600		48	49.50		97.50	129	
1800	Metal, high temp. steel jacket, factory lining, 24" diam.	E-2	65	.862		186	32	20.50	238.50	281	
1900	60" diameter	"	30	1.867		675	69	44	788	910	
2100	Poured concrete, brick lining, 200' high x 10' diam.					5,600			5,600	6,150	
2800	500' x 20' diameter				↓	9,800			9,800	10,800	
172	0010	**CONCRETE BLOCK, BACK-UP,** C90, 2000 psi										**172**
0020	Normal weight, 8" x 16" units, tooled joint 1 side										
0050	Not-reinforced, 2000 psi, 2" thick	D-8	475	.084	S.F.	.83	2.69		3.52	5	
0200	4" thick		460	.087		.98	2.77		3.75	5.30	
0300	6" thick		440	.091		1.44	2.90		4.34	6	
0350	8" thick		400	.100		1.56	3.19		4.75	6.60	
0400	10" thick	↓	330	.121		2.19	3.87		6.06	8.30	
0450	12" thick	D-9	310	.155		2.26	4.81		7.07	9.85	
1000	Reinforced, alternate courses, 4" thick	D-8	450	.089		1.08	2.84		3.92	5.50	
1100	6" thick		430	.093		1.55	2.97		4.52	6.20	
1150	8" thick		395	.101		1.67	3.23		4.90	6.75	
1200	10" thick	↓	320	.125		2.30	3.99		6.29	8.65	
1250	12" thick	D-9	300	.160		2.38	4.97		7.35	10.20	
2000	Lightweight, not reinforced, 4" thick	D-8	460	.087		1.16	2.77		3.93	5.50	
2100	6" thick		445	.090		1.58	2.87		4.45	6.10	
2150	8" thick		435	.092		1.93	2.93		4.86	6.60	
2200	10" thick	↓	410	.098		2.53	3.11		5.64	7.50	
2250	12" thick	D-9	390	.123		2.96	3.83		6.79	9.10	
3000	Lightweight, reinforced, alternate courses, 4" thick	D-8	450	.089		1.26	2.84		4.10	5.70	
3100	6" thick		430	.093		1.68	2.97		4.65	6.35	
3150	8" thick		420	.095		2.04	3.04		5.08	6.85	
3200	10" thick	↓	400	.100		2.64	3.19		5.83	7.75	
3250	12" thick	D-9	380	.126	↓	3.08	3.93		7.01	9.40	
175	0010	**CONCRETE BLOCK BOND BEAM** C90, 2000 psi										**175**
0020	Not including grout or reinforcing										
0130	8" high, 8" thick	D-8	565	.071	L.F.	1.64	2.26		3.90	5.25	
0150	12" thick	D-9	510	.094		2.28	2.93		5.21	6.95	
0530	8" high, 8" thick	D-8	575	.070		1.53	2.22		3.75	5.05	
0550	12" thick	D-9	520	.092	↓	2.57	2.87		5.44	7.20	
2000	Including grout and 2 #5 bars										
2100	Regular block, 8" high, 8" thick	D-8	300	.133	L.F.	3.27	4.25		7.52	10.10	

MASONRY 4

	04810	**Unit Masonry Assemblies**	CREW	DAILY OUTPUT	LABOR-HOURS	UNIT	2005 BARE COSTS				TOTAL INCL O&P	
							MAT.	LABOR	EQUIP.	TOTAL		
175	2150	12" thick	D-9	250	.192	L.F.	4.33	5.95		10.28	13.85	**175**
	2500	Lightweight, 8" high, 8" thick	D-8	305	.131		3.51	4.19		7.70	10.25	
	2550	12" thick	D-9	255	.188	↓	4.62	5.85		10.47	14	
180	0010	**CONCRETE BLOCK COLUMN** or pilaster, scaffolding not included										**180**
	0050	Including vertical reinforcing (4-#4 bars) and grout										
	0160	1 piece unit, 16" x 16"	D-1	26	.615	V.L.F.	14.45	19.15		33.60	45	
	0170	2 piece units, 16" x 20"		24	.667		19.30	20.50		39.80	52.50	
	0180	20" x 20"		22	.727		28.50	22.50		51	65.50	
	0190	22" x 24"		18	.889		39.50	27.50		67	85.50	
	0200	20" x 32"	↓	14	1.143	↓	43	35.50		78.50	101	
184	0010	**CONCRETE BLOCK, EXTERIOR** C90, 2000 psi										**184**
	0020	Reinforced alt courses, tooled joints 2 sides										
	0100	Normal weight, 8" x 16" x 6" thick	D-8	395	.101	S.F.	1.67	3.23		4.90	6.75	
	0200	8" thick		360	.111		2.48	3.55		6.03	8.15	
	0250	10" thick	↓	290	.138		3	4.40		7.40	10	
	0300	12" thick	D-9	250	.192		3.01	5.95		8.96	12.40	
	0500	Lightweight, 8" x 16" x 6" thick	D-8	450	.089		1.96	2.84		4.80	6.45	
	0600	8" thick		430	.093		2.65	2.97		5.62	7.45	
	0650	10" thick	↓	395	.101		2.99	3.23		6.22	8.20	
	0700	12" thick	D-9	350	.137	↓	4.31	4.26		8.57	11.25	
186	0010	**CONCRETE BLOCK FOUNDATION WALL** C90/C145										**186**
	0050	Normal-weight, cut joints, horiz joint reinf, no vert reinf										
	0200	Hollow, 8" x 16" x 6" thick	D-8	455	.088	S.F.	1.67	2.81		4.48	6.10	
	0250	8" thick		425	.094		1.80	3		4.80	6.55	
	0300	10" thick	↓	350	.114		2.43	3.65		6.08	8.25	
	0350	12" thick	D-9	300	.160		2.51	4.97		7.48	10.35	
	0500	Solid, 8" x 16" block, 6" thick	D-8	440	.091		1.81	2.90		4.71	6.40	
	0550	8" thick	"	415	.096		2.59	3.08		5.67	7.55	
	0600	12" thick	D-9	350	.137	↓	3.76	4.26		8.02	10.65	
	1000	Reinforced, #4 vert @ 48"										
	1125	6" thick	D-8	445	.090	S.F.	2.33	2.87		5.20	6.95	
	1150	8" thick		415	.096		2.73	3.08		5.81	7.70	
	1200	10" thick	↓	340	.118		3.64	3.75		7.39	9.70	
	1250	12" thick	D-9	290	.166		3.98	5.15		9.13	12.25	
	1500	Solid, 8" x 16" block, 6" thick	D-8	430	.093		1.81	2.97		4.78	6.50	
	1600	8" thick	"	405	.099		2.59	3.15		5.74	7.65	
	1650	12" thick	D-9	340	.141	↓	3.76	4.39		8.15	10.85	
187	0010	**CONCRETE BLOCK, HIGH STRENGTH** Normal weight										**187**
	0050	Hollow, reinforced alternate courses, 8" x 16" units										
	0200	3500 psi, 4" thick	D-8	440	.091	S.F.	1.24	2.90		4.14	5.80	
	0250	6" thick		395	.101		1.64	3.23		4.87	6.70	
	0300	8" thick	↓	360	.111		2.44	3.55		5.99	8.10	
	0350	12" thick	D-9	250	.192		2.96	5.95		8.91	12.35	
	0500	5000 psi, 4" thick	D-8	440	.091		1.40	2.90		4.30	5.95	
	0550	6" thick		395	.101		2.05	3.23		5.28	7.15	
	0600	8" thick	↓	360	.111		2.77	3.55		6.32	8.45	
	0650	12" thick	D-9	300	.160	↓	4.09	4.97		9.06	12.10	
	1000	For 75% solid block, add					30%					
	1050	For 100% solid block, add					50%					
188	0010	**CONCRETE BLOCK INSULATION INSERTS**										**188**
	0100	Inserts, styrofoam, plant installed, add to block prices										
	0200	8" x 16" units, 6" thick				S.F.	.85			.85	.94	
	0250	8" thick				↓	.85			.85	.94	

Important: See the Reference Section for critical supporting data - Reference Nos., Crews, & City Cost Indexes

MASONRY **4**

		04810 | Unit Masonry Assemblies	CREW	DAILY OUTPUT	LABOR-HOURS	UNIT	2005 BARE COSTS MAT.	LABOR	EQUIP.	TOTAL	TOTAL INCL O&P	
188	0300	10" thick				S.F.	1			1	1.10	188
	0350	12" thick					1.05			1.05	1.16	
	0500	8" x 8" units, 8" thick					.70			.70	.77	
	0550	12" thick					.85			.85	.94	
225	0010	**CONCRETE BRICK** C55, grade N, type I										225
	0100	Regular, 4 x 2-1/4 x 8	D-8	220	.182	Ea.	.34	5.80		6.14	9.20	
	0125	Rusticated, 4 x 2-1/4 x 8		220	.182		.38	5.80		6.18	9.25	
	0150	Frog, 4 x 2-1/4 x 8		220	.182		.37	5.80		6.17	9.25	
	0200	Double, 4 x 4-7/8 x 8		180	.222		.60	7.10		7.70	11.45	
350	0010	**GLAZED CONCRETE BLOCK** C744										350
	0100	Single face, 8" x 16" units, 2" thick	D-8	360	.111	S.F.	5.90	3.55		9.45	11.90	
	0200	4" thick		345	.116		6	3.70		9.70	12.25	
	0250	6" thick		330	.121		6.35	3.87		10.22	12.90	
	0300	8" thick		310	.129		7.25	4.12		11.37	14.20	
	0350	10" thick		295	.136		8	4.33		12.33	15.40	
	0400	12" thick	D-9	280	.171		8.60	5.35		13.95	17.55	
	0700	Double face, 8" x 16" units, 4" thick	D-8	340	.118		9.70	3.75		13.45	16.35	
	0750	6" thick		320	.125		10.35	3.99		14.34	17.45	
	0800	8" thick		300	.133		10.70	4.25		14.95	18.30	
	1000	Jambs, bullnose or square, single face, 8" x 16", 2" thick		315	.127	Ea.	10.45	4.05		14.50	17.65	
	1050	4" thick		285	.140	"	11.65	4.48		16.13	19.65	
	1200	Caps, bullnose or square, 8" x 16", 2" thick		420	.095	L.F.	10.10	3.04		13.14	15.80	
	1250	4" thick		380	.105		13.10	3.36		16.46	19.50	
	1500	Cove base, 8" x 16", 2" thick		315	.127		5.95	4.05		10	12.70	
	1550	4" thick		285	.140		5.95	4.48		10.43	13.35	
	1600	6" thick		265	.151		6.35	4.82		11.17	14.35	
	1650	8" thick		245	.163		6.95	5.20		12.15	15.60	
650	0010	**WALLS** Building brick, including mortar										650
	0140	4" thick, facing, 4" x 2-2/3" x 8"	D-8	1.45	27.586	M	345	880		1,225	1,725	
	0150	4" thick, as back-up, 6.75 bricks per S.F.		1.60	25		345	800		1,145	1,600	
	0204	8" thick, 13.50 bricks per S.F.		1.80	22.222		350	710		1,060	1,450	
	0250	12" thick, 20.25 bricks per S.F.		1.90	21.053		350	670		1,020	1,400	
	0304	16" thick, 27.00 bricks per S.F.		2	20		355	640		995	1,350	
	0500	Reinforced, 4" wall, 4" x 2-2/3" x 8"		1.40	28.571		345	910		1,255	1,775	
	0550	8" thick, 13.50 bricks per S.F.		1.75	22.857		410	730		1,140	1,550	
	0600	12" thick, 20.25 bricks per S.F.		1.85	21.622		410	690		1,100	1,500	
	0650	16" thick, 27.00 bricks per S.F.		1.95	20.513		415	655		1,070	1,450	
	0660	4" thick, select common, face, 4" x 2-2/3" x 8"		1.45	27.586		400	880		1,280	1,800	
	0790	Alternate method of figuring by square foot										
	0800	4" wall, face, 4" x 2-2/3" x 8"	D-8	215	.186	S.F.	2.83	5.95		8.78	12.15	
	0850	4" thick, as back up, 6.75 bricks per S.F.		240	.167		2.31	5.30		7.61	10.65	
	0900	8" thick wall, 13.50 brick per S.F.		135	.296		4.73	9.45		14.18	19.60	
	1000	12" thick wall, 20.25 bricks per S.F.		95	.421		7.10	13.45		20.55	28.50	
	1050	16" thick wall, 27.00 bricks per S.F.		75	.533		9.60	17		26.60	36.50	
	1200	Reinforced, 4" x 2-2/3" x 8", 4" wall		205	.195		2.31	6.25		8.56	12.05	
	1250	8" thick wall, 13.50 brick per S.F.		130	.308		4.73	9.80		14.53	20	
	1300	12" thick wall, 20.25 bricks per S.F.		90	.444		7.10	14.20		21.30	29.50	
	1350	16" thick wall, 27.00 bricks per S.F.		70	.571		9.60	18.25		27.85	38.50	
670	0010	**STEPS** With select common brick	D-1	.30	53.333	M	355	1,650		2,005	2,925	670

04840 | Prefabricated Masonry Panels

			CREW	DAILY OUTPUT	LABOR-HOURS	UNIT	MAT.	LABOR	EQUIP.	TOTAL	TOTAL INCL O&P	
900	0010	**WALL PANELS** Prefabricated, 4" thick, minimum	C-11	775	.093	S.F.	7	3.46	1.86	12.32	15.75	900
	0100	Maximum	"	500	.144		9	5.35	2.89	17.24	22.50	

04840	Prefabricated Masonry Panels	CREW	DAILY OUTPUT	LABOR-HOURS	UNIT	2005 BARE COSTS				TOTAL INCL O&P	
						MAT.	LABOR	EQUIP.	TOTAL		
900	0200	4" brick & 2" concrete back-up, add				S.F.	50%					**900**
	0300	4" brick & 1" urethane & 3" concrete back-up, add				↓	70%					
	04850	**Stone Assemblies**										
050	0011	**ASHLAR VENEER** 4" + or - thk, random or random rectangular										**050**
	0150	Sawn face, split joints, low priced stone	D-8	140	.286	S.F.	8.40	9.10		17.50	23	
	0200	Medium priced stone		130	.308		10.90	9.80		20.70	27	
	0300	High priced stone		120	.333		13.90	10.65		24.55	31.50	
	0600	Seam face, split joints, medium price stone		125	.320		10.45	10.20		20.65	27	
	0700	High price stone		120	.333		12.75	10.65		23.40	30.50	
	1000	Split or rock face, split joints, medium price stone		125	.320		10.50	10.20		20.70	27	
	1100	High price stone	↓	120	.333	↓	12.85	10.65		23.50	30.50	
100	0010	**BLUESTONE** Cut to size										**100**
	0100	Paving, natural cleft, to 4', 1" thick	D-8	150	.267	S.F.	6.40	8.50		14.90	20	
	0150	1-1/2" thick		145	.276		5.85	8.80		14.65	19.80	
	0200	Smooth finish, 1" thick		150	.267		6.90	8.50		15.40	20.50	
	0250	1-1/2" thick		145	.276		7.35	8.80		16.15	21.50	
	0300	Thermal finish, 1" thick		150	.267		8.40	8.50		16.90	22	
	0350	1-1/2" thick	↓	145	.276	↓	9.45	8.80		18.25	24	
	1000	Stair treads, natural cleft, 12" wide, 6' long, 1-1/2" thick	D-10	115	.348	L.F.	16.55	11.30	4.40	32.25	40	
	1050	2" thick		105	.381		17.40	12.35	4.82	34.57	43	
	1100	Smooth finish, 1-1/2" thick		115	.348		20	11.30	4.40	35.70	44	
	1150	2" thick		105	.381		20	12.35	4.82	37.17	46	
	1300	Thermal finish		115	.348		25	11.30	4.40	40.70	49.50	
	1350	2" thick	↓	105	.381	↓	28	12.35	4.82	45.17	55	
300	0010	**GRANITE** Cut to size										**300**
	0050	Veneer, polished face, 3/4" to 1-1/2" thick										
	0150	Low price, gray, light gray, etc.	D-10	130	.308	S.F.	20.50	10	3.89	34.39	42	
	0220	High price, red, black, etc.	"	130	.308	"	34	10	3.89	47.89	57	
	0300	1-1/2" to 2-1/2" thick, veneer										
	0350	Low price, gray, light gray, etc.	D-10	130	.308	S.F.	22.50	10	3.89	36.39	44	
	0550	High price, red, black, etc.	"	130	.308	"	39	10	3.89	52.89	62	
	0700	2-1/2" to 4" thick, veneer										
	0750	Low price, gray, light gray, etc.	D-10	110	.364	S.F.	28	11.80	4.60	44.40	54	
	0950	High price, red, black, etc.	"	110	.364	↓	45	11.80	4.60	61.40	72.50	
	1000	For bush hammered finish, deduct					5%					
	1050	Coarse rubbed finish, deduct					10%					
	1100	Honed finish, deduct					5%					
	1150	Thermal finish, deduct				↓	18%					
	1800	Carving or bas-relief, from templates or plaster molds										
	1850	Minimum	D-10	80	.500	C.F.	128	16.20	6.35	150.55	171	
	1900	Maximum	"	80	.500	"	385	16.20	6.35	407.55	450	
	2000	Intricate or hand finished pieces										
	2010	Mouldings, radius cuts, bullnose edges, etc.										
	2050	Add, minimum					30%					
	2100	Add, maximum					300%					
	2500	Steps, copings, etc., finished on more than one surface										
	2550	Minimum	D-10	50	.800	C.F.	76.50	26	10.10	112.60	135	
	2600	Maximum	"	50	.800	"	122	26	10.10	158.10	186	
	2700	Pavers, Belgian block, 8"-13" long, 4"-6" wide, 4"-6" deep	D-11	120	.200	S.F.	19.40	6.65		26.05	31.50	
	2800	Pavers, 4" x 4" x 4" blocks, split face and joints										
	2850	Minimum	D-11	80	.300	S.F.	11.20	9.95		21.15	27.50	
	2900	Maximum	"	80	.300	"	22.50	9.95		32.45	39.50	
	3000	Pavers, 4" x 4" x 4", thermal face, sawn joints										
	3050	Minimum	D-11	65	.369	S.F.	20.50	12.25		32.75	41	

Important: See the Reference Section for critical supporting data - Reference Nos., Crews, & City Cost Indexes

04800 | Masonry Assemblies

		04850	Stone Assemblies	CREW	DAILY OUTPUT	LABOR-HOURS	UNIT	2005 BARE COSTS				TOTAL INCL O&P	
								MAT.	LABOR	EQUIP.	TOTAL		
300	3100		Maximum	D-11	65	.369	S.F.	27	12.25		39.25	48	300
	3500		Curbing, city street type, See Division 02770-300										
	3700		Slope, 4-1/2" x 12", split face,										
	3710		sawn top, 2' to 6' lengths	D-10	300	.133	L.F.	8.65	4.33	1.69	14.67	17.95	
	3800		Radius curbs, over 5' radius, add					50%					
	3850		Under 5' radius, add					100%					
	4000		Soffits, 2" thick, minimum	D-13	35	1.371	S.F.	31.50	45	14.45	90.95	119	
	4100		Maximum		35	1.371		66.50	45	14.45	125.95	157	
	4200		4" thick, minimum		35	1.371		46	45	14.45	105.45	135	
	4300		Maximum		35	1.371		85.50	45	14.45	144.95	179	
350	0011		LIGHTWEIGHT NATURAL STONE Lava type										350
	0100		Veneer, rubble face, sawed back, irregular shapes	D-10	130	.308	S.F.	5.50	10	3.89	19.39	25.50	
	0200		Sawed face and back, irregular shapes	"	130	.308	"	5.50	10	3.89	19.39	25.50	
600	0011		ROUGH STONE WALL, Dry										600
	0150		Over 18" thick	D-12	63	.508	C.F.	9.70	16.05		25.75	35	
800	0010		SLATE Pennsylvania, blue gray to gray black; Vermont,										800
	0050		Unfading green, mottled green & purple, gray & purple										
	0100		Virginia, blue black										
	0200		Exterior paving, natural cleft, 1" thick										
	0250		6" x 6" Pennsylvania	D-12	100	.320	S.F.	6.85	10.10		16.95	23	
	0300		Vermont		100	.320		6.55	10.10		16.65	22.50	
	0350		Virginia		100	.320		10.65	10.10		20.75	27	
	0500		24" x 24", Pennsylvania		120	.267		9.70	8.40		18.10	23.50	
	0550		Vermont		120	.267		12.60	8.40		21	26.50	
	0600		Virginia		120	.267		14.20	8.40		22.60	28.50	
	0700		18" x 30" Pennsylvania		120	.267		10.75	8.40		19.15	24.50	
	0750		Vermont		120	.267		11.80	8.40		20.20	26	
	0800		Virginia		120	.267		12.60	8.40		21	26.50	
	3500		Stair treads, sand finish, 1" thick x 12" wide										
	3550		Under 3 L.F.	D-10	85	.471	L.F.	15.25	15.25	5.95	36.45	46.50	
	3600		3 L.F. to 6 L.F.	"	120	.333	"	16.30	10.80	4.22	31.32	39	
900	0010		WINDOW SILL Bluestone, thermal top, 10" wide, 1-1/2" thick	D-1	85	.188	S.F.	13.50	5.85		19.35	24	900
	0050		2" thick		75	.213	"	15.75	6.65		22.40	27.50	
	0100		Cut stone, 5" x 8" plain		48	.333	L.F.	10.20	10.35		20.55	27	
	0200		Face brick on edge, brick, 8" wide		80	.200		2.15	6.20		8.35	11.80	
	0400		Marble, 9" wide, 1" thick		85	.188		7.50	5.85		13.35	17.15	
	0900		Slate, colored, unfading, honed, 12" wide, 1" thick		85	.188		15.25	5.85		21.10	25.50	
	0950		2" thick		70	.229		21.50	7.10		28.60	34.50	

04900 | Masonry Restoration and Cleaning

		04910	Unit Masonry Restoration	CREW	DAILY OUTPUT	LABOR-HOURS	UNIT	2005 BARE COSTS				TOTAL INCL O&P	
								MAT.	LABOR	EQUIP.	TOTAL		
200	0010		CAULKING MASONRY 1/2" x 1/2" joint										200
	0050		Re-caulk only, oil base	1 Bric	225	.036	L.F.	.22	1.25		1.47	2.15	
	0100		Acrylic latex		205	.039		.25	1.38		1.63	2.37	
	0200		Polyurethane		200	.040		.32	1.41		1.73	2.50	
	0300		Silicone		195	.041		.47	1.45		1.92	2.72	
	1000		Cut out and re-caulk, oil base		145	.055		.22	1.94		2.16	3.20	

04910 | Unit Masonry Restoration

			CREW	DAILY OUTPUT	LABOR-HOURS	UNIT	MAT.	LABOR	EQUIP.	TOTAL	TOTAL INCL O&P	
200	1050	Acrylic latex	1 Bric	130	.062	L.F.	.25	2.17		2.42	3.57	**200**
	1100	Polyurethane		125	.064		.32	2.26		2.58	3.79	
	1150	Silicone	▼	120	.067	▼	.47	2.35		2.82	4.10	
600	0010	**NEEDLE BEAM MASONRY** Incl. wood shoring 10′ x 10′ opening										**600**
	0400	Block, concrete, 8″ thick	B-9	7.10	5.634	Ea.	37	153	19.95	209.95	300	
	0420	12″ thick		6.70	5.970		43.50	162	21	226.50	325	
	0800	Brick, 4″ thick with 8″ backup block		5.70	7.018		43.50	190	25	258.50	370	
	1000	Brick, solid, 8″ thick		6.20	6.452		37	175	23	235	340	
	1040	12″ thick		4.90	8.163		43.50	221	29	293.50	425	
	1080	16″ thick	▼	4.50	8.889		57.50	241	31.50	330	475	
	2000	Add for additional floors of shoring	B-1	6	4	▼	37	109		146	211	
720	0010	**POINTING MASONRY**										**720**
	0300	Cut and repoint brick, hard mortar, running bond	1 Bric	80	.100	S.F.	.27	3.53		3.80	5.65	
	0320	Common bond		77	.104		.27	3.66		3.93	5.90	
	0360	Flemish bond		70	.114		.28	4.03		4.31	6.45	
	0400	English bond		65	.123		.28	4.34		4.62	6.90	
	0600	Soft old mortar, running bond		100	.080		.27	2.82		3.09	4.60	
	0620	Common bond		96	.083		.27	2.94		3.21	4.76	
	0640	Flemish bond		90	.089		.28	3.13		3.41	5.10	
	0680	English bond		82	.098	▼	.28	3.44		3.72	5.55	
	0700	Stonework, hard mortar		140	.057	L.F.	.35	2.01		2.36	3.46	
	0720	Soft old mortar		160	.050	″	.35	1.76		2.11	3.08	
	1000	Repoint, mask and grout method, running bond		95	.084	S.F.	.35	2.97		3.32	4.91	
	1020	Common bond		90	.089		.35	3.13		3.48	5.15	
	1040	Flemish bond		86	.093		.35	3.28		3.63	5.40	
	1060	English bond		77	.104		.35	3.66		4.01	6	
	2000	Scrub coat, sand grout on walls, minimum		120	.067		2.82	2.35		5.17	6.70	
	2020	Maximum	▼	98	.082	▼	2.03	2.88		4.91	6.60	
750	0010	**SAWING**										**750**
	0050	Brick or block by hand, per inch depth	D-5	300	.027	L.F.		.94		.94	1.43	
800	0010	**TOOTHING MASONRY**										**800**
	0500	Brickwork, soft old mortar	1 Clab	40	.200	V.L.F.		5.35		5.35	8.30	
	0520	Hard mortar		30	.267			7.10		7.10	11.10	
	0700	Blockwork, soft old mortar		70	.114			3.05		3.05	4.75	
	0720	Hard mortar	▼	50	.160	▼		4.27		4.27	6.65	

04930 | Unit Masonry Cleaning

			CREW	DAILY OUTPUT	LABOR-HOURS	UNIT	MAT.	LABOR	EQUIP.	TOTAL	TOTAL INCL O&P	
220	0010	**MASONRY CLEANING**										**220**
	0200	Chemical cleaning, new construction, brush and wash, minimum	D-1	1,000	.016	S.F.	.04	.50		.54	.80	
	0220	Average		800	.020		.05	.62		.67	1.01	
	0240	Maximum		600	.027		.07	.83		.90	1.34	
	0260	Light restoration, minimum		800	.020		.07	.62		.69	1.02	
	0270	Average		400	.040		.10	1.24		1.34	2	
	0280	Maximum		330	.048		.13	1.51		1.64	2.45	
	0300	Heavy restoration, minimum		600	.027		.07	.83		.90	1.34	
	0310	Average		400	.040		.11	1.24		1.35	2.01	
	0320	Maximum	▼	250	.064		.15	1.99		2.14	3.19	
	0400	High pressure water only, minimum	B-9	2,000	.020			.54	.07	.61	.92	
	0420	Average		1,500	.027			.72	.09	.81	1.22	
	0440	Maximum		1,000	.040			1.08	.14	1.22	1.85	
	0800	High pressure water and chemical, minimum		1,800	.022		.08	.60	.08	.76	1.12	
	0820	Average		1,200	.033		.12	.90	.12	1.14	1.67	
	0840	Maximum	▼	800	.050	▼	.16	1.36	.18	1.70	2.48	

04930	Unit Masonry Cleaning	CREW	DAILY OUTPUT	LABOR-HOURS	UNIT	2005 BARE COSTS				TOTAL INCL O&P	
						MAT.	LABOR	EQUIP.	TOTAL		
220	1200	Sandblast, wet system, minimum	B-9	1,750	.023	S.F.	.15	.62	.08	.85	1.22	**220**
	1220	Average		1,100	.036		.23	.99	.13	1.35	1.92	
	1240	Maximum		700	.057		.31	1.55	.20	2.06	2.97	
	1400	Dry system, minimum		2,500	.016		.15	.43	.06	.64	.90	
	1420	Average		1,750	.023		.23	.62	.08	.93	1.30	
	1440	Maximum	↓	1,000	.040		.31	1.08	.14	1.53	2.19	
	1800	For walnut shells, add					.38			.38	.42	
	1820	For corn chips, add					.38			.38	.42	
	2000	Steam cleaning, minimum	B-9	3,000	.013			.36	.05	.41	.61	
	2020	Average		2,500	.016			.43	.06	.49	.73	
	2040	Maximum	↓	1,500	.027			.72	.09	.81	1.22	
	4000	Add for masking doors and windows				↓					.80	
	4200	Add for pedestrian protection				Job					10%	
	4300	Add for common face brick				S.F.					.21	
	4400	Add for wire cut face brick				"	.13			.13	.15	
900	0010	**BRICK WASHING** Acid, smooth brick	1 Bric	560	.014	S.F.	.02	.50		.52	.80	**900**
	0050	Rough brick		400	.020		.03	.71		.74	1.11	
	0060	Stone, acid wash	↓	600	.013	↓	.04	.47		.51	.76	
	1000	Muriatic acid, price per gallon in 5 gallon lots				Gal.	4.90			4.90	5.40	

For information about Means Estimating Seminars, see yellow pages 12 and 13 in back of book

MASONRY **4**

Division Notes

	CREW	DAILY OUTPUT	LABOR-HOURS	UNIT	2005 BARE COSTS				TOTAL INCL O&P
					MAT.	LABOR	EQUIP.	TOTAL	

Division 5
Metals

Estimating Tips

05050 Basic Metal Materials & Methods

- Nuts, bolts, washers, connection angles and plates can add a significant amount to both the tonnage of a structural steel job as well as the estimated cost. As a rule of thumb add 10% to the total weight to account for these accessories.
- Type 2 steel construction, commonly referred to as "simple construction," consists generally of field bolted connections with lateral bracing supplied by other elements of the building, such as masonry walls or x-bracing. The estimator should be aware, however, that shop connections may be accomplished by welding or bolting. The method may be particular to the fabrication shop and may have an impact on the estimated cost.

05200 Metal Joists

- In any given project the total weight of open web steel joists is determined by the loads to be supported and the design. However, economies can be realized in minimizing the amount of labor used to place the joists. This is done by maximizing the joist spacing and therefore minimizing the number of joists required to be installed on the job. Certain spacings and locations may be required by the design, but in other cases maximizing the spacing and keeping it as uniform as possible will keep the costs down.

05300 Metal Deck

- The takeoff and estimating of metal deck involves more than simply the area of the floor or roof and the type of deck specified or shown on the drawings. Many different sizes and types of openings may exist. Small openings for individual pipes or conduits may be drilled after the floor/roof is installed, but larger openings may require special deck lengths as well as reinforcing or structural support. The estimator should determine who will be supplying this reinforcing. Additionally, some deck terminations are part of the deck package, such as screed angles and pour stops, and others will be part of the steel contract, such as angles attached to structural members and cast-in-place angles and plates. The estimator must ensure that all pieces are accounted for in the complete estimate.

05500 Metal Fabrications

- The most economical steel stairs are those that use common materials, standard details and most importantly, a uniform and relatively simple method of field assembly. Commonly available A36 channels and plates are very good choices for the main stringers of the stairs, as are angles and tees for the carrier members. Risers and treads are usually made by specialty shops, and it is most economical to use a typical detail in as many places as possible. The stairs should be pre-assembled and shipped directly to the site. The field connections should be simple and straightforward to be accomplished efficiently and with a minimum of equipment and labor.

Reference Numbers

Reference numbers are shown in bold squares at the beginning of some major classifications. These numbers refer to related items in the Reference Section. The reference information may be an estimating procedure, an alternate pricing method or technical information.

Note: Not all subdivisions listed here necessarily appear in this publication.

05060 | Selective Demolition

		CREW	DAILY OUTPUT	LABOR-HOURS	UNIT	2005 BARE COSTS MAT.	2005 BARE COSTS LABOR	2005 BARE COSTS EQUIP.	2005 BARE COSTS TOTAL	TOTAL INCL O&P	
110	0010	**SELECTIVE METALS DEMOLITION**										110
	0015	Excludes shores, bracing, cutting, loading, hauling, dumping	R02220 -510									
	0020	Remove nuts only up to 3/4" diameter	1 Sswk	480	.017	Ea.		.64		.64	1.14	
	0030	7/8" to 1-1/4" diameter		240	.033			1.27		1.27	2.27	
	0040	1-3/8" to 2" diameter		160	.050			1.91		1.91	3.41	
	0060	Unbolt and remove structural bolts up to 3/4" diameter		240	.033			1.27		1.27	2.27	
	0070	7/8" to 2" diameter		160	.050			1.91		1.91	3.41	
	0140	Light weight framing members, remove whole or cut up, up to 20 lb		240	.033			1.27		1.27	2.27	
	0150	21 - 40 lb	2 Sswk	210	.076			2.91		2.91	5.20	
	0160	41 - 80 lb	3 Sswk	180	.133			5.10		5.10	9.10	
	0170	81 - 120 lb	4 Sswk	150	.213			8.15		8.15	14.55	
	0230	Structural members, remove whole or cut up, up to 500 lb	E-19	48	.500			18.55	16.70	35.25	50	
	0240	1/4 - 2 tons	E-18	36	1.111			42	22	64	97.50	
	0250	2 - 5 tons	E-24	30	1.067			40	20.50	60.50	91	
	0260	5 - 10 tons	E-20	24	2.667			99	40	139	215	
	0270	10 - 15 tons	E-2	18	3.111			115	73.50	188.50	278	
	0340	Fabricated item, remove whole or cut up, up to 20 lb	1 Sswk	96	.083			3.18		3.18	5.70	
	0350	21 - 40 lb	2 Sswk	84	.190			7.25		7.25	13	
	0360	41 - 80 lb	3 Sswk	72	.333			12.70		12.70	23	
	0370	81 - 120 lb	4 Sswk	60	.533			20.50		20.50	36.50	
	0380	121 - 500 lb	E-19	48	.500			18.55	16.70	35.25	50	
	2000	Steel framing, beams, 4" x 6"	B-13	500	.112	L.F.		3.22	1.23	4.45	6.35	

05090 | Metal Fastenings

		CREW	DAILY OUTPUT	LABOR-HOURS	UNIT	MAT.	LABOR	EQUIP.	TOTAL	TOTAL INCL O&P	
080	0010	**ANCHOR BOLTS**										080
	0020	See also divisions 03150-080 and 04080-070										
	0100	J-type, incl. hex nut & washer, 1/2" diameter x 6" long	2 Carp	70	.229	Ea.	.94	7.85		8.79	13.25	
	0110	12" long		65	.246		1.17	8.45		9.62	14.45	
	0120	18" long		60	.267		1.52	9.15		10.67	15.90	
	0130	3/4" diameter x 8" long		50	.320		1.39	10.95		12.34	18.60	
	0140	12" long		45	.356		1.74	12.20		13.94	21	
	0150	18" long		40	.400		2.26	13.70		15.96	24	
	0160	1" diameter x 12" long		35	.457		3.32	15.65		18.97	28	
	0170	18" long		30	.533		4	18.25		22.25	33	
	0180	24" long		25	.640		4.88	22		26.88	39.50	
	0190	36" long		20	.800		6.70	27.50		34.20	50	
	0200	1-1/2" diameter x 18" long		22	.727		11.85	25		36.85	52	
	0210	24" long		16	1		14.10	34.50		48.60	69	
	0300	L-type, incl. hex nut & washer, 3/4" diameter x 12" long		45	.356		1.31	12.20		13.51	20.50	
	0310	18" long		40	.400		1.67	13.70		15.37	23.50	
	0320	24" long		35	.457		2.04	15.65		17.69	26.50	
	0330	30" long		30	.533		2.59	18.25		20.84	31.50	
	0340	36" long		25	.640		2.96	22		24.96	37.50	
	0350	1" diameter x 12" long		35	.457		2.28	15.65		17.93	27	
	0360	18" long		30	.533		2.84	18.25		21.09	31.50	
	0370	24" long		25	.640		3.50	22		25.50	38	
	0380	30" long		23	.696		4.13	24		28.13	41.50	
	0390	36" long		20	.800		4.73	27.50		32.23	47.50	
	0400	42" long		18	.889		5.75	30.50		36.25	54	
	0410	48" long		15	1.067		6.45	36.50		42.95	64	
	0420	1-1/4" diameter x 18" long		25	.640		5.10	22		27.10	39.50	
	0430	24" long		22	.727		6.05	25		31.05	45.50	
	0440	30" long		20	.800		7	27.50		34.50	50	
	0450	36" long		18	.889		8	30.50		38.50	56.50	
	0460	42" long		16	1		9	34.50		43.50	63.50	
	0470	48" long		14	1.143		10.30	39		49.30	72.50	

5 METALS

05090	Metal Fastenings	CREW	DAILY OUTPUT	LABOR-HOURS	UNIT	2005 BARE COSTS				TOTAL INCL O&P
						MAT.	LABOR	EQUIP.	TOTAL	
080 0480	54" long	2 Carp	12	1.333	Ea.	12.15	45.50		57.65	84.50 **080**
0490	60" long		10	1.600		13.35	55		68.35	100
0500	1-1/2" diameter x 18" long		22	.727		7.85	25		32.85	47.50
0510	24" long		19	.842		9.15	29		38.15	55
0520	30" long		17	.941		10.35	32		42.35	61.50
0530	36" long		16	1		11.90	34.50		46.40	66.50
0540	42" long		15	1.067		13.55	36.50		50.05	72
0550	48" long		13	1.231		15.25	42		57.25	82.50
0560	54" long		11	1.455		18.55	50		68.55	98
0570	60" long		9	1.778		20.50	61		81.50	118
0580	1-3/4" diameter x 18" long		20	.800		11.85	27.50		39.35	55.50
0590	24" long		18	.889		13.90	30.50		44.40	63
0600	30" long		17	.941		16.20	32		48.20	68
0610	36" long		16	1		18.45	34.50		52.95	74
0620	42" long		14	1.143		20.50	39		59.50	84
0630	48" long		12	1.333		23	45.50		68.50	96
0640	54" long		10	1.600		28.50	55		83.50	117
0650	60" long		8	2		30.50	68.50		99	141
0660	2" diameter x 24" long		17	.941		17.70	32		49.70	69.50
0670	30" long		15	1.067		19.95	36.50		56.45	79
0680	36" long		13	1.231		22	42		64	89.50
0690	42" long		11	1.455		24.50	50		74.50	105
0700	48" long		10	1.600		28	55		83	117
0710	54" long		9	1.778		33.50	61		94.50	132
0720	60" long		8	2		36	68.50		104.50	147
0730	66" long		7	2.286		38.50	78.50		117	165
0740	72" long	▼	6	2.667		42	91.50		133.50	189
0990	For galvanized, add				▼	75%				
150 0010	**BOLTS & HEX NUTS** Steel, A307									**150**
0100	1/4" diameter, 1/2" long	1 Sswk	140	.057	Ea.	.06	2.18		2.24	3.96
0200	1" long		140	.057		.07	2.18		2.25	3.97
0300	2" long		130	.062		.09	2.35		2.44	4.30
0400	3" long		130	.062		.13	2.35		2.48	4.35
0500	4" long		120	.067		.15	2.54		2.69	4.71
0600	3/8" diameter, 1" long		130	.062		.11	2.35		2.46	4.32
0700	2" long		130	.062		.14	2.35		2.49	4.35
0800	3" long		120	.067		.18	2.54		2.72	4.75
0900	4" long		120	.067		.23	2.54		2.77	4.80
1000	5" long		115	.070		.28	2.65		2.93	5.05
1100	1/2" diameter, 1-1/2" long		120	.067		.21	2.54		2.75	4.78
1200	2" long		120	.067		.24	2.54		2.78	4.81
1300	4" long		115	.070		.36	2.65		3.01	5.15
1400	6" long		110	.073		.49	2.77		3.26	5.50
1500	8" long		105	.076		.64	2.91		3.55	5.90
1600	5/8" diameter, 1-1/2" long		120	.067		.40	2.54		2.94	4.99
1700	2" long		120	.067		.44	2.54		2.98	5.05
1800	4" long		115	.070		.61	2.65		3.26	5.40
1900	6" long		110	.073		.77	2.77		3.54	5.80
2000	8" long		105	.076		1.12	2.91		4.03	6.45
2100	10" long		100	.080		1.39	3.05		4.44	7
2200	3/4" diameter, 2" long		120	.067		.63	2.54		3.17	5.25
2300	4" long		110	.073		.88	2.77		3.65	5.95
2400	6" long		105	.076		1.12	2.91		4.03	6.45
2500	8" long		95	.084		1.66	3.21		4.87	7.60
2600	10" long		85	.094		2.16	3.59		5.75	8.80
2700	12" long	▼	80	.100	▼	2.52	3.82		6.34	9.60

METALS 5

05090 | Metal Fastenings

		CREW	DAILY OUTPUT	LABOR-HOURS	UNIT	2005 BARE COSTS				TOTAL INCL O&P	
						MAT.	LABOR	EQUIP.	TOTAL		
150	2800	1" diameter, 3" long	1 Sswk	105	.076	Ea.	1.65	2.91		4.56	7	**150**
	2900	6" long		90	.089		2.54	3.39		5.93	8.85	
	3000	12" long	↓	75	.107		4.78	4.07		8.85	12.55	
	3100	For galvanized, add					75%					
	3200	For stainless, add				↓	350%					
300	0010	**CHEMICAL ANCHORS**, Includes layout & drilling										**300**
	1430	Chemical anchor, w/rod & epoxy cartridge, 3/4" diam. x 9-1/2" long	B-89A	27	.593	Ea.	11.65	18.25	3.82	33.72	45.50	
	1435	1" diameter x 11-3/4" long		24	.667		22.50	20.50	4.29	47.29	61.50	
	1440	1-1/4" diameter x 14" long		21	.762		43	23.50	4.91	71.41	89	
	1445	1-3/4" diameter x 15" long		20	.800		81	24.50	5.15	110.65	134	
	1450	18" long		17	.941		97.50	29	6.05	132.55	159	
	1455	2" diameter x 18" long		16	1		124	31	6.45	161.45	192	
	1460	24" long	↓	15	1.067	↓	162	33	6.85	201.85	238	
	1500	Chemical anchoring, epoxy cartridge, excludes layout, drilling, fastener										
	1530	For fastener 3/4" dia x 6" embedment	B-89A	27	.593	Ea.	4.58	18.25	3.82	26.65	38	
	1535	1" dia x 8" embedment		24	.667		6.85	20.50	4.29	31.64	44.50	
	1540	1-1/4" dia x 10" embedment		21	.762		13.75	23.50	4.91	42.16	57	
	1545	1-3/4" dia x 12" embedment		20	.800		23	24.50	5.15	52.65	69	
	1550	14" embedment		17	.941		27.50	29	6.05	62.55	81.50	
	1555	2" dia x 12" embedment		16	1		36.50	31	6.45	73.95	95.50	
	1560	18" embedment	↓	15	1.067	↓	46	33	6.85	85.85	109	
340	0010	**DRILLING** For anchors, up to 4" deep, incl. bit and layout										**340**
	0050	in concrete or brick walls and floors, no anchor										
	0100	Holes, 1/4" diameter	1 Carp	75	.107	Ea.	.08	3.65		3.73	5.80	
	0150	For each additional inch of depth, add		430	.019		.02	.64		.66	1.01	
	0200	3/8" diameter		63	.127		.08	4.35		4.43	6.85	
	0250	For each additional inch of depth, add		340	.024		.02	.81		.83	1.28	
	0300	1/2" diameter		50	.160		.08	5.50		5.58	8.65	
	0350	For each additional inch of depth, add		250	.032		.02	1.10		1.12	1.73	
	0400	5/8" diameter		48	.167		.15	5.70		5.85	9.05	
	0450	For each additional inch of depth, add		240	.033		.04	1.14		1.18	1.82	
	0500	3/4" diameter		45	.178		.18	6.10		6.28	9.70	
	0550	For each additional inch of depth, add		220	.036		.04	1.25		1.29	1.99	
	0600	7/8" diameter		43	.186		.21	6.35		6.56	10.20	
	0650	For each additional inch of depth, add		210	.038		.05	1.30		1.35	2.09	
	0700	1" diameter		40	.200		.24	6.85		7.09	10.90	
	0750	For each additional inch of depth, add		190	.042		.06	1.44		1.50	2.32	
	0800	1-1/4" diameter		38	.211		.35	7.20		7.55	11.65	
	0850	For each additional inch of depth, add		180	.044		.09	1.52		1.61	2.47	
	0900	1-1/2" diameter		35	.229		.53	7.85		8.38	12.80	
	0950	For each additional inch of depth, add	↓	165	.048	↓	.13	1.66		1.79	2.73	
	1000	For ceiling installations, add						40%				
	1100	Drilling & layout for drywall/plaster walls, up to 1" deep, no anchor										
	1200	Holes, 1/4" diameter	1 Carp	150	.053	Ea.	.01	1.83		1.84	2.86	
	1300	3/8" diameter		140	.057		.01	1.96		1.97	3.06	
	1400	1/2" diameter		130	.062		.01	2.11		2.12	3.29	
	1500	3/4" diameter		120	.067		.02	2.28		2.30	3.58	
	1600	1" diameter		110	.073		.03	2.49		2.52	3.91	
	1700	1-1/4" diameter		100	.080		.04	2.74		2.78	4.32	
	1800	1-1/2" diameter	↓	90	.089	↓	.07	3.04		3.11	4.81	
	1900	For ceiling installations, add						40%				
	1910	Drilling & layout for steel, up to 1/4" deep, no anchor										
	1920	Holes, 1/4" diameter	1 Sswk	112	.071	Ea.	.11	2.73		2.84	5	
	1925	For each additional 1/4" depth, add		336	.024		.11	.91		1.02	1.75	
	1930	3/8" diameter	↓	104	.077	↓	.13	2.93		3.06	5.40	

Important: See the Reference Section for critical supporting data - Reference Nos., Crews, & City Cost Indexes

	05090	Metal Fastenings	CREW	DAILY OUTPUT	LABOR-HOURS	UNIT	2005 BARE COSTS				TOTAL INCL O&P	
							MAT.	LABOR	EQUIP.	TOTAL		
340	1935	For each additional 1/4" depth, add	1 Sswk	312	.026	Ea.	.13	.98		1.11	1.89	340
	1940	1/2" diameter		96	.083		.14	3.18		3.32	5.85	
	1945	For each additional 1/4" depth, add		288	.028		.14	1.06		1.20	2.06	
	1950	5/8" diameter		88	.091		.24	3.47		3.71	6.45	
	1955	For each additional 1/4" depth, add		264	.030		.24	1.16		1.40	2.33	
	1960	3/4" diameter		80	.100		.26	3.82		4.08	7.15	
	1965	For each additional 1/4" depth, add		240	.033		.26	1.27		1.53	2.56	
	1970	7/8" diameter		72	.111		.31	4.24		4.55	7.95	
	1975	For each additional 1/4" depth, add		216	.037		.31	1.41		1.72	2.87	
	1980	1" diameter		64	.125		.35	4.77		5.12	8.95	
	1985	For each additional 1/4" depth, add	↓	192	.042	↓	.35	1.59		1.94	3.23	
	1990	For drilling up, add						40%				
380	0010	**EXPANSION ANCHORS** & shields										380
	0100	Bolt anchors for concrete, brick or stone, no layout and drilling										
	0200	Expansion shields, zinc, 1/4" diameter, 1-5/16" long, single	1 Carp	90	.089	Ea.	1.01	3.04		4.05	5.85	
	0300	1-3/8" long, double		85	.094		1.11	3.22		4.33	6.20	
	0400	3/8" diameter, 1-1/2" long, single		85	.094		1.66	3.22		4.88	6.85	
	0500	2" long, double		80	.100		2.05	3.43		5.48	7.60	
	0600	1/2" diameter, 2-1/16" long, single		80	.100		2.75	3.43		6.18	8.40	
	0700	2-1/2" long, double		75	.107		2.65	3.65		6.30	8.60	
	0800	5/8" diameter, 2-5/8" long, single		75	.107		3.93	3.65		7.58	10	
	0900	2-3/4" long, double		70	.114		3.93	3.91		7.84	10.40	
	1000	3/4" diameter, 2-3/4" long, single		70	.114		5.85	3.91		9.76	12.50	
	1100	3-15/16" long, double		65	.123		7.80	4.22		12.02	15.10	
	1500	Self drilling anchor, snap-off, for 1/4" diameter bolt		26	.308		.83	10.55		11.38	17.30	
	1600	3/8" diameter bolt		23	.348		1.20	11.90		13.10	19.85	
	1700	1/2" diameter bolt		20	.400		1.84	13.70		15.54	23.50	
	1800	5/8" diameter bolt		18	.444		3.08	15.20		18.28	27	
	1900	3/4" diameter bolt		16	.500		5.20	17.15		22.35	32	
	5700	Lag screw shields, 1/4" diameter, short		90	.089		.44	3.04		3.48	5.20	
	5800	Long		85	.094		.51	3.22		3.73	5.55	
	5900	3/8" diameter, short		85	.094		.80	3.22		4.02	5.90	
	6000	Long		80	.100		.94	3.43		4.37	6.40	
	6100	1/2" diameter, short		80	.100		1.11	3.43		4.54	6.55	
	6200	Long		75	.107		1.39	3.65		5.04	7.25	
	6300	3/4" diameter, short		70	.114		3.12	3.91		7.03	9.55	
	6400	Long		65	.123		3.78	4.22		8	10.70	
	6600	Lead, #6 & #8, 3/4" long		260	.031		.16	1.05		1.21	1.82	
	6700	#10 - #14, 1-1/2" long		200	.040		.23	1.37		1.60	2.38	
	6800	#16 & #18, 1-1/2" long		160	.050		.31	1.71		2.02	3.01	
	6900	Plastic, #6 & #8, 3/4" long		260	.031		.10	1.05		1.15	1.75	
	7000	#8 & #10, 7/8" long		240	.033		.04	1.14		1.18	1.82	
	7100	#10 & #12, 1" long		220	.036		.13	1.25		1.38	2.08	
	7200	#14 & #16, 1-1/2" long	↓	160	.050	↓	.07	1.71		1.78	2.75	
	8000	Wedge anchors, not including layout or drilling										
	8050	Carbon steel, 1/4" diameter, 1-3/4" long	1 Carp	150	.053	Ea.	.42	1.83		2.25	3.31	
	8100	3 1/4" long		140	.057		.55	1.96		2.51	3.66	
	8150	3/8" diameter, 2-1/4" long		145	.055		.63	1.89		2.52	3.63	
	8200	5" long		140	.057		1.10	1.96		3.06	4.26	
	8250	1/2" diameter, 2-3/4" long		140	.057		.96	1.96		2.92	4.11	
	8300	7" long		125	.064		1.65	2.19		3.84	5.20	
	8350	5/8" diameter, 3-1/2" long		130	.062		1.90	2.11		4.01	5.35	
	8400	8-1/2" long		115	.070		4.05	2.38		6.43	8.15	
	8450	3/4" diameter, 4-1/4" long		115	.070		2.29	2.38		4.67	6.25	
	8500	10" long		95	.084		5.20	2.88		8.08	10.25	
	8550	1" diameter, 6" long	↓	100	.080		7.70	2.74		10.44	12.70	

METALS 5

187

05090 | Metal Fastenings

		CREW	DAILY OUTPUT	LABOR-HOURS	UNIT	2005 BARE COSTS MAT.	LABOR	EQUIP.	TOTAL	TOTAL INCL O&P	
380	8575	9" long	1 Carp	85	.094	Ea.	10	3.22		13.22	16	**380**
	8600	12" long		75	.107		10.80	3.65		14.45	17.55	
	8650	1-1/4" diameter, 9" long		70	.114		14	3.91		17.91	21.50	
	8700	12" long	↓	60	.133	↓	17.90	4.57		22.47	27	
	8750	For type 303 stainless steel, add					350%					
	8800	For type 316 stainless steel, add					450%					
	8950	Self-drilling concrete screw, hex washer head, 3/16" dia x 1-3/4" long	1 Carp	300	.027	Ea.	.32	.91		1.23	1.77	
	8960	2-1/4" long		250	.032		.49	1.10		1.59	2.25	
	8970	Phillips flat head, 3/16" dia x 1-3/4" long		300	.027		.33	.91		1.24	1.78	
	8980	2-1/4" long		250	.032	↓	.48	1.10		1.58	2.24	
460	0010	**LAG SCREWS**										**460**
	0020	Steel, 1/4" diameter, 2" long	1 Carp	200	.040	Ea.	.08	1.37		1.45	2.22	
	0100	3/8" diameter, 3" long		150	.053		.23	1.83		2.06	3.10	
	0200	1/2" diameter, 3" long		130	.062		.38	2.11		2.49	3.70	
	0300	5/8" diameter, 3" long	↓	120	.067	↓	.75	2.28		3.03	4.39	
540	0010	**MACHINERY ANCHORS**, heavy duty, incl. sleeve, floating base nut,										**540**
	0020	lower stud & coupling nut, fiber plug, connecting stud, washer & nut.										
	0030	For flush mounted embedment in poured concrete heavy equip. pads.										
	0200	Material only, 1/2" diameter stud & bolt				Ea.	52.50			52.50	58	
	0300	5/8" diameter					58.50			58.50	64	
	0500	3/4" diameter					67.50			67.50	74	
	0600	7/8" diameter					73.50			73.50	81	
	0800	1" diameter					77.50			77.50	85	
	0900	1-1/4" diameter				↓	103			103	113	
580	0010	**POWDER ACTUATED** Tools & fasteners										**580**
	0020	Stud driver, .22 caliber, buy, minimum				Ea.	310			310	340	
	0100	Maximum				"	495			495	545	
	0300	Powder charges for above, low velocity				C	16.60			16.60	18.25	
	0400	Standard velocity					23.50			23.50	26	
	0600	Drive pins & studs, 1/4" & 3/8" diam., to 3" long, minimum					11.85			11.85	13.05	
	0700	Maximum				↓	46.50			46.50	51	
	0800	Pneumatic stud driver for 1/8" diameter studs				Ea.	2,150			2,150	2,350	
	0900	Drive pins for above, 1/2" to 3/4" long				M	490			490	540	
600	0010	**RIVETS**										**600**
	0100	Aluminum rivet & mandrel, 1/2" grip length x 1/8" diameter				C	5			5	5.50	
	0200	3/16" diameter					7.70			7.70	8.45	
	0300	Aluminum rivet, steel mandrel, 1/8" diameter					7.30			7.30	8.05	
	0400	3/16" diameter					6.70			6.70	7.35	
	0500	Copper rivet, steel mandrel, 1/8" diameter					6.65			6.65	7.30	
	0600	Monel rivet, steel mandrel, 1/8" diameter					23.50			23.50	26	
	0700	3/16" diameter					68			68	75	
	0800	Stainless rivet & mandrel, 1/8" diameter					11.90			11.90	13.05	
	0900	3/16" diameter					22			22	24.50	
	1000	Stainless rivet, steel mandrel, 1/8" diameter					9.25			9.25	10.20	
	1100	3/16" diameter					16.75			16.75	18.45	
	1200	Steel rivet and mandrel, 1/8" diameter					5.75			5.75	6.30	
	1300	3/16" diameter				↓	8.60			8.60	9.50	
	1400	Hand riveting tool, minimum				Ea.	111			111	122	
	1500	Maximum					212			212	233	
	1600	Power riveting tool, minimum					800			800	880	
	1700	Maximum				↓	2,025			2,025	2,225	
820	0010	**VIBRATION PADS**										**820**
	0300	Laminated synthetic rubber impregnated cotton duck, 1/2" thick	2 Sswk	24	.667	S.F.	53.50	25.50		79	105	

Important: See the Reference Section for critical supporting data - Reference Nos., Crews, & City Cost Indexes

05090	Metal Fastenings	CREW	DAILY OUTPUT	LABOR-HOURS	UNIT	2005 BARE COSTS				TOTAL INCL O&P	
						MAT.	LABOR	EQUIP.	TOTAL		
820	0400	1" thick	2 Sswk	20	.800	S.F.	108	30.50		138.50	174	**820**
	0600	Neoprene bearing pads, 1/2" thick		24	.667		20	25.50		45.50	68	
	0700	1" thick		20	.800		41.50	30.50		72	100	
	0900	Fabric reinforced neoprene, 5000 psi, 1/2" thick		24	.667		9.10	25.50		34.60	55.50	
	1000	1" thick		20	.800		18.20	30.50		48.70	74.50	
	1200	Felt surfaced vinyl pads, cork and sisal, 5/8" thick		24	.667		23.50	25.50		49	71.50	
	1300	1" thick		20	.800		42.50	30.50		73	102	
	1600	3/32" layer		24	.667		63.50	25.50		89	115	
	1800	Bonded to 10 ga. stainless steel, 1/32" layer		24	.667		75	25.50		100.50	128	
	1900	3/32" layer	▼	24	.667	▼	97.50	25.50		123	153	
	2100	Circular machine leveling pad & stud				Kip	6.15			6.15	6.75	
840	0010	**WELD SHEAR CONNECTORS**										**840**
	0020	3/4" diameter, 3-3/16" long	E-10	960	.017	Ea.	.37	.65	.27	1.29	1.88	
	0030	3-3/8" long		950	.017		.39	.66	.27	1.32	1.91	
	0200	3-7/8" long		945	.017		.42	.66	.27	1.35	1.95	
	0300	4-3/16" long		935	.017		.44	.67	.28	1.39	1.98	
	0500	4-7/8" long		930	.017		.49	.67	.28	1.44	2.04	
	0600	5-3/16" long		920	.017		.51	.68	.28	1.47	2.09	
	0800	5-3/8" long		910	.018		.51	.69	.28	1.48	2.11	
	0900	6-3/16" long		905	.018		.56	.69	.28	1.53	2.17	
	1000	7-3/16" long		895	.018		.70	.70	.29	1.69	2.34	
	1100	8-3/16" long		890	.018		.77	.70	.29	1.76	2.42	
	1500	7/8" diameter, 3-11/16" long		920	.017		.61	.68	.28	1.57	2.20	
	1600	4-3/16" long		910	.018		.65	.69	.28	1.62	2.26	
	1700	5-3/16" long		905	.018		.74	.69	.28	1.71	2.36	
	1800	6-3/16" long		895	.018		.83	.70	.29	1.82	2.48	
	1900	7-3/16" long		890	.018		.92	.70	.29	1.91	2.59	
	2000	8-3/16" long	▼	880	.018	▼	1	.71	.29	2	2.69	
860	0010	**WELD STUDS**										**860**
	0020	1/4" diameter, 2-11/16" long	E-10	1,120	.014	Ea.	.24	.56	.23	1.03	1.52	
	0100	4-1/8" long		1,080	.015		.23	.58	.24	1.05	1.55	
	0200	3/8" diameter, 4-1/8" long		1,080	.015		.26	.58	.24	1.08	1.59	
	0300	6-1/8" long		1,040	.015		.34	.60	.25	1.19	1.72	
	0400	1/2" diameter, 2-1/8" long		1,040	.015		.25	.60	.25	1.10	1.63	
	0500	3-1/8" long		1,025	.016		.31	.61	.25	1.17	1.71	
	0600	4-1/8" long		1,010	.016		.36	.62	.25	1.23	1.78	
	0700	5-5/16" long		990	.016		.44	.63	.26	1.33	1.90	
	0800	6-1/8" long		975	.016		.48	.64	.26	1.38	1.96	
	0900	8-1/8" long		960	.017		.67	.65	.27	1.59	2.21	
	1000	5/8" diameter, 2-11/16" long		1,000	.016		.43	.63	.26	1.32	1.87	
	1010	4-3/16" long		990	.016		.53	.63	.26	1.42	2	
	1100	6-9/16" long		975	.016		.69	.64	.26	1.59	2.20	
	1200	8-3/16" long	▼	960	.017	▼	.92	.65	.27	1.84	2.49	
880	0010	**WELD ROD**										**880**
	0020	Steel, type 6011, 1/8" dia, less than 500#				Lb.	1.72			1.72	1.89	
	0100	500# to 2,000#					1.55			1.55	1.71	
	0200	2,000# to 5,000#					1.46			1.46	1.60	
	0300	5/32" diameter, less than 500#					1.65			1.65	1.82	
	0310	500# to 2,000#					1.49			1.49	1.64	
	0320	2,000# to 5,000#					1.40			1.40	1.54	
	0400	3/16" dia, less than 500#					1.68			1.68	1.84	
	0500	500# to 2,000#					1.51			1.51	1.66	
	0600	2,000# to 5,000#	▼			▼	1.42			1.42	1.56	

METALS 5

05090 | Metal Fastenings

			CREW	DAILY OUTPUT	LABOR-HOURS	UNIT	2005 BARE COSTS				TOTAL INCL O&P	
							MAT.	LABOR	EQUIP.	TOTAL		
880	0620	Steel, type 6010, 1/8" dia, less than 500#				Lb.	1.90			1.90	2.09	880
	0630	500# to 2,000#					1.71			1.71	1.88	
	0640	2,000# to 5,000#					1.61			1.61	1.77	
	0650	Steel, type 7018 Low Hydrogen, 1/8" dia, less than 500#					1.62			1.62	1.78	
	0660	500# to 2,000#					1.46			1.46	1.61	
	0670	2,000# to 5,000#					1.37			1.37	1.51	
	0700	Steel, type 7024 Jet Weld, 1/8" dia, less than 500#					1.72			1.72	1.89	
	0710	500# to 2,000#					1.55			1.55	1.71	
	0720	2,000# to 5,000#					1.46			1.46	1.60	
	1550	Aluminum, type 4043 TIG, 1/8" dia, less than 10#					5.95			5.95	6.55	
	1560	10# to 60#					5.35			5.35	5.90	
	1570	Over 60#					5.05			5.05	5.55	
	1600	Aluminum, type 5356 TIG, 1/8" dia, less than 10#					6.60			6.60	7.25	
	1610	10# to 60#					5.95			5.95	6.55	
	1620	Over 60#					5.60			5.60	6.15	
	1900	Cast iron, type 8 Nickel, 1/8" dia, less than 500#					20.50			20.50	22.50	
	1910	500# to 1,000#					18.35			18.35	20	
	1920	Over 1,000#					17.25			17.25	18.95	
	2000	Stainless steel, type 316/316L, 1/8" dia, less than 500#					9.40			9.40	10.30	
	2100	500# to 1000#					8.45			8.45	9.30	
	2220	Over 1000#					7.95			7.95	8.75	
900	0010	**WELDING STRUCTURAL**										900
	0020	Field welding, 1/8" E6011, cost per welder, no oper. engr R05090 -520	E-14	8	1	Hr.	3.44	40	10.15	53.59	87	
	0200	With 1/2 operating engineer	E-13	8	1.500		3.44	56.50	10.15	70.09	112	
	0300	With 1 operating engineer	E-12	8	2		3.44	73	10.15	86.59	137	
	0500	With no operating engineer, 2# weld rod per ton	E-14	8	1	Ton	3.44	40	10.15	53.59	87	
	0600	8# E6011 per ton	"	2	4		13.75	161	40.50	215.25	345	
	0800	With one operating engineer per welder, 2# E6011 per ton	E-12	8	2		3.44	73	10.15	86.59	137	
	0900	8# E6011 per ton	"	2	8		13.75	293	40.50	347.25	545	
	1200	Continuous fillet, stick welding, incl. equipment										
	1300	Single pass, 1/8" thick, 0.1#/L.F.	E-14	150	.053	L.F.	.17	2.14	.54	2.85	4.62	
	1400	3/16" thick, 0.2#/L.F.		75	.107		.34	4.28	1.08	5.70	9.20	
	1500	1/4" thick, 0.3#/L.F.		50	.160		.52	6.40	1.62	8.54	13.85	
	1610	5/16" thick, 0.4#/L.F.		38	.211		.69	8.45	2.14	11.28	18.25	
	1800	3 passes, 3/8" thick, 0.5#/L.F.		30	.267		.86	10.70	2.71	14.27	23	
	2010	4 passes, 1/2" thick, 0.7#/L.F.		22	.364		1.20	14.60	3.69	19.49	31.50	
	2200	5 to 6 passes, 3/4" thick, 1.3#/L.F.		12	.667		2.24	27	6.75	35.99	58	
	2400	8 to 11 passes, 1" thick, 2.4#/L.F.		6	1.333		4.13	53.50	13.55	71.18	115	
	2600	For all position welding, add, minimum						20%				
	2700	Maximum						300%				
	2900	For semi-automatic welding, deduct, minimum						5%				
	3000	Maximum						15%				
	4000	Cleaning and welding plates, bars, or rods										
	4010	to existing beams, columns, or trusses	E-14	12	.667	L.F.	.86	27	6.75	34.61	56.50	
920	0010	**STEEL CUTTING**										920
	0020	Hand burning, incl. preparation, torch cutting & grinding, no staging										
	0100	Steel to 1/2" thick	E-25	320	.025	L.F.		1	.26	1.26	2.08	
	0150	3/4" thick		260	.031			1.24	.32	1.56	2.56	
	0200	1" thick		200	.040			1.61	.41	2.02	3.33	

05120	Structural Steel	CREW	DAILY OUTPUT	LABOR-HOURS	UNIT	MAT.	LABOR	EQUIP.	TOTAL	TOTAL INCL O&P
							2005 BARE COSTS			
250	**0010 COLUMNS, LIGHTWEIGHT**									**250**
1000	Lightweight units (lally), 3-1/2" diameter	E-2	780	.072	L.F.	2.72	2.65	1.70	7.07	9.40
1050	4" diameter	"	900	.062	"	3.99	2.30	1.47	7.76	9.95
5800	Adjustable jack post, 8' maximum height, 2-3/4" diameter				Ea.	29			29	32
5850	4" diameter				"	46.50			46.50	51
260	**0010 COLUMNS, STRUCTURAL**									**260**
0020	Shop fab'd for 100-ton, 1-2 story project, bolted conn's.									
0800	Steel, concrete filled, extra strong pipe, 3-1/2" diameter	E-2	660	.085	L.F.	29	3.14	2.01	34.15	39.50
0830	4" diameter		780	.072		32	2.65	1.70	36.35	42
0890	5" diameter		1,020	.055		38.50	2.03	1.30	41.83	47
0930	6" diameter		1,200	.047		51	1.72	1.10	53.82	60
1100	For galvanizing, add				Lb.	.21			.21	.24
1300	For web ties, angles, etc., add per added lb.	1 Sswk	945	.008		.88	.32		1.20	1.54
1500	Steel pipe, extra strong, no concrete, 3" to 5" diameter	E-2	16,000	.003		.88	.13	.08	1.09	1.27
1600	6" to 12" diameter		14,000	.004		.88	.15	.09	1.12	1.31
1700	Steel pipe, extra strong, no concrete, 3" diameter x 12'-0"		60	.933	Ea.	108	34.50	22	164.50	202
1750	4" diameter x 12'-0"		58	.966		157	35.50	23	215.50	260
1800	6" diameter x 12'-0"		54	1.037		300	38.50	24.50	363	425
1850	8" diameter x 14'-0"		50	1.120		530	41.50	26.50	598	685
1900	10" diameter x 16'-0"		48	1.167		765	43	27.50	835.50	950
1950	12" diameter x 18'-0"		45	1.244		1,025	46	29.50	1,100.50	1,225
3300	Structural tubing, square, A500GrB, 4" to 6" square, light section		11,270	.005	Lb.	.88	.18	.12	1.18	1.41
3600	Heavy section		32,000	.002	"	.88	.06	.04	.98	1.12
4000	Concrete filled, add				L.F.	3.14			3.14	3.46
4500	Structural tubing, sq, 4" x 4" x 1/4" x 12'-0"	E-2	58	.966	Ea.	144	35.50	23	202.50	246
4550	6" x 6" x 1/4" x 12'-0"		54	1.037		236	38.50	24.50	299	355
4600	8" x 8" x 3/8" x 14'-0"		50	1.120		510	41.50	26.50	578	665
4650	10" x 10" x 1/2" x 16'-0"		48	1.167		950	43	27.50	1,020.50	1,150
5100	Structural tubing, rect, 5" to 6" wide, light section		8,000	.007	Lb.	.88	.26	.17	1.31	1.58
5200	Heavy section		12,000	.005		.88	.17	.11	1.16	1.38
5300	7" to 10" wide, light section		15,000	.004		.88	.14	.09	1.11	1.30
5400	Heavy section		18,000	.003		.88	.11	.07	1.06	1.24
5500	Structural tubing, rect, 5" x 3" x 1/4" x 12'-0"		58	.966	Ea.	140	35.50	23	198.50	241
5550	6" x 4" x 5/16" x 12'-0"		54	1.037		219	38.50	24.50	282	335
5600	8" x 4" x 3/8" x 12'-0"		54	1.037		320	38.50	24.50	383	445
5650	10" x 6" x 3/8" x 14'-0"		50	1.120		510	41.50	26.50	578	665
5700	12" x 8" x 1/2" x 16'-0"		48	1.167		945	43	27.50	1,015.50	1,150
6800	W Shape, A992 steel, 2 tier, W8 x 24		1,080	.052	L.F.	23	1.92	1.23	26.15	30
6850	W8 x 31		1,080	.052		30	1.92	1.23	33.15	37.50
6900	W8 x 48		1,032	.054		46	2.01	1.28	49.29	56
6950	W8 x 67		984	.057		64.50	2.10	1.35	67.95	76
7000	W10 x 45		1,032	.054		43.50	2.01	1.28	46.79	52.50
7050	W10 x 68		984	.057		65.50	2.10	1.35	68.95	77
7100	W10 x 112		960	.058		108	2.16	1.38	111.54	124
7150	W12 x 50		1,032	.054		48	2.01	1.28	51.29	58
7200	W12 x 87		984	.057		83.50	2.10	1.35	86.95	97
7250	W12 x 120		960	.058		116	2.16	1.38	119.54	132
7300	W12 x 190		912	.061		183	2.27	1.45	186.72	207
7350	W14 x 74		984	.057		71	2.10	1.35	74.45	83.50
7400	W14 x 120		960	.058		116	2.16	1.38	119.54	132
7450	W14 x 176		912	.061		169	2.27	1.45	172.72	192
8090	For projects 75 to 99 tons, add				All	10%				
8092	50 to 74 tons, add					20%				
8094	25 to 49 tons, add					30%	10%			
8096	10 to 24 tons, add					50%	25%			

METALS 5

191

		05120	Structural Steel	CREW	DAILY OUTPUT	LABOR-HOURS	UNIT	2005 BARE COSTS				TOTAL INCL O&P	
								MAT.	LABOR	EQUIP.	TOTAL		
260	8098		2 to 9 tons, add				All	75%	50%				260
	8099		Less than 2 tons, add				↓	100%	100%				
300	0010	**CURB EDGING**											300
	0020		Steel angle w/anchors, on forms, 1" x 1", 0.8#/L.F.	E-4	350	.091	L.F.	1.48	3.53	.23	5.24	8.20	
	0100		2" x 2" angles, 3.92#/L.F.		330	.097		4.76	3.75	.25	8.76	12.20	
	0200		3" x 3" angles, 6.1#/L.F.		300	.107		7.50	4.12	.27	11.89	15.95	
	0300		4" x 4" angles, 8.2#/L.F.		275	.116		9.70	4.50	.30	14.50	19.05	
	1000		6" x 4" angles, 12.3#/L.F.		250	.128		14	4.95	.33	19.28	24.50	
	1050		Steel channels with anchors, on forms, 3" channel, 5#/L.F.		290	.110		5.90	4.26	.28	10.44	14.45	
	1100		4" channel, 5.4#/L.F.		270	.119		6.30	4.58	.30	11.18	15.50	
	1200		6" channel, 8.2#/L.F.		255	.125		9.70	4.85	.32	14.87	19.75	
	1300		8" channel, 11.5#/L.F.		225	.142		13.20	5.50	.36	19.06	25	
	1400		10" channel, 15.3#/L.F.		180	.178		17.15	6.85	.45	24.45	31.50	
	1500		12" channel, 20.7#/L.F.	↓	140	.229	↓	23	8.85	.58	32.43	41.50	
	2000		For curved edging, add				↓	35%	10%				
440	0010	**LIGHTWEIGHT FRAMING**											440
	0400		Angle framing, field fabricated, 4" and larger	E-3	440	.055	Lb.	.51	2.12	.18	2.81	4.55	
	0450		Less than 4" angles		265	.091		.53	3.52	.31	4.36	7.20	
	0600		Channel framing, field fabricated, 8" and larger		500	.048		.53	1.86	.16	2.55	4.09	
	0650		Less than 8" channels	↓	335	.072		.53	2.78	.24	3.55	5.85	
	1000		Continuous slotted channel framing system, shop fab, min	2 Sswk	2,400	.007		2.71	.25		2.96	3.44	
	1200		Maximum	"	1,600	.010		3.06	.38		3.44	4.05	
	1250		Plate & bar stock for reinforcing beams and trusses					.96			.96	1.06	
	1300		Cross bracing, rods, shop fabricated, 3/4" diameter	E-3	700	.034		1.05	1.33	.12	2.50	3.67	
	1310		7/8" diameter		850	.028		1.05	1.10	.10	2.25	3.23	
	1320		1" diameter		1,000	.024		1.05	.93	.08	2.06	2.92	
	1330		Angle, 5" x 5" x 3/8"		2,800	.009		1.05	.33	.03	1.41	1.79	
	1350		Hanging lintels, shop fabricated, average	↓	850	.028		1.05	1.10	.10	2.25	3.23	
	1380		Roof frames, shop fabricated, 3'-0" square, 5' span	E-2	4,200	.013		1.05	.49	.32	1.86	2.36	
	1400		Tie rod, not upset, 1-1/2" to 4" diameter, with turnbuckle	2 Sswk	800	.020		1.14	.76		1.90	2.62	
	1420		No turnbuckle		700	.023		1.09	.87		1.96	2.76	
	1500		Upset, 1-3/4" to 4" diameter, with turnbuckle		800	.020		1.14	.76		1.90	2.62	
	1520		No turnbuckle	↓	700	.023	↓	1.09	.87		1.96	2.76	
480	0010	**LINTELS**											480
	0020		Plain steel angles, under 500 lb.	1 Bric	550	.015	Lb.	.67	.51		1.18	1.52	
	0100		500 to 1000 lb.		640	.013		.66	.44		1.10	1.39	
	0200		1,000 to 2,000 lb.		640	.013		.64	.44		1.08	1.37	
	0300		2,000 to 4,000 lb.	↓	640	.013		.62	.44		1.06	1.35	
	0500		For built-up angles and plates, add to above					.22			.22	.24	
	0700		For engineering, add to above					.09			.09	.10	
	0900		For galvanizing, add to above, under 500 lb.					.28			.28	.31	
	0950		500 to 2,000 lb.					.26			.26	.28	
	1000		Over 2,000 lb.				↓	.21			.21	.24	
520	0010	**PIPE SUPPORT FRAMING**											520
	0020		Under 10#/L.F.	E-4	3,900	.008	Lb.	1.17	.32	.02	1.51	1.88	
	0200		10.1 to 15#/L.F.		4,300	.007		1.16	.29	.02	1.47	1.80	
	0400		15.1 to 20#/L.F.		4,800	.007		1.14	.26	.02	1.42	1.73	
	0600		Over 20#/L.F.	↓	5,400	.006	↓	1.12	.23	.02	1.37	1.66	
600	0010	**STRESSED SKIN** Roof & ceiling system											600
	0020		Double panel flat roof, spans to 100'	E-2	1,150	.049	S.F.	7	1.80	1.15	9.95	12.05	
	0100		Double panel convex roof, spans to 200'		960	.058		11.40	2.16	1.38	14.94	17.70	
	0200		Double panel arched roof, spans to 300'	↓	760	.074	↓	17.50	2.72	1.74	21.96	26	
640	0010	**STRUCTURAL STEEL MEMBERS**											640
	0020		Shop fab'd for 100-ton, 1-2 story project, bolted conn's.										

05120	Structural Steel	CREW	DAILY OUTPUT	LABOR-HOURS	UNIT	MAT.	LABOR	EQUIP.	TOTAL	TOTAL INCL O&P	
						2005 BARE COSTS					
640 0100	W 6 x 9	E-2	600	.093	L.F.	8.65	3.45	2.21	14.31	17.90	**640**
0120	x 16		600	.093		15.40	3.45	2.21	21.06	25.50	
0140	x 20		600	.093		19.25	3.45	2.21	24.91	29.50	
0300	W 8 x 10		600	.093		9.65	3.45	2.21	15.31	18.95	
0320	x 15		600	.093		14.45	3.45	2.21	20.11	24	
0350	x 21		600	.093		20	3.45	2.21	25.66	30.50	
0360	x 24		550	.102		23	3.76	2.41	29.17	34.50	
0370	x 28		550	.102		27	3.76	2.41	33.17	38.50	
0500	x 31		550	.102		30	3.76	2.41	36.17	42	
0520	x 35		550	.102		33.50	3.76	2.41	39.67	46	
0540	x 48		550	.102		46	3.76	2.41	52.17	60	
0600	W 10 x 12		600	.093		11.55	3.45	2.21	17.21	21	
0620	x 15		600	.093		14.45	3.45	2.21	20.11	24	
0700	x 22		600	.093		21	3.45	2.21	26.66	32	
0720	x 26		600	.093		25	3.45	2.21	30.66	36	
0740	x 33		550	.102		32	3.76	2.41	38.17	44	
0900	x 49		550	.102		47	3.76	2.41	53.17	61	
1100	W 12 x 14		880	.064		13.50	2.35	1.51	17.36	20.50	
1300	x 22		880	.064		21	2.35	1.51	24.86	29	
1500	x 26		880	.064		25	2.35	1.51	28.86	33	
1520	x 35		810	.069		33.50	2.56	1.64	37.70	43	
1560	x 50		750	.075		48	2.76	1.77	52.53	59.50	
1580	x 58		750	.075		56	2.76	1.77	60.53	68	
1700	x 72		640	.087		69.50	3.23	2.07	74.80	84	
1740	x 87		640	.087		83.50	3.23	2.07	88.80	100	
1900	W 14 x 26		990	.057		25	2.09	1.34	28.43	32.50	
2100	x 30		900	.062		29	2.30	1.47	32.77	37.50	
2300	x 34		810	.069		32.50	2.56	1.64	36.70	42	
2320	x 43		810	.069		41.50	2.56	1.64	45.70	51.50	
2340	x 53		800	.070		51	2.59	1.66	55.25	62.50	
2360	x 74		760	.074		71	2.72	1.74	75.46	85	
2380	x 90		740	.076		86.50	2.80	1.79	91.09	102	
2500	x 120		720	.078		116	2.87	1.84	120.71	134	
2700	W 16 x 26		1,000	.056		25	2.07	1.33	28.40	32.50	
2900	x 31		900	.062		30	2.30	1.47	33.77	38.50	
3100	x 40		800	.070		38.50	2.59	1.66	42.75	49	
3120	x 50		800	.070		48	2.59	1.66	52.25	59.50	
3140	x 67		760	.074		64.50	2.72	1.74	68.96	77.50	
3300	W 18 x 35	E-5	960	.083		33.50	3.13	1.46	38.09	44	
3500	x 40		960	.083		38.50	3.13	1.46	43.09	49.50	
3520	x 46		960	.083		44.50	3.13	1.46	49.09	55.50	
3700	x 50		912	.088		48	3.29	1.54	52.83	60.50	
3900	x 55		912	.088		53	3.29	1.54	57.83	65.50	
3920	x 65		900	.089		62.50	3.34	1.56	67.40	76.50	
3940	x 76		900	.089		73	3.34	1.56	77.90	88	
3960	x 86		900	.089		83	3.34	1.56	87.90	98.50	
3980	x 106		900	.089		102	3.34	1.56	106.90	120	
4100	W 21 x 44		1,064	.075		42.50	2.82	1.32	46.64	53	
4300	x 50		1,064	.075		48	2.82	1.32	52.14	59.50	
4500	x 62		1,036	.077		59.50	2.90	1.36	63.76	72	
4700	x 68		1,036	.077		65.50	2.90	1.36	69.76	78.50	
4720	x 83		1,000	.080		80	3	1.41	84.41	95	
4740	x 93		1,000	.080		89.50	3	1.41	93.91	105	
4760	x 101		1,000	.080		97	3	1.41	101.41	114	
4780	x 122		1,000	.080		117	3	1.41	121.41	136	
4900	W 24 x 55		1,110	.072		53	2.70	1.27	56.97	64	

METALS **5**

	05120	Structural Steel	CREW	DAILY OUTPUT	LABOR-HOURS	UNIT	2005 BARE COSTS				TOTAL INCL O&P	
							MAT.	LABOR	EQUIP.	TOTAL		
640	5100	x 62	E-5	1,110	.072	L.F.	59.50	2.70	1.27	63.47	71.50	**640**
	5300	x 68		1,110	.072		65.50	2.70	1.27	69.47	78	
	5500	x 76		1,110	.072		73	2.70	1.27	76.97	86.50	
	5700	x 84		1,080	.074		81	2.78	1.30	85.08	95.50	
	5720	x 94		1,080	.074		90.50	2.78	1.30	94.58	106	
	5740	x 104		1,050	.076		100	2.86	1.34	104.20	116	
	5760	x 117		1,050	.076		113	2.86	1.34	117.20	130	
	5780	x 146		1,050	.076		141	2.86	1.34	145.20	161	
	5800	W 27 x 84		1,190	.067		81	2.52	1.18	84.70	94.50	
	5900	x 94		1,190	.067		90.50	2.52	1.18	94.20	105	
	5920	x 114		1,150	.070		110	2.61	1.22	113.83	127	
	5940	x 146		1,150	.070		141	2.61	1.22	144.83	161	
	5960	x 161		1,150	.070		155	2.61	1.22	158.83	176	
	6100	W 30 x 99		1,200	.067		95.50	2.50	1.17	99.17	111	
	6300	x 108		1,200	.067		104	2.50	1.17	107.67	120	
	6500	x 116		1,160	.069		112	2.59	1.21	115.80	129	
	6520	x 132		1,160	.069		127	2.59	1.21	130.80	146	
	6540	x 148		1,160	.069		142	2.59	1.21	145.80	163	
	6560	x 173		1,120	.071		167	2.68	1.26	170.94	189	
	6580	x 191		1,120	.071		184	2.68	1.26	187.94	208	
	6700	W 33 x 118		1,176	.068		114	2.55	1.20	117.75	131	
	6900	x 130		1,134	.071		125	2.65	1.24	128.89	144	
	7100	x 141		1,134	.071		136	2.65	1.24	139.89	155	
	7120	x 169		1,100	.073		163	2.73	1.28	167.01	185	
	7140	x 201		1,100	.073		193	2.73	1.28	197.01	219	
	7300	W 36 x 135		1,170	.068		130	2.57	1.20	133.77	149	
	7500	x 150		1,170	.068		144	2.57	1.20	147.77	165	
	7600	x 170		1,150	.070		164	2.61	1.22	167.83	186	
	7700	x 194		1,125	.071		187	2.67	1.25	190.92	211	
	7900	x 230		1,125	.071		221	2.67	1.25	224.92	250	
	7920	x 260		1,035	.077		250	2.90	1.36	254.26	282	
	8100	x 300	▼	1,035	.077		289	2.90	1.36	293.26	325	
	8490	For projects 75 to 99 tons, add					10%					
	8492	50 to 74 tons, add					20%					
	8494	25 to 49 tons, add					30%	10%				
	8496	10 to 24 tons, add					50%	25%				
	8498	2 to 9 tons, add					75%	50%				
	8499	Less than 2 tons, add				▼	100%	100%				
680	0010	**STRUCTURAL STEEL PROJECTS**										**680**
	0020	Shop fab'd for 100-ton, 1-2 story project, bolted conn's.										
	1300	Industrial bldgs., 1 story, beams & girders, steel bearing	E-5	12.90	6.202	Ton	1,750	233	109	2,092	2,450	
	1400	Masonry bearing	"	10	8	"	1,750	300	141	2,191	2,600	
	1500	Industrial bldgs., 1 story, under 10 tons,										
	1510	steel from warehouse, trucked	E-2	7.50	7.467	Ton	2,100	276	177	2,553	2,975	
	1600	1 story with roof trusses, steel bearing	E-5	10.60	7.547		2,075	283	133	2,491	2,925	
	1700	Masonry bearing	"	8.30	9.639		2,075	360	169	2,604	3,100	
	1900	Monumental structures, banks, stores, etc., minimum	E-6	13	9.846		1,750	370	117	2,237	2,700	
	2000	Maximum		9	14.222		2,900	535	169	3,604	4,325	
	2800	Power stations, fossil fuels, minimum		11	11.636		1,750	435	138	2,323	2,825	
	2900	Maximum		5.70	22.456		2,625	845	266	3,736	4,675	
	2950	Nuclear fuels, non-safety steel, minimum		7	18.286		1,750	685	217	2,652	3,375	
	3000	Maximum		5.50	23.273		2,625	875	276	3,776	4,725	
	3040	Safety steel, minimum		2.50	51.200		2,550	1,925	605	5,080	6,825	
	3070	Maximum	▼	1.50	85.333		3,350	3,200	1,000	7,550	10,400	
	3100	Roof trusses, minimum	E-5	13	6.154		2,450	231	108	2,789	3,225	
	3200	Maximum	"	8.30	9.639	▼	2,975	360	169	3,504	4,100	

Important: See the Reference Section for critical supporting data - Reference Nos., Crews, & City Cost Indexes

05100 | Structural Metal Framing

05120 | Structural Steel

			CREW	DAILY OUTPUT	LABOR-HOURS	UNIT	MAT.	LABOR	EQUIP.	TOTAL	TOTAL INCL O&P	
680	3900	High strength steel mill spec extras: A242, A441,										680
	3950	A529, A572 (42 ksi) and A992: same as A36 steel										
	4000	Add to A992 price for A572 (50, 60, 65 ksi)				Ton	25			25	27.50	
	4100	A588 Weathering				"	60			60	66	
	4200	Mill size extras for W-Shapes: 0 to 30 plf: no extra charge										
	4210	Member sizes 31 to 65 plf, add				Ton	5			5	5.50	
	4220	Member sizes 66 to 100 plf, add					20			20	22	
	4230	Member sizes 101 to 387 plf, add				↓	40			40	44	
	4300	Column base plates, light, up to 150 lb	2 Sswk	2,000	.008	Lb.	.96	.31		1.27	1.61	
	4400	Heavy, over 150 lb	E-2	7,500	.007	"	1.01	.28	.18	1.47	1.77	
	5390	For projects 75 to 99 tons, add				Ton	10%					
	5392	50 to 74 tons, add					20%					
	5394	25 to 49 tons, add					30%	10%				
	5396	10 to 24 tons, add					50%	25%				
	5398	2 to 9 tons, add					75%	50%				
	5399	Less than 2 tons, add				↓	100%	100%				

05140 | Structural Aluminum

			CREW	DAILY OUTPUT	LABOR-HOURS	UNIT	MAT.	LABOR	EQUIP.	TOTAL	TOTAL INCL O&P	
080	0010	**ALUMINUM**										080
	0020	Structural shapes, 1" to 10" members, under 1 ton	E-2	1,050	.053	Lb.	2.19	1.97	1.26	5.42	7.20	
	0050	1 to 5 tons		1,330	.042		2.09	1.56	1	4.65	6.05	
	0100	Over 5 tons		1,330	.042		2.03	1.56	1	4.59	6	
	0300	Extrusions, over 5 tons, stock shapes		1,330	.042		2.19	1.56	1	4.75	6.20	
	0400	Custom shapes	↓	1,330	.042	↓	2.23	1.56	1	4.79	6.20	
	0600	For formed aluminum columns, see div. 05580-200										

05150 | Wire Rope Assemblies

			CREW	DAILY OUTPUT	LABOR-HOURS	UNIT	MAT.	LABOR	EQUIP.	TOTAL	TOTAL INCL O&P	
800	0010	**STEEL WIRE ROPE**										800
	0020	6 x 19, bright, fiber core, 5000' rolls, 1/2" diameter				L.F.	.80			.80	.88	
	0050	Steel core					1.06			1.06	1.16	
	0100	Fiber core, 1" diameter					2.71			2.71	2.98	
	0150	Steel core					3.09			3.09	3.40	
	0300	6 x 19, galvanized, fiber core, 1/2" diameter					1.18			1.18	1.30	
	0350	Steel core					1.35			1.35	1.49	
	0400	Fiber core, 1" diameter					3.47			3.47	3.82	
	0450	Steel core					3.64			3.64	4	
	0500	6 x 7, bright, IPS, fiber core, <500 L.F. w/acc., 1/4" diameter	E-17	6,400	.002		.60	.10		.70	.84	
	0510	1/2" diameter		2,100	.008		1.46	.30		1.76	2.14	
	0520	3/4" diameter		960	.017		2.64	.65	.01	3.30	4.09	
	0550	6 x 19, bright, IPS, IWRC, <500 L.F. w/acc., 1/4" diameter		5,760	.003		.89	.11		1	1.17	
	0560	1/2" diameter		1,730	.009		1.44	.36		1.80	2.23	
	0570	3/4" diameter		770	.021		2.50	.81	.01	3.32	4.22	
	0580	1" diameter		420	.038		4.23	1.49	.01	5.73	7.35	
	0590	1-1/4" diameter		290	.055		7	2.16	.02	9.18	11.60	
	0600	1-1/2" diameter	↓	192	.083		8.65	3.26	.03	11.94	15.40	
	0610	1-3/4" diameter	E-18	240	.167		13.75	6.30	3.32	23.37	29.50	
	0620	2" diameter		160	.250		17.65	9.45	4.98	32.08	41.50	
	0630	2-1/4" diameter	↓	160	.250		23.50	9.45	4.98	37.93	48	
	0650	6 x 37, bright, IPS, IWRC, <500 L.F. w/acc., 1/4" diameter	E-17	6,400	.002		1.12	.10		1.22	1.42	
	0660	1/2" diameter		1,730	.009		1.91	.36		2.27	2.75	
	0670	3/4" diameter		770	.021		3.08	.81	.01	3.90	4.86	
	0680	1" diameter		430	.037		4.89	1.46	.01	6.36	8	
	0690	1-1/4" diameter		290	.055		7.40	2.16	.02	9.58	12	
	0700	1-1/2" diameter	↓	190	.084		10.55	3.30	.03	13.88	17.55	
	0710	1-3/4" diameter	E-18	260	.154	↓	16.75	5.80	3.07	25.62	32	

			DAILY	LABOR-		2005 BARE COSTS				TOTAL
05150	**Wire Rope Assemblies**	CREW	OUTPUT	HOURS	UNIT	MAT.	LABOR	EQUIP.	TOTAL	INCL O&P
800	0720	2" diameter	E-18	200	.200	L.F.	22	7.55	3.99	33.54	41.50	800
	0730	2-1/4" diameter	↓	160	.250		29	9.45	4.98	43.43	53.50
	0800	6 x 19 & 6 x 37, swaged, 1/2" diameter	E-17	1,220	.013		3.43	.51		3.94	4.69
	0810	9/16" diameter		1,120	.014		3.98	.56		4.54	5.40
	0820	5/8" diameter		930	.017		4.72	.67	.01	5.40	6.40
	0830	3/4" diameter		640	.025		6	.98	.01	6.99	8.35
	0840	7/8" diameter		480	.033		7.60	1.30	.01	8.91	10.70
	0850	1" diameter		350	.046		9.25	1.79	.02	11.06	13.40
	0860	1-1/8" diameter		288	.056		11.40	2.18	.02	13.60	16.45
	0870	1-1/4" diameter		230	.070		13.80	2.72	.02	16.54	20
	0880	1-3/8" diameter	↓	192	.083		15.95	3.26	.03	19.24	23.50
	0890	1-1/2" diameter	E-18	300	.133	↓	19.35	5.05	2.66	27.06	33

			DAILY	LABOR-		2005 BARE COSTS				TOTAL
05210	**Steel Joists**	CREW	OUTPUT	HOURS	UNIT	MAT.	LABOR	EQUIP.	TOTAL	INCL O&P
600	0010	**OPEN WEB JOISTS**, 40-ton job lots										600
	0020	K series, horizontal bridging, spans up to 30', minimum	E-7	15	5.333	Ton	1,075	200	99	1,374	1,650
	0050	Average		12	6.667		1,225	250	124	1,599	1,925
	0080	Maximum		9	8.889		1,450	335	165	1,950	2,350
	0410	Span 30' to 50', minimum		17	4.706		1,075	177	87.50	1,339.50	1,575
	0440	Average		17	4.706		1,200	177	87.50	1,464.50	1,725
	0460	Maximum	↓	10	8	↓	1,275	300	149	1,724	2,075
	1010	CS series, horizontal bridging									
	1020	Spans to 30', minimum	E-7	15	5.333	Ton	1,125	200	99	1,424	1,700
	1040	Average		12	6.667		1,250	250	124	1,624	1,950
	1060	Maximum	↓	9	8.889	↓	1,475	335	165	1,975	2,375
	2000	LH series, bolted cross bridging									
	2020	Spans to 96', minimum	E-7	16	5	Ton	1,225	188	93	1,506	1,775
	2040	Average		13	6.154		1,350	231	114	1,695	2,000
	2080	Maximum	↓	11	7.273	↓	1,600	273	135	2,008	2,375
	3010	DLH series, bolted cross bridging									
	3020	Spans to 144' (shipped in 2 pieces), minimum	E-7	16	5	Ton	1,325	188	93	1,606	1,900
	3040	Average		13	6.154		1,425	231	114	1,770	2,100
	3100	Maximum	↓	11	7.273	↓	1,725	273	135	2,133	2,525
	4010	SLH series, bolted cross bridging									
	4020	Spans to 200' (shipped in 3 pieces), minimum	E-7	16	5	Ton	1,300	188	93	1,581	1,875
	4040	Average		13	6.154		1,475	231	114	1,820	2,150
	4060	Maximum	↓	11	7.273	↓	1,750	273	135	2,158	2,550
	6000	For welded cross bridging, add						30%			
	6100	For less than 40-ton job lots									
	6102	For 30 to 39 tons, add					10%				
	6104	20 to 29 tons, add					20%				
	6106	10 to 19 tons, add					30%				
	6107	5 to 9 tons, add					50%	25%			
	6108	1 to 4 tons, add					75%	50%			
	6109	Less than 1 ton, add					100%	100%			
	6200	For shop prime paint other than mfrs. standard, add					20%				
	6300	For bottom chord extensions, add per chord				Ea.	26.50			26.50	29
	7000	Joist girders, minimum	E-5	15	5.333	Ton	1,100	200	94	1,394	1,675

05200 | Metal Joists

05210	Steel Joists		CREW	DAILY OUTPUT	LABOR-HOURS	UNIT	2005 BARE COSTS				TOTAL INCL O&P	
							MAT.	LABOR	EQUIP.	TOTAL		
600	7020	Average	E-5	13	6.154	Ton	1,225	231	108	1,564	1,875	600
	7040	Maximum	↓	11	7.273	↓	1,275	273	128	1,676	2,025	
	8000	Trusses, factory fabricated WT chords, average	▼	11	7.273	▼	4,000	273	128	4,401	5,000	

05300 | Metal Deck

05310	Steel Deck		CREW	DAILY OUTPUT	LABOR-HOURS	UNIT	2005 BARE COSTS				TOTAL INCL O&P	
							MAT.	LABOR	EQUIP.	TOTAL		
300	0010	**METAL DECKING** Steel decking R05310-100										300
	0200	Cellular units, galv, 2" deep, 20-20 gauge, over 15 squares	E-4	1,460	.022	S.F.	5.85	.85	.06	6.76	8.05	
	0250	18-20 gauge		1,420	.023		6.65	.87	.06	7.58	8.90	
	0300	18-18 gauge		1,390	.023		6.85	.89	.06	7.80	9.15	
	0320	16-18 gauge		1,360	.024		8.15	.91	.06	9.12	10.65	
	0340	16-16 gauge		1,330	.024		9.05	.93	.06	10.04	11.70	
	0400	3" deep, galvanized, 20-20 gauge		1,375	.023		6.45	.90	.06	7.41	8.75	
	0500	18-20 gauge		1,350	.024		7.80	.92	.06	8.78	10.25	
	0600	18-18 gauge		1,290	.025		7.75	.96	.06	8.77	10.35	
	0700	16-18 gauge		1,230	.026		8.75	1.01	.07	9.83	11.45	
	0800	16-16 gauge		1,150	.028		9.55	1.08	.07	10.70	12.50	
	1000	4-1/2" deep, galvanized, 20-18 gauge		1,100	.029		9	1.12	.07	10.19	12	
	1100	18-18 gauge		1,040	.031		8.95	1.19	.08	10.22	12.05	
	1200	16-18 gauge		980	.033		10.05	1.26	.08	11.39	13.40	
	1300	16-16 gauge	▼	935	.034	▼	10.95	1.32	.09	12.36	14.50	
	1500	For acoustical deck, add					15%					
	1700	For cells used for ventilation, add					15%					
	1900	For multi-story or congested site, add						50%				
	2100	Open type, galv., 1-1/2" deep wide rib, 22 gauge, under 50 squares	E-4	4,500	.007	S.F.	1.47	.27	.02	1.76	2.12	
	2400	Over 500 squares		5,100	.006		1.06	.24	.02	1.32	1.61	
	2600	20 gauge, under 50 squares		3,865	.008		1.73	.32	.02	2.07	2.49	
	2700	Over 500 squares		4,300	.007		1.24	.29	.02	1.55	1.90	
	2900	18 gauge, under 50 squares		3,800	.008		2.24	.33	.02	2.59	3.06	
	3000	Over 500 squares		4,300	.007		1.61	.29	.02	1.92	2.30	
	3050	16 gauge, under 50 squares		3,700	.009		3.01	.33	.02	3.36	3.93	
	3100	Over 500 squares		4,200	.008		2.17	.29	.02	2.48	2.94	
	3200	3" deep, 22 gauge, under 50 squares		3,600	.009		2.02	.34	.02	2.38	2.85	
	3300	20 gauge, under 50 squares		3,400	.009		2.36	.36	.02	2.74	3.27	
	3400	18 gauge, under 50 squares		3,200	.010		3.05	.39	.03	3.47	4.08	
	3500	16 gauge, under 50 squares		3,000	.011		4.02	.41	.03	4.46	5.20	
	3700	4-1/2" deep, long span roof, over 50 squares, 20 gauge		2,700	.012		3.78	.46	.03	4.27	5	
	3800	18 gauge		2,460	.013		4.88	.50	.03	5.41	6.30	
	3900	16 gauge		2,350	.014		3.64	.53	.03	4.20	4.98	
	4100	6" deep, long span, 18 gauge		2,000	.016		7	.62	.04	7.66	8.85	
	4200	16 gauge		1,930	.017		5.20	.64	.04	5.88	6.90	
	4300	14 gauge		1,860	.017		6.70	.66	.04	7.40	8.60	
	4500	7-1/2" deep, long span, 18 gauge		1,690	.019		7.65	.73	.05	8.43	9.80	
	4600	16 gauge		1,590	.020		5.70	.78	.05	6.53	7.75	
	4700	14 gauge	▼	1,490	.021	▼	7.35	.83	.05	8.23	9.65	
	4800	For painted instead of galvanized, deduct					2%					
	5000	For acoustical perforated, with fiberglass, add				S.F.	1			1	1.10	
	5200	Non-cellular composite deck, galv., 2" deep, 22 gauge	E-4	3,860	.008	↓	1.40	.32	.02	1.74	2.13	

METALS 5

197

05310 | Steel Deck

			CREW	DAILY OUTPUT	LABOR-HOURS	UNIT	MAT.	LABOR	EQUIP.	TOTAL	TOTAL INCL O&P	
300	5300	20 gauge	E-4	3,600	.009	S.F.	1.55	.34	.02	1.91	2.34	**300**
	5400	18 gauge		3,380	.009		1.97	.37	.02	2.36	2.85	
	5500	16 gauge		3,200	.010		2.46	.39	.03	2.88	3.43	
	5700	3" deep, galv., 22 gauge		3,200	.010		1.53	.39	.03	1.95	2.40	
	5800	20 gauge		3,000	.011		1.71	.41	.03	2.15	2.65	
	5900	18 gauge		2,850	.011		2.10	.43	.03	2.56	3.12	
	6000	16 gauge		2,700	.012		2.80	.46	.03	3.29	3.93	
	6100	Slab form, steel, 28 gauge, 9/16" deep, uncoated		4,000	.008		.98	.31	.02	1.31	1.65	
	6200	Galvanized		4,000	.008		.87	.31	.02	1.20	1.53	
	6220	24 gauge, 1" deep, uncoated		3,900	.008		1.07	.32	.02	1.41	1.77	
	6240	Galvanized		3,900	.008		1.26	.32	.02	1.60	1.98	
	6300	24 gauge, 1-5/16" deep, uncoated		3,800	.008		1.14	.33	.02	1.49	1.85	
	6400	Galvanized		3,800	.008		1.34	.33	.02	1.69	2.07	
	6500	22 gauge, 1-5/16" deep, uncoated		3,700	.009		1.43	.33	.02	1.78	2.19	
	6600	Galvanized		3,700	.009		1.46	.33	.02	1.81	2.23	
	6700	22 gauge, 2" deep uncoated		3,600	.009		1.89	.34	.02	2.25	2.71	
	6800	Galvanized		3,600	.009		1.85	.34	.02	2.21	2.67	
	7000	Sheet metal edge closure form, 12" wide with 2 bends, galv										
	7100	18 gauge	E-14	360	.022	L.F.	3.04	.89	.23	4.16	5.20	
	7200	16 gauge	"	360	.022	"	4.12	.89	.23	5.24	6.40	
	8000	Metal deck and trench, 2" thick, 20 gauge, combination										
	8010	60% cellular, 40% non-cellular, inserts and trench	R-4	1,100	.036	S.F.	11.70	1.42	.07	13.19	15.40	

Note: Row 5400–5500 references R05310-100.

05514 | Ladders

			CREW	DAILY OUTPUT	LABOR-HOURS	UNIT	MAT.	LABOR	EQUIP.	TOTAL	TOTAL INCL O&P	
500	0010	**LADDER**, shop fabricated										**500**
	0020	Steel, 20" wide, bolted to concrete, with cage	E-4	50	.640	V.L.F.	64.50	24.50	1.63	90.63	117	
	0100	Without cage		85	.376		30	14.55	.96	45.51	60	
	0300	Aluminum, bolted to concrete, with cage		50	.640		87.50	24.50	1.63	113.63	142	
	0400	Without cage		85	.376		50.50	14.55	.96	66.01	82.50	

05517 | Metal Stairs

			CREW	DAILY OUTPUT	LABOR-HOURS	UNIT	MAT.	LABOR	EQUIP.	TOTAL	TOTAL INCL O&P	
700	0010	**STAIR**, shop fabricated, steel stringers, safety nosing on treads										**700**
	0020	Grating tread and pipe railing, 3'-6" wide	E-4	35	.914	Riser	128	35.50	2.32	165.82	207	
	0100	4'-0" wide		30	1.067		167	41	2.71	210.71	260	
	0200	Cement fill metal pan, picket rail, 3'-6" wide		35	.914		192	35.50	2.32	229.82	277	
	0300	4'-0" wide		30	1.067		218	41	2.71	261.71	315	
	0350	Wall rail, both sides, 3'-6" wide		53	.604		147	23.50	1.53	172.03	205	
	0400	Cast iron tread and pipe rail, 3'-6" wide		35	.914		205	35.50	2.32	242.82	292	
	0500	Checkered plate tread, industrial, 3'-6" wide		28	1.143		128	44	2.90	174.90	223	
	0550	Circular, for tanks, 3'-0" wide		33	.970		141	37.50	2.46	180.96	225	
	0600	For isolated stairs, add						100%				
	0800	Custom steel stairs, 3'-6" wide, minimum	E-4	35	.914		192	35.50	2.32	229.82	277	
	0810	Average		30	1.067		256	41	2.71	299.71	360	
	0900	Maximum		20	1.600		320	62	4.06	386.06	465	
	1100	For 4' wide stairs, add					5%	5%				
	1300	For 5' wide stairs, add					10%	10%				
	1500	Landing, steel pan, conventional	E-4	160	.200	S.F.	25.50	7.75	.51	33.76	42.50	

Important: See the Reference Section for critical supporting data - Reference Nos., Crews, & City Cost Indexes

05517	Metal Stairs	CREW	DAILY OUTPUT	LABOR-HOURS	UNIT	2005 BARE COSTS				TOTAL INCL O&P	
						MAT.	LABOR	EQUIP.	TOTAL		
700	1600	Pre-erected	E-4	255	.125	S.F.	45	4.85	.32	50.17	58.50	**700**
	1700	Pre-erected, steel pan tread, 3'-6" wide, 2 line pipe rail	E-2	87	.644	Riser	211	24	15.25	250.25	291	

05520 | Handrails & Railings

						MAT.	LABOR	EQUIP.	TOTAL	INCL O&P	
700	0010	**RAILING, PIPE**, shop fabricated										**700**
	0020	Aluminum, 2 rail, satin finish, 1-1/4" diameter	E-4	160	.200	L.F.	14.60	7.75	.51	22.86	30.50	
	0030	Clear anodized		160	.200		18.10	7.75	.51	26.36	34.50	
	0040	Dark anodized		160	.200		20.50	7.75	.51	28.76	37	
	0080	1-1/2" diameter, satin finish		160	.200		17.50	7.75	.51	25.76	33.50	
	0090	Clear anodized		160	.200		19.50	7.75	.51	27.76	36	
	0100	Dark anodized		160	.200		21.50	7.75	.51	29.76	38.50	
	0140	Aluminum, 3 rail, 1-1/4" diam., satin finish		137	.234		22.50	9.05	.59	32.14	41.50	
	0150	Clear anodized		137	.234		28	9.05	.59	37.64	48	
	0160	Dark anodized		137	.234		31	9.05	.59	40.64	51	
	0200	1-1/2" diameter, satin finish		137	.234		27	9.05	.59	36.64	46.50	
	0210	Clear anodized		137	.234		30.50	9.05	.59	40.14	50.50	
	0220	Dark anodized		137	.234		33.50	9.05	.59	43.14	54	
	0500	Steel, 2 rail, on stairs, primed, 1-1/4" diameter		160	.200		12	7.75	.51	20.26	27.50	
	0520	1-1/2" diameter		160	.200		13.20	7.75	.51	21.46	29	
	0540	Galvanized, 1-1/4" diameter		160	.200		16.65	7.75	.51	24.91	32.50	
	0560	1-1/2" diameter		160	.200		18.65	7.75	.51	26.91	35	
	0580	Steel, 3 rail, primed, 1-1/4" diameter		137	.234		17.90	9.05	.59	27.54	36.50	
	0600	1-1/2" diameter		137	.234		19	9.05	.59	28.64	38	
	0620	Galvanized, 1-1/4" diameter		137	.234		25	9.05	.59	34.64	44.50	
	0640	1-1/2" diameter		137	.234		29.50	9.05	.59	39.14	49	
	0700	Stainless steel, 2 rail, 1-1/4" diam. #4 finish		137	.234		43.50	9.05	.59	53.14	64.50	
	0720	High polish		137	.234		69.50	9.05	.59	79.14	93.50	
	0740	Mirror polish		137	.234		87.50	9.05	.59	97.14	113	
	0760	Stainless steel, 3 rail, 1-1/2" diam., #4 finish		120	.267		65.50	10.30	.68	76.48	91	
	0770	High polish		120	.267		108	10.30	.68	118.98	138	
	0780	Mirror finish		120	.267		131	10.30	.68	141.98	163	
	0900	Wall rail, alum. pipe, 1-1/4" diam., satin finish		213	.150		8.35	5.80	.38	14.53	20	
	0905	Clear anodized		213	.150		10.20	5.80	.38	16.38	22	
	0910	Dark anodized		213	.150		12.35	5.80	.38	18.53	24.50	
	0915	1-1/2" diameter, satin finish		213	.150		9.25	5.80	.38	15.43	21	
	0920	Clear anodized		213	.150		11.65	5.80	.38	17.83	23.50	
	0925	Dark anodized		213	.150		14.35	5.80	.38	20.53	26.50	
	0930	Steel pipe, 1-1/4" diameter, primed		213	.150		7.30	5.80	.38	13.48	18.80	
	0935	Galvanized		213	.150		10.55	5.80	.38	16.73	22.50	
	0940	1-1/2" diameter		176	.182		7.50	7.05	.46	15.01	21.50	
	0945	Galvanized		213	.150		10.60	5.80	.38	16.78	22.50	
	0955	Stainless steel pipe, 1-1/2" diam., #4 finish		107	.299		34.50	11.55	.76	46.81	59.50	
	0960	High polish		107	.299		70.50	11.55	.76	82.81	99	
	0965	Mirror polish		107	.299		83	11.55	.76	95.31	112	
780	0010	**RAILINGS, INDUSTRIAL** Welded, shop fabricated										**780**
	0020	2 rail, 3'-6" high, 1-1/2" pipe	E-4	255	.125	L.F.	17.95	4.85	.32	23.12	29	
	0100	2" angle rail	"	255	.125		16.35	4.85	.32	21.52	27	
	0200	For 4" high kick plate, 10 gauge, add					3.73			3.73	4.10	
	0300	1/4" thick, add					4.80			4.80	5.30	
	0500	For curved rails, add					30%	30%				

05530 | Gratings

300	0010	**FLOOR GRATING, ALUMINUM**, field fabricated from panels										**300**
	0110	Bearing bars @ 1-3/16" O.C., cross bars @ 4" O.C.,										
	0111	Up to 300 S.F., 1" x 1/8" bar	E-4	900	.036	S.F.	9.25	1.37	.09	10.71	12.75	
	0112	Over 300 S.F.		850	.038		8.40	1.46	.10	9.96	11.95	

METALS 5

5 METALS

05530		Gratings	CREW	DAILY OUTPUT	LABOR-HOURS	UNIT	2005 BARE COSTS				TOTAL INCL O&P	
							MAT.	LABOR	EQUIP.	TOTAL		
300	0113	1-1/4" x 1/8" bar, up to 300 S.F.	E-4	800	.040	S.F.	10.50	1.55	.10	12.15	14.45	**300**
	0114	Over 300 S.F.		1,000	.032		9.55	1.24	.08	10.87	12.80	
	0122	1-1/4" x 3/16" bar, up to 300 S.F.		750	.043		13.50	1.65	.11	15.26	17.90	
	0124	Over 300 S.F.		1,000	.032		12.25	1.24	.08	13.57	15.80	
	0132	1-1/2" x 1/8" bar, up to 300 S.F.		700	.046		8.45	1.77	.12	10.34	12.55	
	0134	Over 300 S.F.		1,000	.032		7.65	1.24	.08	8.97	10.75	
	0136	1-3/4" x 3/16" bar, up to 300 S.F.		500	.064		18.10	2.47	.16	20.73	24.50	
	0138	Over 300 S.F.		1,000	.032		16.45	1.24	.08	17.77	20.50	
	0146	2-1/4" x 3/16" bar, up to 300 S.F.		600	.053		21.50	2.06	.14	23.70	27.50	
	0148	Over 300 S.F.		1,000	.032		19.35	1.24	.08	20.67	24	
	0162	Cross bars @ 2" O.C., 1" x 1/8", up to 300 S.F.		600	.053		11.65	2.06	.14	13.85	16.70	
	0164	Over 300 S.F.		1,000	.032		10.60	1.24	.08	11.92	13.95	
	0172	1-1/4" x 3/16" bar, up to 300 S.F.		600	.053		16.15	2.06	.14	18.35	21.50	
	0174	Over 300 S.F.		1,000	.032		14.70	1.24	.08	16.02	18.45	
	0182	1-1/2" x 1/8" bar, up to 300 S.F.		600	.053		11.60	2.06	.14	13.80	16.60	
	0184	Over 300 S.F.		1,000	.032		10.55	1.24	.08	11.87	13.90	
	0186	1-3/4" x 3/16" bar, up to 300 S.F.		600	.053		16.80	2.06	.14	19	22.50	
	0188	Over 300 S.F.	▼	1,000	.032	▼	15.30	1.24	.08	16.62	19.10	
	0200	For straight cuts, add				L.F.	2.38			2.38	2.62	
	0212	Close mesh, 1" x 1/8", up to 300 S.F.	E-4	520	.062	S.F.	29.50	2.38	.16	32.04	37	
	0214	Over 300 S.F.		920	.035		27	1.34	.09	28.43	32	
	0222	1-1/4" x 3/16", up to 300 S.F.		520	.062		36.50	2.38	.16	39.04	45	
	0224	Over 300 S.F.		920	.035		33.50	1.34	.09	34.93	39	
	0232	1-1/2" x 1/8", up to 300 S.F.		520	.062		14.85	2.38	.16	17.39	21	
	0234	Over 300 S.F.	▼	920	.035	▼	13.50	1.34	.09	14.93	17.35	
	0600	For aluminum checkered plate nosings, add				L.F.	3.84			3.84	4.22	
	0700	For straight toe plate, add					6.95			6.95	7.65	
	0800	For curved toe plate, add					8.70			8.70	9.60	
	1000	For cast aluminum abrasive nosings, add				▼	4.67			4.67	5.15	
	1200	Expanded aluminum, .65# per S.F.	E-4	1,050	.030	S.F.	4.07	1.18	.08	5.33	6.70	
	1600	Heavy duty, all extruded plank, 3/4" deep, 1.8 # per S.F.		1,100	.029		11.40	1.12	.07	12.59	14.65	
	1700	1-1/4" deep, 2.9# per S.F.		1,000	.032		12.15	1.24	.08	13.47	15.65	
	1800	1-3/4" deep, 4.2# per S.F.		925	.035		14.50	1.34	.09	15.93	18.45	
	1900	2-1/4" deep, 5.0# per S.F.	▼	875	.037	▼	16.60	1.41	.09	18.10	21	
	2100	For safety serrated surface, add					15%					
320	0010	**FLOOR GRATING PLANKS**, field fabricated from planks										**320**
	0020	Aluminum, 9-1/2" wide, 14 ga., 2" rib	E-4	950	.034	L.F.	13.40	1.30	.09	14.79	17.15	
	0200	Galvanized steel, 9-1/2" wide, 14 ga., 2-1/2" rib		950	.034		8.65	1.30	.09	10.04	11.95	
	0300	4" rib		950	.034		10.40	1.30	.09	11.79	13.85	
	0500	12 gauge, 2-1/2" rib		950	.034		11.50	1.30	.09	12.89	15.05	
	0600	3" rib		950	.034		13.50	1.30	.09	14.89	17.25	
	0800	Stainless steel, type 304, 16 ga., 2" rib		950	.034		23	1.30	.09	24.39	28	
	0900	Type 316	▼	950	.034	▼	28	1.30	.09	29.39	33.50	
340	0010	**FLOOR GRATING, STEEL**, field fabricated from panels										**340**
	0050	Labor for installing, from ground/floor	E-4	845	.038	S.F.		1.46	.10	1.56	2.73	
	0100	Elevated		460	.070	"		2.69	.18	2.87	5	
	0300	Platforms, to 12' high, rectangular		3,150	.010	Lb.	1.32	.39	.03	1.74	2.18	
	0400	Circular	▼	2,300	.014	"	1.46	.54	.04	2.04	2.61	
	0410	Painted bearing bars @ 1-3/16"										
	0412	Cross bars @ 4" O.C., 3/4" x 1/8" bar, up to 300 S.F.	E-2	500	.112	S.F.	5.60	4.14	2.65	12.39	16.15	
	0414	Over 300 S.F.		750	.075		5.05	2.76	1.77	9.58	12.30	
	0422	1-1/4" x 3/16", up to 300 S.F.		400	.140		7.50	5.15	3.31	15.96	21	
	0424	Over 300 S.F.		600	.093		6.85	3.45	2.21	12.51	15.85	
	0432	1-1/2" x 1/8", up to 300 S.F.		400	.140		6.75	5.15	3.31	15.21	19.95	
	0434	Over 300 S.F.	▼	600	.093	▼	6.10	3.45	2.21	11.76	15.10	

Important: See the Reference Section for critical supporting data - Reference Nos., Crews, & City Cost Indexes

05530 | Gratings

		Description	CREW	DAILY OUTPUT	LABOR-HOURS	UNIT	MAT.	LABOR	EQUIP.	TOTAL	TOTAL INCL O&P	
340	0436	1-3/4" x 3/16", up to 300 S.F.	E-2	400	.140	S.F.	9.80	5.15	3.31	18.26	23.50	340
	0438	Over 300 S.F.		600	.093		8.90	3.45	2.21	14.56	18.15	
	0452	2-1/4" x 3/16", up to 300 S.F.		300	.187		11.60	6.90	4.42	22.92	29.50	
	0454	Over 300 S.F.		450	.124		10.55	4.60	2.94	18.09	22.50	
	0462	Cross bars @ 2" O.C., 3/4" x 1/8", up to 300 S.F.		500	.112		10.75	4.14	2.65	17.54	22	
	0464	Over 300 S.F.		750	.075		8.95	2.76	1.77	13.48	16.55	
	0472	1-1/4" x 3/16", up to 300 S.F.		400	.140		16.10	5.15	3.31	24.56	30	
	0474	Over 300 S.F.		600	.093		13.40	3.45	2.21	19.06	23	
	0482	1-1/2" x 1/8", up to 300 S.F.		400	.140		12.45	5.15	3.31	20.91	26	
	0484	Over 300 S.F.		600	.093		10.35	3.45	2.21	16.01	19.75	
	0486	1-3/4" x 3/16", up to 300 S.F.		400	.140		16.75	5.15	3.31	25.21	31	
	0488	Over 300 S.F.		600	.093		13.95	3.45	2.21	19.61	23.50	
	0502	2-1/4" x 3/16", up to 300 S.F.		300	.187		20	6.90	4.42	31.32	38.50	
	0504	Over 300 S.F.	↓	450	.124	↓	16.70	4.60	2.94	24.24	29.50	
	0601	Painted bearing bars @ 15/16" O.C., cross bars @ 4" O.C.,										
	0612	Up to 300 S.F., 3/4" x 1/8" bars	E-4	850	.038	S.F.	6.95	1.46	.10	8.51	10.35	
	0622	1-1/4" x 3/16" bars		600	.053		9.80	2.06	.14	12	14.65	
	0632	1-1/2" x 1/8" bars		550	.058		8.95	2.25	.15	11.35	14.05	
	0636	1-3/4" x 3/16" bars	↓	450	.071		12.90	2.75	.18	15.83	19.30	
	0652	2-1/4" x 3/16" bars	E-2	300	.187		16.90	6.90	4.42	28.22	35.50	
	0662	Cross bars @ 2" O.C., up to 300 S.F., 3/4" x 1/8"		500	.112		11.80	4.14	2.65	18.59	23	
	0672	1-1/4" x 3/16" bars		400	.140		12.75	5.15	3.31	21.21	26.50	
	0682	1-1/2" x 1/8" bars		400	.140		12.15	5.15	3.31	20.61	26	
	0686	1-3/4" x 3/16" bars	↓	300	.187		16.50	6.90	4.42	27.82	35	
	0690	For galvanized grating, add				↓	25%					
	0800	For straight cuts, add				L.F.	2.53			2.53	2.78	
	0900	For curved cuts, add					3.48			3.48	3.83	
	1000	For straight banding, add					2.90			2.90	3.19	
	1100	For curved banding, add					4.48			4.48	4.93	
	1200	For checkered plate nosings, add					5			5	5.50	
	1300	For straight toe or kick plate, add					7.05			7.05	7.75	
	1400	For curved toe or kick plate, add					8.60			8.60	9.45	
	1500	For abrasive nosings, add				↓	5.20			5.20	5.70	
	1600	For safety serrated surface, minimum, add					15%					
	1700	Maximum, add					25%					
	2000	Stainless steel gratings, close spaced, 1" x 1/8" bars, up to 300 S.F.	E-4	450	.071	S.F.	31.50	2.75	.18	34.43	40	
	2100	Standard spacing, 3/4" x 1/8" bars		500	.064		25	2.47	.16	27.63	32	
	2200	1-1/4" x 3/16" bars		400	.080		39.50	3.09	.20	42.79	49	
	2400	Expanded steel grating, at ground, 3.0# per S.F.		900	.036		3.11	1.37	.09	4.57	6	
	2500	3.14# per S.F.		900	.036		3.29	1.37	.09	4.75	6.20	
	2600	4.0# per S.F.		850	.038		4.06	1.46	.10	5.62	7.20	
	2650	4.27# per S.F.		850	.038		4.40	1.46	.10	5.96	7.55	
	2700	5.0# per S.F.		800	.040		5.20	1.55	.10	6.85	8.65	
	2800	6.25# per S.F.		750	.043		6.60	1.65	.11	8.36	10.30	
	2900	7.0# per S.F.	↓	700	.046		7.45	1.77	.12	9.34	11.45	
	3100	For flattened expanded steel grating, add					8%					
	3300	For elevated installation above 15', add				↓		15%				
360	0010	**GRATING FRAME**, field fabricated										360
	0020	Aluminum, for gratings 1" to 1-1/2" deep	1 Sswk	70	.114	L.F.	7.05	4.36		11.41	15.60	
	0100	For each corner, add				Ea.	5.55			5.55	6.15	

05540 | Floor Plates

		Description	CREW	DAILY OUTPUT	LABOR-HOURS	UNIT	MAT.	LABOR	EQUIP.	TOTAL	TOTAL INCL O&P	
200	0010	**CHECKERED PLATE**, field fabricated										200
	0020	1/4" & 3/8", 2000 to 5000 S.F., bolted	E-4	2,900	.011	Lb.	.76	.43	.03	1.22	1.63	

METALS 5

5 METALS

05540		Floor Plates	CREW	DAILY OUTPUT	LABOR-HOURS	UNIT	2005 BARE COSTS				TOTAL INCL O&P	
							MAT.	LABOR	EQUIP.	TOTAL		
200	0100	Welded	E-4	4,400	.007	Lb.	.71	.28	.02	1.01	1.30	**200**
	0300	Pit or trench cover and frame, 1/4" plate, 2' to 3' wide	↓	100	.320	S.F.	20	12.35	.81	33.16	45	
	0400	For galvanizing, add				Lb.	.41			.41	.45	
	0500	Platforms, 1/4" plate, no handrails included, rectangular	E-4	4,200	.008		1.21	.29	.02	1.52	1.88	
	0600	Circular	"	2,500	.013	↓	1.69	.49	.03	2.21	2.79	
700	0010	**TRENCH COVER**, field fabricated										**700**
	0020	Cast iron grating with bar stops and angle frame, to 18" wide	1 Sswk	20	.400	L.F.	53	15.25		68.25	86	
	0100	Frame only (both sides of trench), 1" grating	↓	45	.178	↓	11.65	6.80		18.45	25	
	0150	2" grating		35	.229		18.10	8.70		26.80	35.50	
	0200	Aluminum, stock units, including frames and										
	0210	3/8" plain cover plate, 4" opening	E-4	205	.156	L.F.	33.50	6.05	.40	39.95	48	
	0300	6" opening		185	.173		41.50	6.70	.44	48.64	58	
	0400	10" opening		170	.188		57	7.30	.48	64.78	76	
	0500	16" opening	↓	155	.206		78	8	.52	86.52	101	
	0700	Add per inch for additional widths to 24"					3.11			3.11	3.42	
	0900	For custom fabrication, add					50%					
	1100	For 1/4" plain cover plate, deduct					12%					
	1500	For cover recessed for tile, 1/4" thick, deduct					12%					
	1600	3/8" thick, add					5%					
	1800	For checkered plate cover, 1/4" thick, deduct					12%					
	1900	3/8" thick, add					2%					
	2100	For slotted or round holes in cover, 1/4" thick, add					3%					
	2200	3/8" thick, add					4%					
	2300	For abrasive cover, add				↓	12%					

05560		Metal Castings	CREW	DAILY OUTPUT	LABOR-HOURS	UNIT	MAT.	LABOR	EQUIP.	TOTAL	TOTAL INCL O&P	
200	0010	**CONSTRUCTION CASTINGS**										**200**
	0020	Manhole covers and frames see Division 02630-110										
	0100	Column bases, cast iron, 16" x 16", approx. 65 lbs.	E-4	46	.696	Ea.	102	27	1.77	130.77	163	
	0200	32" x 32", approx. 256 lbs.	"	23	1.391		380	54	3.53	437.53	520	
	0400	Cast aluminum for wood columns, 8" x 8"	1 Carp	32	.250		33	8.55		41.55	49.50	
	0500	12" x 12"	"	32	.250	↓	71	8.55		79.55	91.50	
	0600	Miscellaneous C.I. castings, light sections, less than 150 lbs	E-4	3,200	.010	Lb.	1.58	.39	.03	2	2.46	
	1100	Heavy sections, more than 150 lb	↓	4,200	.008		.77	.29	.02	1.08	1.40	
	1300	Special low volume items	↓	3,200	.010		2.75	.39	.03	3.17	3.75	
	1500	For ductile iron, add				↓	100%					

05580		Formed Metal Fabrications	CREW	DAILY OUTPUT	LABOR-HOURS	UNIT	MAT.	LABOR	EQUIP.	TOTAL	TOTAL INCL O&P	
150	0010	**ALLOY STEEL CHAIN**, Grade 80										**150**
	0015	Self-colored, cut lengths, w/accessories, 1/4"	E-17	4	4	C.L.F.	445	157	1.32	603.32	770	
	0020	3/8"		2	8		575	315	2.64	892.64	1,200	
	0030	1/2"		1.20	13.333		905	520	4.40	1,429.40	1,925	
	0040	5/8"		.72	22.222		1,425	870	7.35	2,302.35	3,125	
	0050	3/4"		.48	33.333		2,075	1,300	11	3,386	4,625	
	0060	7/8"		.40	40		3,075	1,575	13.20	4,663.20	6,200	
	0070	1"		.35	45.714		5,050	1,800	15.10	6,865.10	8,775	
	0080	1-1/4"	↓	.24	66.667	↓	9,425	2,600	22	12,047	15,100	
	0110	Clevis slip hook, 1/4"				Ea.	12.95			12.95	14.25	
	0120	3/8"					19.40			19.40	21.50	
	0130	1/2"					32.50			32.50	36	
	0140	5/8"					49.50			49.50	54	
	0150	3/4"					103			103	113	
	0160	Eye/sling hook w/ hammerlock coupling, 7/8"					289			289	320	
	0170	1"				↓	405			405	450	

Important: See the Reference Section for critical supporting data - Reference Nos., Crews, & City Cost Indexes

05580 | Formed Metal Fabrications

			CREW	DAILY OUTPUT	LABOR-HOURS	UNIT	2005 BARE COSTS				TOTAL INCL O&P	
							MAT.	LABOR	EQUIP.	TOTAL		
150	0180	1-1/4"				Ea.	615			615	675	150
200	0010	**ALUMINUM COLUMNS**										200
	0020	Aluminum, extruded, stock units, no cap or base, 6" diameter	E-4	240	.133	L.F.	8.25	5.15	.34	13.74	18.65	
	0100	8" diameter		170	.188		11.15	7.30	.48	18.93	26	
	0200	10" diameter		150	.213		14.70	8.25	.54	23.49	31.50	
	0300	12" diameter		140	.229		25.50	8.85	.58	34.93	44.50	
	0400	15" diameter		120	.267		31	10.30	.68	41.98	53	
	0410	Caps and bases, plain, 6" diameter				Set	18.05			18.05	19.85	
	0420	8" diameter					22			22	24.50	
	0430	10" diameter					29.50			29.50	32.50	
	0440	12" diameter					54.50			54.50	60	
	0450	15" diameter					82			82	90.50	
	0460	Caps, ornamental, minimum					207			207	228	
	0470	Maximum					905			905	995	
	0500	For square columns, add to column prices above				L.F.	50%					

METALS 5

05650 | Railroad Track & Accessories

05655 | Railroad Trackwork

			CREW	DAILY OUTPUT	LABOR-HOURS	UNIT	2005 BARE COSTS				TOTAL INCL O&P	
							MAT.	LABOR	EQUIP.	TOTAL		
700	0010	**RAILROAD SIDINGS** R05650-100										700
	0020	Car bumpers, standard	B-14	2	24	Ea.	2,525	675	108	3,308	3,950	
	0100	Heavy duty R05650-200		2	24		4,775	675	108	5,558	6,425	
	0200	Derails hand throw (sliding)		10	4.800		795	135	21.50	951.50	1,100	
	0300	Hand throw with standard timbers, open stand & target		8	6		865	169	27	1,061	1,250	
	0400	Resurface and realign existing track		200	.240	L.F.		6.75	1.08	7.83	11.65	
	0600	For crushed stone ballast, add		500	.096	"	12.55	2.70	.43	15.68	18.45	
	0800	Siding, yard spur, level grade										
	0808	Wood ties and ballast, 80 lb. new rail	B-14	57	.842	L.F.	78	23.50	3.79	105.29	127	
	0809	80 lb. relay rail		57	.842		57.50	23.50	3.79	84.79	104	
	0812	90 lb. new rail		57	.842		83.50	23.50	3.79	110.79	133	
	0813	90 lb. relay rail		57	.842		58	23.50	3.79	85.29	104	
	0820	100 lb. new rail		57	.842		88.50	23.50	3.79	115.79	138	
	0822	100 lb. relay rail		57	.842		60.50	23.50	3.79	87.79	107	
	0830	110 lb. new rail		57	.842		95	23.50	3.79	122.29	146	
	0832	110 lb. relay rail		57	.842		64	23.50	3.79	91.29	111	
	1002	Steel ties in concrete, incl. fasteners & plates										
	1003	80 lb. new rail	B-14	22	2.182	L.F.	114	61.50	9.80	185.30	231	
	1005	80 lb. relay rail		22	2.182		94.50	61.50	9.80	165.80	210	
	1012	90 lb. new rail		22	2.182		118	61.50	9.80	189.30	236	
	1015	90 lb. relay rail		22	2.182		97	61.50	9.80	168.30	213	
	1020	100 lb. new rail		22	2.182		122	61.50	9.80	193.30	240	
	1025	100 lb. relay rail		22	2.182		99	61.50	9.80	170.30	215	
	1030	110 lb. new rail		22	2.182		128	61.50	9.80	199.30	247	
	1035	110 lb. relay rail		22	2.182		102	61.50	9.80	173.30	218	
	1200	Switch timber, for a #8 switch, pressure treated		3.70	12.973	M.B.F.	2,125	365	58.50	2,548.50	2,975	
	1300	Complete set of timbers, 3.7 M.B.F. for #8 switch		1	48	Total	8,225	1,350	216	9,791	11,400	
	1400	Ties, concrete, 8'-6" long, 30" O.C.		80	.600	Ea.	85	16.85	2.70	104.55	122	
	1600	Wood, pressure treated, 6" x 8" x 8'-6", C.L. lots		90	.533		35.50	15	2.40	52.90	64.50	
	1700	L.C.L. lots		90	.533		37.50	15	2.40	54.90	66.50	

05655 | Railroad Trackwork

			CREW	DAILY OUTPUT	LABOR-HOURS	UNIT	2005 BARE COSTS				TOTAL INCL O&P	
							MAT.	LABOR	EQUIP.	TOTAL		
700	1900	Heavy duty, 7" x 9" x 8'-6", C.L. lots	B-14	70	.686	Ea.	39	19.25	3.09	61.34	76.50	700
	2000	L.C.L. lots	↓	70	.686	↓	39	19.25	3.09	61.34	76.50	
	2200	Turnouts, #8, incl. 100 lb. rails, plates, bars, frog, switch pt.	R05650-200									
	2300	Timbers and ballast 6" below bottom of tie	B-14	.50	96	Ea.	9,600	2,700	430	12,730	15,300	
	2400	Wheel stops, fixed		18	2.667	Pr.	565	75	12	652	750	
	2450	Hinged	↓	14	3.429	"	765	96.50	15.45	876.95	1,000	
750	0010	**RAILROAD TRACK** R05650-100										750
	0020	Track bolts				Ea.	1.98			1.98	2.18	
	0100	Joint bars				Pr.	60			60	66	
	0200	Spikes				Ea.	.46			.46	.51	
	0300	Tie plates				"	10			10	11	
	1000	Rail, 100 lb. prime grade				L.F.	22.50			22.50	25	
	1500	Relay rail	↓			"	11.25			11.25	12.40	

05730 | Ornamental Formed Metal

			CREW	DAILY OUTPUT	LABOR-HOURS	UNIT	2005 BARE COSTS				TOTAL INCL O&P	
							MAT.	LABOR	EQUIP.	TOTAL		
200	0010	**COLUMNS, ORNAMENTAL**, shop fabricated										200
	6400	Mild steel, flat, 9" wide, stock units, painted, plain	E-4	160	.200	L.F.	5.80	7.75	.51	14.06	21	
	6450	Fancy		160	.200		11.75	7.75	.51	20.01	27.50	
	6500	Corner columns, painted, plain		160	.200		10.30	7.75	.51	18.56	26	
	6550	Fancy	↓	160	.200	↓	21	7.75	.51	29.26	37.50	

05810 | Exp. Joint Cover Assemblies

			CREW	DAILY OUTPUT	LABOR-HOURS	UNIT	2005 BARE COSTS				TOTAL INCL O&P	
							MAT.	LABOR	EQUIP.	TOTAL		
350	0010	**EXPANSION JOINT ASSEMBLIES** Custom units										350
	0200	Floor cover assemblies, 1" space, aluminum	1 Sswk	38	.211	L.F.	15.45	8.05		23.50	31.50	
	0300	Bronze		38	.211		31.50	8.05		39.55	49	
	0500	2" space, aluminum		38	.211		18.65	8.05		26.70	35	
	0600	Bronze		38	.211		33.50	8.05		41.55	51.50	
	0800	Wall and ceiling assemblies, 1" space, aluminum		38	.211		9.20	8.05		17.25	24.50	
	0900	Bronze		38	.211		29.50	8.05		37.55	47	
	1100	2" space, aluminum		38	.211		15.65	8.05		23.70	31.50	
	1200	Bronze		38	.211		36	8.05		44.05	54	
	1400	Floor to wall assemblies, 1" space, aluminum		38	.211		14.05	8.05		22.10	30	
	1500	Bronze or stainless		38	.211		35.50	8.05		43.55	53.50	
	1700	Gym floor angle covers, aluminum, 3" x 3" angle		46	.174		12.15	6.65		18.80	25.50	
	1800	3" x 4" angle		46	.174		14.35	6.65		21	27.50	
	2000	Roof closures, aluminum, flat roof, low profile, 1" space		57	.140		28	5.35		33.35	40.50	
	2100	High profile		57	.140		34.50	5.35		39.85	47.50	
	2300	Roof to wall, low profile, 1" space		57	.140		15.55	5.35		20.90	26.50	
	2400	High profile	↓	57	.140	↓	20	5.35		25.35	31.50	

05910	Metal Cleaning	CREW	DAILY OUTPUT	LABOR-HOURS	UNIT	2005 BARE COSTS				TOTAL INCL O&P	
						MAT.	LABOR	EQUIP.	TOTAL		
500	**0010**	**METAL CLEANING**									**500**
6125	Steel surface treatments										
6170	Wire brush, hand (SSPC-SP2)	1 Psst	240	.033	S.F.	.02	1.03		1.05	1.93	
6180	Power tool (SSPC-SP3)	"	600	.013		.05	.41		.46	.81	
6215	Pressure washing, 2800-6000 S.F./day	1 Pord	3,500	.002			.07		.07	.11	
6220	Steam cleaning, 2800-4000 S.F./day		2,400	.003			.10		.10	.15	
6225	Water blasting	↓	3,200	.002			.08		.08	.12	
6230	Brush-off blast (SSPC-SP7)	E-11	6,000	.005		.08	.16	.02	.26	.38	
6235	Com'l blast (SSPC-SP6), loose scale, fine pwder rust, 2.0 #/S.F. sand		4,000	.008		.15	.24	.03	.42	.62	
6240	Tight mill scale, little/no rust, 3.0 #/S.F. sand		3,200	.010		.23	.30	.04	.57	.81	
6245	Exist coat blistered/pitted, 4.0 #/S.F. sand		2,100	.015		.31	.46	.07	.84	1.19	
6250	Exist coat badly pitted/nodules, 6.7 #/S.F. sand		1,300	.025		.52	.75	.11	1.38	1.95	
6255	Near white blast (SSPC-SP10), loose scale, fine rust, 5.6 #/S.F. sand		1,700	.019		.43	.57	.08	1.08	1.53	
6260	Tight mill scale, little/no rust, 6.9 #/S.F. sand		1,400	.023		.53	.70	.10	1.33	1.86	
6265	Exist coat blistered/pitted, 9.0 #/S.F. sand		1,100	.029		.69	.89	.13	1.71	2.39	
6270	Exist coat badly pitted/nodules, 11.3 #/S.F. sand	↓	800	.040	↓	.87	1.22	.17	2.26	3.20	

05950	Paints & Protective Coatings										
650	**0010**	**PAINTS & PROTECTIVE COATINGS**									**650**
5900	Galvanizing structural steel in shop, under 1 ton				Ton	515			515	565	
5950	1 ton to 20 tons					430			430	470	
6000	Over 20 tons				↓	375			375	415	
6100	Cold galvanizing, brush in field	1 Psst	1,100	.007	S.F.	.07	.23		.30	.48	
6510	Paints & protective coatings, sprayed in field										
6520	Alkyds, primer	2 Psst	3,600	.004	S.F.	.06	.14		.20	.31	
6540	Gloss topcoats		3,200	.005		.05	.16		.21	.34	
6560	Silicone alkyd		3,200	.005		.10	.16		.26	.40	
6610	Epoxy, primer		3,000	.005		.17	.17		.34	.49	
6630	Intermediate or topcoat		2,800	.006		.14	.18		.32	.48	
6650	Enamel coat		2,800	.006		.17	.18		.35	.52	
6700	Epoxy ester, primer		2,800	.006		.34	.18		.52	.71	
6720	Topcoats		2,800	.006		.11	.18		.29	.46	
6810	Latex primer		3,600	.004		.05	.14		.19	.31	
6830	Topcoats		3,200	.005		.06	.16		.22	.35	
6910	Universal primers, one part, phenolic, modified alkyd		2,000	.008		.08	.25		.33	.55	
6940	Two part, epoxy spray		2,000	.008		.20	.25		.45	.68	
7000	Zinc rich primers, self cure, spray, inorganic		1,800	.009		.51	.28		.79	1.07	
7010	Epoxy, spray, organic	↓	1,800	.009	↓	.14	.28		.42	.66	
7020	Above one story, spray painting simple structures, add						25%				
7030	Intricate structures, add						50%				

For information about Means Estimating Seminars, see yellow pages 12 and 13 in back of book

METALS 5

Division Notes

	CREW	DAILY OUTPUT	LABOR-HOURS	UNIT	2005 BARE COSTS				TOTAL INCL O&P
					MAT.	LABOR	EQUIP.	TOTAL	

Division 6
Wood & Plastics

Estimating Tips

06050 Basic Wood & Plastic Materials & Methods

- Common to any wood framed structure are the accessory connector items such as screws, nails, adhesives, hangers, connector plates, straps, angles and holdowns. For typical wood framed buildings, such as residential projects, the aggregate total for these items can be significant, especially in areas where seismic loading is a concern. For floor and wall framing, the material cost is based on 10 to 25 lbs. per MBF. Holdowns, hangers and other connectors should be taken off by the piece.

06100 Rough Carpentry

- Lumber is a traded commodity and therefore sensitive to supply and demand in the marketplace. Even in "budgetary" estimating of wood framed projects, it is advisable to call local suppliers for the latest market pricing.

- Common quantity units for wood framed projects are "thousand board feet" (MBF). A board foot is a volume of wood, 1″ x 1′ x 1′,

or 144 cubic inches. Board foot quantities are generally calculated using nominal material dimensions—dressed sizes are ignored. Board foot per lineal foot of any stick of lumber can be calculated by dividing the nominal cross sectional area by 12. As an example, 2,000 lineal feet of 2 x 12 equates to 4 MBF by dividing the nominal area, 2 x 12, by 12, which equals 2, and multiplying by 2,000 to give 4,000 board feet. This simple rule applies to all nominal dimensioned lumber.

- Waste is an issue of concern at the quantity takeoff for any area of construction. Framing lumber is sold in even foot lengths, i.e., 10′, 12′, 14′, 16′, and depending on spans, wall heights and the grade of lumber, waste is inevitable. A rule of thumb for lumber waste is 5% to 10% depending on material quality and the complexity of the framing.

- Wood in various forms and shapes is used in many projects, even where the main structural framing is steel, concrete or masonry. Plywood as a back-up partition material and 2x boards used as blocking and cant strips around roof edges are two common examples. The estimator should

ensure that the costs of all wood materials are included in the final estimate.

06200 Finish Carpentry

- It is necessary to consider the grade of workmanship when estimating labor costs for erecting millwork and interior finish. In practice, there are three grades: premium, custom and economy. The Means daily output for base and case moldings is in the range of 200 to 250 L.F. per carpenter per day. This is appropriate for most average custom grade projects. For premium projects an adjustment to productivity of 25% to 50% should be made depending on the complexity of the job.

Reference Numbers

Reference numbers are shown in bold squares at the beginning of some major classifications. These numbers refer to related items in the Reference Section. The reference information may be an estimating procedure, an alternate pricing method or technical information.

Note: Not all subdivisions listed here necessarily appear in this publication.

06052	Selective Demolition		CREW	DAILY OUTPUT	LABOR-HOURS	UNIT	2005 BARE COSTS				TOTAL INCL O&P
							MAT.	LABOR	EQUIP.	TOTAL	
110	0010	**SELECTIVE DEMOLITION, WOOD FRAMING** R02220 -510									110
	0100	Timber connector, nailed, small	1 Clab	96	.083	Ea.		2.22		2.22	3.46
	0110	Medium		60	.133			3.56		3.56	5.55
	0120	Large		48	.167			4.45		4.45	6.95
	0130	Bolted, small		48	.167			4.45		4.45	6.95
	0140	Medium		32	.250			6.70		6.70	10.40
	0150	Large		24	.333			8.90		8.90	13.85
	3000	6″ x 8″	B-1	275	.087	L.F.		2.39		2.39	3.72
	3040	6″ x 10″		220	.109			2.99		2.99	4.65
	3080	6″ x 12″		185	.130			3.55		3.55	5.55
	3120	8″ x 12″		140	.171			4.69		4.69	7.30
	3160	10″ x 12″		110	.218			5.95		5.95	9.30
	3162	Alternate pricing method		1.10	21.818	M.B.F.		595		595	930
	3400	Fascia boards, 1″ x 6″	1 Clab	500	.016	L.F.		.43		.43	.66
	3440	1″ x 8″		450	.018			.47		.47	.74
	3480	1″ x 10″		400	.020			.53		.53	.83
	3490	2″ x 6″		450	.018			.47		.47	.74
	3500	2″ x 8″		400	.020			.53		.53	.83
	3510	2″ x 10″		350	.023			.61		.61	.95
	3800	Headers over openings, 2 @ 2″ x 6″		110	.073			1.94		1.94	3.02
	3840	2 @ 2″ x 8″		100	.080			2.14		2.14	3.32
	4230	2″ x 6″	2 Clab	970	.016			.44		.44	.69
	4240	2″ x 8″		940	.017			.45		.45	.71
	4250	2″ x 10″		910	.018			.47		.47	.73
	4280	2″ x 12″		880	.018			.49		.49	.76
	4281	2″ x 14″		850	.019			.50		.50	.78
	4290	Wood joists, alternate pricing method		1.50	10.667	M.B.F.		285		285	445
	4300	Plates, single		1,000	.016	L.F.		.43		.43	.66
	4320	Double		800	.020			.53		.53	.83
	5400	Posts, 4″ x 4″		800	.020			.53		.53	.83
	5440	6″ x 6″		400	.040			1.07		1.07	1.66
	5480	8″ x 8″		300	.053			1.42		1.42	2.22
	5500	10″ x 10″		240	.067			1.78		1.78	2.77
	5800	2″ x 6″ (alternate method)		850	.019			.50		.50	.78
	5840	2″ x 8″ (alternate method)		837	.019			.51		.51	.79
	6200	Stairs and stringers, minimum		40	.400	Riser		10.70		10.70	16.60
	6240	Maximum		26	.615	″		16.45		16.45	25.50
	6590	Wood studs, 2″ x 3″		3,076	.005	L.F.		.14		.14	.22
	6600	2″ x 4″		2,000	.008			.21		.21	.33
	6640	2″ x 6″		1,600	.010			.27		.27	.42
	7000	Trusses									
	7050	12′ span	2 Clab	74	.216	Ea.		5.75		5.75	9
	7150	24′ span	F-3	66	.606			21	9.15	30.15	42.50
	7350	32′ span		56	.714			24.50	10.75	35.25	50
	7450	36′ span		52	.769			26.50	11.60	38.10	54
	9500	See Div. 02220-350 for rubbish handling									
120	0010	**SELECTIVE DEMOLITION, MILLWORK AND TRIM** R02220 -510									120
	1000	Cabinets, wood, base cabinets, per L.F.	2 Clab	80	.200	L.F.		5.35		5.35	8.30
	1020	Wall cabinets, per L.F.	″	80	.200	″		5.35		5.35	8.30
	1060	Remove and reset, base cabinets	2 Carp	18	.889	Ea.		30.50		30.50	47.50
	1070	Wall cabinets		20	.800			27.50		27.50	42.50
	1072	Oven cabinet, 7′ high		11	1.455			50		50	77.50
	1074	Cabinet door, up to 2′ high	1 Clab	66	.121			3.24		3.24	5.05
	1076	2′ - 4′ high	″	46	.174			4.64		4.64	7.25
	1100	Steel, painted, base cabinets	2 Clab	60	.267	L.F.		7.10		7.10	11.10
	1120	Wall cabinets		60	.267	″		7.10		7.10	11.10

6 WOOD & PLASTICS

Important: See the Reference Section for critical supporting data - Reference Nos., Crews, & City Cost Indexes

06050 | Basic Wood / Plastic Materials / Methods

06052 | Selective Demolition

			CREW	DAILY OUTPUT	LABOR-HOURS	UNIT	2005 BARE COSTS				TOTAL INCL O&P	
							MAT.	LABOR	EQUIP.	TOTAL		
120	1200	Casework, large area	2 Clab	320	.050	S.F.		1.34		1.34	2.08	120
	1220	Selective [R02220-510]		200	.080	"		2.14		2.14	3.32	
	1500	Counter top, minimum		200	.080	L.F.		2.14		2.14	3.32	
	1510	Maximum		120	.133			3.56		3.56	5.55	
	1550	Remove and reset, minimum	2 Carp	50	.320			10.95		10.95	17.05	
	1560	Maximum	"	40	.400			13.70		13.70	21.50	
	2000	Paneling, 4' x 8' sheets	2 Clab	2,000	.008	S.F.		.21		.21	.33	
	2100	Boards, 1" x 4"		700	.023			.61		.61	.95	
	2120	1" x 6"		750	.021			.57		.57	.89	
	2140	1" x 8"		800	.020			.53		.53	.83	
	3000	Trim, baseboard, to 6" wide		1,200	.013	L.F.		.36		.36	.55	
	3040	Greater than 6" and up to 12" wide		1,000	.016			.43		.43	.66	
	3080	Remove and reset, minimum	2 Carp	400	.040			1.37		1.37	2.13	
	3090	Maximum	"	300	.053			1.83		1.83	2.85	
	3100	Ceiling trim	2 Clab	1,000	.016			.43		.43	.66	
	3120	Chair rail		1,200	.013			.36		.36	.55	
	3140	Railings with balusters		240	.067			1.78		1.78	2.77	
	3160	Wainscoting		700	.023	S.F.		.61		.61	.95	

06070 | Lumber Treatment

			CREW	DAILY OUTPUT	LABOR-HOURS	UNIT	MAT.	LABOR	EQUIP.	TOTAL	TOTAL INCL O&P	
400	0011	**LUMBER TREATMENT**										400
	0400	Fire retardant, wet				M.B.F.	291			291	320	
	0500	KDAT					277			277	305	
	0700	Salt treated, water borne, .40 lb. retention					133			133	146	
	0800	Oil borne, 8 lb. retention					156			156	171	
	1000	Kiln dried lumber, 1" & 2" thick, softwoods					88.50			88.50	97.50	
	1100	Hardwoods					94.50			94.50	104	
	1500	For small size 1" stock, add					11.95			11.95	13.15	
	1700	For full size rough lumber, add					20%					
600	0010	**PLYWOOD TREATMENT** Fire retardant, 1/4" thick				M.S.F.	222			222	244	600
	0030	3/8" thick					244			244	268	
	0050	1/2" thick					261			261	287	
	0070	5/8" thick					277			277	305	
	0100	3/4" thick					305			305	335	
	0200	For KDAT, add					66.50			66.50	73	
	0500	Salt treated water borne, .25 lb., wet, 1/4" thick					122			122	134	
	0530	3/8" thick					127			127	140	
	0550	1/2" thick					133			133	146	
	0570	5/8" thick					145			145	159	
	0600	3/4" thick					150			150	165	
	0800	For KDAT add					66.50			66.50	73	
	0900	For .40 lb., per C.F. retention, add					55.50			55.50	61	
	1000	For certification stamp, add					32.50			32.50	36	

06090 | Wood & Plastic Fastenings

			CREW	DAILY OUTPUT	LABOR-HOURS	UNIT	MAT.	LABOR	EQUIP.	TOTAL	TOTAL INCL O&P	
600	0010	**NAILS** Prices of material only, based on 50# box purchase, copper, plain				Lb.	4.99			4.99	5.50	600
	0400	Stainless steel, plain					5.75			5.75	6.30	
	0500	Box, 3d to 20d, bright					1.32			1.32	1.45	
	0520	Galvanized					1.44			1.44	1.58	
	0600	Common, 3d to 60d, plain					1.02			1.02	1.12	
	0700	Galvanized					1.39			1.39	1.53	
	0800	Aluminum					3.38			3.38	3.72	
	1000	Annular or spiral thread, 4d to 60d, plain					1.60			1.60	1.76	
	1200	Galvanized					1.80			1.80	1.98	
	1400	Drywall nails, plain					.79			.79	.87	

WOOD & PLASTICS 6

For expanded coverage of these items see *Means Interior Cost Data 2005*

06090	Wood & Plastic Fastenings	CREW	DAILY OUTPUT	LABOR-HOURS	UNIT	2005 BARE COSTS				TOTAL INCL O&P	
						MAT.	LABOR	EQUIP.	TOTAL		
600	1600	Galvanized				Lb.	1.54			1.54	1.69	**600**
	1800	Finish nails, 4d to 10d, plain					.94			.94	1.03	
	2000	Galvanized					1.32			1.32	1.45	
	2100	Aluminum					4.10			4.10	4.51	
	2300	Flooring nails, hardened steel, 2d to 10d, plain					1.40			1.40	1.54	
	2400	Galvanized					2.25			2.25	2.48	
	2500	Gypsum lath nails, 1-1/8", 13 ga. flathead, blued					1.54			1.54	1.69	
	2600	Masonry nails, hardened steel, 3/4" to 3" long, plain					1.48			1.48	1.63	
	2700	Galvanized					1.87			1.87	2.06	
	2900	Roofing nails, threaded, galvanized					1.35			1.35	1.49	
	3100	Aluminum					4.80			4.80	5.30	
	3300	Compressed lead head, threaded, galvanized					1.50			1.50	1.65	
	3600	Siding nails, plain shank, galvanized					1.45			1.45	1.60	
	3800	Aluminum					4.11			4.11	4.52	
	5000	Add to prices above for cement coating					.10			.10	.11	
	5200	Zinc or tin plating					.13			.13	.14	
	5500	Vinyl coated sinkers, 8d to 16d					.57			.57	.63	
650	0010	**NAILS** mat. only, for pneumatic tools, framing, per carton of 5000, 2"				Ea.	37			37	41	**650**
	0100	2-3/8"					42.50			42.50	46.50	
	0200	Per carton of 4000, 3"					37.50			37.50	41.50	
	0300	3-1/4"					40			40	44	
	0400	Per carton of 5000, 2-3/8", galv.					57.50			57.50	63.50	
	0500	Per carton of 4000, 3", galv.					65			65	71.50	
	0600	3-1/4", galv.					80.50			80.50	88.50	
	0700	Roofing, per carton of 7200, 1"					35			35	38.50	
	0800	1-1/4"					32.50			32.50	36	
	0900	1-1/2"					37.50			37.50	41.50	
	1000	1-3/4"					45.50			45.50	50	
700	0010	**SHEET METAL SCREWS** Steel, standard, #8 x 3/4", plain				C	2.81			2.81	3.09	**700**
	0100	Galvanized					3.44			3.44	3.78	
	0300	#10 x 1", plain					3.76			3.76	4.14	
	0400	Galvanized					4.35			4.35	4.79	
	0600	With washers, #14 x 1", plain					10.20			10.20	11.20	
	0700	Galvanized					10.20			10.20	11.20	
	0900	#14 x 2", plain					10.20			10.20	11.20	
	1000	Galvanized					18.15			18.15	19.95	
	1500	Self-drilling, with washers, (pinch point) #8 x 3/4", plain					6.10			6.10	6.70	
	1600	Galvanized					6.10			6.10	6.70	
	1800	#10 x 3/4", plain					6.10			6.10	6.70	
	1900	Galvanized					6.10			6.10	6.70	
	3000	Stainless steel w/aluminum or neoprene washers, #14 x 1", plain					18.35			18.35	20	
	3100	#14 x 2", plain					25			25	27.50	
750	0010	**WOOD SCREWS** #8, 1" long, steel				C	3.36			3.36	3.70	**750**
	0100	Brass					11.20			11.20	12.35	
	0200	#8, 2" long, steel					3.76			3.76	4.14	
	0300	Brass					11.70			11.70	12.90	
	0400	#10, 1" long, steel					4.30			4.30	4.73	
	0500	Brass					23			23	25	
	0600	#10, 2" long, steel					7.65			7.65	8.45	
	0700	Brass					40.50			40.50	44.50	
	0800	#10, 3" long, steel					11.95			11.95	13.15	
	1000	#12, 2" long, steel					4.89			4.89	5.40	
	1100	Brass					16.30			16.30	17.95	
	1500	#12, 3" long, steel					16.20			16.20	17.85	

6 WOOD & PLASTICS

Important: See the Reference Section for critical supporting data - Reference Nos., Crews, & City Cost Indexes

06090	Wood & Plastic Fastenings	CREW	DAILY OUTPUT	LABOR-HOURS	UNIT	2005 BARE COSTS MAT.	LABOR	EQUIP.	TOTAL	TOTAL INCL O&P	
750	2000	#12, 4" long, steel				C	29			29	32	**750**
800	0010	**TIMBER CONNECTORS** Add up cost of each part for total										**800**
	0020	cost of connection										
	0200	Bolts, machine, sq. hd. with nut & washer, 1/2" diameter, 4" long	1 Carp	140	.057	Ea.	.36	1.96		2.32	3.45	
	0300	7-1/2" long		130	.062		.63	2.11		2.74	3.97	
	0500	3/4" diameter, 7-1/2" long		130	.062		1.66	2.11		3.77	5.10	
	0610	Machine bolts, w/ nut, washer, 3/4" dia, 15" L, HD's & beam hangers		95	.084		3.05	2.88		5.93	7.85	
	0800	Drilling bolt holes in timber, 1/2" diameter		450	.018	Inch		.61		.61	.95	
	0900	1" diameter		350	.023	"		.78		.78	1.22	
	1100	Framing anchors, 2 or 3 dimensional, 10 gauge, no nails incl.		175	.046	Ea.	.46	1.57		2.03	2.95	
	1150	Framing anchors, 18 gauge, 4 1/2" x 2 3/4"		175	.046		.46	1.57		2.03	2.95	
	1160	Framing anchors, 18 gauge, 4 1/2" x 3"		175	.046		.46	1.57		2.03	2.95	
	1170	Clip anchors plates, 18 gauge, 12" x 1 1/8"		175	.046		.46	1.57		2.03	2.95	
	1250	Holdowns, 3 gauge base, 10 gauge body		8	1		15.60	34.50		50.10	70.50	
	1260	Holdowns, 7 gauge 11 1/16" x 3 1/4"		8	1		15.60	34.50		50.10	70.50	
	1270	Holdowns, 7 gauge 14 3/8" x 3 1/8"		8	1		15.60	34.50		50.10	70.50	
	1275	Holdowns, 12 gauge 8" x 2 1/2"		8	1		15.60	34.50		50.10	70.50	
	1300	Joist and beam hangers, 18 ga. galv., for 2" x 4" joist		175	.046		.58	1.57		2.15	3.08	
	1400	2" x 6" to 2" x 10" joist		165	.048		.67	1.66		2.33	3.33	
	1600	16 ga. galv., 3" x 6" to 3" x 10" joist		160	.050		2.64	1.71		4.35	5.55	
	1700	3" x 10" to 3" x 14" joist		160	.050		3.06	1.71		4.77	6.05	
	1800	4" x 6" to 4" x 10" joist		155	.052		2.32	1.77		4.09	5.30	
	1900	4" x 10" to 4" x 14" joist		155	.052		3.14	1.77		4.91	6.20	
	2000	Two-2" x 6" to two-2" x 10" joists		150	.053		2.53	1.83		4.36	5.65	
	2100	Two-2" x 10" to two-2" x 14" joists		150	.053		2.53	1.83		4.36	5.65	
	2300	3/16" thick, 6" x 8" joist		145	.055		5.50	1.89		7.39	9	
	2400	6" x 10" joist		140	.057		6.50	1.96		8.46	10.20	
	2500	6" x 12" joist		135	.059		7.80	2.03		9.83	11.75	
	2700	1/4" thick, 6" x 14" joist		130	.062		9.70	2.11		11.81	14	
	2900	Plywood clips, extruded aluminum H clip, for 3/4" panels					.15			.15	.17	
	3000	Galvanized 18 ga. back-up clip					.14			.14	.15	
	3200	Post framing, 16 ga. galv. for 4" x 4" base, 2 piece	1 Carp	130	.062		5.45	2.11		7.56	9.30	
	3300	Cap		130	.062		2.67	2.11		4.78	6.20	
	3500	Rafter anchors, 18 ga. galv., 1-1/2" wide, 5-1/4" long		145	.055		.46	1.89		2.35	3.45	
	3600	10-3/4" long		145	.055		.91	1.89		2.80	3.94	
	3800	Shear plates, 2-5/8" diameter		120	.067		1.63	2.28		3.91	5.35	
	3900	4" diameter		115	.070		3.75	2.38		6.13	7.85	
	4000	Sill anchors, embedded in concrete or block, 18-5/8" long		115	.070		1.10	2.38		3.48	4.92	
	4100	Spike grids, 4" x 4", flat or curved		120	.067		.41	2.28		2.69	4.01	
	4400	Split rings, 2-1/2" diameter		120	.067		1.32	2.28		3.60	5	
	4500	4" diameter		110	.073		2.04	2.49		4.53	6.10	
	4550	Tie plate, 20 gauge, 7" x 3 1/8"		110	.073		2.04	2.49		4.53	6.10	
	4560	Tie plate, 20 gauge, 5" x 4 1/8"		110	.073		2.04	2.49		4.53	6.10	
	4575	Twist straps, 18 gauge, 12" x 1 1/4"		110	.073		2.04	2.49		4.53	6.10	
	4580	Twist straps, 18 gauge, 16" x 1 1/4"		110	.073		2.04	2.49		4.53	6.10	
	4600	Strap ties, 20 ga., 2 -1/16" wide, 12 13/16" long		180	.044		1.19	1.52		2.71	3.68	
	4700	Strap ties, 16 ga., 1-3/8" wide, 12" long		180	.044		1.19	1.52		2.71	3.68	
	4800	24" long		160	.050		1.64	1.71		3.35	4.47	
	5000	Toothed rings, 2-5/8" or 4" diameter		90	.089		1.13	3.04		4.17	6	
	5200	Truss plates, nailed, 20 gauge, up to 32' span		17	.471	Truss	8.15	16.10		24.25	34	
	5400	Washers, 2" x 2" x 1/8"				Ea.	.26			.26	.29	
	5500	3" x 3" x 3/16"				"	.67			.67	.74	
825	0010	**ROUGH HARDWARE** Average % of carpentry material, minimum					.50%					**825**
	0200	Maximum					1.50%					

WOOD & PLASTICS 6

06090	Wood & Plastic Fastenings	CREW	DAILY OUTPUT	LABOR-HOURS	UNIT	2005 BARE COSTS				TOTAL INCL O&P	
						MAT.	LABOR	EQUIP.	TOTAL		
825	0210	In seismic or hurricane areas, up to					10%					825

06100 | Rough Carpentry

	06110	Wood Framing	CREW	DAILY OUTPUT	LABOR-HOURS	UNIT	2005 BARE COSTS				TOTAL INCL O&P	
							MAT.	LABOR	EQUIP.	TOTAL		
100	0010	**BLOCKING**										100
	2600	Miscellaneous, to wood construction										
	2620	2" x 4"	1 Carp	.17	47.059	M.B.F.	510	1,600		2,110	3,050	
	2625	Pneumatic nailed		.21	38.095		510	1,300		1,810	2,575	
	2660	2" x 8"		.27	29.630		620	1,025		1,645	2,250	
	2665	Pneumatic nailed	▼	.33	24.242	▼	620	830		1,450	1,975	
	2720	To steel construction										
	2740	2" x 4"	1 Carp	.14	57.143	M.B.F.	510	1,950		2,460	3,600	
	2780	2" x 8"	"	.21	38.095	"	620	1,300		1,920	2,700	
200	0010	**BRIDGING** Wood, for joists 16" O.C., 1" x 3"	1 Carp	1.30	6.154	C.Pr.	42.50	211		253.50	375	200
	0015	Pneumatic nailed		1.70	4.706		42.50	161		203.50	298	
	0100	2" x 3" bridging		1.30	6.154		39.50	211		250.50	375	
	0105	Pneumatic nailed		1.70	4.706		39.50	161		200.50	295	
	0300	Steel, galvanized, 18 ga., for 2" x 10" joists at 12" O.C.		1.30	6.154		98	211		309	440	
	0400	24" O.C.		1.40	5.714		118	196		314	435	
	0900	Compression type, 16" O.C., 2" x 8" joists		2	4		115	137		252	340	
	1000	2" x 12" joists	▼	2	4	▼	115	137		252	340	
505	0010	**FRAMING, BEAMS & GIRDERS** R06100 -010										505
	1000	Single, 2" x 6"	2 Carp	700	.023	L.F.	.54	.78		1.32	1.81	
	1005	Pneumatic nailed		812	.020		.54	.67		1.21	1.64	
	1020	2" x 8"		650	.025		.83	.84		1.67	2.22	
	1025	Pneumatic nailed		754	.021		.83	.73		1.56	2.04	
	1040	2" x 10"		600	.027		1.18	.91		2.09	2.71	
	1045	Pneumatic nailed		696	.023		1.18	.79		1.97	2.52	
	1060	2" x 12"		550	.029		1.60	1		2.60	3.31	
	1065	Pneumatic nailed		638	.025		1.60	.86		2.46	3.10	
	1080	2" x 14"		500	.032		2.16	1.10		3.26	4.08	
	1085	Pneumatic nailed		580	.028		2.16	.95		3.11	3.84	
	1100	3" x 8"		550	.029		2.59	1		3.59	4.40	
	1120	3" x 10"		500	.032		3.26	1.10		4.36	5.30	
	1140	3" x 12"		450	.036		3.92	1.22		5.14	6.20	
	1160	3" x 14"	▼	400	.040		4.73	1.37		6.10	7.35	
	1180	4" x 8"	F-3	1,000	.040		2.69	1.38	.60	4.67	5.75	
	1200	4" x 10"		950	.042		3.37	1.46	.63	5.46	6.65	
	1220	4" x 12"		900	.044		4.40	1.54	.67	6.61	7.95	
	1240	4" x 14"	▼	850	.047		4.95	1.63	.71	7.29	8.75	
	2000	Double, 2" x 6"	2 Carp	625	.026		1.07	.88		1.95	2.55	
	2005	Pneumatic nailed		725	.022		1.07	.76		1.83	2.36	
	2020	2" x 8"		575	.028		1.65	.95		2.60	3.30	
	2025	Pneumatic nailed		667	.024		1.65	.82		2.47	3.10	
	2040	2" x 10"		550	.029		2.35	1		3.35	4.14	
	2045	Pneumatic nailed		638	.025		2.35	.86		3.21	3.93	
	2060	2" x 12"	▼	525	.030	▼	3.20	1.04		4.24	5.15	

Important: See the Reference Section for critical supporting data - Reference Nos., Crews, & City Cost Indexes

06110 | Wood Framing

		CREW	DAILY OUTPUT	LABOR-HOURS	UNIT	2005 BARE COSTS MAT.	LABOR	EQUIP.	TOTAL	TOTAL INCL O&P	
505	2065	Pneumatic nailed	2 Carp	610	.026	L.F.	3.20	.90		4.10	4.92	**505**
	2080	2" x 14"		475	.034		4.32	1.15		5.47	6.55	
	2085	Pneumatic nailed		551	.029		4.32	.99		5.31	6.30	
	3000	Triple, 2" x 6"		550	.029		1.61	1		2.61	3.32	
	3005	Pneumatic nailed		638	.025		1.61	.86		2.47	3.11	
	3020	2" x 8"		525	.030		2.48	1.04		3.52	4.36	
	3025	Pneumatic nailed		609	.026		2.48	.90		3.38	4.13	
	3040	2" x 10"		500	.032		3.53	1.10		4.63	5.60	
	3045	Pneumatic nailed		580	.028		3.53	.95		4.48	5.35	
	3060	2" x 12"		475	.034		4.80	1.15		5.95	7.10	
	3065	Pneumatic nailed		551	.029		4.80	.99		5.79	6.85	
	3080	2" x 14"		450	.036		6.50	1.22		7.72	9	
	3085	Pneumatic nailed		522	.031		6.50	1.05		7.55	8.75	
515	0010	**FRAMING, COLUMNS**										**515**
	0100	4" x 4"	2 Carp	390	.041	L.F.	1.25	1.41		2.66	3.57	
	0150	4" x 6"		275	.058		2	1.99		3.99	5.30	
	0200	4" x 8"		220	.073		2.69	2.49		5.18	6.85	
	0250	6" x 6"		215	.074		4.45	2.55		7	8.85	
	0300	6" x 8"		175	.091		5.85	3.13		8.98	11.35	
	0350	6" x 10"		150	.107		9.50	3.65		13.15	16.10	
520	0010	**FRAMING, HEAVY** Mill timber, beams, single 6" x 10"	2 Carp	1.10	14.545	M.B.F.	2,125	500		2,625	3,125	**520**
	0100	Single 8" x 16"		1.20	13.333		2,675	455		3,130	3,650	
	0200	Built from 2" lumber, multiple 2" x 14"		.90	17.778		925	610		1,535	1,975	
	0210	Built from 3" lumber, multiple 3" x 6"		.70	22.857		1,275	785		2,060	2,625	
	0220	Multiple 3" x 8"		.80	20		1,300	685		1,985	2,500	
	0230	Multiple 3" x 10"		.90	17.778		1,300	610		1,910	2,375	
	0240	Multiple 3" x 12"		1	16		1,300	550		1,850	2,300	
	0250	Built from 4" lumber, multiple 4" x 6"		.80	20		1,000	685		1,685	2,175	
	0260	Multiple 4" x 8"		.90	17.778		1,000	610		1,610	2,050	
	0270	Multiple 4" x 10"		1	16		1,000	550		1,550	1,975	
	0280	Multiple 4" x 12"		1.10	14.545		1,100	500		1,600	1,975	
	0290	Columns, structural grade, 1500f, 4" x 4"		.60	26.667		1,725	915		2,640	3,325	
	0300	6" x 6"		.65	24.615		2,125	845		2,970	3,650	
	0400	8" x 8"		.70	22.857		2,300	785		3,085	3,775	
	0500	10" x 10"		.75	21.333		2,350	730		3,080	3,750	
	0600	12" x 12"		.80	20		2,375	685		3,060	3,675	
	0800	Floor planks, 2" thick, T & G, 2" x 6"		1.05	15.238		1,850	520		2,370	2,875	
	0900	2" x 10"		1.10	14.545		1,875	500		2,375	2,825	
	1100	3" thick, 3" x 6"		1.05	15.238		1,325	520		1,845	2,300	
	1200	3" x 10"		1.10	14.545		1,325	500		1,825	2,250	
	1400	Girders, structural grade, 12" x 12"		.80	20		1,925	685		2,610	3,175	
	1500	10" x 16"		1	16		1,875	550		2,425	2,900	
	2050	Roof planks, see division 06150-600										
	2300	Roof purlins, 4" thick, structural grade	2 Carp	1.05	15.238	M.B.F.	1,300	520		1,820	2,275	
	2500	Roof trusses, add timber connectors, division 06090-800	"	.45	35.556	"	1,275	1,225		2,500	3,300	
530	0010	**FRAMING, JOISTS**										**530**
	2000	Joists, 2" x 4"	2 Carp	1,250	.013	L.F.	.34	.44		.78	1.05	
	2005	Pneumatic nailed		1,438	.011		.34	.38		.72	.96	
	2100	2" x 6"		1,250	.013		.54	.44		.98	1.27	
	2105	Pneumatic nailed		1,438	.011		.54	.38		.92	1.18	
	2150	2" x 8"		1,100	.015		.83	.50		1.33	1.69	
	2155	Pneumatic nailed		1,265	.013		.83	.43		1.26	1.58	
	2200	2" x 10"		900	.018		1.18	.61		1.79	2.24	

Reference codes: R06100-010, R06110-030

WOOD & PLASTICS 6

For expanded coverage of these items see *Means Interior Cost Data 2005*

06110	Wood Framing	CREW	DAILY OUTPUT	LABOR-HOURS	UNIT	MAT.	LABOR	EQUIP.	TOTAL	TOTAL INCL O&P	
530	2205	Pneumatic nailed	2 Carp	1,035	.015	L.F.	1.18	.53		1.71	2.11	**530**
	2250	2" x 12"		875	.018		1.60	.63		2.23	2.74	
	2255	Pneumatic nailed		1,006	.016		1.60	.54		2.14	2.61	
	2300	2" x 14"		770	.021		2.16	.71		2.87	3.48	
	2305	Pneumatic nailed		886	.018		2.16	.62		2.78	3.33	
	2350	3" x 6"		925	.017		1.92	.59		2.51	3.03	
	2400	3" x 10"		780	.021		3.26	.70		3.96	4.68	
	2450	3" x 12"		600	.027		3.92	.91		4.83	5.75	
	2500	4" x 6"		800	.020		2	.69		2.69	3.27	
	2550	4" x 10"		600	.027		3.37	.91		4.28	5.15	
	2600	4" x 12"		450	.036		4.40	1.22		5.62	6.75	
	2605	Sister joist, 2" x 6"		800	.020		.54	.69		1.23	1.66	
	2606	Pneumatic nailed		960	.017		.54	.57		1.11	1.48	
	2610	2" x 8"		640	.025		.83	.86		1.69	2.24	
	2611	Pneumatic nailed		768	.021		.83	.71		1.54	2.02	
	2615	2" x 10"		535	.030		1.18	1.02		2.20	2.89	
	2616	Pneumatic nailed		642	.025		1.18	.85		2.03	2.62	
	2620	2" x 12"		455	.035		1.60	1.20		2.80	3.64	
	2625	Pneumatic nailed	▼	546	.029	▼	1.60	1		2.60	3.32	
550	0010	**PARTITIONS** Wood stud with single bottom plate and										**550**
	0020	double top plate, no waste, std. & better lumber										
	0180	2" x 4" studs, 8' high, studs 12" O.C.	2 Carp	80	.200	L.F.	3.73	6.85		10.58	14.75	
	0185	12" O.C., pneumatic nailed		96	.167		3.73	5.70		9.43	13	
	0200	16" O.C.		100	.160		3.05	5.50		8.55	11.90	
	0205	16" O.C., pneumatic nailed		120	.133		3.05	4.57		7.62	10.45	
	0300	24" O.C.		125	.128		2.37	4.38		6.75	9.45	
	0305	24" O.C., pneumatic nailed		150	.107		2.37	3.65		6.02	8.30	
	0380	10' high, studs 12" O.C.		80	.200		4.40	6.85		11.25	15.50	
	0385	12" O.C., pneumatic nailed		96	.167		4.40	5.70		10.10	13.75	
	0400	16" O.C.		100	.160		3.56	5.50		9.06	12.45	
	0405	16" O.C., pneumatic nailed		120	.133		3.56	4.57		8.13	11	
	0500	24" O.C.		125	.128		2.71	4.38		7.09	9.85	
	0505	24" O.C., pneumatic nailed		150	.107		2.71	3.65		6.36	8.70	
	0580	12' high, studs 12" O.C.		65	.246		5.10	8.45		13.55	18.75	
	0585	12" O.C., pneumatic nailed		78	.205		5.10	7.05		12.15	16.55	
	0600	16" O.C.		80	.200		4.06	6.85		10.91	15.10	
	0605	16" O.C., pneumatic nailed		96	.167		4.06	5.70		9.76	13.35	
	0700	24" O.C.		100	.160		3.05	5.50		8.55	11.90	
	0705	24" O.C., pneumatic nailed		120	.133		3.05	4.57		7.62	10.45	
	0780	2" x 6" studs, 8' high, studs 12" O.C.		70	.229		5.90	7.85		13.75	18.70	
	0785	12" O.C., pneumatic nailed		84	.190		5.90	6.50		12.40	16.65	
	0800	16" O.C.		90	.178		4.83	6.10		10.93	14.80	
	0805	16" O.C., pneumatic nailed		108	.148		4.83	5.05		9.88	13.20	
	0900	24" O.C.		115	.139		3.76	4.77		8.53	11.55	
	0905	24" O.C., pneumatic nailed		138	.116		3.76	3.97		7.73	10.35	
	0980	10' high, studs 12" O.C.		70	.229		7	7.85		14.85	19.90	
	0985	12" O.C., pneumatic nailed		84	.190		7	6.50		13.50	17.85	
	1000	16" O.C.		90	.178		5.65	6.10		11.75	15.70	
	1005	16" O.C., pneumatic nailed		108	.148		5.65	5.05		10.70	14.10	
	1100	24" O.C.		115	.139		4.30	4.77		9.07	12.15	
	1105	24" O.C., pneumatic nailed		138	.116		4.30	3.97		8.27	10.95	
	1180	12' high, studs 12" O.C.		55	.291		8.05	9.95		18	24.50	
	1185	12" O.C., pneumatic nailed		66	.242		8.05	8.30		16.35	22	
	1200	16" O.C.		70	.229		6.45	7.85		14.30	19.30	
	1205	16" O.C., pneumatic nailed	▼	84	.190	▼	6.45	6.50		12.95	17.25	

Important: See the Reference Section for critical supporting data - Reference Nos., Crews, & City Cost Indexes

	06110	Wood Framing	CREW	DAILY OUTPUT	LABOR-HOURS	UNIT	MAT.	LABOR	EQUIP.	TOTAL	TOTAL INCL O&P	
550	1300	24" O.C.	2 Carp	90	.178	L.F.	4.83	6.10		10.93	14.80	550
	1305	24" O.C., pneumatic nailed		108	.148		4.83	5.05		9.88	13.20	
	1400	For horizontal blocking, 2" x 4", add		600	.027		.34	.91		1.25	1.79	
	1500	2" x 6", add		600	.027		.54	.91		1.45	2.01	
	1600	For openings, add		250	.064			2.19		2.19	3.41	
	1700	Headers for above openings, material only, add				M.B.F.	620			620	680	
555	0010	**FRAMING, ROOFS** R06100-010										555
	6070	Fascia boards, 2" x 8"	2 Carp	.30	53.333	M.B.F.	620	1,825		2,445	3,525	
	6080	2" x 10"		.30	53.333		705	1,825		2,530	3,625	
	7000	Rafters, to 4 in 12 pitch, 2" x 6"		1	16		535	550		1,085	1,450	
	7060	2" x 8"		1.26	12.698		620	435		1,055	1,350	
	7300	Hip and valley rafters, 2" x 6"		.76	21.053		535	720		1,255	1,725	
	7360	2" x 8"		.96	16.667		620	570		1,190	1,575	
	7540	Hip and valley jacks, 2" x 6"		.60	26.667		535	915		1,450	2,025	
	7600	2" x 8"		.65	24.615		620	845		1,465	2,000	
	7780	For slopes steeper than 4 in 12, add						30%				
	7790	For dormers or complex roofs, add						50%				
	7800	Rafter tie, 1" x 4", #3	2 Carp	.27	59.259	M.B.F.	1,225	2,025		3,250	4,475	
	7820	Ridge board, #2 or better, 1" x 6"		.30	53.333		1,825	1,825		3,650	4,850	
	7840	1" x 8"		.37	43.243		1,825	1,475		3,300	4,325	
	7860	1" x 10"		.42	38.095		1,825	1,300		3,125	4,025	
	7880	2" x 6"		.50	32		535	1,100		1,635	2,300	
	7900	2" x 8"		.60	26.667		620	915		1,535	2,100	
	7920	2" x 10"		.66	24.242		705	830		1,535	2,075	
	7940	Roof cants, split, 4" x 4"		.86	18.605		940	635		1,575	2,025	
	7960	6" x 6"		1.80	8.889		1,475	305		1,780	2,100	
	7980	Roof curbs, untreated, 2" x 6"		.52	30.769		535	1,050		1,585	2,250	
	8000	2" x 12"		.80	20		800	685		1,485	1,950	
	8020	Sister rafters, 2" x 6"		.80	20		535	685		1,220	1,675	
	8040	2" x 8"		.85	18.824		620	645		1,265	1,675	
	8060	2" x 10"		.89	17.978		705	615		1,320	1,725	
	8080	2" x 12"		.91	17.582		800	600		1,400	1,825	
560	0010	**FRAMING, SILLS**										560
	2000	Ledgers, nailed, 2" x 4"	2 Carp	755	.021	L.F.	.34	.73		1.07	1.50	
	2050	2" x 6"		600	.027		.54	.91		1.45	2.01	
	2100	Bolted, not including bolts, 3" x 6"		325	.049		1.92	1.69		3.61	4.74	
	2150	3" x 12"		233	.069		3.92	2.35		6.27	7.95	
	2600	Mud sills, redwood, construction grade, 2" x 4"		895	.018		2.20	.61		2.81	3.37	
	2620	2" x 6"		780	.021		3.30	.70		4	4.72	
	4000	Sills, 2" x 4"		600	.027		.34	.91		1.25	1.79	
	4050	2" x 6"		550	.029		.54	1		1.54	2.14	
	4080	2" x 8"		500	.032		.83	1.10		1.93	2.62	
	4200	Treated, 2" x 4"		550	.029		.49	1		1.49	2.09	
	4220	2" x 6"		500	.032		.77	1.10		1.87	2.56	
	4240	2" x 8"		450	.036		1	1.22		2.22	3	
	4400	4" x 4"		450	.036		1.35	1.22		2.57	3.39	
	4420	4" x 6"		350	.046		2.99	1.57		4.56	5.75	
	4460	4" x 8"		300	.053		5.95	1.83		7.78	9.40	
	4480	4" x 10"		260	.062		7.40	2.11		9.51	11.45	
565	0010	**FRAMING, SLEEPERS**										565
	0100	On concrete, treated, 1" x 2"	2 Carp	2,350	.007	L.F.	.10	.23		.33	.47	
	0150	1" x 3"		2,000	.008		.19	.27		.46	.64	
	0200	2" x 4"		1,500	.011		.49	.37		.86	1.11	
	0250	2" x 6"		1,300	.012		.77	.42		1.19	1.51	

WOOD & PLASTICS 6

		06110	Wood Framing		CREW	DAILY OUTPUT	LABOR-HOURS	UNIT	2005 BARE COSTS				TOTAL INCL O&P	
									MAT.	LABOR	EQUIP.	TOTAL		
575	0010	**FRAMING, TREATED LUMBER**												575
	0100		2" x 4"					M.B.F.	730			730	805	
	0110		2" x 6"						770			770	850	
	0120		2" x 8"						750			750	825	
	0130		2" x 10"						890			890	980	
	0140		2" x 12"						955			955	1,050	
	0200		4" x 4"						1,025			1,025	1,125	
	0210		4" x 6"						1,500			1,500	1,650	
	0220		4" x 8"						2,225			2,225	2,450	
590	0010	**FRAMING, WALLS**		R06100 -010										590
	2000	Headers over openings, 2" x 6"			2 Carp	360	.044	L.F.	.54	1.52		2.06	2.96	
	2005		2" x 6", pneumatic nailed			432	.037		.54	1.27		1.81	2.57	
	2050		2" x 8"			340	.047		.83	1.61		2.44	3.42	
	2055		2" x 8", pneumatic nailed			408	.039		.83	1.34		2.17	3	
	2100		2" x 10"			320	.050		1.18	1.71		2.89	3.96	
	2105		2" x 10", pneumatic nailed			384	.042		1.18	1.43		2.61	3.51	
	2150		2" x 12"			300	.053		1.60	1.83		3.43	4.61	
	2155		2" x 12", pneumatic nailed			360	.044		1.60	1.52		3.12	4.13	
	2200		4" x 12"			190	.084		4.40	2.88		7.28	9.35	
	2205		4" x 12", pneumatic nailed			228	.070		4.40	2.40		6.80	8.60	
	2250		6" x 12"			140	.114		9.10	3.91		13.01	16.10	
	2255		6" x 12", pneumatic nailed			168	.095		9.10	3.26		12.36	15.10	
	2710	Headers over openings, 2" x 6"				270	.059		.54	2.03		2.57	3.75	
	5000	Plates, untreated, 2" x 3"				850	.019		.26	.64		.90	1.29	
	5005		2" x 3", pneumatic nailed			1,020	.016		.26	.54		.80	1.13	
	5020		2" x 4"			800	.020		.34	.69		1.03	1.44	
	5025		2" x 4", pneumatic nailed			960	.017		.34	.57		.91	1.26	
	5040		2" x 6"			750	.021		.54	.73		1.27	1.73	
	5045		2" x 6", pneumatic nailed			900	.018		.54	.61		1.15	1.54	
	5120	Studs, 8' high wall, 2" x 3"				1,200	.013		.26	.46		.72	1	
	5125		2" x 3", pneumatic nailed			1,440	.011		.26	.38		.64	.88	
	5140		2" x 4"			1,100	.015		.34	.50		.84	1.15	
	5146		2" x 4", pneumatic nailed			1,320	.012		.34	.42		.76	1.02	
	5160		2" x 6"			1,000	.016		.54	.55		1.09	1.44	
	5166		2" x 6", pneumatic nailed			1,200	.013		.54	.46		1	1.30	
	5180		3" x 4"			800	.020		1.09	.69		1.78	2.27	
	5185		3" x 4", pneumatic nailed			960	.017		1.09	.57		1.66	2.09	
	8200	For 12' high walls, deduct								5%				
	8220	For stub wall, 6' high, add								20%				
	8240	3' high, add								40%				
	8250	For second story & above, add								5%				
	8300	For dormer & gable, add								15%				
600	0010	**FURRING** Wood strips, 1" x 2", on walls, on wood			1 Carp	550	.015	L.F.	.19	.50		.69	.99	600
	0015		On wood, pneumatic nailed			710	.011		.19	.39		.58	.81	
	0300		On masonry			495	.016		.19	.55		.74	1.07	
	0400		On concrete			260	.031		.19	1.05		1.24	1.85	
	0600		1" x 3", on walls, on wood			550	.015		.28	.50		.78	1.09	
	0605		On wood, pneumatic nailed			710	.011		.28	.39		.67	.91	
	0700		On masonry			495	.016		.28	.55		.83	1.17	
	0800		On concrete			260	.031		.28	1.05		1.33	1.95	
	0850		On ceilings, on wood			350	.023		.28	.78		1.06	1.53	
	0855		On wood, pneumatic nailed			450	.018		.28	.61		.89	1.26	
	0900		On masonry			320	.025		.28	.86		1.14	1.64	
	0950		On concrete			210	.038		.28	1.30		1.58	2.34	

6

WOOD & PLASTICS

Important: See the Reference Section for critical supporting data - Reference Nos., Crews, & City Cost Indexes

06120 | Structural Panels

		CREW	DAILY OUTPUT	LABOR-HOURS	UNIT	2005 BARE COSTS				TOTAL INCL O&P
						MAT.	LABOR	EQUIP.	TOTAL	
900	**0010**	**STRUCTURAL INSULATED PANELS**									**900**
	0100	Structural insul. panels, 7/16" OSB both faces, EPS insul, 3-5/8" T	F-3	2,075	.019	S.F.	3.72	.67	.29	4.68	5.45
	0110	5-5/8" thick		1,725	.023		4.06	.80	.35	5.21	6.10
	0120	7-3/8" thick		1,425	.028		4.34	.97	.42	5.73	6.75
	0130	9-3/8" thick		1,125	.036		4.68	1.23	.54	6.45	7.65
	0140	7/16" OSB one face, EPS insul, 3-5/8" thick		2,175	.018		1.87	.64	.28	2.79	3.34
	0150	5-5/8" thick		1,825	.022		2.12	.76	.33	3.21	3.86
	0160	7-3/8" thick		1,525	.026		2.35	.91	.40	3.66	4.42
	0170	9-3/8" thick		1,225	.033		2.56	1.13	.49	4.18	5.10
	0180	11-3/8" thick		925	.043		2.75	1.50	.65	4.90	6.05
	0190	7/16" OSB - 1/2" GWB faces , EPS insul, 3-5/8" T		2,075	.019		3.06	.67	.29	4.02	4.72
	0200	5-5/8" thick		1,725	.023		3.50	.80	.35	4.65	5.45
	0210	7-3/8" thick		1,425	.028		4.12	.97	.42	5.51	6.50
	0220	9-3/8" thick		1,125	.036		4.78	1.23	.54	6.55	7.75
	0230	11-3/8" thick		825	.048		5.10	1.68	.73	7.51	9
	0240	7/16" OSB - 1/2" MRGWB faces , EPS insul, 3-5/8" T		2,075	.019		3.06	.67	.29	4.02	4.72
	0250	5-5/8" thick	▼	1,725	.023		3.50	.80	.35	4.65	5.45
	0300	For 1/2" GWB added to OSB skin, add					.60			.60	.66
	0310	For 1/2" MRGWB added to OSB skin, add					.75			.75	.83
	0320	For one T1-11 skin, add to OSB-OSB					.75			.75	.83
	0330	For one 19/32" CDX skin, add to OSB-OSB				▼	.72			.72	.79

06150 | Wood Decking

		CREW	DAILY OUTPUT	LABOR-HOURS	UNIT	2005 BARE COSTS				TOTAL INCL O&P
						MAT.	LABOR	EQUIP.	TOTAL	
600	**0010**	**ROOF DECKS**									**600**
	0400	Cedar planks, 3" thick	2 Carp	320	.050	S.F.	6.40	1.71		8.11	9.70
	0500	4" thick		250	.064		8.65	2.19		10.84	12.90
	0700	Douglas fir, 3" thick		320	.050		2.40	1.71		4.11	5.30
	0800	4" thick		250	.064		3.21	2.19		5.40	6.95
	1000	Hemlock, 3" thick		320	.050		2.40	1.71		4.11	5.30
	1100	4" thick		250	.064		3.20	2.19		5.39	6.95
	1300	Western white spruce, 3" thick		320	.050		2.31	1.71		4.02	5.20
	1400	4" thick	▼	250	.064	▼	3.08	2.19		5.27	6.80

06160 | Sheathing

		CREW	DAILY OUTPUT	LABOR-HOURS	UNIT	2005 BARE COSTS				TOTAL INCL O&P
						MAT.	LABOR	EQUIP.	TOTAL	
800	**0010**	**SHEATHING** Plywood on roof, CDX									**800**
	0030	5/16" thick [R06110-030]	2 Carp	1,600	.010	S.F.	.63	.34		.97	1.22
	0035	Pneumatic nailed		1,952	.008		.63	.28		.91	1.13
	0050	3/8" thick		1,525	.010		.65	.36		1.01	1.28
	0055	Pneumatic nailed		1,860	.009		.65	.29		.94	1.18
	0100	1/2" thick		1,400	.011		.62	.39		1.01	1.29
	0105	Pneumatic nailed		1,708	.009		.62	.32		.94	1.18
	0200	5/8" thick		1,300	.012		.78	.42		1.20	1.52
	0205	Pneumatic nailed		1,586	.010		.78	.35		1.13	1.40
	0300	3/4" thick		1,200	.013		.94	.46		1.40	1.74
	0305	Pneumatic nailed		1,464	.011		.94	.37		1.31	1.61
	0500	Plywood on walls with exterior CDX, 3/8" thick		1,200	.013		.65	.46		1.11	1.43
	0505	Pneumatic nailed		1,488	.011		.65	.37		1.02	1.29
	0600	1/2" thick		1,125	.014		.62	.49		1.11	1.44
	0605	Pneumatic nailed		1,395	.011		.62	.39		1.01	1.29
	0700	5/8" thick		1,050	.015		.78	.52		1.30	1.67
	0705	Pneumatic nailed		1,302	.012		.78	.42		1.20	1.52
	0800	3/4" thick		975	.016		.94	.56		1.50	1.91
	0805	Pneumatic nailed	▼	1,209	.013	▼	.94	.45		1.39	1.74
	1000	For shear wall construction, add						20%			

WOOD & PLASTICS 6

06160 | Sheathing

			CREW	DAILY OUTPUT	LABOR-HOURS	UNIT	2005 BARE COSTS				TOTAL INCL O&P	
							MAT.	LABOR	EQUIP.	TOTAL		
800	1200	For structural 1 exterior plywood, add				S.F.	10%					800
	1400	With boards, on roof 1″ x 6″ boards, laid horizontal	2 Carp	725	.022		1.25	.76		2.01	2.56	
	1500	Laid diagonal		650	.025		1.25	.84		2.09	2.69	
	1700	1″ x 8″ boards, laid horizontal		875	.018		1.11	.63		1.74	2.21	
	1800	Laid diagonal		725	.022		1.11	.76		1.87	2.41	
	2000	For steep roofs, add						40%				
	2200	For dormers, hips and valleys, add					5%	50%				
	2400	Boards on walls, 1″ x 6″ boards, laid regular	2 Carp	650	.025		1.25	.84		2.09	2.69	
	2500	Laid diagonal		585	.027		1.25	.94		2.19	2.84	
	2700	1″ x 8″ boards, laid regular		765	.021		1.11	.72		1.83	2.35	
	2800	Laid diagonal		650	.025		1.11	.84		1.95	2.54	
	2850	Gypsum, weatherproof, 1/2″ thick		1,125	.014		.29	.49		.78	1.08	
	2900	Sealed, 4/10″ thick		1,100	.015		.45	.50		.95	1.28	
	3000	Wood fiber, regular, no vapor barrier, 1/2″ thick		1,200	.013		.55	.46		1.01	1.32	
	3100	5/8″ thick		1,200	.013		.72	.46		1.18	1.50	
	3300	No vapor barrier, in colors, 1/2″ thick		1,200	.013		.78	.46		1.24	1.57	
	3400	5/8″ thick		1,200	.013		.96	.46		1.42	1.77	
	3600	With vapor barrier one side, white, 1/2″ thick		1,200	.013		.55	.46		1.01	1.32	
	3700	Vapor barrier 2 sides, 1/2″ thick		1,200	.013		.77	.46		1.23	1.56	
	3800	Asphalt impregnated, 25/32″ thick		1,200	.013		.28	.46		.74	1.02	
	3850	Intermediate, 1/2″ thick		1,200	.013		.18	.46		.64	.91	
850	0010	**SUBFLOOR** Plywood, CDX, 1/2″ thick	2 Carp	1,500	.011	SF Flr.	.62	.37		.99	1.25	850
	0015	Pneumatic nailed		1,860	.009		.62	.29		.91	1.14	
	0100	5/8″ thick		1,350	.012		.78	.41		1.19	1.49	
	0105	Pneumatic nailed		1,674	.010		.78	.33		1.11	1.37	
	0200	3/4″ thick		1,250	.013		.94	.44		1.38	1.71	
	0205	Pneumatic nailed		1,550	.010		.94	.35		1.29	1.58	
	0300	1-1/8″ thick, 2-4-1 including underlayment		1,050	.015		1.19	.52		1.71	2.12	
	0450	1″ x 8″ S4S, laid regular		1,000	.016		1.11	.55		1.66	2.08	
	0460	Laid diagonal		850	.019		1.11	.64		1.75	2.23	
	0500	1″ x 10″ S4S, laid regular		1,100	.015		1.36	.50		1.86	2.28	
	0600	Laid diagonal		900	.018		1.36	.61		1.97	2.45	
900	0011	**UNDERLAYMENT** Plywood, underlayment grade, 3/8″ thick	2 Carp	1,500	.011	SF Flr.	.87	.37		1.24	1.53	900
	0016	Pneumatic nailed		1,860	.009		.87	.29		1.16	1.42	
	0100	1/2″ thick		1,450	.011		1.06	.38		1.44	1.76	
	0105	Pneumatic nailed		1,798	.009		1.06	.30		1.36	1.64	
	0200	5/8″ thick		1,400	.011		1.22	.39		1.61	1.95	
	0205	Pneumatic nailed		1,736	.009		1.22	.32		1.54	1.83	
	0300	3/4″ thick		1,300	.012		1.40	.42		1.82	2.20	
	0305	Pneumatic nailed		1,612	.010		1.40	.34		1.74	2.07	
	0500	Particle board, 3/8″ thick		1,500	.011		.37	.37		.74	.98	
	0505	Pneumatic nailed		1,860	.009		.37	.29		.66	.87	
	0600	1/2″ thick		1,450	.011		.39	.38		.77	1.02	
	0605	Pneumatic nailed		1,798	.009		.39	.30		.69	.90	
	0800	5/8″ thick		1,400	.011		.49	.39		.88	1.15	
	0805	Pneumatic nailed		1,736	.009		.49	.32		.81	1.03	
	0900	3/4″ thick		1,300	.012		.55	.42		.97	1.27	
	0905	Pneumatic nailed		1,612	.010		.55	.34		.89	1.14	
	1100	Hardboard, underlayment grade, 4′ x 4′, .215″ thick		1,500	.011		.39	.37		.76	1	

R06110
-030

06170 | Prefabricated Structural Wood

			CREW	DAILY OUTPUT	LABOR-HOURS	UNIT	MAT.	LABOR	EQUIP.	TOTAL	TOTAL INCL O&P	
600	0010	**STRUCTURAL JOISTS** Fabricated "I" joists with wood flanges,										600
	0100	Plywood webs, incl. bridging & blocking, panels 24″ O.C.										

6
WOOD & PLASTICS

06100 | Rough Carpentry

06170	Prefabricated Structural Wood	CREW	DAILY OUTPUT	LABOR-HOURS	UNIT	2005 BARE COSTS				TOTAL INCL O&P	
						MAT.	LABOR	EQUIP.	TOTAL		
600 1200	15' to 24' span, 50 psf live load	F-5	2,400	.013	SF Flr.	1.67	.46		2.13	2.56	**600**
1300	55 psf live load		2,250	.014		1.78	.49		2.27	2.72	
1400	24' to 30' span, 45 psf live load		2,600	.012		2.03	.43		2.46	2.91	
1500	55 psf live load		2,400	.013		2.90	.46		3.36	3.91	
1600	Tubular steel open webs, 45 psf, 24" O.C., 40' span	F-3	6,250	.006		1.80	.22	.10	2.12	2.43	
1700	55' span		7,750	.005		1.75	.18	.08	2.01	2.30	
1800	70' span		9,250	.004		2.27	.15	.07	2.49	2.80	
1900	85 psf live load, 26' span		2,300	.017		2.11	.60	.26	2.97	3.54	
980 0010	**ROOF TRUSSES**										**980**
0020	For timber connectors, see div. 06090-800										
0100	Fink (W) or King post type, 2'-0" O.C.										
0200	Metal plate connected, 4 in 12 slope										
0210	24' to 29' span	F-3	3,000	.013	SF Flr.	1.60	.46	.20	2.26	2.69	
0300	30' to 43' span		3,000	.013		1.77	.46	.20	2.43	2.88	
0400	44' to 60' span		3,000	.013		1.96	.46	.20	2.62	3.09	
0700	Glued and nailed, add					50%					
0800	Flat wood truss 16' to 29' span	F-3	3,000	.013		1.91	.46	.20	2.57	3.03	

06180	Glued-Laminated Construction	CREW	DAILY OUTPUT	LABOR-HOURS	UNIT	MAT.	LABOR	EQUIP.	TOTAL	TOTAL INCL O&P	
400 0010	**LAMINATED FRAMING** Not including decking										**400**
0200	Straight roof beams, 20' clear span, beams 8' O.C.	F-3	2,560	.016	SF Flr.	1.63	.54	.24	2.41	2.89	
0300	Beams 16' O.C.		3,200	.013		1.17	.43	.19	1.79	2.17	
0500	40' clear span, beams 8' O.C.		3,200	.013		3.12	.43	.19	3.74	4.31	
0600	Beams 16' O.C.		3,840	.010		2.54	.36	.16	3.06	3.52	
0800	60' clear span, beams 8' O.C.	F-4	2,880	.017		5.35	.56	.31	6.22	7.10	
0900	Beams 16' O.C.	"	3,840	.013		3.99	.42	.23	4.64	5.30	
1100	Tudor arches, 30' to 40' clear span, frames 8' O.C.	F-3	1,680	.024		7	.82	.36	8.18	9.35	
1200	Frames 16' O.C.	"	2,240	.018		5.45	.62	.27	6.34	7.25	
1400	50' to 60' clear span, frames 8' O.C.	F-4	2,200	.022		7.55	.74	.41	8.70	9.90	
1500	Frames 16' O.C.		2,640	.018		6.40	.62	.34	7.36	8.35	
1700	Radial arches, 60' clear span, frames 8' O.C.		1,920	.025		7.05	.85	.47	8.37	9.55	
1800	Frames 16' O.C.		2,880	.017		5.40	.56	.31	6.27	7.15	
2000	100' clear span, frames 8' O.C.		1,600	.030		7.30	1.01	.56	8.87	10.20	
2100	Frames 16' O.C.		2,400	.020		6.40	.68	.37	7.45	8.50	
2300	120' clear span, frames 8' O.C.		1,440	.033		9.70	1.13	.62	11.45	13.05	
2400	Frames 16' O.C.		1,920	.025		8.85	.85	.47	10.17	11.55	
2600	Bowstring trusses, 20' O.C., 40' clear span	F-3	2,400	.017		4.37	.58	.25	5.20	6	
2700	60' clear span	F-4	3,600	.013		3.92	.45	.25	4.62	5.30	
2800	100' clear span		4,000	.012		5.55	.41	.22	6.18	7	
2900	120' clear span		3,600	.013		5.95	.45	.25	6.65	7.50	
3100	For premium appearance, add to S.F. prices					5%					
3300	For industrial type, deduct					15%					
3500	For stain and varnish, add					5%					
3900	For 3/4" laminations, add to straight					25%					
4100	Add to curved					15%					
4300	Alternate pricing method: (use nominal footage of										
4310	components). Straight beams, camber less than 6"	F-3	3.50	11.429	M.B.F.	2,425	395	172	2,992	3,450	
4400	Columns, including hardware		2	20		2,600	690	300	3,590	4,250	
4600	Curved members, radius over 32'		2.50	16		2,650	555	241	3,446	4,050	
4700	Radius 10' to 32'		3	13.333		2,625	460	201	3,286	3,825	
4900	For complicated shapes, add maximum					100%					
5100	For pressure treating, add to straight					35%					
5200	Add to curved					45%					
6000	Laminated veneer members, southern pine or western species										
6050	1-3/4" wide x 5-1/2" deep	2 Carp	480	.033	L.F.	2.77	1.14		3.91	4.83	

WOOD & PLASTICS 6

06180	Glued-Laminated Construction	CREW	DAILY OUTPUT	LABOR-HOURS	UNIT	2005 BARE COSTS				TOTAL INCL O&P	
						MAT.	LABOR	EQUIP.	TOTAL		
400 6100	9-1/2" deep	2 Carp	480	.033	L.F.	3.45	1.14		4.59	5.60	**400**
6150	14" deep		450	.036		5.15	1.22		6.37	7.55	
6200	18" deep	▼	450	.036	▼	6.95	1.22		8.17	9.55	
6300	Parallel strand members, southern pine or western species										
6350	1-3/4" wide x 9-1/4" deep	2 Carp	480	.033	L.F.	3.27	1.14		4.41	5.40	
6400	11-1/4" deep		450	.036		4.02	1.22		5.24	6.30	
6450	14" deep		400	.040		4.79	1.37		6.16	7.40	
6500	3-1/2" wide x 9-1/4" deep		480	.033		7.95	1.14		9.09	10.55	
6550	11-1/4" deep		450	.036		9.85	1.22		11.07	12.70	
6600	14" deep		400	.040		11.65	1.37		13.02	15	
6650	7" wide x 9-1/4" deep		450	.036		16.55	1.22		17.77	20	
6700	11-1/4" deep		420	.038		20.50	1.30		21.80	25	
6750	14" deep	▼	400	.040	▼	25	1.37		26.37	29.50	

For information about Means Estimating Seminars, see yellow pages 12 and 13 in back of book

6 WOOD & PLASTICS

Division 7
Thermal & Moisture Protection

Estimating Tips

07100 Dampproofing & Waterproofing

- Be sure of the job specifications before pricing this subdivision. The difference in cost between waterproofing and dampproofing can be great. Waterproofing will hold back standing water. Dampproofing prevents the transmission of water vapor. Also included in this section are vapor retarding membranes.

07200 Thermal Protection

- Insulation and fireproofing products are measured by area, thickness, volume or R value. Specifications may only give what the specific R value should be in a certain situation. The estimator may need to choose the type of insulation to meet that R value.

07300 Shingles, Roof Tiles & Roof Coverings
07400 Roofing & Siding Panels

- Many roofing and siding products are bought and sold by the square. One square is equal to an area that measures 100 square feet.

This simple change in unit of measure could create a large error if the estimator is not observant. Accessories necessary for a complete installation must be figured into any calculations for both material and labor.

07500 Membrane Roofing
07600 Flashing & Sheet Metal
07700 Roof Specialties & Accessories

- The items in these subdivisions compose a roofing system. No one component completes the installation and all must be estimated. Built-up or single ply membrane roofing systems are made up of many products and installation trades. Wood blocking at roof perimeters or penetrations, parapet coverings, reglets, roof drains, gutters, downspouts, sheet metal flashing, skylights, smoke vents or roof hatches all need to be considered along with the roofing material. Several different installation trades will need to work together on the roofing system. Inherent difficulties in the scheduling and coordination of various trades must be accounted for when estimating labor costs.

07900 Joint Sealers

- To complete the weather-tight shell the sealants and caulkings must be estimated. Where different materials meet—at expansion joints, at flashing penetrations, and at hundreds of other locations throughout a construction project—they provide another line of defense against water penetration. Often, an entire system is based on the proper location and placement of caulking or sealants. The detail drawings that are included as part of a set of architectural plans, show typical locations for these materials. When caulking or sealants are shown at typical locations, this means the estimator must include them for all the locations where this detail is applicable. Be careful to keep different types of sealants separate, and remember to consider backer rods and primers if necessary.

Reference Numbers

Reference numbers are shown in bold squares at the beginning of some major classifications. These numbers refer to related items in the Reference Section. The reference information may be an estimating procedure, an alternate pricing method or technical information.

Note: Not all subdivisions listed here necessarily appear in this publication.

07060	Selective Demolition	CREW	DAILY OUTPUT	LABOR-HOURS	UNIT	2005 BARE COSTS				TOTAL INCL O&P
						MAT.	LABOR	EQUIP.	TOTAL	
110 0010	**SELECTIVE DEMOLITION, ROOFING AND SIDING** R02220-510									**110**
1000	Deck, roof, concrete plank	B-13	1,680	.033	S.F.		.96	.37	1.33	1.88
1100	Gypsum plank		3,900	.014			.41	.16	.57	.81
1150	Metal decking	↓	3,500	.016			.46	.18	.64	.90
1200	Wood, boards, tongue and groove, 2" x 6"	2 Clab	960	.017			.45		.45	.69
1220	2" x 10"		1,040	.015			.41		.41	.64
1280	Standard planks, 1" x 6"		1,080	.015			.40		.40	.62
1320	1" x 8"		1,160	.014			.37		.37	.57
1340	1" x 12"		1,200	.013			.36		.36	.55
1350	Plywood, to 1" thick	↓	2,000	.008			.21		.21	.33
1360	Flashing, aluminum	1 Clab	290	.028	↓		.74		.74	1.15
2000	Gutters, aluminum or wood, edge hung		240	.033	L.F.		.89		.89	1.38
2100	Built-in		100	.080	"		2.14		2.14	3.32
2500	Roof accessories, plumbing vent flashing		14	.571	Ea.		15.25		15.25	23.50
2600	Adjustable metal chimney flashing	↓	9	.889	"		23.50		23.50	37
3000	Roofing, built-up, 5 ply roof, no gravel	B-2	1,600	.025	S.F.		.68		.68	1.05
3001	Including gravel		890	.045			1.22		1.22	1.90
3100	Gravel removal, minimum		5,000	.008			.22		.22	.34
3120	Maximum		2,000	.020			.54		.54	.84
3400	Roof insulation board, up to 2" thick		3,900	.010			.28		.28	.43
4000	Shingles, asphalt strip, 1 layer		3,500	.011			.31		.31	.48
4100	Slate		2,500	.016			.43		.43	.67
4300	Wood	↓	2,200	.018	↓		.49		.49	.77
4500	Skylight to 10 S.F.	1 Clab	8	1	Ea.		26.50		26.50	41.50
5000	Siding, metal, horizontal		444	.018	S.F.		.48		.48	.75
5020	Vertical		400	.020			.53		.53	.83
5200	Wood, boards, vertical		400	.020			.53		.53	.83
5220	Clapboards, horizontal		380	.021			.56		.56	.87
5240	Shingles		350	.023			.61		.61	.95
5260	Textured plywood	↓ ↓	725	.011	↓		.29		.29	.46

07100 | Dampproofing and Waterproofing

07110	Dampproofing	CREW	DAILY OUTPUT	LABOR-HOURS	UNIT	2005 BARE COSTS				TOTAL INCL O&P
						MAT.	LABOR	EQUIP.	TOTAL	
100 0010	**BITUMINOUS ASPHALT COATING** For foundation									**100**
0030	Brushed on, below grade, 1 coat	1 Rofc	665	.012	S.F.	.07	.35		.42	.68
0100	2 coat		500	.016		.10	.47		.57	.91
0300	Sprayed on, below grade, 1 coat, 25.6 S.F./gal.		830	.010		.07	.28		.35	.56
0400	2 coat, 20.5 S.F./gal.	↓	500	.016	↓	.14	.47		.61	.96
0500	Asphalt coating, with fibers				Gal.	3.74			3.74	4.11
0600	Troweled on, asphalt with fibers, 1/16" thick	1 Rofc	500	.016	S.F.	.16	.47		.63	.98
0700	1/8" thick		400	.020		.29	.59		.88	1.32
1000	1/2" thick	↓	350	.023	↓	.94	.67		1.61	2.18
200 0010	**CEMENT PARGING** 2 coats, 1/2" thick, regular P.C.	D-1	250	.064	S.F.	.17	1.99		2.16	3.22	**200**
0100	Waterproofed Portland cement	"	250	.064	"	.18	1.99		2.17	3.23

07130	Sheet Waterproofing									
200 0010	**ELASTOMERIC WATERPROOFING**									**200**
0050	Acrylic rubber, fluid applied, 20 mils thick	3 Rofc	1,000	.024	S.F.	1.59	.71		2.30	2.95

07130 | Sheet Waterproofing

			DAILY OUTPUT	LABOR-HOURS	UNIT	2005 BARE COSTS				TOTAL INCL O&P	
			CREW			MAT.	LABOR	EQUIP.	TOTAL		
200	0060	50 mil, reinforced, stucco texture	3 Rofc	600	.040	S.F.	2.55	1.18		3.73	4.82	**200**
	0090	EPDM, plain, 45 mils thick	2 Rofc	580	.028		.85	.81		1.66	2.33	
	0100	60 mils thick		570	.028		.92	.83		1.75	2.42	
	0300	Nylon reinforced sheets, 45 mils thick		580	.028		1.32	.81		2.13	2.84	
	0400	60 mils thick		570	.028		1.68	.83		2.51	3.26	
	0600	Vulcanizing splicing tape for above, 2" wide				C.L.F.	33.50			33.50	36.50	
	0700	4" wide				"	67			67	73.50	
	0900	Adhesive, bonding, 60 SF per gal				Gal.	13.60			13.60	14.95	
	1000	Splicing, 75 SF per gal				"	21			21	23	
	1200	Neoprene sheets, plain, 45 mils thick	2 Rofc	580	.028	S.F.	.93	.81		1.74	2.41	
	1300	60 mils thick		570	.028		1.48	.83		2.31	3.04	
	1500	Nylon reinforced, 45 mils thick		580	.028		1.20	.81		2.01	2.71	
	1600	60 mils thick		570	.028		1.36	.83		2.19	2.91	
	1800	120 mils thick		500	.032		2.25	.94		3.19	4.09	
	1900	Adhesive, splicing, 150 S.F. per gal. per coat				Gal.	18.15			18.15	19.95	
	2100	Fiberglass reinforced, fluid applied, 1/8" thick	2 Rofc	500	.032	S.F.	1.50	.94		2.44	3.26	
	2200	Polyethylene and rubberized asphalt sheets, 1/8" thick		550	.029		.54	.86		1.40	2.05	
	2210	Asphaltic hardboard protection board, 1/8" thick		500	.032		.30	.94		1.24	1.94	
	2220	Asphaltic hardboard protection board, 1/4" thick		450	.036		.52	1.05		1.57	2.36	
	2400	Polyvinyl chloride sheets, plain, 10 mils thick		580	.028		.14	.81		.95	1.54	
	2500	20 mils thick		570	.028		.22	.83		1.05	1.65	
	2700	30 mils thick		560	.029		.30	.84		1.14	1.76	
	3000	Adhesives, trowel grade, 40-100 SF per gal				Gal.	20			20	22	
	3100	Brush grade, 100-250 SF per gal.				"	20			20	22	
	3300	Bitumen modified polyurethane, fluid applied, 55 mils thick	2 Rofc	665	.024	S.F.	.65	.71		1.36	1.93	
	3600	Vinyl plastic, sprayed on, 25 to 40 mils thick	"	475	.034	"	.98	.99		1.97	2.77	
500	0010	**MEMBRANE WATERPROOFING** On slabs, 1 ply, felt	G-1	3,000	.019	S.F.	.19	.51	.10	.80	1.19	**500**
	0015	Membrane waterproofing on walls, 1 ply, felt, mopped		3,000	.019		.19	.51	.10	.80	1.19	
	0100	Glass fiber fabric		2,100	.027		.20	.73	.14	1.07	1.63	
	0300	2 ply, felt		2,500	.022		.38	.62	.12	1.12	1.60	
	0400	Glass fiber fabric		1,650	.034		.44	.93	.18	1.55	2.28	
	0600	3 ply, felt		2,100	.027		.57	.73	.14	1.44	2.04	
	0700	Glass fiber fabric		1,550	.036		.59	.99	.19	1.77	2.55	
	0900	For installation on walls, add						15%				
	1050	3/8" thick, add	2 Rofc	3,500	.005		.16	.13		.29	.41	
	1060	1/2" thick, add		3,500	.005		.18	.13		.31	.43	
	1070	Fiberglass fabric, black, 20/10 mesh		116	.138	Sq.	9.95	4.06		14.01	17.85	
	1080	White, 20/10 mesh		116	.138	"	10.25	4.06		14.31	18.20	
	1100	1/16" urethane, troweled		200	.080	S.F.	.66	2.36		3.02	4.75	
	1200	Roller applied		120	.133	"	.60	3.93		4.53	7.35	

07160 | Cement. & Reactive Waterproofing

			CREW	DAILY OUTPUT	LABOR-HOURS	UNIT	MAT.	LABOR	EQUIP.	TOTAL	TOTAL INCL O&P	
150	0010	**CEMENTITIOUS WATERPROOFING** One coat cement base										**150**
	0020	1/8" application, sprayed on	G-2	1,000	.024	S.F.	1.50	.68	.11	2.29	2.81	
	0030	2 coat, cementitious/metallic slurry, troweled, 1/4" thick	1 Cefi	2.48	3.226	C.S.F.	35	106		141	195	
	0040	3 coat, 3/8" thick		1.84	4.348		52.50	143		195.50	268	
	0050	4 coat, 1/2" thick		1.20	6.667		70	219		289	395	

07170 | Bentonite Waterproofing

			CREW	DAILY OUTPUT	LABOR-HOURS	UNIT	MAT.	LABOR	EQUIP.	TOTAL	TOTAL INCL O&P	
700	0010	**BENTONITE**, Panels, 4' x 4', 3/16" thick	1 Rofc	625	.013	S.F.	.74	.38		1.12	1.45	**700**
	0100	Rolls, 3/8" thick, with geotextile fabric both sides	"	550	.015	"	.94	.43		1.37	1.76	
	0300	Granular bentonite, 50 lb. bags (.625 C.F.)				Bag	18.65			18.65	20.50	
	0400	3/8" thick, troweled on	1 Rofc	475	.017	S.F.	.93	.50		1.43	1.88	
	0500	Drain board, expanded polystyrene, binder encapsulated, 1-1/2" thick	1 Rohe	1,600	.005		1.04	.11		1.15	1.32	
	0510	2" thick		1,600	.005		1.38	.11		1.49	1.70	

THERMAL & MOISTURE PROTECTION 7

223

07100 | Dampproofing and Waterproofing

07170	Bentonite Waterproofing	CREW	DAILY OUTPUT	LABOR-HOURS	UNIT	2005 BARE COSTS				TOTAL INCL O&P	
						MAT.	LABOR	EQUIP.	TOTAL		
700 0520	3" thick	1 Rohe	1,600	.005	S.F.	2.07	.11		2.18	2.46	**700**
0530	4" thick		1,600	.005		2.75	.11		2.86	3.21	
0600	With filter fabric, 1-1/2" thick		1,600	.005		1.25	.11		1.36	1.56	
0625	2" thick		1,600	.005		1.65	.11		1.76	2	
0650	3" thick		1,600	.005		2.34	.11		2.45	2.75	
0675	4" thick	↓	1,600	.005	↓	3.03	.11		3.14	3.51	
0700	Vapor retarder, see polyethelene, 07260-100										

07190	Water Repellents										
700 0010	**RUBBER COATING** Water base liquid, roller applied	2 Rofc	7,000	.002	S.F.	.55	.07		.62	.73	**700**
0200	Silicone or stearate, sprayed on CMU, 1 coat	1 Rofc	4,000	.002		.31	.06		.37	.44	
0300	2 coats	"	3,000	.003	↓	.63	.08		.71	.82	

07200 | Thermal Protection

07210	Building Insulation	CREW	DAILY OUTPUT	LABOR-HOURS	UNIT	2005 BARE COSTS				TOTAL INCL O&P	
						MAT.	LABOR	EQUIP.	TOTAL		
150 0010	**BLOWN-IN INSULATION** Ceilings, with open access										**150**
0020	Cellulose, 3-1/2" thick, R13	G-4	5,000	.005	S.F.	.13	.13	.04	.30	.39	
0030	5-3/16" thick, R19		3,800	.006		.21	.17	.06	.44	.56	
0050	6-1/2" thick, R22		3,000	.008		.26	.22	.07	.55	.71	
0100	8-11/16" thick, R30		2,600	.009		.34	.25	.08	.67	.85	
0120	10-7/8" thick, R38		1,800	.013		.43	.36	.12	.91	1.17	
1000	Fiberglass, 5" thick, R11		3,800	.006		.16	.17	.06	.39	.51	
1050	6" thick, R13		3,000	.008		.22	.22	.07	.51	.66	
1100	8-1/2" thick, R19		2,200	.011		.31	.30	.10	.71	.91	
1200	10" thick, R22		1,800	.013		.36	.36	.12	.84	1.10	
1300	12" thick, R26		1,500	.016		.43	.44	.14	1.01	1.31	
2000	Mineral wool, 4" thick, R12		3,500	.007		.18	.19	.06	.43	.56	
2050	6" thick, R17		2,500	.010		.20	.26	.09	.55	.73	
2100	9" thick, R23	↓	1,750	.014	↓	.29	.38	.12	.79	1.04	
2500	Wall installation, incl. drilling & patching from outside, two 1"										
2510	diam. holes @ 16" O.C., top & mid-point of wall, add to above										
2700	For masonry	G-4	415	.058	S.F.	.06	1.58	.52	2.16	3.10	
2800	For wood siding		840	.029		.06	.78	.26	1.10	1.57	
2900	For stucco/plaster	↓	665	.036	↓	.06	.99	.33	1.38	1.97	
500 0010	**POURED INSULATION** Cellulose fiber, R3.8 per inch	1 Carp	200	.040	C.F.	.48	1.37		1.85	2.66	**500**
0040	Ceramic type (perlite), R3.2 per inch		200	.040		1.65	1.37		3.02	3.95	
0080	Fiberglass wool, R4 per inch		200	.040		.39	1.37		1.76	2.56	
0100	Mineral wool, R3 per inch		200	.040		.33	1.37		1.70	2.49	
0300	Polystyrene, R4 per inch		200	.040		2.23	1.37		3.60	4.58	
0400	Vermiculite or perlite, R2.7 per inch		200	.040		1.65	1.37		3.02	3.95	
0700	Wood fiber, R3.85 per inch	↓	200	.040	↓	.61	1.37		1.98	2.80	
550 0010	**MASONRY INSULATION** Vermiculite or perlite, poured										**550**
0100	In cores of concrete block, 4" thick wall, .115 CF/SF	D-1	4,800	.003	S.F.	.19	.10		.29	.37	
0200	6" thick wall, .175 CF/SF		3,000	.005		.29	.17		.46	.57	
0300	8" thick wall, .258 CF/SF	↓	2,400	.007	↓	.43	.21		.64	.79	

Important: See the Reference Section for critical supporting data - Reference Nos., Crews, & City Cost Indexes

	07210	Building Insulation	CREW	DAILY OUTPUT	LABOR-HOURS	UNIT	2005 BARE COSTS				TOTAL INCL O&P	
							MAT.	LABOR	EQUIP.	TOTAL		
550	0400	10" thick wall, .340 CF/SF	D-1	1,850	.009	S.F.	.56	.27		.83	1.03	**550**
	0500	12" thick wall, .422 CF/SF	↓	1,200	.013	↓	.70	.41		1.11	1.40	
	0550	For sand fill, deduct from above					70%					
	0600	Poured cavity wall, vermiculite or perlite, water repellant	D-1	250	.064	C.F.	1.65	1.99		3.64	4.85	
	0700	Foamed in place, urethane in 2-5/8" cavity	G-2	1,035	.023	S.F.	.40	.66	.10	1.16	1.56	
	0800	For each 1" added thickness, add	"	2,372	.010	"	.12	.29	.05	.46	.62	
600	0010	**PERIMETER INSULATION**										**600**
	0600	Polystyrene, expanded, 1" thick, R4	1 Carp	680	.012	S.F.	.19	.40		.59	.84	
	0700	2" thick, R8	"	675	.012	"	.41	.41		.82	1.08	
700	0010	**REFLECTIVE INSULATION**, aluminum foil on reinforced scrim	1 Carp	19	.421	C.S.F.	13.90	14.40		28.30	38	**700**
	0100	Reinforced with woven polyolefin		19	.421		16.90	14.40		31.30	41	
	0500	With single bubble air space, R8.8		15	.533		27	18.25		45.25	58	
	0600	With double bubble air space, R9.8	↓	15	.533	↓	29	18.25		47.25	60.50	
900	0010	**WALL INSULATION, RIGID**										**900**
	0040	Fiberglass, 1.5#/CF, unfaced, 1" thick, R4.1	1 Carp	1,000	.008	S.F.	.31	.27		.58	.77	
	0060	1-1/2" thick, R6.2		1,000	.008		.41	.27		.68	.88	
	0080	2" thick, R8.3		1,000	.008		.47	.27		.74	.95	
	0120	3" thick, R12.4		800	.010		.57	.34		.91	1.16	
	0370	3#/CF, unfaced, 1" thick, R4.3		1,000	.008		.36	.27		.63	.83	
	0390	1-1/2" thick, R6.5		1,000	.008		.69	.27		.96	1.19	
	0400	2" thick, R8.7		890	.009		.83	.31		1.14	1.39	
	0420	2-1/2" thick, R10.9		800	.010		1.02	.34		1.36	1.65	
	0440	3" thick, R13		800	.010		1.21	.34		1.55	1.86	
	0520	Foil faced, 1" thick, R4.3		1,000	.008		.80	.27		1.07	1.31	
	0540	1-1/2" thick, R6.5		1,000	.008		1.08	.27		1.35	1.62	
	0560	2" thick, R8.7		890	.009		1.35	.31		1.66	1.97	
	0580	2-1/2" thick, R10.9		800	.010		1.60	.34		1.94	2.29	
	0600	3" thick, R13		800	.010		1.74	.34		2.08	2.44	
	0670	6#/CF, unfaced, 1" thick, R4.3		1,000	.008		.77	.27		1.04	1.28	
	0690	1-1/2" thick, R6.5		890	.009		1.19	.31		1.50	1.79	
	0700	2" thick, R8.7		800	.010		1.68	.34		2.02	2.38	
	0721	2-1/2" thick, R10.9		800	.010		1.84	.34		2.18	2.55	
	0741	3" thick, R13		730	.011		2.20	.38		2.58	3	
	0821	Foil faced, 1" thick, R4.3		1,000	.008		1.09	.27		1.36	1.63	
	0840	1-1/2" thick, R6.5		890	.009		1.57	.31		1.88	2.21	
	0850	2" thick, R8.7		800	.010		2.05	.34		2.39	2.79	
	0880	2-1/2" thick, R10.9		800	.010		2.46	.34		2.80	3.24	
	0900	3" thick, R13		730	.011		2.94	.38		3.32	3.81	
	1500	Foamglass, 1-1/2" thick, R4.5		800	.010		1.21	.34		1.55	1.86	
	1550	3" thick, R9		730	.011		2.92	.38		3.30	3.79	
	1700	Perlite, 1" thick, R2.77		800	.010		.26	.34		.60	.82	
	1750	2" thick, R5.55		730	.011		.50	.38		.88	1.13	
	1900	Extruded polystyrene, 25 PSI compressive strength, 1" thick, R5		800	.010		.38	.34		.72	.95	
	1940	2" thick R10		730	.011		.75	.38		1.13	1.41	
	1960	3" thick, R15		730	.011		1.04	.38		1.42	1.72	
	2100	Expanded polystyrene, 1" thick, R3.85		800	.010		.17	.34		.51	.72	
	2120	2" thick, R7.69		730	.011		.42	.38		.80	1.04	
	2140	3" thick, R11.49	↓	730	.011	↓	.52	.38		.90	1.15	
950	0010	**WALL OR CEILING INSUL., NON-RIGID**										**950**
	0040	Fiberglass, kraft faced, batts or blankets										
	0060	3-1/2" thick, R11, 11" wide	1 Carp	1,150	.007	S.F.	.27	.24		.51	.67	
	0080	15" wide		1,600	.005		.27	.17		.44	.57	
	0100	23" wide		1,600	.005		.27	.17		.44	.57	
	0140	6" thick, R19, 11" wide	↓	1,000	.008	↓	.35	.27		.62	.82	

	07210	Building Insulation	CREW	DAILY OUTPUT	LABOR-HOURS	UNIT	2005 BARE COSTS				TOTAL INCL O&P	
							MAT.	LABOR	EQUIP.	TOTAL		
950	0160	15" wide	1 Carp	1,350	.006	S.F.	.35	.20		.55	.71	**950**
	0180	23" wide		1,600	.005		.35	.17		.52	.66	
	0200	9" thick, R30, 15" wide		1,150	.007		.60	.24		.84	1.03	
	0220	23" wide		1,350	.006		.60	.20		.80	.98	
	0240	12" thick, R38, 15" wide		1,000	.008		.76	.27		1.03	1.27	
	0260	23" wide	↓	1,350	.006	↓	.76	.20		.96	1.16	
	0400	Fiberglass, foil faced, batts or blankets										
	0420	3-1/2" thick, R11, 15" wide	1 Carp	1,600	.005	S.F.	.40	.17		.57	.71	
	0440	23" wide		1,600	.005		.40	.17		.57	.71	
	0460	6" thick, R19, 15" wide		1,350	.006		.45	.20		.65	.82	
	0480	23" wide		1,600	.005		.45	.17		.62	.77	
	0500	9" thick, R30, 15" wide		1,150	.007		.71	.24		.95	1.15	
	0550	23" wide	↓	1,350	.006	↓	.71	.20		.91	1.10	
	0800	Fiberglass, unfaced, batts or blankets										
	0820	3-1/2" thick, R11, 15" wide	1 Carp	1,350	.006	S.F.	.24	.20		.44	.58	
	0830	23" wide		1,600	.005		.24	.17		.41	.53	
	0860	6" thick, R19, 15" wide		1,150	.007		.46	.24		.70	.88	
	0880	23" wide		1,350	.006		.46	.20		.66	.83	
	0900	9" thick, R30, 15" wide		1,000	.008		.60	.27		.87	1.09	
	0920	23" wide		1,150	.007		.60	.24		.84	1.03	
	0940	12" thick, R38, 15" wide		1,000	.008		.76	.27		1.03	1.27	
	0960	23" wide	↓	1,150	.007	↓	.76	.24		1	1.21	
	1300	Mineral fiber batts, kraft faced										
	1320	3-1/2" thick, R12	1 Carp	1,600	.005	S.F.	.28	.17		.45	.58	
	1340	6" thick, R19		1,600	.005		.38	.17		.55	.69	
	1380	10" thick, R30		1,350	.006	↓	.57	.20		.77	.95	
	1850	Friction fit wire insulation supports, 16" O.C.	↓	960	.008	Ea.	.05	.29		.34	.50	
	1900	For foil backing, add				S.F.	.04			.04	.04	

	07220	Roof and Deck Insulation										
700	0010	**ROOF DECK INSULATION**										**700**
	0020	Fiberboard low density, 1/2" thick R1.39	1 Rofc	1,000	.008	S.F.	.19	.24		.43	.61	
	0030	1" thick R2.78		800	.010		.34	.29		.63	.87	
	0080	1 1/2" thick R4.17		800	.010		.50	.29		.79	1.05	
	0100	2" thick R5.56		800	.010		.68	.29		.97	1.25	
	0110	Fiberboard high density, 1/2" thick R1.3		1,000	.008		.20	.24		.44	.62	
	0120	1" thick R2.5		800	.010		.36	.29		.65	.90	
	0130	1-1/2" thick R3.8		800	.010		.59	.29		.88	1.15	
	0200	Fiberglass, 3/4" thick R2.78		1,000	.008		.46	.24		.70	.91	
	0400	15/16" thick R3.70		1,000	.008		.61	.24		.85	1.07	
	0460	1-1/16" thick R4.17		1,000	.008		.76	.24		1	1.24	
	0600	1-5/16" thick R5.26		1,000	.008		1.05	.24		1.29	1.56	
	0650	2-1/16" thick R8.33		800	.010		1.12	.29		1.41	1.73	
	0700	2-7/16" thick R10		800	.010		1.28	.29		1.57	1.91	
	1500	Foamglass, 1-1/2" thick R4.5		800	.010		1.26	.29		1.55	1.89	
	1530	3" thick R9		700	.011	↓	2.57	.34		2.91	3.40	
	1600	Tapered for drainage		600	.013	B.F.	1.10	.39		1.49	1.88	
	1650	Perlite, 1/2" thick R1.32		1,050	.008	S.F.	.27	.22		.49	.68	
	1655	3/4" thick R2.08		800	.010		.29	.29		.58	.82	
	1660	1" thick R2.78		800	.010		.31	.29		.60	.84	
	1670	1-1/2" thick R4.17		800	.010		.39	.29		.68	.93	
	1680	2" thick R5.56		700	.011		.62	.34		.96	1.25	
	1685	2-1/2" thick R6.67		700	.011	↓	.77	.34		1.11	1.42	
	1690	Tapered for drainage		800	.010	B.F.	.59	.29		.88	1.15	
	1700	Polyisocyanurate, 2#/CF density, 3/4" thick, R5.1		1,500	.005	S.F.	.30	.16		.46	.60	
	1705	1" thick R7.14	↓	1,400	.006	↓	.33	.17		.50	.65	

		07220 | Roof and Deck Insulation	CREW	DAILY OUTPUT	LABOR-HOURS	UNIT	2005 BARE COSTS				TOTAL INCL O&P	
							MAT.	LABOR	EQUIP.	TOTAL		
700	1715	1-1/2" thick R10.87	1 Rofc	1,250	.006	S.F.	.35	.19		.54	.71	**700**
	1725	2" thick R14.29		1,100	.007		.45	.21		.66	.87	
	1735	2-1/2" thick R16.67		1,050	.008		.50	.22		.72	.93	
	1745	3" thick R21.74		1,000	.008		.73	.24		.97	1.20	
	1755	3-1/2" thick R25		1,000	.008	▼	.74	.24		.98	1.21	
	1765	Tapered for drainage	▼	1,400	.006	B.F.	.38	.17		.55	.71	
	1900	Extruded Polystyrene										
	1910	15 PSI compressive strength, 1" thick, R5	1 Rofc	1,500	.005	S.F.	.23	.16		.39	.52	
	1920	2" thick, R10		1,250	.006		.35	.19		.54	.71	
	1930	3" thick R15		1,000	.008		.68	.24		.92	1.15	
	1932	4" thick R20		1,000	.008	▼	1.07	.24		1.31	1.58	
	1934	Tapered for drainage		1,500	.005	B.F.	.35	.16		.51	.66	
	1940	25 PSI compressive strength, 1" thick R5		1,500	.005	S.F.	.50	.16		.66	.82	
	1942	2" thick R10		1,250	.006		.97	.19		1.16	1.39	
	1944	3" thick R15		1,000	.008		1.47	.24		1.71	2.02	
	1946	4" thick R20		1,000	.008	▼	1.96	.24		2.20	2.56	
	1948	Tapered for drainage		1,500	.005	B.F.	.40	.16		.56	.71	
	1950	40 psi compressive strength, 1" thick R5		1,500	.005	S.F.	.35	.16		.51	.66	
	1952	2" thick R10		1,250	.006		.69	.19		.88	1.08	
	1954	3" thick R15		1,000	.008		1.01	.24		1.25	1.51	
	1956	4" thick R20		1,000	.008	▼	1.35	.24		1.59	1.89	
	1958	Tapered for drainage		1,400	.006	B.F.	.50	.17		.67	.84	
	1960	60 PSI compressive strength, 1" thick R5		1,450	.006	S.F.	.42	.16		.58	.74	
	1962	2" thick R10		1,200	.007		.75	.20		.95	1.16	
	1964	3" thick R15		975	.008		1.12	.24		1.36	1.64	
	1966	4" thick R20		950	.008	▼	1.56	.25		1.81	2.14	
	1968	Tapered for drainage		1,400	.006	B.F.	.61	.17		.78	.96	
	2010	Expanded polystyrene, 1#/CF density, 3/4" thick R2.89		1,500	.005	S.F.	.19	.16		.35	.48	
	2020	1" thick R3.85		1,500	.005		.19	.16		.35	.48	
	2100	2" thick R7.69		1,250	.006		.41	.19		.60	.77	
	2110	3" thick R11.49		1,250	.006		.70	.19		.89	1.09	
	2120	4" thick R15.38		1,200	.007		.57	.20		.77	.96	
	2130	5" thick R19.23		1,150	.007		.71	.21		.92	1.13	
	2140	6" thick R23.26		1,150	.007	▼	.83	.21		1.04	1.26	
	2150	Tapered for drainage	▼	1,500	.005	B.F.	.34	.16		.50	.64	
	2400	Composites with 2" EPS										
	2410	1" fiberboard	1 Rofc	950	.008	S.F.	.78	.25		1.03	1.28	
	2420	7/16" oriented strand board		800	.010		.93	.29		1.22	1.52	
	2430	1/2" plywood		800	.010		1	.29		1.29	1.60	
	2440	1" perlite	▼	800	.010	▼	.82	.29		1.11	1.40	
	2450	Composites with 1 1/2" polyisocyanurate										
	2460	1" fiberboard	1 Rofc	800	.010	S.F.	.84	.29		1.13	1.42	
	2470	1" perlite		850	.009		.88	.28		1.16	1.44	
	2480	7/16" oriented strand board	▼	800	.010	▼	1.01	.29		1.30	1.61	

		07260 | Vapor Retarders	CREW	DAILY OUTPUT	LABOR-HOURS	UNIT	MAT.	LABOR	EQUIP.	TOTAL	TOTAL INCL O&P	
100	0010	**BUILDING PAPER** Aluminum and kraft laminated, foil 1 side	1 Carp	37	.216	Sq.	3.64	7.40		11.04	15.55	**100**
	0100	Foil 2 sides		37	.216		6.40	7.40		13.80	18.55	
	0300	Asphalt, two ply, 30#, for subfloors		19	.421		13.95	14.40		28.35	38	
	0400	Asphalt felt sheathing paper, 15#	▼	37	.216	▼	3.41	7.40		10.81	15.30	
	0450	Housewrap, exterior, spun bonded polypropylene										
	0470	Small roll	1 Carp	3,800	.002	S.F.	.16	.07		.23	.29	
	0480	Large roll	"	4,000	.002	"	.10	.07		.17	.22	
	0500	Material only, 3' x 111.1' roll				Ea.	55			55	60.50	
	0520	9' x 111.1' roll				"	100			100	110	
	0600	Polyethylene vapor barrier, standard, .002" thick	1 Carp	37	.216	Sq.	.96	7.40		8.36	12.60	

07200 | Thermal Protection

07260 | Vapor Retarders

		CREW	DAILY OUTPUT	LABOR-HOURS	UNIT	2005 BARE COSTS MAT.	LABOR	EQUIP.	TOTAL	TOTAL INCL O&P	
100	0700	.004" thick	1 Carp	37	.216	Sq.	1.88	7.40		9.28	13.60	100
	0900	.006" thick		37	.216		2.75	7.40		10.15	14.60	
	1200	.010" thick		37	.216		5.10	7.40		12.50	17.15	
	1300	Clear reinforced, fire retardant, .008" thick		37	.216		8.55	7.40		15.95	21	
	1350	Cross laminated type, .003" thick		37	.216		6.65	7.40		14.05	18.85	
	1400	.004" thick		37	.216		7.40	7.40		14.80	19.70	
	1500	Red rosin paper, 5 sq rolls, 4 lb per square		37	.216		1.58	7.40		8.98	13.30	
	1600	5 lbs. per square		37	.216		2.04	7.40		9.44	13.80	
	1800	Reinf. waterproof, .002" polyethylene backing, 1 side		37	.216		5	7.40		12.40	17.05	
	1900	2 sides		37	.216		6.65	7.40		14.05	18.85	
	2100	Roof deck vapor barrier, class 1 metal decks	1 Rofc	37	.216		11.60	6.35		17.95	23.50	
	2200	For all other decks	"	37	.216		8.20	6.35		14.55	19.90	
	2400	Waterproofed kraft with sisal or fiberglass fibers, minimum	1 Carp	37	.216		5.40	7.40		12.80	17.50	
	2500	Maximum	"	37	.216		13.50	7.40		20.90	26.50	

07500 | Membrane Roofing

07590 | Roof Maintenance and Repairs

		CREW	DAILY OUTPUT	LABOR-HOURS	UNIT	2005 BARE COSTS MAT.	LABOR	EQUIP.	TOTAL	TOTAL INCL O&P	
300	0010	ROOF COATINGS Asphalt				Gal.	2.95			2.95	3.25	300
	0200	Asphalt base, fibered aluminum coating					9.60			9.60	10.55	
	0300	Asphalt primer, 5 gallon					3.92			3.92	4.31	
	0600	Coal tar pitch, 200 lb. barrels				Ton	580			580	640	
	0700	Tar roof cement, 5 gal. lots				Gal.	6.10			6.10	6.70	
	0800	Glass fibered roof & patching cement, 5 gallon				"	4.43			4.43	4.87	
	0900	Reinforcing glass membrane, 450 S.F./roll				Ea.	45.50			45.50	50	
	1000	Neoprene roof coating, 5 gal, 2 gal/sq				Gal.	22			22	24	
	1100	Roof patch & flashing cement, 5 gallon					18.05			18.05	19.85	
	1200	Roof resurtant, glass fibered, 3 gal/sq					7.30			7.30	8.05	
	1300	Mineral rubber, 3 gal/sq					4.71			4.71	5.20	

07700 | Roof Specialties and Accessories

07720 | Roof Accessories

		CREW	DAILY OUTPUT	LABOR-HOURS	UNIT	2005 BARE COSTS MAT.	LABOR	EQUIP.	TOTAL	TOTAL INCL O&P	
550	0010	RIDGE VENT										550
	0100	Aluminum strips, mill finish	1 Rofc	160	.050	L.F.	1.20	1.47		2.67	3.83	
	0150	Painted finish		160	.050	"	2.12	1.47		3.59	4.85	
	0200	Connectors		48	.167	Ea.	1.91	4.91		6.82	10.45	
	0300	End caps		48	.167	"	.93	4.91		5.84	9.35	
	0400	Galvanized strips		160	.050	L.F.	2.07	1.47		3.54	4.79	
560	0010	SNOW GUARDS										560
	0100	Slate & asphalt shingle roofs	1 Rofc	160	.050	Ea.	8	1.47		9.47	11.30	
	0200	Standing seam metal roofs		48	.167		12.25	4.91		17.16	22	
	0300	Surface mount for metal roofs		48	.167		6.75	4.91		11.66	15.80	

07920	Joint Sealants	CREW	DAILY OUTPUT	LABOR-HOURS	UNIT	2005 BARE COSTS MAT.	LABOR	EQUIP.	TOTAL	TOTAL INCL O&P	
800	**0010** **CAULKING AND SEALANTS**										**800**
0020	Acoustical sealant, elastomeric, cartridges				Ea.	2.21			2.21	2.43	
0030	Backer rod, polyethylene, 1/4" diameter	1 Bric	4.60	1.739	C.L.F.	2.12	61.50		63.62	96	
0050	1/2" diameter		4.60	1.739		3.39	61.50		64.89	97	
0070	3/4" diameter		4.60	1.739		5.80	61.50		67.30	100	
0090	1" diameter	↓	4.60	1.739	↓	9.35	61.50		70.85	104	
0100	Acrylic latex caulk, white										
0200	11 fl. oz cartridge				Ea.	1.88			1.88	2.07	
0500	1/4" x 1/2"	1 Bric	248	.032	L.F.	.15	1.14		1.29	1.90	
0600	1/2" x 1/2"		250	.032		.31	1.13		1.44	2.06	
0800	3/4" x 3/4"		230	.035		.69	1.23		1.92	2.63	
0900	3/4" x 1"		200	.040		.92	1.41		2.33	3.16	
1000	1" x 1"	↓	180	.044	↓	1.15	1.57		2.72	3.66	
1400	Butyl based, bulk				Gal.	22			22	24.50	
1500	Cartridges				"	27			27	29.50	
1700	Bulk, in place 1/4" x 1/2", 154 L.F./gal.	1 Bric	230	.035	L.F.	.14	1.23		1.37	2.03	
1800	1/2" x 1/2", 77 L.F./gal.	"	180	.044	"	.29	1.57		1.86	2.71	
2000	Latex acrylic based, bulk				Gal.	23			23	25.50	
2100	Cartridges					26			26	28.50	
2300	Polysulfide compounds, 1 component, bulk				↓	44			44	48	
2600	1 or 2 component, in place, 1/4" x 1/4", 308 L.F./gal.	1 Bric	145	.055	L.F.	.14	1.94		2.08	3.12	
2700	1/2" x 1/4", 154 L.F./gal.		135	.059		.28	2.09		2.37	3.49	
2900	3/4" x 3/8", 68 L.F./gal.		130	.062		.64	2.17		2.81	4.01	
3000	1" x 1/2", 38 L.F./gal.	↓	130	.062	↓	1.15	2.17		3.32	4.57	
3200	Polyurethane, 1 or 2 component				Gal.	47.50			47.50	52.50	
3500	Bulk, in place, 1/4" x 1/4"	1 Bric	150	.053	L.F.	.15	1.88		2.03	3.03	
3600	1/2" x 1/4"		145	.055		.31	1.94		2.25	3.30	
3800	3/4" x 3/8", 68 L.F./gal.		130	.062		.70	2.17		2.87	4.07	
3900	1" x 1/2"	↓	110	.073	↓	1.24	2.56		3.80	5.25	
4100	Silicone rubber, bulk				Gal.	34.50			34.50	38	
4200	Cartridges				"	40.50			40.50	44.50	
4400	Neoprene gaskets, closed cell, adhesive, 1/8" x 3/8"	1 Bric	240	.033	L.F.	.20	1.17		1.37	2.01	
4500	1/4" x 3/4"		215	.037		.48	1.31		1.79	2.53	
4700	1/2" x 1"		200	.040		1.40	1.41		2.81	3.69	
4800	3/4" x 1-1/2"	↓	165	.048	↓	2.91	1.71		4.62	5.80	
5500	Resin epoxy coating, 2 component, heavy duty				Gal.	26			26	28.50	
5800	Tapes, sealant, P.V.C. foam adhesive, 1/16" x 1/4"				C.L.F.	4.60			4.60	5.05	
5900	1/16" x 1/2"					6.80			6.80	7.50	
5950	1/16" x 1"					11.30			11.30	12.45	
6000	1/8" x 1/2"				↓	7.65			7.65	8.40	
6200	Urethane foam, 2 component, handy pack, 1 C.F.				Ea.	27.50			27.50	30.50	
6300	50.0 C.F. pack				C.F.	14.05			14.05	15.45	

For information about Means Estimating Seminars, see yellow pages 12 and 13 in back of book

THERMAL & MOISTURE PROTECTION **7**

Division Notes

	CREW	DAILY OUTPUT	LABOR-HOURS	UNIT	2005 BARE COSTS				TOTAL INCL O&P
					MAT.	LABOR	EQUIP.	TOTAL	

Division 8
Doors & Windows

Estimating Tips

08100 Metal Doors & Frames
- Most metal doors and frames look alike, but there may be significant differences among them. When estimating these items be sure to choose the line item that most closely compares to the specification or door schedule requirements regarding:
  - type of metal
  - metal gauge
  - door core material
  - fire rating
  - finish

08200 Wood & Plastic Doors
- Wood and plastic doors vary considerably in price. The primary determinant is the veneer material. Lauan, birch and oak are the most common veneers. Other variables include the following:
  - hollow or solid core
  - fire rating
  - flush or raised panel
  - finish
- If the specifications require compliance with AWI (Architectural Woodwork Institute) standards or acoustical standards, the cost of the door may increase substantially. All wood doors are priced pre-mortised for hinges and predrilled for cylindrical locksets.

- Frequently doors, frames, and windows are unique in old buildings. Specified replacement units could be stock, custom (similar to the original) or exact reproduction. The estimator should work closely with a window consultant to determine any extra costs that may be associated with the unusual installation requirements.

08300 Specialty Doors
- There are many varieties of special doors, and they are usually priced per each. Add frames, hardware or operators required for a complete installation.

08510 Steel Windows
- Most metal windows are delivered preglazed. However, some metal windows are priced without glass. Refer to 08800 Glazing for glass pricing. The grade C indicates commercial grade windows, usually ASTM C-35.

08550 Wood Windows
- All wood windows are priced preglazed. The two glazing options priced are single pane float glass and insulating glass 1/2" thick. Add the cost of screens and grills if required.

08700 Hardware
- Hardware costs add considerably to the cost of a door. The most efficient method to determine the hardware requirements for a project is to review the door schedule. This schedule, in conjunction with the specifications, is all you should need to take off the door hardware.

- Door hinges are priced by the pair, with most doors requiring 1-1/2 pairs per door. The hinge prices do not include installation labor because it is included in door installation. Hinges are classified according to the frequency of use.

08800 Glazing
- Different openings require different types of glass. The three most common types are:
  - float
  - tempered
  - insulating
- Most exterior windows are glazed with insulating glass. Entrance doors and window walls, where the glass is less than 18" from the floor, are generally glazed with tempered glass. Interior windows and some residential windows are glazed with float glass.
- Energy efficient coatings are also available

08900 Glazed Curtain Wall
- Glazed curtain walls consist of the metal tube framing and the glazing material. The cost data in this subdivision is presented for the metal tube framing alone or the composite wall. If your estimate requires a detailed takeoff of the framing, be sure to add the glazing cost.

Reference Numbers
Reference numbers are shown in bold squares at the beginning of some major classifications. These numbers refer to related items in the Reference Section. The reference information may be an estimating procedure, an alternate pricing method or technical information.

Note: Not all subdivisions listed here necessarily appear in this publication.

08060	Selective Demolition		CREW	DAILY OUTPUT	LABOR-HOURS	UNIT	MAT.	LABOR	EQUIP.	TOTAL	TOTAL INCL O&P	
110	0010	**SELECTIVE DEMOLITION, DOORS**	R02220 -510										110
	0200	Doors, exterior, 1-3/4" thick, single, 3' x 7' high		1 Clab	16	.500	Ea.		13.35		13.35	21
	0220	Double, 6' x 7' high			12	.667			17.80		17.80	27.50
	0500	Interior, 1-3/8" thick, single, 3' x 7' high			20	.400			10.70		10.70	16.60
	0520	Double, 6' x 7' high			16	.500			13.35		13.35	21
	0700	Bi-folding, 3' x 6'-8" high			20	.400			10.70		10.70	16.60
	0720	6' x 6'-8" high			18	.444			11.85		11.85	18.45
	0900	Bi-passing, 3' x 6'-8" high			16	.500			13.35		13.35	21
	0940	6' x 6'-8" high		▼	14	.571			15.25		15.25	23.50
	1500	Remove and reset, minimum		1 Carp	8	1			34.50		34.50	53.50
	1520	Maximum			6	1.333			45.50		45.50	71
	2000	Frames, including trim, metal		▼	8	1			34.50		34.50	53.50
	2200	Wood		2 Carp	32	.500			17.15		17.15	26.50
	3000	Special doors, counter doors			6	2.667			91.50		91.50	142
	3100	Double acting			10	1.600			55		55	85.50
	3200	Floor door (trap type), or access type			8	2			68.50		68.50	107
	3300	Glass, sliding, including frames			12	1.333			45.50		45.50	71
	3400	Overhead, commercial, 12' x 12' high			4	4			137		137	213
	3440	up to 20' x 16' high			3	5.333			183		183	285
	3445	up to 35' x 30' high			1	16			550		550	855
	3500	Residential, 9' x 7' high			8	2			68.50		68.50	107
	3540	16' x 7' high			7	2.286			78.50		78.50	122
	3600	Remove and reset, minimum			4	4			137		137	213
	3620	Maximum			2.50	6.400			219		219	340
	3700	Roll-up grille			5	3.200			110		110	171
	3800	Revolving door			2	8			274		274	425
	3900	Storefront swing door		▼	3	5.333	▼		183		183	285
120	0010	**SELECTIVE DEMOLITION, WINDOWS**	R02220 -510										120
	0200	Aluminum, including trim, to 12 S.F.		1 Clab	16	.500	Ea.		13.35		13.35	21
	0240	To 25 S.F.			11	.727			19.40		19.40	30
	0280	To 50 S.F.			5	1.600			42.50		42.50	66.50
	0320	Storm windows, to 12 S.F.			27	.296			7.90		7.90	12.30
	0360	To 25 S.F.			21	.381			10.15		10.15	15.85
	0400	To 50 S.F.			16	.500	▼		13.35		13.35	21
	0600	Glass, minimum			200	.040	S.F.		1.07		1.07	1.66
	0620	Maximum			150	.053	"		1.42		1.42	2.22
	1000	Steel, including trim, to 12 S.F.			13	.615	Ea.		16.45		16.45	25.50
	1020	To 25 S.F.			9	.889			23.50		23.50	37
	1040	To 50 S.F.			4	2			53.50		53.50	83
	2000	Wood, including trim, to 12 S.F.			22	.364			9.70		9.70	15.10
	2020	To 25 S.F.			18	.444			11.85		11.85	18.45
	2060	To 50 S.F.			13	.615			16.45		16.45	25.50
	2065	To 180 S.F.		▼	8	1			26.50		26.50	41.50
	5020	Remove and reset window, minimum		1 Carp	6	1.333			45.50		45.50	71
	5040	Average			4	2			68.50		68.50	107
	5080	Maximum		▼	2	4	▼		137		137	213

8
DOORS & WINDOWS

08770	Door/Window Accessories	CREW	DAILY OUTPUT	LABOR-HOURS	UNIT	2005 BARE COSTS				TOTAL INCL O&P	
						MAT.	LABOR	EQUIP.	TOTAL		
100	0010	**AREA WINDOW WELL**, Galvanized steel, 20 ga., 3'-2" wide, 1' deep	1 Sswk	29	.276	Ea.	12.95	10.50		23.45	33	100
	0100	2' deep		23	.348		23	13.25		36.25	49	
	0300	16 ga., galv., 3'-2" wide, 1' deep		29	.276		18.25	10.50		28.75	39	
	0400	3' deep		23	.348		37	13.25		50.25	64	
	0600	Welded grating for above, 15 lbs., painted		45	.178		33	6.80		39.80	48	
	0700	Galvanized		45	.178		78.50	6.80		85.30	98	
	0900	Translucent plastic cap for above	▼	60	.133	▼	13.40	5.10		18.50	24	

For information about Means Estimating Seminars, see yellow pages 12 and 13 in back of book

DOORS & WINDOWS **8**

233

Division Notes

	CREW	DAILY OUTPUT	LABOR-HOURS	UNIT	2005 BARE COSTS MAT.	LABOR	EQUIP.	TOTAL	TOTAL INCL O&P

Division 9
Finishes

Estimating Tips
General
- Room Finish Schedule: A complete set of plans should contain a room finish schedule. If one is not available, it would be well worth the time and effort to put one together. A room finish schedule should contain the room number, room name (for clarity), floor materials, base materials, wainscot materials, wainscot height, wall materials (for each wall), ceiling materials, ceiling height and special instructions.
- Surplus Finishes: Review the specifications to determine if there is any requirement to provide certain amounts of extra materials for the owner's maintenance department. In some cases the owner may require a substantial amount of materials, especially when it is a special order item or long lead time item.

09200 Plaster & Gypsum Board
- Lath is estimated by the square yard for both gypsum and metal lath, plus usually 5% allowance for waste. Furring, channels and accessories are measured by the linear foot. An extra foot should be allowed for each accessory miter or stop.
- Plaster is also estimated by the square yard. Deductions for openings vary by preference, from zero deduction to 50% of all openings over 2 feet in width. Some estimators deduct a percentage of the total yardage for openings. The estimator should allow one extra square foot for each linear foot of horizontal interior or exterior angle located below the ceiling level. Also, double the areas of small radius work.
- Each room should be measured, perimeter times maximum wall height. Floors and ceiling areas are equal to length times width.
- Drywall accessories, studs, track, and acoustical caulking are all measured by the linear foot. Drywall taping is figured by the square foot. Gypsum wallboard is estimated by the square foot. No material deductions should be made for door or window openings under 32 S.F. Coreboard can be obtained in a 1″ thickness for solid wall and shaft work. Additions should be made to price out the inside or outside corners.
- Different types of partition construction should be listed separately on the quantity sheets. There may be walls with studs of various widths, double studded, and similar or dissimilar surface materials. Shaft work is usually different construction from surrounding partitions requiring separate quantities and pricing of the work.

09300 Tile
09400 Terrazzo
- Tile and terrazzo areas are taken off on a square foot basis. Trim and base materials are measured by the linear foot. Accent tiles are listed per each. Two basic methods of installation are used. Mud set is approximately 30% more expensive than the thin set. In terrazzo work, be sure to include the linear footage of embedded decorative strips, grounds, machine rubbing and power cleanup.

09600 Flooring
- Wood flooring is available in strip, parquet, or block configuration. The latter two types are set in adhesives with quantities estimated by the square foot. The laying pattern will influence labor costs and material waste. In addition to the material and labor for laying wood floors, the estimator must make allowances for sanding and finishing these areas unless the flooring is prefinished.
- Most of the various types of flooring are all measured on a square foot basis. Base is measured by the linear foot. If adhesive materials are to be quantified, they are estimated at a specified coverage rate by the gallon depending upon the specified type and the manufacturer's recommendations.
- Sheet flooring is measured by the square yard. Roll widths vary, so consideration should be given to use the most economical width, as waste must be figured into the total quantity. Consider also the installation methods available, direct glue down or stretched.

09700 Wall Finishes
- Wall coverings are estimated by the square foot. The area to be covered is measured, length by height of wall above baseboards, to calculate the square footage of each wall. This figure is divided by the number of square feet in the single roll which is being used. Deduct, in full, the areas of openings such as doors and windows. Where a pattern match is required allow 25%-30% waste. One gallon of paste should be sufficient to hang 12 single rolls of light to medium weight paper.

09800 Acoustical Treatment
- Acoustical systems fall into several categories. The takeoff of these materials should be by the square foot of area with a 5% allowance for waste. Do not forget about scaffolding, if applicable, when estimating these systems.

09900 Paints & Coatings
- A major portion of the work in painting involves surface preparation. Be sure to include cleaning, sanding, filling and masking costs in the estimate.
- Painting is one area where bids vary to a greater extent than almost any other section of a project. This arises from the many methods of measuring surfaces to be painted. The estimator should check the plans and specifications carefully to be sure of the required number of coats.
- Protection of adjacent surfaces is not included in painting costs. When considering the method of paint application, an important factor is the amount of protection and masking required. These must be estimated separately and may be the determining factor in choosing the method of application.

Reference Numbers
Reference numbers are shown in bold squares at the beginning of some major classifications. These numbers refer to related items in the Reference Section. The reference information may be an estimating procedure, an alternate pricing method or technical information.

Note: Not all subdivisions listed here necessarily appear in this publication.

			DAILY	LABOR-			2005 BARE COSTS			TOTAL
09060	**Selective Demolition**	CREW	OUTPUT	HOURS	UNIT	MAT.	LABOR	EQUIP.	TOTAL	INCL O&P
110	**0010 SELECTIVE DEMOLITION, CEILINGS** R02220 -510									**110**
0200	Ceiling, drywall, furred and nailed or screwed	2 Clab	800	.020	S.F.		.53		.53	.83
0220	On metal frame		760	.021			.56		.56	.87
0240	On suspension system, including system		720	.022			.59		.59	.92
1000	Plaster, lime and horse hair, on wood lath, incl. lath		700	.023			.61		.61	.95
1020	On metal lath		570	.028			.75		.75	1.17
1100	Gypsum, on gypsum lath		720	.022			.59		.59	.92
1120	On metal lath		500	.032			.85		.85	1.33
1500	Tile, wood fiber, 12" x 12", glued		900	.018			.47		.47	.74
1540	Stapled		1,500	.011			.28		.28	.44
1580	On suspension system, incl. system		760	.021			.56		.56	.87
2000	Wood, tongue and groove, 1" x 4"		1,000	.016			.43		.43	.66
2040	1" x 8"		1,100	.015			.39		.39	.60
2400	Plywood or wood fiberboard, 4' x 8' sheets		1,200	.013			.36		.36	.55
120	**0010 SELECTIVE DEMOLITION, FLOORING** R02220 -510									**120**
0200	Brick with mortar	2 Clab	475	.034	S.F.		.90		.90	1.40
0400	Carpet, bonded, including surface scraping		2,000	.008			.21		.21	.33
0440	Scrim applied		8,000	.002			.05		.05	.08
0480	Tackless		9,000	.002			.05		.05	.07
0600	Composition, acrylic or epoxy		400	.040			1.07		1.07	1.66
0700	Concrete, scarify skin	A-1A	225	.036			1.24	1.50	2.74	3.58
0800	Resilient, sheet goods	2 Clab	1,400	.011			.31		.31	.47
0820	For gym floors		900	.018			.47		.47	.74
0900	Vinyl composition tile, 12" x 12"		1,000	.016			.43		.43	.66
2000	Tile, ceramic, thin set		675	.024			.63		.63	.98
2020	Mud set		625	.026			.68		.68	1.06
2200	Marble, slate, thin set		675	.024			.63		.63	.98
2220	Mud set		625	.026			.68		.68	1.06
2600	Terrazzo, thin set		450	.036			.95		.95	1.48
2620	Mud set		425	.038			1.01		1.01	1.56
2640	Cast in place		300	.053			1.42		1.42	2.22
3000	Wood, block, on end	1 Carp	400	.020			.69		.69	1.07
3200	Parquet		450	.018			.61		.61	.95
3400	Strip flooring, interior, 2-1/4" x 25/32" thick		325	.025			.84		.84	1.31
3500	Exterior, porch flooring, 1" x 4"		220	.036			1.25		1.25	1.94
3800	Subfloor, tongue and groove, 1" x 6"		325	.025			.84		.84	1.31
3820	1" x 8"		430	.019			.64		.64	.99
3840	1" x 10"		520	.015			.53		.53	.82
4000	Plywood, nailed		600	.013			.46		.46	.71
4100	Glued and nailed		400	.020			.69		.69	1.07
130	**0010 SELECTIVE DEMOLITION, WALLS AND PARTITIONS** R02220 -510									**130**
0100	Brick, 4" to 12" thick	B-9C	220	.182	C.F.		4.93	.64	5.57	8.35
0200	Concrete block, 4" thick		1,000	.040	S.F.		1.08	.14	1.22	1.85
0280	8" thick		810	.049			1.34	.17	1.51	2.27
0300	Exterior stucco 1" thick over mesh	B-9	3,200	.013			.34	.04	.38	.58
1000	Drywall, nailed or screwed	1 Clab	1,000	.008			.21		.21	.33
1020	Glued and nailed		900	.009			.24		.24	.37
1500	Fiberboard, nailed		900	.009			.24		.24	.37
1520	Glued and nailed		800	.010			.27		.27	.42
1568	Plenum barrier, sheet lead		300	.027			.71		.71	1.11
2000	Movable walls, metal, 5' high		300	.027			.71		.71	1.11
2020	8' high		400	.020			.53		.53	.83
2200	Metal or wood studs, finish 2 sides, fiberboard	B-1	520	.046			1.26		1.26	1.97
2250	Lath and plaster		260	.092			2.53		2.53	3.93
2350	Plywood		450	.053			1.46		1.46	2.27
3000	Plaster, lime and horsehair, on wood lath	1 Clab	400	.020			.53		.53	.83

9

FINISHES

Important: See the Reference Section for critical supporting data - Reference Nos., Crews, & City Cost Indexes

09050 | Basic Material Finishes and Methods

09060 | Selective Demolition

			CREW	DAILY OUTPUT	LABOR-HOURS	UNIT	MAT.	LABOR	EQUIP.	TOTAL	TOTAL INCL O&P	
								2005 BARE COSTS				
130	3020	On metal lath	1 Clab	335	.024	S.F.		.64		.64	.99	**130**
	3400	Gypsum or perlite, on gypsum lath	R02220 -510	410	.020			.52		.52	.81	
	3420	On metal lath	↓	300	.027			.71		.71	1.11	
	3600	Plywood, one side	B-1	1,500	.016			.44		.44	.68	
	3750	Terra cotta block and plaster, to 6" thick	"	175	.137	↓		3.75		3.75	5.85	
	3800	Toilet partitions, slate or marble	1 Clab	5	1.600	Ea.		42.50		42.50	66.50	
	3820	Hollow metal	"	8	1	"		26.50		26.50	41.50	

09600 | Flooring

09631 | Brick Flooring

			CREW	DAILY OUTPUT	LABOR-HOURS	UNIT	MAT.	LABOR	EQUIP.	TOTAL	TOTAL INCL O&P	
								2005 BARE COSTS				
100	0010	**BRICK FLOORING**										**100**
	0400	Heavy duty industrial, cement mortar bed, 2" thick, not incl. brick	D-1	80	.200	S.F.	.67	6.20		6.87	10.20	
	0450	Acid proof joints, 1/4" wide	"	65	.246		1.13	7.65		8.78	12.90	
	0500	Pavers, 8" x 4", 1" to 1-1/4" thick, red	D-7	95	.168		2.88	4.89		7.77	10.35	
	0510	Ironspot	"	95	.168		4.07	4.89		8.96	11.70	
	0540	1-3/8" to 1-3/4" thick, red	D-1	95	.168		2.78	5.25		8.03	11	
	0560	Ironspot		95	.168		4.02	5.25		9.27	12.35	
	0580	2-1/4" thick, red		90	.178		2.83	5.55		8.38	11.50	
	0590	Ironspot	↓	90	.178	↓	4.38	5.55		9.93	13.20	
	0800	For sidewalks and patios with pavers, see division 02780-200										
	0870	For epoxy joints, add	D-1	600	.027	S.F.	2.15	.83		2.98	3.63	
	0880	For Furan underlayment, add	"	600	.027		1.78	.83		2.61	3.22	
	0890	For waxed surface, steam cleaned, add	D-5	1,000	.008	↓	.15	.28		.43	.60	

09900 | Paints & Coatings

09910 | Paints

			CREW	DAILY OUTPUT	LABOR-HOURS	UNIT	MAT.	LABOR	EQUIP.	TOTAL	TOTAL INCL O&P	
								2005 BARE COSTS				
630	0010	**MISCELLANEOUS, INTERIOR**										**630**
	5000	Pipe, to 4" diameter, primer or sealer coat, oil base, brushwork	2 Pord	1,250	.013	L.F.	.06	.39		.45	.66	
	5100	Spray		2,165	.007		.06	.23		.29	.40	
	5200	Paint 1 coat, brushwork		1,250	.013		.06	.39		.45	.66	
	5300	Spray		2,165	.007		.06	.23		.29	.40	
	5350	Paint 2 coats, brushwork		775	.021		.11	.63		.74	1.07	
	5400	Spray		1,240	.013		.13	.39		.52	.73	
	5450	To 8" diameter, primer or sealer coat, brushwork		620	.026		.12	.79		.91	1.32	
	5500	Spray		1,085	.015		.20	.45		.65	.90	
	5550	Paint 1 coat, brushwork		620	.026		.17	.79		.96	1.38	
	5600	Spray		1,085	.015		.19	.45		.64	.89	
	5650	Paint 2 coats, brushwork		385	.042		.23	1.27		1.50	2.16	
	5700	Spray		620	.026		.25	.79		1.04	1.47	
	5750	To 12" diameter, primer or sealer coat, brushwork	↓	415	.039	↓	.18	1.18		1.36	1.97	

09910	Paints	CREW	DAILY OUTPUT	LABOR-HOURS	UNIT	2005 BARE COSTS				TOTAL INCL O&P	
						MAT.	LABOR	EQUIP.	TOTAL		
630	5800	Spray	2 Pord	725	.022	L.F.	.21	.68		.89	1.25	**630**
	5850	Paint 1 coat, brushwork		415	.039		.17	1.18		1.35	1.96	
	6000	Spray		725	.022		.19	.68		.87	1.23	
	6200	Paint 2 coats, brushwork		260	.062		.34	1.88		2.22	3.20	
	6250	Spray		415	.039		.38	1.18		1.56	2.18	
	6300	To 16" diameter, primer or sealer coat, brushwork		310	.052		.24	1.58		1.82	2.64	
	6350	Spray		540	.030		.27	.91		1.18	1.66	
	6400	Paint 1 coat, brushwork		310	.052		.23	1.58		1.81	2.63	
	6450	Spray		540	.030		.26	.91		1.17	1.65	
	6500	Paint 2 coats, brushwork		195	.082		.45	2.51		2.96	4.27	
	6550	Spray	▼	310	.052	▼	.50	1.58		2.08	2.92	
	8900	Trusses and wood frames, primer coat, oil base, brushwork	1 Pord	800	.010	S.F.	.05	.31		.36	.51	
	8950	Spray		1,200	.007		.05	.20		.25	.37	
	9000	Paint 1 coat, brushwork		750	.011		.05	.33		.38	.55	
	9200	Spray		1,200	.007		.06	.20		.26	.38	
	9220	Paint 2 coats, brushwork		500	.016		.10	.49		.59	.85	
	9240	Spray		600	.013		.12	.41		.53	.74	
	9260	Stain, brushwork, wipe off		600	.013		.05	.41		.46	.66	
	9280	Varnish, 3 coats, brushwork	▼	275	.029	▼	.17	.89		1.06	1.53	
	9350	For latex paint, deduct					10%					
650	0010	**COATINGS & PAINTS** In 5 gallon lots										**650**
	0050	For 100 gallons or more, deduct					10%				10%	
	0100	Paint, Exterior alkyd (oil base)										
	0200	Flat				Gal.	24			24	26.50	
	0300	Gloss					23.50			23.50	26	
	0400	Primer				▼	23			23	25.50	
	0500	Latex (water base)										
	0600	Acrylic stain				Gal.	20			20	22	
	0700	Gloss enamel					24.50			24.50	27	
	0800	Flat					18.85			18.85	20.50	
	0900	Primer					20.50			20.50	22.50	
	1000	Semi-gloss				▼	23.50			23.50	26	
	2400	Masonry, Exterior										
	2500	Alkali resistant primer				Gal.	20			20	22	
	2600	Block filler, epoxy					22			22	24	
	2700	Latex					10.60			10.60	11.65	
	2800	Latex, flat					19.55			19.55	21.50	
	2900	Semi-gloss				▼	21.50			21.50	24	
	4000	Metal										
	4100	Galvanizing paint				Gal.	26			26	28.50	
	4200	High heat					46.50			46.50	51	
	4300	Heat resistant					28			28	31	
	4400	Machinery enamel, alkyd					27.50			27.50	30	
	4500	Metal pretreatment (polyvinyl butyral)					25			25	27.50	
	4600	Rust inhibitor, ferrous metal					24			24	26.50	
	4700	Zinc chromate					21			21	23.50	
	4800	Zinc rich primer				▼	101			101	112	
	5500	Coatings										
	5600	Heavy duty										
	5700	Acrylic urethane				Gal.	47.50			47.50	52	
	5800	Chlorinated rubber					32			32	35.50	
	5900	Coal tar epoxy					40.50			40.50	44.50	
	6000	Polyamide epoxy, finish					34			34	37.50	
	6100	Primer					33.50			33.50	37	
	6200	Silicone alkyd					34			34	37.50	
	6300	2 component solvent based acrylic epoxy				▼	49			49	54	

9 FINISHES

09910 | Paints

		CREW	DAILY OUTPUT	LABOR-HOURS	UNIT	2005 BARE COSTS				TOTAL INCL O&P	
						MAT.	LABOR	EQUIP.	TOTAL		
650	6400	Polyester epoxy				Gal.	69			69	76	650
	6500	Vinyl				↓	29			29	31.50	
	6600	Special/Miscellaneous										
	6700	Aluminum				Gal.	24.50			24.50	27	
	6900	Dry fall out, flat					13.05			13.05	14.35	
	7000	Fire retardant, intumescent					39			39	43	
	7100	Linseed oil					12.15			12.15	13.40	
	7200	Shellac					22.50			22.50	25	
	7300	Swimming pool, epoxy or urethane base					39			39	42.50	
	7400	Rubber base					29			29	32	
	7500	Texture paint					14.30			14.30	15.70	
	7600	Turpentine					11.40			11.40	12.55	
	7700	Water repellent, 5% silicone				↓	19			19	21	

09963 | Glazed Coatings

		CREW	DAILY OUTPUT	LABOR-HOURS	UNIT	2005 BARE COSTS				TOTAL INCL O&P	
						MAT.	LABOR	EQUIP.	TOTAL		
200	0010	**WALL COATINGS**										200
	0100	Acrylic glazed coatings, minimum	1 Pord	525	.015	S.F.	.26	.47		.73	.99	
	0200	Maximum		305	.026		.54	.80		1.34	1.80	
	0300	Epoxy coatings, minimum		525	.015		.33	.47		.80	1.06	
	0400	Maximum		170	.047		1.02	1.44		2.46	3.28	
	0600	Exposed aggregate, troweled on, 1/16" to 1/4", minimum		235	.034		.50	1.04		1.54	2.12	
	0700	Maximum (epoxy or polyacrylate)		130	.062		1.09	1.88		2.97	4.03	
	0900	1/2" to 5/8" aggregate, minimum		130	.062		1.01	1.88		2.89	3.94	
	1000	Maximum		80	.100		1.72	3.06		4.78	6.50	
	1200	1" aggregate size, minimum		90	.089		1.75	2.72		4.47	6	
	1300	Maximum		55	.145		2.68	4.45		7.13	9.65	
	1500	Exposed aggregate, sprayed on, 1/8" aggregate, minimum		295	.027		.47	.83		1.30	1.77	
	1600	Maximum	↓	145	.055	↓	.87	1.69		2.56	3.50	

For information about Means Estimating Seminars, see yellow pages 12 and 13 in back of book

FINISHES 9

For expanded coverage of these items see *Means Interior Cost Data 2005*

	CREW	DAILY OUTPUT	LABOR-HOURS	UNIT	2005 BARE COSTS				TOTAL INCL O&P
					MAT.	LABOR	EQUIP.	TOTAL	

Division 11
Equipment

Estimating Tips

General

- The items in this division are usually priced per square foot or each. Many of these items are purchased by the owner for installation by the contractor. Check the specifications for responsibilities, and include time for receiving, storage, installation and mechanical and electrical hook-ups in the appropriate divisions.

- Many items in Division 11 require some type of support system that is not usually furnished with the item. Examples of these systems include blocking for the attachment of casework and support angles for ceiling hung projection screens. The required blocking or supports must be added to the estimate in the appropriate division.

- Some items in Division 11 may require assembly or electrical hook-ups. Verify the amount of assembly required or the need for a hard electrical connection and add the appropriate costs.

Reference Numbers

Reference numbers are shown in bold squares at the beginning of some major classifications. These numbers refer to related items in the Reference Section. The reference information may be an estimating procedure, an alternate pricing method or technical information.

Note: Not all subdivisions listed here necessarily appear in this publication.

11285	**Hydraulic Gates**	CREW	DAILY OUTPUT	LABOR-HOURS	UNIT	2005 BARE COSTS				TOTAL INCL O&P	
						MAT.	LABOR	EQUIP.	TOTAL		
150	0010 **CANAL GATES** Cast iron body, fabricated frame										150
	0100 12" diameter	L-5A	4.60	6.957	Ea.	560	265	125	950	1,225	
	0110 18" diameter		4	8		925	305	144	1,374	1,700	
	0120 24" diameter		3.50	9.143		1,325	350	164	1,839	2,250	
	0130 30" diameter		2.80	11.429		2,150	435	205	2,790	3,350	
	0140 36" diameter		2.30	13.913		2,650	530	250	3,430	4,125	
	0150 42" diameter		1.70	18.824		4,675	715	340	5,730	6,750	
	0160 48" diameter		1.20	26.667		5,900	1,025	480	7,405	8,775	
	0170 54" diameter		.90	35.556		9,975	1,350	640	11,965	14,000	
	0180 60" diameter		.50	64		11,900	2,450	1,150	15,500	18,600	
	0190 66" diameter		.50	64		13,000	2,450	1,150	16,600	19,800	
	0200 72" diameter		.40	80		18,000	3,050	1,450	22,500	26,600	
190	0010 **FLAP GATES**										190
	0100 Aluminum, 18" diameter	L-5A	5	6.400	Ea.	2,200	244	115	2,559	2,975	
	0110 24" diameter		4	8		2,775	305	144	3,224	3,725	
	0120 30" diameter		3.50	9.143		3,325	350	164	3,839	4,425	
	0130 36" diameter		2.80	11.429		4,100	435	205	4,740	5,500	
	0140 42" diameter		2.30	13.913		5,525	530	250	6,305	7,275	
	0150 48" diameter		1.70	18.824		6,725	715	340	7,780	9,000	
	0160 54" diameter		1.20	26.667		9,275	1,025	480	10,780	12,500	
	0170 60" diameter		.80	40		11,700	1,525	720	13,945	16,300	
	0180 66" diameter		.50	64		13,400	2,450	1,150	17,000	20,200	
	0190 72" diameter		.40	80		15,600	3,050	1,450	20,100	23,900	
400	0010 **KNIFE GATES**										400
	0100 Incl. handwheel operator for hub, 6" diameter	Q-23	7.70	3.117	Ea.	665	123	104	892	1,025	
	0110 8" diameter		7.20	3.333		920	132	111	1,163	1,350	
	0120 10" diameter		4.80	5		1,175	197	167	1,539	1,775	
	0130 12" diameter		3.60	6.667		1,675	263	223	2,161	2,500	
	0140 14" diameter		3.40	7.059		3,025	278	236	3,539	4,000	
	0150 16" diameter		3.20	7.500		3,450	296	251	3,997	4,500	
	0160 18" diameter		3	8		4,400	315	267	4,982	5,625	
	0170 20" diameter		2.70	8.889		5,350	350	297	5,997	6,725	
	0180 24" diameter		2.40	10		6,425	395	335	7,155	8,050	
	0190 30" diameter		1.80	13.333		11,700	525	445	12,670	14,100	
	0200 36" diameter		1.20	20		15,700	790	670	17,160	19,100	
600	0010 **SLIDE GATES**										600
	0100 Steel, self contained incl. anchor bolts and grout, 12" x 12"	L-5A	4.60	6.957	Ea.	2,400	265	125	2,790	3,250	
	0110 18" x 18"		4	8		2,825	305	144	3,274	3,800	
	0120 24" x 24"		3.50	9.143		3,050	350	164	3,564	4,125	
	0130 30" x 30"		2.80	11.429		3,450	435	205	4,090	4,775	
	0140 36" x 36"		2.30	13.913		3,775	530	250	4,555	5,375	
	0150 42" x 42"		1.70	18.824		4,875	715	340	5,930	6,950	
	0160 48" x 48"		1.20	26.667		5,100	1,025	480	6,605	7,875	
	0170 54" x 54"		.90	35.556		6,150	1,350	640	8,140	9,800	
	0180 60" x 60"		.55	58.182		6,825	2,225	1,050	10,100	12,500	
	0190 72" x 72"		.36	88.889		9,425	3,375	1,600	14,400	18,000	
700	0010 **SLUICE GATES** Cast iron, AWWA C501										700
	0100 Heavy duty, self contained w/crank oper. gate, 18" x 18"	L-5A	1.70	18.824	Ea.	5,650	715	340	6,705	7,825	
	0110 24" x 24"		1.20	26.667		7,225	1,025	480	8,730	10,200	
	0120 30" x 30"		1	32		8,675	1,225	575	10,475	12,300	
	0130 36" x 36"		.90	35.556		10,400	1,350	640	12,390	14,400	
	0140 42" x 42"		.80	40		13,500	1,525	720	15,745	18,200	
	0150 48" x 48"		.50	64		14,900	2,450	1,150	18,500	21,900	
	0160 54" x 54"		.40	80		19,500	3,050	1,450	24,000	28,200	

Important: See the Reference Section for critical supporting data - Reference Nos., Crews, & City Cost Indexes

11 EQUIPMENT

11280 | Hydraulic Gates & Valves

		11285	Hydraulic Gates	CREW	DAILY OUTPUT	LABOR-HOURS	UNIT	2005 BARE COSTS				TOTAL INCL O&P	
								MAT.	LABOR	EQUIP.	TOTAL		
700	0170		60" x 60"	L-5A	.30	106	Ea.	25,100	4,075	1,925	31,100	36,700	700
	0180		66" x 66"		.30	106		29,100	4,075	1,925	35,100	41,100	
	0190		72" x 72"		.20	160		31,800	6,100	2,875	40,775	48,700	
	0200		78" x 78"		.20	160		36,000	6,100	2,875	44,975	53,500	
	0210		84" x 84"		.10	320		40,800	12,200	5,750	58,750	72,000	
	0220		90" x 90"	E-20	.30	213		46,500	7,925	3,200	57,625	68,000	
	0230		96" x 96"		.30	213		51,500	7,925	3,200	62,625	73,500	
	0240		108" x 108"		.20	320		60,500	11,900	4,775	77,175	92,500	
	0250		120" x 120"		.10	640		70,500	23,800	9,575	103,875	129,000	
	0260		132" x 132"		.10	640		109,000	23,800	9,575	142,375	171,500	

11300 | Fluid Waste Treatment & Disposal Equipment

EQUIPMENT 11

		11310 Sewage & Sludge Pumps	CREW	DAILY OUTPUT	LABOR-HOURS	UNIT	2005 BARE COSTS				TOTAL INCL O&P	
							MAT.	LABOR	EQUIP.	TOTAL		
100	0010	**OIL/WATER SEPARATORS**										100
	0020	Complete system, not including excavation and backfill										
	0100	Treated capacity 0.2 cubic feet per second	B-22	1	30	Ea.	6,675	940	238	7,853	9,025	
	0200	0.5 cubic foot per second	B-13	.75	74.667		10,100	2,150	820	13,070	15,400	
	0300	1.0 cubic foot per second		.60	93.333		14,500	2,675	1,025	18,200	21,300	
	0400	2.4 - 3 cubic feet per second		.30	186		21,600	5,375	2,050	29,025	34,300	
	0500	11 cubic feet per second		.17	329		24,400	9,475	3,625	37,500	45,400	
	0600	22 cubic feet per second		.10	560		33,100	16,100	6,175	55,375	68,000	
700	0010	**SEWAGE PUMPING STATIONS** Prefabricated steel, concrete										700
	0020	or fiberglass, 200 GPM	C-17D	.17	494	Total	34,100	17,400	2,925	54,425	68,000	
	0200	1,000 GPM		.07	1,200		48,100	42,300	7,100	97,500	126,500	
	0500	Add for generator unit, 200 GPM, steel		.34	247		24,600	8,725	1,475	34,800	42,200	
	0600	Concrete		.51	164		15,600	5,800	975	22,375	27,300	
	1000	Add for generator unit, 1,000 GPM, steel		.30	280		26,200	9,875	1,650	37,725	45,900	
	1200	Concrete		.38	221		22,100	7,800	1,300	31,200	37,900	
	1500	For drilled water well, if required, add	B-23	.50	80		6,375	2,175	6,700	15,250	17,800	

		11390 Pkg Sewage Treat Plants	CREW	DAILY OUTPUT	LABOR-HOURS	UNIT	MAT.	LABOR	EQUIP.	TOTAL	TOTAL INCL O&P	
200	0010	**SEWAGE TREATMENT PLANTS,** not incl fencing or external piping										200
	0020	Steel packaged, blown air aeration plants										
	0100	1,000 GPD				Gal.				16.85	19.40	
	0200	5,000 GPD								11.25	12.90	
	0300	15,000 GPD								6.20	7.10	
	0400	30,000 GPD								5.85	6.70	
	0500	50,000 GPD								4.50	5.05	
	0600	100,000 GPD								3.95	4.50	
	0700	200,000 GPD								2.81	3.23	
	0800	500,000 GPD								2.75	3.15	
	1000	Concrete, extended aeration, primary and secondary treatment										
	1010	10,000 GPD				Gal.				12.35	14.65	
	1100	30,000 GPD								6.20	7.15	
	1200	50,000 GPD								5.05	5.85	
	1400	100,000 GPD								3.94	4.55	
	1500	500,000 GPD								2.81	3.26	
	1700	Municipal wastewater treatment facility										
	1720	1.0 MGD				Gal.				4.83	5.55	

243

		11390	Pkg Sewage Treat Plants	CREW	DAILY OUTPUT	LABOR-HOURS	UNIT	2005 BARE COSTS				TOTAL INCL O&P	
								MAT.	LABOR	EQUIP.	TOTAL		
200	1740		1.5 MGD				Gal.				4.78	5.50	200
	1760		2.0 MGD								4.10	4.73	
	1780		3.0 MGD								3.20	3.71	
	1800		5.0 MGD				↓				2.92	3.37	
	2000		Holding tank system, not incl. excavation or backfill										
	2010		Recirculating chemical water closet	2 Plum	4	4	Ea.	750	163		913	1,075	
	2100		For voltage converter, add	"	16	1		199	41		240	281	
	2200		For high level alarm, add	1 Plum	7.80	1.026	↓	114	42		156	188	
900	0010		**WASTEWATER TREATMENT SYSTEM** Fiberglass, 1,000 gallon	B-21	1.29	21.705	Ea.	2,925	670	123	3,718	4,400	900
	0100		1,500 gallon	"	1.03	27.184	"	6,400	840	154	7,394	8,525	

		11488	Shooting Ranges	CREW	DAILY OUTPUT	LABOR-HOURS	UNIT	2005 BARE COSTS				TOTAL INCL O&P	
								MAT.	LABOR	EQUIP.	TOTAL		
700	0010		**SHOOTING RANGE** Incl. bullet traps, target provisions, controls,										700
	0100		separators, ceiling system, etc. Not incl. structural shell										
	0200		Commercial	L-9	.64	56.250	Point	11,500	1,750		13,250	15,500	
	0300		Law enforcement		.28	128		26,000	3,975		29,975	35,000	
	0400		National Guard armories		.71	50.704		12,500	1,575		14,075	16,300	
	0500		Reserve training centers		.71	50.704		10,500	1,575		12,075	14,100	
	0600		Schools and colleges		.32	112		22,000	3,475		25,475	29,800	
	0700		Major acadamies	↓	.19	189		40,000	5,850		45,850	53,500	
	0800		For acoustical treatment, add					10%	10%				
	0900		For lighting, add					28%	25%				
	1000		For plumbing, add					5%	5%				
	1100		For ventilating system, add, minimum					40%	40%				
	1200		Add, average					25%	25%				
	1300		Add, maximum				↓	35%	35%				

For information about Means Estimating Seminars, see yellow pages 12 and 13 in back of book

Division 13
Special Construction

Estimating Tips

General

- The items and systems in this division are usually estimated, purchased, supplied and installed as a unit by one or more subcontractors. The estimator must ensure that all parties are operating from the same set of specifications and assumptions and that all necessary items are estimated and will be provided. Many times the complex items and systems are covered but the more common ones such as excavation or a crane are overlooked for the very reason that everyone assumes nobody could miss them. The estimator should be the central focus and be able to ensure that all systems are complete.

- Another area where problems can develop in this division is at the interface between systems. The estimator must ensure, for instance, that anchor bolts, nuts and washers are estimated and included for the air-supported structures and pre-engineered buildings to be bolted to their foundations.

Utility supply is a common area where essential items or pieces of equipment can be missed or overlooked due to the fact that each subcontractor may feel it is the others' responsibility. The estimator should also be aware of certain items which may be supplied as part of a package but installed by others, and ensure that the installing contractor's estimate includes the cost of installation. Conversely, the estimator must also ensure that items are not costed by two different subcontractors, resulting in an inflated overall estimate.

13120 Pre-Engineered Structures

- The foundations and floor slab, as well as rough mechanical and electrical, should be estimated, as this work is required for the assembly and erection of the structure. Generally, as noted in the book, the pre-engineered building comes as a shell and additional features, such as windows and doors, must be included by the estimator. Here again, the estimator must have a clear understanding of the scope of each portion of the work and all the necessary interfaces.

13200 Storage Tanks

- The prices in this subdivision for above and below ground storage tanks do not include foundations or hold-down slabs. The estimator should refer to Divisions 2 and 3 for foundation system pricing. In addition to the foundations, required tank accessories such as tank gauges, leak detection devices, and additional manholes and piping must be added to the tank prices.

Reference Numbers

Reference numbers are shown in bold squares at the beginning of some major classifications. These numbers refer to related items in the Reference Section. The reference information may be an estimating procedure, an alternate pricing method or technical information.

Note: Not all subdivisions listed here necessarily appear in this publication.

13005		Selective Demolition	CREW	DAILY OUTPUT	LABOR-HOURS	UNIT	MAT.	2005 BARE COSTS LABOR	EQUIP.	TOTAL	TOTAL INCL O&P	
011	0010	**SELECTIVE DEMOLITION, AIR SUPPORTED STRUCTURES**										**011**
	0020	Tank covers, scrim, dbl layer, vinyl poly w/ hdw, blower & controls										
	0050	Round and rectangular R02220-510	B-2	9,000	.004	S.F.		.12		.12	.19	
	0100	Warehouse structures										
	0120	Whse, poly/vinyl fabric, 28 oz., w/tension cables & inflation sys	4 Clab	9,000	.004	SF Flr.		.10		.10	.15	
	0150	Warehouse, reinforced vinyl, 12 oz., 3000 S.F.	"	5,000	.006			.17		.17	.27	
	0200	12,000 to 24,000 S.F.	8 Clab	20,000	.003			.09		.09	.13	
	0250	Warehouse, tedlar vinyl fabric, 28 oz. w/liner, to 3000 S.F.	4 Clab	5,000	.006			.17		.17	.27	
	0300	12,000 to 24,000 S.F.	8 Clab	20,000	.003	▼		.09		.09	.13	
	0350	Greenhouse/shelter, woven polyethylene with liner										
	0400	3000 S.F.	4 Clab	5,000	.006	SF Flr.		.17		.17	.27	
	0450	12,000 to 24,000 S.F.	8 Clab	20,000	.003			.09		.09	.13	
	0500	Tennis/gymnasium, poly/vinyl fabric, 28 oz., incl thermal liner	4 Clab	9,000	.004			.10		.10	.15	
	0600	Stadiun/Convention Ctr, teflon coated fibergalss, incl thermal liner	9 Clab	40,000	.002	▼		.05		.05	.07	
	0700	Doors, air lock, 15' long, 10' x 10'	2 Carp	1.50	10.667	Ea.		365		365	570	
	0720	15' x 15'		.80	20			685		685	1,075	
	0750	Revolving personnel door, 6' dia. x 6'-6" high	▼	1.50	10.667	▼		365		365	570	
128	0010	**PRE-ENGD STEEL BLDGS DEMOLITION**										**128**
	0500	Pre-engd steel bldgs, rigid frame, clear span & multi post, excl salvage										
	0550	3500 to 7500 S.F	L-11	1,350	.024	SF Flr.		.73	1.05	1.78	2.33	
	0600	7501 to 12,500 S.F		2,000	.016			.49	.71	1.20	1.57	
	0650	12,500 S.F or greater	▼	2,200	.015	▼		.45	.64	1.09	1.43	
	0700	Pre-engd steel building components										
	0710	Entrance canopy, including frame 4' x 4'	E-24	8	4	Ea.		149	77	226	340	
	0720	4' x 8'	"	7	4.571			170	88	258	390	
	0730	H.M doors, self framing, single leaf	2 Skwk	8	2			69.50		69.50	108	
	0740	double leaf		5	3.200	▼		112		112	173	
	0760	Gutter, eave type		600	.027	L.F.		.93		.93	1.45	
	0770	Sash, single slide, double slide or fixed		24	.667	Ea.		23		23	36	
	0780	Skylight, fiberglass, to 30 S.F.		16	1			35		35	54	
	0785	Roof vents, circular, 12" to 24" diameter		12	1.333			46.50		46.50	72.50	
	0790	continuous, 10' long	▼	8	2	▼		69.50		69.50	108	
	0900	Shelters, aluminum frame										
	0910	Aluminum frame, acrylic glazing, 3' x 9' x 8' high	2 Skwk	2	8	Ea.		279		279	435	
	0920	9' x 12' x 8' high	"	1.50	10.667	"		370		370	580	
	0930	Silos, concrete stave industrial, not incl foundations										
201	0010	**STORAGE TANK DEMOLITION**										**201**
	0500	Steel tank, single wall, aboveground, not incl fdn, pumps or piping										
	0510	Single wall, 275 gallon R02220-510	Q-1	3	5.333	Ea.		196		196	295	
	0520	550 thru 2000 gallon	B-34P	2	12			410	196	606	835	
	0530	5000 thru 10000 gallon	B-34Q	2	12			415	515	930	1,200	
	0540	15000 thru 30000 gallon	B-34N	2	4	▼		110	223	333	415	
	0600	Steel tank, double wall, above ground not incl fdn, pumps & piping										
	0620	500 thru 2000 gallon	B-34P	2	12	Ea.		410	196	606	835	

13011	Air Supported Structures	CREW	DAILY OUTPUT	LABOR-HOURS	UNIT	2005 BARE COSTS				TOTAL INCL O&P
						MAT.	LABOR	EQUIP.	TOTAL	
100	**0010**	**AIR SUPPORTED TANK COVERS**, vinyl polyester									**100**
	0100	scrim, double layer, with hardware, blower, standby & controls									
	0200	Round, 75' diameter	B-2	4,500	.009	S.F.	6.50	.24		6.74	7.55
	0300	100' diameter		5,000	.008		6	.22		6.22	6.95
	0400	150' diameter		5,000	.008		4.51	.22		4.73	5.30
	0500	Rectangular, 20' x 20'		4,500	.009		14.65	.24		14.89	16.55
	0600	30' x 40'		4,500	.009		14.65	.24		14.89	16.55
	0700	50' x 60'		4,500	.009		14.65	.24		14.89	16.55
	0800	For single wall construction, deduct, minimum					.42			.42	.46
	0900	Maximum					1.39			1.39	1.53
	1000	For maximum resistance to atmosphere or cold, add					.58			.58	.64
	1100	For average shipping charges, add				Total	1,150			1,150	1,275
200	**0010**	**AIR SUPPORTED STRUCTURES**									**200**
	0020	Site preparation, incl. anchor placement and utilities	B-11B	1,000	.016	SF Flr.	.77	.48	.18	1.43	1.78
	0030	For concrete curb, see division 03310-240									
	0050	Warehouse, polyester/vinyl fabric, 28 oz., over 10 yr. life, welded									
	0060	Seams, tension cables, primary & auxiliary inflation system,									
	0070	airlock, personnel doors and liner									
	0100	5000 S.F.	4 Clab	5,000	.006	SF Flr.	16.25	.17		16.42	18.15
	0250	12,000 S.F.	"	6,000	.005		11.75	.14		11.89	13.15
	0400	24,000 S.F.	8 Clab	12,000	.005		8.60	.14		8.74	9.70
	0500	50,000 S.F.	"	12,500	.005		7.45	.14		7.59	8.40
	0700	12 oz. reinforced vinyl fabric, 5 yr. life, sewn seams,									
	0710	accordian door, including liner									
	0750	3000 S.F.	4 Clab	3,000	.011	SF Flr.	8.50	.28		8.78	9.80
	0800	12,000 S.F.	"	6,000	.005		6.30	.14		6.44	7.10
	0850	24,000 S.F.	8 Clab	12,000	.005		6.25	.14		6.39	7.05
	0950	Deduct for single layer					.68			.68	.75
	1000	Add for welded seams					.98			.98	1.08
	1050	Add for double layer, welded seams included					2.02			2.02	2.22
	1250	Tedlar/vinyl fabric, 28 oz., with liner, over 10 yr. life,									
	1260	incl. overhead and personnel doors									
	1300	3000 S.F.	4 Clab	3,000	.011	SF Flr.	15.80	.28		16.08	17.80
	1450	12,000 S.F.	"	6,000	.005		11.10	.14		11.24	12.40
	1550	24,000 S.F.	8 Clab	12,000	.005		8.60	.14		8.74	9.70
	1700	Deduct for single layer					1.40			1.40	1.54
	2860	For low temperature conditions, add					.69			.69	.76
	2870	For average shipping charges, add				Total	3,025			3,025	3,325
	2900	Thermal liner, translucent reinforced vinyl				SF Flr.	.95			.95	1.05
	2950	Metalized mylar fabric and mesh, double liner				"	1.51			1.51	1.66
	3050	Stadium/convention center, teflon coated fiberglass, heavy weight,									
	3060	over 20 yr. life, incl. thermal liner and heating system									
	3100	Minimum	9 Clab	26,000	.003	SF Flr.	39	.07		39.07	42.50
	3110	Maximum	"	19,000	.004	"	45.50	.10		45.60	50
	3400	Doors, air lock, 15' long, 10' x 10'	2 Carp	.80	20	Ea.	14,200	685		14,885	16,800
	3600	15' x 15'	"	.50	32		20,600	1,100		21,700	24,300
	3700	For each added 5' length, add					2,950			2,950	3,250
	3900	Revolving personnel door, 6' diameter, 6'-6" high	2 Carp	.80	20		10,400	685		11,085	12,600
	4200	Double wall, self supporting, shell only, minimum				SF Flr.					21
	4300	Maximum				"					39

		13128	Pre-Engineered Structures	CREW	DAILY OUTPUT	LABOR-HOURS	UNIT	MAT.	LABOR	EQUIP.	TOTAL	TOTAL INCL O&P	
								2005 BARE COSTS					
070	0010		**CONTROL TOWERS**, 12' x 10', incl. instruments, min.				Ea.					510,000	070
	0100		Maximum									1,000,000	
	0500		With standard 40' tower, average				↓					710,500	
	1000		Temporary portable control towers, 8' x 12',										
	1010		complete with one position communications, minimum				Ea.					275,000	
	2000		For fixed facilities, depending on height, minimum									50,000	
	2010		Maximum				↓					100,000	
160	0010		**TENSION STRUCTURES** Rigid steel/alum. frame, vyl. coated polyester										160
	0100		fabric shell, 60' clear span, not incl. foundations or floors										
	0200		6,000 S.F.	B-41	1,000	.044	SF Flr.	10.90	1.22	.22	12.34	14.15	
	0300		12,000 S.F.		1,100	.040		9.55	1.11	.20	10.86	12.45	
	0400		80' clear span, 20,800 S.F.	↓	1,220	.036		9.75	1	.18	10.93	12.50	
	0410		100' clear span, 10,000 S.F.	L-5	2,175	.026		10.85	.98	.28	12.11	13.95	
	0430		26,000 S.F.		2,300	.024		9.60	.93	.27	10.80	12.50	
	0450		36,000 S.F.		2,500	.022		9.40	.85	.25	10.50	12.10	
	0460		120' clear span, 24,000 S.F.		3,000	.019		11.35	.71	.21	12.27	14	
	0470		150' clear span, 30,000 S.F.	↓	6,000	.009		11.65	.36	.10	12.11	13.60	
	0480		200' clear span, 40,000 S.F.	E-6	8,000	.016	↓	14.15	.60	.19	14.94	16.85	
	0500		For roll-up door, 12' x 14', add	L-2	1	16	Ea.	4,275	480		4,755	5,450	
	0600		For personnel doors, add, minimum				SF Flr.	5%					
	0700		Add, maximum					15%					
	0800		For site work, simple foundation, etc., add, minimum								1.25	1.95	
	0900		Add, maximum				↓				2.75	3.05	
200	0010		**COMFORT STATIONS** Prefab., stock, w/doors, windows & fixt.										200
	0100		Not incl. interior finish or electrical										
	0300		Mobile, on steel frame, minimum				S.F.	36.50			36.50	40	
	0350		Maximum					57.50			57.50	63.50	
	0400		Permanent, including concrete slab, minimum	B-12J	50	.320		142	10	16.75	168.75	190	
	0500		Maximum	"	43	.372	↓	206	11.65	19.50	237.15	266	
	0600		Alternate pricing method, mobile, minimum				Fixture	1,775			1,775	1,950	
	0650		Maximum					2,650			2,650	2,925	
	0700		Permanent, minimum	B-12J	.70	22.857	↓	10,200	715	1,200	12,115	13,600	
	0750		Maximum	"	.50	32		17,000	1,000	1,675	19,675	22,100	
300	0010		**DOMES**										300
	1500		Bulk storage, shell only, dual radius hemispher. arch, steel										
	1600		framing, corrugated steel covering, 150' diameter	E-2	550	.102	SF Flr.	28	3.76	2.41	34.17	39.50	
	1700		400' diameter	"	720	.078		22.50	2.87	1.84	27.21	32	
	1800		Wood framing, wood decking, to 400' diameter	F-4	400	.120	↓	20.50	4.06	2.24	26.80	31	
	1900		Radial framed wood (2" x 6"), 1/2" thick										
	2000		plywood, asphalt shingles, 50' diameter	F-3	2,000	.020	SF Flr.	38.50	.69	.30	39.49	44	
	2100		60' diameter		1,900	.021		29.50	.73	.32	30.55	34	
	2200		72' diameter		1,800	.022		26.50	.77	.33	27.60	30.50	
	2300		116' diameter		1,730	.023		23	.80	.35	24.15	27	
	2400		150' diameter	↓	1,500	.027	↓	20	.92	.40	21.32	24	
510	0010		**BLEACHERS**										510
	0020		Bleachers, outdoor, portable, 3 to 5 tiers, to 300' long, min	2 Sswk	120	.133	Seat	33.50	5.10		38.60	46	
	0100		Maximum, less than 15' long, prefabricated		80	.200		44	7.65		51.65	62	
	0200		6 to 20 tiers, minimum, up to 300' long		120	.133		40.50	5.10		45.60	53.50	
	0300		Max., under 15', (highly prefabricated, on wheels)	↓	80	.200	↓	58.50	7.65		66.15	78	
	0500		Permanent grandstands, wood seat, steel frame, 24" row										
	0600		3 to 15 tiers, minimum	2 Sswk	60	.267	Seat	100	10.15		110.15	128	
	0700		Maximum		48	.333		110	12.70		122.70	144	
	0900		16 to 30 tiers, minimum		60	.267		116	10.15		126.15	145	
	0950		Average	↓	55	.291		145	11.10		156.10	179	

13 SPECIAL CONSTRUCTION

		13128	Pre-Engineered Structures	CREW	DAILY OUTPUT	LABOR-HOURS	UNIT	2005 BARE COSTS				TOTAL INCL O&P	
								MAT.	LABOR	EQUIP.	TOTAL		
510	1000		Maximum	2 Sswk	48	.333	Seat	173	12.70		185.70	214	**510**
	1200		Seat backs only, 30" row, fiberglass		160	.100		20.50	3.82		24.32	29.50	
	1300		Steel and wood	↓	160	.100	↓	25.50	3.82		29.32	35	
	1400	NOTE: average seating is 1.5' in width											
580	0010	**HANGARS** Prefabricated steel T hangars, Galv. steel roof &											**580**
	0100	walls, incl. electric bi-folding doors, 4 or more units,											
	0110	not including floors or foundations, minimum		E-2	1,275	.044	SF Flr.	10.25	1.62	1.04	12.91	15.25	
	0130	Maximum			1,063	.053		10.65	1.95	1.25	13.85	16.40	
	0900	With bottom rolling doors, minimum			1,386	.040		9.80	1.49	.96	12.25	14.35	
	1000	Maximum		↓	966	.058	↓	10.65	2.14	1.37	14.16	16.90	
	1200	Alternate pricing method:											
	1300	Galv. roof and walls, electric bi-folding doors, minimum		E-2	1.06	52.830	Plane	12,700	1,950	1,250	15,900	18,600	
	1500	Maximum			.91	61.538		14,100	2,275	1,450	17,825	21,000	
	1600	With bottom rolling doors, minimum			1.25	44.800		10,100	1,650	1,050	12,800	15,100	
	1800	Maximum		↓	.97	57.732	↓	12,000	2,125	1,375	15,500	18,500	
	2000	Circular type, prefab., steel frame, plastic skin, electric											
	2010	door, including foundations, 80' diameter,											
	2020	for up to 5 light planes, minimum		E-2	.50	112	Total	63,000	4,150	2,650	69,800	79,500	
	2200	Maximum		"	.25	224	"	70,500	8,275	5,300	84,075	97,500	
700	0010	**PRE-ENGINEERED STEEL BUILDINGS** R13128 -210											**700**
	0100	Clear span rigid frame, 26 ga. colored roofing and siding											
	0150	20 wide, 10' eave height		E-2	425	.132	SF Flr.	7.60	4.87	3.12	15.59	20	
	0160	14' eave height			350	.160		8.05	5.90	3.79	17.74	23	
	0170	16' eave height			320	.175		8.45	6.45	4.14	19.04	25	
	0180	20' eave height			275	.204		9.35	7.55	4.82	21.72	28.50	
	0190	24' eave height			240	.233		10.65	8.60	5.50	24.75	32.50	
	0200	30' to 40' wide, 10' eave height			535	.105		6.30	3.87	2.48	12.65	16.30	
	0300	14' eave height			450	.124		6.70	4.60	2.94	14.24	18.50	
	0400	16' eave height			415	.135		7.05	4.99	3.19	15.23	19.80	
	0500	20' eave height			360	.156		7.70	5.75	3.68	17.13	22.50	
	0600	24' eave height			320	.175		8.65	6.45	4.14	19.24	25	
	0700	50' to 100' wide, 10' eave height			865	.065		5.45	2.39	1.53	9.37	11.80	
	0800	14' eave height			770	.073		5.80	2.69	1.72	10.21	12.85	
	0900	16' eave height			730	.077		6.15	2.84	1.82	10.81	13.60	
	1000	20' eave height			660	.085		6.60	3.14	2.01	11.75	14.85	
	1100	24' eave height		↓	605	.093	↓	7.25	3.42	2.19	12.86	16.20	
	1200	Clear span tapered beam frame, 26 ga. colored roofing/siding											
	1300	30' wide, 10' eave height		E-2	535	.105	SF Flr.	7	3.87	2.48	13.35	17.05	
	1400	14' eave height			450	.124		7.70	4.60	2.94	15.24	19.60	
	1500	16' eave height			415	.135		8.25	4.99	3.19	16.43	21	
	1600	20' eave height			360	.156		9.10	5.75	3.68	18.53	24	
	1700	40' wide, 10' eave height			600	.093		6.40	3.45	2.21	12.06	15.40	
	1800	14' eave height			510	.110		7	4.06	2.60	13.66	17.50	
	1900	16' eave height			475	.118		7.30	4.36	2.79	14.45	18.55	
	2000	20' eave height			415	.135		8	4.99	3.19	16.18	21	
	2100	50' to 80' wide, 10' eave height			770	.073		6.10	2.69	1.72	10.51	13.20	
	2200	14' eave height			675	.083		6.50	3.07	1.96	11.53	14.55	
	2300	16' eave height			635	.088		6.75	3.26	2.09	12.10	15.35	
	2400	20' eave height		↓	565	.099	↓	7.55	3.66	2.35	13.56	17.20	
	2500	Single post 2-span frame, 26 ga. colored roofing and siding											
	2600	80' wide, 14' eave height		E-2	740	.076	SF Flr.	5.20	2.80	1.79	9.79	12.50	
	2700	16' eave height			695	.081		5.55	2.98	1.91	10.44	13.30	
	2800	20' eave height			625	.090		6.05	3.31	2.12	11.48	14.70	
	2900	24' eave height			570	.098		6.55	3.63	2.32	12.50	16	
	3000	100' wide, 14' eave height		↓	835	.067	↓	5.05	2.48	1.59	9.12	11.55	

SPECIAL CONSTRUCTION 13

		CREW	DAILY OUTPUT	LABOR-HOURS	UNIT	2005 BARE COSTS				TOTAL INCL O&P	
13128	**Pre-Engineered Structures**					MAT.	LABOR	EQUIP.	TOTAL		
700 3100	16' eave height R13128 -210	E-2	795	.070	SF Flr.	5	2.60	1.67	9.27	11.80	**700**
3200	20' eave height		730	.077		5.85	2.84	1.82	10.51	13.25	
3300	24' eave height		670	.084		6.30	3.09	1.98	11.37	14.45	
3400	120' wide, 14' eave height		870	.064		5	2.38	1.52	8.90	11.25	
3500	16' eave height		830	.067		5.25	2.49	1.60	9.34	11.80	
3600	20' eave height		765	.073		5.65	2.71	1.73	10.09	12.80	
3700	24' eave height	▼	705	.079	▼	6.15	2.94	1.88	10.97	13.90	
3800	Double post 3-span frame, 26 ga. colored roofing and siding										
3900	150' wide, 14' eave height	E-2	925	.061	SF Flr.	4.65	2.24	1.43	8.32	10.50	
4000	16' eave height		890	.063		4.80	2.33	1.49	8.62	10.95	
4100	20' eave height		820	.068		5.10	2.52	1.62	9.24	11.70	
4200	24' eave height	▼	765	.073	▼	6	2.71	1.73	10.44	13.15	
4300	Triple post 4-span frame, 26 ga. colored roofing and siding										
4400	160' wide, 14' eave height	E-2	970	.058	SF Flr.	4.52	2.13	1.37	8.02	10.15	
4500	16' eave height		930	.060		4.73	2.23	1.42	8.38	10.60	
4600	20' eave height		870	.064		4.99	2.38	1.52	8.89	11.25	
4700	24' eave height		815	.069		5.90	2.54	1.63	10.07	12.65	
4800	200' wide, 14' eave height		1,030	.054		4.55	2.01	1.29	7.85	9.85	
4900	16' eave height		995	.056		4.69	2.08	1.33	8.10	10.20	
5000	20' eave height		935	.060		5.10	2.21	1.42	8.73	10.95	
5100	24' eave height	▼	885	.063	▼	5.95	2.34	1.50	9.79	12.15	
5200	Accessory items: add to the basic building cost above										
5250	Eave overhang, 2' wide, 26 ga., with soffit	E-2	360	.156	L.F.	22.50	5.75	3.68	31.93	38.50	
5300	4' wide, without soffit		300	.187		21.50	6.90	4.42	32.82	40.50	
5350	With soffit		250	.224		30	8.30	5.30	43.60	53	
5400	6' wide, without soffit		250	.224		29.50	8.30	5.30	43.10	52	
5450	With soffit		200	.280	▼	38	10.35	6.60	54.95	67	
5500	Entrance canopy, incl. frame, 4' x 4'		25	2.240	Ea.	288	83	53	424	515	
5550	4' x 8'		19	2.947	"	395	109	69.50	573.50	700	
5600	End wall roof overhang, 4' wide, without soffit		850	.066	L.F.	14.55	2.43	1.56	18.54	22	
5650	With soffit	▼	500	.112	"	21	4.14	2.65	27.79	33.50	
5700	Doors, H.M. self-framing, incl. butts, lockset and trim										
5750	Single leaf, 3070 (3' x 7'), economy	2 Sswk	5	3.200	Opng.	280	122		402	530	
5800	Deluxe		4	4		294	153		447	600	
5825	Glazed		4	4		335	153		488	645	
5850	3670 (3'-6" x 7')		4	4		345	153		498	655	
5900	4070 (4' x 7')		3	5.333		375	203		578	780	
5950	Double leaf, 6070 (6' x 7')		2	8		495	305		800	1,100	
6000	Glazed	▼	2	8		605	305		910	1,225	
6050	Framing only, for openings, 3' x 7'	E-2	25	2.240		122	83	53	258	335	
6100	10' x 10'		21	2.667		400	98.50	63	561.50	680	
6150	For windows below, 2020 (2' x 2')		25	2.240		129	83	53	265	345	
6200	4030 (4' x 3')	▼	22	2.545	▼	158	94	60	312	405	
6250	Flashings, 26 ga., corner or eave, painted	2 Sswk	240	.067	L.F.	3.53	2.54		6.07	8.45	
6300	Galvanized		240	.067		2.90	2.54		5.44	7.75	
6350	Rake flashing, painted		240	.067		3.78	2.54		6.32	8.70	
6400	Galvanized		240	.067		3.18	2.54		5.72	8.05	
6450	Ridge flashing, 18" wide, painted		240	.067		4.67	2.54		7.21	9.70	
6500	Galvanized		240	.067		4.09	2.54		6.63	9.05	
6550	Gutter, eave type, 26 ga., painted		320	.050		4.85	1.91		6.76	8.75	
6600	Galvanized		320	.050		2.97	1.91		4.88	6.70	
6650	Valley type, between buildings, painted		120	.133		7.70	5.10		12.80	17.60	
6700	Galvanized	▼	120	.133	▼	8.25	5.10		13.35	18.15	
6750	Insulation, rated .6 lb density, vinyl faced										
6800	1-1/2" thick, R5	2 Carp	2,300	.007	S.F.	.24	.24		.48	.63	
6850	3" thick, R10	▼	2,300	.007	▼	.30	.24		.54	.70	

Important: See the Reference Section for critical supporting data - Reference Nos., Crews, & City Cost Indexes

13 SPECIAL CONSTRUCTION

			CREW	DAILY OUTPUT	LABOR-HOURS	UNIT	2005 BARE COSTS MAT.	LABOR	EQUIP.	TOTAL	TOTAL INCL O&P	
		13128 Pre-Engineered Structures										
700	6900	4" thick, R13	2 Carp	2,300	.007	S.F.	.41	.24		.65	.82	**700**
	6950	Foil/scrim/kraft (FSK) faced, 1-1/2" thick, R5 [R13128-210]		2,300	.007		.26	.24		.50	.66	
	7000	2" thick, R6		2,300	.007		.32	.24		.56	.72	
	7050	3" thick, R10		2,300	.007		.38	.24		.62	.79	
	7100	4" thick, R13		2,300	.007		.47	.24		.71	.89	
	7150	Met polyester/scrim/kraft (PSK) facing,1-1/2" thk,R5		2,300	.007		.42	.24		.66	.83	
	7200	2" thick, R6		2,300	.007		.48	.24		.72	.90	
	7250	3" thick, R11		2,300	.007		.49	.24		.73	.91	
	7300	4" thick, R13		2,300	.007		.58	.24		.82	1.01	
	7350	Vinyl/scrim/foil (VSF), 1-1/2" thick, R5		2,300	.007		.37	.24		.61	.78	
	7400	2" thick, R6		2,300	.007		.46	.24		.70	.88	
	7450	3" thick, R10		2,300	.007		.55	.24		.79	.98	
	7500	4" thick, R13		2,300	.007		.61	.24		.85	1.04	
	7650	Sash, single slide, glazed, with screens, 2020 (2'x 2')	E-1	22	1.091	Opng.	76	40.50	3.69	120.19	157	
	7700	3030 (3' x 3')		14	1.714		171	63.50	5.80	240.30	305	
	7750	4030 (4' x 3')		13	1.846		228	68.50	6.25	302.75	375	
	7800	6040 (6' x 4')		12	2		455	74	6.75	535.75	635	
	7850	Double slide sash, 3030 (3' x 3')		14	1.714		143	63.50	5.80	212.30	272	
	7900	6040 (6' x 4')		12	2		380	74	6.75	460.75	555	
	7950	Fixed glass, no screens, 3030 (3' x 3')		14	1.714		185	63.50	5.80	254.30	320	
	8000	6040 (6' x 4')		12	2		495	74	6.75	575.75	680	
	8050	Prefinished storm sash, 3030 (3' x 3')		70	.343		61.50	12.75	1.16	75.41	90.50	
	8100	Siding and roofing, see division 07300 & 07400										
	8200	Skylight, fiberglass panels, to 30 S.F.	E-1	10	2.400	Ea.	97	89	8.10	194.10	268	
	8250	Larger sizes, add for excess over 30 S.F.	"	300	.080	S.F.	3.23	2.97	.27	6.47	8.90	
	8300	Roof vents, circular with damper, birdscreen										
	8350	and operator hardware, painted										
	8400	26 ga., 12" diameter	1 Sswk	4	2	Ea.	72	76.50		148.50	216	
	8450	20" diameter		3	2.667		148	102		250	345	
	8500	24 ga., 24" diameter		2	4		293	153		446	595	
	8550	Galvanized		2	4		245	153		398	545	
	8600	Continuous, 26 ga., 10' long, 9" wide	2 Sswk	4	4		20.50	153		173.50	296	
	8650	12" wide	"	4	4		25	153		178	300	
840	0010	**SILOS** Concrete stave industrial, not incl. foundations, conical or										**840**
	0100	sloping bottoms, 12' diameter, 35' high	D-8	.11	363	Ea.	18,100	11,600		29,700	37,600	
	0200	16' diameter, 45' high		.08	500		24,200	16,000		40,200	51,000	
	0400	25' diameter, 75' high		.05	800		59,500	25,500		85,000	104,500	
	0500	Steel, factory fab., 30,000 gallon cap., painted, minimum	L-5	1	56		17,200	2,125	615	19,940	23,400	
	0700	Maximum		.50	112		27,400	4,275	1,225	32,900	39,000	
	0800	Epoxy lined, minimum		1	56		28,200	2,125	615	30,940	35,400	
	1000	Maximum		.50	112		35,700	4,275	1,225	41,200	48,100	

13200 | Storage Tanks

		13201 Storage Tanks	CREW	DAILY OUTPUT	LABOR-HOURS	UNIT	2005 BARE COSTS MAT.	LABOR	EQUIP.	TOTAL	TOTAL INCL O&P	
200	0010	**ELEVATED STORAGE TANKS**, not incl pipe, pumps or foundation										**200**
	3000	Elevated water tanks, 100' to bottom capacity line, incl painting										
	3010	50,000 gallons				Ea.					193,764	
	3300	100,000 gallons									267,300	

SPECIAL CONSTRUCTION 13

13201	Storage Tanks	CREW	DAILY OUTPUT	LABOR-HOURS	UNIT	2005 BARE COSTS				TOTAL INCL O&P	
						MAT.	LABOR	EQUIP.	TOTAL		
200 3400	250,000 gallons				Ea.					368,851	**200**
3600	500,000 gallons									595,298	
3700	750,000 gallons									820,577	
3900	1,000,000 gallons									824,180	
300 0010	**GROUND TANKS** Not incl. pipe or pumps, prestress conc., 250,000 gal.				Ea.					284,200	**300**
0100	500,000 gallons									385,700	
0300	1,000,000 gallons									548,100	
0400	2,000,000 gallons									828,240	
0600	4,000,000 gallons									1,309,350	
0700	6,000,000 gallons									1,786,400	
0750	8,000,000 gallons									2,273,600	
0800	10,000,000 gallons									2,740,500	
0900	Steel, ground level, ht/dia less than 1, not incl. fdn, 100,000 gallons									141,500	
1000	250,000 gallons									159,910	
1200	500,000 gallons									243,955	
1250	750,000 gallons									290,645	
1300	1,000,000 gallons									444,360	
1500	2,000,000 gallons									717,858	
1600	4,000,000 gallons									1,114,723	
1800	6,000,000 gallons									1,546,606	
1850	8,000,000 gallons									2,025,179	
1900	10,000,000 gallons									2,784,500	
2100	Steel standpipes, hgt/dia more than 1, 100' to overflow, no fdn										
2200	500,000 gallons				Ea.					320,993	
2400	750,000 gallons									391,029	
2500	1,000,000 gallons									476,238	
2700	1,500,000 gallons									665,333	
2800	2,000,000 gallons									817,075	
4000	Fixed roof oil storage tanks, steel, (1 BBL=42 GAL w/ fdn 3'D x 1'W)										
4200	5,000 barrels				Ea.					165,600	
4300	25,000 barrels									232,300	
4500	55,000 barrels									393,300	
4600	100,000 barrels									632,500	
4800	150,000 barrels									891,250	
4900	225,000 barrels									1,343,200	
5100	Floating roof gasoline tanks, steel, 5,000 barrels									140,300	
5200	25,000 barrels									313,950	
5400	55,000 barrels									474,950	
5500	100,000 barrels									718,750	
5700	150,000 barrels									972,900	
5800	225,000 barrels									1,348,950	
6000	Wood tanks, ground level, 2" cypress, 3,000 gallons	C-1	.19	168		6,800	5,450		12,250	16,000	
6100	2-1/2" cypress, 10,000 gallons		.12	266		18,000	8,625		26,625	33,200	
6300	3" redwood or 3" fir, 20,000 gallons		.10	320		27,500	10,400		37,900	46,400	
6400	30,000 gallons		.08	400		34,000	12,900		46,900	57,500	
6600	45,000 gallons		.07	457		51,500	14,800		66,300	79,500	
6700	Larger sizes, minimum				Gal.					.66	
6900	Maximum				"					.81	
7000	Vinyl coated fabric pillow tanks, freestanding, 5,000 gallons	4 Clab	4	8	Ea.	3,625	214		3,839	4,325	
7100	Supporting embankment not included, 25,000 gallons	6 Clab	2	24		6,900	640		7,540	8,575	
7200	50,000 gallons	8 Clab	1.50	42.667		13,600	1,150		14,750	16,700	
7300	100,000 gallons	9 Clab	.90	80		28,200	2,125		30,325	34,300	
7400	150,000 gallons		.50	144		34,500	3,850		38,350	44,000	
7500	200,000 gallons		.40	180		40,400	4,800		45,200	52,000	
7600	250,000 gallons		.30	240		45,500	6,400		51,900	60,000	

13 SPECIAL CONSTRUCTION

For information about Means Estimating Seminars, see yellow pages 12 and 13 in back of book

Important: See the Reference Section for critical supporting data - Reference Nos., Crews, & City Cost Indexes

Division 14
Conveying Systems

Estimating Tips

General

- Many products in Division 14 will require some type of support or blocking for installation not included with the item itself. Examples are supports for conveyors or tube systems, attachment points for lifts, and footings for hoists or cranes. Add these supports in the appropriate division.

14100 Dumbwaiters
14200 Elevators

- Dumbwaiters and elevators are estimated and purchased in a method similar to buying a car. The manufacturer has a base unit with standard features. Added to this base unit price will be whatever options the owner or specifications require. Increased load capacity, additional vertical travel, additional stops, higher speed, and cab finish options are items to be considered. When developing an estimate for dumbwaiters and elevators, remember that some items needed by the installers may have to be included as part of the general contract.

Examples are:
  - shaftway
  - rail support brackets
  - machine room
  - electrical supply
  - sill angles
  - electrical connections
  - pits
  - roof penthouses
  - pit ladders

  Check the job specifications and drawings before pricing.

- Installation of elevators and handicapped lifts in historic structures can require significant additional costs. The associated structural requirements may involve cutting into and repairing finishes, mouldings, flooring, etc. The estimator must account for these special conditions.

14300 Escalators & Moving Walks

- Escalators and moving walks are specialty items installed by specialty contractors. There are numerous options associated with these items. For specific options contact a manufacturer or contractor. In a method similar to estimating dumbwaiters and elevators, you should verify the extent of general contract work and add items as necessary.

14400 Lifts
14500 Material Handling
14600 Hoists & Cranes

- Products such as correspondence lifts, conveyors, chutes, pneumatic tube systems, material handling cranes and hoists as well as other items specified in this subdivision may require trained installers. The general contractor might not have any choice as to who will perform the installation or when it will be performed. Long lead times are often required for these products, making early decisions in scheduling necessary.

Reference Numbers

Reference numbers are shown in bold squares at the beginning of some major classifications. These numbers refer to related items in the Reference Section. The reference information may be an estimating procedure, an alternate pricing method or technical information.

Note: Not all subdivisions listed here necessarily appear in this publication.

		14550	Conveyors	CREW	DAILY OUTPUT	LABOR-HOURS	UNIT	2005 BARE COSTS				TOTAL INCL O&P	
								MAT.	LABOR	EQUIP.	TOTAL		
350	0010		MATERIAL HANDLING Conveyors, gravity type 2" rollers, 3" O.C.										350
	0050		10' sections with 2 supports, 600 lb. capacity, 18" wide				Ea.	375			375	410	
	0100		24" wide					430			430	475	
	0150		1400 lb. capacity, 18" wide					345			345	380	
	0200		24" wide				↓	380			380	420	
	0350		Horizontal belt, center drive and takeup, 60 fpm										
	0400		16" belt, 26.5' length	2 Mill	.50	32	Ea.	2,625	1,125		3,750	4,575	
	0450		24" belt, 41.5' length	↓	.40	40		4,075	1,425		5,500	6,600	
	0500		61.5' length	↓	.30	53.333	↓	5,450	1,900		7,350	8,775	
	0600		Inclined belt, 10' rise with horizontal loader and										
	0620		End idler assembly, 27.5' length, 18" belt	2 Mill	.30	53.333	Ea.	7,675	1,900		9,575	11,200	
	0700		24" belt	"	.15	106	"	8,600	3,775		12,375	15,100	
	3600		Monorail, overhead, manual, channel type										
	3700		125 lb. per L.F.	1 Mill	26	.308	L.F.	8.50	10.90		19.40	25.50	
	3900		500 lb. per L.F.	"	21	.381	"	7.70	13.50		21.20	28.50	
	4000		Trolleys for above, 2 wheel, 125 lb. capacity				Ea.	53.50			53.50	59	
	4200		4 wheel, 250 lb. capacity					163			163	179	
	4300		8 wheel, 1,000 lb. capacity				↓	315			315	345	

		14610	Fixed Hoists	CREW	DAILY OUTPUT	LABOR-HOURS	UNIT	2005 BARE COSTS				TOTAL INCL O&P	
								MAT.	LABOR	EQUIP.	TOTAL		
500	0010		MATERIAL HANDLING										500
	1500		Cranes, portable hydraulic, floor type, 2,000 lb capacity				Ea.	1,800			1,800	1,975	
	1600		4,000 lb capacity					3,100			3,100	3,425	
	1800		Movable gantry type, 12' to 15' range, 2,000 lb capacity					2,775			2,775	3,050	
	1900		6,000 lb capacity					4,025			4,025	4,425	
	2100		Hoists, electric overhead, chain, hook hung, 15' lift, 1 ton cap.					1,825			1,825	2,000	
	2200		3 ton capacity					2,275			2,275	2,500	
	2500		5 ton capacity				↓	5,000			5,000	5,500	
	2600		For hand-pushed trolley, add					15%					
	2700		For geared trolley, add					30%					
	2800		For motor trolley, add					75%					
	3000		For lifts over 15', 1 ton, add				L.F.	17.10			17.10	18.80	
	3100		5 ton, add				"	40			40	44	
	3300		Lifts, scissor type, portable, electric, 36" high, 2,000 lb				Ea.	3,275			3,275	3,600	
	3400		48" high, 4,000 lb				"	3,750			3,750	4,125	

		14630	Bridge Cranes	CREW	DAILY OUTPUT	LABOR-HOURS	UNIT	2005 BARE COSTS				TOTAL INCL O&P	
								MAT.	LABOR	EQUIP.	TOTAL		
300	0010		CRANE RAIL Box beam bridge, no equipment included	E-4	3,400	.009	Lb.	.91	.36	.02	1.29	1.68	300
	0210		Running track only, 104 lb per yard, 20' piece	"	160	.200	L.F.	16.70	7.75	.51	24.96	33	

For information about Means Estimating Seminars, see yellow pages 12 and 13 in back of book

Important: See the Reference Section for critical supporting data - Reference Nos., Crews, & City Cost Indexes

Division 15
Mechanical

Estimating Tips

15100 Building Services Piping

This subdivision is primarily basic pipe and related materials. The pipe may be used by any of the mechanical disciplines, i.e., plumbing, fire protection, heating, and air conditioning.

- The piping section lists the add to labor for elevated pipe installation. These adds apply to all elevated pipe, fittings, valves, insulation, etc., that are placed above 10' high. CAUTION: the correct percentage may vary for the same pipe. For example, the percentage add for the basic pipe installation should be based on the maximum height that the craftsman must install for that particular section. If the pipe is to be located 14' above the floor but it is suspended on threaded rod from beams, the bottom flange of which is 18' high (4' rods), then the height is actually 18' and the add is 20%. The pipe coverer, however, does not have to go above the 14' and so his add should be 10%.

- Most pipe is priced first as straight pipe with a joint (coupling, weld, etc.) every 10' and a hanger usually every 10'. There are exceptions with hanger spacing such as: for cast iron pipe (5') and plastic pipe (3 per 10'). Following each type of pipe there are several lines listing sizes and the amount to be subtracted to delete couplings and hangers. This is for pipe that is to be buried or supported together on trapeze hangers. The reason that the couplings are deleted is that these runs are usually long and frequently longer lengths of pipe are used. By deleting the couplings the estimator is expected to look up and add back the correct reduced number of couplings.

- When preparing an estimate it may be necessary to approximate the fittings. Fittings usually run between 25% and 50% of the cost of the pipe. The lower percentage is for simpler runs, and the higher number is for complex areas like mechanical rooms.

- For historic restoration projects, the systems must be as invisible as possible, and pathways must be sought for pipes, conduits, and ductwork. While installations in accessible spaces (such as basements and attics) are relatively straightforward to estimate, labor costs may be more difficult to determine when delivery systems must be concealed.

15400 Plumbing Fixtures & Equipment

- Plumbing fixture costs usually require two lines, the fixture itself and its "rough-in, supply and waste".

- In the Assemblies Section (Plumbing D2010) for the desired fixture, the System Components Group in the center of the page shows the fixture itself on the first line while the rest of the list (fittings, pipe, tubing, etc.) will total up to what we refer to in the Unit Price section as "Rough-in, supply, waste and vent". Note that for most fixtures we allow a nominal 5' of tubing to reach from the fixture to a main or riser.

- Remember that gas and oil fired units need venting.

15500 Heat Generation Equipment

- When estimating the cost of an HVAC system, check to see who is responsible for providing and installing the temperature control system. It is possible to overlook controls, assuming that they would be included in the electrical estimate.

- When looking up a boiler be careful on specified capacity. Some manufacturers rate their products on output while others use input.

- Include HVAC insulation for pipe, boiler and duct (wrap and liner).

- Be careful when looking up mechanical items to get the correct pressure rating and connection type (thread, weld, flange).

15700 Heating/Ventilation/ Air Conditioning Equipment

- Combination heating and cooling units are sized by the air conditioning requirements. (See Reference No. R15710-020 for preliminary sizing guide.)

- A ton of air conditioning is nominally 400 CFM.

- Rectangular duct is taken off by the linear foot for each size, but its cost is usually estimated by the pound. Remember that SMACNA standards now base duct on internal pressure.

- Prefabricated duct is estimated and purchased like pipe: straight sections and fittings.

- Note that cranes or other lifting equipment are not included on any lines in Division 15. For example, if a crane is required to lift a heavy piece of pipe into place high above a gym floor, or to put a rooftop unit on the roof of a four-story building, etc., it must be added. Due to the potential for extreme variation—from nothing additional required, to a major crane or helicopter—we feel that including a nominal amount for "lifting contingency" would be useless and detract from the accuracy of the estimate. When using equipment rental from Means do not forget to include the cost of the operator(s).

Reference Numbers

Reference numbers are shown in bold squares at the beginning of some major classifications. These numbers refer to related items in the Reference Section. The reference information may be an estimating procedure, an alternate pricing method or technical information.

Note: Not all subdivisions listed here necessarily appear in this publication.

Note: **i2 Trade Service,** *in part, has been used as a reference source for some of the material prices used in Division 15.*

15051	Mechanical General	CREW	DAILY OUTPUT	LABOR-HOURS	UNIT	MAT.	2005 BARE COSTS LABOR	EQUIP.	TOTAL	TOTAL INCL O&P	
700	0010	**PIPING** See also divisions 02500 & 02600 for site work										700
	1000	Add to labor for elevated installation										
	1080	10' to 15' high						10%				
	1100	15' to 20' high						20%				
	1120	20' to 25' high						25%				
	1140	25' to 30' high						35%				
	1160	30' to 35' high						40%				
	1180	35' to 40' high						50%				
	1200	Over 40' high						55%				

15055	Selective Mech Demolition	CREW	DAILY OUTPUT	LABOR-HOURS	UNIT	MAT.	LABOR	EQUIP.	TOTAL	TOTAL INCL O&P	
300	0010	**HVAC DEMOLITION**										300
	0100	Air conditioner, split unit, 3 ton	Q-5	2	8	Ea.		295		295	445	
	0150	Package unit, 3 ton	Q-6	3	8	"		305		305	460	
	0298	Boilers										
	0300	Electric, up thru 148 kW	Q-19	2	12	Ea.		460		460	685	
	0310	150 thru 518 kW	"	1	24			915		915	1,375	
	0320	550 thru 2000 kW	Q-21	.40	80			3,100		3,100	4,650	
	0330	2070 kW and up	"	.30	106			4,150		4,150	6,200	
	0340	Gas and/or oil, up thru 150 MBH	Q-7	2.20	14.545			570		570	855	
	0350	160 thru 2000 MBH		.80	40			1,550		1,550	2,350	
	0360	2100 thru 4500 MBH		.50	64			2,500		2,500	3,750	
	0370	4600 thru 7000 MBH		.30	106			4,175		4,175	6,250	
	0380	7100 thru 12,000 MBH		.16	200			7,800		7,800	11,700	
	0390	12,200 thru 25,000 MBH	↓	.12	266	↓		10,400		10,400	15,600	
	1000	Ductwork, 4" high, 8" wide	1 Clab	200	.040	L.F.		1.07		1.07	1.66	
	1100	6" high, 8" wide		165	.048			1.29		1.29	2.01	
	1200	10" high, 12" wide		125	.064			1.71		1.71	2.66	
	1300	12"-14" high, 16"-18" wide		85	.094			2.51		2.51	3.91	
	1400	18" high, 24" wide		67	.119			3.19		3.19	4.96	
	1500	30" high, 36" wide		56	.143			3.81		3.81	5.95	
	1540	72" wide	↓	50	.160	↓		4.27		4.27	6.65	
	3000	Mechanical equipment, light items. Unit is weight, not cooling.	Q-5	.90	17.778	Ton		655		655	985	
	3600	Heavy items	"	1.10	14.545	"		535		535	805	
600	0010	**PLUMBING DEMOLITION**										600
	1020	Fixtures, including 10' piping										
	1100	Bath tubs, cast iron	1 Plum	4	2	Ea.		81.50		81.50	123	
	1120	Fiberglass		6	1.333			54.50		54.50	82	
	1140	Steel		5	1.600			65.50		65.50	98	
	1200	Lavatory, wall hung		10	.800			32.50		32.50	49	
	1220	Counter top		8	1			41		41	61.50	
	1250	Shower pan, minimum		24	.333			13.60		13.60	20.50	
	1260	Shower pan, maximum		16	.500			20.50		20.50	30.50	
	1300	Sink, steel or cast iron, single		8	1			41		41	61.50	
	1320	Double		7	1.143			46.50		46.50	70	
	1400	Water closet, floor mounted		8	1			41		41	61.50	
	1420	Wall mounted		7	1.143			46.50		46.50	70	
	1500	Urinal, floor mounted		4	2			81.50		81.50	123	
	1520	Wall mounted		7	1.143			46.50		46.50	70	
	1600	Water fountains, free standing		8	1			41		41	61.50	
	1620	Recessed		6	1.333	↓		54.50		54.50	82	
	2000	Piping, metal, up thru 1-1/2" diameter		200	.040	L.F.		1.63		1.63	2.46	
	2050	2" thru 3-1/2" diameter	↓	150	.053			2.18		2.18	3.27	
	2100	4" thru 6" diameter	2 Plum	100	.160	↓		6.55		6.55	9.80	

Important: See the Reference Section for critical supporting data - Reference Nos., Crews, & City Cost Indexes

15050 | Basic Materials & Methods

15055	Selective Mech Demolition	CREW	DAILY OUTPUT	LABOR-HOURS	UNIT	2005 BARE COSTS				TOTAL INCL O&P
						MAT.	LABOR	EQUIP.	TOTAL	
600 2150	8" thru 14" diameter	2 Plum	60	.267	L.F.		10.90		10.90	16.35	**600**
2153	16" thru 20" diameter	Q-18	70	.343			13.10	1.15	14.25	21
2155	24" thru 26" diameter		55	.436			16.70	1.47	18.17	26.50
2156	30" thru 36" diameter	↓	40	.600	↓		23	2.02	25.02	36.50
2212	Deduct for salvage, aluminum scrap				Ton					525
2214	Brass scrap									420
2216	Copper scrap									1,025
2218	Lead scrap									235
2220	Steel scrap				↓					65
2250	Water heater, 40 gal.	1 Plum	6	1.333	Ea.		54.50		54.50	82

15100 | Building Services Piping

15106	Glass Pipe & Fittings	CREW	DAILY OUTPUT	LABOR-HOURS	UNIT	2005 BARE COSTS				TOTAL INCL O&P
						MAT.	LABOR	EQUIP.	TOTAL	
120 0010	**PIPE, GLASS** Borosilicate, couplings & hangers 10' O.C.										**120**
0020	Drainage									
1100	1-1/2" diameter	Q-1	52	.308	L.F.	9.15	11.30		20.45	27
1120	2" diameter		44	.364		12.20	13.35		25.55	33.50
1140	3" diameter		39	.410		16.30	15.10		31.40	40.50
1160	4" diameter		30	.533		29.50	19.60		49.10	62
1180	6" diameter	↓	26	.615	↓	53.50	22.50		76	93
1870	To delete coupling & hanger, subtract									
1880	1-1/2" diam. to 2" diam.					19%	22%			
1890	3" diam. to 6" diam.					20%	17%			
2000	Process supply (pressure), beaded joints									
2040	1/2" diameter	1 Plum	36	.222	L.F.	3.05	9.10		12.15	17
2060	3/4" diameter		31	.258		3.53	10.55		14.08	19.75
2080	1" diameter	↓	27	.296		12.85	12.10		24.95	32.50
2100	1-1/2" diameter	Q-1	47	.340		15.40	12.50		27.90	35.50
2120	2" diameter		39	.410		20.50	15.10		35.60	45.50
2140	3" diameter		34	.471		30	17.30		47.30	58.50
2160	4" diameter		25	.640		41.50	23.50		65	81.50
2180	6" diameter	↓	21	.762	↓	82.50	28		110.50	133
2860	To delete coupling & hanger, subtract									
2870	1/2" diam. to 1" diam.					25%	33%			
2880	1-1/2" diam. to 3" diam.					22%	21%			
2890	4" diam. to 6" diam.					23%	15%			
3000	Beaded joint, armored, translucent									
3040	1/2" diameter	1 Plum	36	.222	L.F.	13.55	9.10		22.65	28.50
3060	3/4" diameter		31	.258		16.10	10.55		26.65	33.50
3080	1" diameter	↓	27	.296		20.50	12.10		32.60	40.50
3100	1-1/2" diameter	Q-1	47	.340		24	12.50		36.50	45
3120	2" diameter		39	.410		31	15.10		46.10	56.50
3140	3" diameter		34	.471		48.50	17.30		65.80	79.50
3160	4" diameter		25	.640		64.50	23.50		88	107
3180	6" diameter	↓	21	.762	↓	125	28		153	180
3710	To delete coupling & hanger, subtract									
3720	1/2" diam. to 3/4" diam.					30%	27%			
3730	1" diam. to 3" diam.					28%	22%			
3740	4" diam. to 6" diam.					24%	15%			

		15106	Glass Pipe & Fittings	CREW	DAILY OUTPUT	LABOR-HOURS	UNIT	2005 BARE COSTS				TOTAL INCL O&P	
								MAT.	LABOR	EQUIP.	TOTAL		
120	3800		Conical joint, transparent										**120**
	3980		6" diameter	Q-1	21	.762	L.F.	570	28		598	670	
	4500		To delete couplings & hangers, subtract										
	4530		6" diam.					22%	26%				
160	0010	**PIPE, GLASS, FITTINGS**											**160**
	0020		Drainage, beaded ends										
	0040		Coupling & labor required at joints not incl. in fitting										
	0050		price. Add 1 per joint for installed price										
	0070		90° Bend or sweep, 1-1/2"				Ea.	19.70			19.70	21.50	
	0090		2"					25			25	27.50	
	0100		3"					41			41	45	
	0110		4"					65.50			65.50	72	
	0120		6" (sweep only)				↓	200			200	219	
	0200		45° Bend or sweep same as 90°										
	0350		Tee, single sanitary, 1-1/2"				Ea.	32			32	35	
	0370		2"					32			32	35	
	0380		3"					47.50			47.50	52	
	0390		4"					85			85	93.50	
	0400		6"					228			228	251	
	0410		Tee, straight, 1-1/2"					39.50			39.50	43.50	
	0430		2"					39.50			39.50	43.50	
	0440		3"					57			57	62.50	
	0450		4"					100			100	110	
	0460		6"				↓	246			246	270	
	0500		Coupling, stainless steel, TFE seal ring										
	0520		1-1/2"	Q-1	32	.500	Ea.	13.45	18.40		31.85	42.50	
	0530		2"		30	.533		17.20	19.60		36.80	48.50	
	0540		3"		25	.640		23	23.50		46.50	61	
	0550		4"		23	.696		39.50	25.50		65	82	
	0560		6"	↓	20	.800	↓	92	29.50		121.50	145	
	0600		Coupling, stainless steel, bead to plain end										
	0610		1-1/2"	Q-1	36	.444	Ea.	17.40	16.35		33.75	43.50	
	0620		2"		34	.471		21.50	17.30		38.80	50	
	0630		3"		29	.552		37	20.50		57.50	71	
	0640		4"		27	.593		55	22		77	94	
	0650		6"	↓	24	.667	↓	188	24.50		212.50	244	
	2000		Process supply (pressure), beaded ends										
	2050		90° Sweep elbow, 1/2"				Ea.	19.90			19.90	22	
	2070		3/4"					23.50			23.50	25.50	
	2080		1"					31			31	34	
	2090		1-1/2"					32.50			32.50	35.50	
	2100		2"					44			44	48	
	2120		3"					69			69	76	
	2130		4"					97			97	107	
	2140		6"				↓	243			243	267	
	2150		45° Sweep elbow, same as 90°										
	2250		Tee, straight, 1/2"				Ea.	34			34	37	
	2270		3/4"					34			34	37	
	2280		1"					37.50			37.50	41.50	
	2290		1-1/2"					55.50			55.50	61	
	2300		2"					65.50			65.50	72	
	2310		3"					80			80	88	
	2320		4"					135			135	148	
	2330		6"				↓	250			250	275	
	2350		Coupling, Viton liner, for temperatures to 400°F										
	2370		1/2"	Q-1	40	.400	Ea.	28	14.70		42.70	53	

15 MECHANICAL

		15106	Glass Pipe & Fittings	CREW	DAILY OUTPUT	LABOR-HOURS	UNIT	2005 BARE COSTS				TOTAL INCL O&P	
								MAT.	LABOR	EQUIP.	TOTAL		
160	2380		3/4"	Q-1	37	.432	Ea.	32	15.90		47.90	59	**160**
	2390		1"		35	.457		38	16.80		54.80	67.50	
	2400		1-1/2"		32	.500		42	18.40		60.40	74	
	2410		2"		30	.533		51.50	19.60		71.10	86.50	
	2420		3"		25	.640		95	23.50		118.50	140	
	2430		4"		23	.696		114	25.50		139.50	165	
	2440		6"	↓	20	.800	↓	245	29.50		274.50	315	
	2550		For beaded joint armored fittings, add					200%					
	2600		Conical ends. Flange set, gasket & labor not incl. in fitting										
	2620		price. Add 1 per joint for installed price.										
	2650		90° Sweep elbow, 1"				Ea.	53			53	58.50	
	2670		1-1/2"					77			77	84.50	
	2680		2"					89			89	98	
	2690		3"					151			151	166	
	2700		4"					340			340	375	
	2710		6"					645			645	710	
	2750		Cross (straight), add					55%					
	2850		Tee, add				↓	20%					

		15107	Metal Pipe & Fittings										
320	0010		**PIPE, CAST IRON** Soil, on hangers 5' O.C.										**320**
	0020		Single hub, service wt., lead & oakum joints 10' O.C. R15100 -050										
	2120		2" diameter	Q-1	63	.254	L.F.	4.63	9.35		13.98	19.15	
	2140		3" diameter		60	.267		6.35	9.80		16.15	21.50	
	2160		4" diameter	↓	55	.291		8.10	10.70		18.80	25	
	2180		5" diameter	Q-2	76	.316		8.20	12.05		20.25	27	
	2200		6" diameter	"	73	.329		13.75	12.55		26.30	34	
	2220		8" diameter	Q-3	59	.542		21.50	21		42.50	55	
	2240		10" diameter		54	.593		39	23		62	77.50	
	2260		12" diameter	↓	48	.667		55.50	26		81.50	100	
	2320		For service weight, double hub, add					10%					
	2340		For extra heavy, single hub, add					48%	4%				
	2360		For extra heavy, double hub, add				↓	71%	4%				
	2400		Lead for caulking, (1#/dia in)	Q-1	160	.100	Lb.	.96	3.68		4.64	6.60	
	2420		Oakum for caulking, (1/8#/dia in)	"	40	.400	"	2.36	14.70		17.06	24.50	
	2960		To delete hangers, subtract										
	2970		2" diam. to 4" diam.					16%	19%				
	2980		5" diam. to 8" diam.					14%	14%				
	2990		10" diam. to 15" diam.					13%	19%				
	3000		Single hub, service wt, push-on gasket joints 10' O.C.										
	3010		2" diameter	Q-1	66	.242	L.F.	4.94	8.90		13.84	18.85	
	3020		3" diameter		63	.254		6.75	9.35		16.10	21.50	
	3030		4" diameter	↓	57	.281		8.65	10.30		18.95	25	
	3040		5" diameter	Q-2	79	.304		9	11.60		20.60	27.50	
	3050		6" diameter	"	75	.320		14.60	12.20		26.80	34.50	
	3060		8" diameter	Q-3	62	.516		23	20		43	55.50	
	3070		10" diameter		56	.571		42	22.50		64.50	79.50	
	3080		12" diameter	↓	49	.653	↓	59	25.50		84.50	103	
	3100		For service weight, double hub, add					65%					
	3110		For extra heavy, single hub, add					48%	4%				
	3120		For extra heavy, double hub, add					29%	4%				
	3130		To delete hangers, subtract										
	3140		2" diam. to 4" diam.					12%	21%				
	3150		5" diam. to 8" diam.					10%	16%				
	3160		10" diam. to 15" diam.					9%	21%				
	4000		No hub, couplings 10' O.C.	↓									

MECHANICAL 15

15107 | Metal Pipe & Fittings

		CREW	DAILY OUTPUT	LABOR-HOURS	UNIT	2005 BARE COSTS MAT.	LABOR	EQUIP.	TOTAL	TOTAL INCL O&P	
320	4100	1-1/2" diameter R15100 -050	Q-1	71	.225	L.F.	5.30	8.30		13.60	18.30	320
	4120	2" diameter		67	.239		5.45	8.80		14.25	19.20	
	4140	3" diameter		64	.250		7.15	9.20		16.35	21.50	
	4160	4" diameter		58	.276		9.10	10.15		19.25	25.50	
	4180	5" diameter	Q-2	83	.289		13.10	11.05		24.15	31	
	4200	6" diameter	"	79	.304		16.30	11.60		27.90	35.50	
	4220	8" diameter	Q-3	69	.464		25.50	18.05		43.55	55.50	
	4240	10" diameter	"	61	.525		46	20.50		66.50	81	
	4280	To delete hangers, subtract										
	4290	1-1/2" diam. to 6" diam.					22%	47%				
	4300	8" diam. to 10" diam.					21%	44%				
360	0010	**PIPE, CAST IRON, FITTINGS** Soil R15100 -050										360
	0040	Hub and spigot, service weight, lead & oakum joints										
	0080	1/4 bend, 2"	Q-1	16	1	Ea.	9.75	37		46.75	66.50	
	0120	3"		14	1.143		13	42		55	77.50	
	0140	4"		13	1.231		20.50	45.50		66	90.50	
	0160	5"	Q-2	18	1.333		28.50	51		79.50	108	
	0180	6"	"	17	1.412		35.50	54		89.50	120	
	0200	8"	Q-3	11	2.909		107	113		220	287	
	0220	10"		10	3.200		156	125		281	360	
	0224	12"		9	3.556		211	138		349	440	
	0266	Closet bend, 3" diameter with flange 10" x 16"	Q-1	14	1.143		53	42		95	121	
	0268	16"x16"		12	1.333		73	49		122	154	
	0270	Closet bend, 4" diameter, 1" x 4" ring, 6" x 16"		13	1.231		59	45.50		104.50	133	
	0280	8" x 16"		13	1.231		52.50	45.50		98	126	
	0290	10" x 12"		12	1.333		49.50	49		98.50	128	
	0300	10" x 18"		11	1.455		75	53.50		128.50	163	
	0310	12" x 16"		11	1.455		59	53.50		112.50	146	
	0330	16" x 16"		10	1.600		74.50	59		133.50	171	
	0340	1/8 bend, 2"		16	1		6.95	37		43.95	63	
	0350	3"		14	1.143		10.90	42		52.90	75	
	0360	4"		13	1.231		15.85	45.50		61.35	85.50	
	0380	5"	Q-2	18	1.333		22	51		73	101	
	0400	6"	"	17	1.412		27	54		81	111	
	0420	8"	Q-3	11	2.909		80.50	113		193.50	259	
	0440	10"		10	3.200		116	125		241	315	
	0460	12"		9	3.556		219	138		357	450	
	0500	Sanitary tee, 2"	Q-1	10	1.600		12.60	59		71.60	102	
	0540	3"		9	1.778		22	65.50		87.50	123	
	0620	4"		8	2		27	73.50		100.50	141	
	0700	5"	Q-2	12	2		53.50	76.50		130	174	
	0800	6"	"	11	2.182		61	83		144	192	
	0880	8"	Q-3	7	4.571		161	178		339	445	
	1000	Tee, 2"	Q-1	10	1.600		17.05	59		76.05	107	
	1060	3"		9	1.778		27.50	65.50		93	129	
	1120	4"		8	2		35.50	73.50		109	150	
	1200	5"	Q-2	12	2		53.50	76.50		130	174	
	1300	6"	"	11	2.182		61	83		144	192	
	1380	8"	Q-3	7	4.571		161	178		339	445	
	1400	Combination Y and 1/8 bend										
	1420	2"	Q-1	10	1.600	Ea.	17.05	59		76.05	107	
	1460	3"		9	1.778		26	65.50		91.50	127	
	1520	4"		8	2		35.50	73.50		109	151	
	1540	5"	Q-2	12	2		68	76.50		144.50	190	
	1560	6"		11	2.182		85.50	83		168.50	220	

Important: See the Reference Section for critical supporting data - Reference Nos., Crews, & City Cost Indexes

			CREW	DAILY OUTPUT	LABOR-HOURS	UNIT	2005 BARE COSTS				TOTAL INCL O&P
15107	**Metal Pipe & Fittings**						MAT.	LABOR	EQUIP.	TOTAL	
360 1580	8"	R15100 -050	Q-2	7	3.429	Ea.	211	131		342	430	360
1582	12"		Q-3	6	5.333		480	208		688	835
1600	Double Y, 2"		Q-1	8	2		20.50	73.50		94	134
1610	3"			7	2.286		35.50	84		119.50	165
1620	4"		▼	6.50	2.462		49.50	90.50		140	190
1630	5"		Q-2	9	2.667		83.50	102		185.50	245
1640	6"		"	8	3		127	114		241	310
1650	8"		Q-3	5.50	5.818		310	227		537	680
1660	10"			5	6.400		470	249		719	895
1670	12"		▼	4.50	7.111		630	277		907	1,100
1740	Reducer, 3" x 2"		Q-1	15	1.067		9.50	39		48.50	69.50
1750	4" x 2"			14.50	1.103		10	40.50		50.50	72
1760	4" x 3"			14	1.143		12.40	42		54.40	76.50
1770	5" x 2"			14	1.143		16.40	42		58.40	81
1780	5" x 3"			13.50	1.185		16.40	43.50		59.90	83.50
1790	5" x 4"			13	1.231		16.40	45.50		61.90	86
1800	6" x 2"			13.50	1.185		22	43.50		65.50	90
1810	6" x 3"			13	1.231		24	45.50		69.50	94.50
1840	6" x 5"		▼	11	1.455		27	53.50		80.50	110
1880	8" x 3"		Q-2	13.50	1.778		43	68		111	150
1900	8" x 4"			13	1.846		42	70.50		112.50	152
1920	8" x 5"			12	2		43	76.50		119.50	163
1940	8" x 6"		▼	12	2		43	76.50		119.50	162
1960	Increaser, 2" x 3"		Q-1	15	1.067		21.50	39		60.50	82.50
1980	2" x 4"			14	1.143		21.50	42		63.50	86.50
2000	2" x 5"			13	1.231		26	45.50		71.50	97
2020	3" x 4"			13	1.231		22.50	45.50		68	93
2040	3" x 5"			13	1.231		26	45.50		71.50	96.50
2060	3" x 6"			12	1.333		32.50	49		81.50	109
2070	4" x 5"			13	1.231		26	45.50		71.50	97
2080	4" x 6"		▼	12	1.333		32.50	49		81.50	109
2090	4" x 8"		Q-2	13	1.846		66	70.50		136.50	179
2100	5" x 6"		Q-1	11	1.455		48	53.50		101.50	134
2110	5" x 8"		Q-2	12	2		77.50	76.50		154	201
2120	6" x 8"			12	2		77.50	76.50		154	201
2130	6" x 10"			8	3		133	114		247	320
2140	8" x 10"			6.50	3.692		133	141		274	360
2150	10" x 12"		▼	5.50	4.364		251	166		417	525
2500	Y, 2"		Q-1	10	1.600		11.85	59		70.85	102
2510	3"			9	1.778		22	65.50		87.50	123
2520	4"		▼	8	2		29.50	73.50		103	144
2530	5"		Q-2	12	2		49.50	76.50		126	170
2540	6"		"	11	2.182		67.50	83		150.50	200
2550	8"		Q-3	7	4.571		165	178		343	450
2560	10"			6	5.333		267	208		475	605
2570	12"			5	6.400		460	249		709	880
2580	15"		▼	4	8		975	310		1,285	1,550
3000	For extra heavy, add					▼	44%	4%			
3600	Hub and spigot, service weight gasket joint										
3605	Note: gaskets and joint labor have										
3606	been included with all listed fittings.										
3610	1/4 bend, 2"		Q-1	20	.800	Ea.	12.80	29.50		42.30	58
3620	3"			17	.941		17.05	34.50		51.55	71
3630	4"		▼	15	1.067		25.50	39		64.50	87
3640	5"		Q-2	21	1.143		36.50	43.50		80	106
3650	6"		"	19	1.263		43.50	48		91.50	121

MECHANICAL 15

	15107	Metal Pipe & Fittings	CREW	DAILY OUTPUT	LABOR-HOURS	UNIT	MAT.	LABOR	EQUIP.	TOTAL	TOTAL INCL O&P	
								2005 BARE COSTS				
360	3660	8"	Q-3	12	2.667	Ea.	125	104		229	293	360
	3670	10"		11	2.909		183	113		296	370	
	3680	12"	▼	10	3.200		246	125		371	460	
	3700	Closet bend, 3" diameter with ring 10" x 16"	Q-1	17	.941		57	34.50		91.50	115	
	3710	16" x 16"		15	1.067		77	39		116	144	
	3730	Closet bend, 4" diameter, 1" x 4" ring, 6" x 16"		15	1.067		64	39		103	130	
	3740	8" x 16"		15	1.067		57.50	39		96.50	123	
	3750	10" x 12"		14	1.143		54.50	42		96.50	123	
	3760	10" x 18"		13	1.231		80.50	45.50		126	157	
	3770	12" x 16"		13	1.231		64.50	45.50		110	139	
	3780	16" x 16"		12	1.333		79.50	49		128.50	161	
	3800	1/8 bend, 2"		20	.800		10	29.50		39.50	55	
	3810	3"		17	.941		14.95	34.50		49.45	68.50	
	3820	4"	▼	15	1.067		21	39		60	82	
	3830	5"	Q-2	21	1.143		30	43.50		73.50	98.50	
	3840	6"	"	19	1.263		35	48		83	111	
	3850	8"	Q-3	12	2.667		98.50	104		202.50	264	
	3860	10"		11	2.909		143	113		256	325	
	3870	12"	▼	10	3.200		254	125		379	465	
	3900	Sanitary Tee, 2"	Q-1	12	1.333		18.70	49		67.70	94	
	3910	3"		10	1.600		30	59		89	122	
	3920	4"	▼	9	1.778		37	65.50		102.50	140	
	3930	5"	Q-2	13	1.846		69	70.50		139.50	182	
	3940	6"	"	11	2.182		77	83		160	210	
	3950	8"	Q-3	8	4		197	156		353	450	
	3980	Tee, 2"	Q-1	12	1.333		23	49		72	99	
	3990	3"		10	1.600		35.50	59		94.50	128	
	4000	4"	▼	9	1.778		45.50	65.50		111	149	
	4010	5"	Q-2	13	1.846		69	70.50		139.50	182	
	4020	6"	"	11	2.182		77	83		160	210	
	4030	8"	Q-3	8	4	▼	197	156		353	450	
	4060	Combination Y and 1/8 bend										
	4070	2"	Q-1	12	1.333	Ea.	23	49		72	99	
	4080	3"		10	1.600		34	59		93	126	
	4090	4"	▼	9	1.778		46	65.50		111.50	149	
	4100	5"	Q-2	13	1.846		83.50	70.50		154	198	
	4110	6"	"	11	2.182		102	83		185	237	
	4120	8"	Q-3	8	4		247	156		403	505	
	4160	Double Y, 2"	Q-1	10	1.600		29.50	59		88.50	121	
	4170	3"		8	2		47.50	73.50		121	164	
	4180	4"	▼	7	2.286		64.50	84		148.50	197	
	4190	5"	Q-2	10	2.400		107	91.50		198.50	256	
	4200	6"	"	9	2.667		151	102		253	320	
	4210	8"	Q-3	6	5.333		365	208		573	710	
	4220	10"		5	6.400		555	249		804	985	
	4230	12"	▼	4.50	7.111		735	277		1,012	1,225	
	4260	Reducer, 3" x 2"	Q-1	17	.941		16.60	34.50		51.10	70.50	
	4270	4" x 2"		16.50	.970		18.15	35.50		53.65	73.50	
	4280	4" x 3"		16	1		21.50	37		58.50	79	
	4290	5" x 2"		16	1		27.50	37		64.50	85.50	
	4300	5" x 3"		15.50	1.032		28.50	38		66.50	88	
	4310	5" x 4"		15	1.067		29.50	39		68.50	91.50	
	4320	6" x 2"		15.50	1.032		33.50	38		71.50	93.50	
	4330	6" x 3"		15	1.067		36.50	39		75.50	99	
	4340	6" x 5"	▼	13	1.231		43	45.50		88.50	115	
	4360	8" x 3"	Q-2	15	1.600	▼	65	61		126	163	

R15100 -050

15107	Metal Pipe & Fittings		CREW	DAILY OUTPUT	LABOR-HOURS	UNIT	2005 BARE COSTS				TOTAL INCL O&P
							MAT.	LABOR	EQUIP.	TOTAL	
360	4370	8" x 4"	R15100 -050	Q-2	15	1.600	Ea.	65	61		126	163	360
	4380	8" x 5"			14	1.714		69	65.50		134.50	175
	4390	8" x 6"		▼	14	1.714		69	65.50		134.50	175
	4430	Increaser, 2" x 3"		Q-1	17	.941		25.50	34.50		60	80
	4440	2" x 4"			16	1		26.50	37		63.50	85
	4450	2" x 5"			15	1.067		34	39		73	96.50
	4460	3" x 4"			15	1.067		27.50	39		66.50	89.50
	4470	3" x 5"			15	1.067		34	39		73	96.50
	4480	3" x 6"			14	1.143		40.50	42		82.50	108
	4490	4" x 5"			15	1.067		34	39		73	96.50
	4500	4" x 6"		▼	14	1.143		40.50	42		82.50	108
	4510	4" x 8"		Q-2	15	1.600		84	61		145	184
	4520	5" x 6"		Q-1	13	1.231		56	45.50		101.50	130
	4530	5" x 8"		Q-2	14	1.714		95.50	65.50		161	204
	4540	6" x 8"			14	1.714		95.50	65.50		161	204
	4550	6" x 10"			10	2.400		160	91.50		251.50	315
	4560	8" x 10"			8.50	2.824		160	108		268	340
	4570	10" x 12"		▼	7.50	3.200		286	122		408	500
	4600	Y, 2"		Q-1	12	1.333		17.90	49		66.90	93
	4610	3"			10	1.600		30	59		89	122
	4620	4"		▼	9	1.778		39.50	65.50		105	142
	4630	5"		Q-2	13	1.846		65	70.50		135.50	178
	4640	6"		"	11	2.182		84	83		167	217
	4650	8"		Q-3	8	4		201	156		357	455
	4660	10"			7	4.571		320	178		498	625
	4670	12"		▼	6	5.333	▼	530	208		738	890
	4900	For extra heavy, add						44%	4%			
	4940	Gasket and making push-on joint										
	4950	2"		Q-1	40	.400	Ea.	3.04	14.70		17.74	25.50
	4960	3"			35	.457		4.05	16.80		20.85	30
	4970	4"		▼	32	.500		5.10	18.40		23.50	33
	4980	5"		Q-2	43	.558		7.85	21.50		29.35	40.50
	4990	6"		"	40	.600		8.15	23		31.15	43.50
	5000	8"		Q-3	32	1		17.95	39		56.95	78.50
	5010	10"			29	1.103		27.50	43		70.50	94.50
	5020	12"			25	1.280		35	50		85	114
	5022	15"		▼	21	1.524	▼	42	59.50		101.50	135
	5030	Note: gaskets and joint labor have										
	5040	Been included with all listed fittings.										
	5990	No hub										
	6000	Cplg. & labor required at joints not incl. in fitting										
	6010	price. Add 1 coupling per joint for installed price										
	6020	1/4 Bend, 1-1/2"					Ea.	5.50			5.50	6.05
	6060	2"						5.95			5.95	6.55
	6080	3"						8.30			8.30	9.10
	6120	4"						11.90			11.90	13.10
	6140	5"						26.50			26.50	29.50
	6160	6"						30			30	33
	6180	8"						83			83	91
	6184	1/4 Bend, long sweep, 1-1/2"						12.85			12.85	14.10
	6186	2"						12.85			12.85	14.10
	6188	3"						15.30			15.30	16.85
	6189	4"						24.50			24.50	27
	6190	5"						45			45	49.50
	6191	6"						25			25	27.50
	6192	8"		▼			▼	133			133	147

MECHANICAL 15

15107	Metal Pipe & Fittings						2005 BARE COSTS				TOTAL	
			DAILY	LABOR-			MAT.	LABOR	EQUIP.	TOTAL	INCL O&P	
		CREW	OUTPUT	HOURS	UNIT							
360	6193	10"				Ea.	239			239	263	**360**
6200	1/8 Bend, 1-1/2"	R15100 -050					4.55			4.55	5
6210	2"						5.10			5.10	5.60
6212	3"						6.85			6.85	7.55
6214	4"						8.70			8.70	9.60
6216	5"						18.60			18.60	20.50
6218	6"						20			20	22
6220	8"						58			58	64
6222	10"						108			108	119
6380	Sanitary Tee, tapped, 1-1/2"						10.05			10.05	11.05
6382	2" x 1-1/2"						9.15			9.15	10.05
6384	2"						10.10			10.10	11.10
6386	3" x 2"						14.15			14.15	15.60
6388	3"						26			26	29
6390	4" x 1-1/2"						12.65			12.65	13.90
6392	4" x 2"						14.15			14.15	15.60
6394	6" x 1-1/2"						29.50			29.50	32.50
6396	6" x 2"						30			30	33
6459	Sanitary Tee, 1-1/2"						7.45			7.45	8.15
6460	2"						8.30			8.30	9.10
6470	3"						10.55			10.55	11.60
6472	4"						15.60			15.60	17.15
6474	5"						44.50			44.50	49
6476	6"						45.50			45.50	50
6478	8"						184			184	202
6730	Y, 1-1/2"						7.75			7.75	8.55
6740	2"						7.75			7.75	8.55
6750	3"						11			11	12.10
6760	4"						17.90			17.90	19.70
6762	5"						42			42	46.50
6764	6"						47.50			47.50	52.50
6768	8"						112			112	124
6791	Y, reducing, 3" x 2"						8.30			8.30	9.10
6792	4" x 2"						11.45			11.45	12.60
6793	5" x 2"						26			26	28.50
6794	6" x 2"						28.50			28.50	31.50
6795	6" x 4"						37			37	41
6796	8" x 4"						64			64	70.50
6797	8" x 6"						79			79	86.50
6798	10" x 6"						177			177	195
6799	10" x 8"						213			213	234
6800	Double Y, 2"						11.90			11.90	13.10
6920	3"						22			22	24
7000	4"						45			45	49.50
7100	6"						79			79	87
7120	8"					↓	226			226	248
7200	Combination Y and 1/8 Bend										
7220	1-1/2"					Ea.	8.20			8.20	9
7260	2"						8.70			8.70	9.60
7320	3"						13.25			13.25	14.60
7400	4"						24.50			24.50	26.50
7480	5"						53.50			53.50	58.50
7500	6"						67.50			67.50	74.50
7520	8"						167			167	184
7800	Reducer, 3" x 2"						4.13			4.13	4.54
7820	4" x 2"	↓			↓		6.45			6.45	7.10

15107	Metal Pipe & Fittings		CREW	DAILY OUTPUT	LABOR-HOURS	UNIT	2005 BARE COSTS				TOTAL INCL O&P
							MAT.	LABOR	EQUIP.	TOTAL	
360	7840	4" x 3"				Ea.	6.45			6.45	7.10	**360**
	8000	Coupling, standard (by CISPI Mfrs.)	R15100 -050								
	8020	1-1/2"	Q-1	48	.333	Ea.	5.25	12.25		17.50	24
	8040	2"		44	.364		5.25	13.35		18.60	26
	8080	3"		38	.421		6.20	15.50		21.70	30.50
	8120	4"	▼	33	.485		7.35	17.85		25.20	35
	8160	5"	Q-2	44	.545		17.95	21		38.95	51
	8180	6"	"	40	.600		18.70	23		41.70	55
	8200	8"	Q-3	33	.970		35.50	38		73.50	96
	8220	10"	"	26	1.231	▼	48.50	48		96.50	126
	8300	Coupling, cast iron clamp & neoprene gasket (by MG)									
	8310	1-1/2"	Q-1	48	.333	Ea.	7.75	12.25		20	27
	8320	2"		44	.364		8	13.35		21.35	29
	8330	3"		38	.421		8.35	15.50		23.85	32.50
	8340	4"	▼	33	.485		9.80	17.85		27.65	38
	8350	5"	Q-2	44	.545		18.55	21		39.55	52
	8360	6"	"	40	.600		25.50	23		48.50	62.50
	8380	8"	Q-3	33	.970		44.50	38		82.50	106
	8400	10"	"	26	1.231	▼	68	48		116	147
	8600	Coupling, Stainless steel, heavy duty									
	8620	1-1/2"	Q-1	48	.333	Ea.	11.10	12.25		23.35	30.50
	8630	2"		44	.364		11.85	13.35		25.20	33
	8640	2" x 1-1/2"		44	.364		15.60	13.35		28.95	37
	8650	3"		38	.421		13.65	15.50		29.15	38.50
	8660	4"		33	.485		15.90	17.85		33.75	44.50
	8670	4" x 3"	▼	33	.485		24	17.85		41.85	53
	8680	5"	Q-2	44	.545		32.50	21		53.50	67.50
	8690	6"	"	40	.600		37	23		60	75.50
	8700	8"	Q-3	33	.970		76.50	38		114.50	142
	8710	10"	"	26	1.231	▼	99	48		147	181
420	0010	**PIPE, COPPER** Solder joints	R15100 -050									**420**
	1000	Type K tubing, couplings & clevis hangers 10' O.C.									
	1100	1/4" diameter	1 Plum	84	.095	L.F.	1.37	3.89		5.26	7.35
	1120	3/8" diameter		82	.098		1.57	3.99		5.56	7.70
	1140	1/2" diameter		78	.103		1.79	4.19		5.98	8.25
	1160	5/8" diameter		77	.104		3.05	4.24		7.29	9.75
	1180	3/4" diameter		74	.108		3.02	4.42		7.44	9.95
	1200	1" diameter		66	.121		3.97	4.95		8.92	11.80
	1220	1-1/4" diameter		56	.143		4.91	5.85		10.76	14.15
	1240	1-1/2" diameter		50	.160		6.50	6.55		13.05	16.95
	1260	2" diameter	▼	40	.200		9.95	8.15		18.10	23.50
	1280	2-1/2" diameter	Q-1	60	.267		14.85	9.80		24.65	31
	1300	3" diameter		54	.296		20.50	10.90		31.40	39.50
	1320	3-1/2" diameter		42	.381		28	14		42	52
	1330	4" diameter		38	.421		35	15.50		50.50	62
	1340	5" diameter	▼	32	.500		91	18.40		109.40	128
	1360	6" diameter	Q-2	38	.632		133	24		157	182
	1380	8" diameter	"	34	.706		234	27		261	299
	1390	For other than full hard temper, add				▼	13%				
	1440	For silver solder, add						15%			
	1800	For medical clean, (oxygen class), add					12%				
	1950	To delete cplgs. & hngrs., 1/4"-1" pipe, subtract					27%	60%			
	1960	1-1/4" -3" pipe, subtract					14%	52%			
	1970	3-1/2"-5" pipe, subtract					10%	60%			
	1980	6"-8" pipe, subtract					19%	53%			

15107	Metal Pipe & Fittings	CREW	DAILY OUTPUT	LABOR-HOURS	UNIT	2005 BARE COSTS				TOTAL INCL O&P
						MAT.	LABOR	EQUIP.	TOTAL	
460	0010	**PIPE, COPPER, FITTINGS** Wrought unless otherwise noted	R15100 -050									460
	0040	Solder joints, copper x copper									
	0070	90° elbow, 1/4"	1 Plum	22	.364	Ea.	1.57	14.85		16.42	24
	0090	3/8"		22	.364		1.49	14.85		16.34	24
	0100	1/2"		20	.400		.48	16.35		16.83	25
	0110	5/8"		19	.421		2.99	17.20		20.19	29.50
	0120	3/4"		19	.421		1.06	17.20		18.26	27
	0130	1"		16	.500		2.61	20.50		23.11	33.50
	0140	1-1/4"		15	.533		3.94	22		25.94	37.50
	0150	1-1/2"		13	.615		5.55	25		30.55	44
	0160	2"		11	.727		11.20	29.50		40.70	57
	0170	2-1/2"	Q-1	13	1.231		23.50	45.50		69	94
	0180	3"		11	1.455		31.50	53.50		85	115
	0190	3-1/2"		10	1.600		110	59		169	210
	0200	4"		9	1.778		72.50	65.50		138	179
	0210	5"		6	2.667		405	98		503	590
	0220	6"	Q-2	9	2.667		450	102		552	650
	0230	8"	"	8	3		1,675	114		1,789	2,000
	0250	45° elbow, 1/4"	1 Plum	22	.364		2.65	14.85		17.50	25.50
	0270	3/8"		22	.364		2.16	14.85		17.01	25
	0280	1/2"		20	.400		.86	16.35		17.21	25.50
	0290	5/8"		19	.421		4.56	17.20		21.76	31
	0300	3/4"		19	.421		1.52	17.20		18.72	27.50
	0310	1"		16	.500		3.82	20.50		24.32	34.50
	0320	1-1/4"		15	.533		5.45	22		27.45	39
	0330	1-1/2"		13	.615		6.55	25		31.55	45
	0340	2"		11	.727		10.95	29.50		40.45	56.50
	0350	2-1/2"	Q-1	13	1.231		21	45.50		66.50	91
	0360	3"		13	1.231		34.50	45.50		80	106
	0370	3-1/2"		10	1.600		55	59		114	149
	0380	4"		9	1.778		73.50	65.50		139	180
	0390	5"		6	2.667		285	98		383	460
	0400	6"	Q-2	9	2.667		450	102		552	650
	0410	8"	"	8	3		1,525	114		1,639	1,850
	0450	Tee, 1/4"	1 Plum	14	.571		3.30	23.50		26.80	38.50
	0470	3/8"		14	.571		2.53	23.50		26.03	38
	0480	1/2"		13	.615		.81	25		25.81	39
	0490	5/8"		12	.667		5.50	27		32.50	47
	0500	3/4"		12	.667		1.95	27		28.95	43
	0510	1"		10	.800		6	32.50		38.50	55.50
	0520	1-1/4"		9	.889		8.65	36.50		45.15	64
	0530	1-1/2"		8	1		12	41		53	74.50
	0540	2"		7	1.143		18.75	46.50		65.25	90.50
	0550	2-1/2"	Q-1	8	2		42	73.50		115.50	158
	0560	3"		7	2.286		64	84		148	197
	0570	3-1/2"		6	2.667		186	98		284	350
	0580	4"		5	3.200		155	118		273	350
	0590	5"		4	4		510	147		657	780
	0600	6"	Q-2	6	4		695	153		848	995
	0610	8"	"	5	4.800		2,675	183		2,858	3,225
	0612	Tee, reducing on the outlet, 1/4"	1 Plum	15	.533		5.70	22		27.70	39.50
	0613	3/8"		15	.533		4.99	22		26.99	38.50
	0614	1/2"		14	.571		4.38	23.50		27.88	40
	0615	5/8"		13	.615		8.85	25		33.85	47.50
	0616	3/4"		12	.667		2.82	27		29.82	44
	0617	1"		11	.727		5.75	29.50		35.25	51

15 MECHANICAL

15107	Metal Pipe & Fittings		CREW	DAILY OUTPUT	LABOR-HOURS	UNIT	2005 BARE COSTS				TOTAL INCL O&P
							MAT.	LABOR	EQUIP.	TOTAL	
460												460
0618	1-1/4"	R15100 -050	1 Plum	10	.800	Ea.	9.35	32.50		41.85	59.50
0619	1-1/2"			9	.889		9.95	36.50		46.45	65.50
0620	2"			8	1		16.25	41		57.25	79.50
0621	2-1/2"		Q-1	9	1.778		44	65.50		109.50	147
0622	3"			8	2		54	73.50		127.50	171
0623	4"			6	2.667		110	98		208	268
0624	5"			5	3.200		510	118		628	735
0625	6"		Q-2	7	3.429		695	131		826	960
0626	8"		"	6	4		2,675	153		2,828	3,175
0630	Tee, reducing on the run, 1/4"		1 Plum	15	.533		6.65	22		28.65	40.50
0631	3/8"			15	.533		7.55	22		29.55	41.50
0632	1/2"			14	.571		8.90	23.50		32.40	45
0633	5/8"			13	.615		9.40	25		34.40	48.50
0634	3/4"			12	.667		2.30	27		29.30	43.50
0635	1"			11	.727		7.55	29.50		37.05	53
0636	1-1/4"			10	.800		12	32.50		44.50	62
0637	1-1/2"			9	.889		21	36.50		57.50	78
0638	2"			8	1		27	41		68	91.50
0639	2-1/2"		Q-1	9	1.778		66	65.50		131.50	171
0640	3"			8	2		93.50	73.50		167	214
0641	4"			6	2.667		207	98		305	375
0642	5"			5	3.200		480	118		598	705
0643	6"		Q-2	7	3.429		735	131		866	1,000
0644	8"		"	6	4		2,675	153		2,828	3,175
0650	Coupling, 1/4"		1 Plum	24	.333		.37	13.60		13.97	21
0670	3/8"			24	.333		.48	13.60		14.08	21
0680	1/2"			22	.364		.33	14.85		15.18	23
0690	5/8"			21	.381		1.02	15.55		16.57	24.50
0700	3/4"			21	.381		.71	15.55		16.26	24.50
0710	1"			18	.444		1.44	18.15		19.59	29
0715	1-1/4"			17	.471		2.69	19.20		21.89	32
0716	1-1/2"			15	.533		3.55	22		25.55	37
0718	2"			13	.615		5.95	25		30.95	44.50
0721	2-1/2"		Q-1	15	1.067		12.60	39		51.60	73
0722	3"			13	1.231		18.90	45.50		64.40	89
0724	3-1/2"			8	2		36.50	73.50		110	151
0726	4"			7	2.286		40	84		124	170
0728	5"			6	2.667		98	98		196	255
0731	6"		Q-2	8	3		162	114		276	350
0732	8"		"	7	3.429		525	131		656	770
0850	Unions, 1/4"		1 Plum	21	.381		9.50	15.55		25.05	34
0870	3/8"			21	.381		9.60	15.55		25.15	34
0880	1/2"			19	.421		4.89	17.20		22.09	31.50
0890	5/8"			18	.444		22	18.15		40.15	51.50
0900	3/4"			18	.444		6.40	18.15		24.55	34.50
0910	1"			15	.533		11	22		33	45
0920	1-1/4"			14	.571		19.05	23.50		42.55	56
0930	1-1/2"			12	.667		25	27		52	68.50
0940	2"			10	.800		43	32.50		75.50	96
0950	2-1/2"		Q-1	12	1.333		93.50	49		142.50	177
0960	3"		"	10	1.600		242	59		301	355
0980	Adapter, copper x male IPS, 1/4"		1 Plum	20	.400		5.20	16.35		21.55	30
0990	3/8"			20	.400		2.58	16.35		18.93	27.50
1000	1/2"			18	.444		.99	18.15		19.14	28.50
1010	3/4"			17	.471		1.66	19.20		20.86	31
1020	1"			15	.533		4.30	22		26.30	37.50

MECHANICAL 15

			CREW	DAILY OUTPUT	LABOR-HOURS	UNIT	2005 BARE COSTS				TOTAL INCL O&P	
	15107	**Metal Pipe & Fittings**					MAT.	LABOR	EQUIP.	TOTAL		
460	1030	1-1/4" R15100-050	1 Plum	13	.615	Ea.	6.70	25		31.70	45.50	**460**
	1040	1-1/2"		12	.667		7.70	27		34.70	49.50	
	1050	2"		11	.727		13	29.50		42.50	59	
	1060	2-1/2"	Q-1	10.50	1.524		50	56		106	139	
	1070	3"		10	1.600		66	59		125	161	
	1080	3-1/2"		9	1.778		86.50	65.50		152	194	
	1090	4"		8	2		86.50	73.50		160	206	
	1200	5"		6	2.667		500	98		598	695	
	1210	6"	Q-2	8.50	2.824		500	108		608	710	
	1250	Cross, 1/2"	1 Plum	10	.800		6.90	32.50		39.40	56.50	
	1260	3/4"		9.50	.842		13.40	34.50		47.90	66.50	
	1270	1"		8	1		23	41		64	86.50	
	1280	1-1/4"		7.50	1.067		32.50	43.50		76	102	
	1290	1-1/2"		6.50	1.231		46.50	50.50		97	127	
	1300	2"		5.50	1.455		88	59.50		147.50	187	
	1310	2-1/2"	Q-1	6.50	2.462		203	90.50		293.50	360	
	1320	3"	"	5.50	2.909		256	107		363	440	
	1500	Tee fitting, mechanically formed, (Type 1).										
	1520	1/2" run size, 3/8" to 1/2" branch size	1 Plum	80	.100	Ea.		4.09		4.09	6.15	
	1530	3/4" run size, 3/8" to 3/4" branch size		60	.133			5.45		5.45	8.20	
	1540	1" run size, 3/8" to 1" branch size		54	.148			6.05		6.05	9.10	
	1550	1-1/4" run size, 3/8" to 1-1/4" branch size		48	.167			6.80		6.80	10.25	
	1560	1-1/2" run size, 3/8" to 1-1/2" branch size		40	.200			8.15		8.15	12.30	
	1570	2" run size, 3/8" to 2" branch size		35	.229			9.35		9.35	14.05	
	1580	2-1/2" run size, 1/2" to 2" branch size		32	.250			10.20		10.20	15.35	
	1590	3" run size, 1" to 2" branch size		26	.308			12.55		12.55	18.90	
	1600	4" run size, 1" to 2" branch size		24	.333			13.60		13.60	20.50	
	1640	Tee fitting, mechanically formed, (Type 2)										
	1650	2-1/2" run size, 2-1/2" branch size	1 Plum	12.50	.640	Ea.		26		26	39.50	
	1660	3" run size, 2-1/2" to 3" branch size		12	.667			27		27	41	
	1670	3-1/2" run size, 2-1/2" to 3-1/2" branch size		11	.727			29.50		29.50	44.50	
	1680	4" run size, 2-1/2" to 4" branch size		10.50	.762			31		31	47	
	1698	5" run size, 2" to 4" branch size		9.50	.842			34.50		34.50	51.50	
	1700	6" run size, 2" to 4" branch size		8.50	.941			38.50		38.50	58	
	1710	8" run size, 2" to 4" branch size		7	1.143			46.50		46.50	70	
	2000	DWV, solder joints, copper x copper										
	2030	90° Elbow, 1-1/4"	1 Plum	13	.615	Ea.	4.82	25		29.82	43.50	
	2050	1-1/2"		12	.667		7.20	27		34.20	49	
	2070	2"		10	.800		9.40	32.50		41.90	59.50	
	2090	3"	Q-1	10	1.600		23.50	59		82.50	114	
	2100	4"	"	9	1.778		100	65.50		165.50	209	
	2150	45° Elbow, 1-1/4"	1 Plum	13	.615		3.99	25		28.99	42.50	
	2170	1-1/2"		12	.667		3.30	27		30.30	44.50	
	2180	2"		10	.800		7.60	32.50		40.10	57.50	
	2190	3"	Q-1	10	1.600		16.10	59		75.10	106	
	2200	4"	"	9	1.778		76.50	65.50		142	183	
	2250	Tee, Sanitary, 1-1/4"	1 Plum	9	.889		8.55	36.50		45.05	64	
	2270	1-1/2"		8	1		10.65	41		51.65	73.50	
	2290	2"		7	1.143		12.45	46.50		58.95	83.50	
	2310	3"	Q-1	7	2.286		50	84		134	181	
	2330	4"	"	6	2.667		127	98		225	287	
	2400	Coupling, 1-1/4"	1 Plum	14	.571		2.03	23.50		25.53	37	
	2420	1-1/2"		13	.615		2.53	25		27.53	41	
	2440	2"		11	.727		3.50	29.50		33	48.50	
	2460	3"	Q-1	11	1.455		6.80	53.50		60.30	88	
	2480	4"	"	10	1.600		21.50	59		80.50	113	

Important: See the Reference Section for critical supporting data - Reference Nos., Crews, & City Cost Indexes

15107 | Metal Pipe & Fittings

		CREW	DAILY OUTPUT	LABOR-HOURS	UNIT	2005 BARE COSTS				TOTAL INCL O&P
						MAT.	LABOR	EQUIP.	TOTAL	
500	0010	**PIPE, CORROSION RESISTANT** No couplings or hangers									**500**
	0020	Iron alloy, drain, mechanical joint									
	1000	1-1/2" diameter	Q-1	70	.229	L.F.	22.50	8.40		30.90	37.50
	1100	2" diameter		66	.242		26	8.90		34.90	42
	1120	3" diameter		60	.267		36	9.80		45.80	55
	1140	4" diameter		52	.308		46.50	11.30		57.80	68
	1980	Iron alloy, drain, B&S joint									
	2000	2" diameter	Q-1	54	.296	L.F.	28	10.90		38.90	47.50
	2100	3" diameter		52	.308		40.50	11.30		51.80	61.50
	2120	4" diameter		48	.333		56	12.25		68.25	80
	2140	6" diameter	Q-2	59	.407		93.50	15.50		109	127
	2160	8" diameter	"	54	.444		169	16.95		185.95	212
	2980	Plastic, epoxy, fiberglass filament wound									
	3000	2" diameter	Q-1	62	.258	L.F.	5.20	9.50		14.70	19.95
	3100	3" diameter		51	.314		7.10	11.55		18.65	25
	3120	4" diameter		45	.356		9.40	13.10		22.50	30
	3140	6" diameter		32	.500		14.40	18.40		32.80	43.50
	3160	8" diameter	Q-2	38	.632		19.80	24		43.80	58
	3180	10" diameter		32	.750		34	28.50		62.50	80.50
	3200	12" diameter		28	.857		43	32.50		75.50	96.50
	3980	Polyester, fiberglass filament wound									
	4000	2" diameter	Q-1	62	.258	L.F.	9.85	9.50		19.35	25
	4100	3" diameter		51	.314		13.20	11.55		24.75	32
	4120	4" diameter		45	.356		14.90	13.10		28	36
	4140	6" diameter		32	.500		23.50	18.40		41.90	53.50
	4160	8" diameter	Q-2	38	.632		32	24		56	71
	4180	10" diameter		32	.750		41	28.50		69.50	88
	4200	12" diameter		28	.857		47	32.50		79.50	101
	4980	Polypropylene, acid resistant, fire retardant, schedule 40									
	5000	1-1/2" diameter	Q-1	68	.235	L.F.	1.13	8.65		9.78	14.25
	5100	2" diameter		62	.258		1.53	9.50		11.03	15.95
	5120	3" diameter		51	.314		4.28	11.55		15.83	22
	5140	4" diameter		45	.356		4.55	13.10		17.65	24.50
	5160	6" diameter		32	.500		7.65	18.40		26.05	36
	5980	Proxylene, fire retardant, Schedule 40									
	6000	1-1/2" diameter	Q-1	68	.235	L.F.	3.87	8.65		12.52	17.25
	6100	2" diameter		62	.258		5.30	9.50		14.80	20
	6120	3" diameter		51	.314		9.60	11.55		21.15	28
	6140	4" diameter		45	.356		13.65	13.10		26.75	34.50
	6160	6" diameter		32	.500		24.50	18.40		42.90	54.50
	6820	For Schedule 80, add					35%	2%			
560	0010	**PIPE, CORROSION RESISTANT, FITTINGS**									**560**
	0030	Iron alloy									
	0050	Mechanical joint									
	0060	1/4 Bend, 1-1/2"	Q-1	12	1.333	Ea.	28	49		77	105
	0080	2"		10	1.600		36.50	59		95.50	129
	0090	3"		9	1.778		51	65.50		116.50	155
	0100	4"		8	2		70.50	73.50		144	189
	0110	1/8 Bend, 1-1/2"		12	1.333		28	49		77	105
	0130	2"		10	1.600		36.50	59		95.50	129
	0140	3"		9	1.778		51	65.50		116.50	155
	0150	4"		8	2		70.50	73.50		144	189
	0160	Tee and Y, sanitary, straight									
	0170	1-1/2"	Q-1	8	2	Ea.	37.50	73.50		111	152
	0180	2"		7	2.286		49.50	84		133.50	180

MECHANICAL 15

15 MECHANICAL

		CREW	DAILY OUTPUT	LABOR-HOURS	UNIT	MAT.	LABOR	EQUIP.	TOTAL	TOTAL INCL O&P
15107	**Metal Pipe & Fittings**						2005 BARE COSTS			
560											560
0190	3"	Q-1	6	2.667	Ea.	74	98		172	229
0200	4"		5	3.200		111	118		229	299
0360	Coupling, 1-1/2"		14	1.143		26.50	42		68.50	92
0380	2"		12	1.333		30.50	49		79.50	107
0390	3"		11	1.455		32.50	53.50		86	116
0400	4"		10	1.600		36.50	59		95.50	129
0500	Bell & Spigot									
0510	1/4 and 1/16 bend, 2"	Q-1	16	1	Ea.	39	37		76	98.50
0520	3"		14	1.143		58	42		100	127
0530	4"		13	1.231		75	45.50		120.50	151
0540	6"	Q-2	17	1.412		269	54		323	375
0550	8"	"	12	2		925	76.50		1,001.50	1,150
0620	1/8 bend, 2"	Q-1	16	1		39	37		76	98.50
0640	3"		14	1.143		58.50	42		100.50	128
0650	4"		13	1.231		75	45.50		120.50	151
0660	6"	Q-2	17	1.412		181	54		235	280
0680	8"	"	12	2		535	76.50		611.50	705
0700	Tee, sanitary, 2"	Q-1	10	1.600		51	59		110	145
0710	3"		9	1.778		168	65.50		233.50	284
0720	4"		8	2		187	73.50		260.50	315
0730	6"	Q-2	11	2.182		310	83		393	465
0740	8"	"	8	3		925	114		1,039	1,200
1800	Y, sanitary, 2"	Q-1	10	1.600		50	59		109	144
1820	3"		9	1.778		85	65.50		150.50	192
1830	4"		8	2		139	73.50		212.50	264
1840	6"	Q-2	11	2.182		226	83		309	375
1850	8"	"	8	3		1,375	114		1,489	1,675
3000	Epoxy, filament wound									
3030	Quick-lock joint									
3040	90° Elbow, 2"	Q-1	28	.571	Ea.	58.50	21		79.50	96
3060	3"		16	1		67.50	37		104.50	130
3070	4"		13	1.231		92	45.50		137.50	169
3080	6"		8	2		134	73.50		207.50	259
3090	8"	Q-2	9	2.667		248	102		350	425
3100	10"		7	3.429		310	131		441	535
3110	12"		6	4		445	153		598	720
3120	45° Elbow, 2"	Q-1	28	.571		45.50	21		66.50	81.50
3130	3"		16	1		64.50	37		101.50	126
3140	4"		13	1.231		88	45.50		133.50	165
3150	6"		8	2		134	73.50		207.50	259
3160	8"	Q-2	9	2.667		248	102		350	425
3170	10"		7	3.429		310	131		441	535
3180	12"		6	4		445	153		598	720
3190	Tee, 2"	Q-1	19	.842		140	31		171	200
3200	3"		11	1.455		169	53.50		222.50	267
3210	4"		9	1.778		203	65.50		268.50	320
3220	6"		5	3.200		310	118		428	515
3230	8"	Q-2	6	4		400	153		553	670
3240	10"		5	4.800		520	183		703	845
3250	12"		4	6		755	229		984	1,175
4000	Polypropylene, acid resistant									
4020	Non-pressure, electrofusion joints									
4050	1/4 bend, 1-1/2"	1 Plum	16	.500	Ea.	14	20.50		34.50	46
4060	2"	Q-1	28	.571		18	21		39	51.50
4080	3"		17	.941		29.50	34.50		64	84.50
4090	4"		14	1.143		47	42		89	115

Important: See the Reference Section for critical supporting data - Reference Nos., Crews, & City Cost Indexes

15107	Metal Pipe & Fittings	CREW	DAILY OUTPUT	LABOR-HOURS	UNIT	2005 BARE COSTS				TOTAL INCL O&P
						MAT.	LABOR	EQUIP.	TOTAL	
560 4110	6"	Q-1	8	2	Ea.	115	73.50		188.50	238	**560**
4150	1/4 Bend, long sweep									
4170	1-1/2"	1 Plum	16	.500	Ea.	14.45	20.50		34.95	46.50
4180	2"	Q-1	28	.571		19.75	21		40.75	53
4200	3"		17	.941		34	34.50		68.50	89
4210	4"	↓	14	1.143		48.50	42		90.50	117
4250	1/8 bend, 1-1/2"	1 Plum	16	.500		13.75	20.50		34.25	45.50
4260	2"	Q-1	28	.571		16.75	21		37.75	50
4280	3"		17	.941		30.50	34.50		65	86
4290	4"		14	1.143		35	42		77	102
4310	6"	↓	8	2	↓	96.50	73.50		170	217
4400	Tee, sanitary									
4420	1-1/2"	1 Plum	10	.800	Ea.	17.40	32.50		49.90	68
4430	2"	Q-1	17	.941		21	34.50		55.50	75
4450	3"		11	1.455		41.50	53.50		95	127
4460	4"		9	1.778		62.50	65.50		128	167
4480	6"		5	3.200		106	118		224	293
4490	Tee, sanitary reducing, 2" x 2" x 1-1/2"		17	.941		21	34.50		55.50	75
4492	3" x 3" x 2"		11	1.455		39	53.50		92.50	124
4494	4" x 4" x 3"		9	1.778		58.50	65.50		124	163
4496	6" x 6" x 4"	↓	5	3.200	↓	106	118		224	293
4500	Tee/wye, long turn									
4520	1-1/2"	1 Plum	10	.800	Ea.	23	32.50		55.50	74.50
4530	2"	Q-1	17	.941		31	34.50		65.50	86
4550	3"		11	1.455		49.50	53.50		103	135
4570	4"	↓	9	1.778		70	65.50		135.50	176
4650	Wye 45°, 1-1/2"	1 Plum	10	.800		17.70	32.50		50.20	68.50
4652	2"	Q-1	17	.941		25	34.50		59.50	80
4653	3"		11	1.455		44.50	53.50		98	129
4654	4"		9	1.778		64.50	65.50		130	170
4656	6"	↓	5	3.200	↓	164	118		282	355
620 0010	**PIPE, STEEL**									**620**
0020	All pipe sizes are to Spec. A-53 unless noted otherwise	R15100 -050								
0050	Schedule 40, threaded, with couplings, and clevis type									
0060	hangers sized for covering, 10' O.C.									
0540	Black, 1/4" diameter	1 Plum	66	.121	L.F.	2.09	4.95		7.04	9.75
0550	3/8" diameter		65	.123		2.02	5.05		7.07	9.80
0560	1/2" diameter		63	.127		1.28	5.20		6.48	9.20
0570	3/4" diameter		61	.131		1.54	5.35		6.89	9.75
0580	1" diameter	↓	53	.151		2.16	6.15		8.31	11.65
0590	1-1/4" diameter	Q-1	89	.180		2.75	6.60		9.35	12.95
0600	1-1/2" diameter		80	.200		3.16	7.35		10.51	14.55
0610	2" diameter		64	.250		4.13	9.20		13.33	18.35
0620	2-1/2" diameter		50	.320		7	11.75		18.75	25.50
0630	3" diameter		43	.372		9.05	13.70		22.75	30.50
0640	3-1/2" diameter		40	.400		11.95	14.70		26.65	35
0650	4" diameter		36	.444		14.25	16.35		30.60	40
0660	5" diameter	↓	26	.615		26	22.50		48.50	62.50
0670	6" diameter	Q-2	31	.774		32.50	29.50		62	80.50
0680	8" diameter		27	.889		51	34		85	107
0690	10" diameter		23	1.043		78	40		118	146
0700	12" diameter	↓	18	1.333	↓	86	51		137	171
0809	A-106, gr. A/B, seamless w/cplgs. & hangers									
0811	1/4" diameter	1 Plum	66	.121	L.F.	4.93	4.95		9.88	12.85
0812	3/8" diameter	↓	65	.123	↓	3.66	5.05		8.71	11.55

MECHANICAL 15

For expanded coverage of these items see *Means Mechanical or Plumbing Cost Data 2005*

15107	Metal Pipe & Fittings		CREW	DAILY OUTPUT	LABOR-HOURS	UNIT	2005 BARE COSTS				TOTAL INCL O&P	
							MAT.	LABOR	EQUIP.	TOTAL		
620	0813	1/2" diameter	R15100 -050	1 Plum	63	.127	L.F.	2.75	5.20		7.95	10.85	**620**
	0814	3/4" diameter			61	.131		2.25	5.35		7.60	10.50	
	0815	1" diameter		↓	53	.151		2.95	6.15		9.10	12.50	
	0816	1-1/4" diameter		Q-1	89	.180		3.67	6.60		10.27	14	
	0817	1-1/2" diameter			80	.200		4.88	7.35		12.23	16.40	
	0819	A-53, 2" diameter			64	.250		5.15	9.20		14.35	19.45	
	0821	2-1/2" diameter			50	.320		9.95	11.75		21.70	28.50	
	0822	3" diameter			43	.372		12.05	13.70		25.75	34	
	0823	4" diameter		↓	36	.444	↓	20	16.35		36.35	46.50	
	1220	To delete coupling & hanger, subtract											
	1230	1/4" diam. to 3/4" diam.						31%	56%				
	1240	1" diam. to 1-1/2" diam.						23%	51%				
	1250	2" diam. to 4" diam.						23%	41%				
	1260	5" diam. to 12" diam.						21%	45%				
	1280	All pipe sizes are to Spec. A-53 unless noted otherwise											
	1281	Schedule 40, threaded, with couplings and clevis type											
	1282	hangers sized for covering, 10' O. C.											
	1290	Galvanized, 1/4" diameter		1 Plum	66	.121	L.F.	2.36	4.95		7.31	10.05	
	1300	3/8" diameter			65	.123		2.69	5.05		7.74	10.50	
	1310	1/2" diameter			63	.127		1.52	5.20		6.72	9.45	
	1320	3/4" diameter			61	.131		1.76	5.35		7.11	10	
	1330	1" diameter		↓	53	.151		2.46	6.15		8.61	11.95	
	1340	1-1/4" diameter		Q-1	89	.180		3.20	6.60		9.80	13.45	
	1350	1-1/2" diameter			80	.200		3.71	7.35		11.06	15.15	
	1360	2" diameter			64	.250		4.85	9.20		14.05	19.15	
	1370	2-1/2" diameter			50	.320		7.90	11.75		19.65	26.50	
	1380	3" diameter			43	.372		10.20	13.70		23.90	32	
	1390	3-1/2" diameter			40	.400		16.50	14.70		31.20	40	
	1400	4" diameter			36	.444		16.65	16.35		33	43	
	1410	5" diameter		↓	26	.615		27.50	22.50		50	64.50	
	1420	6" diameter		Q-2	31	.774		38	29.50		67.50	86	
	1430	8" diameter			27	.889		57	34		91	114	
	1440	10" diameter			23	1.043		84	40		124	153	
	1450	12" diameter		↓	18	1.333	↓	115	51		166	204	
	1750	To delete coupling & hanger, subtract											
	1760	1/4" diam. to 3/4" diam.						31%	56%				
	1770	1" diam. to 1-1/2" diam.						23%	51%				
	1780	2" diam. to 4" diam.						23%	41%				
	1790	5" diam. to 12" diam.		↓				21%	45%				
640	0010	**PIPE, STEEL, FITTINGS** Threaded	R15100 -050										**640**
	0020	Cast Iron											
	1300	Extra heavy weight, black											
	1310	Couplings, steel straight											
	1320	1/4"		1 Plum	19	.421	Ea.	1.76	17.20		18.96	28	
	1330	3/8"			19	.421		1.92	17.20		19.12	28	
	1340	1/2"			19	.421		2.59	17.20		19.79	29	
	1350	3/4"			18	.444		2.77	18.15		20.92	30.50	
	1360	1"		↓	15	.533		3.53	22		25.53	37	
	1370	1-1/4"		Q-1	26	.615		5.65	22.50		28.15	40.50	
	1380	1-1/2"			24	.667		5.65	24.50		30.15	43.50	
	1390	2"			21	.762		8.65	28		36.65	51.50	
	1400	2-1/2"			18	.889		12.80	32.50		45.30	63	
	1410	3"			14	1.143		15.20	42		57.20	80	
	1420	3-1/2"			12	1.333		20.50	49		69.50	96	
	1430	4"		↓	10	1.600	↓	24	59		83	115	

			DAILY	LABOR-			2005 BARE COSTS			TOTAL	
	15107 | **Metal Pipe & Fittings**	CREW	OUTPUT	HOURS	UNIT	MAT.	LABOR	EQUIP.	TOTAL	INCL O&P	
640	1440	5"	R15100 -050	Q-1	6	2.667	Ea.	34.50	98		132.50	185	**640**
	1450	6"		Q-2	8	3		42	114		156	219	
	1460	8"			7	3.429		76.50	131		207.50	281	
	1470	10"			6	4		120	153		273	360	
	1480	12"			4	6		161	229		390	520	
	1510	90° Elbow, straight											
	1520	1/2"		1 Plum	15	.533	Ea.	12.60	22		34.60	47	
	1530	3/4"			14	.571		13.35	23.50		36.85	49.50	
	1540	1"			13	.615		16.15	25		41.15	56	
	1550	1-1/4"		Q-1	22	.727		24	27		51	66.50	
	1560	1-1/2"			20	.800		30	29.50		59.50	77	
	1580	2"			18	.889		37	32.50		69.50	89.50	
	1590	2-1/2"			14	1.143		86	42		128	158	
	1600	3"			10	1.600		114	59		173	214	
	1610	4"			6	2.667		238	98		336	410	
	1620	6"		Q-2	7	3.429		620	131		751	875	
	1650	45° Elbow, straight											
	1660	1/2"		1 Plum	15	.533	Ea.	18	22		40	53	
	1670	3/4"			14	.571		17.35	23.50		40.85	54	
	1680	1"			13	.615		21	25		46	61	
	1690	1-1/4"		Q-1	22	.727		34.50	27		61.50	78	
	1700	1-1/2"			20	.800		38	29.50		67.50	85.50	
	1710	2"			18	.889		54	32.50		86.50	108	
	1720	2-1/2"			14	1.143		93	42		135	165	
	1800	Tee, straight											
	1810	1/2"		1 Plum	9	.889	Ea.	19.90	36.50		56.40	76.50	
	1820	3/4"			9	.889		19.90	36.50		56.40	76.50	
	1830	1"			8	1		24	41		65	88	
	1840	1-1/4"		Q-1	14	1.143		36	42		78	103	
	1850	1-1/2"			13	1.231		46	45.50		91.50	119	
	1860	2"			11	1.455		57	53.50		110.50	144	
	1870	2-1/2"			9	1.778		122	65.50		187.50	233	
	1880	3"			6	2.667		166	98		264	330	
	1890	4"			4	4		325	147		472	575	
	1900	6"		Q-2	4	6		715	229		944	1,125	
	4000	Standard weight, black											
	4010	Couplings, steel straight, merchants											
	4030	1/4"		1 Plum	19	.421	Ea.	.47	17.20		17.67	26.50	
	4040	3/8"			19	.421		.57	17.20		17.77	26.50	
	4050	1/2"			19	.421		.61	17.20		17.81	26.50	
	4060	3/4"			18	.444		.77	18.15		18.92	28.50	
	4070	1"			15	.533		1.08	22		23.08	34	
	4080	1-1/4"		Q-1	26	.615		1.38	22.50		23.88	35.50	
	4090	1-1/2"			24	.667		1.75	24.50		26.25	39	
	4100	2"			21	.762		2.50	28		30.50	45	
	4110	2-1/2"			18	.889		7.10	32.50		39.60	57	
	4120	3"			14	1.143		10	42		52	74	
	4130	3-1/2"			12	1.333		17.70	49		66.70	93	
	4140	4"			10	1.600		17.70	59		76.70	108	
	4150	5"			6	2.667		32.50	98		130.50	183	
	4160	6"		Q-2	8	3		39	114		153	215	
	4200	Standard weight, galvanized											
	4210	Couplings, steel straight, merchants											
	4230	1/4"		1 Plum	19	.421	Ea.	.55	17.20		17.75	26.50	
	4240	3/8"			19	.421		.70	17.20		17.90	27	
	4250	1/2"			19	.421		.74	17.20		17.94	27	

MECHANICAL **15**

15107	Metal Pipe & Fittings		CREW	DAILY OUTPUT	LABOR-HOURS	UNIT	2005 BARE COSTS				TOTAL INCL O&P	
							MAT.	LABOR	EQUIP.	TOTAL		
640	4260	3/4"	R15100 -050	1 Plum	18	.444	Ea.	.93	18.15		19.08	28.50	**640**
	4270	1"		↓	15	.533		1.30	22		23.30	34.50	
	4280	1-1/4"		Q-1	26	.615		1.66	22.50		24.16	36	
	4290	1-1/2"			24	.667		2.07	24.50		26.57	39.50	
	4300	2"			21	.762		3.08	28		31.08	45.50	
	4310	2-1/2"			18	.889		8.75	32.50		41.25	58.50	
	4320	3"			14	1.143		11.60	42		53.60	76	
	4330	3-1/2"			12	1.333		20.50	49		69.50	96	
	4340	4"			10	1.600		20.50	59		79.50	111	
	4350	5"		↓	6	2.667		38	98		136	189	
	4360	6"		Q-2	8	3	↓	47	114		161	224	
690	0010	**PIPE, GROOVED-JOINT STEEL FITTINGS & VALVES**											**690**
	0012	Fittings are ductile iron. Steel fittings noted.											
	0020	Pipe includes coupling & clevis type hanger 10' O.C.											
	1000	Schedule 40, black											
	1040	3/4" diameter		1 Plum	71	.113	L.F.	1.98	4.60		6.58	9.10	
	1050	1" diameter			63	.127		1.85	5.20		7.05	9.85	
	1060	1-1/4" diameter			58	.138		2.37	5.65		8.02	11.05	
	1070	1-1/2" diameter			51	.157		2.21	6.40		8.61	12.10	
	1080	2" diameter		↓	40	.200		3.30	8.15		11.45	15.95	
	1090	2-1/2" diameter		Q-1	57	.281		4.80	10.30		15.10	21	
	1100	3" diameter			50	.320		6	11.75		17.75	24.50	
	1110	4" diameter			45	.356		10.95	13.10		24.05	31.50	
	1120	5" diameter		↓	37	.432		19.95	15.90		35.85	46	
	1130	6" diameter		Q-2	42	.571		25.50	22		47.50	61	
	1140	8" diameter			37	.649		38	24.50		62.50	79	
	1150	10" diameter			31	.774		58.50	29.50		88	109	
	1160	12" diameter			27	.889		79.50	34		113.50	139	
	1170	14" diameter			20	1.200		91	46		137	169	
	1180	16" diameter			17	1.412		99.50	54		153.50	191	
	1190	18" diameter			14	1.714		160	65.50		225.50	275	
	1200	20" diameter			12	2	↓	190	76.50		266.50	325	
	1210	24" diameter		↓	10	2.400		224	91.50		315.50	385	
	1740	To delete coupling & hanger, subtract											
	1750	3/4" diam. to 2" diam.						65%	27%				
	1760	2-1/2" diam. to 5" diam.						41%	18%				
	1770	6" diam. to 12" diam.						31%	13%				
	1780	14" diam. to 24" diam.						35%	10%				
	1800	Galvanized											
	1840	3/4" diameter		1 Plum	71	.113	L.F.	2.19	4.60		6.79	9.30	
	1850	1" diameter			63	.127		2.13	5.20		7.33	10.15	
	1860	1-1/4" diameter			58	.138		2.76	5.65		8.41	11.50	
	1870	1-1/2" diameter			51	.157		3.18	6.40		9.58	13.15	
	1880	2" diameter		↓	40	.200		3.91	8.15		12.06	16.60	
	1890	2-1/2" diameter		Q-1	57	.281		5.75	10.30		16.05	22	
	1900	3" diameter			50	.320		7.25	11.75		19	25.50	
	1910	4" diameter			45	.356		10	13.10		23.10	30.50	
	1920	5" diameter		↓	37	.432		26	15.90		41.90	52.50	
	1930	6" diameter		Q-2	42	.571		33.50	22		55.50	70	
	1940	8" diameter			37	.649		49.50	24.50		74	91.50	
	1950	10" diameter			31	.774		75	29.50		104.50	127	
	1960	12" diameter		↓	27	.889	↓	103	34		137	164	
	2540	To delete coupling & hanger, subtract											
	2550	3/4" diam. to 2" diam.						36%	27%				
	2560	2-1/2" diam. to 5" diam.						19%	18%				

15 MECHANICAL

15107	Metal Pipe & Fittings	CREW	DAILY OUTPUT	LABOR-HOURS	UNIT	2005 BARE COSTS				TOTAL INCL O&P
						MAT.	LABOR	EQUIP.	TOTAL	
690 2570	6" diam. to 12" diam.					14%	13%			690
4690	Tee, painted									
4700	3/4" diameter	1 Plum	38	.211	Ea.	20.50	8.60		29.10	35.50
4740	1" diameter		33	.242		15.60	9.90		25.50	32
4750	1-1/4" diameter		27	.296		15.60	12.10		27.70	35.50
4760	1-1/2" diameter		22	.364		15.60	14.85		30.45	39.50
4770	2" diameter		17	.471		15.60	19.20		34.80	46
4780	2-1/2" diameter	Q-1	27	.593		15.60	22		37.60	50
4790	3" diameter		22	.727		22	27		49	64
4800	4" diameter		17	.941		33	34.50		67.50	88.50
4810	5" diameter		13	1.231		76	45.50		121.50	152
4820	6" diameter	Q-2	17	1.412		90	54		144	180
4830	8" diameter		14	1.714		198	65.50		263.50	315
4840	10" diameter		12	2		410	76.50		486.50	570
4850	12" diameter		10	2.400		575	91.50		666.50	770
4851	14" diameter		9	2.667		610	102		712	825
4852	16" diameter		8	3		685	114		799	925
4853	18" diameter	Q-3	11	2.909		860	113		973	1,125
4854	20" diameter		10	3.200		1,225	125		1,350	1,525
4855	24" diameter		8	4		1,875	156		2,031	2,300
4900	For galvanized tees, add					24%				
4939	Couplings									
4940	Flexible, standard, painted									
4950	3/4" diameter	1 Plum	100	.080	Ea.	5.40	3.27		8.67	10.80
4960	1" diameter		100	.080		5.40	3.27		8.67	10.80
4970	1-1/4" diameter		80	.100		7.15	4.09		11.24	14
4980	1-1/2" diameter		67	.119		7.80	4.88		12.68	15.95
4990	2" diameter		50	.160		8.25	6.55		14.80	18.90
5000	2-1/2" diameter	Q-1	80	.200		9.85	7.35		17.20	22
5010	3" diameter		67	.239		10.90	8.80		19.70	25
5020	3-1/2" diameter		57	.281		15.85	10.30		26.15	33
5030	4" diameter		50	.320		12.95	11.75		24.70	32
5040	5" diameter		40	.400		24.50	14.70		39.20	48.50
5050	6" diameter	Q-2	50	.480		28.50	18.30		46.80	59
5070	8" diameter		42	.571		46.50	22		68.50	84.50
5090	10" diameter		35	.686		77.50	26		103.50	125
5110	12" diameter		32	.750		88.50	28.50		117	140
5120	14" diameter		24	1		107	38		145	176
5130	16" diameter		20	1.200		141	46		187	224
5140	18" diameter		18	1.333		165	51		216	258
5150	20" diameter		16	1.500		259	57		316	370
5160	24" diameter		13	1.846		276	70.50		346.50	410
5200	For galvanized couplings, add					33%				
5750	Flange, w/groove gasket, black steel									
5780	2" pipe size	1 Plum	23	.348	Ea.	34	14.20		48.20	59
5790	2-1/2" pipe size	Q-1	37	.432		42.50	15.90		58.40	70.50
5800	3" pipe size		31	.516		45.50	19		64.50	79
5820	4" pipe size		23	.696		61.50	25.50		87	106
5830	5" pipe size		19	.842		71	31		102	125
5840	6" pipe size	Q-2	23	1.043		77.50	40		117.50	146
5850	8" pipe size		17	1.412		87.50	54		141.50	178
5860	10" pipe size		14	1.714		139	65.50		204.50	252
5870	12" pipe size		12	2		182	76.50		258.50	315
5880	14" pipe size		10	2.400		415	91.50		506.50	600
5890	16" pipe size		9	2.667		485	102		587	690
5900	18" pipe size		6	4		595	153		748	885

MECHANICAL 15

15107	Metal Pipe & Fittings	CREW	DAILY OUTPUT	LABOR-HOURS	UNIT	2005 BARE COSTS				TOTAL INCL O&P	
						MAT.	LABOR	EQUIP.	TOTAL		
690											**690**
5910	20" pipe size	Q-2	5	4.800	Ea.	720	183		903	1,075	
5920	24" pipe size	↓	4.50	5.333	↓	920	203		1,123	1,300	
8000	Butterfly valve, 2 position handle, with standard trim										
8010	1-1/2" pipe size	1 Plum	50	.160	Ea.	96	6.55		102.55	116	
8020	2" pipe size	"	38	.211		96	8.60		104.60	119	
8030	3" pipe size	Q-1	50	.320		138	11.75		149.75	170	
8050	4" pipe size	"	38	.421		171	15.50		186.50	212	
8070	6" pipe size	Q-2	38	.632		305	24		329	370	
8080	8" pipe size	↓	27	.889		435	34		469	525	
8090	10" pipe size	↓	20	1.200	↓	625	46		671	760	
8200	With stainless steel trim										
8240	1-1/2" pipe size	1 Plum	50	.160	Ea.	122	6.55		128.55	144	
8250	2" pipe size	"	38	.211		122	8.60		130.60	147	
8270	3" pipe size	Q-1	50	.320		164	11.75		175.75	198	
8280	4" pipe size	"	38	.421		178	15.50		193.50	220	
8300	6" pipe size	Q-2	38	.632		330	24		354	400	
8310	8" pipe size	↓	27	.889		505	34		539	610	
8320	10" pipe size	↓	20	1.200	↓	725	46		771	870	
9000	Cut one groove, labor										
9010	3/4" pipe size	Q-1	152	.105	Ea.		3.87		3.87	5.80	
9020	1" pipe size		140	.114			4.20		4.20	6.30	
9030	1-1/4" pipe size		124	.129			4.75		4.75	7.15	
9040	1-1/2" pipe size		114	.140			5.15		5.15	7.75	
9050	2" pipe size		104	.154			5.65		5.65	8.50	
9060	2-1/2" pipe size		96	.167			6.15		6.15	9.20	
9070	3" pipe size		88	.182			6.70		6.70	10.05	
9080	3-1/2" pipe size		83	.193			7.10		7.10	10.65	
9090	4" pipe size		78	.205			7.55		7.55	11.35	
9100	5" pipe size		72	.222			8.15		8.15	12.30	
9110	6" pipe size		70	.229			8.40		8.40	12.65	
9120	8" pipe size		54	.296			10.90		10.90	16.40	
9130	10" pipe size		38	.421			15.50		15.50	23.50	
9140	12" pipe size		30	.533			19.60		19.60	29.50	
9150	14" pipe size		20	.800			29.50		29.50	44	
9160	16" pipe size		19	.842			31		31	46.50	
9170	18" pipe size		18	.889			32.50		32.50	49	
9180	20" pipe size	↓	17	.941	↓		34.50		34.50	52	
9190	24" pipe size	↓	15	1.067	↓		39		39	59	
9210	Roll one groove										
9220	3/4" pipe size	Q-1	266	.060	Ea.		2.21		2.21	3.33	
9230	1" pipe size		228	.070			2.58		2.58	3.88	
9240	1-1/4" pipe size		200	.080			2.94		2.94	4.42	
9250	1-1/2" pipe size		178	.090			3.31		3.31	4.97	
9260	2" pipe size		116	.138			5.05		5.05	7.60	
9270	2-1/2" pipe size		110	.145			5.35		5.35	8.05	
9280	3" pipe size		100	.160			5.90		5.90	8.85	
9290	3-1/2" pipe size		94	.170			6.25		6.25	9.40	
9300	4" pipe size		86	.186			6.85		6.85	10.30	
9310	5" pipe size		84	.190			7		7	10.55	
9320	6" pipe size		80	.200			7.35		7.35	11.05	
9330	8" pipe size		66	.242			8.90		8.90	13.40	
9340	10" pipe size		58	.276			10.15		10.15	15.25	
9350	12" pipe size		46	.348			12.80		12.80	19.25	
9360	14" pipe size		30	.533			19.60		19.60	29.50	
9370	16" pipe size		28	.571			21		21	31.50	
9380	18" pipe size	↓	27	.593	↓		22		22	33	

MECHANICAL 15

15100 | Building Services Piping

15107 | Metal Pipe & Fittings

			CREW	DAILY OUTPUT	LABOR-HOURS	UNIT	2005 BARE COSTS				TOTAL INCL O&P	
							MAT.	LABOR	EQUIP.	TOTAL		
690	9390	20" pipe size	Q-1	25	.640	Ea.		23.50		23.50	35.50	**690**
	9400	24" pipe size	↓	23	.696	↓		25.50		25.50	38.50	

15108 | Plastic Pipe & Fittings

			CREW	DAILY OUTPUT	LABOR-HOURS	UNIT	MAT.	LABOR	EQUIP.	TOTAL	TOTAL INCL O&P	
520	0010	**PIPE, PLASTIC**										**520**
	0020	Fiberglass reinforced, couplings 10' O.C., hangers 3 per 10'										
	0200	High strength										
	0240	2" diameter	Q-1	58	.276	L.F.	9.20	10.15		19.35	25.50	
	0260	3" diameter		51	.314		13.10	11.55		24.65	32	
	0280	4" diameter		47	.340		15.85	12.50		28.35	36.50	
	0300	6" diameter	↓	38	.421		24.50	15.50		40	50.50	
	0320	8" diameter	Q-2	48	.500		39	19.05		58.05	71	
	0340	10" diameter		40	.600		61.50	23		84.50	103	
	0360	12" diameter	↓	36	.667	↓	76	25.50		101.50	122	
	0550	To delete coupling & hangers, subtract										
	0560	2" diam. to 6" diam.					33%	56%				
	0570	8" diam. to 12" diam.					31%	52%				
	0600	PVC, high impact/pressure, cplgs. 10' O.C., hangers 3 per 10'										
	4100	DWV type, schedule 40, couplings 10' O.C., hangers 3 per 10'										
	4120	ABS										
	4140	1-1/4" diameter	1 Plum	42	.190	L.F.	1.08	7.80		8.88	12.90	
	4150	1-1/2" diameter	"	36	.222		1.09	9.10		10.19	14.85	
	4160	2" diameter	Q-1	59	.271		1.19	9.95		11.14	16.30	
	4170	3" diameter		53	.302		2.08	11.10		13.18	19	
	4180	4" diameter		48	.333		2.78	12.25		15.03	21.50	
	4190	6" diameter	↓	39	.410	↓	6.80	15.10		21.90	30	
	4360	To delete coupling & hangers, subtract										
	4370	1-1/4" diam.					64%	68%				
	4380	1-1/2" diam. to 6" diam.					54%	57%				
	4400	PVC										
	4410	1-1/4" diameter	1 Plum	42	.190	L.F.	1.19	7.80		8.99	13	
	4420	1-1/2" diameter	"	36	.222		1.20	9.10		10.30	14.95	
	4460	2" diameter	Q-1	59	.271		1.35	9.95		11.30	16.50	
	4470	3" diameter		53	.302		2.40	11.10		13.50	19.35	
	4480	4" diameter		48	.333		3.13	12.25		15.38	22	
	4490	6" diameter	↓	39	.410		5.95	15.10		21.05	29	
	4500	8" diameter	Q-2	48	.500	↓	15.10	19.05		34.15	45	
	4750	To delete coupling & hangers, subtract										
	4760	1-1/4" diam. to 1-1/2" diam.					71%	64%				
	4770	2" diam. to 8" diam.					60%	57%				
	5360	CPVC, couplings 10' O.C., hangers 3 per 10'										
	5380	Schedule 40										
	5460	1/2" diameter	1 Plum	54	.148	L.F.	2.31	6.05		8.36	11.65	
	5470	3/4" diameter		51	.157		3.05	6.40		9.45	13	
	5480	1" diameter		46	.174		3.74	7.10		10.84	14.80	
	5490	1-1/4" diameter		42	.190		4.35	7.80		12.15	16.50	
	5500	1-1/2" diameter	↓	36	.222		4.86	9.10		13.96	19	
	5510	2" diameter	Q-1	59	.271		6	9.95		15.95	21.50	
	5520	2-1/2" diameter		56	.286		9.55	10.50		20.05	26.50	
	5530	3" diameter		53	.302		11.75	11.10		22.85	29.50	
	5540	4" diameter		48	.333		19.05	12.25		31.30	39.50	
	5550	6" diameter	↓	43	.372	↓	31	13.70		44.70	55	
	5730	To delete coupling & hangers, subtract										
	5740	1/2" diam. to 3/4" diam.					37%	77%				
	5750	1" diam. to 1-1/4" diam.					27%	70%				
	5760	1-1/2" diam. to 3" diam.					21%	57%				

MECHANICAL 15

15108	Plastic Pipe & Fittings	CREW	DAILY OUTPUT	LABOR-HOURS	UNIT	2005 BARE COSTS				TOTAL INCL O&P
						MAT.	LABOR	EQUIP.	TOTAL	
520	5770	4″ diam. to 6″ diam.					16%	57%				**520**
	7280	Polyethylene, flexible, no couplings or hangers									
	7282	Note: For labor costs add 25% to the couplings and fittings labor total.									
	7300	SDR 15, 100 psi									
	7310	3/4″ diameter				L.F.	.24			.24	.26
	7350	1″ diameter					.32			.32	.35
	7360	1-1/4″ diameter					.57			.57	.63
	7370	1-1/2″ diameter					.69			.69	.76
	7380	2″ diameter				↓	1.13			1.13	1.24
	8120	SDR 9, 200 psi									
	8150	3/4″ diameter				L.F.	.28			.28	.31
	8160	1″ diameter					.46			.46	.51
	8170	1-1/4″ diameter				↓	.69			.69	.76
560	0010	**PIPE, PLASTIC, FITTINGS**										**560**
	0030	Epoxy resin, fiberglass reinforced, general service									
	0090	Elbow, 90°, 2″	Q-1	33.10	.483	Ea.	45.50	17.80		63.30	76.50
	0100	3″		20.80	.769		64.50	28.50		93	113
	0110	4″		16.50	.970		88	35.50		123.50	151
	0120	6″	↓	10.10	1.584		134	58.50		192.50	236
	0130	8″	Q-2	9.30	2.581		248	98.50		346.50	420
	0140	10″		8.50	2.824		310	108		418	500
	0150	12″	↓	7.60	3.158	↓	445	120		565	670
	0160	45° Elbow, same as 90°									
	0290	Tee, 2″	Q-1	20	.800	Ea.	55	29.50		84.50	105
	0300	3″		13.90	1.151		64.50	42.50		107	134
	0310	4″		11	1.455		76.50	53.50		130	165
	0320	6″	↓	6.70	2.388		205	88		293	360
	0330	8″	Q-2	6.20	3.871		237	148		385	480
	0340	10″		5.70	4.211		380	161		541	655
	0350	12″	↓	5.10	4.706	↓	460	179		639	775
	0380	Couplings									
	0410	2″	Q-1	33.10	.483	Ea.	9.85	17.80		27.65	37.50
	0420	3″		20.80	.769		16.95	28.50		45.45	61
	0430	4″		16.50	.970		22	35.50		57.50	78
	0440	6″	↓	10.10	1.584		42	58.50		100.50	134
	0450	8″	Q-2	9.30	2.581		64	98.50		162.50	219
	0460	10″		8.50	2.824		96	108		204	268
	0470	12″	↓	7.60	3.158	↓	129	120		249	325
	0473	High corrosion resistant couplings, add					30%				
	2100	PVC schedule 80, socket joint									
	2110	90° elbow, 1/2″	1 Plum	30.30	.264	Ea.	.59	10.80		11.39	16.85
	2130	3/4″		26	.308		.76	12.55		13.31	19.75
	2140	1″		22.70	.352		1.72	14.40		16.12	23.50
	2150	1-1/4″		20.20	.396		1.63	16.20		17.83	26.50
	2160	1-1/2″	↓	18.20	.440		1.74	17.95		19.69	29
	2170	2″	Q-1	33.10	.483		2.11	17.80		19.91	29
	2180	3″		20.80	.769		5.55	28.50		34.05	48.50
	2190	4″		16.50	.970		6.30	35.50		41.80	60.50
	2200	6″	↓	10.10	1.584		63.50	58.50		122	158
	2210	8″	Q-2	9.30	2.581		175	98.50		273.50	340
	2250	45° elbow, 1/2″	1 Plum	30.30	.264		1.12	10.80		11.92	17.45
	2270	3/4″		26	.308		1.70	12.55		14.25	21
	2280	1″		22.70	.352		2.55	14.40		16.95	24.50
	2290	1-1/4″		20.20	.396		3.25	16.20		19.45	28
	2300	1-1/2″	↓	18.20	.440	↓	3.84	17.95		21.79	31

15108 | Plastic Pipe & Fittings

		CREW	DAILY OUTPUT	LABOR-HOURS	UNIT	2005 BARE COSTS MAT.	LABOR	EQUIP.	TOTAL	TOTAL INCL O&P
560	2310	2"	Q-1	33.10	.483	Ea.	4.97	17.80		22.77	32
	2320	3"		20.80	.769		12.70	28.50		41.20	56.50
	2330	4"		16.50	.970		23	35.50		58.50	78.50
	2340	6"	▼	10.10	1.584		29	58.50		87.50	119
	2350	8"	Q-2	9.30	2.581		166	98.50		264.50	330
	2400	Tee, 1/2"	1 Plum	20.20	.396		1.67	16.20		17.87	26.50
	2420	3/4"		17.30	.462		1.75	18.90		20.65	30.50
	2430	1"		15.20	.526		2.18	21.50		23.68	35
	2440	1-1/4"		13.50	.593		6	24		30	43
	2450	1-1/2"	▼	12.10	.661		6	27		33	47
	2460	2"	Q-1	20	.800		7.50	29.50		37	52.50
	2470	3"		13.90	1.151		8.15	42.50		50.65	72.50
	2480	4"		11	1.455		11.80	53.50		65.30	93.50
	2490	6"	▼	6.70	2.388		40.50	88		128.50	177
	2500	8"	Q-2	6.20	3.871		237	148		385	480
	2510	Flange, socket, 150 lb., 1/2"	1 Plum	55.60	.144		3.23	5.90		9.13	12.40
	2514	3/4"		47.60	.168		3.46	6.85		10.31	14.10
	2518	1"		41.70	.192		3.85	7.85		11.70	16.05
	2522	1-1/2"	▼	33.30	.240		4.05	9.80		13.85	19.20
	2526	2"	Q-1	60.60	.264		5.40	9.70		15.10	20.50
	2530	4"		30.30	.528		11.65	19.40		31.05	42
	2534	6"	▼	18.50	.865		18.30	32		50.30	68
	2538	8"	Q-2	17.10	1.404		87	53.50		140.50	176
	2550	Coupling, 1/2"	1 Plum	30.30	.264		1.07	10.80		11.87	17.40
	2570	3/4"		26	.308		1.45	12.55		14	20.50
	2580	1"		22.70	.352		1.49	14.40		15.89	23
	2590	1-1/4"		20.20	.396		2.27	16.20		18.47	27
	2600	1-1/2"	▼	18.20	.440		2.44	17.95		20.39	29.50
	2610	2"	Q-1	33.10	.483		2.62	17.80		20.42	29.50
	2620	3"		20.80	.769		7.40	28.50		35.90	50.50
	2630	4"		16.50	.970		9.25	35.50		44.75	63.50
	2640	6"	▼	10.10	1.584		19.95	58.50		78.45	110
	2650	8"	Q-2	9.30	2.581		37.50	98.50		136	189
	2660	10"		8.50	2.824		40.50	108		148.50	207
	2670	12"	▼	7.60	3.158	▼	47	120		167	233
	4500	DWV, ABS, non pressure, socket joints									
	4540	1/4 Bend, 1-1/4"	1 Plum	20.20	.396	Ea.	2.57	16.20		18.77	27.50
	4560	1-1/2"	"	18.20	.440		1.97	17.95		19.92	29
	4570	2"	Q-1	33.10	.483		3.04	17.80		20.84	30
	4580	3"		20.80	.769		7.30	28.50		35.80	50.50
	4590	4"		16.50	.970		10.95	35.50		46.45	65.50
	4600	6"	▼	10.10	1.584	▼	58.50	58.50		117	152
	4650	1/8 Bend, same as 1/4 Bend									
	4800	Tee, sanitary									
	4820	1-1/4"	1 Plum	13.50	.593	Ea.	2.91	24		26.91	39.50
	4830	1-1/2"	"	12.10	.661		2.68	27		29.68	43.50
	4840	2"	Q-1	20	.800		3.92	29.50		33.42	48.50
	4850	3"		13.90	1.151		8.80	42.50		51.30	73
	4860	4"		11	1.455		16.10	53.50		69.60	98
	4862	Tee, sanitary, reducing, 2" x 1-1/2"		22	.727		3.78	27		30.78	44
	4864	3" x 2"		15.30	1.046		6.80	38.50		45.30	65.50
	4868	4" x 3"	▼	12.10	1.322		18	48.50		66.50	93
	4870	Combination Y and 1/8 bend									
	4872	1-1/2"	1 Plum	12.10	.661	Ea.	5.45	27		32.45	46.50
	4874	2"	Q-1	20	.800		6.60	29.50		36.10	51.50
	4876	3"	▼	13.90	1.151		14.65	42.50		57.15	79.50

MECHANICAL 15

15108	Plastic Pipe & Fittings	CREW	DAILY OUTPUT	LABOR-HOURS	UNIT	2005 BARE COSTS				TOTAL INCL O&P	
						MAT.	LABOR	EQUIP.	TOTAL		
560	4878	4"	Q-1	11	1.455	Ea.	27.50	53.50		81	111	560
	4880	3" x 1-1/2"		15.50	1.032		13.60	38		51.60	72	
	4882	4" x 3"	▼	12.10	1.322		20.50	48.50		69	95.50	
	4900	Wye, 1-1/4"	1 Plum	13.50	.593		3.36	24		27.36	40	
	4902	1-1/2"	"	12.10	.661		3.43	27		30.43	44.50	
	4904	2"	Q-1	20	.800		4.82	29.50		34.32	49.50	
	4906	3"		13.90	1.151		11.95	42.50		54.45	76.50	
	4908	4"		11	1.455		21	53.50		74.50	104	
	4910	6"		6.70	2.388		72.50	88		160.50	212	
	4918	3" x 1-1/2"		15.50	1.032		8	38		46	66	
	4920	4" x 3"		12.10	1.322		15.30	48.50		63.80	90	
	4922	6" x 4"	▼	6.90	2.319		60	85.50		145.50	194	
	4930	Double Wye, 1-1/2"	1 Plum	9.10	.879		7.95	36		43.95	63	
	4932	2"	Q-1	16.60	.964		9.35	35.50		44.85	64	
	4934	3"		10.40	1.538		24	56.50		80.50	112	
	4936	4"		8.25	1.939		49	71.50		120.50	161	
	4940	2" x 1-1/2"		16.80	.952		9.35	35		44.35	63	
	4942	3" x 2"		10.60	1.509		17.95	55.50		73.45	103	
	4944	4" x 3"		8.45	1.893		38.50	69.50		108	148	
	4946	6" x 4"		7.25	2.207		81	81		162	211	
	4950	Reducer bushing, 2" x 1-1/2"		36.40	.440		1.29	16.15		17.44	26	
	4952	3" x 1-1/2"		27.30	.586		4.33	21.50		25.83	37.50	
	4954	4" x 2"		18.20	.879		9.95	32.50		42.45	59.50	
	4956	6" x 4"	▼	11.10	1.441		26.50	53		79.50	109	
	4960	Couplings, 1-1/2"	1 Plum	18.20	.440		.82	17.95		18.77	28	
	4962	2"	Q-1	33.10	.483		1.22	17.80		19.02	28	
	4963	3"		20.80	.769		3.13	28.50		31.63	46	
	4964	4"		16.50	.970		4.90	35.50		40.40	59	
	4966	6"		10.10	1.584		22	58.50		80.50	112	
	4970	2" x 1-1/2"		33.30	.480		2.12	17.65		19.77	29	
	4972	3" x 1-1/2"		21	.762		5.95	28		33.95	48.50	
	4974	4" x 3"	▼	16.70	.958		10.55	35		45.55	64.50	
	4978	Closet flange, 4"	1 Plum	32	.250		5.20	10.20		15.40	21	
	4980	4" x 3"	"	34	.235	▼	5.20	9.60		14.80	20	
	5000	DWV, PVC, schedule 40, socket joints										
	5040	1/4 bend, 1-1/4"	1 Plum	20.20	.396	Ea.	3.25	16.20		19.45	28	
	5060	1-1/2"	"	18.20	.440		1.24	17.95		19.19	28.50	
	5070	2"	Q-1	33.10	.483		1.93	17.80		19.73	28.50	
	5080	3"		20.80	.769		5.55	28.50		34.05	48.50	
	5090	4"		16.50	.970		9.05	35.50		44.55	63.50	
	5100	6"	▼	10.10	1.584		40.50	58.50		99	133	
	5105	8"	Q-2	9.30	2.581		82.50	98.50		181	239	
	5110	1/4 bend, long sweep, 1-1/2"	1 Plum	18.20	.440		3.05	17.95		21	30.50	
	5112	2"	Q-1	33.10	.483		2.93	17.80		20.73	29.50	
	5114	3"		20.80	.769		6.85	28.50		35.35	50	
	5116	4"	▼	16.50	.970		12.80	35.50		48.30	67.50	
	5150	1/8 bend, 1-1/4"	1 Plum	20.20	.396		1.95	16.20		18.15	26.50	
	5170	1-1/2"	"	18.20	.440		1.21	17.95		19.16	28.50	
	5180	2"	Q-1	33.10	.483		1.76	17.80		19.56	28.50	
	5190	3"		20.80	.769		4.81	28.50		33.31	48	
	5200	4"		16.50	.970		7.85	35.50		43.35	62	
	5210	6"	▼	10.10	1.584		37.50	58.50		96	129	
	5215	8"	Q-2	9.30	2.581		66.50	98.50		165	221	
	5250	Tee, sanitary 1-1/4"	1 Plum	13.50	.593		3.36	24		27.36	40	
	5254	1-1/2"	"	12.10	.661	▼	2.13	27		29.13	43	
	5255	2"	Q-1	20	.800		3.18	29.50		32.68	47.50	

Important: See the Reference Section for critical supporting data - Reference Nos., Crews, & City Cost Indexes

			DAILY	LABOR-		2005 BARE COSTS				TOTAL	
15108	**Plastic Pipe & Fittings**	CREW	OUTPUT	HOURS	UNIT	MAT.	LABOR	EQUIP.	TOTAL	INCL O&P	
560	5256	3"	Q-1	13.90	1.151	Ea.	7.05	42.50		49.55	71.50	560
	5257	4"		11	1.455		12.30	53.50		65.80	94	
	5259	6"	▼	6.70	2.388		60.50	88		148.50	199	
	5261	8"	Q-2	6.20	3.871		181	148		329	420	
	5264	2" x 1-1/2"	Q-1	22	.727		2.94	27		29.94	43	
	5266	3" x 1-1/2"		15.50	1.032		4.47	38		42.47	62	
	5268	4" x 3"		12.10	1.322		19.20	48.50		67.70	94	
	5271	6" x 4"	▼	6.90	2.319		65	85.50		150.50	200	
	5314	Combination Y & 1/8 bend, 1-1/2"	1 Plum	12.10	.661		4.59	27		31.59	45.50	
	5315	2"	Q-1	20	.800		5.95	29.50		35.45	50.50	
	5317	3"		13.90	1.151		11.70	42.50		54.20	76.50	
	5318	4"	▼	11	1.455	▼	21	53.50		74.50	104	
	5324	Combination Y & 1/8 bend, reducing										
	5325	2" x 2" x 1-1/2"	Q-1	22	.727	Ea.	6.55	27		33.55	47.50	
	5327	3" x 3" x 1-1/2"		15.50	1.032		11.40	38		49.40	69.50	
	5328	3" x 3" x 2"		15.30	1.046		8.45	38.50		46.95	67.50	
	5329	4" x 4" x 2"	▼	12.20	1.311		15.85	48		63.85	90	
	5331	Wye, 1-1/4"	1 Plum	13.50	.593		4.29	24		28.29	41	
	5332	1-1/2"	"	12.10	.661		2.86	27		29.86	43.50	
	5333	2"	Q-1	20	.800		3.65	29.50		33.15	48	
	5334	3"		13.90	1.151		9.90	42.50		52.40	74.50	
	5335	4"		11	1.455		16	53.50		69.50	98	
	5336	6"	▼	6.70	2.388		62	88		150	200	
	5337	8"	Q-2	6.20	3.871		73.50	148		221.50	305	
	5341	2" x 1-1/2"	Q-1	22	.727		5.05	27		32.05	45.50	
	5342	3" x 1-1/2"		15.50	1.032		6.70	38		44.70	64.50	
	5343	4" x 3"		12.10	1.322		12.75	48.50		61.25	87	
	5344	6" x 4"	▼	6.90	2.319		44.50	85.50		130	177	
	5345	8" x 6"	Q-2	6.40	3.750		124	143		267	350	
	5347	Double wye, 1-1/2"	1 Plum	9.10	.879		6.10	36		42.10	60.50	
	5348	2"	Q-1	16.60	.964		7.85	35.50		43.35	62	
	5349	3"		10.40	1.538		20	56.50		76.50	107	
	5350	4"		8.25	1.939		41	71.50		112.50	152	
	5354	2" x 1-1/2"		16.80	.952		7.15	35		42.15	60.50	
	5355	3" x 2"		10.60	1.509		15.05	55.50		70.55	100	
	5356	4" x 3"		8.45	1.893		32.50	69.50		102	141	
	5357	6" x 4"	▼	7.25	2.207		67.50	81		148.50	197	
	5374	Coupling, 1-1/4"	1 Plum	20.20	.396		1.80	16.20		18	26.50	
	5376	1-1/2"	"	18.20	.440		.63	17.95		18.58	27.50	
	5378	2"	Q-1	33.10	.483		.78	17.80		18.58	27.50	
	5380	3"		20.80	.769		2.71	28.50		31.21	45.50	
	5390	4"		16.50	.970		4.45	35.50		39.95	58.50	
	5400	6"	▼	10.10	1.584		16.30	58.50		74.80	105	
	5402	8"	Q-2	9.30	2.581		36.50	98.50		135	189	
	5404	2" x 1-1/2"	Q-1	33.30	.480		1.60	17.65		19.25	28.50	
	5406	3" x 1-1/2"		21	.762		5.10	28		33.10	47.50	
	5408	4" x 3"		16.70	.958		9	35		44	63	
	5410	Reducer bushing, 2" x 1-1/4"		36.50	.438		.97	16.10		17.07	25	
	5412	3" x 1-1/2"		27.30	.586		4.75	21.50		26.25	38	
	5414	4" x 2"		18.20	.879		9.25	32.50		41.75	58.50	
	5416	6" x 4"	▼	11.10	1.441		26.50	53		79.50	109	
	5418	8" x 6"	Q-2	10.20	2.353		53	89.50		142.50	194	
	5425	Closet flange 4"	Q-1	32	.500		5.40	18.40		23.80	33.50	
	5426	4" x 3"	"	34	.471	▼	8.25	17.30		25.55	35	
	5450	Solvent cement for PVC, industrial grade, per quart				Qt.	11.65			11.65	12.80	
	7340	PVC flange, slip-on, Sch 80 std., 1/2"	1 Plum	22	.364	Ea.	8.20	14.85		23.05	31.50	

MECHANICAL **T5**

			DAILY	LABOR-				2005 BARE COSTS			TOTAL
15108	**Plastic Pipe & Fittings**	CREW	OUTPUT	HOURS	UNIT	MAT.	LABOR	EQUIP.	TOTAL	**INCL O&P**	

560	7350	3/4"	1 Plum	21	.381	Ea.	8.75	15.55		24.30	33	**560**
	7360	1"		18	.444		9.75	18.15		27.90	38	
	7370	1-1/4"		17	.471		10.05	19.20		29.25	40	
	7380	1-1/2"		16	.500		10.25	20.50		30.75	42	
	7390	2"	Q-1	26	.615		13.65	22.50		36.15	49	
	7400	2-1/2"		24	.667		21	24.50		45.50	60	
	7410	3"		18	.889		23.50	32.50		56	74.50	
	7420	4"		15	1.067		29.50	39		68.50	91.50	
	7430	6"		10	1.600		48	59		107	142	
	7440	8"	Q-2	11	2.182		80.50	83		163.50	214	
	7550	Union, schedule 40, socket joints, 1/2"	1 Plum	19	.421		2.85	17.20		20.05	29	
	7560	3/4"		18	.444		3.25	18.15		21.40	31	
	7570	1"		15	.533		3.34	22		25.34	36.50	
	7580	1-1/4"		14	.571		10.75	23.50		34.25	47	
	7590	1-1/2"		13	.615		11.20	25		36.20	50.50	
	7600	2"	Q-1	20	.800		15.10	29.50		44.60	60.50	

15110	**Valves**										

160	0010	**VALVES, BRONZE**										**160**
	1020	Angle, 150 lb., rising stem, threaded	R15100 -050									
	1030	1/8"	1 Plum	24	.333	Ea.	49.50	13.60		63.10	75	
	1040	1/4"		24	.333		49.50	13.60		63.10	75	
	1050	3/8"		24	.333		49.50	13.60		63.10	75	
	1060	1/2"		22	.364		49.50	14.85		64.35	77	
	1070	3/4"		20	.400		67	16.35		83.35	98	
	1080	1"		19	.421		95.50	17.20		112.70	131	
	1100	1-1/2"		13	.615		161	25		186	215	
	1110	2"		11	.727		260	29.50		289.50	330	
	1510	2-1/2"		9	.889		156	36.50		192.50	227	
	1520	3"		8	1		255	41		296	345	
	1750	Check, swing, class 150, regrinding disc, threaded										
	1800	1/8"	1 Plum	24	.333	Ea.	25	13.60		38.60	48	
	1830	1/4"		24	.333		25	13.60		38.60	48	
	1840	3/8"		24	.333		26	13.60		39.60	49	
	1850	1/2"		24	.333		26	13.60		39.60	49	
	1860	3/4"		20	.400		35	16.35		51.35	63	
	1870	1"		19	.421		52	17.20		69.20	83	
	1880	1-1/4"		15	.533		73	22		95	113	
	1890	1-1/2"		13	.615		85.50	25		110.50	132	
	1900	2"		11	.727		125	29.50		154.50	183	
	1910	2-1/2"	Q-1	15	1.067		265	39		304	350	
	2000	For 200 lb, add					5%	10%				
	2040	For 300 lb, add					15%	15%				
	2850	Gate, N.R.S., soldered, 125 psi										
	2900	3/8"	1 Plum	24	.333	Ea.	18.30	13.60		31.90	40.50	
	2920	1/2"		24	.333		16	13.60		29.60	38	
	2940	3/4"		20	.400		25	16.35		41.35	52	
	2950	1"		19	.421		34.50	17.20		51.70	64	
	2960	1-1/4"		15	.533		46.50	22		68.50	84	
	2970	1-1/2"		13	.615		52.50	25		77.50	95.50	
	2980	2"		11	.727		83.50	29.50		113	137	
	2990	2-1/2"	Q-1	15	1.067		195	39		234	274	
	3000	3"	"	13	1.231		256	45.50		301.50	350	
	3850	Rising stem, soldered, 300 psi										
	3950	1"	1 Plum	19	.421	Ea.	59.50	17.20		76.70	91	
	3980	2"	"	11	.727		161	29.50		190.50	222	

Important: See the Reference Section for critical supporting data - Reference Nos., Crews, & City Cost Indexes

			CREW	DAILY OUTPUT	LABOR-HOURS	UNIT	2005 BARE COSTS				TOTAL INCL O&P
							MAT.	LABOR	EQUIP.	TOTAL	
160	4000	3"	Q-1	13	1.231	Ea.	455	45.50		500.50	570
	4250	Threaded, class 150	R15100 -050								
	4310	1/4"	1 Plum	24	.333	Ea.	27	13.60		40.60	50.50
	4320	3/8"		24	.333		27	13.60		40.60	50.50
	4330	1/2"		24	.333		25.50	13.60		39.10	48.50
	4340	3/4"		20	.400		30	16.35		46.35	57.50
	4350	1"		19	.421		40.50	17.20		57.70	70.50
	4360	1-1/4"		15	.533		53.50	22		75.50	91.50
	4370	1-1/2"		13	.615		67.50	25		92.50	113
	4380	2"		11	.727		91.50	29.50		121	146
	4390	2-1/2"	Q-1	15	1.067		214	39		253	294
	4400	3"	"	13	1.231		299	45.50		344.50	400
	4500	For 300 psi, threaded, add					100%	15%			
	4540	For chain operated type, add					15%				
	4850	Globe, class 150, rising stem, threaded									
	4920	1/4"	1 Plum	24	.333	Ea.	35.50	13.60		49.10	60
	4940	3/8" size		24	.333		39	13.60		52.60	63.50
	4950	1/2"		24	.333		39	13.60		52.60	63.50
	4960	3/4"		20	.400		53.50	16.35		69.85	83
	4970	1"		19	.421		83.50	17.20		100.70	118
	4980	1-1/4"		15	.533		131	22		153	177
	4990	1-1/2"		13	.615		160	25		185	214
	5000	2"		11	.727		239	29.50		268.50	310
	5010	2-1/2"	Q-1	15	1.067		480	39		519	590
	5020	3"	"	13	1.231		605	45.50		650.50	735
	5120	For 300 lb threaded, add					50%	15%			
	5600	Relief, pressure & temperature, self-closing, ASME, threaded									
	5640	3/4"	1 Plum	28	.286	Ea.	81.50	11.65		93.15	107
	5650	1"		24	.333		118	13.60		131.60	151
	5660	1-1/4"		20	.400		237	16.35		253.35	286
	5670	1-1/2"		18	.444		455	18.15		473.15	530
	5680	2"		16	.500		495	20.50		515.50	575
	5950	Pressure, poppet type, threaded									
	6000	1/2"	1 Plum	30	.267	Ea.	13.55	10.90		24.45	31.50
	6040	3/4"	"	28	.286	"	45	11.65		56.65	67
	6400	Pressure, water, ASME, threaded									
	6440	3/4"	1 Plum	28	.286	Ea.	47	11.65		58.65	69.50
	6450	1"		24	.333		93.50	13.60		107.10	124
	6460	1-1/4"		20	.400		153	16.35		169.35	193
	6470	1-1/2"		18	.444		211	18.15		229.15	260
	6480	2"		16	.500		305	20.50		325.50	365
	6490	2-1/2"		15	.533		1,550	22		1,572	1,725
	6900	Reducing, water pressure									
	6920	300 psi to 25-75 psi, threaded or sweat									
	6940	1/2"	1 Plum	24	.333	Ea.	147	13.60		160.60	182
	6950	3/4"		20	.400		147	16.35		163.35	186
	6960	1"		19	.421		227	17.20		244.20	276
	6970	1-1/4"		15	.533		405	22		427	480
	6980	1-1/2"		13	.615		615	25		640	715
	8350	Tempering, water, sweat connections									
	8400	1/2"	1 Plum	24	.333	Ea.	52.50	13.60		66.10	78
	8440	3/4"	"	20	.400	"	64	16.35		80.35	95
	8650	Threaded connections									
	8700	1/2"	1 Plum	24	.333	Ea.	79	13.60		92.60	107
	8740	3/4"		20	.400		244	16.35		260.35	294
	8750	1"		19	.421		269	17.20		286.20	320

MECHANICAL 15

			DAILY	LABOR-		2005 BARE COSTS				TOTAL	
15110	**Valves**	CREW	OUTPUT	HOURS	UNIT	MAT.	LABOR	EQUIP.	TOTAL	INCL O&P	
160 8760	1-1/4"	1 Plum	15	.533	Ea.	430	22		452	505	**160**
8770	1-1/2" R15100 -050		13	.615		465	25		490	555	
8780	2"		11	.727		700	29.50		729.50	815	
200 0010	**VALVES, IRON BODY** R15100 -050										**200**
1020	Butterfly, wafer type, gear actuator, 200 lb.										
1030	2"	1 Plum	14	.571	Ea.	113	23.50		136.50	159	
1040	2-1/2"	Q-1	9	1.778		116	65.50		181.50	227	
1050	3"		8	2		121	73.50		194.50	244	
1060	4"		5	3.200		151	118		269	345	
1070	5"	Q-2	5	4.800		182	183		365	475	
1080	6"	"	5	4.800		205	183		388	500	
1650	Gate, 125 lb., N.R.S.										
2150	Flanged										
2200	2"	1 Plum	5	1.600	Ea.	320	65.50		385.50	450	
2240	2-1/2"	Q-1	5	3.200		330	118		448	535	
2260	3"		4.50	3.556		365	131		496	600	
2280	4"		3	5.333		525	196		721	875	
2300	6"	Q-2	3	8		895	305		1,200	1,450	
3550	OS&Y, 125 lb., flanged										
3600	2"	1 Plum	5	1.600	Ea.	226	65.50		291.50	345	
3660	3"	Q-1	4.50	3.556		263	131		394	485	
3680	4"	"	3	5.333		375	196		571	705	
3700	6"	Q-2	3	8		615	305		920	1,150	
3900	For 175 lb, flanged, add					200%	10%				
4350	Globe, OS&Y										
5450	Swing check, 125 lb., threaded										
5470	1"	1 Plum	13	.615	Ea.	310	25		335	380	
5500	2"	"	11	.727		400	29.50		429.50	485	
5540	2-1/2"	Q-1	15	1.067		455	39		494	560	
5550	3"		13	1.231		510	45.50		555.50	630	
5560	4"		10	1.600		815	59		874	985	
5950	Flanged										
6000	2"	1 Plum	5	1.600	Ea.	156	65.50		221.50	269	
6040	2-1/2"	Q-1	5	3.200		189	118		307	385	
6050	3"		4.50	3.556		202	131		333	420	
6060	4"		3	5.333		320	196		516	645	
6070	6"	Q-2	3	8		545	305		850	1,050	
500 0010	**VALVES, PLASTIC**										**500**
1150	Ball, PVC, socket or threaded, single union										
1230	1/2"	1 Plum	26	.308	Ea.	24	12.55		36.55	45.50	
1240	3/4"		25	.320		28.50	13.05		41.55	51	
1250	1"		23	.348		34.50	14.20		48.70	59.50	
1260	1-1/4"		21	.381		57	15.55		72.55	86	
1270	1-1/2"		20	.400		57	16.35		73.35	87	
1280	2"		17	.471		82	19.20		101.20	119	
1290	2-1/2"	Q-1	26	.615		204	22.50		226.50	258	
1300	3"		24	.667		204	24.50		228.50	261	
1310	4"		20	.800		278	29.50		307.50	350	
1360	For PVC, flanged, add					100%	15%				
3150	Ball check, PVC, socket or threaded										
3200	1/4"	1 Plum	26	.308	Ea.	30.50	12.55		43.05	52.50	
3220	3/8"		26	.308		30.50	12.55		43.05	52.50	
3240	1/2"		26	.308		30.50	12.55		43.05	52.50	
3250	3/4"		25	.320		34	13.05		47.05	57	
3260	1"		23	.348		43	14.20		57.20	68.50	

Important: See the Reference Section for critical supporting data - Reference Nos., Crews, & City Cost Indexes

15 MECHANICAL

15110 | Valves

		CREW	DAILY OUTPUT	LABOR-HOURS	UNIT	MAT.	LABOR	EQUIP.	TOTAL	TOTAL INCL O&P	
500	3270	1-1/4"	1 Plum	21	.381	Ea.	71.50	15.55		87.05	103	**500**
	3280	1-1/2"		20	.400		71.50	16.35		87.85	104	
	3290	2"		17	.471		170	19.20		189.20	216	
	3310	3"	Q-1	24	.667		271	24.50		295.50	335	
	3320	4"	"	20	.800		385	29.50		414.50	465	
	3360	For PVC, flanged, add					50%	15%				
700	0010	**VALVES, STEEL** R15100-050										**700**
	0800	Cast										
	1350	Check valve, swing type, 150 lb., flanged										
	1370	1"	1 Plum	10	.800	Ea.	365	32.50		397.50	455	
	1400	2"	"	8	1		515	41		556	625	
	1440	2-1/2"	Q-1	5	3.200		505	118		623	730	
	1450	3"		4.50	3.556		595	131		726	850	
	1460	4"		3	5.333		875	196		1,071	1,250	
	1540	For 300 lb., flanged, add					50%	15%				
	1548	For 600 lb., flanged, add					110%	20%				
	1950	Gate valve, 150 lb., flanged										
	2000	2"	1 Plum	8	1	Ea.	540	41		581	655	
	2040	2-1/2"	Q-1	5	3.200		765	118		883	1,025	
	2050	3"		4.50	3.556		765	131		896	1,025	
	2060	4"		3	5.333		930	196		1,126	1,325	
	2070	6"	Q-2	3	8		1,475	305		1,780	2,075	
	3650	Globe valve, 150 lb., flanged										
	3700	2"	1 Plum	8	1	Ea.	710	41		751	840	
	3740	2-1/2"	Q-1	5	3.200		910	118		1,028	1,175	
	3750	3"		4.50	3.556		910	131		1,041	1,200	
	3760	4"		3	5.333		1,325	196		1,521	1,750	
	3770	6"	Q-2	3	8		2,100	305		2,405	2,750	
	5150	Forged										
	5650	Check valve, class 800, horizontal, socket										
	5698	Threaded										
	5700	1/4"	1 Plum	24	.333	Ea.	61	13.60		74.60	88	
	5720	3/8"		24	.333		61	13.60		74.60	88	
	5730	1/2"		24	.333		61	13.60		74.60	88	
	5740	3/4"		20	.400		67	16.35		83.35	98	
	5750	1"		19	.421		78	17.20		95.20	112	
	5760	1-1/4"		15	.533		153	22		175	202	

15120 | Piping Specialties

		CREW	DAILY OUTPUT	LABOR-HOURS	UNIT	MAT.	LABOR	EQUIP.	TOTAL	TOTAL INCL O&P	
350	0010	**FLEXIBLE CONNECTORS**, Corrugated, 7/8" O.D., 1/2" I.D.										**350**
	0050	Gas, seamless brass, steel fittings										
	0200	12" long	1 Plum	36	.222	Ea.	11.95	9.10		21.05	27	
	0220	18" long		36	.222		14.85	9.10		23.95	30	
	0240	24" long		34	.235		17.55	9.60		27.15	34	
	0260	30" long		34	.235		18.95	9.60		28.55	35.50	
	0280	36" long		32	.250		21	10.20		31.20	38.50	
	0320	48" long		30	.267		26.50	10.90		37.40	46	
	0340	60" long		30	.267		31.50	10.90		42.40	51.50	
	0360	72" long		30	.267		36.50	10.90		47.40	56.50	
	2000	Water, copper tubing, dielectric separators										
	2100	12" long	1 Plum	36	.222	Ea.	9.40	9.10		18.50	24	
	2220	15" long		36	.222		10.45	9.10		19.55	25	
	2240	18" long		36	.222		11.40	9.10		20.50	26	
	2260	24" long		34	.235		13.95	9.60		23.55	30	

MECHANICAL 15

15120	Piping Specialties	CREW	DAILY OUTPUT	LABOR-HOURS	UNIT	2005 BARE COSTS				TOTAL INCL O&P
						MAT.	LABOR	EQUIP.	TOTAL	
940	**0010**	**WATER SUPPLY METERS**									**940**
1000	Detector, serves dual systems such as fire and domestic or									
1020	process water, wide range cap., UL and FM approved									
1100	3″ mainline x 2″ by-pass, 400 GPM	Q-1	3.60	4.444	Ea.	5,150	163		5,313	5,900
1140	4″ mainline x 2″ by-pass, 700 GPM	″	2.50	6.400		5,150	235		5,385	6,000
1180	6″ mainline x 3″ by-pass, 1600 GPM	Q-2	2.60	9.231		7,775	350		8,125	9,075
1220	8″ mainline x 4″ by-pass, 2800 GPM		2.10	11.429		11,500	435		11,935	13,300
1260	10″ mainline x 6″ by-pass, 4400 GPM		2	12		15,700	460		16,160	18,000
1300	10″x12″ mainlines x 6″ by-pass,5400 GPM		1.70	14.118		20,700	540		21,240	23,600
2000	Domestic/commercial, bronze									
2020	Threaded									
2060	5/8″ diameter, to 20 GPM	1 Plum	16	.500	Ea.	40	20.50		60.50	74.50
2080	3/4″ diameter, to 30 GPM		14	.571		67.50	23.50		91	110
2100	1″ diameter, to 50 GPM		12	.667		94	27		121	144
2300	Threaded/flanged									
2340	1-1/2″ diameter, to 100 GPM	1 Plum	8	1	Ea.	330	41		371	425
2360	2″ diameter, to 160 GPM	″	6	1.333	″	415	54.50		469.50	535
2600	Flanged, compound									
2640	3″ diameter, 320 GPM	Q-1	3	5.333	Ea.	1,950	196		2,146	2,425
2660	4″ diameter, to 500 GPM		1.50	10.667		3,025	390		3,415	3,925
2680	6″ diameter, to 1,000 GPM		1	16		4,350	590		4,940	5,650
2700	8″ diameter, to 1,800 GPM		.80	20		8,600	735		9,335	10,600
7000	Turbine									
7260	Flanged									
7300	2″ diameter, to 160 GPM	1 Plum	7	1.143	Ea.	455	46.50		501.50	570
7320	3″ diameter, to 450 GPM	Q-1	3.60	4.444		745	163		908	1,050
7340	4″ diameter, to 650 GPM	″	2.50	6.400		1,350	235		1,585	1,825
7360	6″ diameter, to 1800 GPM	Q-2	2.60	9.231		2,550	350		2,900	3,325
7380	8″ diameter, to 2500 GPM		2.10	11.429		4,050	435		4,485	5,100
7400	10″ diameter, to 5500 GPM		1.70	14.118		5,400	540		5,940	6,750

15140 | Domestic Water Piping

		CREW	DAILY OUTPUT	LABOR-HOURS	UNIT	MAT.	LABOR	EQUIP.	TOTAL	TOTAL INCL O&P
100	**0010**	**BACKFLOW PREVENTER** Includes valves									**100**
0020	and four test cocks, corrosion resistant, automatic operation									
1000	Double check principle									
1010	Threaded, with ball valves									
1020	3/4″ pipe size	1 Plum	16	.500	Ea.	127	20.50		147.50	170
1030	1″ pipe size		14	.571		140	23.50		163.50	189
1040	1-1/2″ pipe size		10	.800		271	32.50		303.50	350
1050	2″ pipe size		7	1.143		315	46.50		361.50	415
1080	Threaded, with gate valves									
1100	3/4″ pipe size	1 Plum	16	.500	Ea.	500	20.50		520.50	575
1120	1″ pipe size		14	.571		500	23.50		523.50	590
1140	1-1/2″ pipe size		10	.800		645	32.50		677.50	760
1160	2″ pipe size		7	1.143		790	46.50		836.50	940
1200	Flanged, valves are gate									
1210	3″ pipe size	Q-1	4.50	3.556	Ea.	1,375	131		1,506	1,700
1220	4″ pipe size	″	3	5.333		1,675	196		1,871	2,150
1230	6″ pipe size	Q-2	3	8		2,475	305		2,780	3,175
1240	8″ pipe size		2	12		4,500	460		4,960	5,650
1250	10″ pipe size		1	24		6,675	915		7,590	8,725
1300	Flanged, valves are OS&Y									
1380	3″ pipe size	Q-1	4.50	3.556	Ea.	1,550	131		1,681	1,925
1400	4″ pipe size	″	3	5.333		2,175	196		2,371	2,700
1420	6″ pipe size	Q-2	3	8		3,425	305		3,730	4,225
4000	Reduced pressure principle									

15 MECHANICAL

Important: See the Reference Section for critical supporting data - Reference Nos., Crews, & City Cost Indexes

15140	Domestic Water Piping	CREW	DAILY OUTPUT	LABOR-HOURS	UNIT	2005 BARE COSTS				TOTAL INCL O&P	
						MAT.	LABOR	EQUIP.	TOTAL		
100	4100	Threaded, bronze, valves are ball										100
	4120	3/4" pipe size	1 Plum	16	.500	Ea.	188	20.50		208.50	238
	4140	1" pipe size		14	.571		203	23.50		226.50	258
	4150	1-1/4" pipe size		12	.667		345	27		372	420
	4160	1-1/2" pipe size		10	.800		380	32.50		412.50	470
	4180	2" pipe size	▼	7	1.143	▼	425	46.50		471.50	540
	5000	Flanged, valves are OS&Y									
	5060	2-1/2" pipe size	Q-1	5	3.200	Ea.	2,175	118		2,293	2,575
	5080	3" pipe size		4.50	3.556		2,300	131		2,431	2,725
	5100	4" pipe size	▼	3	5.333		2,900	196		3,096	3,475
	5120	6" pipe size	Q-2	3	8	▼	4,175	305		4,480	5,050
	5200	Flanged, iron, valves are gate									
	5210	2-1/2" pipe size	Q-1	5	3.200	Ea.	1,550	118		1,668	1,875
	5220	3" pipe size		4.50	3.556		1,600	131		1,731	1,975
	5230	4" pipe size	▼	3	5.333		2,175	196		2,371	2,700
	5240	6" pipe size	Q-2	3	8		3,075	305		3,380	3,825
	5250	8" pipe size		2	12		5,500	460		5,960	6,750
	5260	10" pipe size	▼	1	24	▼	7,750	915		8,665	9,900
	5600	Flanged, iron, valves are OS&Y									
	5660	2-1/2" pipe size	Q-1	5	3.200	Ea.	1,750	118		1,868	2,100
	5680	3" pipe size		4.50	3.556		1,850	131		1,981	2,225
	5700	4" pipe size	▼	3	5.333		2,375	196		2,571	2,925
	5720	6" pipe size	Q-2	3	8		3,350	305		3,655	4,125
	5740	8" pipe size		2	12		5,875	460		6,335	7,175
	5760	10" pipe size	▼	1	24	▼	7,875	915		8,790	10,000

15150	Sanitary Waste and Vent Piping										
100	0010	**BACKWATER VALVES**, C.I. body										100
	6980	Bronze gate and automatic flapper valves									
	7000	3" and 4" pipe size	Q-1	13	1.231	Ea.	800	45.50		845.50	950
	7100	5" and 6" pipe size	"	13	1.231	"	1,225	45.50		1,270.50	1,425
	7240	Bronze flapper valve, bolted cover									
	7260	2" pipe size	Q-1	16	1	Ea.	235	37		272	315
	7300	4" pipe size	"	13	1.231	▼	450	45.50		495.50	565
	7340	6" pipe size	Q-2	17	1.412		650	54		704	795
200	0010	**CLEANOUTS**										200
	0060	Floor type									
	0080	Round or square, scoriated nickel bronze top									
	0100	2" pipe size	1 Plum	10	.800	Ea.	77	32.50		109.50	134
	0120	3" pipe size		8	1		116	41		157	189
	0140	4" pipe size		6	1.333		116	54.50		170.50	209
	0160	5" pipe size	▼	4	2		147	81.50		228.50	285
	0180	6" pipe size	Q-1	6	2.667		147	98		245	310
	0200	8" pipe size	"	4	4	▼	260	147		407	505
	0340	Recessed for tile, same price									
	0980	Round top, recessed for terrazzo									
	1000	2" pipe size	1 Plum	9	.889	Ea.	77	36.50		113.50	139
	1080	3" pipe size		6	1.333		116	54.50		170.50	209
	1100	4" pipe size	▼	4	2		116	81.50		197.50	250
	1120	5" pipe size	Q-1	6	2.667		147	98		245	310
	1140	6" pipe size		5	3.200		147	118		265	340
	1160	8" pipe size	▼	4	4	▼	260	147		407	505
	2000	Round scoriated nickel bronze top, extra heavy duty									
	2060	2" pipe size	1 Plum	9	.889	Ea.	105	36.50		141.50	171
	2080	3" pipe size	▼	6	1.333		144	54.50		198.50	240

MECHANICAL 15

	15150	Sanitary Waste and Vent Piping	CREW	DAILY OUTPUT	LABOR-HOURS	UNIT	MAT.	LABOR	EQUIP.	TOTAL	TOTAL INCL O&P	
200	2100	4" pipe size	1 Plum	4	2	Ea.	144	81.50		225.50	281	**200**
	2120	5" pipe size	Q-1	6	2.667		175	98		273	340	
	2140	6" pipe size		5	3.200		175	118		293	370	
	2160	8" pipe size	↓	4	4	↓	288	147		435	535	
	4000	Wall type, square smooth cover, over wall frame										
	4060	2" pipe size	1 Plum	14	.571	Ea.	126	23.50		149.50	174	
	4080	3" pipe size		12	.667		136	27		163	190	
	4100	4" pipe size		10	.800		146	32.50		178.50	209	
	4120	5" pipe size		9	.889		211	36.50		247.50	288	
	4140	6" pipe size	↓	8	1		235	41		276	320	
	4160	8" pipe size	Q-1	11	1.455	↓	315	53.50		368.50	425	
	5000	Extension, C.I.;bronze countersunk plug, 8" long										
	5040	2" pipe size	1 Plum	16	.500	Ea.	66	20.50		86.50	103	
	5060	3" pipe size		14	.571		76.50	23.50		100	119	
	5080	4" pipe size		13	.615		78.50	25		103.50	124	
	5100	5" pipe size		12	.667		116	27		143	169	
	5120	6" pipe size	↓	11	.727	↓	148	29.50		177.50	208	
250	0010	**CLEANOUT TEE**										**250**
	0100	Cast iron, B&S, with countersunk plug										
	0200	2" pipe size	1 Plum	4	2	Ea.	105	81.50		186.50	239	
	0220	3" pipe size		3.60	2.222		115	91		206	262	
	0240	4" pipe size	↓	3.30	2.424		143	99		242	305	
	0260	5" pipe size	Q-1	5.50	2.909		310	107		417	500	
	0280	6" pipe size	"	5	3.200		385	118		503	600	
	0300	8" pipe size	Q-3	5	6.400		440	249		689	860	
	0500	For round smooth access cover, same price										
	0600	For round scoriated access cover, same price										
	0700	For square smooth access cover, add				↓	60%					
	4000	Plastic, tees and adapters. Add plugs										
	4010	ABS, DWV										
	4020	Cleanout tee, 1-1/2" pipe size	1 Plum	15	.533	Ea.	3.41	22		25.41	37	
	4030	2" pipe size	Q-1	27	.593		4.50	22		26.50	38	
	4040	3" pipe size		21	.762		9	28		37	52	
	4050	4" pipe size	↓	16	1		16.70	37		53.70	74	
	4100	Cleanout plug, 1-1/2" pipe size	1 Plum	32	.250		.95	10.20		11.15	16.40	
	4110	2" pipe size	Q-1	56	.286		1.33	10.50		11.83	17.25	
	4120	3" pipe size		36	.444		3.77	16.35		20.12	28.50	
	4130	4" pipe size	↓	30	.533		2.48	19.60		22.08	32	
	4180	Cleanout adapter fitting, 1-1/2" pipe size	1 Plum	32	.250		.95	10.20		11.15	16.40	
	4190	2" pipe size	Q-1	56	.286		1.33	10.50		11.83	17.25	
	4200	3" pipe size		36	.444		3.77	16.35		20.12	28.50	
	4210	4" pipe size	↓	30	.533	↓	6.05	19.60		25.65	36	
	5000	PVC, DWV										
	5010	Cleanout tee, 1-1/2" pipe size	1 Plum	15	.533	Ea.	3.42	22		25.42	37	
	5020	2" pipe size	Q-1	27	.593		4.60	22		26.60	38	
	5030	3" pipe size		21	.762		9.20	28		37.20	52	
	5040	4" pipe size	↓	16	1		17.10	37		54.10	74.50	
	5090	Cleanout plug, 1-1/2" pipe size	1 Plum	32	.250		.77	10.20		10.97	16.20	
	5100	2" pipe size	Q-1	56	.286		.95	10.50		11.45	16.85	
	5110	3" pipe size		36	.444		1.98	16.35		18.33	26.50	
	5120	4" pipe size		30	.533		2.94	19.60		22.54	32.50	
	5130	6" pipe size	↓	24	.667		9	24.50		33.50	47	
	5170	Cleanout adapter fitting, 1-1/2" pipe size	1 Plum	32	.250		1.11	10.20		11.31	16.55	
	5180	2" pipe size	Q-1	56	.286		1.57	10.50		12.07	17.55	
	5190	3" pipe size	↓	36	.444	↓	4.47	16.35		20.82	29.50	

15
MECHANICAL

15100 | Building Services Piping

		15150	Sanitary Waste and Vent Piping	CREW	DAILY OUTPUT	LABOR-HOURS	UNIT	MAT.	LABOR	EQUIP.	TOTAL	TOTAL INCL O&P	
								2005 BARE COSTS					
250	5200		4" pipe size	Q-1	30	.533	Ea.	7.30	19.60		26.90	37.50	250
	5210		6" pipe size	↓	24	.667	↓	16.45	24.50		40.95	55	
300	0010	**FLOOR AND AREA DRAINS**											300
	0400	Deck, auto park, C.I., 13" top											
	0440	3", 4", 5", and 6" pipe size	Q-1	8	2	Ea.	565	73.50		638.50	730	
	0480	For galvanized body, add					"	268			268	295	
	2000	Floor, medium duty, C.I., deep flange, 7" dia top											
	2040	2" and 3" pipe size	Q-1	12	1.333	Ea.	79	49		128	161	
	2080	For galvanized body, add					↓	33.50			33.50	37	
	2120	For polished bronze top, add						37			37	41	
	2400	Heavy duty, with sediment bucket, C.I., 12" dia. loose grate											
	2420	2", 3", 4", 5", and 6" pipe size	Q-1	9	1.778	Ea.	272	65.50		337.50	400	
	2460	For polished bronze top, add					"	112			112	123	
	2500	Heavy duty, cleanout & trap w/bucket, C.I., 15" top											
	2540	2", 3", and 4" pipe size	Q-1	6	2.667	Ea.	2,525	98		2,623	2,925	
	2560	For galvanized body, add						580			580	640	
	2580	For polished bronze top, add					↓	204			204	225	

		15160	Storm Drainage Piping										
500	0010	**STORM AREA DRAINS**											500
	5980	Trench, floor, hvy duty, modular, C.I., 12" x 12" top											
	6000	2", 3", 4", 5", & 6" pipe size	Q-1	8	2	Ea.	350	73.50		423.50	495	
	6100	For polished bronze top, add					"	172			172	189	

MECHANICAL 15

15200 | Process Piping

		15230	Industrial Process Piping	CREW	DAILY OUTPUT	LABOR-HOURS	UNIT	MAT.	LABOR	EQUIP.	TOTAL	TOTAL INCL O&P	
								2005 BARE COSTS					
500	0010	**PUMPS, GENERAL UTILITY** With motor											500
	0200	Multi-stage, horizontal split, for boiler feed applications											
	0300	Two stage, 3" discharge x 4" suction, 75 HP	Q-7	.30	106	Ea.	15,700	4,175		19,875	23,500	
	0340	Four stage, 3" discharge x 4" suction, 150 HP	"	.18	177	"	25,800	6,950		32,750	38,700	
	2000	Single stage											
	2060	End suction, 1"D. x 2"S., 3 HP	Q-1	.50	32	Ea.	3,825	1,175		5,000	5,975	
	2100	1-1/2"D. x 3"S., 10 HP	"	.40	40		4,125	1,475		5,600	6,725	
	2140	2"D. x 3"S., 15 HP	Q-2	.60	40		4,475	1,525		6,000	7,225	
	2180	3"D. x 4"S., 20 HP		.50	48		4,775	1,825		6,600	8,000	
	2220	4"D. x 6"S., 30 HP	↓	.40	60		5,175	2,300		7,475	9,150	
	3000	Double suction, 2"D. x 2-1/2"S., 10 HP	Q-1	.30	53.333		5,300	1,950		7,250	8,775	
	3060	3"D. x 4"S., 15 HP	Q-2	.46	52.174		5,475	2,000		7,475	9,000	
	3100	4"D. x 5"S., 30 HP		.40	60		7,425	2,300		9,725	11,600	
	3140	5"D. x 6"S., 50 HP	↓	.33	72.727		7,950	2,775		10,725	12,900	
	3180	6"D. x 8"S., 60 HP	Q-3	.30	106		8,900	4,150		13,050	16,100	
	3190	75 HP, to 2500 GPM		.28	114		11,100	4,450		15,550	18,900	
	3220	100 HP, to 3000 GPM		.26	123		14,100	4,800		18,900	22,700	
	3240	150 HP, to 4000 GPM	↓	.24	133	↓	18,900	5,200		24,100	28,600	
	8000	Vertical submerged, with non-submerged motor											
	8060	1"D., 3 HP	Q-1	.50	32	Ea.	5,275	1,175		6,450	7,575	
	8100	1-1/2"D., 10 HP	"	.30	53.333		5,525	1,950		7,475	9,025	
	8140	2"D., 15 HP	Q-2	.40	60	↓	5,675	2,300		7,975	9,700	

15230	Industrial Process Piping	CREW	DAILY OUTPUT	LABOR-HOURS	UNIT	2005 BARE COSTS				TOTAL INCL O&P	
						MAT.	LABOR	EQUIP.	TOTAL		
500 8180	3"D., 25 HP	Q-2	.30	80	Ea.	5,800	3,050		8,850	11,000	**500**
8220	4"D., 30 HP	↓	.20	120	↓	6,500	4,575		11,075	14,000	

15400 | Plumbing Fixtures & Equipment

15411	Commercial/Indust Fixtures	CREW	DAILY OUTPUT	LABOR-HOURS	UNIT	2005 BARE COSTS				TOTAL INCL O&P	
						MAT.	LABOR	EQUIP.	TOTAL		
500 0010	**HYDRANTS**										**500**
0050	Wall type, moderate climate, bronze, encased										
0200	3/4" IPS connection	1 Plum	16	.500	Ea.	300	20.50		320.50	360	
0300	1" IPS connection	"	14	.571		299	23.50		322.50	365	
0500	Anti-siphon type				↓	259			259	285	
1000	Non-freeze, bronze, exposed										
1100	3/4" IPS connection, 4" to 9" thick wall	1 Plum	14	.571	Ea.	203	23.50		226.50	258	
1120	10" to 14" thick wall	"	12	.667		220	27		247	283	
1280	For anti-siphon type, add				↓	54			54	59.50	
2000	Non-freeze bronze, encased, anti-siphon type										
2100	3/4" IPS connection, 5" to 9" thick wall	1 Plum	14	.571	Ea.	271	23.50		294.50	335	
2140	15" to 19" thick wall	"	12	.667	"	395	27		422	470	
3000	Ground box type, bronze frame, 3/4" IPS connection										
3080	Non-freeze, all bronze, polished face, set flush										
3100	2 feet depth of bury	1 Plum	8	1	Ea.	375	41		416	470	
3180	6 feet depth of bury		7	1.143		490	46.50		536.50	605	
3220	8 feet depth of bury	↓	5	1.600		545	65.50		610.50	700	
3400	For 1" IPS connection, add					15%	10%				
3550	For 2" connection, add					445%	24%				
3600	For tapped drain port in box, add				↓	34.50			34.50	37.50	
5000	Moderate climate, all bronze, polished face										
5020	and scoriated cover, set flush										
5100	3/4" IPS connection	1 Plum	16	.500	Ea.	265	20.50		285.50	325	
5120	1" IPS connection	"	14	.571		265	23.50		288.50	325	
5200	For tapped drain port in box, add				↓	34.50			34.50	37.50	

15440	Plumbing Pumps	CREW	DAILY OUTPUT	LABOR-HOURS	UNIT	MAT.	LABOR	EQUIP.	TOTAL	INCL O&P	
240 0010	**PUMPS, PRESSURE BOOSTER SYSTEM**										**240**
0200	Pump system, with diaphragm tank, control, press. switch										
0300	1 HP pump	Q-1	1.30	12.308	Ea.	4,000	455		4,455	5,075	
0400	1-1/2 HP pump		1.25	12.800		4,050	470		4,520	5,150	
0420	2 HP pump		1.20	13.333		4,150	490		4,640	5,300	
0440	3 HP pump	↓	1.10	14.545		4,200	535		4,735	5,425	
0460	5 HP pump	Q-2	1.50	16		4,650	610		5,260	6,050	
0480	7-1/2 HP pump		1.42	16.901		5,175	645		5,820	6,675	
0500	10 HP pump	↓	1.34	17.910	↓	5,425	685		6,110	7,000	
1000	Pump/ energy storage system, diaphragm tank, 3 HP pump										
1100	motor, PRV, switch, gauge, control center, flow switch										
1200	125 lb. working pressure	Q-2	.70	34.286	Ea.	11,700	1,300		13,000	14,800	
1300	250 lb. working pressure	"	.64	37.500	"	12,800	1,425		14,225	16,200	
400 0010	**PUMPS, GRINDER SYSTEM** Complete, incl. check valve, tank, std.										**400**
0020	controls incl. alarm/disconnect panel w/wire. Excavation not included										

Important: See the Reference Section for critical supporting data - Reference Nos., Crews, & City Cost Indexes

15440	Plumbing Pumps	CREW	DAILY OUTPUT	LABOR-HOURS	UNIT	2005 BARE COSTS				TOTAL INCL O&P	
						MAT.	LABOR	EQUIP.	TOTAL		
400	0260	Simplex, 9 GPM at 60 PSIG, 70 gal. tank				Ea.	2,425			2,425	2,675	**400**
	0300	For manway, 26" I.D., 18" high, add					310			310	340	
	0340	26" I.D., 36" high, add					425			425	470	
	0380	43" I.D., 4' high, add					475			475	525	
	0600	Simplex, 9 GPM at 60 PSIG, 150 gal. tank				▼	2,625			2,625	2,875	
	0660	For manway, add										
	0700	26" I.D., 36" high, add				Ea.	565			565	625	
	0740	26" I.D., 4' high, add					645			645	710	
	2000	Duplex, 18 GPM at 60 PSIG, 150 gal. tank					1,725			1,725	1,900	
	2060	For manway 43" I.D., 4' high, add					1,425			1,425	1,550	
	2400	For core only				▼	1,300			1,300	1,425	
800	0010	**PUMPS, SEWAGE EJECTOR** With operating and level controls										**800**
	0100	Simplex system incl. tank, cover, pump 15' head										
	0500	37 gal PE tank, 12 GPM, 1/2 HP, 2" discharge	Q-1	3.20	5	Ea.	375	184		559	690	
	0510	3" discharge		3.10	5.161		405	190		595	730	
	0530	87 GPM, .7 HP, 2" discharge		3.20	5		570	184		754	905	
	0540	3" discharge		3.10	5.161		620	190		810	965	
	0600	45 gal. coated stl tank, 12 GPM, 1/2 HP, 2" discharge		3	5.333		670	196		866	1,025	
	0610	3" discharge		2.90	5.517		695	203		898	1,075	
	0630	87 GPM, .7 HP, 2" discharge		3	5.333		855	196		1,051	1,225	
	0640	3" discharge		2.90	5.517		905	203		1,108	1,300	
	0660	134 GPM, 1 HP, 2" discharge		2.80	5.714		925	210		1,135	1,350	
	0680	3" discharge		2.70	5.926		975	218		1,193	1,400	
	0700	70 gal. PE tank, 12 GPM, 1/2 HP, 2" discharge		2.60	6.154		735	226		961	1,150	
	0710	3" discharge		2.40	6.667		785	245		1,030	1,225	
	0730	87 GPM, 0.7 HP, 2" discharge		2.50	6.400		940	235		1,175	1,375	
	0740	3" discharge		2.30	6.957		1,000	256		1,256	1,475	
	0760	134 GPM, 1 HP, 2" discharge		2.20	7.273		1,025	267		1,292	1,525	
	0770	3" discharge		2	8		1,100	294		1,394	1,650	
	0800	75 gal. coated stl. tank, 12 GPM, 1/2 HP, 2 " discharge		2.40	6.667		760	245		1,005	1,200	
	0810	3" discharge		2.20	7.273		790	267		1,057	1,275	
	0830	87 GPM, .7 HP, 2" discharge		2.30	6.957		955	256		1,211	1,425	
	0840	3" discharge		2.10	7.619		1,000	280		1,280	1,525	
	0860	134 GPM, 1 HP, 2" discharge		2	8		1,025	294		1,319	1,575	
	0880	3" discharge	▼	1.80	8.889	▼	1,075	325		1,400	1,675	
	1040	Duplex system incl. tank, covers, pumps										
	1060	110 gal. fiberglass tank, 24 GPM, 1/2 HP, 2" discharge	Q-1	1.60	10	Ea.	1,425	370		1,795	2,125	
	1080	3" discharge		1.40	11.429		1,500	420		1,920	2,275	
	1100	174 GPM, .7 HP, 2" discharge		1.50	10.667		1,875	390		2,265	2,650	
	1120	3" discharge		1.30	12.308		1,950	455		2,405	2,825	
	1140	268 GPM, 1 HP, 2" discharge		1.20	13.333		2,050	490		2,540	2,975	
	1160	3" discharge	▼	1	16		2,100	590		2,690	3,200	
	1260	135 gal. coated stl. tank, 24 GPM, 1/2 HP, 2" discharge	Q-2	1.70	14.118		1,500	540		2,040	2,450	
	2000	3" discharge		1.60	15		1,575	570		2,145	2,575	
	2640	174 GPM, .7 HP, 2" discharge		1.60	15		1,950	570		2,520	3,000	
	2660	3" discharge		1.50	16		2,050	610		2,660	3,175	
	2700	268 GPM, 1 HP, 2" discharge		1.30	18.462		2,125	705		2,830	3,375	
	3040	3" discharge		1.10	21.818		2,250	830		3,080	3,725	
	3060	275 gal. coated stl. tank, 24 GPM, 1/2 HP, 2" discharge		1.50	16		1,850	610		2,460	2,975	
	3080	3" discharge		1.40	17.143		1,875	655		2,530	3,050	
	3100	174 GPM, .7 HP, 2" discharge		1.40	17.143		2,400	655		3,055	3,625	
	3120	3" discharge		1.30	18.462		2,425	705		3,130	3,700	
	3140	268 GPM, 1 HP, 2" discharge		1.10	21.818		2,625	830		3,455	4,150	
	3160	3" discharge	▼	.90	26.667	▼	2,725	1,025		3,750	4,525	
	3260	Pump system accessories, add										

MECHANICAL 15

15440 | Plumbing Pumps

			CREW	DAILY OUTPUT	LABOR-HOURS	UNIT	MAT.	LABOR	EQUIP.	TOTAL	TOTAL INCL O&P	
800	3300	Alarm horn and lights, 115V mercury switch	Q-1	8	2	Ea.	82.50	73.50		156	202	800
	3340	Switch, mag. contactor, alarm bell, light, 3 level control		5	3.200		590	118		708	825	
	3380	Alternator, mercury switch activated	↓	4	4	↓	750	147		897	1,050	
900	0010	**PUMPS, PEDESTAL SUMP** With float control										900
	0400	Molded PVC base, 21 GPM at 15' head, 1/3 HP	1 Plum	5	1.600	Ea.	87	65.50		152.50	194	
	0800	Iron base, 21 GPM at 15' head, 1/3 HP		5	1.600		110	65.50		175.50	219	
	1200	Solid brass, 21 GPM at 15' head, 1/3 HP	↓	5	1.600	↓	179	65.50		244.50	295	
940	0010	**PUMPS, SUBMERSIBLE** Sump										940
	7000	Sump pump, automatic										
	7100	Plastic, 1-1/4" discharge, 1/4 HP	1 Plum	6	1.333	Ea.	105	54.50		159.50	198	
	7140	1/3 HP		5	1.600		127	65.50		192.50	238	
	7160	1/2 HP		5	1.600		153	65.50		218.50	266	
	7180	1-1/2" discharge, 1/2 HP		4	2		171	81.50		252.50	310	
	7500	Cast iron, 1-1/4" discharge, 1/4 HP		6	1.333		122	54.50		176.50	216	
	7540	1/3 HP		6	1.333		144	54.50		198.50	240	
	7560	1/2 HP	↓	5	1.600	↓	170	65.50		235.50	285	

15460 | Domestic Water Cond Equipment

			CREW	DAILY OUTPUT	LABOR-HOURS	UNIT	MAT.	LABOR	EQUIP.	TOTAL	TOTAL INCL O&P	
900	0010	**WATER SOFTENER**										900
	5800	Softener systems, automatic, intermediate sizes										
	5820	available, may be used in multiples.										
	6000	Hardness capacity between regenerations and flow										
	6100	150,000 grains, 37 GPM cont., 51 GPM peak	Q-1	1.20	13.333	Ea.	3,775	490		4,265	4,875	
	6200	300,000 grains, 81 GPM cont., 113 GPM peak		1	16		5,600	590		6,190	7,050	
	6300	750,000 grains, 160 GPM cont., 230 GPM peak		.80	20		7,300	735		8,035	9,150	
	6400	900,000 grains, 185 GPM cont., 270 GPM peak	↓	.70	22.857	↓	11,800	840		12,640	14,200	

15490 | Pool and Fountain Equipment

			CREW	DAILY OUTPUT	LABOR-HOURS	UNIT	MAT.	LABOR	EQUIP.	TOTAL	TOTAL INCL O&P	
100	0010	**FOUNTAINS/AERATORS**										100
	0100	Pump w/controls										
	0200	Single phase, 100' cord, 1/2 H.P. pump	2 Skwk	4.40	3.636	Ea.	2,675	127		2,802	3,150	
	0300	3/4 H.P. pump		4.30	3.721		3,075	130		3,205	3,575	
	0400	1 H.P. pump		4.20	3.810		3,100	133		3,233	3,600	
	0500	1-1/2 H.P. pump		4.10	3.902		3,200	136		3,336	3,725	
	0600	2 H.P. pump		4	4		3,250	139		3,389	3,800	
	0700	Three phase, 200' cord, 5 H.P. pump		3.90	4.103		7,850	143		7,993	8,875	
	0800	7-1/2 H.P. pump		3.80	4.211		8,875	147		9,022	9,975	
	0900	10 H.P. pump		3.70	4.324		9,900	151		10,051	11,100	
	1000	15 H.P. pump		3.60	4.444		11,700	155		11,855	13,100	
	1100	Nozzles, minimum		8	2		197	69.50		266.50	325	
	1200	Maximum		8	2		259	69.50		328.50	395	
	1300	Lights w/mounting kits, 200 watt		18	.889		325	31		356	410	
	1400	300 watt		18	.889		360	31		391	445	
	1500	500 watt		18	.889		395	31		426	485	
	1600	Color blender	↓	12	1.333	↓	305	46.50		351.50	410	

15 MECHANICAL

Important: See the Reference Section for critical supporting data - Reference Nos., Crews, & City Cost Indexes

		15955	HVAC Test/Adjust/Balance	CREW	DAILY OUTPUT	LABOR-HOURS	UNIT	2005 BARE COSTS				TOTAL INCL O&P	
								MAT.	LABOR	EQUIP.	TOTAL		
700	0010		**PIPING, TESTING**										700
	0100		Nondestructive testing										
	0110		Nondestructive hydraulic pressure test, isolate & 1 hr. hold										
	0120		1" - 4" pipe										
	0140		0 - 250 L.F.	1 Stpi	1.33	6.015	Ea.		246		246	370	
	0160		250 - 500 L.F.	"	.80	10			410		410	615	
	0180		500 - 1000 L.F.	Q-5	1.14	14.035			515		515	775	
	0200		1000 - 2000 L.F.	"	.80	20	↓		735		735	1,100	
	0300		6" - 10" pipe										
	0320		0 - 250 L.F.	Q-5	1	16	Ea.		590		590	885	
	0340		250 - 500 L.F.		.73	21.918			810		810	1,225	
	0360		500 - 1000 L.F.		.53	30.189			1,100		1,100	1,675	
	0380		1000 - 2000 L.F.	↓	.38	42.105	↓		1,550		1,550	2,325	
	1000		Pneumatic pressure test, includes soaping joints										
	1120		1" - 4" pipe										
	1140		0 - 250 L.F.	Q-5	2.67	5.993	Ea.	5.10	221		226.10	335	
	1160		250 - 500 L.F.		1.33	12.030		10.25	445		455.25	675	
	1180		500 - 1000 L.F.		.80	20		15.35	735		750.35	1,125	
	1200		1000 - 2000 L.F.	↓	.50	32	↓	20.50	1,175		1,195.50	1,800	
	1300		6" - 10" pipe										
	1320		0 - 250 L.F.	Q-5	1.33	12.030	Ea.	5.10	445		450.10	670	
	1340		250 - 500 L.F.		.67	23.881		10.25	880		890.25	1,325	
	1360		500 - 1000 L.F.		.40	40		20.50	1,475		1,495.50	2,250	
	1380		1000 - 2000 L.F.	↓	.25	64	↓	25.50	2,350		2,375.50	3,575	
	2000		X-Ray of welds										
	2110		2" diam.	1 Stpi	8	1	Ea.	8.20	41		49.20	70.50	
	2120		3" diam.		8	1		8.20	41		49.20	70.50	
	2130		4" diam.		8	1		12.30	41		53.30	75	
	2140		6" diam.		8	1		12.30	41		53.30	75	
	2150		8" diam.		6.60	1.212		12.30	49.50		61.80	88	
	2160		10" diam.	↓	6	1.333	↓	16.40	54.50		70.90	100	
	3000		Liquid penetration of welds										
	3110		2" diam.	1 Stpi	14	.571	Ea.	2.30	23.50		25.80	37.50	
	3120		3" diam.		13.60	.588		2.30	24		26.30	38.50	
	3130		4" diam.		13.40	.597		2.30	24.50		26.80	39.50	
	3140		6" diam.		13.20	.606		2.30	25		27.30	40	
	3150		8" diam.		13	.615		3.45	25		28.45	42	
	3160		10" diam.	↓	12.80	.625	↓	3.45	25.50		28.95	42.50	

For information about Means Estimating Seminars, see yellow pages 12 and 13 in back of book

MECHANICAL 15

Division Notes

	CREW	DAILY OUTPUT	LABOR-HOURS	UNIT	2005 BARE COSTS				TOTAL INCL O&P
					MAT.	LABOR	EQUIP.	TOTAL	

Division 16 Electrical

Estimating Tips

16060 Grounding & Bonding
- When taking off grounding system, identify separately the type and size of wire and list each unique type of ground connection.

16100 Wiring Methods
- Conduit should be taken off in three main categories: power distribution, branch power, and branch lighting, so the estimator can concentrate on systems and components, therefore making it easier to ensure all items have been accounted for.
- For cost modifications for elevated conduit installation, add the percentages to labor according to the height of installation and only the quantities exceeding the different height levels, not to the total conduit quantities.
- Remember that aluminum wiring of equal ampacity is larger in diameter than copper and may require larger conduit.
- If more than three wires at a time are being pulled, deduct percentages from the labor hours of that grouping of wires.

- The estimator should take the weights of materials into consideration when completing a takeoff. Topics to consider include: How will the materials be supported? What methods of support are available? How high will the support structure have to reach? Will the final support structure be able to withstand the total burden? Is the support material included or separate from the fixture, equipment and material specified?

16200 Electrical Power
- Do not overlook the costs for equipment used in the installation. If scaffolding or highlifts are available in the field, contractors may use them in lieu of the proposed ladders and rolling staging.

16400 Low-Voltage Distribution
- Supports and concrete pads may be shown on drawings for the larger equipment, or the support system may be just a piece of plywood for the back of a panelboard. In either case, it must be included in the costs.

16500 Lighting
- Fixtures should be taken off room by room, using the fixture schedule, specifications, and the ceiling plan. For large concentrations of lighting fixtures in the same area deduct the percentages from labor hours.

16700 Communications
16800 Sound & Video
- When estimating material costs for special systems, it is always prudent to obtain manufacturers' quotations for equipment prices and special installation requirements which will affect the total costs.

Reference Numbers
Reference numbers are shown in bold squares at the beginning of some major classifications. These numbers refer to related items in the Reference Section. The reference information may be an estimating procedure, an alternate pricing method or technical information.

Note: Not all subdivisions listed here necessarily appear in this publication.

Note: **i2 Trade Service,** *in part, has been used as a reference source for some of the material prices used in Division 16.*

			DAILY	LABOR-			2005 BARE COSTS			TOTAL	
16055	**Selective Demolition**	CREW	OUTPUT	HOURS	UNIT	MAT.	LABOR	EQUIP.	TOTAL	INCL O&P	
300	0010	**ELECTRICAL DEMOLITION**									300
	0020	Conduit to 15' high, including fittings & hangers									
	0100	Rigid galvanized steel, 1/2" to 1" diameter	1 Elec	242	.033	L.F.		1.35		1.35	2.01
	0120	1-1/4" to 2"	"	200	.040			1.63		1.63	2.43
	0140	2-1/2" to 3-1/2"	2 Elec	302	.053			2.16		2.16	3.21
	0160	4" to 6"	"	160	.100			4.08		4.08	6.05
	0200	Electric metallic tubing (EMT), 1/2" to 1"	1 Elec	394	.020			.83		.83	1.23
	0220	1-1/4" to 1-1/2"		326	.025			1		1	1.49
	0240	2" to 3"	↓	236	.034			1.38		1.38	2.06
	0260	3-1/2" to 4"	2 Elec	310	.052	↓		2.10		2.10	3.13
	0270	Armored cable, (BX) avg. 50' runs									
	0280	#14, 2 wire	1 Elec	690	.012	L.F.		.47		.47	.70
	0290	#14, 3 wire		571	.014			.57		.57	.85
	0300	#12, 2 wire		605	.013			.54		.54	.80
	0310	#12, 3 wire		514	.016			.63		.63	.94
	0320	#10, 2 wire		514	.016			.63		.63	.94
	0330	#10, 3 wire		425	.019			.77		.77	1.14
	0340	#8, 3 wire	↓	342	.023	↓		.95		.95	1.42
	0350	Non metallic sheathed cable (Romex)									
	0360	#14, 2 wire	1 Elec	720	.011	L.F.		.45		.45	.67
	0370	#14, 3 wire		657	.012			.50		.50	.74
	0380	#12, 2 wire		629	.013			.52		.52	.77
	0390	#10, 3 wire	↓	450	.018	↓		.72		.72	1.08
	0400	Wiremold raceway, including fittings & hangers									
	0420	No. 3000	1 Elec	250	.032	L.F.		1.30		1.30	1.94
	0440	No. 4000		217	.037			1.50		1.50	2.24
	0460	No. 6000	↓	166	.048	↓		1.96		1.96	2.92
	0500	Channels, steel, including fittings & hangers									
	0520	3/4" x 1-1/2"	1 Elec	308	.026	L.F.		1.06		1.06	1.58
	0540	1-1/2" x 1-1/2"		269	.030			1.21		1.21	1.80
	0560	1-1/2" x 1-7/8"	↓	229	.035	↓		1.42		1.42	2.12
	0600	Copper bus duct, indoor, 3 phase									
	0610	Including hangers & supports									
	0620	225 amp	2 Elec	135	.119	L.F.		4.83		4.83	7.20
	0640	400 amp		106	.151			6.15		6.15	9.15
	0660	600 amp		86	.186			7.60		7.60	11.30
	0680	1000 amp		60	.267			10.85		10.85	16.15
	0700	1600 amp		40	.400			16.30		16.30	24.50
	0720	3000 amp	↓	10	1.600	↓		65		65	97
	1300	Transformer, dry type, 1 ph, incl. removal of									
	1320	supports, wire & conduit terminations									
	1340	1 kVA	1 Elec	7.70	1.039	Ea.		42.50		42.50	63
	1420	75 kVA	2 Elec	2.50	6.400	"		261		261	390
	1440	3 Phase to 600V, primary									
	1460	3 kVA	1 Elec	3.85	2.078	Ea.		84.50		84.50	126
	1520	75 kVA	2 Elec	2.70	5.926			241		241	360
	1550	300 kVA	R-3	1.80	11.111			445	88	533	760
	1570	750 kVA	"	1.10	18.182	↓		725	144	869	1,225
	1800	Wire, THW-THWN-THHN, removed from									
	1810	in place conduit, to 15' high									
	1830	#14	1 Elec	65	.123	C.L.F.		5		5	7.45
	1840	#12		55	.145			5.95		5.95	8.80
	1850	#10		45.50	.176			7.15		7.15	10.65
	1860	#8		40.40	.198			8.05		8.05	12
	1870	#6	↓	32.60	.245			10		10	14.90
	1880	#4	2 Elec	53	.302	↓		12.30		12.30	18.30

Important: See the Reference Section for critical supporting data - Reference Nos., Crews, & City Cost Indexes

16050 | Basic Electrical Materials & Methods

16055 | Selective Demolition

			CREW	DAILY OUTPUT	LABOR-HOURS	UNIT	2005 BARE COSTS MAT.	LABOR	EQUIP.	TOTAL	TOTAL INCL O&P	
300	1890	#3	2 Elec	50	.320	C.L.F.		13.05		13.05	19.40	300
	1900	#2		44.60	.359			14.60		14.60	22	
	1910	1/0		33.20	.482			19.65		19.65	29	
	1920	2/0		29.20	.548			22.50		22.50	33	
	1930	3/0		25	.640			26		26	39	
	1940	4/0		22	.727			29.50		29.50	44	
	1950	250 kcmil		20	.800			32.50		32.50	48.50	
	1960	300 kcmil		19	.842			34.50		34.50	51	
	1970	350 kcmil		18	.889			36		36	54	
	1980	400 kcmil		17	.941			38.50		38.50	57	
	1990	500 kcmil	▼	16.20	.988	▼		40.50		40.50	60	
	2000	Interior fluorescent fixtures, incl. supports										
	2010	& whips, to 15' high										
	2100	Recessed drop-in 2' x 2', 2 lamp	2 Elec	35	.457	Ea.		18.65		18.65	27.50	
	2120	2' x 4', 2 lamp		33	.485			19.75		19.75	29.50	
	2140	2' x 4', 4 lamp		30	.533			21.50		21.50	32.50	
	2160	4' x 4', 4 lamp	▼	20	.800	▼		32.50		32.50	48.50	
	2180	Surface mount, acrylic lens & hinged frame										
	2200	1' x 4', 2 lamp	2 Elec	44	.364	Ea.		14.80		14.80	22	
	2220	2' x 2', 2 lamp		44	.364			14.80		14.80	22	
	2260	2' x 4', 4 lamp		33	.485			19.75		19.75	29.50	
	2280	4' x 4', 4 lamp	▼	23	.696	▼		28.50		28.50	42	
	2300	Strip fixtures, surface mount										
	2320	4' long, 1 lamp	2 Elec	53	.302	Ea.		12.30		12.30	18.30	
	2340	4' long, 2 lamp		50	.320			13.05		13.05	19.40	
	2360	8' long, 1 lamp		42	.381			15.50		15.50	23	
	2380	8' long, 2 lamp	▼	40	.400	▼		16.30		16.30	24.50	
	2400	Pendant mount, industrial, incl. removal										
	2410	of chain or rod hangers, to 15' high										
	2420	4' long, 2 lamp	2 Elec	35	.457	Ea.		18.65		18.65	27.50	
	2440	8' long, 2 lamp	"	27	.593	"		24		24	36	

16100 | Wiring Methods

16120 | Conductors & Cables

			CREW	DAILY OUTPUT	LABOR-HOURS	UNIT	2005 BARE COSTS MAT.	LABOR	EQUIP.	TOTAL	TOTAL INCL O&P	
550	0010	**NON-METALLIC SHEATHED CABLE** 600 volt										550
	1450	UF underground feeder cable, copper with ground, #14, 2 conductor	1 Elec	4	2	C.L.F.	21.50	81.50		103	145	
	1500	#12, 2 conductor		3.50	2.286		30.50	93		123.50	173	
	1550	#10, 2 conductor		3	2.667		53	109		162	220	
	1600	#14, 3 conductor		3.50	2.286		34	93		127	177	
	1650	#12, 3 conductor		3	2.667		48	109		157	215	
	1700	#10, 3 conductor	▼	2.50	3.200	▼	73.50	130		203.50	275	
700	0010	**SHIELDED CABLE** Splicing & terminations not included										700
	0040	Copper, XLP shielding, 5 kV, #6	2 Elec	4.40	3.636	C.L.F.	109	148		257	340	
	0050	#4		4.40	3.636		142	148	.	290	375	
	0100	#2		4	4		168	163		331	425	
	0200	#1		4	4		192	163		355	455	
	0400	1/0	▼	3.80	4.211	▼	212	172		384	490	

		16120	Conductors & Cables	CREW	DAILY OUTPUT	LABOR-HOURS	UNIT	2005 BARE COSTS				TOTAL INCL O&P	
								MAT.	LABOR	EQUIP.	TOTAL		
700	0600		2/0	2 Elec	3.60	4.444	C.L.F.	267	181		448	565	700
	0800		4/0	↓	3.20	5		345	204		549	685	
	1000		250 kcmil	3 Elec	4.50	5.333		430	217		647	795	
	1200		350 kcmil		3.90	6.154		555	251		806	985	
	1400		500 kcmil	↓	3.60	6.667		675	272		947	1,150	
	1600	15 kV, ungrounded neutral, #1		2 Elec	4	4		234	163		397	500	
	1800		1/0		3.80	4.211		280	172		452	560	
	2000		2/0		3.60	4.444		320	181		501	620	
	2200		4/0	↓	3.20	5		425	204		629	770	
	2400		250 kcmil	3 Elec	4.50	5.333		470	217		687	840	
	2600		350 kcmil		3.90	6.154		590	251		841	1,025	
	2800		500 kcmil	↓	3.60	6.667		730	272		1,002	1,200	
	3000	25 kV, grounded neutral, #1/0		2 Elec	3.60	4.444		385	181		566	695	
	3200		2/0		3.40	4.706		425	192		617	755	
	3400		4/0	↓	3	5.333		535	217		752	910	
	3600		250 kcmil	3 Elec	4.20	5.714		665	233		898	1,075	
	3800		350 kcmil		3.60	6.667		775	272		1,047	1,250	
	3900		500 kcmil	↓	3.30	7.273		910	296		1,206	1,450	
	4000	35 kV, grounded neutral, #1/0		2 Elec	3.40	4.706		410	192		602	735	
	4200		2/0		3.20	5		480	204		684	835	
	4400		4/0	↓	2.80	5.714		605	233		838	1,000	
	4600		250 kcmil	3 Elec	3.90	6.154		710	251		961	1,150	
	4800		350 kcmil		3.30	7.273		850	296		1,146	1,375	
	5000		500 kcmil	↓	3	8		1,000	325		1,325	1,575	
	5050	Aluminum, XLP shielding, 5 kV, #2		2 Elec	5	3.200		135	130		265	340	
	5070		#1		4.40	3.636		139	148		287	375	
	5090		1/0		4	4		161	163		324	420	
	5100		2/0		3.80	4.211		181	172		353	455	
	5150		4/0	↓	3.60	4.444		214	181		395	505	
	5200		250 kcmil	3 Elec	4.80	5		259	204		463	590	
	5220		350 kcmil		4.50	5.333		305	217		522	660	
	5240		500 kcmil		3.90	6.154		390	251		641	805	
	5260		750 kcmil	↓	3.60	6.667		520	272		792	975	
	5300	15 kV aluminum, XLP, #1		2 Elec	4.40	3.636		173	148		321	410	
	5320		1/0		4	4		178	163		341	440	
	5340		2/0		3.80	4.211		213	172		385	490	
	5360		4/0	↓	3.60	4.444		236	181		417	530	
	5380		250 kcmil	3 Elec	4.80	5		282	204		486	615	
	5400		350 kcmil		4.50	5.333		315	217		532	675	
	5420		500 kcmil		3.90	6.154		435	251		686	855	
	5440		750 kcmil	↓	3.60	6.667	↓	655	272		927	1,125	

		16132	Conduit & Tubing										
230	0010	**CONDUIT IN CONCRETE SLAB** Including terminations,											230
	0020	fittings and supports											
	3230	PVC, schedule 40, 1/2" diameter		1 Elec	270	.030	L.F.	.49	1.21		1.70	2.34	
	3250	3/4" diameter			230	.035		.58	1.42		2	2.75	
	3270	1" diameter			200	.040		.77	1.63		2.40	3.28	
	3300	1-1/4" diameter			170	.047		1.09	1.92		3.01	4.05	
	3330	1-1/2" diameter			140	.057		1.34	2.33		3.67	4.95	
	3350	2" diameter			120	.067		1.70	2.72		4.42	5.90	
	3370	2-1/2" diameter		↓	90	.089		2.81	3.62		6.43	8.50	
	3400	3" diameter		2 Elec	160	.100		3.68	4.08		7.76	10.10	
	3430	3-1/2" diameter			120	.133		4.81	5.45		10.26	13.40	
	3440	4" diameter		↓	100	.160	↓	5.45	6.50		11.95	15.70	

16 ELECTRICAL

Important: See the Reference Section for critical supporting data - Reference Nos., Crews, & City Cost Indexes

16132 | Conduit & Tubing

		CREW	DAILY OUTPUT	LABOR-HOURS	UNIT	2005 BARE COSTS				TOTAL INCL O&P	
						MAT.	LABOR	EQUIP.	TOTAL		
230	3450	5" diameter	2 Elec	80	.200	L.F.	8.20	8.15		16.35	21	230
	3460	6" diameter	↓	60	.267	↓	11.95	10.85		22.80	29.50	
	3530	Sweeps, 1" diameter, 30" radius	1 Elec	32	.250	Ea.	15.70	10.20		25.90	32.50	
	3550	1-1/4" diameter		24	.333		19.85	13.60		33.45	42	
	3570	1-1/2" diameter		21	.381		20.50	15.50		36	45.50	
	3600	2" diameter		18	.444		21.50	18.10		39.60	50.50	
	3630	2-1/2" diameter		14	.571		36.50	23.50		60	75	
	3650	3" diameter		10	.800		42	32.50		74.50	94.50	
	3670	3-1/2" diameter		8	1		62.50	41		103.50	130	
	3700	4" diameter		7	1.143		60	46.50		106.50	135	
	3710	5" diameter	↓	6	1.333		91	54.50		145.50	181	
	3730	Couplings, 1/2" diameter					.41			.41	.45	
	3750	3/4" diameter					.50			.50	.55	
	3770	1" diameter					.77			.77	.85	
	3800	1-1/4" diameter					1.02			1.02	1.12	
	3830	1-1/2" diameter					1.40			1.40	1.54	
	3850	2" diameter					1.86			1.86	2.05	
	3870	2-1/2" diameter					3.30			3.30	3.63	
	3900	3" diameter					5.40			5.40	5.95	
	3930	3-1/2" diameter					6			6	6.60	
	3950	4" diameter					8.35			8.35	9.20	
	3960	5" diameter					21			21	23.50	
	3970	6" diameter					27			27	30	
	4030	End bells 1" diameter, PVC	1 Elec	60	.133		4	5.45		9.45	12.50	
	4050	1-1/4" diameter		53	.151		4.95	6.15		11.10	14.60	
	4100	1-1/2" diameter		48	.167		4.95	6.80		11.75	15.55	
	4150	2" diameter		34	.235		7.30	9.60		16.90	22.50	
	4170	2-1/2" diameter		27	.296		8.10	12.05		20.15	27	
	4200	3" diameter		20	.400		8.60	16.30		24.90	34	
	4250	3-1/2" diameter		16	.500		9.40	20.50		29.90	41	
	4300	4" diameter		14	.571		10.20	23.50		33.70	45.50	
	4310	5" diameter		12	.667		16.05	27		43.05	58	
	4320	6" diameter	↓	9	.889	↓	17.60	36		53.60	73.50	
	4350	Rigid galvanized steel, 1/2" diameter		200	.040	L.F.	2.36	1.63		3.99	5	
	4400	3/4" diameter		170	.047		2.72	1.92		4.64	5.85	
	4450	1" diameter		130	.062		4.15	2.51		6.66	8.30	
	4500	1-1/4" diameter		110	.073		5.55	2.96		8.51	10.50	
	4600	1-1/2" diameter		100	.080		6.45	3.26		9.71	11.95	
	4800	2" diameter	↓	90	.089	↓	8.25	3.62		11.87	14.45	
240	0010	**CONDUIT IN TRENCH** Includes terminations and fittings										240
	0020	Does not include excavation or backfill, see div. 02315										
	0200	Rigid galvanized steel, 2" diameter	1 Elec	150	.053	L.F.	8	2.17		10.17	12.05	
	0400	2-1/2" diameter	"	100	.080		15.45	3.26		18.71	22	
	0600	3" diameter	2 Elec	160	.100		18.90	4.08		22.98	27	
	0800	3-1/2" diameter		140	.114		23.50	4.66		28.16	33	
	1000	4" diameter		100	.160		26.50	6.50		33	38.50	
	1200	5" diameter		80	.200		55.50	8.15		63.65	73.50	
	1400	6" diameter	↓	60	.267	↓	81.50	10.85		92.35	106	

16136 | Boxes

		CREW	DAILY OUTPUT	LABOR-HOURS	UNIT	MAT.	LABOR	EQUIP.	TOTAL	TOTAL INCL O&P	
700	0010	**PULL BOXES & CABINETS**										700
	2100	Pull box, NEMA 3R, type SC, raintight & weatherproof R16136 -700										
	2150	6" L x 6" W x 6" D	1 Elec	10	.800	Ea.	16.70	32.50		49.20	67	
	2200	8" L x 6" W x 6" D	↓	8	1	↓	21	41		62	83.50	

16136 | Boxes

		CREW	DAILY OUTPUT	LABOR-HOURS	UNIT	2005 BARE COSTS				TOTAL INCL O&P	
						MAT.	LABOR	EQUIP.	TOTAL		
700	2250	10" L x 6" W x 6" D [R16136 -700]	1 Elec	7	1.143	Ea.	27.50	46.50		74	100	**700**
	2300	12" L x 12" W x 6" D		5	1.600		39	65		104	140	
	2350	16" L x 16" W x 6" D		4.50	1.778		78	72.50		150.50	194	
	2400	20" L x 20" W x 6" D		4	2		107	81.50		188.50	239	
	2450	24" L x 18" W x 8" D		3	2.667		117	109		226	291	
	2500	24" L x 24" W x 10" D		2.50	3.200		157	130		287	365	
	2550	30" L x 24" W x 12" D		2	4		286	163		449	560	
	2600	36" L x 36" W x 12" D	↓	1.50	5.333	↓	375	217		592	740	
	2800	Cast iron, pull boxes for surface mounting										
	3000	NEMA 4, watertight & dust tight										
	3050	6" L x 6" W x 6" D	1 Elec	4	2	Ea.	158	81.50		239.50	295	
	3100	8" L x 6" W x 6" D		3.20	2.500		214	102		316	390	
	3150	10" L x 6" W x 6" D		2.50	3.200		266	130		396	485	
	3200	12" L x 12" W x 6" D		2.30	3.478		450	142		592	705	
	3250	16" L x 16" W x 6" D		1.30	6.154		920	251		1,171	1,400	
	3300	20" L x 20" W x 6" D		.80	10		1,725	410		2,135	2,475	
	3350	24" L x 18" W x 8" D		.70	11.429		1,800	465		2,265	2,675	
	3400	24" L x 24" W x 10" D		.50	16		2,750	650		3,400	4,000	
	3450	30" L x 24" W x 12" D		.40	20		4,625	815		5,440	6,325	
	3500	36" L x 36" W x 12" D	↓	.20	40	↓	5,725	1,625		7,350	8,725	

16 ELECTRICAL

16210 | Electrical Utility Services

		CREW	DAILY OUTPUT	LABOR-HOURS	UNIT	2005 BARE COSTS				TOTAL INCL O&P	
						MAT.	LABOR	EQUIP.	TOTAL		
600	0010	**METER CENTERS AND SOCKETS**										**600**
	0100	Sockets, single position, 4 terminal, 100 amp	1 Elec	3.20	2.500	Ea.	34.50	102		136.50	190	
	0200	150 amp		2.30	3.478		38.50	142		180.50	253	
	0300	200 amp	↓	1.90	4.211	↓	51	172		223	310	
	2000	Meter center, main fusible switch, 1P 3W 120/240 volt										
	2030	400 amp	2 Elec	1.60	10	Ea.	1,275	410		1,685	2,000	
	2040	600 amp		1.10	14.545		2,200	595		2,795	3,300	
	2050	800 amp		.90	17.778		3,450	725		4,175	4,875	
	2060	Rainproof 1P 3W 120/240 volt, 400 amp		1.60	10		1,275	410		1,685	2,000	
	2070	600 amp		1.10	14.545		2,200	595		2,795	3,300	
	2080	800 amp		.90	17.778		3,450	725		4,175	4,875	
	2100	3P 4W 120/208V, 400 amp		1.60	10		1,450	410		1,860	2,200	
	2110	600 amp		1.10	14.545		2,725	595		3,320	3,850	
	2120	800 amp		.90	17.778		5,050	725		5,775	6,625	
	2130	Rainproof 3P 4W 120/208V, 400 amp		1.60	10		1,450	410		1,860	2,200	
	2140	600 amp		1.10	14.545		2,725	595		3,320	3,850	
	2150	800 amp	↓	.90	17.778	↓	5,050	725		5,775	6,625	
	2170	Main circuit breaker, 1P 3W 120/240V										
	2180	400 amp	2 Elec	1.60	10	Ea.	2,275	410		2,685	3,100	
	2190	600 amp		1.10	14.545		3,050	595		3,645	4,250	
	2200	800 amp		.90	17.778		3,575	725		4,300	5,000	
	2210	1000 amp		.80	20		4,925	815		5,740	6,625	
	2220	1200 amp		.76	21.053		6,625	860		7,485	8,575	
	2230	1600 amp		.68	23.529		10,800	960		11,760	13,300	
	2240	Rainproof 1P 3W 120/240V, 400 amp		1.60	10		2,275	410		2,685	3,100	
	2250	600 amp	↓	1.10	14.545	↓	3,050	595		3,645	4,250	

Important: See the Reference Section for critical supporting data - Reference Nos., Crews, & City Cost Indexes

16210	Electrical Utility Services		CREW	DAILY OUTPUT	LABOR-HOURS	UNIT	2005 BARE COSTS				TOTAL INCL O&P	
							MAT.	LABOR	EQUIP.	TOTAL		
600	2260	800 amp	2 Elec	.90	17.778	Ea.	3,575	725		4,300	5,000	**600**
	2270	1000 amp		.80	20		4,925	815		5,740	6,625	
	2280	1200 amp		.76	21.053		6,625	860		7,485	8,575	
	2300	3P 4W 120/208V, 400 amp		1.60	10		2,600	410		3,010	3,450	
	2310	600 amp		1.10	14.545		3,650	595		4,245	4,875	
	2320	800 amp		.90	17.778		4,325	725		5,050	5,850	
	2330	1000 amp		.80	20		5,700	815		6,515	7,475	
	2340	1200 amp		.76	21.053		7,275	860		8,135	9,275	
	2350	1600 amp		.68	23.529		10,800	960		11,760	13,300	
	2360	Rainproof 3P 4W 120/208V, 400 amp		1.60	10		2,600	410		3,010	3,450	
	2370	600 amp		1.10	14.545		3,650	595		4,245	4,875	
	2380	800 amp		.90	17.778		4,325	725		5,050	5,850	
	2390	1000 amp		.76	21.053		5,700	860		6,560	7,525	
	2400	1200 amp		.68	23.529		7,275	960		8,235	9,425	

16230	Generator Assemblies											
450	0010	**GENERATOR SET**										**450**
	0020	Gas or gasoline operated, includes battery,										
	0050	charger, muffler & transfer switch										
	0200	3 phase 4 wire, 277/480 volt, 7.5 kW	R-3	.83	24.096	Ea.	6,000	965	191	7,156	8,225	
	0300	11.5 kW		.71	28.169		8,500	1,125	223	9,848	11,300	
	0400	20 kW		.63	31.746		10,000	1,275	252	11,527	13,200	
	0500	35 kW		.55	36.364		12,000	1,450	288	13,738	15,700	
	0520	60 kW		.50	40		16,500	1,600	315	18,415	20,900	
	0600	80 kW		.40	50		20,500	2,000	395	22,895	26,000	
	0700	100 kW		.33	60.606		22,500	2,425	480	25,405	29,000	
	0800	125 kW		.28	71.429		46,000	2,850	565	49,415	55,500	
	0900	185 kW		.25	80		61,000	3,200	635	64,835	72,500	
	2000	Diesel engine, including battery, charger,										
	2010	muffler, automatic transfer switch & day tank, 30 kW	R-3	.55	36.364	Ea.	16,000	1,450	288	17,738	20,100	
	2100	50 kW		.42	47.619		19,700	1,900	380	21,980	25,000	
	2200	75 kW		.35	57.143		25,700	2,275	455	28,430	32,200	
	2300	100 kW		.31	64.516		28,500	2,575	510	31,585	35,800	
	2400	125 kW		.29	68.966		30,000	2,750	545	33,295	37,700	
	2500	150 kW		.26	76.923		34,400	3,075	610	38,085	43,100	
	2600	175 kW		.25	80		37,500	3,200	635	41,335	46,800	
	2700	200 kW		.24	83.333		38,700	3,325	660	42,685	48,300	
	2800	250 kW		.23	86.957		45,600	3,475	690	49,765	56,000	
	2900	300 kW		.22	90.909		49,300	3,625	720	53,645	60,000	
	3000	350 kW		.20	100		55,500	4,000	795	60,295	68,500	
	3100	400 kW		.19	105		69,000	4,200	835	74,035	82,500	
	3200	500 kW		.18	111		86,500	4,450	880	91,830	102,500	
	3220	600 kW		.17	117		113,500	4,700	935	119,135	133,000	
	3240	750 kW	R-13	.38	110		142,500	4,275	420	147,195	164,000	

16270	Transformers											
600	0010	**OIL FILLED TRANSFORMER** Pad mounted, primary delta or Y,										**600**
	0050	5 kV or 15 kV, with taps, 277/480 V secondary, 3 phase [R16270-600]										
	0100	150 kVA	R-3	.65	30.769	Ea.	7,025	1,225	244	8,494	9,825	
	0110	225 kVA		.55	36.364		7,425	1,450	288	9,163	10,700	
	0200	300 kVA		.45	44.444		9,300	1,775	350	11,425	13,200	
	0300	500 kVA		.40	50		14,000	2,000	395	16,395	18,800	
	0400	750 kVA		.38	52.632		16,700	2,100	415	19,215	22,000	
	0500	1000 kVA		.26	76.923		19,800	3,075	610	23,485	27,100	
610	0010	**TRANSFORMER, LIQUID-FILLED** Pad mounted										**610**
	0020	5 kV or 15 kV primary, 277/480 volt secondary, 3 phase										

ELECTRICAL 16

16200 | Electrical Power

		16270	Transformers	CREW	DAILY OUTPUT	LABOR-HOURS	UNIT	MAT.	LABOR	EQUIP.	TOTAL	TOTAL INCL O&P	
									2005 BARE COSTS				
610	0050		225 kVA	R-3	.55	36.364	Ea.	9,800	1,450	288	11,538	13,300	610
	0100		300 kVA		.45	44.444		11,700	1,775	350	13,825	15,800	
	0200		500 kVA		.40	50		14,700	2,000	395	17,095	19,600	
	0250		750 kVA		.38	52.632		19,000	2,100	415	21,515	24,500	
	0300		1000 kVA		.26	76.923		22,100	3,075	610	25,785	29,600	

16300 | Transmission & Distribution

		16310	Transmission & Dist Accessories	CREW	DAILY OUTPUT	LABOR-HOURS	UNIT	MAT.	LABOR	EQUIP.	TOTAL	TOTAL INCL O&P	
									2005 BARE COSTS				
600	0010	**LINE POLES & FIXTURES**											600
	0100		Digging holes in earth, average	R-5	25.14	3.500	Ea.		123	59.50	182.50	252	
	0105		In rock, average	"	4.51	19.512			690	330	1,020	1,400	
	0200		Wood poles, material handling and spotting	R-7	6.49	7.396			208	15.25	223.25	335	
	0220		Erect wood poles & backfill holes, in earth	R-5	6.77	12.999		1,100	460	221	1,781	2,125	
	0250		In rock	"	5.87	14.991		1,100	530	255	1,885	2,275	
	0260		Disposal of surplus material	R-7	20.87	2.300	Mile		65	4.74	69.74	105	
	0300		Crossarms for wood pole structure										
	0310		Material handling and spotting	R-7	14.55	3.299	Ea.		93	6.80	99.80	150	
	0320		Install crossarms	R-5	11	8	"	360	282	136	778	970	
	0330		Disposal of surplus material	R-7	40	1.200	Mile		34	2.47	36.47	54.50	
	0400		Formed plate pole structure										
	0410		Material handling and spotting	R-7	2.40	20	Ea.		565	41	606	910	
	0420		Erect steel plate pole	R-5	1.95	45.128		6,775	1,600	765	9,140	10,700	
	0500		Guys, anchors and hardware for pole, in earth		7.04	12.500		410	440	212	1,062	1,350	
	0510		In rock		17.96	4.900		490	173	83	746	890	
	0900		Foundations for line poles										
	0920		Excavation, in earth	R-5	135.38	.650	C.Y.		23	11.05	34.05	46.50	
	0940		In rock	"	20	4.400	"		155	74.50	229.50	315	
	0950		See also Division 02400										
	0960		Concrete foundations	R-5	11	8	C.Y.	95.50	282	136	513.50	680	
	0970		See also Division 03300										
610	0010	**LINE TOWERS & FIXTURES**											610
	0100		Excavation and backfill, earth	R-5	135.38	.650	C.Y.		23	11.05	34.05	46.50	
	0105		Rock		21.46	4.101	"		145	69.50	214.50	295	
	0200		Steel footings (grillage) in earth		3.91	22.506	Ton	1,525	795	380	2,700	3,300	
	0205		In rock		3.20	27.500	"	1,525	970	465	2,960	3,650	
	0290		See also Division 02400										
	0300		Rock anchors	R-5	5.87	14.991	Ea.	360	530	255	1,145	1,475	
	0400		Concrete foundations	"	12.85	6.848	C.Y.	90.50	242	116	448.50	595	
	0490		See also Division 03300										
	0500		Towers-material handling and spotting	R-7	22.56	2.128	Ton		60	4.38	64.38	97.50	
	0540		Steel tower erection	R-5	7.65	11.503		1,475	405	195	2,075	2,425	
	0550		Lace and box		7.10	12.394		1,475	435	210	2,120	2,500	
	0560		Painting total structure		1.47	59.864	Ea.	284	2,100	1,025	3,409	4,600	
	0570		Disposal of surplus material	R-7	20.87	2.300	Mile		65	4.74	69.74	105	
	0600		Special towers-material handling and spotting	"	12.31	3.899	Ton		110	8.05	118.05	178	
	0640		Special steel structure erection	R-6	6.52	13.497		1,800	475	470	2,745	3,225	
	0650		Special steel lace and box	"	6.29	13.990		1,800	495	490	2,785	3,275	
	0670		Disposal of surplus material	R-7	7.87	6.099	Mile		172	12.55	184.55	278	

Important: See the Reference Section for critical supporting data - Reference Nos., Crews, & City Cost Indexes

16310 | Transmission & Dist Accessories	CREW	DAILY OUTPUT	LABOR-HOURS	UNIT	2005 BARE COSTS				TOTAL INCL O&P
					MAT.	LABOR	EQUIP.	TOTAL	
700 0010 **OVERHEAD LINE CONDUCTORS & DEVICES**									**700**
0100 Conductors, primary circuits									
0110 Material handling and spotting	R-5	9.78	8.998	W.Mile		315	153	468	650
0120 For river crossing, add		11	8			282	136	418	575
0150 Conductors, per wire, 210 to 636 kcmil		1.96	44.898		7,950	1,575	760	10,285	12,000
0160 795 to 954 kcmil		1.87	47.059		11,700	1,650	800	14,150	16,300
0170 1000 to 1600 kcmil		1.47	59.864		19,200	2,100	1,025	22,325	25,400
0180 Over 1600 kcmil		1.35	65.185		25,600	2,300	1,100	29,000	32,800
0200 For river crossing, add, 210 to 636 kcmil		1.24	70.968			2,500	1,200	3,700	5,100
0220 795 to 954 kcmil		1.09	80.734			2,850	1,375	4,225	5,800
0230 1000 to 1600 kcmil		.97	90.722			3,200	1,550	4,750	6,525
0240 Over 1600 kcmil	↓	.87	101	↓		3,575	1,725	5,300	7,275
0300 Joints and dead ends	R-8	6	8	Ea.	1,050	286	29.50	1,365.50	1,650
0400 Sagging	R-5	7.33	12.001	W.Mile		425	204	629	860
0500 Clipping, per structure, 69 kV	R-10	9.60	5	Ea.		192	46.50	238.50	335
0510 161 kV		5.33	9.006			345	83.50	428.50	605
0520 345 to 500 kV	↓	2.53	18.972			725	176	901	1,275
0600 Make and install jumpers, per structure, 69 kV	R-8	3.20	15		287	535	55	877	1,175
0620 161 kV		1.20	40		575	1,425	147	2,147	2,950
0640 345 to 500 kV	↓	.32	150		965	5,375	550	6,890	9,725
0700 Spacers	R-10	68.57	.700		57	27	6.50	90.50	110
0720 For river crossings, add	"	60	.800	↓		30.50	7.40	37.90	54
0800 Installing pulling line (500 kV only)	R-9	1.45	44.138	W.Mile	580	1,475	122	2,177	3,000
0810 Disposal of surplus material	R-7	6.96	6.897	Mile		194	14.20	208.20	315
0820 With trailer mounted reel stands	"	13.71	3.501	"		98.50	7.20	105.70	160
0900 Insulators and hardware, primary circuits									
0920 Material handling and spotting, 69 kV	R-7	480	.100	Ea.		2.82	.21	3.03	4.57
0930 161 kV		685.71	.070			1.97	.14	2.11	3.20
0950 345 to 500 kV	↓	960	.050			1.41	.10	1.51	2.28
1000 Disk insulators, 69 kV	R-5	880	.100		58.50	3.53	1.70	63.73	71.50
1020 161 kV		977.78	.090		67	3.17	1.53	71.70	80
1040 345 to 500 kV	↓	1,100	.080	↓	67	2.82	1.36	71.18	79
1060 See Div. 16360-800-7400 for pin or pedestal insulator									
1100 Install disk insulator at river crossing, add									
1110 69 kV	R-5	586.67	.150	Ea.		5.30	2.55	7.85	10.75
1120 161 kV		880	.100			3.53	1.70	5.23	7.15
1140 345 to 500 kV	↓	880	.100	↓		3.53	1.70	5.23	7.15
1150 Disposal of surplus material	R-7	41.74	1.150	Mile		32.50	2.37	34.87	52.50
1300 Overhead ground wire installation									
1320 Material handling and spotting	R-7	5.65	8.496	W.Mile		239	17.50	256.50	390
1340 Overhead ground wire	R-5	1.76	50		3,500	1,775	850	6,125	7,400
1350 At river crossing, add		1.17	75.214	↓		2,650	1,275	3,925	5,400
1360 Disposal of surplus material	↓	41.74	2.108	Mile		74.50	36	110.50	152
1400 Installing conductors, underbuilt circuits									
1420 Material handling and spotting	R-7	5.65	8.496	W.Mile		239	17.50	256.50	390
1440 Conductors, per wire, 210 to 636 kcmil	R-5	1.96	44.898		7,950	1,575	760	10,285	12,000
1450 795 to 954 kcmil		1.87	47.059		11,700	1,650	800	14,150	16,300
1460 1000 to 1600 kcmil		1.47	59.864		19,200	2,100	1,025	22,325	25,400
1470 Over 1600 kcmil	↓	1.35	65.185	↓	25,600	2,300	1,100	29,000	32,800
1500 Joints and dead ends	R-8	6	8	Ea.	1,050	286	29.50	1,365.50	1,650
1550 Sagging	R-5	8.80	10	W.Mile		355	170	525	715
1600 Clipping, per structure, 69 kV	R-10	9.60	5	Ea.		192	46.50	238.50	335
1620 161 kV		5.33	9.006			345	83.50	428.50	605
1640 345 to 500 kV	↓	2.53	18.972			725	176	901	1,275
1700 Making and installing jumpers, per structure, 69 kV	R-8	5.87	8.177		287	293	30	610	790
1720 161 kV	↓	.96	50	↓	575	1,800	184	2,559	3,525

ELECTRICAL 16

16310	Transmission & Dist Accessories	CREW	DAILY OUTPUT	LABOR-HOURS	UNIT	*2005 BARE COSTS MAT.	LABOR	EQUIP.	TOTAL	TOTAL INCL O&P	
700 1740	345 to 500 kV	R-8	.32	150	Ea.	965	5,375	550	6,890	9,725	700
1800	Spacers	R-10	96	.500	↓	57	19.15	4.63	80.78	96.50	
1810	Disposal of surplus material	R-7	6.96	6.897	Mile		194	14.20	208.20	315	
2000	Insulators and hardware for underbuilt circuits										
2100	Material handling and spotting	R-7	1,200	.040	Ea.		1.13	.08	1.21	1.82	
2150	Disk insulators, 69 kV	R-8	600	.080		58.50	2.86	.29	61.65	69	
2160	161 kV		686	.070		67	2.50	.26	69.76	77.50	
2170	345 to 500 kV	↓	800	.060	↓	67	2.15	.22	69.37	77	
2180	Disposal of surplus material	R-7	41.74	1.150	Mile		32.50	2.37	34.87	52.50	
2300	Sectionalizing switches, 69 kV	R-5	1.26	69.841	Ea.	15,100	2,475	1,175	18,750	21,600	
2310	161 kV		.80	110		17,000	3,875	1,875	22,750	26,600	
2500	Protective devices	↓	5.50	16	↓	4,925	565	272	5,762	6,575	
2600	Clearance poles, 8 poles per mile										
2650	In earth, 69 kV	R-5	1.16	75.862	Mile	4,050	2,675	1,300	8,025	9,900	
2660	161 kV	"	.64	137		6,650	4,850	2,325	13,825	17,200	
2670	345 to 500 kV	R-6	.48	183		8,000	6,475	6,425	20,900	25,600	
2800	In rock, 69 kV	R-5	.69	127		4,050	4,500	2,175	10,725	13,600	
2820	161 kV	"	.35	251		6,650	8,875	4,275	19,800	25,400	
2840	345 to 500 kV	R-6	.24	366	↓	8,000	12,900	12,800	33,700	42,400	

850 0010	**TRANSMISSION LINE RIGHT OF WAY**										850
0100	Clearing right of way	B-87	6.67	5.997	Acre		198	360	558	700	
0200	Restoration & seeding	B-10D	4	3	"	920	96	380	1,396	1,575	

16330	**Med-Voltage Sw & Prot**										
760 0010	**SWITCHGEAR**, Incorporate switch with cable connections, transformer,										760
0100	& Low Voltage section										
0200	Load interrupter switch, 600 amp, 2 position										
0300	NEMA 1, 4.8 KV, 300 kVA & below w/CLF fuses	R-3	.40	50	Ea.	14,700	2,000	395	17,095	19,500	
0400	400 kVA & above w/CLF fuses		.38	52.632		16,400	2,100	415	18,915	21,600	
0500	Non fusible		.41	48.780		11,900	1,950	385	14,235	16,400	
0600	13.8 kV, 300 kVA & below		.38	52.632		18,400	2,100	415	20,915	23,900	
0700	400 kVA & above		.36	55.556		18,400	2,225	440	21,065	24,100	
0800	Non fusible	↓	.40	50		14,000	2,000	395	16,395	18,800	
0900	Cable lugs for 2 feeders 4.8 kV or 13.8 kV	1 Elec	8	1		445	41		486	550	
1000	Pothead, one 3 conductor or three 1 conductor		4	2		2,150	81.50		2,231.50	2,475	
1100	Two 3 conductor or six 1 conductor		2	4		4,225	163		4,388	4,900	
1200	Key interlocks	↓	8	1	↓	495	41		536	605	
1300	Lightning arresters, Distribution class (no charge)										
1400	Intermediate class or line type 4.8 kV	1 Elec	2.70	2.963	Ea.	2,400	121		2,521	2,825	
1500	13.8 kV		2	4		3,200	163		3,363	3,750	
1600	Station class, 4.8 kV		2.70	2.963		4,125	121		4,246	4,725	
1700	13.8 kV	↓	2	4		7,100	163		7,263	8,050	
1800	Transformers, 4800 volts to 480/277 volts, 75 kVA	R-3	.68	29.412		12,500	1,175	233	13,908	15,800	
1900	112.5 kVA		.65	30.769		15,300	1,225	244	16,769	18,900	
2000	150 kVA		.57	35.088		17,400	1,400	278	19,078	21,600	
2100	225 kVA		.48	41.667		20,000	1,675	330	22,005	24,800	
2200	300 kVA		.41	48.780		22,400	1,950	385	24,735	27,900	
2300	500 kVA		.36	55.556		29,500	2,225	440	32,165	36,200	
2400	750 kVA		.29	68.966		33,400	2,750	545	36,695	41,500	
2500	13,800 volts to 480/277 volts, 75 kVA		.61	32.787		17,700	1,300	260	19,260	21,600	
2600	112.5 kVA		.55	36.364		23,400	1,450	288	25,138	28,300	
2700	150 kVA		.49	40.816		23,700	1,625	325	25,650	28,800	
2800	225 kVA		.41	48.780		27,300	1,950	385	29,635	33,400	
2900	300 kVA	↓	.37	54.054	↓	27,900	2,150	430	30,480	34,400	

Important: See the Reference Section for critical supporting data - Reference Nos., Crews, & City Cost Indexes

16 ELECTRICAL

16330	Med-Voltage Sw & Prot	CREW	DAILY OUTPUT	LABOR-HOURS	UNIT	2005 BARE COSTS				TOTAL INCL O&P
						MAT.	LABOR	EQUIP.	TOTAL	
760 3000	500 kVA	R-3	.31	64.516	Ea.	30,900	2,575	510	33,985	38,400 **760**
3100	750 kVA	↓	.26	76.923		34,000	3,075	610	37,685	42,700
3200	Forced air cooling & temperature alarm	1 Elec	1	8	↓	2,725	325		3,050	3,475
3300	Low voltage components									
3400	Maximum panel height 49-1/2", single or twin row									
3500	Breaker heights, type FA or FH, 6"									
3600	type KA or KH, 8"									
3700	type LA, 11"									
3800	type MA, 14"									
3900	Breakers, 2 pole, 15 to 60 amp, type FA	1 Elec	5.60	1.429	Ea.	305	58		363	420
4000	70 to 100 amp, type FA		4.20	1.905		370	77.50		447.50	520
4100	15 to 60 amp, type FH		5.60	1.429		475	58		533	605
4200	70 to 100 amp, type FH		4.20	1.905		565	77.50		642.50	735
4300	125 to 225 amp, type KA		3.40	2.353		855	96		951	1,075
4400	125 to 225 amp, type KH		3.40	2.353		1,975	96		2,071	2,325
4500	125 to 400 amp, type LA		2.50	3.200		1,575	130		1,705	1,950
4600	125 to 600 amp, type MA		1.80	4.444		2,575	181		2,756	3,125
4700	700 & 800 amp, type MA		1.50	5.333		3,250	217		3,467	3,900
4800	3 pole, 15 to 60 amp, type FA		5.30	1.509		375	61.50		436.50	505
4900	70 to 100 amp, type FA		4	2		450	81.50		531.50	620
5000	15 to 60 amp, type FH		5.30	1.509		560	61.50		621.50	710
5100	70 to 100 amp, type FH		4	2		635	81.50		716.50	820
5200	125 to 225 amp, type KA		3.20	2.500		1,075	102		1,177	1,325
5300	125 to 225 amp, type KH		3.20	2.500		2,400	102		2,502	2,800
5400	125 to 400 amp, type LA		2.30	3.478		1,925	142		2,067	2,325
5500	125 to 600 amp, type MA		1.60	5		3,175	204		3,379	3,800
5600	700 & 800 amp, type MA	↓	1.30	6.154	↓	4,150	251		4,401	4,925

16360	Unit Substations									
800 0010	**SUBSTATION EQUIPMENT**									**800**
1000	Main conversion equipment									
1050	Power transformers, 13 to 26 kV	R-11	1.72	32.558	MVA	16,800	1,250	410	18,460	20,700
1060	46 kV		3.50	16		15,700	610	200	16,510	18,400
1070	69 kV		3.11	18.006		13,400	685	226	14,311	16,000
1080	110 kV		3.29	17.021		12,700	650	213	13,563	15,200
1090	161 kV		4.31	12.993		11,800	495	163	12,458	13,900
1100	500 kV		7	8	↓	11,800	305	100	12,205	13,600
1200	Grounding transformers	↓	3.11	18.006	Ea.	74,000	685	226	74,911	83,000
1300	Station capacitors									
1350	Synchronous, 13 to 26 kV	R-11	3.11	18.006	MVAR	4,675	685	226	5,586	6,425
1360	46 kV		3.33	16.817		5,975	640	211	6,826	7,775
1370	69 kV		3.81	14.698		5,875	560	184	6,619	7,500
1380	161 kV		6.51	8.602		5,500	330	108	5,938	6,650
1390	500 kV		10.37	5.400		4,775	206	67.50	5,048.50	5,625
1450	Static, 13 to 26 kV		3.11	18.006		3,975	685	226	4,886	5,650
1460	46 kV		3.01	18.605		5,000	710	233	5,943	6,825
1470	69 kV		3.81	14.698		4,875	560	184	5,619	6,400
1480	161 kV		6.51	8.602		4,525	330	108	4,963	5,575
1490	500 kV		10.37	5.400	↓	4,125	206	67.50	4,398.50	4,925
1600	Voltage regulators, 13 to 26 kV	↓	.75	74.667	Ea.	177,500	2,850	935	181,285	201,000
2000	Power circuit breakers									
2050	Oil circuit breakers, 13 to 26 kV	R-11	1.12	50	Ea.	39,600	1,900	625	42,125	47,100
2060	46 kV		.75	74.667		58,000	2,850	935	61,785	69,000
2070	69 kV		.45	124		134,500	4,750	1,550	140,800	157,000
2080	161 kV	↓	.16	350	↓	205,500	13,300	4,375	223,175	251,000

ELECTRICAL T6

		16360	Unit Substations	CREW	DAILY OUTPUT	LABOR-HOURS	UNIT	2005 BARE COSTS				TOTAL INCL O&P	
								MAT.	LABOR	EQUIP.	TOTAL		
800	2090		500 kV	R-11	.06	933	Ea.	769,000	35,600	11,700	816,300	912,500	**800**
	2100		Air circuit breakers, 13 to 26 kV		.56	100		41,400	3,800	1,250	46,450	52,500	
	2110		161 kV		.24	233		178,000	8,900	2,925	189,825	212,500	
	2150		Gas circuit breakers, 13 to 26 kV		.56	100		155,500	3,800	1,250	160,550	178,500	
	2160		161 kV		.08	700		204,500	26,700	8,775	239,975	274,500	
	2170		500 kV		.04	1,400		720,000	53,500	17,500	791,000	891,500	
	2200		Vacuum circuit breakers, 13 to 26 kV	↓	.56	100	↓	33,900	3,800	1,250	38,950	44,400	
	3000		Disconnecting switches										
	3050		Gang operated switches										
	3060		Manual operation, 13 to 26 kV	R-11	1.65	33.939	Ea.	9,550	1,300	425	11,275	12,900	
	3070		46 kV		1.12	50		15,900	1,900	625	18,425	21,000	
	3080		69 kV		.80	70		17,800	2,675	875	21,350	24,600	
	3090		161 kV		.56	100		21,400	3,800	1,250	26,450	30,700	
	3100		500 kV		.14	400		58,500	15,200	5,000	78,700	92,500	
	3110		Motor operation, 161 kV		.51	109		32,200	4,175	1,375	37,750	43,300	
	3120		500 kV		.28	200		89,000	7,625	2,500	99,125	112,000	
	3250		Circuit switches, 161 kV	↓	.41	136	↓	64,500	5,200	1,700	71,400	80,000	
	3300		Single pole switches										
	3350		Disconnecting switches, 13 to 26 kV	R-11	28	2	Ea.	8,375	76	25	8,476	9,375	
	3360		46 kV		8	7		14,800	267	87.50	15,154.50	16,800	
	3370		69 kV		5.60	10		16,400	380	125	16,905	18,700	
	3380		161 kV		2.80	20		62,500	760	251	63,511	70,000	
	3390		500 kV		.22	254		178,000	9,700	3,200	190,900	213,500	
	3450		Grounding switches, 46 kV		5.60	10		22,500	380	125	23,005	25,500	
	3460		69 kV		3.73	15.013		23,400	570	188	24,158	26,900	
	3470		161 kV		2.24	25		24,800	955	315	26,070	29,100	
	3480		500 kV	↓	.62	90.323	↓	30,800	3,450	1,125	35,375	40,300	
	4000		Instrument transformers										
	4050		Current transformers, 13 to 26 kV	R-11	14	4	Ea.	2,175	152	50	2,377	2,675	
	4060		46 kV		9.33	6.002		6,375	229	75	6,679	7,450	
	4070		69 kV		7	8		6,600	305	100	7,005	7,850	
	4080		161 kV		1.87	29.947		21,400	1,150	375	22,925	25,700	
	4100		Potential transformers, 13 to 26 kV		11.20	5		3,125	191	62.50	3,378.50	3,775	
	4110		46 kV		8	7		6,425	267	87.50	6,779.50	7,550	
	4120		69 kV		6.22	9.003		6,800	345	113	7,258	8,125	
	4130		161 kV		2.24	25		14,700	955	315	15,970	18,000	
	4140		500 kV	↓	1.40	40	↓	43,900	1,525	500	45,925	51,000	
	7000		Conduit, conductors, and insulators										
	7100		Conduit, metallic	R-11	560	.100	Lb.	2.65	3.81	1.25	7.71	10	
	7110		Non-metallic		800	.070		2.01	2.67	.88	5.56	7.15	
	7200		Wire and cable	↓	700	.080	↓	1.87	3.05	1	5.92	7.75	
	7290		See also Division 16100										
	7300		Bus	R-11	590	.095	Lb.	2.70	3.62	1.19	7.51	9.70	
	7400		Insulators, pedestal type	"	112	.500	Ea.		19.05	6.25	25.30	35.50	
	7490		See also Line 16310-700-1000										
	7500		Grounding systems	R-11	280	.200	Lb.	9.25	7.60	2.51	19.36	24.50	
	7600		Manholes	R-11	4.15	13.494	Ea.	2,925	515	169	3,609	4,175	
	7690		See also Division 02080										
	7700		Cable tray	R-11	40	1.400	L.F.	13.30	53.50	17.55	84.35	114	
	8000		Protective equipment										
	8050		Lightning arresters, 13 to 26 kV	R-11	18.67	2.999	Ea.	1,075	114	37.50	1,226.50	1,400	
	8060		46 kV		14	4		2,900	152	50	3,102	3,475	
	8070		69 kV		11.20	5		3,725	191	62.50	3,978.50	4,450	
	8080		161 kV		5.60	10		5,150	380	125	5,655	6,350	
	8090		500 kV		1.40	40		17,400	1,525	500	19,425	21,900	
	8150		Reactors and resistors, 13 to 26 kV		28	2		2,425	76	25	2,526	2,825	

16 ELECTRICAL

Important: See the Reference Section for critical supporting data - Reference Nos., Crews, & City Cost Indexes

16360	Unit Substations	CREW	DAILY OUTPUT	LABOR-HOURS	UNIT	2005 BARE COSTS				TOTAL INCL O&P	
						MAT.	LABOR	EQUIP.	TOTAL		
800	8160	46 kV	R-11	4.31	12.993	Ea.	7,325	495	163	7,983	9,000	800
	8170	69 kV		2.80	20		12,000	760	251	13,011	14,500
	8180	161 kV		2.24	25		13,600	955	315	14,870	16,800
	8190	500 kV		.08	700		54,500	26,700	8,775	89,975	109,500
	8250	Fuses, 13 to 26 kV		18.67	2.999		1,450	114	37.50	1,601.50	1,800
	8260	46 kV		11.20	5		1,650	191	62.50	1,903.50	2,150
	8270	69 kV		8	7		1,700	267	87.50	2,054.50	2,375
	8280	161 kV		4.67	11.991		2,150	455	150	2,755	3,225
	9000	Station service equipment									
	9100	Conversion equipment									
	9110	Station service transformers	R-11	5.60	10	Ea.	64,500	380	125	65,005	71,500
	9120	Battery chargers		11.20	5	"	2,650	191	62.50	2,903.50	3,275
	9200	Control batteries		14	4	K.A.H.	61	152	50	263	350

16410	Encl Switches & Circuit Breakers	CREW	DAILY OUTPUT	LABOR-HOURS	UNIT	2005 BARE COSTS				TOTAL INCL O&P	
						MAT.	LABOR	EQUIP.	TOTAL		
800	0010	**SAFETY SWITCHES**										800
	0100	General duty 240 volt, 3 pole NEMA 1, fusible, 30 amp	1 Elec	3.20	2.500	Ea.	78	102		180	238
	0200	60 amp		2.30	3.478		132	142		274	355
	0300	100 amp		1.90	4.211		227	172		399	505
	0400	200 amp		1.30	6.154		490	251		741	910
	0500	400 amp	2 Elec	1.80	8.889		1,225	360		1,585	1,900
	0600	600 amp	"	1.20	13.333		2,300	545		2,845	3,325
	5510	Heavy duty, 600 volt, 3 pole 3ph. NEMA 3R fusible, 30 amp	1 Elec	3.10	2.581		360	105		465	550
	5520	60 amp		2.20	3.636		420	148		568	680
	5530	100 amp		1.80	4.444		655	181		836	990
	5540	200 amp		1.20	6.667		905	272		1,177	1,400
	5550	400 amp	2 Elec	1.60	10		2,125	410		2,535	2,925

16440	Swbds, Panels & Control Centers	CREW	DAILY OUTPUT	LABOR-HOURS	UNIT	2005 BARE COSTS				TOTAL INCL O&P	
						MAT.	LABOR	EQUIP.	TOTAL		
720	0010	**PANELBOARDS** (Commercial use)										720
	0050	NQOD, w/20 amp 1 pole bolt-on circuit breakers									
	0100	3 wire, 120/240 volts, 100 amp main lugs									
	0150	10 circuits	1 Elec	1	8	Ea.	410	325		735	935
	0200	14 circuits		.88	9.091		480	370		850	1,075
	0250	18 circuits		.75	10.667		525	435		960	1,225
	0300	20 circuits		.65	12.308		595	500		1,095	1,400
	0350	225 amp main lugs, 24 circuits	2 Elec	1.20	13.333		675	545		1,220	1,550
	0400	30 circuits		.90	17.778		785	725		1,510	1,925
	0450	36 circuits		.80	20		895	815		1,710	2,200
	0500	38 circuits		.72	22.222		960	905		1,865	2,400
	0550	42 circuits		.66	24.242		1,000	990		1,990	2,575
	0600	4 wire, 120/208 volts, 100 amp main lugs, 12 circuits	1 Elec	1	8		465	325		790	995
	0650	16 circuits		.75	10.667		535	435		970	1,225
	0700	20 circuits		.65	12.308		625	500		1,125	1,425
	0750	24 circuits		.60	13.333		675	545		1,220	1,550
	0800	30 circuits		.53	15.094		780	615		1,395	1,775
	0850	225 amp main lugs, 32 circuits	2 Elec	.90	17.778		880	725		1,605	2,050

ELECTRICAL 16

16440	Swbds, Panels & Control Centers	CREW	DAILY OUTPUT	LABOR-HOURS	UNIT	2005 BARE COSTS				TOTAL INCL O&P	
						MAT.	LABOR	EQUIP.	TOTAL		
720	0900	34 circuits	2 Elec	.84	19.048	Ea.	900	775		1,675	2,150	**720**
	0950	36 circuits		.80	20		920	815		1,735	2,250	
	1000	42 circuits		.68	23.529		1,025	960		1,985	2,550	

16520	Exterior Luminaires	CREW	DAILY OUTPUT	LABOR-HOURS	UNIT	2005 BARE COSTS				TOTAL INCL O&P	
						MAT.	LABOR	EQUIP.	TOTAL		
300	0010	**EXTERIOR FIXTURES** With lamps										**300**
	0200	Wall mounted, incandescent, 100 watt	1 Elec	8	1	Ea.	28	41		69	91.50	
	0400	Quartz, 500 watt		5.30	1.509		54	61.50		115.50	151	
	0420	1500 watt		4.20	1.905		99	77.50		176.50	225	
	1100	Wall pack, low pressure sodium, 35 watt		4	2		214	81.50		295.50	355	
	1150	55 watt		4	2		255	81.50		336.50	400	
	1160	High pressure sodium, 70 watt		4	2		200	81.50		281.50	340	
	1170	150 watt		4	2		230	81.50		311.50	375	
	1180	Metal Halide, 175 watt		4	2		255	81.50		336.50	400	
	1190	250 watt		4	2		265	81.50		346.50	415	
	1200	Floodlights with ballast and lamp,										
	1400	pole mounted, pole not included										
	1950	Metal halide, 175 watt	1 Elec	2.70	2.963	Ea.	300	121		421	510	
	2000	400 watt	2 Elec	4.40	3.636		350	148		498	605	
	2200	1000 watt		4	4		505	163		668	800	
	2210	1500 watt		3.70	4.324		530	176		706	845	
	2250	Low pressure sodium, 55 watt	1 Elec	2.70	2.963		485	121		606	715	
	2270	90 watt		2	4		535	163		698	835	
	2290	180 watt		2	4		680	163		843	995	
	2340	High pressure sodium, 70 watt		2.70	2.963		207	121		328	405	
	2360	100 watt		2.70	2.963		212	121		333	415	
	2380	150 watt		2.70	2.963		235	121		356	440	
	2400	400 watt	2 Elec	4.40	3.636		340	148		488	590	
	2600	1000 watt	"	4	4		500	163		663	795	
	2610	Incandescent, 300 watt	1 Elec	4	2		76.50	81.50		158	205	
	2620	500 watt	"	4	2		122	81.50		203.50	255	
	2630	1000 watt	2 Elec	6	2.667		131	109		240	305	
	2640	1500 watt	"	6	2.667		144	109		253	320	
	2650	Roadway area luminaire, low pressure sodium, 135 watt	1 Elec	2	4		535	163		698	835	
	2700	180 watt	"	2	4		565	163		728	865	
	2750	Metal halide, 400 watt	2 Elec	4.40	3.636		445	148		593	705	
	2760	1000 watt		4	4		500	163		663	795	
	2780	High pressure sodium, 400 watt		4.40	3.636		460	148		608	725	
	2790	1000 watt		4	4		525	163		688	825	
	2800	Light poles, anchor base										
	2820	not including concrete bases										
	2840	Aluminum pole, 8' high	1 Elec	4	2	Ea.	520	81.50		601.50	690	
	2850	10' high		4	2		540	81.50		621.50	715	
	2860	12' high		3.80	2.105		565	86		651	750	
	2870	14' high		3.40	2.353		590	96		686	790	
	2880	16' high		3	2.667		645	109		754	870	
	3000	20' high	R-3	2.90	6.897		675	276	54.50	1,005.50	1,225	

16 ELECTRICAL

Important: See the Reference Section for critical supporting data - Reference Nos., Crews, & City Cost Indexes

			DAILY	LABOR-			2005 BARE COSTS				TOTAL	
16520		**Exterior Luminaires**	CREW	OUTPUT	HOURS	UNIT	MAT.	LABOR	EQUIP.	TOTAL	INCL O&P	
300	3200	30' high	R-3	2.60	7.692	Ea.	1,325	310	61	1,696	1,975	**300**
	3400	35' high		2.30	8.696		1,425	350	69	1,844	2,175	
	3600	40' high	▼	2	10		1,625	400	79.50	2,104.50	2,475	
	3800	Bracket arms, 1 arm	1 Elec	8	1		88.50	41		129.50	158	
	4000	2 arms		8	1		178	41		219	257	
	4200	3 arms		5.30	1.509		267	61.50		328.50	385	
	4400	4 arms		5.30	1.509		355	61.50		416.50	480	
	4500	Steel pole, galvanized, 8' high		3.80	2.105		460	86		546	640	
	4510	10' high		3.70	2.162		485	88		573	660	
	4520	12' high		3.40	2.353		525	96		621	720	
	4530	14' high		3.10	2.581		555	105		660	765	
	4540	16' high		2.90	2.759		590	112		702	810	
	4550	18' high		2.70	2.963		620	121		741	865	
	4600	20' high	R-3	2.60	7.692		820	310	61	1,191	1,425	
	4800	30' high		2.30	8.696		965	350	69	1,384	1,650	
	5000	35' high		2.20	9.091		1,050	365	72	1,487	1,775	
	5200	40' high	▼	1.70	11.765		1,300	470	93.50	1,863.50	2,225	
	5400	Bracket arms, 1 arm	1 Elec	8	1		134	41		175	209	
	5600	2 arms		8	1		208	41		249	290	
	5800	3 arms		5.30	1.509		226	61.50		287.50	340	
	6000	4 arms	▼	5.30	1.509		315	61.50		376.50	435	
	6100	Fiberglass pole, 1 or 2 fixtures, 20' high	R-3	4	5		465	200	39.50	704.50	850	
	6200	30' high		3.60	5.556		575	222	44	841	1,025	
	6300	35' high		3.20	6.250		900	250	49.50	1,199.50	1,425	
	6400	40' high	▼	2.80	7.143		1,050	286	56.50	1,392.50	1,650	
	6420	Wood pole, 4-1/2" x 5-1/8", 8' high	1 Elec	6	1.333		265	54.50		319.50	370	
	6430	10' high		6	1.333		299	54.50		353.50	410	
	6440	12' high		5.70	1.404		380	57		437	500	
	6450	15' high		5	1.600		445	65		510	580	
	6460	20' high	▼	4	2	▼	535	81.50		616.50	710	
	6500	Bollard light, lamp & ballast, 42" high with polycarbonate lens										
	6800	Metal halide, 175 watt	1 Elec	3	2.667	Ea.	645	109		754	865	
	6900	High pressure sodium, 70 watt		3	2.667		660	109		769	885	
	7000	100 watt		3	2.667		660	109		769	885	
	7100	150 watt		3	2.667		645	109		754	865	
	7200	Incandescent, 150 watt	▼	3	2.667	▼	465	109		574	670	
	7300	Transformer bases, not including concrete bases										
	7320	Maximum pole size, steel, 40' high	1 Elec	2	4	Ea.	1,100	163		1,263	1,475	
	7340	Cast aluminum, 30' high		3	2.667		600	109		709	820	
	7350	40' high	▼	2.50	3.200	▼	905	130		1,035	1,200	
	7380	Landscape recessed uplight, incl. housing, ballast, transformer										
	7390	& reflector										
	7420	Incandescent, 250 watt	1 Elec	5	1.600	Ea.	445	65		510	585	
	7440	Quartz, 250 watt		5	1.600		425	65		490	560	
	7460	500 watt	▼	4	2	▼	435	81.50		516.50	600	
	7500	Replacement (H.I.D.) ballasts,										
	7510	Multi-tap 120/208/240/277 volt										
	7550	High pressure sodium, 70 watt	1 Elec	10	.800	Ea.	124	32.50		156.50	185	
	7560	100 watt		9.40	.851		130	34.50		164.50	195	
	7570	150 watt		9	.889		140	36		176	208	
	7580	250 watt		8.50	.941		208	38.50		246.50	286	
	7590	400 watt		7	1.143		235	46.50		281.50	330	
	7600	1000 watt		6	1.333		325	54.50		379.50	435	
	7610	Metal halide, 175 watt		8	1		76.50	41		117.50	145	
	7620	250 watt		8	1		98.50	41		139.50	169	
	7630	400 watt	▼	7	1.143		123	46.50		169.50	205	

ELECTRICAL 16

For expanded coverage of these items see *Means Electrical Cost Data 2005*

			DAILY	LABOR-		2005 BARE COSTS				TOTAL	
16520	**Exterior Luminaires**	CREW	OUTPUT	HOURS	UNIT	MAT.	LABOR	EQUIP.	TOTAL	INCL O&P	
300 7640	1000 watt	1 Elec	6	1.333	Ea.	211	54.50		265.50	315	**300**
7650	1500 watt		5	1.600		263	65		328	385	
7810	Walkway luminaire, square 16", metal halide 250 watt		2.70	2.963		520	121		641	750	
7820	High pressure sodium, 70 watt		3	2.667		595	109		704	815	
7830	100 watt		3	2.667		605	109		714	825	
7840	150 watt		3	2.667		605	109		714	825	
7850	200 watt		3	2.667		610	109		719	830	
7910	Round 19", metal halide, 250 watt		2.70	2.963		755	121		876	1,000	
7920	High pressure sodium, 70 watt		3	2.667		830	109		939	1,075	
7930	100 watt		3	2.667		830	109		939	1,075	
7940	150 watt		3	2.667		835	109		944	1,075	
7950	250 watt		2.70	2.963		875	121		996	1,150	
8000	Sphere 14" opal, incandescent, 200 watt		4	2		244	81.50		325.50	390	
8020	Sphere 18" opal, incandescent, 300 watt		3.50	2.286		295	93		388	465	
8040	Sphere 16" clear, high pressure sodium, 70 watt		3	2.667		510	109		619	720	
8050	100 watt		3	2.667		545	109		654	760	
8100	Cube 16" opal, incandescent, 300 watt		3.50	2.286		325	93		418	500	
8120	High pressure sodium, 70 watt		3	2.667		475	109		584	685	
8130	100 watt		3	2.667		490	109		599	695	
8230	Lantern, high pressure sodium, 70 watt		3	2.667		420	109		529	625	
8240	100 watt		3	2.667		455	109		564	660	
8250	150 watt		3	2.667		425	109		534	630	
8260	250 watt		2.70	2.963		595	121		716	835	
8270	Incandescent, 300 watt		3.50	2.286		315	93		408	485	
8330	Reflector 22" w/globe, high pressure sodium, 70 watt		3	2.667		395	109		504	595	
8340	100 watt		3	2.667		405	109		514	605	
8350	150 watt		3	2.667		410	109		519	610	
8360	250 watt		2.70	2.963		520	121		641	755	
500 0010	**LANDSCAPE FIXTURES** Incl. conduit, wire, trench										**500**
0030	Bollards										
0040	Incandescent, 24"	1 Elec	2.50	3.200	Ea.	296	130		426	520	
0050	36"		2	4		355	163		518	635	
0060	42"		2	4		375	163		538	660	
0070	H.I.D., 24"		2.50	3.200		293	130		423	515	
0080	36"		2	4		325	163		488	605	
0090	42"		2	4		425	163		588	715	
0100	Concrete, 18" dia.		1.20	6.667		1,125	272		1,397	1,650	
0110	24" dia.		.75	10.667		1,425	435		1,860	2,200	
0120	Dry niche										
0130	300 W 120 Volt	1 Elec	4	2	Ea.	745	81.50		826.50	940	
0140	1000 W 120 Volt		2	4		1,000	163		1,163	1,350	
0150	300 W 12 Volt		4	2		745	81.50		826.50	940	
0160	Low voltage										
0170	Recessed uplight	1 Elec	2.20	3.636	Ea.	243	148		391	490	
0180	Walkway		4	2		222	81.50		303.50	365	
0190	Malibu - 5 light set		3	2.667		178	109		287	360	
0200	Mushroom 24" pier		4	2		182	81.50		263.50	320	
0210	Recessed, adjustable										
0220	Incandescent, 150 W	1 Elec	2.50	3.200	Ea.	370	130		500	605	
0230	300 W	"	2.50	3.200	"	430	130		560	665	
0250	Recessed uplight										
0260	Incandescent, 50 W	1 Elec	2.50	3.200	Ea.	340	130		470	570	
0270	150 W		2.50	3.200		355	130		485	585	
0280	300 W		2.50	3.200		395	130		525	630	
0310	Quartz 500 W		2.50	3.200		380	130		510	610	
0400	Recessed wall light										

Important: See the Reference Section for critical supporting data - Reference Nos., Crews, & City Cost Indexes

16520 | Exterior Luminaires

			CREW	DAILY OUTPUT	LABOR-HOURS	UNIT	2005 BARE COSTS				TOTAL INCL O&P	
							MAT.	LABOR	EQUIP.	TOTAL		
500	0410	Incandescent 100 W	1 Elec	4	2	Ea.	186	81.50		267.50	325	**500**
	0420	Fluorescent		4	2		169	81.50		250.50	305	
	0430	H.I.D. 100 W	▼	3	2.667	▼	345	109		454	540	
	0500	Step lights										
	0510	Incadescent	1 Elec	5	1.600	Ea.	171	65		236	285	
	0520	Fluorescent	"	5	1.600	"	182	65		247	297	
	0600	Tree lights, surface adjustable										
	0610	Incandescent 50 W	1 Elec	3	2.667	Ea.	405	109		514	605	
	0620	Incandescent 100 W		3	2.667		425	109		534	625	
	0630	Incandescent 150 W	▼	2	4	▼	450	163		613	740	
	0700	Underwater lights										
	0710	150 W 120 Volt	1 Elec	6	1.333	Ea.	675	54.50		729.50	820	
	0720	300 W 120 Volt		6	1.333		805	54.50		859.50	965	
	0730	1000 W 120 Volt		4	2		1,025	81.50		1,106.50	1,250	
	0740	50 W 12 Volt		6	1.333		765	54.50		819.50	920	
	0750	300 W 12 Volt	▼	6	1.333	▼	690	54.50		744.50	840	
	0800	Walkway, adjustable										
	0810	Fluorescent, 2'	1 Elec	3	2.667	Ea.	289	109		398	480	
	0820	Fluorescent, 4'		3	2.667		305	109		414	500	
	0830	Fluorescent 8'		2	4		600	163		763	905	
	0840	Incandescent, 50 W		4	2		273	81.50		354.50	420	
	0850	150 W	▼	4	2	▼	294	81.50		375.50	445	
	0900	Wet niche										
	0910	300 W 120 Volt	1 Elec	2.50	3.200	Ea.	600	130		730	855	
	0920	1000 W 120 Volt		1.50	5.333		720	217		937	1,125	
	0930	300 W 12 Volt	▼	2.50	3.200	▼	685	130		815	950	

16525 | Aviation Lighting

			CREW	DAILY OUTPUT	LABOR-HOURS	UNIT	MAT.	LABOR	EQUIP.	TOTAL	TOTAL INCL O&P	
100	0010	**AIRPORT LIGHTING**										**100**
	0100	Runway centerline, bidir., semi-flush, 200 W, w/shallow insert base	R-22	12.40	3.006	Ea.	1,175	103		1,278	1,450	
	0120	Flush, 200 W, w/shallow insert base		12.40	3.006		1,175	103		1,278	1,450	
	0130	for mounting in base housing		18.64	2		625	68.50		693.50	790	
	0150	Touchdown zone light, unidirectional, 200 W, w/shallow insert base		12.40	3.006		970	103		1,073	1,225	
	0160	115 W		12.40	3.006		1,025	103		1,128	1,275	
	0180	Unidirectional, 200 W, for mounting in base housing		18.60	2.004		475	69		544	625	
	0190	115 W		18.60	2.004		510	69		579	665	
	0210	Runway edge & threshold light, bidir., 200 W, for base housing		9.36	3.983		670	137		807	940	
	0240	Threshold & approach light, unidir., 200 W, for base housing		9.36	3.983		390	137		527	635	
	0260	Runway edge, bi-directional, 2-115 W, for base housing		12.40	3.006		1,025	103		1,128	1,300	
	0280	Runway threshold & end, bidir., 2-115 W, for base housing		12.40	3.006		1,150	103		1,253	1,400	
	0370	45 W, flush, for mounting in base housing		18.64	2		560	68.50		628.50	720	
	0380	115 W	▼	18.64	2		575	68.50		643.50	735	
	1200	Wind cone, 12' lighted assembly, rigid, w/obstruction light	R-21	1.36	24.118		6,575	980	42	7,597	8,725	
	1210	Without obstruction light		1.52	21.579		5,500	880	38	6,418	7,400	
	1220	Unlighted assembly, w/obstruction light		1.68	19.524		6,000	795	34	6,829	7,825	
	1230	Without obstruction light		1.84	17.826		5,500	725	31	6,256	7,150	
	1240	Wind cone slip fitter, 2-1/2" pipe		21.84	1.502		72.50	61	2.63	136.13	173	
	1250	Wind cone sock, 12' x 3', cotton		6.56	5		400	204	8.75	612.75	755	
	1260	Nylon	▼	6.56	5	▼	405	204	8.75	617.75	760	

16585 | Lamps

			CREW	DAILY OUTPUT	LABOR-HOURS	UNIT	MAT.	LABOR	EQUIP.	TOTAL	TOTAL INCL O&P	
600	0010	**LAMPS**										**600**
	0600	Mercury vapor, mogul base, deluxe white, 100 watt	1 Elec	.30	26.667	C	2,575	1,075		3,650	4,475	
	0650	175 watt		.30	26.667		1,925	1,075		3,000	3,725	
	0700	250 watt	▼	.30	26.667	▼	3,425	1,075		4,500	5,375	

ELECTRICAL 16

16585 | Lamps

		CREW	DAILY OUTPUT	LABOR-HOURS	UNIT	2005 BARE COSTS				TOTAL INCL O&P
						MAT.	LABOR	EQUIP.	TOTAL	
0800	400 watt	1 Elec	.30	26.667	C	2,750	1,075		3,825	4,650
0900	1000 watt		.20	40		6,425	1,625		8,050	9,475
1000	Metal halide, mogul base, 175 watt		.30	26.667		3,225	1,075		4,300	5,175
1100	250 watt		.30	26.667		3,650	1,075		4,725	5,650
1200	400 watt		.30	26.667		3,475	1,075		4,550	5,425
1300	1000 watt		.20	40		9,100	1,625		10,725	12,400
1320	1000 watt, 125,000 initial lumens		.20	40		12,700	1,625		14,325	16,300
1330	1500 watt		.20	40		12,300	1,625		13,925	15,900
1350	High pressure sodium, 70 watt		.30	26.667		3,750	1,075		4,825	5,750
1360	100 watt		.30	26.667		4,275	1,075		5,350	6,325
1370	150 watt		.30	26.667		4,025	1,075		5,100	6,050
1380	250 watt		.30	26.667		4,275	1,075		5,350	6,350
1400	400 watt		.30	26.667		4,400	1,075		5,475	6,475
1450	1000 watt		.20	40		13,200	1,625		14,825	16,900
1500	Low pressure sodium, 35 watt		.30	26.667		5,700	1,075		6,775	7,900
1550	55 watt		.30	26.667		6,275	1,075		7,350	8,525
1600	90 watt		.30	26.667		7,250	1,075		8,325	9,600
1650	135 watt		.20	40		9,275	1,625		10,900	12,600
1700	180 watt		.20	40		10,200	1,625		11,825	13,600
1750	Quartz line, clear, 500 watt		1.10	7.273		790	296		1,086	1,300
1760	1500 watt		.20	40		3,475	1,625		5,100	6,225
1762	Spot, MR 16, 50 watt		1.30	6.154		750	251		1,001	1,200

16710 | Communication Circuits

		CREW	DAILY OUTPUT	LABOR-HOURS	UNIT	2005 BARE COSTS				TOTAL INCL O&P
						MAT.	LABOR	EQUIP.	TOTAL	
0010	**FIBER OPTICS**									
0020	Fiber optics cable only. Added costs depend on the type of fiber									
0030	special connectors, optical modems, and networking parts.									
0040	Specialized tools & techniques cause installation costs to vary.									
0070	Cable, minimum, bulk simplex	1 Elec	8	1	C.L.F.	24	41		65	87
0080	Cable, maximum, bulk plenum quad	"	2.29	3.493	"	117	142		259	340
0150	Fiber optic jumper				Ea.	55.50			55.50	61
0200	Fiber optic pigtail					30			30	33
0300	Fiber optic connector	1 Elec	24	.333		16.90	13.60		30.50	38.50
0350	Fiber optic finger splice		32	.250		32	10.20		42.20	50
0400	Transceiver (low cost bi-directional)		8	1		425	41		466	530
0450	Multi-channel rack enclosure (10 modules)		2	4		475	163		638	770
0500	Fiber optic patch panel (12 ports)		6	1.333		178	54.50		232.50	277
1000	Cable, 62.5 microns, direct burial, 4 fiber	R-15	1,200	.040	L.F.	.74	1.58	.21	2.53	3.40
1020	Indoor, 2 fiber	R-19	1,000	.020		.34	.82		1.16	1.59
1040	Outdoor, aerial/duct	"	1,670	.012		1.89	.49		2.38	2.81
1060	50 microns, direct burial, 8 fiber	R-22	4,000	.009		1.55	.32		1.87	2.19
1080	12 fiber		4,000	.009		2.10	.32		2.42	2.79
1100	Indoor, 12 fiber		759	.049		1.65	1.68		3.33	4.36
1120	Connectors, 62.5 micron cable, transmission	R-19	40	.500	Ea.	9.40	20.50		29.90	41
1140	Cable splice		40	.500		6.45	20.50		26.95	37.50
1160	125 micron cable, transmission		16	1.250		9.25	51		60.25	86
1180	Receiver, 1.2 mile range		20	1		256	41		297	340
1200	1.9 mile range		20	1		485	41		526	590
1220	6.2 mile range		5	4		685	163		848	995
1240	Transmitter, 1.2 mile range		20	1		255	41		296	340
1260	1.9 mile range		20	1		485	41		526	590
1280	6.2 mile range		5	4		685	163		848	995
1300	Modem, 1.2 mile range		5	4		325	163		488	605
1320	6.2 mile range		5	4		900	163		1,063	1,225
1340	1.9 mile range, 12 channel		5	4		1,800	163		1,963	2,225
1360	Repeater, 1.2 mile range		10	2		430	81.50		511.50	590

Important: See the Reference Section for critical supporting data - Reference Nos., Crews, & City Cost Indexes

16 ELECTRICAL

16710	Communication Circuits	CREW	DAILY OUTPUT	LABOR-HOURS	UNIT	2005 BARE COSTS				TOTAL INCL O&P	
						MAT.	LABOR	EQUIP.	TOTAL		
400											400
1380	1.9 mile range	R-19	10	2	Ea.	430	81.50		511.50	590	
1400	6.2 mile range		5	4		1,100	163		1,263	1,450	
1420	1.2 mile range, digital		5	4		430	163		593	715	
1440	Transceiver, 1.9 mile range		5	4		745	163		908	1,050	
1460	1.2 mile range, digital		5	4		255	163		418	525	
1480	Cable enclosure, interior NEMA 13		7	2.857		143	117		260	330	
1500	Splice w/enclosure encapsulant		16	1.250		214	51		265	310	

For information about Means Estimating Seminars, see yellow pages 12 and 13 in back of book

ELECTRICAL 16

Division Notes

		CREW	DAILY OUTPUT	LABOR-HOURS	UNIT	2005 BARE COSTS				TOTAL INCL O&P
						MAT.	LABOR	EQUIP.	TOTAL	

Assemblies
Section

Table of Contents

Table No.	Page
A SUBSTRUCTURE	319
A1010 Standard Foundations	
A1010 100 Wall Foundations	320
A1010 200 Column Foundations & Pile Caps	321
A1010 300 Perimeter Drainage	325
A1020 Special Foundations	
A1020 100 Pile Foundations	327
A1020 200 Grade Beams	338

Table No.	Page
A1020 300 Caissons	340
A1020 700 Pressure Injected Footings	341
A2020 Basement Walls	
A2020 100 Basement Wall Construction	342
G BUILDING SITEWORK	347
G1030 Site Earthwork	
G1030 800 Utilities Trenching	348

Table No.	Page
G2030 Pedestrian Paving	
G2030 100 Paving & Surfacing	353
G2040 Site Development	
G2040 200 Paving & Surfacing	355
G3030 Storm Sewer	
G3030 200 Manholes & Cleanouts	358
G3030 300 Headwalls & Catch Basins	360

How to Use the Assemblies Cost Tables

The following is a detailed explanation of a sample Assemblies Cost Table. Most Assembly Tables are separated into three parts: 1) an illustration of the system to be estimated; 2) the components and related costs of a typical system; and 3) the costs for similar systems with dimensional and/or size variations. For costs of the components that comprise these systems or "assemblies," refer to the Unit Price Section. Next to each bold number below is the item being described with the appropriate component of the sample entry following in parenthesis. In most cases, if the work is to be subcontracted, the general contractor will need to add an additional markup (RSMeans suggests using 10%) to the "Total" figures.

System/Line Numbers (G2040 210)

Each Assemblies Cost Line has been assigned a unique identification number based on the UNIFORMAT II classification system.

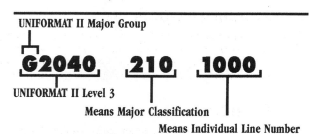

UNIFORMAT II Major Group

G2040   210   1000

UNIFORMAT II Level 3

Means Major Classification

Means Individual Line Number

G20 Site Improvements
G2040 Site Development

There are four basic types of Concrete Retaining Wall Systems: reinforced concrete with level backfill; reinforced concrete with sloped backfill or surcharge; unreinforced with level backfill; and unreinforced with sloped backfill or surcharge. System elements include: all necessary forms (4 uses); 3,000 p.s.i. concrete with an 8" chute; all necessary reinforcing steel; and underdrain. Exposed concrete is patched and rubbed.

The Expanded System Listing shows walls that range in thickness from 10" to 18" for reinforced concrete walls with level backfill and 12" to 24" for reinforced walls with sloped backfill. Walls range from a height of 4' to 20'. Unreinforced level and sloped backfill walls range from a height of 3' to 10'.

System Components	QUANTITY	UNIT	COST PER L.F. MAT.	COST PER L.F. INST.	COST PER L.F. TOTAL
SYSTEM G2040 210 1000					
CONC.RETAIN. WALL REINFORCED, LEVEL BACKFILL 4' HIGH					
Forms in place, cont. wall footing & keyway, 4 uses	2.000	S.F.	1.66	6.66	8.32
Forms in place, retaining wall forms, battered to 8' high, 4 uses	8.000	SFCA	4.88	51.20	56.08
Reinforcing in place, walls, #3 to #7	.004	Ton	3.34	2.68	6.02
Concrete ready mix, regular weight, 3000 psi	.204	C.Y.	18.16		18.16
Placing concrete and vibrating footing con., shallow direct chute	.074	C.Y.		1.31	1.31
Placing concrete and vibrating walls, 8" thick, direct chute	.130	C.Y.		3.07	3.07
Pipe bedding, crushed or screened bank run gravel	1.000	L.F.	2.42	.95	3.37
Pipe, subdrainage, corrugated plastic, 4" diameter	1.000	L.F.	.63	.55	1.18
Finish walls and break ties, patch walls	4.000	S.F.	.12	2.88	3
TOTAL			31.21	69.30	100.51

G2040 210	Concrete Retaining Walls	COST PER L.F. MAT.	COST PER L.F. INST.	COST PER L.F. TOTAL
1000	Conc.retain.wall, reinforced, level backfill, 4' high x 2'-2" base,10"thick	31	69.50	100.50
1200	6' high x 3'-3" base, 10" thick	45.50	100	145.50
1400	8' high x 4'-3" base, 10" thick	60	132	192
1600	10' high x 5'-4" base, 13" thick	79	193	272
1800	12' high x 6'-6" base, 14" thick	98	231	329
2200	16' high x 8'-6" base, 16" thick	149	310	459
2600	20' high x 10'-5" base, 18" thick	218	405	623
3000	Sloped backfill, 4' high x 3'-2" base, 12" thick	38	72	110

2 Illustration

At the top of most assembly tables is an illustration, a brief description, and the design criteria used to develop the cost.

3 System Components

The components of a typical system are listed separately to show what has been included in the development of the total system price. The table below contains prices for other similar systems with dimensional and/or size variations.

4 Quantity

This is the number of line item units required for one system unit. For example, we assume that it will take 2 S.F. of footing forms per L.F. of wall.

5 Unit of Measure for Each Item

The abbreviated designation indicates the unit of measure, as defined by industry standards, upon which the price of the component is based. For example, reinforcing is priced per ton and concrete is priced per C.Y.

6 Unit of Measure for Each System (Cost per L.F.)

Costs shown in the three right hand columns have been adjusted by the component quantity and unit of measure for the entire system. In this example, "Cost per L.F." is the unit of measure for this system or "assembly."

7 Materials (31.21)

This column contains the Materials Cost of each component. These cost figures are bare costs plus 10% for profit.

8 Installation (69.30)

Installation includes labor and equipment plus the installing contractor's overhead and profit. Equipment costs are the bare rental costs plus 10% for profit. The labor overhead and profit is defined on the inside back cover of this book.

9 Total (100.51)

The figure in this column is the sum of the material and installation costs.

Material Cost	+	Installation Cost	=	Total
$31.21	+	$69.30	=	$100.51

A SUBSTRUCTURE

A1010 Standard Foundations

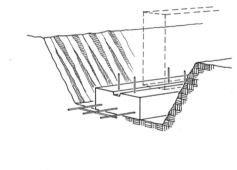

The Strip Footing System includes: excavation; hand trim; all forms needed for footing placement; forms for 2″ x 6″ keyway (four uses); dowels; and 3,000 p.s.i. concrete.

The footing size required varies for different soils. Soil bearing capacities are listed for 3 KSF and 6 KSF. Depths of the system range from 8″ to 24″. Widths range from 16″ to 96″. Smaller strip footings may not require reinforcement.

Please see the reference section for further design and cost information.

System Components	QUANTITY	UNIT	COST PER L.F. MAT.	COST PER L.F. INST.	COST PER L.F. TOTAL
SYSTEM A1010 110 2500					
STRIP FOOTING, LOAD 5.1KLF, SOIL CAP. 3 KSF, 24″WIDE X12″DEEP, REINF.					
Trench excavation	.148	C.Y.		.97	.97
Hand trim	2.000	S.F.		1.42	1.42
Compacted backfill	.074	C.Y.		.24	.24
Formwork, 4 uses	2.000	S.F.	1.66	6.66	8.32
Keyway form, 4 uses	1.000	L.F.	.30	.85	1.15
Reinforcing, fy = 60000 psi	3.000	Lb.	1.38	1.44	2.82
Dowels	2.000	Ea.	1.30	4.18	5.48
Concrete, f'c = 3000 psi	.074	C.Y.	6.59		6.59
Place concrete, direct chute	.074	C.Y.		1.31	1.31
Screed finish	2.000	S.F.		1.06	1.06
TOTAL			**11.23**	**18.13**	**29.36**

A1010 110	Strip Footings	COST PER L.F. MAT.	COST PER L.F. INST.	COST PER L.F. TOTAL
2100	Strip footing, load 2.6KLF, soil capacity 3KSF, 16″wide x 8″deep plain	5.45	10.90	16.35
2300	Load 3.9 KLF, soil capacity, 3 KSF, 24″wide x 8″deep, plain	6.95	12.05	19
2500	Load 5.1KLF, soil capacity 3 KSF, 24″wide x 12″deep, reinf.	11.25	18.15	29.40
2700	Load 11.1KLF, soil capacity 6 KSF, 24″wide x 12″deep, reinf.	11.25	18.15	29.40
2900	Load 6.8 KLF, soil capacity 3 KSF, 32″wide x 12″deep, reinf.	13.95	19.85	33.80
3100	Load 14.8 KLF, soil capacity 6 KSF, 32″wide x 12″deep, reinf.	13.95	19.85	33.80
3300	Load 9.3 KLF, soil capacity 3 KSF, 40″wide x 12″deep, reinf.	16.65	21.50	38.15
3500	Load 18.4 KLF, soil capacity 6 KSF, 40″wide x 12″deep, reinf.	16.80	22	38.80
3700	Load 10.1KLF, soil capacity 3 KSF, 48″wide x 12″deep, reinf.	18.75	23	41.75
3900	Load 22.1KLF, soil capacity 6 KSF, 48″wide x 12″deep, reinf.	20	24.50	44.50
4100	Load 11.8KLF, soil capacity 3 KSF, 56″wide x 12″deep, reinf.	22	25.50	47.50
4300	Load 25.8KLF, soil capacity 6 KSF, 56″wide x 12″deep, reinf.	24	28	52
4500	Load 10KLF, soil capacity 3 KSF, 48″wide x 16″deep, reinf.	23.50	27	50.50
4700	Load 22KLF, soil capacity 6 KSF, 48″wide, 16″deep, reinf.	24	27.50	51.50
4900	Load 11.6KLF, soil capacity 3 KSF, 56″wide x 16″deep, reinf.	27	38	65
5100	Load 25.6KLF, soil capacity 6 KSF, 56″wide x 16″deep, reinf.	28.50	40	68.50
5300	Load 13.3KLF, soil capacity 3 KSF, 64″wide x 16″deep, reinf.	31.50	32	63.50
5500	Load 29.3KLF, soil capacity 6 KSF, 64″wide x 16″deep, reinf.	33.50	34.50	68
5700	Load 15KLF, soil capacity 3 KSF, 72″wide x 20″deep, reinf.	42	38.50	80.50
5900	Load 33KLF, soil capacity 6 KSF, 72″wide x 20″deep, reinf.	44.50	41.50	86
6100	Load 18.3KLF, soil capacity 3 KSF, 88″wide x 24″deep, reinf.	59.50	49	108.50
6300	Load 40.3KLF, soil capacity 6 KSF, 88″wide x 24″deep, reinf.	65	54.50	119.50
6500	Load 20KLF, soil capacity 3 KSF, 96″wide x 24″deep, reinf.	65	51.50	116.50
6700	Load 44 KLF, soil capacity 6 KSF, 96″ wide x 24″ deep, reinf.	69	56	125

SUBSTRUCTURE

A

Important: See the Reference Section for critical supporting data - Reference Numbers and City Cost Indexes

A1010 Standard Foundations

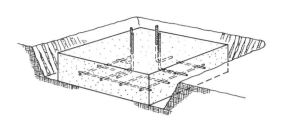

The Spread Footing System includes: excavation; backfill; forms (four uses); all reinforcement; 3,000 p.s.i. concrete (chute placed); and screed finish.

Footing systems are priced per individual unit. The Expanded System Listing at the bottom shows footings that range from 3' square x 12" deep, to 18' square x 52" deep. It is assumed that excavation is done by a truck mounted hydraulic excavator with an operator and oiler.

Backfill is with a dozer, and compaction by air tamp. The excavation and backfill equipment is assumed to operate at 30 C.Y. per hour.

Please see the reference section for further design and cost information.

System Components	QUANTITY	UNIT	COST EACH MAT.	COST EACH INST.	COST EACH TOTAL
SYSTEM A1010 210 7100					
SPREAD FOOTINGS, LOAD 25K, SOIL CAPACITY 3 KSF, 3' SQ X 12" DEEP					
Bulk excavation	.590	C.Y.		4.15	4.15
Hand trim	9.000	S.F.		6.39	6.39
Compacted backfill	.260	C.Y.		.84	.84
Formwork, 4 uses	12.000	S.F.	7.08	46.80	53.88
Reinforcing, fy = 60,000 psi	.006	Ton	5.01	5.73	10.74
Dowel or anchor bolt templates	6.000	L.F.	5.04	19.38	24.42
Concrete, f'c = 3,000 psi	.330	C.Y.	29.37		29.37
Place concrete, direct chute	.330	C.Y.		5.84	5.84
Screed finish	9.000	S.F.		3.87	3.87
TOTAL			46.50	93	139.50

A1010 210	Spread Footings	COST EACH MAT.	COST EACH INST.	COST EACH TOTAL
7090	Spread footings, 3000 psi concrete, chute delivered			
7100	Load 25K, soil capacity 3 KSF, 3'-0" sq. x 12" deep	46.50	93	139.50
7150	Load 50K, soil capacity 3 KSF, 4'-6" sq. x 12" deep	99	161	260
7200	Load 50K, soil capacity 6 KSF, 3'-0" sq. x 12" deep	46.50	93	139.50
7250	Load 75K, soil capacity 3 KSF, 5'-6" sq. x 13" deep	157	227	384
7300	Load 75K, soil capacity 6 KSF, 4'-0" sq. x 12" deep	80.50	138	218.50
7350	Load 100K, soil capacity 3 KSF, 6'-0" sq. x 14" deep	199	273	472
7410	Load 100K, soil capacity 6 KSF, 4'-6" sq. x 15" deep	122	189	311
7450	Load 125K, soil capacity 3 KSF, 7'-0" sq. x 17" deep	315	390	705
7500	Load 125K, soil capacity 6 KSF, 5'-0" sq. x 16" deep	157	227	384
7550	Load 150K, soil capacity 3 KSF 7'-6" sq. x 18" deep	380	460	840
7610	Load 150K, soil capacity 6 KSF, 5'-6" sq. x 18" deep	209	287	496
7650	Load 200K, soil capacity 3 KSF, 8'-6" sq. x 20" deep	540	605	1,145
7700	Load 200K, soil capacity 6 KSF, 6'-0" sq. x 20" deep	274	355	629
7750	Load 300K, soil capacity 3 KSF, 10'-6" sq. x 25" deep	1,000	985	1,985
7810	Load 300K, soil capacity 6 KSF, 7'-6" sq. x 25" deep	520	595	1,115
7850	Load 400K, soil capacity 3 KSF, 12'-6" sq. x 28" deep	1,575	1,450	3,025
7900	Load 400K, soil capacity 6 KSF, 8'-6" sq. x 27" deep	725	770	1,495
7950	Load 500K, soil capacity 3 KSF, 14'-0" sq. x 31" deep	2,200	1,925	4,125
8010	Load 500K, soil capacity 6 KSF, 9'-6" sq. x 30" deep	995	1,000	1,995
8050	Load 600K, soil capacity 3 KSF, 16'-0" sq. x 35" deep	3,225	2,650	5,875
8100	Load 600K, soil capacity 6 KSF, 10'-6" sq. x 33" deep	1,350	1,300	2,650

A SUBSTRUCTURE

A1010 Standard Foundations

A1010 210	Spread Footings	COST EACH		
		MAT.	INST.	TOTAL
8150	Load 700K, soil capacity 3 KSF, 17'-0" sq. x 37" deep	3,800	3,000	6,800
8200	Load 700K, soil capacity 6 KSF, 11'-6" sq. x 36" deep	1,725	1,575	3,300
8250	Load 800K, soil capacity 3 KSF, 18'-0" sq. x 39" deep	4,475	3,500	7,975
8300	Load 800K, soil capacity 6 KSF, 12'-0" sq. x 37" deep	1,925	1,725	3,650
8350	Load 900K, soil capacity 3 KSF, 19'-0" sq. x 40" deep	5,200	4,000	9,200
8400	Load 900K, soil capacity 6 KSF, 13'-0" sq. x 39" deep	2,375	2,050	4,425
8450	Load 1000K, soil capacity 3 KSF, 20'-0" sq. x 42" deep	6,000	4,500	10,500
8500	Load 1000K, soil capacity 6 KSF, 13'-6" sq. x 41" deep	2,700	2,300	5,000
8550	Load 1200K, soil capacity 6 KSF, 15'-6" sq. x 48" deep	3,575	2,925	6,500
8600	Load 1400K, soil capacity 6 KSF, 16'-0" sq. x 47" deep	4,375	3,475	7,850
8650	Load 1600K, soil capacity 6 KSF, 18'-0" sq. x 52" deep	6,075	4,600	10,675

Important: See the Reference Section for critical supporting data - Reference Numbers and City Cost Indexes

A1010 Standard Foundations

These pile cap systems include excavation with a truck mounted hydraulic excavator, hand trimming, compacted backfill, forms for concrete, templates for dowels or anchor bolts, reinforcing steel and concrete placed and screeded.

Pile embedment is assumed as 6″. Design is consistent with the Concrete Reinforcing Steel Institute Handbook f'c = 3000 psi, fy = 60,000.

Please see the reference section for further design and cost information.

System Components			COST EACH		
	QUANTITY	UNIT	MAT.	INST.	TOTAL
SYSTEM A1010 250 5100					
CAP FOR 2 PILES,6'-6″X3'-6″X20″,15 TON PILE,8″ MIN. COL., 45K COL. LOAD					
Excavation, bulk, hyd excavator, truck mtd. 30″ bucket 1/2 CY	2.890	C.Y.		20.32	20.32
Trim sides and bottom of trench, regular soil	23.000	S.F.		16.33	16.33
Dozer backfill & roller compaction	1.500	C.Y.		4.81	4.81
Forms in place pile cap, square or rectangular, 4 uses	33.000	SFCA	27.72	138.93	166.65
Templates for dowels or anchor bolts	8.000	Ea.	6.72	25.84	32.56
Reinforcing in place footings, #8 to #14	.025	Ton	19.75	13.88	33.63
Concrete ready mix, regular weight, 3000 psi	1.400	C.Y.	124.60		124.60
Place and vibrate concrete for pile caps, under 5CY, direct chute	1.400	C.Y.		33.03	33.03
Monolithic screed finish	23.000	S.F.		9.89	9.89
TOTAL			178.79	263.03	441.82

A1010 250	Pile Caps							
	NO. PILES	SIZE FT-IN X FT-IN X IN	PILE CAPACITY (TON)	COLUMN SIZE (IN)		COST EACH		
						MAT.	INST.	TOTAL
5100	2	6-6x3-6x20	15	8	45	179	263	442
5150		26	40	8	155	220	320	540
5200		34	80	11	314	297	410	707
5250		37	120	14	473	320	440	760
5300	3	5-6x5-1x23	15	8	75	212	305	517
5350		28	40	10	232	238	340	578
5400		32	80	14	471	274	390	664
5450		38	120	17	709	315	445	760
5500	4	5-6x5-6x18	15	10	103	239	305	544
5550		30	40	11	308	350	440	790
5600		36	80	16	626	410	510	920
5650		38	120	19	945	435	530	965
5700	6	8-6x5-6x18	15	12	156	400	445	845
5750		37	40	14	458	650	650	1,300
5800		40	80	19	936	740	725	1,465
5850		45	120	24	1413	825	800	1,625
5900	8	8-6x7-9x19	15	12	205	600	605	1,205
5950		36	40	16	610	870	785	1,655
6000		44	80	22	1243	1,075	950	2,025
6050		47	120	27	1881	1,175	1,025	2,200

SUBSTRUCTURE

A

A1010 Standard Foundations

A1010 250	Pile Caps

	NO. PILES	SIZE FT-IN X FT-IN X IN	PILE CAPACITY (TON)	COLUMN SIZE (IN)		COST EACH		
						MAT.	INST.	TOTAL
6100	10	11-6x7-9x21	15	14	250	850	740	1,590
6150		39	40	17	756	1,300	1,050	2,350
6200		47	80	25	1547	1,575	1,250	2,825
6250		49	120	31	2345	1,675	1,325	3,000
6300	12	11-6x8-6x22	15	15	316	1,075	895	1,970
6350		49	40	19	900	1,725	1,325	3,050
6400		52	80	27	1856	1,900	1,450	3,350
6450		55	120	34	2812	2,050	1,575	3,625
6500	14	11-6x10-9x24	15	16	345	1,400	1,100	2,500
6550		41	40	21	1056	1,850	1,350	3,200
6600		55	80	29	2155	2,425	1,725	4,150
6700	16	11-6x11-6x26	15	18	400	1,675	1,275	2,950
6750		48	40	22	1200	2,275	1,600	3,875
6800		60	80	31	2460	2,825	1,950	4,775
6900	18	13-0x11-6x28	15	20	450	1,925	1,375	3,300
6950		49	40	23	1349	2,625	1,800	4,425
7000		56	80	33	2776	3,100	2,125	5,225
7100	20	14-6x11-6x30	15	20	510	2,350	1,675	4,025
7150		52	40	24	1491	3,175	2,125	5,300

Important: See the Reference Section for critical supporting data - Reference Numbers and City Cost Indexes

A1010 Standard Foundations

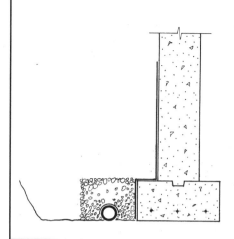

General: Footing drains can be placed either inside or outside of foundation walls depending upon the source of water to be intercepted. If the source of subsurface water is principally from grade or a subsurface stream above the bottom of the footing, outside drains should be used. For high water tables, use inside drains or both inside and outside.

The effectiveness of underdrains depends on good waterproofing. This must be carefully installed and protected during construction.

Costs below include the labor and materials for the pipe and 6″ only of gravel or crushed stone around pipe. Excavation and backfill are not included.

System Components	QUANTITY	UNIT	COST PER L.F.		
			MAT.	INST.	TOTAL
SYSTEM A1010 310 1000					
FOUNDATION UNDERDRAIN, OUTSIDE ONLY, PVC 4″ DIAM.					
PVC pipe 4″ diam. S.D.R. 35	1.000	L.F.	2.11	3	5.11
Pipe bedding, graded gravel 3/4″ to 1/2″	.070	C.Y.	2.10	.61	2.71
TOTAL			4.21	3.61	7.82

A1010 310	Foundation Underdrain	COST PER L.F.		
		MAT.	INST.	TOTAL
1000	Foundation underdrain, outside only, PVC, 4″ diameter	4.21	3.61	7.82
1100	6″ diameter	6.50	3.99	10.49
1200	Bituminous fiber, 4″ diameter	4.21	3.61	7.82
1300	6″ diameter	6.50	3.99	10.49
1400	Porous concrete, 6″ diameter	5.50	4.35	9.85
1450	8″ diameter	6.75	5.85	12.60
1500	12″ diameter	12.10	6.70	18.80
1600	Corrugated metal, 16 ga. asphalt coated, 6″ diameter	5.75	6.90	12.65
1650	8″ diameter	7.65	7.25	14.90
1700	10″ diameter	9.30	7.55	16.85
2000	Vitrified clay, C-211, 4″ diameter	4.30	6.40	10.70
2050	6″ diameter	6.35	8.15	14.50
2100	12″ diameter	15.70	9.85	25.55
3000	Outside and inside, PVC, 4″ diameter	8.40	7.20	15.60
3100	6″ diameter	12.95	8	20.95
3200	Bituminous fiber, 4″ diameter	8.40	7.20	15.60
3300	6″ diameter	12.95	8	20.95
3400	Porous concrete, 6″ diameter	11	8.70	19.70
3450	8″ diameter	13.50	11.65	25.15
3500	12″ diameter	24	13.40	37.40
3600	Corrugated metal, 16 ga., asphalt coated, 6″ diameter	11.50	13.85	25.35
3650	8″ diameter	15.25	14.50	29.75
3700	10″ diameter	18.60	15.20	33.80
4000	Vitrified clay, C-211, 4″ diameter	8.60	12.80	21.40
4050	6″ diameter	12.70	16.30	29
4100	12″ diameter	31.50	19.70	51.20

SUBSTRUCTURE

A

A1010 Standard Foundations

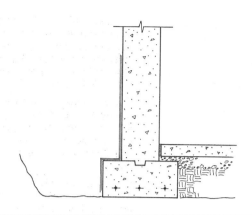

General: Apply foundation wall dampproofing over clean concrete giving particular attention to the joint between the wall and the footing. Use care in backfilling to prevent damage to the dampproofing.

Costs for four types of dampproofing are listed below.

System Components				COST PER L.F.		
	QUANTITY	UNIT		MAT.	INST.	TOTAL
SYSTEM A1010 320 1000						
FOUNDATION DAMPPROOFING, BITUMINOUS, 1 COAT, 4' HIGH						
Bituminous asphalt dampproofing brushed on below grade, 1 coat	4.000	S.F.		.32	2.40	2.72
Labor for protection of dampproofing during backfilling	4.000	S.F.			.99	.99
TOTAL				.32	3.39	3.71

A1010 320	Foundation Dampproofing		COST PER L.F.		
			MAT.	INST.	TOTAL
1000	Foundation dampproofing, bituminous, 1 coat, 4' high		.32	3.39	3.71
1400	8' high		.64	6.80	7.44
1800	12' high		.96	10.50	11.46
2000	2 coats, 4' high		.44	4.19	4.63
2400	8' high		.88	8.40	9.28
2800	12' high		1.32	12.90	14.22
3000	Asphalt with fibers, 1/16" thick, 4' high		.72	4.19	4.91
3400	8' high		1.44	8.40	9.84
3800	12' high		2.16	12.90	15.06
4000	1/8" thick, 4' high		1.28	4.99	6.27
4400	8' high		2.56	10	12.56
4800	12' high		3.84	15.30	19.14
5000	Asphalt coated board and mastic, 1/4" thick, 4' high		2.72	4.55	7.27
5400	8' high		5.45	9.10	14.55
5800	12' high		8.15	14	22.15
6000	1/2" thick, 4' high		4.08	6.35	10.43
6400	8' high		8.15	12.65	20.80
6800	12' high		12.25	19.30	31.55
7000	Cementitious coating, on walls, 1/8" thick coating, 4' high		6.60	4.64	11.24
7400	8' high		13.20	9.25	22.45
7800	12' high		19.80	13.95	33.75
8000	Cementitious/metallic slurry, 2 coat, 1/4" thick, 2' high		38.50	156	194.50
8400	4' high		77	310	387
8800	6' high		116	470	586

SUBSTRUCTURE A

Important: See the Reference Section for critical supporting data - Reference Numbers and City Cost Indexes

A1020   Special Foundations

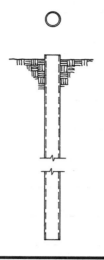

The Cast-in-Place Concrete Pile System includes: a defined number of 4,000 p.s.i. concrete piles with thin-wall, straight-sided, steel shells that have a standard steel plate driving point. An allowance for cutoffs is included.

The Expanded System Listing shows costs per cluster of piles. Clusters range from one pile to twenty piles. Loads vary from 50 Kips to 1,600 Kips. Both end-bearing and friction-type piles are shown.

Please see the reference section for cost of mobilization of the pile driving equipment and other design and cost information.

System Components	QUANTITY	UNIT	COST EACH		
			MAT.	INST.	TOTAL
SYSTEM A1020 110 2220					
CIP SHELL CONCRETE PILE, 25′ LONG, 50K LOAD, END BEARING, 1 PILE					
Cast in place piles, end bearing, no mobil, 7 Ga. shell, 12″ diam.	27.000	V.L.F.	526.50	231.12	757.62
Steel pipe pile standard point, 12″ or 14″ diameter pile	1.000	Ea.	49.25	67.50	116.75
Pile cutoff, conc. pile with thin steel shell	1.000	Ea.		11.55	11.55
TOTAL			575.75	310.17	885.92

A1020 110	C.I.P. Concrete Piles	COST EACH		
		MAT.	INST.	TOTAL
2220	CIP shell concrete pile, 25′ long, 50K load, end bearing, 1 pile	575	310	885
2240	100K load, end bearing, 2 pile cluster	1,150	620	1,770
2260	200K load, end bearing, 4 pile cluster	2,300	1,250	3,550
2280	400K load, end bearing, 7 pile cluster	4,025	2,175	6,200
2300	10 pile cluster	5,750	3,100	8,850
2320	800K load, end bearing, 13 pile cluster	11,400	5,650	17,050
2340	17 pile cluster	14,900	7,400	22,300
2360	1200K load, end bearing, 14 pile cluster	12,300	6,100	18,400
2380	19 pile cluster	16,600	8,275	24,875
2400	1600K load, end bearing, 19 pile cluster	16,600	8,275	24,875
2420	50′ long, 50K load, end bearing, 1 pile	1,075	535	1,610
2440	Friction type, 2 pile cluster	2,075	1,075	3,150
2460	3 pile cluster	3,125	1,600	4,725
2480	100K load, end bearing, 2 pile cluster	2,175	1,075	3,250
2500	Friction type, 4 pile cluster	4,150	2,125	6,275
2520	6 pile cluster	6,250	3,200	9,450
2540	200K load, end bearing, 4 pile cluster	4,325	2,125	6,450
2560	Friction type, 8 pile cluster	8,325	4,275	12,600
2580	10 pile cluster	10,400	5,350	15,750
2600	400K load, end bearing, 7 pile cluster	7,575	3,725	11,300
2620	Friction type, 16 pile cluster	16,600	8,525	25,125
2640	19 pile cluster	19,800	10,100	29,900
2660	800K load, end bearing, 14 pile cluster	23,600	10,000	33,600
2680	20 pile cluster	33,600	14,300	47,900
2700	1200K load, end bearing, 15 pile cluster	25,200	10,700	35,900
2720	1600K load, end bearing, 20 pile cluster	33,600	14,300	47,900

SUBSTRUCTURE

A

A1020 Special Foundations

A1020 110	C.I.P. Concrete Piles	COST EACH		
		MAT.	INST.	TOTAL
3740	75' long, 50K load, end bearing, 1 pile	1,675	895	2,570
3760	Friction type, 2 pile cluster	3,225	1,800	5,025
3780	3 pile cluster	4,825	2,700	7,525
3800	100K load, end bearing, 2 pile cluster	3,350	1,800	5,150
3820	Friction type, 3 pile cluster	4,825	2,700	7,525
3840	5 pile cluster	8,050	4,475	12,525
3860	200K load, end bearing, 4 pile cluster	6,700	3,600	10,300
3880	6 pile cluster	10,000	5,400	15,400
3900	Friction type, 6 pile cluster	9,650	5,400	15,050
3910	7 pile cluster	11,300	6,300	17,600
3920	400K load, end bearing, 7 pile cluster	11,700	6,300	18,000
3930	11 pile cluster	18,400	9,900	28,300
3940	Friction type, 12 pile cluster	19,300	10,800	30,100
3950	14 pile cluster	22,500	12,600	35,100
3960	800K load, end bearing, 15 pile cluster	38,900	17,600	56,500
3970	20 pile cluster ·	52,000	23,500	75,500
3980	1200K load, end bearing, 17 pile cluster	44,000	19,900	63,900

SUBSTRUCTURE A

Important: See the Reference Section for critical supporting data - Reference Numbers and City Cost Indexes

A1020 Special Foundations

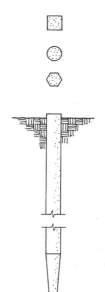

The Precast Concrete Pile System includes: pre-stressed concrete piles; standard steel driving point; and an allowance for cutoffs.

The Expanded System Listing shows costs per cluster of piles. Clusters range from one pile to twenty piles. Loads vary from 50 Kips to 1,600 Kips. Both end-bearing and friction type piles are listed.

Please see the reference section for cost of mobilization of the pile driving equipment and other design and cost information.

SUBSTRUCTURE

A

System Components	QUANTITY	UNIT	COST EACH		
			MAT.	INST.	TOTAL
SYSTEM A1020 120 2220					
PRECAST CONCRETE PILE, 50' LONG, 50K LOAD, END BEARING, 1 PILE					
Precast, prestressed conc. piles, 10" square, no mobil.	53.000	V.L.F.	506.15	388.49	894.64
Steel pipe pile standard point, 8" to 10" diameter	1.000	Ea.	35.25	59	94.25
Piling special costs cutoffs concrete piles plain	1.000	Ea.		80	80
TOTAL			541.40	527.49	1,068.89

A1020 120	Precast Concrete Piles	COST EACH		
		MAT.	INST.	TOTAL
2220	Precast conc pile, 50' long, 50K load, end bearing, 1 pile	540	525	1,065
2240	Friction type, 2 pile cluster	1,375	1,100	2,475
2260	4 pile cluster	2,750	2,175	4,925
2280	100K load, end bearing, 2 pile cluster	1,075	1,050	2,125
2300	Friction type, 2 pile cluster	1,375	1,100	2,475
2320	4 pile cluster	2,750	2,175	4,925
2340	7 pile cluster	4,800	3,825	8,625
2360	200K load, end bearing, 3 pile cluster	1,625	1,575	3,200
2380	4 pile cluster	2,175	2,100	4,275
2400	Friction type, 8 pile cluster	5,475	4,375	9,850
2420	9 pile cluster	6,175	4,900	11,075
2440	14 pile cluster	9,600	7,650	17,250
2460	400K load, end bearing, 6 pile cluster	3,250	3,175	6,425
2480	8 pile cluster	4,325	4,225	8,550
2500	Friction type, 14 pile cluster	9,600	7,650	17,250
2520	16 pile cluster	11,000	8,750	19,750
2540	18 pile cluster	12,300	9,850	22,150
2560	800K load, end bearing, 12 pile cluster	7,050	6,850	13,900
2580	16 pile cluster	8,650	8,425	17,075
2600	1200K load, end bearing, 19 pile cluster	29,300	12,300	41,600
2620	20 pile cluster	30,900	13,000	43,900
2640	1600K load, end bearing, 19 pile cluster	29,300	12,300	41,600
4660	100' long, 50K load, end bearing, 1 pile	1,050	905	1,955
4680	Friction type, 1 pile	1,300	940	2,240

A1020 Special Foundations

A1020 120	Precast Concrete Piles	COST EACH		
		MAT.	INST.	TOTAL
4700	2 pile cluster	2,625	1,875	4,500
4720	100K load, end bearing, 2 pile cluster	2,075	1,825	3,900
4740	Friction type, 2 pile cluster	2,625	1,875	4,500
4760	3 pile cluster	3,925	2,800	6,725
4780	4 pile cluster	5,225	3,750	8,975
4800	200K load, end bearing, 3 pile cluster	3,125	2,725	5,850
4820	4 pile cluster	4,150	3,625	7,775
4840	Friction type, 3 pile cluster	3,925	2,800	6,725
4860	5 pile cluster	6,550	4,700	11,250
4880	400K load, end bearing, 6 pile cluster	6,225	5,450	11,675
4900	8 pile cluster	8,300	7,275	15,575
4910	Friction type, 8 pile cluster	10,500	7,500	18,000
4920	10 pile cluster	13,100	9,375	22,475
4930	800K load, end bearing, 13 pile cluster	13,500	11,800	25,300
4940	16 pile cluster	16,600	14,500	31,100
4950	1200K load, end bearing, 19 pile cluster	57,000	21,300	78,300
4960	20 pile cluster	60,000	22,400	82,400
4970	1600K load, end bearing, 19 pile cluster	57,000	21,300	78,300

Important: See the Reference Section for critical supporting data - Reference Numbers and City Cost Indexes

A10 Foundations

A1020 Special Foundations

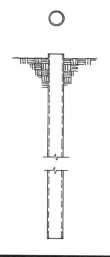

The Steel Pipe Pile System includes: steel pipe sections filled with 4,000 p.s.i. concrete; a standard steel driving point; splices when required and an allowance for cutoffs.

The Expanded System Listing shows costs per cluster of piles. Clusters range from one pile to twenty piles. Loads vary from 50 Kips to 1,600 Kips. Both end-bearing and friction-type piles are shown.

Please see the reference section for cost of mobilization of the pile driving equipment and other design and cost information.

System Components	QUANTITY	UNIT	COST EACH		
			MAT.	INST.	TOTAL
SYSTEM A1020 130 2220 **CONC. FILL STEEL PIPE PILE, 50' LONG, 50K LOAD, END BEARING, 1 PILE**					
Piles, steel, pipe, conc. filled, 12" diameter	53.000	V.L.F.	1,192.50	655.08	1,847.58
Steel pipe pile, standard point, for 12" or 14" diameter pipe	1.000	Ea.	98.50	135	233.50
Pile cut off, concrete pile, thin steel shell	1.000	Ea.		11.55	11.55
TOTAL			1,291	801.63	2,092.63

A1020 130	Steel Pipe Piles	COST EACH		
		MAT.	INST.	TOTAL
2220	Conc. fill steel pipe pile, 50' long, 50K load, end bearing, 1 pile	1,300	800	2,100
2240	Friction type, 2 pile cluster	2,575	1,600	4,175
2250	100K load, end bearing, 2 pile cluster	2,575	1,600	4,175
2260	3 pile cluster	3,875	2,400	6,275
2300	Friction type, 4 pile cluster	5,175	3,200	8,375
2320	5 pile cluster	6,450	4,000	10,450
2340	10 pile cluster	12,900	8,025	20,925
2360	200K load, end bearing, 3 pile cluster	3,875	2,400	6,275
2380	4 pile cluster	5,175	3,200	8,375
2400	Friction type, 4 pile cluster	5,175	3,200	8,375
2420	8 pile cluster	10,300	6,425	16,725
2440	9 pile cluster	11,600	7,200	18,800
2460	400K load, end bearing, 6 pile cluster	7,750	4,825	12,575
2480	7 pile cluster	9,025	5,600	14,625
2500	Friction type, 9 pile cluster	11,600	7,200	18,800
2520	16 pile cluster	20,700	12,800	33,500
2540	19 pile cluster	24,500	15,200	39,700
2560	800K load, end bearing, 11 pile cluster	14,200	8,825	23,025
2580	14 pile cluster	18,100	11,200	29,300
2600	15 pile cluster	19,400	12,000	31,400
2620	Friction type, 17 pile cluster	21,900	13,600	35,500
2640	1200K load, end bearing, 16 pile cluster	20,700	12,800	33,500
2660	20 pile cluster	25,800	16,000	41,800
2680	1600K load, end bearing, 17 pile cluster	21,900	13,600	35,500
3700	100' long, 50K load, end bearing, 1 pile	2,550	1,575	4,125
3720	Friction type, 1 pile	2,550	1,575	4,125

SUBSTRUCTURE

A

A1020 Special Foundations

A1020 130	Steel Pipe Piles	COST EACH		
		MAT.	INST.	TOTAL
3740	2 pile cluster	5,100	3,175	8,275
3760	100K load, end bearing, 2 pile cluster	5,100	3,175	8,275
3780	Friction type, 2 pile cluster	5,100	3,175	8,275
3800	3 pile cluster	7,625	4,750	12,375
3820	200K load, end bearing, 3 pile cluster	7,625	4,750	12,375
3840	4 pile cluster	10,200	6,350	16,550
3860	Friction type, 3 pile cluster	7,625	4,750	12,375
3880	4 pile cluster	10,200	6,350	16,550
3900	400K load, end bearing, 6 pile cluster	15,300	9,525	24,825
3910	7 pile cluster	17,800	11,100	28,900
3920	Friction type, 5 pile cluster	12,700	7,925	20,625
3930	8 pile cluster	20,400	12,700	33,100
3940	800K load, end bearing, 11 pile cluster	28,000	17,500	45,500
3950	14 pile cluster	35,600	22,300	57,900
3960	15 pile cluster	38,200	23,800	62,000
3970	1200K load, end bearing, 16 pile cluster	40,700	25,400	66,100
3980	20 pile cluster	51,000	31,700	82,700
3990	1600K load, end bearing, 17 pile cluster	43,200	27,100	70,300

Important: See the Reference Section for critical supporting data - Reference Numbers and City Cost Indexes

A1020 Special Foundations

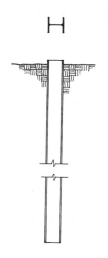

A Steel "H" Pile System includes: steel H sections; heavy duty driving point; splices where applicable and allowance for cutoffs.

The Expanded System Listing shows costs per cluster of piles. Clusters range from one pile to seventeen piles. Loads vary from 50 Kips to 2,000 Kips. All loads for Steel H Pile systems are given in terms of end bearing capacity.

Steel sections range from 10" x 10" to 14" x 14" in the Expanded System Listing. The 14" x 14" steel section is used for all H piles used in applications requiring a working load over 800 Kips.

Please see the reference section for cost of mobilization of the pile driving equipment and other design and cost information.

System Components			COST EACH		
	QUANTITY	UNIT	MAT.	INST.	TOTAL
SYSTEM A1020 140 2220					
STEEL H PILES, 50' LONG, 100K LOAD, END BEARING, 1 PILE					
Steel H piles 10" x 10", 42 #/L.F.	53.000	V.L.F.	805.60	445.73	1,251.33
Heavy duty point, 10"	1.000	Ea.	132	137	269
Pile cut off, steel pipe or H piles	1.000	Ea.		23	23
TOTAL			937.60	605.73	1,543.33

A1020 140	Steel H Piles	COST EACH		
		MAT.	INST.	TOTAL
2220	Steel H piles, 50' long, 100K load, end bearing, 1 pile	940	605	1,545
2260	2 pile cluster	1,875	1,225	3,100
2280	200K load, end bearing, 2 pile cluster	1,875	1,225	3,100
2300	3 pile cluster	2,825	1,825	4,650
2320	400K load, end bearing, 3 pile cluster	2,825	1,825	4,650
2340	4 pile cluster	3,750	2,425	6,175
2360	6 pile cluster	5,625	3,625	9,250
2380	800K load, end bearing, 5 pile cluster	4,700	3,025	7,725
2400	7 pile cluster	6,575	4,250	10,825
2420	12 pile cluster	11,300	7,275	18,575
2440	1200K load, end bearing, 8 pile cluster	7,500	4,850	12,350
2460	11 pile cluster	10,300	6,650	16,950
2480	17 pile cluster	15,900	10,300	26,200
2500	1600K load, end bearing, 10 pile cluster	11,800	6,350	18,150
2520	14 pile cluster	16,500	8,900	25,400
2540	2000K load, end bearing, 12 pile cluster	14,100	7,625	21,725
2560	18 pile cluster	21,200	11,400	32,600
3580	100' long, 50K load, end bearing, 1 pile	3,125	1,400	4,525
3600	100K load, end bearing, 1 pile	3,125	1,400	4,525
3620	2 pile cluster	6,250	2,800	9,050
3640	200K load, end bearing, 2 pile cluster	6,250	2,800	9,050
3660	3 pile cluster	9,400	4,225	13,625
3680	400K load, end bearing, 3 pile cluster	9,400	4,225	13,625
3700	4 pile cluster	12,500	5,625	18,125
3720	6 pile cluster	18,800	8,425	27,225
3740	800K load, end bearing, 5 pile cluster	15,700	7,050	22,750

SUBSTRUCTURE

A

A1020 Special Foundations

A1020 140	Steel H Piles	COST EACH		
		MAT.	INST.	TOTAL
3760	7 pile cluster	21,900	9,850	31,750
3780	12 pile cluster	37,600	16,900	54,500
3800	1200K load, end bearing, 8 pile cluster	25,000	11,300	36,300
3820	11 pile cluster	34,400	15,500	49,900
3840	17 pile cluster	53,000	23,900	76,900
3860	1600K load, end bearing, 10 pile cluster	31,300	14,100	45,400
3880	14 pile cluster	43,800	19,700	63,500
3900	2000K load, end bearing, 12 pile cluster	37,600	16,900	54,500
3920	18 pile cluster	56,500	25,300	81,800

SUBSTRUCTURE A

A1020 Special Foundations

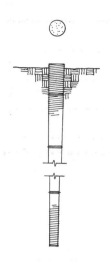

The Step Tapered Steel Pile System includes: step tapered piles filled with 4,000 p.s.i. concrete. The cost for splices and pile cutoffs is included.

The Expanded System Listing shows costs per cluster of piles. Clusters range from one pile to twenty-four piles. Both end bearing piles and friction piles are listed. Loads vary from 50 Kips to 1,600 Kips.

Please see the reference section for cost of mobilization of the pile driving equipment and other design and cost information.

System Components	QUANTITY	UNIT	COST EACH		
			MAT.	INST.	TOTAL
SYSTEM A1020 150 1000					
STEEL PILE, STEP TAPERED, 50' LONG, 50K LOAD, END BEARING, 1 PILE					
Steel shell step tapered conc. filled piles, 8" tip 60 ton capacity to 60'	53.000	V.L.F.	453.15	367.29	820.44
Pile cutoff, steel pipe or H piles	1.000	Ea.		23	23
TOTAL			453.15	390.29	843.44

A1020 150	Step-Tapered Steel Piles	COST EACH		
		MAT.	INST.	TOTAL
1000	Steel pile, step tapered, 50' long, 50K load, end bearing, 1 pile	455	390	845
1200	Friction type, 3 pile cluster	1,350	1,175	2,525
1400	100K load, end bearing, 2 pile cluster	905	780	1,685
1600	Friction type, 4 pile cluster	1,825	1,550	3,375
1800	200K load, end bearing, 4 pile cluster	1,825	1,550	3,375
2000	Friction type, 6 pile cluster	2,725	2,350	5,075
2200	400K load, end bearing, 7 pile cluster	3,175	2,750	5,925
2400	Friction type, 10 pile cluster	4,525	3,900	8,425
2600	800K load, end bearing, 14 pile cluster	6,350	5,450	11,800
2800	Friction type, 18 pile cluster	8,150	7,025	15,175
3000	1200K load, end bearing, 16 pile cluster	8,150	6,675	14,825
3200	Friction type, 21 pile cluster	10,700	8,775	19,475
3400	1600K load, end bearing, 18 pile cluster	9,150	7,500	16,650
3600	Friction type, 24 pile cluster	12,200	10,000	22,200
5000	100' long, 50K load, end bearing, 1 pile	910	775	1,685
5200	Friction type, 2 pile cluster	1,825	1,550	3,375
5400	100K load, end bearing, 2 pile cluster	1,825	1,550	3,375
5600	Friction type, 3 pile cluster	2,725	2,350	5,075
5800	200K load, end bearing, 4 pile cluster	3,650	3,125	6,775
6000	Friction type, 5 pile cluster	4,550	3,875	8,425
6200	400K load, end bearing, 7 pile cluster	6,375	5,450	11,825
6400	Friction type, 8 pile cluster	7,300	6,225	13,525
6600	800K load, end bearing, 15 pile cluster	13,700	11,700	25,400
6800	Friction type, 16 pile cluster	14,600	12,500	27,100
7000	1200K load, end bearing, 17 pile cluster	26,900	20,300	47,200
7200	Friction type, 19 pile cluster	20,300	15,400	35,700
7400	1600K load, end bearing, 20 pile cluster	21,300	16,300	37,600
7600	Friction type, 22 pile cluster	23,500	17,900	41,400

A SUBSTRUCTURE

A1020 Special Foundations

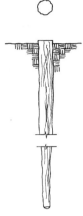

The Treated Wood Pile System includes: creosoted wood piles; a standard steel driving point; and an allowance for cutoffs.

The Expanded System Listing shows costs per cluster of piles. Clusters range from three piles to twenty piles. Loads vary from 50 Kips to 400 Kips. Both end-bearing and friction type piles are listed.

Please see the reference section for cost of mobilization of the pile driving equipment and other design and cost information.

SUBSTRUCTURE

A

System Components	QUANTITY	UNIT	COST EACH MAT.	COST EACH INST.	COST EACH TOTAL
SYSTEM A1020 160 2220					
WOOD PILES, 25' LONG, 50K LOAD, END BEARING, 3 PILE CLUSTER					
Wood piles, treated, 12" butt, 8" tip, up to 30' long	81.000	V.L.F.	830.25	663.39	1,493.64
Point for driving wood piles	3.000	Ea.	63	66	129
Pile cutoff, wood piles	3.000	Ea.		34.65	34.65
TOTAL			893.25	764.04	1,657.29

A1020 160	Treated Wood Piles	COST EACH MAT.	COST EACH INST.	COST EACH TOTAL
2220	Wood piles, 25' long, 50K load, end bearing, 3 pile cluster	895	765	1,660
2240	Friction type, 3 pile cluster	830	700	1,530
2260	5 pile cluster	1,375	1,175	2,550
2280	100K load, end bearing, 4 pile cluster	1,200	1,025	2,225
2300	5 pile cluster	1,500	1,275	2,775
2320	6 pile cluster	1,775	1,525	3,300
2340	Friction type, 5 pile cluster	1,375	1,175	2,550
2360	6 pile cluster	1,650	1,400	3,050
2380	10 pile cluster	2,775	2,325	5,100
2400	200K load, end bearing, 8 pile cluster	2,375	2,025	4,400
2420	10 pile cluster	2,975	2,550	5,525
2440	12 pile cluster	3,575	3,050	6,625
2460	Friction type, 10 pile cluster	2,775	2,325	5,100
2480	400K load, end bearing, 16 pile cluster	4,775	4,075	8,850
2500	20 pile cluster	5,950	5,100	11,050
4520	50' long, 50K load, end bearing, 3 pile cluster	1,950	1,125	3,075
4540	4 pile cluster	2,600	1,500	4,100
4560	Friction type, 2 pile cluster	1,225	690	1,915
4580	3 pile cluster	1,850	1,025	2,875
4600	100K load, end bearing, 5 pile cluster	3,250	1,900	5,150
4620	8 pile cluster	5,175	3,025	8,200
4640	Friction type, 3 pile cluster	1,850	1,025	2,875
4660	5 pile cluster	3,075	1,725	4,800
4680	200K load, end bearing, 9 pile cluster	5,825	3,400	9,225
4700	10 pile cluster	6,475	3,775	10,250
4720	15 pile cluster	9,725	5,675	15,400
4740	Friction type, 5 pile cluster	3,075	1,725	4,800
4760	6 pile cluster	3,700	2,075	5,775

Important: See the Reference Section for critical supporting data - Reference Numbers and City Cost Indexes

A1020 Special Foundations

A1020 160	Treated Wood Piles	COST EACH		
		MAT.	INST.	TOTAL
4780	10 pile cluster	6,150	3,450	9,600
4800	400K load, end bearing, 18 pile cluster	11,700	6,800	18,500
4820	20 pile cluster	13,000	7,550	20,550
4840	Friction type, 9 pile cluster	5,550	3,100	8,650
4860	10 pile cluster	6,150	3,450	9,600
9000	Add for boot for driving tip, each pile	21	16.25	37.25

SUBSTRUCTURE

A

A1020 Special Foundations

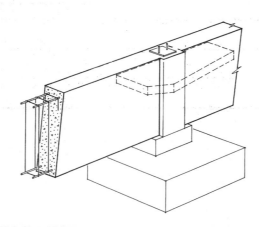

The Grade Beam System includes: excavation with a truck mounted backhoe; hand trim; backfill; forms (four uses); reinforcing steel; and 3,000 p.s.i. concrete placed from chute.

Superimposed loads vary in the listing from 8 Kips per linear foot (KLF) to 50 KLF. In the Expanded System Listing, the span of the beams varies from 15' to 40'. Depth varies from 28" to 52". Width varies from 12" to 28".

Please see the reference section for further design and cost information.

System Components	QUANTITY	UNIT	COST PER L.F.		
			MAT.	INST.	TOTAL
SYSTEM A1020 210 2220					
GRADE BEAM, 15' SPAN, 28" DEEP, 12" WIDE, 8 KLF LOAD					
Excavation, trench, hydraulic backhoe, 3/8 CY bucket	.260	C.Y.		1.71	1.71
Trim sides and bottom of trench, regular soil	2.000	S.F.		1.42	1.42
Backfill, by hand, compaction in 6" layers, using vibrating plate	.170	C.Y.		1.02	1.02
Forms in place, grade beam, 4 uses	4.700	SFCA	2.63	19.36	21.99
Reinforcing in place, beams & girders, #8 to #14	.019	Ton	16.72	14.06	30.78
Concrete ready mix, regular weight, 3000 psi	.090	C.Y.	8.01		8.01
Place and vibrate conc. for grade beam, direct chute	.090	C.Y.		1.27	1.27
TOTAL			27.36	38.84	66.20

A1020 210	Grade Beams	COST PER L.F.		
		MAT.	INST.	TOTAL
2220	Grade beam, 15' span, 28" deep, 12" wide, 8 KLF load	27.50	39	66.50
2240	14" wide, 12 KLF load	28.50	39	67.50
2260	40" deep, 12" wide, 16 KLF load	27.50	48	75.50
2280	20 KLF load	32	51.50	83.50
2300	52" deep, 12" wide, 30 KLF load	36.50	65	101.50
2320	40 KLF load	45.50	72.50	118
2340	50 KLF load	54.50	79	133.50
3360	20' span, 28" deep, 12" wide, 2 KLF load	16.80	30	46.80
3380	16" wide, 4 KLF load	22.50	33.50	56
3400	40" deep, 12" wide, 8 KLF load	27.50	48	75.50
3420	12 KLF load	35.50	54	89.50
3440	14" wide, 16 KLF load	43.50	60	103.50
3460	52" deep, 12" wide, 20 KLF load	46.50	73	119.50
3480	14" wide, 30 KLF load	65	87	152
3500	20" wide, 40 KLF load	76	91.50	167.50
3520	24" wide, 50 KLF load	95	105	200
4540	30' span, 28" deep, 12" wide, 1 KLF load	17.70	30.50	48.20
4560	14" wide, 2 KLF load	28.50	42.50	71
4580	40" deep, 12" wide, 4 KLF load	34	50.50	84.50
4600	18" wide, 8 KLF load	49	62.50	111.50
4620	52" deep, 14" wide, 12 KLF load	62.50	84.50	147
4640	20" wide, 16 KLF load	76.50	92.50	169
4660	24" wide, 20 KLF load	95	105	200
4680	36" wide, 30 KLF load	137	131	268
4700	48" wide, 40 KLF load	182	161	343
5720	40' span, 40" deep, 12" wide, 1 KLF load	24	45	69

Important: See the Reference Section for critical supporting data - Reference Numbers and City Cost Indexes

A10 Foundations

A1020 Special Foundations

A1020 210	Grade Beams	COST PER L.F.		
		MAT.	INST.	TOTAL
5740	2 KLF load	30.50	50	80.50
5760	52" deep, 12" wide, 4 KLF load	45.50	72.50	118
5780	20" wide, 8 KLF load	74	90	164
5800	28" wide, 12 KLF load	102	108	210
5820	38" wide, 16 KLF load	141	133	274
5840	46" wide, 20 KLF load	185	165	350

A1020 Special Foundations

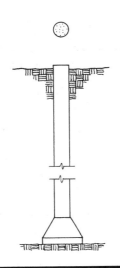

Caisson Systems are listed for three applications: stable ground, wet ground and soft rock. Concrete used is 3,000 p.s.i. placed from chute. Included are a bell at the bottom of the caisson shaft (if applicable) along with required excavation and disposal of excess excavated material up to two miles from job site.

The Expanded System lists cost per caisson. End-bearing loads vary from 200 Kips to 3,200 Kips. The dimensions of the caissons range from 2' x 50' to 7' x 200'.

Please see the reference section for further design and cost information.

System Components	QUANTITY	UNIT	COST EACH		
			MAT.	INST.	TOTAL
SYSTEM A1020 310 2200					
CAISSON, STABLE GROUND, 3000 PSI CONC., 10KSF BRNG, 200K LOAD, 2'X50'					
Caissons, drilled, to 50', 24" shaft diameter, .116 C.Y./L.F.	50.000	V.L.F.	645	1,485	2,130
Reinforcing in place, columns, #3 to #7	.060	Ton	52.80	79.50	132.30
Caisson bell excavation and concrete, 4' diameter .444 CY	1.000	Ea.	39.50	283	322.50
Load & haul excess excavation, 2 miles	6.240	C.Y.		30.14	30.14
TOTAL			737.30	1,877.64	2,614.94

A1020 310	Caissons	COST EACH		
		MAT.	INST.	TOTAL
2200	Caisson, stable ground, 3000 PSI conc, 10KSF brng, 200K load, 2'x50'	735	1,900	2,635
2400	400K load, 2'-6"x50'-0"	1,200	3,025	4,225
2600	800K load, 3'-0"x100'-0"	3,325	7,175	10,500
2800	1200K load, 4'-0"x100'-0"	5,750	9,000	14,750
3000	1600K load, 5'-0"x150'-0"	13,000	13,900	26,900
3200	2400K load, 6'-0"x150'-0"	18,900	18,000	36,900
3400	3200K load, 7'-0"x200'-0"	34,000	26,000	60,000
5000	Wet ground, 3000 PSI conc., 10 KSF brng, 200K load, 2'-0"x50'-0"	640	2,575	3,215
5200	400K load, 2'-6"x50'-0"	1,050	4,400	5,450
5400	800K load, 3'-0"x100'-0"	2,900	11,800	14,700
5600	1200K load, 4'-0"x100'-0"	4,975	16,600	21,575
5800	1600K load, 5'-0"x150'-0"	11,200	35,800	47,000
6000	2400K load, 6'-0"x150'-0"	16,300	44,000	60,300
6200	3200K load, 7'-0"x200'-0"	29,200	70,500	99,700
7800	Soft rock, 3000 PSI conc., 10 KSF brng, 200K load, 2'-0"x50'-0"	640	13,200	13,840
8000	400K load, 2'-6"x50'-0"	1,050	21,300	22,350
8200	800K load, 3'-0"x100'-0"	2,900	56,000	58,900
8400	1200K load, 4'-0"x100'-0"	4,975	82,000	86,975
8600	1600K load, 5'-0"x150'-0"	11,200	168,500	179,700
8800	2400K load, 6'-0"x150'-0"	16,300	200,000	216,300
9000	3200K load, 7'-0"x200'-0"	29,200	320,000	349,200

A1020 Special Foundations

Pressure Injected Piles are usually uncased up to 25' and cased over 25' depending on soil conditions.

These costs include excavation and hauling of excess materials; steel casing over 25'; reinforcement; 3,000 p.s.i. concrete; plus mobilization and demobilization of equipment for a distance of up to fifty miles to and from the job site.

The Expanded System lists cost per cluster of piles. Clusters range from one pile to eight piles. End-bearing loads range from 50 Kips to 1,600 Kips.

Please see the reference section for further design and cost information.

System Components			COST EACH		
	QUANTITY	UNIT	MAT.	INST.	TOTAL
SYSTEM A1020 710 4200					
PRESSURE INJECTED FOOTING, END BEARING, 50' LONG, 50K LOAD, 1 PILE					
Pressure injected footings, cased, 30-60 ton cap., 12" diameter	50.000	V.L.F.	685	969	1,654
Pile cutoff, concrete pile with thin steel shell	1.000	Ea.		11.55	11.55
TOTAL			685	980.55	1,665.55

A1020 710	Pressure Injected Footings	COST EACH		
		MAT.	INST.	TOTAL
2200	Pressure injected footing, end bearing, 25' long, 50K load, 1 pile	320	685	1,005
2400	100K load, 1 pile	320	685	1,005
2600	2 pile cluster	640	1,375	2,015
2800	200K load, 2 pile cluster	640	1,375	2,015
3200	400K load, 4 pile cluster	1,275	2,725	4,000
3400	7 pile cluster	2,225	4,775	7,000
3800	1200K load, 6 pile cluster	2,675	5,250	7,925
4000	1600K load, 7 pile cluster	3,125	6,125	9,250
4200	50' long, 50K load, 1 pile	685	980	1,665
4400	100K load, 1 pile	1,200	980	2,180
4600	2 pile cluster	2,400	1,950	4,350
4800	200K load, 2 pile cluster	2,400	1,950	4,350
5000	4 pile cluster	4,800	3,925	8,725
5200	400K load, 4 pile cluster	4,800	3,925	8,725
5400	8 pile cluster	9,600	7,825	17,425
5600	800K load, 7 pile cluster	8,400	6,850	15,250
5800	1200K load, 6 pile cluster	7,650	5,875	13,525
6000	1600K load, 7 pile cluster	8,925	6,850	15,775

A SUBSTRUCTURE

A2020 Basement Walls

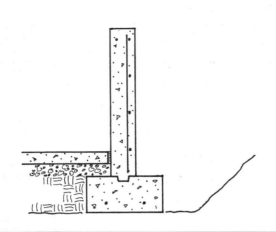

The Foundation Bearing Wall System includes: forms up to 16' high (four uses); 3,000 p.s.i. concrete placed and vibrated; and form removal with breaking form ties and patching walls. The wall systems list walls from 6" to 16" thick and are designed with minimum reinforcement.

Excavation and backfill are not included.

Please see the reference section for further design and cost information.

System Components		QUANTITY	UNIT	COST PER L.F.		
				MAT.	INST.	TOTAL
SYSTEM A2020 110 1500						
FOUNDATION WALL, CAST IN PLACE, DIRECT CHUTE, 4' HIGH, 6" THICK						
Formwork		8.000	SFCA	3.28	33.20	36.48
Reinforcing		3.300	Lb.	1.38	1.11	2.49
Unloading & sorting reinforcing		3.300	Lb.		.07	.07
Concrete, 3,000 psi		.074	C.Y.	6.59		6.59
Place concrete, direct chute		.074	C.Y.		1.74	1.74
Finish walls, break ties and patch voids, one side		4.000	S.F.	.12	2.88	3
	TOTAL			11.37	39	50.37

A2020 110	Walls, Cast in Place							
	WALL HEIGHT (FT.)	PLACING METHOD	CONCRETE (C.Y. per L.F.)	REINFORCING (LBS. per L.F.)	WALL THICKNESS (IN.)	COST PER L.F.		
						MAT.	INST.	TOTAL
1500	4'	direct chute	.074	3.3	6	11.35	39	50.35
1520			.099	4.8	8	14.20	40	54.20
1540			.123	6.0	10	16.85	40.50	57.35
1560			.148	7.2	12	19.60	41.50	61.10
1580			.173	8.1	14	22	42.50	64.50
1600			.197	9.44	16	25	43.50	68.50
1700	4'	pumped	.074	3.3	6	11.35	40	51.35
1720			.099	4.8	8	14.20	41.50	55.70
1740			.123	6.0	10	16.85	42.50	59.35
1760			.148	7.2	12	19.60	43.50	63.10
1780			.173	8.1	14	22	44	66
1800			.197	9.44	16	25	45.50	70.50
3000	6'	direct chute	.111	4.95	6	17.05	58.50	75.55
3020			.149	7.20	8	21.50	60	81.50
3040			.184	9.00	10	25	61	86
3060			.222	10.8	12	29.50	62.50	92
3080			.260	12.15	14	33.50	63.50	97
3100			.300	14.39	16	38	65	103

Important: See the Reference Section for critical supporting data - Reference Numbers and City Cost Indexes

SUBSTRUCTURE A

A2020 Basement Walls

A2020 110	Walls, Cast in Place

	WALL HEIGHT (FT.)	PLACING METHOD	CONCRETE (C.Y. per L.F.)	REINFORCING (LBS. per L.F.)	WALL THICKNESS (IN.)	COST PER L.F.		
						MAT.	INST.	TOTAL
3200	6'	pumped	.111	4.95	6	17.05	60	77.05
3220			.149	7.20	8	21.50	62.50	84
3240			.184	9.00	10	25	63.50	88.50
3260			.222	10.8	12	29.50	65	94.50
3280			.260	12.15	14	33.50	66.50	100
3300			.300	14.39	16	38	68	106
5000	8'	direct chute	.148	6.6	6	22.50	78	100.50
5020			.199	9.6	8	28.50	80	108.50
5040			.250	12	10	34	81.50	115.50
5060			.296	14.39	12	39	83.50	122.50
5080			.347	16.19	14	41	84	125
5100			.394	19.19	16	50	87	137
5200	8'	pumped	.148	6.6	6	22.50	80	102.50
5220			1.99	9.6	8	28.50	82.50	111
5240			.250	12	10	34	85	119
5260			.296	14.39	12	39	87.50	126.50
5280			.347	16.19	14	41	87	128
5300			.394	19.19	16	50	91	141
6020	10'	direct chute	.248	12	8	35.50	100	135.50
6040			.307	14.99	10	42	102	144
6060			.370	17.99	12	49	104	153
6080			.433	20.24	14	55.50	106	161.50
6100			.493	23.99	16	62.50	108	170.50
6220	10'	pumped	.248	12	8	35.50	103	138.50
6240			.307	14.99	10	42	105	147
6260			.370	17.99	12	49	109	158
6280			.433	20.24	14	55.50	111	166.50
6300			.493	23.99	16	62.50	113	175.50
7220	12'	pumped	.298	14.39	8	42.50	125	167.50
7240			.369	17.99	10	50.50	127	177.50
7260			.444	21.59	12	58.50	130	188.50
7280			.52	24.29	14	66.50	133	199.50
7300			.591	28.79	16	75	136	211
7420	12'	crane & bucket	.298	14.39	8	42.50	129	171.50
7440			.369	17.99	10	50.50	132	182.50
7460			.444	21.59	12	58.50	137	195.50
7480			.52	24.29	14	66.50	140	206.50
7500			.591	28.79	16	75	146	221
8220	14'	pumped	.347	16.79	8	50	145	195
8240			.43	20.99	10	59	148	207
8260			.519	25.19	12	68.50	152	220.50
8280			.607	28.33	14	78	155	233
8300			.69	33.59	16	87.50	159	246.50
8420	14'	crane & bucket	.347	16.79	8	50	151	201
8440			.43	20.99	10	59	154	213
8460			.519	25.19	12	68.50	160	228.50
8480			.607	28.33	14	78	164	242
8500			.69	33.59	16	87.50	169	256.50
9220	16'	pumped	.397	19.19	8	57	165	222
9240			.492	23.99	10	67.50	169	236.50
9260			.593	28.79	12	78.50	175	253.50
9280			.693	32.39	14	89	177	266
9300			.788	38.38	16	100	182	282

SUBSTRUCTURE

A

A2020 Basement Walls

A2020 110	Walls, Cast in Place							
	WALL HEIGHT (FT.)	PLACING METHOD	CONCRETE (C.Y. per L.F.)	REINFORCING (LBS. per L.F.)	WALL THICKNESS (IN.)	COST PER L.F.		
						MAT.	INST.	TOTAL
9420	16'	crane & bucket	.397	19.19	8	57	172	229
9440			.492	23.99	10	67.50	176	243.50
9460			.593	28.79	12	78.50	182	260.50
9480			.693	32.39	14	89	187	276
9500			.788	38.38	16	100	193	293

Important: See the Reference Section for critical supporting data - Reference Numbers and City Cost Indexes

A2020 Basement Walls

The Wood Wall Foundations System includes: all required pressure treated wood framing members; sheathing (treated and bonded with CDX-grade glue); asphalt paper; polyethylene vapor barrier; and 3-1/2" R11 fiberglass insulation.

The Expanded System Listing shows various walls fabricated with 2" x 4", 2" x 6", 2" x 8", or 2" x 10" studs. Wall height varies from 2' to 8'. The insulation is standard, fiberglass batt type for all walls listed.

System Components	QUANTITY	UNIT	COST PER S.F. MAT.	COST PER S.F. INST.	COST PER S.F. TOTAL
SYSTEM A2020 150 2000					
WOOD WALL FOUNDATION, 2" X 4" STUDS, 12" O.C., 1/2" SHEATHING, 2' HIGH					
Framing plates & studs, 2" x 4" treated	.550	B.F.	1.26	1.90	3.16
Sheathing, 1/2" thick CDX, treated	1.005	S.F.	.88	.76	1.64
Building paper, asphalt felt sheathing paper 15 lb	1.100	S.F.	.04	.13	.17
Polyethylene vapor barrier, standard, .006" thick	1.000	S.F.	.03	.12	.15
Fiberglass insulation, 3-1/2" thick, 11" wide	1.000	S.F.	.30	.37	.67
TOTAL			2.51	3.28	5.79

A2020 150	Wood Wall Foundations	COST PER S.F. MAT.	COST PER S.F. INST.	COST PER S.F. TOTAL
2000	Wood wall foundation, 2" x 4" studs, 12" O.C., 1/2" sheathing, 2' high	2.51	3.28	5.79
2008	4' high	2.15	2.73	4.88
2012	6' high	2.03	2.56	4.59
2020	8' high	1.97	2.47	4.44
2500	2" x 6" studs, 12" O.C., 1/2" sheathing, 2' high	3.21	3.90	7.11
2508	4' high	2.66	3.19	5.85
2512	6' high	2.47	2.95	5.42
2520	8' high	2.38	2.84	5.22
3000	2" x 8" studs, 12" O.C., 1/2" sheathing, 2' high	3.34	3.47	6.81
3008	4' high	2.75	2.88	5.63
3016	6' high	2.55	2.68	5.23
3024	8' high	2.45	2.58	5.03
3032	16" O.C., 3/4" sheathing, 2' high	3.79	3.41	7.20
3040	4' high	3.18	2.80	5.98
3048	6' high	2.97	2.59	5.56
3056	8' high	2.87	2.49	5.36
4000	2" x 10" studs, 12" O.C., 1/2" sheathing, 2' high	5.05	3.58	8.63
4008	4' high	3.95	2.95	6.90
4012	6' high	3.60	2.75	6.35
4040	8' high	3.43	2.65	6.08
4048	16" O.C., 3/4" sheathing, 2' high	5.35	3.51	8.86
4056	4' high	4.21	2.86	7.07
4070	6' high	3.85	2.65	6.50
4080	8' high	3.66	2.53	6.19

For information about Means Estimating Seminars, see yellow pages 12 and 13 in back of book

SUBSTRUCTURE

A

G BUILDING SITEWORK

G1030 Site Earthwork

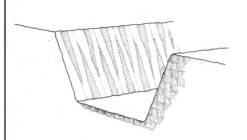

Trenching Systems are shown on a cost per linear foot basis. The systems include: excavation; backfill and removal of spoil; and compaction for various depths and trench bottom widths. The backfill has been reduced to accommodate a pipe of suitable diameter and bedding.

The slope for trench sides varies from 0:1 to 2:1.

The Expanded System Listing shows Trenching Systems that range from 2' to 12' in width. Depths range from 2' to 25'.

System Components			COST PER L.F.		
	QUANTITY	UNIT	EQUIP.	LABOR	TOTAL
SYSTEM G1030 805 1310					
TRENCHING, BACKHOE, 0 TO 1 SLOPE, 2' WIDE, 2' DP, 3/8 C.Y. BUCKET					
Excavation, trench, hyd. backhoe, track mtd., 3/8 C.Y. bucket	.174	C.Y.	.27	.87	1.14
Backfill and load spoil, from stockpile	.174	C.Y.	.09	.25	.34
Compaction by rammer tamper, 8" lifts, 4 passes	.014	C.Y.		.04	.04
Remove excess spoil, 6 C.Y. dump truck, 2 mile roundtrip	.160	C.Y.	.57	.54	1.11
TOTAL			.93	1.70	2.63

G1030 805	Trenching	COST PER L.F.		
		EQUIP.	LABOR	TOTAL
1310	Trenching, backhoe, 0 to 1 slope, 2' wide, 2' deep, 3/8 C.Y. bucket	.93	1.70	2.63
1320	3' deep, 3/8 C.Y. bucket	1.15	2.49	3.64
1330	4' deep, 3/8 C.Y. bucket	1.36	3.28	4.64
1340	6' deep, 3/8 C.Y. bucket	1.69	4.18	5.87
1350	8' deep, 1/2 C.Y. bucket	1.98	5.30	7.28
1360	10' deep, 1 C.Y. bucket	3.01	6.30	9.31
1400	4' wide, 2' deep, 3/8 C.Y. bucket	1.98	3.42	5.40
1410	3' deep, 3/8 C.Y. bucket	2.80	5.10	7.90
1420	4' deep, 1/2 C.Y. bucket	2.79	5.10	7.89
1430	6' deep, 1/2 C.Y. bucket	3.56	7.75	11.31
1440	8' deep, 1/2 C.Y. bucket	5.60	10	15.60
1450	10' deep, 1 C.Y. bucket	6.25	12.40	18.65
1460	12' deep, 1 C.Y. bucket	7.95	15.90	23.85
1470	15' deep, 1-1/2 C.Y. bucket	6.75	13.70	20.45
1480	18' deep, 2-1/2 C.Y. bucket	9.30	19.50	28.80
1520	6' wide, 6' deep, 5/8 C.Y. bucket	7.25	10.75	18
1530	8' deep, 3/4 C.Y. bucket	8.85	12.90	21.75
1540	10' deep, 1 C.Y. bucket	9.50	15.50	25
1550	12' deep, 1-1/4 C.Y. bucket	8.35	12.30	20.65
1560	16' deep, 2 C.Y. bucket	10.95	14.05	25
1570	20' deep, 3-1/2 C.Y. bucket	14.90	25	39.90
1580	24' deep, 3-1/2 C.Y. bucket	24	41.50	65.50
1640	8' wide, 12' deep, 1-1/4 C.Y. bucket	12.40	16.80	29.20
1650	15' deep, 1-1/2 C.Y. bucket	14.60	20.50	35.10
1660	18' deep, 2-1/2 C.Y. bucket	17	21	38
1680	24' deep, 3-1/2 C.Y. bucket	33	56	89
1730	10' wide, 20' deep, 3-1/2 C.Y. bucket	27	42	69
1740	24' deep, 3-1/2 C.Y. bucket	42	70	112
1800	1/2 to 1 slope, 2' wide, 2' deep, 3/8 C.Y. bucket	1.27	2.18	3.45
1810	3' deep, 3/8 C.Y. bucket	1.75	3.60	5.35
1820	4' deep, 3/8 C.Y. bucket	2.22	5.25	7.47
1840	6' deep, 3/8 C.Y. bucket	3.23	8.10	11.33
1860	8' deep, 1/2 C.Y. bucket	4.49	12.35	16.84
1880	10' deep, 1 C.Y. bucket	8.35	17	25.35

G1030 Site Earthwork

G1030 805	Trenching	COST PER L.F.		
		EQUIP.	LABOR	TOTAL
2300	4' wide, 2' deep, 3/8 C.Y. bucket	2.11	3.63	5.74
2310	3' deep, 3/8 C.Y. bucket	3.48	6	9.48
2320	4' deep, 1/2 C.Y. bucket	3.66	6.25	9.91
2340	6' deep, 1/2 C.Y. bucket	5.10	10.85	15.95
2360	8' deep, 1/2 C.Y. bucket	8.65	14.70	23.35
2380	10' deep, 1 C.Y. bucket	11.35	21.50	32.85
2400	12' deep, 1 C.Y. bucket	11.95	23.50	35.45
2430	15' deep, 1-1/2 C.Y. bucket	15.20	29	44.20
2460	18' deep, 2-1/2 C.Y. bucket	21	35	56
2840	6' wide, 6' deep, 5/8 C.Y. bucket	8.60	11.85	20.45
2860	8' deep, 3/4 C.Y. bucket	12.25	17.75	30
2880	10' deep, 1 C.Y. bucket	14.30	23.50	37.80
2900	12' deep, 1-1/4 C.Y. bucket	13.45	19.55	33
2940	16' deep, 2 C.Y. bucket	21	31	52
2980	20' deep, 3-1/2 C.Y. bucket	32.50	53.50	86
3020	24' deep, 3-1/2 C.Y. bucket	58.50	100	158.50
3100	8' wide, 12' deep, 1-1/4 C.Y. bucket	18.05	24	42.05
3120	15' deep, 1-1/2 C.Y. bucket	23	32	55
3140	18' deep, 2-1/2 C.Y. bucket	29	35	64
3180	24' deep, 3-1/2 C.Y. bucket	67.50	112	179.50
3270	10' wide, 20' deep, 3-1/2 C.Y. bucket	42	67	109
3280	24' deep, 3-1/2 C.Y. bucket	77	126	203
3500	1 to 1 slope, 2' wide, 2' deep, 3/8 C.Y. bucket	1.48	2.55	4.03
3520	3' deep, 3/8 C.Y. bucket	2.27	4.46	6.73
3540	4' deep, 3/8 C.Y. bucket	2.92	6.70	9.62
3560	6' deep, 3/8 C.Y. bucket	4.40	10.70	15.10
3580	8' deep, 1/2 C.Y. bucket	6.35	16.85	23.20
3600	10' deep, 1 C.Y. bucket	13.90	27	40.90
3800	4' wide, 2' deep, 3/8 C.Y. bucket	2.52	4.35	6.87
3820	3' deep, 3/8 C.Y. bucket	4.44	7.65	12.09
3840	4' deep, 1/2 C.Y. bucket	4.45	7	11.45
3860	6' deep, 1/2 C.Y. bucket	6.35	13	19.35
3880	8' deep, 1/2 C.Y. bucket	11.65	19.40	31.05
3900	10' deep, 1 C.Y. bucket	15.15	28	43.15
3920	12' deep, 1 C.Y. bucket	22	40.50	62.50
3940	15' deep, 1-1/2 C.Y. bucket	21.50	38	59.50
3960	18' deep, 2-1/2 C.Y. bucket	30	45.50	75.50
4030	6' wide, 6' deep, 5/8 C.Y. bucket	10.45	14.40	24.85
4040	8' deep, 3/4 C.Y. bucket	15.30	23	38.30
4050	10' deep, 1 C.Y. bucket	18.45	32	50.45
4060	12' deep, 1-1/4 C.Y. bucket	17.95	29	46.95
4070	16' deep, 2 C.Y. bucket	28.50	41	69.50
4080	20' deep, 3-1/2 C.Y. bucket	46.50	85	131.50
4090	24' deep, 3-1/2 C.Y. bucket	87	165	252
4500	8' wide, 12' deep, 1-1/4 C.Y. bucket	23	32.50	55.50
4550	15' deep, 1-1/2 C.Y. bucket	30.50	46.50	77
4600	18' deep, 2-1/2 C.Y. bucket	39.50	54	93.50
4650	24' deep, 3-1/2 C.Y. bucket	96	176	272
4800	10' wide, 20' deep, 3-1/2 C.Y. bucket	58.50	98.50	157
4850	24' deep, 3-1/2 C.Y. bucket	104	187	291
5000	1-1/2 to 1 slope, 2' wide, 2' deep, 3/8 C.Y. bucket	1.71	2.95	4.66
5020	3' deep, 3/8 C.Y. bucket	2.77	5.25	8.02
5040	4' deep, 3/8 C.Y. bucket	3.58	7.85	11.43
5060	6' deep, 3/8 C.Y. bucket	5.45	12.50	17.95
5080	8' deep, 1/2 C.Y. bucket	7.90	19.80	27.70
5100	10' deep, 1 C.Y. bucket	15.55	28	43.55
5300	4' wide, 2' deep, 3/8 C.Y. bucket	2.39	4.10	6.49
5320	3' deep, 3/8 C.Y. bucket	4.32	7.45	11.77

BUILDING SITEWORK

G

G1030 805	Trenching	COST PER L.F.		
		EQUIP.	LABOR	TOTAL
5340	4' deep, 1/2 C.Y. bucket	5.30	8.20	13.50
5360	6' deep, 1/2 C.Y. bucket	7.55	14.75	22.30
5380	8' deep, 1/2 C.Y. bucket	14	22	36
5400	10' deep, 1 C.Y. bucket	18.40	31.50	49.90
5420	12' deep, 1 C.Y. bucket	27	47	74
5450	15' deep, 1-1/2 C.Y. bucket	26.50	42	68.50
5480	18' deep, 2-1/2 C.Y. bucket	37.50	49.50	87
5660	6' wide, 6' deep, 5/8 C.Y. bucket	12.10	16.10	28.20
5680	8' deep, 3/4 C.Y. bucket	17.90	25.50	43.40
5700	10' deep, 1 C.Y. bucket	22	35.50	57.50
5720	12' deep, 1-1/4 C.Y. bucket	21	30.50	51.50
5760	16' deep, 2 C.Y. bucket	34.50	43.50	78
5800	20' deep, 3-1/2 C.Y. bucket	57	96	153
6050	15' deep, 1-1/2 C.Y. bucket	36	51	87
6080	18' deep, 2-1/2 C.Y. bucket	48	58	106
6140	24' deep, 3-1/2 C.Y. bucket	118	198	316
6300	10' wide, 20' deep, 3-1/2 C.Y. bucket	70	109	179
6350	24' deep, 3-1/2 C.Y. bucket	125	207	332
6600	2 to 1 slope, 2' wide, 2' deep, 3/8 C.Y. bucket	2.70	2.68	5.38
6620	3' deep, 3/8 C.Y. bucket	3.19	5.85	9.04
6640	4' deep, 3/8 C.Y. bucket	4.10	8.75	12.85
6660	6' deep, 3/8 C.Y. bucket	5.85	13.25	19.10
6680	8' deep, 1/2 C.Y. bucket	9	22.50	31.50
6700	10' deep, 1 C.Y. bucket	17.65	31.50	49.15
6900	4' wide, 2' deep, 3/8 C.Y. bucket	2.38	4.08	6.46
6920	3' deep, 3/8 C.Y. bucket	4.44	7.65	12.09
6940	4' deep, 1/2 C.Y. bucket	5.95	8.75	14.70
6960	6' deep, 1/2 C.Y. bucket	8.45	16	24.45
6980	8' deep, 1/2 C.Y. bucket	15.55	24	39.55
7000	10' deep, 1 C.Y. bucket	20.50	35	55.50
7020	12' deep, 1 C.Y. bucket	30.50	52.50	83
7050	15' deep, 1-1/2 C.Y. bucket	29.50	47	76.50
7080	18' deep, 2-1/2 C.Y. bucket	42.50	56	98.50
7260	6' wide, 6' deep, 5/8 C.Y. bucket	13.35	17.20	30.55
7280	8' deep, 3/4 C.Y. bucket	19.70	27.50	47.20
7300	10' deep, 3/4 C.Y. bucket	26	39.50	65.50
7320	12' deep, 1-1/4 C.Y. bucket	23.50	34.50	58
7360	16' deep, 2 C.Y. bucket	38.50	48.50	87
7400	20' deep, 3-1/2 C.Y. bucket	64	107	171
7440	24' deep, 3-1/2 C.Y. bucket	122	210	332
7620	8' wide, 12' deep, 1-1/4 C.Y. bucket	30.50	39	69.50
7650	15' deep, 1-1/2 C.Y. bucket	40.50	56	96.50
7680	18' deep, 2-1/2 C.Y. bucket	53	64	117
7740	24' deep, 3-1/2 C.Y. bucket	132	220	352
7920	10' wide, 20' deep, 3-1/2 C.Y. bucket	77	120	197
7940	24' deep, 3-1/2 C.Y. bucket	139	229	368

G1030 Site Earthwork

The Pipe Bedding System is shown for various pipe diameters. Compacted bank sand is used for pipe bedding and to fill 12″ over the pipe. No backfill is included. Various side slopes are shown to accommodate different soil conditions. Pipe sizes vary from 6″ to 84″ diameter.

System Components			COST PER L.F.		
	QUANTITY	UNIT	MAT.	INST.	TOTAL
SYSTEM G1030 815 1440					
PIPE BEDDING, SIDE SLOPE 0 TO 1, 1′ WIDE, PIPE SIZE 6″ DIAMETER					
Borrow, bank sand, 2 mile haul, machine spread	.067	C.Y.	.30	.37	.67
Compaction, vibrating plate	.067	C.Y.		.12	.12
TOTAL			.30	.49	.79

G1030 815	Pipe Bedding	COST PER L.F.		
		MAT.	INST.	TOTAL
1440	Pipe bedding, side slope 0 to 1, 1′ wide, pipe size 6″ diameter	.30	.49	.79
1460	2′ wide, pipe size 8″ diameter	.67	1.09	1.76
1480	Pipe size 10″ diameter	.69	1.11	1.80
1500	Pipe size 12″ diameter	.71	1.15	1.86
1520	3′ wide, pipe size 14″ diameter	1.17	1.89	3.06
1540	Pipe size 15″ diameter	1.18	1.90	3.08
1560	Pipe size 16″ diameter	1.19	1.93	3.12
1580	Pipe size 18″ diameter	1.22	1.97	3.19
1600	4′ wide, pipe size 20″ diameter	1.76	2.83	4.59
1620	Pipe size 21″ diameter	1.78	2.86	4.64
1640	Pipe size 24″ diameter	1.83	2.95	4.78
1660	Pipe size 30″ diameter	1.87	3.01	4.88
1680	6′ wide, pipe size 32″ diameter	3.26	5.25	8.51
1700	Pipe size 36″ diameter	3.35	5.40	8.75
1720	7′ wide, pipe size 48″ diameter	4.36	7.05	11.41
1740	8′ wide, pipe size 60″ diameter	5.45	8.80	14.25
1760	10′ wide, pipe size 72″ diameter	7.85	12.70	20.55
1780	12′ wide, pipe size 84″ diameter	10.70	17.25	27.95
2140	Side slope 1/2 to 1, 1′ wide, pipe size 6″ diameter	.64	1.02	1.66
2160	2′ wide, pipe size 8″ diameter	1.06	1.71	2.77
2180	Pipe size 10″ diameter	1.15	1.86	3.01
2200	Pipe size 12″ diameter	1.23	1.99	3.22
2220	3′ wide, pipe size 14″ diameter	1.77	2.84	4.61
2240	Pipe size 15″ diameter	1.81	2.92	4.73
2260	Pipe size 16″ diameter	1.87	3.01	4.88
2280	Pipe size 18″ diameter	1.97	3.18	5.15
2300	4′ wide, pipe size 20″ diameter	2.61	4.20	6.81
2320	Pipe size 21″ diameter	2.67	4.31	6.98
2340	Pipe size 24″ diameter	2.86	4.62	7.48
2360	Pipe size 30″ diameter	3.21	5.20	8.41
2380	6′ wide, pipe size 32″ diameter	4.73	7.65	12.38
2400	Pipe size 36″ diameter	5.05	8.15	13.20
2420	7′ wide, pipe size 48″ diameter	6.90	11.15	18.05
2440	8′ wide, pipe size 60″ diameter	9	14.50	23.50
2460	10′ wide, pipe size 72″ diameter	12.60	20.50	33.10
2480	12′ wide, pipe size 84″ diameter	16.75	27	43.75
2620	Side slope 1 to 1, 1′ wide, pipe size 6″ diameter	.98	1.58	2.56
2640	2′ wide, pipe size 8″ diameter	1.47	2.38	3.85

BUILDING SITEWORK

G

G1030 Site Earthwork

G1030 815	Pipe Bedding	COST PER L.F.		
		MAT.	INST.	TOTAL
2660	Pipe size 10" diameter	1.61	2.59	4.20
2680	Pipe size 12" diameter	1.77	2.84	4.61
2700	3' wide, pipe size 14" diameter	2.37	3.81	6.18
2720	Pipe size 15" diameter	2.45	3.95	6.40
2740	Pipe size 16" diameter	2.54	4.10	6.64
2760	Pipe size 18" diameter	2.73	4.40	7.13
2780	4' wide, pipe size 20" diameter	3.45	5.55	9
2800	Pipe size 21" diameter	3.55	5.75	9.30
2820	Pipe size 24" diameter	3.89	6.25	10.14
2840	Pipe size 30" diameter	4.54	7.30	11.84
2860	6' wide, pipe size 32" diameter	6.20	9.95	16.15
2880	Pipe size 36" diameter	6.75	10.90	17.65
2900	7' wide, pipe size 48" diameter	9.45	15.25	24.70
2920	8' wide, pipe size 60" diameter	12.55	20.50	33.05
2940	10' wide, pipe size 72" diameter	17.30	28	45.30
2960	12' wide, pipe size 84" diameter	23	37	60
3000	Side slope 1-1/2 to 1, 1' wide, pipe size 6" diameter	1.31	2.11	3.42
3020	2' wide, pipe size 8" diameter	1.86	3.01	4.87
3040	Pipe size 10" diameter	2.06	3.32	5.38
3060	Pipe size 12" diameter	2.29	3.69	5.98
3080	3' wide, pipe size 14" diameter	2.96	4.77	7.73
3100	Pipe size 15" diameter	3.08	4.98	8.06
3120	Pipe size 16" diameter	3.21	5.20	8.41
3140	Pipe size 18" diameter	3.49	5.65	9.14
3160	4' wide, pipe size 20" diameter	4.30	6.95	11.25
3180	Pipe size 21" diameter	4.44	7.15	11.59
3200	Pipe size 24" diameter	4.92	7.90	12.82
3220	Pipe size 30" diameter	5.90	9.50	15.40
3240	6' wide, pipe size 32" diameter	7.65	12.30	19.95
3260	Pipe size 36" diameter	8.45	13.65	22.10
3280	7' wide, pipe size 48" diameter	12	19.35	31.35
3300	8' wide, pipe size 60" diameter	16.10	26	42.10
3320	10' wide, pipe size 72" diameter	22	35.50	57.50
3340	12' wide, pipe size 84" diameter	29	46.50	75.50
3400	Side slope 2 to 1, 1' wide, pipe size 6" diameter	1.68	2.70	4.38
3420	2' wide, pipe size 8" diameter	2.26	3.64	5.90
3440	Pipe size 10" diameter	2.52	4.06	6.58
3460	Pipe size 12" diameter	2.81	4.53	7.34
3480	3' wide, pipe size 14" diameter	3.55	5.75	9.30
3500	Pipe size 15" diameter	3.70	5.95	9.65
3520	Pipe size 16" diameter	3.88	6.25	10.13
3540	Pipe size 18" diameter	4.24	6.85	11.09
3560	4' wide, pipe size 20" diameter	5.15	8.30	13.45
3580	Pipe size 21" diameter	5.35	8.60	13.95
3600	Pipe size 24" diameter	5.95	9.60	15.55
3620	Pipe size 30" diameter	7.25	11.70	18.95
3640	6' wide, pipe size 32" diameter	9.10	14.65	23.75
3660	Pipe size 36" diameter	10.15	16.40	26.55
3680	7' wide, pipe size 48" diameter	14.55	23.50	38.05
3700	8' wide, pipe size 60" diameter	19.65	31.50	51.15
3720	10' wide, pipe size 72" diameter	27	43	70
3740	12' wide, pipe size 84" diameter	35	56.50	91.50

Important: See the Reference Section for critical supporting data - Reference Numbers and City Cost Indexes

G2030 Pedestrian Paving

The Bituminous Sidewalk System includes: excavation; compacted gravel base (hand graded), bituminous surface; and hand grading along the edge of the completed walk.

The Expanded System Listing shows Bituminous Sidewalk systems with wearing course depths ranging from 1″ to 2-1/2″ of bituminous material. The gravel base ranges from 4″ to 8″. Sidewalk widths are shown ranging from 3′ to 5′. Costs are on a linear foot basis.

System Components	QUANTITY	UNIT	COST PER L.F.		
			MAT.	INST.	TOTAL
SYSTEM G2030 110 1580					
BITUMINOUS SIDEWALK, 1″ THICK PAVING, 4″ GRAVEL BASE, 3′ WIDTH					
Excavation, bulk, dozer, push 50′	.046	C.Y.		.07	.07
Borrow, bank run gravel, haul 2 mi., spread w/dozer, no compaction	.037	C.Y.	.72	.21	.93
Compact w/ vib plate, 8″ lifts	.037	C.Y.		.07	.07
Fine grade,area to be paved,small area	.333	S.Y.		2.17	2.17
Sidewalk, bituminous, no base, 1″ thick	.333	S.Y.	.67	.51	1.18
Backfill by hand, no compaction, light soil	.006	C.Y.		.14	.14
TOTAL			1.39	3.17	4.56

G2030 110	Bituminous Sidewalks	COST PER L.F.		
		MAT.	INST.	TOTAL
1580	Bituminous sidewalk, 1″ thick paving, 4″ gravel base, 3′ width	1.39	3.17	4.56
1600	4′ width	1.85	3.44	5.29
1620	5′ width	2.35	3.60	5.95
1640	6″ gravel base, 3′ width	1.76	3.33	5.09
1660	4′ width	2.34	3.65	5.99
1680	5′ width	2.93	4	6.93
1700	8″ gravel base, 3′ width	2.12	3.48	5.60
1720	4′ width	2.83	3.88	6.71
1740	5′ width	3.53	4.30	7.83
1800	1-1/2″ thick paving, 4″ gravel base, 3′ width	1.76	3.32	5.08
1820	4′ width	2.33	3.68	6.01
1840	5′ width	2.88	4.24	7.12
1860	6″ gravel base, 3′ width	2.13	3.48	5.61
1880	4′ width	2.79	4.08	6.87
1900	5′ width	3.53	4.33	7.86
1920	8″ gravel base, 3′ width	2.45	3.81	6.26
1940	4′ width	3.31	4.11	7.42
1960	5′ width	4.09	4.80	8.89
2120	2″ thick paving, 4″ gravel base, 3′ width	2.06	3.84	5.90
2140	4′ width	2.74	4.28	7.02
2160	5′ width	3.44	4.75	8.19
2180	6″ gravel base, 3′ width	2.43	3.99	6.42
2200	4′ width	3.23	4.49	7.72
2220	5′ width	4.05	5	9.05
2240	8″ gravel base, 3′ width	2.79	4.15	6.94
2260	4′ width	3.72	4.72	8.44
2280	5′ width	4.65	5.30	9.95
2400	2-1/2″ thick paving, 4″gravel base, 3′ width	2.85	4.29	7.14
2420	4′ width	3.79	4.86	8.65
2440	5′ width	4.78	5.45	10.23
2460	6″ gravel base, 3′ width	3.22	4.46	7.68
2480	4′ width	4.28	5.05	9.33
2500	5′ width	5.40	5.75	11.15
2520	8″ gravel base, 3′ width	3.58	4.60	8.18
2540	4′ width	4.77	5.30	10.07
2560	5′ width	6	6.05	12.05

BUILDING SITEWORK

G

G2030 Pedestrian Paving

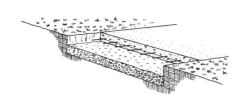

The Concrete Sidewalk System includes: excavation; compacted gravel base (hand graded); forms; welded wire fabric; and 3,000 p.s.i. air-entrained concrete (broom finish).

The Expanded System Listing shows Concrete Sidewalk systems with wearing course depths ranging from 4″ to 6″. The gravel base ranges from 4″ to 8″. Sidewalk widths are shown ranging from 3′ to 5′. Costs are on a linear foot basis.

System Components			COST PER L.F.		
	QUANTITY	UNIT	MAT.	INST.	TOTAL
SYSTEM G2030 120 1580					
CONCRETE, SIDEWALK 4″ THICK, 4″ GRAVEL BASE, 3′ WIDE					
Excavation, box out with dozer	.100	C.Y.		.15	.15
Gravel base, haul 2 miles, spread with dozer	.037	C.Y.	.72	.21	.93
Compaction with vibrating plate	.037	C.Y.		.07	.07
Fine grade by hand	.333	S.Y.		2.20	2.20
Concrete in place including forms and reinforcing	.037	C.Y.	4.47	5.73	10.20
Backfill edges by hand	.010	C.Y.		.24	.24
TOTAL			5.19	8.60	13.79

G2030 120	Concrete Sidewalks	COST PER L.F.		
		MAT.	INST.	TOTAL
1580	Concrete sidewalk, 4″ thick, 4″ gravel base, 3′ wide	5.20	8.60	13.80
1600	4′ wide	6.90	10.65	17.55
1620	5′ wide	8.65	12.65	21.30
1640	6″ gravel base, 3′ wide	5.55	8.75	14.30
1660	4′ wide	7.40	10.85	18.25
1680	5′ wide	9.25	12.95	22.20
1700	8″ gravel base, 3′ wide	5.90	8.90	14.80
1720	4′ wide	7.90	11.05	18.95
1740	5′ wide	9.85	13.20	23.05
1800	5″ thick concrete, 4″ gravel base, 3′ wide	6.65	9.30	15.95
1820	4′ wide	8.90	11.50	20.40
1840	5′ wide	11.10	13.75	24.85
1860	6″ gravel base, 3′ wide	7.05	9.45	16.50
1900	5′ wide	11.70	13.55	25.25
1920	8″ gravel base, 3′ wide	7.40	9.60	17
1940	4′ wide	9.85	12.55	22.40
1960	5′ wide	12.30	13.85	26.15
2120	6″ thick concrete, 4″ gravel base, 3′ wide	7.65	9.80	17.45
2140	4′ wide	10.20	12.15	22.35
2160	5′ wide	12.75	14.60	27.35
2180	6″ gravel base, 3′ wide	8	10	18
2200	4′ wide	10.70	12.35	23.05
2220	5′ wide	13.35	14.80	28.15
2240	8″ gravel base, 3′ wide	8.40	10.10	18.50
2260	4′ wide	11.20	12.65	23.85
2280	5′ wide	13.95	15.10	29.05

G2040   Site Development

There are four basic types of Concrete Retaining Wall Systems: reinforced concrete with level backfill; reinforced concrete with sloped backfill or surcharge; unreinforced with level backfill; and unreinforced with sloped backfill or surcharge. System elements include: all necessary forms (4 uses); 3,000 p.s.i. concrete with an 8" chute; all necessary reinforcing steel; and underdrain. Exposed concrete is patched and rubbed.

The Expanded System Listing shows walls that range in thickness from 10" to 18" for reinforced concrete walls with level backfill and 12" to 24" for reinforced walls with sloped backfill. Walls range from a height of 4' to 20'. Unreinforced level and sloped backfill walls range from a height of 3' to 10'.

System Components	QUANTITY	UNIT	COST PER L.F. MAT.	COST PER L.F. INST.	COST PER L.F. TOTAL
SYSTEM G2040 210 1000					
CONC.RETAIN. WALL REINFORCED, LEVEL BACKFILL 4' HIGH					
Forms in place, cont. wall footing & keyway, 4 uses	2.000	S.F.	1.66	6.66	8.32
Forms in place, retaining wall forms, battered to 8' high, 4 uses	8.000	SFCA	4.88	51.20	56.08
Reinforcing in place, walls, #3 to #7	.004	Ton	3.34	2.68	6.02
Concrete ready mix, regular weight, 3000 psi	.204	C.Y.	18.16		18.16
Placing concrete and vibrating footing con., shallow direct chute	.074	C.Y.		1.31	1.31
Placing concrete and vibrating walls, 8" thick, direct chute	.130	C.Y.		3.07	3.07
Pipe bedding, crushed or screened bank run gravel	1.000	L.F.	2.42	.95	3.37
Pipe, subdrainage, corrugated plastic, 4" diameter	1.000	L.F.	.63	.55	1.18
Finish walls and break ties, patch walls	4.000	S.F.	.12	2.88	3
TOTAL			31.21	69.30	100.51

G2040 210	Concrete Retaining Walls	COST PER L.F. MAT.	COST PER L.F. INST.	COST PER L.F. TOTAL
1000	Conc.retain.wall, reinforced, level backfill, 4' high x 2'-2" base, 10"thick	31	69.50	100.50
1200	6' high x 3'-3" base, 10" thick	45.50	100	145.50
1400	8' high x 4'-3" base, 10" thick	60	132	192
1600	10' high x 5'-4" base, 13" thick	79	193	272
1800	12' high x 6'-6" base, 14" thick	98	231	329
2200	16' high x 8'-6" base, 16" thick	149	310	459
2600	20' high x 10'-5" base, 18" thick	218	405	623
3000	Sloped backfill, 4' high x 3'-2" base, 12" thick	38	72	110
3200	6' high x 4'-6" base, 12" thick	54	103	157
3400	8' high x 5'-11" base, 12" thick	72.50	136	208.50
3600	10' high x 7'-5" base, 16" thick	104	202	306
3800	12' high x 8'-10" base, 18" thick	137	246	383
4200	16' high x 11'-10" base, 21" thick	229	350	579
4600	20' high x 15'-0" base, 24" thick	355	470	825
5000	Unreinforced, level backfill, 3'-0" high x 1'-6" base	17.45	46.50	63.95
5200	4'-0" high x 2'-0" base	27	62	89
5400	6'-0" high x 3'-0" base	50.50	96	146.50
5600	8'-0" high x 4'-0" base	80	128	208
5800	10'-0" high x 5'-0" base	119	198	317
7000	Sloped backfill, 3'-0" high x 2'-0" base	21	48.50	69.50
7200	4'-0" high x 3'-0" base	35	64.50	99.50
7400	6'-0" high x 5'-0" base	72	102	174
7600	8'-0" high x 7'-0" base	122	138	260
7800	10'-0" high x 9'-0" base	188	214	402

BUILDING SITEWORK

G

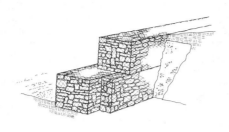

The Gabion Retaining Wall Systems list three types of surcharge conditions: level, sloped and highway, for two different facing configurations: stepped and straight with batter. Costs are expressed per L.F. for heights ranging from 6' to 18' for retaining sandy or clay soil. For protection against sloughing in wet clay materials counterforts have been added for these systems. Drainage stone has been added behind the walls to avoid any additional lateral pressures.

System Components				COST PER L.F.		
	QUANTITY	UNIT		MAT.	INST.	TOTAL
SYSTEM G2040 270 1000						
GABION RET. WALL, LEVEL BACKFILL, STEPPED FACE, 4' BASE, 6' HIGH						
3' x 3' cross section gabion	2.000	Ea.		47.33	117.33	164.66
3' x 1' cross section gabion	1.000	Ea.		10.50	7.01	17.51
Crushed stone drainage	.220	C.Y.		7.50	1.38	8.88
TOTAL				65.33	125.72	191.05

G2040 270	Gabion Retaining Walls	COST PER L.F.		
		MAT.	INST.	TOTAL
1000	Gabion ret. wall, level backfill, stepped face, 4' base, 6' high, sandy soil	65.50	126	191.50
1040	Clay soil with counterforts @ 16' O.C.	100	258	358
1100	5' base, 9' high, sandy soil	113	200	313
1140	Clay soil with counterforts @ 16' O.C.	193	395	588
1200	6' base, 12' high, sandy soil	165	315	480
1240	Clay soil with counterforts @ 16' O.C.	305	845	1,150
1300	7'-6" base, 15' high, sandy soil	231	445	676
1340	Clay soil with counterforts @ 16' O.C.	555	1,175	1,730
1400	9' base, 18' high, sandy soil	305	620	925
1440	Clay soil with counterforts @ 16' O.C.	625	1,400	2,025
2000	Straight face w/1:6 batter, 3' base, 6' high, sandy soil	62.50	121	183.50
2040	Clay soil with counterforts @ 16' O.C.	97.50	252	349.50
2100	4'-6" base, 9' high, sandy soil	106	196	302
2140	Clay soil with counterforts @ 16' O.C.	186	395	581
2200	6' base, 12' high, sandy soil	159	315	474
2240	Clay soil with counterforts @ 16' O.C.	300	840	1,140
2300	7'-6" base, 15' high, sandy soil	226	450	676
2340	Clay soil with counterforts @ 16' O.C.	445	930	1,375
2400	18' high, sandy soil	292	585	877
2440	Clay soil with counterforts @ 16' O.C.	435	680	1,115
3000	Backfill sloped 1-1/2:1, stepped face, 4'-6" base, 6' high, sandy soil	68.50	135	203.50
3040	Clay soil with counterforts @ 16' O.C.	104	267	371
3100	6' base, 9' high, sandy soil	119	253	372
3140	Clay soil with counterforts @ 16' O.C.	199	450	649
3200	7'-6" base, 12' high, sandy soil	185	390	575
3240	Clay soil with counterforts @ 16' O.C.	325	915	1,240
3300	9' base, 15' high, sandy soil	259	560	819
3340	Clay soil with counterforts @ 16' O.C.	585	1,300	1,885
3400	10'-6" base, 18' high, sandy soil	350	760	1,110
3440	Clay soil with counterforts @ 16' O.C.	665	1,550	2,215
4000	Straight face with 1:6 batter, 4'-6" base, 6' high, sandy soil	76	137	213
4040	Clay soil with counterforts @ 16' O.C.	111	269	380
4100	6' base, 9' high, sandy soil	129	254	383
4140	Clay soil with counterforts @ 16' O.C.	209	455	664

G2040 Site Development

G2040 270	Gabion Retaining Walls	COST PER L.F.		
		MAT.	INST.	TOTAL
4200	7'-6" base, 12' high, sandy soil	197	390	587
4240	Clay soil with counterforts @ 16' O.C.	335	920	1,255
4300	9' base, 15' high, sandy soil	274	565	839
4340	Clay soil with counterforts @ 16' O.C.	600	1,300	1,900
4400	18' high, sandy soil	350	740	1,090
4440	Clay soil with counterforts @ 16' O.C.	670	1,525	2,195
5000	Highway surcharge, straight face, 6' base, 6' high, sandy soil	102	236	338
5040	Clay soil with counterforts @ 16' O.C.	137	370	507
5100	9' base, 9' high, sandy soil	176	415	591
5140	Clay soil with counterforts @ 16' O.C.	256	610	866
5200	12' high, sandy soil	385	615	1,000
5240	Clay soil with counterforts @ 16' O.C.	500	1,525	2,025
5300	12' base, 15' high, sandy soil	350	825	1,175
5340	Clay soil with counterforts @ 16' O.C.	570	1,300	1,870
5400	18' high, sandy soil	450	1,050	1,500
5440	Clay soil with counterforts @ 16' O.C.	770	1,850	2,620

BUILDING SITEWORK

G

G3030 Storm Sewer

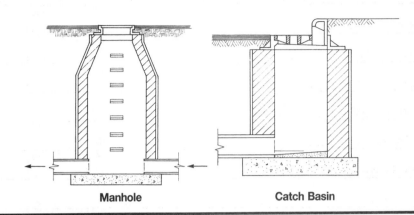

Manhole   Catch Basin

The Manhole and Catch Basin System includes: excavation with a backhoe; a formed concrete footing; frame and cover; cast iron steps and compacted backfill.

The Expanded System Listing shows manholes that have a 4', 5' and 6' inside diameter riser. Depths range from 4' to 14'. Construction material shown is either concrete, concrete block, precast concrete, or brick.

BUILDING SITEWORK G

System Components	QUANTITY	UNIT	COST PER EACH		
			MAT.	INST.	TOTAL
SYSTEM G3030 210 1920					
MANHOLE/CATCH BASIN, BRICK, 4' I.D. RISER, 4' DEEP					
Excavation, hydraulic backhoe, 3/8 C.Y. bucket	14.815	C.Y.		84.74	84.74
Trim sides and bottom of excavation	64.000	S.F.		45.44	45.44
Forms in place, manhole base, 4 uses	20.000	SFCA	11.80	78	89.80
Reinforcing in place footings, #4 to #7	.019	Ton	15.87	18.15	34.02
Concrete, 3000 psi	.925	C.Y.	82.33		82.33
Place and vibrate concrete, footing, direct chute	.925	C.Y.		35.58	35.58
Catch basin or MH, brick, 4' ID, 4' deep	1.000	Ea.	350	760	1,110
Catch basin or MH steps; heavy galvanized cast iron	1.000	Ea.	13.45	10.75	24.20
Catch basin or MH frame and cover	1.000	Ea.	212	166.50	378.50
Fill, granular	12.954	C.Y.	112.05		112.05
Backfill, spread with wheeled front end loader	12.954	C.Y.		22.93	22.93
Backfill compaction, 12" lifts, air tamp	12.954	C.Y.		83.55	83.55
TOTAL			797.50	1,305.64	2,103.14

G3030 210	Manholes & Catch Basins	COST PER EACH		
		MAT.	INST.	TOTAL
1920	Manhole/catch basin, brick, 4' I.D. riser, 4' deep	800	1,325	2,125
1940	6' deep	1,050	1,800	2,850
1960	8' deep	1,350	2,500	3,850
1980	10' deep	1,850	3,075	4,925
3000	12' deep	2,425	3,350	5,775
3020	14' deep	3,075	4,675	7,750
3200	Block, 4' I.D. riser, 4' deep	730	1,050	1,780
3220	6' deep	940	1,475	2,415
3240	8' deep	1,175	2,050	3,225
3260	10' deep	1,375	2,525	3,900
3280	12' deep	1,675	3,200	4,875
3300	14' deep	2,025	3,900	5,925
4620	Concrete, cast-in-place, 4' I.D. riser, 4' deep	940	1,800	2,740
4640	6' deep	1,275	2,425	3,700
4660	8' deep	1,750	3,525	5,275
4680	10' deep	2,100	4,375	6,475
4700	12' deep	2,550	5,400	7,950
4720	14' deep	3,025	6,475	9,500
5820	Concrete, precast, 4' I.D. riser, 4' deep	1,200	965	2,165
5840	6' deep	1,525	1,300	2,825

G3030 Storm Sewer

G3030 210	Manholes & Catch Basins	COST PER EACH		
		MAT.	INST.	TOTAL
5860	8' deep	1,850	1,825	3,675
5880	10' deep	2,275	2,250	4,525
5900	12' deep	2,800	2,750	5,550
5920	14' deep	3,375	3,500	6,875
6000	5' I.D. riser, 4' deep	1,275	1,050	2,325
6020	6' deep	1,700	1,475	3,175
6040	8' deep	2,150	1,925	4,075
6060	10' deep	2,700	2,475	5,175
6080	12' deep	3,275	3,150	6,425
6100	14' deep	3,925	3,850	7,775
6200	6' I.D. riser, 4' deep	1,875	1,375	3,250
6220	6' deep	2,400	1,825	4,225
6240	8' deep	3,000	2,575	5,575
6260	10' deep	3,750	3,250	7,000
6280	12' deep	4,575	4,075	8,650
6300	14' deep	5,425	4,950	10,375

BUILDING SITEWORK

G

G3030 Storm Sewer

The Headwall Systems are listed in concrete and two different stone wall materials for two different backfill slope conditions. The backfill slope directly affects the length of the wing walls. Walls are listed for three different culvert sizes 30″, 48″ and 60″ diameter. Excavation and backfill are included in the system components, and are figured from an elevation 2′ below the bottom of the pipe.

System Components	QUANTITY	UNIT	COST PER EACH		
			MAT.	INST.	TOTAL
SYSTEM G3030 310 2000					
HEADWALL C.I.P. CONCRETE FOR 30″ PIPE, 3′ LONG WING WALLS					
Excavation, hydraulic backhoe, 3/8 C.Y. bucket	2.500	C.Y.		160.16	160.16
Formwork, 2 uses	157.000	SFCA	294.69	1,402.91	1,697.60
Reinforcing in place, A615 Gr 60, longer and heavier dowels, add	45.000	Lb.	21.60	62.10	83.70
Concrete, 3000 psi	2.600	C.Y.	231.40		231.40
Place concrete, spread footings, direct chute	2.600	C.Y.		100	100
Backfill, dozer	2.500	C.Y.		37.24	37.24
TOTAL			547.69	1,762.41	2,310.10

G3030 310	Headwalls	COST PER EACH		
		MAT.	INST.	TOTAL
2000	Headwall, 1-1/2 to 1 slope soil, C.I.P. conc, 30″pipe, 3′long wing walls	550	1,775	2,325
2020	Pipe size 36″, 3′-6″ long wing walls	685	2,125	2,810
2040	Pipe size 42″, 4′ long wing walls	830	2,475	3,305
2060	Pipe size 48″, 4′-6″ long wing walls	1,000	2,900	3,900
2080	Pipe size 54″, 5′-0″ long wing walls	1,175	3,375	4,550
2100	Pipe size 60″, 5′-6″ long wing walls	1,375	3,825	5,200
2120	Pipe size 72″, 6′-6″ long wing walls	1,850	4,925	6,775
2140	Pipe size 84″, 7′-6″ long wing walls	2,350	6,075	8,425
3000	$350/ton stone, pipe size 30″, 3′ long wing walls	1,800	640	2,440
3020	Pipe size 36″, 3′-6″ long wing walls	2,325	770	3,095
3040	Pipe size 42″, 4′ long wing walls	2,925	915	3,840
3060	Pipe size 48″, 4′-6″ long wing walls	3,600	1,075	4,675
3080	Pipe size 54″, 5′ long wing walls	4,350	1,275	5,625
3100	Pipe size 60″, 5′-6″ long wing walls	5,175	1,475	6,650
3120	Pipe size 72″, 6′-6″ long wing walls	7,125	1,950	9,075
3140	Pipe size 84″, 7′-6″ long wing walls	9,375	2,500	11,875
4500	2 to 1 slope soil, C.I.P. concrete, pipe size 30″, 4′-3″ long wing walls	655	2,050	2,705
4520	Pipe size 36″, 5′ long wing walls	830	2,500	3,330
4540	Pipe size 42″, 5′-9″ long wing walls	1,025	3,000	4,025
4560	Pipe size 48″, 6′-6″ long wing walls	1,225	3,475	4,700
4580	Pipe size 54″, 7′-3″ long wing walls	1,475	4,050	5,525
4600	Pipe size 60″, 8′-0″ long wing walls	1,700	4,650	6,350
4620	Pipe size 72″, 9′-6″ long wing walls	2,300	6,025	8,325
4640	Pipe size 84″, 11′-0″ long wing walls	2,975	7,525	10,500
5500	$350/ton stone, pipe size 30″, 4′-3″ long wing walls	2,200	735	2,935
5520	Pipe size 36″, 5′ long wing walls	2,875	905	3,780
5540	Pipe size 42″, 5′-9″ long wing walls	3,650	1,100	4,750
5560	Pipe size 48″, 6′-6″ long wing walls	4,525	1,325	5,850
5580	Pipe size 54″, 7′-3″ long wing walls	5,475	1,550	7,025
5600	Pipe size 60″, 8′-0″ long wing walls	6,500	1,800	8,300
5620	Pipe size 72″, 9′-6″ long wing walls	9,025	2,425	11,450
5640	Pipe size 84″, 11′-0″ long wing walls	12,000	3,150	15,150

Reference Section

All the reference information is in one section making it easy to find what you need to know . . . and easy to use the book on a daily basis. This section is visually identified by a vertical gray bar on the edge of pages.

In the reference number information that follows, you'll see the background that relates to the "reference numbers" that appeared in the Unit Price Sections. You'll find reference tables, explanations and estimating information that support how we arrived at the unit price data. Also included are alternate pricing methods, technical data and estimating procedures along with information on design and economy in construction.

Also in this Reference Section, we've included Crew Listings, a full listing of all the crews, equipment and their costs; Historical Cost Indexes for cost comparisons over time; City Cost Indexes and Location Factors for adjusting costs to the region you are in; and an explanation of all Abbreviations used in the book.

Table of Contents

Reference Numbers

R011	Overhead & Misc. Data	362
R015	Construction Aids	373
R020	Paving & Surfacing	375
R021	Subsurface Investigation & Demolition	376
R022	Site Preparation & Excavation Support	376
R023	Earthwork	380
R024	Foundation & L.B. Elements	383
R025	Sewerage & Drainage	389
R027	Paving & Surfacing	391
R029	Landscaping	392
R032	Concrete Reinforcement	395
R033	Cast-in-Place Concrete	397
R050	Materials, Coatings, & Fastenings	399
R053	Metal Deck	399
R056	Railroad & Marine Work	400
R061	Rough Carpentry	401
R131	Pre-Eng. Struct. & Aquatic Facil.	402

Reference Numbers (cont.)

R151	Plumbing	402
R160	Basic Electrical Materials & Methods	404
R161	Wiring Methods	406
R162	Electrical Power	407
R164	Low-Voltage Distribution	408

Crew Listings	409
Historical Cost Indexes	437
City Cost Indexes	438
Location Factors	458
Abbreviations	464

R01100-005 Tips for Accurate Estimating

1. Use pre-printed or columnar forms for orderly sequence of dimensions and locations and for recording telephone quotations.
2. Use only the front side of each paper or form except for certain pre-printed summary forms.
3. Be consistent in listing dimensions: For example, length x width x height. This helps in rechecking to ensure that, the total length of partitions is appropriate for the building area.
4. Use printed (rather than measured) dimensions where given.
5. Add up multiple printed dimensions for a single entry where possible.
6. Measure all other dimensions carefully.
7. Use each set of dimensions to calculate multiple related quantities.
8. Convert foot and inch measurements to decimal feet when listing. Memorize decimal equivalents to .01 parts of a foot (1/8″ equals approximately .01′).
9. Do not "round off" quantities until the final summary.
10. Mark drawings with different colors as items are taken off.
11. Keep similar items together, different items separate.
12. Identify location and drawing numbers to aid in future checking for completeness.
13. Measure or list everything on the drawings or mentioned in the specifications.
14. It may be necessary to list items not called for to make the job complete.
15. Be alert for: Notes on plans such as N.T.S. (not to scale); changes in scale throughout the drawings; reduced size drawings; discrepancies between the specifications and the drawings.
16. Develop a consistent pattern of performing an estimate. For example:
    a. Start the quantity takeoff at the lower floor and move to the next higher floor.
    b. Proceed from the main section of the building to the wings.
    c. Proceed from south to north or vice versa, clockwise or counterclockwise.
    d. Take off floor plan quantities first, elevations next, then detail drawings.
17. List all gross dimensions that can be either used again for different quantities, or used as a rough check of other quantities for verification (exterior perimeter, gross floor area, individual floor areas, etc.).
18. Utilize design symmetry or repetition (repetitive floors, repetitive wings, symmetrical design around a center line, similar room layouts, etc.). Note: Extreme caution is needed here so as not to omit or duplicate an area.
19. Do not convert units until the final total is obtained. For instance, when estimating concrete work, keep all units to the nearest cubic foot, then summarize and convert to cubic yards.
20. When figuring alternatives, it is best to total all items involved in the basic system, then total all items involved in the alternates. Therefore you work with positive numbers in all cases. When adds and deducts are used, it is often confusing whether to add or subtract a portion of an item; especially on a complicated or involved alternate.

R01100-040 Builder's Risk Insurance

Builder's Risk Insurance is insurance on a building during construction. Premiums are paid by the owner or the contractor. Blasting, collapse and underground insurance would raise total insurance costs above those listed. Floater policy for materials delivered to the job runs $.75 to $1.25 per $100 value. Contractor equipment insurance runs $.50 to $1.50 per $100 value. Insurance for miscellaneous tools to $1,500 value runs from $3.00 to $7.50 per $100 value.

Tabulated below are New England Builder's Risk insurance rates in dollars per $100 value for $1,000 deductible. For $25,000 deductible, rates can be reduced 13% to 34%. On contracts over $1,000,000, rates may be lower than those tabulated. Policies are written annually for the total completed value in place. For "all risk" insurance (excluding flood, earthquake and certain other perils) add $.025 to total rates below.

Coverage	Frame Construction (Class 1) Range			Average	Brick Construction (Class 4) Range			Average	Fire Resistive (Class 6) Range			Average
Fire Insurance	$.350	to	$.850	$.600	$.158	to	$.189	$.174	$.052	to	$.080	$.070
Extended Coverage	.115	to	.200	.158	.080	to	.105	.101	.081	to	.105	.100
Vandalism	.012	to	.016	.014	.008	to	.011	.011	.008	to	.011	.010
Total Annual Rate	$.477	to	$1.066	$.772	$.246	to	$.305	$.286	$.141	to	$.196	$.180

R01100-050   General Contractor's Overhead

There are two distinct types of overhead on a construction project: Project Overhead and Main Office Overhead. Project Overhead includes those costs at a construction site not directly associated with the installation of construction materials. Examples of Project Overhead costs include the following:

1. Superintendent
2. Construction office and storage trailers
3. Temporary sanitary facilities
4. Temporary utilities
5. Security fencing
6. Photographs
7. Clean up
8. Performance and payment bonds

The above Project Overhead items are also referred to as General Requirements and therefore are estimated in Division 1. Division 1 is the first division listed in the CSI MasterFormat but it is usually the last division estimated. The sum of the costs in Divisions 1 through 16 is referred to as the sum of the direct costs.

All construction projects also include indirect costs. The primary components of indirect costs are the contractor's Main Office Overhead and profit. The amount of the Main Office Overhead expense varies depending on the the following:

1. Owner's compensation
2. Project managers and estimator's wages
3. Clerical support wages
4. Office rent and utilities
5. Corporate legal and accounting costs
6. Advertising
7. Automobile expenses
8. Association dues
9. Travel and entertainment expenses

These costs are usually calculated as a percentage of annual sales volume. This percentage can range from 35% for a small contractor doing less than $500,000 to 5% for a large contractor with sales in excess of $100 million.

R01100-060   Workers' Compensation Insurance Rates by Trade

The table below tabulates the national averages for Workers' Compensation insurance rates by trade and type of building. The average "Insurance Rate" is multiplied by the "% of Building Cost" for each trade. This produces the "Workers' Compensation Cost" by % of total labor cost, to be added for each trade by building type to determine the weighted average Workers' Compensation rate for the building types analyzed.

Trade	Insurance Rate (% Labor Cost)			% of Building Cost			Workers' Compensation		
	Range		Average	Office Bldgs.	Schools & Apts.	Mfg.	Office Bldgs.	Schools & Apts.	Mfg.
Excavation, Grading, etc.	4.2 % to	19.5%	10.4%	4.8%	4.9%	4.5%	.50%	.51%	.47%
Piles & Foundations	7.7 to	76.7	22.5	7.1	5.2	8.7	1.60	1.17	1.96
Concrete	5.2 to	35.8	16.1	5.0	14.8	3.7	.81	2.38	.60
Masonry	5.1 to	28.3	15.1	6.9	7.5	1.9	1.04	1.13	.29
Structural Steel	7.8 to	103.8	38.6	10.7	3.9	17.6	4.13	1.51	6.79
Miscellaneous & Ornamental Metals	4.9 to	24.8	12.6	2.8	4.0	3.6	.35	.50	.45
Carpentry & Millwork	7.0 to	53.2	18.4	3.7	4.0	0.5	.68	.74	.09
Metal or Composition Siding	5.3 to	32.1	17.0	2.3	0.3	4.3	.39	.05	.73
Roofing	7.8 to	78.9	33.1	2.3	2.6	3.1	.76	.86	1.03
Doors & Hardware	4.5 to	24.9	11.3	0.9	1.4	0.4	.10	.16	.05
Sash & Glazing	4.6 to	33.7	13.9	3.5	4.0	1.0	.49	.56	.14
Lath & Plaster	4.3 to	40.2	14.9	3.3	6.9	0.8	.49	1.03	.12
Tile, Marble & Floors	2.4 to	31.8	9.7	2.6	3.0	0.5	.25	.29	.05
Acoustical Ceilings	2.4 to	29.7	10.9	2.4	0.2	0.3	.26	.02	.03
Painting	4.5 to	33.7	13.1	1.5	1.6	1.6	.20	.21	.21
Interior Partitions	7.0 to	53.2	18.4	3.9	4.3	4.4	.72	.79	.81
Miscellaneous Items	2.6 to	137.6	16.9	5.2	3.7	9.7	.88	.63	1.64
Elevators	2.8 to	42.8	7.6	2.1	1.1	2.2	.16	.08	.17
Sprinklers	3.2 to	22.1	8.8	0.5	—	2.0	.04	—	.18
Plumbing	2.7 to	12.4	8.0	4.9	7.2	5.2	.39	.58	.42
Heat., Vent., Air Conditioning	4.0 to	27.2	11.5	13.5	11.0	12.9	1.55	1.27	1.48
Electrical	2.6 to	12.7	6.5	10.1	8.4	11.1	.66	.55	.72
Total	2.4 % to	137.6%	—	100.0%	100.0%	100.0%	16.45%	15.02%	18.43%

Overall Weighted Average 16.63%

Workers' Compensation Insurance Rates by States

The table below lists the weighted average Workers' Compensation base rate for each state with a factor comparing this with the national average of 16.2%.

State	Weighted Average	Factor	State	Weighted Average	Factor	State	Weighted Average	Factor
Alabama	25.2%	156	Kentucky	17.4%	107	North Dakota	11.4%	70
Alaska	19.7	122	Louisiana	26.2	162	Ohio	12.8	79
Arizona	7.4	46	Maine	19.3	119	Oklahoma	13.7	85
Arkansas	13.7	85	Maryland	17.7	109	Oregon	12.6	78
California	17.4	107	Massachusetts	12.4	77	Pennsylvania	13.4	83
Colorado	12.3	76	Michigan	17.6	109	Rhode Island	19.6	121
Connecticut	23.1	143	Minnesota	25.1	155	South Carolina	12.9	80
Delaware	14.2	88	Mississippi	16.1	99	South Dakota	14.2	88
District of Columbia	17.3	107	Missouri	17.6	109	Tennessee	17.9	110
Florida	25.1	155	Montana	18.9	117	Texas	13.5	83
Georgia	21.3	131	Nebraska	18.8	116	Utah	11.4	70
Hawaii	15.7	97	Nevada	13.9	86	Vermont	18.6	115
Idaho	10.4	64	New Hampshire	21.3	131	Virginia	10.8	67
Illinois	17.2	106	New Jersey	10.5	65	Washington	10.4	64
Indiana	5.4	33	New Mexico	16.3	101	West Virginia	12.4	77
Iowa	11.8	73	New York	13.4	83	Wisconsin	15.2	94
Kansas	8.5	52	North Carolina	13.2	81	Wyoming	7.9	49

Weighted Average for U.S. is   15.5% of payroll = 100%

Rates in the following table are the base or manual costs per $100 of payroll for Workers' Compensation in each state. Rates are usually applied to straight time wages only and not to premium time wages and bonuses.

The weighted average skilled worker rate for 35 trades is 16.2%. For bidding purposes, apply the full value of Workers' Compensation directly to total labor costs, or if labor is 38%, materials 42% and overhead and profit 20% of total cost, carry 38/80 x 16.2% =7.7% of cost (before overhead and profit) into overhead. Rates vary not only from state to state but also with the experience rating of the contractor.

Rates are the most current available at the time of publication.

R01100-060 Workers' Compensation Insurance Rates by Trade and State (cont.)

State	Carpentry — 3 stories or less	Carpentry — interior cab. work	Carpentry — general	Concrete Work — NOC	Concrete Work — flat (flr., sdwk.)	Electrical Wiring — inside	Excavation — earth NOC	Excavation — rock	Glaziers	Insulation Work	Lathing	Masonry	Painting & Decorating	Pile Driving	Plastering	Plumbing	Roofing	Sheet Metal Work (HVAC)	Steel Erection — door & sash	Steel Erection — inter, ornam.	Steel Erection — structure	Steel Erection — NOC	Tile Work — (interior ceramic)	Waterproofing	Wrecking
	5651	5437	5403	5213	5221	5190	6217	6217	5462	5479	5443	5022	5474	6003	5480	5183	5551	5538	5102	5102	5040	5057	5348	9014	5701
AL	26.93	15.28	34.89	12.39	9.90	8.95	13.69	13.69	33.66	17.12	12.78	27.57	33.72	34.63	40.23	11.91	60.85	27.15	23.11	23.11	51.16	39.09	13.16	6.65	51.16
AK	17.02	14.74	13.84	12.07	11.18	12.70	19.54	19.54	19.86	25.41	9.78	24.13	16.28	59.80	19.42	11.79	38.40	12.13	12.24	12.24	42.33	18.60	8.48	10.79	42.33
AZ	7.71	5.37	12.63	6.33	3.44	4.18	4.83	4.83	7.59	9.87	4.70	6.40	5.23	9.47	15.76	3.77	11.96	4.71	7.92	7.92	16.58	7.75	2.37	2.55	50.31
AR	13.04	11.01	16.99	13.77	6.64	6.11	9.42	9.42	15.49	33.95	8.65	11.34	10.50	12.82	12.93	5.69	21.95	10.28	7.70	7.70	37.63	28.19	7.10	3.75	37.63
CA	28.28	9.07	28.28	13.18	13.18	9.93	7.88	7.88	16.37	23.34	10.90	14.55	19.45	21.18	18.12	11.59	39.79	15.33	13.55	13.55	23.93	21.91	7.74	19.45	21.91
CO	17.36	9.17	12.24	14.36	7.69	5.32	10.44	10.44	9.28	13.68	5.57	13.41	10.77	17.70	10.81	8.51	24.47	12.11	7.39	7.39	31.30	15.53	7.99	5.78	31.30
CT	20.96	18.42	31.08	29.77	13.28	8.63	11.18	11.18	15.10	24.22	17.24	25.81	19.56	27.54	27.61	10.37	51.40	14.91	18.42	18.42	70.81	35.84	12.65	6.69	50.86
DE	15.39	15.39	12.92	13.03	7.13	6.79	10.03	10.03	11.96	12.92	13.83	12.33	16.85	21.38	13.83	7.94	27.84	10.69	13.02	13.02	29.92	13.02	10.51	12.33	29.92
DC	13.05	10.16	15.52	17.41	20.87	6.52	10.32	10.32	27.59	11.07	8.44	22.34	10.52	18.70	13.70	12.43	21.23	9.29	17.97	17.97	54.79	20.38	31.80	4.05	54.79
FL	30.55	21.73	31.10	30.26	15.05	10.74	13.79	13.79	23.24	22.05	13.08	23.21	21.38	58.93	36.28	10.71	46.17	18.10	14.78	14.78	58.02	39.86	11.26	9.25	58.02
GA	32.05	17.69	24.16	15.16	10.44	8.97	16.86	16.86	15.69	22.02	17.16	20.30	17.13	31.16	19.94	10.77	46.74	17.52	13.97	13.97	44.46	48.01	9.64	8.55	44.46
HI	17.38	11.62	28.20	13.74	12.27	7.10	7.73	7.73	20.55	21.19	10.94	17.87	11.33	20.15	15.61	6.06	33.24	8.02	11.11	11.11	31.71	21.70	9.78	11.54	31.71
ID	11.48	6.50	13.42	8.40	5.87	5.01	6.24	6.24	9.48	7.60	5.27	9.01	7.40	13.14	9.74	4.67	29.72	9.37	9.80	9.80	30.78	13.76	6.47	4.09	30.78
IL	16.38	13.48	17.63	29.67	11.05	8.51	10.49	10.49	15.50	11.58	9.74	16.53	11.34	28.71	11.92	9.79	28.32	14.36	14.87	14.87	52.98	26.69	14.11	4.36	52.98
IN	5.31	4.50	7.02	5.18	3.17	2.64	4.17	4.17	4.59	4.15	2.43	5.07	4.48	7.71	4.30	2.70	10.93	4.03	4.94	4.94	16.12	9.00	3.08	2.55	16.12
IA	11.67	5.80	10.47	12.55	6.82	4.18	5.98	5.98	13.66	9.22	5.45	8.79	7.73	10.70	9.26	5.92	18.71	7.17	9.92	9.92	51.91	36.28	5.29	4.20	30.67
KS	10.58	8.42	10.47	7.12	6.36	3.72	4.73	4.73	6.49	7.80	4.37	7.45	6.06	9.36	8.73	5.40	20.70	6.76	7.83	7.83	23.76	12.28	4.75	3.48	23.76
KY	15.79	12.91	19.21	20.50	6.16	7.64	17.08	17.08	21.66	24.14	11.20	10.81	11.40	18.49	14.90	6.85	48.37	13.64	11.61	11.61	40.21	30.64	12.56	4.72	40.21
LA	23.14	24.93	53.17	26.43	15.61	9.88	17.49	17.49	20.14	21.43	24.51	28.33	29.58	31.25	22.37	8.64	77.12	21.20	18.98	18.98	51.76	24.21	13.77	13.78	66.41
ME	13.43	10.42	43.59	24.29	11.12	5.03	12.19	12.19	13.35	15.39	14.03	17.26	16.07	31.44	17.95	7.32	32.07	8.82	15.08	15.08	37.84	61.81	11.66	6.06	37.84
MD	13.71	5.95	10.55	18.82	8.70	5.15	12.91	12.91	25.52	20.07	11.37	13.31	9.37	26.22	14.21	8.58	48.72	12.74	13.66	13.66	56.25	37.81	8.03	7.14	26.80
MA	9.93	6.07	16.09	17.95	8.10	3.69	6.49	6.49	7.55	13.48	5.76	13.04	7.54	14.31	5.34	4.55	38.25	7.47	12.35	12.35	35.13	27.39	9.24	3.32	28.82
MI	21.68	13.59	19.80	22.12	9.35	5.71	11.69	11.69	13.87	12.38	13.59	19.33	14.93	39.06	15.59	8.24	41.33	10.41	11.36	11.36	39.06	29.15	11.31	6.40	39.06
MN	21.84	21.96	43.39	16.70	14.64	6.94	16.68	16.68	17.42	11.95	21.02	20.28	17.69	26.83	21.02	10.70	78.89	12.72	14.64	14.64	103.82	38.11	14.59	6.96	132.92
MS	16.92	15.26	19.71	11.11	8.98	6.80	12.22	12.22	12.02	14.11	8.02	13.18	12.86	19.38	22.03	8.15	37.48	21.88	11.67	11.67	33.42	32.27	9.89	6.93	33.42
MO	21.84	12.24	16.70	15.66	11.37	7.09	10.05	10.05	11.75	18.49	12.47	16.14	13.64	21.93	16.25	9.37	33.06	14.20	12.51	12.51	59.20	40.85	9.59	7.02	59.20
MT	22.03	10.79	17.11	12.89	9.98	5.54	17.20	17.20	11.51	16.19	19.02	14.58	12.03	76.71	13.82	8.93	57.20	9.44	9.73	9.73	41.23	17.93	6.54	5.33	41.23
NE	23.15	10.97	20.95	28.60	9.65	8.32	12.20	12.20	15.50	24.47	11.30	18.50	12.32	23.35	14.75	11.17	36.72	13.05	13.15	13.15	52.27	38.35	9.97	6.15	50.40
NV	19.78	7.76	13.18	10.57	11.57	6.18	11.21	11.21	12.58	13.22	7.97	11.38	10.62	14.55	12.30	8.93	21.55	18.17	12.87	12.87	31.33	33.90	6.64	6.15	44.29
NH	21.98	14.43	24.90	35.84	11.10	6.02	17.16	17.16	14.70	33.70	8.60	24.98	15.72	17.80	18.65	10.22	60.89	14.43	14.01	14.01	54.76	25.50	17.84	6.36	54.76
NJ	11.76	8.69	11.76	9.82	7.77	4.22	7.74	7.74	6.91	11.95	10.39	12.07	9.66	13.10	10.39	6.14	29.59	6.69	9.94	9.94	19.67	10.61	4.42	4.65	23.42
NM	28.63	8.03	18.67	17.07	9.38	6.85	9.33	9.33	14.38	12.83	7.20	13.69	12.57	18.59	11.22	8.64	32.00	11.70	24.78	24.78	43.20	25.65	7.98	6.81	43.20
NY	17.66	6.98	15.91	17.16	13.57	6.32	10.02	10.02	12.02	9.09	5.11	19.20	13.75	16.81	8.43	8.53	33.44	17.47	9.63	9.63	18.80	19.14	9.43	6.09	11.29
NC	16.34	11.24	15.16	11.77	7.00	8.63	8.65	8.65	10.30	11.88	8.28	10.32	9.75	13.68	15.22	7.78	26.83	11.26	7.71	7.71	52.96	19.19	5.79	4.27	52.96
ND	9.60	9.60	9.60	5.66	5.66	4.00	5.39	5.39	9.60	9.60	9.04	8.10	6.27	20.12	9.04	5.35	20.74	5.35	20.12	20.12	20.12	20.12	9.60	20.24	14.48
OH	6.32	11.48	9.82	11.66	10.37	5.54	8.42	8.42	9.22	14.76	29.68	12.20	15.20	25.34	4.41	7.14	22.43	8.85	10.53	10.53	25.27	18.05	9.73	5.02	25.27
OK	13.44	9.11	12.70	12.65	6.63	5.82	10.25	10.25	16.10	19.50	8.46	10.98	9.09	20.29	12.21	7.24	22.32	9.49	13.78	13.78	39.78	24.01	6.95	5.88	39.78
OR	19.10	9.61	15.51	13.55	8.43	4.65	9.70	9.70	15.02	8.08	7.48	15.36	12.79	15.81	10.75	5.89	23.26	10.49	9.53	9.53	32.00	13.92	11.23	4.97	32.00
PA	13.85	13.85	12.80	14.90	9.67	6.74	8.44	8.44	9.60	12.80	11.67	12.18	14.33	17.72	11.67	7.51	27.83	8.19	15.17	15.17	26.43	15.17	8.22	12.18	26.43
RI	19.53	11.65	18.07	18.23	16.24	4.43	10.38	10.38	12.85	22.78	11.97	25.11	24.13	37.66	17.25	8.52	33.92	10.29	14.07	14.07	59.49	37.50	14.36	7.70	78.79
SC	19.10	13.03	19.24	12.88	6.19	7.02	8.32	8.32	12.22	10.69	7.27	9.55	11.19	16.64	18.89	6.88	28.98	11.42	9.49	9.49	21.80	23.66	6.23	4.65	21.80
SD	17.83	7.27	14.55	18.00	6.34	5.44	13.24	13.24	10.58	13.24	8.39	8.01	12.63	24.87	11.78	8.53	21.26	9.64	11.53	11.53	53.25	20.15	6.68	4.20	53.25
TN	31.79	14.64	24.26	19.40	9.95	8.72	14.06	14.06	12.09	14.84	13.05	17.02	14.59	17.43	17.47	11.33	38.38	14.51	13.38	13.38	45.26	25.19	9.73	6.57	45.26
TX	16.89	11.33	13.25	12.10	8.85	7.36	10.49	10.49	11.61	15.68	9.53	13.68	9.89	13.95	20.99	8.06	22.80	14.58	11.23	11.23	33.39	14.78	7.29	8.09	17.86
UT	9.45	6.23	10.52	7.68	7.99	7.05	6.38	6.38	9.55	10.95	12.32	13.75	15.29	17.24	10.60	6.41	26.27	6.30	8.99	8.99	23.51	23.51	6.06	5.83	26.53
VT	19.90	11.34	19.46	34.13	12.01	6.10	10.61	10.61	21.25	25.54	10.85	19.53	10.41	23.91	18.23	10.10	28.73	12.62	13.94	13.94	47.05	37.23	8.52	9.59	47.05
VA	11.97	7.81	9.72	10.47	5.32	4.43	6.53	6.53	7.60	7.96	13.79	8.17	9.95	13.55	7.16	5.52	21.62	8.30	11.34	11.34	39.01	16.86	6.32	3.09	39.01
WA	9.13	9.13	9.13	8.17	8.17	3.12	8.54	8.54	13.74	8.92	9.13	13.27	9.39	20.04	11.57	5.04	19.54	4.60	12.96	12.96	9.39	9.59	10.76	10.23	9.59
WV	12.30	12.30	12.30	27.37	27.37	5.86	8.90	8.90	6.99	13.15	13.15	11.94	12.35	8.90	12.35	5.48	14.63	6.99	10.89	10.89	10.89	10.89	13.15	13.15	48.62
WI	12.40	10.86	19.86	13.12	11.66	5.32	6.74	6.74	16.53	14.04	10.60	18.99	14.82	17.24	13.95	6.31	43.57	8.01	14.30	14.30	34.84	21.34	14.42	5.80	34.84
WY	7.77	7.77	7.77	7.77	7.77	7.77	7.77	7.77	7.77	7.77	7.77	7.77	7.77	7.77	7.77	7.77	7.77	7.77	7.77	7.77	7.77	7.77	7.77	7.77	7.77
AVG.	16.96	11.33	18.42	16.07	9.94	6.46	10.43	10.43	13.91	15.53	10.87	15.06	13.12	22.45	14.88	7.96	33.14	11.46	12.57	12.57	38.60	24.71	9.66	7.12	39.48

R01100-060 Workers' Compensation (cont.) (Canada in Canadian dollars)

Province		Alberta	British Columbia	Manitoba	Ontario	New Brunswick	Newfndld. & Labrador	Northwest Territories	Nova Scotia	Prince Edward Island	Quebec	Saskatchewan	Yukon
Carpentry—3 stories or less	Rate	10.80	6.66	4.57	4.83	4.63	10.05	4.25	7.67	6.15	14.69	7.01	4.36
	Code	42143	721028	40102	723	422	4226	4-41	4226	401	80110	B1317	202
Carpentry—interior cab. work	Rate	2.59	5.94	4.57	4.83	4.21	6.35	4.25	5.74	3.41	14.69	3.65	4.36
	Code	42133	721021	40102	723	427	4270	4-41	4274	402	80110	B11-27	202
CARPENTRY—general	Rate	10.80	6.66	4.57	4.83	4.63	6.35	4.25	7.67	6.15	14.69	7.01	4.36
	Code	42143	721028	40102	723	422	4299	4-41	4226	401	80110	B1317	202
CONCRETE WORK—NOC	Rate	7.26	8.48	7.60	16.47	4.63	10.05	4.25	4.83	6.15	15.84	7.01	4.36
	Code	42104	721010	40110	748	422	4224	4-41	4224	401	80100	B13-14	203
CONCRETE WORK—flat (flr. sidewalk)	Rate	7.26	8.48	7.60	16.47	4.63	10.05	4.25	4.83	6.15	15.84	7.01	4.36
	Code	42104	721010	40110	748	422	4224	4-41	4224	401	80100	B13-14	203
ELECTRICAL Wiring—inside	Rate	2.86	2.67	2.52	3.03	2.43	3.19	3.00	2.23	3.41	7.29	3.65	4.36
	Code	42124	721019	40203	704	426	4261	4-46	4261	402	80170	B11-05	206
EXCAVATION—earth NOC	Rate	3.59	4.29	3.84	4.20	3.69	5.55	3.50	4.11	3.58	8.80	4.02	4.36
	Code	40604	721031	40706	711	421	4214	4-43	4214	404	80030	R11-06	207
EXCAVATION—rock	Rate	3.59	4.29	3.84	4.20	3.69	5.55	3.50	4.11	3.58	8.80	4.02	4.36
	Code	40604	721031	40706	711	421	4214	4-43	4214	404	80030	R11-06	207
GLAZIERS	Rate	4.84	3.00	4.57	8.12	6.12	5.97	4.25	7.67	3.41	14.59	7.01	4.36
	Code	42121	715020	40109	751	423	4233	4-41	4233	402	80150	B13-04	212
INSULATION WORK	Rate	3.67	6.94	4.57	8.12	6.12	5.97	4.25	7.67	6.15	14.69	6.20	4.36
	Code	42184	721029	40102	751	423	4234	4-41	4234	401	80110	B12-07	202
LATHING	Rate	8.30	11.20	4.57	4.83	4.21	6.35	4.25	5.74	3.41	14.69	7.01	4.36
	Code	42135	721033	40102	723	427	4279	4-41	4271	402	80110	B13-16	202
MASONRY	Rate	7.26	11.20	4.57	12.21	6.12	5.97	4.25	7.67	6.15	15.84	7.01	4.36
	Code	42102	721037	40102	741	423	4231	4-41	4231	401	80100	B13-18	202
PAINTING & DECORATING	Rate	5.64	6.94	3.45	6.83	4.21	6.35	4.25	5.74	3.41	14.69	6.20	4.36
	Code	42111	721041	40105	719	427	4275	4-41	4275	402	80110	B12-01	202
PILE DRIVING	Rate	7.26	13.07	3.84	5.84	4.63	10.05	3.50	4.83	6.15	8.80	7.01	4.36
	Code	42159	722004	40706	732	422	4221	4-43	4221	401	80030	B13-10	202
PLASTERING	Rate	8.30	11.20	5.13	6.83	4.21	6.35	4.25	5.74	3.41	14.69	6.20	4.36
	Code	42135	721042	40108	719	427	4271	4-41	4271	402	80110	B12-21	202
PLUMBING	Rate	2.86	5.48	2.88	3.83	3.74	3.89	3.00	2.23	3.41	7.75	3.65	4.36
	Code	42122	721043	40204	707	424	4241	4-46	4241	402	80160	B11-01	214
ROOFING	Rate	11.09	10.45	6.52	12.34	8.00	10.05	4.25	9.49	6.15	23.29	7.01	4.36
	Code	42118	721036	40403	728	430	4236	4-41	4236	401	80130	B13-20	202
SHEET METAL WORK (HVAC)	Rate	2.86	5.48	6.52	3.83	3.74	3.89	3.50	3.23	3.41	7.75	3.65	6.60
	Code	42117	721043	40402	707	424	4244	4-46	4244	402	80160	B11-07	208
STEEL ERECTION—door & sash	Rate	3.67	13.07	9.89	16.47	4.63	10.05	4.25	7.67	6.15	29.23	7.01	4.36
	Code	42106	722005	40502	748	422	4227	4-41	4227	401	80080	B13-22	202
STEEL ERECTION—inter., ornam.	Rate	3.67	13.07	9.89	16.47	4.63	10.05	4.25	7.67	6.15	29.23	7.01	4.36
	Code	42106	722005	40502	748	422	4227	4-41	4227	401	80080	B13-22	202
STEEL ERECTION—structure	Rate	3.67	13.07	9.89	16.47	4.63	10.05	4.25	7.67	6.15	29.23	7.01	4.36
	Code	42106	722005	40502	748	422	4227	4-41	4227	401	80080	B13-22	202
STEEL ERECTION—NOC	Rate	3.67	13.07	9.89	16.47	4.63	10.05	4.25	7.67	6.15	29.23	7.01	4.36
	Code	42106	722005	40502	748	422	4227	4-41	4227	401	80080	B13-22	202
TILE WORK—inter. (ceramic)	Rate	4.86	4.70	2.17	6.83	4.21	6.35	4.25	5.74	3.41	14.69	7.01	4.36
	Code	42113	721054	40103	719	427	4276	4-41	4276	402	80110	B13-01	202
WATERPROOFING	Rate	5.64	6.94	4.57	4.83	6.12	6.35	4.25	7.67	3.41	23.29	6.20	4.36
	Code	42139	721016	40102	723	423	4299	4-41	4239	402	80130	B12-17	202
WRECKING	Rate	3.59	6.53	6.46	16.47	3.69	5.55	3.50	4.11	6.15	14.69	7.01	4.36
	Code	40604	721005	40106	748	421	4211	4-43	4211	401	80110	B13-09	202

1

GENERAL REQUIREMENTS

REFERENCE NOS.

R01100-070 Contractor's Overhead & Profit

Below are the **average** installing contractor's percentage mark-ups applied to base labor rates to arrive at typical billing rates.

Column A: Labor rates are based on union wages averaged for 30 major U.S. cities. Base rates including fringe benefits are listed hourly and daily. These figures are the sum of the wage rate and employer-paid fringe benefits such as vacation pay, employer-paid health and welfare costs, pension costs, plus appropriate training and industry advancement funds costs.

Column B: Workers' Compensation rates are the national average of state rates established for each trade.

Column C: Column C lists average fixed overhead figures for all trades. Included are Federal and State Unemployment costs set at 6.2%; Social Security Taxes (FICA) set at 7.65%; Builder's Risk Insurance costs set at 0.44%; and Public Liability costs set at 2.02%. All the percentages except those for Social Security Taxes vary from state to state as well as from company to company.

Columns D and E: Percentages in Columns D and E are based on the presumption that the installing contractor has annual billing of $4,000,000 and up. Overhead percentages may increase with smaller annual billing. The overhead percentages for any given contractor may vary greatly and depend on a number of factors, such as the contractor's annual volume, engineering and logistical support costs, and staff requirements. The figures for overhead and profit will also vary depending on the type of job, the job location, and the prevailing economic conditions. All factors should be examined very carefully for each job.

Column F: Column F lists the total of Columns B, C, D, and E.

Column G: Column G is Column A (hourly base labor rate) multiplied by the percentage in Column F (O&P percentage).

Column H: Column H is the total of Column A (hourly base labor rate) plus Column G (Total O&P).

Column I: Column I is Column H multiplied by eight hours.

Abbr.	Trade	A Base Rate Incl. Fringes Hourly	A Base Rate Incl. Fringes Daily	B Workers' Comp. Ins.	C Average Fixed Over-head	D Over-head	E Profit	F Total Overhead & Profit %	F Total Overhead & Profit Amount	H Rate with O & P Hourly	I Rate with O & P Daily
Skwk	Skilled Workers Average (35 trades)	$34.85	$278.80	16.2%	16.3%	13.0%	10.0%	55.5%	$19.35	$54.20	$433.60
	Helpers Average (5 trades)	25.55	204.40	18.2		11.0		55.5	$14.20	39.75	318.00
	Foreman Average, Inside ($.50 over trade)	35.35	282.80	16.2		13.0		55.5	19.60	54.95	439.60
	Foreman Average, Outside ($2.00 over trade)	36.85	294.80	16.2		13.0		55.5	20.45	57.30	458.40
Clab	Common Building Laborers	26.70	213.60	18.4		11.0		55.7	14.85	41.55	332.40
Asbe	Asbestos/Insulation Workers/Pipe Coverers	37.10	296.80	15.5		16.0		57.8	21.45	58.55	468.40
Boil	Boilermakers	42.90	343.20	13.7		16.0		56.0	24.00	66.90	535.20
Bric	Bricklayers	35.25	282.00	15.1		11.0		52.4	18.45	53.70	429.60
Brhe	Bricklayer Helpers	26.90	215.20	15.1		11.0		52.4	14.10	41.00	328.00
Carp	Carpenters	34.25	274.00	18.4		11.0		55.7	19.10	53.35	426.80
Cefi	Cement Finishers	32.85	262.80	9.9		11.0		47.2	15.50	48.35	386.80
Elec	Electricians	40.75	326.00	6.5		16.0		48.8	19.90	60.65	485.20
Elev	Elevator Constructors	45.05	360.40	7.6		16.0		49.9	22.50	67.55	540.40
Eqhv	Equipment Operators, Crane or Shovel	35.90	287.20	10.4		14.0		50.7	18.20	54.10	432.80
Eqmd	Equipment Operators, Medium Equipment	34.65	277.20	10.4		14.0		50.7	17.55	52.20	417.60
Eqlt	Equipment Operators, Light Equipment	33.05	264.40	10.4		14.0		50.7	16.75	49.80	398.40
Eqol	Equipment Operators, Oilers	30.10	240.80	10.4		14.0		50.7	15.25	45.35	362.80
Eqmm	Equipment Operators, Master Mechanics	36.20	289.60	10.4		14.0		50.7	18.35	54.55	436.40
Glaz	Glaziers	33.00	264.00	13.9		11.0		51.2	16.90	49.90	399.20
Lath	Lathers	31.45	251.60	10.9		11.0		48.2	15.15	46.60	372.80
Marb	Marble Setters	33.20	265.60	15.1		11.0		52.4	17.40	50.60	404.80
Mill	Millwrights	35.50	284.00	10.4		11.0		47.7	16.95	52.45	419.60
Mstz	Mosaic and Terrazzo Workers	32.65	261.20	9.7		11.0		47.0	15.35	48.00	384.00
Pord	Painters, Ordinary	30.60	244.80	13.1		11.0		50.4	15.40	46.00	368.00
Psst	Painters, Structural Steel	31.00	248.00	46.8		11.0		84.1	26.05	57.05	456.40
Pape	Paper Hangers	30.30	242.40	13.1		11.0		50.4	15.25	45.55	364.40
Pile	Pile Drivers	33.30	266.40	22.5		16.0		64.8	21.60	54.90	439.20
Plas	Plasterers	31.35	250.80	14.9		11.0		52.2	16.35	47.70	381.60
Plah	Plasterer Helpers	27.10	216.80	14.9		11.0		52.2	14.15	41.25	330.00
Plum	Plumbers	40.85	326.80	8.0		16.0		50.3	20.55	61.40	491.20
Rodm	Rodmen (Reinforcing)	37.95	303.60	24.7		14.0		65.0	24.65	62.60	500.80
Rofc	Roofers, Composition	29.45	235.60	33.1		11.0		70.4	20.75	50.20	401.60
Rots	Roofers, Tile and Slate	29.70	237.60	33.1		11.0		70.4	20.90	50.60	404.80
Rohe	Roofer Helpers (Composition)	21.60	172.80	33.1		11.0		70.4	15.20	36.80	294.40
Shee	Sheet Metal Workers	40.20	321.60	11.5		16.0		53.8	21.65	61.85	494.80
Spri	Sprinkler Installers	40.75	326.00	8.8		16.0		51.1	20.80	61.55	492.40
Stpi	Steamfitters or Pipefitters	40.95	327.60	8.0		16.0		50.3	20.60	61.55	492.40
Ston	Stone Masons	35.30	282.40	15.1		11.0		52.4	18.50	53.80	430.40
Sswk	Structural Steel Workers	38.15	305.20	38.6		14.0		78.9	30.10	68.25	546.00
Tilf	Tile Layers	32.70	261.60	9.7		11.0		47.0	15.35	48.05	384.40
Tilh	Tile Layer Helpers	25.35	202.80	9.7		11.0		47.0	11.90	37.25	298.00
Trlt	Truck Drivers, Light	26.80	214.40	15.2		11.0		52.5	14.05	40.85	326.80
Trhv	Truck Drivers, Heavy	27.55	220.40	15.2		11.0		52.5	14.45	42.00	336.00
Sswl	Welders, Structural Steel	38.15	305.20	38.6		14.0		78.9	30.10	68.25	546.00
Wrck	*Wrecking	26.70	213.60	39.5		11.0		76.8	20.50	47.20	377.60

*Not included in Averages.

R01100-080　Performance Bond

This table shows the cost of a Performance Bond for a construction job scheduled to be completed in 12 months. Add 1% of the premium cost per month for jobs requiring more than 12 months to complete. The rates are "standard" rates offered to contractors that the bonding company considers financially sound and capable of doing the work. Preferred rates are offered by some bonding companies based upon financial strength of the contractor. Actual rates vary from contractor to contractor and from bonding company to bonding company. Contractors should prequalify through a bonding agency before submitting a bid on a contract that requires a bond.

Contract Amount	Building Construction Class B Projects			Highways & Bridges					
				Class A New Construction			Class A-1 Highway Resurfacing		
First $ 100,000 bid	\$25.00 per M			\$15.00 per M			\$9.40 per M		
Next 400,000 bid	$ 2,500	plus	\$15.00 per M	$ 1,500	plus	\$10.00 per M	$ 940	plus	\$7.20 per M
Next 2,000,000 bid	8,500	plus	10.00 per M	5,500	plus	7.00 per M	3,820	plus	5.00 per M
Next 2,500,000 bid	28,500	plus	7.50 per M	19,500	plus	5.50 per M	15,820	plus	4.50 per M
Next 2,500,000 bid	47,250	plus	7.00 per M	33,250	plus	5.00 per M	28,320	plus	4.50 per M
Over 7,500,000 bid	64,750	plus	6.00 per M	45,750	plus	4.50 per M	39,570	plus	4.00 per M

R01100-090　Sales Tax by State

State sales tax on materials is tabulated below (5 states have no sales tax). Many states allow local jurisdictions, such as a county or city, to levy additional sales tax.

Some projects may be sales tax exempt, particularly those constructed with public funds.

State	Tax (%)	State	Tax (%)	State	Tax (%)	State	Tax (%)
Alabama	4	Illinois	6.25	Montana	0	Rhode Island	7
Alaska	0	Indiana	6	Nebraska	5.5	South Carolina	5
Arizona	5.6	Iowa	5	Nevada	6.5	South Dakota	4
Arkansas	5.125	Kansas	5.3	New Hampshire	0	Tennessee	7
California	7.25	Kentucky	6	New Jersey	6	Texas	6.25
Colorado	2.9	Louisiana	4	New Mexico	5	Utah	4.75
Connecticut	6	Maine	5	New York	4.25	Vermont	6
Delaware	0	Maryland	5	North Carolina	4.5	Virginia	5
District of Columbia	5.75	Massachusetts	5	North Dakota	5	Washington	6.5
Florida	6	Michigan	6	Ohio	6	West Virginia	6
Georgia	4	Minnesota	6.5	Oklahoma	4.5	Wisconsin	5
Hawaii	4	Mississippi	7	Oregon	0	Wyoming	4
Idaho	6	Missouri	4.225	Pennsylvania	6	Average	4.86 %

Sales Tax by Province (Canada)

GST - a value-added tax, which the government imposes on most goods and services provided in or imported into Canada. PST - a retail sales tax, which five of the provinces impose on the price of most goods and some services. QST - a value-added tax, similar to the federal GST, which Quebec imposes. HST - Three provinces have combined their retail sales tax with the federal GST into one harmonized tax.

Province	PST (%)	QST (%)	GST(%)	HST(%)
Alberta	0	0	7	0
British Columbia	7.5	0	7	0
Manitoba	7	0	7	0
New Brunswick	0	0	0	15
Newfoundland	0	0	0	15
Northwest Territories	0	0	7	0
Nova Scotia	0	0	0	15
Ontario	8	0	7	0
Prince Edward Island	10	0	7	0
Quebec	0	7.5	7	0
Saskatchewan	6	0	7	0
Yukon	0	0	7	0

R01100-100　Unemployment Taxes and Social Security Taxes

Mass. State Unemployment tax ranges from 1.12% to 10.96% plus an experience rating assessment the following year, on the first \$14,000 of wages. Federal Unemployment tax is 6.2% of the first \$7,000 of wages. This is reduced by a credit for payment to the state. The minimum Federal Unemployment tax is 0.8% after all credits.

Combined rates in Mass. thus vary from 1.92% to 11.76% of the first \$14,000 of wages. Combined average U.S. rate is about 6.2% of the first \$7,000. Contractors with permanent workers will pay less since the average annual wages for skilled workers is \$34.85 x 2,000 hours or about \$69,700 per year. The average combined rate for U.S. would thus be 6.2% x \$7,000 ÷ \$69,700 = 0.6% of total wages for permanent employees.

Rates vary not only from state to state but also with the experience rating of the contractor.

Social Security (FICA) for 2005 is estimated at time of publication to be 7.65% of wages up to \$87,900.

R01100-110   Overtime

One way to improve the completion date of a project or eliminate negative float from a schedule is to compress activity duration times. This can be achieved by increasing the crew size or working overtime with the proposed crew.

To determine the costs of working overtime to compress activity duration times, consider the following examples. Below is an overtime efficiency and cost chart based on a five, six, or seven day week with an eight through twelve hour day. Payroll percentage increases for time and one half and double time are shown for the various working days.

Days per Week	Hours per Day	Production Efficiency					Payroll Cost Factors	
		1 Week	2 Weeks	3 Weeks	4 Weeks	Average 4 Weeks	@ 1-1/2 Times	@ 2 Times
5	8	100%	100%	100%	100%	100 %	100 %	100 %
	9	100	100	95	90	96.25	105.6	111.1
	10	100	95	90	85	91.25	110.0	120.0
	11	95	90	75	65	81.25	113.6	127.3
	12	90	85	70	60	76.25	116.7	133.3
6	8	100	100	95	90	96.25	108.3	116.7
	9	100	95	90	85	92.50	113.0	125.9
	10	95	90	85	80	87.50	116.7	133.3
	11	95	85	70	65	78.75	119.7	139.4
	12	90	80	65	60	73.75	122.2	144.4
7	8	100	95	85	75	88.75	114.3	128.6
	9	95	90	80	70	83.75	118.3	136.5
	10	90	85	75	65	78.75	121.4	142.9
	11	85	80	65	60	72.50	124.0	148.1
	12	85	75	60	55	68.75	126.2	152.4

R01100-740   Weather Data and Design Conditions

City	Latitude (1)		Winter Temperatures (1)			Winter Degree Days (2)	Summer (Design Dry Bulb) Temperatures and Relative Humidity		
	0	1'	Med. of Annual Extremes	99%	97½%		1%	2½%	5%
UNITED STATES									
Albuquerque, NM	35	0	5.1	12	16	4,400	96/61	94/61	92/61
Atlanta, GA	33	4	11.9	17	22	3,000	94/74	92/74	90/73
Baltimore, MD	39	2	7	14	17	4,600	94/75	91/75	89/74
Birmingham, AL	33	3	13	17	21	2,600	96/74	94/75	92/74
Bismarck, ND	46	5	-32	-23	-19	8,800	95/68	91/68	88/67
Boise, ID	43	3	1	3	10	5,800	96/65	94/64	91/64
Boston, MA	42	2	-1	6	9	5,600	91/73	88/71	85/70
Burlington, VT	44	3	-17	-12	-7	8,200	88/72	85/70	82/69
Charleston, WV	38	2	3	7	11	4,400	92/74	90/73	87/72
Charlotte, NC	35	1	13	18	22	3,200	95/74	93/74	91/74
Casper, WY	42	5	-21	-11	-5	7,400	92/58	90/57	87/57
Chicago, IL	41	5	-8	-3	2	6,600	94/75	91/74	88/73
Cincinnati, OH	39	1	0	1	6	4,400	92/73	90/72	88/72
Cleveland, OH	41	2	-3	1	5	6,400	91/73	88/72	86/71
Columbia, SC	34	0	16	20	24	2,400	97/76	95/75	93/75
Dallas, TX	32	5	14	18	22	2,400	102/75	100/75	97/75
Denver, CO	39	5	-10	-5	1	6,200	93/59	91/59	89/59
Des Moines, IA	41	3	-14	-10	-5	6,600	94/75	91/74	88/73
Detroit, MI	42	2	-3	3	6	6,200	91/73	88/72	86/71
Great Falls, MT	47	3	-25	-21	-15	7,800	91/60	88/60	85/59
Hartford, CT	41	5	-4	3	7	6,200	91/74	88/73	85/72
Houston, TX	29	5	24	28	33	1,400	97/77	95/77	93/77
Indianapolis, IN	39	4	-7	-2	2	5,600	92/74	90/74	87/73
Jackson, MS	32	2	16	21	25	2,200	97/76	95/76	93/76
Kansas City, MO	39	1	-4	2	6	4,800	99/75	96/74	93/74
Las Vegas, NV	36	1	18	25	28	2,800	108/66	106/65	104/65
Lexington, KY	38	0	-1	3	8	4,600	93/73	91/73	88/72
Little Rock, AR	34	4	11	15	20	3,200	99/76	96/77	94/77
Los Angeles, CA	34	0	36	41	43	2,000	93/70	89/70	86/69
Memphis, TN	35	0	10	13	18	3,200	98/77	95/76	93/76
Miami, FL	25	5	39	44	47	200	91/77	90/77	89/77
Milwaukee, WI	43	0	-11	-8	-4	7,600	90/74	87/73	84/71
Minneapolis, MN	44	5	-22	-16	-12	8,400	92/75	89/73	86/71
New Orleans, LA	30	0	28	29	33	1,400	93/78	92/77	90/77
New York, NY	40	5	6	11	15	5,000	92/74	89/73	87/72
Norfolk, VA	36	5	15	20	22	3,400	93/77	91/76	89/76
Oklahoma City, OK	35	2	4	9	13	3,200	100/74	97/74	95/73
Omaha, NE	41	2	-13	-8	-3	6,600	94/76	91/75	88/74
Philadelphia, PA	39	5	6	10	14	4,400	93/75	90/74	87/72
Phoenix, AZ	33	3	27	31	34	1,800	109/71	107/71	105/71
Pittsburgh, PA	40	3	-1	3	7	6,000	91/72	88/71	86/70
Portland, ME	43	4	-10	-6	-1	7,600	87/72	84/71	81/69
Portland, OR	45	4	18	17	23	4,600	89/68	85/67	81/65
Portsmouth, NH	43	1	-8	-2	2	7,200	89/73	85/71	83/70
Providence, RI	41	4	-1	5	9	6,000	89/73	86/72	83/70
Rochester, NY	43	1	-5	1	5	6,800	91/73	88/71	85/70
Salt Lake City, UT	40	5	0	3	8	6,000	97/62	95/62	92/61
San Francisco, CA	37	5	36	38	40	3,000	74/63	71/62	69/61
Seattle, WA	47	4	22	22	27	5,200	85/68	82/66	78/65
Sioux Falls, SD	43	4	-21	-15	-11	7,800	94/73	91/72	88/71
St. Louis, MO	38	4	-3	3	8	5,000	98/75	94/75	91/75
Tampa, FL	28	0	32	36	40	680	92/77	91/77	90/76
Trenton, NJ	40	1	4	11	14	5,000	91/75	88/74	85/73
Washington, DC	38	5	7	14	17	4,200	93/75	91/74	89/74
Wichita, KS	37	4	-3	3	7	4,600	101/72	98/73	96/73
Wilmington, DE	39	4	5	10	14	5,000	92/74	89/74	87/73
ALASKA									
Anchorage	61	1	-29	-23	-18	10,800	71/59	68/58	66/56
Fairbanks	64	5	-59	-51	-47	14,280	82/62	78/60	75/59
CANADA									
Edmonton, Alta.	53	3	-30	-29	-25	11,000	85/66	82/65	79/63
Halifax, N.S.	44	4	-4	1	5	8,000	79/66	76/65	74/64
Montreal, Que.	45	3	-20	-16	-10	9,000	88/73	85/72	83/71
Saskatoon, Sask.	52	1	-35	-35	-31	11,000	89/68	86/66	83/65
St. John's, N.F.	47	4	1	3	7	8,600	77/66	75/65	73/64
Saint John, N.B.	45	2	-15	-12	-8	8,200	80/67	77/65	75/64
Toronto, Ont.	43	4	-10	-5	-1	7,000	90/73	87/72	85/71
Vancouver, B.C.	49	1	13	15	19	6,000	79/67	77/66	74/65
Winnipeg, Man.	49	5	-31	-30	-27	10,800	89/73	86/71	84/70

(1) Handbook of Fundamentals, ASHRAE, Inc., NY 1989
(2) Local Climatological Annual Survey, USDC Env. Science Services
Administration, Asheville, NC

R01100-750 Metric Conversion Factors

Description: This table is primarily for converting customary U.S. units in the left hand column to SI metric units in the right hand column. In addition, conversion factors for some commonly encountered Canadian and non-SI metric units are included.

	If You Know		Multiply By		To Find
Length	Inches	x	25.4[a]	=	Millimeters
	Feet	x	0.3048[a]	=	Meters
	Yards	x	0.9144[a]	=	Meters
	Miles (statute)	x	1.609	=	Kilometers
Area	Square inches	x	645.2	=	Square millimeters
	Square feet	x	0.0929	=	Square meters
	Square yards	x	0.8361	=	Square meters
Volume (Capacity)	Cubic inches	x	16,387	=	Cubic millimeters
	Cubic feet	x	0.02832	=	Cubic meters
	Cubic yards	x	0.7646	=	Cubic meters
	Gallons (U.S. liquids)[b]	x	0.003785	=	Cubic meters[c]
	Gallons (Canadian liquid)[b]	x	0.004546	=	Cubic meters[c]
	Ounces (U.S. liquid)[b]	x	29.57	=	Milliliters[c, d]
	Quarts (U.S. liquid)[b]	x	0.9464	=	Liters[c, d]
	Gallons (U.S. liquid)[b]	x	3.785	=	Liters[c, d]
Force	Kilograms force[d]	x	9.807	=	Newtons
	Pounds force	x	4.448	=	Newtons
	Pounds force	x	0.4536	=	Kilograms force[d]
	Kips	x	4448	=	Newtons
	Kips	x	453.6	=	Kilograms force[d]
Pressure, Stress, Strength (Force per unit area)	Kilograms force per square centimeter[d]	x	0.09807	=	Megapascals
	Pounds force per square inch (psi)	x	0.006895	=	Megapascals
	Kips per square inch	x	6.895	=	Megapascals
	Pounds force per square inch (psi)	x	0.07031	=	Kilograms force per square centimeter[d]
	Pounds force per square foot	x	47.88	=	Pascals
	Pounds force per square foot	x	4.882	=	Kilograms force per square meter[d]
Bending Moment Or Torque	Inch-pounds force	x	0.01152	=	Meter-kilograms force[d]
	Inch-pounds force	x	0.1130	=	Newton-meters
	Foot-pounds force	x	0.1383	=	Meter-kilograms force[d]
	Foot-pounds force	x	1.356	=	Newton-meters
	Meter-kilograms force[d]	x	9.807	=	Newton-meters
Mass	Ounces (avoirdupois)	x	28.35	=	Grams
	Pounds (avoirdupois)	x	0.4536	=	Kilograms
	Tons (metric)	x	1000	=	Kilograms
	Tons, short (2000 pounds)	x	907.2	=	Kilograms
	Tons, short (2000 pounds)	x	0.9072	=	Megagrams[e]
Mass per Unit Volume	Pounds mass per cubic foot	x	16.02	=	Kilograms per cubic meter
	Pounds mass per cubic yard	x	0.5933	=	Kilograms per cubic meter
	Pounds mass per gallon (U.S. liquid)[b]	x	119.8	=	Kilograms per cubic meter
	Pounds mass per gallon (Canadian liquid)[b]	x	99.78	=	Kilograms per cubic meter
Temperature	Degrees Fahrenheit	(F-32)/1.8		=	Degrees Celsius
	Degrees Fahrenheit	(F+459.67)/1.8		=	Degrees Kelvin
	Degrees Celsius	C+273.15		=	Degrees Kelvin

[a]The factor given is exact
[b]One U.S. gallon = 0.8327 Canadian gallon
[c]1 liter = 1000 milliliters = 1000 cubic centimeters
1 cubic decimeter = 0.001 cubic meter

[d]Metric but not SI unit
[e]Called "tonne" in England and "metric ton" in other metric countries

R01100-750 Equivalents of Cement Content for Concrete Mixes

94-Pound Bags per Cubic Yard	Kilograms per Cubic Meter	94-Pound Bags per Cubic Yard	Kilograms per Cubic Meter
1.0	55.77	7.0	390.4
1.5	83.65	7.5	418.3
2.0	111.5	8.0	446.2
2.5	139.4	8.5	474.0
3.0	167.3	9.0	501.9
3.5	195.2	9.5	529.8
4.0	223.1	10.0	557.7
4.5	251.0	10.5	585.6
5.0	278.8	11.0	613.5
5.5	306.7	11.5	641.3
6.0	334.6	12.0	669.2
6.5	362.5	12.5	697.1

(a) If you know cement content in pounds per cubic yard,
  multiply by 0.5933 to obtain kilograms per cubic meter.
(b) If you know cement content in 94-pound bags per cubic yard,
  multiply by 55.77 to obtain kilograms per cubic meter.

R01100-750 Equivalents of Concrete Strengths Commonly Encountered (To convert other psi values to megapascals multiply by 0.006895.)

U.S. Value, PSI	SI Value, Megapascals	Non-SI Metric Value, kgf/cm^2*
2,000	14	140
2,500	17	175
3,000	21	210
3,500	24	245
4,000	28	280
4,500	31	315
5,000	34	350
6,000	41	420
7,000	48	490
8,000	55	560
9,000	62	630
10,000	69	705

* Kilograms force per square centimeter

R01540-100 Steel Tubular Scaffolding

On new construction, tubular scaffolding is efficient up to 60' high or five stories. Above this it is usually better to use a hung scaffolding if construction permits. Swing scaffolding operations may interfere with tenants. In this case, the tubular is more practical at all heights.

In repairing or cleaning the front of an existing building the cost of tubular scaffolding per S.F. of building front increases as the height increases above the first tier. The first tier cost is relatively high due to leveling and alignment.

The minimum efficient crew for erection is three workers. For heights over 50', a crew of four is more efficient. Use two or more on top and two at the bottom for handing up or hoisting. Four workers can erect and

dismantle about nine frames per hour up to five stories. From five to eight stories they will average six frames per hour. With 7' horizontal spacing this will run about 400 S.F. and 265 S.F. of wall surface, respectively. Time for placing planks must be added to the above. On heights above 50', five planks can be placed per labor-hour.

The table below shows the number of pieces required to erect tubular steel scaffolding for 1000 S.F. of building frontage. This area is made up of a scaffolding system that is 12 frames (11 bays) long by 2 frames high.

For jobs under twenty-five frames, add 50% to rental cost. Rental rates will be lower for jobs over three months duration. Large quantities for long periods can reduce rental rates by 20%.

Description of Component	CSI Line Item	Number of Pieces for 1000 S.F. of Building Front	Unit
5' Wide Standard Frame, 6'-4" High	01540-750-2200	24	Ea.
Leveling Jack & Plate	01540-750-2650	24	
Cross Brace	01540-750-2500	44	
Side Arm Bracket, 21"	01540-750-2700	12	
Guardrail Post	01540-750-2550	12	
Guardrail, 7' section	01540-750-2600	22	
Stairway Section	01540-750-2900	2	
Stairway Starter Bar	01540-750-2910	1	
Stairway Inside Handrail	01540-750-2920	2	
Stairway Outside Handrail	01540-750-2930	2	
Walk-Thru Frame Guardrail	01540-750-2940	2	

Scaffolding is often used as falsework over 15' high during construction of cast-in-place concrete beams and slabs. Two foot wide scaffolding is generally used for heavy beam construction. The span between frames depends upon the load to be carried with a maximum span of 5'.

Heavy duty shoring frames with a capacity of 10,000#/leg can be spaced up to 10' O.C. depending upon form support design and loading.

Scaffolding used as horizontal shoring requires less than half the material required with conventional shoring.

On new construction, erection is done by carpenters.

Rolling towers supporting horizontal shores can reduce labor and speed the job. For maintenance work, catwalks with spans up to 70' can be supported by the rolling towers.

R01590-100   Contractor Equipment

Rental Rates shown in Division 01590 pertain to late model high quality machines in excellent working condition, rented from equipment dealers. Rental rates from contractors may be substantially lower than the rental rates from equipment dealers depending upon economic conditions; for older, less productive machines, reduce rates by a maximum of 15%. Any overtime must be added to the base rates. For shift work, rates are lower. Usual rule of thumb is 150% of one shift rate for two shifts; 200% for three shifts.

For periods of less than one week, operated equipment is usually more economical to rent than renting bare equipment and hiring an operator.

Costs to move equipment to a job site (mobilization) or from a job site (demobilization) are not included in rental rates in Division 01590, nor in any Equipment costs on any Unit Price line items or crew listings. These costs can be found in section 02305-250. If a piece of equipment is already at a job site, it is not appropriate to utilize mob/demob costs in an estimate again.

Rental rates vary throughout the country with larger cities generally having lower rates. Lease plans for new equipment are available for periods in excess of six months with a percentage of payments applying toward purchase.

Monthly rental rates vary from 2% to 5% of the cost of the equipment depending on the anticipated life of the equipment and its wearing parts. Weekly rates are about 1/3 the monthly rates and daily rental rates about 1/3 the weekly rate.

The hourly operating costs for each piece of equipment include costs to the user such as fuel, oil, lubrication, normal expendables for the equipment, and a percentage of mechanic's wages chargeable to maintenance. The hourly operating costs listed do not include the operator's wages.

The daily cost for equipment used in the standard crews is figured by dividing the weekly rate by five, then adding eight times the hourly operating cost to give the total daily equipment cost, not including the operator. This figure is in the right hand column of Division 01590 under Crew Equipment Cost/Day.

Pile Driving rates shown for pile hammer and extractor do not include leads, crane, boiler or compressor. Vibratory pile driving requires an added field specialist during set-up and pile driving operation for the electric model. The hydraulic model requires a field specialist for set-up only. Up to 125 reuses of sheet piling are possible using vibratory drivers. For normal conditions, crane capacity for hammer type and size are as follows.

Crane Capacity	Hammer Type and Size		
	Air or Steam	Diesel	Vibratory
25 ton	to 8,750 ft.-lb.		70 H.P.
40 ton	15,000 ft.-lb.	to 32,000 ft.-lb.	170 H.P.
60 ton	25,000 ft.-lb.		300 H.P.
100 ton		112,000 ft.-lb.	

Cranes should be specified for the job by size, building and site characteristics, availability, performance characteristics, and duration of time required.

Backhoes & Shovels rent for about the same as equivalent size cranes but maintenance and operating expense is higher. Crane operators rate must be adjusted for high boom heights. Average adjustments: for 150' boom add 2% per hour; over 185', add 4% per hour; over 210', add 6% per hour; over 250', add 8% per hour and over 295', add 12% per hour.

Tower Cranes of the climbing or static type have jibs from 50' to 200' and capacities at maximum reach range from 4,000 to 14,000 pounds. Lifting capacities increase up to maximum load as the hook radius decreases.

Typical rental rates, based on purchase price are about 2% to 3% per month.

Erection and dismantling runs between 500 and 2000 labor hours. Climbing operation takes 10 labor hours per 20' climb. Crane dead time is about 5 hours per 40' climb. If crane is bolted to side of the building add cost of ties and extra mast sections. Climbing cranes have from 80' to 180' of mast while static cranes have 80' to 800' of mast.

Truck Cranes can be converted to tower cranes by using tower attachments. Mast heights over 400' have been used. See Division 01590-600 for rental rates of high boom cranes.

A single 100' high material **Hoist and Tower** can be erected and dismantled in about 400 labor hours; a double 100' high hoist and tower in about 600 labor hours. Erection times for additional heights are 3 and 4 labor hours per vertical foot respectively up to 150', and 4 to 5 labor hours per vertical foot over 150' high. A 40' high portable Buck hoist takes about 160 labor hours to erect and dismantle. Additional heights take 2 labor hours per vertical foot to 80' and 3 labor hours per vertical foot for the next 100'. Most material hoists do not meet local code requirements for carrying personnel.

A 150' high **Personnel Hoist** requires about 500 to 800 labor hours to erect and dismantle. Budget erection time at 5 labor hours per vertical foot for all trades. Local code requirements or labor scarcity requiring overtime can add up to 50% to any of the above erection costs.

Earthmoving Equipment: The selection of earthmoving equipment depends upon the type and quantity of material, moisture content, haul distance, haul road, time available, and equipment available. Short haul cut and fill operations may require dozers only, while another operation may require excavators, a fleet of trucks, and spreading and compaction equipment. Stockpiled material and granular material are easily excavated with front end loaders. Scrapers are most economically used with hauls between 300' and 1-1/2 miles if adequate haul roads can be maintained. Shovels are often used for blasted rock and any material where a vertical face of 8' or more can be excavated. Special conditions may dictate the use of draglines, clamshells, or backhoes. Spreading and compaction equipment must be matched to the soil characteristics, the compaction required and the rate the fill is being supplied.

R01590-150   Heavy Lifting

Hydraulic Climbing Jacks

The use of hydraulic heavy lift systems is an alternative to conventional type crane equipment. The lifting, lowering, pushing, or pulling mechanism is a hydraulic climbing jack moving on a square steel jackrod from 1-5/8" to 4" square, or a steel cable. The jackrod or cable can be vertical or horizontal, stationary or movable, depending on the individual application. When the jackrod is stationary, the climbing jack will climb the rod and push or pull the load along with itself. When the climbing jack is stationary, the jackrod is movable with the load attached to the end and the climbing jack will lift or lower the jackrod with the attached load. The heavy lift system is normally operated by a single control lever located at the hydraulic pump.

The system is flexible in that one or more climbing jacks can be applied wherever a load support point is required, and the rate of lift synchronized.

Economic benefits have been demonstrated on projects such as: erection of ground assembled roofs and floors, complete bridge spans, girders and trusses, towers, chimney liners and steel vessels, storage tanks, and heavy machinery. Other uses are raising and lowering offshore work platforms, caissons, tunnel sections and pipelines.

R02065-300 Bituminous Paving

City	Sidewalk Mix Bituminous Asphalt per Ton*	Sidewalks (2") 9.2 S.Y./ton				Pavement (3") 6.13 S.Y./ton			
		Cost per S.Y.			Per Ton	Cost per S.Y.			Per Ton
		Material*	Installation	Total	Total	Material*	Installation	Total	Total
Atlanta	$30.00	$3.26	$1.94	$5.20	$47.84	$4.89	$ .98	$5.87	$35.98
Baltimore	34.73	3.78	1.89	5.67	52.16	5.67	.96	6.63	40.64
Boston	39.50	4.29	2.01	6.30	57.96	6.44	1.02	7.46	45.73
Buffalo	34.60	3.76	1.89	5.65	51.98	5.64	.96	6.60	40.46
Chicago	31.75	3.45	2.08	5.53	50.88	5.18	1.06	6.24	38.25
Cincinnati	41.00	4.46	2.20	6.66	61.27	6.69	1.12	7.81	47.88
Cleveland	29.50	3.21	2.16	5.37	49.40	4.81	1.10	5.91	36.23
Columbus	30.50	3.32	2.07	5.39	49.59	4.98	1.05	6.03	36.96
Dallas	29.25	3.18	1.75	4.93	45.36	4.77	.89	5.66	34.70
Denver	30.00	3.26	2.16	5.42	49.86	4.89	1.09	5.98	36.66
Detroit	32.00	3.48	1.97	5.45	50.14	5.22	1.00	6.22	38.13
Houston	34.50	3.75	1.80	5.55	51.06	5.63	.91	6.54	40.09
Indianapolis	29.75	3.23	2.03	5.26	48.39	4.85	1.03	5.88	36.04
Kansas City	29.50	3.21	1.98	5.19	47.75	4.81	1.00	5.81	35.62
Los Angeles	36.00	3.91	2.18	6.09	56.03	5.87	1.11	6.98	42.79
Memphis	38.75	4.21	1.86	6.07	55.84	6.32	.94	7.26	44.50
Milwaukee	32.50	3.53	1.89	5.42	49.86	5.30	.96	6.26	38.37
Minneapolis	31.38	3.41	2.21	5.62	51.70	5.12	1.12	6.24	38.25
Nashville	27.75	3.02	2.02	5.04	46.37	4.53	1.02	5.55	34.02
New Orleans	33.75	3.67	1.78	5.45	50.14	5.51	.90	6.41	39.29
New York City	49.50	5.38	2.33	7.71	70.93	8.08	1.18	9.26	56.76
Philadelphia	32.50	3.53	1.92	5.45	50.14	5.30	.97	6.27	38.44
Phoenix	30.00	3.26	2.13	5.39	49.59	4.89	1.08	5.97	36.60
Pittsburgh	35.00	3.80	2.21	6.01	55.29	5.71	1.12	6.83	41.87
St. Louis	32.00	3.48	1.96	5.44	50.05	5.22	.99	6.21	38.07
San Antonio	33.25	3.61	1.80	5.41	49.77	5.42	.91	6.33	38.80
San Diego	37.00	4.02	2.04	6.06	55.75	6.04	1.04	7.08	43.40
San Francisco	40.25	4.38	2.25	6.63	61.00	6.57	1.14	7.71	47.26
Seattle	39.50	4.29	2.30	6.59	60.63	6.44	1.16	7.60	46.59
Washington, D.C.	35.41	3.85	1.82	5.67	52.16	5.78	.92	6.70	41.07
Average	$34.00	$3.70	$2.02	$5.72	$52.62	$5.55	$1.02	$6.57	$40.27

Assumed density is 145 lb. per C.F.

*Includes delivery within 20 miles for quantities over 300 tons only.

Table below shows quantities and bare costs for 1000 S.Y. of Bituminous Paving.

Item	Sidewalks, 2" Thick (02775-275-0020)		Roads and Parking Areas, 3" Thick (02740-310-0460)	
	Quantities	Cost	Quantities	Cost
Bituminous asphalt	108.7 tons @ $34.00 per ton	$3,696.00	163.1 tons@ $34.00 per ton	$5,546.00
Installation using	Crew B-37 @ $1,455.80 /720 SY/day x 1000	2,021.94	Crew B-25B @ $5,023.80 /4900SY/ day x 1000	1,025.30
Total per 1000 S.Y.		$5,717.94		$6,571.30
Total per S.Y.		$ 5.72		$ 6.57
Total per Ton		$ 52.62		$ 40.27

R02115-200 Underground Storage Tank Removal

Underground Storage Tank Removal can be divided into two categories: Non-Leaking and Leaking. Prior to removing an underground storage tank, tests should be made, with the proper authorities present, to determine whether a tank has been leaking or the surrounding soil has been contaminated.

To safely remove Liquid Underground Storage Tanks:

1. Excavate to the top of the tank.
2. Disconnect all piping.
3. Open all tank vents and access ports.
4. Remove all liquids and/or sludge.
5. Purge the tank with an inert gas.
6. Provide access to the inside of the tank and clean out the interior using proper personal protective equipment (PPE).
7. Excavate soil surrounding the tank using proper PPE for on-site personnel.
8. Pull and properly dispose of the tank.
9. Clean up the site of all contaminated material.
10. Install new tanks or close the excavation.

R02220-510 Demolition Defined

Whole Building Demolition (Division 02220-110) - Demolition of the whole building with no concern for any particular building element, component, or material type being demolished. This type of demolition is accomplished with large pieces of construction equipment that break up the structure, load it into trucks and haul it to a disposal site, but disposal or dump fees are not included (Divisions 02220-320 and 02220-330). Demolition of below-grade foundation elements, such as footings, foundation walls, grade beams, slabs on grade, etc. (Division 02220-130), is not included. Certain mechanical equipment containing flammable liquids or ozone-depleting refrigerants, electric lighting elements, communication equipment components, and other building elements may contain hazardous waste, and must be removed, either selectively or carefully, as hazardous waste before the building can be demolished (Division 13281).

Foundation Demolition (Division 02220-130) - Demolition of below-grade foundation footings, foundation walls, grade beams, and slabs on grade. This type of demolition is accomplished by hand or pneumatic hand tools, and does not include saw cutting (Division 02220-360), or handling, loading, hauling, or disposal of the debris (Divisions 02220-320, 02220-330 and 02220-350).

Gutting (Division 02220-340) - Removal of building interior finishes and electrical/mechanical systems down to the load-bearing and sub-floor elements of the rough building frame, with no concern for any particular building element, component, or material type being demolished. This type of demolition is accomplished by hand or pneumatic hand tools, and includes loading into trucks, but not hauling, disposal or dump fees (Divisions 02220-320 and 02220-330); scaffolding (01540-750); or shoring (03150-600). Certain mechanical equipment containing flammable liquids or ozone-depleting refrigerants, electric lighting elements, communication equipment components, and other building elements may contain hazardous waste, and must be removed, either selectively or carefully, as hazardous waste, before the building is gutted (Division 13281).

Selective Demolition - Demolition of a selected building element, component, or finish, with some concern for surrounding or adjacent elements, components, or finishes (see the first Subdivision (s) at the beginning of Divisions 3 through 16). This type of demolition is accomplished by hand or pneumatic hand tools, and does not include handling, loading, storing, hauling, or disposal of the debris (Divisions 02220-320, 02220-330 and 02220-350); scaffolding (Division 01540-750); or shoring (Division 03150-600). "Gutting" methods may be used in order to save time, but damage that was caused to surrounding or adjacent elements, components, or finishes may have to be repaired at a later time.

Careful Removal - Removal of a piece of service equipment, building element or component, or material type, with great concern for both the removed item and surrounding or adjacent elements, components or finishes. The purpose of careful removal may be to protect the removed item for later re-use, preserve a higher salvage value of the removed item, or replace an item while taking care to protect surrounding or adjacent elements, components, connections, or finishes from cosmetic and/or structural damage. An approximation of the time required to perform this type of removal is 1/3 to 1/2 the time it would take to install a new item of like kind (see Reference Numbers R15050-720 and R16055-300). This type of removal is accomplished by hand or pneumatic hand tools, and does not include loading, hauling, or storing the removed item (Divisions 01840-100 and 02220-350); scaffolding (Division 01540-750); shoring (Division 03150-600); or lifting equipment (Division 01590-600).

Cutout Demolition (Division 02220-310) - Demolition of a small quantity of floor, wall, roof, or other assembly, with concern for the appearance and structural integrity of the surrounding materials. This type of demolition is accomplished by hand or pneumatic hand tools, and does not include saw cutting (Division 02220-360); handling, loading, hauling, or disposal of debris (Divisions 02220-320, 02220-330 and 02220-350); scaffolding (Division 01540-750); or shoring (Division 03150-600).

Rubbish Handling (Division 02220-350) - Work activities that involve handling, loading or hauling of debris. Generally, the cost of rubbish handling must be added to the cost of all types of demolition, with the exception of whole building demolition.

Minor Site Demolition (Division 02220-240) - Demolition of site elements outside the footprint of a building. This type of demolition is accomplished by hand or pneumatic hand tools, or with larger pieces of construction equipment, and may include loading a removed item onto a truck (check the Crew for equipment used). It does not include saw cutting (Division 02220-360), hauling or disposal of debris (Divisions 02220-320 and 02220-330), and, sometimes, handling or loading (Division 02220-350).

R02240-900   Wellpoints

A single stage wellpoint system is usually limited to dewatering an average 15' depth below normal ground water level. Multi-stage systems are employed for greater depth with the pumping equipment installed only at the lowest header level. Ejectors, with unlimited lift capacity, can be economical when two or more stages of wellpoints can be replaced or when horizontal clearance is restricted, such as in deep trenches or tunneling projects, and where low water flows are expected. Wellpoints are usually spaced on 2-1/2' to 10' centers along a header pipe. Wellpoint spacing, header size, and pump size are all determined by the expected flow, as dictated by soil conditions.

In almost all soils encountered in wellpoint dewatering, the wellpoints may be jetted into place. Cemented soils and stiff clays may require sand wicks about 12" in diameter around each wellpoint to increase efficiency and eliminate weeping into the excavation. These sand wicks require 1/2 to 3 C.Y. of washed filter sand and are installed by using a 12" diameter steel casing and hole puncher jetted into the ground 2' deeper than the wellpoint. Rock may require predrilled holes.

Labor required for the complete installation and removal of a single stage wellpoint system is in the range of 3/4 to 2 labor-hours per linear foot of header, depending upon jetting conditions, wellpoint spacing, etc.

Continuous pumping is necessary except in some free draining soil where temporary flooding is permissible (as in trenches which are backfilled after each day's work). Good practice requires provision of a stand-by pump during the continuous pumping operation.

Systems for continuous trenching below the water table should be installed three to four times the length of expected daily progress to insure uninterrupted digging, and header pipe size should not be changed during the job.

For pervious free draining soils, deep wells in place of wellpoints may be economical because of lower installation and maintenance costs. Daily production ranges between two to three wells per day, for 25' to 40' depths, to one well per day for depths over 50'.

Detailed analysis and estimating for any dewatering problem is available at no cost from wellpoint manufacturers. Major firms will quote "sufficient equipment" quotes or their affiliates offer lump sum proposals to cover complete dewatering responsibility.

Description for 200' System with 8" Header		Quantities	1st Month		Thereafter	
			Unit	Total	Unit	Total
Equipment & Material	Wellpoints 25' long, 2" diameter @ 5' O.C.	40 Each	$ 31.00	$ 1,240.00	$ 23.25	$ 930.00
	Header pipe, 8" diameter	200 L.F.	5.90	1,180.00	2.95	590.00
	Discharge pipe, 8" diameter	100 L.F.	3.96	396.00	2.57	257.00
	8" valves	3 Each	151.00	453.00	98.15	294.45
	Combination Jetting & Wellpoint pump (standby)	1 Each	2,450.00	2,450.00	1,715.00	1,715.00
	Wellpoint pump, 8" diameter	1 Each	2,400.00	2,400.00	1,680.00	1,680.00
	Transportation to and from site	1 Day	697.00	697.00	—	—
	Fuel 30 days x 60 gal./day	1800 Gallons	2.00	3,600.00	2.00	3,600.00
	Lubricants for 30 days x 16 lbs./day	480 Lbs.	1.50	720.00	1.50	720.00
	Sand for points	40 C.Y.	21.35	854.00	—	—
	Equipment & Materials Subtotal			$13,990.00		$ 9,786.45
Labor	Technician to supervise installation	1 Week	$ 36.20	$ 1,448.00	—	—
	Labor for installation and removal of system	300 Labor-hours	26.70	8,010.00	—	—
	4 Operators straight time 40 hrs./wk. for 4.33 wks.	693 Hrs.	33.05	22,903.65	$ 33.05	$22,903.65
	4 Operators overtime 2 hrs./wk. for 4.33 wks.	35 Hrs.	66.10	2,313.50	66.10	2,313.50
	Bare Labor Subtotal			$34,675.15		$25,217.15
	Total Bare Cost			$48,665.15		$35,003.60
	Monthly Bare Cost/L.F. Header			$ 243.33		$ 175.02

R02250-400 Wood Sheet Piling

Wood sheet piling may be used for depths to 20' where there is no ground water. If moderate ground water is encountered Tongue & Groove sheeting will help to keep it out. When considerable ground water is present, steel sheeting must be used.

For estimating purposes on trench excavation, sizes are as follows:

Depth	Sheeting	Wales	Braces	B.F. per S.F.
To 8'	3 x 12's	6 x 8's, 2 line	6 x 8's, @ 10'	4.0 @ 8'
8' x 12'	3 x 12's	10 x 10's, 2 line	10 x 10's, @ 9'	5.0 average
12' to 20'	3 x 12's	12 x 12's, 3 line	12 x 12's, @ 8'	7.0 average

Sheeting to be toed in at least 2' depending upon soil conditions. A five person crew with an air compressor and sheeting driver can drive and brace 440 SF/day at 8' deep, 360 SF/day at 12' deep, and 320 SF/day at 16' deep. For normal soils, piling can be pulled in 1/3 the time to install. Pulling difficulty increases with the time in the ground. Production can be increased by high pressure jetting. Figures below assume 50% of lumber is salvaged and includes pulling costs. Some jurisdictions require an equipment operator in addition to Crew B-31.

Sheeting Pulled

Daily Cost Crew B-31	L.H./ Day	Hourly Cost	Daily Cost	8' Depth, 440 S.F./Day		16' Depth, 320 S.F./Day	
				To Drive (1 Day)	To Pull (1/3 Day)	To Drive (1 Day)	To Pull (1/3 Day)
1 Foreman	8	$28.70	$ 229.60	$ 229.60	$ 76.46	$ 229.60	$ 76.46
3 Laborers	24	26.70	640.80	640.80	213.39	640.80	213.39
1 Carpenter	8	34.25	274.00	274.00	91.24	274.00	91.24
1 Air Compressor			112.40	112.40	37.43	112.40	37.43
1 Sheeting Driver			7.20	7.20	2.40	7.20	2.40
2 -50 Ft. Air Hoses, 1-1/2" Diam.			25.50	25.50	8.49	25.50	8.49
Lumber (50% salvage)				1.76 MBF	616.00	2.24 MBF	784.00
Bare Total			$1,289.50	$1,905.50	$429.41	$2,073.50	$429.41
Bare Total/S.F.				$ 4.33	$ .98	$ 4.71	$ 1.34
Bare Total (Drive and Pull)/S.F.				$ 5.31		$ 6.05	

Sheeting Left in Place

Daily Cost	8' Depth 440 S.F./Day		10' Depth 400 S.F./Day		12' Depth 360 S.F./Day		16' Depth 320 S.F./Day		18' Depth 305 S.F./Day		20' Depth 280 S.F./Day	
Crew B-31		$1,273.40		$1,273.40		$1,273.40		$1,273.40		$1,273.40		$1,273.40
Lumber	1.76 MBF	1,232.00	1.8 MBF	1,260.00	1.8 MBF	1,260.00	2.24 MBF	1,568.00	2.1 MBF	1,470.00	1.9 MBF	1,330.00
Bare Total in Place		$2,505.40		$2,533.40		$2,533.40		$2,841.40		$2,743.40		$2,603.40
Bare Total/S.F.		$ 5.69		$ 6.33		$ 7.04		$ 8.88		$ 8.99		$ 9.30
Bare Total/M.B.F.		$1,423.52		$1,407.44		$1,407.44		$1,268.48		$1,306.38		$1,370.21

R02250-450   Steel Sheet Piling

Limiting weights are 22 to 38#/S.F. of wall surface with 27#/S.F. average for usual types and sizes. (Weights of piles themselves are from 30.7#/L.F. to 57#/L.F. but they are 15″ to 21″ wide.) Lightweight sections 12″ to 28″ wide from 3 ga. to 12 ga. thick are also available for shallow excavations. Piles may be driven two at a time with an impact or vibratory hammer (use vibratory to pull) hung from a crane without leads. A reasonable estimate of the life of steel sheet piling is 10 uses with up to 125 uses possible if a vibratory hammer is used. Used piling costs from 50% to 80% of new piling depending on location and market conditions. Sheet piling and H piles

can be rented for about 30% of the delivered mill price for the first month and 5% per month thereafter. Allow 1 labor-hour per pile for cleaning and trimming after driving. These costs increase with depth and hydrostatic head. Vibratory drivers are faster in wet granular soils and are excellent for pile extraction. Pulling difficulty increases with the time in the ground and may cost more than driving. It is often economical to abandon the sheet piling, especially if it can be used as the outer wall form. Allow about 1/3 additional length or more, for toeing into ground. Add bracing, waler and strut costs. Waler costs can equal the cost per ton of sheeting.

Cost of Sheet Piling & Production Rate by Ton & S.F.

Depth of Excavation	15′ Depth		20′ Depth		25′ Depth	
Description of Pile	**22 psf = 90.9 S.F./Ton**		**27 psf = 74 S.F./Ton**		**38 psf = 52.6 S.F./Ton**	
Type of operation:	Drive &	Drive &	Drive &	Drive &	Drive &	Drive &
Left in Place or Removed	Left	Extract*	Left	Extract*	Left	Extract*
Labor & Equip. to Drive 1 ton	$ 412.06	$742.40	$ 343.97	$680.06	$ 234.44	$424.23
Piling (75% Salvage)	880.00	220.00	880.00	220.00	880.00	220.00
Bare Cost/Ton in Place	$1,292.06	$962.40	$1,223.97	$900.06	$1,114.44	$644.23
Production Rate, Tons/Day	10.81	6.00	12.95	6.55	19.00	10.50
Bare Cost, S.F. in Place (incl. 33% toe in)	$ 14.21	$ 10.59	$ 16.54	$ 12.16	$ 21.19	$ 12.25
Production Rate, S.F./Day (incl. 33% toe in)	983	545	959	485	1000	553

Crew B-40	Bare Costs		Inc. Subs O&P		
	Hr.	**Daily**	**Hr.**	**Daily**	
1 Pile Driver Foreman (Out)	$35.30	$ 282.40	$58.15	$ 465.20	
4 Pile Drivers	33.30	1,065.60	54.90	1,756.80	
2 Equip. Oper. (crane)	35.90	574.40	54.10	865.60	
1 Equip. Oper. Oiler	30.10	240.80	45.35	362.80	
1 Crane, 40 Ton		888.20		977.00	
Vibratory Hammer & Gen.		1,403.00		1,543.30	
64 L.H., Daily Totals		$4,454.40		$5,970.70	

Bare Installation Cost per Ton
for 15′ Deep Excavation plus 33% toe-in
$4,454 per day/10.81 tons = $412.03 per ton

Bare Installation Cost per S.F.

$$\frac{2000\#}{22\#/S.F.} = 90.9 \text{ S.F. x } 10.81 \text{ tons} = 983 \text{ S.F. per day}$$

$4,454 / 983 S.F. per day = $4.53 per S.F.

*For driving & extracting, two mobilizations & demobilizations will be necessary

R02250-900 Vibroflotation and Vibro Replacement Soil Compaction

Vibroflotation is a proprietary system of compacting sandy soils in place to increase relative density to about 70%. Typical bearing capacities attained will be 6000 psf for saturated sand and 12,000 psf for dry sand. Usual range is 4000 to 8000 psf capacity. Costs in the front of the book are for a vertical foot of compacted cylinder 6' to 10' in diameter.

Vibro replacement is a proprietary system of improving cohesive soils in place to increase bearing capacity. Most silts and clays above or below the water table can be strengthened by installation of stone columns.

The process consists of radial displacement of the soil by vibration. The created hole is then backfilled in stages with coarse granular fill which is thoroughly compacted and displaced into the surrounding soil in the form of a column.

The total project cost would depend on the number and depth of the compacted cylinders. The installing company guarantees relative soil density of the sand cylinders after compaction and the bearing capacity of the soil after the replacement process. Detailed estimating information is available from the installer at no cost.

Site Construction | **R023** | Earthwork

R02315-260 Trench Excavation and Pipe Bedding

						6' Wide Bottom of Trench			
Example: Find the cost per LF for excavating, bedding, and backfilling a 36" diameter pipe 10' deep in damp sandy loam. **Method:** For trench bottom widths for various size pipe, Assembly G1030-815, a 36" diameter pipe uses a 6' wide trench bottom.						Slope 1/2 to 1	Slope 1 to 1	Slope 0	Slope 2 to 1
Excavation & Backfill System	For 1/2 to 1 slope see: 1 to 1 slope 1 2 to 1 slope	Trench Excavation	System	G1030-805	-2880 -4050 -5700 -7300	$37.80	$50.45	$57.50	$65.50
Pipe Bedding System	For 1/2 to 1 slope see: 1 to 1 slope 1 2 to 1 slope	Trench Pipe Bedding	System	G1030-815	-2400 -2880 -3260 -3660	$13.20	$17.65	$22.10	$26.55
Total Cost, including O&P, per LF (not incl. pipe)						$51.00	$68.10	$79.60	$92.05

R02315-300 Compacting Backfill

Compaction of fill in embankments, around structures, in trenches, and under slabs is important to control settlement. Factors affecting compaction are:

1. Soil gradation
2. Moisture content
3. Equipment used
4. Depth of fill per lift
5. Density required

The costs for testing and soil analyses are listed in Division 01450-500. Also, see Division 02315 for further backfill, borrow, and compaction costs.

Example:

Compact granular fill around a building foundation using a 21" wide x 24" vibratory plate in 8" lifts. Operator moves at 50 FPM working a 50 minute hour to develop 95% Modified Proctor Density with 4 passes.

Production Rate:

$$\frac{1.75' \text{ plate width x 50 F.P.M. x 50 min./hr. x .67' lift}}{27 \text{ C.F. per C.Y.}} = 108.5 \text{ C.Y./hr.}$$

Production Rate for 4 Passes:

$$\frac{108.5 \text{ C.Y.}}{4 \text{ passes}} = 27.125 \text{ C.Y./hr. x 8 hrs.} = 217 \text{ C.Y./day}$$

Compacting 217 C.Y. with 21" Wide Vibratory Plate		L.H./ Day	Hourly Cost	Daily Cost	C.Y. Cost
1	Laborer	8	$26.70	$213.60	$ .98
1	Vibratory Plate Compactor			36.80	.17
Bare Total for 217 C.Y./day				$250.40	$1.15

R02315-400   Excavating

The selection of equipment used for structural excavation and bulk excavation or for grading is determined by the following factors.

1. Quantity of material.
2. Type of material.
3. Depth or height of cut.
4. Length of haul.
5. Condition of haul road.
6. Accessibility of site.
7. Moisture content and dewatering requirements.
8. Availability of excavating and hauling equipment.

Some additional costs must be allowed for hand trimming the sides and bottom of concrete pours and other excavation below the general excavation.

Number of B.C.Y. per truck = 1.5 C.Y. bucket x 8 passes = 12 loose C.Y.

$$= 12 \times \frac{100}{118} = 10.2 \text{ B.C.Y. per truck}$$

Truck Haul Cycle:

Load truck 8 passes	=	4 minutes
Haul distance 1 mile	=	9 minutes
Dump time	=	2 minutes
Return 1 mile	=	7 minutes
Spot under machine	=	1 minute
		23 minute cycle

When planning excavation and fill, the following should also be considered.

1. Swell factor.
2. Compaction factor.
3. Moisture content.
4. Density requirements.

A typical example for scheduling and estimating the cost of excavation of a 15' deep basement on a dry site when the material must be hauled off the site, is outlined below.

Assumptions:

1. Swell factor, 18%.
2. No mobilization or demobilization.
3. Allowance included for idle time and moving on job.
4. No dewatering, sheeting, or bracing.
5. No truck spotter or hand trimming.

Fleet Haul Production per day in B.C.Y.

$$4 \text{ trucks} \times \frac{50 \text{ min. hour}}{23 \text{ min. haul cycle}} \times 8 \text{ hrs.} \times 10.2 \text{ B.C.Y.}$$

$$= 4 \times 2.2 \times 8 \times 10.2 = 718 \text{ B.C.Y./day}$$

Excavating Cost with a 1-1/2 C.Y. Hydraulic Excavator 15' Deep, 2 Mile Round Trip Haul	L.H./Day	Hourly Cost	Daily Cost	Subtotal	Unit Price
1 Equipment Operator	8	$35.90	$ 287.20		
1 Oiler	8	26.70	213.60		
4 Truck Drivers	32	27.55	881.60	$1,382.40	$1.92
1 Hydraulic Excavator			720.40		
4 Dump Trucks			1,906.40	2,626.80	$3.65
Bare Total for 720 B.C.Y.			$4,009.20	$4,009.20	$5.57

Description		1-1/2 C.Y. Hyd. Backhoe 15' Deep		1-1/2 C.Y. Power Shovel 7' Bank		1-1/2 C.Y. Dragline 7' Deep		2-1/2 C.Y. Trackloader Stockpile
Operator (and Oiler, if required)		$ 500.80		$ 500.80		$ 500.80		$ 500.80
Truck Drivers	3 Ea.	661.20	4 Ea.	881.60	3 Ea.	661.20	4 Ea.	881.60
Equipment Rental		720.40		956.00		1,236.60		551.80
20 C.Y. Trailer Dump Trucks	3 Ea.	1,429.80	4 Ea.	1,906.40	3 Ea.	1,429.80	4 Ea.	1,906.40
Bare Total Cost per Day		$3,312.20		$4,244.80		$3,828.40		$3,840.60
Daily Production, C.Y. Bank Measure		720.00		960.00		640.00		1,000.00
Bare Cost per C.Y.		$ 4.60		$ 4.42		$ 5.98		$ 3.84

Add the mobilization and demobilization costs to the total excavation costs. When equipment is rented for more than three days, there is often no mobilization charge by the equipment dealer. On larger jobs outside of urban areas, scrapers can move earth economically provided a dump site or fill area and adequate haul roads are available. Excavation within sheeting bracing or cofferdam bracing is usually done with a clamshell and production

is low, since the clamshell may have to be guided by hand between the bracing. When excavating or filling an area enclosed with a wellpoint system, add 10% to 15% to the cost to allow for restricted access. When estimating earth excavation quantities for structures, allow work space outside the building footprint for construction of the foundation, and a slope of 1:1 unless sheeting is used.

R02315-450 Excavating Equipment

The table below lists THEORETICAL hourly production in C.Y./hr. bank measure for some typical excavation equipment. Figures assume 50 minute hours, 83% job efficiency, 100% operator efficiency, 90° swing and properly sized hauling units, which must be modified for adverse digging and loading conditions. Actual production costs in the front of the book average about 50% of the theoretical values listed here.

Equipment	Soil Type	B.C.Y. Weight	% Swell	1 C.Y.	1-1/2 C.Y.	2 C.Y.	2-1/2 C.Y.	3 C.Y.	3-1/2 C.Y.	4 C.Y.
Hydraulic Excavator	Moist loam, sandy clay	3400 lb.	40%	85	125	175	220	275	330	380
"Backhoe"	Sand and gravel	3100	18	80	120	160	205	260	310	365
15' Deep Cut	Common earth	2800	30	70	105	150	190	240	280	330
	Clay, hard, dense	3000	33	65	100	130	170	210	255	300
	Moist loam, sandy clay	3400	40	170 (6.0)	245 (7.0)	295 (7.8)	335 (8.4)	385 (8.8)	435 (9.1)	475 (9.4)
Power Shovel	Sand and gravel	3100	18	165 (6.0)	225 (7.0)	275 (7.8)	325 (8.4)	375 (8.8)	420 (9.1)	460 (9.4)
Optimum Cut (Ft.)	Common earth	2800	30	145 (7.8)	200 (9.2)	250 (10.2)	295 (11.2)	335 (12.1)	375 (13.0)	425 (13.8)
	Clay, hard, dense	3000	33	120 (9.0)	175 (10.7)	220 (12.2)	255 (13.3)	300 (14.2)	335 (15.1)	375 (16.0)
	Moist loam, sandy clay	3400	40	130 (6.6)	180 (7.4)	220 (8.0)	250 (8.5)	290 (9.0)	325 (9.5)	385 (10.0)
Drag Line	Sand and gravel	3100	18	130 (6.6)	175 (7.4)	210 (8.0)	245 (8.5)	280 (9.0)	315 (9.5)	375 (10.0)
Optimum Cut (Ft.)	Common earth	2800	30	110 (8.0)	160 (9.0)	190 (9.9)	220 (10.5)	250 (11.0)	280 (11.5)	310 (12.0)
	Clay, hard, dense	3000	33	90 (9.3)	130 (10.7)	160 (11.8)	190 (12.3)	225 (12.8)	250 (13.3)	280 (12.0)

				Wheel Loaders				Track Loaders		
				3 C.Y.	4 C.Y.	6 C.Y.	8 C.Y.	2-1/4 C.Y.	3 C.Y.	4 C.Y.
	Moist loam, sandy clay	3400	40	260	340	510	690	135	180	250
	Sand and gravel	3100	18	245	320	480	650	130	170	235
Loading Tractors	Common earth	2800	30	230	300	460	620	120	155	220
	Clay, hard, dense	3000	33	200	270	415	560	110	145	200
	Rock, well-blasted	4000	50	180	245	380	520	100	130	180

R02400-110 Tunnel Excavation

Bored tunnel excavation is common in rock for diameters from 4 feet for sewer and utilities, to 60 feet for vehicles. Production varies from a few linear feet per day to over 200 linear feet per day. In the smaller diameters, the productivity is limited by the restricted area for mucking or the removal of excavated material.

Most of the tunnels in rock today are excavated by boring machines called moles. Preparation for starting the excavation or setting up the mole is very costly. Shafts must be excavated to the invert of the proposed tunnel and the mole must be lowered into the shaft. If excavating a portal tunnel, that is starting at an open face, the cost is reduced considerably both for mobilization and mucking.

In soft ground and mixed material, special bucket excavators and rotary excavators are used inside a shield. Tunnel liners must follow directly behind the shield to support the earth and prevent cave-ins.

Traditional muck haulage operations are performed by rail with locomotives and muck cars. Sometimes conveyors are more economical and require less ventilation of the tunnel.

Ventilation and air compression are other important cost factors to consider in tunnel excavation. Continuous ventilation ducts are sometimes fabricated at the tunnel site.

Tunnel linings are steel, cast in place reinforced concrete, shotcrete, or a combination of these. When required, contact grouting is performed by pumping grout between the lining and the excavation. Intermittent holes are drilled into the lining and separate costs are determined for drilling per hole, grout pump connecting per hole, and grout per cubic foot.

Consolidation grouting and roof bolts may also be required where the excavation is unstable or faulting occurs.

Tunnel boring is usually done 24 hours per day. A typical crew for rock boring is:

Tunneling Crew based on three 8 hour shifts

1 Shifter
1 Walker
1 Machine Operator for mole
1 Oiler
1 Mechanic
3 Locomotives with operators
5 Miners for rails, vent ducts, and roof bolts
1 Electrician
2 Pumps
2 Laborers for hoisting
1 Hoist operator for muck removal
1 Oiler

Surface Crew Based on normal 8 hour shift

2 Shop Mechanics
1 Electrician
1 Shifter
2 Laborers
1 Operator with 18 ton cherry picker
1 Operator with front end loader

For tunneling costs, please see Division 02400.

R02450-110 Spread Footings

General: A spread footing is used to convert a concentrated load (from one superstructure column, or substructure grade beams) into an allowable area load on supporting soil.

Because of punching action from the column load, a spread footing is usually thicker than strip footings which support wall loads. One or two story commercial or residential buildings should have no less than 1' thick spread footings. Heavier loads require no less than 2' thick. Spread footings may be square, rectangular or octagonal in plan.

Spread footings tend to minimize excavation and foundation materials, as well as labor and equipment. Another advantage is that footings and soil conditions can be readily examined. They are the most widely used type of footing, especially in mild climates and for buildings of four stories or under. This is because they are usually more economical than other types, if suitable soil and site conditions exist.

They are used when suitable supporting soil is located within several feet of the surface or line of subsurface excavation. Suitable soil types include sands and gravels, gravels with a small amount of clay or silt, hardpan, chalk, and rock. Pedestals may be used to bring the column base load down to the top of footing. Alternately, undesirable soil between underside of footing and top of bearing level can be removed and replaced with lean concrete mix or compacted granular material.

Depth of footing should be below topsoil, uncompacted fill, muck, etc. It must be lower than frost penetration but should be above the water table. It must not be at the ground surface because of potential surface erosion. If the ground slopes, approximately three horizontal feet of edge protection must remain. Differential footing elevations may overlap soil stresses or cause excavation problems if clear spacing between footings is less than the difference in depth.

Other footing types are usually used for the following reasons:

A. Bearing capacity of soil is low.
B. Very large footings are required, at a cost disadvantage.
C. Soil under footing (shallow or deep) is very compressible, with probability of causing excessive or differential settlement.
D. Good bearing soil is deep.
E. Potential for scour action exists.
F. Varying subsoil conditions within building perimeter.

Cost of spread footings for a building is determined by:
1. The soil bearing capacity.
2. Typical bay size.
3. Total load (live plus dead) per S.F. for roof and elevated floor levels.
4. The size and shape of the building.
5. Footing configuration. Does the building utilize outer spread footings or are there continuous perimeter footings only or a combination of spread footings plus continuous footings?

Soil Bearing Capacity in Kips per S.F.

Bearing Material	Typical Allowable Bearing Capacity
Hard sound rock	120 KSF
Medium hard rock	80
Hardpan overlaying rock	24
Compact gravel and boulder-gravel; very compact sandy gravel	20
Soft rock	16
Loose gravel; sandy gravel; compact sand; very compact sand-inorganic silt	12
Hard dry consolidated clay	10
Loose coarse to medium sand; medium compact fine sand	8
Compact sand-clay	6
Loose fine sand; medium compact sand-inorganic silts	4
Firm or stiff clay	3
Loose saturated sand-clay; medium soft clay	2

R02450-660 Maximum Depth of Frost Penetration in Inches

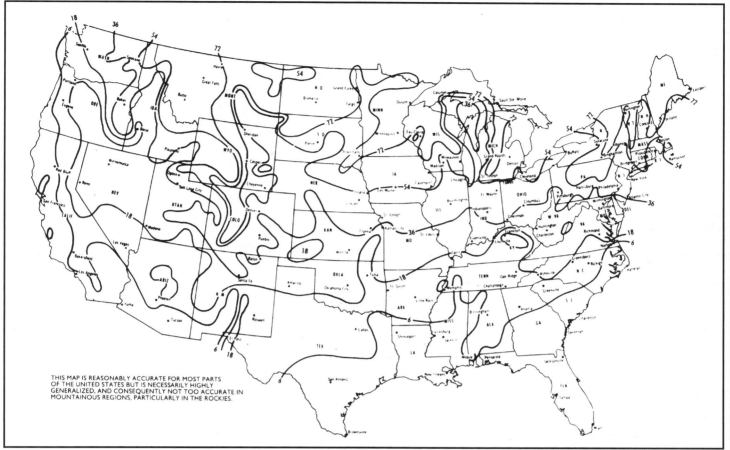

THIS MAP IS REASONABLY ACCURATE FOR MOST PARTS
OF THE UNITED STATES BUT IS NECESSARILY HIGHLY
GENERALIZED, AND CONSEQUENTLY NOT TOO ACCURATE IN
MOUNTAINOUS REGIONS, PARTICULARLY IN THE ROCKIES.

R02450-820 Pile Caps, Piles and Caissons

General: The function of a reinforced concrete pile cap is to transfer superstructure load from isolated column or pier to each pile in its supporting cluster. To do this, the cap must be thick and rigid, with all piles securely embedded into and bonded to it.

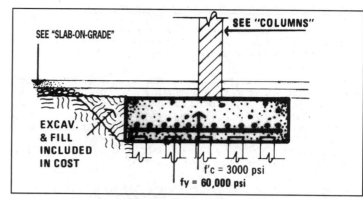

Figure 1.1-331 Section Through Pile Cap

Table 1.1-332   Concrete Quantities for Pile Caps

Load Working (K)	Number of Piles @ 3'-0" O.C. Per Footing Cluster									
	2 (CY)	4 (CY)	6 (CY)	8 (CY)	10 (CY)	12 (CY)	14 (CY)	16 (CY)	18 (CY)	20 (CY)
50	(.9)	(1.9)	(3.3)	(4.9)	(5.7)	(7.8)	(9.9)	(11.1)	(14.4)	(16.5)
100	(1.0)	(2.2)	(3.3)	(4.9)	(5.7)	(7.8)	(9.9)	(11.1)	(14.4)	(16.5)
200	(1.0)	(2.2)	(4.0)	(4.9)	(5.7)	(7.8)	(9.9)	(11.1)	(14.4)	(16.5)
400	(1.1)	(2.6)	(5.2)	(6.3)	(7.4)	(8.2)	(13.7)	(11.1)	(14.4)	(16.5)
800		(2.9)	(5.8)	(7.5)	(9.2)	(13.6)	(17.6)	(15.9)	(19.7)	(22.1)
1200			(5.8)	(8.3)	(9.7)	(14.2)	(18.3)	(20.4)	(21.2)	(22.7)
1600				(9.8)	(11.4)	(14.5)	(19.5)	(20.4)	(24.6)	(27.2)
2000				(9.8)	(11.4)	(16.6)	(24.1)	(21.7)	(26.0)	(28.8)
3000						(17.5)		(26.5)	(30.3)	(32.9)
4000								(30.2)	(30.7)	(36.5)

Table 1.1-333   Concrete Quantities for Pile Caps

Load Working (K)	Number of Piles @ 4'-6" O.C. Per Footing					
	2 (CY)	3 (CY)	4 (CY)	5 (CY)	6 (CY)	7 (CY)
50	(2.3)	(3.6)	(5.6)	(11.0)	(13.7)	(12.9)
100	(2.3)	(3.6)	(5.6)	(11.0)	(13.7)	(12.9)
200	(2.3)	(3.6)	(5.6)	(11.0)	(13.7)	(12.9)
400	(3.0)	(3.6)	(5.6)	(11.0)	(13.7)	(12.9)
800			(6.2)	(11.5)	(14.0)	(12.9)
1200				(13.0)	(13.7)	(13.4)
1600						(14.0)

R02450-820 Pile Caps, Piles and Caissons (cont.)

General: Piles are column-like shafts which receive superstructure loads, overturning forces, or uplift forces. They receive these loads from isolated column or pier foundations (pile caps), foundation walls, grade beams, or foundation mats. The piles then transfer these loads through shallower poor soil strata to deeper soil of adequate support strength and acceptable settlement with load.

Be sure that other foundation types aren't better suited to the job. Consider ground and settlement, as well as loading, when reviewing. Piles usually are associated with difficult foundation problems and substructure condition. Ground conditions determine type of pile (different pile types have been developed to suit ground conditions.) Decide each case by technical study, experience and sound engineering judgment—not rules of thumb. A full investigation of ground conditions, early, is essential to provide maximum information for professional foundation engineering and an acceptable structure.

Piles support loads by end bearing and friction. Both are generally present; however, piles are designated by their principal method of load transfer to soil.

Boring should be taken at expected pile locations. Ground strata (to bedrock or depth of 1-1/2 building width) must be located and identified with appropriate strengths and compressibilities. The sequence of strata determines if end bearing or friction piles are best suited. See **Table 1.4-105** for site investigation costs.

End bearing piles have shafts which pass through soft strata or thin hard strata and tip bear on bedrock or penetrate some distance into a dense, adequate soil (sand or gravel.)

Friction piles have shafts which may be entirely embedded in cohesive soil (moist clay), and develop required support mainly by adhesion or "skin-friction" between soil and shaft area.

Piles pass through soil by either one of two ways:
1. Displacement piles force soil out of the way. This may cause compaction, ground heaving, remolding of sensitive soils, damage to adjacent structures, or hard driving.
2. Non-displacement piles have either a hole bored and the pile cast or placed in hole, or open ended pipe (casing) driven and the soil core removed. They tend to eliminate heaving or lateral pressure damage to adjacent structures of piles. Steel "HP" piles are considered of small displacement.

Placement of piles (attitude) is most often vertical; however, they are sometimes battered (placed at a small angel from vertical) to advantageously resist lateral loads. Seldom are piles installed singly but rather in clusters (or groups). Codes require a minimum of three piles per major column load or two per foundation wall or grade beam. Single pile capacity is limited by pile structural strength or support strength of soil. Support capacity of a pile cluster is almost always less than the sum of its individual pile capacities due to overlapping of bearing the friction stresses. See **Table 1.4-101** for minimum pile spacing requirements.

Large rigs for heavy, long piles create large soil surface loads and additional expense on weak ground. See **Table 1.4-103** for percent of cost increases.

Fewer piles create higher costs per pile. See **Table 1.4-102** for effect of mobilization.

Pile load tests are frequently required by code, ground situation, or pile type. See **Table 1.4-104** for costs. Test load is twice the design load.

Table 1.4-101 Min. Pile Spacing by Pile Type

Type Support		Min. Pile Spacing	
		X's Butt Diameter	Foot
End	Bedrock	2	2'-0"
Bearing	Hard Strata	2.5	2'-6"
Friction		3 to 5	3'-6"

Table 1.4-102 Add cost for Mobilization (Set Up and Removal)

Job Size	Cost	
	$/Job	$/L.F. Pile
Small (12,000 L.F.)	$11,400	$.95
Large (25,000 L.F.)	19,000	$.76

Table 1.4-103 Add Costs for Special Soil Conditions

Special Conditions	Add%ofTotal
Soft, damp ground	40%
Swampy, wet ground	40%
Barge mounted drilling rig	30%

Table 1.4-104 Testing Costs, if Required, Any Pile Type

Test Weight (Tons)	$/Test
50-100T	$16,050
100-200T	20,300
150-300T	25,700
200-400T	28,800

Table 1.4-105 Cost/L.F. of Boring Types (4"), and Per Job Cost of Other Items

Type Soil	Type Boring	Sample	$/L.F. Boring	Total $
Earth	Auger	None	$23.50	
	Cased	Yes	49.00	
Rock, "BX"	Core	None	52.00	
	Cased	Yes	72.50	
Filed survey, mobilization, demobilization & engineering report $2,230				

R02450-820 Pile Caps, Piles and Caissons (cont.)

General: Caissons, as covered in this section, are drilled cylindrical foundation shafts which function primarily as short column-like compression members. They transfer superstructure loads through inadequate soils to bedrock or hard stratum. They may be either reinforced or unreinforced and either straight or belled out at the bearing level.

Shaft diameters range in size from 20″ to 84″ with the most usual sizes beginning at 34″. If inspection of bottom is required, the minimum diameter practical is 30″. If handwork is required (in addition to mechanical belling, etc.) the minimum diameter is 32″. The most frequently used shaft diameter is probably 36″ with a 5′ or 6′ bell diameter. The maximum bell diameter practical is three times the shaft diameter.

Plain concrete is commonly used, poured directly against the excavated face of soil. Permanent casings add to cost and economically should be avoided. Wet or loose strata are undesirable. The associated installation sometimes involves a mudding operation with bentonite clay slurry to keep walls of excavation stable (costs not included here).

Reinforcement is sometimes used, especially for heavy loads. It is required if uplift, bending moment, or lateral loads exist. A small amount of reinforcement is desirable at the top portion of each caisson, even if the above conditions theoretically are not present. This will provide for construction eccentricities and other possibilities. Reinforcement, if present, should extend below the soft strata. Horizontal reinforcement is not required for belled bottoms.

There are three basic types of caisson bearing details:

1. Belled, which are generally recommended to provide reduced bearing pressure on soil. These are not for shallow depths or poor soils. Good soils for belling include most clays, hardpan, soft shale, and decomposed rock.

Soils requiring handwork include hard shale, limestone, and sandstone.

Soils not recommended include sand, gravel, silt, and igneous rock. Compact sand and gravel above water table may stand. Water in the bearing strata is undesirable.

2. Straight shafted, which have no bell but the entire length is enlarged to permit safe bearing pressures. They are most economical for light loads on high bearing capacity soil.

3. Socketed (or keyed), which are used for extremely heavy loads. They involve sinking the shaft into rock for combined friction and bearing support action. Reinforcement of shaft is usually necessary. Wide flange cores are frequently used here.

Advantages include:

A. Shafts can pass through soils that piles cannot

B. No soil heaving or displacement during installation

C. No vibration during installation

D. Less noise than pile driving

E. Bearing strata can be visually inspected & tested

Uses include:

A. Situations where unsuitable soil exists to moderate depth

B. Tall structures

C. Heavy structures

D. Underpinning (extensive use)

See R02450-110 for Soil Bearing Capacities.

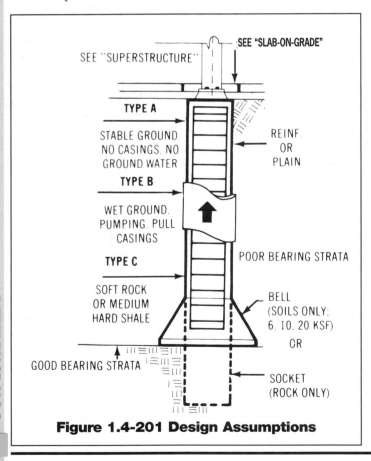

Figure 1.4-201 Design Assumptions

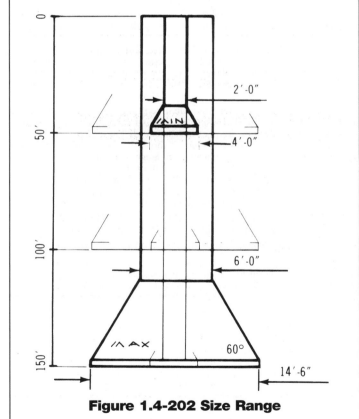

Figure 1.4-202 Size Range

R02455-900   Wood Bearing Piles

Untreated Southern Yellow Pine is most generally used for pile foundations cut off below the low water line. These are driven with the bark on. All

piles cut off above the low water line should be treated with a preservative or encased in concrete for the section above water.

Item	Unit Cost	Units	Quantity	Total Cost	Cost/Pile	Cost/L.F.
50' Treated Piles, 13" Butt, 7" Tip	$ 8.75	L.F.	200	$ 87,500.00	$437.50	$ 8.75
Installation, Crew B-19	3,690.40	Day	13	47,975.20	239.88	4.80
Mobilization & Demobilization, Crew B-19	3,690.40	Day	3	11,071.20	55.36	1.11
Transportation of Equip. One Way	744.20	Day	1	744.20	3.72	.07
Bare Totals				$147,290.60	$736.46	$14.73

The above figures are based on driving 800 L.F. daily which can be considered average. Time is included for moving rig, cutoff & ordinary delays. A general observation is that the cost of a pile in place complete is about

two times the cost of pile only. See also equipment rental division 01590 and R01590-100 for equipment capacities.

R02465-800   Pressure Injected Footings

Pressure Injected Footings are end bearing foundation units consisting of expanded bulbous bases formed by high energy blows, and concrete shafts either cased or uncased, to transfer the load.

The bulb must be formed in granular, rock or hardpan soils. Bearing is achieved at a predetermined depth so that the length of the shaft is usually

less than for friction piles. High load capacity reduces the number of units and caps required.

Mobilization and demobilization costs below assume maximum distance of 50 miles.

100 Uncased Pressure Injected Footings 25' Long	Quantity Unit	Unit Cost	Total	Cost/Pile	Cost/L.F.
Crew B-44, 8 piles/day	13 Days	$3,080.45	$40,045.85	$400.46	$16.02
Mobilization & Demobilization	2 Days	3,080.45	6,160.90	61.61	2.46
Transportation one way	1.5 Days	744.20	1,116.30	11.16	.45
Total Installation			$47,323.05	$473.23	$18.93
Shafts: 100 ea. 17" diam., 25' deep	155 C.Y.	$ 117.97	$18,285.35	$182.85	$ 7.31
Pressure Bulbs: 100 ea. @ .75 C.Y.	75 C.Y.	117.97	8,847.75	88.48	3.54
Waste: 15% of above	35 C.Y.	117.97	4,128.95	41.29	1.65
Total Bare Material Cost			$31,262.05	$312.62	$12.50
Total Bare Cost in Place			$78,585.10	$785.85	$31.43

R02510-800   Piping Designations

There are several systems currently in use to describe pipe and fittings. The following paragraphs will help to identify and clarify classifications of piping systems used for water distribution.

Piping may be classified by schedule. Piping schedules include 5S, 10S, 10, 20, 30, Standard, 40, 60, Extra Strong, 80, 100, 120, 140, 160 and Double Extra Strong. These schedules are dependent upon the pipe wall thickness. The wall thickness of a particular schedule may vary with pipe size.

Ductile iron pipe for water distribution is classified by Pressure Classes such as Class 150, 200, 250, 300 and 350. These classes are actually the rated water working pressure of the pipe in pounds per square inch (psi). The pipe in these pressure classes is designed to withstand the rated water working pressure plus a surge allowance of 100 psi.

The American Water Works Association (AWWA) provides standards for various types of **plastic pipe.** C-900 is the specification for polyvinyl chloride (PVC) piping used for water distribution in sizes ranging from 4" through 12". C-901 is the specification for polyethylene (PE) pressure pipe, tubing and fittings used for water distribution in sizes ranging from 1/2" through 3". C-905 is the specification for PVC piping sizes 14" and greater.

PVC pressure-rated pipe is identified using the standard dimensional ratio (SDR) method. This method is defined by the American Society for Testing and Materials (ASTM) Standard D 2241. This pipe is available in SDR numbers 64, 41, 32.5, 26, 21, 17, and 13.5. Pipe with an SDR of 64 will have the thinnest wall while pipe with an SDR of 13.5 will have the thickest wall. When the pressure rating (PR) of a pipe is given in psi, it is based on a line supplying water at 73 degrees F.

The National Sanitation Foundation (NSF) seal of approval is applied to products that can be used with potable water. These products have been tested to ANSI/NSF Standard 14.

Valves and strainers are classified by American National Standards Institute (ANSI) Classes. These Classes are 125, 150, 200, 250, 300, 400, 600, 900, 1500 and 2500. Within each class there is an operating pressure range dependent upon temperature. Design parameters should be compared to the appropriate material dependent, pressure-temperature rating chart for accurate valve selection.

R02510-810 Concrete Pipe

Prices given are for inside 20 mile delivery zone. Add $2.15 per ton of pipe for each additional 10 miles. Minimum truckload is 10 tons. The non-reinforced pipe listed in the front of the book is designation ASTM C14-59 extra strength. The reinforced pipe listed is ASTM C76-65T class 3, no gaskets. The installation cost given includes shaping bottom of the trench, placing the pipe, and backfilling and tamping to the top of the pipe only.

R02580-300 Concrete for Conduit Encasement

Table below lists C.Y. of concrete for 100 L.F. of trench. Conduits separation center to center should meet 7.5″ (N.E.C.).

Number of Conduits	1	2	3	4	6	8	9	Number of Conduits
Trench Dimension	11.5″ x 11.5″	11.5″ x 19″	11.5″ x 27″	19″ x 19″	19″ x 27″	19″ x 38″	27″ x 27″	Trench Dimension
Conduit Diameter 2.0″	3.29	5.39	7.64	8.83	12.51	17.66	17.72	Conduit Diameter 2.0″
2.5″	3.23	5.29	7.49	8.62	12.19	17.23	17.25	2.5″
3.0″	3.15	5.13	7.24	8.29	11.71	16.59	16.52	3.0″
3.5″	3.08	4.97	7.02	7.99	11.26	15.98	15.84	3.5″
4.0″	2.99	4.80	6.76	7.65	10.74	15.30	15.07	4.0″
5.0″	2.78	4.37	6.11	6.78	9.44	13.57	13.12	5.0″
6.0″	2.52	3.84	5.33	5.74	7.87	11.48	10.77	6.0″

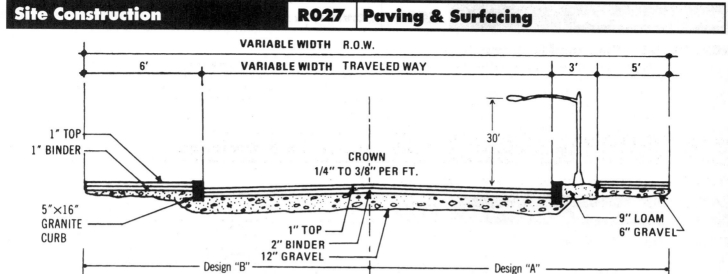

R02700-120 Options to be Added for the Roadway Total Cost, Design "A"

Description	Quantity		w/O&P Cost per L.F.
2" Bituminous concrete mix for sidewalk, incl. fine grade & roll	1.167	S.Y.	$ 8.23
5" x 16" Granite vertical curbing	2.000	L.F.	32.60
9" Loam for tree lawn	.069	C.Y.	1.93
1" Sod, in the East	.003	M.S.F.	1.14
Paint 4" centerline stripe	1.000	L.F.	.29
Lighting 30' aluminum, 140' on center, 400 watt metal halide	.007	Ea.	23.31

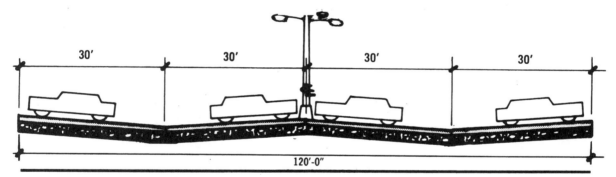

R02700-130 Options to be Added for the Parking Lot Total Cost

Description	Unit	Installed Cost w/O&P
8" x 6" Asphalt berm curbing	L.F.	$ 2.27
5" x 16" Vertical granite curbing	L.F.	16.30
4 1/2" x 12" Sloped granite edging	L.F.	16.55
6" x 18" Precast concrete curb	L.F.	12.95
24" Wide 6" high concrete curb and gutter	L.F.	21.50
4'-0" Wide bituminous concrete walk, 2" thick, 6" gravel base	L.F.	7.72
4'-0" Wide concrete walk, 4" thick, 6" gravel base	L.F.	23.50
2 Coat sealcoating petroleum resistant, under 1000 S.Y.	S.Y.	1.48
1/4 to 3/8 C.Y. rap slope protection, 18" thick, non grouted	S.Y.	77.00
40'-0" High aluminum pole with (3) - 1,000 watt metal halide roadway type fixtures	Ea.	6,060.00
6'-0" Long concrete car bumpers, 6" x 10"	Ea.	90.00
6'-0" High chain link fence, 1 5/8" top rail, 2" line posts, 10'-0" O.C., 6 ga.	L.F.	24.50
3'-0" High corrugated steel guardrail, post @ 6'-3" O.C.	L.F.	19.10
18" Deep perimeter planter bed with 5' Arborvitae shrub @ 3' O.C.	S.Y.	234.05
6" Topsoil including fine grading and seeding	S.Y.	10.56

R02700-710   Pavement Maintenance

Routine pavement maintenance should be performed to keep a paved surface from deteriorating under the normal forces of nature and traffic.

The msot important maintenance function is the early detection and repair of minor pavement defects. Cracks and other surface breaks can develop into serious defects if not repaired in their earliest stages. For these reasons a pavement preventive maintenance program should include frequent close inspections of pavement surfaces. When suspicious areas are detected, a detailed investigation should be undertaken to determine the appropriate repair. Where subsurface or pavement deterioration is detected, the Benkelman Beam can be used to make deflection measurements under normal traffic stresses. This is done to determine the extent of the affected area.

Patching or resurfacing work should be done during warm (10°C and above) and dry weather. Adequate compaction is dificult to achieve when hot or warm mixtures are placed on cold pavements. Asphalt and asphalt mixtures usually do not bond well to damp surfaces.

R02920-500   Seeding

The type of grass is determined by light, shade and moisture content of soil plus intended use. Fertilizer should be disked 4″ before seeding. For steep slopes disk five tons of mulch and lay two tons of hay or straw on surface per acre after seeding. Surface mulch can be staked, lightly disked or tar emulsion sprayed. Material for mulch can be wood chips, peat moss, partially rotted hay or straw, wood fibers and sprayed emulsions. Hemp seed blankets with fertilizer are also available. For spring seeding, watering is necessary. Late fall seeding may have to be reseeded in the spring. Hydraulic seeding, power mulching, and aerial seeding can be used on large areas.

R02920-520 Trees and Plants by Environment and Purposes

Dry, Windy, Exposed Areas
Barberry
Junipers, all varieties
Locust
Maple
Oak
Pines, all varieties
Poplar, Hybrid
Privet
Spruce, all varieties
Sumac, Staghorn

Lightly Wooded Areas
Dogwood
Hemlock
Larch
Pine, White
Rhododendron
Spruce, Norway
Redbud

Total Shade Areas
Hemlock
Ivy, English
Myrtle
Pachysandra
Privet
Spice Bush
Yews, Japanese

Cold Temperatures of Northern U.S. and Canada
Arborvitae, American
Birch, White
Dogwood, Silky
Fir, Balsam
Fir, Douglas
Hemlock
Juniper, Andorra
Juniper, Blue Rug
Linden, Little Leaf
Maple, Sugar
Mountain Ash
Myrtle
Olive, Russian

Pine, Mugho
Pine, Ponderosa
Pine, Red
Pine, Scotch
Poplar, Hybrid
Privet
Rosa Rugosa
Spruce, Dwarf Alberta
Spruce, Black Hills
Spruce, Blue
Spruce, Norway
Spruce, White, Engelman
Yellow Wood

Wet, Swampy Areas
American Arborvitae
Birch, White
Black Gum
Hemlock
Maple, Red
Pine, White
Willow

Poor, Dry, Rocky Soil
Barberry
Crownvetch
Eastern Red Cedar
Juniper, Virginiana
Locust, Black
Locust, Bristly
Locust, Honey
Olive, Russian
Pines, all varieties
Privet
Rosa Rugosa
Sumac, Staghorn

Seashore Planting
Arborvitae, American
Juniper, Tamarix
Locust, Black
Oak, White
Olive, Russian
Pine, Austrian
Pine, Japanese Black

Pine, Mugho
Pine, Scotch
Privet, Amur River
Rosa Rugosa
Yew, Japanese

City Planting
Barberry
Fir, Concolor
Forsythia
Hemlock
Holly, Japanese
Ivy, English
Juniper, Andorra
Linden, Little Leaf
Locust, Honey
Maple, Norway, Silver
Oak, Pin, Red
Olive, Russian
Pachysandra
Pine, Austrian
Pine, White
Privet
Rosa Rugosa
Sumac, Staghorn
Yew, Japanese

Bonsal Planting
Azaleas
Birch, White
Ginkgo
Junipers
Pine, Bristlecone
Pine, Mugho
Spruce,k Engleman
Spruce, Dwarf Alberta

Street Planting
Linden, Little Leaf
Oak, Pin
Ginkgo

Fast Growth
Birch, White
Crownvetch
Dogwood, Silky

Fir, Douglas
Juniper, Blue Pfitzer
Juniper, Blue Rug
Maple, Silver
Olive, Autumn
Pines, Austrian, Ponderosa, Red
   Scotch and White
Poplar, Hybrid
Privet
Spruce, Norway
Spruce, Serbian
Texus, Cuspidata, Hicksi
Willow

Dense, Impenetrable Hedges
Field Plantings:
   Locust, Bristly,
   Olive, Autumn
   Sumac
Residential Area:
   Barberry, Red or Green
   Juniper, Blue Pfitzer
   Rosa Rugosa

Food for Birds
Ash, Mountain
Barberry
Bittersweet
Cherry, Manchu
Dogwood, Silky
Honesuckle, Rem Red
Hawthorn
Oaks
Olive, Autumn, Russian
Privet
Rosa Rugosa
Sumac

Erosion Control
Crownvetch
Locust, Bristly
Willow

R02920-530 Zones of Plant Hardiness

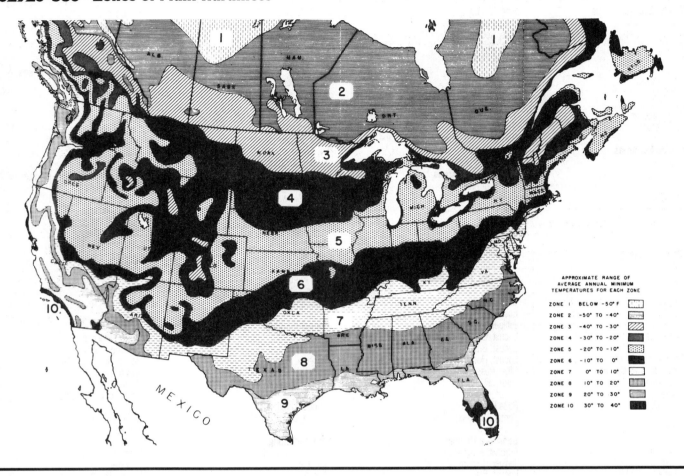

APPROXIMATE RANGE OF
AVERAGE ANNUAL MINIMUM
TEMPERATURES FOR EACH ZONE

ZONE 1   BELOW −50° F

ZONE 2   −50° TO −40°

ZONE 3   −40° TO −30°

ZONE 4   −30° TO −20°

ZONE 5   −20° TO −10°

ZONE 6   −10° TO 0°

ZONE 7   0° TO 10°

ZONE 8   10° TO 20°

ZONE 9   20° TO 30°

ZONE 10 30° TO 40°

2 SITE CONSTRUCTION

REFERENCE NOS.

R03210-020   Metric Rebar Specification - ASTM A615-81

Grade 300 (300 MPa* = 43,560 psi; +8.7% vs. Grade 40)				
Grade 400 (400 MPa* = 58,000 psi; −3.4% vs. Grade 60)				
Bar No.	Diameter mm	Area mm²	Equivalent in.²	Comparison with U.S. Customary Bars
10M	11.3	100	.16	Between #3 & #4
15M	16.0	200	.31	#5 (.31 in.²)
20M	19.5	300	.47	#6 (.44 in.²)
25M	25.2	500	.78	#8 (.79 in.²)
30M	29.9	700	1.09	#9 (1.00 in.²)
35M	35.7	1000	1.55	#11 (1.56 in.²)
45M	43.7	1500	2.33	#14 (2.25 in.²)
55M	56.4	2500	3.88	#18 (4.00 in.²)

* MPa = megapascals

R03210-025   Comparison of U.S. Customary Units and SI Units for Reinforcing Bars

U.S. Customary Units

Bar Designation No.[b]	Nominal Weight, lb/ft	Nominal Dimensions[a]			Deformation Requirements, in.		
		Diameter in.	Cross Sectional Area, in.²	Perimeter in.	Maximum Average Spacing	Minimum Average Height	Maximum Gap (Chord of 12-1/2% of Nominal Perimeter)
3	0.376	0.375	0.11	1.178	0.262	0.015	0.143
4	0.668	0.500	0.20	1.571	0.350	0.020	0.191
5	1.043	0.625	0.31	1.963	0.437	0.028	0.239
6	1.502	0.750	0.44	2.356	0.525	0.038	0.286
7	2.044	0.875	0.60	2.749	0.612	0.044	0.334
8	2.670	1.000	0.79	3.142	0.700	0.050	0.383
9	3.400	1.128	1.00	3.544	0.790	0.056	0.431
10	4.303	1.270	1.27	3.990	0.889	0.064	0.487
11	5.313	1.410	1.56	4.430	0.987	0.071	0.540
14	7.65	1.693	2.25	5.32	1.185	0.085	0.648
18	13.60	2.257	4.00	7.09	1.58	0.102	0.864

SI UNITS

Bar Designation No.[b]	Nominal Weight kg/m	Nominal Dimensions[a]			Deformation Requirements, mm		
		Diameter, mm	Cross Sectional Area, cm²	Perimeter, mm	Maximum Average Spacing	Minimum Average Height	Maximum Gap (Chord of 12-1/2% of Nominal Perimeter)
3	0.560	9.52	0.71	29.9	6.7	0.38	3.5
4	0.994	12.70	1.29	39.9	8.9	0.51	4.9
5	1.552	15.88	2.00	49.9	11.1	0.71	6.1
6	2.235	19.05	2.84	59.8	13.3	0.96	7.3
7	3.042	22.22	3.87	69.8	15.5	1.11	8.5
8	3.973	25.40	5.10	79.8	17.8	1.27	9.7
9	5.059	28.65	6.45	90.0	20.1	1.42	10.9
10	6.403	32.26	8.19	101.4	22.6	1.62	11.4
11	7.906	35.81	10.06	112.5	25.1	1.80	13.6
14	11.384	43.00	14.52	135.1	30.1	2.16	16.5
18	20.238	57.33	25.81	180.1	40.1	2.59	21.9

[a]Nominal dimensions of a deformed bar are equivalent to those of a plain round bar having the same weight per foot as the deformed bar.

[b]Bar numbers are based on the number of eighths of an inch included in the nominal diameter of the bars.

CONCRETE

REFERENCE NOS.

R03210-080 Reinforcing Key Prices

Costs given below are for over 50 tons delivered and bent from fabricators in metropolitan area. Over 500 tons, subtract $16 per ton. Figures are for domestic A615, grade 60. Economic conditions can alter the prices substantially.

City	Bare Costs Per Ton					Costs Incl. Subs O & P		
	Material		Installation			Material	Installation	Total
	50 Tons Fabricated	Allow for Accessories	Unload, Sort and Pile	Place & Tie	in Place	Incl. 10% Mark-Up	Include Mark-up	per Ton
Atlanta	$830	$125	$21	$ 588	$1,563	$1,050	$1,001	$2,051
Baltimore	905	136	25	707	1,773	1,145	1,204	2,348
Boston	905	136	33	933	2,006	1,145	1,588	2,733
Buffalo	805	121	28	807	1,761	1,018	1,374	2,393
Chicago	775	116	34	977	1,903	980	1,663	2,644
Cincinnati	830	125	25	713	1,693	1,050	1,214	2,264
Cleveland	750	113	29	831	1,723	949	1,415	2,364
Columbus	705	106	25	711	1,547	892	1,211	2,103
Dallas	855	128	16	466	1,466	1,082	794	1,875
Denver	805	121	20	559	1,505	1,018	953	1,971
Detroit	855	128	31	885	1,899	1,082	1,506	2,588
Houston	750	113	18	520	1,401	949	885	1,834
Indianapolis	750	113	25	717	1,605	949	1,221	2,170
Kansas City	980	147	28	798	1,953	1,240	1,359	2,599
Los Angeles	805	121	31	883	1,840	1,018	1,504	2,522
Memphis	705	106	20	555	1,386	892	945	1,837
Milwaukee	725	109	28	803	1,665	917	1,367	2,285
Minneapolis	815	122	32	911	1,880	1,031	1,551	2,582
Nashville	705	106	19	530	1,360	892	903	1,795
New Orleans	805	121	16	450	1,392	1,018	767	1,785
New York	805	121	51	1,445	2,421	1,018	2,460	3,478
Philadelphia	805	121	37	1,040	2,002	1,018	1,771	2,789
Phoenix	775	116	22	635	1,549	980	1,081	2,062
Pittsburgh	775	116	29	813	1,733	980	1,384	2,364
St. Louis	665	100	29	813	1,606	841	1,384	2,226
San Antonio	805	121	15	430	1,371	1,018	732	1,751
San Diego	725	109	31	883	1,748	917	1,504	2,421
San Francisco	855	128	31	883	1,897	1,082	1,504	2,585
Seattle	775	116	29	820	1,741	980	1,397	2,377
Washington, D.C.	955	143	23	660	1,781	1,208	1,123	2,331
Average per ton	$800	$120	$27	$ 759	$1,706	$1,012	$1,292	$2,304

Note: For field fabrication, add $85 to $110 per ton to shop fabrication costs. Place and tie costs do not include crane charge.

Bare Material Costs		Total Installed Costs		
Item	Cost per Ton	Description	Bare Costs	Cost Incl. Subs O&P
Base price at mill	$588	Unload, sort & pile, C-5 crew @ .56 Hours	$ 27	$ 40
Size and length extras	59	Place & tie, 4 Rodmen @ 20 Hours	759	1,252
Freight from mill	27			
Warehouse handling & storage	22	Total Installation Cost	$ 786	$1,292
Shearing and shop bending	44			
Drafting, detailing and listing	35	Bare Material Cost	$ 920	
Trucking to job site	25	Material Cost + 10%		$1,012
		Total Installed Cost	$1,706	$2,304
Reinforcing Delivered	$800			
Accessories Delivered	120			
Total Material Delivered	$920			

Engineering or design fees are not included above.

Material and erection costs can be considerably higher than those above for small jobs consisting primarily of smaller bars.

R03310-030 Metric Equivalents of Cement Content for Concrete Mixes

94 Pound Bags per Cubic Yard	Kilograms per Cubic Meter	94 Pound Bags per Cubic Yard	Kilograms per Cubic Meter
1.0	55.77	7.0	390.4
1.5	83.65	7.5	418.3
2.0	111.5	8.0	446.2
2.5	139.4	8.5	474.0
3.0	167.3	9.0	501.9
3.5	195.2	9.5	529.8
4.0	223.1	10.0	557.7
4.5	251.0	10.5	585.6
5.0	278.8	11.0	613.5
5.5	306.7	11.5	641.3
6.0	334.6	12.0	669.2
6.5	362.5	12.5	697.1

a. If you know the cement content in pounds per cubic yard,
   multiply by .5933 to obtain kilograms per cubic meter.

b. If you know the cement content in 94 pound bags per cubic yard,
   multiply by 55.77 to obtain kilograms per cubic meter.

R03310-040 Metric Equivalents of Common Concrete Strengths
(to convert other psi values to megapascals, multiply by 0.006895)

U.S. Values psi	SI Value Megapascals	Non-SI Metric Value kgf/cm²*
2000	14	140
2500	17	175
3000	21	210
3500	24	245
4000	28	280
4500	31	315
5000	34	350
6000	41	420
7000	48	490
8000	55	560
9000	62	630
10,000	69	705

* kilograms force per square centimeter

CONCRETE 3

REFERENCE NOS.

R03310-060 Concrete Material Net Prices

Costs below are C.Y. of concrete delivered; per ton of bulk cement; per bag cement delivered T.L.L.; per ton for stone and sand aggregates loaded at plant (no trucking included) and per 4 C.F. bag for perlite or vermiculite aggregate delivered T.L.L.

City	Ready Mix Concrete Regular Weight per C.Y. 3000 psi	Ready Mix Concrete Regular Weight per C.Y. 5000 psi	Cement T.L. Lots Bulk per Ton	Cement T.L. Lots Bags per Bag	Aggregates per Ton Crushed Stone 1-1/2"	Aggregates per Ton Crushed Stone 3/4"	Aggregates per Ton Sand	Vermiculite or Perlite 4 C.F. Bag
Atlanta	$ 81.50	$ 90.00	$ 88.95	$ 7.60	$18.20	$18.75	$17.75	$6.15
Baltimore	80.25	88.00	81.95	7.00	13.10	13.50	15.75	5.65
Boston	86.00	96.00	105.35	9.00	14.55	15.00	13.25	7.30
Buffalo	96.25	104.00	87.20	7.45	13.90	14.35	16.75	6.05
Chicago	91.75	100.00	83.70	7.15	18.65	19.25	16.25	5.80
Cincinnati	67.75	77.00	101.25	8.65	9.70	10.00	9.75	7.00
Cleveland	78.60	88.00	97.75	8.35	16.50	17.00	16.25	6.75
Columbus	73.25	83.00	107.70	9.20	13.10	13.50	11.25	7.45
Dallas	78.00	88.00	108.25	9.25	23.30	24.00	23.25	7.50
Denver	74.50	85.00	114.70	9.80	14.05	14.50	15.25	7.95
Detroit	82.50	91.00	88.35	7.55	12.75	13.15	11.75	6.10
Houston	71.00	79.00	88.95	7.60	20.60	21.25	19.75	6.15
Indianapolis	80.25	90.00	107.10	9.15	11.95	12.30	11.25	7.40
Kansas City	76.75	85.00	93.05	7.95	11.95	12.30	11.75	6.45
Los Angeles	71.00	81.00	110.00	9.40	12.95	13.35	15.25	7.60
Memphis	72.25	81.00	96.55	8.25	17.45	18.00	17.25	6.70
Milwaukee	83.75	94.00	105.35	9.00	13.10	13.50	16.25	7.30
Minneapolis	88.25	98.00	107.10	9.15	14.55	15.00	12.50	7.40
Nashville	75.50	84.00	89.55	7.65	12.35	12.75	14.75	6.20
New Orleans	73.25	83.00	98.90	8.45	19.40	20.00	16.50	6.85
New York City	98.50	107.00	90.10	7.70	23.30	24.00	24.25	6.25
Philadelphia	78.00	86.00	88.95	7.60	16.50	17.00	16.75	6.15
Phoenix	80.25	89.00	92.45	7.90	12.15	12.50	13.75	6.40
Pittsburgh	80.25	89.00	98.30	8.40	16.00	16.50	15.75	6.80
St. Louis	68.75	78.00	94.80	8.10	12.35	12.75	13.25	6.55
San Antonio	64.25	72.00	86.00	7.35	12.60	13.00	13.25	5.95
San Diego	90.50	100.00	103.00	8.80	15.75	16.25	19.25	7.15
San Francisco	111.25	122.00	117.65	10.05	21.35	22.00	23.75	8.15
Seattle	82.50	89.00	69.05	5.90	18.45	19.00	19.75	4.80
Washington, D.C.	94.00	103.00	93.65	8.00	15.05	15.50	17.75	6.50
Average	$ 81.00	$ 90.00	$ 96.50	$ 8.25	$15.50	$16.00	$16.00	$6.68

For information about Means Estimating Seminars, see yellow pages 12 and 13 in back of book

R05090-520   Welded Structural Steel

Usual weight reductions with welded design run 10% to 20% compared with bolted or riveted connections. This amounts to about the same total cost compared with bolted structures since field welding is more expensive than bolts. For normal spans of 18' to 24' figure 6 to 7 connections per ton.

Trusses — For welded trusses add 4% to weight of main members for connections. Up to 15% less steel can be expected in a welded truss compared to one that is shop bolted. Cost of erection is the same whether shop bolted or welded.

General — Typical electrodes for structural steel welding are E6010, E6011, E60T and E70T. Typical buildings vary between 2# to 8# of weld rod per

ton of steel. Buildings utilizing continuous design require about three times as much welding as conventional welded structures. In estimating field erection by welding, it is best to use the average linear feet of weld per ton to arrive at the welding cost per ton. The type, size and position of the weld will have a direct bearing on the cost per linear foot. A typical field welder will deposit 1.8# to 2# of weld rod per hour manually. Using semiautomatic methods can increase production by as much as 50% to 75%.

R05310-100   Decking Descriptions

General - All Deck Products

Steel deck is made by cold forming structural grade sheet steel into a repeating pattern of parallel ribs. The strength and stiffness of the panels are the result of the ribs and the material properties of the steel. Deck lengths can be varied to suit job conditions, but because of shipping considerations, are usually less than 40 feet. Standard deck width varies with the product used but full sheets are usually 12", 18", 24", 30", or 36". Deck is typically furnished in a standard width with the ends cut square. Any cutting for width, such as at openings or for angular fit, is done at the job site.

Deck is typically attached to the building frame with arc puddle welds, self-drilling screws, or powder or pneumatically driven pins. Sheet to sheet fastening is done with screws, button punching (crimping), or welds.

Composite Floor Deck

After installation and adequate fastening, floor deck serves several purposes. It (a) acts as a working platform, (b) stabilizes the frame, (c) serves as a concrete form for the slab, and (d) reinforces the slab to carry the design loads applied during the life of the building. Composite decks are distinguished by the presence of shear connector devices as part of the deck. These devices are designed to mechanically lock the concrete and deck together so that the concrete and the deck work together to carry subsequent floor loads. These shear connector devices can be rolled-in embossments, lugs, holes, or wires welded to the panels. The deck profile can also be used to interlock concrete and steel.

Composite deck finishes are either galvanized (zinc coated) or phosphatized/painted. Galvanized deck has a zinc coating on both the top and bottom surfaces. The phosphatized/painted deck has a bare (phosphatized) top surface that will come into contact with the concrete. This bare top surface can be expected to develop rust before the concrete is placed. The bottom side of the deck has a primer coat of paint.

Composite floor deck is normally installed so the panel ends do not overlap on the supporting beams. Shear lugs or panel profile shape often prevent a tight metal to metal fit if the panel ends overlap; the air gap caused by overlapping will prevent proper fusion with the structural steel supports when the panel end laps are shear stud welded.

Adequate end bearing of the deck must be obtained as shown on the drawings. If bearing is actually less in the field than shown on the drawings, further investigation is required.

Roof Deck

Roof deck is not designed to act compositely with other materials. Roof deck acts alone in transferring horizontal and vertical loads into the building frame. Roof deck rib openings are usually narrower than floor deck rib openings. This provides adequate support of rigid thermal insulation board.

Roof deck is typically installed to endlap approximately 2" over supports. However, it can be butted (or lapped more than 2") to solve field fit problems. Since designers frequently use the installed deck system as part of the horizontal bracing system (the deck as a diaphragm), any fastening substitution or change should be approved by the designer. Continuous perimeter support of the deck is necessary to limit edge deflection in the finished roof and may be required for diaphragm shear transfer.

Standard roof deck finishes are galvanized or primer painted. The standard factory applied paint for roof deck is a primer paint and is not intended to weather for extended periods of time. Field painting or touching up of abrasions and deterioration of the primer coat or other protective finishes is the responsibility of the contractor.

Cellular Deck

Cellular deck is made by attaching a bottom steel sheet to a roof deck or composite floor deck panel. Cellular deck can be used in the same manner as floor deck. Electrical, telephone, and data wires are easily run through the chase created between the deck panel and the bottom sheet.

When used as part of the electrical distribution system, the cellular deck must be installed so that the ribs line up and create a smooth cell transition at abutting ends. The joint that occurs at butting cell ends must be taped or otherwise sealed to prevent wet concrete from seeping into the cell. Cell interiors must be free of welding burrs, or other sharp intrusions, to prevent damage to wires.

When used as a roof deck, the bottom flat plate is usually left exposed to view. Care must be maintained during erection to keep good alignment and prevent damage.

Cellular deck is sometimes used with the flat plate on the top side to provide a flat working surface. Installation of the deck for this purpose requires special methods for attachment to the frame because the flat plate, now on the top, can prevent direct access to the deck material that is bearing on the structural steel. It may be advisable to treat the flat top surface to prevent slipping.

Cellular deck is always furnished galvanized or painted over galvanized.

Form Deck

Form deck can be any floor or roof deck product used as a concrete form. Connections to the frame are by the same methods used to anchor floor and roof deck. Welding washers are recommended when welding deck that is less than 20 gauge thickness.

Form deck is furnished galvanized, prime painted, or uncoated. Galvanized deck must be used for those roof deck systems where form deck is used to carry a lightweight insulating concrete fill.

R05650-100 Single Track R.R. Siding

The costs for a single track RR siding in section 05655-700 include the components shown in the table below.

Description of Component	CSI Line Item	Qty. per L.F. of Track	Unit
Ballast, 1-1/2" crushed stone	02060-150-0320	.667	C.Y.
6" x 8" x 8'-6" Treated timber ties, 22" O.C.	05655-700-1600	.545	Ea.
Tie plates, 2 per tie	05655-750-0300	1.091	Ea.
Track rail	05655-750-1000	2.000	L.F.
Spikes, 6", 4 per tie	05655-750-0200	2.182	Ea.
Splice bars w/ bolts, lock washers & nuts, @ 33' O.C.	05655-750-0100	.061	Pair
Crew B-14 @ 57 L.F./Day		.018	Day

R05650-200 Single Track, Steel Ties, Concrete Bed

The costs for a R.R. siding with steel ties and a concrete bed in section 05655-700 include the components shown in the table below.

Description of Component	CSI Line Item	Qty. per L.F. of Track	Unit
Concrete bed, 9' wide, 10" thick	03310-240-3950	.278	C.Y.
Ties, W6x16 x 6'-6" long, @ 30" O.C.	05120-640-0120	.400	Ea.
Tie plates, 4 per tie	05655-750-0300	1.600	Ea.
Track rail	05655-750-1000	2.000	L.F.
Tie plate bolts, 1", 8 per tie	05655-750-0020	3.200	Ea.
Splice bars w/bolts, lock washers & nuts, @ 33' O.C.	05655-750-0100	.061	Pair
Crew B-14 @ 22 L.F./Day		.045	Day

R06100-010 Thirty City Lumber Prices

Prices for boards are for #2 or better or sterling, whichever is in best supply. Dimension lumber is "Standard or Better" either Southern Yellow Pine (S.Y.P.), Spruce-Pine-Fir (S.P.F.), Hem-Fir (H.F.) or Douglas Fir (D.F.). The species of lumber used in a geographic area is listed by city. Plyform is 3/4" BB oil sealed fir or S.Y.P. whichever prevails locally, 3/4" CDX is S.Y.P. or Fir.

These are prices at the time of publication and should be checked against the current market price. Relative differences between cities will stay approximately constant.

City	Species	Contractor Purchases per M.B.F. S4S								Contractor Purchases per M.S.F.	
		Dimensions						Boards		3/4" Ext. Plyform	3/4" Thick CDX T&G
		2"x4"	2"x6"	2"x8"	2"x10"	2"x12"	4"x4"	1"x6"	1"x12"		
Atlanta	S.Y.P.	$428	$453	$523	$595	$675	$793	$970	$1,316	$1,223	$1,067
Baltimore	S.P.F.	536	566	654	743	843	991	1212	1645	1529	1334
Boston	S.P.F.	517	547	631	718	814	957	1171	1588	1476	1288
Buffalo	S.P.F.	519	549	634	721	818	961	1175	1594	1482	1293
Chicago	H.F.	519	549	634	721	818	961	1175	1594	1482	1293
Cincinnati	S.P.F.	536	566	654	743	843	991	1212	1645	1529	1334
Cleveland	S.P.F.	520	550	635	722	819	963	1178	1598	1485	1295
Columbus	S.P.F.	530	561	647	736	835	982	1201	1629	1514	1321
Dallas	S.Y.P.	431	456	527	599	680	798	977	1325	1232	1074
Denver	H.F.	517	547	631	718	814	957	1171	1588	1476	1288
Detroit	H.F.	531	562	649	738	837	983	1203	1632	1517	1323
Houston	S.Y.P.	427	452	521	593	673	790	967	1312	1219	1064
Indianapolis	S.P.F.	524	554	640	728	826	970	1187	1610	1497	1306
Kansas City	D.F.	517	547	631	718	814	957	1171	1588	1476	1288
Los Angeles	D.F.	527	557	644	732	831	976	1194	1620	1505	1313
Memphis	S.Y.P.	508	537	620	705	800	940	1150	1560	1450	1265
Milwaukee	H.F.	519	549	634	721	818	961	1175	1594	1482	1293
Minneapolis	H.F.	517	547	631	718	814	957	1171	1588	1476	1288
Nashville	S.Y.P.	508	537	620	705	800	940	1150	1560	1450	1265
New Orleans	S.Y.P.	440	465	537	611	693	814	996	1352	1256	1096
New York City	S.P.F.	532	563	650	739	839	985	1205	1635	1520	1326
Philadelphia	S.P.F.	532	563	650	739	839	985	1205	1635	1520	1326
Phoenix	D.F.	536	566	654	743	843	991	1212	1645	1529	1334
Pittsburgh	S.P.F.	530	561	647	736	835	982	1201	1629	1514	1321
St. Louis	H.F.	524	554	640	728	826	970	1187	1610	1497	1306
San Antonio	S.Y.P.	436	461	532	605	686	806	987	1338	1244	1085
San Diego	D.F.	533	563	650	739	839	986	1206	1636	1520	1326
San Francisco	D.F.	508	537	620	705	800	940	1150	1560	1450	1265
Seattle	D.F.	503	532	614	698	792	931	1139	1544	1436	1252
Washington, DC	S.P.F.	541	572	660	751	852	1001	1224	1661	1544	1347
Average		$508	$537	$620	$705	$800	$940	$1,150	$1,560	$1,450	$1,265

To convert square feet of surface to board feet, 4% waste included.

S4S Size	Multiply S.F. by	T & G Size	Multiply S.F. by	Flooring Size	Multiply S.F. by
1 x 4	1.18	1 x 4	1.27	25/32" x 2-1/4"	1.37
1 x 6	1.13	1 x 6	1.18	25/32" x 3-1/4"	1.29
1 x 8	1.11	1 x 8	1.14	15/32" x 1-1/2"	1.54
1 x 10	1.09	2 x 6	2.36	1" x 3"	1.28
				1" x 4"	1.24

R06110-030 Lumber Product Material Prices

The price of forest products fluctuates widely from location to location and from season to season depending upon economic conditions. The table below indicates National Average material prices in effect Jan. 1 of this book year. The table shows relative differences between various sizes, grades and species. These percentage differentials remain fairly constant even though lumber prices in general may change significantly during the year.

Availability of certain items depends upon geographic location and must be checked prior to firm price bidding.

	National Average Contractor Price, Quantity Purchase						Heavy Timbers, Fir	
	Dimension Lumber, S4S, #2 & Better, KD							
	Species	2"x4"	2"x6"	2"x8"	2"x10"	2"x12"		
Framing	Douglas Fir	$ 608	$ 540	$ 567	$ 527	$ 594	3" x 4" thru 3" x 12"	$872
Lumber	Spruce	543	473	527	533	578	4" x 4" thru 4" x 12"	872
per MBF	Southern Yellow Pine	540	358	459	493	709	6" x 6" thru 6" x 12"	1,060
	Hem-Fir	486	473	493	520	473	8" x 8" thru 8" x 12"	1,222
							10" x 10" and 10" x 12"	1,285

	S4S "D" Quality or Clear, KD					S4S # 2 & Better or Sterling, KD				
	Species	1"x4"	1"x6"	1"x8"	1"x10"	1"x12"	Species	1"x4"	1"x6"	1"x8"	1"x10"	1"x12"
Boards per MBF	Sugar Pine	$1,485	$1,667	$1,688	$2,329	$3,071	Sugar Pine	$648	$770	$ 790	$ 797	$1,073
*See also	Idaho Pine	979	986	1,033	1,013	1,087	Idaho Pine	628	986	1,033	1,013	1,019
Cedar Siding	Engleman Spruce	905	1,168	1,127	1,202	1,715	Engleman Spruce	702	736	682	675	992
	So. Yellow Pine	662	878	965	898	1,121	So. Yellow Pine	513	608	689	554	1,121
	Ponderosa Pine	1,033	1,175	1,046	1,249	2,025	Ponderosa Pine	689	648	655	662	945

R13128-210 Pre-engineered Steel Buildings

These buildings are manufactured by many companies and normally erected by franchised dealers throughout the U.S. The four basic types are: Rigid Frames, Truss type, Post and Beam and the Sloped Beam type. Most popular roof slope is low pitch of 1" in 12". The minimum economical area of these buildings is about 3000 S.F. of floor area. Bay sizes are usually 20' to 24' but can go as high as 30' with heavier girts and purlins. Eave heights are usually 12' to 24' with 18' to 20' most typical.

Material prices shown in Division 13128-700 are bare costs for the building shell only and do not include floors, foundations, anchor bolts, interior finishes or utilities. Costs assume at least three bays of 24' each, a 1" in 12" roof slope, and they are based on 30 psf roof load and 20 psf wind load

(wind load is a function of wind speed, building height, and terrain characteristics; this should be determined by a Registered Structural Engineer) and no unusual requirements. Costs include the structural frame, 26 ga. non-insulated colored corrugated or ribbed roofing and siding panels, fasteners, closures, trim and flashing but no allowance for insulation, doors, windows, skylights, gutters or downspouts. Very large projects would generally cost less for materials than the prices shown. For roof panel substitutions, see Section 07410-700. For wall panel substitutions, see Section 07460-800.

Conditions at the site, weather, shape and size of the building, and labor availability will affect the erection cost of the building.

R15100-050 Pipe Material Considerations

1. Malleable fittings should be used for gas service.
2. Malleable fittings are used where there are stresses/strains due to expansion and vibration.
3. Cast fittings may be broken as an aid to disassembling of heating lines frozen by long use, temperature and minerals.
4. Cast iron pipe is extensively used for underground and submerged service.
5. Type M (light wall) copper tubing is available in hard temper only and is used for nonpressure and less severe applications than K and L.
6. Type L (medium wall) copper tubing, available hard or soft for interior service.
7. Type K (heavy wall) copper tubing, available in hard or soft temper for use where conditions are severe. For underground and interior service.
8. Hard drawn tubing requires fewer hangers or supports but should not be bent. Silver brazed fittings are recommended, however soft solder is normally used.
9. Type DMV (very light wall) copper tubing designed for drainage, waste and vent plus other non-critical pressure services.

Domestic/Imported Pipe and Fittings Cost

The prices shown in this publication for steel/cast iron pipe and steel, cast iron, malleable iron fittings are based on domestic production sold at the normal trade discounts. The above listed items of foreign manufacture may be available at prices of 1/3 to 1/2 those shown. Some imported items after minor machining or finishing operations are being sold as domestic to further complicate the system.

Caution: Most pipe prices in this book also include a coupling and pipe hangers which for the larger sizes can add significantly to the per foot cost and should be taken into account when comparing "book cost" with quoted supplier's cost.

R15100-080 Valves

VALVE MATERIALS

Bronze:
Bronze is one of the oldest materials used to make valves. It is most commonly used in hot and cold water systems and other non-corrosive services. It is often used as a seating surface in larger iron body valves to ensure tight closure.

Carbon Steel:
Carbon steel is a high strength material. Therefore, valves made from this metal are used in higher pressure services, such as steam lines up to 600 psi at 850°F. Many steel valves are available with butt-weld ends for economy and are generally used in high pressure steam service as well as other higher pressure non-corrosive services.

Forged Steel:
Valves from tough carbon steel are used in service up to 2000 psi and temperatures up to 1000°F in Gate, Globe and Check valves.

Iron:
Valves are normally used in medium to large pipe lines to control non-corrosive fluid and gases, where pressures do not exceed 250 psi at 450° or 500 psi cold water, oil or gas.

Stainless Steel:
Developed steel alloys can be used in over 90% corrosive services.

Plastic PVC:
This is used in a great variety of valves generally in high corrosive service with lower temperatures and pressures.

VALVE SERVICE PRESSURES

Pressure ratings on valves provide an indication of the safe operating pressure for a valve at some elevated temperature. This temperature is dependent upon the materials used and the fabrication of the valve. When specific data is not available, a good "rule-of-thumb" to follow is the temperature of saturated steam on the primary rating indicated on the valve body. Example: The valve has the number 150S printed on the side indicating 150 psi and hence, a maximum operating temperature of 367°F (temperature of saturated steam and 150 psi).

DEFINITIONS

1. "WOG" – Water, oil, gas (cold working pressures).
2. "SWP" – Steam working pressure.
3. 100% area (full port) – means the area through the valve is equal to or greater than the area of standard pipe.
4. "Standard Opening" – means that the area through the valve is less than the area of standard pipe and therefore these valves should be used only where restriction of flow is unimportant.
5. "Round Port" – means the valve has a full roung opening through the plug and body, of the same size and area as standard pipe.
6. "Rectangular Port" – valves have rectangular shaped ports through the plug body. The area of the port is either equal to 100% of the area of standard pipe, or restricted (standard opening). In either case it is clearly marked.
7. "ANSI" – American National Standards Institute.

R15100-090 Valve Selection Considerations

INTRODUCTION: In any piping application, valve performance is critical. Valves should be selected to give the best performance at the lowest cost.

The following is a list of performance characteristics generally expected of valves.

1. Stopping flow or starting it.
2. Throttling flow (Modulation).
3. Flow direction changing.
4. Checking backflow (Permitting flow in only one direction).
5. Relieving or regulating pressure.

In order to properly select the right valve, some facts must be determined.

A. What liquid or gas will flow through the valve?
B. Does the fluid contain suspended particles?
C. Does the fluid remain in liquid form at all times?
D. Which metals does fluid corrode?

E. What are the pressure and temperature limits? (As temperature and pressure rise, so will the price of the valve.)
F. Is there constant line pressure?
G. Is the valve merely an on-off valve?
H. Will checking of backflow be required?
I. Will the valve operate frequently or infrequently?

Valves are classified by design type into such classifications as Gate, Globe, Angle, Check, Ball, Butterfly and Plug. They are also classified by end connection, stem, pressure restrictions and material such as bronze, cast iron, etc. Each valve has a specific use. A quality valve used correctly will provide a lifetime of trouble free service, but a high quality valve installed in the wrong service may require frequent attention.

MECHANICAL 15

REFERENCE NOS.

R16050-005 Typical Overhead Service Entrance

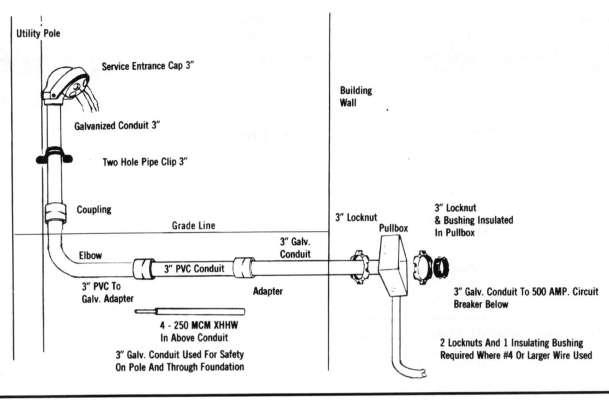

Utility Pole

Service Entrance Cap 3"

Galvanized Conduit 3"

Two Hole Pipe Clip 3"

Coupling

Grade Line

Elbow

3" PVC To
Galv. Adapter

3" PVC Conduit

Adapter

4 - 250 MCM XHHW
In Above Conduit

3" Galv. Conduit Used For Safety
On Pole And Through Foundation

Building
Wall

3" Locknut

Pullbox

3" Galv.
Conduit

3" Locknut
& Bushing Insulated
In Pullbox

3" Galv. Conduit To 500 AMP. Circuit
Breaker Below

2 Locknuts And 1 Insulating Bushing
Required Where #4 Or Larger Wire Used

R16050-031 Ampere Values Determined by Horsepower, Voltage and Phase Values

H.P.	Amperes							
	Single Phase			Three Phase				
	115V	208V	230V	200V	208V	230V	460V	575V
1/6	4.4A	2.4A	2.2A					
1/4	5.8	3.2	2.9					
1/3	7.2	4.0	3.6					
1/2	9.8	5.4	4.9	2.5A	2.4A	2.2A	1.1A	0.9A
3/4	13.8	7.6	6.9	3.7	3.5	3.2	1.6	1.3
1	16	8.8	8	4.8	4.6	4.2	2.1	1.7
1-1/2	20	11	10	6.9	6.6	6.0	3.0	2.4
2	24	13.2	12	7.8	7.5	6.8	3.4	2.7
3	34	18.7	17	11.0	10.6	9.6	4.8	3.9
5	56	30.8	28	17.5	16.7	15.2	7.6	6.1
7-1/2	80	44	40	25.3	24.2	22	11	9
10	100	55	50	32.2	30.8	28	14	11
15				48.3	46.2	42	21	17
20				62.1	59.4	54	27	22
25				78.2	74.8	68	34	27
30				92	88	80	40	32
40				120	114	104	52	41
50				150	143	130	65	52
60				177	169	154	77	62
75				221	211	192	96	77
100				285	273	248	124	99
125				359	343	312	156	125
150				414	396	360	180	144
200				552	528	480	240	192
250							302	242
300							361	289
350							414	336
400							477	382

ELECTRICAL 16

REFERENCE NOS.

R16120-920 Size Required and Weight (Lbs./1000 L.F.) of Aluminum and Copper THW Wire by Ampere Load

Amperes	Copper Size	Aluminum Size	Copper Weight	Aluminum Weight
15	14	12	24	11
20	12	10	33	17
30	10	8	48	39
45	8	6	77	52
65	6	4	112	72
85	4	2	167	101
100	3	1	205	136
115	2	1/0	252	162
130	1	2/0	324	194
150	1/0	3/0	397	233
175	2/0	4/0	491	282
200	3/0	250	608	347
230	4/0	300	753	403
255	250	400	899	512
285	300	500	1068	620
310	350	500	1233	620
335	400	600	1396	772
380	500	750	1732	951

R16136-700 Pull Boxes and Cabinets

List cabinets and pull boxes by NEMA type and size.

Example:	TYPE	SIZE
	NEMA 1	6"W x 6"H x 4"D
	NEMA 3R	6"W x 6"H x 4"D

Labor-hours for wall mount (indoor or outdoor) installations include:

1. Unloading and uncrating
2. Handling of enclosures up to 200' from loading dock using a dolly or pipe rollers
3. Measuring and marking
4. Drilling (4) anchor type lead fasteners using a hammer drill
5. Mounting and leveling boxes

Note: A plywood backboard is not included.

Labor-hours for ceiling mounting include:

1. Unloading and uncrating
2. Handling boxes up to 100' from loading dock

3. Measuring and marking
4. Drilling (4) anchor type lead fasteners using a hammer drill
5. Installing and leveling boxes to a height of 15' using rolling staging

Labor-hours for free standing cabinets include:

1. Unloading and uncrating
2. Handling of cabinets up to 200' from loading dock using a dolly or pipe rollers
3. Marking of floor
4. Drilling (4) anchor type lead fasteners using a hammer drill
5. Leveling and shimming

Labor-hours for telephone cabinets include:

1. Unloading and uncrating
2. Handling cabinets up to 200' using a dolly or pipe rollers
3. Measuring and marking
4. Mounting and leveling, using (4) lead anchor type fasteners

R16136-705 Weight Comparisons of Common Size Cast Boxes in Lbs.

Size NEMA 4 or 9	Cast Iron	Cast Aluminum	Size NEMA 7	Cast Iron	Cast Aluminum
6" x 6" x 6"	17	7	6" x 6" x 6"	40	15
8" x 6" x 6"	21	8	8" x 6" x 6"	50	19
10" x 6" x 6"	23	9	10" x 6" x 6"	55	21
12" x 12" x 6"	52	20	12" x 6" x 6"	100	37
16" x 16" x 6"	97	36	16" x 16" x 6"	140	52
20" x 20" x 6"	133	50	20" x 20" x 6"	180	67
24" x 18" x 8"	149	56	24" x 18" x 8"	250	93
24" x 24" x 10"	238	88	24" x 24" x 10"	358	133
30" x 24" x 12"	324	120	30" x 24" x 10"	475	176
36" x 36" x 12"	500	185	30" x 24" x 12"	510	189

R16230-450 Generator Weight (Lbs.) by kW

3 Phase 4 Wire 277/480 Volt			
Gas		Diesel	
kW	Lbs.	kW	Lbs.
7.5	600	30	1800
10	630	50	2230
15	960	75	2250
30	1500	100	3840
65	2350	125	4030
85	2570	150	5500
115	4310	175	5650
170	6530	200	5930
		250	6320
		300	7840
		350	8220
		400	10750
		500	11900

R16270-600 Oil Filled Transformers

Transformers in this section include:
1. Rigging (as required)
2. Rental of crane and operator
3. Setting of oil filled transformer
4. (4) Anchor bolts, nuts and washers in concrete pad

Price does not include:
1. Primary and secondary terminations
2. Transformer pad
3. Equipment grounding
4. Cable
5. Conduit locknuts or bushings

DRY TYPE TRANSFORMERS Section 16270-200

BUCK-BOOST TRANSFORMERS Section 16270-100

ISOLATING TRANSFORMERS Section 16280-340

Transformers in these sections include:
1. Unloading and uncrating
2. Hauling transformer to within 200' of loading dock
3. Setting in place
4. Wall mounting hardware
5. Testing

Price does not include:
1. Structural supports
2. Suspension systems
3. Welding or fabrication
4. Primary & secondary terminations

Add the following percentages to the labor for ceiling mounted transformers:

10' to 15'	= + 15%
15' to 25'	= + 30%
Over 25'	= + 35%

Job Conditions: Productivities are based on new construction. Installation is assumed to be on the first floor, in an obstructed area to a height of 10'. Material staging area is within 100' of final transformer location.

ELECTRICAL 16

REFERENCE NOS.

R16270-605 Transformer Weight (Lbs.) by kVA

Oil Filled 3 Phase 5/15 KV To 480/277			
kVA	Lbs.	kVA	Lbs.
150	1800	1000	6200
300	2900	1500	8400
500	4700	2000	9700
750	5300	3000	15000

Dry 240/480 To 120/240 Volt			
1 Phase		3 Phase	
kVA	Lbs.	kVA	Lbs.
1	23	3	90
2	36	6	135
3	59	9	170
5	73	15	220
7.5	131	30	310
10	149	45	400
15	205	75	600
25	255	112.5	950
37.5	295	150	1140
50	340	225	1575
75	550	300	1870
100	670	500	2850
167	900	750	4300

R16450-105 Weight (Lbs./L.F.) of 4 Pole Aluminum and Copper Bus Duct by Ampere Load

Amperes	Aluminum Feeder	Copper Feeder	Aluminum Plug-In	Copper Plug-In
225			7	7
400			8	13
600	10	10	11	14
800	10	19	13	18
1000	11	19	16	22
1350	14	24	20	30
1600	17	26	25	39
2000	19	30	29	46
2500	27	43	36	56
3000	30	48	42	73
4000	39	67		
5000		78		

Crew No.	Bare Costs		Incl. Subs O & P		Cost Per Labor-Hour	
Crew A-1	Hr.	Daily	Hr.	Daily	Bare Costs	Incl. O&P
1 Building Laborer	$26.70	$213.60	$41.55	$332.40	$26.70	$41.55
1 Concrete saw, gas manual		49.80		54.80	6.23	6.85
8 L.H., Daily Totals		$263.40		$387.20	$32.93	$48.40

Crew No.	Bare Costs		Incl. Subs O & P		Cost Per Labor-Hour	
Crew A-1A	Hr.	Daily	Hr.	Daily	Bare Costs	Incl. O&P
1 Skilled Worker	$34.85	$278.80	$54.20	$433.60	$34.85	$54.20
1 Shot Blaster, 20"		338.35		372.20	42.29	46.52
8 L.H., Daily Totals		$617.15		$805.80	$77.14	$100.72

Crew No.	Bare Costs		Incl. Subs O & P		Cost Per Labor-Hour	
Crew A-1B	Hr.	Daily	Hr.	Daily	Bare Costs	Incl. O&P
1 Building Laborer	$26.70	$213.60	$41.55	$332.40	$26.70	$41.55
1 Concr. saw, gas, self-prop.		114.40		125.85	14.30	15.73
8 L.H., Daily Totals		$328.00		$458.25	$41.00	$57.28

Crew No.	Bare Costs		Incl. Subs O & P		Cost Per Labor-Hour	
Crew A-1C	Hr.	Daily	Hr.	Daily	Bare Costs	Incl. O&P
1 Building Laborer	$26.70	$213.60	$41.55	$332.40	$26.70	$41.55
1 Brush saw		19.20		21.10	2.40	2.64
8 L.H., Daily Totals		$232.80		$353.50	$29.10	$44.19

Crew No.	Bare Costs		Incl. Subs O & P		Cost Per Labor-Hour	
Crew A-1D	Hr.	Daily	Hr.	Daily	Bare Costs	Incl. O&P
1 Building Laborer	$26.70	$213.60	$41.55	$332.40	$26.70	$41.55
1 Vibrating plate, gas, 18"		25.20		27.70	3.15	3.47
8 L.H., Daily Totals		$238.80		$360.10	$29.85	$45.02

Crew No.	Bare Costs		Incl. Subs O & P		Cost Per Labor-Hour	
Crew A-1E	Hr.	Daily	Hr.	Daily	Bare Costs	Incl. O&P
1 Building Laborer	$26.70	$213.60	$41.55	$332.40	$26.70	$41.55
1 Vibrating plate, gas, 21"		36.80		40.50	4.60	5.06
8 L.H., Daily Totals		$250.40		$372.90	$31.30	$46.61

Crew No.	Bare Costs		Incl. Subs O & P		Cost Per Labor-Hour	
Crew A-1F	Hr.	Daily	Hr.	Daily	Bare Costs	Incl. O&P
1 Building Laborer	$26.70	$213.60	$41.55	$332.40	$26.70	$41.55
1 Rammer/tamper, gas, 8"		34.80		38.30	4.35	4.79
8 L.H., Daily Totals		$248.40		$370.70	$31.05	$46.34

Crew No.	Bare Costs		Incl. Subs O & P		Cost Per Labor-Hour	
Crew A-1G	Hr.	Daily	Hr.	Daily	Bare Costs	Incl. O&P
1 Building Laborer	$26.70	$213.60	$41.55	$332.40	$26.70	$41.55
1 Rammer/tamper, gas, 8"		34.80		38.30	4.35	4.79
8 L.H., Daily Totals		$248.40		$370.70	$31.05	$46.34

Crew No.	Bare Costs		Incl. Subs O & P		Cost Per Labor-Hour	
Crew A-1H	Hr.	Daily	Hr.	Daily	Bare Costs	Incl. O&P
1 Building Laborer	$26.70	$213.60	$41.55	$332.40	$26.70	$41.55
1 Pressure washer		55.00		60.50	6.88	7.56
8 L.H., Daily Totals		$268.60		$392.90	$33.58	$49.11

Crew No.	Bare Costs		Incl. Subs O & P		Cost Per Labor-Hour	
Crew A-1J	Hr.	Daily	Hr.	Daily	Bare Costs	Incl. O&P
1 Building Laborer	$26.70	$213.60	$41.55	$332.40	$26.70	$41.55
1 Rototiller		84.50		92.95	10.56	11.61
8 L.H., Daily Totals		$298.10		$425.35	$37.26	$53.16

Crew No.	Bare Costs		Incl. Subs O & P		Cost Per Labor-Hour	
Crew A-1K	Hr.	Daily	Hr.	Daily	Bare Costs	Incl. O&P
1 Building Laborer	$26.70	$213.60	$41.55	$332.40	$26.70	$41.55
1 Lawn aerator		23.95		26.35	2.99	3.29
8 L.H., Daily Totals		$237.55		$358.75	$29.69	$44.84

Crew No.	Bare Costs		Incl. Subs O & P		Cost Per Labor-Hour	
Crew A-1L	Hr.	Daily	Hr.	Daily	Bare Costs	Incl. O&P
1 Building Laborer	$26.70	$213.60	$41.55	$332.40	$26.70	$41.55
1 Power blower/vacuum		23.95		26.35	2.99	3.29
8 L.H., Daily Totals		$237.55		$358.75	$29.69	$44.84

Crew No.	Bare Costs		Incl. Subs O & P		Cost Per Labor-Hour	
Crew A-1M	Hr.	Daily	Hr.	Daily	Bare Costs	Incl. O&P
1 Building Laborer	$26.70	$213.60	$41.55	$332.40	$26.70	$41.55
1 Snow blower		84.50		92.95	10.56	11.61
8 L.H., Daily Totals		$298.10		$425.35	$37.26	$53.16

Crew No.	Bare Costs		Incl. Subs O & P		Cost Per Labor-Hour	
Crew A-2	Hr.	Daily	Hr.	Daily	Bare Costs	Incl. O&P
2 Laborers	$26.70	$427.20	$41.55	$664.80	$26.73	$41.32
1 Truck Driver (light)	26.80	214.40	40.85	326.80		
1 Light Truck, 1.5 Ton		128.80		141.70	5.37	5.90
24 L.H., Daily Totals		$770.40		$1133.30	$32.10	$47.22

Crew No.	Bare Costs		Incl. Subs O & P		Cost Per Labor-Hour	
Crew A-2A	Hr.	Daily	Hr.	Daily	Bare Costs	Incl. O&P
2 Laborers	$26.70	$427.20	$41.55	$664.80	$26.73	$41.32
1 Truck Driver (light)	26.80	214.40	40.85	326.80		
1 Light Truck, 1.5 Ton		128.80		141.70		
1 Concrete Saw		114.40		125.85	10.13	11.15
24 L.H., Daily Totals		$884.80		$1259.15	$36.86	$52.47

Crew No.	Bare Costs		Incl. Subs O & P		Cost Per Labor-Hour	
Crew A-3	Hr.	Daily	Hr.	Daily	Bare Costs	Incl. O&P
1 Truck Driver (heavy)	$27.55	$220.40	$42.00	$336.00	$27.55	$42.00
1 Dump Truck, 12 Ton		326.60		359.25	40.83	44.91
8 L.H., Daily Totals		$547.00		$695.25	$68.38	$86.91

Crew No.	Bare Costs		Incl. Subs O & P		Cost Per Labor-Hour	
Crew A-3A	Hr.	Daily	Hr.	Daily	Bare Costs	Incl. O&P
1 Truck Driver (light)	$26.80	$214.40	$40.85	$326.80	$26.80	$40.85
1 Pickup Truck (4x4)		85.20		93.70	10.65	11.72
8 L.H., Daily Totals		$299.60		$420.50	$37.45	$52.57

Crew No.	Bare Costs		Incl. Subs O & P		Cost Per Labor-Hour	
Crew A-3B	Hr.	Daily	Hr.	Daily	Bare Costs	Incl. O&P
1 Equip. Oper. (medium)	$34.65	$277.20	$52.20	$417.60	$31.10	$47.10
1 Truck Driver (heavy)	27.55	220.40	42.00	336.00		
1 Dump Truck, 16 Ton		476.60		524.25		
1 F.E. Loader, 3 C.Y.		305.00		335.50	48.85	53.74
16 L.H., Daily Totals		$1279.20		$1613.35	$79.95	$100.84

Crew No.	Bare Costs		Incl. Subs O & P		Cost Per Labor-Hour	
Crew A-3C	Hr.	Daily	Hr.	Daily	Bare Costs	Incl. O&P
1 Equip. Oper. (light)	$33.05	$264.40	$49.80	$398.40	$33.05	$49.80
1 Wheeled Skid Steer Loader		197.80		217.60	24.73	27.20
8 L.H., Daily Totals		$462.20		$616.00	$57.78	$77.00

Crew No.	Bare Costs		Incl. Subs O & P		Cost Per Labor-Hour	
Crew A-3D	Hr.	Daily	Hr.	Daily	Bare Costs	Incl. O&P
1 Truck Driver, Light	$26.80	$214.40	$40.85	$326.80	$26.80	$40.85
1 Pickup Truck (4x4)		85.20		93.70		
1 Flatbed Trailer, 25 Ton		88.00		96.80	21.65	23.82
8 L.H., Daily Totals		$387.60		$517.30	$48.45	$64.67

Crew No.	Bare Costs		Incl. Subs O & P		Cost Per Labor-Hour	
Crew A-3E	Hr.	Daily	Hr.	Daily	Bare Costs	Incl. O&P
1 Equip. Oper. (crane)	$35.90	$287.20	$54.10	$432.80	$31.73	$48.05
1 Truck Driver (heavy)	27.55	220.40	42.00	336.00		
1 Pickup Truck (4x4)		85.20		93.70	5.33	5.86
16 L.H., Daily Totals		$592.80		$862.50	$37.06	$53.91

Crew No.	Bare Costs		Incl. Subs O & P		Cost Per Labor-Hour	
Crew A-3F	Hr.	Daily	Hr.	Daily	Bare Costs	Incl. O&P
1 Equip. Oper. (crane)	$35.90	$287.20	$54.10	$432.80	$31.73	$48.05
1 Truck Driver (heavy)	27.55	220.40	42.00	336.00		
1 Pickup Truck (4x4)		85.20		93.70		
1 Tractor, 6x2, 40 Ton Cap.		301.60		331.75		
1 Lowbed Trailer, 75 Ton		170.20		187.20	34.81	38.29
16 L.H., Daily Totals		$1064.60		$1381.45	$66.54	$86.34

Crews

Crew No.	Bare Costs Hr.	Bare Costs Daily	Incl. Subs O & P Hr.	Incl. Subs O & P Daily	Cost Per Labor-Hour Bare Costs	Cost Per Labor-Hour Incl. O&P
Crew A-3G						
1 Equip. Oper. (crane)	$35.90	$287.20	$54.10	$432.80	$31.73	$48.05
1 Truck Driver (heavy)	27.55	220.40	42.00	336.00		
1 Pickup Truck (4x4)		85.20		93.70		
1 Tractor, 6x4, 45 Ton Cap.		353.60		388.95		
1 Lowbed Trailer, 75 Ton		170.20		187.20	38.06	41.87
16 L.H., Daily Totals		$1116.60		$1438.65	$69.79	$89.92

Crew A-4	Hr.	Daily	Hr.	Daily	Bare Costs	Incl. O&P
2 Carpenters	$34.25	$548.00	$53.35	$853.60	$33.03	$50.90
1 Painter, Ordinary	30.60	244.80	46.00	368.00		
24 L.H., Daily Totals		$792.80		$1221.60	$33.03	$50.90

Crew A-5	Hr.	Daily	Hr.	Daily	Bare Costs	Incl. O&P
2 Laborers	$26.70	$427.20	$41.55	$664.80	$26.71	$41.47
.25 Truck Driver (light)	26.80	53.60	40.85	81.70		
.25 Light Truck, 1.5 Ton		32.20		35.40	1.79	1.97
18 L.H., Daily Totals		$513.00		$781.90	$28.50	$43.44

Crew A-6	Hr.	Daily	Hr.	Daily	Bare Costs	Incl. O&P
1 Instrument Man	$34.85	$278.80	$54.20	$433.60	$33.93	$52.05
1 Rodman/Chainman	33.00	264.00	49.90	399.20		
1 Laser Transit/Level		60.75		66.85	3.79	4.17
16 L.H., Daily Totals		$603.55		$899.65	$37.72	$56.22

Crew A-7	Hr.	Daily	Hr.	Daily	Bare Costs	Incl. O&P
1 Chief Of Party	$42.90	$343.20	$66.90	$535.20	$36.92	$57.00
1 Instrument Man	34.85	278.80	54.20	433.60		
1 Rodman/Chainman	33.00	264.00	49.90	399.20		
1 Laser Transit/Level		60.75		66.85	2.53	2.78
24 L.H., Daily Totals		$946.75		$1434.85	$39.45	$59.78

Crew A-8	Hr.	Daily	Hr.	Daily	Bare Costs	Incl. O&P
1 Chief Of Party	$42.90	$343.20	$66.90	$535.20	$35.94	$55.23
1 Instrument Man	34.85	278.80	54.20	433.60		
2 Rodmen/Chainmen	33.00	528.00	49.90	798.40		
1 Laser Transit/Level		60.75		66.85	1.90	2.09
32 L.H., Daily Totals		$1210.75		$1834.05	$37.84	$57.32

Crew A-9	Hr.	Daily	Hr.	Daily	Bare Costs	Incl. O&P
1 Asbestos Foreman	$37.60	$300.80	$59.35	$474.80	$37.16	$58.65
7 Asbestos Workers	37.10	2077.60	58.55	3278.80		
64 L.H., Daily Totals		$2378.40		$3753.60	$37.16	$58.65

Crew A-10	Hr.	Daily	Hr.	Daily	Bare Costs	Incl. O&P
1 Asbestos Foreman	$37.60	$300.80	$59.35	$474.80	$37.16	$58.65
7 Asbestos Workers	37.10	2077.60	58.55	3278.80		
64 L.H., Daily Totals		$2378.40		$3753.60	$37.16	$58.65

Crew A-10A	Hr.	Daily	Hr.	Daily	Bare Costs	Incl. O&P
1 Asbestos Foreman	$37.60	$300.80	$59.35	$474.80	$37.27	$58.82
2 Asbestos Workers	37.10	593.60	58.55	936.80		
24 L.H., Daily Totals		$894.40		$1411.60	$37.27	$58.82

Crew A-10B	Hr.	Daily	Hr.	Daily	Bare Costs	Incl. O&P
1 Asbestos Foreman	$37.60	$300.80	$59.35	$474.80	$37.23	$58.75
3 Asbestos Workers	37.10	890.40	58.55	1405.20		
32 L.H., Daily Totals		$1191.20		$1880.00	$37.23	$58.75

Crew A-10C	Hr.	Daily	Hr.	Daily	Bare Costs	Incl. O&P
3 Asbestos Workers	$37.10	$890.40	$58.55	$1405.20	$37.10	$58.55
1 Flatbed Truck		128.80		141.70	5.37	5.90
24 L.H., Daily Totals		$1019.20		$1546.90	$42.47	$64.45

Crew A-10D	Hr.	Daily	Hr.	Daily	Bare Costs	Incl. O&P
2 Asbestos Workers	$37.10	$593.60	$58.55	$936.80	$35.05	$54.14
1 Equip. Oper. (crane)	35.90	287.20	54.10	432.80		
1 Equip. Oper. Oiler	30.10	240.80	45.35	362.80		
1 Hydraulic Crane, 33 Ton		641.60		705.75	20.05	22.06
32 L.H., Daily Totals		$1763.20		$2438.15	$55.10	$76.20

Crew A-11	Hr.	Daily	Hr.	Daily	Bare Costs	Incl. O&P
1 Asbestos Foreman	$37.60	$300.80	$59.35	$474.80	$37.16	$58.65
7 Asbestos Workers	37.10	2077.60	58.55	3278.80		
2 Chipping Hammers		32.00		35.20	.50	.55
64 L.H., Daily Totals		$2410.40		$3788.80	$37.66	$59.20

Crew A-12	Hr.	Daily	Hr.	Daily	Bare Costs	Incl. O&P
1 Asbestos Foreman	$37.60	$300.80	$59.35	$474.80	$37.16	$58.65
7 Asbestos Workers	37.10	2077.60	58.55	3278.80		
1 Large Prod. Vac. Loader		480.00		528.00	7.50	8.25
64 L.H., Daily Totals		$2858.40		$4281.60	$44.66	$66.90

Crew A-13	Hr.	Daily	Hr.	Daily	Bare Costs	Incl. O&P
1 Equip. Oper. (light)	$33.05	$264.40	$49.80	$398.40	$33.05	$49.80
1 Large Prod. Vac. Loader		480.00		528.00	60.00	66.00
8 L.H., Daily Totals		$744.40		$926.40	$93.05	$115.80

Crew B-1	Hr.	Daily	Hr.	Daily	Bare Costs	Incl. O&P
1 Labor Foreman (outside)	$28.70	$229.60	$44.70	$357.60	$27.37	$42.60
2 Laborers	26.70	427.20	41.55	664.80		
24 L.H., Daily Totals		$656.80		$1022.40	$27.37	$42.60

Crew B-1A	Hr.	Daily	Hr.	Daily	Bare Costs	Incl. O&P
1 Laborer Foreman	$28.70	$229.60	$44.70	$357.60	$27.37	$42.60
2 Laborers	26.70	427.20	41.55	664.80		
2 Cutting Torches		36.00		39.60		
2 Gases		129.60		142.55	6.90	7.59
24 L.H., Daily Totals		$822.40		$1204.55	$34.27	$50.19

Crew B-1B	Hr.	Daily	Hr.	Daily	Bare Costs	Incl. O&P
1 Laborer Foreman	$28.70	$229.60	$44.70	$357.60	$29.50	$45.48
2 Laborers	26.70	427.20	41.55	664.80		
1 Equip. Oper. (crane)	35.90	287.20	54.10	432.80		
2 Cutting Torches		36.00		39.60		
2 Gases		129.60		142.55		
1 Hyd. Crane, 12 Ton		602.60		662.85	24.01	26.41
32 L.H., Daily Totals		$1712.20		$2300.20	$53.51	$71.89

Crew B-2	Hr.	Daily	Hr.	Daily	Bare Costs	Incl. O&P
1 Labor Foreman (outside)	$28.70	$229.60	$44.70	$357.60	$27.10	$42.18
4 Laborers	26.70	854.40	41.55	1329.60		
40 L.H., Daily Totals		$1084.00		$1687.20	$27.10	$42.18

Crew B-3

Crew B-3	Bare Costs Hr.	Daily	Incl. Subs O & P Hr.	Daily	Cost Per Labor-Hour Bare Costs	Incl. O&P
1 Labor Foreman (outside)	$28.70	$229.60	$44.70	$357.60	$28.64	$44.00
2 Laborers	26.70	427.20	41.55	664.80		
1 Equip. Oper. (med.)	34.65	277.20	52.20	417.60		
2 Truck Drivers (heavy)	27.55	440.80	42.00	672.00		
1 F.E. Loader, T.M., 2.5 C.Y.		788.00		866.80		
2 Dump Trucks, 16 Ton		953.20		1048.50	36.28	39.90
48 L.H., Daily Totals		$3116.00		$4027.30	$64.92	$83.90

Crew B-3A	Hr.	Daily	Hr.	Daily	Bare Costs	Incl. O&P
4 Laborers	$26.70	$854.40	$41.55	$1329.60	$28.29	$43.68
1 Equip. Oper. (med.)	34.65	277.20	52.20	417.60		
1 Hyd. Excavator, 1.5 C.Y.		720.40		792.45	18.01	19.81
40 L.H., Daily Totals		$1852.00		$2539.65	$46.30	$63.49

Crew B-3B	Hr.	Daily	Hr.	Daily	Bare Costs	Incl. O&P
2 Laborers	$26.70	$427.20	$41.55	$664.80	$28.90	$44.33
1 Equip. Oper. (med.)	34.65	277.20	52.20	417.60		
1 Truck Driver (heavy)	27.55	220.40	42.00	336.00		
1 Backhoe Loader, 80 H.P.		256.40		282.05		
1 Dump Truck, 16 Ton		476.60		524.25	22.91	25.20
32 L.H., Daily Totals		$1657.80		$2224.70	$51.81	$69.53

Crew B-3C	Hr.	Daily	Hr.	Daily	Bare Costs	Incl. O&P
3 Laborers	$26.70	$640.80	$41.55	$997.20	$28.69	$44.21
1 Equip. Oper. (med.)	34.65	277.20	52.20	417.60		
1 F.E. Crawler Ldr, 4 C.Y.		1094.00		1203.40	34.19	37.61
32 L.H., Daily Totals		$2012.00		$2618.20	$62.88	$81.82

Crew B-4	Hr.	Daily	Hr.	Daily	Bare Costs	Incl. O&P
1 Labor Foreman (outside)	$28.70	$229.60	$44.70	$357.60	$27.18	$42.15
4 Laborers	26.70	854.40	41.55	1329.60		
1 Truck Driver (heavy)	27.55	220.40	42.00	336.00		
1 Tractor, 4 x 2, 195 H.P.		208.20		229.00		
1 Platform Trailer		119.40		131.35	6.83	7.51
48 L.H., Daily Totals		$1632.00		$2383.55	$34.01	$49.66

Crew B-5	Hr.	Daily	Hr.	Daily	Bare Costs	Incl. O&P
1 Labor Foreman (outside)	$28.70	$229.60	$44.70	$357.60	$29.26	$45.04
4 Laborers	26.70	854.40	41.55	1329.60		
2 Equip. Oper. (med.)	34.65	554.40	52.20	835.20		
1 Air Compr., 250 C.F.M.		112.40		123.65		
2 Air Tools & Accessories		19.60		21.55		
2-50 Ft. Air Hoses, 1.5" Dia.		9.40		10.35		
1 F.E. Loader, T.M., 2.5 C.Y.		788.00		866.80	16.60	18.26
56 L.H., Daily Totals		$2567.80		$3544.75	$45.86	$63.30

Crew B-5A	Hr.	Daily	Hr.	Daily	Bare Costs	Incl. O&P
1 Foreman	$28.70	$229.60	$44.70	$357.60	$28.86	$44.35
6 Laborers	26.70	1281.60	41.55	1994.40		
2 Equip. Oper. (med.)	34.65	554.40	52.20	835.20		
1 Equip. Oper. (light)	33.05	264.40	49.80	398.40		
2 Truck Drivers (heavy)	27.55	440.80	42.00	672.00		
1 Air Compr. 365 C.F.M.		146.80		161.50		
2 Pavement Breakers		19.60		21.55		
8 Air Hoses w/Coup.,1"		29.20		32.10		
2 Dump Trucks, 12 Ton		653.20		718.50	8.84	9.73
96 L.H., Daily Totals		$3619.60		$5191.25	$37.70	$54.08

Crew B-5B	Hr.	Daily	Hr.	Daily	Bare Costs	Incl. O&P
1 Powderman	$34.85	$278.80	$54.20	$433.60	$31.13	$47.43
2 Equip. Oper. (med.)	34.65	554.40	52.20	835.20		
3 Truck Drivers (heavy)	27.55	661.20	42.00	1008.00		
1 F.E. Ldr. 2-1/2 CY		305.00		335.50		
3 Dump Trucks, 16 Ton		1429.80		1572.80		
1 Air Compr. 365 C.F.M.		146.80		161.50	39.20	43.12
48 L.H., Daily Totals		$3376.00		$4346.60	$70.33	$90.55

Crew B-5C	Hr.	Daily	Hr.	Daily	Bare Costs	Incl. O&P
3 Laborers	$26.70	$640.80	$41.55	$997.20	$29.48	$45.04
1 Equip. Oper. (medium)	34.65	277.20	52.20	417.60		
2 Truck Drivers (heavy)	27.55	440.80	42.00	672.00		
1 Equip. Oper. (crane)	35.90	287.20	54.10	432.80		
1 Equip. Oper. Oiler	30.10	240.80	45.35	362.80		
2 Dump Trucks, 16 Ton		953.20		1048.50		
1 F.E. Crawler Ldr, 4 C.Y.		1094.00		1203.40		
1 Hyd. Crane, 25 Ton		575.00		632.50	40.97	45.07
64 L.H., Daily Totals		$4509.00		$5766.80	$70.45	$90.11

Crew B-6	Hr.	Daily	Hr.	Daily	Bare Costs	Incl. O&P
2 Laborers	$26.70	$427.20	$41.55	$664.80	$28.82	$44.30
1 Equip. Oper. (light)	33.05	264.40	49.80	398.40		
1 Backhoe Loader, 48 H.P.		215.80		237.40	8.99	9.89
24 L.H., Daily Totals		$907.40		$1300.60	$37.81	$54.19

Crew B-6A	Hr.	Daily	Hr.	Daily	Bare Costs	Incl. O&P
.5 Labor Foreman (outside)	$28.70	$114.80	$44.70	$178.80	$30.28	$46.44
1 Laborer	26.70	213.60	41.55	332.40		
1 Equip. Oper. (med.)	34.65	277.20	52.20	417.60		
1 Vacuum Trk.,5000 Gal.		331.50		364.65	16.58	18.23
20 L.H., Daily Totals		$937.10		$1293.45	$46.86	$64.67

Crew B-6B	Hr.	Daily	Hr.	Daily	Bare Costs	Incl. O&P
2 Labor Foreman (out)	$28.70	$459.20	$44.70	$715.20	$27.37	$42.60
4 Laborers	26.70	854.40	41.55	1329.60		
1 Winch Truck		317.00		348.70		
1 Flatbed Truck		128.80		141.70		
1 Butt Fusion Machine		432.80		476.10	18.30	20.13
48 L.H., Daily Totals		$2192.20		$3011.30	$45.67	$62.73

Crew B-7	Hr.	Daily	Hr.	Daily	Bare Costs	Incl. O&P
1 Labor Foreman (outside)	$28.70	$229.60	$44.70	$357.60	$28.36	$43.85
4 Laborers	26.70	854.40	41.55	1329.60		
1 Equip. Oper. (med.)	34.65	277.20	52.20	417.60		
1 Chipping Machine		164.80		181.30		
1 F.E. Loader, T.M., 2.5 C.Y.		788.00		866.80		
2 Chain Saws, 36"		66.80		73.50	21.24	23.37
48 L.H., Daily Totals		$2380.80		$3226.40	$49.60	$67.22

Crew B-7A	Hr.	Daily	Hr.	Daily	Bare Costs	Incl. O&P
2 Laborers	$26.70	$427.20	$41.55	$664.80	$28.82	$44.30
1 Equip. Oper. (light)	33.05	264.40	49.80	398.40		
1 Rake w/Tractor		191.50		210.65		
2 Chain Saws, 18"		38.40		42.25	9.58	10.54
24 L.H., Daily Totals		$921.50		$1316.10	$38.40	$54.84

Crew No.	Bare Costs Hr.	Bare Costs Daily	Incl. Subs O & P Hr.	Incl. Subs O & P Daily	Cost Per Labor-Hour Bare Costs	Cost Per Labor-Hour Incl. O&P
Crew B-8	**Hr.**	**Daily**	**Hr.**	**Daily**	**Bare Costs**	**Incl. O&P**
1 Labor Foreman (outside)	$28.70	$229.60	$44.70	$357.60	$29.58	$45.19
2 Laborers	26.70	427.20	41.55	664.80		
2 Equip. Oper. (med.)	34.65	554.40	52.20	835.20		
1 Equip. Oper. Oiler	30.10	240.80	45.35	362.80		
2 Truck Drivers (heavy)	27.55	440.80	42.00	672.00		
1 Hyd. Crane, 25 Ton		616.80		678.50		
1 F.E. Loader, T.M., 2.5 C.Y.		788.00		866.80		
2 Dump Trucks, 16 Ton		953.20		1048.50	36.84	40.53
64 L.H., Daily Totals		$4250.80		$5486.20	$66.42	$85.72
Crew B-9	**Hr.**	**Daily**	**Hr.**	**Daily**	**Bare Costs**	**Incl. O&P**
1 Labor Foreman (outside)	$28.70	$229.60	$44.70	$357.60	$27.10	$42.18
4 Laborers	26.70	854.40	41.55	1329.60		
1 Air Compr., 250 C.F.M.		112.40		123.65		
2 Air Tools & Accessories		19.60		21.55		
2-50 Ft. Air Hoses, 1.5″ Dia.		9.40		10.35	3.54	3.89
40 L.H., Daily Totals		$1225.40		$1842.75	$30.64	$46.07
Crew B-9A	**Hr.**	**Daily**	**Hr.**	**Daily**	**Bare Costs**	**Incl. O&P**
2 Laborers	$26.70	$427.20	$41.55	$664.80	$26.98	$41.70
1 Truck Driver (heavy)	27.55	220.40	42.00	336.00		
1 Water Tanker		116.80		128.50		
1 Tractor		208.20		229.00		
2-50 Ft. Disch. Hoses		5.00		5.50	13.75	15.13
24 L.H., Daily Totals		$977.60		$1363.80	$40.73	$56.83
Crew B-9B	**Hr.**	**Daily**	**Hr.**	**Daily**	**Bare Costs**	**Incl. O&P**
2 Laborers	$26.70	$427.20	$41.55	$664.80	$26.98	$41.70
1 Truck Driver (heavy)	27.55	220.40	42.00	336.00		
2-50 Ft. Disch. Hoses		5.00		5.50		
1 Water Tanker		116.80		128.50		
1 Tractor		208.20		229.00		
1 Pressure Washer		46.20		50.80	15.68	17.24
24 L.H., Daily Totals		$1023.80		$1414.60	$42.66	$58.94
Crew B-9C	**Hr.**	**Daily**	**Hr.**	**Daily**	**Bare Costs**	**Incl. O&P**
1 Labor Foreman (outside)	$28.70	$229.60	$44.70	$357.60	$27.10	$42.18
4 Laborers	26.70	854.40	41.55	1329.60		
1 Air Compr., 250 C.F.M.		112.40		123.65		
2-50 Ft. Air Hoses, 1.5″ Dia.		9.40		10.35		
2 Breaker, Pavement, 60 lb.		19.60		21.55	3.54	3.89
40 L.H., Daily Totals		$1225.40		$1842.75	$30.64	$46.07
Crew B-9D	**Hr.**	**Daily**	**Hr.**	**Daily**	**Bare Costs**	**Incl. O&P**
1 Labor Foreman (Outside)	$28.70	$229.60	$44.70	$357.60	$27.10	$42.18
4 Common Laborers	26.70	854.40	41.55	1329.60		
1 Air Compressor, 250 CFM		112.40		123.65		
2 Air hoses, 1.5″ x 50′		9.40		10.35		
2 Air tamper		55.30		60.85	4.43	4.87
40 L.H., Daily Totals		$1261.10		$1882.05	$31.53	$47.05
Crew B-10	**Hr.**	**Daily**	**Hr.**	**Daily**	**Bare Costs**	**Incl. O&P**
1 Equip. Oper. (med.)	$34.65	$277.20	$52.20	$417.60	$32.00	$48.65
.5 Laborer	26.70	106.80	41.55	166.20		
12 L.H., Daily Totals		$384.00		$583.80	$32.00	$48.65
Crew B-10A	**Hr.**	**Daily**	**Hr.**	**Daily**	**Bare Costs**	**Incl. O&P**
1 Equip. Oper. (med.)	$34.65	$277.20	$52.20	$417.60	$32.00	$48.65
.5 Laborer	26.70	106.80	41.55	166.20		
1 Walk behind compactor, 7.5 HP		122.80		135.10	10.23	11.26
12 L.H., Daily Totals		$506.80		$718.90	$42.23	$59.91

Crew No.	Bare Costs Hr.	Bare Costs Daily	Incl. Subs O & P Hr.	Incl. Subs O & P Daily	Cost Per Labor-Hour Bare Costs	Cost Per Labor-Hour Incl. O&P
Crew B-10B	**Hr.**	**Daily**	**Hr.**	**Daily**	**Bare Costs**	**Incl. O&P**
1 Equip. Oper. (med.)	$34.65	$277.20	$52.20	$417.60	$32.00	$48.65
.5 Laborer	26.70	106.80	41.55	166.20		
1 Dozer, 200 H.P.		919.60		1011.55	76.63	84.30
12 L.H., Daily Totals		$1303.60		$1595.35	$108.63	$132.95
Crew B-10C	**Hr.**	**Daily**	**Hr.**	**Daily**	**Bare Costs**	**Incl. O&P**
1 Equip. Oper. (med.)	$34.65	$277.20	$52.20	$417.60	$32.00	$48.65
.5 Laborer	26.70	106.80	41.55	166.20		
1 Dozer, 200 H.P.		919.60		1011.55		
1 Vibratory Roller, Towed		591.60		650.75	125.93	138.53
12 L.H., Daily Totals		$1895.20		$2246.10	$157.93	$187.18
Crew B-10D	**Hr.**	**Daily**	**Hr.**	**Daily**	**Bare Costs**	**Incl. O&P**
1 Equip. Oper. (med.)	$34.65	$277.20	$52.20	$417.60	$32.00	$48.65
.5 Laborer	26.70	106.80	41.55	166.20		
1 Dozer, 200 H.P.		919.60		1011.55		
1 Sheepsft. Roller, Towed		610.20		671.20	127.48	140.23
12 L.H., Daily Totals		$1913.80		$2266.55	$159.48	$188.88
Crew B-10E	**Hr.**	**Daily**	**Hr.**	**Daily**	**Bare Costs**	**Incl. O&P**
1 Equip. Oper. (med.)	$34.65	$277.20	$52.20	$417.60	$32.00	$48.65
.5 Laborer	26.70	106.80	41.55	166.20		
1 Tandem Roller, 5 Ton		107.40		118.15	8.95	9.85
12 L.H., Daily Totals		$491.40		$701.95	$40.95	$58.50
Crew B-10F	**Hr.**	**Daily**	**Hr.**	**Daily**	**Bare Costs**	**Incl. O&P**
1 Equip. Oper. (med.)	$34.65	$277.20	$52.20	$417.60	$32.00	$48.65
.5 Laborer	26.70	106.80	41.55	166.20		
1 Tandem Roller, 10 Ton		179.20		197.10	14.93	16.43
12 L.H., Daily Totals		$563.20		$780.90	$46.93	$65.08
Crew B-10G	**Hr.**	**Daily**	**Hr.**	**Daily**	**Bare Costs**	**Incl. O&P**
1 Equip. Oper. (med.)	$34.65	$277.20	$52.20	$417.60	$32.00	$48.65
.5 Laborer	26.70	106.80	41.55	166.20		
1 Sheepsft. Roll., 130 H.P.		794.80		874.30	66.23	72.86
12 L.H., Daily Totals		$1178.80		$1458.10	$98.23	$121.51
Crew B-10H	**Hr.**	**Daily**	**Hr.**	**Daily**	**Bare Costs**	**Incl. O&P**
1 Equip. Oper. (med.)	$34.65	$277.20	$52.20	$417.60	$32.00	$48.65
.5 Laborer	26.70	106.80	41.55	166.20		
1 Diaphr. Water Pump, 2″		50.60		55.65		
1-20 Ft. Suction Hose, 2″		2.55		2.80		
2-50 Ft. Disch. Hoses, 2″		3.80		4.20	4.75	5.22
12 L.H., Daily Totals		$440.95		$646.45	$36.75	$53.87
Crew B-10I	**Hr.**	**Daily**	**Hr.**	**Daily**	**Bare Costs**	**Incl. O&P**
1 Equip. Oper. (med.)	$34.65	$277.20	$52.20	$417.60	$32.00	$48.65
.5 Laborer	26.70	106.80	41.55	166.20		
1 Diaphr. Water Pump, 4″		74.40		81.85		
1-20 Ft. Suction Hose, 4″		5.05		5.55		
2-50 Ft. Disch. Hoses, 4″		7.10		7.80	7.21	7.93
12 L.H., Daily Totals		$470.55		$679.00	$39.21	$56.58
Crew B-10J	**Hr.**	**Daily**	**Hr.**	**Daily**	**Bare Costs**	**Incl. O&P**
1 Equip. Oper. (med.)	$34.65	$277.20	$52.20	$417.60	$32.00	$48.65
.5 Laborer	26.70	106.80	41.55	166.20		
1 Centr. Water Pump, 3″		53.60		58.95		
1-20 Ft. Suction Hose, 3″		4.25		4.70		
2-50 Ft. Disch. Hoses, 3″		5.00		5.50	5.24	5.76
12 L.H., Daily Totals		$446.85		$652.95	$37.24	$54.41

Crews

Crew No.	Bare Costs		Incl. Subs O & P		Cost Per Labor-Hour	
Crew B-10K	Hr.	Daily	Hr.	Daily	Bare Costs	Incl. O&P
1 Equip. Oper. (med.)	$34.65	$277.20	$52.20	$417.60	$32.00	$48.65
.5 Laborer	26.70	106.80	41.55	166.20		
1 Centr. Water Pump, 6"		217.00		238.70		
1-20 Ft. Suction Hose, 6"		12.80		14.10		
2-50 Ft. Disch. Hoses, 6"		17.80		19.60	20.63	22.70
12 L.H., Daily Totals		$631.60		$856.20	$52.63	$71.35
Crew B-10L	Hr.	Daily	Hr.	Daily	Bare Costs	Incl. O&P
1 Equip. Oper. (med.)	$34.65	$277.20	$52.20	$417.60	$32.00	$48.65
.5 Laborer	26.70	106.80	41.55	166.20		
1 Dozer, 80 H.P.		314.80		346.30	26.23	28.86
12 L.H., Daily Totals		$698.80		$930.10	$58.23	$77.51
Crew B-10M	Hr.	Daily	Hr.	Daily	Bare Costs	Incl. O&P
1 Equip. Oper. (med.)	$34.65	$277.20	$52.20	$417.60	$32.00	$48.65
.5 Laborer	26.70	106.80	41.55	166.20		
1 Dozer, 300 H.P.		1195.00		1314.50	99.58	109.54
12 L.H., Daily Totals		$1579.00		$1898.30	$131.58	$158.19
Crew B-10N	Hr.	Daily	Hr.	Daily	Bare Costs	Incl. O&P
1 Equip. Oper. (med.)	$34.65	$277.20	$52.20	$417.60	$32.00	$48.65
.5 Laborer	26.70	106.80	41.55	166.20		
1 F.E. Loader, T.M., 1.5 C.Y		312.60		343.85	26.05	28.66
12 L.H., Daily Totals		$696.60		$927.65	$58.05	$77.31
Crew B-10O	Hr.	Daily	Hr.	Daily	Bare Costs	Incl. O&P
1 Equip. Oper. (med.)	$34.65	$277.20	$52.20	$417.60	$32.00	$48.65
.5 Laborer	26.70	106.80	41.55	166.20		
1 F.E. Loader, T.M., 2.25 C.Y.		551.80		607.00	45.98	50.58
12 L.H., Daily Totals		$935.80		$1190.80	$77.98	$99.23
Crew B-10P	Hr.	Daily	Hr.	Daily	Bare Costs	Incl. O&P
1 Equip. Oper. (med.)	$34.65	$277.20	$52.20	$417.60	$32.00	$48.65
.5 Laborer	26.70	106.80	41.55	166.20		
1 F.E. Loader, T.M., 2.5 C.Y.		788.00		866.80	65.67	72.23
12 L.H., Daily Totals		$1172.00		$1450.60	$97.67	$120.88
Crew B-10Q	Hr.	Daily	Hr.	Daily	Bare Costs	Incl. O&P
1 Equip. Oper. (med.)	$34.65	$277.20	$52.20	$417.60	$32.00	$48.65
.5 Laborer	26.70	106.80	41.55	166.20		
1 F.E. Loader, T.M., 5 C.Y.		1094.00		1203.40	91.17	100.28
12 L.H., Daily Totals		$1478.00		$1787.20	$123.17	$148.93
Crew B-10R	Hr.	Daily	Hr.	Daily	Bare Costs	Incl. O&P
1 Equip. Oper. (med.)	$34.65	$277.20	$52.20	$417.60	$32.00	$48.65
.5 Laborer	26.70	106.80	41.55	166.20		
1 F.E. Loader, W.M., 1 C.Y.		193.80		213.20	16.15	17.77
12 L.H., Daily Totals		$577.80		$797.00	$48.15	$66.42
Crew B-10S	Hr.	Daily	Hr.	Daily	Bare Costs	Incl. O&P
1 Equip. Oper. (med.)	$34.65	$277.20	$52.20	$417.60	$32.00	$48.65
.5 Laborer	26.70	106.80	41.55	166.20		
1 F.E. Loader, W.M., 1.5 C.Y.		241.00		265.10	20.08	22.09
12 L.H., Daily Totals		$625.00		$848.90	$52.08	$70.74
Crew B-10T	Hr.	Daily	Hr.	Daily	Bare Costs	Incl. O&P
1 Equip. Oper. (med.)	$34.65	$277.20	$52.20	$417.60	$32.00	$48.65
.5 Laborer	26.70	106.80	41.55	166.20		
1 F.E. Loader, W.M., 2.5 C.Y.		305.00		335.50	25.42	27.96
12 L.H., Daily Totals		$689.00		$919.30	$57.42	$76.61

Crew No.	Bare Costs		Incl. Subs O & P		Cost Per Labor-Hour	
Crew B-10U	Hr.	Daily	Hr.	Daily	Bare Costs	Incl. O&P
1 Equip. Oper. (med.)	$34.65	$277.20	$52.20	$417.60	$32.00	$48.65
.5 Laborer	26.70	106.80	41.55	166.20		
1 F.E. Loader, W.M., 5.5 C.Y.		700.40		770.45	58.37	64.20
12 L.H., Daily Totals		$1084.40		$1354.25	$90.37	$112.85
Crew B-10V	Hr.	Daily	Hr.	Daily	Bare Costs	Incl. O&P
1 Equip. Oper. (med.)	$34.65	$277.20	$52.20	$417.60	$32.00	$48.65
.5 Laborer	26.70	106.80	41.55	166.20		
1 Dozer, 700 H.P.		3139.00		3452.90	261.58	287.74
12 L.H., Daily Totals		$3523.00		$4036.70	$293.58	$336.39
Crew B-10W	Hr.	Daily	Hr.	Daily	Bare Costs	Incl. O&P
1 Equip. Oper. (med.)	$34.65	$277.20	$52.20	$417.60	$32.00	$48.65
.5 Laborer	26.70	106.80	41.55	166.20		
1 Dozer, 105 H.P.		453.80		499.20	37.82	41.60
12 L.H., Daily Totals		$837.80		$1083.00	$69.82	$90.25
Crew B-10X	Hr.	Daily	Hr.	Daily	Bare Costs	Incl. O&P
1 Equip. Oper. (med.)	$34.65	$277.20	$52.20	$417.60	$32.00	$48.65
.5 Laborer	26.70	106.80	41.55	166.20		
1 Dozer, 410 H.P.		1524.00		1676.40	127.00	139.70
12 L.H., Daily Totals		$1908.00		$2260.20	$159.00	$188.35
Crew B-10Y	Hr.	Daily	Hr.	Daily	Bare Costs	Incl. O&P
1 Equip. Oper. (med.)	$34.65	$277.20	$52.20	$417.60	$32.00	$48.65
.5 Laborer	26.70	106.80	41.55	166.20		
1 Vibratory Drum Roller		344.80		379.30	28.73	31.61
12 L.H., Daily Totals		$728.80		$963.10	$60.73	$80.26
Crew B-11A	Hr.	Daily	Hr.	Daily	Bare Costs	Incl. O&P
1 Equipment Oper. (med.)	$34.65	$277.20	$52.20	$417.60	$30.68	$46.88
1 Laborer	26.70	213.60	41.55	332.40		
1 Dozer, 200 H.P.		919.60		1011.55	57.48	63.22
16 L.H., Daily Totals		$1410.40		$1761.55	$88.16	$110.10
Crew B-11B	Hr.	Daily	Hr.	Daily	Bare Costs	Incl. O&P
1 Equipment Oper. (light)	$33.05	$264.40	$49.80	$398.40	$29.88	$45.68
1 Laborer	26.70	213.60	41.55	332.40		
1 Air Powered Tamper		27.65		30.40		
1 Air Compr. 365 C.F.M.		146.80		161.50		
2-50 Ft. Air Hoses, 1.5" Dia.		9.40		10.35	11.49	12.64
16 L.H., Daily Totals		$661.85		$933.05	$41.37	$58.32
Crew B-11C	Hr.	Daily	Hr.	Daily	Bare Costs	Incl. O&P
1 Equipment Oper. (med.)	$34.65	$277.20	$52.20	$417.60	$30.68	$46.88
1 Laborer	26.70	213.60	41.55	332.40		
1 Backhoe Loader, 48 H.P.		215.80		237.40	13.49	14.84
16 L.H., Daily Totals		$706.60		$987.40	$44.17	$61.72
Crew B-11J	Hr.	Daily	Hr.	Daily	Bare Costs	Incl. O&P
1 Equipment Oper. (med.)	$34.65	$277.20	$52.20	$417.60	$30.68	$46.88
1 Laborer	26.70	213.60	41.55	332.40		
1 Grader, 30,000 Lbs.		456.60		502.25		
1 Ripper, beam & 1 shank		71.40		78.55	33.00	36.30
16 L.H., Daily Totals		$1018.80		$1330.80	$63.68	$83.18

Crew No.	Bare Costs		Incl. Subs O & P		Cost Per Labor-Hour	
Crew B-11K	Hr.	Daily	Hr.	Daily	Bare Costs	Incl. O&P
1 Equipment Oper. (med.)	$34.65	$277.20	$52.20	$417.60	$30.68	$46.88
1 Laborer	26.70	213.60	41.55	332.40		
1 Trencher, 8' D., 16" W.		1420.00		1562.00	88.75	97.63
16 L.H., Daily Totals		$1910.80		$2312.00	$119.43	$144.51
Crew B-11L	Hr.	Daily	Hr.	Daily	Bare Costs	Incl. O&P
1 Equipment Oper. (med.)	$34.65	$277.20	$52.20	$417.60	$30.68	$46.88
1 Laborer	26.70	213.60	41.55	332.40		
1 Grader, 30,000 Lbs.		456.60		502.25	28.54	31.39
16 L.H., Daily Totals		$947.40		$1252.25	$59.22	$78.27
Crew B-11M	Hr.	Daily	Hr.	Daily	Bare Costs	Incl. O&P
1 Equipment Oper. (med.)	$34.65	$277.20	$52.20	$417.60	$30.68	$46.88
1 Laborer	26.70	213.60	41.55	332.40		
1 Backhoe Loader, 80 H.P.		256.40		282.05	16.03	17.63
16 L.H., Daily Totals		$747.20		$1032.05	$46.71	$64.51
Crew B-11N	Hr.	Daily	Hr.	Daily	Bare Costs	Incl. O&P
1 Labor Foreman	$28.70	$229.60	$44.70	$357.60	$29.26	$44.57
2 Equipment Operators (med.)	34.65	554.40	52.20	835.20		
6 Truck Drivers (hvy.)	27.55	1322.40	42.00	2016.00		
1 F.E. Loader, 5.5 C.Y.		700.40		770.45		
1 Dozer, 400 H.P.		1524.00		1676.40		
6 Off Hwy. Tks. 50 Ton		7086.00		7794.60	129.31	142.24
72 L.H., Daily Totals		$11416.80		$13450.25	$158.57	$186.81
Crew B-11Q	Hr.	Daily	Hr.	Daily	Bare Costs	Incl. O&P
1 Equipment Operator (med.)	$34.65	$277.20	$52.20	$417.60	$32.00	$48.65
.5 Laborer	26.70	106.80	41.55	166.20		
1 Dozer, 140 H.P.		572.00		629.20	47.67	52.43
12 L.H., Daily Totals		$956.00		$1213.00	$79.67	$101.08
Crew B-11R	Hr.	Daily	Hr.	Daily	Bare Costs	Incl. O&P
1 Equipment Operator (med.)	$34.65	$277.20	$52.20	$417.60	$32.00	$48.65
.5 Laborer	26.70	106.80	41.55	166.20		
1 Dozer, 215 H.P.		919.60		1011.55	76.63	84.30
12 L.H., Daily Totals		$1303.60		$1595.35	$108.63	$132.95
Crew B-11S	Hr.	Daily	Hr.	Daily	Bare Costs	Incl. O&P
1 Equipment Operator (med.)	$34.65	$277.20	$52.20	$417.60	$32.00	$48.65
.5 Laborer	26.70	106.80	41.55	166.20		
1 Dozer, 300 H.P.		1195.00		1314.50		
1 Ripper, Beam & 1 Shank		71.40		78.55	105.53	116.09
12 L.H., Daily Totals		$1650.40		$1976.85	$137.53	$164.74
Crew B-11T	Hr.	Daily	Hr.	Daily	Bare Costs	Incl. O&P
1 Equipment Operator (med.)	$34.65	$277.20	$52.20	$417.60	$32.00	$48.65
.5 Laborer	26.70	106.80	41.55	166.20		
1 Dozer, 410 H.P.		1524.00		1676.40		
1 Ripper, Beam & 2 Shanks		80.20		88.20	133.68	147.05
12 L.H., Daily Totals		$1988.20		$2348.40	$165.68	$195.70
Crew B-11U	Hr.	Daily	Hr.	Daily	Bare Costs	Incl. O&P
1 Equipment Operator (med.)	$34.65	$277.20	$52.20	$417.60	$32.00	$48.65
.5 Laborer	26.70	106.80	41.55	166.20		
1 Dozer, 520 H.P.		2035.00		2238.50	169.58	186.54
12 L.H., Daily Totals		$2419.00		$2822.30	$201.58	$235.19

Crew No.	Bare Costs		Incl. Subs O & P		Cost Per Labor-Hour	
Crew B-11V	Hr.	Daily	Hr.	Daily	Bare Costs	Incl. O&P
3 Laborers	$26.70	$640.80	$41.55	$997.20	$26.70	$41.55
1 Walk behind compactor, 7.5 HP		122.80		135.10	5.12	5.63
24 L.H., Daily Totals		$763.60		$1132.30	$31.82	$47.18
Crew B-11W	Hr.	Daily	Hr.	Daily	Bare Costs	Incl. O&P
1 Equipment Operator (med.)	$34.65	$277.20	$52.20	$417.60	$28.07	$42.81
1 Common Laborer	26.70	213.60	41.55	332.40		
10 Truck Drivers, Heavy	27.55	2204.00	42.00	3360.00		
1 Dozer, 200 H.P.		919.60		1011.55		
1 Vib. roller, smth. towed, 23 Ton		591.60		650.75		
10 Dump Truck, 10 Ton		3266.00		3592.60	49.76	54.74
96 L.H., Daily Totals		$7472.00		$9364.90	$77.83	$97.55
Crew B-11Y	Hr.	Daily	Hr.	Daily	Bare Costs	Incl. O&P
1 Labor Foreman (Outside)	$28.70	$229.60	$44.70	$357.60	$29.57	$45.45
5 Common Laborers	26.70	1068.00	41.55	1662.00		
3 Equipment Operator (med.)	34.65	831.60	52.20	1252.80		
1 Dozer, 80 H.P.		314.80		346.30		
2 Walk behind compactor, 7.5 HP		245.60		270.15		
4 Vibratory plate, gas, 21"		147.20		161.90	9.83	10.81
72 L.H., Daily Totals		$2836.80		$4050.75	$39.40	$56.26
Crew B-12A	Hr.	Daily	Hr.	Daily	Bare Costs	Incl. O&P
1 Equip. Oper. (crane)	$35.90	$287.20	$54.10	$432.80	$31.30	$47.83
1 Laborer	26.70	213.60	41.55	332.40		
1 Hyd. Excavator, 1 C.Y.		557.80		613.60	34.86	38.35
16 L.H., Daily Totals		$1058.60		$1378.80	$66.16	$86.18
Crew B-12B	Hr.	Daily	Hr.	Daily	Bare Costs	Incl. O&P
1 Equip. Oper. (crane)	$35.90	$287.20	$54.10	$432.80	$31.30	$47.83
1 Laborer	26.70	213.60	41.55	332.40		
1 Hyd. Excavator, 1.5 C.Y.		720.40		792.45	45.03	49.53
16 L.H., Daily Totals		$1221.20		$1557.65	$76.33	$97.36
Crew B-12C	Hr.	Daily	Hr.	Daily	Bare Costs	Incl. O&P
1 Equip. Oper. (crane)	$35.90	$287.20	$54.10	$432.80	$31.30	$47.83
1 Laborer	26.70	213.60	41.55	332.40		
1 Hyd. Excavator, 2 C.Y.		909.80		1000.80	56.86	62.55
16 L.H., Daily Totals		$1410.60		$1766.00	$88.16	$110.38
Crew B-12D	Hr.	Daily	Hr.	Daily	Bare Costs	Incl. O&P
1 Equip. Oper. (crane)	$35.90	$287.20	$54.10	$432.80	$31.30	$47.83
1 Laborer	26.70	213.60	41.55	332.40		
1 Hyd. Excavator, 3.5 C.Y.		2020.00		2222.00	126.25	138.88
16 L.H., Daily Totals		$2520.80		$2987.20	$157.55	$186.71
Crew B-12E	Hr.	Daily	Hr.	Daily	Bare Costs	Incl. O&P
1 Equip. Oper. (crane)	$35.90	$287.20	$54.10	$432.80	$31.30	$47.83
1 Laborer	26.70	213.60	41.55	332.40		
1 Hyd. Excavator, .5 C.Y.		332.40		365.65	20.78	22.85
16 L.H., Daily Totals		$833.20		$1130.85	$52.08	$70.68
Crew B-12F	Hr.	Daily	Hr.	Daily	Bare Costs	Incl. O&P
1 Equip. Oper. (crane)	$35.90	$287.20	$54.10	$432.80	$31.30	$47.83
1 Laborer	26.70	213.60	41.55	332.40		
1 Hyd. Excavator, .75 C.Y.		474.00		521.40	29.63	32.59
16 L.H., Daily Totals		$974.80		$1286.60	$60.93	$80.42

Crew No.	Bare Costs		Incl. Subs O & P		Cost Per Labor-Hour	
Crew B-12G	Hr.	Daily	Hr.	Daily	Bare Costs	Incl. O&P
1 Equip. Oper. (crane)	$35.90	$287.20	$54.10	$432.80	$31.30	$47.83
1 Laborer	26.70	213.60	41.55	332.40		
1 Power Shovel, .5 C.Y.		508.10		558.90		
1 Clamshell Bucket, .5 C.Y.		33.80		37.20	33.87	37.25
16 L.H., Daily Totals		$1042.70		$1361.30	$65.17	$85.08

Crew No.	Bare Costs		Incl. Subs O & P		Cost Per Labor-Hour	
Crew B-12H	Hr.	Daily	Hr.	Daily	Bare Costs	Incl. O&P
1 Equip. Oper. (crane)	$35.90	$287.20	$54.10	$432.80	$31.30	$47.83
1 Laborer	26.70	213.60	41.55	332.40		
1 Power Shovel, 1 C.Y.		833.20		916.50		
1 Clamshell Bucket, 1 C.Y.		43.40		47.75	54.79	60.27
16 L.H., Daily Totals		$1377.40		$1729.45	$86.09	$108.10

Crew No.	Bare Costs		Incl. Subs O & P		Cost Per Labor-Hour	
Crew B-12I	Hr.	Daily	Hr.	Daily	Bare Costs	Incl. O&P
1 Equip. Oper. (crane)	$35.90	$287.20	$54.10	$432.80	$31.30	$47.83
1 Laborer	26.70	213.60	41.55	332.40		
1 Power Shovel, .75 C.Y.		630.45		693.50		
1 Dragline Bucket, .75 C.Y.		18.80		20.70	40.58	44.63
16 L.H., Daily Totals		$1150.05		$1479.40	$71.88	$92.46

Crew No.	Bare Costs		Incl. Subs O & P		Cost Per Labor-Hour	
Crew B-12J	Hr.	Daily	Hr.	Daily	Bare Costs	Incl. O&P
1 Equip. Oper. (crane)	$35.90	$287.20	$54.10	$432.80	$31.30	$47.83
1 Laborer	26.70	213.60	41.55	332.40		
1 Gradall, 3 Ton, .5 C.Y.		837.80		921.60	52.36	57.60
16 L.H., Daily Totals		$1338.60		$1686.80	$83.66	$105.43

Crew No.	Bare Costs		Incl. Subs O & P		Cost Per Labor-Hour	
Crew B-12K	Hr.	Daily	Hr.	Daily	Bare Costs	Incl. O&P
1 Equip. Oper. (crane)	$35.90	$287.20	$54.10	$432.80	$31.30	$47.83
1 Laborer	26.70	213.60	41.55	332.40		
1 Gradall, 3 Ton, 1 C.Y.		971.00		1068.10	60.69	66.76
16 L.H., Daily Totals		$1471.80		$1833.30	$91.99	$114.59

Crew No.	Bare Costs		Incl. Subs O & P		Cost Per Labor-Hour	
Crew B-12L	Hr.	Daily	Hr.	Daily	Bare Costs	Incl. O&P
1 Equip. Oper. (crane)	$35.90	$287.20	$54.10	$432.80	$31.30	$47.83
1 Laborer	26.70	213.60	41.55	332.40		
1 Power Shovel, .5 C.Y.		508.10		558.90		
1 F.E. Attachment, .5 C.Y.		47.40		52.15	34.72	38.19
16 L.H., Daily Totals		$1056.30		$1376.25	$66.02	$86.02

Crew No.	Bare Costs		Incl. Subs O & P		Cost Per Labor-Hour	
Crew B-12M	Hr.	Daily	Hr.	Daily	Bare Costs	Incl. O&P
1 Equip. Oper. (crane)	$35.90	$287.20	$54.10	$432.80	$31.30	$47.83
1 Laborer	26.70	213.60	41.55	332.40		
1 Power Shovel, .75 C.Y.		630.45		693.50		
1 F.E. Attachment, .75 C.Y.		51.80		57.00	42.64	46.90
16 L.H., Daily Totals		$1183.05		$1515.70	$73.94	$94.73

Crew No.	Bare Costs		Incl. Subs O & P		Cost Per Labor-Hour	
Crew B-12N	Hr.	Daily	Hr.	Daily	Bare Costs	Incl. O&P
1 Equip. Oper. (crane)	$35.90	$287.20	$54.10	$432.80	$31.30	$47.83
1 Laborer	26.70	213.60	41.55	332.40		
1 Power Shovel, 1 C.Y.		833.20		916.50		
1 F.E. Attachment, 1 C.Y.		58.60		64.45	55.74	61.31
16 L.H., Daily Totals		$1392.60		$1746.15	$87.04	$109.14

Crew No.	Bare Costs		Incl. Subs O & P		Cost Per Labor-Hour	
Crew B-12O	Hr.	Daily	Hr.	Daily	Bare Costs	Incl. O&P
1 Equip. Oper. (crane)	$35.90	$287.20	$54.10	$432.80	$31.30	$47.83
1 Laborer	26.70	213.60	41.55	332.40		
1 Power Shovel, 1.5 C.Y.		888.20		977.00		
1 F.E. Attachment, 1.5 C.Y.		67.80		74.60	59.75	65.73
16 L.H., Daily Totals		$1456.80		$1816.80	$91.05	$113.56

Crew No.	Bare Costs		Incl. Subs O & P		Cost Per Labor-Hour	
Crew B-12P	Hr.	Daily	Hr.	Daily	Bare Costs	Incl. O&P
1 Equip. Oper. (crane)	$35.90	$287.20	$54.10	$432.80	$31.30	$47.83
1 Laborer	26.70	213.60	41.55	332.40		
1 Crawler Crane, 40 Ton		888.20		977.00		
1 Dragline Bucket, 1.5 C.Y.		30.60		33.65	57.43	63.17
16 L.H., Daily Totals		$1419.60		$1775.85	$88.73	$111.00

Crew No.	Bare Costs		Incl. Subs O & P		Cost Per Labor-Hour	
Crew B-12Q	Hr.	Daily	Hr.	Daily	Bare Costs	Incl. O&P
1 Equip. Oper. (crane)	$35.90	$287.20	$54.10	$432.80	$31.30	$47.83
1 Laborer	26.70	213.60	41.55	332.40		
1 Hyd. Excavator, 5/8 C.Y.		432.40		475.65	27.03	29.73
16 L.H., Daily Totals		$933.20		$1240.85	$58.33	$77.56

Crew No.	Bare Costs		Incl. Subs O & P		Cost Per Labor-Hour	
Crew B-12R	Hr.	Daily	Hr.	Daily	Bare Costs	Incl. O&P
1 Equip. Oper. (crane)	$35.90	$287.20	$54.10	$432.80	$31.30	$47.83
1 Laborer	26.70	213.60	41.55	332.40		
1 Hyd. Excavator, 1.5 C.Y.		720.40		792.45	45.03	49.53
16 L.H., Daily Totals		$1221.20		$1557.65	$76.33	$97.36

Crew No.	Bare Costs		Incl. Subs O & P		Cost Per Labor-Hour	
Crew B-12S	Hr.	Daily	Hr.	Daily	Bare Costs	Incl. O&P
1 Equip. Oper. (crane)	$35.90	$287.20	$54.10	$432.80	$31.30	$47.83
1 Laborer	26.70	213.60	41.55	332.40		
1 Hyd. Excavator, 2.5 C.Y.		1215.00		1336.50	75.94	83.53
16 L.H., Daily Totals		$1715.80		$2101.70	$107.24	$131.36

Crew No.	Bare Costs		Incl. Subs O & P		Cost Per Labor-Hour	
Crew B-12T	Hr.	Daily	Hr.	Daily	Bare Costs	Incl. O&P
1 Equip. Oper. (crane)	$35.90	$287.20	$54.10	$432.80	$31.30	$47.83
1 Laborer	26.70	213.60	41.55	332.40		
1 Crawler Crane, 75 Ton		1170.00		1287.00		
1 F.E. Attachment, 3 C.Y.		90.20		99.20	78.76	86.64
16 L.H., Daily Totals		$1761.00		$2151.40	$110.06	$134.47

Crew No.	Bare Costs		Incl. Subs O & P		Cost Per Labor-Hour	
Crew B-12V	Hr.	Daily	Hr.	Daily	Bare Costs	Incl. O&P
1 Equip. Oper. (crane)	$35.90	$287.20	$54.10	$432.80	$31.30	$47.83
1 Laborer	26.70	213.60	41.55	332.40		
1 Crawler Crane, 75 Ton		1170.00		1287.00		
1 Dragline Bucket, 3 C.Y.		47.40		52.15	76.09	83.70
16 L.H., Daily Totals		$1718.20		$2104.35	$107.39	$131.53

Crew No.	Bare Costs		Incl. Subs O & P		Cost Per Labor-Hour	
Crew B-13	Hr.	Daily	Hr.	Daily	Bare Costs	Incl. O&P
1 Labor Foreman (outside)	$28.70	$229.60	$44.70	$357.60	$28.79	$44.34
4 Laborers	26.70	854.40	41.55	1329.60		
1 Equip. Oper. (crane)	35.90	287.20	54.10	432.80		
1 Equip. Oper. Oiler	30.10	240.80	45.35	362.80		
1 Hyd. Crane, 25 Ton		616.80		678.50	11.01	12.12
56 L.H., Daily Totals		$2228.80		$3161.30	$39.80	$56.46

Crew No.	Bare Costs		Incl. Subs O & P		Cost Per Labor-Hour	
Crew B-13A	Hr.	Daily	Hr.	Daily	Bare Costs	Incl. O&P
1 Foreman	$28.70	$229.60	$44.70	$357.60	$29.50	$45.17
2 Laborers	26.70	427.20	41.55	664.80		
2 Equipment Operators	34.65	554.40	52.20	835.20		
2 Truck Drivers (heavy)	27.55	440.80	42.00	672.00		
1 Crane, 75 Ton		1170.00		1287.00		
1 F.E. Lder, 3.75 C.Y.		1094.00		1203.40		
2 Dump Trucks, 12 Ton		653.20		718.50	52.09	57.30
56 L.H., Daily Totals		$4569.20		$5738.50	$81.59	$102.47

Crews

Crew B-13B	Hr.	Daily	Hr.	Daily	Bare Costs	Incl. O&P
1 Labor Foreman (outside)	$28.70	$229.60	$44.70	$357.60	$28.79	$44.34
4 Laborers	26.70	854.40	41.55	1329.60		
1 Equip. Oper. (crane)	35.90	287.20	54.10	432.80		
1 Equip. Oper. Oiler	30.10	240.80	45.35	362.80		
1 Hyd. Crane, 55 Ton		895.40		984.95	15.99	17.59
56 L.H., Daily Totals		$2507.40		$3467.75	$44.78	$61.93

Crew B-13C	Hr.	Daily	Hr.	Daily	Bare Costs	Incl. O&P
1 Labor Foreman (outside)	$28.70	$229.60	$44.70	$357.60	$28.79	$44.34
4 Laborers	26.70	854.40	41.55	1329.60		
1 Equip. Oper. (crane)	35.90	287.20	54.10	432.80		
1 Equip. Oper. Oiler	30.10	240.80	45.35	362.80		
1 Crawler Crane, 100 Ton		1519.00		1670.90	27.13	29.84
56 L.H., Daily Totals		$3131.00		$4153.70	$55.92	$74.18

Crew B-14	Hr.	Daily	Hr.	Daily	Bare Costs	Incl. O&P
1 Labor Foreman (outside)	$28.70	$229.60	$44.70	$357.60	$28.09	$43.45
4 Laborers	26.70	854.40	41.55	1329.60		
1 Equip. Oper. (light)	33.05	264.40	49.80	398.40		
1 Backhoe Loader, 48 H.P.		215.80		237.40	4.50	4.95
48 L.H., Daily Totals		$1564.20		$2323.00	$32.59	$48.40

Crew B-15	Hr.	Daily	Hr.	Daily	Bare Costs	Incl. O&P
1 Equipment Oper. (med)	$34.65	$277.20	$52.20	$417.60	$29.46	$44.85
.5 Laborer	26.70	106.80	41.55	166.20		
2 Truck Drivers (heavy)	27.55	440.80	42.00	672.00		
2 Dump Trucks, 16 Ton		953.20		1048.50		
1 Dozer, 200 H.P.		919.60		1011.55	66.89	73.57
28 L.H., Daily Totals		$2697.60		$3315.85	$96.35	$118.42

Crew B-16	Hr.	Daily	Hr.	Daily	Bare Costs	Incl. O&P
1 Labor Foreman (outside)	$28.70	$229.60	$44.70	$357.60	$27.41	$42.45
2 Laborers	26.70	427.20	41.55	664.80		
1 Truck Driver (heavy)	27.55	220.40	42.00	336.00		
1 Dump Truck, 16 Ton		476.60		524.25	14.89	16.38
32 L.H., Daily Totals		$1353.80		$1882.65	$42.30	$58.83

Crew B-17	Hr.	Daily	Hr.	Daily	Bare Costs	Incl. O&P
2 Laborers	$26.70	$427.20	$41.55	$664.80	$28.50	$43.73
1 Equip. Oper. (light)	33.05	264.40	49.80	398.40		
1 Truck Driver (heavy)	27.55	220.40	42.00	336.00		
1 Backhoe Loader, 48 H.P.		215.80		237.40		
1 Dump Truck, 12 Ton		326.60		359.25	16.95	18.65
32 L.H., Daily Totals		$1454.40		$1995.85	$45.45	$62.38

Crew B-17A	Hr.	Daily	Hr.	Daily	Bare Costs	Incl. O&P
2 Laborer Foremen	$28.70	$459.20	$44.70	$715.20	$28.93	$45.02
6 Laborers	26.70	1281.60	41.55	1994.40		
1 Skilled Worker Foreman	36.85	294.80	57.30	458.40		
1 Skilled Worker	34.85	278.80	54.20	433.60		
80 L.H., Daily Totals		$2314.40		$3601.60	$28.93	$45.02

Crew B-18	Hr.	Daily	Hr.	Daily	Bare Costs	Incl. O&P
1 Labor Foreman (outside)	$28.70	$229.60	$44.70	$357.60	$27.37	$42.60
2 Laborers	26.70	427.20	41.55	664.80		
1 Vibrating Compactor		36.80		40.50	1.53	1.69
24 L.H., Daily Totals		$693.60		$1062.90	$28.90	$44.29

Crew B-19	Hr.	Daily	Hr.	Daily	Bare Costs	Incl. O&P
1 Pile Driver Foreman	$35.30	$282.40	$58.15	$465.20	$33.80	$53.91
4 Pile Drivers	33.30	1065.60	54.90	1756.80		
2 Equip. Oper. (crane)	35.90	574.40	54.10	865.60		
1 Equip. Oper. Oiler	30.10	240.80	45.35	362.80		
1 Crane, 40 Ton & Access.		888.20		977.00		
60 L.F. Pile Leads		75.00		82.50		
1 Hammer, Diesel, 22k Ft-Lb		564.00		620.40	23.91	26.30
64 L.H., Daily Totals		$3690.40		$5130.30	$57.71	$80.21

Crew B-19A	Hr.	Daily	Hr.	Daily	Bare Costs	Incl. O&P
1 Pile Driver Foreman	$35.30	$282.40	$58.15	$465.20	$33.80	$53.91
4 Pile Drivers	33.30	1065.60	54.90	1756.80		
2 Equip. Oper. (crane)	35.90	574.40	54.10	865.60		
1 Equip. Oper. Oiler	30.10	240.80	45.35	362.80		
1 Crawler Crane, 75 Ton		1170.00		1287.00		
60 L.F. Leads, 25K Ft. Lbs.		120.00		132.00		
1 Hammer, Diesel, 41k Ft-Lb		662.60		728.85	30.51	33.56
64 L.H., Daily Totals		$4115.80		$5598.25	$64.31	$87.47

Crew B-20	Hr.	Daily	Hr.	Daily	Bare Costs	Incl. O&P
1 Labor Foreman (out)	$28.70	$229.60	$44.70	$357.60	$30.08	$46.82
1 Skilled Worker	34.85	278.80	54.20	433.60		
1 Laborer	26.70	213.60	41.55	332.40		
24 L.H., Daily Totals		$722.00		$1123.60	$30.08	$46.82

Crew B-20A	Hr.	Daily	Hr.	Daily	Bare Costs	Incl. O&P
1 Labor Foreman	$28.70	$229.60	$44.70	$357.60	$32.24	$49.20
1 Laborer	26.70	213.60	41.55	332.40		
1 Plumber	40.85	326.80	61.40	491.20		
1 Plumber Apprentice	32.70	261.60	49.15	393.20		
32 L.H., Daily Totals		$1031.60		$1574.40	$32.24	$49.20

Crew B-21	Hr.	Daily	Hr.	Daily	Bare Costs	Incl. O&P
1 Labor Foreman (out)	$28.70	$229.60	$44.70	$357.60	$30.91	$47.86
1 Skilled Worker	34.85	278.80	54.20	433.60		
1 Laborer	26.70	213.60	41.55	332.40		
.5 Equip. Oper. (crane)	35.90	143.60	54.10	216.40		
.5 S.P. Crane, 5 Ton		158.50		174.35	5.66	6.23
28 L.H., Daily Totals		$1024.10		$1514.35	$36.57	$54.09

Crew B-21A	Hr.	Daily	Hr.	Daily	Bare Costs	Incl. O&P
1 Labor Foreman	$28.70	$229.60	$44.70	$357.60	$32.97	$50.18
1 Laborer	26.70	213.60	41.55	332.40		
1 Plumber	40.85	326.80	61.40	491.20		
1 Plumber Apprentice	32.70	261.60	49.15	393.20		
1 Equip. Oper. (crane)	35.90	287.20	54.10	432.80		
1 S.P. Crane, 12 Ton		505.80		556.40	12.65	13.91
40 L.H., Daily Totals		$1824.60		$2563.60	$45.62	$64.09

Crew B-21B	Hr.	Daily	Hr.	Daily	Bare Costs	Incl. O&P
1 Laborer Foreman	$28.70	$229.60	$44.70	$357.60	$28.94	$44.69
3 Laborers	26.70	640.80	41.55	997.20		
1 Equip. Oper. (crane)	35.90	287.20	54.10	432.80		
1 Hyd. Crane, 12 Ton		602.60		662.85	15.07	16.57
40 L.H., Daily Totals		$1760.20		$2450.45	$44.01	$61.26

Crew B-21C

Crew B-21C	Hr.	Daily	Hr.	Daily	Bare Costs	Incl. O&P
1 Laborer Foreman	$28.70	$229.60	$44.70	$357.60	$28.79	$44.34
4 Laborers	26.70	854.40	41.55	1329.60		
1 Equip. Oper. (crane)	35.90	287.20	54.10	432.80		
1 Equip. Oper. Oiler	30.10	240.80	45.35	362.80		
2 Cutting Torches		36.00		39.60		
2 Gases		129.60		142.55		
1 Crane, 90 Ton		1325.00		1457.50	26.62	29.28
56 L.H., Daily Totals		$3102.60		$4122.45	$55.41	$73.62

Crew B-22

Crew B-22	Hr.	Daily	Hr.	Daily	Bare Costs	Incl. O&P
1 Labor Foreman (out)	$28.70	$229.60	$44.70	$357.60	$31.25	$48.27
1 Skilled Worker	34.85	278.80	54.20	433.60		
1 Laborer	26.70	213.60	41.55	332.40		
.75 Equip. Oper. (crane)	35.90	215.40	54.10	324.60		
.75 S.P. Crane, 5 Ton		237.75		261.50	7.93	8.72
30 L.H., Daily Totals		$1175.15		$1709.70	$39.18	$56.99

Crew B-22A

Crew B-22A	Hr.	Daily	Hr.	Daily	Bare Costs	Incl. O&P
1 Labor Foreman (out)	$28.70	$229.60	$44.70	$357.60	$30.29	$46.86
1 Skilled Worker	34.85	278.80	54.20	433.60		
2 Laborers	26.70	427.20	41.55	664.80		
.75 Equipment Oper. (crane)	35.90	215.40	54.10	324.60		
.75 Crane, 5 Ton		237.75		261.50		
1 Generator, 5 KW		32.20		35.40		
1 Butt Fusion Machine		432.80		476.10	18.49	20.34
38 L.H., Daily Totals		$1853.75		$2553.60	$48.78	$67.20

Crew B-22B

Crew B-22B	Hr.	Daily	Hr.	Daily	Bare Costs	Incl. O&P
1 Skilled Worker	$34.85	$278.80	$54.20	$433.60	$30.78	$47.88
1 Laborer	26.70	213.60	41.55	332.40		
1 Electro Fusion Machine		171.80		189.00	10.74	11.81
16 L.H., Daily Totals		$664.20		$955.00	$41.52	$59.69

Crew B-23

Crew B-23	Hr.	Daily	Hr.	Daily	Bare Costs	Incl. O&P
1 Labor Foreman (outside)	$28.70	$229.60	$44.70	$357.60	$27.10	$42.18
4 Laborers	26.70	854.40	41.55	1329.60		
1 Drill Rig, Wells		3179.00		3496.90		
1 Light Truck, 3 Ton		176.20		193.80	83.88	92.27
40 L.H., Daily Totals		$4439.20		$5377.90	$110.98	$134.45

Crew B-23A

Crew B-23A	Hr.	Daily	Hr.	Daily	Bare Costs	Incl. O&P
1 Labor Foreman (outside)	$28.70	$229.60	$44.70	$357.60	$30.02	$46.15
1 Laborer	26.70	213.60	41.55	332.40		
1 Equip. Operator (medium)	34.65	277.20	52.20	417.60		
1 Drill Rig, Wells		3179.00		3496.90		
1 Pickup Truck, 3/4 Ton		78.00		85.80	135.71	149.28
24 L.H., Daily Totals		$3977.40		$4690.30	$165.73	$195.43

Crew B-23B

Crew B-23B	Hr.	Daily	Hr.	Daily	Bare Costs	Incl. O&P
1 Labor Foreman (outside)	$28.70	$229.60	$44.70	$357.60	$30.02	$46.15
1 Laborer	26.70	213.60	41.55	332.40		
1 Equip. Operator (medium)	34.65	277.20	52.20	417.60		
1 Drill Rig, Wells		3179.00		3496.90		
1 Pickup Truck, 3/4 Ton		78.00		85.80		
1 Pump, Cntfgl, 6"		217.00		238.70	144.75	159.23
24 L.H., Daily Totals		$4194.40		$4929.00	$174.77	$205.38

Crew B-24

Crew B-24	Hr.	Daily	Hr.	Daily	Bare Costs	Incl. O&P
1 Cement Finisher	$32.85	$262.80	$48.35	$386.80	$31.27	$47.63
1 Laborer	26.70	213.60	41.55	332.40		
1 Carpenter	34.25	274.00	53.35	426.80		
24 L.H., Daily Totals		$750.40		$1146.00	$31.27	$47.63

Crew B-25

Crew B-25	Hr.	Daily	Hr.	Daily	Bare Costs	Incl. O&P
1 Labor Foreman	$28.70	$229.60	$44.70	$357.60	$29.05	$44.74
7 Laborers	26.70	1495.20	41.55	2326.80		
3 Equip. Oper. (med.)	34.65	831.60	52.20	1252.80		
1 Asphalt Paver, 130 H.P.		1590.00		1749.00		
1 Tandem Roller, 10 Ton		179.20		197.10		
1 Roller, Pneumatic Wheel		241.80		266.00	22.85	25.14
88 L.H., Daily Totals		$4567.40		$6149.30	$51.90	$69.88

Crew B-25B

Crew B-25B	Hr.	Daily	Hr.	Daily	Bare Costs	Incl. O&P
1 Labor Foreman	$28.70	$229.60	$44.70	$357.60	$29.52	$45.36
7 Laborers	26.70	1495.20	41.55	2326.80		
4 Equip. Oper. (medium)	34.65	1108.80	52.20	1670.80		
1 Asphalt Paver, 130 H.P.		1590.00		1749.00		
2 Rollers, Steel Wheel		358.40		394.25		
1 Roller, Pneumatic Wheel		241.80		266.00	22.81	25.10
96 L.H., Daily Totals		$5023.80		$6764.05	$52.33	$70.46

Crew B-25C

Crew B-25C	Hr.	Daily	Hr.	Daily	Bare Costs	Incl. O&P
1 Labor Foreman	$28.70	$229.60	$44.70	$357.60	$29.68	$45.63
3 Laborers	26.70	640.80	41.55	997.20		
2 Equip. Oper. (medium)	34.65	554.40	52.20	835.20		
1 Asphalt Paver, 130 H.P.		1590.00		1749.00		
1 Rollers, Steel Wheel		179.20		197.10	36.86	40.54
48 L.H., Daily Totals		$3194.00		$4136.10	$66.54	$86.17

Crew B-26

Crew B-26	Hr.	Daily	Hr.	Daily	Bare Costs	Incl. O&P
1 Labor Foreman (outside)	$28.70	$229.60	$44.70	$357.60	$29.91	$46.27
6 Laborers	26.70	1281.60	41.55	1994.40		
2 Equip. Oper. (med.)	34.65	554.40	52.20	835.20		
1 Rodman (reinf.)	37.95	303.60	62.60	500.80		
1 Cement Finisher	32.85	262.80	48.35	386.80		
1 Grader, 30,000 Lbs.		456.60		502.25		
1 Paving Mach. & Equip.		1910.00		2101.00	26.89	29.58
88 L.H., Daily Totals		$4998.60		$6678.05	$56.80	$75.85

Crew B-27

Crew B-27	Hr.	Daily	Hr.	Daily	Bare Costs	Incl. O&P
1 Labor Foreman (outside)	$28.70	$229.60	$44.70	$357.60	$27.20	$42.34
3 Laborers	26.70	640.80	41.55	997.20		
1 Berm Machine		216.20		237.80	6.76	7.43
32 L.H., Daily Totals		$1086.60		$1592.60	$33.96	$49.77

Crew B-28

Crew B-28	Hr.	Daily	Hr.	Daily	Bare Costs	Incl. O&P
2 Carpenters	$34.25	$548.00	$53.35	$853.60	$31.73	$49.42
1 Laborer	26.70	213.60	41.55	332.40		
24 L.H., Daily Totals		$761.60		$1186.00	$31.73	$49.42

Crew B-29

Crew B-29	Hr.	Daily	Hr.	Daily	Bare Costs	Incl. O&P
1 Labor Foreman (outside)	$28.70	$229.60	$44.70	$357.60	$28.79	$44.34
4 Laborers	26.70	854.40	41.55	1329.60		
1 Equip. Oper. (crane)	35.90	287.20	54.10	432.80		
1 Equip. Oper. Oiler	30.10	240.80	45.35	362.80		
1 Gradall, 3 Ton, 1/2 C.Y.		837.80		921.60	14.96	16.46
56 L.H., Daily Totals		$2449.80		$3404.40	$43.75	$60.80

Crews

Crew B-30

Crew No.	Bare Costs Hr.	Daily	Incl. Subs O & P Hr.	Daily	Cost Per Labor-Hour Bare Costs	Incl. O&P
1 Equip. Oper. (med.)	$34.65	$277.20	$52.20	$417.60	$29.92	$45.40
2 Truck Drivers (heavy)	27.55	440.80	42.00	672.00		
1 Hyd. Excavator, 1.5 C.Y.		720.40		792.45		
2 Dump Trucks, 16 Ton		953.20		1048.50	69.73	76.71
24 L.H., Daily Totals		$2391.60		$2930.55	$99.65	$122.11

Crew B-31

	Hr.	Daily	Hr.	Daily	Bare Costs	Incl. O&P
1 Labor Foreman (outside)	$28.70	$229.60	$44.70	$357.60	$28.61	$44.54
3 Laborers	26.70	640.80	41.55	997.20		
1 Carpenter	34.25	274.00	53.35	426.80		
1 Air Compr., 250 C.F.M.		112.40		123.65		
1 Sheeting Driver		7.20		7.90		
2-50 Ft. Air Hoses, 1.5" Dia.		9.40		10.35	3.23	3.55
40 L.H., Daily Totals		$1273.40		$1923.50	$31.84	$48.09

Crew B-32

	Hr.	Daily	Hr.	Daily	Bare Costs	Incl. O&P
1 Laborer	$26.70	$213.60	$41.55	$332.40	$32.66	$49.54
3 Equip. Oper. (med.)	34.65	831.60	52.20	1252.80		
1 Grader, 30,000 Lbs.		456.60		502.25		
1 Tandem Roller, 10 Ton		179.20		197.10		
1 Dozer, 200 H.P.		919.60		1011.55	48.61	53.47
32 L.H., Daily Totals		$2600.60		$3296.10	$81.27	$103.01

Crew B-32A

	Hr.	Daily	Hr.	Daily	Bare Costs	Incl. O&P
1 Laborer	$26.70	$213.60	$41.55	$332.40	$32.00	$48.65
2 Equip. Oper. (medium)	34.65	554.40	52.20	835.20		
1 Grader, 30,000 Lbs.		456.60		502.25		
1 Roller, Vibratory, 29,000 Lbs.		433.20		476.50	37.08	40.78
24 L.H., Daily Totals		$1657.80		$2146.35	$69.08	$89.43

Crew B-32B

	Hr.	Daily	Hr.	Daily	Bare Costs	Incl. O&P
1 Laborer	$26.70	$213.60	$41.55	$332.40	$32.00	$48.65
2 Equip. Oper. (medium)	34.65	554.40	52.20	835.20		
1 Dozer, 200 H.P.		919.60		1011.55		
1 Roller, Vibratory, 29,000 Lbs.		433.20		476.50	56.37	62.00
24 L.H., Daily Totals		$2120.80		$2655.65	$88.37	$110.65

Crew B-32C

	Hr.	Daily	Hr.	Daily	Bare Costs	Incl. O&P
1 Labor Foreman	$28.70	$229.60	$44.70	$357.60	$31.01	$47.40
2 Laborers	26.70	427.20	41.55	664.80		
3 Equip. Oper. (medium)	34.65	831.60	52.20	1252.80		
1 Grader, 30,000 Lbs.		456.60		502.25		
1 Roller, Steel Wheel		179.20		197.10		
1 Dozer, 200 H.P.		919.60		1011.55	32.40	35.64
48 L.H., Daily Totals		$3043.80		$3986.10	$63.41	$83.04

Crew B-33A

	Hr.	Daily	Hr.	Daily	Bare Costs	Incl. O&P
1 Equip. Oper. (med.)	$34.65	$277.20	$52.20	$417.60	$32.38	$49.16
.5 Laborer	26.70	106.80	41.55	166.20		
.25 Equip. Oper. (med.)	34.65	69.30	52.20	104.40		
1 Scraper, Towed, 7 C.Y.		150.75		165.80		
1.25 Dozer, 300 H.P.		1493.75		1643.15	117.47	129.21
14 L.H., Daily Totals		$2097.80		$2497.15	$149.85	$178.37

Crew B-33B

	Hr.	Daily	Hr.	Daily	Bare Costs	Incl. O&P
1 Equip. Oper. (med.)	$34.65	$277.20	$52.20	$417.60	$32.38	$49.16
.5 Laborer	26.70	106.80	41.55	166.20		
.25 Equip. Oper. (med.)	34.65	69.30	52.20	104.40		
1 Scraper, Towed, 10 C.Y.		169.20		186.10		
1.25 Dozer, 300 H.P.		1493.75		1643.15	118.78	130.66
14 L.H., Daily Totals		$2116.25		$2517.45	$151.16	$179.82

Crew B-33C

	Hr.	Daily	Hr.	Daily	Bare Costs	Incl. O&P
1 Equip. Oper. (med.)	$34.65	$277.20	$52.20	$417.60	$32.38	$49.16
.5 Laborer	26.70	106.80	41.55	166.20		
.25 Equip. Oper. (med.)	34.65	69.30	52.20	104.40		
1 Scraper, Towed, 12 C.Y.		169.20		186.10		
1.25 Dozer, 300 H.P.		1493.75		1643.15	118.78	130.66
14 L.H., Daily Totals		$2116.25		$2517.45	$151.16	$179.82

Crew B-33D

	Hr.	Daily	Hr.	Daily	Bare Costs	Incl. O&P
1 Equip. Oper. (med.)	$34.65	$277.20	$52.20	$417.60	$32.38	$49.16
.5 Laborer	26.70	106.80	41.55	166.20		
.25 Equip. Oper. (med.)	34.65	69.30	52.20	104.40		
1 S.P. Scraper, 14 C.Y.		1552.00		1707.20		
.25 Dozer, 300 H.P.		298.75		328.65	132.20	145.42
14 L.H., Daily Totals		$2304.05		$2724.05	$164.58	$194.58

Crew B-33E

	Hr.	Daily	Hr.	Daily	Bare Costs	Incl. O&P
1 Equip. Oper. (med.)	$34.65	$277.20	$52.20	$417.60	$32.38	$49.16
.5 Laborer	26.70	106.80	41.55	166.20		
.25 Equip. Oper. (med.)	34.65	69.30	52.20	104.40		
1 S.P. Scraper, 24 C.Y.		2317.00		2548.70		
.25 Dozer, 300 H.P.		298.75		328.65	186.84	205.52
14 L.H., Daily Totals		$3069.05		$3565.55	$219.22	$254.68

Crew B-33F

	Hr.	Daily	Hr.	Daily	Bare Costs	Incl. O&P
1 Equip. Oper. (med.)	$34.65	$277.20	$52.20	$417.60	$32.38	$49.16
.5 Laborer	26.70	106.80	41.55	166.20		
.25 Equip. Oper. (med.)	34.65	69.30	52.20	104.40		
1 Elev. Scraper, 11 C.Y.		819.40		901.35		
.25 Dozer, 300 H.P.		298.75		328.65	79.87	87.85
14 L.H., Daily Totals		$1571.45		$1918.20	$112.25	$137.01

Crew B-33G

	Hr.	Daily	Hr.	Daily	Bare Costs	Incl. O&P
1 Equip. Oper. (med.)	$34.65	$277.20	$52.20	$417.60	$32.38	$49.16
.5 Laborer	26.70	106.80	41.55	166.20		
.25 Equip. Oper. (med.)	34.65	69.30	52.20	104.40		
1 Elev. Scraper, 20 C.Y.		1648.00		1812.80		
.25 Dozer, 300 H.P.		298.75		328.65	139.05	152.96
14 L.H., Daily Totals		$2400.05		$2829.65	$171.43	$202.12

Crew B-33H

	Hr.	Daily	Hr.	Daily	Bare Costs	Incl. O&P
.25 Laborer	$26.70	$53.40	$41.55	$83.10	$33.28	$50.36
1 Equipment Operator (med.)	34.65	277.20	52.20	417.60		
.2 Equipment Operator (med.)	34.65	55.44	52.20	83.52		
1 Scraper, 32-44 C.Y.		2728.00		3000.80		
.2 Dozer, 400 H.P.		304.80		335.30	261.45	287.59
11. L.H., Daily Totals		$3418.84		$3920.32	$294.73	$337.95

Crew B-33J

	Hr.	Daily	Hr.	Daily	Bare Costs	Incl. O&P
1 Equipment Operator (med.)	$34.65	$277.20	$52.20	$417.60	$34.65	$52.20
1 Scraper 17 C.Y.		1552.00		1707.20	194.00	213.40
8 L.H., Daily Totals		$1829.20		$2124.80	$228.65	$265.60

Crew B-34A

	Hr.	Daily	Hr.	Daily	Bare Costs	Incl. O&P
1 Truck Driver (heavy)	$27.55	$220.40	$42.00	$336.00	$27.55	$42.00
1 Dump Truck, 12 Ton		326.60		359.25	40.83	44.91
8 L.H., Daily Totals		$547.00		$695.25	$68.38	$86.91

CREWS

Crew B-34B

Crew No.	Bare Costs Hr.	Daily	Incl. Subs O & P Hr.	Daily	Cost Per Labor-Hour Bare Costs	Incl. O&P
Crew B-34B	Hr.	Daily	Hr.	Daily	Bare Costs	Incl. O&P
1 Truck Driver (heavy)	$27.55	$220.40	$42.00	$336.00	$27.55	$42.00
1 Dump Truck, 16 Ton		476.60		524.25	59.58	65.53
8 L.H., Daily Totals		$697.00		$860.25	$87.13	$107.53

Crew B-34C	Hr.	Daily	Hr.	Daily	Bare Costs	Incl. O&P
1 Truck Driver (heavy)	$27.55	$220.40	$42.00	$336.00	$27.55	$42.00
1 Truck Tractor, 40 Ton		301.60		331.75		
1 Dump Trailer, 16.5 C.Y.		103.00		113.30	50.58	55.63
8 L.H., Daily Totals		$625.00		$781.05	$78.13	$97.63

Crew B-34D	Hr.	Daily	Hr.	Daily	Bare Costs	Incl. O&P
1 Truck Driver (heavy)	$27.55	$220.40	$42.00	$336.00	$27.55	$42.00
1 Truck Tractor, 40 Ton		301.60		331.75		
1 Dump Trailer, 20 C.Y.		115.80		127.40	52.18	57.39
8 L.H., Daily Totals		$637.80		$795.15	$79.73	$99.39

Crew B-34E	Hr.	Daily	Hr.	Daily	Bare Costs	Incl. O&P
1 Truck Driver (heavy)	$27.55	$220.40	$42.00	$336.00	$27.55	$42.00
1 Truck, Off Hwy., 25 Ton		912.20		1003.40	114.03	125.43
8 L.H., Daily Totals		$1132.60		$1339.40	$141.58	$167.43

Crew B-34F	Hr.	Daily	Hr.	Daily	Bare Costs	Incl. O&P
1 Truck Driver (heavy)	$27.55	$220.40	$42.00	$336.00	$27.55	$42.00
1 Truck, Off Hwy., 22 C.Y.		933.40		1026.75	116.68	128.34
8 L.H., Daily Totals		$1153.80		$1362.75	$144.23	$170.34

Crew B-34G	Hr.	Daily	Hr.	Daily	Bare Costs	Incl. O&P
1 Truck Driver (heavy)	$27.55	$220.40	$42.00	$336.00	$27.55	$42.00
1 Truck, Off Hwy., 34 C.Y.		1181.00		1299.10	147.63	162.39
8 L.H., Daily Totals		$1401.40		$1635.10	$175.18	$204.39

Crew B-34H	Hr.	Daily	Hr.	Daily	Bare Costs	Incl. O&P
1 Truck Driver (heavy)	$27.55	$220.40	$42.00	$336.00	$27.55	$42.00
1 Truck, Off Hwy., 42 C.Y.		1276.00		1403.60	159.50	175.45
8 L.H., Daily Totals		$1496.40		$1739.60	$187.05	$217.45

Crew B-34J	Hr.	Daily	Hr.	Daily	Bare Costs	Incl. O&P
1 Truck Driver (heavy)	$27.55	$220.40	$42.00	$336.00	$27.55	$42.00
1 Truck, Off Hwy., 60 C.Y.		1649.00		1813.90	206.13	226.74
8 L.H., Daily Totals		$1869.40		$2149.90	$233.68	$268.74

Crew B-34K	Hr.	Daily	Hr.	Daily	Bare Costs	Incl. O&P
1 Truck Driver (heavy)	$27.55	$220.40	$42.00	$336.00	$27.55	$42.00
1 Truck Tractor, 240 H.P.		353.60		388.95		
1 Low Bed Trailer		170.20		187.20	65.48	72.02
8 L.H., Daily Totals		$744.20		$912.15	$93.03	$114.02

Crew B-34N	Hr.	Daily	Hr.	Daily	Bare Costs	Incl. O&P
1 Truck Driver (heavy)	$27.55	$220.40	$42.00	$336.00	$27.55	$42.00
1 Dump Truck, 12 Ton		326.60		359.25		
1 Flatbed Trailer, 40 Ton		119.40		131.35	55.75	61.33
8 L.H., Daily Totals		$666.40		$826.60	$83.30	$103.33

Crew B-34P	Hr.	Daily	Hr.	Daily	Bare Costs	Incl. O&P
1 Pipe Fitter	$40.95	$327.60	$61.55	$492.40	$34.13	$51.53
1 Truck Driver (light)	26.80	214.40	40.85	326.80		
1 Equip. Oper. (medium)	34.65	277.20	52.20	417.60		
1 Flatbed Truck, 3 Ton		176.20		193.80		
1 Backhoe Loader, 48 H.P.		215.80		237.40	16.33	17.97
24 L.H., Daily Totals		$1211.20		$1668.00	$50.46	$69.50

Crew B-34Q	Hr.	Daily	Hr.	Daily	Bare Costs	Incl. O&P
1 Pipe Fitter	$40.95	$327.60	$61.55	$492.40	$34.55	$52.17
1 Truck Driver (light)	26.80	214.40	40.85	326.80		
1 Eqip. Oper. (crane)	35.90	287.20	54.10	432.80		
1 Flatbed Trailer, 25 Ton		88.00		96.80		
1 Dump Truck, 12 Ton		326.60		359.25		
1 Hyd. Crane, 25 Ton		616.80		678.50	42.98	47.27
24 L.H., Daily Totals		$1860.60		$2386.55	$77.53	$99.44

Crew B-34R	Hr.	Daily	Hr.	Daily	Bare Costs	Incl. O&P
1 Pipe Fitter	$40.95	$327.60	$61.55	$492.40	$34.55	$52.17
1 Truck Driver (light)	26.80	214.40	40.85	326.80		
1 Eqip. Oper. (crane)	35.90	287.20	54.10	432.80		
1 Flatbed Trailer, 25 Ton		88.00		96.80		
1 Dump Truck, 12 Ton		326.60		359.25		
1 Hyd. Crane, 25 Ton		616.80		678.50		
1 Hyd. Excavator, 1 C.Y.		557.80		613.60	66.22	72.84
24 L.H., Daily Totals		$2418.40		$3000.15	$100.77	$125.01

Crew B-34S	Hr.	Daily	Hr.	Daily	Bare Costs	Incl. O&P
2 Pipe Fitters	$40.95	$655.20	$61.55	$984.80	$36.34	$54.80
1 Truck Driver (heavy)	27.55	220.40	42.00	336.00		
1 Eqip. Oper. (crane)	35.90	287.20	54.10	432.80		
1 Flatbed Trailer, 40 Ton		119.40		131.35		
1 Truck Tractor, 40 Ton		301.60		331.75		
1 Truck Crane, 80 Ton		995.00		1094.50		
1 Hyd. Excavator, 2 C.Y.		909.80		1000.80	72.68	79.95
32 L.H., Daily Totals		$3488.60		$4312.00	$109.02	$134.75

Crew B-34T	Hr.	Daily	Hr.	Daily	Bare Costs	Incl. O&P
2 Pipe Fitters	$40.95	$655.20	$61.55	$984.80	$36.34	$54.80
1 Truck Driver (heavy)	27.55	220.40	42.00	336.00		
1 Eqip. Oper. (crane)	35.90	287.20	54.10	432.80		
1 Flatbed Trailer, 40 Ton		119.40		131.35		
1 Truck Tractor, 40 Ton		301.60		331.75		
1 Truck Crane, 80 Ton		995.00		1094.50	44.25	48.68
32 L.H., Daily Totals		$2578.80		$3311.20	$80.59	$103.48

Crew B-35	Hr.	Daily	Hr.	Daily	Bare Costs	Incl. O&P
1 Laborer Foreman (out)	$28.70	$229.60	$44.70	$357.60	$32.85	$50.22
1 Skilled Worker	34.85	278.80	54.20	433.60		
1 Welder (plumber)	40.85	326.80	61.40	491.20		
1 Laborer	26.70	213.60	41.55	332.40		
1 Equip. Oper. (crane)	35.90	287.20	54.10	432.80		
1 Equip. Oper. Oiler	30.10	240.80	45.35	362.80		
1 Electric Welding Mach.		80.65		88.70		
1 Hyd. Excavator, .75 C.Y.		474.00		521.40	11.56	12.71
48 L.H., Daily Totals		$2131.45		$3020.50	$44.41	$62.93

CREWS

Crew No.	Bare Costs		Incl. Subs O & P		Cost Per Labor-Hour	

Crew B-35A	Hr.	Daily	Hr.	Daily	Bare Costs	Incl. O&P
1 Laborer Foreman (out)	$28.70	$229.60	$44.70	$357.60	$31.97	$48.98
2 Laborers	26.70	427.20	41.55	664.80		
1 Skilled Worker	34.85	278.80	54.20	433.60		
1 Welder (plumber)	40.85	326.80	61.40	491.20		
1 Equip. Oper. (crane)	35.90	287.20	54.10	432.80		
1 Equip. Oper. Oiler	30.10	240.80	45.35	362.80		
1 Welder, 300 amp		81.20		89.30		
1 Crane, 75 Ton		1170.00		1287.00	22.34	24.58
56 L.H., Daily Totals		$3041.60		$4119.10	$54.31	$73.56

Crew B-36	Hr.	Daily	Hr.	Daily	Bare Costs	Incl. O&P
1 Labor Foreman (outside)	$28.70	$229.60	$44.70	$357.60	$30.28	$46.44
2 Laborers	26.70	427.20	41.55	664.80		
2 Equip. Oper. (med.)	34.65	554.40	52.20	835.20		
1 Dozer, 200 H.P.		919.60		1011.55		
1 Aggregate Spreader		35.00		38.50		
1 Tandem Roller, 10 Ton		179.20		197.10	28.35	31.18
40 L.H., Daily Totals		$2345.00		$3104.75	$58.63	$77.62

Crew B-36A	Hr.	Daily	Hr.	Daily	Bare Costs	Incl. O&P
1 Labor Foreman (outside)	$28.70	$229.60	$44.70	$357.60	$31.53	$48.09
2 Laborers	26.70	427.20	41.55	664.80		
4 Equip. Oper. (med.)	34.65	1108.80	52.20	1670.40		
1 Dozer, 200 H.P.		919.60		1011.55		
1 Aggregate Spreader		35.00		38.50		
1 Roller, Steel Wheel		179.20		197.10		
1 Roller, Pneumatic Wheel		241.80		266.00	24.56	27.02
56 L.H., Daily Totals		$3141.20		$4205.95	$56.09	$75.11

Crew B-36B	Hr.	Daily	Hr.	Daily	Bare Costs	Incl. O&P
1 Labor Foreman (outside)	$28.70	$229.60	$44.70	$357.60	$31.03	$47.33
2 Laborers	26.70	427.20	41.55	664.80		
4 Equip. Oper. (medium)	34.65	1108.80	52.20	1670.40		
1 Truck Driver, Heavy	27.55	220.40	42.00	336.00		
1 Grader, 30,000 Lbs.		456.60		502.25		
1 F.E. Loader, crl, 1.5 C.Y.		368.80		405.70		
1 Dozer, 300 H.P.		1195.00		1314.50		
1 Roller, Vibratory		433.20		476.50		
1 Truck, Tractor, 240 H.P.		353.60		388.95		
1 Water Tanker, 5000 Gal.		116.80		128.50	45.69	50.26
64 L.H., Daily Totals		$4910.00		$6245.20	$76.72	$97.59

Crew B-36C	Hr.	Daily	Hr.	Daily	Bare Costs	Incl. O&P
1 Labor Foreman (outside)	$28.70	$229.60	$44.70	$357.60	$32.04	$48.66
3 Equip. Oper. (medium)	34.65	831.60	52.20	1252.80		
1 Truck Driver, Heavy	27.55	220.40	42.00	336.00		
1 Grader, 30,000 Lbs.		456.60		502.25		
1 Dozer, 300 H.P.		1195.00		1314.50		
1 Roller, Vibratory		433.20		476.50		
1 Truck, Tractor, 240 H.P.		353.60		388.95		
1 Water Tanker, 5000 Gal.		116.80		128.50	63.88	70.27
40 L.H., Daily Totals		$3836.80		$4757.10	$95.92	$118.93

Crew B-37	Hr.	Daily	Hr.	Daily	Bare Costs	Incl. O&P
1 Labor Foreman (outside)	$28.70	$229.60	$44.70	$357.60	$28.09	$43.45
4 Laborers	26.70	854.40	41.55	1329.60		
1 Equip. Oper. (light)	33.05	264.40	49.80	398.40		
1 Tandem Roller, 5 Ton		107.40		118.15	2.24	2.46
48 L.H., Daily Totals		$1455.80		$2203.75	$30.33	$45.91

Crew B-38	Hr.	Daily	Hr.	Daily	Bare Costs	Incl. O&P
1 Labor Foreman (outside)	$28.70	$229.60	$44.70	$357.60	$29.96	$45.96
2 Laborers	26.70	427.20	41.55	664.80		
1 Equip. Oper. (light)	33.05	264.40	49.80	398.40		
1 Equip. Oper. (medium)	34.65	277.20	52.20	417.60		
1 Backhoe Loader, 48 H.P.		215.80		237.40		
1 Hyd.Hammer, (1200 lb)		112.00		123.20		
1 F.E. Loader (170 H.P.)		445.60		490.15		
1 Pavt. Rem. Bucket		51.80		57.00	20.63	22.69
40 L.H., Daily Totals		$2023.60		$2746.15	$50.59	$68.65

Crew B-39	Hr.	Daily	Hr.	Daily	Bare Costs	Incl. O&P
1 Labor Foreman (outside)	$28.70	$229.60	$44.70	$357.60	$28.09	$43.45
4 Laborers	26.70	854.40	41.55	1329.60		
1 Equip. Oper. (light)	33.05	264.40	49.80	398.40		
1 Air Compr., 250 C.F.M.		112.40		123.65		
2 Air Tools & Accessories		19.60		21.55		
2-50 Ft. Air Hoses, 1.5" Dia.		9.40		10.35	2.95	3.24
48 L.H., Daily Totals		$1489.80		$2241.15	$31.04	$46.69

Crew B-40	Hr.	Daily	Hr.	Daily	Bare Costs	Incl. O&P
1 Pile Driver Foreman (out)	$35.30	$282.40	$58.15	$465.20	$33.80	$53.91
4 Pile Drivers	33.30	1065.60	54.90	1756.80		
2 Equip. Oper. (crane)	35.90	574.40	54.10	865.60		
1 Equip. Oper. Oiler	30.10	240.80	45.35	362.80		
1 Crane, 40 Ton		888.20		977.00		
1 Vibratory Hammer & Gen.		1403.00		1543.30	35.80	39.38
64 L.H., Daily Totals		$4454.40		$5970.70	$69.60	$93.29

Crew B-40B	Hr.	Daily	Hr.	Daily	Bare Costs	Incl. O&P
1 Laborer Foreman	$28.70	$229.60	$44.70	$357.60	$29.13	$44.80
3 Laborers	26.70	640.80	41.55	997.20		
1 Equip. Oper. (crane)	35.90	287.20	54.10	432.80		
1 Equip. Oper. Oiler	30.10	240.80	45.35	362.80		
1 Crane, 40 Ton		951.90		1047.10	19.83	21.81
48 L.H., Daily Totals		$2350.30		$3197.50	$48.96	$66.61

Crew B-41	Hr.	Daily	Hr.	Daily	Bare Costs	Incl. O&P
1 Labor Foreman (outside)	$28.70	$229.60	$44.70	$357.60	$27.64	$42.87
4 Laborers	26.70	854.40	41.55	1329.60		
.25 Equip. Oper. (crane)	35.90	71.80	54.10	108.20		
.25 Equip. Oper. Oiler	30.10	60.20	45.35	90.70		
.25 Crawler Crane, 40 Ton		222.05		244.25	5.05	5.55
44 L.H., Daily Totals		$1438.05		$2130.35	$32.69	$48.42

Crew B-42	Hr.	Daily	Hr.	Daily	Bare Costs	Incl. O&P
1 Labor Foreman (outside)	$28.70	$229.60	$44.70	$357.60	$29.96	$47.33
4 Laborers	26.70	854.40	41.55	1329.60		
1 Equip. Oper. (crane)	35.90	287.20	54.10	432.80		
1 Equip. Oper. Oiler	30.10	240.80	45.35	362.80		
1 Welder	38.15	305.20	68.25	546.00		
1 Hyd. Crane, 25 Ton		616.80		678.50		
1 Gas Welding Machine		81.20		89.30		
1 Horz. Boring Csg. Mch.		372.40		409.65	16.73	18.40
64 L.H., Daily Totals		$2987.60		$4206.25	$46.69	$65.73

| Crew No. | Bare Costs | | Incl. Subs O & P | | Cost Per Labor-Hour | |

Crew B-43

Crew B-43	Hr.	Daily	Hr.	Daily	Bare Costs	Incl. O&P
1 Labor Foreman (outside)	$28.70	$229.60	$44.70	$357.60	$29.13	$44.80
3 Laborers	26.70	640.80	41.55	997.20		
1 Equip. Oper. (crane)	35.90	287.20	54.10	432.80		
1 Equip. Oper. Oiler	30.10	240.80	45.35	362.80		
1 Drill Rig & Augers		3179.00		3496.90	66.23	72.85
48 L.H., Daily Totals		$4577.40		$5647.30	$95.36	$117.65

Crew B-44	Hr.	Daily	Hr.	Daily	Bare Costs	Incl. O&P
1 Pile Driver Foreman	$35.30	$282.40	$58.15	$465.20	$33.38	$53.44
4 Pile Drivers	33.30	1065.60	54.90	1756.80		
2 Equip. Oper. (crane)	35.90	574.40	54.10	865.60		
1 Laborer	26.70	213.60	41.55	332.40		
1 Crane, 40 Ton, & Access.		888.20		977.00		
45 L.F. Leads, 15K Ft. Lbs.		56.25		61.90	14.79	16.27
64 L.H., Daily Totals		$3080.45		$4458.90	$48.17	$69.71

Crew B-45	Hr.	Daily	Hr.	Daily	Bare Costs	Incl. O&P
1 Equip. Oper. (med.)	$34.65	$277.20	$52.20	$417.60	$31.10	$47.10
1 Truck Driver (heavy)	27.55	220.40	42.00	336.00		
1 Dist. Tank Truck, 3K Gal.		233.80		257.20		
1 Tractor, 4 x 2, 250 H.P.		291.60		320.75	32.84	36.12
16 L.H., Daily Totals		$1023.00		$1331.55	$63.94	$83.22

Crew B-46	Hr.	Daily	Hr.	Daily	Bare Costs	Incl. O&P
1 Pile Driver Foreman	$35.30	$282.40	$58.15	$465.20	$30.33	$48.77
2 Pile Drivers	33.30	532.80	54.90	878.40		
3 Laborers	26.70	640.80	41.55	997.20		
1 Chain Saw, 36" Long		33.40		36.75	.70	.77
48 L.H., Daily Totals		$1489.40		$2377.55	$31.03	$49.54

Crew B-47	Hr.	Daily	Hr.	Daily	Bare Costs	Incl. O&P
1 Blast Foreman	$28.70	$229.60	$44.70	$357.60	$29.48	$45.35
1 Driller	26.70	213.60	41.55	332.40		
1 Equip. Oper. (light)	33.05	264.40	49.80	398.40		
1 Crawler Type Drill, 4"		664.40		730.85		
1 Air Compr., 600 C.F.M.		297.80		327.60		
2-50 Ft. Air Hoses, 3" Dia.		35.40		38.95	41.57	45.72
24 L.H., Daily Totals		$1705.20		$2185.80	$71.05	$91.07

Crew B-47A	Hr.	Daily	Hr.	Daily	Bare Costs	Incl. O&P
1 Drilling Foreman	$28.70	$229.60	$44.70	$357.60	$31.57	$48.05
1 Equip. Oper. (heavy)	35.90	287.20	54.10	432.80		
1 Oiler	30.10	240.80	45.35	362.80		
1 Quarry Drill		822.60		904.85	34.28	37.70
24 L.H., Daily Totals		$1580.20		$2058.05	$65.85	$85.75

Crew B-47C	Hr.	Daily	Hr.	Daily	Bare Costs	Incl. O&P
1 Laborer	$26.70	$213.60	$41.55	$332.40	$29.88	$45.68
1 Equip. Oper. (light)	33.05	264.40	49.80	398.40		
1 Air Compressor, 750 CFM		317.40		349.15		
2-50' Air Hose, 3"		35.40		38.95		
1 Air Track Drill, 4"		664.40		730.85	63.58	69.93
16 L.H., Daily Totals		$1495.20		$1849.75	$93.46	$115.61

Crew B-47E	Hr.	Daily	Hr.	Daily	Bare Costs	Incl. O&P
1 Laborer Foreman	$28.70	$229.60	$44.70	$357.60	$27.20	$42.34
3 Laborers	26.70	640.80	41.55	997.20		
1 Truck, Flatbed, 3 Ton		176.20		193.80	5.51	6.06
32 L.H., Daily Totals		$1046.60		$1548.60	$32.71	$48.40

Crew B-47G	Hr.	Daily	Hr.	Daily	Bare Costs	Incl. O&P
1 Laborer Foreman	$28.70	$229.60	$44.70	$357.60	$28.79	$44.40
2 Laborers	26.70	427.20	41.55	664.80		
1 Equip. Oper. (light)	33.05	264.40	49.80	398.40		
1 Air Track Drill, 4"		664.40		730.85		
1 Air Compr., 600 C.F.M.		297.80		327.60		
2-50 Ft. Air Hoses, 3" Dia.		35.40		38.95		
1 Grout Pump		101.40		111.55	34.34	37.78
32 L.H., Daily Totals		$2020.20		$2629.75	$63.13	$82.18

Crew B-48	Hr.	Daily	Hr.	Daily	Bare Costs	Incl. O&P
1 Labor Foreman (outside)	$28.70	$229.60	$44.70	$357.60	$29.69	$45.51
3 Laborers	26.70	640.80	41.55	997.20		
1 Equip. Oper. (crane)	35.90	287.20	54.10	432.80		
1 Equip. Oper. Oiler	30.10	240.80	45.35	362.80		
1 Equip. Oper. (light)	33.05	264.40	49.80	398.40		
1 Centr. Water Pump, 6"		217.00		238.70		
1-20 Ft. Suction Hose, 6"		12.80		14.10		
1-50 Ft. Disch. Hose, 6"		8.90		9.80		
1 Drill Rig & Augers		3179.00		3496.90	61.03	67.13
56 L.H., Daily Totals		$5080.50		$6308.30	$90.72	$112.64

Crew B-49	Hr.	Daily	Hr.	Daily	Bare Costs	Incl. O&P
1 Labor Foreman (outside)	$28.70	$229.60	$44.70	$357.60	$30.95	$47.99
3 Laborers	26.70	640.80	41.55	997.20		
2 Equip. Oper. (crane)	35.90	574.40	54.10	865.60		
2 Equip. Oper. Oilers	30.10	481.60	45.35	725.60		
1 Equip. Oper. (light)	33.05	264.40	49.80	398.40		
2 Pile Drivers	33.30	532.80	54.90	878.40		
1 Hyd. Crane, 25 Ton		616.80		678.50		
1 Centr. Water Pump, 6"		217.00		238.70		
1-20 Ft. Suction Hose, 6"		12.80		14.10		
1-50 Ft. Disch. Hose, 6"		8.90		9.80		
1 Drill Rig & Augers		3179.00		3496.90	45.85	50.43
88 L.H., Daily Totals		$6758.10		$8660.80	$76.80	$98.42

Crew B-50	Hr.	Daily	Hr.	Daily	Bare Costs	Incl. O&P
2 Pile Driver Foremen	$35.30	$564.80	$58.15	$930.40	$32.31	$51.71
6 Pile Drivers	33.30	1598.40	54.90	2635.20		
2 Equip. Oper. (crane)	35.90	574.40	54.10	865.60		
1 Equip. Oper. Oiler	30.10	240.80	45.35	362.80		
3 Laborers	26.70	640.80	41.55	997.20		
1 Crane, 40 Ton		888.20		977.00		
60 L.F. Leads, 15K Ft. Lbs.		75.00		82.50		
1 Hammer, 15K Ft. Lbs.		359.80		395.80		
1 Air Compr., 600 C.F.M.		297.80		327.60		
2-50 Ft. Air Hoses, 3" Dia.		35.40		38.95		
1 Chain Saw, 36" Long		33.40		36.75	15.11	16.62
112 L.H., Daily Totals		$5308.80		$7649.80	$47.42	$68.33

Crew B-51	Hr.	Daily	Hr.	Daily	Bare Costs	Incl. O&P
1 Labor Foreman (outside)	$28.70	$229.60	$44.70	$357.60	$27.05	$41.96
4 Laborers	26.70	854.40	41.55	1329.60		
1 Truck Driver (light)	26.80	214.40	40.85	326.80		
1 Light Truck, 1.5 Ton		128.80		141.70	2.68	2.95
48 L.H., Daily Totals		$1427.20		$2155.70	$29.73	$44.91

CREWS

Crews

Crew B-52

Crew B-52	Hr.	Daily	Hr.	Daily	Bare Costs	Incl. O&P
1 Carpenter Foreman	$36.25	$290.00	$56.45	$451.60	$31.39	$48.55
1 Carpenter	34.25	274.00	53.35	426.80		
3 Laborers	26.70	640.80	41.55	997.20		
1 Cement Finisher	32.85	262.80	48.35	386.80		
.5 Rodman (reinf.)	37.95	151.80	62.60	250.40		
.5 Equip. Oper. (med.)	34.65	138.60	52.20	208.80		
.5 F.E. Ldr., T.M., 2.5 C.Y.		394.00		433.40	7.04	7.74
56 L.H., Daily Totals		$2152.00		$3155.00	$38.43	$56.29

Crew B-53

Crew B-53	Hr.	Daily	Hr.	Daily	Bare Costs	Incl. O&P
1 Equip. Oper. (light)	$33.05	$264.40	$49.80	$398.40	$33.05	$49.80
1 Trencher, Chain, 12 H.P.		48.00		52.80	6.00	6.60
8 L.H., Daily Totals		$312.40		$451.20	$39.05	$56.40

Crew B-54

Crew B-54	Hr.	Daily	Hr.	Daily	Bare Costs	Incl. O&P
1 Equip. Oper. (light)	$33.05	$264.40	$49.80	$398.40	$33.05	$49.80
1 Trencher, Chain, 40 H.P.		213.40		234.75	26.68	29.34
8 L.H., Daily Totals		$477.80		$633.15	$59.73	$79.14

Crew B-54A

Crew B-54A	Hr.	Daily	Hr.	Daily	Bare Costs	Incl. O&P
.17 Labor Foreman (outside)	$28.70	$39.03	$44.70	$60.79	$33.79	$51.11
1 Equipment Operator (med.)	34.65	277.20	52.20	417.60		
1 Wheel Trencher, 67 H.P.		816.20		897.80	87.20	95.92
9.36 L.H., Daily Totals		$1132.43		$1376.19	$120.99	$147.03

Crew B-54B

Crew B-54B	Hr.	Daily	Hr.	Daily	Bare Costs	Incl. O&P
.25 Labor Foreman (outside)	$28.70	$57.40	$44.70	$89.40	$33.46	$50.70
1 Equipment Operator (med.)	34.65	277.20	52.20	417.60		
1 Wheel Trencher, 150 H.P.		1460.00		1606.00	146.00	160.60
10 L.H., Daily Totals		$1794.60		$2113.00	$179.46	$211.30

Crew B-55

Crew B-55	Hr.	Daily	Hr.	Daily	Bare Costs	Incl. O&P
2 Laborers	$26.70	$427.20	$41.55	$664.80	$26.73	$41.32
1 Truck Driver (light)	26.80	214.40	40.85	326.80		
1 Auger, 4" to 36" Dia		604.60		665.05		
1 Flatbed 3 Ton Truck		176.20		193.80	32.53	35.79
24 L.H., Daily Totals		$1422.40		$1850.45	$59.26	$77.11

Crew B-56

Crew B-56	Hr.	Daily	Hr.	Daily	Bare Costs	Incl. O&P
1 Laborer	$26.70	$213.60	$41.55	$332.40	$29.88	$45.68
1 Equip. Oper. (light)	33.05	264.40	49.80	398.40		
1 Crawler Type Drill, 4"		664.40		730.85		
1 Air Compr., 600 C.F.M.		297.80		327.60		
1-50 Ft. Air Hose, 3" Dia.		17.70		19.45	61.24	67.37
16 L.H., Daily Totals		$1457.90		$1808.70	$91.12	$113.05

Crew B-57

Crew B-57	Hr.	Daily	Hr.	Daily	Bare Costs	Incl. O&P
1 Labor Foreman (outside)	$28.70	$229.60	$44.70	$357.60	$30.19	$46.18
2 Laborers	26.70	427.20	41.55	664.80		
1 Equip. Oper. (crane)	35.90	287.20	54.10	432.80		
1 Equip. Oper. (light)	33.05	264.40	49.80	398.40		
1 Equip. Oper. Oiler	30.10	240.80	45.35	362.80		
1 Power Shovel, 1 C.Y.		833.20		916.50		
1 Clamshell Bucket, 1 C.Y.		43.40		47.75		
1 Centr. Water Pump, 6"		217.00		238.70		
1-20 Ft. Suction Hose, 6"		12.80		14.10		
20-50 Ft. Disch. Hoses, 6"		178.00		195.80	26.76	29.43
48 L.H., Daily Totals		$2733.60		$3629.25	$56.95	$75.61

Crew B-58

Crew B-58	Hr.	Daily	Hr.	Daily	Bare Costs	Incl. O&P
2 Laborers	$26.70	$427.20	$41.55	$664.80	$28.82	$44.30
1 Equip. Oper. (light)	33.05	264.40	49.80	398.40		
1 Backhoe Loader, 48 H.P.		215.80		237.40		
1 Small Helicopter, w/pilot		2153.00		2368.30	98.70	108.57
24 L.H., Daily Totals		$3060.40		$3668.90	$127.52	$152.87

Crew B-59

Crew B-59	Hr.	Daily	Hr.	Daily	Bare Costs	Incl. O&P
1 Truck Driver (heavy)	$27.55	$220.40	$42.00	$336.00	$27.55	$42.00
1 Truck, 30 Ton		208.20		229.00		
1 Water tank, 6000 Gal.		116.80		128.50	40.63	44.69
8 L.H., Daily Totals		$545.40		$693.50	$68.18	$86.69

Crew B-59A

Crew B-59A	Hr.	Daily	Hr.	Daily	Bare Costs	Incl. O&P
2 Laborers	$26.70	$427.20	$41.55	$664.80	$26.98	$41.70
1 Truck Driver (heavy)	27.55	220.40	42.00	336.00		
1 Water tank, 5K w/pum		116.80		128.50		
1 Truck, 30 Ton		208.20		229.00	13.54	14.90
24 L.H., Daily Totals		$972.60		$1358.30	$40.52	$56.60

Crew B-60

Crew B-60	Hr.	Daily	Hr.	Daily	Bare Costs	Incl. O&P
1 Labor Foreman (outside)	$28.70	$229.60	$44.70	$357.60	$30.60	$46.69
2 Laborers	26.70	427.20	41.55	664.80		
1 Equip. Oper. (crane)	35.90	287.20	54.10	432.80		
2 Equip. Oper. (light)	33.05	528.80	49.80	796.80		
1 Equip. Oper. Oiler	30.10	240.80	45.35	362.80		
1 Crawler Crane, 40 Ton		888.20		977.00		
45 L.F. Leads, 15K Ft. Lbs.		56.25		61.90		
1 Backhoe Loader, 48 H.P.		215.80		237.40	20.76	22.83
56 L.H., Daily Totals		$2873.85		$3891.10	$51.36	$69.52

Crew B-61

Crew B-61	Hr.	Daily	Hr.	Daily	Bare Costs	Incl. O&P
1 Labor Foreman (outside)	$28.70	$229.60	$44.70	$357.60	$28.37	$43.83
3 Laborers	26.70	640.80	41.55	997.20		
1 Equip. Oper. (light)	33.05	264.40	49.80	398.40		
1 Cement Mixer, 2 C.Y.		162.00		178.20		
1 Air Compr., 160 C.F.M.		87.40		96.15	6.24	6.86
40 L.H., Daily Totals		$1384.20		$2027.55	$34.61	$50.69

Crew B-62

Crew B-62	Hr.	Daily	Hr.	Daily	Bare Costs	Incl. O&P
2 Laborers	$26.70	$427.20	$41.55	$664.80	$28.82	$44.30
1 Equip. Oper. (light)	33.05	264.40	49.80	398.40		
1 Loader, Skid Steer		153.40		168.75	6.39	7.03
24 L.H., Daily Totals		$845.00		$1231.95	$35.21	$51.33

Crew B-63

Crew B-63	Hr.	Daily	Hr.	Daily	Bare Costs	Incl. O&P
4 Laborers	$26.70	$854.40	$41.55	$1329.60	$27.97	$43.20
1 Equip. Oper. (light)	33.05	264.40	49.80	398.40		
1 Loader, Skid Steer		153.40		168.75	3.84	4.22
40 L.H., Daily Totals		$1272.20		$1896.75	$31.81	$47.42

Crew B-64

Crew B-64	Hr.	Daily	Hr.	Daily	Bare Costs	Incl. O&P
1 Laborer	$26.70	$213.60	$41.55	$332.40	$26.75	$41.20
1 Truck Driver (light)	26.80	214.40	40.85	326.80		
1 Power Mulcher (small)		103.20		113.50		
1 Light Truck, 1.5 Ton		128.80		141.70	14.50	15.95
16 L.H., Daily Totals		$660.00		$914.40	$41.25	$57.15

CREWS

Crew No.	Bare Costs		Incl. Subs O & P		Cost Per Labor-Hour	
Crew B-65	Hr.	Daily	Hr.	Daily	Bare Costs	Incl. O&P
1 Laborer	$26.70	$213.60	$41.55	$332.40	$26.75	$41.20
1 Truck Driver (light)	26.80	214.40	40.85	326.80		
1 Power Mulcher (large)		184.00		202.40		
1 Light Truck, 1.5 Ton		128.80		141.70	19.55	21.51
16 L.H., Daily Totals		$740.80		$1003.30	$46.30	$62.71
Crew B-66	Hr.	Daily	Hr.	Daily	Bare Costs	Incl. O&P
1 Equip. Oper. (light)	$33.05	$264.40	$49.80	$398.40	$33.05	$49.80
1 Backhoe Ldr. w/Attchmt.		173.20		190.50	21.65	23.82
8 L.H., Daily Totals		$437.60		$588.90	$54.70	$73.62
Crew B-67	Hr.	Daily	Hr.	Daily	Bare Costs	Incl. O&P
1 Millwright	$35.50	$284.00	$52.45	$419.60	$34.28	$51.13
1 Equip. Oper. (light)	33.05	264.40	49.80	398.40		
1 Forklift		233.00		256.30	14.56	16.02
16 L.H., Daily Totals		$781.40		$1074.30	$48.84	$67.15
Crew B-68	Hr.	Daily	Hr.	Daily	Bare Costs	Incl. O&P
2 Millwrights	$35.50	$568.00	$52.45	$839.20	$34.68	$51.57
1 Equip. Oper. (light)	33.05	264.40	49.80	398.40		
1 Forklift		233.00		256.30	9.71	10.68
24 L.H., Daily Totals		$1065.40		$1493.90	$44.39	$62.25
Crew B-69	Hr.	Daily	Hr.	Daily	Bare Costs	Incl. O&P
1 Labor Foreman (outside)	$28.70	$229.60	$44.70	$357.60	$29.13	$44.80
3 Laborers	26.70	640.80	41.55	997.20		
1 Equip Oper. (crane)	35.90	287.20	54.10	432.80		
1 Equip Oper. Oiler	30.10	240.80	45.35	362.80		
1 Truck Crane, 80 Ton		995.00		1094.50	20.73	22.80
48 L.H., Daily Totals		$2393.40		$3244.90	$49.86	$67.60
Crew B-69A	Hr.	Daily	Hr.	Daily	Bare Costs	Incl. O&P
1 Labor Foreman	$28.70	$229.60	$44.70	$357.60	$29.38	$44.93
3 Laborers	26.70	640.80	41.55	997.20		
1 Equip. Oper. (medium)	34.65	277.20	52.20	417.60		
1 Concrete Finisher	32.85	262.80	48.35	386.80		
1 Curb Paver		582.80		641.10	12.14	13.36
48 L.H., Daily Totals		$1993.20		$2800.30	$41.52	$58.29
Crew B-69B	Hr.	Daily	Hr.	Daily	Bare Costs	Incl. O&P
1 Labor Foreman	$28.70	$229.60	$44.70	$357.60	$29.38	$44.93
3 Laborers	26.70	640.80	41.55	997.20		
1 Equip. Oper. (medium)	34.65	277.20	52.20	417.60		
1 Cement Finisher	32.85	262.80	48.35	386.80		
1 Curb/Gutter Paver		725.00		797.50	15.10	16.61
48 L.H., Daily Totals		$2135.40		$2956.70	$44.48	$61.54
Crew B-70	Hr.	Daily	Hr.	Daily	Bare Costs	Incl. O&P
1 Labor Foreman (outside)	$28.70	$229.60	$44.70	$357.60	$30.39	$46.56
3 Laborers	26.70	640.80	41.55	997.20		
3 Equip. Oper. (med.)	34.65	831.60	52.20	1252.80		
1 Motor Grader, 30,000 Lb.		456.60		502.25		
1 Grader Attach., Ripper		71.40		78.55		
1 Road Sweeper, S.P.		445.00		489.50		
1 F.E. Loader, 1-3/4 C.Y.		241.00		265.10	21.68	23.85
56 L.H., Daily Totals		$2916.00		$3943.00	$52.07	$70.41

Crew No.	Bare Costs		Incl. Subs O & P		Cost Per Labor-Hour	
Crew B-71	Hr.	Daily	Hr.	Daily	Bare Costs	Incl. O&P
1 Labor Foreman (outside)	$28.70	$229.60	$44.70	$357.60	$30.39	$46.56
3 Laborers	26.70	640.80	41.55	997.20		
3 Equip. Oper. (med.)	34.65	831.60	52.20	1252.80		
1 Pvmt. Profiler, 750 H.P.		4266.00		4692.60		
1 Road Sweeper, S.P.		445.00		489.50		
1 F.E. Loader, 1-3/4 C.Y.		241.00		265.10	88.43	97.27
56 L.H., Daily Totals		$6654.00		$8054.80	$118.82	$143.83
Crew B-72	Hr.	Daily	Hr.	Daily	Bare Costs	Incl. O&P
1 Labor Foreman (outside)	$28.70	$229.60	$44.70	$357.60	$30.93	$47.27
3 Laborers	26.70	640.80	41.55	997.20		
4 Equip. Oper. (med.)	34.65	1108.80	52.20	1670.40		
1 Pvmt. Profiler, 750 H.P.		4266.00		4692.60		
1 Hammermill, 250 H.P.		1344.00		1478.40		
1 Windrow Loader		791.20		870.30		
1 Mix Paver 165 H.P.		1656.00		1821.60		
1 Roller, Pneu. Tire, 12 T.		241.80		266.00	129.67	142.64
64 L.H., Daily Totals		$10278.20		$12154.10	$160.60	$189.91
Crew B-73	Hr.	Daily	Hr.	Daily	Bare Costs	Incl. O&P
1 Labor Foreman (outside)	$28.70	$229.60	$44.70	$357.60	$31.92	$48.60
2 Laborers	26.70	427.20	41.55	664.80		
5 Equip. Oper. (med.)	34.65	1386.00	52.20	2088.00		
1 Road Mixer, 310 H.P.		1546.00		1700.60		
1 Roller, Tandem, 12 Ton		179.20		197.10		
1 Hammermill, 250 H.P.		1344.00		1478.40		
1 Motor Grader, 30,000 Lb.		456.60		502.25		
.5 F.E. Loader, 1-3/4 C.Y.		120.50		132.55		
.5 Truck, 30 Ton		104.10		114.50		
.5 Water Tank 5000 Gal.		58.40		64.25	59.51	65.46
64 L.H., Daily Totals		$5851.60		$7300.05	$91.43	$114.06
Crew B-74	Hr.	Daily	Hr.	Daily	Bare Costs	Incl. O&P
1 Labor Foreman (outside)	$28.70	$229.60	$44.70	$357.60	$31.14	$47.38
1 Laborer	26.70	213.60	41.55	332.40		
4 Equip. Oper. (med.)	34.65	1108.80	52.20	1670.40		
2 Truck Drivers (heavy)	27.55	440.80	42.00	672.00		
1 Motor Grader, 30,000 Lb.		456.60		502.25		
1 Grader Attach., Ripper		71.40		78.55		
2 Stabilizers, 310 H.P.		2090.00		2299.00		
1 Flatbed Truck, 3 Ton		176.20		193.80		
1 Chem. Spreader, Towed		54.60		60.05		
1 Vibr. Roller, 29,000 Lb.		433.20		476.50		
1 Water Tank 5000 Gal.		116.80		128.50		
1 Truck, 30 Ton		208.20		229.00	56.36	62.00
64 L.H., Daily Totals		$5599.80		$7000.05	$87.50	$109.38
Crew B-75	Hr.	Daily	Hr.	Daily	Bare Costs	Incl. O&P
1 Labor Foreman (outside)	$28.70	$229.60	$44.70	$357.60	$31.65	$48.15
1 Laborer	26.70	213.60	41.55	332.40		
4 Equip. Oper. (med.)	34.65	1108.80	52.20	1670.40		
1 Truck Driver (heavy)	27.55	220.40	42.00	336.00		
1 Motor Grader, 30,000 Lb.		456.60		502.25		
1 Grader Attach., Ripper		71.40		78.55		
2 Stabilizers, 310 H.P.		2090.00		2299.00		
1 Dist. Truck, 3000 Gal.		233.80		257.20		
1 Vibr. Roller, 29,000 Lb.		433.20		476.50	58.66	64.53
56 L.H., Daily Totals		$5057.40		$6309.90	$90.31	$112.68

Crew No.	Bare Costs		Incl. Subs O & P		Cost Per Labor-Hour	
Crew B-76	**Hr.**	**Daily**	**Hr.**	**Daily**	**Bare Costs**	**Incl. O&P**
1 Dock Builder Foreman	$35.30	$282.40	$58.15	$465.20	$33.74	$54.02
5 Dock Builders	33.30	1332.00	54.90	2196.00		
2 Equip. Oper. (crane)	35.90	574.40	54.10	865.60		
1 Equip. Oper. Oiler	30.10	240.80	45.35	362.80		
1 Crawler Crane, 50 Ton		1206.00		1326.60		
1 Barge, 400 Ton		273.20		300.50		
1 Hammer, 15K Ft. Lbs.		359.80		395.80		
60 L.F. Leads, 15K Ft. Lbs.		75.00		82.50		
1 Air Compr., 600 C.F.M.		297.80		327.60		
2-50 Ft. Air Hoses, 3" Dia.		35.40		38.95	31.25	34.38
72 L.H., Daily Totals		$4676.80		$6361.55	$64.99	$88.40
Crew B-76A	**Hr.**	**Daily**	**Hr.**	**Daily**	**Bare Costs**	**Incl. O&P**
1 Laborer Foreman	$28.70	$229.60	$44.70	$357.60	$28.53	$43.99
5 Laborers	26.70	1068.00	41.55	1662.00		
1 Equip. Oper. (crane)	35.90	287.20	54.10	432.80		
1 Equip. Oper. Oiler	30.10	240.80	45.35	362.80		
1 Crawler Crane, 50 Ton		1206.00		1326.60		
1 Barge, 400 Ton		273.20		300.50	23.11	25.42
64 L.H., Daily Totals		$3304.80		$4442.30	$51.64	$69.41
Crew B-77	**Hr.**	**Daily**	**Hr.**	**Daily**	**Bare Costs**	**Incl. O&P**
1 Labor Foreman	$28.70	$229.60	$44.70	$357.60	$27.12	$42.04
3 Laborers	26.70	640.80	41.55	997.20		
1 Truck Driver (light)	26.80	214.40	40.85	326.80		
1 Crack Cleaner, 25 H.P.		44.40		48.85		
1 Crack Filler, Trailer Mtd.		157.00		172.70		
1 Flatbed Truck, 3 Ton		176.20		193.80	9.44	10.38
40 L.H., Daily Totals		$1462.40		$2096.95	$36.56	$52.42
Crew B-78	**Hr.**	**Daily**	**Hr.**	**Daily**	**Bare Costs**	**Incl. O&P**
1 Labor Foreman	$28.70	$229.60	$44.70	$357.60	$27.05	$41.96
4 Laborers	26.70	854.40	41.55	1329.60		
1 Truck Driver (light)	26.80	214.40	40.85	326.80		
1 Paint Striper, S.P.		134.20		147.60		
1 Flatbed Truck, 3 Ton		176.20		193.80		
1 Pickup Truck, 3/4 Ton		78.00		85.80	8.09	8.90
48 L.H., Daily Totals		$1686.80		$2441.20	$35.14	$50.86
Crew B-79	**Hr.**	**Daily**	**Hr.**	**Daily**	**Bare Costs**	**Incl. O&P**
1 Labor Foreman	$28.70	$229.60	$44.70	$357.60	$27.12	$42.04
3 Laborers	26.70	640.80	41.55	997.20		
1 Truck Driver (light)	26.80	214.40	40.85	326.80		
1 Thermo. Striper, T.M.		549.20		604.10		
1 Flatbed Truck, 3 Ton		176.20		193.80		
2 Pickup Truck, 3/4 Ton		156.00		171.60	22.04	24.24
40 L.H., Daily Totals		$1966.20		$2651.10	$49.16	$66.28
Crew B-80	**Hr.**	**Daily**	**Hr.**	**Daily**	**Bare Costs**	**Incl. O&P**
1 Labor Foreman	$28.70	$229.60	$44.70	$357.60	$28.81	$44.23
1 Laborer	26.70	213.60	41.55	332.40		
1 Truck Driver (light)	26.80	214.40	40.85	326.80		
1 Equip. Oper. (light)	33.05	264.40	49.80	398.40		
1 Flatbed Truck, 3 Ton		176.20		193.80		
1 Fence Post Auger, T.M.		357.40		393.15	16.68	18.34
32 L.H., Daily Totals		$1455.60		$2002.15	$45.49	$62.57
Crew B-80A	**Hr.**	**Daily**	**Hr.**	**Daily**	**Bare Costs**	**Incl. O&P**
3 Laborers	$26.70	$640.80	$41.55	$997.20	$26.70	$41.55
1 Flatbed Truck, 3 Ton		176.20		193.80	7.34	8.08
24 L.H., Daily Totals		$817.00		$1191.00	$34.04	$49.63

Crew No.	Bare Costs		Incl. Subs O & P		Cost Per Labor-Hour	
Crew B-80B	**Hr.**	**Daily**	**Hr.**	**Daily**	**Bare Costs**	**Incl. O&P**
3 Laborers	$26.70	$640.80	$41.55	$997.20	$28.29	$43.61
1 Equip. Oper. (light)	33.05	264.40	49.80	398.40		
1 Crane, Flatbed Mnt.		209.60		230.55	6.55	7.21
32 L.H., Daily Totals		$1114.80		$1626.15	$34.84	$50.82
Crew B-80C	**Hr.**	**Daily**	**Hr.**	**Daily**	**Bare Costs**	**Incl. O&P**
2 Laborers	$26.70	$427.20	$41.55	$664.80	$26.73	$41.32
1 Truck Driver (light)	26.80	214.40	40.85	326.80		
1 Light Truck, 1.5 Ton		128.80		141.70		
1 Manual fence post auger, gas		5.00		5.50	5.58	6.13
24 L.H., Daily Totals		$775.40		$1138.80	$32.31	$47.45
Crew B-81	**Hr.**	**Daily**	**Hr.**	**Daily**	**Bare Costs**	**Incl. O&P**
1 Laborer	$26.70	$213.60	$41.55	$332.40	$29.63	$45.25
1 Equip. Oper. (med.)	34.65	277.20	52.20	417.60		
1 Truck Driver (heavy)	27.55	220.40	42.00	336.00		
1 Hydromulcher, T.M.		196.00		215.60		
1 Tractor Truck, 4x2		208.20		229.00	16.84	18.53
24 L.H., Daily Totals		$1115.40		$1530.60	$46.47	$63.78
Crew B-82	**Hr.**	**Daily**	**Hr.**	**Daily**	**Bare Costs**	**Incl. O&P**
1 Laborer	$26.70	$213.60	$41.55	$332.40	$29.88	$45.68
1 Equip. Oper. (light)	33.05	264.40	49.80	398.40		
1 Horiz. Borer, 6 H.P.		62.60		68.85	3.91	4.30
16 L.H., Daily Totals		$540.60		$799.65	$33.79	$49.98
Crew B-83	**Hr.**	**Daily**	**Hr.**	**Daily**	**Bare Costs**	**Incl. O&P**
1 Tugboat Captain	$34.65	$277.20	$52.20	$417.60	$30.68	$46.88
1 Tugboat Hand	26.70	213.60	41.55	332.40		
1 Tugboat, 250 H.P.		440.60		484.65	27.54	30.29
16 L.H., Daily Totals		$931.40		$1234.65	$58.22	$77.17
Crew B-84	**Hr.**	**Daily**	**Hr.**	**Daily**	**Bare Costs**	**Incl. O&P**
1 Equip. Oper. (med.)	$34.65	$277.20	$52.20	$417.60	$34.65	$52.20
1 Rotary Mower/Tractor		228.80		251.70	28.60	31.46
8 L.H., Daily Totals		$506.00		$669.30	$63.25	$83.66
Crew B-85	**Hr.**	**Daily**	**Hr.**	**Daily**	**Bare Costs**	**Incl. O&P**
3 Laborers	$26.70	$640.80	$41.55	$997.20	$28.46	$43.77
1 Equip. Oper. (med.)	34.65	277.20	52.20	417.60		
1 Truck Driver (heavy)	27.55	220.40	42.00	336.00		
1 Aerial Lift Truck, 80'		496.80		546.50		
1 Brush Chipper, 130 H.P.		164.80		181.30		
1 Pruning Saw, Rotary		8.75		9.65	16.76	18.43
40 L.H., Daily Totals		$1808.75		$2488.25	$45.22	$62.20
Crew B-86	**Hr.**	**Daily**	**Hr.**	**Daily**	**Bare Costs**	**Incl. O&P**
1 Equip. Oper. (med.)	$34.65	$277.20	$52.20	$417.60	$34.65	$52.20
1 Stump Chipper, S.P.		70.70		77.75	8.84	9.72
8 L.H., Daily Totals		$347.90		$495.35	$43.49	$61.92
Crew B-86A	**Hr.**	**Daily**	**Hr.**	**Daily**	**Bare Costs**	**Incl. O&P**
1 Equip. Oper. (medium)	$34.65	$277.20	$52.20	$417.60	$34.65	$52.20
1 Grader, 30,000 Lbs.		456.60		502.25	57.08	62.78
8 L.H., Daily Totals		$733.80		$919.85	$91.73	$114.98
Crew B-86B	**Hr.**	**Daily**	**Hr.**	**Daily**	**Bare Costs**	**Incl. O&P**
1 Equip. Oper. (medium)	$34.65	$277.20	$52.20	$417.60	$34.65	$52.20
1 Dozer, 200 H.P.		919.60		1011.55	114.95	126.45
8 L.H., Daily Totals		$1196.80		$1429.15	$149.60	$178.65

Crew No.	Bare Costs		Incl. Subs O & P		Cost Per Labor-Hour	
Crew B-87	**Hr.**	**Daily**	**Hr.**	**Daily**	**Bare Costs**	**Incl. O&P**
1 Laborer	$26.70	$213.60	$41.55	$332.40	$33.06	$50.07
4 Equip. Oper. (med.)	34.65	1108.80	52.20	1670.40		
2 Feller Bunchers, 50 H.P.		864.80		951.30		
1 Log Chipper, 22" Tree		1062.00		1168.20		
1 Dozer, 105 H.P.		453.80		499.20		
1 Chainsaw, Gas, 36" Long		33.40		36.75	60.35	66.39
40 L.H., Daily Totals		$3736.40		$4658.25	$93.41	$116.46
Crew B-88	**Hr.**	**Daily**	**Hr.**	**Daily**	**Bare Costs**	**Incl. O&P**
1 Laborer	$26.70	$213.60	$41.55	$332.40	$33.51	$50.68
6 Equip. Oper. (med.)	34.65	1663.20	52.20	2505.60		
2 Feller Bunchers, 50 H.P.		864.80		951.30		
1 Log Chipper, 22" Tree		1062.00		1168.20		
2 Log Skidders, 50 H.P.		1490.80		1639.90		
1 Dozer, 105 H.P.		453.80		499.20		
1 Chainsaw, Gas, 36" Long		33.40		36.75	69.73	76.70
56 L.H., Daily Totals		$5781.60		$7133.35	$103.24	$127.38
Crew B-89	**Hr.**	**Daily**	**Hr.**	**Daily**	**Bare Costs**	**Incl. O&P**
1 Equip. Oper. (light)	$33.05	$264.40	$49.80	$398.40	$29.93	$45.33
1 Truck Driver (light)	26.80	214.40	40.85	326.80		
1 Truck, Stake Body, 3 Ton		176.20		193.80		
1 Concrete Saw		114.40		125.85		
1 Water Tank, 65 Gal.		13.60		14.95	19.01	20.91
16 L.H., Daily Totals		$783.00		$1059.80	$48.94	$66.24
Crew B-89A	**Hr.**	**Daily**	**Hr.**	**Daily**	**Bare Costs**	**Incl. O&P**
1 Skilled Worker	$34.85	$278.80	$54.20	$433.60	$30.78	$47.88
1 Laborer	26.70	213.60	41.55	332.40		
1 Core Drill (large)		103.05		113.35	6.44	7.08
16 L.H., Daily Totals		$595.45		$879.35	$37.22	$54.96
Crew B-89B	**Hr.**	**Daily**	**Hr.**	**Daily**	**Bare Costs**	**Incl. O&P**
1 Equip. Oper. (light)	$33.05	$264.40	$49.80	$398.40	$29.93	$45.33
1 Truck Driver, Light	26.80	214.40	40.85	326.80		
1 Wall Saw, Hydraulic, 10 H.P.		73.30		80.65		
1 Generator, Diesel, 100 KW		176.60		194.25		
1 Water Tank, 65 Gal.		13.60		14.95		
1 Flatbed Truck, 3 Ton		176.20		193.80	27.48	30.23
16 L.H., Daily Totals		$918.50		$1208.85	$57.41	$75.56
Crew B-90	**Hr.**	**Daily**	**Hr.**	**Daily**	**Bare Costs**	**Incl. O&P**
1 Labor Foreman (outside)	$28.70	$229.60	$44.70	$357.60	$28.75	$44.12
3 Laborers	26.70	640.80	41.55	997.20		
2 Equip. Oper. (light)	33.05	528.80	49.80	796.80		
2 Truck Drivers (heavy)	27.55	440.80	42.00	672.00		
1 Road Mixer, 310 H.P.		1546.00		1700.60		
1 Dist. Truck, 2000 Gal.		204.80		225.30	27.36	30.09
64 L.H., Daily Totals		$3590.80		$4749.50	$56.11	$74.21
Crew B-90A	**Hr.**	**Daily**	**Hr.**	**Daily**	**Bare Costs**	**Incl. O&P**
1 Labor Foreman	$28.70	$229.60	$44.70	$357.60	$31.53	$48.09
2 Laborers	26.70	427.20	41.55	664.80		
4 Equip. Oper. (medium)	34.65	1108.80	52.20	1670.40		
2 Graders, 30,000 Lbs.		913.20		1004.50		
1 Roller, Steel Wheel		179.20		197.10		
1 Roller, Pneumatic Wheel		241.80		266.00	23.83	26.21
56 L.H., Daily Totals		$3099.80		$4160.40	$55.36	$74.30

Crew No.	Bare Costs		Incl. Subs O & P		Cost Per Labor-Hour	
Crew B-90B	**Hr.**	**Daily**	**Hr.**	**Daily**	**Bare Costs**	**Incl. O&P**
1 Labor Foreman	$28.70	$229.60	$44.70	$357.60	$31.01	$47.40
2 Laborers	26.70	427.20	41.55	664.80		
3 Equip. Oper. (medium)	34.65	831.60	52.20	1252.80		
1 Roller, Steel Wheel		179.20		197.10		
1 Roller, Pneumatic Wheel		241.80		266.00		
1 Road Mixer, 310 H.P.		1546.00		1700.60	40.98	45.08
48 L.H., Daily Totals		$3455.40		$4438.90	$71.99	$92.48
Crew B-91	**Hr.**	**Daily**	**Hr.**	**Daily**	**Bare Costs**	**Incl. O&P**
1 Labor Foreman (outside)	$28.70	$229.60	$44.70	$357.60	$31.03	$47.33
2 Laborers	26.70	427.20	41.55	664.80		
4 Equip. Oper. (med.)	34.65	1108.80	52.20	1670.40		
1 Truck Driver (heavy)	27.55	220.40	42.00	336.00		
1 Dist. Truck, 3000 Gal.		233.80		257.20		
1 Aggreg. Spreader, S.P.		771.20		848.30		
1 Roller, Pneu. Tire, 12 Ton		241.80		266.00		
1 Roller, Steel, 10 Ton		179.20		197.10	22.28	24.51
64 L.H., Daily Totals		$3412.00		$4597.40	$53.31	$71.84
Crew B-92	**Hr.**	**Daily**	**Hr.**	**Daily**	**Bare Costs**	**Incl. O&P**
1 Labor Foreman (outside)	$28.70	$229.60	$44.70	$357.60	$27.20	$42.34
3 Laborers	26.70	640.80	41.55	997.20		
1 Crack Cleaner, 25 H.P.		44.40		48.85		
1 Air Compressor		55.40		60.95		
1 Tar Kettle, T.M.		40.95		45.05		
1 Flatbed Truck, 3 Ton		176.20		193.80	9.90	10.90
32 L.H., Daily Totals		$1187.35		$1703.45	$37.10	$53.24
Crew B-93	**Hr.**	**Daily**	**Hr.**	**Daily**	**Bare Costs**	**Incl. O&P**
1 Equip. Oper. (med.)	$34.65	$277.20	$52.20	$417.60	$34.65	$52.20
1 Feller Buncher, 50 H.P.		432.40		475.65	54.05	59.46
8 L.H., Daily Totals		$709.60		$893.25	$88.70	$111.66
Crew B-94A	**Hr.**	**Daily**	**Hr.**	**Daily**	**Bare Costs**	**Incl. O&P**
1 Laborer	$26.70	$213.60	$41.55	$332.40	$26.70	$41.55
1 Diaph. Water Pump, 2"		50.60		55.65		
1-20 Ft. Suction Hose, 2"		2.55		2.80		
2-50 Ft. Disch. Hoses, 2"		3.80		4.20	7.12	7.83
8 L.H., Daily Totals		$270.55		$395.05	$33.82	$49.38
Crew B-94B	**Hr.**	**Daily**	**Hr.**	**Daily**	**Bare Costs**	**Incl. O&P**
1 Laborer	$26.70	$213.60	$41.55	$332.40	$26.70	$41.55
1 Diaph. Water Pump, 4"		74.40		81.85		
1-20 Ft. Suction Hose, 4"		5.05		5.55		
2-50 Ft. Disch. Hoses, 4"		7.10		7.80	10.82	11.90
8 L.H., Daily Totals		$300.15		$427.60	$37.52	$53.45
Crew B-94C	**Hr.**	**Daily**	**Hr.**	**Daily**	**Bare Costs**	**Incl. O&P**
1 Laborer	$26.70	$213.60	$41.55	$332.40	$26.70	$41.55
1 Centr. Water Pump, 3"		53.60		58.95		
1-20 Ft. Suction Hose, 3"		4.25		4.70		
2-50 Ft. Disch. Hoses, 3"		5.00		5.50	7.86	8.64
8 L.H., Daily Totals		$276.45		$401.55	$34.56	$50.19
Crew B-94D	**Hr.**	**Daily**	**Hr.**	**Daily**	**Bare Costs**	**Incl. O&P**
1 Laborer	$26.70	$213.60	$41.55	$332.40	$26.70	$41.55
1 Centr. Water Pump, 6"		217.00		238.70		
1-20 Ft. Suction Hose, 6"		12.80		14.10		
2-50 Ft. Disch. Hoses, 6"		17.80		19.60	30.95	34.05
8 L.H., Daily Totals		$461.20		$604.80	$57.65	$75.60

Crew No.	Bare Costs		Incl. Subs O & P		Cost Per Labor-Hour	
	Hr.	Daily	Hr.	Daily	Bare Costs	Incl. O&P
Crew B-95A	Hr.	Daily	Hr.	Daily	Bare Costs	Incl. O&P
1 Equip. Oper. (crane)	$35.90	$287.20	$54.10	$432.80	$31.30	$47.83
1 Laborer	26.70	213.60	41.55	332.40		
1 Hyd. Excavator, 5/8 C.Y.		432.40		475.65	27.03	29.73
16 L.H., Daily Totals		$933.20		$1240.85	$58.33	$77.56
Crew B-95B	Hr.	Daily	Hr.	Daily	Bare Costs	Incl. O&P
1 Equip. Oper. (crane)	$35.90	$287.20	$54.10	$432.80	$31.30	$47.83
1 Laborer	26.70	213.60	41.55	332.40		
1 Hyd. Excavator, 1.5 C.Y.		720.40		792.45	45.03	49.53
16 L.H., Daily Totals		$1221.20		$1557.65	$76.33	$97.36
Crew B-95C	Hr.	Daily	Hr.	Daily	Bare Costs	Incl. O&P
1 Equip. Oper. (crane)	$35.90	$287.20	$54.10	$432.80	$31.30	$47.83
1 Laborer	26.70	213.60	41.55	332.40		
1 Hyd. Excavator, 2.5 C.Y.		1215.00		1336.50	75.94	83.53
16 L.H., Daily Totals		$1715.80		$2101.70	$107.24	$131.36
Crew C-1	Hr.	Daily	Hr.	Daily	Bare Costs	Incl. O&P
3 Carpenters	$34.25	$822.00	$53.35	$1280.40	$32.36	$50.40
1 Laborer	26.70	213.60	41.55	332.40		
32 L.H., Daily Totals		$1035.60		$1612.80	$32.36	$50.40
Crew C-2	Hr.	Daily	Hr.	Daily	Bare Costs	Incl. O&P
1 Carpenter Foreman (out)	$36.25	$290.00	$56.45	$451.60	$33.33	$51.90
4 Carpenters	34.25	1096.00	53.35	1707.20		
1 Laborer	26.70	213.60	41.55	332.40		
48 L.H., Daily Totals		$1599.60		$2491.20	$33.33	$51.90
Crew C-2A	Hr.	Daily	Hr.	Daily	Bare Costs	Incl. O&P
1 Carpenter Foreman (out)	$36.25	$290.00	$56.45	$451.60	$33.09	$51.01
3 Carpenters	34.25	822.00	53.35	1280.40		
1 Cement Finisher	32.85	262.80	48.35	386.80		
1 Laborer	26.70	213.60	41.55	332.40		
48 L.H., Daily Totals		$1588.40		$2451.20	$33.09	$51.01
Crew C-3	Hr.	Daily	Hr.	Daily	Bare Costs	Incl. O&P
1 Rodman Foreman	$39.95	$319.60	$65.90	$527.20	$34.78	$56.15
4 Rodmen (reinf.)	37.95	1214.40	62.60	2003.20		
1 Equip. Oper. (light)	33.05	264.40	49.80	398.40		
2 Laborers	26.70	427.20	41.55	664.80		
3 Stressing Equipment		40.80		44.90		
.5 Grouting Equipment		76.40		84.05	1.83	2.01
64 L.H., Daily Totals		$2342.80		$3722.55	$36.61	$58.16
Crew C-4	Hr.	Daily	Hr.	Daily	Bare Costs	Incl. O&P
1 Rodman Foreman	$39.95	$319.60	$65.90	$527.20	$38.45	$63.43
3 Rodmen (reinf.)	37.95	910.80	62.60	1502.40		
3 Stressing Equipment		40.80		44.90	1.28	1.40
32 L.H., Daily Totals		$1271.20		$2074.50	$39.73	$64.83
Crew C-5	Hr.	Daily	Hr.	Daily	Bare Costs	Incl. O&P
1 Rodman Foreman	$39.95	$319.60	$65.90	$527.20	$36.82	$59.39
4 Rodmen (reinf.)	37.95	1214.40	62.60	2003.20		
1 Equip. Oper. (crane)	35.90	287.20	54.10	432.80		
1 Equip. Oper. Oiler	30.10	240.80	45.35	362.80		
1 Hyd. Crane, 25 Ton		616.80		678.50	11.01	12.12
56 L.H., Daily Totals		$2678.80		$4004.50	$47.83	$71.51

Crew No.	Bare Costs		Incl. Subs O & P		Cost Per Labor-Hour	
	Hr.	Daily	Hr.	Daily	Bare Costs	Incl. O&P
Crew C-6	Hr.	Daily	Hr.	Daily	Bare Costs	Incl. O&P
1 Labor Foreman (outside)	$28.70	$229.60	$44.70	$357.60	$28.06	$43.15
4 Laborers	26.70	854.40	41.55	1329.60		
1 Cement Finisher	32.85	262.80	48.35	386.80		
2 Gas Engine Vibrators		48.00		52.80	1.00	1.10
48 L.H., Daily Totals		$1394.80		$2126.80	$29.06	$44.25
Crew C-7	Hr.	Daily	Hr.	Daily	Bare Costs	Incl. O&P
1 Labor Foreman (outside)	$28.70	$229.60	$44.70	$357.60	$28.87	$44.22
5 Laborers	26.70	1068.00	41.55	1662.00		
1 Cement Finisher	32.85	262.80	48.35	386.80		
1 Equip. Oper. (med.)	34.65	277.20	52.20	417.60		
1 Equip. Oper. (oiler)	30.10	240.80	45.35	362.80		
2 Gas Engine Vibrators		48.00		52.80		
1 Concrete Bucket, 1 C.Y.		16.00		17.60		
1 Hyd. Crane, 55 Ton		895.40		984.95	13.33	14.66
72 L.H., Daily Totals		$3037.80		$4242.15	$42.20	$58.88
Crew C-7A	Hr.	Daily	Hr.	Daily	Bare Costs	Incl. O&P
1 Labor Foreman (outside)	$28.70	$229.60	$44.70	$357.60	$27.16	$42.06
5 Laborers	26.70	1068.00	41.55	1662.00		
2 Truck Drivers (Heavy)	27.55	440.80	42.00	672.00		
2 Conc. Transit Mixers		1480.80		1628.90	23.14	25.45
64 L.H., Daily Totals		$3219.20		$4320.50	$50.30	$67.51
Crew C-7B	Hr.	Daily	Hr.	Daily	Bare Costs	Incl. O&P
1 Labor Foreman (outside)	$28.70	$229.60	$44.70	$357.60	$28.53	$43.99
5 Laborers	26.70	1068.00	41.55	1662.00		
1 Equipment Operator (heavy)	35.90	287.20	54.10	432.80		
1 Equipment Oiler	30.10	240.80	45.35	362.80		
1 Conc. Bucket, 2 C.Y.		24.60		27.05		
1 Truck Crane, 165 Ton		1912.00		2103.20	30.26	33.29
64 L.H., Daily Totals		$3762.20		$4945.45	$58.79	$77.28
Crew C-7C	Hr.	Daily	Hr.	Daily	Bare Costs	Incl. O&P
1 Labor Foreman (outside)	$28.70	$229.60	$44.70	$357.60	$28.94	$44.61
5 Laborers	26.70	1068.00	41.55	1662.00		
2 Equipment Operator (medium)	34.65	554.40	52.20	835.20		
2 Wheel Loader, 4 C.Y.		891.20		980.30	13.93	15.32
64 L.H., Daily Totals		$2743.20		$3835.10	$42.87	$59.93
Crew C-7D	Hr.	Daily	Hr.	Daily	Bare Costs	Incl. O&P
1 Labor Foreman (outside)	$28.70	$229.60	$44.70	$357.60	$28.12	$43.52
5 Laborers	26.70	1068.00	41.55	1662.00		
1 Equipment Operator (med.)	34.65	277.20	52.20	417.60		
1 Concrete Conveyer		152.80		168.10	2.73	3.00
56 L.H., Daily Totals		$1727.60		$2605.30	$30.85	$46.52
Crew C-8	Hr.	Daily	Hr.	Daily	Bare Costs	Incl. O&P
1 Labor Foreman (outside)	$28.70	$229.60	$44.70	$357.60	$29.88	$45.36
3 Laborers	26.70	640.80	41.55	997.20		
2 Cement Finishers	32.85	525.60	48.35	773.60		
1 Equip. Oper. (med.)	34.65	277.20	52.20	417.60		
1 Concrete Pump (small)		704.20		774.60	12.58	13.83
56 L.H., Daily Totals		$2377.40		$3320.60	$42.46	$59.19
Crew C-8A	Hr.	Daily	Hr.	Daily	Bare Costs	Incl. O&P
1 Labor Foreman (outside)	$28.70	$229.60	$44.70	$357.60	$29.08	$44.23
3 Laborers	26.70	640.80	41.55	997.20		
2 Cement Finishers	32.85	525.60	48.35	773.60		
48 L.H., Daily Totals		$1396.00		$2128.40	$29.08	$44.23

Crews

Crew C-8B	Hr.	Daily	Hr.	Daily	Bare Costs	Incl. O&P
1 Labor Foreman (outside)	$28.70	$229.60	$44.70	$357.60	$28.69	$44.31
3 Laborers	26.70	640.80	41.55	997.20		
1 Equipment Operator	34.65	277.20	52.20	417.60		
1 Vibrating Screed		43.05		47.35		
1 Vibratory Roller		433.20		476.50		
1 Dozer, 200 H.P.		919.60		1011.55	34.90	38.38
40 L.H., Daily Totals		$2543.45		$3307.80	$63.59	$82.69

Crew C-8C	Hr.	Daily	Hr.	Daily	Bare Costs	Incl. O&P
1 Labor Foreman (outside)	$28.70	$229.60	$44.70	$357.60	$29.38	$44.93
3 Laborers	26.70	640.80	41.55	997.20		
1 Cement Finisher	32.85	262.80	48.35	386.80		
1 Equipment Operator (med.)	34.65	277.20	52.20	417.60		
1 Shotcrete Rig, 12 CY/hr		223.80		246.20	4.66	5.13
48 L.H., Daily Totals		$1634.20		$2405.40	$34.04	$50.06

Crew C-8D	Hr.	Daily	Hr.	Daily	Bare Costs	Incl. O&P
1 Labor Foreman (outside)	$28.70	$229.60	$44.70	$357.60	$30.33	$46.01
1 Laborer	26.70	213.60	41.55	332.40		
1 Cement Finisher	32.85	262.80	48.35	386.80		
1 Equipment Operator (light)	33.05	264.40	49.80	398.40		
1 Compressor, 250 CFM		112.40		123.65		
2 Hoses, 1", 50'		7.30		8.05	3.74	4.11
32 L.H., Daily Totals		$1090.10		$1606.90	$34.07	$50.12

Crew C-8E	Hr.	Daily	Hr.	Daily	Bare Costs	Incl. O&P
1 Labor Foreman (outside)	$28.70	$229.60	$44.70	$357.60	$30.33	$46.01
1 Laborer	26.70	213.60	41.55	332.40		
1 Cement Finisher	32.85	262.80	48.35	386.80		
1 Equipment Operator (light)	33.05	264.40	49.80	398.40		
1 Compressor, 250 CFM		112.40		123.65		
2 Hoses, 1", 50'		7.30		8.05		
1 Concrete Pump (small)		704.20		774.60	25.75	28.32
32 L.H., Daily Totals		$1794.30		$2381.50	$56.08	$74.33

Crew C-10	Hr.	Daily	Hr.	Daily	Bare Costs	Incl. O&P
1 Laborer	$26.70	$213.60	$41.55	$332.40	$30.80	$45.85
2 Cement Finishers	32.85	525.60	48.35	773.60		
24 L.H., Daily Totals		$739.20		$1106.00	$30.80	$45.85

Crew C-10B	Hr.	Daily	Hr.	Daily	Bare Costs	Incl. O&P
3 Laborers	$26.70	$640.80	$41.55	$997.20	$29.16	$44.13
2 Cement Finishers	32.85	525.60	48.35	773.60		
1 Concrete mixer, 10 CF		123.20		135.50		
2 Concrete finisher, 48" dia		49.20		54.10	4.31	4.74
40 L.H., Daily Totals		$1338.80		$1960.40	$33.47	$48.87

Crew C-11	Hr.	Daily	Hr.	Daily	Bare Costs	Incl. O&P
1 Struc. Steel Foreman	$40.15	$321.20	$71.85	$574.80	$37.23	$64.53
6 Struc. Steel Workers	38.15	1831.20	68.25	3276.00		
1 Equip. Oper. (crane)	35.90	287.20	54.10	432.80		
1 Equip. Oper. Oiler	30.10	240.80	45.35	362.80		
1 Truck Crane, 150 Ton		1445.00		1589.50	20.07	22.08
72 L.H., Daily Totals		$4125.40		$6235.90	$57.30	$86.61

Crew C-12	Hr.	Daily	Hr.	Daily	Bare Costs	Incl. O&P
1 Carpenter Foreman (out)	$36.25	$290.00	$56.45	$451.60	$33.60	$52.03
3 Carpenters	34.25	822.00	53.35	1280.40		
1 Laborer	26.70	213.60	41.55	332.40		
1 Equip. Oper. (crane)	35.90	287.20	54.10	432.80		
1 Hyd. Crane, 12 Ton		602.60		662.85	12.55	13.81
48 L.H., Daily Totals		$2215.40		$3160.05	$46.15	$65.84

Crew C-13	Hr.	Daily	Hr.	Daily	Bare Costs	Incl. O&P
1 Struc. Steel Worker	$38.15	$305.20	$68.25	$546.00	$36.85	$63.28
1 Welder	38.15	305.20	68.25	546.00		
1 Carpenter	34.25	274.00	53.35	426.80		
1 Gas Welding Machine		81.20		89.30	3.38	3.72
24 L.H., Daily Totals		$965.60		$1608.10	$40.23	$67.00

Crew C-14	Hr.	Daily	Hr.	Daily	Bare Costs	Incl. O&P
1 Carpenter Foreman (out)	$36.25	$290.00	$56.45	$451.60	$33.21	$51.96
5 Carpenters	34.25	1370.00	53.35	2134.00		
4 Laborers	26.70	854.40	41.55	1329.60		
4 Rodmen (reinf.)	37.95	1214.40	62.60	2003.20		
2 Cement Finishers	32.85	525.60	48.35	773.60		
1 Equip. Oper. (crane)	35.90	287.20	54.10	432.80		
1 Equip. Oper. Oiler	30.10	240.80	45.35	362.80		
1 Crane, 80 Ton, & Tools		995.00		1094.50	6.91	7.60
144 L.H., Daily Totals		$5777.40		$8582.10	$40.12	$59.56

Crew C-14A	Hr.	Daily	Hr.	Daily	Bare Costs	Incl. O&P
1 Carpenter Foreman (out)	$36.25	$290.00	$56.45	$451.60	$34.28	$53.75
16 Carpenters	34.25	4384.00	53.35	6828.80		
4 Rodmen (reinf.)	37.95	1214.40	62.60	2003.20		
2 Laborers	26.70	427.20	41.55	664.80		
1 Cement Finisher	32.85	262.80	48.35	386.80		
1 Equip. Oper. (med.)	34.65	277.20	52.20	417.60		
1 Gas Engine Vibrator		24.00		26.40		
1 Concrete Pump (small)		704.20		774.60	3.64	4.01
200 L.H., Daily Totals		$7583.80		$11553.80	$37.92	$57.76

Crew C-14B	Hr.	Daily	Hr.	Daily	Bare Costs	Incl. O&P
1 Carpenter Foreman (out)	$36.25	$290.00	$56.45	$451.60	$34.22	$53.53
16 Carpenters	34.25	4384.00	53.35	6828.80		
4 Rodmen (reinf.)	37.95	1214.40	62.60	2003.20		
2 Laborers	26.70	427.20	41.55	664.80		
2 Cement Finishers	32.85	525.60	48.35	773.60		
1 Equip. Oper. (med.)	34.65	277.20	52.20	417.60		
1 Gas Engine Vibrator		24.00		26.40		
1 Concrete Pump (small)		704.20		774.60	3.50	3.85
208 L.H., Daily Totals		$7846.60		$11940.60	$37.72	$57.38

Crew C-14C	Hr.	Daily	Hr.	Daily	Bare Costs	Incl. O&P
1 Carpenter Foreman (out)	$36.25	$290.00	$56.45	$451.60	$32.66	$51.14
6 Carpenters	34.25	1644.00	53.35	2560.80		
2 Rodmen (reinf.)	37.95	607.20	62.60	1001.60		
4 Laborers	26.70	854.40	41.55	1329.60		
1 Cement Finisher	32.85	262.80	48.35	386.80		
1 Gas Engine Vibrator		24.00		26.40	.21	.24
112 L.H., Daily Totals		$3682.40		$5756.80	$32.87	$51.38

Crew C-14D

Crew No.	Bare Costs Hr.	Bare Costs Daily	Incl. Subs O & P Hr.	Incl. Subs O & P Daily	Cost Per Labor-Hour Bare Costs	Cost Per Labor-Hour Incl. O&P
1 Carpenter Foreman (out)	$36.25	$290.00	$56.45	$451.60	$33.98	$53.01
18 Carpenters	34.25	4932.00	53.35	7682.40		
2 Rodmen (reinf.)	37.95	607.20	62.60	1001.60		
2 Laborers	26.70	427.20	41.55	664.80		
1 Cement Finisher	32.85	262.80	48.35	386.80		
1 Equip. Oper. (med.)	34.65	277.20	52.20	417.60		
1 Gas Engine Vibrator		24.00		26.40		
1 Concrete Pump (small)		704.20		774.60	3.64	4.01
200 L.H., Daily Totals		$7524.60		$11405.80	$37.62	$57.02

Crew C-14E

Crew No.	Bare Costs Hr.	Bare Costs Daily	Incl. Subs O & P Hr.	Incl. Subs O & P Daily	Cost Per Labor-Hour Bare Costs	Cost Per Labor-Hour Incl. O&P
1 Carpenter Foreman (out)	$36.25	$290.00	$56.45	$451.60	$33.59	$53.29
2 Carpenters	34.25	548.00	53.35	853.60		
4 Rodmen (reinf.)	37.95	1214.40	62.60	2003.20		
3 Laborers	26.70	640.80	41.55	997.20		
1 Cement Finisher	32.85	262.80	48.35	386.80		
1 Gas Engine Vibrator		24.00		26.40	.27	.30
88 L.H., Daily Totals		$2980.00		$4718.80	$33.86	$53.59

Crew C-14F

Crew No.	Bare Costs Hr.	Bare Costs Daily	Incl. Subs O & P Hr.	Incl. Subs O & P Daily	Cost Per Labor-Hour Bare Costs	Cost Per Labor-Hour Incl. O&P
1 Laborer Foreman (out)	$28.70	$229.60	$44.70	$357.60	$31.02	$46.20
2 Laborers	26.70	427.20	41.55	664.80		
6 Cement Finishers	32.85	1576.80	48.35	2320.80		
1 Gas Engine Vibrator		24.00		26.40	.33	.37
72 L.H., Daily Totals		$2257.60		$3369.60	$31.35	$46.57

Crew C-14G

Crew No.	Bare Costs Hr.	Bare Costs Daily	Incl. Subs O & P Hr.	Incl. Subs O & P Daily	Cost Per Labor-Hour Bare Costs	Cost Per Labor-Hour Incl. O&P
1 Laborer Foreman (out)	$28.70	$229.60	$44.70	$357.60	$30.50	$45.69
2 Laborers	26.70	427.20	41.55	664.80		
4 Cement Finishers	32.85	1051.20	48.35	1547.20		
1 Gas Engine Vibrator		24.00		26.40	.43	.47
56 L.H., Daily Totals		$1732.00		$2596.00	$30.93	$46.16

Crew C-14H

Crew No.	Bare Costs Hr.	Bare Costs Daily	Incl. Subs O & P Hr.	Incl. Subs O & P Daily	Cost Per Labor-Hour Bare Costs	Cost Per Labor-Hour Incl. O&P
1 Carpenter Foreman (out)	$36.25	$290.00	$56.45	$451.60	$33.71	$52.55
2 Carpenters	34.25	548.00	53.35	853.60		
1 Rodman (reinf.)	37.95	303.60	62.60	500.80		
1 Laborer	26.70	213.60	41.55	332.40		
1 Cement Finisher	32.85	262.80	48.35	386.80		
1 Gas Engine Vibrator		24.00		26.40	.50	.55
48 L.H., Daily Totals		$1642.00		$2551.60	$34.21	$53.10

Crew C-15

Crew No.	Bare Costs Hr.	Bare Costs Daily	Incl. Subs O & P Hr.	Incl. Subs O & P Daily	Cost Per Labor-Hour Bare Costs	Cost Per Labor-Hour Incl. O&P
1 Carpenter Foreman (out)	$36.25	$290.00	$56.45	$451.60	$32.06	$49.60
2 Carpenters	34.25	548.00	53.35	853.60		
3 Laborers	26.70	640.80	41.55	997.20		
2 Cement Finishers	32.85	525.60	48.35	773.60		
1 Rodman (reinf.)	37.95	303.60	62.60	500.80		
72 L.H., Daily Totals		$2308.00		$3576.80	$32.06	$49.60

Crew C-16

Crew No.	Bare Costs Hr.	Bare Costs Daily	Incl. Subs O & P Hr.	Incl. Subs O & P Daily	Cost Per Labor-Hour Bare Costs	Cost Per Labor-Hour Incl. O&P
1 Labor Foreman (outside)	$28.70	$229.60	$44.70	$357.60	$31.67	$49.19
3 Laborers	26.70	640.80	41.55	997.20		
2 Cement Finishers	32.85	525.60	48.35	773.60		
1 Equip. Oper. (med.)	34.65	277.20	52.20	417.60		
2 Rodmen (reinf.)	37.95	607.20	62.60	1001.60		
1 Concrete Pump (small)		704.20		774.60	9.78	10.76
72 L.H., Daily Totals		$2984.60		$4322.20	$41.45	$59.95

Crew C-17

Crew No.	Bare Costs Hr.	Bare Costs Daily	Incl. Subs O & P Hr.	Incl. Subs O & P Daily	Cost Per Labor-Hour Bare Costs	Cost Per Labor-Hour Incl. O&P
2 Skilled Worker Foremen	$36.85	$589.60	$57.30	$916.80	$35.25	$54.82
8 Skilled Workers	34.85	2230.40	54.20	3468.80		
80 L.H., Daily Totals		$2820.00		$4385.60	$35.25	$54.82

Crew C-17A

Crew No.	Bare Costs Hr.	Bare Costs Daily	Incl. Subs O & P Hr.	Incl. Subs O & P Daily	Cost Per Labor-Hour Bare Costs	Cost Per Labor-Hour Incl. O&P
2 Skilled Worker Foremen	$36.85	$589.60	$57.30	$916.80	$35.26	$54.81
8 Skilled Workers	34.85	2230.40	54.20	3468.80		
.125 Equip. Oper. (crane)	35.90	35.90	54.10	54.10		
.125 Crane, 80 Ton, & Tools		124.38		136.80	1.54	1.69
81 L.H., Daily Totals		$2980.28		$4576.50	$36.80	$56.50

Crew C-17B

Crew No.	Bare Costs Hr.	Bare Costs Daily	Incl. Subs O & P Hr.	Incl. Subs O & P Daily	Cost Per Labor-Hour Bare Costs	Cost Per Labor-Hour Incl. O&P
2 Skilled Worker Foremen	$36.85	$589.60	$57.30	$916.80	$35.27	$54.80
8 Skilled Workers	34.85	2230.40	54.20	3468.80		
.25 Equip. Oper. (crane)	35.90	71.80	54.10	108.20		
.25 Crane, 80 Ton, & Tools		248.75		273.65		
.25 Walk Behind Power Tools		6.15		6.75	3.11	3.42
82 L.H., Daily Totals		$3146.70		$4774.20	$38.38	$58.22

Crew C-17C

Crew No.	Bare Costs Hr.	Bare Costs Daily	Incl. Subs O & P Hr.	Incl. Subs O & P Daily	Cost Per Labor-Hour Bare Costs	Cost Per Labor-Hour Incl. O&P
2 Skilled Worker Foremen	$36.85	$589.60	$57.30	$916.80	$35.27	$54.79
8 Skilled Workers	34.85	2230.40	54.20	3468.80		
.375 Equip. Oper. (crane)	35.90	107.70	54.10	162.30		
.375 Crane, 80 Ton & Tools		373.13		410.45	4.50	4.95
83 L.H., Daily Totals		$3300.83		$4958.35	$39.77	$59.74

Crew C-17D

Crew No.	Bare Costs Hr.	Bare Costs Daily	Incl. Subs O & P Hr.	Incl. Subs O & P Daily	Cost Per Labor-Hour Bare Costs	Cost Per Labor-Hour Incl. O&P
2 Skilled Worker Foremen	$36.85	$589.60	$57.30	$916.80	$35.28	$54.79
8 Skilled Workers	34.85	2230.40	54.20	3468.80		
.5 Equip. Oper. (crane)	35.90	143.60	54.10	216.40		
.5 Crane, 80 Ton & Tools		497.50		547.25	5.92	6.51
84 L.H., Daily Totals		$3461.10		$5149.25	$41.20	$61.30

Crew C-17E

Crew No.	Bare Costs Hr.	Bare Costs Daily	Incl. Subs O & P Hr.	Incl. Subs O & P Daily	Cost Per Labor-Hour Bare Costs	Cost Per Labor-Hour Incl. O&P
2 Skilled Worker Foremen	$36.85	$589.60	$57.30	$916.80	$35.25	$54.82
8 Skilled Workers	34.85	2230.40	54.20	3468.80		
1 Hyd. Jack with Rods		79.00		86.90	.99	1.09
80 L.H., Daily Totals		$2899.00		$4472.50	$36.24	$55.91

Crew C-18

Crew No.	Bare Costs Hr.	Bare Costs Daily	Incl. Subs O & P Hr.	Incl. Subs O & P Daily	Cost Per Labor-Hour Bare Costs	Cost Per Labor-Hour Incl. O&P
.125 Labor Foreman (out)	$28.70	$28.70	$44.70	$44.70	$26.92	$41.90
1 Laborer	26.70	213.60	41.55	332.40		
1 Concrete Cart, 10 C.F.		49.80		54.80	5.53	6.09
9 L.H., Daily Totals		$292.10		$431.90	$32.45	$47.99

Crew C-19

Crew No.	Bare Costs Hr.	Bare Costs Daily	Incl. Subs O & P Hr.	Incl. Subs O & P Daily	Cost Per Labor-Hour Bare Costs	Cost Per Labor-Hour Incl. O&P
.125 Labor Foreman (out)	$28.70	$28.70	$44.70	$44.70	$26.92	$41.90
1 Laborer	26.70	213.60	41.55	332.40		
1 Concrete Cart, 18 C.F.		76.80		84.50	8.53	9.39
9 L.H., Daily Totals		$319.10		$461.60	$35.45	$51.29

Crew C-20

Crew No.	Bare Costs Hr.	Bare Costs Daily	Incl. Subs O & P Hr.	Incl. Subs O & P Daily	Cost Per Labor-Hour Bare Costs	Cost Per Labor-Hour Incl. O&P
1 Labor Foreman (outside)	$28.70	$229.60	$44.70	$357.60	$28.71	$44.08
5 Laborers	26.70	1068.00	41.55	1662.00		
1 Cement Finisher	32.85	262.80	48.35	386.80		
1 Equip. Oper. (med.)	34.65	277.20	52.20	417.60		
2 Gas Engine Vibrators		48.00		52.80		
1 Concrete Pump (small)		704.20		774.60	11.75	12.93
64 L.H., Daily Totals		$2589.80		$3651.40	$40.46	$57.01

Crew No.	Bare Costs		Incl. Subs O & P		Cost Per Labor-Hour	
Crew C-21	**Hr.**	**Daily**	**Hr.**	**Daily**	**Bare Costs**	**Incl. O&P**
1 Labor Foreman (outside)	$28.70	$229.60	$44.70	$357.60	$28.71	$44.08
5 Laborers	26.70	1068.00	41.55	1662.00		
1 Cement Finisher	32.85	262.80	48.35	386.80		
1 Equip. Oper. (med.)	34.65	277.20	52.20	417.60		
2 Gas Engine Vibrators		48.00		52.80		
1 Concrete Conveyer		152.80		168.10	3.14	3.45
64 L.H., Daily Totals		$2038.40		$3044.90	$31.85	$47.53

Crew C-22	Hr.	Daily	Hr.	Daily	Bare Costs	Incl. O&P
1 Rodman Foreman	$39.95	$319.60	$65.90	$527.20	$38.10	$62.62
4 Rodmen (reinf.)	37.95	1214.40	62.60	2003.20		
.125 Equip. Oper. (crane)	35.90	35.90	54.10	54.10		
.125 Equip. Oper. Oiler	30.10	30.10	45.35	45.35		
.125 Hyd. Crane, 25 Ton		77.10		84.80	1.84	2.02
42 L.H., Daily Totals		$1677.10		$2714.65	$39.94	$64.64

Crew C-23	Hr.	Daily	Hr.	Daily	Bare Costs	Incl. O&P
2 Skilled Worker Foremen	$36.85	$589.60	$57.30	$916.80	$34.88	$53.93
6 Skilled Workers	34.85	1672.80	54.20	2601.60		
1 Equip. Oper. (crane)	35.90	287.20	54.10	432.80		
1 Equip. Oper. Oiler	30.10	240.80	45.35	362.80		
1 Crane, 90 Ton		1325.00		1457.50	16.56	18.22
80 L.H., Daily Totals		$4115.40		$5771.50	$51.44	$72.15

Crew C-23A	Hr.	Daily	Hr.	Daily	Bare Costs	Incl. O&P
1 Labor Foreman (outside)	$28.70	$229.60	$44.70	$357.60	$29.62	$45.45
2 Laborers	26.70	427.20	41.55	664.80		
1 Equip. Oper. (crane)	35.90	287.20	54.10	432.80		
1 Equip. Oper. Oiler	30.10	240.80	45.35	362.80		
1 Crane, 100 ton capacity		1519.00		1670.90		
3 Conc. bucket, 8 C.Y.		465.60		512.15	49.62	54.58
40 L.H., Daily Totals		$3169.40		$4001.05	$79.24	$100.03

Crew C-24	Hr.	Daily	Hr.	Daily	Bare Costs	Incl. O&P
2 Skilled Worker Foremen	$36.85	$589.60	$57.30	$916.80	$34.88	$53.93
6 Skilled Workers	34.85	1672.80	54.20	2601.60		
1 Equip. Oper. (crane)	35.90	287.20	54.10	432.80		
1 Equip. Oper. Oiler	30.10	240.80	45.35	362.80		
1 Truck Crane, 150 Ton		1445.00		1589.50	18.06	19.87
80 L.H., Daily Totals		$4235.40		$5903.50	$52.94	$73.80

Crew C-25	Hr.	Daily	Hr.	Daily	Bare Costs	Incl. O&P
2 Rodmen (reinf.)	$37.95	$607.20	$62.60	$1001.60	$29.78	$49.70
2 Rodmen Helpers	21.60	345.60	36.80	588.80		
32 L.H., Daily Totals		$952.80		$1590.40	$29.78	$49.70

Crew C-27	Hr.	Daily	Hr.	Daily	Bare Costs	Incl. O&P
2 Cement Finishers	$32.85	$525.60	$48.35	$773.60	$32.85	$48.00
1 Concrete Saw		114.40		125.85	7.15	7.87
16 L.H., Daily Totals		$640.00		$899.45	$40.00	$55.87

Crew C-28	Hr.	Daily	Hr.	Daily	Bare Costs	Incl. O&P
1 Cement Finisher	$32.85	$262.80	$48.35	$386.80	$32.85	$48.00
1 Portable Air Compressor		17.30		19.05	2.16	2.37
8 L.H., Daily Totals		$280.10		$405.85	$35.01	$50.37

Crew D-1	Hr.	Daily	Hr.	Daily	Bare Costs	Incl. O&P
1 Bricklayer	$35.25	$282.00	$53.70	$429.60	$31.08	$47.35
1 Bricklayer Helper	26.90	215.20	41.00	328.00		
16 L.H., Daily Totals		$497.20		$757.60	$31.08	$47.35

Crew No.	Bare Costs		Incl. Subs O & P		Cost Per Labor-Hour	
Crew D-2	**Hr.**	**Daily**	**Hr.**	**Daily**	**Bare Costs**	**Incl. O&P**
3 Bricklayers	$35.25	$846.00	$53.70	$1288.80	$32.12	$49.05
2 Bricklayer Helpers	26.90	430.40	41.00	656.00		
.5 Carpenter	34.25	137.00	53.35	213.40		
44 L.H., Daily Totals		$1413.40		$2158.20	$32.12	$49.05

Crew D-3	Hr.	Daily	Hr.	Daily	Bare Costs	Incl. O&P
3 Bricklayers	$35.25	$846.00	$53.70	$1288.80	$32.02	$48.85
2 Bricklayer Helpers	26.90	430.40	41.00	656.00		
.25 Carpenter	34.25	68.50	53.35	106.70		
42 L.H., Daily Totals		$1344.90		$2051.50	$32.02	$48.85

Crew D-4	Hr.	Daily	Hr.	Daily	Bare Costs	Incl. O&P
1 Bricklayer	$35.25	$282.00	$53.70	$429.60	$30.53	$46.38
2 Bricklayer Helpers	26.90	430.40	41.00	656.00		
1 Equip. Oper. (light)	33.05	264.40	49.80	398.40		
1 Grout Pump, 50 C.F./hr		107.75		118.55		
1 Hoses & Hopper		15.20		16.70		
1 Accessories		11.90		13.10	4.22	4.64
32 L.H., Daily Totals		$1111.65		$1632.35	$34.75	$51.02

Crew D-5	Hr.	Daily	Hr.	Daily	Bare Costs	Incl. O&P
1 Bricklayer	$35.25	$282.00	$53.70	$429.60	$35.25	$53.70
8 L.H., Daily Totals		$282.00		$429.60	$35.25	$53.70

Crew D-6	Hr.	Daily	Hr.	Daily	Bare Costs	Incl. O&P
3 Bricklayers	$35.25	$846.00	$53.70	$1288.80	$31.20	$47.59
3 Bricklayer Helpers	26.90	645.60	41.00	984.00		
.25 Carpenter	34.25	68.50	53.35	106.70		
50 L.H., Daily Totals		$1560.10		$2379.50	$31.20	$47.59

Crew D-7	Hr.	Daily	Hr.	Daily	Bare Costs	Incl. O&P
1 Tile Layer	$32.70	$261.60	$48.05	$384.40	$29.03	$42.65
1 Tile Layer Helper	25.35	202.80	37.25	298.00		
16 L.H., Daily Totals		$464.40		$682.40	$29.03	$42.65

Crew D-8	Hr.	Daily	Hr.	Daily	Bare Costs	Incl. O&P
3 Bricklayers	$35.25	$846.00	$53.70	$1288.80	$31.91	$48.62
2 Bricklayer Helpers	26.90	430.40	41.00	656.00		
40 L.H., Daily Totals		$1276.40		$1944.80	$31.91	$48.62

Crew D-9	Hr.	Daily	Hr.	Daily	Bare Costs	Incl. O&P
3 Bricklayers	$35.25	$846.00	$53.70	$1288.80	$31.08	$47.35
3 Bricklayer Helpers	26.90	645.60	41.00	984.00		
48 L.H., Daily Totals		$1491.60		$2272.80	$31.08	$47.35

Crew D-10	Hr.	Daily	Hr.	Daily	Bare Costs	Incl. O&P
1 Bricklayer Foreman	$37.25	$298.00	$56.75	$454.00	$32.44	$49.31
1 Bricklayer	35.25	282.00	53.70	429.60		
2 Bricklayer Helpers	26.90	430.40	41.00	656.00		
1 Equip. Oper. (crane)	35.90	287.20	54.10	432.80		
1 Truck Crane, 12.5 Ton		505.80		556.40	12.65	13.91
40 L.H., Daily Totals		$1803.40		$2528.80	$45.09	$63.22

Crew D-11	Hr.	Daily	Hr.	Daily	Bare Costs	Incl. O&P
1 Bricklayer Foreman	$37.25	$298.00	$56.75	$454.00	$33.13	$50.48
1 Bricklayer	35.25	282.00	53.70	429.60		
1 Bricklayer Helper	26.90	215.20	41.00	328.00		
24 L.H., Daily Totals		$795.20		$1211.60	$33.13	$50.48

Crew No.	Bare Costs		Incl. Subs O & P		Cost Per Labor-Hour	
Crew D-12	Hr.	Daily	Hr.	Daily	Bare Costs	Incl. O&P
1 Bricklayer Foreman	$37.25	$298.00	$56.75	$454.00	$31.58	$48.11
1 Bricklayer	35.25	282.00	53.70	429.60		
2 Bricklayer Helpers	26.90	430.40	41.00	656.00		
32 L.H., Daily Totals		$1010.40		$1539.60	$31.58	$48.11
Crew D-13	Hr.	Daily	Hr.	Daily	Bare Costs	Incl. O&P
1 Bricklayer Foreman	$37.25	$298.00	$56.75	$454.00	$32.74	$49.98
1 Bricklayer	35.25	282.00	53.70	429.60		
2 Bricklayer Helpers	26.90	430.40	41.00	656.00		
1 Carpenter	34.25	274.00	53.35	426.80		
1 Equip. Oper. (crane)	35.90	287.20	54.10	432.80		
1 Truck Crane, 12.5 Ton		505.80		556.40	10.54	11.59
48 L.H., Daily Totals		$2077.40		$2955.60	$43.28	$61.57
Crew E-1	Hr.	Daily	Hr.	Daily	Bare Costs	Incl. O&P
1 Welder Foreman	$40.15	$321.20	$71.85	$574.80	$37.12	$63.30
1 Welder	38.15	305.20	68.25	546.00		
1 Equip. Oper. (light)	33.05	264.40	49.80	398.40		
1 Gas Welding Machine		81.20		89.30	3.38	3.72
24 L.H., Daily Totals		$972.00		$1608.50	$40.50	$67.02
Crew E-2	Hr.	Daily	Hr.	Daily	Bare Costs	Incl. O&P
1 Struc. Steel Foreman	$40.15	$321.20	$71.85	$574.80	$36.96	$63.47
4 Struc. Steel Workers	38.15	1220.80	68.25	2184.00		
1 Equip. Oper. (crane)	35.90	287.20	54.10	432.80		
1 Equip. Oper. Oiler	30.10	240.80	45.35	362.80		
1 Crane, 90 Ton		1325.00		1457.50	23.66	26.03
56 L.H., Daily Totals		$3395.00		$5011.90	$60.62	$89.50
Crew E-3	Hr.	Daily	Hr.	Daily	Bare Costs	Incl. O&P
1 Struc. Steel Foreman	$40.15	$321.20	$71.85	$574.80	$38.82	$69.45
1 Struc. Steel Worker	38.15	305.20	68.25	546.00		
1 Welder	38.15	305.20	68.25	546.00		
1 Gas Welding Machine		81.20		89.30	3.38	3.72
24 L.H., Daily Totals		$1012.80		$1756.10	$42.20	$73.17
Crew E-4	Hr.	Daily	Hr.	Daily	Bare Costs	Incl. O&P
1 Struc. Steel Foreman	$40.15	$321.20	$71.85	$574.80	$38.65	$69.15
3 Struc. Steel Workers	38.15	915.60	68.25	1638.00		
1 Gas Welding Machine		81.20		89.30	2.54	2.79
32 L.H., Daily Totals		$1318.00		$2302.10	$41.19	$71.94
Crew E-5	Hr.	Daily	Hr.	Daily	Bare Costs	Incl. O&P
2 Struc. Steel Foremen	$40.15	$642.40	$71.85	$1149.60	$37.52	$65.27
5 Struc. Steel Workers	38.15	1526.00	68.25	2730.00		
1 Equip. Oper. (crane)	35.90	287.20	54.10	432.80		
1 Welder	38.15	305.20	68.25	546.00		
1 Equip. Oper. Oiler	30.10	240.80	45.35	362.80		
1 Crane, 90 Ton		1325.00		1457.50		
1 Gas Welding Machine		81.20		89.30	17.58	19.34
80 L.H., Daily Totals		$4407.80		$6768.00	$55.10	$84.61

Crew No.	Bare Costs		Incl. Subs O & P		Cost Per Labor-Hour	
Crew E-6	Hr.	Daily	Hr.	Daily	Bare Costs	Incl. O&P
3 Struc. Steel Foremen	$40.15	$963.60	$71.85	$1724.40	$37.56	$65.46
9 Struc. Steel Workers	38.15	2746.80	68.25	4914.00		
1 Equip. Oper. (crane)	35.90	287.20	54.10	432.80		
1 Welder	38.15	305.20	68.25	546.00		
1 Equip. Oper. Oiler	30.10	240.80	45.35	362.80		
1 Equip. Oper. (light)	33.05	264.40	49.80	398.40		
1 Crane, 90 Ton		1325.00		1457.50		
1 Gas Welding Machine		81.20		89.30		
1 Air Compr., 160 C.F.M.		87.40		96.15		
2 Impact Wrenches		24.00		26.40	11.86	13.04
128 L.H., Daily Totals		$6325.60		$10047.75	$49.42	$78.50
Crew E-7	Hr.	Daily	Hr.	Daily	Bare Costs	Incl. O&P
1 Struc. Steel Foreman	$40.15	$321.20	$71.85	$574.80	$37.52	$65.27
4 Struc. Steel Workers	38.15	1220.80	68.25	2184.00		
1 Equip. Oper. (crane)	35.90	287.20	54.10	432.80		
1 Equip. Oper. Oiler	30.10	240.80	45.35	362.80		
1 Welder Foreman	40.15	321.20	71.85	574.80		
2 Welders	38.15	610.40	68.25	1092.00		
1 Crane, 90 Ton		1325.00		1457.50		
2 Gas Welding Machines		162.40		178.65	18.59	20.45
80 L.H., Daily Totals		$4489.00		$6857.35	$56.11	$85.72
Crew E-8	Hr.	Daily	Hr.	Daily	Bare Costs	Incl. O&P
1 Struc. Steel Foreman	$40.15	$321.20	$71.85	$574.80	$37.27	$64.53
4 Struc. Steel Workers	38.15	1220.80	68.25	2184.00		
1 Welder Foreman	40.15	321.20	71.85	574.80		
4 Welders	38.15	1220.80	68.25	2184.00		
1 Equip. Oper. (crane)	35.90	287.20	54.10	432.80		
1 Equip. Oper. Oiler	30.10	240.80	45.35	362.80		
1 Equip. Oper. (light)	33.05	264.40	49.80	398.40		
1 Crane, 90 Ton		1325.00		1457.50		
4 Gas Welding Machines		324.80		357.30	15.86	17.45
104 L.H., Daily Totals		$5526.20		$8526.40	$53.13	$81.98
Crew E-9	Hr.	Daily	Hr.	Daily	Bare Costs	Incl. O&P
2 Struc. Steel Foremen	$40.15	$642.40	$71.85	$1149.60	$37.56	$65.46
5 Struc. Steel Workers	38.15	1526.00	68.25	2730.00		
1 Welder Foreman	40.15	321.20	71.85	574.80		
5 Welders	38.15	1526.00	68.25	2730.00		
1 Equip. Oper. (crane)	35.90	287.20	54.10	432.80		
1 Equip. Oper. Oiler	30.10	240.80	45.35	362.80		
1 Equip. Oper. (light)	33.05	264.40	49.80	398.40		
1 Crane, 90 Ton		1325.00		1457.50		
5 Gas Welding Machines		406.00		446.60	13.52	14.88
128 L.H., Daily Totals		$6539.00		$10282.50	$51.08	$80.34
Crew E-10	Hr.	Daily	Hr.	Daily	Bare Costs	Incl. O&P
1 Welder Foreman	$40.15	$321.20	$71.85	$574.80	$39.15	$70.05
1 Welder	38.15	305.20	68.25	546.00		
1 Gas Welding Machines		81.20		89.30		
1 Truck, 3 Ton		176.20		193.80	16.09	17.70
16 L.H., Daily Totals		$883.80		$1403.90	$55.24	$87.75
Crew E-11	Hr.	Daily	Hr.	Daily	Bare Costs	Incl. O&P
2 Painters, Struc. Steel	$31.00	$496.00	$57.05	$912.80	$30.44	$51.36
1 Building Laborer	26.70	213.60	41.55	332.40		
1 Equip. Oper. (light)	33.05	264.40	49.80	398.40		
1 Air Compressor 250 C.F.M.		112.40		123.65		
1 Sand Blaster		15.20		16.70		
1 Sand Blasting Accessories		11.90		13.10	4.36	4.80
32 L.H., Daily Totals		$1113.50		$1797.05	$34.80	$56.16

Crews

Crew No.	Bare Costs		Incl. Subs O & P		Cost Per Labor-Hour	
Crew E-12	**Hr.**	**Daily**	**Hr.**	**Daily**	**Bare Costs**	**Incl. O&P**
1 Welder Foreman	$40.15	$321.20	$71.85	$574.80	$36.60	$60.83
1 Equip. Oper. (light)	33.05	264.40	49.80	398.40		
1 Gas Welding Machine		81.20		89.30	5.08	5.58
16 L.H., Daily Totals		$666.80		$1062.50	$41.68	$66.41

Crew E-13	Hr.	Daily	Hr.	Daily	Bare Costs	Incl. O&P
1 Welder Foreman	$40.15	$321.20	$71.85	$574.80	$37.78	$64.50
.5 Equip. Oper. (light)	33.05	132.20	49.80	199.20		
1 Gas Welding Machine		81.20		89.30	6.77	7.44
12 L.H., Daily Totals		$534.60		$863.30	$44.55	$71.94

Crew E-14	Hr.	Daily	Hr.	Daily	Bare Costs	Incl. O&P
1 Welder Foreman	$40.15	$321.20	$71.85	$574.80	$40.15	$71.85
1 Gas Welding Machine		81.20		89.30	10.15	11.17
8 L.H., Daily Totals		$402.40		$664.10	$50.30	$83.02

Crew E-16	Hr.	Daily	Hr.	Daily	Bare Costs	Incl. O&P
1 Welder Foreman	$40.15	$321.20	$71.85	$574.80	$39.15	$70.05
1 Welder	38.15	305.20	68.25	546.00		
1 Gas Welding Machine		81.20		89.30	5.08	5.58
16 L.H., Daily Totals		$707.60		$1210.10	$44.23	$75.63

Crew E-17	Hr.	Daily	Hr.	Daily	Bare Costs	Incl. O&P
1 Structural Steel Foreman	$40.15	$321.20	$71.85	$574.80	$39.15	$70.05
1 Structural Steel Worker	38.15	305.20	68.25	546.00		
1 Power Tool		5.20		5.70	.33	.36
16 L.H., Daily Totals		$631.60		$1126.50	$39.48	$70.41

Crew E-18	Hr.	Daily	Hr.	Daily	Bare Costs	Incl. O&P
1 Structural Steel Foreman	$40.15	$321.20	$71.85	$574.80	$37.85	$65.76
3 Structural Steel Workers	38.15	915.60	68.25	1638.00		
1 Equipment Operator (med.)	34.65	277.20	52.20	417.60		
1 Crane, 20 Ton		797.10		876.80	19.93	21.92
40 L.H., Daily Totals		$2311.10		$3507.20	$57.78	$87.68

Crew E-19	Hr.	Daily	Hr.	Daily	Bare Costs	Incl. O&P
1 Structural Steel Worker	$38.15	$305.20	$68.25	$546.00	$37.12	$63.30
1 Structural Steel Foreman	40.15	321.20	71.85	574.80		
1 Equip. Oper. (light)	33.05	264.40	49.80	398.40		
1 Power Tool		5.20		5.70		
1 Crane, 20 Ton		797.10		876.80	33.43	36.77
24 L.H., Daily Totals		$1693.10		$2401.70	$70.55	$100.07

Crew E-20	Hr.	Daily	Hr.	Daily	Bare Costs	Incl. O&P
1 Structural Steel Foreman	$40.15	$321.20	$71.85	$574.80	$37.11	$64.07
5 Structural Steel Workers	38.15	1526.00	68.25	2730.00		
1 Equip. Oper. (crane)	35.90	287.20	54.10	432.80		
1 Oiler	30.10	240.80	45.35	362.80		
1 Power Tool		5.20		5.70		
1 Crane, 40 Ton		951.90		1047.10	14.95	16.45
64 L.H., Daily Totals		$3332.30		$5153.20	$52.06	$80.52

Crew E-22	Hr.	Daily	Hr.	Daily	Bare Costs	Incl. O&P
1 Skilled Worker Foreman	$36.85	$294.80	$57.30	$458.40	$35.52	$55.23
2 Skilled Workers	34.85	557.60	54.20	867.20		
24 L.H., Daily Totals		$852.40		$1325.60	$35.52	$55.23

Crew No.	Bare Costs		Incl. Subs O & P		Cost Per Labor-Hour	

Crew E-24	Hr.	Daily	Hr.	Daily	Bare Costs	Incl. O&P
3 Structural Steel Workers	$38.15	$915.60	$68.25	$1638.00	$37.28	$64.24
1 Equipment Operator (medium)	34.65	277.20	52.20	417.60		
1-25 Ton Crane		616.80		678.50	19.28	21.20
32 L.H., Daily Totals		$1809.60		$2734.10	$56.56	$85.44

Crew E-25	Hr.	Daily	Hr.	Daily	Bare Costs	Incl. O&P
1 Welder Foreman	$40.15	$321.20	$71.85	$574.80	$40.15	$71.85
1 Cutting Torch		18.00		19.80		
1 Gases		64.80		71.30	10.35	11.39
8 L.H., Daily Totals		$404.00		$665.90	$50.50	$83.24

Crew F-3	Hr.	Daily	Hr.	Daily	Bare Costs	Incl. O&P
4 Carpenters	$34.25	$1096.00	$53.35	$1707.20	$34.58	$53.50
1 Equip. Oper. (crane)	35.90	287.20	54.10	432.80		
1 Hyd. Crane, 12 Ton		602.60		662.85	15.07	16.57
40 L.H., Daily Totals		$1985.80		$2802.85	$49.65	$70.07

Crew F-4	Hr.	Daily	Hr.	Daily	Bare Costs	Incl. O&P
4 Carpenters	$34.25	$1096.00	$53.35	$1707.20	$33.83	$52.14
1 Equip. Oper. (crane)	35.90	287.20	54.10	432.80		
1 Equip. Oper. Oiler	30.10	240.80	45.35	362.80		
1 Hyd. Crane, 55 Ton		895.40		984.95	18.65	20.52
48 L.H., Daily Totals		$2519.40		$3487.75	$52.48	$72.66

Crew F-5	Hr.	Daily	Hr.	Daily	Bare Costs	Incl. O&P
1 Carpenter Foreman	$36.25	$290.00	$56.45	$451.60	$34.75	$54.13
3 Carpenters	34.25	822.00	53.35	1280.40		
32 L.H., Daily Totals		$1112.00		$1732.00	$34.75	$54.13

Crew F-6	Hr.	Daily	Hr.	Daily	Bare Costs	Incl. O&P
2 Carpenters	$34.25	$548.00	$53.35	$853.60	$31.56	$48.78
2 Building Laborers	26.70	427.20	41.55	664.80		
1 Equip. Oper. (crane)	35.90	287.20	54.10	432.80		
1 Hyd. Crane, 12 Ton		602.60		662.85	15.07	16.57
40 L.H., Daily Totals		$1865.00		$2614.05	$46.63	$65.35

Crew F-7	Hr.	Daily	Hr.	Daily	Bare Costs	Incl. O&P
2 Carpenters	$34.25	$548.00	$53.35	$853.60	$30.48	$47.45
2 Building Laborers	26.70	427.20	41.55	664.80		
32 L.H., Daily Totals		$975.20		$1518.40	$30.48	$47.45

Crew G-1	Hr.	Daily	Hr.	Daily	Bare Costs	Incl. O&P
1 Roofer Foreman	$31.45	$251.60	$53.60	$428.80	$27.49	$46.86
4 Roofers, Composition	29.45	942.40	50.20	1606.40		
2 Roofer Helpers	21.60	345.60	36.80	588.80		
1 Application Equipment		146.60		161.25		
1 Tar Kettle/Pot		52.50		57.75		
1 Crew Truck		98.80		108.70	5.32	5.85
56 L.H., Daily Totals		$1837.50		$2951.70	$32.81	$52.71

Crew G-2	Hr.	Daily	Hr.	Daily	Bare Costs	Incl. O&P
1 Plasterer	$31.35	$250.80	$47.70	$381.60	$28.38	$43.50
1 Plasterer Helper	27.10	216.80	41.25	330.00		
1 Building Laborer	26.70	213.60	41.55	332.40		
1 Grouting Equipment		107.75		118.55	4.49	4.94
24 L.H., Daily Totals		$788.95		$1162.55	$32.87	$48.44

Crews

Crew G-3	Hr.	Daily	Hr.	Daily	Bare Costs	Incl. O&P
2 Sheet Metal Workers	$40.20	$643.20	$61.85	$989.60	$33.45	$51.70
2 Building Laborers	26.70	427.20	41.55	664.80		
32 L.H., Daily Totals		$1070.40		$1654.40	$33.45	$51.70

Crew G-4	Hr.	Daily	Hr.	Daily	Bare Costs	Incl. O&P
1 Labor Foreman (outside)	$28.70	$229.60	$44.70	$357.60	$27.37	$42.60
2 Building Laborers	26.70	427.20	41.55	664.80		
1 Light Truck, 1.5 Ton		128.80		141.70		
1 Air Compr., 160 C.F.M.		87.40		96.15	9.01	9.91
24 L.H., Daily Totals		$873.00		$1260.25	$36.38	$52.51

Crew G-5	Hr.	Daily	Hr.	Daily	Bare Costs	Incl. O&P
1 Roofer Foreman	$31.45	$251.60	$53.60	$428.80	$26.71	$45.52
2 Roofers, Composition	29.45	471.20	50.20	803.20		
2 Roofer Helpers	21.60	345.60	36.80	588.80		
1 Application Equipment		146.60		161.25	3.67	4.03
40 L.H., Daily Totals		$1215.00		$1982.05	$30.38	$49.55

Crew G-6A	Hr.	Daily	Hr.	Daily	Bare Costs	Incl. O&P
2 Roofers Composition	$29.45	$471.20	$50.20	$803.20	$29.45	$50.20
1 Small Compressor		12.15		13.35		
2 Pneumatic Nailers		39.00		42.90	3.19	3.51
16 L.H., Daily Totals		$522.35		$859.45	$32.64	$53.71

Crew G-7	Hr.	Daily	Hr.	Daily	Bare Costs	Incl. O&P
1 Carpenter	$34.25	$274.00	$53.35	$426.80	$34.25	$53.35
1 Small Compressor		12.15		13.35		
1 Pneumatic Nailer		19.50		21.45	3.94	4.34
8 L.H., Daily Totals		$305.65		$461.60	$38.19	$57.69

Crew H-1	Hr.	Daily	Hr.	Daily	Bare Costs	Incl. O&P
2 Glaziers	$33.00	$528.00	$49.90	$798.40	$35.58	$59.08
2 Struc. Steel Workers	38.15	610.40	68.25	1092.00		
32 L.H., Daily Totals		$1138.40		$1890.40	$35.58	$59.08

Crew H-2	Hr.	Daily	Hr.	Daily	Bare Costs	Incl. O&P
2 Glaziers	$33.00	$528.00	$49.90	$798.40	$30.90	$47.12
1 Building Laborer	26.70	213.60	41.55	332.40		
24 L.H., Daily Totals		$741.60		$1130.80	$30.90	$47.12

Crew H-3	Hr.	Daily	Hr.	Daily	Bare Costs	Incl. O&P
1 Glazier	$33.00	$264.00	$49.90	$399.20	$29.28	$44.83
1 Helper	25.55	204.40	39.75	318.00		
16 L.H., Daily Totals		$468.40		$717.20	$29.28	$44.83

Crew J-1	Hr.	Daily	Hr.	Daily	Bare Costs	Incl. O&P
3 Plasterers	$31.35	$752.40	$47.70	$1144.80	$29.65	$45.12
2 Plasterer Helpers	27.10	433.60	41.25	660.00		
1 Mixing Machine, 6 C.F.		103.00		113.30	2.58	2.83
40 L.H., Daily Totals		$1289.00		$1918.10	$32.23	$47.95

Crew J-2	Hr.	Daily	Hr.	Daily	Bare Costs	Incl. O&P
3 Plasterers	$31.35	$752.40	$47.70	$1144.80	$29.95	$45.37
2 Plasterer Helpers	27.10	433.60	41.25	660.00		
1 Lather	31.45	251.60	46.60	372.80		
1 Mixing Machine, 6 C.F.		103.00		113.30	2.15	2.36
48 L.H., Daily Totals		$1540.60		$2290.90	$32.10	$47.73

Crew J-3	Hr.	Daily	Hr.	Daily	Bare Costs	Incl. O&P
1 Terrazzo Worker	$32.65	$261.20	$48.00	$384.00	$29.38	$43.18
1 Terrazzo Helper	26.10	208.80	38.35	306.80		
1 Terrazzo Grinder, Electric		70.00		77.00		
1 Terrazzo Mixer		139.60		153.55	13.10	14.41
16 L.H., Daily Totals		$679.60		$921.35	$42.48	$57.59

Crew J-4	Hr.	Daily	Hr.	Daily	Bare Costs	Incl. O&P
1 Tile Layer	$32.70	$261.60	$48.05	$384.40	$29.03	$42.65
1 Tile Layer Helper	25.35	202.80	37.25	298.00		
16 L.H., Daily Totals		$464.40		$682.40	$29.03	$42.65

Crew K-1	Hr.	Daily	Hr.	Daily	Bare Costs	Incl. O&P
1 Carpenter	$34.25	$274.00	$53.35	$426.80	$30.53	$47.10
1 Truck Driver (light)	26.80	214.40	40.85	326.80		
1 Truck w/Power Equip.		176.20		193.80	11.01	12.11
16 L.H., Daily Totals		$664.60		$947.40	$41.54	$59.21

Crew K-2	Hr.	Daily	Hr.	Daily	Bare Costs	Incl. O&P
1 Struc. Steel Foreman	$40.15	$321.20	$71.85	$574.80	$35.03	$60.32
1 Struc. Steel Worker	38.15	305.20	68.25	546.00		
1 Truck Driver (light)	26.80	214.40	40.85	326.80		
1 Truck w/Power Equip.		176.20		193.80	7.34	8.08
24 L.H., Daily Totals		$1017.00		$1641.40	$42.37	$68.40

Crew L-1	Hr.	Daily	Hr.	Daily	Bare Costs	Incl. O&P
1 Electrician	$40.75	$326.00	$60.65	$485.20	$40.80	$61.03
1 Plumber	40.85	326.80	61.40	491.20		
16 L.H., Daily Totals		$652.80		$976.40	$40.80	$61.03

Crew L-2	Hr.	Daily	Hr.	Daily	Bare Costs	Incl. O&P
1 Carpenter	$34.25	$274.00	$53.35	$426.80	$29.90	$46.55
1 Carpenter Helper	25.55	204.40	39.75	318.00		
16 L.H., Daily Totals		$478.40		$744.80	$29.90	$46.55

Crew L-3	Hr.	Daily	Hr.	Daily	Bare Costs	Incl. O&P
1 Carpenter	$34.25	$274.00	$53.35	$426.80	$37.36	$57.30
.5 Electrician	40.75	163.00	60.65	242.60		
.5 Sheet Metal Worker	40.20	160.80	61.85	247.40		
16 L.H., Daily Totals		$597.80		$916.80	$37.36	$57.30

Crew L-3A	Hr.	Daily	Hr.	Daily	Bare Costs	Incl. O&P
1 Carpenter Foreman (outside)	$36.25	$290.00	$56.45	$451.60	$37.57	$58.25
.5 Sheet Metal Worker	40.20	160.80	61.85	247.40		
12 L.H., Daily Totals		$450.80		$699.00	$37.57	$58.25

Crew L-4	Hr.	Daily	Hr.	Daily	Bare Costs	Incl. O&P
2 Skilled Workers	$34.85	$557.60	$54.20	$867.20	$31.75	$49.38
1 Helper	25.55	204.40	39.75	318.00		
24 L.H., Daily Totals		$762.00		$1185.20	$31.75	$49.38

Crew L-5	Hr.	Daily	Hr.	Daily	Bare Costs	Incl. O&P
1 Struc. Steel Foreman	$40.15	$321.20	$71.85	$574.80	$38.11	$66.74
5 Struc. Steel Workers	38.15	1526.00	68.25	2730.00		
1 Equip. Oper. (crane)	35.90	287.20	54.10	432.80		
1 Hyd. Crane, 25 Ton		616.80		678.50	11.01	12.12
56 L.H., Daily Totals		$2751.20		$4416.10	$49.12	$78.86

Crew L-5A

Crew No.	Bare Costs Hr.	Daily	Incl. Subs O & P Hr.	Daily	Cost Per Labor-Hour Bare Costs	Incl. O&P
1 Structural Steel Foreman	$40.15	$321.20	$71.85	$574.80	$38.09	$65.61
2 Structural Steel Workers	38.15	610.40	68.25	1092.00		
1 Equip. Oper. (crane)	35.90	287.20	54.10	432.80		
1 Crane,SP, 25 Ton		575.00		632.50	17.97	19.77
32 L.H., Daily Totals		$1793.80		$2732.10	$56.06	$85.38

Crew L-6

Crew No.	Bare Costs Hr.	Daily	Incl. Subs O & P Hr.	Daily	Cost Per Labor-Hour Bare Costs	Incl. O&P
1 Plumber	$40.85	$326.80	$61.40	$491.20	$40.82	$61.15
.5 Electrician	40.75	163.00	60.65	242.60		
12 L.H., Daily Totals		$489.80		$733.80	$40.82	$61.15

Crew L-7

Crew No.	Bare Costs Hr.	Daily	Incl. Subs O & P Hr.	Daily	Cost Per Labor-Hour Bare Costs	Incl. O&P
2 Carpenters	$34.25	$548.00	$53.35	$853.60	$33.02	$51.02
1 Building Laborer	26.70	213.60	41.55	332.40		
.5 Electrician	40.75	163.00	60.65	242.60		
28 L.H., Daily Totals		$924.60		$1428.60	$33.02	$51.02

Crew L-8

Crew No.	Bare Costs Hr.	Daily	Incl. Subs O & P Hr.	Daily	Cost Per Labor-Hour Bare Costs	Incl. O&P
2 Carpenters	$34.25	$548.00	$53.35	$853.60	$35.57	$54.96
.5 Plumber	40.85	163.40	61.40	245.60		
20 L.H., Daily Totals		$711.40		$1099.20	$35.57	$54.96

Crew L-9

Crew No.	Bare Costs Hr.	Daily	Incl. Subs O & P Hr.	Daily	Cost Per Labor-Hour Bare Costs	Incl. O&P
1 Labor Foreman (inside)	$27.20	$217.60	$42.35	$338.80	$30.92	$49.78
2 Building Laborers	26.70	427.20	41.55	664.80		
1 Struc. Steel Worker	38.15	305.20	68.25	546.00		
.5 Electrician	40.75	163.00	60.65	242.60		
36 L.H., Daily Totals		$1113.00		$1792.20	$30.92	$49.78

Crew L-10

Crew No.	Bare Costs Hr.	Daily	Incl. Subs O & P Hr.	Daily	Cost Per Labor-Hour Bare Costs	Incl. O&P
1 Structural Steel Foreman	$40.15	$321.20	$71.85	$574.80	$38.07	$64.73
1 Structural Steel Worker	38.15	305.20	68.25	546.00		
1 Equip. Oper. (crane)	35.90	287.20	54.10	432.80		
1 Hyd. Crane, 12 Ton		602.60		662.85	25.11	27.62
24 L.H., Daily Totals		$1516.20		$2216.45	$63.18	$92.35

Crew L-11

Crew No.	Bare Costs Hr.	Daily	Incl. Subs O & P Hr.	Daily	Cost Per Labor-Hour Bare Costs	Incl. O&P
2 Wreckers	$26.70	$427.20	$47.20	$755.20	$30.59	$49.58
1 Equip. Oper. (crane)	35.90	287.20	54.10	432.80		
1 Equip. Oper. (light)	33.05	264.40	49.80	398.40		
1 Hyd. Excavator, 2.5 C.Y.		1215.00		1336.50		
1 Skid steer loader		197.80		217.60	44.15	48.57
32 L.H., Daily Totals		$2391.60		$3140.50	$74.74	$98.15

Crew M-1

Crew No.	Bare Costs Hr.	Daily	Incl. Subs O & P Hr.	Daily	Cost Per Labor-Hour Bare Costs	Incl. O&P
3 Elevator Constructors	$45.05	$1081.20	$67.55	$1621.20	$42.80	$64.18
1 Elevator Apprentice	36.05	288.40	54.05	432.40		
5 Hand Tools		63.00		69.30	1.97	2.17
32 L.H., Daily Totals		$1432.60		$2122.90	$44.77	$66.35

Crew M-3

Crew No.	Bare Costs Hr.	Daily	Incl. Subs O & P Hr.	Daily	Cost Per Labor-Hour Bare Costs	Incl. O&P
1 Electrician Foreman (out)	$42.75	$342.00	$63.60	$508.80	$37.46	$56.42
1 Common Laborer	26.70	213.60	41.55	332.40		
.25 Equipment Operator, Medium	34.65	69.30	52.20	104.40		
1 Elevator Constructor	45.05	360.40	67.55	540.40		
1 Elevator Apprentice	36.05	288.40	54.05	432.40		
.25 Crane, SP, 4 x 4, 20 ton		145.35		159.90	4.28	4.70
34 L.H., Daily Totals		$1419.05		$2078.30	$41.74	$61.12

Crew M-4

Crew No.	Bare Costs Hr.	Daily	Incl. Subs O & P Hr.	Daily	Cost Per Labor-Hour Bare Costs	Incl. O&P
1 Electrician Foreman (out)	$42.75	$342.00	$63.60	$508.80	$37.12	$55.91
1 Common Laborer	26.70	213.60	41.55	332.40		
.25 Equipment Operator, Crane	35.90	71.80	54.10	108.20		
.25 Equipment Operator, Oiler	30.10	60.20	45.35	90.70		
1 Elevator Constructor	45.05	360.40	67.55	540.40		
1 Elevator Apprentice	36.05	288.40	54.05	432.40		
.25 Crane, Hyd, SP, 4WD, 40 Ton		217.40		239.15	6.04	6.64
36 L.H., Daily Totals		$1553.80		$2252.05	$43.16	$62.55

Crew Q-1

Crew No.	Bare Costs Hr.	Daily	Incl. Subs O & P Hr.	Daily	Cost Per Labor-Hour Bare Costs	Incl. O&P
1 Plumber	$40.85	$326.80	$61.40	$491.20	$36.78	$55.28
1 Plumber Apprentice	32.70	261.60	49.15	393.20		
16 L.H., Daily Totals		$588.40		$884.40	$36.78	$55.28

Crew Q-1C

Crew No.	Bare Costs Hr.	Daily	Incl. Subs O & P Hr.	Daily	Cost Per Labor-Hour Bare Costs	Incl. O&P
1 Plumber	$40.85	$326.80	$61.40	$491.20	$36.07	$54.25
1 Plumber Apprentice	32.70	261.60	49.15	393.20		
1 Equip. Oper. (medium)	34.65	277.20	52.20	417.60		
1 Trencher, Chain		1420.00		1562.00	59.17	65.08
24 L.H., Daily Totals		$2285.60		$2864.00	$95.24	$119.33

Crew Q-2

Crew No.	Bare Costs Hr.	Daily	Incl. Subs O & P Hr.	Daily	Cost Per Labor-Hour Bare Costs	Incl. O&P
2 Plumbers	$40.85	$653.60	$61.40	$982.40	$38.13	$57.32
1 Plumber Apprentice	32.70	261.60	49.15	393.20		
24 L.H., Daily Totals		$915.20		$1375.60	$38.13	$57.32

Crew Q-3

Crew No.	Bare Costs Hr.	Daily	Incl. Subs O & P Hr.	Daily	Cost Per Labor-Hour Bare Costs	Incl. O&P
1 Plumber Foreman (inside)	$41.35	$330.80	$62.15	$497.20	$38.94	$58.53
2 Plumbers	40.85	653.60	61.40	982.40		
1 Plumber Apprentice	32.70	261.60	49.15	393.20		
32 L.H., Daily Totals		$1246.00		$1872.80	$38.94	$58.53

Crew Q-4

Crew No.	Bare Costs Hr.	Daily	Incl. Subs O & P Hr.	Daily	Cost Per Labor-Hour Bare Costs	Incl. O&P
1 Plumber Foreman (inside)	$41.35	$330.80	$62.15	$497.20	$38.94	$58.53
1 Plumber	40.85	326.80	61.40	491.20		
1 Welder (plumber)	40.85	326.80	61.40	491.20		
1 Plumber Apprentice	32.70	261.60	49.15	393.20		
1 Electric Welding Mach.		80.65		88.70	2.52	2.77
32 L.H., Daily Totals		$1326.65		$1961.50	$41.46	$61.30

Crew Q-5

Crew No.	Bare Costs Hr.	Daily	Incl. Subs O & P Hr.	Daily	Cost Per Labor-Hour Bare Costs	Incl. O&P
1 Steamfitter	$40.95	$327.60	$61.55	$492.40	$36.85	$55.38
1 Steamfitter Apprentice	32.75	262.00	49.20	393.60		
16 L.H., Daily Totals		$589.60		$886.00	$36.85	$55.38

Crew Q-6

Crew No.	Bare Costs Hr.	Daily	Incl. Subs O & P Hr.	Daily	Cost Per Labor-Hour Bare Costs	Incl. O&P
2 Steamfitters	$40.95	$655.20	$61.55	$984.80	$38.22	$57.43
1 Steamfitter Apprentice	32.75	262.00	49.20	393.60		
24 L.H., Daily Totals		$917.20		$1378.40	$38.22	$57.43

Crew Q-7

Crew No.	Bare Costs Hr.	Daily	Incl. Subs O & P Hr.	Daily	Cost Per Labor-Hour Bare Costs	Incl. O&P
1 Steamfitter Foreman (inside)	$41.45	$331.60	$62.30	$498.40	$39.03	$58.65
2 Steamfitters	40.95	655.20	61.55	984.80		
1 Steamfitter Apprentice	32.75	262.00	49.20	393.60		
32 L.H., Daily Totals		$1248.80		$1876.80	$39.03	$58.65

Crew No.	Bare Costs		Incl. Subs O & P		Cost Per Labor-Hour	
Crew Q-8	Hr.	Daily	Hr.	Daily	Bare Costs	Incl. O&P
1 Steamfitter Foreman (inside)	$41.45	$331.60	$62.30	$498.40	$39.03	$58.65
1 Steamfitter	40.95	327.60	61.55	492.40		
1 Welder (steamfitter)	40.95	327.60	61.55	492.40		
1 Steamfitter Apprentice	32.75	262.00	49.20	393.60		
1 Electric Welding Mach.		80.65		88.70	2.52	2.77
32 L.H., Daily Totals		$1329.45		$1965.50	$41.55	$61.42

Crew No.	Bare Costs		Incl. Subs O & P		Cost Per Labor-Hour	
Crew Q-9	Hr.	Daily	Hr.	Daily	Bare Costs	Incl. O&P
1 Sheet Metal Worker	$40.20	$321.60	$61.85	$494.80	$36.18	$55.65
1 Sheet Metal Apprentice	32.15	257.20	49.45	395.60		
16 L.H., Daily Totals		$578.80		$890.40	$36.18	$55.65

Crew No.	Bare Costs		Incl. Subs O & P		Cost Per Labor-Hour	
Crew Q-10	Hr.	Daily	Hr.	Daily	Bare Costs	Incl. O&P
2 Sheet Metal Workers	$40.20	$643.20	$61.85	$989.60	$37.52	$57.72
1 Sheet Metal Apprentice	32.15	257.20	49.45	395.60		
24 L.H., Daily Totals		$900.40		$1385.20	$37.52	$57.72

Crew No.	Bare Costs		Incl. Subs O & P		Cost Per Labor-Hour	
Crew Q-11	Hr.	Daily	Hr.	Daily	Bare Costs	Incl. O&P
1 Sheet Metal Foreman (inside)	$40.70	$325.60	$62.60	$500.80	$38.31	$58.94
2 Sheet Metal Workers	40.20	643.20	61.85	989.60		
1 Sheet Metal Apprentice	32.15	257.20	49.45	395.60		
32 L.H., Daily Totals		$1226.00		$1886.00	$38.31	$58.94

Crew No.	Bare Costs		Incl. Subs O & P		Cost Per Labor-Hour	
Crew Q-12	Hr.	Daily	Hr.	Daily	Bare Costs	Incl. O&P
1 Sprinkler Installer	$40.75	$326.00	$61.55	$492.40	$36.68	$55.40
1 Sprinkler Apprentice	32.60	260.80	49.25	394.00		
16 L.H., Daily Totals		$586.80		$886.40	$36.68	$55.40

Crew No.	Bare Costs		Incl. Subs O & P		Cost Per Labor-Hour	
Crew Q-13	Hr.	Daily	Hr.	Daily	Bare Costs	Incl. O&P
1 Sprinkler Foreman (inside)	$41.25	$330.00	$62.35	$498.80	$38.84	$58.68
2 Sprinkler Installers	40.75	652.00	61.55	984.80		
1 Sprinkler Apprentice	32.60	260.80	49.25	394.00		
32 L.H., Daily Totals		$1242.80		$1877.60	$38.84	$58.68

Crew No.	Bare Costs		Incl. Subs O & P		Cost Per Labor-Hour	
Crew Q-14	Hr.	Daily	Hr.	Daily	Bare Costs	Incl. O&P
1 Asbestos Worker	$37.10	$296.80	$58.55	$468.40	$33.40	$52.70
1 Asbestos Apprentice	29.70	237.60	46.85	374.80		
16 L.H., Daily Totals		$534.40		$843.20	$33.40	$52.70

Crew No.	Bare Costs		Incl. Subs O & P		Cost Per Labor-Hour	
Crew Q-15	Hr.	Daily	Hr.	Daily	Bare Costs	Incl. O&P
1 Plumber	$40.85	$326.80	$61.40	$491.20	$36.78	$55.28
1 Plumber Apprentice	32.70	261.60	49.15	393.20		
1 Electric Welding Mach.		80.65		88.70	5.04	5.54
16 L.H., Daily Totals		$669.05		$973.10	$41.82	$60.82

Crew No.	Bare Costs		Incl. Subs O & P		Cost Per Labor-Hour	
Crew Q-16	Hr.	Daily	Hr.	Daily	Bare Costs	Incl. O&P
2 Plumbers	$40.85	$653.60	$61.40	$982.40	$38.13	$57.32
1 Plumber Apprentice	32.70	261.60	49.15	393.20		
1 Electric Welding Mach.		80.65		88.70	3.36	3.70
24 L.H., Daily Totals		$995.85		$1464.30	$41.49	$61.02

Crew No.	Bare Costs		Incl. Subs O & P		Cost Per Labor-Hour	
Crew Q-17	Hr.	Daily	Hr.	Daily	Bare Costs	Incl. O&P
1 Steamfitter	$40.95	$327.60	$61.55	$492.40	$36.85	$55.38
1 Steamfitter Apprentice	32.75	262.00	49.20	393.60		
1 Electric Welding Mach.		80.65		88.70	5.04	5.54
16 L.H., Daily Totals		$670.25		$974.70	$41.89	$60.92

Crew No.	Bare Costs		Incl. Subs O & P		Cost Per Labor-Hour	
Crew Q-17A	Hr.	Daily	Hr.	Daily	Bare Costs	Incl. O&P
1 Steamfitter	$40.95	$327.60	$61.55	$492.40	$36.53	$54.95
1 Steamfitter Apprentice	32.75	262.00	49.20	393.60		
1 Equip. Oper. (crane)	35.90	287.20	54.10	432.80		
1 Truck Crane, 12 Ton		602.60		662.85		
1 Electric Welding Mach.		80.65		88.70	28.47	31.32
24 L.H., Daily Totals		$1560.05		$2070.35	$65.00	$86.27

Crew No.	Bare Costs		Incl. Subs O & P		Cost Per Labor-Hour	
Crew Q-18	Hr.	Daily	Hr.	Daily	Bare Costs	Incl. O&P
2 Steamfitters	$40.95	$655.20	$61.55	$984.80	$38.22	$57.43
1 Steamfitter Apprentice	32.75	262.00	49.20	393.60		
1 Electric Welding Mach.		80.65		88.70	3.36	3.70
24 L.H., Daily Totals		$997.85		$1467.10	$41.58	$61.13

Crew No.	Bare Costs		Incl. Subs O & P		Cost Per Labor-Hour	
Crew Q-19	Hr.	Daily	Hr.	Daily	Bare Costs	Incl. O&P
1 Steamfitter	$40.95	$327.60	$61.55	$492.40	$38.15	$57.13
1 Steamfitter Apprentice	32.75	262.00	49.20	393.60		
1 Electrician	40.75	326.00	60.65	485.20		
24 L.H., Daily Totals		$915.60		$1371.20	$38.15	$57.13

Crew No.	Bare Costs		Incl. Subs O & P		Cost Per Labor-Hour	
Crew Q-20	Hr.	Daily	Hr.	Daily	Bare Costs	Incl. O&P
1 Sheet Metal Worker	$40.20	$321.60	$61.85	$494.80	$37.09	$56.65
1 Sheet Metal Apprentice	32.15	257.20	49.45	395.60		
.5 Electrician	40.75	163.00	60.65	242.60		
20 L.H., Daily Totals		$741.80		$1133.00	$37.09	$56.65

Crew No.	Bare Costs		Incl. Subs O & P		Cost Per Labor-Hour	
Crew Q-21	Hr.	Daily	Hr.	Daily	Bare Costs	Incl. O&P
2 Steamfitters	$40.95	$655.20	$61.55	$984.80	$38.85	$58.24
1 Steamfitter Apprentice	32.75	262.00	49.20	393.60		
1 Electrician	40.75	326.00	60.65	485.20		
32 L.H., Daily Totals		$1243.20		$1863.60	$38.85	$58.24

Crew No.	Bare Costs		Incl. Subs O & P		Cost Per Labor-Hour	
Crew Q-22	Hr.	Daily	Hr.	Daily	Bare Costs	Incl. O&P
1 Plumber	$40.85	$326.80	$61.40	$491.20	$36.78	$55.28
1 Plumber Apprentice	32.70	261.60	49.15	393.20		
1 Truck Crane, 12 Ton		602.60		662.85	37.66	41.43
16 L.H., Daily Totals		$1191.00		$1547.25	$74.44	$96.71

Crew No.	Bare Costs		Incl. Subs O & P		Cost Per Labor-Hour	
Crew Q-22A	Hr.	Daily	Hr.	Daily	Bare Costs	Incl. O&P
1 Plumber	$40.85	$326.80	$61.40	$491.20	$34.04	$51.55
1 Plumber Apprentice	32.70	261.60	49.15	393.20		
1 Laborer	26.70	213.60	41.55	332.40		
1 Equip. Oper. (crane)	35.90	287.20	54.10	432.80		
1 Truck Crane, 12 Ton		602.60		662.85	18.83	20.71
32 L.H., Daily Totals		$1691.80		$2312.45	$52.87	$72.26

Crew No.	Bare Costs		Incl. Subs O & P		Cost Per Labor-Hour	
Crew Q-23	Hr.	Daily	Hr.	Daily	Bare Costs	Incl. O&P
1 Plumber Foreman	$42.85	$342.80	$64.40	$515.20	$39.45	$59.33
1 Plumber	40.85	326.80	61.40	491.20		
1 Equip. Oper. (medium)	34.65	277.20	52.20	417.60		
1 Power Tools		5.20		5.70		
1 Crane, 20 Ton		797.10		876.80	33.43	36.77
24 L.H., Daily Totals		$1749.10		$2306.50	$72.88	$96.10

Crew No.	Bare Costs		Incl. Subs O & P		Cost Per Labor-Hour	
Crew R-1	Hr.	Daily	Hr.	Daily	Bare Costs	Incl. O&P
1 Electrician Foreman	$41.25	$330.00	$61.40	$491.20	$35.77	$53.81
3 Electricians	40.75	978.00	60.65	1455.60		
2 Helpers	25.55	408.80	39.75	636.00		
48 L.H., Daily Totals		$1716.80		$2582.80	$35.77	$53.81

Crew No.	Bare Costs		Incl. Subs O & P		Cost Per Labor-Hour	
Crew R-1A	Hr.	Daily	Hr.	Daily	Bare Costs	Incl. O&P
1 Electrician	$40.75	$326.00	$60.65	$485.20	$33.15	$50.20
1 Helper	25.55	204.40	39.75	318.00		
16 L.H., Daily Totals		$530.40		$803.20	$33.15	$50.20
Crew R-2	Hr.	Daily	Hr.	Daily	Bare Costs	Incl. O&P
1 Electrician Foreman	$41.25	$330.00	$61.40	$491.20	$35.79	$53.85
3 Electricians	40.75	978.00	60.65	1455.60		
2 Helpers	25.55	408.80	39.75	636.00		
1 Equip. Oper. (crane)	35.90	287.20	54.10	432.80		
1 S.P. Crane, 5 Ton		317.00		348.70	5.66	6.23
56 L.H., Daily Totals		$2321.00		$3364.30	$41.45	$60.08
Crew R-3	Hr.	Daily	Hr.	Daily	Bare Costs	Incl. O&P
1 Electrician Foreman	$41.25	$330.00	$61.40	$491.20	$39.98	$59.64
1 Electrician	40.75	326.00	60.65	485.20		
.5 Equip. Oper. (crane)	35.90	143.60	54.10	216.40		
.5 S.P. Crane, 5 Ton		158.50		174.35	7.93	8.72
20 L.H., Daily Totals		$958.10		$1367.15	$47.91	$68.36
Crew R-4	Hr.	Daily	Hr.	Daily	Bare Costs	Incl. O&P
1 Struc. Steel Foreman	$40.15	$321.20	$71.85	$574.80	$39.07	$67.45
3 Struc. Steel Workers	38.15	915.60	68.25	1638.00		
1 Electrician	40.75	326.00	60.65	485.20		
1 Gas Welding Machine		81.20		89.30	2.03	2.23
40 L.H., Daily Totals		$1644.00		$2787.30	$41.10	$69.68
Crew R-5	Hr.	Daily	Hr.	Daily	Bare Costs	Incl. O&P
1 Electrician Foreman	$41.25	$330.00	$61.40	$491.20	$35.27	$53.12
4 Electrician Linemen	40.75	1304.00	60.65	1940.80		
2 Electrician Operators	40.75	652.00	60.65	970.40		
4 Electrician Groundmen	25.55	817.60	39.75	1272.00		
1 Crew Truck		98.80		108.70		
1 Tool Van		122.15		134.35		
1 Pickup Truck, 3/4 Ton		78.00		85.80		
.2 Crane, 55 Ton		179.08		197.00		
.2 Crane, 12 Ton		120.52		132.55		
.2 Auger, Truck Mtd.		635.80		699.40		
1 Tractor w/Winch		260.00		286.00	16.98	18.68
88 L.H., Daily Totals		$4597.95		$6318.20	$52.25	$71.80
Crew R-6	Hr.	Daily	Hr.	Daily	Bare Costs	Incl. O&P
1 Electrician Foreman	$41.25	$330.00	$61.40	$491.20	$35.27	$53.12
4 Electrician Linemen	40.75	1304.00	60.65	1940.80		
2 Electrician Operators	40.75	652.00	60.65	970.40		
4 Electrician Groundmen	25.55	817.60	39.75	1272.00		
1 Crew Truck		98.80		108.70		
1 Tool Van		122.15		134.35		
1 Pickup Truck, 3/4 Ton		78.00		85.80		
.2 Crane, 55 Ton		179.08		197.00		
.2 Crane, 12 Ton		120.52		132.55		
.2 Auger, Truck Mtd.		635.80		699.40		
1 Tractor w/Winch		260.00		286.00		
3 Cable Trailers		490.65		539.70		
.5 Tensioning Rig		160.00		176.00		
.5 Cable Pulling Rig		934.00		1027.40	34.99	38.49
88 L.H., Daily Totals		$6182.60		$8061.30	$70.26	$91.61
Crew R-7	Hr.	Daily	Hr.	Daily	Bare Costs	Incl. O&P
1 Electrician Foreman	$41.25	$330.00	$61.40	$491.20	$28.17	$43.36
5 Electrician Groundmen	25.55	1022.00	39.75	1590.00		
1 Crew Truck		98.80		108.70	2.06	2.26
48 L.H., Daily Totals		$1450.80		$2189.90	$30.23	$45.62
Crew R-8	Hr.	Daily	Hr.	Daily	Bare Costs	Incl. O&P
1 Electrician Foreman	$41.25	$330.00	$61.40	$491.20	$35.77	$53.81
3 Electrician Linemen	40.75	978.00	60.65	1455.60		
2 Electrician Groundmen	25.55	408.80	39.75	636.00		
1 Pickup Truck, 3/4 Ton		78.00		85.80		
1 Crew Truck		98.80		108.70	3.68	4.05
48 L.H., Daily Totals		$1893.60		$2777.30	$39.45	$57.86
Crew R-9	Hr.	Daily	Hr.	Daily	Bare Costs	Incl. O&P
1 Electrician Foreman	$41.25	$330.00	$61.40	$491.20	$33.21	$50.29
1 Electrician Lineman	40.75	326.00	60.65	485.20		
2 Electrician Operators	40.75	652.00	60.65	970.40		
4 Electrician Groundmen	25.55	817.60	39.75	1272.00		
1 Pickup Truck, 3/4 Ton		78.00		85.80		
1 Crew Truck		98.80		108.70	2.76	3.04
64 L.H., Daily Totals		$2302.40		$3413.30	$35.97	$53.33
Crew R-10	Hr.	Daily	Hr.	Daily	Bare Costs	Incl. O&P
1 Electrician Foreman	$41.25	$330.00	$61.40	$491.20	$38.30	$57.29
4 Electrician Linemen	40.75	1304.00	60.65	1940.80		
1 Electrician Groundman	25.55	204.40	39.75	318.00		
1 Crew Truck		98.80		108.70		
3 Tram Cars		345.75		380.35	9.26	10.19
48 L.H., Daily Totals		$2282.95		$3239.05	$47.56	$67.48
Crew R-11	Hr.	Daily	Hr.	Daily	Bare Costs	Incl. O&P
1 Electrician Foreman	$41.25	$330.00	$61.40	$491.20	$38.12	$57.09
4 Electricians	40.75	1304.00	60.65	1940.80		
1 Equip. Oper. (crane)	35.90	287.20	54.10	432.80		
1 Common Laborer	26.70	213.60	41.55	332.40		
1 Crew Truck		98.80		108.70		
1 Crane, 12 Ton		602.60		662.85	12.53	13.78
56 L.H., Daily Totals		$2836.20		$3968.75	$50.65	$70.87
Crew R-12	Hr.	Daily	Hr.	Daily	Bare Costs	Incl. O&P
1 Carpenter Foreman	$34.75	$278.00	$54.10	$432.80	$31.94	$50.38
4 Carpenters	34.25	1096.00	53.35	1707.20		
4 Common Laborers	26.70	854.40	41.55	1329.60		
1 Equip. Oper. (med.)	34.65	277.20	52.20	417.60		
1 Steel Worker	38.15	305.20	68.25	546.00		
1 Dozer, 200 H.P.		919.60		1011.55		
1 Pickup Truck, 3/4 Ton		78.00		85.80	11.34	12.47
88 L.H., Daily Totals		$3808.40		$5530.55	$43.28	$62.85
Crew R-13	Hr.	Daily	Hr.	Daily	Bare Costs	Incl. O&P
1 Electrician Foreman	$41.25	$330.00	$61.40	$491.20	$38.59	$57.57
3 Electricians	40.75	978.00	60.65	1455.60		
.25 Equip. Oper. (crane)	35.90	71.80	54.10	108.20		
1 Equipment Oiler	30.10	240.80	45.35	362.80		
.25-1 Hyd. Crane, 33 Ton		160.40		176.45	3.82	4.20
42 L.H., Daily Totals		$1781.00		$2594.25	$42.41	$61.77

Crew No.	Bare Costs		Incl. Sub O & P		Cost Per Labor-Hour	

Crew R-15	Hr.	Daily	Hr.	Daily	Bare Costs	Incl. O&P
1 Electrician Foreman	$41.25	$330.00	$61.40	$491.20	$39.55	$58.97
4 Electricians	40.75	1304.00	60.65	1940.80		
1 Equipment Operator	33.05	264.40	49.80	398.40		
1 Aerial Lift Truck		252.60		277.85	5.26	5.79
48 L.H., Daily Totals		$2151.00		$3108.25	$44.81	$64.76

Crew R-18	Hr.	Daily	Hr.	Daily	Bare Costs	Incl. O&P
.25 Electrician Foreman	$41.25	$82.50	$61.40	$122.80	$31.43	$47.85
1 Electrician	40.75	326.00	60.65	485.20		
2 Helpers	25.55	408.80	39.75	636.00		
26 L.H., Daily Totals		$817.30		$1244.00	$31.43	$47.85

Crew R-19	Hr.	Daily	Hr.	Daily	Bare Costs	Incl. O&P
.5 Electrician Foreman	$41.25	$165.00	$61.40	$245.60	$40.85	$60.80
2 Electricians	40.75	652.00	60.65	970.40		
20 L.H., Daily Totals		$817.00		$1216.00	$40.85	$60.80

Crew R-21	Hr.	Daily	Hr.	Daily	Bare Costs	Incl. O&P
1 Electrician Foreman	$41.25	$330.00	$61.40	$491.20	$40.72	$60.63
3 Electricians	40.75	978.00	60.65	1455.60		
.1 Equip. Oper. (med.)	34.65	27.72	52.20	41.76		
.1 Hyd. Crane 25 Ton		57.50		63.25	1.75	1.93
32. L.H., Daily Totals		$1393.22		$2051.81	$42.47	$62.56

Crew R-22	Hr.	Daily	Hr.	Daily	Bare Costs	Incl. O&P
.66 Electrician Foreman	$41.25	$217.80	$61.40	$324.19	$34.30	$51.79
2 Helpers	25.55	408.80	39.75	636.00		
2 Electricians	40.75	652.00	60.65	970.40		
37.28 L.H., Daily Totals		$1278.60		$1930.59	$34.30	$51.79

Crew R-30	Hr.	Daily	Hr.	Daily	Bare Costs	Incl. O&P
.25 Electrician	$42.75	$85.50	$63.60	$127.20	$32.26	$49.12
1 Electrician	40.75	326.00	60.65	485.20		
2 Laborers, (Semi-Skilled)	26.70	427.20	41.55	664.80		
26 L.H., Daily Totals		$838.70		$1277.20	$32.26	$49.12

Crew R-31	Hr.	Daily	Hr.	Daily	Bare Costs	Incl. O&P
1 Electrician	$40.75	$326.00	$60.65	$485.20	$40.75	$60.65
1 Core Drill, Elec, 2.5 HP		50.30		55.35	6.28	6.91
8 L.H., Daily Totals		$376.30		$540.55	$47.03	$67.56

Crew W-41E	Hr.	Daily	Hr.	Daily	Bare Costs	Incl. O&P
1 Laborers, (Semi-Skilled)	$26.70	$213.60	$41.55	$332.40	$35.59	$54.06
1 Plumber	40.85	326.80	61.40	491.20		
.5 Plumber	42.85	171.40	64.40	257.60		
20 L.H., Daily Totals		$711.80		$1081.20	$35.59	$54.06

The table below lists both the Means Historical Cost Index based on Jan. 1, 1993 = 100 as well as the computed value of an index based on Jan. 1, 2005 costs. Since the Jan. 1, 2005 figure is estimated, space is left to write in the actual index figures as they become available through either the quarterly "Means Construction Cost Indexes" or as printed in the "Engineering News-Record." To compute the actual index based on Jan. 1, 2005 = 100, divide the Historical Cost Index for a particular year by the actual Jan. 1, 2005 Construction Cost Index. Space has been left to advance the index figures as the year progresses.

Year	Historical Cost Index Jan. 1, 1993 = 100		Current Index Based on Jan. 1, 2005 = 100		Year	Historical Cost Index Jan. 1, 1993 = 100	Current Index Based on Jan. 1, 2005 = 100		Year	Historical Cost Index Jan. 1, 1993 = 100	Current Index Based on Jan. 1, 2005 = 100	
	Est.	Actual	Est.	Actual		Actual	Est.	Actual		Actual	Est.	Actual
Oct 2005					July 1990	94.3	63.5		July 1972	34.8	23.4	
July 2005					1989	92.1	62.0		1971	32.1	21.6	
April 2005					1988	89.9	60.5		1970	28.7	19.3	
Jan 2005	148.5		100.0	100.0	1987	87.7	59.0		1969	26.9	18.1	
July 2004		143.7	96.8		1986	84.2	56.7		1968	24.9	16.8	
2003		132.0	88.9		1985	82.6	55.6		1967	23.5	15.8	
2002		128.7	86.7		1984	82.0	55.2		1966	22.7	15.3	
2001		125.1	84.2		1983	80.2	54.0		1965	21.7	14.6	
2000		120.9	81.4		1982	76.1	51.3		1964	21.2	14.3	
1999		117.6	79.2		1981	70.0	47.1		1963	20.7	13.9	
1998		115.1	77.5		1980	62.9	42.4		1962	20.2	13.6	
1997		112.8	76.0		1979	57.8	38.9		1961	19.8	13.3	
1996		110.2	74.2		1978	53.5	36.0		1960	19.7	13.3	
1995		107.6	72.5		1977	49.5	33.3		1959	19.3	13.0	
1994		104.4	70.3		1976	46.9	31.6		1958	18.8	12.7	
1993		101.7	68.5		1975	44.8	30.2		1957	18.4	12.4	
1992		99.4	67.0		1974	41.4	27.9		1956	17.6	11.9	
1991		96.8	65.2		1973	37.7	25.4		1955	16.6	11.2	

Adjustments to Costs

The Historical Cost Index can be used to convert National Average building costs at a particular time to the approximate building costs for some other time.

Example:

Estimate and compare construction costs for different years in the same city.

To estimate the National Average construction cost of a building in 1970, knowing that it cost $900,000 in 2005:

INDEX in 1970 = 28.7

INDEX in 2005 = 148.5

Note: The City Cost Indexes for Canada can be used to convert U.S. National averages to local costs in Canadian dollars.

Time Adjustment using the Historical Cost Indexes:

$$\frac{\text{Index for Year A}}{\text{Index for Year B}} \times \text{Cost in Year B} = \text{Cost in Year A}$$

$$\frac{\text{INDEX 1970}}{\text{INDEX 2005}} \times \text{Cost 2005} = \text{Cost 1970}$$

$$\frac{28.7}{148.5} \times \$900,000 = .193 \times \$900,000 = \$173,700$$

The construction cost of the building in 1970 is $173,700.

How to Use the City Cost Indexes

What you should know before you begin

Means City Cost Indexes (CCI) are an extremely useful tool to use when you want to compare costs from city to city and region to region.

This publication contains average construction cost indexes for 316 major U.S. and Canadian cities and Location Factors covering over 930 three-digit zip code locations.

Keep in mind that a City Cost Index number is a *percentage ratio* of a specific city's cost to the national average cost of the same item at a stated time period.

In other words, these index figures represent relative construction *factors* (or, if you prefer, multipliers) for Material and Installation costs, as well as the weighted average for Total In Place costs for each CSI MasterFormat division. Installation costs include both labor and equipment rental costs. When estimating equipment rental rates only, for a specific location, use 01590 EQUIPMENT RENTAL index.

The 30 City Average Index is the average of 30 major U.S. cities and serves as a National Average.

Index figures for both material and installation are based on the 30 major city average of 100 and represent the cost relationship as of July 1, 2003. The index for each division is computed from representative material and labor quantities for that division. The weighted average for each city is a weighted total of the components listed above it, but does not include relative productivity between trades or cities.

As changes occur in local material prices, labor rates and equipment rental rates, the impact of these changes should be accurately measured by the change in the City Cost Index for each particular city (as compared to the 30 City Average).

Therefore, if you know (or have estimated) building costs in one city today, you can easily convert those costs to expected building costs in another city.

In addition, by using the Historical Cost Index, you can easily convert National Average building costs at a particular time to the approximate building costs for some other time. The City Cost Indexes can then be applied to calculate the costs for a particular city.

Quick Calculations

Location Adjustment Using the City Cost Indexes:

$$\frac{\text{Index for City A}}{\text{Index for City B}} \times \text{Cost in City B} = \text{Cost in City A}$$

Time Adjustment for the National Average Using the Historical Cost Index:

$$\frac{\text{Index for Year A}}{\text{Index for Year B}} \times \text{Cost in Year B} = \text{Cost in Year A}$$

Adjustment from the National Average:

$$\frac{\text{Index for City A}}{100} \times \text{National Average Cost} = \text{Cost in City A}$$

Since each of the other RSMeans publications contains many different items, any *one* item multiplied by the particular city index may give incorrect results. However, the larger the number of items compiled, the closer the results should be to actual costs for that particular city.

The City Cost Indexes for Canadian cities are calculated using Canadian material and equipment prices and labor rates, in Canadian dollars. Therefore, indexes for Canadian cities can be used to convert U.S. National Average prices to local costs in Canadian dollars.

How to use this section

1. Compare costs from city to city.

In using the Means Indexes, remember that an index number is not a fixed number but a *ratio:* It's a percentage ratio of a building component's cost at any stated time to the National Average cost of that same component at the same time period. Put in the form of an equation:

$$\frac{\text{Specific City Cost}}{\text{National Average Cost}} \times 100 = \text{City Index Number}$$

Therefore, when making cost comparisons between cities, do not subtract one city's index number from the index number of another city and read the result as a percentage difference. Instead, divide one city's index number by that of the other city. The resulting number may then be used as a multiplier to calculate cost differences from city to city.

The formula used to find cost differences between cities for the purpose of comparison is as follows:

$$\frac{\text{City A Index}}{\text{City B Index}} \times \text{City B Cost (Known)} = \text{City A Cost (Unknown)}$$

In addition, you can use *Means CCI* to calculate and compare costs division by division between cities using the same basic formula. (Just be sure that you're comparing similar divisions.)

2. Compare a specific city's construction costs with the National Average.

When you're studying construction location feasibility, it's advisable to compare a prospective project's cost index with an index of the National Average cost.

For example, divide the weighted average index of construction costs of a specific city by that of the 30 City Average, which = 100.

$$\frac{\text{City Index}}{100} = \text{\% of National Average}$$

As a result, you get a ratio that indicates the relative cost of construction in that city in comparison with the National Average.

3. Convert U.S. National Average to actual costs in Canadian City.

$$\frac{\text{Index for Canadian City}}{100} \times \text{National Average Cost} = \text{Cost in Canadian City in \$ CAN}$$

4. Adjust construction cost data based on a National Average.

When you use a source of construction cost data which is based on a National Average (such as *Means cost data publications*), it is necessary to adjust those costs to a specific location.

$$\frac{\text{City Index}}{100} \times \frac{\text{"Book" Cost Based on}}{\text{National Average Costs}} = \frac{\text{City Cost}}{\text{(Unknown)}}$$

5. When applying the City Cost Indexes to demolition projects, use the appropriate division index. For example, for removal of existing doors and windows, use the Division 8 index.

What you might like to know about how we developed the Indexes

To create a reliable index, RSMeans researched the building type most often constructed in the United States and Canada. Because it was concluded that no one type of building completely represented the building construction industry, nine different types of buildings were combined to create a composite model.

The exact material, labor and equipment quantities are based on detailed analysis of these nine building types, then each quantity is weighted in proportion to expected usage. These various material items, labor hours, and equipment rental rates are thus combined to form a composite building representing as closely as possible the actual usage of materials, labor and equipment used in the North American Building Construction Industry.

The following structures were chosen to make up that composite model:

1. Factory, 1 story
2. Office, 2–4 story
3. Store, Retail
4. Town Hall, 2–3 story
5. High School, 2–3 story
6. Hospital, 4–8 story
7. Garage, Parking
8. Apartment, 1–3 story
9. Hotel/Motel, 2–3 story

For the purposes of ensuring the timeliness of the data, the components of the index for the composite model have been streamlined. They currently consist of:

- specific quantities of 66 commonly used construction materials;
- specific labor-hours for 21 building construction trades; and
- specific days of equipment rental for 6 types of construction equipment (normally used to install the 66 material items by the 21 trades.)

A sophisticated computer program handles the updating of all costs for each city on a quarterly basis. Material and equipment price quotations are gathered quarterly from over 316 cities in the United States and Canada. These prices and the latest negotiated labor wage rates for 21 different building trades are used to compile the quarterly update of the City Cost Index.

The 30 major U.S. cities used to calculate the National Average are:

Atlanta, GA	Memphis, TN
Baltimore, MD	Milwaukee, WI
Boston, MA	Minneapolis, MN
Buffalo, NY	Nashville, TN
Chicago, IL	New Orleans, LA
Cincinnati, OH	New York, NY
Cleveland, OH	Philadelphia, PA
Columbus, OH	Phoenix, AZ
Dallas, TX	Pittsburgh, PA
Denver, CO	St. Louis, MO
Detroit, MI	San Antonio, TX
Houston, TX	San Diego, CA
Indianapolis, IN	San Francisco, CA
Kansas City, MO	Seattle, WA
Los Angeles, CA	Washington, DC

F.Y.I.: The CSI MasterFormat Divisions

1. General Requirements
2. Site Construction
3. Concrete
4. Masonry
5. Metals
6. Wood & Plastics
7. Thermal & Moisture Protection
8. Doors & Windows
9. Finishes
10. Specialties
11. Equipment
12. Furnishings
13. Special Construction
14. Conveying Systems
15. Mechanical
16. Electrical

The information presented in the CCI is organized according to the Construction Specifications Institute (CSI) MasterFormat.

What the CCI does not *indicate*

The weighted average for each city is a total of the components listed above weighted to reflect typical usage, but it does *not* include the productivity variations between trades or cities.

In addition, the CCI does not take into consideration factors such as the following:

- managerial efficiency
- competitive conditions
- automation
- restrictive union practices
- unique local requirements
- regional variations due to specific building codes

DIVISION		UNITED STATES 30 CITY AVERAGE			BIRMINGHAM			HUNTSVILLE			MOBILE			MONTGOMERY			TUSCALOOSA		
		MAT.	INST.	TOTAL	MAT.	INST.	TOTAL	MAT.	INST.	TOTAL	MAT.	INST.	TOTAL	MAT.	INST.	TOTAL	MAT.	INST.	TOTAL
01590	EQUIPMENT RENTAL	.0	100.0	100.0	.0	101.3	101.3	.0	101.2	101.2	.0	97.7	97.7	.0	97.7	97.7	.0	101.2	101.2
02	SITE CONSTRUCTION	100.0	100.0	100.0	84.6	93.1	90.9	82.7	92.9	90.2	93.7	86.1	88.0	94.1	86.6	88.6	83.2	91.9	89.6
03100	CONCRETE FORMS & ACCESSORIES	100.0	100.0	100.0	95.0	75.6	78.2	97.1	68.6	72.4	97.1	53.0	58.9	95.9	48.1	54.5	97.0	39.7	47.3
03200	CONCRETE REINFORCEMENT	100.0	100.0	100.0	92.1	88.2	90.1	92.1	77.4	84.7	95.0	53.9	74.2	95.0	86.7	90.8	92.1	87.0	89.5
03300	CAST-IN-PLACE CONCRETE	100.0	100.0	100.0	92.8	68.1	82.5	87.8	66.2	78.8	92.7	54.7	76.8	94.3	49.4	75.6	91.4	47.1	72.9
03	CONCRETE	100.0	100.0	100.0	90.9	76.4	83.7	88.7	70.7	79.7	91.5	55.4	73.5	92.2	57.5	74.9	90.4	53.0	71.8
04	MASONRY	100.0	100.0	100.0	85.9	76.0	79.7	85.8	65.2	72.8	86.3	52.3	65.0	86.9	36.3	55.2	86.0	38.6	56.3
05	METALS	100.0	100.0	100.0	94.6	95.5	94.9	96.0	91.0	94.4	94.8	79.6	89.8	94.6	92.1	93.8	95.2	92.7	94.4
06	WOOD & PLASTICS	100.0	100.0	100.0	96.7	76.0	85.9	96.7	68.4	81.9	96.7	52.7	73.7	95.3	48.3	70.7	96.7	38.3	66.2
07	THERMAL & MOISTURE PROTECTION	100.0	100.0	100.0	95.7	80.9	88.4	95.4	75.1	85.4	95.4	69.5	82.7	95.1	63.0	79.3	95.3	61.8	78.8
08	DOORS & WINDOWS	100.0	100.0	100.0	98.5	78.5	93.3	98.5	65.5	89.9	98.5	53.2	86.6	98.5	58.2	88.0	98.5	58.0	88.0
09200	PLASTER & GYPSUM BOARD	100.0	100.0	100.0	103.1	75.9	85.9	100.5	68.1	80.0	100.5	52.0	69.7	100.5	47.4	66.9	100.5	37.2	60.4
095,098	CEILINGS & ACOUSTICAL TREATMENT	100.0	100.0	100.0	102.7	75.9	86.6	102.7	68.1	81.9	102.7	52.0	72.3	102.7	47.4	69.6	102.7	37.2	63.4
09600	FLOORING	100.0	100.0	100.0	104.4	50.5	90.1	104.4	54.2	91.1	114.7	54.7	98.8	112.9	28.3	90.5	104.4	41.5	87.7
097,099	WALL FINISHES, PAINTS & COATINGS	100.0	100.0	100.0	95.3	70.8	80.4	95.3	64.2	76.4	100.1	56.9	73.9	95.3	56.0	71.4	95.3	48.2	66.7
09	FINISHES	100.0	100.0	100.0	100.8	70.1	84.8	100.3	65.0	81.9	104.8	53.1	77.9	104.0	43.9	72.8	100.2	39.2	68.5
10 - 14	TOTAL DIV. 10000 - 14000	100.0	100.0	100.0	100.0	86.6	97.1	100.0	84.5	96.7	100.0	69.0	93.4	100.0	77.1	95.1	100.0	74.5	94.6
15	MECHANICAL	100.0	100.0	100.0	100.3	69.6	86.9	100.3	69.8	86.9	99.8	70.8	87.2	99.8	41.7	74.4	99.8	36.7	72.2
16	ELECTRICAL	100.0	100.0	100.0	96.3	66.6	81.8	96.4	70.5	83.8	96.4	51.3	74.4	96.3	69.1	83.0	96.4	66.6	81.9
01 - 16	WEIGHTED AVERAGE	100.0	100.0	100.0	96.4	76.5	87.4	96.3	73.3	85.8	97.1	63.0	81.6	97.0	58.4	79.5	96.3	56.4	78.2

DIVISION		ANCHORAGE			FAIRBANKS			JUNEAU			FLAGSTAFF			MESA/TEMPE			PHOENIX		
		MAT.	INST.	TOTAL	MAT.	INST.	TOTAL	MAT.	INST.	TOTAL	MAT.	INST.	TOTAL	MAT.	INST.	TOTAL	MAT.	INST.	TOTAL
01590	EQUIPMENT RENTAL	.0	118.4	118.4	.0	118.4	118.4	.0	118.4	118.4	.0	94.9	94.9	.0	98.2	98.2	.0	98.8	98.8
02	SITE CONSTRUCTION	143.5	133.8	136.3	127.1	133.8	132.1	139.1	133.8	135.1	83.5	100.9	96.4	88.3	103.9	99.9	88.8	104.7	100.6
03100	CONCRETE FORMS & ACCESSORIES	132.4	113.9	116.4	134.1	119.7	121.6	133.8	113.9	116.6	102.8	61.5	67.1	97.3	63.1	67.7	98.4	69.8	73.6
03200	CONCRETE REINFORCEMENT	146.2	106.6	126.2	122.9	106.7	114.7	108.7	106.6	107.7	105.5	73.5	89.3	104.8	72.9	88.7	103.2	74.2	88.5
03300	CAST-IN-PLACE CONCRETE	195.7	114.6	161.8	163.1	115.1	143.0	196.5	114.6	162.3	91.8	78.3	86.2	98.3	71.4	87.1	98.4	78.9	90.3
03	CONCRETE	153.5	112.2	132.9	129.8	115.0	122.4	147.9	112.2	130.1	116.7	69.5	93.1	99.2	67.8	83.5	98.8	73.6	86.2
04	MASONRY	217.0	121.0	156.7	210.3	121.0	154.3	219.4	121.0	157.7	102.9	56.2	73.6	108.2	52.8	73.4	96.0	68.1	78.5
05	METALS	130.1	102.2	121.0	130.1	102.5	121.1	130.1	102.2	121.0	93.8	67.9	85.4	94.5	69.2	86.2	96.0	70.9	87.8
06	WOOD & PLASTICS	117.8	111.5	114.5	118.0	118.9	118.5	117.8	111.5	114.5	108.6	60.1	83.3	99.6	68.7	83.5	100.6	70.5	84.9
07	THERMAL & MOISTURE PROTECTION	198.5	115.8	157.8	194.6	118.6	157.2	195.2	115.8	156.2	107.3	66.9	87.4	105.4	63.8	85.0	105.3	70.8	88.4
08	DOORS & WINDOWS	126.2	108.8	121.7	123.3	112.9	120.6	123.3	108.8	119.6	102.2	64.2	92.3	98.9	65.2	90.2	100.0	69.9	92.2
09200	PLASTER & GYPSUM BOARD	132.6	111.6	119.3	132.6	119.3	124.1	132.6	111.6	119.3	90.8	59.0	70.6	91.6	67.7	76.4	93.9	69.6	78.5
095,098	CEILINGS & ACOUSTICAL TREATMENT	132.1	111.6	119.8	132.1	119.3	124.4	132.1	111.6	119.8	101.3	59.0	75.9	95.6	67.7	78.8	102.7	69.6	82.8
09600	FLOORING	164.9	127.0	154.9	164.9	127.0	154.9	164.9	127.0	154.9	95.2	51.8	83.8	97.3	64.6	88.7	97.6	67.5	89.7
097,099	WALL FINISHES, PAINTS & COATINGS	168.1	111.4	133.7	168.1	125.0	141.9	168.1	111.4	133.7	94.6	48.8	66.8	105.0	53.1	73.5	105.0	60.4	77.9
09	FINISHES	154.8	116.0	134.6	152.6	121.9	136.6	153.4	116.0	134.0	96.7	57.6	76.4	97.2	62.0	78.9	99.5	68.0	83.1
10 - 14	TOTAL DIV. 10000 - 14000	100.0	112.8	102.7	100.0	113.7	102.9	100.0	112.8	102.7	100.0	72.6	94.2	100.0	69.0	93.4	100.0	74.4	94.6
15	MECHANICAL	100.9	108.3	104.2	100.9	115.9	107.4	100.9	108.3	104.2	100.2	77.4	90.2	100.2	67.9	86.1	100.2	77.5	90.2
16	ELECTRICAL	149.1	112.3	131.2	150.9	112.3	132.1	150.9	112.3	132.1	96.9	48.0	73.1	90.6	60.3	75.8	99.6	65.7	83.1
01 - 16	WEIGHTED AVERAGE	134.5	113.8	125.1	130.7	116.9	124.4	133.6	113.8	124.6	100.5	67.6	85.5	97.8	67.7	84.1	98.8	74.3	87.7

DIVISION		PRESCOTT			TUCSON			FORT SMITH			JONESBORO			LITTLE ROCK			PINE BLUFF		
		MAT.	INST.	TOTAL	MAT.	INST.	TOTAL	MAT.	INST.	TOTAL	MAT.	INST.	TOTAL	MAT.	INST.	TOTAL	MAT.	INST.	TOTAL
01590	EQUIPMENT RENTAL	.0	94.9	94.9	.0	98.2	98.2	.0	86.0	86.0	.0	107.8	107.8	.0	86.0	86.0	.0	86.0	86.0
02	SITE CONSTRUCTION	71.2	100.3	92.7	85.1	104.4	99.4	77.6	83.8	82.2	100.5	99.2	99.6	77.4	83.8	82.1	79.6	83.8	82.7
03100	CONCRETE FORMS & ACCESSORIES	98.2	56.0	61.6	97.9	69.2	73.1	97.7	42.0	49.4	84.5	47.8	52.7	91.8	59.2	63.5	77.2	58.9	61.4
03200	CONCRETE REINFORCEMENT	105.5	72.5	88.8	86.6	73.5	80.0	96.9	74.6	85.6	92.7	49.6	70.9	97.1	72.2	84.5	97.1	72.1	84.5
03300	CAST-IN-PLACE CONCRETE	91.7	64.2	80.2	101.1	78.7	91.8	89.8	66.8	80.2	85.6	57.1	73.7	89.8	67.0	80.3	82.4	66.9	75.9
03	CONCRETE	102.6	62.0	82.3	97.4	73.1	85.3	87.4	57.4	72.4	84.0	53.0	68.6	87.0	64.7	75.9	85.0	64.5	74.8
04	MASONRY	103.3	58.8	75.4	97.8	56.2	71.7	97.5	53.3	69.8	92.8	46.9	64.0	95.9	53.3	69.2	118.2	53.3	77.5
05	METALS	93.8	66.7	85.0	95.3	68.7	86.6	95.7	71.3	87.7	90.0	75.9	85.4	91.6	70.8	84.8	94.3	70.6	86.6
06	WOOD & PLASTICS	103.8	54.6	78.1	99.8	70.5	84.5	100.4	40.6	69.2	86.4	48.8	66.8	97.4	63.3	79.6	78.9	63.3	70.7
07	THERMAL & MOISTURE PROTECTION	105.6	61.7	84.0	106.6	64.2	85.8	99.4	48.4	74.3	109.1	51.6	80.8	97.9	50.7	74.7	98.0	50.7	74.8
08	DOORS & WINDOWS	102.2	58.3	90.8	96.0	69.9	89.2	97.2	45.8	83.9	98.8	47.9	85.6	97.2	58.4	87.1	92.5	58.4	83.7
09200	PLASTER & GYPSUM BOARD	87.9	53.3	66.0	92.0	69.6	77.8	85.4	39.6	56.4	92.3	47.7	64.0	85.4	62.9	71.1	77.7	62.9	68.3
095,098	CEILINGS & ACOUSTICAL TREATMENT	99.6	53.3	71.8	95.6	69.6	80.0	91.2	39.6	60.3	89.0	47.7	64.2	91.2	62.9	74.2	86.8	62.9	72.4
09600	FLOORING	93.6	51.6	82.5	96.8	51.8	84.9	115.8	70.0	103.7	77.3	45.6	68.9	117.1	70.0	104.6	104.8	70.0	95.6
097,099	WALL FINISHES, PAINTS & COATINGS	94.6	48.8	66.8	102.6	48.8	70.0	96.4	58.3	73.3	84.9	52.4	65.2	96.4	59.9	74.2	96.4	59.9	74.2
09	FINISHES	94.3	53.8	73.2	96.9	63.8	79.6	96.4	48.4	71.4	87.7	47.9	67.0	96.7	61.9	78.6	91.1	61.9	75.9
10 - 14	TOTAL DIV. 10000 - 14000	100.0	71.1	93.9	100.0	74.4	94.6	100.0	63.4	92.2	100.0	54.9	90.4	100.0	66.4	92.8	100.0	66.4	92.8
15	MECHANICAL	100.2	73.6	88.6	100.1	69.2	86.6	100.1	43.2	75.2	100.2	43.1	75.2	100.1	58.4	81.8	100.1	45.6	76.3
16	ELECTRICAL	96.7	46.2	72.1	92.5	60.9	77.1	95.6	64.4	80.4	102.7	47.8	75.9	95.3	70.0	83.0	95.0	70.0	82.8
01 - 16	WEIGHTED AVERAGE	98.2	64.7	83.0	97.0	69.8	84.6	96.0	56.8	78.2	95.5	55.0	77.1	95.2	64.3	81.2	95.3	61.7	80.0

ALABAMA (spanning Birmingham, Huntsville, Mobile, Montgomery, Tuscaloosa)
ALASKA (spanning Anchorage, Fairbanks, Juneau)
ARIZONA (spanning Flagstaff, Mesa/Tempe, Phoenix, Prescott, Tucson)
ARKANSAS (spanning Fort Smith, Jonesboro, Little Rock, Pine Bluff)

DIVISION		ARKANSAS TEXARKANA MAT.	INST.	TOTAL	CALIFORNIA ANAHEIM MAT.	INST.	TOTAL	BAKERSFIELD MAT.	INST.	TOTAL	FRESNO MAT.	INST.	TOTAL	LOS ANGELES MAT.	INST.	TOTAL	OAKLAND MAT.	INST.	TOTAL
01590	EQUIPMENT RENTAL	.0	86.8	86.8	.0	102.1	102.1	.0	99.5	99.5	.0	99.5	99.5	.0	98.5	98.5	.0	103.3	103.3
02	SITE CONSTRUCTION	95.3	84.6	87.4	103.2	109.0	107.5	108.1	106.0	106.6	109.6	106.0	106.9	98.1	107.8	105.3	155.8	104.6	117.9
03100	CONCRETE FORMS & ACCESSORIES	85.0	39.6	45.7	105.0	121.5	119.3	97.5	121.1	117.9	101.2	127.2	123.7	105.5	121.4	119.3	107.4	133.5	130.0
03200	CONCRETE REINFORCEMENT	96.6	48.1	72.1	95.6	112.5	104.1	108.9	112.5	110.7	92.2	112.8	102.6	113.0	112.7	112.9	103.1	113.4	108.3
03300	CAST-IN-PLACE CONCRETE	89.8	44.6	70.9	107.2	117.2	111.4	102.3	116.1	108.1	112.0	113.5	112.6	87.3	115.4	99.1	139.5	118.9	130.9
03	CONCRETE	84.3	43.9	64.1	107.8	117.4	112.6	107.1	116.8	112.0	109.3	118.7	114.0	103.1	116.7	109.9	126.4	123.6	125.0
04	MASONRY	97.5	32.8	56.9	89.0	111.4	103.1	107.5	111.2	109.8	111.1	111.2	111.2	97.7	115.3	108.8	148.4	122.5	132.1
05	METALS	86.9	59.4	77.9	110.9	99.8	107.2	105.4	99.0	103.3	111.1	100.2	107.6	112.8	98.4	108.1	105.1	104.9	105.0
06	WOOD & PLASTICS	88.2	42.3	64.3	94.9	121.2	108.7	86.0	121.3	104.4	99.1	129.8	115.1	91.8	120.9	107.0	101.9	134.8	119.1
07	THERMAL & MOISTURE PROTECTION	98.9	42.2	71.0	121.2	115.0	118.1	105.5	108.9	107.2	101.3	110.5	105.8	111.6	117.0	114.3	113.0	124.0	118.4
08	DOORS & WINDOWS	97.6	41.1	83.0	102.8	116.2	106.2	101.6	113.3	104.6	101.9	118.6	106.2	96.7	116.0	101.7	106.4	125.7	111.4
09200	PLASTER & GYPSUM BOARD	83.3	41.4	56.7	98.1	121.7	113.1	98.1	121.7	113.1	96.1	130.4	117.9	101.6	121.7	114.4	94.7	135.1	120.3
095,098	CEILINGS & ACOUSTICAL TREATMENT	93.9	41.4	62.4	119.6	121.7	120.9	119.6	121.7	120.9	119.6	130.4	126.1	119.8	121.7	121.0	114.8	135.1	127.0
09600	FLOORING	107.6	43.5	90.6	123.1	103.7	118.0	117.3	93.4	111.0	133.8	138.6	135.1	114.0	103.7	111.3	116.1	122.8	117.9
097,099	WALL FINISHES, PAINTS & COATINGS	96.4	36.8	60.2	111.4	110.7	111.0	110.9	92.5	99.7	134.6	100.1	113.7	99.6	110.7	106.4	115.6	136.3	128.1
09	FINISHES	95.0	40.1	66.4	114.3	117.4	115.9	114.5	113.6	114.0	120.9	127.5	124.3	110.8	117.1	114.1	115.9	132.6	124.6
10-14	TOTAL DIV. 10000 - 14000	100.0	39.3	87.1	100.0	114.8	103.2	100.0	119.2	104.1	100.0	128.8	106.1	100.0	114.1	103.0	100.0	132.3	106.9
15	MECHANICAL	100.1	38.1	73.0	100.2	108.4	103.8	100.2	99.6	99.9	100.2	111.6	105.2	100.1	108.4	103.7	100.2	127.0	111.9
16	ELECTRICAL	96.8	39.8	69.0	88.6	107.9	98.0	89.5	93.6	91.5	87.5	97.1	92.2	96.5	112.9	104.5	102.6	134.4	118.1
01-16	WEIGHTED AVERAGE	94.7	45.3	72.3	102.9	111.1	106.6	102.6	106.2	104.2	104.3	111.9	107.8	102.7	111.9	106.8	110.6	123.9	116.6

DIVISION		CALIFORNIA OXNARD MAT.	INST.	TOTAL	REDDING MAT.	INST.	TOTAL	RIVERSIDE MAT.	INST.	TOTAL	SACRAMENTO MAT.	INST.	TOTAL	SAN DIEGO MAT.	INST.	TOTAL	SAN FRANCISCO MAT.	INST.	TOTAL
01590	EQUIPMENT RENTAL	.0	98.1	98.1	.0	99.2	99.2	.0	100.6	100.6	.0	102.9	102.9	.0	97.1	97.1	.0	108.6	108.6
02	SITE CONSTRUCTION	109.7	103.8	105.3	114.4	105.2	107.6	100.8	106.7	105.1	119.5	110.9	113.2	102.2	100.6	101.0	158.2	111.2	123.4
03100	CONCRETE FORMS & ACCESSORIES	103.4	121.5	119.1	103.1	126.9	123.7	105.4	121.4	119.3	105.9	127.6	124.7	106.2	111.0	110.3	107.8	134.8	131.2
03200	CONCRETE REINFORCEMENT	108.9	112.2	110.6	105.6	112.6	109.1	107.9	112.5	110.2	95.3	112.9	104.2	105.6	112.4	109.0	117.5	113.9	115.7
03300	CAST-IN-PLACE CONCRETE	108.6	116.4	111.9	123.0	113.3	119.0	106.1	117.2	110.7	120.2	114.3	117.8	110.8	104.9	108.3	139.4	120.7	131.6
03	CONCRETE	110.5	117.0	113.8	120.2	118.4	119.3	109.3	117.4	113.3	115.8	119.1	117.5	113.0	108.4	110.7	128.7	124.9	126.8
04	MASONRY	112.3	105.9	108.2	114.4	108.2	110.5	85.6	111.0	101.5	119.0	112.2	114.7	98.1	106.4	103.3	148.8	129.0	136.4
05	METALS	105.2	99.0	103.1	110.3	99.4	106.7	111.1	99.6	107.4	100.4	99.8	100.2	108.8	98.8	105.6	111.2	107.1	109.9
06	WOOD & PLASTICS	94.0	121.3	108.3	94.2	129.8	112.8	94.9	121.2	108.7	96.7	130.0	114.1	100.3	109.2	104.9	101.9	135.0	119.2
07	THERMAL & MOISTURE PROTECTION	110.5	110.0	110.3	111.3	108.5	109.9	118.9	112.0	115.5	123.8	112.8	118.4	113.8	104.1	109.0	113.0	130.8	121.7
08	DOORS & WINDOWS	100.4	116.2	104.5	103.2	119.8	107.5	102.8	116.2	106.2	117.7	120.8	118.5	103.9	107.7	104.9	110.8	128.8	115.5
09200	PLASTER & GYPSUM BOARD	98.1	121.7	113.1	96.8	130.4	118.1	97.0	121.7	112.7	92.1	130.4	116.4	97.6	109.2	104.9	97.6	135.1	121.4
095,098	CEILINGS & ACOUSTICAL TREATMENT	119.6	121.7	120.9	126.7	130.4	128.9	115.1	121.7	119.1	114.8	130.4	124.2	103.7	109.2	107.0	123.5	135.1	130.5
09600	FLOORING	117.3	103.7	113.7	116.6	106.8	114.0	121.8	103.7	117.0	119.0	115.6	118.1	110.7	103.5	108.8	116.1	128.7	119.5
097,099	WALL FINISHES, PAINTS & COATINGS	110.3	102.7	105.7	110.3	112.3	111.5	107.8	110.7	109.6	112.9	116.2	114.9	109.4	105.0	106.7	115.6	151.7	137.5
09	FINISHES	114.3	116.5	115.5	116.1	123.1	119.8	112.1	117.4	114.8	114.5	125.5	120.2	107.8	109.1	108.5	118.6	136.0	127.7
10-14	TOTAL DIV. 10000 - 14000	100.0	115.1	103.2	100.0	128.9	106.1	100.0	114.8	103.1	100.0	129.6	106.3	100.0	112.5	102.7	100.0	133.1	107.0
15	MECHANICAL	100.2	108.4	103.8	100.2	105.7	102.6	100.1	108.4	103.7	100.2	111.2	105.0	100.3	106.8	103.1	100.2	151.6	122.7
16	ELECTRICAL	94.6	106.1	100.2	97.2	102.7	99.9	88.7	102.0	95.3	97.9	102.8	100.3	96.0	98.7	97.4	102.6	152.1	126.8
01-16	WEIGHTED AVERAGE	103.9	109.5	106.4	106.8	110.4	108.4	102.6	109.9	105.9	106.9	113.1	109.7	103.9	105.0	104.4	112.6	133.8	122.2

DIVISION		CALIFORNIA SAN JOSE MAT.	INST.	TOTAL	SANTA BARBARA MAT.	INST.	TOTAL	STOCKTON MAT.	INST.	TOTAL	VALLEJO MAT.	INST.	TOTAL	COLORADO COLORADO SPRINGS MAT.	INST.	TOTAL	DENVER MAT.	INST.	TOTAL
01590	EQUIPMENT RENTAL	.0	100.0	100.0	.0	99.5	99.5	.0	99.2	99.2	.0	103.5	103.5	.0	95.2	95.2	.0	100.4	100.4
02	SITE CONSTRUCTION	148.5	99.7	112.4	109.6	106.1	107.0	107.6	105.3	105.9	116.2	111.1	112.4	95.0	97.1	96.5	93.6	106.4	103.0
03100	CONCRETE FORMS & ACCESSORIES	105.7	133.8	130.0	104.1	121.4	119.1	103.8	126.9	123.8	107.0	131.7	128.4	92.6	85.9	86.8	101.9	87.1	89.1
03200	CONCRETE REINFORCEMENT	96.2	113.6	105.0	108.9	112.5	110.7	109.5	112.6	111.1	102.7	113.5	108.2	113.0	84.6	98.6	113.0	84.7	98.7
03300	CAST-IN-PLACE CONCRETE	128.5	118.6	124.4	108.2	116.3	111.6	108.2	113.4	110.3	125.1	115.4	121.1	98.0	88.1	93.9	91.7	88.5	90.3
03	CONCRETE	118.1	123.8	120.9	110.4	117.0	113.7	110.4	118.5	114.5	119.4	121.4	120.4	108.6	86.5	97.5	102.6	87.2	94.9
04	MASONRY	147.2	125.5	133.6	107.9	111.1	109.9	111.9	108.2	109.5	84.4	121.6	107.7	108.7	82.6	92.3	108.4	85.8	94.2
05	METALS	104.5	106.9	105.3	105.7	99.3	103.6	107.1	99.8	104.7	103.8	101.4	103.0	97.9	88.1	94.7	100.2	88.4	96.3
06	WOOD & PLASTICS	101.0	134.5	118.5	94.0	121.3	108.3	95.9	129.8	113.6	95.1	134.5	115.6	93.9	88.3	91.0	103.3	88.8	95.7
07	THERMAL & MOISTURE PROTECTION	107.2	129.5	118.2	106.8	108.9	107.8	110.8	107.5	109.2	126.1	121.6	123.9	103.8	85.8	95.0	103.2	84.3	93.9
08	DOORS & WINDOWS	92.8	126.7	101.6	101.6	116.2	105.4	100.8	120.6	106.0	119.2	128.6	121.6	98.5	90.6	96.5	99.9	91.5	97.7
09200	PLASTER & GYPSUM BOARD	97.9	135.1	121.5	98.1	121.7	113.1	98.5	130.4	118.7	94.9	135.1	120.4	87.6	88.1	87.9	97.6	88.8	92.0
095,098	CEILINGS & ACOUSTICAL TREATMENT	106.2	135.1	123.5	119.6	121.7	120.9	119.6	130.4	126.1	123.7	135.1	130.5	99.2	88.1	92.5	98.3	88.8	92.6
09600	FLOORING	114.3	128.7	118.1	117.3	100.4	112.9	117.3	110.9	115.6	123.7	128.7	125.0	107.8	89.1	102.9	108.8	101.0	106.7
097,099	WALL FINISHES, PAINTS & COATINGS	112.0	139.4	128.6	110.3	102.7	105.7	110.3	101.4	104.9	112.1	136.0	126.6	106.7	53.6	74.5	106.7	77.7	89.1
09	FINISHES	111.8	134.0	123.3	114.6	116.0	115.3	114.4	122.6	118.7	116.0	132.8	124.8	98.2	83.0	90.3	99.6	88.8	94.0
10-14	TOTAL DIV. 10000 - 14000	100.0	131.7	106.7	100.0	115.1	103.2	100.0	128.9	106.1	100.0	129.6	106.4	100.0	91.3	98.1	100.0	91.5	98.2
15	MECHANICAL	100.2	141.3	118.2	100.2	108.4	103.8	100.2	101.3	100.7	100.2	118.4	108.2	100.2	81.3	91.9	100.2	86.4	94.1
16	ELECTRICAL	102.5	142.9	122.2	85.8	107.2	96.3	97.0	109.6	103.2	92.2	123.1	107.2	95.3	90.8	93.1	97.1	95.4	96.3
01-16	WEIGHTED AVERAGE	107.3	128.4	116.9	102.7	110.3	106.1	104.5	110.5	107.2	105.6	120.4	112.3	100.1	86.6	94.0	100.3	90.3	95.8

COST INDEXES

COLORADO / CONNECTICUT

DIVISION		FORT COLLINS MAT.	INST.	TOTAL	GRAND JUNCTION MAT.	INST.	TOTAL	GREELEY MAT.	INST.	TOTAL	PUEBLO MAT.	INST.	TOTAL	BRIDGEPORT MAT.	INST.	TOTAL	BRISTOL MAT.	INST.	TOTAL
01590	EQUIPMENT RENTAL	.0	96.8	96.8	.0	100.0	100.0	.0	96.8	96.8	.0	96.9	96.9	.0	101.6	101.6	.0	101.6	101.6
02	SITE CONSTRUCTION	104.3	99.5	100.8	124.7	102.1	108.0	91.7	98.2	96.5	116.8	96.5	101.7	102.3	104.4	103.8	101.4	104.3	103.6
03100	CONCRETE FORMS & ACCESSORIES	101.3	80.4	83.2	106.7	81.0	84.4	98.7	49.5	56.1	103.3	86.1	88.4	101.0	110.7	109.4	101.0	110.3	109.1
03200	CONCRETE REINFORCEMENT	113.9	83.0	98.3	113.7	83.0	98.2	113.8	82.4	97.9	109.5	84.6	96.9	108.2	122.7	115.5	108.2	122.6	115.5
03300	CAST-IN-PLACE CONCRETE	106.4	83.5	96.9	113.2	84.6	101.2	89.0	60.8	77.2	101.5	89.3	96.4	105.2	117.7	110.4	98.6	117.6	106.5
03	CONCRETE	115.0	82.1	98.6	115.3	82.6	99.0	100.0	60.5	80.3	105.2	87.1	96.1	108.8	115.3	112.0	105.6	115.1	110.3
04	MASONRY	122.8	60.9	84.0	148.3	63.4	95.0	116.4	40.4	68.7	107.1	82.6	91.7	108.2	117.3	113.9	99.9	117.3	110.8
05	METALS	96.1	83.4	92.0	96.9	82.6	92.3	96.1	81.7	91.4	98.4	89.2	95.4	101.2	120.2	107.4	101.2	119.9	107.3
06	WOOD & PLASTICS	103.0	83.6	92.9	106.4	83.8	94.6	100.2	48.3	73.1	103.1	88.7	95.6	99.4	108.0	103.8	99.4	108.0	103.8
07	THERMAL & MOISTURE PROTECTION	103.8	72.5	88.4	104.9	68.6	87.0	102.9	60.7	82.2	104.2	85.5	95.0	101.6	117.5	109.4	101.8	113.6	107.6
08	DOORS & WINDOWS	95.9	87.9	93.8	100.8	88.0	97.5	95.9	68.7	88.8	95.2	90.8	94.1	107.1	121.0	110.7	107.1	113.0	108.6
09200	PLASTER & GYPSUM BOARD	96.0	83.4	88.0	106.7	83.4	91.9	94.7	47.1	64.5	85.2	88.1	87.0	104.4	107.2	106.2	104.4	107.2	106.2
095,098	CEILINGS & ACOUSTICAL TREATMENT	92.1	83.4	86.9	96.0	83.4	88.4	92.1	65.1	65.1	107.6	88.1	95.9	101.1	107.2	104.8	101.1	107.2	104.8
09600	FLOORING	108.6	71.5	98.8	116.1	71.5	104.4	107.2	71.5	97.8	112.5	101.0	109.5	99.8	115.1	103.9	99.8	115.1	103.9
097,099	WALL FINISHES, PAINTS & COATINGS	106.7	53.2	74.2	117.0	77.7	93.2	106.7	33.1	62.1	117.1	49.6	76.2	92.8	109.6	103.0	92.8	109.6	103.0
09	FINISHES	98.3	75.9	86.7	106.3	79.5	92.4	97.1	50.2	72.7	104.6	85.3	94.6	102.7	110.7	106.8	102.7	110.7	106.9
10 - 14	TOTAL DIV. 10000 - 14000	100.0	89.1	97.7	100.0	90.5	98.0	100.0	80.3	95.8	100.0	92.3	98.4	100.0	110.1	102.1	100.0	110.1	102.1
15	MECHANICAL	100.2	82.5	92.4	100.1	72.5	88.0	100.2	76.7	89.9	100.1	71.3	87.5	100.1	104.0	101.8	100.1	103.9	101.8
16	ELECTRICAL	92.3	93.0	92.6	88.5	65.4	77.3	92.3	93.0	92.6	89.3	82.6	86.0	100.3	105.8	103.0	100.3	108.4	104.2
01 - 16	WEIGHTED AVERAGE	101.0	82.7	92.7	103.8	77.6	91.9	98.5	71.1	86.0	100.0	83.9	92.6	102.7	110.8	106.4	101.9	110.6	105.9

CONNECTICUT

DIVISION		HARTFORD MAT.	INST.	TOTAL	NEW BRITAIN MAT.	INST.	TOTAL	NEW HAVEN MAT.	INST.	TOTAL	NORWALK MAT.	INST.	TOTAL	STAMFORD MAT.	INST.	TOTAL	WATERBURY MAT.	INST.	TOTAL
01590	EQUIPMENT RENTAL	.0	101.6	101.6	.0	101.6	101.6	.0	102.1	102.1	.0	101.6	101.6	.0	101.6	101.6	.0	101.6	101.6
02	SITE CONSTRUCTION	101.8	104.3	103.7	101.6	104.3	103.6	101.5	105.2	104.2	102.0	104.4	103.8	102.7	104.4	103.9	101.8	104.4	103.7
03100	CONCRETE FORMS & ACCESSORIES	99.9	110.3	108.9	101.2	110.4	109.1	100.8	110.6	109.3	101.0	110.7	109.4	101.0	110.9	109.6	101.0	110.6	109.3
03200	CONCRETE REINFORCEMENT	108.2	122.6	115.5	108.2	122.6	115.5	108.2	122.7	115.5	108.2	122.8	115.6	108.2	122.8	115.6	108.2	122.7	115.5
03300	CAST-IN-PLACE CONCRETE	99.7	117.6	107.2	100.1	117.6	107.4	101.9	117.7	108.5	103.5	119.3	110.1	105.2	119.4	111.1	105.2	117.7	110.4
03	CONCRETE	106.1	115.1	110.6	106.4	115.1	110.7	121.6	115.2	118.4	107.9	115.9	111.9	108.8	116.0	112.4	108.8	115.2	112.0
04	MASONRY	100.0	117.3	110.9	100.0	117.3	110.9	100.2	117.3	110.9	100.3	119.0	112.1	100.4	119.0	112.1	100.4	117.3	111.0
05	METALS	106.3	119.9	110.7	97.5	119.9	104.8	97.7	120.2	105.1	101.2	120.4	107.5	101.2	120.7	107.5	101.2	120.1	107.4
06	WOOD & PLASTICS	99.4	108.0	103.8	99.4	108.0	103.8	99.4	108.0	103.8	99.4	108.0	103.8	99.4	108.0	103.8	99.4	108.0	103.8
07	THERMAL & MOISTURE PROTECTION	100.3	113.6	106.9	101.8	115.2	108.4	101.9	115.2	108.5	101.8	118.2	109.9	101.8	118.2	109.9	101.8	115.2	108.4
08	DOORS & WINDOWS	107.1	113.0	108.6	107.1	113.0	108.6	107.1	121.0	110.7	107.1	121.0	110.7	107.1	121.0	110.7	107.1	121.0	110.7
09200	PLASTER & GYPSUM BOARD	104.4	107.2	106.2	104.4	107.2	106.2	104.4	107.2	106.2	104.4	107.2	106.2	104.4	107.2	106.2	104.4	107.2	106.2
095,098	CEILINGS & ACOUSTICAL TREATMENT	101.1	107.2	104.8	101.1	107.2	104.8	101.1	107.2	104.8	101.1	107.2	104.8	101.1	107.2	104.8	101.1	107.2	104.8
09600	FLOORING	99.8	115.1	103.9	99.8	115.1	103.9	99.8	115.1	103.9	99.8	115.1	103.9	99.8	115.1	103.9	99.8	115.1	103.9
097,099	WALL FINISHES, PAINTS & COATINGS	92.8	109.6	103.0	92.8	109.6	103.0	92.8	109.6	103.0	92.8	109.6	103.0	92.8	109.6	103.0	92.8	109.6	103.0
09	FINISHES	102.7	110.7	106.9	102.7	110.7	106.9	102.8	110.7	106.9	102.7	110.7	106.9	102.8	110.7	106.9	102.6	110.7	106.8
10 - 14	TOTAL DIV. 10000 - 14000	100.0	110.1	102.1	100.0	110.1	102.1	100.0	110.1	102.1	100.0	110.1	102.1	100.0	110.2	102.2	100.0	110.1	102.1
15	MECHANICAL	100.1	103.9	101.8	100.1	103.9	101.8	100.1	104.0	101.8	100.1	104.0	101.8	100.1	104.0	101.8	100.1	104.0	101.8
16	ELECTRICAL	99.9	103.1	101.5	100.3	108.4	104.2	100.2	108.4	104.2	100.3	108.4	104.2	100.3	137.7	118.5	99.8	105.8	102.7
01 - 16	WEIGHTED AVERAGE	102.7	109.9	106.0	101.4	110.7	105.6	103.2	111.2	106.8	102.2	111.5	106.4	102.3	115.6	108.4	102.2	110.7	106.1

D.C. / DELAWARE / FLORIDA

DIVISION		WASHINGTON MAT.	INST.	TOTAL	WILMINGTON MAT.	INST.	TOTAL	DAYTONA BEACH MAT.	INST.	TOTAL	FORT LAUDERDALE MAT.	INST.	TOTAL	JACKSONVILLE MAT.	INST.	TOTAL	MELBOURNE MAT.	INST.	TOTAL
01590	EQUIPMENT RENTAL	.0	102.3	102.3	.0	118.4	118.4	.0	97.7	97.7	.0	89.3	89.3	.0	97.7	97.7	.0	97.7	97.7
02	SITE CONSTRUCTION	100.3	89.5	92.3	87.5	111.4	105.1	115.6	86.7	94.2	101.3	73.8	81.0	115.7	87.2	94.6	123.2	87.2	96.6
03100	CONCRETE FORMS & ACCESSORIES	98.0	80.7	83.0	96.7	101.4	100.7	98.2	69.8	73.6	95.6	68.4	72.0	98.0	53.3	59.3	92.5	73.7	76.2
03200	CONCRETE REINFORCEMENT	105.6	89.5	97.5	102.2	99.0	100.6	95.0	85.5	90.2	95.0	72.9	83.8	95.0	51.2	72.9	96.1	85.5	90.8
03300	CAST-IN-PLACE CONCRETE	114.7	88.1	103.6	78.8	94.4	85.3	89.9	70.5	81.8	94.3	69.4	83.9	90.8	58.7	77.4	104.5	75.8	92.5
03	CONCRETE	111.2	86.2	98.7	97.8	99.5	98.6	90.4	74.2	82.3	92.3	70.9	81.6	90.8	56.5	73.7	99.6	77.8	88.7
04	MASONRY	92.2	82.0	85.8	108.5	88.8	96.1	86.7	64.7	72.9	86.9	67.6	74.8	86.2	51.4	64.4	83.9	73.6	77.4
05	METALS	107.4	107.8	107.5	101.8	115.1	106.1	96.4	95.9	96.2	96.3	89.9	94.2	96.0	80.2	90.8	105.3	96.4	102.4
06	WOOD & PLASTICS	95.0	80.9	87.7	95.3	103.2	99.4	97.8	72.5	84.6	93.3	65.8	79.0	97.8	52.1	74.0	91.6	72.5	81.7
07	THERMAL & MOISTURE PROTECTION	98.3	81.7	90.2	100.6	104.3	102.4	95.4	69.7	82.7	95.4	74.7	85.2	95.7	58.0	77.2	95.6	76.3	86.1
08	DOORS & WINDOWS	102.6	90.7	99.5	94.4	106.2	97.4	100.9	69.9	92.8	98.5	63.3	89.4	100.9	49.7	87.6	100.0	75.4	93.7
09200	PLASTER & GYPSUM BOARD	101.1	80.5	88.0	112.3	103.1	106.5	100.5	72.3	82.6	100.1	65.5	78.1	100.5	51.3	69.3	96.2	72.3	81.1
095,098	CEILINGS & ACOUSTICAL TREATMENT	99.3	80.5	88.0	103.6	103.1	103.3	102.7	72.3	84.5	102.7	65.5	80.4	102.7	51.3	71.9	97.3	72.3	82.3
09600	FLOORING	113.7	102.8	110.8	83.7	98.3	87.5	118.6	73.5	106.7	118.6	61.0	103.4	118.6	50.1	100.5	115.3	73.5	104.2
097,099	WALL FINISHES, PAINTS & COATINGS	113.0	87.8	97.7	90.7	97.0	94.5	111.2	72.1	87.5	107.4	49.9	72.5	111.2	48.1	72.9	111.2	92.6	99.9
09	FINISHES	101.2	85.3	92.9	102.2	100.2	101.2	108.4	70.8	88.9	106.4	64.5	84.6	108.4	51.5	78.8	106.0	75.4	90.0
10 - 14	TOTAL DIV. 10000 - 14000	100.0	97.5	99.5	100.0	100.0	100.0	100.0	82.5	96.3	100.0	89.4	97.8	100.0	77.5	95.2	100.0	86.0	97.0
15	MECHANICAL	100.0	90.9	96.0	100.3	116.6	107.4	99.8	72.6	87.9	99.8	69.7	86.7	99.8	51.4	78.6	99.8	77.7	90.2
16	ELECTRICAL	97.7	97.6	97.6	97.2	104.8	100.9	96.3	67.0	82.0	96.3	75.2	86.0	96.1	67.2	82.0	96.2	73.9	85.3
01 - 16	WEIGHTED AVERAGE	102.0	90.8	96.9	99.5	105.7	102.3	98.4	74.5	87.5	97.7	72.3	86.2	98.3	61.0	81.4	100.5	79.0	90.7

COST INDEXES

442

City Cost Indexes

		FLORIDA																
DIVISION		MIAMI			ORLANDO			PANAMA CITY			PENSACOLA			ST. PETERSBURG			TALLAHASSEE		
		MAT.	INST.	TOTAL	MAT.	INST.	TOTAL	MAT.	INST.	TOTAL	MAT.	INST.	TOTAL	MAT.	INST.	TOTAL	MAT.	INST.	TOTAL
01590	EQUIPMENT RENTAL	.0	89.3	89.3	.0	97.7	97.7	.0	97.7	97.7	.0	97.7	97.7	.0	97.7	97.7	.0	97.7	97.7
02	SITE CONSTRUCTION	100.6	73.6	80.6	116.2	86.4	94.1	130.2	83.9	95.9	127.8	86.3	97.1	116.5	85.7	93.7	117.0	85.8	93.9
03100	CONCRETE FORMS & ACCESSORIES	95.8	68.4	72.0	98.0	71.3	74.9	97.0	27.7	37.0	87.4	51.1	55.9	95.5	47.2	53.7	97.9	39.0	46.9
03200	CONCRETE REINFORCEMENT	95.0	72.9	83.8	95.0	81.8	88.3	99.2	50.1	74.4	101.7	50.4	75.8	98.5	57.6	77.8	95.0	50.6	72.6
03300	CAST-IN-PLACE CONCRETE	91.9	69.6	82.5	97.7	70.4	86.3	95.4	34.6	70.0	95.4	54.9	78.5	101.7	55.3	82.3	94.1	48.4	75.0
03	CONCRETE	91.1	70.9	81.0	92.0	74.2	83.1	99.0	36.1	67.6	98.0	54.0	76.0	96.4	53.7	75.1	92.3	46.5	69.5
04	MASONRY	84.8	68.0	74.3	90.9	64.7	74.5	91.0	27.6	51.2	88.1	49.9	64.1	134.0	49.3	80.8	89.1	38.8	57.5
05	METALS	101.7	89.3	97.6	102.5	94.1	99.8	96.2	65.6	86.2	96.2	79.6	90.8	99.6	80.6	93.4	89.4	78.5	85.9
06	WOOD & PLASTICS	93.3	65.8	79.0	97.8	75.2	86.0	96.6	27.8	60.7	86.3	51.4	68.1	95.0	47.2	70.0	96.2	37.4	65.5
07	THERMAL & MOISTURE PROTECTION	99.3	71.4	85.6	95.7	69.9	83.1	96.0	31.4	64.2	95.7	52.8	74.6	95.3	49.6	72.8	95.7	46.5	71.5
08	DOORS & WINDOWS	98.5	63.9	89.5	100.9	69.6	92.8	98.4	26.7	79.9	98.4	50.2	85.9	99.5	46.5	85.8	99.5	41.0	84.3
09200	PLASTER & GYPSUM BOARD	100.1	65.5	78.1	103.1	75.0	85.3	98.4	26.4	52.7	94.0	50.6	66.5	97.6	46.3	65.1	100.5	36.2	59.7
095,098	CEILINGS & ACOUSTICAL TREATMENT	102.7	65.5	80.4	102.7	75.0	86.1	96.5	26.4	54.4	96.5	50.6	69.0	96.5	46.3	66.4	102.7	36.2	62.8
09600	FLOORING	126.0	61.7	109.0	118.6	73.5	106.7	118.0	19.2	91.9	112.2	53.2	96.6	116.9	53.0	100.0	118.6	38.9	97.5
097,099	WALL FINISHES, PAINTS & COATINGS	107.4	49.9	72.5	111.2	55.2	77.2	111.2	24.6	58.6	111.2	55.9	77.6	111.2	46.5	71.9	111.2	39.6	67.8
09	FINISHES	108.6	64.6	85.7	108.8	70.4	88.9	107.5	25.4	64.8	105.0	51.8	77.3	106.0	48.0	75.8	108.6	38.2	72.0
10 - 14	TOTAL DIV. 10000 - 14000	100.0	89.4	97.8	100.0	82.7	96.3	100.0	46.0	88.5	100.0	51.5	89.7	100.0	53.9	90.2	100.0	63.9	92.3
15	MECHANICAL	99.8	74.2	88.6	99.8	63.8	84.1	99.8	25.2	67.2	99.8	50.0	78.1	99.8	49.9	78.0	99.8	39.9	73.6
16	ELECTRICAL	96.7	73.3	85.3	96.7	44.6	71.3	95.3	33.4	65.1	98.4	58.8	79.1	96.6	48.1	73.0	96.7	41.3	69.7
01 - 16	WEIGHTED AVERAGE	98.7	72.9	87.0	99.8	69.3	86.0	99.5	37.9	71.5	99.2	58.1	80.5	101.6	55.8	80.8	97.6	49.2	75.6

		FLORIDA			GEORGIA														
DIVISION		TAMPA			ALBANY			ATLANTA			AUGUSTA			COLUMBUS			MACON		
		MAT.	INST.	TOTAL	MAT.	INST.	TOTAL	MAT.	INST.	TOTAL	MAT.	INST.	TOTAL	MAT.	INST.	TOTAL	MAT.	INST.	TOTAL
01590	EQUIPMENT RENTAL	.0	97.7	97.7	.0	90.1	90.1	.0	92.8	92.8	.0	92.1	92.1	.0	90.1	90.1	.0	102.6	102.6
02	SITE CONSTRUCTION	116.7	86.4	94.3	101.6	76.2	82.8	101.2	95.6	97.0	97.8	92.6	94.0	101.5	76.6	83.1	102.3	93.8	96.0
03100	CONCRETE FORMS & ACCESSORIES	99.4	74.9	78.2	97.1	45.9	52.8	96.0	78.8	81.1	93.1	51.9	57.4	97.0	58.7	63.8	96.4	55.5	61.0
03200	CONCRETE REINFORCEMENT	95.0	90.6	92.8	94.6	91.7	93.1	98.7	93.5	96.1	99.8	77.3	88.4	95.0	92.7	93.9	96.3	91.9	94.1
03300	CAST-IN-PLACE CONCRETE	99.5	61.6	83.7	96.0	46.2	75.2	99.6	74.8	89.2	94.1	48.9	75.2	95.7	55.0	78.7	94.4	45.8	74.1
03	CONCRETE	95.0	74.3	84.7	93.0	56.1	74.6	98.5	80.0	89.3	94.2	56.1	75.2	92.9	65.1	79.0	92.5	60.4	76.4
04	MASONRY	87.7	75.7	80.2	88.9	39.9	58.1	90.9	70.6	78.1	91.2	40.2	59.2	88.9	57.2	69.0	102.8	38.4	62.4
05	METALS	99.7	95.6	98.4	95.6	88.8	93.4	93.3	80.4	89.1	92.2	69.6	84.8	95.2	92.3	94.3	90.9	90.8	90.9
06	WOOD & PLASTICS	99.3	77.4	87.9	96.7	42.5	68.4	96.4	80.9	88.3	93.4	53.3	72.5	96.7	58.6	76.8	104.4	57.5	79.9
07	THERMAL & MOISTURE PROTECTION	95.7	62.3	79.3	95.5	54.1	75.1	92.3	75.5	84.0	91.7	49.8	71.1	95.1	61.9	78.8	93.9	56.8	75.7
08	DOORS & WINDOWS	100.9	71.2	93.2	98.5	50.1	85.9	100.0	77.5	94.2	94.1	54.7	83.9	98.5	60.9	88.7	96.9	59.3	87.2
09200	PLASTER & GYPSUM BOARD	100.5	77.4	85.8	100.2	41.5	63.0	113.1	80.8	92.7	111.9	52.5	74.2	100.2	58.0	73.5	106.2	56.9	75.0
095,098	CEILINGS & ACOUSTICAL TREATMENT	102.7	77.4	87.5	101.8	41.5	65.6	107.9	80.8	91.7	108.9	52.5	75.1	101.8	58.0	75.6	97.7	56.9	73.2
09600	FLOORING	118.6	53.0	101.3	118.6	32.5	95.8	84.0	78.0	82.4	82.8	41.4	71.9	118.6	54.8	101.7	92.5	38.2	78.1
097,099	WALL FINISHES, PAINTS & COATINGS	111.2	46.5	71.9	107.4	41.4	67.3	90.3	84.5	86.8	90.3	39.3	59.4	107.4	49.1	72.0	109.1	48.4	72.3
09	FINISHES	108.5	68.3	87.6	106.2	41.5	72.6	96.2	79.3	87.4	95.7	48.2	71.0	106.1	56.2	80.2	94.1	51.1	71.7
10 - 14	TOTAL DIV. 10000 - 14000	100.0	85.3	96.9	100.0	77.5	95.2	100.0	84.6	96.7	100.0	67.6	93.1	100.0	79.6	95.7	100.0	77.6	95.2
15	MECHANICAL	99.8	73.9	88.5	99.8	49.7	77.9	100.0	80.4	91.5	100.0	66.3	85.3	99.8	50.1	78.1	99.8	41.9	74.5
16	ELECTRICAL	95.7	48.1	72.5	91.9	60.4	76.5	96.6	82.1	89.5	96.7	49.3	73.6	92.3	50.9	72.1	90.0	59.6	75.2
01 - 16	WEIGHTED AVERAGE	99.4	72.8	87.3	97.3	56.8	78.9	97.4	80.7	89.8	96.0	59.2	79.3	97.2	61.7	81.1	95.7	59.2	79.1

		GEORGIA						HAWAII			IDAHO								
DIVISION		SAVANNAH			VALDOSTA			HONOLULU			BOISE			LEWISTON			POCATELLO		
		MAT.	INST.	TOTAL	MAT.	INST.	TOTAL	MAT.	INST.	TOTAL	MAT.	INST.	TOTAL	MAT.	INST.	TOTAL	MAT.	INST.	TOTAL
01590	EQUIPMENT RENTAL	.0	91.2	91.2	.0	90.1	90.1	.0	99.3	99.3	.0	101.7	101.7	.0	94.6	94.6	.0	101.7	101.7
02	SITE CONSTRUCTION	102.1	77.1	83.6	111.8	76.4	85.6	139.1	106.5	115.0	79.9	102.7	96.8	84.1	96.9	93.6	81.1	102.6	97.0
03100	CONCRETE FORMS & ACCESSORIES	97.4	52.8	58.8	84.2	45.7	50.8	106.8	144.6	139.5	97.7	82.3	84.4	111.5	67.5	73.4	97.8	81.8	84.0
03200	CONCRETE REINFORCEMENT	96.0	77.5	86.6	97.0	44.0	70.2	108.9	126.2	117.6	108.2	78.0	92.9	115.6	97.7	106.5	108.5	77.5	92.8
03300	CAST-IN-PLACE CONCRETE	92.7	51.6	75.5	94.0	52.2	76.5	199.7	129.5	170.3	95.9	91.0	93.9	104.0	86.7	96.8	95.2	90.8	93.3
03	CONCRETE	91.7	58.4	75.1	97.0	49.3	73.2	154.0	134.6	144.3	104.2	84.4	94.3	111.9	80.0	96.0	100.9	84.0	92.4
04	MASONRY	91.9	53.9	68.0	94.8	49.0	66.1	132.4	129.8	130.8	137.5	74.5	98.0	138.3	86.6	105.8	132.9	66.5	91.2
05	METALS	91.2	84.3	89.0	94.9	72.5	87.6	145.3	111.2	134.1	99.7	77.6	92.5	92.6	88.0	91.1	107.8	76.4	97.6
06	WOOD & PLASTICS	110.2	50.2	78.9	82.3	41.5	61.0	97.7	148.7	124.3	95.7	81.9	88.5	103.9	60.5	81.2	95.7	81.9	88.5
07	THERMAL & MOISTURE PROTECTION	95.5	53.0	74.6	95.2	57.4	76.7	113.0	130.0	121.4	96.9	80.0	88.6	167.0	80.7	124.6	97.1	70.5	84.1
08	DOORS & WINDOWS	99.6	50.9	87.0	93.8	38.7	79.5	105.8	140.1	114.7	94.7	77.8	90.3	114.8	67.5	102.6	94.7	71.0	88.6
09200	PLASTER & GYPSUM BOARD	100.5	49.4	68.1	91.2	40.4	59.0	116.7	149.8	137.7	83.2	81.2	81.9	134.0	59.3	86.7	83.2	81.2	81.9
095,098	CEILINGS & ACOUSTICAL TREATMENT	102.7	49.4	70.7	96.5	40.4	62.8	119.6	149.8	137.7	107.6	81.2	91.7	128.4	59.3	87.0	107.6	81.2	91.7
09600	FLOORING	118.6	48.8	100.1	110.4	39.0	91.6	168.4	134.9	159.5	99.8	56.3	88.3	137.7	93.5	126.1	100.1	56.3	88.5
097,099	WALL FINISHES, PAINTS & COATINGS	108.1	49.1	72.3	107.4	35.8	64.0	104.9	148.5	131.4	103.7	52.2	72.5	129.6	74.3	96.0	103.7	54.3	73.7
09	FINISHES	106.7	51.4	77.9	101.8	42.7	71.1	134.0	144.6	139.5	98.5	74.5	86.0	157.0	71.7	112.6	98.6	74.7	86.2
10 - 14	TOTAL DIV. 10000 - 14000	100.0	70.6	93.8	100.0	69.4	93.5	100.0	121.1	104.5	100.0	86.4	97.1	100.0	88.3	97.5	100.0	86.4	97.1
15	MECHANICAL	99.8	47.8	77.1	99.8	42.1	74.6	100.2	117.8	107.9	100.1	77.1	90.1	101.3	85.9	94.6	100.1	77.1	90.0
16	ELECTRICAL	93.8	58.9	76.8	90.6	30.2	61.1	108.0	124.1	115.9	93.5	78.3	86.1	79.1	85.1	82.0	88.7	73.9	81.5
01 - 16	WEIGHTED AVERAGE	97.1	58.8	79.7	97.0	49.1	75.2	121.0	125.7	123.1	100.2	80.5	91.2	108.1	83.3	96.8	100.3	78.3	90.3

443

DIVISION		IDAHO TWIN FALLS			ILLINOIS CHICAGO			DECATUR			EAST ST. LOUIS			JOLIET			PEORIA		
		MAT.	INST.	TOTAL	MAT.	INST.	TOTAL	MAT.	INST.	TOTAL	MAT.	INST.	TOTAL	MAT.	INST.	TOTAL	MAT.	INST.	TOTAL
01590	EQUIPMENT RENTAL	.0	101.7	101.7	.0	91.5	91.5	.0	102.4	102.4	.0	109.0	109.0	.0	89.3	89.3	.0	101.6	101.6
02	SITE CONSTRUCTION	87.8	100.6	97.2	96.9	90.9	92.5	87.2	96.0	93.7	103.5	96.3	98.6	96.8	89.9	91.7	95.9	95.0	95.3
03100	CONCRETE FORMS & ACCESSORIES	98.5	35.2	43.7	101.6	140.3	135.2	97.1	103.2	102.4	91.0	102.4	100.9	102.5	131.8	127.8	95.4	108.0	106.3
03200	CONCRETE REINFORCEMENT	110.4	49.1	79.4	100.7	139.3	120.2	97.0	99.8	98.4	98.9	102.5	100.7	100.7	128.0	114.5	97.0	99.8	98.5
03300	CAST-IN-PLACE CONCRETE	97.7	45.2	75.7	113.2	136.6	123.0	98.9	103.5	100.8	92.6	108.3	99.2	113.1	121.2	116.4	96.6	106.9	100.9
03	CONCRETE	108.4	42.3	75.4	111.4	137.9	124.6	97.7	103.1	100.4	88.6	105.5	97.1	111.4	126.6	118.9	96.6	106.4	101.5
04	MASONRY	136.3	39.0	75.2	93.4	134.3	119.1	68.6	91.8	83.2	73.9	107.9	95.2	94.9	121.5	111.6	112.3	108.7	110.1
05	METALS	107.8	62.2	92.9	96.2	122.1	104.7	97.9	106.4	100.7	95.6	118.6	103.1	94.1	114.3	100.7	97.9	107.4	101.1
06	WOOD & PLASTICS	96.4	35.1	64.4	103.3	139.5	122.2	98.8	102.1	100.5	94.9	99.9	97.5	104.9	132.2	119.2	98.8	104.3	101.7
07	THERMAL & MOISTURE PROTECTION	98.2	46.7	72.9	100.0	131.3	115.4	97.2	94.0	95.6	92.3	101.5	96.8	99.8	123.0	111.2	97.1	105.3	101.1
08	DOORS & WINDOWS	98.0	36.0	82.0	103.4	139.3	112.7	96.3	101.4	97.6	86.1	107.7	91.7	101.4	132.5	109.4	96.3	102.6	98.0
09200	PLASTER & GYPSUM BOARD	82.9	33.1	51.4	102.8	140.5	126.7	104.4	102.0	102.9	100.5	99.8	100.1	106.3	133.1	121.0	104.4	104.2	104.3
095,098	CEILINGS & ACOUSTICAL TREATMENT	100.4	33.1	60.1	106.3	140.5	126.8	97.1	102.0	100.1	90.0	99.8	95.9	106.3	133.1	122.3	97.1	104.2	101.4
09600	FLOORING	101.5	56.3	89.5	83.2	127.6	94.9	96.4	84.6	93.3	105.1	104.5	104.9	82.7	116.2	91.6	96.4	99.7	97.3
097,099	WALL FINISHES, PAINTS & COATINGS	103.7	32.0	60.2	80.5	129.5	110.2	91.2	97.6	95.1	99.3	90.4	93.9	78.3	115.5	100.9	91.2	96.3	94.3
09	FINISHES	97.7	38.4	66.9	93.1	137.4	116.1	95.8	99.4	97.7	94.5	100.2	97.5	92.5	127.9	110.9	95.9	105.4	100.8
10-14	TOTAL DIV. 10000-14000	100.0	48.6	89.1	100.0	126.9	105.7	100.0	93.5	98.6	100.0	94.8	98.9	100.0	123.2	104.9	100.0	95.3	99.0
15	MECHANICAL	100.1	36.4	72.2	100.0	123.2	110.1	100.0	101.4	100.6	100.0	95.8	98.2	100.1	117.7	107.8	100.0	102.1	101.0
16	ELECTRICAL	81.0	31.5	56.9	98.0	119.9	108.7	98.8	91.3	95.1	96.1	91.7	94.0	96.3	106.3	101.2	98.1	88.7	93.5
01-16	WEIGHTED AVERAGE	100.8	45.8	75.8	99.7	125.8	111.6	96.5	98.5	97.4	94.2	101.1	97.3	99.0	117.2	107.3	98.7	101.7	100.1

DIVISION		ILLINOIS ROCKFORD			SPRINGFIELD			INDIANA ANDERSON			BLOOMINGTON			EVANSVILLE			FORT WAYNE		
		MAT.	INST.	TOTAL	MAT.	INST.	TOTAL	MAT.	INST.	TOTAL	MAT.	INST.	TOTAL	MAT.	INST.	TOTAL	MAT.	INST.	TOTAL
01590	EQUIPMENT RENTAL	.0	101.6	101.6	.0	102.4	102.4	.0	96.7	96.7	.0	85.9	85.9	.0	121.3	121.3	.0	96.7	96.7
02	SITE CONSTRUCTION	95.0	95.6	95.4	93.3	96.0	95.3	87.6	96.6	94.2	77.1	95.6	90.8	82.1	130.1	117.6	88.6	96.5	94.4
03100	CONCRETE FORMS & ACCESSORIES	99.7	113.7	111.9	99.0	103.8	103.2	96.1	75.6	78.3	99.9	76.1	79.3	92.6	75.8	78.0	94.7	75.0	77.6
03200	CONCRETE REINFORCEMENT	97.0	130.2	113.8	97.0	99.8	98.4	93.8	81.2	87.4	83.9	81.0	82.5	92.0	81.2	86.6	93.8	78.0	85.8
03300	CAST-IN-PLACE CONCRETE	98.9	98.8	98.9	91.9	104.2	97.0	102.4	80.8	93.4	101.6	77.3	91.5	97.0	90.4	94.2	108.8	76.2	95.2
03	CONCRETE	97.9	112.0	104.9	94.5	103.6	99.0	96.7	79.1	87.9	103.1	77.2	90.2	103.4	82.1	92.8	99.6	76.6	88.2
04	MASONRY	83.1	110.8	100.5	69.8	92.8	84.3	91.3	77.5	82.6	90.6	75.9	81.4	86.8	81.2	83.3	91.6	77.7	82.9
05	METALS	97.9	123.4	106.2	100.5	106.6	102.5	90.9	89.1	90.3	93.4	75.1	87.4	86.2	86.3	86.2	90.9	87.1	89.7
06	WOOD & PLASTICS	98.8	112.0	105.7	101.5	102.1	101.8	113.2	74.1	92.8	118.9	74.7	95.8	97.7	72.2	84.4	112.9	73.6	92.4
07	THERMAL & MOISTURE PROTECTION	97.1	111.2	104.1	96.7	98.5	97.6	101.0	76.1	88.7	91.5	80.6	86.1	94.8	84.1	89.5	100.7	82.4	91.7
08	DOORS & WINDOWS	96.3	114.8	101.1	96.3	101.4	97.6	99.3	78.1	93.8	103.2	78.4	96.8	95.6	75.8	90.5	99.3	74.7	92.9
09200	PLASTER & GYPSUM BOARD	104.4	112.1	109.3	104.4	102.0	102.9	96.7	74.0	82.3	98.2	74.7	83.3	91.8	70.6	78.4	95.9	73.5	81.7
095,098	CEILINGS & ACOUSTICAL TREATMENT	97.1	112.1	106.1	97.1	102.0	100.1	89.4	74.0	80.2	81.7	74.7	77.5	86.4	70.6	77.0	89.4	73.5	79.9
09600	FLOORING	96.4	93.3	95.6	96.6	84.7	93.4	87.3	88.6	87.6	103.9	88.6	99.9	97.0	83.6	93.5	87.3	79.9	85.3
097,099	WALL FINISHES, PAINTS & COATINGS	91.2	108.8	101.9	91.2	90.2	90.6	92.1	67.6	77.3	89.8	86.2	87.7	96.5	88.1	91.4	92.1	74.7	81.6
09	FINISHES	95.9	109.3	102.9	95.9	98.8	97.4	88.9	76.8	82.6	94.7	79.2	86.7	92.9	77.9	85.1	88.7	75.6	81.8
10-14	TOTAL DIV. 10000-14000	100.0	103.9	100.8	100.0	93.9	98.7	100.0	81.7	96.1	100.0	81.2	96.0	100.0	84.4	96.7	100.0	83.6	96.5
15	MECHANICAL	100.0	105.7	102.5	100.0	103.7	101.7	99.9	81.0	91.7	99.6	82.9	92.3	99.8	85.0	93.3	99.9	78.6	90.6
16	ELECTRICAL	98.1	110.0	103.9	98.7	89.1	94.0	85.2	92.8	88.9	102.4	80.5	91.7	95.5	83.7	89.7	86.3	80.8	83.6
01-16	WEIGHTED AVERAGE	97.4	109.4	102.8	96.8	98.9	97.8	94.6	83.3	89.5	98.2	80.7	90.3	95.2	86.8	91.4	95.1	80.6	88.5

DIVISION		INDIANA GARY			INDIANAPOLIS			MUNCIE			SOUTH BEND			TERRE HAUTE			IOWA CEDAR RAPIDS		
		MAT.	INST.	TOTAL	MAT.	INST.	TOTAL	MAT.	INST.	TOTAL	MAT.	INST.	TOTAL	MAT.	INST.	TOTAL	MAT.	INST.	TOTAL
01590	EQUIPMENT RENTAL	.0	96.7	96.7	.0	91.5	91.5	.0	96.2	96.2	.0	106.8	106.8	.0	121.3	121.3	.0	95.5	95.5
02	SITE CONSTRUCTION	88.2	98.3	95.7	87.5	100.4	97.0	77.2	95.8	90.9	87.7	96.1	93.9	83.4	130.0	117.9	87.6	94.6	92.8
03100	CONCRETE FORMS & ACCESSORIES	96.2	100.6	100.0	96.6	86.8	88.1	90.5	75.3	77.3	100.3	77.6	80.6	94.1	77.7	79.9	103.6	78.2	81.6
03200	CONCRETE REINFORCEMENT	93.8	99.6	96.7	93.2	85.4	89.2	92.9	81.1	86.9	93.8	76.4	85.0	92.0	80.7	86.3	93.3	82.6	87.9
03300	CAST-IN-PLACE CONCRETE	107.0	101.5	104.7	98.5	87.6	93.9	106.8	80.1	95.7	99.6	82.4	92.4	94.0	90.4	92.5	107.8	80.1	96.2
03	CONCRETE	98.9	100.9	99.9	98.8	86.4	92.6	101.8	78.7	90.3	91.4	80.5	86.0	106.3	82.9	94.6	99.7	80.4	90.1
04	MASONRY	92.7	93.1	93.0	98.2	86.6	90.9	92.8	77.5	83.2	89.8	77.4	82.0	93.7	78.2	84.0	106.2	79.6	89.5
05	METALS	90.9	101.1	94.2	92.8	80.1	88.7	95.1	88.8	93.0	90.9	100.9	94.2	86.9	86.4	86.8	87.2	92.2	88.8
06	WOOD & PLASTICS	111.2	100.9	105.8	110.1	86.4	97.7	111.0	73.9	91.6	113.1	77.0	94.3	99.9	76.2	87.5	109.6	76.9	92.5
07	THERMAL & MOISTURE PROTECTION	100.0	98.0	99.0	99.9	85.9	93.0	93.1	76.4	84.9	99.7	82.3	91.1	94.9	81.5	88.3	100.0	80.3	90.3
08	DOORS & WINDOWS	99.3	102.5	100.1	106.9	86.3	101.6	98.2	78.0	93.0	92.7	77.0	88.6	96.2	79.0	91.7	99.3	80.2	94.4
09200	PLASTER & GYPSUM BOARD	91.4	101.5	97.8	94.8	86.5	89.5	91.5	74.0	80.4	96.3	77.0	84.0	91.8	74.7	81.0	104.4	76.6	86.8
095,098	CEILINGS & ACOUSTICAL TREATMENT	89.4	101.5	96.7	93.9	86.5	89.4	82.5	74.0	77.4	89.4	77.0	82.0	86.4	74.7	79.4	112.4	76.6	90.9
09600	FLOORING	87.3	99.4	90.5	87.0	93.4	88.7	96.2	88.6	94.2	87.3	77.4	84.7	97.0	87.3	94.5	123.0	58.2	105.9
097,099	WALL FINISHES, PAINTS & COATINGS	92.1	101.8	98.0	92.1	89.3	90.4	89.8	67.6	76.4	92.1	81.0	85.4	96.5	87.7	91.2	107.7	77.4	89.3
09	FINISHES	88.1	100.5	94.6	90.0	88.4	89.1	91.4	76.7	83.7	88.8	77.8	83.1	92.9	80.2	86.3	111.8	73.5	91.8
10-14	TOTAL DIV. 10000-14000	100.0	98.5	99.7	100.0	85.1	96.8	100.0	81.2	96.0	100.0	77.6	95.2	100.0	85.4	96.9	100.0	81.8	96.1
15	MECHANICAL	99.9	94.7	97.6	99.9	88.6	95.0	99.6	81.0	91.4	99.9	76.9	89.8	99.8	78.6	90.5	100.4	83.6	93.0
16	ELECTRICAL	93.4	102.6	97.9	105.1	92.8	99.1	90.1	76.4	83.4	99.8	83.4	91.8	94.1	80.6	87.5	97.0	82.1	89.7
01-16	WEIGHTED AVERAGE	95.8	98.7	97.1	98.8	88.7	94.2	96.1	80.8	89.2	95.0	82.6	89.4	96.0	85.3	91.1	98.8	82.7	91.5

COST INDEXES

City Cost Indexes

DIVISION		IOWA																	
		COUNCIL BLUFFS			DAVENPORT			DES MOINES			DUBUQUE			SIOUX CITY			WATERLOO		
		MAT.	INST.	TOTAL	MAT.	INST.	TOTAL	MAT.	INST.	TOTAL	MAT.	INST.	TOTAL	MAT.	INST.	TOTAL	MAT.	INST.	TOTAL
01590	EQUIPMENT RENTAL	.0	95.2	95.2	.0	99.5	99.5	.0	101.5	101.5	.0	94.1	94.1	.0	99.5	99.5	.0	99.5	99.5
02	SITE CONSTRUCTION	92.8	91.2	91.6	85.8	98.4	95.1	79.3	99.9	94.6	85.7	91.5	90.0	94.7	95.4	95.2	86.3	94.5	92.4
03100	CONCRETE FORMS & ACCESSORIES	81.5	58.4	61.5	103.3	90.8	92.5	105.3	78.1	81.8	83.3	67.3	69.5	103.6	62.2	67.7	104.2	49.3	56.6
03200	CONCRETE REINFORCEMENT	95.2	78.5	86.8	93.3	95.7	94.5	93.3	79.3	86.2	92.0	82.3	87.1	93.3	67.3	80.2	93.3	82.1	87.6
03300	CAST-IN-PLACE CONCRETE	112.1	74.3	96.3	103.8	100.9	102.6	108.1	78.8	95.8	105.5	87.9	98.2	106.9	57.2	86.1	107.8	48.4	83.0
03	CONCRETE	101.5	68.8	85.2	97.8	95.7	96.7	98.8	79.4	89.1	96.1	78.2	87.1	99.3	62.5	80.9	99.7	56.6	78.2
04	MASONRY	105.8	73.9	85.8	103.8	87.6	93.7	100.0	80.3	87.6	107.2	72.0	85.1	99.1	59.5	74.3	100.7	59.5	74.8
05	METALS	92.4	89.5	91.5	87.2	101.6	91.9	87.3	92.3	89.0	85.8	91.4	87.6	87.2	83.7	86.0	87.2	90.2	88.2
06	WOOD & PLASTICS	85.2	54.5	69.2	109.6	89.1	98.9	111.1	77.0	93.3	87.2	64.7	75.5	109.6	60.7	84.1	110.3	49.3	78.5
07	THERMAL & MOISTURE PROTECTION	99.2	66.8	83.2	99.4	88.3	93.9	100.3	78.5	89.6	99.5	69.5	84.7	99.4	57.4	78.8	99.1	53.6	76.7
08	DOORS & WINDOWS	98.3	61.6	88.8	99.3	91.9	97.4	99.3	81.1	94.6	98.4	75.4	92.4	99.3	62.3	89.7	94.9	61.5	86.3
09200	PLASTER & GYPSUM BOARD	93.7	53.5	68.3	104.4	88.9	94.6	101.3	76.4	85.5	94.1	64.1	75.1	104.4	59.6	76.0	104.4	48.0	68.6
095,098	CEILINGS & ACOUSTICAL TREATMENT	107.0	53.5	74.9	112.4	88.9	98.3	110.6	76.4	90.1	107.0	64.1	81.3	112.4	59.6	80.7	112.4	48.0	73.7
09600	FLOORING	98.1	47.5	84.7	109.4	82.3	102.2	109.2	43.4	91.8	112.0	40.5	93.1	109.9	61.3	97.0	111.1	61.6	98.0
097,099	WALL FINISHES, PAINTS & COATINGS	99.2	72.1	82.8	103.5	94.9	98.3	103.5	78.5	88.3	106.5	77.4	88.9	104.7	63.0	79.4	104.7	36.8	63.5
09	FINISHES	100.8	56.4	77.7	107.4	89.4	98.0	105.6	71.0	87.6	105.5	62.2	83.0	108.6	61.7	84.2	108.0	49.5	77.6
10-14	TOTAL DIV. 10000 - 14000	100.0	74.3	94.5	100.0	86.3	97.1	100.0	82.5	96.3	100.0	78.7	95.5	100.0	77.7	95.2	100.0	72.1	94.1
15	MECHANICAL	100.4	76.9	90.1	100.4	94.9	98.0	100.4	80.7	91.7	100.4	78.9	91.0	100.4	82.0	92.3	100.4	46.9	77.0
16	ELECTRICAL	101.7	81.3	91.7	95.5	89.1	92.4	97.5	83.2	90.5	100.4	70.3	85.7	97.0	77.6	87.6	97.0	63.8	80.8
01-16	WEIGHTED AVERAGE	99.1	74.7	88.0	97.8	93.1	95.7	97.7	82.4	90.7	97.7	76.5	88.1	98.3	73.3	87.0	97.7	62.0	81.5

DIVISION		KANSAS															KENTUCKY		
		DODGE CITY			KANSAS CITY			SALINA			TOPEKA			WICHITA			BOWLING GREEN		
		MAT.	INST.	TOTAL	MAT.	INST.	TOTAL	MAT.	INST.	TOTAL	MAT.	INST.	TOTAL	MAT.	INST.	TOTAL	MAT.	INST.	TOTAL
01590	EQUIPMENT RENTAL	.0	103.5	103.5	.0	100.1	100.1	.0	103.5	103.5	.0	101.6	101.6	.0	103.5	103.5	.0	96.0	96.0
02	SITE CONSTRUCTION	109.9	92.7	97.1	91.0	90.9	90.9	99.9	93.0	94.8	92.9	90.4	91.1	94.1	93.4	93.6	69.5	100.3	92.3
03100	CONCRETE FORMS & ACCESSORIES	93.4	43.4	50.1	99.0	91.9	92.8	89.2	46.0	51.8	98.5	45.2	52.3	95.3	58.1	63.1	83.8	82.9	83.0
03200	CONCRETE REINFORCEMENT	105.1	58.5	81.5	99.8	82.4	91.0	104.5	80.0	92.1	97.0	94.5	95.8	97.0	81.1	89.0	83.3	74.5	78.9
03300	CAST-IN-PLACE CONCRETE	114.8	57.4	90.8	90.0	91.6	90.7	99.6	49.1	78.5	91.0	53.6	75.4	86.8	57.6	74.6	88.6	82.0	85.8
03	CONCRETE	114.8	52.6	83.8	96.2	90.5	93.3	102.0	55.1	78.6	94.1	59.0	76.6	91.9	63.5	77.7	94.9	81.1	88.0
04	MASONRY	104.2	42.2	65.3	103.5	87.5	93.5	118.5	40.6	69.7	98.1	64.0	76.7	91.8	65.8	75.5	93.9	79.6	84.9
05	METALS	95.3	78.2	89.7	100.1	94.5	98.3	95.1	88.2	92.8	100.5	96.9	99.3	100.5	90.3	97.2	91.2	80.0	87.5
06	WOOD & PLASTICS	91.7	43.4	66.5	98.5	92.3	95.2	87.9	46.7	66.4	95.7	41.3	67.3	93.7	56.8	74.4	93.3	82.3	87.5
07	THERMAL & MOISTURE PROTECTION	98.2	49.2	74.1	96.1	90.1	93.2	97.5	54.7	76.5	97.4	71.3	84.6	97.1	65.8	81.7	84.7	80.2	82.5
08	DOORS & WINDOWS	96.2	47.0	83.5	95.1	89.4	93.6	96.1	52.5	84.8	96.3	60.2	87.0	96.3	63.5	87.8	95.3	76.0	90.3
09200	PLASTER & GYPSUM BOARD	100.8	41.7	63.3	97.0	91.9	93.8	100.0	45.1	65.2	102.3	39.5	62.5	102.3	55.4	72.6	87.4	82.2	84.1
095,098	CEILINGS & ACOUSTICAL TREATMENT	88.2	41.7	60.3	89.1	91.9	90.8	88.2	45.1	62.3	89.1	39.5	59.4	89.1	55.4	68.9	86.4	82.2	83.9
09600	FLOORING	95.7	51.2	83.9	86.7	93.1	88.4	93.7	37.4	78.8	97.2	47.5	84.1	96.4	79.3	91.9	92.9	63.0	84.9
097,099	WALL FINISHES, PAINTS & COATINGS	91.2	49.9	66.2	99.3	89.5	93.4	91.2	39.3	59.7	91.2	69.5	78.0	91.2	63.4	74.3	96.5	75.6	83.8
09	FINISHES	93.9	45.1	68.5	92.2	91.2	91.7	92.6	42.9	66.7	93.8	46.0	68.9	93.6	62.2	77.3	91.1	78.5	84.5
10-14	TOTAL DIV. 10000 - 14000	100.0	50.2	89.4	100.0	84.4	96.7	100.0	66.1	92.8	100.0	70.0	93.6	100.0	70.7	93.8	100.0	64.2	92.4
15	MECHANICAL	100.0	47.9	77.2	99.9	88.2	94.8	100.0	36.3	72.1	100.0	70.6	87.1	100.0	67.9	86.0	99.8	88.6	94.9
16	ELECTRICAL	99.7	50.3	75.6	104.4	102.1	103.3	99.4	72.6	86.3	103.3	75.1	89.5	101.5	72.6	87.4	94.5	84.2	89.5
01-16	WEIGHTED AVERAGE	100.3	54.8	79.6	98.7	91.6	95.5	99.0	57.2	80.1	98.5	69.3	85.2	97.7	71.1	85.6	94.4	83.6	89.5

DIVISION		KENTUCKY									LOUISIANA								
		LEXINGTON			LOUISVILLE			OWENSBORO			ALEXANDRIA			BATON ROUGE			LAKE CHARLES		
		MAT.	INST.	TOTAL	MAT.	INST.	TOTAL	MAT.	INST.	TOTAL	MAT.	INST.	TOTAL	MAT.	INST.	TOTAL	MAT.	INST.	TOTAL
01590	EQUIPMENT RENTAL	.0	103.5	103.5	.0	96.0	96.0	.0	121.3	121.3	.0	86.8	86.8	.0	86.2	86.2	.0	86.2	86.2
02	SITE CONSTRUCTION	77.3	102.4	95.8	67.0	100.2	91.5	82.1	129.6	117.2	102.8	85.7	90.1	109.3	85.3	91.5	110.9	85.0	91.8
03100	CONCRETE FORMS & ACCESSORIES	97.6	70.3	74.0	92.5	82.6	83.9	88.7	64.7	67.9	80.0	44.3	49.1	96.9	48.4	54.9	97.4	49.2	55.7
03200	CONCRETE REINFORCEMENT	92.0	96.6	94.3	92.0	97.8	94.9	83.4	93.9	88.7	98.4	63.8	80.9	93.6	51.2	72.2	93.6	51.2	72.2
03300	CAST-IN-PLACE CONCRETE	95.7	80.1	89.2	92.7	80.5	87.6	91.4	76.7	85.3	93.9	44.2	73.2	85.9	52.3	71.9	91.0	53.8	75.5
03	CONCRETE	97.3	79.0	88.2	95.5	84.8	90.1	104.3	74.8	89.5	89.0	49.1	69.1	93.5	51.1	72.3	96.0	52.0	74.0
04	MASONRY	90.8	48.1	64.0	92.4	81.8	85.8	90.8	52.0	66.4	115.3	54.0	76.9	96.6	55.6	70.9	94.4	56.6	70.7
05	METALS	92.6	89.3	91.5	100.0	90.3	96.8	83.1	87.4	84.5	85.7	74.4	82.0	95.7	65.5	85.8	90.1	65.7	82.1
06	WOOD & PLASTICS	104.3	73.3	88.1	102.8	82.3	92.1	93.5	63.6	77.9	82.5	43.9	62.3	103.8	48.3	74.8	102.0	49.0	74.3
07	THERMAL & MOISTURE PROTECTION	95.3	78.7	87.2	85.0	82.0	83.5	94.8	67.5	81.4	99.4	54.2	77.2	97.4	55.2	76.7	100.7	55.3	78.4
08	DOORS & WINDOWS	96.2	80.1	92.1	96.2	87.1	93.8	93.6	75.3	88.8	99.3	50.0	86.6	104.3	48.6	89.9	104.3	50.6	90.4
09200	PLASTER & GYPSUM BOARD	94.8	71.8	80.2	95.9	82.2	87.2	86.5	61.8	70.9	79.1	43.0	56.2	99.1	47.5	66.4	99.1	48.2	66.8
095,098	CEILINGS & ACOUSTICAL TREATMENT	91.8	71.8	79.8	91.8	82.2	86.0	75.7	61.8	67.4	90.3	43.0	61.9	96.7	47.5	67.2	96.7	48.2	67.6
09600	FLOORING	97.9	43.8	83.6	96.4	74.8	90.7	95.3	63.0	86.8	105.0	68.9	95.5	109.4	67.6	98.4	109.7	52.8	94.6
097,099	WALL FINISHES, PAINTS & COATINGS	96.5	59.7	74.2	96.5	84.5	89.2	96.5	99.6	98.4	96.4	42.1	63.5	98.7	48.3	68.1	98.7	44.5	65.8
09	FINISHES	95.0	63.7	78.7	94.4	81.1	87.5	89.0	67.4	77.8	93.2	48.1	69.7	102.1	51.8	75.9	102.2	49.0	74.5
10-14	TOTAL DIV. 10000 - 14000	100.0	81.9	96.2	100.0	85.3	96.9	100.0	84.3	96.7	100.0	71.6	94.0	100.0	75.3	94.7	100.0	75.5	94.8
15	MECHANICAL	99.8	48.8	77.5	99.8	87.3	94.3	99.8	53.4	79.5	100.1	38.5	73.1	100.0	43.7	75.4	100.0	59.3	82.2
16	ELECTRICAL	94.9	50.8	73.4	94.9	84.2	89.7	94.0	82.3	88.3	92.4	56.4	74.8	92.8	50.4	72.1	92.8	63.1	78.4
01-16	WEIGHTED AVERAGE	95.9	66.9	82.7	96.3	86.4	91.8	94.2	74.3	85.2	95.6	54.7	77.0	98.4	55.1	78.7	97.8	60.0	80.7

LOUISIANA / MAINE

DIVISION		LOUISIANA									MAINE								
		MONROE			NEW ORLEANS			SHREVEPORT			AUGUSTA			BANGOR			LEWISTON		
		MAT.	INST.	TOTAL	MAT.	INST.	TOTAL	MAT.	INST.	TOTAL	MAT.	INST.	TOTAL	MAT.	INST.	TOTAL	MAT.	INST.	TOTAL
01590	EQUIPMENT RENTAL	.0	86.8	86.8	.0	87.4	87.4	.0	86.8	86.8	.0	101.6	101.6	.0	101.6	101.6	.0	101.6	101.6
02	SITE CONSTRUCTION	102.8	85.2	89.8	113.4	87.8	94.5	101.5	85.1	89.3	83.4	102.3	97.4	83.3	100.7	96.2	82.0	100.7	95.8
03100	CONCRETE FORMS & ACCESSORIES	79.3	44.9	49.5	97.2	69.5	73.2	98.6	47.6	54.4	99.6	64.9	69.6	93.9	80.3	82.1	99.7	80.3	82.9
03200	CONCRETE REINFORCEMENT	97.4	64.0	80.5	93.6	65.0	79.2	96.9	63.7	80.1	88.1	112.7	100.5	88.1	113.0	100.7	108.2	113.0	110.7
03300	CAST-IN-PLACE CONCRETE	93.9	48.1	74.8	90.1	60.0	77.5	92.6	51.8	75.5	81.5	63.2	73.9	81.5	65.1	74.7	92.8	65.1	81.2
03	CONCRETE	88.8	50.6	69.7	95.5	65.8	80.7	88.8	53.0	70.9	98.6	73.3	86.0	97.7	80.7	89.2	102.8	80.7	91.7
04	MASONRY	110.2	58.1	77.5	96.3	64.3	76.2	102.4	51.1	70.2	96.1	52.3	68.6	113.7	60.2	80.1	97.9	60.2	74.2
05	METALS	85.7	73.0	81.5	103.1	75.1	94.0	82.1	72.2	78.8	87.9	84.9	86.9	87.5	83.1	86.1	90.7	83.1	88.2
06	WOOD & PLASTICS	81.7	44.3	62.2	98.4	72.7	85.0	102.2	48.2	74.0	98.5	63.4	80.2	92.3	82.6	87.2	98.5	82.6	90.2
07	THERMAL & MOISTURE PROTECTION	99.4	57.0	78.5	101.6	65.2	83.7	98.3	55.6	77.3	102.1	52.5	77.7	102.0	59.5	81.1	101.8	59.5	81.0
08	DOORS & WINDOWS	99.3	55.0	87.9	104.9	69.4	95.7	97.2	52.5	85.6	103.9	62.1	93.1	103.8	74.3	96.2	107.1	74.3	98.6
09200	PLASTER & GYPSUM BOARD	78.7	43.4	56.3	97.9	72.5	81.8	85.4	47.4	61.3	102.8	61.4	76.6	98.9	81.2	87.6	105.2	81.2	89.9
095,098	CEILINGS & ACOUSTICAL TREATMENT	90.3	43.4	62.2	95.7	72.5	81.8	91.2	47.4	65.0	92.2	61.4	73.7	89.5	81.2	84.5	101.1	81.2	89.1
09600	FLOORING	104.6	43.6	88.5	110.1	51.8	94.7	115.6	64.6	102.1	99.8	43.9	85.0	97.3	54.7	86.1	99.8	54.7	87.9
097,099	WALL FINISHES, PAINTS & COATINGS	96.4	50.3	68.4	100.5	63.9	78.3	96.4	42.1	63.5	92.8	36.4	58.6	92.8	33.5	56.8	92.8	33.5	56.8
09	FINISHES	93.0	44.3	67.7	102.3	65.8	83.3	97.4	49.7	72.6	99.1	57.1	77.3	97.2	70.9	83.5	101.4	70.9	85.5
10 - 14	TOTAL DIV. 10000 - 14000	100.0	65.4	92.6	100.0	80.3	95.8	100.0	65.7	92.7	100.0	63.9	92.3	100.0	83.0	96.4	100.0	83.0	96.4
15	MECHANICAL	100.1	52.0	79.1	100.0	65.9	85.1	100.1	54.6	80.2	100.1	72.3	88.0	100.1	71.5	87.6	100.1	71.5	87.6
16	ELECTRICAL	94.1	55.9	75.4	94.0	70.5	82.6	92.9	67.0	80.3	100.2	80.9	90.8	98.5	80.9	89.9	100.3	80.9	90.8
01 - 16	WEIGHTED AVERAGE	95.5	57.5	78.2	100.2	69.8	86.4	94.7	59.7	78.8	97.7	72.3	86.2	98.0	76.8	88.3	99.2	76.8	89.0

MAINE / MARYLAND / MASSACHUSETTS

DIVISION		MAINE			MARYLAND						MASSACHUSETTS								
		PORTLAND			BALTIMORE			HAGERSTOWN			BOSTON			BROCKTON			FALL RIVER		
		MAT.	INST.	TOTAL	MAT.	INST.	TOTAL	MAT.	INST.	TOTAL	MAT.	INST.	TOTAL	MAT.	INST.	TOTAL	MAT.	INST.	TOTAL
01590	EQUIPMENT RENTAL	.0	101.6	101.6	.0	103.4	103.4	.0	99.2	99.2	.0	108.4	108.4	.0	103.8	103.8	.0	105.0	105.0
02	SITE CONSTRUCTION	80.8	100.7	95.5	97.6	93.3	94.4	87.4	88.9	88.5	88.4	108.9	103.6	86.1	104.8	100.0	85.2	105.0	99.8
03100	CONCRETE FORMS & ACCESSORIES	98.8	80.3	82.7	101.7	74.8	78.4	90.3	76.1	78.0	105.1	135.6	131.5	104.7	120.9	118.8	104.7	121.2	119.0
03200	CONCRETE REINFORCEMENT	108.2	113.0	110.7	95.9	87.8	91.8	84.8	76.3	80.5	107.1	143.7	125.6	108.2	142.8	125.7	108.2	125.6	117.0
03300	CAST-IN-PLACE CONCRETE	84.5	65.1	76.4	98.3	78.7	90.1	83.6	61.7	74.4	105.2	143.5	121.2	100.2	129.2	112.3	96.9	141.6	115.5
03	CONCRETE	98.8	80.7	89.7	99.5	79.7	89.6	85.3	72.2	78.8	111.7	138.9	125.3	109.0	127.2	118.1	107.4	128.4	117.9
04	MASONRY	95.4	60.2	73.3	92.8	74.6	81.4	97.0	75.9	83.8	116.7	144.8	134.3	112.6	127.3	121.9	112.4	140.0	129.7
05	METALS	92.0	83.1	89.1	100.3	99.3	99.9	96.5	91.4	94.8	103.3	125.9	110.7	100.0	121.6	107.1	100.0	115.7	105.2
06	WOOD & PLASTICS	98.5	82.6	90.2	97.9	75.7	86.3	85.9	75.6	80.5	102.3	136.3	120.0	101.3	120.4	111.3	101.3	120.8	111.4
07	THERMAL & MOISTURE PROTECTION	101.6	59.5	80.9	94.2	81.7	88.1	93.4	71.6	82.7	102.4	141.9	121.8	102.1	130.1	115.9	102.1	127.9	114.8
08	DOORS & WINDOWS	107.1	74.3	98.6	93.8	83.3	91.1	91.1	77.1	87.5	103.1	135.4	111.4	101.0	123.2	106.8	101.0	119.9	105.9
09200	PLASTER & GYPSUM BOARD	105.2	81.2	89.9	108.7	75.3	87.5	103.5	75.3	85.6	107.8	136.5	126.0	102.5	120.1	113.6	102.5	120.1	113.6
095,098	CEILINGS & ACOUSTICAL TREATMENT	101.1	81.2	89.1	104.1	75.3	86.8	105.1	75.3	87.2	100.1	136.5	121.9	102.1	120.1	112.9	102.1	120.1	112.9
09600	FLOORING	99.8	54.7	87.9	92.8	82.2	90.0	87.9	78.8	85.5	100.4	153.6	114.5	100.9	153.6	114.8	100.7	153.6	114.7
097,099	WALL FINISHES, PAINTS & COATINGS	92.8	33.5	56.8	96.0	71.1	80.9	96.4	41.1	62.7	95.9	150.7	129.1	95.3	119.2	109.8	95.3	119.2	109.8
09	FINISHES	101.4	70.9	85.6	97.3	75.2	85.8	94.7	72.9	83.4	101.7	141.0	122.1	101.5	127.0	114.8	101.5	127.2	114.9
10 - 14	TOTAL DIV. 10000 - 14000	100.0	83.0	96.4	100.0	87.2	97.3	100.0	89.7	97.8	100.0	122.5	104.8	100.0	118.7	104.0	100.0	119.6	104.2
15	MECHANICAL	100.1	71.5	87.6	100.0	85.7	93.7	100.0	85.7	93.7	100.1	125.8	111.4	100.1	105.5	102.5	100.1	105.4	102.4
16	ELECTRICAL	100.4	80.9	90.9	107.1	90.7	99.1	103.6	82.1	93.1	99.3	123.3	111.0	98.3	96.3	97.3	97.8	102.2	99.9
01 - 16	WEIGHTED AVERAGE	98.8	76.8	88.8	99.4	84.8	92.7	96.0	80.8	89.1	102.9	130.5	115.4	101.4	115.5	107.8	101.1	117.0	108.4

MASSACHUSETTS

DIVISION		HYANNIS			LAWRENCE			LOWELL			NEW BEDFORD			PITTSFIELD			SPRINGFIELD		
		MAT.	INST.	TOTAL	MAT.	INST.	TOTAL	MAT.	INST.	TOTAL	MAT.	INST.	TOTAL	MAT.	INST.	TOTAL	MAT.	INST.	TOTAL
01590	EQUIPMENT RENTAL	.0	103.8	103.8	.0	103.8	103.8	.0	101.6	101.6	.0	105.0	105.0	.0	101.6	101.6	.0	101.6	101.6
02	SITE CONSTRUCTION	82.7	104.8	99.1	86.7	104.8	100.1	85.8	104.8	99.9	84.8	105.0	99.7	86.9	103.6	99.2	86.3	103.7	99.2
03100	CONCRETE FORMS & ACCESSORIES	95.4	120.9	117.5	104.5	121.4	119.1	101.2	121.6	118.9	104.7	121.2	119.0	101.2	101.2	101.2	101.4	109.8	108.7
03200	CONCRETE REINFORCEMENT	86.8	125.6	106.4	107.4	130.6	119.1	108.2	130.7	119.6	108.2	125.6	117.0	90.2	105.9	98.1	108.2	124.8	116.6
03300	CAST-IN-PLACE CONCRETE	91.4	140.8	112.0	101.0	139.5	117.1	91.8	139.6	111.8	94.0	141.6	113.9	100.2	116.0	106.8	95.5	117.3	104.6
03	CONCRETE	98.6	127.9	113.3	109.2	128.7	118.9	100.0	128.6	114.3	106.1	128.4	117.2	101.0	106.7	103.9	101.8	114.5	108.2
04	MASONRY	111.5	140.1	129.4	111.9	138.4	128.5	97.8	138.5	123.3	112.2	140.0	129.6	98.4	112.1	107.0	98.1	114.1	108.2
05	METALS	96.3	115.2	102.5	97.3	117.6	103.9	97.3	114.9	103.0	100.0	115.7	105.2	97.1	99.4	97.8	99.9	107.5	102.4
06	WOOD & PLASTICS	91.3	120.4	106.5	101.3	120.4	111.3	100.4	120.4	110.9	101.3	120.8	111.4	100.4	100.4	100.4	100.4	110.6	105.7
07	THERMAL & MOISTURE PROTECTION	101.6	128.8	115.0	102.1	134.4	118.0	101.8	134.4	117.8	102.1	127.9	114.8	101.9	109.1	105.4	101.8	111.0	106.3
08	DOORS & WINDOWS	97.2	119.7	103.0	101.0	119.8	105.9	107.1	119.9	110.4	101.0	119.9	105.9	107.1	103.9	106.2	107.1	114.5	109.0
09200	PLASTER & GYPSUM BOARD	94.2	120.1	110.6	105.2	120.1	114.6	105.2	120.1	114.6	102.5	120.1	113.6	105.2	99.4	101.5	105.2	110.0	108.2
095,098	CEILINGS & ACOUSTICAL TREATMENT	90.5	120.1	108.2	101.1	120.1	112.5	101.1	120.1	112.5	102.1	120.1	112.9	101.1	99.4	100.1	101.1	110.0	106.4
09600	FLOORING	96.9	153.6	111.9	99.8	153.6	114.0	99.8	153.6	114.0	100.7	153.6	114.7	100.0	126.7	107.1	99.7	126.7	106.9
097,099	WALL FINISHES, PAINTS & COATINGS	95.3	119.2	109.8	92.9	119.2	108.8	92.8	119.2	108.8	95.3	119.2	109.8	92.8	99.9	96.6	94.6	99.9	97.3
09	FINISHES	96.1	127.0	112.1	101.2	127.0	114.6	101.2	127.0	114.6	101.4	127.2	114.8	101.2	105.6	103.5	101.3	112.1	106.9
10 - 14	TOTAL DIV. 10000 - 14000	100.0	118.7	104.0	100.0	118.8	104.0	100.0	118.8	104.0	100.0	119.6	104.2	100.0	102.7	100.6	100.0	104.8	101.0
15	MECHANICAL	100.1	105.3	102.4	100.1	109.6	104.3	100.1	122.8	110.0	100.1	105.4	102.4	100.1	95.3	98.0	100.1	98.4	99.4
16	ELECTRICAL	96.5	102.2	99.3	99.8	123.3	111.3	100.3	123.3	111.5	99.3	102.2	100.7	100.3	88.7	94.7	100.3	88.0	94.3
01 - 16	WEIGHTED AVERAGE	98.3	116.9	106.7	101.1	121.0	110.2	100.0	123.5	110.7	101.1	117.0	108.4	100.2	101.1	100.6	100.7	105.1	102.7

COST INDEXES

MASSACHUSETTS / MICHIGAN

DIVISION		WORCESTER MAT.	INST.	TOTAL	ANN ARBOR MAT.	INST.	TOTAL	DEARBORN MAT.	INST.	TOTAL	DETROIT MAT.	INST.	TOTAL	FLINT MAT.	INST.	TOTAL	GRAND RAPIDS MAT.	INST.	TOTAL
01590	EQUIPMENT RENTAL	.0	101.6	101.6	.0	112.2	112.2	.0	112.2	112.2	.0	97.7	97.7	.0	112.2	112.2	.0	105.1	105.1
02	SITE CONSTRUCTION	85.8	104.8	99.8	78.1	95.2	90.7	77.9	95.5	90.9	91.5	97.2	95.8	69.0	94.2	87.6	82.2	87.8	86.3
03100	CONCRETE FORMS & ACCESSORIES	101.8	121.0	118.5	97.6	110.4	108.7	97.4	123.0	119.6	99.0	123.1	119.9	101.1	93.5	94.5	97.2	79.1	81.5
03200	CONCRETE REINFORCEMENT	108.2	142.5	125.6	97.1	124.3	110.9	97.1	125.1	111.3	96.5	125.1	110.9	97.1	124.0	110.7	92.0	86.7	89.3
03300	CAST-IN-PLACE CONCRETE	91.8	139.0	111.5	94.3	111.2	101.4	92.1	122.0	104.6	101.2	122.0	109.9	94.9	93.6	94.4	95.9	99.9	97.6
03	CONCRETE	100.1	130.4	115.2	93.0	113.9	103.4	91.9	123.3	107.6	96.2	121.9	109.1	93.5	100.2	96.9	97.3	87.1	92.2
04	MASONRY	97.8	138.0	123.0	98.5	106.2	103.3	98.3	122.1	113.2	97.3	122.1	112.9	98.5	92.6	94.8	94.9	56.0	70.5
05	METALS	100.0	119.1	106.2	100.6	126.0	108.9	100.7	127.9	109.6	105.2	105.7	105.4	100.7	123.3	108.1	92.9	80.6	88.9
06	WOOD & PLASTICS	100.9	120.4	111.1	101.4	109.6	105.7	101.4	123.0	112.7	102.3	123.0	113.1	104.9	94.1	99.3	98.2	79.7	88.5
07	THERMAL & MOISTURE PROTECTION	101.8	129.0	115.2	97.7	109.2	103.4	96.6	124.2	110.2	95.2	124.2	109.5	95.9	92.5	94.2	91.7	62.4	77.3
08	DOORS & WINDOWS	107.1	126.5	112.1	96.5	111.1	100.3	96.5	118.4	102.2	98.0	120.3	103.8	96.5	100.2	97.4	94.2	71.0	88.2
09200	PLASTER & GYPSUM BOARD	105.2	120.1	114.6	102.4	108.6	106.3	102.4	122.3	115.1	102.4	122.3	115.1	104.0	92.6	96.8	93.2	74.2	81.2
095,098	CEILINGS & ACOUSTICAL TREATMENT	101.1	120.1	112.5	95.6	108.6	103.4	95.6	122.3	111.6	96.6	122.3	112.0	95.6	92.6	93.8	91.8	74.2	81.2
09600	FLOORING	99.8	153.6	114.0	87.7	109.8	93.5	87.2	123.8	96.9	87.3	123.8	97.0	87.4	74.9	84.1	96.6	44.2	82.8
097,099	WALL FINISHES, PAINTS & COATINGS	92.8	119.2	108.8	86.5	103.3	96.7	86.5	114.0	103.1	88.3	114.0	103.9	86.5	88.1	87.5	96.5	42.7	63.8
09	FINISHES	101.2	127.0	114.6	93.9	109.5	102.0	93.8	122.5	108.7	94.8	122.5	109.2	93.6	89.0	91.2	94.3	68.5	80.9
10-14	TOTAL DIV. 10000 - 14000	100.0	108.3	101.8	100.0	111.3	102.4	100.0	114.9	103.2	100.0	114.9	103.2	100.0	96.0	99.2	100.0	101.4	100.3
15	MECHANICAL	100.1	108.8	103.9	99.9	103.3	101.4	99.9	116.9	107.3	99.9	119.7	108.6	99.9	93.9	97.1	99.8	57.0	81.1
16	ELECTRICAL	100.3	104.5	102.3	94.1	87.9	91.1	94.1	116.4	105.0	95.9	116.3	105.8	94.1	88.9	91.5	94.6	61.3	78.4
01-16	WEIGHTED AVERAGE	100.5	118.4	108.6	96.9	105.7	100.9	96.7	118.3	106.5	98.7	116.9	107.0	96.7	96.2	96.5	95.8	70.2	84.2

MICHIGAN / MINNESOTA

DIVISION		KALAMAZOO MAT.	INST.	TOTAL	LANSING MAT.	INST.	TOTAL	MUSKEGON MAT.	INST.	TOTAL	SAGINAW MAT.	INST.	TOTAL	DULUTH MAT.	INST.	TOTAL	MINNEAPOLIS MAT.	INST.	TOTAL
01590	EQUIPMENT RENTAL	.0	105.1	105.1	.0	112.2	112.2	.0	105.1	105.1	.0	112.2	112.2	.0	101.8	101.8	.0	106.1	106.1
02	SITE CONSTRUCTION	82.7	88.1	86.7	84.1	94.1	91.5	80.5	88.0	86.1	71.2	93.9	88.0	82.7	102.5	97.4	82.6	108.8	102.0
03100	CONCRETE FORMS & ACCESSORIES	96.7	88.5	89.6	101.1	91.5	92.8	97.4	85.8	87.3	97.6	92.2	92.9	103.5	123.7	121.0	104.2	137.5	133.0
03200	CONCRETE REINFORCEMENT	92.0	88.3	90.1	97.1	123.8	110.6	92.5	88.1	90.3	97.1	123.4	110.4	95.3	109.0	102.2	95.4	130.0	112.9
03300	CAST-IN-PLACE CONCRETE	97.6	100.4	98.8	94.3	94.7	94.5	95.3	98.3	96.5	93.1	93.5	93.3	109.6	104.2	107.3	107.9	126.9	115.8
03	CONCRETE	100.5	91.8	96.2	93.2	99.7	96.4	95.4	89.8	92.6	92.4	99.5	95.9	100.2	114.7	107.4	101.0	132.8	115.8
04	MASONRY	97.9	84.3	89.4	91.9	94.6	93.6	95.1	80.1	85.7	100.0	86.1	91.2	107.9	113.8	111.6	107.9	136.5	125.9
05	METALS	91.5	84.5	89.2	99.1	122.9	106.8	89.2	84.1	87.5	100.7	121.9	107.7	93.9	125.5	104.2	96.7	139.9	110.8
06	WOOD & PLASTICS	99.9	88.2	93.8	104.3	89.6	96.6	96.7	85.2	90.7	97.4	92.4	94.8	114.7	127.0	121.2	115.2	136.5	126.3
07	THERMAL & MOISTURE PROTECTION	91.3	84.4	87.9	96.7	92.5	94.6	90.4	78.7	84.7	96.5	88.9	92.8	101.5	116.6	108.9	101.3	136.3	118.5
08	DOORS & WINDOWS	90.9	83.1	88.9	96.5	97.7	96.8	90.1	82.0	88.0	96.0	98.2	96.6	96.8	124.6	104.0	99.9	143.7	111.2
09200	PLASTER & GYPSUM BOARD	93.2	83.0	86.8	105.1	87.9	94.2	84.1	79.9	81.4	102.4	90.9	95.1	102.3	128.3	118.8	102.5	137.8	124.9
095,098	CEILINGS & ACOUSTICAL TREATMENT	91.8	83.0	86.5	95.6	87.9	91.0	93.6	79.9	85.3	95.6	90.9	92.8	93.4	128.3	114.3	94.2	137.8	120.4
09600	FLOORING	96.6	70.1	89.6	95.8	83.0	92.4	95.8	81.5	92.0	87.7	59.5	80.2	105.8	121.7	110.0	103.2	126.4	109.3
097,099	WALL FINISHES, PAINTS & COATINGS	96.5	86.6	90.5	100.4	94.4	96.8	95.7	61.9	75.2	86.5	81.1	83.2	93.3	103.0	99.2	99.6	127.1	116.3
09	FINISHES	94.3	84.7	89.3	98.2	89.7	93.8	92.7	82.1	87.2	93.7	84.4	88.9	100.6	122.4	111.9	100.4	134.8	118.3
10-14	TOTAL DIV. 10000 - 14000	100.0	103.7	100.8	100.0	97.1	99.4	100.0	103.0	100.6	100.0	96.5	99.2	100.0	99.3	99.8	100.0	109.4	102.0
15	MECHANICAL	99.8	86.3	93.9	99.9	93.2	97.0	99.7	89.7	95.4	99.9	91.0	96.0	100.2	94.8	97.8	100.1	122.2	109.8
16	ELECTRICAL	94.2	71.8	83.2	92.3	88.2	90.3	94.7	67.3	81.4	93.5	83.2	88.5	105.8	95.1	100.6	107.7	112.4	110.0
01-16	WEIGHTED AVERAGE	95.7	84.9	90.8	96.7	96.1	96.4	94.3	83.6	89.5	96.5	93.2	95.0	99.7	108.9	103.8	100.7	126.8	112.6

MINNESOTA / MISSISSIPPI

DIVISION		ROCHESTER MAT.	INST.	TOTAL	SAINT PAUL MAT.	INST.	TOTAL	ST. CLOUD MAT.	INST.	TOTAL	BILOXI MAT.	INST.	TOTAL	GREENVILLE MAT.	INST.	TOTAL	JACKSON MAT.	INST.	TOTAL
01590	EQUIPMENT RENTAL	.0	101.8	101.8	.0	101.8	101.8	.0	101.5	101.5	.0	98.3	98.3	.0	98.3	98.3	.0	98.3	98.3
02	SITE CONSTRUCTION	81.8	101.7	96.5	84.7	103.3	98.5	78.3	105.2	98.2	103.3	86.2	90.6	107.4	86.0	91.5	99.9	86.0	89.6
03100	CONCRETE FORMS & ACCESSORIES	104.0	104.5	104.4	94.5	130.9	126.0	84.5	126.8	121.1	95.5	42.5	49.6	80.3	35.1	41.1	95.3	40.0	47.4
03200	CONCRETE REINFORCEMENT	95.3	129.2	112.4	91.9	129.9	111.1	94.3	129.2	111.9	95.0	62.4	78.6	103.0	44.9	73.6	95.0	48.7	71.6
03300	CAST-IN-PLACE CONCRETE	105.8	98.2	102.6	108.9	126.0	116.0	97.8	124.0	108.7	104.5	46.3	80.2	104.9	40.5	78.0	102.5	43.0	77.7
03	CONCRETE	98.4	108.1	103.2	102.5	129.5	116.0	90.0	126.7	108.4	97.0	49.5	73.3	100.2	40.9	70.6	96.0	44.7	70.4
04	MASONRY	107.2	109.5	108.6	117.7	136.5	129.5	109.2	122.2	117.4	89.3	38.1	57.2	131.8	39.5	73.9	91.3	39.5	58.8
05	METALS	93.7	137.1	107.9	93.2	139.2	108.3	90.0	136.2	105.1	92.1	83.2	89.2	90.4	74.9	85.4	92.1	76.8	87.1
06	WOOD & PLASTICS	115.2	103.7	109.2	104.4	128.1	116.8	91.7	124.5	108.8	96.6	42.4	68.3	79.0	34.4	55.7	96.7	40.8	67.5
07	THERMAL & MOISTURE PROTECTION	101.3	103.7	102.5	101.2	133.9	117.3	98.6	117.3	107.8	95.2	47.3	71.6	95.3	42.4	69.3	95.0	43.6	69.8
08	DOORS & WINDOWS	96.8	125.9	104.3	94.1	139.1	105.8	91.6	137.2	103.4	98.5	48.4	85.5	97.8	39.2	82.7	98.9	43.8	84.6
09200	PLASTER & GYPSUM BOARD	102.0	104.3	103.5	96.3	129.4	117.3	87.0	125.8	111.6	103.7	41.3	64.2	91.2	33.1	54.4	103.7	39.8	63.2
095,098	CEILINGS & ACOUSTICAL TREATMENT	92.5	104.3	99.6	90.7	129.4	113.9	72.1	125.8	104.3	102.7	41.3	65.9	96.5	33.1	58.5	102.7	39.8	65.0
09600	FLOORING	105.6	82.1	99.4	98.6	126.4	106.0	97.4	126.4	105.1	118.6	41.4	98.2	109.1	35.4	89.6	118.6	40.9	98.1
097,099	WALL FINISHES, PAINTS & COATINGS	95.7	103.5	100.4	99.6	120.5	112.3	103.4	127.1	117.8	107.4	36.5	64.4	107.4	35.9	64.0	107.4	35.9	64.0
09	FINISHES	100.3	100.6	100.5	97.5	129.1	113.9	90.9	127.0	109.7	106.9	41.1	72.7	101.1	34.9	66.7	106.9	39.9	72.0
10-14	TOTAL DIV. 10000 - 14000	100.0	97.4	99.4	100.0	107.9	101.7	100.0	102.8	100.6	100.0	55.2	90.5	100.0	53.4	90.1	100.0	54.3	90.3
15	MECHANICAL	100.1	99.2	99.7	100.1	113.2	105.8	99.7	120.5	108.8	99.8	55.9	80.6	99.8	35.6	71.7	99.8	35.3	71.6
16	ELECTRICAL	105.7	89.4	97.8	106.5	106.1	106.3	102.5	106.1	104.3	96.7	61.1	79.3	96.1	38.1	67.8	97.5	38.1	68.5
01-16	WEIGHTED AVERAGE	99.4	105.1	102.0	99.8	122.0	109.8	95.6	120.5	106.9	97.9	56.6	79.1	99.4	45.9	75.1	97.9	47.5	75.0

COST INDEXES

DIVISION		MISSISSIPPI			MISSOURI														
		MERIDIAN			CAPE GIRARDEAU			COLUMBIA			JOPLIN			KANSAS CITY			SPRINGFIELD		
		MAT.	INST.	TOTAL	MAT.	INST.	TOTAL	MAT.	INST.	TOTAL	MAT.	INST.	TOTAL	MAT.	INST.	TOTAL	MAT.	INST.	TOTAL
01590	EQUIPMENT RENTAL	.0	98.3	98.3	.0	108.3	108.3	.0	109.0	109.0	.0	106.9	106.9	.0	104.8	104.8	.0	102.7	102.7
02	SITE CONSTRUCTION	98.7	86.0	89.3	93.6	92.5	92.8	101.5	94.5	96.3	100.7	98.6	99.1	94.3	98.0	97.0	95.3	93.2	93.7
03100	CONCRETE FORMS & ACCESSORIES	78.8	34.1	40.1	84.4	74.9	76.2	85.0	72.4	74.0	101.2	65.4	70.2	100.2	105.2	104.5	100.2	65.6	70.2
03200	CONCRETE REINFORCEMENT	101.7	48.1	74.6	93.9	84.7	89.2	100.7	112.5	106.7	111.3	73.6	92.2	105.8	113.4	109.6	97.0	111.7	104.4
03300	CAST-IN-PLACE CONCRETE	99.0	43.0	75.6	84.6	84.5	84.6	88.5	68.9	80.3	101.2	73.1	89.5	94.1	107.5	99.7	101.5	63.6	85.6
03	CONCRETE	93.0	41.9	67.5	85.5	81.8	83.7	85.9	80.5	83.2	99.9	70.8	85.4	95.9	107.9	101.9	99.2	74.7	87.0
04	MASONRY	89.0	29.4	51.6	114.0	75.8	90.0	131.1	83.4	101.2	98.6	60.4	74.6	103.0	105.7	104.7	87.4	65.4	73.6
05	METALS	90.4	75.8	85.6	92.1	109.3	97.7	92.7	121.1	102.0	99.8	89.8	96.5	107.5	115.0	109.9	96.6	105.6	99.5
06	WOOD & PLASTICS	77.7	33.0	54.4	80.2	72.3	76.1	88.7	68.3	78.1	100.7	63.6	81.3	100.1	104.4	102.4	99.1	66.6	82.1
07	THERMAL & MOISTURE PROTECTION	94.7	40.2	67.9	95.5	79.8	87.7	92.4	81.4	87.0	98.6	63.8	81.5	97.8	107.1	102.4	96.5	70.1	83.5
08	DOORS & WINDOWS	97.8	37.9	82.3	93.9	75.0	89.1	90.9	91.7	91.1	91.9	71.0	86.5	98.0	107.0	100.3	96.3	78.2	91.6
09200	PLASTER & GYPSUM BOARD	91.2	31.7	53.5	96.1	71.4	80.4	98.1	67.3	78.6	103.9	62.3	77.5	100.3	104.3	102.8	102.5	65.5	79.0
095,098	CEILINGS & ACOUSTICAL TREATMENT	96.5	31.7	57.6	85.5	71.4	77.0	90.0	67.3	76.4	87.6	62.3	72.4	93.9	104.3	100.1	90.0	65.5	75.3
09600	FLOORING	108.0	28.7	87.0	92.4	78.2	88.7	101.9	90.5	98.9	111.7	52.6	96.1	91.7	90.5	91.4	104.9	52.6	91.1
097,099	WALL FINISHES, PAINTS & COATINGS	107.4	35.4	63.7	100.9	78.5	87.3	99.3	74.8	84.5	92.3	46.4	64.4	97.2	118.9	110.3	93.6	66.4	77.1
09	FINISHES	100.2	32.6	65.0	90.0	74.9	82.1	93.1	73.3	82.8	99.0	60.6	79.0	95.3	103.6	99.6	96.6	62.4	78.9
10 - 14	TOTAL DIV. 10000 - 14000	100.0	53.4	90.1	100.0	70.2	93.7	100.0	91.0	98.1	100.0	79.8	95.7	100.0	93.8	98.7	100.0	80.5	95.9
15	MECHANICAL	99.8	35.3	71.6	100.1	93.8	97.4	100.0	92.2	96.6	100.1	57.9	81.6	100.0	107.1	103.1	100.0	67.7	85.9
16	ELECTRICAL	95.7	60.2	78.4	102.4	108.8	105.5	99.4	82.4	91.1	97.6	63.6	81.1	109.0	108.1	108.6	102.6	68.7	86.1
01 - 16	WEIGHTED AVERAGE	96.0	47.8	74.1	96.1	89.3	93.0	96.9	88.1	92.9	98.7	69.0	85.2	101.1	106.3	103.4	98.2	74.5	87.4

DIVISION		MISSOURI						MONTANA											
		ST. JOSEPH			ST. LOUIS			BILLINGS			BUTTE			GREAT FALLS			HELENA		
		MAT.	INST.	TOTAL	MAT.	INST.	TOTAL	MAT.	INST.	TOTAL	MAT.	INST.	TOTAL	MAT.	INST.	TOTAL	MAT.	INST.	TOTAL
01590	EQUIPMENT RENTAL	.0	103.3	103.3	.0	109.5	109.5	.0	98.5	98.5	.0	98.2	98.2	.0	98.2	98.2	.0	98.2	98.2
02	SITE CONSTRUCTION	95.8	93.5	94.1	93.3	96.5	95.7	85.2	96.1	93.3	92.2	94.7	94.0	95.7	95.5	95.5	97.5	95.3	95.9
03100	CONCRETE FORMS & ACCESSORIES	100.1	86.0	87.9	99.6	105.7	104.9	98.1	65.3	69.7	85.4	62.9	65.9	99.4	66.5	70.9	99.5	65.9	70.4
03200	CONCRETE REINFORCEMENT	104.5	97.3	100.9	85.8	115.6	100.9	93.2	75.9	84.5	101.2	76.0	88.4	93.3	75.9	84.5	96.6	65.9	81.8
03300	CAST-IN-PLACE CONCRETE	94.0	100.6	96.8	84.6	108.2	94.5	119.0	68.6	97.9	122.5	68.7	100.0	129.4	58.6	99.9	111.4	67.2	89.3
03	CONCRETE	95.6	93.9	94.8	85.3	109.4	97.3	104.6	69.4	87.0	104.6	68.4	86.5	109.7	66.5	88.1	111.4	67.2	89.3
04	MASONRY	102.4	81.0	89.0	95.2	111.2	105.2	120.1	66.4	86.4	120.1	67.4	87.0	123.7	71.5	90.9	119.2	65.7	85.6
05	METALS	103.3	104.0	103.6	96.7	126.0	106.3	105.7	86.8	99.5	98.5	86.5	94.6	102.2	86.6	97.1	101.3	81.2	94.8
06	WOOD & PLASTICS	100.7	85.9	93.0	96.2	103.6	100.0	101.8	65.1	82.6	88.9	63.6	75.7	104.5	65.3	84.0	104.6	65.3	84.1
07	THERMAL & MOISTURE PROTECTION	98.4	90.3	94.4	95.4	106.2	100.7	99.9	68.5	84.5	99.2	67.4	83.6	100.1	68.5	84.5	100.0	68.2	84.4
08	DOORS & WINDOWS	96.8	92.9	95.8	92.9	112.3	97.9	99.1	64.1	90.0	95.8	63.2	87.4	99.3	64.2	90.2	98.8	61.6	89.2
09200	PLASTER & GYPSUM BOARD	105.3	85.2	92.6	105.2	103.5	104.1	104.4	64.5	79.1	94.8	63.0	74.6	104.4	64.7	79.2	103.7	64.7	79.0
095,098	CEILINGS & ACOUSTICAL TREATMENT	93.0	85.2	88.3	90.9	103.5	98.4	112.4	64.5	83.6	109.7	63.0	81.7	112.4	64.7	83.8	109.7	64.7	82.7
09600	FLOORING	94.0	84.2	91.4	99.7	98.5	99.4	107.8	51.2	92.9	99.1	40.8	83.7	107.8	55.3	94.0	107.8	53.6	93.5
097,099	WALL FINISHES, PAINTS & COATINGS	92.9	80.2	85.2	100.9	108.7	105.6	99.2	61.9	76.6	99.2	42.1	64.6	99.2	47.4	67.8	99.2	52.6	71.0
09	FINISHES	96.3	84.5	90.2	94.9	104.1	99.7	106.6	61.9	83.4	101.8	56.0	78.0	106.7	61.9	83.4	106.0	62.0	83.1
10 - 14	TOTAL DIV. 10000 - 14000	100.0	87.3	97.3	100.0	101.1	100.2	100.0	72.0	94.0	100.0	70.8	93.8	100.0	73.3	94.3	100.0	59.2	91.3
15	MECHANICAL	100.1	90.2	95.8	100.1	104.4	102.0	100.4	77.8	90.5	100.4	70.5	87.3	100.4	79.6	91.3	100.4	74.7	89.1
16	ELECTRICAL	105.9	83.4	94.9	105.3	114.1	109.6	95.0	74.7	85.1	101.6	69.6	86.0	95.0	75.1	85.3	95.0	74.1	84.8
01 - 16	WEIGHTED AVERAGE	100.1	89.6	95.3	96.7	108.6	102.1	102.0	74.4	89.5	101.0	71.1	87.4	102.6	74.9	90.0	102.4	72.3	88.7

DIVISION		MONTANA			NEBRASKA												NEVADA		
		MISSOULA			GRAND ISLAND			LINCOLN			NORTH PLATTE			OMAHA			CARSON CITY		
		MAT.	INST.	TOTAL	MAT.	INST.	TOTAL	MAT.	INST.	TOTAL	MAT.	INST.	TOTAL	MAT.	INST.	TOTAL	MAT.	INST.	TOTAL
01590	EQUIPMENT RENTAL	.0	98.2	98.2	.0	101.6	101.6	.0	101.6	101.6	.0	101.6	101.6	.0	90.8	90.8	.0	101.7	101.7
02	SITE CONSTRUCTION	75.6	94.6	89.6	98.1	91.0	92.9	89.5	91.1	90.6	98.8	89.8	92.2	78.9	89.7	86.9	63.0	104.1	93.4
03100	CONCRETE FORMS & ACCESSORIES	89.1	59.0	63.1	96.6	50.7	56.9	101.3	46.8	54.1	96.5	47.3	53.8	96.1	72.4	75.5	94.3	103.0	101.8
03200	CONCRETE REINFORCEMENT	102.9	78.4	90.5	105.6	74.5	89.9	97.0	75.4	86.1	107.1	74.2	90.5	101.6	76.0	88.7	113.7	118.6	116.2
03300	CAST-IN-PLACE CONCRETE	90.2	65.1	79.7	115.5	58.7	91.8	104.3	61.2	86.3	115.5	52.5	89.2	108.3	75.9	94.8	111.6	85.8	100.8
03	CONCRETE	83.6	65.8	74.7	108.5	59.3	83.9	100.6	58.5	79.6	108.7	55.6	82.2	102.0	74.4	88.2	109.3	99.8	104.5
04	MASONRY	145.0	60.5	91.9	105.5	49.9	70.6	95.8	66.7	77.5	91.1	40.5	59.4	100.9	75.4	84.9	135.0	80.6	100.9
05	METALS	95.3	86.6	92.5	92.8	83.3	89.7	97.9	84.8	93.6	93.0	83.2	89.8	97.3	76.1	90.3	93.4	104.5	97.0
06	WOOD & PLASTICS	93.4	59.8	75.9	95.5	45.5	69.4	100.1	39.8	68.6	95.4	45.5	69.3	94.7	73.3	83.5	91.2	106.4	99.1
07	THERMAL & MOISTURE PROTECTION	98.3	70.9	84.8	96.9	57.5	77.5	97.3	60.3	79.1	96.8	47.4	72.5	92.5	70.4	81.6	102.6	89.4	96.1
08	DOORS & WINDOWS	95.9	62.5	87.2	90.0	50.1	79.7	95.5	49.7	83.6	89.3	53.2	80.0	98.8	67.0	90.6	94.4	112.7	99.1
09200	PLASTER & GYPSUM BOARD	96.9	59.1	72.9	100.0	43.8	64.4	104.4	37.9	62.3	100.5	43.8	64.5	106.9	73.0	85.4	81.3	106.4	97.2
095,098	CEILINGS & ACOUSTICAL TREATMENT	109.7	59.1	79.3	88.2	43.8	61.6	97.1	37.9	61.6	90.0	43.8	62.3	107.9	73.0	87.0	100.4	106.4	104.0
09600	FLOORING	101.3	63.7	91.3	94.6	36.3	79.2	96.4	43.7	82.5	94.5	37.8	79.5	122.7	48.4	103.1	101.2	66.7	92.1
097,099	WALL FINISHES, PAINTS & COATINGS	99.2	47.4	67.8	91.2	42.8	61.9	91.2	43.2	62.1	91.2	51.8	67.3	148.5	73.6	103.0	103.7	81.6	90.3
09	FINISHES	102.0	58.2	79.2	93.1	45.7	68.4	96.3	43.9	69.0	93.6	45.3	68.4	114.6	67.9	90.3	96.1	94.0	95.0
10 - 14	TOTAL DIV. 10000 - 14000	100.0	55.2	90.5	100.0	73.9	94.4	100.0	73.2	94.3	100.0	55.2	90.5	100.0	75.9	94.9	100.0	114.9	103.2
15	MECHANICAL	100.4	70.2	87.2	100.0	77.5	90.2	100.0	77.6	90.2	100.0	73.4	88.4	99.8	78.4	90.4	100.1	92.9	96.9
16	ELECTRICAL	99.7	72.3	86.3	91.2	66.6	79.2	96.6	66.6	81.9	93.0	44.0	69.1	90.7	82.1	86.5	88.6	88.0	88.3
01 - 16	WEIGHTED AVERAGE	98.7	70.3	85.8	97.2	66.2	83.1	98.0	67.7	84.2	96.8	59.9	80.0	99.1	76.7	89.0	98.4	95.6	97.2

City Cost Indexes

NEVADA / NEW HAMPSHIRE / NEW JERSEY

DIVISION		LAS VEGAS MAT.	INST.	TOTAL	RENO MAT.	INST.	TOTAL	MANCHESTER MAT.	INST.	TOTAL	NASHUA MAT.	INST.	TOTAL	PORTSMOUTH MAT.	INST.	TOTAL	CAMDEN MAT.	INST.	TOTAL
01590	EQUIPMENT RENTAL	.0	101.7	101.7	.0	101.7	101.7	.0	101.6	101.6	.0	101.6	101.6	.0	101.6	101.6	.0	99.4	99.4
02	SITE CONSTRUCTION	62.5	106.3	94.9	62.7	104.1	93.3	86.7	100.2	96.7	88.0	100.2	97.0	82.2	99.2	94.8	91.3	105.0	101.4
03100	CONCRETE FORMS & ACCESSORIES	93.2	107.9	105.9	94.3	102.9	101.8	100.9	63.9	68.9	101.4	63.9	68.9	88.8	59.4	63.3	101.4	123.2	120.3
03200	CONCRETE REINFORCEMENT	108.1	118.7	113.4	108.1	118.6	113.4	108.2	91.9	100.0	108.2	91.9	100.0	86.6	91.8	89.2	108.2	104.9	106.6
03300	CAST-IN-PLACE CONCRETE	105.3	108.6	106.7	113.7	85.8	102.1	101.0	97.4	99.5	92.8	97.4	94.7	88.0	90.9	89.2	80.7	128.7	100.7
03	CONCRETE	105.3	109.8	107.6	109.4	99.7	104.6	106.8	81.0	93.9	102.9	81.0	91.9	93.8	76.6	85.2	97.1	120.6	108.8
04	MASONRY	125.8	89.8	103.2	134.3	80.6	100.6	97.9	92.8	94.7	98.1	92.8	94.8	93.5	82.2	86.4	90.3	125.0	112.1
05	METALS	94.0	105.9	97.9	93.8	104.5	97.3	100.0	87.1	95.8	99.9	87.1	95.7	96.3	85.0	92.6	99.8	97.9	99.2
06	WOOD & PLASTICS	90.9	106.4	99.0	91.2	106.4	99.1	100.4	55.9	77.2	100.4	55.9	77.2	86.2	55.9	70.4	100.4	122.9	112.1
07	THERMAL & MOISTURE PROTECTION	105.1	98.7	102.0	102.6	89.4	96.1	101.6	89.0	95.4	102.1	89.0	95.6	101.5	92.1	96.9	101.4	122.1	111.6
08	DOORS & WINDOWS	94.7	112.7	99.4	94.7	111.1	99.0	107.1	65.4	96.3	107.1	65.4	96.3	108.0	59.4	95.4	107.1	113.8	108.8
09200	PLASTER & GYPSUM BOARD	81.9	106.4	97.5	83.2	106.4	97.9	105.2	53.8	72.6	105.2	53.8	72.6	94.0	53.8	68.5	105.2	122.6	116.2
095,098	CEILINGS & ACOUSTICAL TREATMENT	105.8	106.4	106.2	107.6	106.4	106.9	101.1	53.8	72.7	101.1	53.8	72.7	89.5	53.8	68.1	101.1	122.6	114.0
09600	FLOORING	101.2	79.5	95.4	101.2	66.7	92.1	100.0	103.7	101.0	99.8	103.7	100.8	94.4	103.7	96.8	99.8	132.6	108.5
097,099	WALL FINISHES, PAINTS & COATINGS	103.7	105.1	104.6	103.7	81.6	90.3	92.8	90.6	91.4	92.8	90.6	91.4	92.8	40.3	61.0	92.8	119.7	109.1
09	FINISHES	97.5	101.9	99.8	98.2	94.0	96.0	101.3	72.5	86.4	101.6	72.5	86.5	95.1	64.2	79.0	101.8	125.0	113.9
10-14	TOTAL DIV. 10000-14000	100.0	110.2	102.2	100.0	114.9	103.2	100.0	80.0	95.7	100.0	80.0	95.7	100.0	75.8	94.9	100.0	117.8	103.8
15	MECHANICAL	100.1	111.1	104.9	100.1	92.9	96.9	100.1	86.1	94.0	100.1	86.1	94.0	100.1	80.4	91.5	100.1	109.6	104.3
16	ELECTRICAL	90.5	105.5	97.8	88.6	88.0	88.3	100.4	74.5	87.8	100.2	74.5	87.7	98.5	74.5	86.8	100.6	117.6	108.9
01-16	WEIGHTED AVERAGE	98.1	105.5	101.5	98.7	95.5	97.3	101.2	82.8	92.9	100.9	82.8	92.7	98.0	78.4	89.1	100.0	115.2	106.9

NEW JERSEY / NEW MEXICO

DIVISION		ELIZABETH MAT.	INST.	TOTAL	JERSEY CITY MAT.	INST.	TOTAL	NEWARK MAT.	INST.	TOTAL	PATERSON MAT.	INST.	TOTAL	TRENTON MAT.	INST.	TOTAL	ALBUQUERQUE MAT.	INST.	TOTAL
01590	EQUIPMENT RENTAL	.0	101.6	101.6	.0	99.4	99.4	.0	101.6	101.6	.0	101.6	101.6	.0	98.9	98.9	.0	116.3	116.3
02	SITE CONSTRUCTION	105.3	105.3	105.3	91.3	104.9	101.3	109.8	105.3	106.5	103.7	104.8	104.5	93.0	105.4	102.1	80.2	111.5	103.3
03100	CONCRETE FORMS & ACCESSORIES	111.5	122.7	121.2	101.4	123.7	120.7	99.8	122.7	119.7	100.4	122.6	119.6	100.3	123.4	120.3	96.1	69.0	72.6
03200	CONCRETE REINFORCEMENT	83.4	118.6	101.2	108.2	118.6	113.5	108.2	118.6	113.5	108.2	118.6	113.5	108.2	114.7	111.5	108.8	67.2	87.8
03300	CAST-IN-PLACE CONCRETE	88.9	128.5	105.4	80.7	130.0	101.3	94.2	128.6	108.5	102.8	128.5	113.6	95.5	119.9	105.7	104.5	76.4	92.8
03	CONCRETE	99.4	123.1	111.2	97.1	123.8	110.5	103.4	123.1	113.2	107.6	123.0	115.3	104.1	119.4	111.8	105.2	72.2	88.7
04	MASONRY	112.4	123.0	119.0	90.3	125.5	112.4	100.1	123.0	114.5	95.8	123.0	112.9	89.9	116.0	106.3	122.3	66.3	87.2
05	METALS	96.0	108.6	100.1	99.9	105.6	101.7	99.8	108.6	102.7	94.7	108.5	99.2	94.6	101.2	96.8	100.2	88.0	96.2
06	WOOD & PLASTICS	115.3	122.9	119.3	100.4	122.9	112.1	102.2	122.9	113.0	102.2	122.9	113.0	100.4	122.8	112.1	95.7	70.1	82.4
07	THERMAL & MOISTURE PROTECTION	102.4	126.0	114.0	101.4	127.1	114.1	101.7	126.0	113.7	102.2	119.7	110.8	100.0	116.0	107.9	101.6	72.6	87.3
08	DOORS & WINDOWS	108.0	118.2	110.7	107.1	118.2	109.9	113.1	118.2	114.4	113.1	118.2	114.4	107.1	116.4	109.5	94.7	72.5	89.0
09200	PLASTER & GYPSUM BOARD	108.9	122.6	117.6	105.2	122.6	116.2	105.2	122.6	116.2	105.2	122.6	116.2	105.2	122.6	116.2	85.4	68.8	74.8
095,098	CEILINGS & ACOUSTICAL TREATMENT	89.5	122.6	109.3	101.1	122.6	114.0	101.1	122.6	114.0	101.1	122.6	114.0	101.1	122.6	114.0	105.8	68.8	83.6
09600	FLOORING	104.5	132.6	111.9	99.8	132.6	108.5	100.0	132.6	108.6	99.8	132.6	108.5	100.0	117.0	104.5	101.2	65.0	91.6
097,099	WALL FINISHES, PAINTS & COATINGS	92.7	119.7	109.1	92.8	119.7	109.1	92.7	119.7	109.1	92.7	119.7	109.1	92.8	119.7	109.1	103.7	60.7	77.6
09	FINISHES	101.6	124.1	113.3	101.8	124.8	113.8	102.4	124.1	113.7	102.0	124.1	113.5	101.9	121.8	112.3	98.5	67.2	82.2
10-14	TOTAL DIV. 10000-14000	100.0	124.0	105.1	100.0	125.1	105.3	100.0	124.1	105.1	100.0	124.0	105.1	100.0	117.7	103.8	100.0	76.4	95.0
15	MECHANICAL	100.1	121.4	109.4	100.1	126.8	111.8	100.1	122.5	109.9	100.1	125.4	111.2	100.1	126.1	111.5	100.1	72.9	88.2
16	ELECTRICAL	98.5	128.9	113.3	102.2	132.5	117.0	102.1	132.5	116.9	102.2	128.9	115.2	100.8	129.8	114.9	81.8	76.3	79.1
01-16	WEIGHTED AVERAGE	101.1	120.7	110.0	100.2	122.5	110.3	102.5	121.4	111.1	101.8	121.2	110.6	99.9	119.1	108.6	98.3	76.8	88.6

NEW MEXICO / NEW YORK

DIVISION		FARMINGTON MAT.	INST.	TOTAL	LAS CRUCES MAT.	INST.	TOTAL	ROSWELL MAT.	INST.	TOTAL	SANTA FE MAT.	INST.	TOTAL	ALBANY MAT.	INST.	TOTAL	BINGHAMTON MAT.	INST.	TOTAL
01590	EQUIPMENT RENTAL	.0	116.3	116.3	.0	86.7	86.7	.0	116.3	116.3	.0	116.3	116.3	.0	115.7	115.7	.0	115.8	115.8
02	SITE CONSTRUCTION	85.7	111.5	104.8	91.2	86.7	87.9	90.3	111.4	105.9	79.9	111.5	103.3	71.8	106.5	97.5	93.8	89.8	90.9
03100	CONCRETE FORMS & ACCESSORIES	96.1	69.0	72.6	93.5	67.2	70.7	96.1	68.9	72.5	96.1	69.0	72.6	98.7	92.1	93.0	103.9	76.5	80.2
03200	CONCRETE REINFORCEMENT	118.7	67.2	92.7	112.8	58.8	85.5	117.9	59.3	88.3	116.6	67.2	91.6	103.1	89.8	96.4	102.0	87.8	94.8
03300	CAST-IN-PLACE CONCRETE	105.1	76.4	93.1	91.4	66.7	81.1	96.8	76.4	88.3	98.8	76.4	89.4	88.7	100.5	93.6	105.9	86.7	97.9
03	CONCRETE	109.0	72.2	90.6	87.3	66.2	76.7	109.3	70.7	90.0	103.8	72.2	88.0	101.3	95.5	98.4	101.1	84.2	92.6
04	MASONRY	127.3	66.3	89.1	110.9	60.7	79.4	127.4	66.3	89.1	119.4	66.3	86.1	89.3	97.0	94.1	106.0	79.0	89.1
05	METALS	97.9	88.0	94.7	97.8	75.2	90.4	97.9	84.1	93.4	97.9	88.0	94.6	96.9	104.9	99.5	93.4	114.1	100.2
06	WOOD & PLASTICS	95.7	70.1	82.4	87.4	68.7	77.6	95.7	70.1	82.4	95.7	70.1	82.4	98.3	91.3	94.6	107.3	75.0	90.4
07	THERMAL & MOISTURE PROTECTION	102.5	72.6	87.8	88.7	65.5	77.3	102.8	72.6	87.9	101.9	72.6	87.5	90.8	92.8	91.8	101.5	80.3	91.1
08	DOORS & WINDOWS	97.6	72.5	91.1	87.1	69.2	82.5	93.6	70.0	87.5	93.7	72.5	88.2	95.7	85.0	92.9	90.7	73.4	86.2
09200	PLASTER & GYPSUM BOARD	80.4	68.8	73.0	82.7	68.8	73.9	80.4	68.8	73.0	80.4	68.8	73.0	111.0	90.8	98.2	116.3	73.7	89.3
095,098	CEILINGS & ACOUSTICAL TREATMENT	96.9	68.8	80.0	96.8	68.8	80.0	96.9	68.8	80.0	96.9	68.8	80.0	99.8	90.8	94.4	99.8	73.7	84.2
09600	FLOORING	101.2	65.0	91.6	133.4	65.2	115.4	101.2	65.0	91.6	101.2	65.0	91.6	86.8	94.1	88.7	97.6	80.2	93.0
097,099	WALL FINISHES, PAINTS & COATINGS	103.7	60.7	77.6	97.1	60.7	75.0	103.7	60.7	77.6	103.7	60.7	77.6	84.9	78.0	80.7	90.1	79.1	83.4
09	FINISHES	95.9	67.2	81.0	111.0	66.2	87.7	96.3	67.2	81.2	95.7	67.2	80.9	97.5	91.0	94.1	98.8	77.0	87.5
10-14	TOTAL DIV. 10000-14000	100.0	76.4	95.0	100.0	72.3	94.1	100.0	76.4	95.0	100.0	76.4	95.0	100.0	87.3	97.3	100.0	89.3	97.7
15	MECHANICAL	100.1	72.9	88.2	100.3	72.4	88.1	100.1	72.8	88.2	100.1	72.9	88.2	100.4	88.9	95.3	100.6	82.4	92.6
16	ELECTRICAL	80.2	76.3	78.3	82.5	60.4	71.7	80.9	76.3	78.7	81.8	76.3	79.1	103.1	89.4	96.4	101.1	80.2	90.9
01-16	WEIGHTED AVERAGE	98.6	76.8	88.7	95.8	69.1	83.7	98.5	76.1	88.4	97.3	76.8	88.0	97.9	93.9	96.1	98.6	84.5	92.2

NEW YORK

DIVISION		BUFFALO			HICKSVILLE			NEW YORK			RIVERHEAD			ROCHESTER			SCHENECTADY		
		MAT.	INST.	TOTAL	MAT.	INST.	TOTAL	MAT.	INST.	TOTAL	MAT.	INST.	TOTAL	MAT.	INST.	TOTAL	MAT.	INST.	TOTAL
01590	EQUIPMENT RENTAL	.0	93.1	93.1	.0	117.2	117.2	.0	117.6	117.6	.0	117.2	117.2	.0	116.5	116.5	.0	115.7	115.7
02	SITE CONSTRUCTION	96.1	93.4	94.1	112.1	130.3	125.6	134.2	128.5	129.9	112.3	130.4	125.7	73.9	107.2	98.6	71.5	106.3	97.3
03100	CONCRETE FORMS & ACCESSORIES	100.3	113.7	111.9	90.5	149.8	141.9	113.8	180.5	171.6	94.6	149.9	142.5	100.2	94.2	95.0	103.5	91.0	92.7
03200	CONCRETE REINFORCEMENT	104.0	105.9	105.0	103.3	184.3	144.2	111.2	187.4	149.7	105.3	184.3	145.2	105.0	86.5	95.7	101.8	89.8	95.7
03300	CAST-IN-PLACE CONCRETE	118.7	119.3	118.9	99.1	151.6	121.0	121.6	160.8	137.9	97.2	151.7	120.0	112.9	100.5	107.8	97.3	99.0	98.0
03	CONCRETE	108.2	113.3	110.7	104.6	155.5	130.0	120.2	172.8	146.5	103.7	155.5	129.6	113.6	96.0	104.8	105.5	94.5	100.0
04	MASONRY	104.9	119.8	114.2	109.5	152.4	136.4	105.3	161.1	140.3	116.0	152.4	138.9	103.0	98.0	99.8	90.8	94.1	92.9
05	METALS	100.4	94.8	98.5	103.5	137.5	114.6	108.1	140.9	118.8	103.6	137.9	114.8	96.8	107.3	100.2	97.0	104.9	99.6
06	WOOD & PLASTICS	104.3	112.9	108.8	88.5	151.2	121.2	111.4	187.9	151.4	92.4	151.2	123.1	101.1	93.8	97.3	103.9	91.3	97.3
07	THERMAL & MOISTURE PROTECTION	98.9	110.7	104.7	104.0	145.1	124.2	105.5	160.3	132.4	104.0	145.1	124.2	95.6	96.9	96.2	91.0	91.6	91.3
08	DOORS & WINDOWS	92.7	103.3	95.5	88.3	153.6	105.2	97.7	176.6	118.1	88.4	153.6	105.2	96.0	86.2	93.5	95.7	85.0	92.9
09200	PLASTER & GYPSUM BOARD	102.4	113.0	109.1	102.1	152.7	134.2	115.8	189.9	162.8	103.0	152.7	134.5	104.6	93.6	97.6	111.0	90.8	98.2
095,098	CEILINGS & ACOUSTICAL TREATMENT	94.6	113.0	105.6	79.3	152.7	123.3	109.4	189.9	157.7	79.3	152.7	123.3	99.1	93.6	95.8	99.8	90.8	94.4
09600	FLOORING	97.7	121.0	103.9	97.4	90.1	95.5	97.4	171.3	117.0	98.7	90.1	96.5	86.8	85.8	86.6	86.8	94.1	88.7
097,099	WALL FINISHES, PAINTS & COATINGS	94.8	114.8	106.9	112.6	146.7	133.3	97.2	150.2	129.3	112.6	146.7	133.3	92.5	98.4	96.1	84.9	78.0	80.7
09	FINISHES	95.7	115.9	106.2	104.0	137.5	121.4	108.9	176.8	144.2	104.4	137.5	121.6	96.8	93.1	94.9	97.3	90.3	93.6
10-14	TOTAL DIV. 10000 - 14000	100.0	108.0	101.7	100.0	120.4	104.3	100.0	151.7	111.0	100.0	120.4	104.3	100.0	91.5	98.2	100.0	86.2	97.1
15	MECHANICAL	99.8	97.4	98.8	99.8	144.9	119.6	100.3	163.2	127.8	99.8	144.9	119.6	99.8	88.7	95.0	100.4	87.4	94.7
16	ELECTRICAL	99.7	97.6	98.7	101.6	152.1	126.3	107.8	178.5	142.3	102.2	152.1	126.6	103.4	89.1	96.4	100.5	89.4	95.1
01-16	WEIGHTED AVERAGE	99.9	104.7	102.1	101.2	145.0	121.1	106.6	163.5	132.4	101.6	145.1	121.4	100.1	94.9	97.7	98.2	93.0	95.8

NEW YORK / NORTH CAROLINA

DIVISION		SYRACUSE			UTICA			WATERTOWN			WHITE PLAINS			YONKERS			ASHEVILLE		
		MAT.	INST.	TOTAL	MAT.	INST.	TOTAL	MAT.	INST.	TOTAL	MAT.	INST.	TOTAL	MAT.	INST.	TOTAL	MAT.	INST.	TOTAL
01590	EQUIPMENT RENTAL	.0	115.7	115.7	.0	115.7	115.7	.0	115.7	115.7	.0	117.3	117.3	.0	117.3	117.3	.0	93.3	93.3
02	SITE CONSTRUCTION	92.6	106.6	103.0	69.6	105.6	96.2	77.5	107.6	99.8	123.8	125.2	124.9	133.5	125.1	127.3	101.6	72.3	79.9
03100	CONCRETE FORMS & ACCESSORIES	102.7	87.4	89.5	104.0	83.3	86.1	85.8	90.6	90.0	113.4	136.1	133.0	113.6	136.3	133.3	94.3	43.0	49.8
03200	CONCRETE REINFORCEMENT	103.1	88.4	95.7	103.1	86.9	94.9	103.7	88.4	96.0	104.0	183.5	144.2	108.1	183.5	146.2	95.1	39.5	67.0
03300	CAST-IN-PLACE CONCRETE	98.3	93.7	96.4	89.9	91.2	90.4	104.7	85.7	96.5	107.1	134.2	118.4	119.9	134.3	125.9	96.3	47.3	75.8
03	CONCRETE	104.4	90.8	97.6	102.3	87.8	95.1	115.2	89.5	102.4	108.1	143.1	125.6	119.0	143.3	131.1	99.5	45.8	72.7
04	MASONRY	96.9	92.4	94.1	89.4	89.1	89.2	90.6	94.0	92.7	100.0	134.1	121.4	104.8	134.1	123.2	79.1	37.1	52.7
05	METALS	96.8	104.2	99.3	94.9	103.2	97.6	94.9	103.9	97.9	94.8	134.3	107.7	104.0	134.6	114.0	93.0	74.5	87.0
06	WOOD & PLASTICS	103.9	86.8	95.0	103.9	82.8	92.9	84.1	91.8	88.2	112.3	135.7	124.5	112.2	135.7	124.5	93.1	44.6	67.8
07	THERMAL & MOISTURE PROTECTION	99.1	93.2	96.2	91.0	92.0	91.5	91.2	93.6	92.4	106.2	137.5	121.6	106.6	137.5	121.8	96.8	42.2	70.0
08	DOORS & WINDOWS	93.6	80.7	90.3	95.7	77.8	91.0	95.7	82.0	92.1	92.2	145.1	105.9	95.8	147.6	109.2	92.2	41.3	79.0
09200	PLASTER & GYPSUM BOARD	111.0	86.2	95.3	111.0	82.0	92.6	102.6	91.4	95.5	108.4	136.3	126.1	115.3	136.3	128.6	100.7	42.9	64.0
095,098	CEILINGS & ACOUSTICAL TREATMENT	99.8	86.2	91.6	99.8	82.0	89.1	99.8	91.4	94.7	79.9	136.3	113.7	107.6	136.3	124.9	91.9	42.9	64.4
09600	FLOORING	88.2	89.4	88.6	86.8	89.5	87.5	79.2	89.5	81.9	93.6	158.9	110.9	93.2	158.9	110.6	101.3	41.4	85.5
097,099	WALL FINISHES, PAINTS & COATINGS	90.3	86.8	88.2	84.9	84.6	84.8	84.9	73.4	78.0	94.8	146.7	126.3	94.8	146.7	126.3	108.5	39.1	66.4
09	FINISHES	98.6	87.6	92.9	97.3	84.3	90.5	94.3	88.3	91.2	98.7	141.6	121.0	107.0	141.6	125.0	96.4	42.3	68.3
10-14	TOTAL DIV. 10000 - 14000	100.0	97.7	99.5	100.0	84.1	96.6	100.0	83.7	96.5	100.0	117.6	103.7	100.0	141.6	108.9	100.0	62.5	92.0
15	MECHANICAL	100.4	84.3	93.3	100.4	80.8	91.8	100.4	77.0	90.1	100.6	127.0	112.1	100.6	127.0	112.1	100.0	43.1	75.1
16	ELECTRICAL	100.5	82.5	91.7	97.1	81.7	89.6	100.5	79.0	90.0	97.3	143.1	119.6	104.3	143.1	123.2	97.9	40.4	69.9
01-16	WEIGHTED AVERAGE	99.0	90.5	95.2	97.0	87.8	92.8	98.7	88.4	94.0	99.9	135.4	116.0	105.1	136.2	119.2	96.3	48.3	74.5

NORTH CAROLINA

DIVISION		CHARLOTTE			DURHAM			FAYETTEVILLE			GREENSBORO			RALEIGH			WILMINGTON		
		MAT.	INST.	TOTAL	MAT.	INST.	TOTAL	MAT.	INST.	TOTAL	MAT.	INST.	TOTAL	MAT.	INST.	TOTAL	MAT.	INST.	TOTAL
01590	EQUIPMENT RENTAL	.0	93.3	93.3	.0	99.7	99.7	.0	99.7	99.7	.0	99.7	99.7	.0	99.7	99.7	.0	93.3	93.3
02	SITE CONSTRUCTION	101.8	72.3	80.0	101.6	82.2	87.2	99.7	82.2	86.7	101.5	82.2	87.2	102.6	82.2	87.5	102.5	72.3	80.2
03100	CONCRETE FORMS & ACCESSORIES	101.7	43.0	50.9	96.2	43.3	50.4	92.0	43.3	49.8	96.3	43.3	50.4	99.0	43.3	50.7	95.8	43.3	50.3
03200	CONCRETE REINFORCEMENT	95.5	39.5	67.2	95.5	54.7	74.9	94.5	54.7	74.4	95.5	54.7	74.9	95.5	54.7	74.9	95.8	54.7	75.0
03300	CAST-IN-PLACE CONCRETE	98.7	47.4	77.3	96.5	47.5	76.0	93.2	47.5	74.1	95.7	47.5	75.6	102.2	47.5	79.3	95.9	47.5	75.6
03	CONCRETE	100.5	45.8	73.2	99.1	48.8	74.0	96.7	48.8	72.8	98.7	48.8	73.8	102.0	48.8	75.4	99.4	48.8	74.2
04	MASONRY	84.0	37.1	54.6	80.8	37.1	53.4	84.4	37.1	54.7	80.6	37.1	53.3	85.5	37.1	55.1	69.7	37.1	49.2
05	METALS	95.5	74.5	88.7	98.1	81.1	92.6	98.1	81.1	92.5	98.7	81.1	92.9	95.9	81.1	91.1	92.5	81.1	88.8
06	WOOD & PLASTICS	102.0	44.6	72.0	95.4	44.6	68.9	90.4	44.6	66.5	95.4	44.6	68.9	98.6	44.6	70.4	94.9	44.6	68.6
07	THERMAL & MOISTURE PROTECTION	96.9	43.2	70.5	97.5	42.8	70.6	97.2	42.8	70.4	97.5	42.8	70.6	97.3	42.8	70.6	96.9	42.8	70.3
08	DOORS & WINDOWS	96.3	41.3	82.1	96.3	47.1	83.6	92.3	47.1	80.6	96.3	47.1	80.6	93.0	47.1	81.1	92.3	47.1	80.6
09200	PLASTER & GYPSUM BOARD	107.2	42.9	66.4	107.2	42.9	66.4	101.0	42.9	64.1	107.2	42.9	66.4	107.2	42.9	66.4	101.8	42.9	64.4
095,098	CEILINGS & ACOUSTICAL TREATMENT	98.1	42.9	65.0	98.1	42.9	65.0	92.8	42.9	62.8	98.1	42.9	65.0	98.1	42.9	65.0	92.8	42.9	62.8
09600	FLOORING	105.1	41.4	88.3	105.3	41.4	88.4	101.4	41.4	85.6	105.3	41.4	88.4	105.3	41.4	88.4	102.3	41.4	86.2
097,099	WALL FINISHES, PAINTS & COATINGS	108.5	39.1	66.4	108.5	39.1	66.4	108.5	39.1	66.4	108.5	39.1	66.4	108.5	39.1	66.4	108.5	39.1	66.4
09	FINISHES	100.0	42.3	70.0	100.1	42.3	70.0	96.7	42.3	68.4	100.1	42.3	70.0	100.1	42.3	70.1	97.0	42.3	68.6
10-14	TOTAL DIV. 10000 - 14000	100.0	62.5	92.0	100.0	65.2	92.6	100.0	65.2	92.6	100.0	62.5	92.0	100.0	65.2	92.6	100.0	65.2	92.6
15	MECHANICAL	100.0	43.1	75.1	100.0	43.3	75.2	100.0	43.3	75.2	100.0	43.3	75.2	100.0	43.3	75.2	100.0	43.3	75.2
16	ELECTRICAL	99.7	36.3	68.8	97.4	38.6	68.8	94.8	38.6	67.4	98.0	38.6	69.1	98.0	38.6	69.1	98.5	38.6	69.3
01-16	WEIGHTED AVERAGE	98.2	47.8	75.3	97.9	50.3	76.3	96.7	50.3	75.6	98.0	50.3	76.3	97.9	50.3	76.3	96.0	49.4	74.8

COST INDEXES

City Cost Indexes

DIVISION		NORTH CAROLINA WINSTON-SALEM			NORTH DAKOTA BISMARCK			NORTH DAKOTA FARGO			NORTH DAKOTA GRAND FORKS			NORTH DAKOTA MINOT			OHIO AKRON		
		MAT.	INST.	TOTAL	MAT.	INST.	TOTAL	MAT.	INST.	TOTAL	MAT.	INST.	TOTAL	MAT.	INST.	TOTAL	MAT.	INST.	TOTAL
01590	EQUIPMENT RENTAL	.0	99.7	99.7	.0	98.2	98.2	.0	98.2	98.2	.0	98.2	98.2	.0	98.2	98.2	.0	97.4	97.4
02	SITE CONSTRUCTION	101.9	82.2	87.3	88.4	96.0	94.0	87.8	96.0	93.9	96.3	94.1	94.7	93.9	96.0	95.5	100.8	106.8	105.2
03100	CONCRETE FORMS & ACCESSORIES	97.6	43.1	50.4	96.4	51.0	57.0	97.4	51.4	57.5	93.1	44.8	51.3	88.1	57.8	61.9	98.1	99.0	98.9
03200	CONCRETE REINFORCEMENT	95.5	39.5	67.2	101.2	75.5	88.2	93.3	74.6	83.8	99.6	81.6	90.5	103.1	75.6	89.2	94.1	98.0	96.1
03300	CAST-IN-PLACE CONCRETE	98.7	47.4	77.3	104.9	56.3	84.6	109.9	58.6	88.5	107.4	54.4	85.2	107.4	54.9	85.5	99.1	106.2	102.0
03	CONCRETE	100.2	45.8	73.1	99.1	58.5	78.9	109.5	59.3	84.4	105.1	56.1	80.6	103.7	61.1	82.4	99.4	100.2	99.8
04	MASONRY	80.8	37.1	53.4	104.0	66.9	80.7	106.3	50.0	71.0	105.6	66.2	80.9	106.7	68.2	82.5	90.0	102.4	97.8
05	METALS	95.9	74.6	88.9	92.0	81.2	88.5	94.4	79.8	89.6	92.0	81.0	88.4	92.3	81.3	88.7	88.2	82.6	86.3
06	WOOD & PLASTICS	95.4	44.6	68.9	88.9	46.2	66.6	88.9	46.8	66.9	85.3	42.4	62.9	80.3	55.4	67.3	92.2	97.8	95.2
07	THERMAL & MOISTURE PROTECTION	97.5	42.8	70.6	100.5	56.9	79.1	101.2	54.2	78.1	101.2	57.1	79.5	100.9	58.1	79.9	104.0	100.4	102.2
08	DOORS & WINDOWS	96.3	43.1	82.5	99.5	50.5	86.8	99.4	50.9	86.9	99.5	47.1	85.9	99.6	55.6	88.2	104.2	98.9	102.8
09200	PLASTER & GYPSUM BOARD	107.2	42.9	66.4	115.0	45.0	70.6	115.0	45.7	71.1	113.0	41.2	67.5	111.4	54.5	75.4	95.2	97.3	96.6
095,098	CEILINGS & ACOUSTICAL TREATMENT	98.1	42.9	65.0	134.3	45.0	80.7	134.3	45.7	81.2	134.3	41.2	78.5	134.3	54.5	86.5	91.9	97.3	95.2
09600	FLOORING	105.3	41.4	88.4	108.4	78.4	100.4	108.2	46.9	91.9	106.6	46.9	90.8	103.9	81.0	97.8	103.2	91.7	100.2
097,099	WALL FINISHES, PAINTS & COATINGS	108.5	39.1	66.4	99.2	38.9	62.6	99.2	73.6	83.7	99.2	29.7	57.1	99.2	32.6	58.8	107.1	122.7	116.5
09	FINISHES	100.1	42.3	70.0	113.9	53.8	82.6	113.7	51.7	81.4	113.6	42.6	76.7	112.3	59.0	84.6	99.5	100.2	99.8
10 - 14	TOTAL DIV. 10000 - 14000	100.0	62.5	92.0	100.0	74.2	94.5	100.0	74.2	94.5	100.0	48.3	89.0	100.0	75.4	94.8	100.0	92.6	98.4
15	MECHANICAL	100.0	43.1	75.1	100.6	64.2	84.7	100.6	69.7	87.0	100.6	46.9	77.1	100.6	61.8	83.6	99.8	101.4	100.5
16	ELECTRICAL	98.0	38.6	69.1	94.4	70.1	82.5	94.2	69.8	82.3	96.8	71.4	84.4	99.7	71.4	85.9	97.9	90.5	94.3
01 - 16	WEIGHTED AVERAGE	97.8	49.1	75.7	99.2	66.9	84.5	100.8	65.9	85.0	100.4	60.7	82.4	100.5	68.2	85.8	97.7	98.0	97.8

DIVISION		OHIO CANTON			OHIO CINCINNATI			OHIO CLEVELAND			OHIO COLUMBUS			OHIO DAYTON			OHIO LORAIN		
		MAT.	INST.	TOTAL	MAT.	INST.	TOTAL	MAT.	INST.	TOTAL	MAT.	INST.	TOTAL	MAT.	INST.	TOTAL	MAT.	INST.	TOTAL
01590	EQUIPMENT RENTAL	.0	97.4	97.4	.0	102.5	102.5	.0	97.7	97.7	.0	96.4	96.4	.0	96.8	96.8	.0	97.4	97.4
02	SITE CONSTRUCTION	100.9	106.9	105.4	74.9	108.7	99.9	100.8	106.8	105.3	86.5	101.8	97.8	73.7	108.2	99.2	100.2	106.0	104.5
03100	CONCRETE FORMS & ACCESSORIES	98.1	86.1	87.7	95.9	90.9	91.6	98.3	105.6	104.6	98.6	86.7	88.3	95.8	80.2	82.3	98.2	106.2	105.1
03200	CONCRETE REINFORCEMENT	94.1	84.1	89.0	98.7	86.7	92.6	94.6	98.3	96.5	76.1	86.5	81.4	98.7	82.1	90.3	94.1	98.1	96.1
03300	CAST-IN-PLACE CONCRETE	100.1	103.4	101.5	84.8	91.5	87.6	97.2	114.1	104.3	89.6	93.7	91.3	77.3	89.2	82.3	94.3	108.9	100.4
03	CONCRETE	99.8	91.0	95.4	92.9	90.4	91.6	98.6	106.1	102.3	92.2	88.7	90.5	89.3	83.4	86.4	97.1	104.4	100.8
04	MASONRY	90.7	93.3	92.4	84.9	95.9	91.8	94.2	110.6	104.5	95.4	93.8	94.4	84.3	85.3	84.9	87.0	100.5	95.5
05	METALS	88.2	76.0	84.2	86.8	89.6	87.7	89.6	85.8	88.3	96.5	82.0	91.8	86.1	77.3	83.3	88.7	83.5	87.0
06	WOOD & PLASTICS	92.6	84.1	88.2	96.4	89.5	92.8	91.5	102.4	97.2	104.3	84.8	94.1	97.5	78.2	87.4	92.2	108.3	100.6
07	THERMAL & MOISTURE PROTECTION	104.7	95.9	100.3	88.9	97.8	93.3	102.9	114.0	108.3	99.4	95.6	97.5	94.5	87.6	91.1	104.6	108.1	106.3
08	DOORS & WINDOWS	98.6	77.9	93.3	95.9	87.6	93.8	95.0	101.3	96.7	99.5	84.1	95.5	96.2	78.1	91.5	98.6	104.6	100.2
09200	PLASTER & GYPSUM BOARD	96.0	83.2	87.9	101.2	89.5	93.8	94.3	102.0	99.2	99.3	84.2	89.8	101.2	77.9	86.5	95.2	108.1	103.4
095,098	CEILINGS & ACOUSTICAL TREATMENT	91.9	83.2	86.7	95.5	89.5	91.9	90.1	102.0	97.2	96.2	84.2	89.0	96.5	77.9	85.3	91.9	108.1	101.6
09600	FLOORING	103.4	83.5	98.1	107.2	101.8	105.8	103.0	112.2	105.4	97.6	91.0	95.9	110.0	84.0	103.1	103.4	109.5	105.0
097,099	WALL FINISHES, PAINTS & COATINGS	107.1	85.8	94.2	104.8	96.1	99.5	107.1	122.4	116.4	99.3	99.6	99.5	104.8	92.2	97.1	107.1	122.4	116.4
09	FINISHES	99.7	84.6	91.8	98.1	93.4	95.6	98.9	108.3	103.8	98.9	88.4	93.5	99.1	81.6	90.0	99.5	109.2	104.6
10 - 14	TOTAL DIV. 10000 - 14000	100.0	80.1	95.8	100.0	94.7	98.9	100.0	108.1	101.7	100.0	94.5	98.8	100.0	91.6	98.2	100.0	105.5	101.2
15	MECHANICAL	99.8	85.4	93.5	99.8	91.6	96.2	99.8	108.1	103.5	99.9	92.6	96.7	100.7	84.4	93.6	99.8	89.4	95.3
16	ELECTRICAL	97.1	90.1	93.7	99.5	83.0	91.4	98.0	109.7	103.7	99.7	87.2	93.6	97.4	80.7	89.3	97.2	90.8	94.1
01 - 16	WEIGHTED AVERAGE	97.2	88.5	93.2	94.7	92.3	93.6	97.1	106.0	101.1	97.8	90.5	94.5	94.4	84.9	90.1	96.8	98.1	97.4

DIVISION		OHIO SPRINGFIELD			OHIO TOLEDO			OHIO YOUNGSTOWN			OKLAHOMA ENID			OKLAHOMA LAWTON			OKLAHOMA MUSKOGEE		
		MAT.	INST.	TOTAL	MAT.	INST.	TOTAL	MAT.	INST.	TOTAL	MAT.	INST.	TOTAL	MAT.	INST.	TOTAL	MAT.	INST.	TOTAL
01590	EQUIPMENT RENTAL	.0	96.8	96.8	.0	99.4	99.4	.0	97.4	97.4	.0	77.1	77.1	.0	78.2	78.2	.0	86.8	86.8
02	SITE CONSTRUCTION	74.2	107.3	98.7	85.8	102.9	98.4	100.7	107.2	105.5	108.8	89.1	94.3	104.2	90.8	94.3	94.8	85.2	87.7
03100	CONCRETE FORMS & ACCESSORIES	95.8	85.1	86.6	98.6	98.5	98.5	98.1	92.4	93.2	93.7	40.4	47.5	97.5	55.4	61.0	99.0	36.6	44.9
03200	CONCRETE REINFORCEMENT	98.7	82.1	90.3	76.1	95.6	85.9	94.1	98.0	96.0	96.7	82.7	89.6	96.9	82.7	89.7	96.7	39.3	67.7
03300	CAST-IN-PLACE CONCRETE	81.0	89.8	84.7	89.6	108.4	97.5	98.2	105.1	101.1	92.6	51.6	75.5	89.6	51.7	73.7	83.3	39.7	65.1
03	CONCRETE	91.1	85.8	88.4	92.2	100.9	96.5	98.9	97.0	97.9	90.6	52.4	71.5	87.3	59.1	73.2	83.1	39.3	61.3
04	MASONRY	84.6	86.4	85.7	105.3	104.7	104.9	90.0	98.1	95.1	103.5	60.2	76.3	97.5	60.2	74.1	113.4	43.2	69.4
05	METALS	86.1	77.2	83.2	96.3	89.5	94.1	88.2	83.0	86.5	89.2	67.3	82.0	92.6	67.4	84.4	89.1	59.1	79.3
06	WOOD & PLASTICS	98.7	84.4	91.2	104.3	97.2	100.6	92.2	90.2	91.2	97.4	37.5	66.1	100.4	57.6	78.1	102.4	37.1	68.3
07	THERMAL & MOISTURE PROTECTION	94.4	88.7	91.6	101.0	109.3	105.1	104.8	100.9	102.9	99.8	60.3	80.4	99.5	62.4	81.2	99.3	45.5	72.9
08	DOORS & WINDOWS	94.2	81.4	90.9	95.7	93.6	96.5	94.8	96.3	98.0	95.6	51.5	84.2	97.2	62.5	88.2	95.6	36.6	80.4
09200	PLASTER & GYPSUM BOARD	101.2	84.2	90.5	99.3	97.0	97.8	95.2	89.5	91.6	82.7	36.6	53.5	85.4	57.3	67.6	86.6	36.1	54.5
095,098	CEILINGS & ACOUSTICAL TREATMENT	96.5	84.2	89.1	96.2	97.0	96.7	91.9	89.5	90.4	81.4	36.6	54.6	91.2	57.3	70.9	91.2	36.1	58.1
09600	FLOORING	110.0	84.0	103.1	96.8	100.8	97.8	103.4	97.9	102.0	112.9	53.7	97.2	115.8	53.7	99.4	117.1	43.3	97.5
097,099	WALL FINISHES, PAINTS & COATINGS	104.8	92.2	97.1	99.4	110.4	106.1	107.1	103.6	105.0	96.4	57.1	72.6	96.4	57.1	72.6	96.4	36.3	59.9
09	FINISHES	99.1	85.5	92.1	98.6	100.1	99.4	99.6	94.0	96.7	94.6	42.8	67.6	97.8	54.6	75.3	97.7	37.7	66.5
10 - 14	TOTAL DIV. 10000 - 14000	100.0	92.9	98.5	100.0	93.4	98.6	100.0	91.9	98.3	100.0	63.9	92.3	100.0	66.5	92.9	100.0	63.3	92.2
15	MECHANICAL	100.7	85.0	93.8	99.9	99.9	99.9	99.8	90.7	95.8	100.1	63.7	84.2	100.1	63.7	84.2	100.1	30.3	69.5
16	ELECTRICAL	97.4	84.0	90.9	99.5	99.4	99.4	97.2	95.4	96.3	95.3	68.5	82.3	96.8	68.5	83.0	95.0	35.1	65.8
01 - 16	WEIGHTED AVERAGE	94.4	86.7	90.9	98.0	99.6	98.7	97.1	94.8	96.0	96.1	61.6	80.4	96.5	65.0	82.2	95.7	43.7	72.1

DIVISION		OKLAHOMA CITY MAT.	INST.	TOTAL	TULSA MAT.	INST.	TOTAL	EUGENE MAT.	INST.	TOTAL	MEDFORD MAT.	INST.	TOTAL	PORTLAND MAT.	INST.	TOTAL	SALEM MAT.	INST.	TOTAL
01590	EQUIPMENT RENTAL	.0	78.5	78.5	.0	86.8	86.8	.0	99.5	99.5	.0	99.5	99.5	.0	99.5	99.5	.0	99.5	99.5
02	SITE CONSTRUCTION	105.5	91.4	95.0	101.2	86.2	90.1	108.8	104.8	105.8	118.4	104.8	108.3	109.9	104.8	106.1	108.4	104.8	105.7
03100	CONCRETE FORMS & ACCESSORIES	98.6	47.7	54.5	98.7	48.2	54.9	103.5	105.8	105.5	98.3	105.6	104.6	104.9	106.3	106.1	104.8	106.0	105.9
03200	CONCRETE REINFORCEMENT	96.9	82.7	89.7	96.9	82.6	89.7	106.0	99.5	102.7	103.4	99.4	101.4	106.9	99.9	103.3	107.0	99.9	103.4
03300	CAST-IN-PLACE CONCRETE	96.9	55.0	79.4	90.8	50.4	73.9	106.0	107.9	106.8	109.5	107.8	108.7	113.8	108.0	111.4	109.9	107.9	109.1
03	CONCRETE	90.9	56.8	73.8	87.9	56.4	72.2	108.8	105.0	106.9	115.6	104.8	110.2	112.8	105.3	109.0	110.9	105.1	108.0
04	MASONRY	99.3	62.4	76.1	96.8	62.4	75.2	112.0	101.4	105.3	108.9	101.4	104.2	113.2	107.8	109.8	116.7	107.8	111.1
05	METALS	94.5	67.3	85.6	92.0	81.9	88.7	94.4	97.0	95.3	94.0	96.8	94.9	95.6	98.0	96.4	94.8	97.7	95.8
06	WOOD & PLASTICS	102.2	46.2	73.0	101.1	47.6	73.1	94.1	105.7	100.1	87.8	105.7	97.1	95.3	105.7	100.7	95.3	105.7	100.7
07	THERMAL & MOISTURE PROTECTION	98.5	62.1	80.6	98.8	59.9	79.7	110.3	92.6	101.6	111.1	89.6	100.5	110.1	102.6	106.4	110.1	98.3	104.3
08	DOORS & WINDOWS	97.2	56.2	86.6	97.2	56.5	86.7	98.8	106.3	100.7	101.5	106.3	102.7	96.3	106.3	98.9	98.1	106.3	100.2
09200	PLASTER & GYPSUM BOARD	85.4	45.5	60.1	85.4	46.8	60.9	94.3	105.7	101.5	92.5	105.7	100.9	92.3	105.7	100.8	92.3	105.7	100.8
095,098	CEILINGS & ACOUSTICAL TREATMENT	91.2	45.5	63.8	91.2	46.8	64.6	105.3	105.7	105.5	115.1	105.7	109.5	105.3	105.7	105.5	105.3	105.7	105.5
09600	FLOORING	115.8	53.7	99.4	115.6	55.9	99.8	115.8	103.1	112.4	113.3	103.1	110.6	115.8	103.1	112.4	115.8	103.1	112.4
097,099	WALL FINISHES, PAINTS & COATINGS	96.4	57.1	72.6	96.4	53.6	70.4	115.8	74.6	90.8	115.8	62.8	83.6	115.8	74.6	90.8	115.8	74.6	90.8
09	FINISHES	97.9	48.4	72.1	97.3	49.3	72.3	110.2	102.1	106.0	112.3	100.7	106.3	110.0	102.1	105.9	109.8	102.1	105.8
10-14	TOTAL DIV. 10000 - 14000	100.0	65.9	92.7	100.0	66.8	92.9	100.0	108.2	101.7	100.0	108.1	101.7	100.0	108.2	101.7	100.0	108.2	101.7
15	MECHANICAL	100.1	64.9	84.7	100.1	62.6	83.7	100.2	103.9	101.8	100.2	103.9	101.8	100.2	109.5	104.2	100.2	104.0	101.8
16	ELECTRICAL	96.3	68.5	82.7	96.8	45.8	71.9	97.3	97.7	97.5	100.6	87.7	94.3	97.8	105.8	101.7	97.4	97.7	97.6
01-16	WEIGHTED AVERAGE	97.3	64.0	82.1	96.3	61.2	80.4	101.7	102.1	101.9	103.4	100.3	102.0	102.2	105.4	103.7	102.2	103.0	102.5

DIVISION		ALLENTOWN MAT.	INST.	TOTAL	ALTOONA MAT.	INST.	TOTAL	ERIE MAT.	INST.	TOTAL	HARRISBURG MAT.	INST.	TOTAL	PHILADELPHIA MAT.	INST.	TOTAL	PITTSBURGH MAT.	INST.	TOTAL
01590	EQUIPMENT RENTAL	.0	115.7	115.7	.0	115.7	115.7	.0	115.7	115.7	.0	114.9	114.9	.0	94.7	94.7	.0	115.5	115.5
02	SITE CONSTRUCTION	91.4	106.4	102.5	95.3	106.6	103.7	92.0	107.1	103.2	82.5	104.9	99.1	101.1	94.8	96.4	98.5	109.1	106.3
03100	CONCRETE FORMS & ACCESSORIES	101.9	105.2	104.8	83.9	86.3	86.0	101.2	96.9	97.4	94.4	85.2	86.4	99.9	134.0	129.5	101.5	98.8	99.1
03200	CONCRETE REINFORCEMENT	103.1	102.0	102.5	100.0	97.1	98.5	102.0	94.3	98.1	103.1	96.9	100.0	104.0	135.7	120.0	99.3	106.2	102.8
03300	CAST-IN-PLACE CONCRETE	89.0	94.9	91.5	99.3	86.9	94.1	97.6	84.9	92.3	98.3	89.2	94.5	97.1	128.4	110.2	99.8	97.3	98.7
03	CONCRETE	98.8	102.1	100.5	94.0	90.0	92.0	93.1	93.5	93.3	100.9	90.3	95.8	97.0	131.9	119.4	95.9	100.9	98.4
04	MASONRY	94.0	90.7	91.9	96.6	80.7	86.6	86.4	92.3	90.1	94.6	84.3	88.1	97.0	129.3	117.2	89.6	97.4	94.5
05	METALS	97.3	120.2	104.8	91.0	117.0	99.5	91.2	116.7	99.5	99.3	117.4	105.2	104.5	127.7	112.1	95.6	123.0	104.6
06	WOOD & PLASTICS	103.2	108.5	105.9	80.4	88.2	84.4	99.6	98.1	98.8	96.8	84.4	90.4	97.9	133.8	116.6	100.1	98.0	99.0
07	THERMAL & MOISTURE PROTECTION	99.1	109.8	104.3	97.6	92.0	94.9	97.8	96.5	97.2	102.8	103.3	103.0	101.2	132.8	116.7	97.5	101.1	99.2
08	DOORS & WINDOWS	93.6	105.3	96.6	88.1	95.2	90.0	88.3	94.3	89.8	93.6	93.1	93.5	96.7	139.5	107.8	93.8	107.6	97.4
09200	PLASTER & GYPSUM BOARD	108.4	108.4	108.4	100.7	87.6	92.4	108.4	97.8	101.7	108.4	83.7	92.8	105.1	134.6	123.8	99.3	97.7	98.3
095,098	CEILINGS & ACOUSTICAL TREATMENT	90.0	108.4	101.1	94.5	87.6	90.3	90.0	97.8	94.7	90.0	83.7	86.2	101.9	134.6	121.5	89.8	97.7	94.5
09600	FLOORING	88.2	83.4	87.0	82.3	59.5	76.3	90.2	88.6	89.8	88.5	78.4	85.8	85.7	134.9	98.7	96.3	106.2	98.9
097,099	WALL FINISHES, PAINTS & COATINGS	90.3	93.3	92.1	85.4	111.7	101.4	96.9	88.6	91.8	90.3	87.0	88.3	92.5	141.7	122.3	96.3	117.6	109.2
09	FINISHES	95.8	100.0	98.0	94.2	83.3	88.5	97.2	94.5	95.8	94.8	83.4	88.9	101.2	135.1	118.9	96.0	101.6	98.9
10-14	TOTAL DIV. 10000 - 14000	100.0	96.8	99.3	100.0	99.7	99.9	100.0	104.0	100.9	100.0	98.0	99.6	100.0	126.5	105.6	100.0	104.8	101.0
15	MECHANICAL	100.4	99.8	100.1	99.8	84.7	93.2	99.8	92.9	96.8	100.4	90.3	95.9	100.1	132.3	114.2	100.1	100.6	100.3
16	ELECTRICAL	100.1	92.4	96.3	89.3	102.0	95.5	90.7	86.4	88.6	99.1	83.7	91.6	97.2	134.2	115.2	95.4	102.0	98.6
01-16	WEIGHTED AVERAGE	98.0	101.2	99.4	94.3	93.3	93.8	94.3	96.2	95.1	98.1	92.3	95.5	100.8	128.9	113.6	96.7	103.8	99.9

PENNSYLVANIA (continued) / PUERTO RICO / RHODE ISLAND / SOUTH CAROLINA

DIVISION		READING MAT.	INST.	TOTAL	SCRANTON MAT.	INST.	TOTAL	YORK MAT.	INST.	TOTAL	SAN JUAN MAT.	INST.	TOTAL	PROVIDENCE MAT.	INST.	TOTAL	CHARLESTON MAT.	INST.	TOTAL
01590	EQUIPMENT RENTAL	.0	118.2	118.2	.0	115.7	115.7	.0	114.9	114.9	.0	89.1	89.1	.0	103.4	103.4	.0	99.3	99.3
02	SITE CONSTRUCTION	98.7	110.7	107.6	91.9	106.3	102.6	81.6	104.8	98.8	115.6	92.1	98.2	80.9	104.0	98.0	95.2	81.2	84.9
03100	CONCRETE FORMS & ACCESSORIES	99.4	86.2	88.0	102.0	84.5	86.9	82.4	79.7	80.1	94.1	20.1	30.0	103.5	115.4	113.8	96.3	38.4	46.1
03200	CONCRETE REINFORCEMENT	102.2	103.1	102.7	103.1	101.9	102.5	102.0	96.8	99.4	194.4	12.5	102.5	108.2	117.3	112.8	95.5	65.6	80.4
03300	CAST-IN-PLACE CONCRETE	71.0	91.3	79.5	93.0	88.7	91.2	87.6	83.9	86.1	105.2	31.4	74.4	86.1	113.2	97.4	85.1	49.3	70.2
03	CONCRETE	94.3	92.7	93.5	100.7	90.8	95.8	98.9	86.0	92.5	113.9	23.3	68.7	102.3	114.6	108.4	93.7	49.1	71.4
04	MASONRY	98.2	87.2	91.3	94.3	92.8	93.4	93.9	81.7	86.2	219.3	17.3	92.6	107.7	126.5	119.5	88.6	36.1	55.7
05	METALS	100.9	120.7	107.4	99.4	119.5	106.0	96.8	117.0	103.4	115.2	31.3	87.8	100.0	110.3	103.3	93.3	81.2	89.4
06	WOOD & PLASTICS	98.1	84.4	90.9	103.2	81.9	92.1	87.4	78.0	82.5	91.1	20.4	54.2	101.2	113.9	107.8	95.4	37.7	65.3
07	THERMAL & MOISTURE PROTECTION	101.2	105.4	103.3	98.9	99.1	99.0	97.2	98.7	97.9	167.7	23.8	97.0	100.6	114.4	107.4	97.1	43.8	70.9
08	DOORS & WINDOWS	94.4	92.5	93.9	93.6	93.2	93.5	90.5	89.6	90.3	153.8	17.1	118.5	101.0	115.1	104.7	96.3	42.6	82.4
09200	PLASTER & GYPSUM BOARD	108.7	83.8	92.9	111.0	81.2	92.1	104.4	77.1	87.1	252.1	18.0	103.7	102.0	113.4	109.2	107.2	35.7	61.9
095,098	CEILINGS & ACOUSTICAL TREATMENT	90.2	83.8	86.4	99.8	81.2	88.6	89.9	77.1	83.0	343.0	18.0	148.1	100.3	113.4	108.2	98.1	35.7	60.7
09600	FLOORING	83.9	87.6	84.9	88.2	82.1	86.6	84.2	83.8	84.1	200.8	17.1	152.3	100.7	133.9	109.5	105.3	41.0	88.3
097,099	WALL FINISHES, PAINTS & COATINGS	90.7	98.8	95.6	90.3	95.3	93.3	90.3	76.4	81.8	201.8	15.8	89.0	95.3	113.4	106.3	108.5	39.5	66.7
09	FINISHES	98.3	86.8	92.3	98.5	84.4	91.2	93.5	79.0	86.0	251.5	19.3	130.7	100.9	119.0	110.3	100.3	38.4	68.1
10-14	TOTAL DIV. 10000 - 14000	100.0	92.1	98.3	100.0	93.0	98.5	100.0	96.6	99.3	100.0	21.0	83.2	100.0	102.8	100.6	100.0	63.8	92.3
15	MECHANICAL	100.3	107.3	103.4	100.4	87.4	94.7	100.4	89.5	95.6	102.4	15.9	64.5	100.1	112.5	105.5	100.0	44.6	75.8
16	ELECTRICAL	98.9	90.9	95.0	100.1	84.5	92.5	94.0	71.7	83.1	133.6	15.7	76.1	99.0	101.4	100.1	97.5	33.2	66.1
01-16	WEIGHTED AVERAGE	98.6	98.5	98.6	98.8	92.9	96.1	96.2	88.6	92.7	136.6	26.3	86.5	100.2	112.4	105.8	96.8	48.9	75.1

SOUTH CAROLINA / SOUTH DAKOTA

DIVISION		COLUMBIA MAT.	COLUMBIA INST.	COLUMBIA TOTAL	FLORENCE MAT.	FLORENCE INST.	FLORENCE TOTAL	GREENVILLE MAT.	GREENVILLE INST.	GREENVILLE TOTAL	SPARTANBURG MAT.	SPARTANBURG INST.	SPARTANBURG TOTAL	ABERDEEN MAT.	ABERDEEN INST.	ABERDEEN TOTAL	PIERRE MAT.	PIERRE INST.	PIERRE TOTAL
01590	EQUIPMENT RENTAL	.0	99.3	99.3	.0	99.3	99.3	.0	99.3	99.3	.0	99.3	99.3	.0	98.2	98.2	.0	98.2	98.2
02	SITE CONSTRUCTION	94.5	81.2	84.7	105.8	81.2	87.6	100.8	80.8	86.0	100.7	80.8	86.0	84.2	93.6	91.1	82.5	93.6	90.7
03100	CONCRETE FORMS & ACCESSORIES	100.9	40.7	48.7	82.0	40.6	46.1	96.1	40.3	47.8	99.8	40.3	48.3	97.0	40.1	47.7	95.3	41.6	48.8
03200	CONCRETE REINFORCEMENT	95.5	65.6	80.4	95.2	65.6	80.2	95.1	40.0	67.2	95.1	40.0	67.2	99.8	49.3	74.3	99.4	61.2	80.1
03300	CAST-IN-PLACE CONCRETE	80.3	49.0	67.2	72.5	49.2	62.8	72.5	49.1	62.8	72.5	49.1	62.8	102.3	51.2	81.0	99.4	46.9	77.5
03	CONCRETE	91.7	50.0	70.9	94.3	50.0	72.2	93.0	45.2	69.1	93.2	45.2	69.2	98.4	47.1	72.8	96.1	48.5	72.4
04	MASONRY	87.3	37.1	55.8	73.3	36.1	50.0	71.2	36.1	49.2	73.3	36.1	50.0	107.7	58.0	76.5	104.2	56.2	74.0
05	METALS	93.3	81.1	89.3	92.3	81.0	88.6	92.3	71.6	85.5	92.3	71.6	85.5	98.3	70.8	89.3	98.3	75.9	91.0
06	WOOD & PLASTICS	101.0	40.9	69.6	79.5	40.9	59.4	95.2	40.9	66.9	99.5	40.9	68.9	101.8	38.2	68.6	99.8	40.3	68.8
07	THERMAL & MOISTURE PROTECTION	97.0	43.7	70.8	97.4	44.2	71.2	97.4	44.2	71.2	97.4	44.2	71.3	99.0	52.2	76.0	99.2	49.9	75.0
08	DOORS & WINDOWS	96.3	44.4	82.8	92.3	44.4	79.9	92.2	38.4	78.3	92.2	38.4	78.3	95.0	41.1	81.1	98.3	45.7	84.7
09200	PLASTER & GYPSUM BOARD	107.2	39.1	64.1	95.8	39.1	59.8	102.4	39.1	62.3	104.4	39.1	63.0	106.3	36.8	62.3	104.4	39.0	63.0
095,098	CEILINGS & ACOUSTICAL TREATMENT	98.1	39.1	62.7	92.8	39.1	60.6	91.9	39.1	60.2	91.9	39.1	60.2	112.0	36.8	66.9	109.4	39.0	67.2
09600	FLOORING	105.1	41.0	88.2	95.4	41.0	81.0	102.7	41.8	86.6	104.4	41.8	87.9	108.7	58.7	95.5	107.8	41.8	90.4
097,099	WALL FINISHES, PAINTS & COATINGS	108.5	39.5	66.7	108.5	39.5	66.7	108.5	39.5	66.7	108.5	39.5	66.7	99.2	35.6	60.6	99.2	40.7	63.7
09	FINISHES	100.3	40.4	69.1	95.3	40.4	66.7	97.9	40.5	68.1	98.7	40.5	68.4	106.8	42.6	73.4	105.5	41.0	72.0
10-14	TOTAL DIV. 10000 - 14000	100.0	64.2	92.4	100.0	64.2	92.4	100.0	64.2	92.4	100.0	64.2	92.4	100.0	57.5	91.0	100.0	68.4	93.3
15	MECHANICAL	100.0	37.6	72.7	100.0	37.6	72.7	100.0	37.5	72.7	100.0	37.5	72.7	100.2	40.1	73.9	100.2	38.4	73.2
16	ELECTRICAL	98.0	35.0	67.3	95.7	20.0	58.8	97.6	31.8	65.5	97.6	31.8	65.5	97.3	60.8	79.5	93.7	51.9	73.3
01-16	WEIGHTED AVERAGE	96.6	48.3	74.7	95.0	46.1	72.8	95.2	46.0	72.9	95.5	46.0	73.0	99.4	54.6	79.0	98.7	53.7	78.2

SOUTH DAKOTA / TENNESSEE

DIVISION		RAPID CITY MAT.	RAPID CITY INST.	RAPID CITY TOTAL	SIOUX FALLS MAT.	SIOUX FALLS INST.	SIOUX FALLS TOTAL	CHATTANOOGA MAT.	CHATTANOOGA INST.	CHATTANOOGA TOTAL	JACKSON MAT.	JACKSON INST.	JACKSON TOTAL	JOHNSON CITY MAT.	JOHNSON CITY INST.	JOHNSON CITY TOTAL	KNOXVILLE MAT.	KNOXVILLE INST.	KNOXVILLE TOTAL
01590	EQUIPMENT RENTAL	.0	98.2	98.2	.0	99.4	99.4	.0	104.5	104.5	.0	104.9	104.9	.0	97.4	97.4	.0	97.4	97.4
02	SITE CONSTRUCTION	82.7	93.3	90.6	83.7	95.5	92.4	103.2	96.5	98.2	100.6	95.7	97.0	113.1	85.2	92.5	89.9	85.3	86.5
03100	CONCRETE FORMS & ACCESSORIES	104.4	37.0	46.0	95.4	42.6	49.6	98.7	47.1	54.0	88.7	38.4	45.1	83.5	47.9	52.7	97.8	47.9	54.6
03200	CONCRETE REINFORCEMENT	93.3	61.3	77.1	93.3	61.3	77.1	87.9	44.6	66.0	89.6	44.3	66.7	88.5	44.8	66.4	87.9	44.8	66.1
03300	CAST-IN-PLACE CONCRETE	98.7	44.3	76.0	102.8	49.8	80.7	100.8	49.4	79.3	100.5	42.7	76.4	81.0	54.1	69.8	94.6	49.6	75.8
03	CONCRETE	95.4	45.6	70.6	95.6	50.0	72.8	92.6	49.3	71.0	93.9	43.1	68.5	97.6	51.3	74.5	89.9	49.8	69.9
04	MASONRY	104.1	52.2	71.6	101.5	56.2	73.0	96.9	44.4	64.0	111.4	34.5	63.2	108.8	44.2	68.3	75.1	44.2	55.7
05	METALS	100.1	76.2	92.3	100.2	76.3	92.4	95.4	77.7	89.6	92.8	75.9	87.3	92.7	77.3	87.7	96.0	77.6	90.0
06	WOOD & PLASTICS	106.1	36.1	69.5	99.8	41.0	69.1	102.3	48.1	74.0	87.2	38.2	61.6	76.1	49.2	62.0	91.5	49.2	69.4
07	THERMAL & MOISTURE PROTECTION	99.6	48.1	74.3	99.1	49.5	74.7	100.3	51.7	76.4	99.3	40.6	70.5	96.1	49.3	73.1	93.7	49.5	72.0
08	DOORS & WINDOWS	99.3	43.4	84.8	99.3	46.0	85.5	101.4	48.2	87.6	102.1	46.0	86.2	97.2	43.9	84.8	93.7	49.3	82.2
09200	PLASTER & GYPSUM BOARD	106.3	34.6	60.9	106.3	39.7	64.1	84.2	47.2	60.8	89.3	37.1	56.2	90.1	48.4	63.6	98.0	48.4	66.6
095,098	CEILINGS & ACOUSTICAL TREATMENT	116.5	34.6	67.4	116.5	39.7	70.4	98.1	47.2	67.6	95.1	37.1	60.3	93.4	48.4	66.4	96.1	48.4	67.5
09600	FLOORING	107.8	73.6	98.8	107.8	72.1	98.4	104.0	51.6	90.1	92.8	25.0	74.8	98.0	53.1	86.1	103.6	53.1	90.2
097,099	WALL FINISHES, PAINTS & COATINGS	99.2	40.7	63.7	99.2	40.7	63.7	109.4	47.4	71.8	96.5	32.1	57.5	107.2	55.7	75.9	107.2	55.7	75.9
09	FINISHES	107.6	43.7	74.4	107.6	47.5	76.4	98.6	47.9	72.3	95.6	35.2	64.2	99.2	49.6	73.4	94.6	49.6	71.2
10-14	TOTAL DIV. 10000 - 14000	100.0	66.0	92.8	100.0	68.6	93.3	100.0	55.4	90.5	100.0	52.3	89.8	100.0	63.7	92.3	100.0	63.7	92.3
15	MECHANICAL	100.2	35.8	72.1	100.2	37.4	72.7	99.8	46.5	76.5	100.0	57.6	81.4	99.8	61.3	83.0	99.8	61.3	83.0
16	ELECTRICAL	94.2	51.9	73.6	93.6	73.7	83.9	103.3	54.1	79.3	100.7	52.3	77.1	91.6	52.4	72.4	96.8	61.2	79.4
01-16	WEIGHTED AVERAGE	99.3	52.5	78.0	99.1	57.8	80.3	98.8	55.8	79.3	98.5	53.3	78.0	97.6	58.2	79.7	94.9	59.2	78.7

TENNESSEE / TEXAS

DIVISION		MEMPHIS MAT.	MEMPHIS INST.	MEMPHIS TOTAL	NASHVILLE MAT.	NASHVILLE INST.	NASHVILLE TOTAL	ABILENE MAT.	ABILENE INST.	ABILENE TOTAL	AMARILLO MAT.	AMARILLO INST.	AMARILLO TOTAL	AUSTIN MAT.	AUSTIN INST.	AUSTIN TOTAL	BEAUMONT MAT.	BEAUMONT INST.	BEAUMONT TOTAL
01590	EQUIPMENT RENTAL	.0	102.2	102.2	.0	105.5	105.5	.0	86.8	86.8	.0	86.8	86.8	.0	85.9	85.9	.0	88.7	88.7
02	SITE CONSTRUCTION	95.7	91.8	92.8	95.0	99.2	98.1	101.5	85.4	89.6	101.9	86.3	90.3	89.8	85.3	86.5	96.5	87.4	89.7
03100	CONCRETE FORMS & ACCESSORIES	99.3	67.1	71.4	97.8	65.2	69.5	95.4	43.9	50.8	97.6	53.2	59.1	94.1	56.6	61.6	103.1	54.5	61.0
03200	CONCRETE REINFORCEMENT	102.2	70.5	86.2	87.1	67.0	76.9	94.1	53.6	73.6	94.1	56.1	74.9	98.7	54.0	76.1	92.5	45.7	68.9
03300	CAST-IN-PLACE CONCRETE	88.9	71.2	81.5	94.4	69.7	84.1	94.2	43.6	73.0	99.0	49.9	78.5	90.0	50.8	73.6	86.2	56.3	73.7
03	CONCRETE	89.6	70.7	80.2	90.0	68.6	79.3	88.9	46.7	67.8	91.3	53.4	72.4	81.8	54.7	68.3	85.6	54.3	70.0
04	MASONRY	84.4	78.0	80.4	85.7	65.9	73.3	100.4	54.4	71.5	102.4	52.9	71.3	94.7	56.4	70.7	104.9	59.1	76.1
05	METALS	97.0	96.1	96.7	96.7	91.9	95.1	95.7	67.7	86.5	95.7	68.9	86.9	94.0	66.7	85.1	97.1	65.7	86.8
06	WOOD & PLASTICS	97.4	67.7	81.9	98.4	66.2	81.6	97.1	43.4	69.0	98.4	55.7	76.1	94.6	59.8	76.4	110.9	55.1	81.7
07	THERMAL & MOISTURE PROTECTION	96.6	74.9	85.9	97.0	66.7	82.1	99.4	51.0	75.6	102.0	50.1	76.5	96.1	55.4	76.1	103.3	59.6	81.8
08	DOORS & WINDOWS	100.9	69.8	92.9	98.5	67.5	90.5	92.9	46.4	80.9	92.9	51.4	82.2	94.5	59.2	85.4	97.6	51.0	85.5
09200	PLASTER & GYPSUM BOARD	95.3	67.3	77.5	92.3	65.7	75.4	85.4	42.5	58.2	85.4	55.1	66.2	90.7	59.3	70.8	90.5	54.5	67.7
095,098	CEILINGS & ACOUSTICAL TREATMENT	106.3	67.3	82.9	96.5	65.7	78.0	91.2	42.5	62.0	91.2	55.1	69.6	88.0	59.3	70.8	93.0	54.5	69.9
09600	FLOORING	98.3	47.1	84.7	106.8	77.5	99.1	115.8	69.1	103.4	115.6	66.7	102.7	101.4	51.2	88.1	110.1	72.6	100.2
097,099	WALL FINISHES, PAINTS & COATINGS	99.0	60.6	75.7	112.7	70.1	86.9	95.2	56.6	71.8	95.2	39.2	61.3	92.2	44.0	63.0	91.4	52.9	68.0
09	FINISHES	97.3	62.1	79.0	104.0	68.0	85.3	97.3	49.6	72.5	97.3	54.3	74.9	92.1	54.2	72.4	91.8	57.5	74.0
10-14	TOTAL DIV. 10000 - 14000	100.0	79.1	95.6	100.0	70.8	93.8	100.0	73.9	94.5	100.0	66.0	92.8	100.0	66.2	92.8	100.0	78.2	95.4
15	MECHANICAL	99.8	76.6	89.7	99.8	73.6	88.3	100.1	46.8	76.8	100.1	52.9	79.5	99.9	56.9	81.1	100.0	64.4	84.4
16	ELECTRICAL	95.8	78.5	87.4	102.9	68.2	86.0	96.9	50.0	74.0	97.4	61.5	79.9	99.0	69.4	84.6	95.0	65.3	80.5
01-16	WEIGHTED AVERAGE	96.6	77.2	87.8	98.0	74.0	87.1	96.8	54.5	77.6	97.3	59.1	79.9	94.9	61.8	79.8	96.6	63.5	81.6

DIVISION		TEXAS																
		CORPUS CHRISTI			DALLAS			EL PASO			FORT WORTH			HOUSTON			LAREDO		
		MAT.	INST.	TOTAL	MAT.	INST.	TOTAL	MAT.	INST.	TOTAL	MAT.	INST.	TOTAL	MAT.	INST.	TOTAL	MAT.	INST.	TOTAL
01590	EQUIPMENT RENTAL	.0	94.7	94.7	.0	98.1	98.1	.0	86.8	86.8	.0	86.8	86.8	.0	98.9	98.9	.0	85.9	85.9
02	SITE CONSTRUCTION	124.9	80.6	92.1	128.3	86.3	97.2	102.0	85.0	89.4	101.4	85.7	89.8	123.8	85.7	95.7	90.3	85.8	86.5
03100	CONCRETE FORMS & ACCESSORIES	97.7	39.9	47.6	93.6	59.3	63.9	94.8	48.5	54.7	96.3	58.6	63.6	91.4	66.7	70.0	91.7	40.2	47.1
03200	CONCRETE REINFORCEMENT	97.8	52.0	74.7	96.5	58.4	77.2	94.1	50.7	72.2	94.1	58.3	76.0	94.2	64.8	79.3	98.7	52.3	75.3
03300	CAST-IN-PLACE CONCRETE	106.7	47.0	81.8	97.1	57.4	80.5	96.5	38.2	72.1	92.6	52.4	75.8	88.8	69.5	80.8	82.1	63.3	74.3
03	CONCRETE	91.6	46.6	69.1	88.2	60.0	74.2	89.9	46.2	68.1	88.2	57.1	72.7	84.9	68.9	76.9	81.6	51.4	66.6
04	MASONRY	86.0	49.4	63.1	102.7	61.3	76.8	99.4	49.4	68.0	95.7	61.3	74.1	100.6	65.3	78.4	95.0	55.4	70.1
05	METALS	93.5	79.9	89.1	93.0	84.3	90.2	95.5	63.7	85.1	92.4	70.0	85.1	101.5	90.9	98.1	94.9	65.6	85.3
06	WOOD & PLASTICS	107.2	39.3	71.8	96.8	60.2	77.7	97.1	52.1	73.6	103.8	60.0	80.9	96.2	66.4	80.6	90.5	39.2	63.7
07	THERMAL & MOISTURE PROTECTION	99.1	47.8	73.9	94.3	60.9	77.9	98.9	52.1	75.9	99.8	55.5	78.0	94.7	67.6	81.3	95.1	52.6	74.2
08	DOORS & WINDOWS	101.6	41.4	86.0	105.6	56.7	93.0	92.9	47.5	81.2	87.5	56.6	79.5	107.1	66.0	96.5	95.1	42.0	81.4
09200	PLASTER & GYPSUM BOARD	90.7	38.2	57.4	91.7	59.5	71.3	85.4	51.4	63.9	85.4	59.5	69.0	88.6	66.0	74.3	87.4	38.2	56.2
095,098	CEILINGS & ACOUSTICAL TREATMENT	88.0	38.2	58.1	94.6	59.5	73.6	91.2	51.4	67.4	91.2	59.5	72.2	94.6	66.0	77.4	88.0	38.2	58.1
09600	FLOORING	114.7	50.6	97.7	105.1	62.0	93.7	115.8	66.4	102.7	150.1	49.6	123.5	98.2	68.2	90.3	92.2	57.5	71.2
097,099	WALL FINISHES, PAINTS & COATINGS	105.7	43.1	67.7	104.7	56.1	75.2	95.2	37.7	60.3	96.4	55.9	71.9	99.8	64.0	78.1	90.8	43.0	65.9
09	FINISHES	98.0	41.4	68.6	99.9	59.3	78.8	97.3	51.2	73.3	107.8	56.6	81.2	96.8	66.5	81.0	90.8	43.0	65.9
10 - 14	TOTAL DIV. 10000 - 14000	100.0	74.7	94.6	100.0	78.8	95.5	100.0	64.3	92.4	100.0	78.3	95.4	100.0	83.2	96.4	100.0	64.4	92.4
15	MECHANICAL	99.9	47.8	77.1	99.8	67.0	85.4	100.1	35.7	71.9	100.1	61.0	83.0	100.0	73.2	88.3	99.8	43.1	75.0
16	ELECTRICAL	97.2	53.6	75.9	98.7	64.7	82.1	96.2	56.9	77.0	96.1	62.4	79.6	97.7	68.6	83.5	98.9	63.9	81.8
01 - 16	WEIGHTED AVERAGE	97.7	54.1	77.8	98.7	67.1	84.3	96.7	52.4	76.6	96.4	63.3	81.4	99.1	72.8	87.2	94.9	55.0	76.8

DIVISION		TEXAS															UTAH		
		LUBBOCK			ODESSA			SAN ANTONIO			WACO			WICHITA FALLS			LOGAN		
		MAT.	INST.	TOTAL	MAT.	INST.	TOTAL	MAT.	INST.	TOTAL	MAT.	INST.	TOTAL	MAT.	INST.	TOTAL	MAT.	INST.	TOTAL
01590	EQUIPMENT RENTAL	.0	96.9	96.9	.0	86.8	86.8	.0	88.8	88.8	.0	86.8	86.8	.0	86.8	86.8	.0	100.7	100.7
02	SITE CONSTRUCTION	131.4	83.5	96.0	101.7	85.8	89.9	90.0	89.8	89.9	100.4	85.7	89.5	101.1	85.4	89.5	92.0	100.8	98.5
03100	CONCRETE FORMS & ACCESSORIES	94.9	43.2	50.1	95.3	41.1	48.3	91.8	57.9	62.4	96.0	41.9	49.2	96.0	44.2	51.1	101.0	61.5	66.8
03200	CONCRETE REINFORCEMENT	95.4	53.5	74.2	94.1	53.4	73.6	105.4	54.3	79.6	94.1	53.8	73.8	94.1	53.8	73.8	108.5	80.7	94.5
03300	CAST-IN-PLACE CONCRETE	94.3	50.0	75.8	94.2	45.8	73.9	80.6	70.6	76.4	85.0	54.0	72.1	90.8	49.5	73.5	88.3	72.8	81.8
03	CONCRETE	88.0	49.5	68.8	88.9	46.1	67.5	82.0	62.3	72.2	84.5	49.4	67.0	87.3	48.8	68.1	108.4	69.3	88.9
04	MASONRY	99.8	51.2	69.3	100.4	51.5	69.7	94.9	65.3	76.3	97.2	59.9	73.8	97.6	58.0	72.7	121.3	62.3	84.3
05	METALS	99.2	82.5	93.8	95.1	66.9	85.8	95.6	69.6	87.1	95.6	67.5	86.4	95.5	68.3	86.6	99.5	77.4	92.3
06	WOOD & PLASTICS	97.4	43.5	69.2	97.1	41.2	67.9	90.5	56.5	72.8	102.8	38.3	69.1	102.8	43.3	71.8	88.1	59.7	73.3
07	THERMAL & MOISTURE PROTECTION	90.6	52.3	71.8	99.4	47.6	73.9	95.1	66.0	80.8	100.1	51.4	76.1	100.1	55.0	77.9	102.2	68.7	85.8
08	DOORS & WINDOWS	104.2	44.6	88.8	92.9	42.9	80.0	97.0	56.7	86.6	87.5	40.2	75.3	87.5	46.3	76.9	89.3	61.8	82.2
09200	PLASTER & GYPSUM BOARD	86.1	42.4	58.4	85.4	40.2	56.8	87.4	55.9	67.5	85.4	37.3	54.9	85.4	42.4	58.1	80.6	58.4	66.5
095,098	CEILINGS & ACOUSTICAL TREATMENT	93.9	42.4	63.0	91.2	40.2	60.6	88.0	55.9	68.8	91.2	37.3	58.8	91.2	42.4	61.9	97.8	58.4	74.1
09600	FLOORING	107.2	42.1	90.0	115.8	41.5	96.2	97.9	57.8	87.3	150.3	38.6	120.7	150.7	75.3	130.8	101.2	62.3	90.9
097,099	WALL FINISHES, PAINTS & COATINGS	106.8	35.2	63.4	95.2	35.2	58.8	92.2	57.5	71.2	96.4	36.3	59.9	98.2	57.5	73.5	103.7	49.6	70.9
09	FINISHES	99.3	41.3	69.1	97.3	39.9	67.4	90.8	57.1	73.3	107.8	39.4	72.2	108.0	50.9	78.3	97.0	59.6	77.5
10 - 14	TOTAL DIV. 10000 - 14000	100.0	73.6	94.4	100.0	63.5	92.2	100.0	70.7	93.8	100.0	75.5	94.8	100.0	64.4	92.4	100.0	75.8	94.9
15	MECHANICAL	99.7	50.1	78.0	100.1	39.1	73.4	99.8	69.5	86.6	100.1	56.6	81.1	100.1	51.0	78.6	100.1	71.3	87.5
16	ELECTRICAL	95.5	47.0	71.9	97.0	43.8	71.0	99.0	63.9	81.9	97.0	63.2	80.5	100.1	57.4	79.3	88.6	76.4	82.7
01 - 16	WEIGHTED AVERAGE	98.8	54.9	78.9	96.7	49.9	75.4	95.2	66.8	82.3	96.6	57.8	79.0	99.8	57.2	79.1	98.9	72.1	86.7

DIVISION		UTAH									VERMONT						VIRGINIA		
		OGDEN			PROVO			SALT LAKE CITY			BURLINGTON			RUTLAND			ALEXANDRIA		
		MAT.	INST.	TOTAL	MAT.	INST.	TOTAL	MAT.	INST.	TOTAL	MAT.	INST.	TOTAL	MAT.	INST.	TOTAL	MAT.	INST.	TOTAL
01590	EQUIPMENT RENTAL	.0	100.7	100.7	.0	99.4	99.4	.0	100.7	100.7	.0	101.6	101.6	.0	101.6	101.6	.0	101.2	101.2
02	SITE CONSTRUCTION	80.1	100.8	95.4	88.4	98.8	96.1	79.8	100.7	95.3	79.6	99.1	94.0	79.6	99.1	94.0	112.3	86.1	92.9
03100	CONCRETE FORMS & ACCESSORIES	101.0	61.5	66.8	102.5	61.5	67.0	99.5	61.5	66.6	91.4	51.9	57.2	101.8	52.0	58.6	91.7	75.0	77.2
03200	CONCRETE REINFORCEMENT	113.4	80.7	96.9	117.4	80.7	98.9	110.7	80.7	95.6	108.2	56.0	81.8	108.2	56.0	81.8	84.8	81.8	83.3
03300	CAST-IN-PLACE CONCRETE	89.6	72.8	82.6	88.4	72.8	81.9	97.2	72.8	87.0	97.2	64.9	83.7	92.6	64.9	81.1	99.6	82.7	92.6
03	CONCRETE	99.2	69.3	84.3	108.4	69.4	88.9	116.9	69.4	93.2	104.3	57.7	81.0	102.8	57.8	80.3	101.8	80.0	90.9
04	MASONRY	114.6	62.3	81.8	127.3	62.3	86.5	132.9	62.3	88.6	98.7	62.0	75.7	86.5	62.0	71.1	85.8	74.6	78.8
05	METALS	99.9	77.4	92.6	97.8	77.5	91.1	103.1	77.5	94.7	101.7	67.6	90.5	99.9	67.7	89.4	102.4	96.0	100.3
06	WOOD & PLASTICS	88.1	59.7	73.3	89.7	59.7	74.0	88.3	59.7	73.4	88.6	50.1	68.5	100.9	50.1	74.4	94.5	75.4	84.5
07	THERMAL & MOISTURE PROTECTION	100.7	68.7	85.0	104.7	68.7	87.0	103.8	68.7	86.6	101.4	56.9	79.5	101.2	58.1	80.0	96.0	78.9	87.6
08	DOORS & WINDOWS	89.3	61.8	82.2	93.7	61.8	85.4	89.3	61.8	82.2	107.1	46.9	91.5	107.1	46.9	91.5	96.3	77.5	91.4
09200	PLASTER & GYPSUM BOARD	80.6	58.4	66.5	81.0	58.4	66.7	80.6	58.4	66.5	103.5	47.8	68.2	103.5	47.8	68.2	107.2	74.6	86.5
095,098	CEILINGS & ACOUSTICAL TREATMENT	97.8	58.4	74.1	97.8	58.4	74.1	97.8	58.4	74.1	94.9	47.8	66.6	94.9	47.8	66.6	98.1	74.6	84.0
09600	FLOORING	101.2	62.3	90.9	102.0	62.3	91.5	100.8	62.3	90.6	99.8	72.0	92.4	99.8	72.0	92.4	105.3	91.2	101.6
097,099	WALL FINISHES, PAINTS & COATINGS	103.7	49.6	70.9	103.7	63.3	79.2	103.7	63.3	79.2	92.8	41.1	61.4	92.8	41.1	61.4	119.6	86.4	99.4
09	FINISHES	95.9	59.6	77.0	97.6	61.1	78.6	96.2	61.1	77.9	98.8	53.8	75.4	98.7	53.8	75.4	100.7	79.1	89.5
10 - 14	TOTAL DIV. 10000 - 14000	100.0	75.8	94.9	100.0	75.8	94.9	100.0	75.8	94.9	100.0	84.8	96.8	100.0	84.8	96.8	100.0	88.0	97.4
15	MECHANICAL	100.1	71.3	87.5	100.1	71.3	87.5	100.2	71.3	87.6	100.1	68.5	86.3	100.1	68.5	86.3	100.0	85.8	93.8
16	ELECTRICAL	88.7	76.4	82.7	88.8	76.4	82.8	92.5	76.4	84.6	100.9	67.4	84.6	100.2	67.4	84.2	98.5	94.4	96.5
01 - 16	WEIGHTED AVERAGE	97.2	72.1	85.8	99.5	72.1	87.1	101.2	72.3	88.1	100.8	66.0	85.0	99.8	66.1	84.5	99.6	84.6	92.8

City Cost Indexes

VIRGINIA

DIVISION		ARLINGTON MAT.	INST.	TOTAL	NEWPORT NEWS MAT.	INST.	TOTAL	NORFOLK MAT.	INST.	TOTAL	PORTSMOUTH MAT.	INST.	TOTAL	RICHMOND MAT.	INST.	TOTAL	ROANOKE MAT.	INST.	TOTAL
01590	EQUIPMENT RENTAL	.0	99.7	99.7	.0	104.9	104.9	.0	105.6	105.6	.0	104.8	104.8	.0	104.8	104.8	.0	99.7	99.7
02	SITE CONSTRUCTION	122.9	82.7	93.2	105.5	84.5	90.0	104.8	85.6	90.6	104.1	84.1	89.3	106.1	84.9	90.4	103.2	81.0	86.8
03100	CONCRETE FORMS & ACCESSORIES	90.5	71.9	74.4	96.4	58.0	63.2	100.7	58.3	64.0	84.5	58.2	61.7	97.3	57.3	62.7	96.2	36.9	44.8
03200	CONCRETE REINFORCEMENT	95.7	65.1	80.3	95.5	72.0	83.6	95.5	72.0	83.6	95.2	72.0	83.5	95.5	71.2	83.2	95.5	64.6	79.9
03300	CAST-IN-PLACE CONCRETE	96.9	74.3	87.5	97.6	56.2	80.3	100.5	56.3	82.0	96.6	56.0	79.7	103.8	56.0	83.8	109.6	47.7	83.8
03	CONCRETE	106.7	72.6	89.7	99.6	61.5	80.6	101.3	61.7	81.5	98.3	61.5	79.9	102.6	61.0	81.8	105.3	47.7	76.6
04	MASONRY	98.5	68.6	79.8	89.6	55.3	68.1	96.2	55.3	70.6	94.6	55.3	69.9	88.5	50.8	64.9	90.0	40.0	58.6
05	METALS	101.1	87.7	96.7	102.4	89.6	98.2	101.5	89.8	97.7	101.4	89.1	97.4	104.4	89.6	99.5	102.2	82.0	95.6
06	WOOD & PLASTICS	91.0	75.4	82.9	95.4	61.5	77.7	100.7	61.5	80.2	82.2	61.5	71.4	96.5	59.5	77.2	95.4	36.4	64.6
07	THERMAL & MOISTURE PROTECTION	98.1	72.7	85.6	97.1	52.3	75.1	96.8	52.4	75.0	97.0	52.4	75.1	96.6	54.3	75.8	97.0	45.9	71.9
08	DOORS & WINDOWS	94.2	72.6	88.6	96.3	58.8	86.6	96.3	61.5	87.3	96.3	61.5	87.3	96.3	56.6	86.0	96.3	43.6	82.6
09200	PLASTER & GYPSUM BOARD	103.4	74.6	85.1	107.2	59.4	76.9	107.2	59.4	76.9	98.8	59.4	73.8	107.2	57.4	75.6	107.2	34.4	61.1
095,098	CEILINGS & ACOUSTICAL TREATMENT	92.8	74.6	81.9	98.1	59.4	74.9	98.1	59.4	74.9	98.1	59.4	74.9	98.1	57.4	73.7	98.1	34.4	59.9
09600	FLOORING	103.5	70.5	94.8	105.3	41.1	88.4	105.1	41.1	88.2	96.5	41.1	81.9	105.1	65.4	94.6	105.3	36.9	87.3
097,099	WALL FINISHES, PAINTS & COATINGS	119.6	86.4	99.4	108.5	43.5	69.1	108.5	64.1	81.6	108.5	64.1	81.6	108.5	56.9	77.2	108.5	34.0	63.3
09	FINISHES	99.3	73.5	85.9	100.1	53.5	75.9	100.1	55.8	77.1	96.3	55.8	75.2	100.1	58.9	78.6	100.0	36.0	66.7
10 - 14	TOTAL DIV. 10000 - 14000	100.0	72.8	94.2	100.0	70.0	93.6	100.0	70.0	93.6	100.0	70.0	93.6	100.0	74.9	94.7	100.0	62.9	92.1
15	MECHANICAL	100.0	82.2	92.2	100.0	58.5	81.9	100.0	62.3	83.5	100.0	62.3	83.5	100.0	62.3	83.5	100.0	40.7	74.1
16	ELECTRICAL	96.0	92.1	94.1	97.9	60.6	79.7	98.0	62.5	80.7	96.3	62.5	79.8	98.5	69.6	84.4	97.9	35.0	67.2
01 - 16	WEIGHTED AVERAGE	100.2	79.4	90.7	99.2	63.6	83.0	99.6	65.2	84.0	98.4	65.0	83.2	99.9	65.9	84.5	99.8	48.4	76.5

WASHINGTON

DIVISION		EVERETT MAT.	INST.	TOTAL	RICHLAND MAT.	INST.	TOTAL	SEATTLE MAT.	INST.	TOTAL	SPOKANE MAT.	INST.	TOTAL	TACOMA MAT.	INST.	TOTAL	VANCOUVER MAT.	INST.	TOTAL
01590	EQUIPMENT RENTAL	.0	104.0	104.0	.0	90.0	90.0	.0	103.8	103.8	.0	90.0	90.0	.0	104.0	104.0	.0	97.2	97.2
02	SITE CONSTRUCTION	94.3	115.2	109.8	104.4	89.5	93.3	98.4	113.3	109.4	105.2	89.5	93.6	97.1	115.5	110.7	109.6	100.9	103.1
03100	CONCRETE FORMS & ACCESSORIES	106.4	101.3	102.0	112.9	81.2	85.4	97.1	103.0	102.2	119.8	81.4	86.5	97.1	102.5	101.8	97.9	97.0	97.1
03200	CONCRETE REINFORCEMENT	107.9	93.6	100.7	103.1	88.0	95.5	106.7	93.9	100.2	103.8	88.1	95.8	106.7	93.7	100.1	107.5	93.5	100.4
03300	CAST-IN-PLACE CONCRETE	97.8	103.4	100.2	117.4	85.4	104.1	102.7	109.9	105.7	121.6	85.5	106.5	100.6	109.7	104.4	112.3	102.6	108.3
03	CONCRETE	98.5	100.0	99.2	108.0	83.8	95.9	101.3	103.2	102.2	110.7	83.9	97.3	100.3	102.7	101.5	110.5	98.0	104.2
04	MASONRY	130.9	99.1	110.9	113.3	84.1	95.0	125.6	103.9	112.0	114.9	84.8	96.0	125.4	101.3	110.3	125.9	98.3	108.6
05	METALS	105.4	87.3	99.4	94.3	82.5	90.4	107.0	90.6	101.6	96.9	82.8	92.3	107.0	88.2	100.8	104.2	89.6	99.5
06	WOOD & PLASTICS	97.2	101.4	99.4	95.3	80.5	87.6	87.7	101.4	94.8	104.6	80.5	92.0	86.7	101.4	94.3	81.0	96.7	89.2
07	THERMAL & MOISTURE PROTECTION	107.2	98.5	102.9	172.6	80.1	127.1	107.2	102.1	104.7	169.8	81.7	126.5	107.0	100.0	103.6	109.6	91.1	100.5
08	DOORS & WINDOWS	98.6	98.2	98.5	112.0	75.9	102.6	100.8	99.1	100.3	111.7	77.3	102.8	99.2	99.1	99.2	96.2	94.5	95.8
09200	PLASTER & GYPSUM BOARD	104.4	101.3	102.5	119.1	79.8	94.2	97.8	101.3	100.0	123.9	79.8	95.9	100.0	101.3	100.8	98.3	96.9	97.4
095,098	CEILINGS & ACOUSTICAL TREATMENT	114.7	101.3	106.7	107.1	79.8	90.7	118.0	101.3	108.0	107.1	79.8	90.7	118.2	101.3	108.1	113.0	96.9	103.4
09600	FLOORING	120.1	101.6	115.2	110.8	45.2	93.5	112.0	105.8	110.4	114.4	76.1	104.3	112.6	101.6	109.7	118.5	83.1	109.2
097,099	WALL FINISHES, PAINTS & COATINGS	113.4	81.4	94.0	113.5	74.1	89.6	113.4	95.6	102.6	113.5	76.8	91.3	113.4	95.6	102.6	121.5	70.5	90.6
09	FINISHES	114.8	99.3	106.8	122.8	73.1	96.9	112.5	102.4	107.2	124.6	79.7	101.2	113.0	101.5	107.0	110.9	91.4	100.8
10 - 14	TOTAL DIV. 10000 - 14000	100.0	95.7	99.1	100.0	88.9	97.6	100.0	105.6	101.2	100.0	88.9	97.6	100.0	105.5	101.2	100.0	85.6	96.9
15	MECHANICAL	99.9	95.7	98.1	100.5	94.5	97.9	99.9	114.1	106.1	100.5	84.6	93.5	100.0	95.1	97.9	100.1	100.2	100.2
16	ELECTRICAL	104.4	95.3	100.0	94.9	86.7	90.9	104.2	109.1	106.6	93.2	83.3	88.4	104.2	97.1	100.8	115.3	103.5	109.5
01 - 16	WEIGHTED AVERAGE	104.0	98.3	101.4	105.5	85.1	96.2	104.3	105.8	105.0	106.2	83.7	96.0	104.1	99.7	102.1	106.0	97.2	102.0

DIVISION		WASHINGTON YAKIMA MAT.	INST.	TOTAL	WEST VIRGINIA CHARLESTON MAT.	INST.	TOTAL	HUNTINGTON MAT.	INST.	TOTAL	PARKERSBURG MAT.	INST.	TOTAL	WHEELING MAT.	INST.	TOTAL	WISCONSIN EAU CLAIRE MAT.	INST.	TOTAL
01590	EQUIPMENT RENTAL	.0	104.0	104.0	.0	99.7	99.7	.0	99.7	99.7	.0	99.7	99.7	.0	99.7	99.7	.0	100.5	100.5
02	SITE CONSTRUCTION	100.5	114.0	110.5	101.0	85.9	89.8	102.4	86.6	90.7	107.1	85.8	91.3	107.6	85.6	91.3	84.7	102.4	97.8
03100	CONCRETE FORMS & ACCESSORIES	97.5	96.2	96.4	103.9	93.8	95.1	97.3	95.2	95.5	87.0	88.2	88.1	88.8	88.2	88.3	99.8	94.8	95.5
03200	CONCRETE REINFORCEMENT	107.2	92.5	99.8	95.5	88.0	91.7	95.5	88.8	92.1	94.1	90.2	92.1	93.5	92.7	93.1	91.8	100.2	96.1
03300	CAST-IN-PLACE CONCRETE	107.5	83.5	97.5	95.5	108.2	100.8	103.0	108.8	105.4	96.2	95.3	95.8	96.2	99.9	97.7	100.1	94.8	97.9
03	CONCRETE	105.2	90.6	97.9	99.1	98.2	98.7	102.2	99.2	100.7	102.3	91.8	97.1	102.3	93.8	98.1	96.4	96.2	96.3
04	MASONRY	117.4	73.2	89.7	87.2	93.7	91.3	89.5	94.6	92.7	76.5	85.5	82.1	99.6	90.5	93.9	92.3	96.2	94.7
05	METALS	105.3	82.9	98.0	102.4	100.2	101.7	102.4	100.5	101.8	101.0	100.3	100.8	101.2	101.9	101.4	91.3	102.5	94.9
06	WOOD & PLASTICS	87.2	101.4	94.6	103.6	93.4	98.3	95.4	94.7	95.0	84.6	86.7	85.7	86.4	86.2	86.3	112.8	93.9	102.9
07	THERMAL & MOISTURE PROTECTION	107.2	78.2	93.0	97.0	90.7	93.9	97.4	91.2	94.4	97.4	86.0	91.8	97.8	89.0	93.4	98.8	88.2	93.5
08	DOORS & WINDOWS	98.8	85.4	95.3	97.5	86.0	94.5	96.3	86.9	93.8	96.9	80.6	92.7	97.8	85.4	94.6	100.7	93.5	98.9
09200	PLASTER & GYPSUM BOARD	99.8	101.3	100.8	106.8	93.0	98.1	105.6	94.4	98.5	99.3	86.2	91.0	99.7	85.6	90.8	106.3	94.1	98.5
095,098	CEILINGS & ACOUSTICAL TREATMENT	112.9	101.3	106.0	96.3	93.0	94.3	91.9	94.4	93.4	91.0	86.2	88.1	91.0	85.6	87.8	97.2	94.1	95.3
09600	FLOORING	114.0	67.9	101.8	105.1	104.2	104.9	105.1	109.0	106.1	99.8	95.1	98.5	100.8	105.0	101.9	90.9	105.2	94.7
097,099	WALL FINISHES, PAINTS & COATINGS	113.4	76.8	91.2	108.5	96.6	101.3	108.5	93.2	99.2	108.5	100.2	103.4	108.5	89.0	96.7	89.5	84.8	86.7
09	FINISHES	112.2	89.5	100.4	99.6	96.3	97.9	98.1	97.9	98.0	95.9	90.7	93.2	96.3	91.1	93.6	97.3	95.8	96.5
10 - 14	TOTAL DIV. 10000 - 14000	100.0	100.7	100.2	100.0	101.4	100.3	100.0	102.0	100.4	100.0	100.3	100.1	100.0	103.6	100.8	100.0	92.8	98.5
15	MECHANICAL	100.0	93.6	97.2	100.0	85.4	93.6	100.0	82.2	92.3	100.0	90.0	95.6	100.0	94.9	97.8	100.2	86.7	94.3
16	ELECTRICAL	106.9	86.7	97.0	97.9	89.4	93.8	97.9	81.9	90.1	98.3	93.5	96.0	95.7	98.7	97.2	102.3	85.2	93.9
01 - 16	WEIGHTED AVERAGE	104.2	90.0	97.8	99.1	92.0	95.9	99.3	90.9	95.5	98.3	90.6	94.8	99.3	93.7	96.8	97.7	93.3	95.7

City Cost Indexes

WISCONSIN

	DIVISION	GREEN BAY MAT.	INST.	TOTAL	KENOSHA MAT.	INST.	TOTAL	LA CROSSE MAT.	INST.	TOTAL	MADISON MAT.	INST.	TOTAL	MILWAUKEE MAT.	INST.	TOTAL	RACINE MAT.	INST.	TOTAL
01590	EQUIPMENT RENTAL	.0	98.2	98.2	.0	97.7	97.7	.0	100.5	100.5	.0	100.0	100.0	.0	86.1	86.1	.0	100.0	100.0
02	SITE CONSTRUCTION	87.2	98.3	95.4	92.8	101.6	99.3	78.9	102.4	96.3	87.5	104.6	100.1	88.8	93.3	92.1	87.9	105.5	100.9
03100	CONCRETE FORMS & ACCESSORIES	110.9	94.0	96.2	108.5	101.2	102.2	83.9	94.3	92.9	99.2	94.9	95.5	101.5	108.6	107.7	98.7	101.3	100.9
03200	CONCRETE REINFORCEMENT	90.1	89.5	89.8	89.3	97.9	93.7	91.5	88.2	89.9	89.5	88.5	89.0	89.5	98.3	93.9	89.5	97.9	93.7
03300	CAST-IN-PLACE CONCRETE	103.5	96.7	100.7	115.2	98.3	108.1	90.0	94.2	91.7	103.8	97.3	101.1	103.8	104.4	104.1	103.8	98.0	101.4
03	CONCRETE	97.6	94.4	96.0	103.3	99.7	101.5	87.9	93.5	90.7	97.3	94.7	96.0	97.7	104.5	101.1	97.2	99.6	98.4
04	MASONRY	123.1	95.2	105.6	98.3	106.4	103.4	91.5	96.2	94.4	102.4	99.1	100.4	101.9	115.1	110.2	101.9	106.4	104.7
05	METALS	93.8	97.5	95.0	93.2	100.8	95.7	91.2	96.8	93.0	94.3	95.2	94.6	96.0	92.0	94.7	94.3	100.8	96.4
06	WOOD & PLASTICS	118.8	93.9	105.8	114.7	99.9	107.0	95.1	93.9	94.5	109.5	93.8	101.3	112.1	108.0	110.0	109.7	99.9	104.6
07	THERMAL & MOISTURE PROTECTION	100.7	86.4	93.7	96.8	101.3	99.0	98.0	87.3	92.7	96.5	95.2	95.9	95.8	105.6	100.6	97.2	100.8	99.0
08	DOORS & WINDOWS	99.1	87.0	96.0	96.9	100.5	97.9	100.7	83.7	96.3	101.8	93.7	99.7	104.0	105.0	104.3	101.8	100.5	101.5
09200	PLASTER & GYPSUM BOARD	97.9	94.1	95.5	94.4	100.4	98.2	99.0	94.1	95.9	103.3	94.1	97.4	104.2	108.5	106.9	103.3	100.4	101.4
095,098	CEILINGS & ACOUSTICAL TREATMENT	91.2	94.1	92.9	84.3	100.4	93.9	95.4	94.1	94.6	90.2	94.1	92.5	93.8	108.5	102.6	90.2	100.4	96.3
09600	FLOORING	108.8	105.2	107.9	110.0	110.6	110.1	83.5	105.2	89.3	93.6	106.0	96.9	96.3	110.2	100.0	93.6	110.6	98.1
097,099	WALL FINISHES, PAINTS & COATINGS	99.2	75.8	85.0	103.6	91.9	96.5	89.5	69.8	77.6	93.3	91.7	92.4	95.7	102.6	99.9	93.3	88.0	90.1
09	FINISHES	100.9	94.5	97.6	98.5	102.2	100.4	93.2	94.1	93.7	95.6	97.0	96.3	97.6	108.8	103.4	95.6	101.7	98.8
10-14	TOTAL DIV. 10000-14000	100.0	92.4	98.4	100.0	91.7	98.2	100.0	92.8	98.5	100.0	92.6	98.4	100.0	94.4	98.8	100.0	91.8	98.2
15	MECHANICAL	100.6	83.2	93.0	100.1	94.2	97.5	100.2	86.2	94.1	99.8	89.7	95.4	99.8	95.7	98.0	99.8	94.2	97.4
16	ELECTRICAL	96.9	84.8	91.0	95.8	85.9	91.0	102.5	85.2	94.1	96.6	85.2	91.1	96.9	101.8	99.3	95.7	93.9	94.8
01-16	WEIGHTED AVERAGE	99.5	90.9	95.6	98.2	97.8	98.0	96.0	91.7	94.0	97.9	93.9	96.1	98.7	101.6	100.0	97.8	99.2	98.4

WYOMING / CANADA

	DIVISION	CASPER MAT.	INST.	TOTAL	CHEYENNE MAT.	INST.	TOTAL	ROCK SPRINGS MAT.	INST.	TOTAL	CALGARY, ALBERTA MAT.	INST.	TOTAL	EDMONTON, ALBERTA MAT.	INST.	TOTAL	HALIFAX, NOVA SCOTIA MAT.	INST.	TOTAL
01590	EQUIPMENT RENTAL	.0	101.7	101.7	.0	101.7	101.7	.0	101.7	101.7	.0	107.0	107.0	.0	107.0	107.0	.0	98.6	98.6
02	SITE CONSTRUCTION	80.9	100.3	95.2	80.7	100.2	95.2	83.8	99.8	95.6	113.8	105.1	107.3	117.5	105.1	108.3	97.2	95.8	96.1
03100	CONCRETE FORMS & ACCESSORIES	97.8	43.6	50.8	98.5	43.4	50.8	99.3	30.6	39.8	122.1	76.4	82.5	119.8	76.4	82.2	91.0	66.3	69.6
03200	CONCRETE REINFORCEMENT	115.6	52.0	83.5	108.5	52.1	80.0	117.9	51.8	84.5	161.3	66.7	113.5	161.3	66.7	113.5	148.1	49.7	98.3
03300	CAST-IN-PLACE CONCRETE	99.4	58.0	82.1	99.4	57.9	82.1	99.9	43.6	76.4	191.6	87.1	147.9	206.9	87.1	156.9	178.4	68.3	132.4
03	CONCRETE	104.0	50.9	77.5	102.9	50.8	76.9	104.7	40.2	72.5	160.8	78.9	119.9	168.0	78.9	123.5	151.7	64.6	108.2
04	MASONRY	109.8	39.3	65.5	107.0	41.4	65.9	174.9	32.1	85.3	180.3	74.1	113.7	181.6	74.1	114.1	172.3	69.6	107.9
05	METALS	97.3	63.0	86.1	98.1	63.0	86.6	95.1	61.6	84.1	135.7	85.6	119.3	136.5	85.6	119.9	120.8	72.5	105.0
06	WOOD & PLASTICS	95.5	41.8	67.5	95.5	41.8	67.5	98.4	29.6	62.5	116.3	76.0	95.3	112.9	76.0	93.6	85.0	65.6	74.8
07	THERMAL & MOISTURE PROTECTION	102.4	54.3	78.8	102.4	54.9	79.1	103.0	41.3	72.7	125.9	79.9	103.3	126.2	79.9	103.4	105.4	67.7	86.9
08	DOORS & WINDOWS	93.5	44.4	80.8	94.3	44.4	81.4	100.6	37.7	84.4	91.8	73.2	87.0	91.8	73.2	87.0	82.5	61.1	77.0
09200	PLASTER & GYPSUM BOARD	80.4	40.0	54.8	80.4	40.0	54.8	86.4	27.5	49.1	151.0	74.8	102.7	153.8	74.8	103.7	166.0	64.7	101.7
095,098	CEILINGS & ACOUSTICAL TREATMENT	96.9	40.0	62.8	96.9	40.0	62.8	96.9	27.5	55.3	107.3	74.8	87.8	125.1	74.8	94.9	108.1	64.7	82.1
09600	FLOORING	101.2	43.1	85.8	101.2	69.2	92.7	103.5	42.2	87.3	132.6	86.2	120.4	132.6	86.2	120.4	110.0	62.6	97.4
097,099	WALL FINISHES, PAINTS & COATINGS	103.7	56.9	75.3	103.6	56.9	75.3	103.7	34.6	61.8	109.1	76.5	89.3	109.2	76.5	89.4	109.2	60.4	79.6
09	FINISHES	95.9	43.9	68.8	95.8	49.2	71.6	97.4	32.3	63.6	121.0	78.0	98.7	125.9	78.0	101.0	115.7	65.1	89.4
10-14	TOTAL DIV. 10000-14000	100.0	75.8	94.8	100.0	75.7	94.8	100.0	50.6	89.5	140.0	85.6	128.4	140.0	85.6	128.4	140.0	62.3	123.5
15	MECHANICAL	100.1	68.4	86.2	100.1	44.8	75.9	100.1	31.6	70.1	101.5	77.6	91.1	101.7	77.6	91.2	101.3	69.0	87.1
16	ELECTRICAL	88.2	62.8	75.8	88.2	52.6	70.8	87.6	45.3	67.0	117.3	86.3	102.2	118.2	86.3	102.6	127.3	70.0	99.3
01-16	WEIGHTED AVERAGE	97.6	60.1	80.5	97.5	54.7	78.0	101.4	44.8	75.7	124.6	82.0	105.2	126.2	82.0	106.2	119.3	70.3	97.0

CANADA

	DIVISION	HAMILTON, ONTARIO MAT.	INST.	TOTAL	KITCHENER, ONTARIO MAT.	INST.	TOTAL	LAVAL, QUEBEC MAT.	INST.	TOTAL	LONDON, ONTARIO MAT.	INST.	TOTAL	MONTREAL, QUEBEC MAT.	INST.	TOTAL	OSHAWA, ONTARIO MAT.	INST.	TOTAL
01590	EQUIPMENT RENTAL	.0	103.1	103.1	.0	103.0	103.0	.0	99.2	99.2	.0	103.2	103.2	.0	101.0	101.0	.0	103.0	103.0
02	SITE CONSTRUCTION	110.0	103.5	105.2	99.2	102.9	101.9	92.2	96.8	95.6	110.0	103.2	105.0	91.9	97.1	95.7	111.5	103.1	105.3
03100	CONCRETE FORMS & ACCESSORIES	121.5	86.8	91.4	115.8	79.5	84.4	130.1	81.3	87.8	127.2	82.4	88.4	130.4	81.6	88.1	122.7	83.2	88.5
03200	CONCRETE REINFORCEMENT	174.6	89.1	131.4	110.3	89.0	99.5	156.8	83.7	119.9	123.3	87.5	105.2	159.4	83.8	121.2	174.6	89.7	131.7
03300	CAST-IN-PLACE CONCRETE	157.2	94.9	131.2	146.7	76.5	117.3	140.1	89.5	119.0	157.2	92.5	130.2	137.5	91.1	118.1	169.7	81.8	133.0
03	CONCRETE	146.6	90.1	118.4	124.4	80.5	102.5	136.2	84.8	110.5	138.6	87.0	112.9	135.4	85.5	110.5	152.6	84.1	118.4
04	MASONRY	167.2	91.2	119.5	162.8	87.1	115.3	160.7	81.5	111.0	167.3	88.4	117.8	163.9	81.5	112.2	165.8	90.5	118.5
05	METALS	139.4	89.2	123.0	116.4	88.7	107.3	105.6	85.4	99.0	124.5	88.8	112.8	125.1	85.9	112.3	107.2	89.0	101.2
06	WOOD & PLASTICS	117.7	85.6	100.9	111.2	77.9	93.8	131.7	81.2	105.3	117.7	80.6	98.3	131.7	81.5	105.5	118.6	81.4	99.2
07	THERMAL & MOISTURE PROTECTION	110.7	88.3	99.7	108.8	83.9	96.5	105.4	85.7	95.7	114.1	86.5	100.5	106.2	86.4	96.5	110.1	84.4	97.5
08	DOORS & WINDOWS	91.8	85.7	90.2	83.2	79.9	82.4	91.8	72.3	86.8	92.7	82.0	90.0	91.8	72.4	86.8	90.9	83.1	88.9
09200	PLASTER & GYPSUM BOARD	190.9	85.2	123.9	157.4	77.3	106.6	156.3	80.7	108.4	191.4	80.2	120.9	159.4	80.7	109.5	159.9	81.0	109.9
095,098	CEILINGS & ACOUSTICAL TREATMENT	112.6	85.2	96.2	107.3	77.3	89.3	95.7	80.7	86.7	114.4	80.2	93.9	107.3	80.7	91.3	100.1	81.0	88.6
09600	FLOORING	132.6	91.9	121.9	128.3	91.9	118.7	132.6	92.6	122.1	134.7	91.9	123.4	132.6	92.6	122.1	132.6	94.3	122.5
097,099	WALL FINISHES, PAINTS & COATINGS	109.2	99.2	103.2	109.2	90.9	98.1	109.2	86.3	95.3	109.2	99.9	103.2	109.2	86.3	95.3	109.2	104.3	106.2
09	FINISHES	127.7	89.2	107.7	119.7	83.0	100.6	118.0	84.3	100.4	128.9	85.9	106.5	121.3	84.5	102.1	120.2	87.6	103.2
10-14	TOTAL DIV. 10000-14000	140.0	91.5	129.7	140.0	89.3	129.2	140.0	79.2	127.1	140.0	90.2	129.4	140.0	80.0	127.2	140.0	90.7	129.5
15	MECHANICAL	101.7	85.5	94.7	101.3	81.9	92.8	101.3	74.6	89.6	101.8	82.1	93.2	101.8	74.7	89.9	101.3	83.5	93.5
16	ELECTRICAL	130.4	93.9	112.6	125.1	90.7	108.3	121.9	73.3	98.2	130.7	90.9	111.3	120.9	73.3	97.7	126.2	91.4	109.2
01-16	WEIGHTED AVERAGE	124.7	90.6	109.2	115.6	86.3	102.3	115.4	81.1	99.8	121.8	88.0	106.5	118.8	81.4	101.8	118.9	88.4	105.1

COST INDEXES

City Cost Indexes

CANADA

DIVISION		OTTAWA, ONTARIO			QUEBEC, QUEBEC			REGINA, SASKATCHEWAN			SASKATOON, SASKATCHEWAN			ST CATHARINES, ONTARIO			ST JOHNS, NEWFOUNDLAND		
		MAT.	INST.	TOTAL	MAT.	INST.	TOTAL	MAT.	INST.	TOTAL	MAT.	INST.	TOTAL	MAT.	INST.	TOTAL	MAT.	INST.	TOTAL
01590	EQUIPMENT RENTAL	.0	103.0	103.0	.0	101.4	101.4	.0	98.6	98.6	.0	98.6	98.6	.0	100.4	100.4	.0	98.6	98.6
02	SITE CONSTRUCTION	110.3	103.1	105.0	93.7	97.2	96.2	110.0	94.7	98.7	105.2	94.8	97.5	99.6	98.8	99.0	112.5	94.5	99.2
03100	CONCRETE FORMS & ACCESSORIES	120.2	83.2	88.2	130.2	81.8	88.3	102.4	56.0	62.2	102.4	55.9	62.1	113.3	81.2	85.5	100.6	56.9	62.8
03200	CONCRETE REINFORCEMENT	174.6	87.5	130.6	148.1	83.8	115.6	113.8	65.8	89.6	118.9	65.8	92.1	111.2	89.0	100.0	148.1	51.5	99.3
03300	CAST-IN-PLACE CONCRETE	159.6	92.4	131.5	151.5	91.6	126.5	162.8	65.7	122.2	148.0	65.6	113.6	140.1	80.9	115.4	179.6	64.7	131.6
03	CONCRETE	147.6	87.4	117.6	140.2	85.7	113.0	131.6	61.9	96.8	125.3	61.8	93.6	121.2	82.8	102.1	164.6	59.5	112.1
04	MASONRY	167.4	88.0	117.6	164.9	81.5	112.6	166.9	60.7	100.3	167.1	60.7	100.3	162.3	84.5	113.5	160.7	60.0	97.5
05	METALS	116.5	88.8	107.4	120.7	86.2	109.4	105.7	72.8	94.9	105.7	72.7	94.9	106.6	88.8	100.8	108.6	70.3	96.0
06	WOOD & PLASTICS	117.2	82.2	99.0	132.3	81.6	105.8	95.6	54.5	74.2	94.2	54.5	73.5	108.5	82.0	94.7	95.0	56.0	74.6
07	THERMAL & MOISTURE PROTECTION	110.9	85.1	98.2	105.7	86.6	96.3	106.5	61.8	84.5	104.9	60.8	83.2	108.8	85.1	97.1	109.9	59.4	85.1
08	DOORS & WINDOWS	91.8	82.8	89.5	91.8	79.6	88.7	86.7	53.7	78.2	85.7	53.7	77.5	82.7	82.5	82.7	98.1	53.8	86.6
09200	PLASTER & GYPSUM BOARD	231.0	81.8	136.4	198.1	80.7	123.7	166.3	53.3	94.7	147.7	53.3	87.9	143.3	81.6	104.2	172.3	54.8	97.8
095,098	CEILINGS & ACOUSTICAL TREATMENT	106.4	81.8	91.6	97.4	80.7	87.4	123.3	53.3	81.3	123.3	53.3	81.3	100.1	81.6	89.0	106.4	54.8	75.4
09600	FLOORING	132.6	90.6	121.5	132.6	92.6	122.1	120.0	58.9	103.9	120.0	58.9	103.9	126.5	91.9	117.4	114.9	53.3	98.6
097,099	WALL FINISHES, PAINTS & COATINGS	109.2	92.9	99.3	109.2	86.3	95.3	109.2	62.9	81.1	109.2	53.6	75.5	109.2	92.9	99.3	109.2	58.2	78.2
09	FINISHES	132.0	85.7	107.9	124.5	84.5	103.7	123.6	57.0	88.9	120.7	55.9	87.0	115.3	84.7	99.4	118.7	56.3	86.2
10 - 14	TOTAL DIV. 10000 - 14000	140.0	88.2	129.0	140.0	80.2	127.3	140.0	60.0	123.0	140.0	60.1	123.0	140.0	68.0	124.7	140.0	60.8	123.2
15	MECHANICAL	101.8	82.4	93.3	101.7	74.7	89.9	101.5	62.0	84.2	101.4	62.0	84.2	101.3	80.3	92.1	101.7	58.7	82.9
16	ELECTRICAL	121.9	91.9	107.3	125.7	73.3	100.2	126.3	63.4	95.6	126.5	63.4	95.7	128.2	91.9	110.5	120.7	58.9	90.6
01 - 16	WEIGHTED AVERAGE	120.7	88.1	105.9	119.6	81.8	102.5	116.0	64.9	92.8	114.7	64.7	92.0	113.6	85.6	100.9	120.0	62.8	94.0

DIVISION		THUNDER BAY, ONTARIO			TORONTO, ONTARIO			VANCOUVER, B C			WINDSOR, ONTARIO			WINNIPEG, MANITOBA					
		MAT.	INST.	TOTAL	MAT.	INST.	TOTAL	MAT.	INST.	TOTAL	MAT.	INST.	TOTAL	MAT.	INST.	TOTAL	MAT.	INST.	TOTAL
01590	EQUIPMENT RENTAL	.0	100.4	100.4	.0	103.2	103.2	.0	111.3	111.3	.0	100.4	100.4	.0	105.3	105.3	.0	.0	.0
02	SITE CONSTRUCTION	105.3	98.9	100.6	111.2	103.9	105.8	115.5	107.5	109.6	95.6	99.1	98.2	109.6	100.3	102.7	.0	.0	.0
03100	CONCRETE FORMS & ACCESSORIES	122.7	82.3	87.7	123.1	93.3	97.3	114.6	84.9	88.9	122.7	82.9	88.2	122.3	67.3	74.7	.0	.0	.0
03200	CONCRETE REINFORCEMENT	99.4	88.4	93.8	170.9	90.0	130.0	161.3	81.7	121.1	109.0	87.6	98.2	161.3	56.5	108.4	.0	.0	.0
03300	CAST-IN-PLACE CONCRETE	154.2	92.6	128.5	163.2	102.1	137.7	153.2	94.6	128.7	143.5	93.7	122.7	157.5	72.5	122.0	.0	.0	.0
03	CONCRETE	130.0	87.1	108.6	149.0	95.7	122.4	146.0	88.0	117.0	123.1	87.7	105.5	144.4	67.8	106.1	.0	.0	.0
04	MASONRY	163.0	85.4	114.3	186.3	99.1	131.6	175.3	88.8	121.0	162.4	91.7	118.1	177.7	63.4	106.0	.0	.0	.0
05	METALS	106.4	87.9	100.4	130.5	91.1	117.6	141.7	90.9	125.1	106.5	89.1	100.8	139.3	75.8	118.5	.0	.0	.0
06	WOOD & PLASTICS	118.6	82.4	99.7	118.6	91.7	104.6	114.3	82.7	97.8	118.6	81.0	99.0	116.3	68.0	91.1	.0	.0	.0
07	THERMAL & MOISTURE PROTECTION	109.1	84.3	96.9	111.0	95.0	103.1	114.4	89.5	102.2	108.8	88.1	98.6	106.1	70.2	88.5	.0	.0	.0
08	DOORS & WINDOWS	81.7	82.1	81.8	90.9	90.8	90.9	94.3	81.2	90.9	81.5	81.8	81.5	91.8	61.3	83.9	.0	.0	.0
09200	PLASTER & GYPSUM BOARD	169.3	82.0	113.9	173.3	91.5	121.5	147.7	81.7	105.9	164.0	80.5	111.1	153.3	66.6	98.4	.0	.0	.0
095,098	CEILINGS & ACOUSTICAL TREATMENT	95.7	82.0	87.5	120.6	91.5	103.2	109.0	81.7	92.7	95.7	80.5	86.6	107.3	66.6	82.9	.0	.0	.0
09600	FLOORING	132.6	52.9	111.6	132.6	97.5	123.3	132.6	89.2	121.2	132.6	92.6	122.1	132.6	65.3	114.8	.0	.0	.0
097,099	WALL FINISHES, PAINTS & COATINGS	109.2	94.9	100.5	112.2	104.3	107.4	109.1	98.7	102.8	109.2	94.5	100.3	109.2	55.3	76.5	.0	.0	.0
09	FINISHES	120.1	79.1	98.8	127.5	95.4	110.8	121.6	86.7	103.4	119.1	85.9	101.8	121.5	65.9	92.6	.0	.0	.0
10 - 14	TOTAL DIV. 10000 - 14000	140.0	69.0	124.9	140.0	93.7	130.2	140.0	89.3	129.2	140.0	70.3	125.2	140.0	63.9	123.8	.0	.0	.0
15	MECHANICAL	101.3	81.0	92.4	101.7	91.4	97.2	101.7	84.6	94.2	101.3	83.6	93.5	101.7	68.5	87.1	.0	.0	.0
16	ELECTRICAL	125.1	90.1	108.0	129.8	94.6	112.6	132.9	82.8	108.5	133.5	91.9	113.2	130.9	70.3	101.3	.0	.0	.0
01 - 16	WEIGHTED AVERAGE	114.8	85.4	101.5	124.4	95.0	111.0	125.6	88.2	108.6	114.7	88.0	102.6	124.3	71.0	100.1	.0	.0	.0

COST INDEXES

457

Costs shown in *Means cost data publications* are based on National Averages for materials and installation. To adjust these costs to a specific location, simply multiply the base cost by the factor and divide by 100 for that city. The data is arranged alphabetically by state and postal zip code numbers. For a city not listed, use the factor for a nearby city with similar economic characteristics.

STATE/ZIP	CITY	MAT.	INST.	TOTAL
ALABAMA				
350-352	Birmingham	96.4	76.5	87.4
354	Tuscaloosa	96.3	56.4	78.2
355	Jasper	96.7	53.5	77.1
356	Decatur	96.3	57.6	78.8
357-358	Huntsville	96.3	73.3	85.8
359	Gadsden	96.3	60.0	79.8
360-361	Montgomery	97.0	58.4	79.5
362	Anniston	96.0	47.7	74.1
363	Dothan	96.5	50.5	75.6
364	Evergreen	95.8	51.8	75.8
365-366	Mobile	97.1	63.0	81.6
367	Selma	96.1	53.0	76.5
368	Phenix City	96.8	56.4	78.5
369	Butler	96.2	50.5	75.5
ALASKA				
995-996	Anchorage	134.5	113.8	125.1
997	Fairbanks	130.7	116.9	124.4
998	Juneau	133.6	113.8	124.6
999	Ketchikan	143.2	113.4	129.6
ARIZONA				
850,853	Phoenix	98.8	74.3	87.7
852	Mesa/Tempe	97.8	67.7	84.1
855	Globe	98.2	62.9	82.2
856-857	Tucson	97.0	69.8	84.6
859	Show Low	98.3	66.9	84.1
860	Flagstaff	100.5	67.6	85.5
863	Prescott	98.2	64.7	83.0
864	Kingman	96.8	65.6	82.6
865	Chambers	96.8	62.5	81.2
ARKANSAS				
716	Pine Bluff	95.3	61.7	80.0
717	Camden	93.4	40.1	69.2
718	Texarkana	94.7	45.3	72.3
719	Hot Springs	92.6	39.6	68.5
720-722	Little Rock	95.2	64.3	81.2
723	West Memphis	95.1	55.0	76.9
724	Jonesboro	95.5	55.0	77.1
725	Batesville	93.5	49.9	73.7
726	Harrison	94.8	49.9	74.4
727	Fayetteville	92.1	48.1	72.1
728	Russellville	93.4	48.0	72.8
729	Fort Smith	96.0	56.8	78.2
CALIFORNIA				
900-902	Los Angeles	102.7	111.9	106.8
903-905	Inglewood	98.5	108.2	102.9
906-908	Long Beach	100.0	108.3	103.8
910-912	Pasadena	100.5	108.3	104.0
913-916	Van Nuys	103.8	108.3	105.8
917-918	Alhambra	102.8	108.2	105.3
919-921	San Diego	103.9	105.0	104.4
922	Palm Springs	100.5	105.8	102.9
923-924	San Bernardino	98.3	106.7	102.2
925	Riverside	102.6	109.9	105.9
926-927	Santa Ana	100.4	106.7	103.2
928	Anaheim	102.9	111.1	106.6
930	Oxnard	103.9	109.5	106.4
931	Santa Barbara	102.7	110.3	106.1
932-933	Bakersfield	102.6	106.2	104.2
934	San Luis Obispo	103.8	106.3	105.0
935	Mojave	100.8	103.7	102.1
936-938	Fresno	104.3	111.9	107.8
939	Salinas	104.5	116.1	109.8
940-941	San Francisco	112.6	133.8	122.2
942,956-958	Sacramento	106.9	113.1	109.7
943	Palo Alto	105.5	125.9	114.8
944	San Mateo	108.4	125.1	116.0
945	Vallejo	105.6	120.4	112.3
946	Oakland	110.6	123.9	116.6
947	Berkeley	110.1	122.5	115.7
948	Richmond	109.5	122.8	115.6
949	San Rafael	110.5	122.3	115.9
950	Santa Cruz	109.6	116.1	112.6

STATE/ZIP	CITY	MAT.	INST.	TOTAL
CALIFORNIA (CONT'D)				
951	San Jose	107.3	128.4	116.9
952	Stockton	104.5	110.5	107.2
953	Modesto	104.4	111.6	107.7
954	Santa Rosa	104.7	121.5	112.4
955	Eureka	106.3	103.7	105.1
959	Marysville	105.3	113.4	109.0
960	Redding	106.8	110.4	108.4
961	Susanville	106.1	111.1	108.4
COLORADO				
800-802	Denver	100.3	90.3	95.8
803	Boulder	97.6	86.7	92.7
804	Golden	99.9	86.2	93.7
805	Fort Collins	101.0	82.7	92.7
806	Greeley	98.5	71.1	86.0
807	Fort Morgan	98.4	85.9	92.7
808-809	Colorado Springs	100.1	86.6	94.0
810	Pueblo	100.0	83.9	92.6
811	Alamosa	101.6	81.3	92.4
812	Salida	101.6	81.1	92.3
813	Durango	102.0	79.2	91.6
814	Montrose	100.6	79.0	90.8
815	Grand Junction	103.8	77.6	91.9
816	Glenwood Springs	101.4	84.6	93.8
CONNECTICUT				
060	New Britain	101.4	110.7	105.6
061	Hartford	102.7	109.9	106.0
062	Willimantic	102.2	109.6	105.6
063	New London	98.6	110.7	104.1
064	Meriden	100.6	111.1	105.4
065	New Haven	103.2	111.2	106.8
066	Bridgeport	102.7	110.8	106.4
067	Waterbury	102.2	110.7	106.1
068	Norwalk	102.2	111.5	106.4
069	Stamford	102.3	115.6	108.4
D.C.				
200-205	Washington	102.0	90.8	96.9
DELAWARE				
197	Newark	100.5	105.7	102.8
198	Wilmington	99.5	105.7	102.3
199	Dover	100.6	105.7	102.9
FLORIDA				
320,322	Jacksonville	98.3	61.0	81.4
321	Daytona Beach	98.4	74.5	87.5
323	Tallahassee	97.6	49.2	75.6
324	Panama City	99.5	37.9	71.5
325	Pensacola	99.2	58.1	80.5
326,344	Gainesville	99.7	58.7	81.1
327-328,347	Orlando	99.8	69.3	86.0
329	Melbourne	100.5	79.0	90.7
330-332,340	Miami	98.7	72.9	87.0
333	Fort Lauderdale	97.7	72.3	86.2
334,349	West Palm Beach	96.6	68.5	83.8
335-336,346	Tampa	99.4	72.8	87.3
337	St. Petersburg	101.6	55.8	80.8
338	Lakeland	98.5	72.3	86.6
339,341	Fort Myers	97.9	64.3	82.6
342	Sarasota	99.6	66.7	84.6
GEORGIA				
300-303,399	Atlanta	97.4	80.7	89.8
304	Statesboro	97.1	45.0	73.4
305	Gainesville	95.8	57.4	78.4
306	Athens	95.2	62.1	80.1
307	Dalton	97.5	50.9	76.3
308-309	Augusta	96.0	59.2	79.3
310-312	Macon	95.7	59.2	79.1
313-314	Savannah	97.1	58.8	79.7
315	Waycross	97.2	51.6	76.5
316	Valdosta	97.0	49.1	75.2
317	Albany	97.3	56.8	78.9
318-319	Columbus	97.2	61.7	81.1

STATE/ZIP	CITY	MAT.	INST.	TOTAL
HAWAII				
967	Hilo	115.3	125.7	120.0
968	Honolulu	121.0	125.7	123.1
STATES & POSS.				
969	Guam	193.9	54.3	130.5
IDAHO				
832	Pocatello	100.3	78.3	90.3
833	Twin Falls	100.8	45.8	75.8
834	Idaho Falls	98.2	55.3	78.7
835	Lewiston	108.1	83.3	96.8
836-837	Boise	100.2	80.5	91.2
838	Coeur d'Alene	107.3	56.8	84.4
ILLINOIS				
600-603	North Suburban	99.1	116.3	106.9
604	Joliet	99.0	117.2	107.3
605	South Suburban	99.1	116.3	106.9
606-608	Chicago	99.7	125.8	111.6
609	Kankakee	95.4	105.3	99.9
610-611	Rockford	97.4	109.4	102.8
612	Rock Island	95.1	97.7	96.3
613	La Salle	96.5	102.7	99.3
614	Galesburg	96.2	100.8	98.3
615-616	Peoria	98.7	101.7	100.1
617	Bloomington	95.5	101.2	98.1
618-619	Champaign	99.0	100.0	99.5
620-622	East St. Louis	94.2	101.1	97.3
623	Quincy	95.6	95.4	95.5
624	Effingham	94.9	97.3	96.0
625	Decatur	96.5	98.5	97.4
626-627	Springfield	96.8	98.9	97.8
628	Centralia	93.1	98.0	95.3
629	Carbondale	92.8	97.8	95.1
INDIANA				
460	Anderson	94.6	83.3	89.5
461-462	Indianapolis	98.8	88.7	94.2
463-464	Gary	95.8	98.7	97.1
465-466	South Bend	95.0	82.6	89.4
467-468	Fort Wayne	95.1	80.6	88.5
469	Kokomo	92.9	82.2	88.0
470	Lawrenceburg	92.2	80.1	86.7
471	New Albany	93.7	74.8	85.1
472	Columbus	96.3	80.2	89.0
473	Muncie	96.1	80.8	89.2
474	Bloomington	98.2	80.7	90.3
475	Washington	94.1	85.2	90.1
476-477	Evansville	95.2	86.8	91.4
478	Terre Haute	96.0	85.3	91.1
479	Lafayette	95.9	79.5	88.4
IOWA				
500-503,509	Des Moines	97.7	82.4	90.7
504	Mason City	96.0	63.3	81.2
505	Fort Dodge	96.2	59.4	79.5
506-507	Waterloo	97.7	62.0	81.5
508	Creston	96.6	66.3	82.8
510-511	Sioux City	98.3	73.3	87.0
512	Sibley	97.0	52.9	77.0
513	Spencer	98.8	51.0	77.1
514	Carroll	95.8	57.0	78.2
515	Council Bluffs	99.1	74.7	88.0
516	Shenandoah	96.1	52.2	76.2
520	Dubuque	97.7	76.5	88.1
521	Decorah	96.8	54.5	77.6
522-524	Cedar Rapids	98.8	82.7	91.5
525	Ottumwa	96.9	70.7	85.0
526	Burlington	96.0	73.7	85.9
527-528	Davenport	97.8	93.1	95.7
KANSAS				
660-662	Kansas City	98.7	91.6	95.5
664-666	Topeka	98.5	69.3	85.2
667	Fort Scott	97.5	66.8	83.6
668	Emporia	97.4	59.4	80.1
669	Belleville	99.2	54.8	79.0
670-672	Wichita	97.7	71.1	85.6
673	Independence	99.0	53.0	78.1
674	Salina	99.0	57.2	80.1
675	Hutchinson	94.5	51.1	74.7
676	Hays	98.6	54.8	78.7
677	Colby	99.2	54.8	79.0

STATE/ZIP	CITY	MAT.	INST.	TOTAL
KANSAS (CONT'D)				
678	Dodge City	100.3	54.8	79.6
679	Liberal	98.4	45.6	74.4
KENTUCKY				
400-402	Louisville	96.3	86.4	91.8
403-405	Lexington	95.9	66.9	82.7
406	Frankfort	96.1	68.5	83.6
407-409	Corbin	93.3	44.6	71.2
410	Covington	94.0	93.6	93.8
411-412	Ashland	92.6	98.6	95.3
413-414	Campton	94.2	45.1	71.9
415-416	Pikeville	95.1	62.1	80.1
417-418	Hazard	93.5	44.8	71.4
420	Paducah	92.2	88.8	90.7
421-422	Bowling Green	94.4	83.6	89.5
423	Owensboro	94.2	74.3	85.2
424	Henderson	91.9	89.2	90.7
425-426	Somerset	91.5	46.5	71.0
427	Elizabethtown	91.1	82.8	87.3
LOUISIANA				
700-701	New Orleans	100.2	69.8	86.4
703	Thibodaux	98.1	64.4	82.8
704	Hammond	95.4	63.9	81.1
705	Lafayette	97.7	55.0	78.3
706	Lake Charles	97.8	60.0	80.7
707-708	Baton Rouge	98.4	55.1	78.7
710-711	Shreveport	94.7	59.7	78.8
712	Monroe	95.5	57.5	78.2
713-714	Alexandria	95.6	54.7	77.0
MAINE				
039	Kittery	95.6	72.4	85.1
040-041	Portland	98.8	76.8	88.8
042	Lewiston	99.2	76.8	89.0
043	Augusta	97.7	72.3	86.2
044	Bangor	98.0	76.8	88.3
045	Bath	96.7	72.4	85.7
046	Machias	96.3	71.8	85.2
047	Houlton	96.5	75.5	87.0
048	Rockland	95.5	70.6	84.2
049	Waterville	96.8	67.3	83.4
MARYLAND				
206	Waldorf	98.0	73.1	86.7
207-208	College Park	98.0	80.6	90.1
209	Silver Spring	97.3	78.2	88.6
210-212	Baltimore	99.4	84.8	92.7
214	Annapolis	98.8	79.9	90.2
215	Cumberland	95.5	80.8	88.8
216	Easton	96.9	45.4	73.5
217	Hagerstown	96.0	80.8	89.1
218	Salisbury	97.4	54.8	78.0
219	Elkton	94.6	67.2	82.2
MASSACHUSETTS				
010-011	Springfield	100.7	105.1	102.7
012	Pittsfield	100.2	101.1	100.6
013	Greenfield	98.3	103.3	100.5
014	Fitchburg	97.0	116.9	106.0
015-016	Worcester	100.5	118.4	108.6
017	Framingham	96.4	122.8	108.4
018	Lowell	100.0	123.5	110.7
019	Lawrence	101.1	121.0	110.2
020-022, 024	Boston	102.9	130.5	115.4
023	Brockton	101.4	115.5	107.8
025	Buzzards Bay	95.8	116.9	105.4
026	Hyannis	98.3	116.9	106.7
027	New Bedford	101.1	117.0	108.4
MICHIGAN				
480,483	Royal Oak	94.9	106.2	100.0
481	Ann Arbor	96.9	105.7	100.9
482	Detroit	98.7	116.9	107.0
484-485	Flint	96.7	96.2	96.5
486	Saginaw	96.5	93.2	95.0
487	Bay City	96.4	93.3	95.0
488-489	Lansing	96.7	96.1	96.4
490	Battle Creek	95.4	88.1	92.1
491	Kalamazoo	95.7	84.9	90.8
492	Jackson	93.9	94.7	94.3
493,495	Grand Rapids	95.8	70.2	84.2
494	Muskegon	94.3	83.6	89.5

STATE/ZIP	CITY	MAT.	INST.	TOTAL
MICHIGAN (CONT'D)				
496	Traverse City	93.4	73.9	84.6
497	Gaylord	94.6	77.9	87.0
498-499	Iron Mountain	96.5	87.1	92.2
MINNESOTA				
550-551	Saint Paul	99.8	122.0	109.8
553-555	Minneapolis	100.7	126.8	112.6
556-558	Duluth	99.7	108.9	103.8
559	Rochester	99.4	105.1	102.0
560	Mankato	96.3	102.1	99.0
561	Windom	95.1	81.4	88.9
562	Willmar	94.5	87.7	91.4
563	St. Cloud	95.6	120.5	106.9
564	Brainerd	96.1	101.0	98.3
565	Detroit Lakes	98.0	99.1	98.5
566	Bemidji	97.3	97.7	97.5
567	Thief River Falls	96.4	94.0	95.3
MISSISSIPPI				
386	Clarksdale	96.0	31.7	66.8
387	Greenville	99.4	45.9	75.1
388	Tupelo	97.4	38.4	70.6
389	Greenwood	97.3	34.3	68.7
390-392	Jackson	97.9	47.5	75.0
393	Meridian	96.0	47.8	74.1
394	Laurel	97.3	34.4	68.7
395	Biloxi	97.9	56.6	79.1
396	McComb	95.8	53.3	76.5
397	Columbus	97.2	37.8	70.2
MISSOURI				
630-631	St. Louis	96.7	108.6	102.1
633	Bowling Green	95.6	89.4	92.8
634	Hannibal	94.4	85.1	90.2
635	Kirksville	96.7	78.3	88.3
636	Flat River	96.6	93.7	95.3
637	Cape Girardeau	96.1	89.3	93.0
638	Sikeston	94.3	84.3	89.8
639	Poplar Bluff	93.9	85.0	89.8
640-641	Kansas City	101.1	106.3	103.4
644-645	St. Joseph	100.1	89.6	95.3
646	Chillicothe	97.0	70.7	85.1
647	Harrisonville	96.7	97.7	97.2
648	Joplin	98.7	69.0	85.2
650-651	Jefferson City	95.9	87.3	92.0
652	Columbia	96.9	88.1	92.9
653	Sedalia	96.1	84.3	90.7
654-655	Rolla	95.0	77.8	87.2
656-658	Springfield	98.2	74.5	87.4
MONTANA				
590-591	Billings	102.0	74.4	89.5
592	Wolf Point	100.8	70.5	87.0
593	Miles City	98.8	71.0	86.2
594	Great Falls	102.6	74.9	90.0
595	Havre	99.6	70.7	86.5
596	Helena	102.4	72.3	88.7
597	Butte	101.0	71.1	87.4
598	Missoula	98.7	70.3	85.8
599	Kalispell	97.8	69.8	85.1
NEBRASKA				
680-681	Omaha	99.1	76.7	89.0
683-685	Lincoln	98.0	67.7	84.2
686	Columbus	96.0	51.2	75.6
687	Norfolk	97.7	62.5	81.7
688	Grand Island	97.2	66.2	83.1
689	Hastings	97.0	58.5	79.5
690	Mccook	96.8	48.8	75.0
691	North Platte	96.8	59.9	80.0
692	Valentine	99.2	39.7	72.2
693	Alliance	98.9	37.0	70.8
NEVADA				
889-891	Las Vegas	98.1	105.5	101.5
893	Ely	98.7	81.2	90.7
894-895	Reno	98.7	95.5	97.3
897	Carson City	98.4	95.6	97.2
898	Elko	97.5	87.9	93.1
NEW HAMPSHIRE				
030	Nashua	100.9	82.8	92.7
031	Manchester	101.2	82.8	92.9

STATE/ZIP	CITY	MAT.	INST.	TOTAL
NEW HAMPSHIRE (CONT'D)				
032-033	Concord	98.6	82.8	91.4
034	Keene	97.6	52.5	77.1
035	Littleton	97.7	62.0	81.5
036	Charleston	97.0	49.3	75.4
037	Claremont	96.3	49.3	75.0
038	Portsmouth	98.0	78.4	89.1
NEW JERSEY				
070-071	Newark	102.5	121.4	111.1
072	Elizabeth	101.1	120.7	110.0
073	Jersey City	100.2	122.5	110.3
074-075	Paterson	101.8	121.2	110.6
076	Hackensack	99.7	121.7	109.7
077	Long Branch	99.4	120.0	108.7
078	Dover	100.0	121.5	109.8
079	Summit	100.1	120.7	109.4
080,083	Vineland	97.8	117.7	106.8
081	Camden	100.0	115.2	106.9
082,084	Atlantic City	98.6	115.7	106.4
085-086	Trenton	99.9	119.1	108.6
087	Point Pleasant	99.8	116.8	107.5
088-089	New Brunswick	100.3	119.4	108.9
NEW MEXICO				
870-872	Albuquerque	98.3	76.8	88.6
873	Gallup	98.3	76.8	88.6
874	Farmington	98.6	76.8	88.7
875	Santa Fe	97.3	76.8	88.0
877	Las Vegas	96.8	76.8	87.7
878	Socorro	96.4	76.8	87.5
879	Truth/Consequences	97.1	71.6	85.5
880	Las Cruces	95.8	69.1	83.7
881	Clovis	96.9	76.1	87.4
882	Roswell	98.5	76.1	88.4
883	Carrizozo	99.0	76.8	88.9
884	Tucumcari	97.8	76.1	87.9
NEW YORK				
100-102	New York	106.6	163.5	132.4
103	Staten Island	102.1	155.2	126.2
104	Bronx	100.2	155.2	125.2
105	Mount Vernon	100.2	135.4	116.2
106	White Plains	99.9	135.4	116.0
107	Yonkers	105.1	136.2	119.2
108	New Rochelle	100.7	135.4	116.5
109	Suffern	100.5	123.7	111.0
110	Queens	101.4	155.0	125.8
111	Long Island City	103.1	155.0	126.7
112	Brooklyn	103.4	156.5	127.5
113	Flushing	103.6	155.0	127.0
114	Jamaica	101.7	155.2	126.0
115,117,118	Hicksville	101.2	145.0	121.1
116	Far Rockaway	103.7	155.0	127.0
119	Riverhead	101.6	145.1	121.4
120-122	Albany	97.9	93.9	96.1
123	Schenectady	98.2	93.0	95.8
124	Kingston	101.1	114.3	107.1
125-126	Poughkeepsie	100.3	119.0	108.8
127	Monticello	99.6	113.8	106.1
128	Glens Falls	92.8	89.9	91.5
129	Plattsburgh	97.0	82.8	90.6
130-132	Syracuse	99.0	90.5	95.2
133-135	Utica	97.0	87.8	92.8
136	Watertown	98.7	88.4	94.0
137-139	Binghamton	98.6	84.5	92.2
140-142	Buffalo	99.9	104.7	102.1
143	Niagara Falls	97.6	102.8	100.0
144-146	Rochester	100.1	94.9	97.7
147	Jamestown	96.6	85.4	91.5
148-149	Elmira	96.5	84.8	91.2
NORTH CAROLINA				
270,272-274	Greensboro	98.0	50.3	76.3
271	Winston-Salem	97.8	49.1	75.7
275-276	Raleigh	97.9	50.3	76.3
277	Durham	97.9	50.3	76.3
278	Rocky Mount	95.4	36.2	68.5
279	Elizabeth City	96.1	38.6	70.0
280	Gastonia	97.2	47.5	74.6
281-282	Charlotte	98.2	47.8	75.3
283	Fayetteville	96.7	50.3	75.6
284	Wilmington	96.0	49.4	74.8
285	Kinston	94.2	35.8	67.6

STATE/ZIP	CITY	MAT.	INST.	TOTAL
NORTH CAROLINA (CONT'D)				
286	Hickory	94.5	34.9	67.5
287-288	Asheville	96.3	48.3	74.5
289	Murphy	95.5	34.0	67.6
NORTH DAKOTA				
580-581	Fargo	100.8	65.9	85.0
582	Grand Forks	100.4	60.7	82.4
583	Devils Lake	100.0	61.0	82.3
584	Jamestown	100.0	54.2	79.2
585	Bismarck	99.2	66.9	84.5
586	Dickinson	100.8	62.3	83.3
587	Minot	100.5	68.2	85.8
588	Williston	99.3	61.7	82.2
OHIO				
430-432	Columbus	97.8	90.5	94.5
433	Marion	94.4	88.3	91.6
434-436	Toledo	98.0	99.6	98.7
437-438	Zanesville	94.8	83.9	89.8
439	Steubenville	95.7	94.3	95.1
440	Lorain	96.8	98.1	97.4
441	Cleveland	97.1	106.0	101.1
442-443	Akron	97.7	98.0	97.8
444-445	Youngstown	97.1	94.8	96.0
446-447	Canton	97.2	88.5	93.2
448-449	Mansfield	94.5	93.4	94.0
450	Hamilton	94.3	90.4	92.5
451-452	Cincinnati	94.7	92.3	93.6
453-454	Dayton	94.4	84.9	90.1
455	Springfield	94.4	86.7	90.9
456	Chillicothe	93.5	92.4	93.0
457	Athens	96.5	82.3	90.1
458	Lima	97.1	87.9	92.9
OKLAHOMA				
730-731	Oklahoma City	97.3	64.0	82.1
734	Ardmore	94.2	63.3	80.2
735	Lawton	96.5	65.0	82.2
736	Clinton	95.7	61.6	80.2
737	Enid	96.1	61.6	80.4
738	Woodward	94.5	61.7	79.6
739	Guymon	95.6	33.9	67.6
740-741	Tulsa	96.3	61.2	80.4
743	Miami	93.3	67.1	81.4
744	Muskogee	95.7	43.7	72.1
745	Mcalester	92.9	56.1	76.2
746	Ponca City	93.4	61.5	78.9
747	Durant	93.4	61.7	79.0
748	Shawnee	94.9	60.1	79.1
749	Poteau	92.5	64.3	79.7
OREGON				
970-972	Portland	102.2	105.4	103.7
973	Salem	102.2	103.0	102.5
974	Eugene	101.7	102.1	101.9
975	Medford	103.4	100.3	102.0
976	Klamath Falls	103.7	100.1	102.1
977	Bend	102.5	102.2	102.4
978	Pendleton	96.6	100.7	98.5
979	Vale	94.4	93.7	94.1
PENNSYLVANIA				
150-152	Pittsburgh	96.7	103.8	99.9
153	Washington	93.8	102.1	97.6
154	Uniontown	94.0	99.6	96.5
155	Bedford	95.0	91.8	93.5
156	Greensburg	95.1	99.6	97.1
157	Indiana	93.9	97.3	95.4
158	Dubois	95.2	95.7	95.5
159	Johnstown	94.9	96.2	95.5
160	Butler	92.3	101.6	96.5
161	New Castle	92.3	100.2	95.9
162	Kittanning	92.8	103.2	97.5
163	Oil City	92.3	95.8	93.9
164-165	Erie	94.3	96.2	95.1
166	Altoona	94.3	93.3	93.8
167	Bradford	95.9	93.3	94.7
168	State College	95.5	93.7	94.7
169	Wellsboro	96.5	87.4	92.4
170-171	Harrisburg	98.1	92.3	95.5
172	Chambersburg	96.1	87.2	92.0
173-174	York	96.2	88.6	92.7
175-176	Lancaster	94.9	87.5	91.5

STATE/ZIP	CITY	MAT.	INST.	TOTAL
PENNSYLVANIA (CONT'D)				
177	Williamsport	93.4	81.8	88.2
178	Sunbury	95.5	87.9	92.0
179	Pottsville	94.6	89.3	92.2
180	Lehigh Valley	96.0	102.7	99.0
181	Allentown	98.0	101.2	99.4
182	Hazleton	95.3	92.4	94.0
183	Stroudsburg	95.3	95.8	95.5
184-185	Scranton	98.8	92.9	96.1
186-187	Wilkes-Barre	95.3	91.4	93.5
188	Montrose	94.9	92.0	93.6
189	Doylestown	94.8	116.5	104.6
190-191	Philadelphia	100.8	128.9	113.6
193	Westchester	97.6	115.6	105.7
194	Norristown	96.5	116.5	105.6
195-196	Reading	98.6	98.5	98.6
PUERTO RICO				
009	San Juan	136.6	26.3	86.5
RHODE ISLAND				
028	Newport	99.9	112.4	105.6
029	Providence	100.2	112.4	105.8
SOUTH CAROLINA				
290-292	Columbia	96.6	48.3	74.7
293	Spartanburg	95.5	46.0	73.0
294	Charleston	96.8	48.9	75.1
295	Florence	95.0	46.1	72.8
296	Greenville	95.2	46.0	72.9
297	Rock Hill	95.0	33.8	67.2
298	Aiken	95.9	70.4	84.3
299	Beaufort	96.7	37.4	69.8
SOUTH DAKOTA				
570-571	Sioux Falls	99.1	57.8	80.3
572	Watertown	97.9	52.0	77.0
573	Mitchell	96.9	51.6	76.3
574	Aberdeen	99.4	54.6	79.0
575	Pierre	98.7	53.7	78.2
576	Mobridge	97.5	51.9	76.8
577	Rapid City	99.3	52.5	78.0
TENNESSEE				
370-372	Nashville	98.0	74.0	87.1
373-374	Chattanooga	98.8	55.8	79.3
375,380-381	Memphis	96.6	77.2	87.8
376	Johnson City	97.6	58.2	79.7
377-379	Knoxville	94.9	59.2	78.7
382	Mckenzie	96.5	49.7	75.3
383	Jackson	98.5	53.3	78.0
384	Columbia	95.0	56.9	77.7
385	Cookeville	96.3	51.4	75.9
TEXAS				
750	Mckinney	98.2	57.1	79.5
751	Waxahackie	98.1	57.8	79.8
752-753	Dallas	98.7	67.1	84.3
754	Greenville	98.3	41.9	72.7
755	Texarkana	97.2	53.0	77.2
756	Longview	97.5	43.9	73.1
757	Tyler	98.1	56.4	79.2
758	Palestine	94.1	44.3	71.5
759	Lufkin	95.1	48.1	73.7
760-761	Fort Worth	96.4	63.3	81.4
762	Denton	96.9	53.2	77.0
763	Wichita Falls	97.4	57.2	79.1
764	Eastland	95.9	43.3	72.0
765	Temple	94.8	51.4	75.1
766-767	Waco	96.6	57.8	79.0
768	Brownwood	96.9	40.7	71.4
769	San Angelo	96.6	49.0	75.0
770-772	Houston	99.1	72.8	87.2
773	Huntsville	97.6	41.6	72.1
774	Wharton	98.8	46.5	75.1
775	Galveston	97.0	70.7	85.0
776-777	Beaumont	96.6	63.5	81.6
778	Bryan	94.5	65.2	81.2
779	Victoria	99.0	49.1	76.4
780	Laredo	94.9	55.0	76.8
781-782	San Antonio	95.2	66.8	82.3
783-784	Corpus Christi	97.7	54.1	77.8
785	Mc Allen	97.8	48.2	75.2
786-787	Austin	94.9	61.8	79.8

Location Factors

STATE/ZIP	CITY	MAT.	INST.	TOTAL
TEXAS (CONT'D)				
788	Del Rio	97.1	34.2	68.6
789	Giddings	94.5	43.4	71.3
790-791	Amarillo	97.3	59.1	79.9
792	Childress	96.5	53.5	76.9
793-794	Lubbock	98.8	54.9	78.9
795-796	Abilene	96.8	54.5	77.6
797	Midland	99.1	51.4	77.4
798-799,885	El Paso	96.7	52.4	76.6
UTAH				
840-841	Salt Lake City	101.2	72.3	88.1
842,844	Ogden	97.2	72.1	85.8
843	Logan	98.9	72.1	86.7
845	Price	99.6	51.1	77.6
846-847	Provo	99.5	72.1	87.1
VERMONT				
050	White River Jct.	99.0	47.9	75.8
051	Bellows Falls	97.5	50.3	76.0
052	Bennington	97.8	50.6	76.4
053	Brattleboro	98.2	50.8	76.7
054	Burlington	100.8	66.0	85.0
056	Montpelier	97.7	66.1	83.3
057	Rutland	99.8	66.1	84.5
058	St. Johnsbury	99.2	50.6	77.1
059	Guildhall	97.8	50.1	76.1
VIRGINIA				
220-221	Fairfax	99.2	81.2	91.0
222	Arlington	100.2	79.4	90.7
223	Alexandria	99.6	84.6	92.8
224-225	Fredericksburg	97.8	68.0	84.3
226	Winchester	98.5	55.4	78.9
227	Culpeper	98.3	56.6	79.4
228	Harrisonburg	98.6	48.1	75.7
229	Charlottesville	98.9	60.5	81.5
230-232	Richmond	99.9	65.9	84.5
233-235	Norfolk	99.6	65.2	84.0
236	Newport News	99.2	63.6	83.0
237	Portsmouth	98.4	65.0	83.2
238	Petersburg	98.6	65.9	83.8
239	Farmville	98.1	42.3	72.8
240-241	Roanoke	99.8	48.4	76.5
242	Bristol	97.8	48.7	75.5
243	Pulaski	97.5	43.7	73.1
244	Staunton	98.2	46.5	74.8
245	Lynchburg	98.4	51.0	76.8
246	Grundy	97.8	45.8	74.2
WASHINGTON				
980-981,987	Seattle	104.3	105.8	105.0
982	Everett	104.0	98.3	101.4
983-984	Tacoma	104.1	99.7	102.1
985	Olympia	102.8	99.7	101.4
986	Vancouver	106.0	97.2	102.0
988	Wenatchee	104.3	84.0	95.1
989	Yakima	104.2	90.0	97.8
990-992	Spokane	106.2	83.7	96.0
993	Richland	105.5	85.1	96.2
994	Clarkston	104.7	82.3	94.6
WEST VIRGINIA				
247-248	Bluefield	96.6	80.2	89.1
249	Lewisburg	98.1	83.9	91.6
250-253	Charleston	99.1	92.0	95.9
254	Martinsburg	98.0	80.0	89.8
255-257	Huntington	99.3	90.9	95.5
258-259	Beckley	96.4	89.5	93.3
260	Wheeling	99.3	93.7	96.8
261	Parkersburg	98.3	90.6	94.8
262	Buckhannon	97.9	93.2	95.8
263-264	Clarksburg	98.3	92.2	95.5
265	Morgantown	98.4	93.2	96.0
266	Gassaway	97.7	92.3	95.3
267	Romney	97.7	86.4	92.6
268	Petersburg	97.7	89.1	93.8
WISCONSIN				
530,532	Milwaukee	98.7	101.6	100.0
531	Kenosha	98.2	97.8	98.0
534	Racine	97.8	99.2	98.4
535	Beloit	97.7	96.4	97.1
537	Madison	97.9	93.9	96.1

STATE/ZIP	CITY	MAT.	INST.	TOTAL
(CONT'D)				
538	Lancaster	95.6	91.4	93.7
539	Portage	94.2	93.3	93.8
540	New Richmond	95.5	94.6	95.1
541-543	Green Bay	99.5	90.9	95.6
544	Wausau	94.9	90.9	93.1
545	Rhinelander	98.0	90.7	94.7
546	La Crosse	96.0	91.7	94.0
547	Eau Claire	97.7	93.3	95.7
548	Superior	95.4	94.8	95.1
549	Oshkosh	95.5	90.5	93.3
WYOMING				
820	Cheyenne	97.5	54.7	78.0
821	Yellowstone Nat'l Park	96.7	49.1	75.1
822	Wheatland	98.1	48.0	75.4
823	Rawlins	99.4	44.1	74.3
824	Worland	97.3	44.8	73.5
825	Riverton	98.4	47.3	75.2
826	Casper	97.6	60.1	80.5
827	Newcastle	97.2	44.1	73.1
828	Sheridan	98.1	52.8	77.5
829-831	Rock Springs	101.4	44.8	75.7
CANADIAN FACTORS (reflect Canadian currency)				
ALBERTA				
	Calgary	124.6	82.0	105.2
	Edmonton	126.2	82.0	106.2
	Fort McMurray	116.8	81.6	100.8
	Lethbridge	117.7	81.0	101.1
	Lloydminster	116.9	81.6	100.8
	Medicine Hat	117.0	81.0	100.6
	Red Deer	117.4	81.0	100.9
BRITISH COLUMBIA				
	Kamloops	117.8	84.3	102.6
	Prince George	119.1	84.3	103.3
	Vancouver	125.6	88.2	108.6
	Victoria	119.1	84.8	103.6
MANITOBA				
	Brandon	117.1	70.7	96.0
	Portage la Prairie	117.1	70.7	96.0
	Winnipeg	124.3	71.0	100.1
NEW BRUNSWICK				
	Bathurst	115.2	62.8	91.4
	Dalhousie	115.2	62.8	91.4
	Fredericton	117.1	66.9	94.3
	Moncton	115.5	62.8	91.6
	Newcastle	115.2	62.8	91.4
	Saint John	118.3	66.9	94.9
NEWFOUNDLAND				
	Corner Brook	120.1	62.8	94.1
	St. John's	120.0	62.8	94.0
NORTHWEST TERRITORIES				
	Yellowknife	114.3	79.8	98.6
NOVA SCOTIA				
	Dartmouth	117.7	69.2	95.7
	Halifax	119.3	70.3	97.0
	New Glasgow	115.8	69.2	94.6
	Sydney	113.3	69.2	93.3
	Yarmouth	115.6	69.2	94.5
ONTARIO				
	Barrie	120.0	86.7	104.9
	Brantford	119.2	90.3	106.1
	Cornwall	119.1	86.9	104.5
	Hamilton	124.7	90.6	109.2
	Kingston	120.1	87.3	105.2
	Kitchener	115.6	86.3	102.3
	London	121.8	88.0	106.5
	North Bay	119.2	85.0	103.6
	Oshawa	118.9	88.4	105.1
	Ottawa	120.7	88.1	105.9
	Owen Sound	120.3	85.0	104.2
	Peterborough	119.2	86.7	104.4
	Sarnia	119.4	90.7	106.3
	St. Catharines	113.6	85.6	100.9
	Sudbury	113.6	85.2	100.7

Location Factors

STATE/ZIP	CITY	MAT.	INST.	TOTAL
ONTARIO (CONT'D)				
	Thunder Bay	114.8	85.4	101.5
	Toronto	124.4	95.0	111.0
	Windsor	114.7	88.0	102.6
PRINCE EDWARD ISLAND				
	Charlottetown	117.8	58.7	91.0
	Summerside	117.5	58.7	90.8
QUEBEC				
	Cap-de-la-Madeleine	116.0	81.4	100.3
	Charlesbourg	116.0	81.4	100.3
	Chicoutimi	115.0	81.1	99.6
	Gatineau	115.4	81.1	99.9
	Laval	115.4	81.1	99.8
	Montreal	118.8	81.4	101.8
	Quebec	119.6	81.8	102.5
	Sherbrooke	115.8	81.1	100.1
	Trois Rivieres	116.3	81.4	100.4
SASKATCHEWAN				
	Moose Jaw	114.2	64.8	91.8
	Prince Albert	113.6	64.7	91.4
	Regina	116.0	64.9	92.8
	Saskatoon	114.7	64.7	92.0
YUKON				
	Whitehorse	114.0	63.8	91.2

A	Area Square Feet; Ampere	Cab.	Cabinet	Demob.	Demobilization
ABS	Acrylonitrile Butadiene Stryrene; Asbestos Bonded Steel	Cair.	Air Tool Laborer	d.f.u.	Drainage Fixture Units
		Calc	Calculated	D.H.	Double Hung
A.C.	Alternating Current; Air-Conditioning; Asbestos Cement; Plywood Grade A & C	Cap.	Capacity	DHW	Domestic Hot Water
		Carp.	Carpenter	Diag.	Diagonal
		C.B.	Circuit Breaker	Diam.	Diameter
		C.C.A.	Chromate Copper Arsenate	Distrib.	Distribution
A.C.I.	American Concrete Institute	C.C.F.	Hundred Cubic Feet	Dk.	Deck
AD	Plywood, Grade A & D	cd	Candela	D.L.	Dead Load; Diesel
Addit.	Additional	cd/sf	Candela per Square Foot	DLH	Deep Long Span Bar Joist
Adj.	Adjustable	CD	Grade of Plywood Face & Back	Do.	Ditto
af	Audio-frequency	CDX	Plywood, Grade C & D, exterior glue	Dp.	Depth
A.G.A.	American Gas Association			D.P.S.T.	Double Pole, Single Throw
Agg.	Aggregate	Cefi.	Cement Finisher	Dr.	Driver
A.H.	Ampere Hours	Cem.	Cement	Drink.	Drinking
A hr.	Ampere-hour	CF	Hundred Feet	D.S.	Double Strength
A.H.U.	Air Handling Unit	C.F.	Cubic Feet	D.S.A.	Double Strength A Grade
A.I.A.	American Institute of Architects	CFM	Cubic Feet per Minute	D.S.B.	Double Strength B Grade
AIC	Ampere Interrupting Capacity	c.g.	Center of Gravity	Dty.	Duty
Allow.	Allowance	CHW	Chilled Water; Commercial Hot Water	DWV	Drain Waste Vent
alt.	Altitude			DX	Deluxe White, Direct Expansion
Alum.	Aluminum	C.I.	Cast Iron	dyn	Dyne
a.m.	Ante Meridiem	C.I.P.	Cast in Place	e	Eccentricity
Amp.	Ampere	Circ.	Circuit	E	Equipment Only; East
Anod.	Anodized	C.L.	Carload Lot	Ea.	Each
Approx.	Approximate	Clab.	Common Laborer	E.B.	Encased Burial
Apt.	Apartment	Clam	Common maintenance laborer	Econ.	Economy
Asb.	Asbestos	C.L.F.	Hundred Linear Feet	E.C.Y	Embankment Cubic Yards
A.S.B.C.	American Standard Building Code	CLF	Current Limiting Fuse	EDP	Electronic Data Processing
Asbe.	Asbestos Worker	CLP	Cross Linked Polyethylene	EIFS	Exterior Insulation Finish System
A.S.H.R.A.E.	American Society of Heating, Refrig. & AC Engineers	cm	Centimeter	E.D.R.	Equiv. Direct Radiation
		CMP	Corr. Metal Pipe	Eq.	Equation
A.S.M.E.	American Society of Mechanical Engineers	C.M.U.	Concrete Masonry Unit	Elec.	Electrician; Electrical
		CN	Change Notice	Elev.	Elevator; Elevating
A.S.T.M.	American Society for Testing and Materials	Col.	Column	EMT	Electrical Metallic Conduit; Thin Wall Conduit
		CO2	Carbon Dioxide		
Attchmt.	Attachment	Comb.	Combination	Eng.	Engine, Engineered
Avg.	Average	Compr.	Compressor	EPDM	Ethylene Propylene Diene Monomer
A.W.G.	American Wire Gauge	Conc.	Concrete		
AWWA	American Water Works Assoc.	Cont.	Continuous; Continued	EPS	Expanded Polystyrene
Bbl.	Barrel	Corr.	Corrugated	Eqhv.	Equip. Oper., Heavy
B&B	Grade B and Better; Balled & Burlapped	Cos	Cosine	Eqlt.	Equip. Oper., Light
		Cot	Cotangent	Eqmd.	Equip. Oper., Medium
B.&S.	Bell and Spigot	Cov.	Cover	Eqmm.	Equip. Oper., Master Mechanic
B.&W.	Black and White	C/P	Cedar on Paneling	Eqol.	Equip. Oper., Oilers
b.c.c.	Body-centered Cubic	CPA	Control Point Adjustment	Equip.	Equipment
B.C.Y.	Bank Cubic Yards	Cplg.	Coupling	ERW	Electric Resistance Welded
BE	Bevel End	C.P.M.	Critical Path Method	E.S.	Energy Saver
B.F.	Board Feet	CPVC	Chlorinated Polyvinyl Chloride	Est.	Estimated
Bg. cem.	Bag of Cement	C.Pr.	Hundred Pair	esu	Electrostatic Units
BHP	Boiler Horsepower; Brake Horsepower	CRC	Cold Rolled Channel	E.W.	Each Way
		Creos.	Creosote	EWT	Entering Water Temperature
B.I.	Black Iron	Crpt.	Carpet & Linoleum Layer	Excav.	Excavation
Bit.; Bitum.	Bituminous	CRT	Cathode-ray Tube	Exp.	Expansion, Exposure
Bk.	Backed	CS	Carbon Steel, Constant Shear Bar Joist	Ext.	Exterior
Bkrs.	Breakers			Extru.	Extrusion
Bldg.	Building	Csc	Cosecant	f.	Fiber stress
Blk.	Block	C.S.F.	Hundred Square Feet	F	Fahrenheit; Female; Fill
Bm.	Beam	CSI	Construction Specifications Institute	Fab.	Fabricated
Boil.	Boilermaker			FBGS	Fiberglass
B.P.M.	Blows per Minute	C.T.	Current Transformer	F.C.	Footcandles
BR	Bedroom	CTS	Copper Tube Size	f.c.c.	Face-centered Cubic
Brg.	Bearing	Cu	Copper, Cubic	f'c.	Compressive Stress in Concrete; Extreme Compressive Stress
Brhe.	Bricklayer Helper	Cu. Ft.	Cubic Foot		
Bric.	Bricklayer	cw	Continuous Wave	F.E.	Front End
Brk.	Brick	C.W.	Cool White; Cold Water	FEP	Fluorinated Ethylene Propylene (Teflon)
Brng.	Bearing	Cwt.	100 Pounds		
Brs.	Brass	C.W.X.	Cool White Deluxe	F.G.	Flat Grain
Brz.	Bronze	C.Y.	Cubic Yard (27 cubic feet)	F.H.A.	Federal Housing Administration
Bsn.	Basin	C.Y./Hr.	Cubic Yard per Hour	Fig.	Figure
Btr.	Better	Cyl.	Cylinder	Fin.	Finished
BTU	British Thermal Unit	d	Penny (nail size)	Fixt.	Fixture
BTUH	BTU per Hour	D	Deep; Depth; Discharge	Fl. Oz.	Fluid Ounces
B.U.R.	Built-up Roofing	Dis.;Disch.	Discharge	Flr.	Floor
BX	Interlocked Armored Cable	Db.	Decibel	F.M.	Frequency Modulation; Factory Mutual
c	Conductivity, Copper Sweat	Dbl.	Double		
C	Hundred; Centigrade	DC	Direct Current	Fmg.	Framing
C/C	Center to Center, Cedar on Cedar	DDC	Direct Digital Control	Fndtn.	Foundation

Fori.	Foreman, Inside
Foro.	Foreman, Outside
Fount.	Fountain
FPM	Feet per Minute
FPT	Female Pipe Thread
Fr.	Frame
F.R.	Fire Rating
FRK	Foil Reinforced Kraft
FRP	Fiberglass Reinforced Plastic
FS	Forged Steel
FSC	Cast Body; Cast Switch Box
Ft.	Foot; Feet
Ftng.	Fitting
Ftg.	Footing
Ft. Lb.	Foot Pound
Furn.	Furniture
FVNR	Full Voltage Non-Reversing
FXM	Female by Male
Fy.	Minimum Yield Stress of Steel
g	Gram
G	Gauss
Ga.	Gauge
Gal.	Gallon
Gal./Min.	Gallon per Minute
Galv.	Galvanized
Gen.	General
G.F.I.	Ground Fault Interrupter
Glaz.	Glazier
GPD	Gallons per Day
GPH	Gallons per Hour
GPM	Gallons per Minute
GR	Grade
Gran.	Granular
Grnd.	Ground
H	High; High Strength Bar Joist; Henry
H.C.	High Capacity
H.D.	Heavy Duty; High Density
H.D.O.	High Density Overlaid
Hdr.	Header
Hdwe.	Hardware
Help.	Helper Average
HEPA	High Efficiency Particulate Air Filter
Hg	Mercury
HIC	High Interrupting Capacity
HM	Hollow Metal
H.O.	High Output
Horiz.	Horizontal
H.P.	Horsepower; High Pressure
H.P.F.	High Power Factor
Hr.	Hour
Hrs./Day	Hours per Day
HSC	High Short Circuit
Ht.	Height
Htg.	Heating
Htrs.	Heaters
HVAC	Heating, Ventilation & Air-Conditioning
Hvy.	Heavy
HW	Hot Water
Hyd.;Hydr.	Hydraulic
Hz.	Hertz (cycles)
I.	Moment of Inertia
I.C.	Interrupting Capacity
ID	Inside Diameter
I.D.	Inside Dimension; Identification
I.F.	Inside Frosted
I.M.C.	Intermediate Metal Conduit
In.	Inch
Incan.	Incandescent
Incl.	Included; Including
Int.	Interior
Inst.	Installation
Insul.	Insulation/Insulated
I.P.	Iron Pipe
I.P.S.	Iron Pipe Size
I.P.T.	Iron Pipe Threaded
I.W.	Indirect Waste
J	Joule
J.I.C.	Joint Industrial Council
K	Thousand; Thousand Pounds; Heavy Wall Copper Tubing, Kelvin
K.A.H.	Thousand Amp. Hours
KCMIL	Thousand Circular Mils
KD	Knock Down
K.D.A.T.	Kiln Dried After Treatment
kg	Kilogram
kG	Kilogauss
kgf	Kilogram Force
kHz	Kilohertz
Kip.	1000 Pounds
KJ	Kiljoule
K.L.	Effective Length Factor
K.L.F.	Kips per Linear Foot
Km	Kilometer
K.S.F.	Kips per Square Foot
K.S.I.	Kips per Square Inch
kV	Kilovolt
kVA	Kilovolt Ampere
K.V.A.R.	Kilovar (Reactance)
KW	Kilowatt
KWh	Kilowatt-hour
L	Labor Only; Length; Long; Medium Wall Copper Tubing
Lab.	Labor
lat	Latitude
Lath.	Lather
Lav.	Lavatory
lb.; #	Pound
L.B.	Load Bearing; L Conduit Body
L. & E.	Labor & Equipment
lb./hr.	Pounds per Hour
lb./L.F.	Pounds per Linear Foot
lbf/sq.in.	Pound-force per Square Inch
L.C.L.	Less than Carload Lot
L.C.Y.	Loose Cubic Yard
Ld.	Load
LE	Lead Equivalent
LED	Light Emitting Diode
L.F.	Linear Foot
Lg.	Long; Length; Large
L & H	Light and Heat
LH	Long Span Bar Joist
L.H.	Labor Hours
L.L.	Live Load
L.L.D.	Lamp Lumen Depreciation
lm	Lumen
lm/sf	Lumen per Square Foot
lm/W	Lumen per Watt
L.O.A.	Length Over All
log	Logarithm
L-O-L	Lateralolet
L.P.	Liquefied Petroleum; Low Pressure
L.P.F.	Low Power Factor
LR	Long Radius
L.S.	Lump Sum
Lt.	Light
Lt. Ga.	Light Gauge
L.T.L.	Less than Truckload Lot
Lt. Wt.	Lightweight
L.V.	Low Voltage
M	Thousand; Material; Male; Light Wall Copper Tubing
M²CA	Meters Squared Contact Area
m/hr; M.H.	Man-hour
mA	Milliampere
Mach.	Machine
Mag. Str.	Magnetic Starter
Maint.	Maintenance
Marb.	Marble Setter
Mat; Mat'l.	Material
Max.	Maximum
MBF	Thousand Board Feet
MBH	Thousand BTU's per hr.
MC	Metal Clad Cable
M.C.F.	Thousand Cubic Feet
M.C.F.M.	Thousand Cubic Feet per Minute
M.C.M.	Thousand Circular Mils
M.C.P.	Motor Circuit Protector
MD	Medium Duty
M.D.O.	Medium Density Overlaid
Med.	Medium
MF	Thousand Feet
M.F.B.M.	Thousand Feet Board Measure
Mfg.	Manufacturing
Mfrs.	Manufacturers
mg	Milligram
MGD	Million Gallons per Day
MGPH	Thousand Gallons per Hour
MH, M.H.	Manhole; Metal Halide; Man-Hour
MHz	Megahertz
Mi.	Mile
MI	Malleable Iron; Mineral Insulated
mm	Millimeter
Mill.	Millwright
Min., min.	Minimum, minute
Misc.	Miscellaneous
ml	Milliliter, Mainline
M.L.F.	Thousand Linear Feet
Mo.	Month
Mobil.	Mobilization
Mog.	Mogul Base
MPH	Miles per Hour
MPT	Male Pipe Thread
MRT	Mile Round Trip
ms	Millisecond
M.S.F.	Thousand Square Feet
Mstz.	Mosaic & Terrazzo Worker
M.S.Y.	Thousand Square Yards
Mtd.	Mounted
Mthe.	Mosaic & Terrazzo Helper
Mtng.	Mounting
Mult.	Multi; Multiply
M.V.A.	Million Volt Amperes
M.V.A.R.	Million Volt Amperes Reactance
MV	Megavolt
MW	Megawatt
MXM	Male by Male
MYD	Thousand Yards
N	Natural; North
nA	Nanoampere
NA	Not Available; Not Applicable
N.B.C.	National Building Code
NC	Normally Closed
N.E.M.A.	National Electrical Manufacturers Assoc.
NEHB	Bolted Circuit Breaker to 600V.
N.L.B.	Non-Load-Bearing
NM	Non-Metallic Cable
nm	Nanometer
No.	Number
NO	Normally Open
N.O.C.	Not Otherwise Classified
Nose.	Nosing
N.P.T.	National Pipe Thread
NQOD	Combination Plug-on/Bolt on Circuit Breaker to 240V.
N.R.C.	Noise Reduction Coefficient
N.R.S.	Non Rising Stem
ns	Nanosecond
nW	Nanowatt
OB	Opposing Blade
OC	On Center
OD	Outside Diameter
O.D.	Outside Dimension
ODS	Overhead Distribution System
O.G.	Ogee
O.H.	Overhead
O&P	Overhead and Profit
Oper.	Operator
Opng.	Opening
Orna.	Ornamental
OSB	Oriented Strand Board

O.S.&Y.	Outside Screw and Yoke	Rsr	Riser	Tilf.	Tile Layer, Floor
Ovhd.	Overhead	RT	Round Trip	Tilh.	Tile Layer, Helper
OWG	Oil, Water or Gas	S.	Suction; Single Entrance; South	THHN	Nylon Jacketed Wire
Oz.	Ounce	SCFM	Standard Cubic Feet per Minute	THW.	Insulated Strand Wire
P.	Pole; Applied Load; Projection	Scaf.	Scaffold	THWN;	Nylon Jacketed Wire
p.	Page	Sch.; Sched.	Schedule	T.L.	Truckload
Pape.	Paperhanger	S.C.R.	Modular Brick	T.M.	Track Mounted
P.A.P.R.	Powered Air Purifying Respirator	S.D.	Sound Deadening	Tot.	Total
PAR	Parabolic Reflector	S.D.R.	Standard Dimension Ratio	T-O-L	Threadolet
Pc., Pcs.	Piece, Pieces	S.E.	Surfaced Edge	T.S.	Trigger Start
P.C.	Portland Cement; Power Connector	Sel.	Select	Tr.	Trade
P.C.F.	Pounds per Cubic Foot	S.E.R.; S.E.U.	Service Entrance Cable	Transf.	Transformer
P.C.M.	Phase Contract Microscopy	S.F.	Square Foot	Trhv.	Truck Driver, Heavy
P.E.	Professional Engineer;	S.F.C.A.	Square Foot Contact Area	Trlr	Trailer
	Porcelain Enamel;	S.F. Flr.	Square Foot of Floor	Trlt.	Truck Driver, Light
	Polyethylene; Plain End	S.F.G.	Square Foot of Ground	TV	Television
Perf.	Perforated	S.F. Hor.	Square Foot Horizontal	T.W.	Thermoplastic Water Resistant
Ph.	Phase	S.F.R.	Square Feet of Radiation		Wire
P.I.	Pressure Injected	S.F. Shlf.	Square Foot of Shelf	UCI	Uniform Construction Index
Pile.	Pile Driver	S4S	Surface 4 Sides	UF	Underground Feeder
Pkg.	Package	Shee.	Sheet Metal Worker	UGND	Underground Feeder
Pl.	Plate	Sin.	Sine	U.H.F.	Ultra High Frequency
Plah.	Plasterer Helper	Skwk.	Skilled Worker	U.L.	Underwriters Laboratory
Plas.	Plasterer	SL	Saran Lined	Unfin.	Unfinished
Pluh.	Plumbers Helper	S.L.	Slimline	URD	Underground Residential
Plum.	Plumber	Sldr.	Solder		Distribution
Ply.	Plywood	SLH	Super Long Span Bar Joist	US	United States
p.m.	Post Meridiem	S.N.	Solid Neutral	USP	United States Primed
Pntd.	Painted	S-O-L	Socketolet	UTP	Unshielded Twisted Pair
Pord.	Painter, Ordinary	sp	Standpipe	V	Volt
pp	Pages	S.P.	Static Pressure; Single Pole; Self-	V.A.	Volt Amperes
PP; PPL	Polypropylene		Propelled	V.C.T.	Vinyl Composition Tile
P.P.M.	Parts per Million	Spri.	Sprinkler Installer	VAV	Variable Air Volume
Pr.	Pair	spwg	Static Pressure Water Gauge	VC	Veneer Core
P.E.S.B.	Pre-engineered Steel Building	S.P.D.T.	Single Pole, Double Throw	Vent.	Ventilation
Prefab.	Prefabricated	SPF	Spruce Pine Fir	Vert.	Vertical
Prefin.	Prefinished	S.P.S.T.	Single Pole, Single Throw	V.F.	Vinyl Faced
Prop.	Propelled	SPT	Standard Pipe Thread	V.G.	Vertical Grain
PSF; psf	Pounds per Square Foot	Sq.	Square; 100 Square Feet	V.H.F.	Very High Frequency
PSI; psi	Pounds per Square Inch	Sq. Hd.	Square Head	VHO	Very High Output
PSIG	Pounds per Square Inch Gauge	Sq. In.	Square Inch	Vib.	Vibrating
PSP	Plastic Sewer Pipe	S.S.	Single Strength; Stainless Steel	V.L.F.	Vertical Linear Foot
Pspr.	Painter, Spray	S.S.B.	Single Strength B Grade	Vol.	Volume
Psst.	Painter, Structural Steel	sst	Stainless Steel	VRP	Vinyl Reinforced Polyester
P.T.	Potential Transformer	Sswk.	Structural Steel Worker	W	Wire; Watt; Wide; West
P. & T.	Pressure & Temperature	Sswl.	Structural Steel Welder	w/	With
Ptd.	Painted	St.; Stl.	Steel	W.C.	Water Column; Water Closet
Ptns.	Partitions	S.T.C.	Sound Transmission Coefficient	W.F.	Wide Flange
Pu	Ultimate Load	Std.	Standard	W.G.	Water Gauge
PVC	Polyvinyl Chloride	STK	Select Tight Knot	Wldg.	Welding
Pvmt.	Pavement	STP	Standard Temperature & Pressure	W. Mile	Wire Mile
Pwr.	Power	Stpi.	Steamfitter, Pipefitter	W-O-L	Weldolet
Q	Quantity Heat Flow	Str.	Strength; Starter; Straight	W.R.	Water Resistant
Quan.; Qty.	Quantity	Strd.	Stranded	Wrck.	Wrecker
Q.C.	Quick Coupling	Struct.	Structural	W.S.P.	Water, Steam, Petroleum
r	Radius of Gyration	Sty.	Story	WT., Wt.	Weight
R	Resistance	Subj.	Subject	WWF	Welded Wire Fabric
R.C.P.	Reinforced Concrete Pipe	Subs.	Subcontractors	XFER	Transfer
Rect.	Rectangle	Surf.	Surface	XFMR	Transformer
Reg.	Regular	Sw.	Switch	XHD	Extra Heavy Duty
Reinf.	Reinforced	Swbd.	Switchboard	XHHW; XLPE	Cross-Linked Polyethylene Wire
Req'd.	Required	S.Y.	Square Yard		Insulation
Res.	Resistant	Syn.	Synthetic	XLP	Cross-linked Polyethylene
Resi.	Residential	S.Y.P.	Southern Yellow Pine	Y	Wye
Rgh.	Rough	Sys.	System	yd	Yard
RGS	Rigid Galvanized Steel	t.	Thickness	yr	Year
R.H.W.	Rubber, Heat & Water Resistant;	T	Temperature; Ton	Δ	Delta
	Residential Hot Water	Tan	Tangent	%	Percent
rms	Root Mean Square	T.C.	Terra Cotta	~	Approximately
Rnd.	Round	T & C	Threaded and Coupled	Ø	Phase
Rodm.	Rodman	T.D.	Temperature Difference	@	At
Rofc.	Roofer, Composition	T.E.M.	Transmission Electron Microscopy	#	Pound; Number
Rofp.	Roofer, Precast	TFE	Tetrafluoroethylene (Teflon)	<	Less Than
Rohe.	Roofer Helpers (Composition)	T. & G.	Tongue & Groove;	>	Greater Than
Rots.	Roofer, Tile & Slate		Tar & Gravel		
R.O.W.	Right of Way	Th.; Thk.	Thick		
RPM	Revolutions per Minute	Thn.	Thin		
R.S.	Rapid Start	Thrded	Threaded		

A

Abrasive aggregate 138
   floor 161
   silicon carbide 161
   stair tread concrete 150
ABS DWV pipe 277
Absorption testing 9
Accelerator set 138
Access door basement 165
   road and parking area 16
Accessories door and window ... 233
   reinforcing steel 151-153
   roof 228
Accessory dock 66
   formwork 147-150
   masonry 168, 172
   reinforcing 151
Acid resistant pipe 269
Acoustical metal deck 197
   sealant 229
Acrylic latex 138
   wall coating 239
Adhesive EPDM 223
   neoprene 223
   PVC 223
Adjustable jack post 191
Admixture cement 138, 161
   concrete 138
Adobe brick 174
Aerate lawn 134
Aeration sewage 243
Aerator 292
Aerial bucket 26
   lift 21
   photography 9
   survey 6
Aggregate abrasive 138
   base course 102
   concrete 139
   exposed 161
   lightweight 139
   marble 139
   masonry 169
   plaster 139
   roof 139
   spreader 25
   testing 9
Air circuit breaker 306
   compressor 21
   conditioner removal 256
   entraining agent 138
   hose 21
   lock 247
   spade 22
   supported building 13, 247
   supported storage tank cover . 246, 247
   supported structures 247
   tool 21
Air-compressor mobilization 73
Airplane hangar 249
Airport lighting 311
Alloy steel chain 202
Aluminium fence 112
Aluminum bar grating 199
   bench 120
   bulkhead 65, 66
   column 203
   column base 202, 203
   column cap 203
   dock 67
   expansion joint 204
   extrusion 195
   fence 111
   foil 225, 227

frame grating 201
framing 195
gangway 66
ladder 66, 198
mesh grating 200
nail 209
pipe 95, 98
plank grating 200
rivet 188
salvage 257
sheeting 65
shielded cable 298
shore 149
stair tread 150
structural 195
trench cover 202
weld rod 190
window demolition 232
Ampere values 405
Analysis petrographic 9
   sieve 9, 10
Anchor bolt 147, 169, 170, 184
   brick 170
   buck 170
   channel slot 171
   chemical 186
   dovetail 148
   epoxy 186
   expansion 187
   framing 211
   lead screw 187
   machinery 188
   masonry 170
   partition 171
   pipe conduit casing 87
   plastic screw 187
   pole 302
   rafter 211
   rigid 171
   self-drilling 187
   sill 211
   steel 171
   stone 171
   super high-tensile 46
   tower 302
   wedge 187
Anchors drilling 186
And cover tunnels cut 69
   shrub removal tree 122
Angle curb edging 192
   framing steel 192
Appliance plumbing 292
Arch culverts oval 101
   laminated 219
   radial 219
Area wall 233
Arrester lightning 306
Arrow 105
Ashlar veneer 178
Asphalt block 107
   block floor 107
   coating 222
   concrete 30
   curb 105
   cutting 39
   distributor 22
   expansion joint 148
   grinding 133
   mix design 9
   paper 227
   paver 23
   primer 228
   recycled 30
   rubberized 109
   sealcoat rubberized 109
   sheathing 218

sidewalk 106
soil stabilization 63
stabilization 63
testing 9
Asphaltic binder 103
   concrete 104
   emulsion 109, 122
   pavement 103
   paving 375
   wearing course 103
Athletic equipment 120
   field seeding 126
   or recreational screening 120
   paving 109
   post 120
Attachment grader 25
Auger 18
   boring 34
   hole 33
Augering 72
Auto park drain 289
Autoclaved concrete block 174

B

B and S pipe gasket 263
Backer rod 148, 229
Backfill 52, 61
   compaction 380
   dozer 52, 53
   general 52
   planting pit 124
   structural 52
   trench 62
Backflow preventer 286
Backhoe 19, 56
   excavation 61
   extension 25
   trenching 348
Backstop baseball 120
   basketball 120
   handball court 120
   squash court 120
   tennis court 120
Backup block 175
Backwater valve 287
Bale hay 65
Ball check valve 284
   valve 284
   wrecking 25
Ballast high intensity discharge .. 309
   railroad 203
   replacement 309
Bankrun gravel 139
Bar grouted 157
   joist painting 196
   tie 151
   ungrouted 157
Barbed wire fence 111
Barge construction 27
   mobilization 68
Bark mulch redwood 122
Barrel 22
Barricade 16, 22, 117
   tape 17
Barrier delineator 118
   impact 117
   median 117
   moisture 223
   parking 118
   slipform 25
   waterstop 148
   weed 122
Barriers and enclosures 16
   crash 118

Barriers, hywy. sound bar traffic . 118
Base column 211
   course 102
   course aggregate 102
   course bituminous 102
   gravel 103
   masonry 177
   plate column 195
   road 103
   stabilization 64
   stabilizer 25
   stone 102
   transformer 309
Baseball backstop 120
Baseboard demolition 209
Basement stair 165
Basic finish materials 236
   thermal materials 222
Basket weave fence, vinyl 113
Basketball backstop 120
Basketweave fence 114
Batch trial 10
Bathtub removal 256
Batt insulation 225, 226
Battery substation 307
Beach stone 107
Beam & girder framing 212
   and girder formwork 140
   bolster 151
   bond 169, 175
   bottom 140
   bridge 119
   concrete 158-160
   formwork 140
   grade 143, 161
   hanger 211
   laminated 219
   precast 164
   reinforcing 154
   side 140
   soldier 46
   spandrel 140
   tee 164
   test 10
   wood 212, 213
Bearing pad 189
Bedding brick 107
   pipe 63
   placement joint 163
Belgian block 106
Bell & spigot pipe 259
   caisson 74
Belt material handling 254
Bench aluminum 120
   fiberglass 120
   park 119
   players 120
   wood 120
Benches 119
Bender duct 94
Bentonite 102, 223
Berm pavement 105
   road 105
Bin retaining walls metal 116
Binder asphaltic 103
Bit drill 34
Bituminous base course 102
   block 107
   coating 131, 222
   concrete curbs 105
   expansion joint 148
   paver 23
   paving 375
   waterproofing 326
Blade saw 39
Blanket curing 163

insulation . . . 225
Blaster shot . . . 24
Blasting . . . 55
cap . . . 55
mat . . . 55
water . . . 180
Bleacher outdoor . . . 248
stadium . . . 248
Bleachers . . . 248
Block asphalt . . . 107
backup . . . 175
belgian . . . 106
bituminous . . . 107
cap . . . 177
column . . . 176
concrete . . . 173, 175, 176
concrete bond beam . . . 175
concrete exterior . . . 176
corner . . . 177
glazed . . . 177
granite . . . 108
high strength . . . 176
insulation . . . 176
manhole . . . 97
manhole/catch basin . . . 358
removal . . . 35
sill . . . 177
wall removal . . . 35
Blocking carpentry . . . 212
steel . . . 212
wood . . . 212, 215
Blockout slab . . . 144
Blower insulation . . . 23
Blown in cellulose . . . 224
in fiberglass . . . 224
in insulation . . . 224
Bluegrass sod . . . 127
Bluestone . . . 178
sidewalk . . . 107
sill . . . 179
step . . . 108
Board drain . . . 223, 224
fence . . . 114
insulation . . . 225
ridge . . . 215
sheathing . . . 218
Boat slip . . . 67
Boiler demolition . . . 256
feed pump . . . 289
mobilization . . . 73
removal . . . 256
Bollard . . . 191
light . . . 309
Bollards pipe . . . 118
Bolster beam . . . 151
slab . . . 151
Bolt anchor . . . 169, 170, 184
steel . . . 211
Bond beam . . . 169, 175
performance . . . 7, 368
Bonding agent . . . 138
Boom lift . . . 21
truck . . . 26
Boot pile . . . 72
Bored pile . . . 73
Borer horizontal . . . 25
Boring and exploratory drilling . . . 33
auger . . . 34
cased . . . 33
horizontal . . . 69
service . . . 69
Borosilicate pipe . . . 257
Borrow clay . . . 53
rock . . . 53
select . . . 53
Bosuns chair . . . 16
Bottom beam . . . 140
Boulder excavation . . . 55, 56
Bowstring truss . . . 219
Box . . . 299
buffalo . . . 81
culvert . . . 100
curb . . . 80
distribution . . . 84
electrical . . . 300
out . . . 144
out opening formwork . . . 142
steel mesh . . . 64
storage . . . 13
trench . . . 46
Boxes & wiring device . . . 300
utility . . . 30
Brace cross . . . 192
Brass salvage . . . 257
screw . . . 210
Breaker low-vol switchgear . . . 305
Brick . . . 172
adobe . . . 174
anchor . . . 170
bedding . . . 107
bedding mortar . . . 106
cart . . . 22
catch basin . . . 97
cleaning . . . 181
concrete . . . 177
demolition . . . 168, 236
economy . . . 172
engineer . . . 172
face . . . 172
floor . . . 107
forklift . . . 22
paving . . . 107
removal . . . 36
saw . . . 180
shelf . . . 144
sidewalk . . . 107
sill . . . 179
stair . . . 177
step . . . 108
testing . . . 10
veneer . . . 174, 175
veneer demolition . . . 168
wall . . . 177
wall panel . . . 177
wash . . . 181
Bridge . . . 119
beam . . . 119
concrete . . . 119
cranes . . . 254
deck . . . 118
foundation . . . 118
girder . . . 118
highway . . . 118
pedestrian . . . 119
railing . . . 119
sidewalk . . . 15
Bridging . . . 212
joist . . . 196
steel . . . 212
wood . . . 212
Broadcast stolen . . . 127
Bronze expansion joint . . . 204
valve . . . 282
Broom finish concrete . . . 161
Brush clearing . . . 42, 134
cutter . . . 19, 20
mowing . . . 42
Brushing wire . . . 205
Buck anchor . . . 170
Bucket aerial . . . 26
concrete . . . 18
crane . . . 19
excavator . . . 25
pavement . . . 25
Buffalo box . . . 81
Buggy concrete . . . 18, 161
Builder's risk insurance . . . 362
Building air supported . . . 247
disposal . . . 38
hangar . . . 249
insulation . . . 224
moving . . . 135
paper . . . 227, 228
permit . . . 7
portable . . . 13
slipform . . . 146
subdrain . . . 325
temporary . . . 12
tension . . . 248
Bulb end pile . . . 341
Bulk bank measure excavating . . . 56
storage dome . . . 248
Bulkhead aluminum . . . 65, 66
canal . . . 65
formwork . . . 142-144
marine . . . 65
residential . . . 65
Bulldozer . . . 20
Bumper car . . . 118
dock . . . 66
parking . . . 118
railroad . . . 203
Buncher feller . . . 19
Burial cable direct . . . 297
Burlap curing . . . 163
rubbing . . . 162
Bus duct . . . 408
substation . . . 306
Bush hammer . . . 162
hammer concrete . . . 162
Butt fusion machine . . . 22
Butterfly valve . . . 31, 77, 284
valve grooved joint . . . 276
Buttress formwork . . . 144
Butyl caulking . . . 229
waterproofing . . . 223

C

Cabinet demolition . . . 208
electrical . . . 299
Cable aluminum shielded . . . 298
copper shielded . . . 297
guide rail . . . 117
jack . . . 27
lug switchgear . . . 304
pulling . . . 24
sheathed nonmetallic . . . 297
shielded . . . 297
tensioning . . . 24
trailer . . . 24
underground feeder . . . 297
Cable-tray substation . . . 306
Caisson . . . 73, 340
bell . . . 74
concrete . . . 73
displacement . . . 74
foundation . . . 73
Calcium chloride . . . 64, 138
Canal bulkhead . . . 65
gate . . . 242
Canopy entrance . . . 250
Cant roof . . . 215
Cantilever retaining wall . . . 115
Cap blasting . . . 55
block . . . 177
pile . . . 143, 160, 161, 323
post . . . 211
Capacitor static . . . 305
station . . . 305
synchronous . . . 305
transmission . . . 305
Capital column . . . 141
Car bumper . . . 117, 118
muck . . . 26
Carbon black . . . 138
Carborundum . . . 162
Carpentry rough . . . 212
Carpet removal . . . 236
Cart brick . . . 22
concrete . . . 18, 161
Cased boring . . . 33
Casework demolition . . . 209
Casing anchor pipe conduit . . . 87
Cast boxes . . . 406
in place concrete . . . 158, 159
in place concrete piles . . . 69
in place pile . . . 70
in place retaining walls . . . 115
iron bench . . . 120
iron fitting . . . 82, 260
iron grate cover . . . 202
iron manhole cover . . . 96
iron pipe . . . 82, 259
iron pipe fitting . . . 260, 261
iron pipe gasket . . . 263
iron pull box . . . 300
iron stair tread . . . 150
Casting . . . 202
Castings construction . . . 202
Cast-iron column base . . . 202
weld rod . . . 190
Catch basin . . . 96
basin and manhole . . . 359
basin brick . . . 97
basin frame and cover . . . 96
basin masonry . . . 97
basin precast . . . 97
basin removal . . . 36
Catwalk . . . 15
Caulking . . . 229
lead . . . 259
masonry . . . 179
oakum . . . 259
polyurethane . . . 229
sealant . . . 229
Cavity truss reinforcing . . . 170
wall grout . . . 169
wall insulation . . . 224, 225
Cedar fence . . . 114, 115
roof plank . . . 217
Ceiling demolition . . . 236
drilling . . . 186
expansion joint . . . 204
furring . . . 216
insulation . . . 224
Cellular concrete . . . 159
fill . . . 166
floor deck . . . 197
Cellulose blown in . . . 224
insulation . . . 224
Cement admixture . . . 138, 161
color . . . 138, 169
concrete curbs . . . 105
content . . . 397
duct . . . 94
flashing . . . 228
grout . . . 44, 166, 169
gunite . . . 162
gypsum . . . 168
liner . . . 130, 131
masonry . . . 168
masonry unit . . . 175, 176

Index

parging . . . 222
Portland . . . 139
soil . . . 64
soil stabilization . . . 64
stabilization . . . 64
testing . . . 9
Cementitious deck . . . 164, 166
roof deck . . . 166
waterproofing . . . 223
Center bulb waterstop . . . 150
meter . . . 300
Centrifugal pump . . . 23
Ceramic mulch . . . 122
tile demolition . . . 236
Certification welding . . . 11
Chain alloy steel . . . 202
hoist . . . 27, 254
link fence . . . 17, 111, 112, 114
link fence removal . . . 36
link gate . . . 112
saw . . . 24
trencher . . . 21
Chair bosuns . . . 16
rail demolition . . . 209
Chamfer strip . . . 147
strip galvanized . . . 148
strip PVC . . . 147
strip wood . . . 148
Channel curb edging . . . 192
framing steel . . . 192
slab concrete . . . 166
slot . . . 171
slot anchor . . . 171
slotted framing . . . 192
Channelizing traffic . . . 118
Charge powder . . . 188
Charges dump . . . 38
Chase formwork . . . 149
Check swing valve . . . 282, 284
valve . . . 31, 75, 285
valve ball . . . 284
Checkered plate . . . 201
plate cover . . . 202
plate platform . . . 202
Chemical anchor . . . 186
cleaning masonry . . . 180
spreader . . . 25
toilet . . . 24
water closet . . . 244
Chilled water distribution . . . 84
Chimney . . . 175
brick . . . 175
demolition . . . 168
flue . . . 174
metal . . . 175
Chipper brush . . . 19
log . . . 20
stump . . . 20
Chipping hammer . . . 21
stump . . . 41
Chips wood . . . 123
Chloride calcium . . . 64
Chute rubbish . . . 38
C.I.P. concrete . . . 157
C.I.P. pile . . . 327, 328
wall . . . 342, 343
Circuit breaker air . . . 306
Circuit-breaker gas . . . 306
oil . . . 305
power . . . 305
transmission . . . 305
vacuum . . . 306
Circular saw . . . 24
Clamp pipe . . . 132, 133
Clamshell . . . 56, 63
bucket . . . 19
Clay borrow . . . 53
fire . . . 174
masonry . . . 172
Clean ditch . . . 43
tank . . . 32
Cleaner crack . . . 25
steam . . . 24
Cleaning brick . . . 181
internal pipe . . . 130
masonry . . . 180
Cleanout floor . . . 287
pipe . . . 287
PVC . . . 288
tee . . . 288
wall type . . . 288
Clear & grub . . . 41, 42
Clearance pole . . . 304
Clearing brush . . . 42, 134
selective . . . 42
site . . . 41
Cleat dock . . . 66
Clevis hooks . . . 202
Climbing crane . . . 26
jack . . . 27
Clip plywood . . . 211
tie . . . 151
Cloth hardware . . . 114
CMU . . . 173
Coal tar pitch . . . 228
Coat tack . . . 109
Coated reinforcing . . . 154
Coating . . . 238
bituminous . . . 131, 222
epoxy . . . 131, 162
polyethylene . . . 131, 132
roof . . . 228
rubber . . . 224
silicone . . . 224
spray . . . 222
trowel . . . 222
wall . . . 239
water repellent . . . 224
waterproofing . . . 222
Cofferdam . . . 46
excavation . . . 57
Cohesion testing . . . 9
Cold laid asphalt . . . 104
mix . . . 104
planing . . . 133
recycling . . . 134
Color concrete . . . 158
floor . . . 162
Column aluminum . . . 203
base . . . 211
base aluminum . . . 202, 203
base cast-iron . . . 202
base plate . . . 195
block . . . 176
cap aluminum . . . 203
capital . . . 141
concrete . . . 158, 160, 164, 165
demolition . . . 168
footing . . . 321, 322
formwork . . . 140
lally . . . 191
laminated wood . . . 219
precast . . . 164
reinforcing . . . 154
removal . . . 168
structural shape . . . 191
tie . . . 171
wood . . . 213
Columns formwork . . . 140
lightweight . . . 191
ornamental . . . 204
structural . . . 191
Comfort station . . . 248
Command dog . . . 17
Commercial gutting . . . 38
Common nail . . . 209, 210
Compact fill . . . 61
Compacted concrete roller . . 158, 163
Compaction . . . 54
backfill . . . 380
earth . . . 54
soil . . . 52
test Proctor . . . 10
vibroflotation . . . 380
Compactor earth . . . 19
landfill . . . 20
plate . . . 54
sheepsfoot . . . 54
tamper . . . 55
Compensation workers' . . . 7
Composite insulation . . . 227
metal deck . . . 197
Composition flooring removal . . . 236
Compressive strength . . . 10
Compressor air . . . 21
Concrete . . . 398
admixture . . . 138
aggregate . . . 139
asphalt . . . 30
asphaltic . . . 104
beam . . . 158-160
block . . . 173, 175, 176
block autoclaved . . . 174
block back-up . . . 175
block bond beam . . . 175
block demolition . . . 37
block exterior . . . 176
block foundation . . . 176
block grout . . . 169
block insulation . . . 224
block paver . . . 108
brick . . . 177
bridge . . . 119
broom finish . . . 161
bucket . . . 18
buggy . . . 161
bush hammer . . . 162
caisson . . . 73, 340
cart . . . 18, 161
cast in place . . . 157, 158
catch basin . . . 97
cellular . . . 159, 166
channel slab . . . 166
chimney . . . 175
C.I.P. . . . 157
color . . . 158
column . . . 158, 160, 165
conversion . . . 397
conveyer . . . 18
core . . . 33
cost . . . 398
cribbing . . . 115
culvert . . . 101
curb . . . 105
curing . . . 105, 138, 163
curtain wall . . . 165
cutout . . . 37
cutting . . . 39
cylinder . . . 10
cylinder pipe . . . 76
demo . . . 37
demolition . . . 37
drill . . . 33, 34
drilling . . . 186
elevated slab . . . 160
encapsulated billet . . . 67
field mix . . . 157
finish . . . 104, 161, 162
float . . . 18
float finish . . . 161
floor . . . 138
footing . . . 159, 160, 320-322
formwork . . . 140
foundation . . . 159
furring . . . 216
gravel . . . 139
grout . . . 169
hand hole . . . 93
hand trowel finish . . . 161
hardener . . . 138, 161
hot weather . . . 158
hydrodemolition . . . 36
inspection . . . 10
insulation . . . 163, 166
joist . . . 159, 164
lightweight . . . 158-160, 166
manhole . . . 93, 97
median . . . 106
mix design . . . 10
mixer . . . 18
mixes . . . 372, 397
monolithic finish . . . 161
patio block . . . 108
pavement . . . 104
paver . . . 23
paving . . . 104
perlite . . . 139
pier . . . 158
piers and shafts drilled . . . 74
pile . . . 327-329
pile cap . . . 323
piles prestressed . . . 70
pipe . . . 76, 95, 100, 101, 390
pipe gasket . . . 101
pipe removal . . . 36
placing . . . 160
post . . . 117
precast . . . 164, 165
protection . . . 139
pump . . . 160
ready mix . . . 157
rehabilitation . . . 166
reinforcement . . . 151
removal . . . 37
restoration . . . 166
retaining wall . . . 115, 355
roller compacted . . . 163
roof . . . 164
sandblasting . . . 162
saw . . . 18, 39
scarify . . . 236
sealer . . . 138
septic tank . . . 84
sidewalk . . . 106, 354
sill . . . 165
silo . . . 251
slab . . . 158, 161
slab sidewalk . . . 354
slabs X-ray . . . 11
spreader . . . 23
stair . . . 160
stair tread . . . 150
stamping . . . 162
stengths . . . 397
structural . . . 157
tank . . . 252
tie railroad . . . 203
trowel . . . 18
truck . . . 18
utility vault . . . 30
vibrator . . . 18
wall . . . 159, 161, 342, 343
wheeling . . . 161
winter . . . 158

INDEX

469

Conductor 297, 312
  &grounding 297
  line 303
  overhead 303
  sagging 303
  substation 306
  underbuilt 303
Conduit direct burial 92
  fitting pipe 86
  in slab 298
  in slab PVC 298
  in trench electrical 299
  in trench steel 299
  pipe prefabricated 88
  prefabricated pipe 85, 87-89
  rigid in slab 299
  substation 306
Cone traffic 16
Connection temporary 12
Connector dock 66
  flexible 285
  joist 211
  timber 211
  water copper tubing 285
Consolidation test 10
Construction aids 13
  barge 27
  castings 202
  cost index 7
  management fee 6
  photography 9
  shaft 69
  sump hole 43
  temporary 13
Contaminated soil 33
Continuous footing 320
Contractor equipment 18
  pump 23
Control crack 109
  erosion 65
  expansion 204
  joint 162, 172
  joint PVC 172
  joint rubber 172
  tower 248
Conveyor 18, 254
  material handling 254
Coping removal 168
Copper fitting 266
  pipe 80, 265
  rivet 188
  salvage 257
  shielded cable 297
  tube pipe 80
Corbel formwork 144
Core drill 10, 18, 33, 34
  testing 10
Cork expansion joint 148
Corner block 177
Corporation stop 80
Corrosion resist. backflow 286
  resistant fitting 269
  resistant pipe 269
  resistant pipe fitting 269, 270
Corrugated HDPE 99
  metal pipe 95, 97
  pipe 95
  tubing 95
  void form 144
Cost control 9
  mark-up 8
Counter top demolition 209
Coupling plastic 281
Course wearing 103
Cover aluminum trench 202
  cast iron grate 202

checkered plate 202
  ground 124
  manhole 96
  stadium 247
  stair tread 13
  tank 247
  trench 202
Covers expansion 204
CPVC pipe 277
Crack cleaner 25
  control 109
  filler 25
  filling 109
  repair 166
Crane 254
  bucket 19
  climbing 26
  crawler 26
  hydraulic 26
  material handling 26, 254
  mobilization 73
  rail 254
  tower 26
Crash barriers 118
Crawler crane 26
Crew survey 6
Cribbing concrete 115
Critical path schedule 9
Cross brace 192
Crosses 75
Crossing pedestrian 121
Crushed gravel 139
  stone sidewalk 107
CT substation 306
Culvert box 100
  concrete 101
  end 100
  formwork 141
  headwall 360
  reinforced 100
Curb 105
  and gutter 105
  asphalt 105
  box 80
  concrete 105
  edging 106, 192
  edging angle 192
  edging channel 192
  formwork 142, 144
  granite 106, 179
  inlet 96, 106
  precast 106
  removal 37
  roof 215
  seal 108
  stop 80
Cured in place pipe 131
Curing blanket 163
  concrete 105, 138, 163
  paper 227
Current transformer 306
Curtain wall concrete 165
Cut and cover tunnels 69
  drainage ditch 43
  in sleeve 75
  in valve 75
  pipe groove 276
Cutoff pile 72
Cutout demolition 37
  slab 37
Cutter brush 19, 20
Cutting asphalt 39
  concrete 39
  saw 39
  steel 39, 190
  torch 24, 190

tree 42
Cylinder concrete 10

D

Dam expansion 119
Dampproofing 223
Deciduous tree 127
Deck bridge 118
  cellular floor 197
  cementitious 164, 166
  drain 289
  edge form 198
  metal 197
  metal floor 197
  metal roof 197
  roof 217
  slab form 198
  steel 197
  wood 217
Decking 399
  floor 197
Dehumidifier 22
Delineator 121
  barrier 118
Delivery charge 52
Demo concrete 37
  concrete selective 138
  steel selective 184
Demolition . 30, 35, 135, 168, 236, 256, 376
  baseboard 209
  boiler 256
  brick 168, 236
  cabinet 208
  casework 209
  ceiling 236
  ceramic tile 236
  chimney 168
  column 168
  concrete 37
  concrete block 37
  cutout 37
  door 232
  drywall 236
  ductwork 256
  electric 296, 297
  explosive 35
  fireplace 168
  flooring 236
  framing 208
  glass 232
  granite 168
  gutter 222
  hammer 19
  house 35
  HVAC 256
  implosive 35
  joist 208
  lath 236
  masonry 37, 168, 236
  metal deck 222
  metal stud 236
  millwork 208
  paneling 209
  partition 236
  pavement 36, 37
  plaster 236
  plenum 236
  plumbing 256
  plywood 222, 236
  post 208
  rafter 208
  railing 209
  roofing 222

selective 208, 232
  siding 222
  site 35
  steel window 232
  stucco 236
  terra cotta 237
  terrazzo 236
  tile 236
  trim 209
  truss 208
  walls and partitions 236
  window 232
  wood 236
Derail railroad 203
Derrick 26
  guyed 26
  stiffleg 27
Detection tape 32
Detector check valve 75
  meter water supply 286
Device & box wiring 300
Dewater 27, 43, 44
  pumping 43
Dewatering 43
Diaphragm pump 23
Diesel generator 301
  hammer 19
  tugboat 27
Direct burial cable 297
  burial conduit 92
Directional sign 121
Disc harrow 19
Discharge hose 23
Disconnect switch 306
Disk insulator 303
Displacement caisson 74
Disposal 35, 38
  building 38
  field 84
Distribution box 84
  pipe water 76
  power 304
  system gas 89
Distributor asphalt 22
Ditch clean 43
Ditching 43, 102
Dock accessory 66
  accessory jetties, ... 66
  bumper 66
  cleat 66
  connector 66
  fixed 67
  floating 67
  galvanized 67
  jetties,floating 67
  ladder 66
  pile guide 66
  utility 66
  wood 67
Dog command 17
Dome fiberglass 142
  metal 142
Domes 248
Door air lock 247
  and window matls 232
  demolition 232
  frame 250
  frame grout 169
  hangar 249
  metal 250
  removal 232
  revolving 247
Dormer gable 216
Double brick 172
  tee beam 164, 165
  wall pipe 91

Index

Dovetail anchor 148
Dowel 171
support 143
Dowels reinforcing 155
Dozer 20
backfill 52, 53
excavating bulk 57
excavation 21, 57, 58
ripping 56
Dozing ripped material 56
Dragline 56, 63
bucket 19
Drag-line exc. bulk drag line 58
Drain 289
board 223, 224
deck 289
floor 289
geotextile 94
pipe 43
sediment bucket 289
trench 289
Drainage ditch cut 43
flexible 95
mat 94
pipe 83, 97, 100, 257, 269
site 97, 243
Dredge 63
hydraulic 63
Dredging 63
Drill & tap main 75
auger 34
bit 34
concrete 33, 34
core 10, 18, 33, 34
earth 33, 73
hammer 22
hole 46
main 75
rig 33
rig mobilization 73
rock 33, 55
rotary 22
steel 22
track 21
wood 211
Drilled concrete piers and shafts . 74
Drilling anchors 186
ceiling 186
concrete 186
horizontal 69
only rock 55
plaster 186
quarry 55
rock bolt 55
steel 186
Driven pile 69
Driver post 25
sheeting 22
Driveway 106
removal 37
Drop panel formwork 142
Drum polyethylene 67
Drywall cutout 38
demolition 236
nail 209
removal 236
Duck tarpaulin 16
Duct bender 94
cement 94
underground 93
utility 93
Ductile iron fitting 77
iron fitting mech joint 77
iron pipe 76
Ductwork demolition 256
Dumbbell waterstop 150
Dump charge 38
charges 38
truck 21, 60
Dumpster 38
Dust partition 38
Dustproofing 138, 162
DWV pipe ABS 277
PVC pipe 277

E

Earth compaction 54
compactor 19
drill 33, 73
retaining 355
vibrator 52, 62
Earthwork 51
equipment 18, 52
Eave overhang 250
Economy brick 172
Edge form 142, 144
Edging curb 106, 192
Efflorescence testing 10
Ejector pump 291
Elastomeric waterproofing 222
Electric curing 163
demolition 296, 297
fixture 308, 309, 312
generator 22
generator set 301
hoist 254
panelboard 307
switch 307
utility 93
Electrical box 300
cabinet 299
fee 6
pole 94
telephone underground 93, 94
Elevated floor 158
installation pipe 256
pipe 256
slab 158
slab concrete 160
slab formwork 141
water tank 251
Elevating scraper 58
Elevator construction 27
fee 6
windrow 26
Employer liability 7
Emulsion asphaltic 109, 122
pavement 108
penetration 102
sprayer 22
Encasement pile 72
End culvert 100
Engineer brick 172
Engineering fee 6
Entrance canopy 250
EPDM adhesive 223
Epoxy 44
anchor 186
coating 131, 162
dustproofing 138
fiberglass wound pipe 269
grout 44
injection 166
lined silo 251
resin 138
resin fitting 278
wall coating 239
Equipment 18
athletic 120
earthwork 18, 52
formwork 143
foundation 143
general 8
hydromulch 25
insurance 7
mobilization 51, 74
playfield 120
playground 120
rental 18, 26, 374
substation 305
Erosion control 65
control synthetic 65
Estimating 362
Excavating bulk bank measure ... 56
bulk dozer 57
equipment 382
Excavation 52, 56, 60, 381, 391
backhoe 61
boulder 55, 56
bulk drag line 58
bulk scraper 58
cofferdam 57
dozer 21, 57, 58
hand 52, 59, 62
machine 52
planting pit 124
rock 55
scraper 58, 59
septic tank 84
structural 59
trench 61-63, 349
Excavation, trench 380
Excavation tunnel 68
Excavator bucket 25
rental 18
Expansion anchor 187
control 204
covers 204
dam 119
joint 148
joint aluminum 204
joint asphalt 148
joint bituminous 148
joint bronze 204
joint ceiling 204
joint cork 148
joint floor 204
joint gymnasium 204
joint neoprene 148
joint polyethylene 148
joint polyurethane 148
joint premolded 148
joint PVC 148
joint roof 204
joint rubber 148
joint wall 204
shield 187
Expense office 8
Explosive demolition 35
Explosives 55
Exposed aggregate 106, 161
aggregate coating 239
Extension backhoe 25
Exterior concrete block 176
floodlamp 308
lighting fixture 308
sprinkler 110
Extra work 8
Extractor 20
Extrusion aluminum 195

F

Fabric fence 113
overlay 109
polypropylene 94
stabilization 103
structure 247
waterproofing 223
welded wire 156
wire 114
Fabrication metal 198
Face brick 172
Fascia board demolition 208
wood 215
Fastener timber 209
wood 210
Fastenings metal 184
Fee engineering 6
Feeder cable underground 297
Feller buncher 19
Felt waterproofing 223
Fence aluminium 112
and gate 111
auger, truck mounted 18
board 114
cedar 114, 115
chain link 17, 111, 114
fabric 113
helical topping 114
mesh 111
metal 111
open rail 114
plywood 17
removal 35
residential 111
security 114
snow 111
temporary 17
tennis court 120
vinyl 113
wire 17, 113
wood 114
Fencing wire 114
Fender 66
Fertilize lawn 128
Fertilizer 126
shrub bed 134
tree 135
Fiber optic cable 312
optics 312
rebar glass 155
reinforcing 157
tube formwork 140
Fiberboard insulation 226
Fiberglass area wall cap 233
bench 120
blown in 224
dome 142
formwork 140
insulation 224, 225, 250
light pole 309
panel 13, 251
waterproofing 223
wool 224
Field disposal 84
mix concrete 157
office 13
office expense 13
personnel 8
Fieldstone 179
Fill 52, 61
cellular 166
floor 159
flowable 158
gravel 61
Filler crack 25
joint 109
Fillet welding 190
Filling crack 109
Film polyethylene 122

INDEX

Filter stone 64
Fine grade 52, 124
Finish concrete 104, 161, 162
  float 162
  grading 52
  lime 168
  nail 210
  wall 162
Finisher floor 18
Fir roof plank 217
Fire brick 174
  clay 174
  hydrant 30
  retardant lumber 209
  retardant pipe 269
  retardant plywood 209
Firebrick 174
Fireplace demolition 168
Fitting adapter pipe 267
  cast iron 260
  cast iron pipe 261, 264, 273
  copper 266
  copper pipe 267
  corrosion resistant 269
  ductile iron 77
  epoxy resin 278
  glass 258
  pipe 78
  pipe conduit 86
  plastic 278
  plastic pipe 279
  polypropylene 270
  PVC 78, 79
  PVC duct 93
  steel 272
  subdrainage 96
  underground duct 93
Fixed dock 67
  dock jetties, 67
  ladder 198
  roof tank 252
Fixture landscape 310
  lighting 308
  plumbing 82, 289
  removal 256
  sodium high pressure 308
Flagging 108
  slate 108
Flange clip 151
  grooved joint 275
  PVC 281
  steel 275
  tie 171
Flap gate 242
Flashing cement 228
  metal 250
Flat plate formwork 141
  slab formwork 142
Flatbed truck 21
Flexible connector 285
  drainage 95
Float concrete 18
  finish 162
  finish concrete 161
Floater equipment 7
Floating dock 67
  roof tank 252
Floodlamp exterior 308
Floodlight 22, 23
  pole mounted 308
Floor 237
  abrasive 161
  asphalt block 107
  brick 107
  cleanout 287
  color 162
  concrete 138
  decking 197
  drain 289
  elevated 158
  expansion joint 204
  fill 159
  finisher 18
  flagging 108
  framing removal 38
  grating 200
  nail 210
  plank 213
  plywood 218
  removal 35
  sander 24
  slab formwork 142
  subfloor 218
  topping 162
  underlayment 218
Flooring 237
  demolition 236
Flotation polystyrene 67
Flowable fill 158
Flue chimney 174
  liner 174, 175
  tile 174
Flying truss 149
Foam glass insulation 225
  insulation 225
  urethane 229
Fog seal 108
Foil aluminum 225, 227
  back insulation 226
Football goalpost 120
Footing column 321, 322
  concrete 159, 160, 320, 322
  formwork 143
  pressure injected 341
  reinforcing 154
  removal 35
  spread 143, 321, 322
  strip 320
  tower 302
Forged valves 285
Forklift brick 22
  rigging 22
Form deck edge 198
  edge 142, 144
  liner 145
  patch 150
  release 138
  slab deck 198
Formwork 140, 141, 144
  accessory 147-150
  beam 140
  beam and girder 140
  box out opening 142
  bulkhead 143, 144
  buttress 144
  chase 149
  columns 140
  concrete 140
  corbel 144
  culvert 141
  curb 142, 144
  drop panel 142
  elevated slab 141
  equipment 143
  fiber tube 140
  flat plate 141
  flat slab 142
  floor slab 142
  footing 143
  foundation 143
  gang wall 145
  girder 140
  interior beam 140
  liner 145
  lintel 146
  oil 150
  pilaster 146
  plywood 141, 144
  retaining wall 145
  sill 146
  sleeve 149
  snap tie 149
  stair 144
  steel frame 141
  steel framed plywood 146
  structural 140
  trench 144
  upstanding beam 140
  void 143
  wall 144
Foundation 331, 333, 335, 336, 338, 340, 345
  beam 338
  caisson 73
  concrete 159
  concrete block 176
  equipment 143
  formwork 143
  mat 159, 160
  pile 70
  pole 302
  tower 302
  underdrain 325
  underpin 45
  wall 176, 342, 343
  waterproofing 326
Fountain 292
  lighting 292
Frame door 250
  trench grating 202
Framing aluminum 195
  anchor 211
  beam & girder 212
  demolition 208
  heavy 213
  laminated 219
  lightweight 192
  metal 191
  opening 250
  removal 208
  roof rafters 215
  roof truss 219
  sleepers 215
  steel angle 192
  steel channel 192
  timber 213
  wall 216
  window 250
  wood 212, 213, 215, 216
Friction pile 70, 72
Front end loader 57
  shovel 20
Frost penetration 385
Fuel storage tank 252
Furnishings site 119
Furring ceiling 216
  wall 216
Fuse substation 307
  transmission 307
Fused pipe 79
Fusion machine butt 22

G

Gabion 65, 116
  retaining walls stone 116
  wall 356
Gable dormer 216
Galley septic 84
Galvanized chamfer strip 148
  dock 67
  plank grating 200
  structural steel 205
Galvanizing 192, 205
Gang operated switch 306
  wall formwork 145
Gantry crane 254
Gas circuit-breaker 306
  connector 285
  distribution system 89
  generator 301
  pipe 89, 90
  station piping 91
  stop 90
Gasket B and S pipe 263
  cast iron pipe 263
  concrete pipe 101
  joint pipe 259
  joint pipe fitting 261-263
  neoprene 229
Gasoline generator 301
  tank 252
Gate box 76
  canal 242
  chain link 112
  fence 111
  flap 242
  knife 242
  slide 242
  sluice 242
  swing 111
  valve 31, 284, 285
  valve grooved-joint 276
  valve soldered 282
General contractor's overhead 363
  equipment 8
  site work maintenance 134
Generator construction 22
  diesel 301
  emergency 301
  gas 301
  gasoline 301
  set 301
  weight 407
Geotextile drain 94
  soil stabilization 64
  stabilization 64
Geotextiles 94
Girder bridge 118
  formwork 140
  joist 196
  reinforcing 154
  wood 212, 213
Gland seal pipe conduit 86
Glass demolition 232
  fiber rebar 155
  fitting 258
  pipe 257
  pipe fitting 258
  process supply pipe 257
Glazed block 177
  brick 172
  wall coating 239
Globe valve 31, 283-285
Glued laminated 219
  laminated construction 219
Goalpost 120
  football 120
  soccer 120
Gore line 105
Gradall 57
Grade beam 143, 161, 338
  fine 52, 124

wall . . . 159
Grader attachment . . . 25
motorized . . . 19
ripping . . . 56
Grading . . . 43, 52, 60
slope . . . 52
Grandstand . . . 248
Granite . . . 178, 179
block . . . 108
building . . . 178
chips . . . 123
curb . . . 106, 179
demolition . . . 168
indian . . . 106
paver . . . 178
paving block . . . 108
sidewalk . . . 108
Grass lawn . . . 126
sprinkler . . . 110
Grate precast . . . 107
tree . . . 107
Grating aluminum bar . . . 199
aluminum frame . . . 201
aluminum mesh . . . 200
aluminum plank . . . 200
area wall . . . 233
area way . . . 233
floor . . . 200
galvanized plank . . . 200
stainless bar . . . 201
stainless plank . . . 200
steel bar . . . 200
steel mesh . . . 201
Gravel base . . . 103
concrete . . . 139
crushed . . . 139
fill . . . 61
pack well . . . 81
pea . . . 123
roof . . . 139
Gravity retaining wall . . . 115
wall . . . 355
Grid spike . . . 211
Grinder pump system . . . 290
terrazzo . . . 18
Grinding asphalt . . . 133
Groove cut labor steel . . . 276
roll labor steel . . . 276
Grooved joint fitting . . . 274
joint flange . . . 275
joint pipe . . . 274
Grooving pavement . . . 134
Ground box hydrant . . . 290
cover . . . 124
water monitoring . . . 11
wire overhead . . . 303
Grounding & conductor . . . 297
switch . . . 306
transformer . . . 305
Grout . . . 169
cavity wall . . . 169
cement . . . 166, 169
column base . . . 166
concrete . . . 169
concrete block . . . 169
door frame . . . 169
epoxy . . . 44
non-shrink . . . 166
wall . . . 169
Grouted bar . . . 157
piles . . . 70
strand . . . 157
Grouting, pressure . . . 44
Grouting tunnel . . . 68
Guard service . . . 17
snow . . . 228
Guardrail temporary . . . 16
Guards snow . . . 228
Guidance inertial . . . 6
Guide rail . . . 117
rail cable . . . 117
rail removal . . . 36
rail timber . . . 117
Guide/guard rail . . . 117
Gunite . . . 162
mesh . . . 163
Gutter . . . 250
demolition . . . 222
monolithic . . . 105
Gutting . . . 38
Guyed derrick . . . 26
tower . . . 92
Guying power pole . . . 302
Gymnasium expansion joint . . . 204
Gypsum block demolition . . . 37
board removal . . . 236
cement . . . 168
lath nail . . . 210
poured . . . 166
roof plank demolition . . . 222
sheathing . . . 218
weatherproof . . . 218

H

H pile . . . 71, 333
pile steel . . . 333
Hammer bush . . . 162
chipping . . . 21
demolition . . . 19
drill . . . 22
hydraulic . . . 22
pile . . . 19
pile mobilization . . . 73
Hammermill . . . 25
Hand clearing . . . 42
excavation . . . 52, 59, 62, 63
hole . . . 31
hole concrete . . . 93
trowel finish concrete . . . 161
Handball court backstop . . . 120
Handicap ramp . . . 159
Handling material . . . 254
rubbish . . . 38
Handrail and railing . . . 199
Hangar . . . 249
Hanger beam . . . 211
joist . . . 211
Hanging lintel . . . 192
Hardboard underlayment . . . 218
Hardener concrete . . . 138, 161
Hardware . . . 233
cloth . . . 114
rough . . . 211
Harrow disc . . . 19
Hauling . . . 60
charge . . . 61
truck . . . 60
Hay . . . 122
bale . . . 65
Hazardous waste cleanup . . . 32
waste disposal . . . 32
HDPE corrugated . . . 99
pipe . . . 79
pipe liner . . . 131
Header pipe . . . 27
wood . . . 216
Headwall sitework . . . 360
Heat temporary . . . 12, 139
Heater contractor . . . 22
Heavy construction . . . 119
duty shoring . . . 13, 14
framing . . . 213
lifting . . . 374
Hex bolt steel . . . 185
High chair . . . 152
early strength cement . . . 139
impact pipe . . . 277
intensity discharge ballast . . . 309
intensity discharge lamp . . . 312
strength block . . . 176
strength concrete . . . 139
strength steel . . . 195
voltage transmission . . . 304
Highway bridge . . . 118
mowing . . . 135
paver . . . 108
paving . . . 118
sign . . . 121
Hip rafter . . . 215
Historical Cost Index . . . 7
Hoist . . . 254
and tower . . . 27
chain . . . 254
contractor . . . 27
electric . . . 254
lift equipment . . . 26
overhead . . . 254
personnel . . . 27
Holding tank . . . 244
Holdown . . . 211
Hole drill . . . 46
Hooks clevis . . . 202
Horizontal auger . . . 18
borer . . . 25
boring . . . 69
drilling . . . 69
shore . . . 149
Hose air . . . 21
discharge . . . 23
suction . . . 23
water . . . 23
Hot water distribution . . . 85
weather concrete . . . 158
House demolition . . . 35
Housewrap . . . 227
Humus peat . . . 122
HVAC demolition . . . 256
Hydrant fire . . . 30
ground box . . . 290
removal . . . 36
wall . . . 290
water . . . 290
Hydrated lime . . . 168
Hydraulic crane . . . 26, 254
dredge . . . 63
hammer . . . 22
jack . . . 27
jacking . . . 374
pressure test . . . 293
structure . . . 242
Hydro seeding . . . 126
Hydrocumulator . . . 290
Hydrodemolition concrete . . . 36
Hydromulch equipment . . . 25

I

Ice melting compound . . . 134
Impact barrier . . . 117
wrench . . . 22
Implosive demolition . . . 35
Improvement site . . . 120
In insulation blown . . . 224
place concrete piles cast . . . 69
place pipe cured . . . 131
place retaining walls cast . . . 115
slab conduit . . . 298
Index construction cost . . . 7
Indian granite . . . 106
Indicator valve post . . . 30
Industrial rail . . . 199
Inert gas . . . 32
Inertial guidance . . . 6
Injection epoxy . . . 166
latex . . . 166
Inlet curb . . . 96, 106
Insecticide . . . 64
Insert slab lifting . . . 153
Inspection technician . . . 10
Instrument transformer . . . 306
Insulated pipe . . . 84
steel pipe . . . 84
Insulating forms . . . 146
Insulation . . . 224, 225
batt . . . 225, 226
blanket . . . 225
blower . . . 23
board . . . 225
building . . . 224
cavity wall . . . 224
cellulose . . . 224
composite . . . 227
concrete . . . 163, 166
fiberglass . . . 224, 225, 250
foam . . . 225
foam glass . . . 225
insert . . . 176
masonry . . . 224
mineral fiber . . . 226
polystyrene . . . 224, 225
roof . . . 226, 250
roof deck . . . 226
vapor barrier . . . 228
vermiculite . . . 224
wall . . . 176, 225, 250
Insulator disk . . . 303
line . . . 303, 304
substation . . . 306
transmission . . . 303, 304
Insurance . . . 7, 362
builder risk . . . 7
equipment . . . 7
Integral waterproofing . . . 139
Interior beam formwork . . . 140
Interlocking retaining walls . . . 115
Interrupter switch load . . . 304
Invert manhole . . . 97
Inverted tee beam . . . 164
Investigation subsurface . . . 33
Iron alloy mechanical joint pipe . . . 269
body valve . . . 284
grate . . . 107
Ironspot brick . . . 237
Irrigation system . . . 110

J

Jack cable . . . 27
hydraulic . . . 27
post adjustable . . . 191
roof . . . 215
screw . . . 45
Jackhammer . . . 21, 55
Jacking . . . 69
Jet water system . . . 82
Jetties, dock accessory . . . 66
fixed dock . . . 67
piers . . . 67
Jetties,floating dock . . . 67
Jetting pump . . . 27

Job condition 6
Joint bar railroad 204
  bedding placement 163
  control 162, 172
  expansion 148
  filler 109
  push-on 77
  reinforcing 170
  sealer 229
  tyton 77
Joist bridging 196
  concrete 159, 164
  connector 211
  demolition 208
  girder 196
  hanger 211
  metal 196
  open web 196
  sister 214
  steel 196
  truss 197
  wood 213, 219
Jumbo brick 172
Jumper transmission 303
Jute mesh 65

K

Kennel fence 114
Kettle tar 25
Kettle/pot tar 24
Keyway 143
Kiln dried lumber 209
King brick 172
Kitchen equipment fee 6
Knife gate 242
Kraft paper 228

L

Labor pipe groove roll 276
Ladder 23
  aluminum 66, 198
  dock 66
  fixed 198
  reinforcing 170
  rolling 15
  steel 198
Lag screw 188
  screw shield 187
Lagging 46
Lally column 191
Laminated beam 219
  construction glued 219
  framing 219
  veneer members 219, 220
  wood 219
Lamp high intensity discharge . . . 312
  incandescent 308
  mercury vapor 311
  metal halide 312
  post 309
  quartz line 312
  sodium high pressure 312
  sodium low pressure 312
Landfill compactor 20
  fees 38
Landing stair 198
Landscape fee 6
  fixture 310
  layout 126
  light 309
  lighting 310
Landscaping 393
Laser level 23
Latex acrylic 138
  caulking 229
  injection 166
  modified emulsion 108
Lath demolition 236
Lava stone 179
Lavatory removal 256
Lawn aerate 134
  bed preparation 123
  edging 134
  fertilize 128
  grass 126
  maintenance 134
  mow 128
  mower, 22" rotary, 5HP 23
  rake 134
  seed 134
  trim 128
  water 129
Layout landscape 126
Leaching field chamber 84
  pit 84
Lead caulking 259
  pile 19
  salvage 257
  screw anchor 187
  wool 148
Level laser 23
Liability employer 7
  insurance 363
Lift aerial 21
  scissor 254
  slab 159
Light bollard 309
  landscape 309
  pole 94, 308
  pole aluminum 308
  pole fiberglass 309
  pole steel 309
  pole wood 309
  portable 22
  runway centerline 311
  shield 117
  temporary 12
  tower 23
  traffic 121
Lighting 309, 312
  fixture 308
  fixture exterior 308
  fountain 292
  landscape 310
  outdoor 22
  roadway 308
  underwater 311
Lightning arrester 304, 306
  arrester switchgear 304
Lightweight aggregate 139
  columns 191
  concrete 158, 159, 166
  floor fill 159
  framing 192
  natural stone 179
Lime finish 168
  hydrated 168
  soil stabilization 64
Limestone paver 107
Line conductor 303
  gore 105
  insulator 303, 304
  pole 302
  pole transmission 302
  spacer 303
  tower 302
  tower special 302
  transmission 304
Liner cement 130, 131
  flue 174
  formwork 145
  pipe 130
Linings tunnel 68
Lintel formwork 146
  hanging 192
  steel 192
Load interrupter switch 304
  test pile 72
Loader front end 57
  tractor 20
Loam 43, 54, 123, 124
  spread 123
Locomotive 25
Log chipper 20
  skidder 20
Lubricated valve 91
Lumber 401, 402
  foundation 345
  kiln dried 209
  treatment 209

M

Macadam 102
  penetration 102
Machine excavation 52
  trowel finish 161
  welding 25
Machinery anchor 188
Magnetic particle test 10
Main office expense 8
Maintenance lawn 134
  shrub 134
  site 134
  walk 134
Management fee construction 6
Manhole 97
  concrete 93, 97
  cover 96
  electric service 93
  invert 97
  precast 97
  raise 96
  removal 36
  step 97
  substation 306
Marble aggregate 139
  chips 123
  sill 179
Marine bulkhead 65
  piling 68
Markings pavement 105
Mark-up cost 8
Masonry accessory 168, 172
  aggregate 169
  anchor 170
  base 177
  brick 175
  catch basin 97
  caulking 179
  cement 168
  clay 172
  cleaning 180
  color 169
  demolition 37, 168, 236
  furring 216
  insulation 224
  manhole 97
  nail 210
  needle 180
  panel 177
  pointing 180
  reinforcing 170
  removal 36, 168
  restoration 179
  saw 24
  sill 179
  step 108
  testing 10
  toothing 37, 180
  unit 172
  wall 116, 175
  wall tie 170
  waterproofing 169
Masons scaffolding 15
Mat blasting 55
  drainage 94
  foundation 143, 159, 160
Material handling 254
  handling belt 254
  handling system 254
  hoist 27
Materials demolition utility 39
Mechanical equip. demo. 256
  fee 6
Median barrier 117
  concrete 106
  precast 117
Medical clean copper tubing 265
Melting compound ice 134
Membrane curing 163
  roofing 228
  waterproof 166
  waterproofing 108, 222, 223
Mercury vapor lamp 311
Mesh fence 111
  gunite 163
  wire 114
Metal bin retaining walls 116
  chimney 175
  deck 197
  deck acoustical 197
  deck composite 197
  deck demolition 222
  deck ventilated 197
  dome 142
  door 250
  door frame 250
  fabrication 198
  fastenings 184
  fence 111
  flashing 250
  floor deck 197
  framing 191
  gutter 250
  halide lamp 312
  joist 196
  miscellaneous 233
  ornamental 233
  paint 238
  pan 142
  pipe 97
  pipe removal 256
  pressure washing 205
  roof deck 197
  roof truss 194
  sandblasting 205
  stair 198
  stair tread 150
  steam cleaning 205
  stud demolition 236
  water blasting 205
  window 251
Metallic coating 326
  foil 225
  hardener 138, 161
  waterproofing 139
Meter center 300
  pit 31

socket 300
   water supply 286
   water supply detector 286
   water supply domestic 286
Metric conversion 397
   conversion factors 371
   rebar specs 395
Microtunneling 68
Mill construction 213
Milling pavement 133, 134
Millwork demolition 208
Mineral fiber insulation 226
   wool blown in 224
Minor site demolition 36
Miscellaneous metal 233
Mix cold 104
   design asphalt 9
   paver 25
   planting pit 124
Mixed bituminous concrete plant . 30
Mixer concrete 18
   mortar 18, 23
   plaster 23
   road 25
   transit 18
Mixes concrete 397
Mixing windrow 134
Mobilization 73, 74
   air-compressor 73
   barge 68
   equipment 45, 74
   or demob. 51
Modification to cost 6
Modulus of elasticity 10
Moil point 22
Moisture barrier 223
   content test 10
Monel rivet 188
Monitoring vibration 11
Monolithic finish concrete 161
   gutter 105
Monorail 254
Monument survey 6
Mortar 169
   mixer 18, 23
   pigments 169
   Portland cement 169
   restoration 169
   sand 169
   testing 10
Moss peat 122
Motor generator 22
Motors 407
Mounting board plywood 217
Moving building 135
   shrub 122
   structure 135
   tree 122
Mow lawn 128
Mower, 22″ rotary, 5HP lawn . . 23
Mowing 128
   brush 42
   highway 135
Muck car 26
Mulch ceramic 122
   stone 122
Mulching 122
Multi-channel rack enclosure . . . 312
Mylar tarpaulin 16

N

Nail 209, 210
   stake 150
Nailer 23

pneumatic 22
Needle masonry 180
Neoprene adhesive 223
   expansion joint 148
   gasket 229
   waterproofing 223
Net safety 13
No hub pipe 260
Nondestructive pipe testing 293
Non-destructive testing 10
Non-metallic hardener 138
Non-shrink grout 166
Norwegian brick 172
Nosing stair 150

O

Oakum caulking 259
Observation well 81
Off highway truck 21
Office expense 8
   field 13
   trailer 12
Oil circuit-breaker 305
   filled transformer 301
   formwork 150
   heater temporary 23
   storage tank 252
Oil/water separator 243
   separators 243
Omitted work 8
Open rail fence 114
   web joist 196
Opening framing 250
   roof frame 192
Operated switch gang 306
Operating cost 18
   cost equipment 18
Ornamental columns 204
   metal 233
Outdoor bleacher 248
   lighting 22
Oval arch culverts 101
Overhaul 38, 60
Overhead and profit 367
   conductor 303
   contractor 363
   ground wire 303
   hoist 254
Overlay fabric 109
   pavement 109
Overpass 119
Overtime 7, 8, 369
Oxygen lance cutting 39

P

Pad bearing 189
   vibration 188
Paint 238
   exterior 238
   metal 238
   miscellaneous 239
   striper 25
Painted traffic lines & markings . . 105
Painting bar joist 196
   parking stall 105
   pavement 105
   pavement letter 105
   pipe 237, 238
   reflective 105
   sprayer 23
   steel 205
   structural steel 205

temporary road 105
   thermoplastic 105
   truss 238
Paints & coatings 237
Pan metal 142
   slab 159
   stair 150
Panel brick wall 177
   fiberglass 13
   masonry 177
   precast 165
   structural 217
   wall 159, 165
Panelboard 307
   electric 307
   w/circuit-breaker 307
Paneling cutout 38
   demolition 209
Paper building 227
   curing 163
   sheathing 228
   tubing 153
Parging cement 222
Park bench 119
Parking barrier 118
   barrier precast 118
   bumper 118
   lots paving 104
   stall painting 105
Particle board underlayment 218
Partition anchor 171
   demolition 236
   dust 38
   wood frame 214
Patch core hole 10
   form 150
   roof 228
Patching asphalt 30
Patio 107, 108
   block concrete 108
Pavement 107
   asphaltic 103
   berm 105
   bucket 25
   concrete 104
   demolition 36, 37
   emulsion 108
   grooving 134
   letter painting 105
   maintenance 392
   markings 105
   milling 133, 134
   overlay 109
   painting 105
   planer 25
   profiler 25
   profiling 134
   pulverization 134
   reclamation 134
   recycled 30
   recycling 133
   replacement 104
   slate 179
   widener 25
Paver bituminous 23
   concrete 23
   concrete block 108
   floor 237
   highway 108
   shoulder 25
   slipform 25
Pavers thinset 107
Paving and surfacing 391
   asphaltic 375
   athletic 109
   block granite 108

brick 107
   concrete 104
   highway 118
   parking lots 104
Pea gravel 123
   stone 139
Peat humus 122
   moss 122
Pedestal pile 341
Pedestrian bridge 119
   crossing 121
Penetration macadam 102
   test 9
Perforated aluminum pipe 95
   pipe 95
   PVC pipe 95
Performance bond 7, 368
Perimeter drain 325
Perlite concrete 139, 166
   insulation 225
Permit building 7
Personnel field 8
   hoist 27
Petrographic analysis 9
Photography 9
   aerial 9
   construction 9
   time lapse 9
Picket fence, vinyl 113
Pickup truck 24
Pier brick 175
   concrete 158
Pigments mortar 169
Pilaster formwork 146
Pile 70, 331, 335, 336, 341
   boot 72
   bored 73
   caisson 340
   cap 143, 160, 161, 323
   caps, piles and caissons . . 386, 387
   cast in place 327, 328
   concrete 327, 328
   concrete filled pipe 331, 332
   cutoff 72
   driven 69
   driver 19
   driving 374
   driving mobilization 73
   encasement 72
   foundation 70
   friction 72
   H 46, 71
   hammer 19
   high strength 44
   lightweight 45
   load test 72
   mobilization hammer 73
   pin 70
   pipe 70, 71
   point 71
   point heavy duty 71
   precast 70
   precast concrete 329
   prestressed 69, 70
   prestressed concrete 329
   sod 124
   splice 71, 72
   steel 71
   steel H 333
   steel pipe 331, 332
   steel sheet 44
   steel shell 328
   step tapered 71
   step-tapered steel 335
   supported dock 67
   testing 72

Index

timber 72
treated 72
treated wood 336
wood 389
wood sheet 45
Piles caps, piles and caissons ... 388
grouted 70
Piling marine 68
sheet 44, 379
Pillow tank 252
Pin pile 70
powder 188
Pipe 97
& fittings .86-88, 257, 258, 260, 261,
264-267, 269-271, 273, 274, 277, 279,
281, 283, 284, 286-288, 290, 402
acid resistant 269
aluminum 95, 98
bedding 63, 351, 352
bedding trench 63
black steel 90
bollards 118
cast iron 259
clamp 132, 133
cleaning internal 130
cleanout 287
concrete 76, 95, 100, 101, 390
concrete cylinder 76
conduit gland seal 86
conduit prefabricated ... 85, 87-89
conduit system 85, 87
copper 80, 265
copper tube 80
corrosion resistant 269
corrugated 95
corrugated metal 95, 97
CPVC 277
double wall 91
drain 43
drainage 83, 97, 100, 269
ductile iron 76
DWV ABS 277
DWV PVC 277
elevated 256
elevated installation 256
epoxy fiberglass wound 269
fire retardant 269
fitting 78
fitting adapter 267
fitting cast iron   260, 261, 263, 264,
272, 273
fitting copper 266, 267
fitting corrosion resistant . 269, 270
fitting DWV 279
fitting gasket joint 261-263
fitting no hub 263
fitting plastic 78, 279
fitting soil 260
freezing 132
fused 79
gas 89, 90
glass 257
groove roll labor 276
grooved joint 274
HDPE 79
header 27
high impact 277
inspection internal 130
insulated 84
insulated steel 84
iron alloy mechanical joint ... 269
liner 130
liner HDPE 131
lining 131
metal 97
no hub 260

painting 237, 238
perforated aluminum 95
pile 70, 71
plastic 82, 277
plastic, FRP 277
polyethylene 78, 89
polypropylene 269
prestressed concrete cylinder .. 76
process pressure 257
proxylene 269
PVC 78, 83
quick coupling 27
rail 199
rail aluminum 199
rail galvanized 199
rail stainless 199
rail steel 199
rail wall 199
reinforced concrete 100
relay 43
removal 36
removal metal 256
repair 132, 133
rodding 130
sewage 83, 97, 100
single hub 259
soil 259
steel 80, 89, 95, 98, 271
subdrainage 95
subdrainage plastic 95
support framing 192
supported dock 67
testing 293
testing nondestructive 293
water 77
water distribution 78, 80
wrapping 131
X-ray 293
Pipebursting 130
Piping designations 389
Piping excavation 348
gas station 91
subdrainage 95
valve 90
Pit excavation 59
leaching 84
meter 31
sump 43
Pitch emulsion tar 108
Pits test 34
Placing concrete 160
Planer pavement 25
Planing cold 133
Plank floor 213
roof 217
scaffolding 14, 15
sheathing 218
Plant bed preparation 124
hardiness 394
mixed bituminous concrete 30
screening 20
stolens 127
Planting 122
shrub 124
tree 124
Plants, planting, transplanting ... 122
Plaster cutout 38
demolition 236
drilling 186
mixer 23
Plastic coupling 281
fitting 278
Plastic, FRP pipe 277
Plastic pipe 82, 277
pipe fitting 78, 278, 279
screw anchor 187

valve 284
Plate checkered 201
compactor 54
roadway 25
shear 211
vibrating 54
Platform checkered plate 202
trailer 24
Plating zinc 210
Players bench 120
Playfield equipment 120
Playground equipment 120
Plenum demolition 236
Plinth 143
Plow vibrator 21
Plug valve 91
wall 172
Plumbing 286
appliance 292
demolition 256
fixture 82, 289
fixtures removal 256
Plywood clip 211
demolition 222, 236
fence 17
floor 218
formwork 141, 144
formwork steel framed 146
joist 219
mounting board 217
sheathing roof & walls 217
sidewalk 16
subfloor 218
treatment 209
underlayment 218
Pneumatic nailer 22
stud driver 188
Point heavy duty pile 71
moil 22
pile 71, 72
Pointing masonry 180
Poisoning soil 64
Pole aluminum light 308
anchor 302
clearance 304
cross arm 94, 302
electrical 94
fiberglass light 309
foundation 302
light 308
steel light 309
transmission 302
utility 94, 308
wood light 309
Polyethylene coating 131, 132
drum 67
expansion joint 148
film 122
pipe 78, 89, 278
septic tank 84
tarpaulin 16
waterproofing 223
Polypropylene fabric 94
fitting 270
pipe 269
Polystyrene insulation 224, 225
Polysulfide caulking 229
Polyurethane caulking 229
expansion joint 148
Polyvinyl tarpaulin 16
Pond and reservoir liners 102
Portable air compressor 21
building 13
heater 22
light 22
Portland cement 139

Post athletic 120
cap 211
concrete 117
demolition 208
driver 25
indicator valve 30
lamp 309
tensioned concrete 164
wood 213
Posts sign 121
Potable water softener 292
water treatment 292
Potential transformer 306
Pothead switchgear 304
Poured gypsum 166
insulation 224
Powder actuated tool 188
charge 188
pin 188
Power circuit-breaker 305
distribution 304
pole guying 302
system & capacitor 301
temporary 11, 12
transformer 305
transmission tower 302
Preblast survey 55
Precast beam 164
bridge 119
catch basin 97
column 164
concrete 164, 165
concrete pile 329
curb 106
grate 107
manhole 97
median 117
panel 165
parking barrier 118
pile 70
roof 166
septic tank 84
sill 165
stair 165
tee 165
tilt-up 165
window sills 165
Pre-engineered steel building ... 249
steel buildings 402
structure 248
Prefabricated comfort station 248
Premolded expansion joint 148
Preparation lawn bed 123
plant bed 124
Pressure booster pump 290
grouting cement 44
injected footing 341
injected footings 389
pipe process 257
regulator valve 90
relief valve 283
test 293
test hydraulic 293
valve relief 283
washing metal 205
Prestressed concrete 164
concrete cylinder pipe 76
concrete piles 70, 329
pile 69, 70
Prestressing steel 157
Pretreatment termite 64
Prime coat 102
Primer asphalt 228
steel 205
Prison fence 114
Privacy fence, vinyl 113

Index

Process supply pipe glass 257
Proctor compaction test 10
Product piping 91
Profiler pavement 25
Profiling pavement 134
Progress schedule 9
Project overhead 8
  sign 17
Protection slope 64, 65
  termite 64
  winter 13
Protective device 304
Protect/mooring structures shore 39
Proxylene pipe 269
Prune shrub 129
  tree 129
Pruning 129
  tree 128
P&T relief valve 283
PT substation 306
Pulverization pavement 134
Pump 82, 289, 290
  concrete 160
  contractor 43
  diaphragm 23
  jetting 27
  operator 44
  pressure booster 290
  rental 23
  sewage ejector 291
  shallow well 82
  submersible 23, 292
  sump 292
  trash 23
  water 23, 27, 81
  wellpoint 27
Pumping 43
  dewater 43
  station 243
Purlin roof 213
Push-on joint 77
Putlog scaffolding 15
PVC adhesive 223
  chamfer strip 147
  cleanout 288
  conduit in slab 298
  control joint 172
  duct fitting 93
  DWV pipe 277
  expansion joint 148
  fitting 78, 79
  flange 281
  pipe 78, 83
  pipe perforated 95
  sheet 223
  underground duct 93
  union 282
  waterstop 150
Pyrex pipe 257

Q

Quarry drilling 55
Quartz 123
  line lamp 312

R

Raceway 299
Radial arch 219
  wall 145
Radio tower 92
Radiography test 11
Rafter anchor 211
  demolition 208
  hip 215
  tie 215
  valley 215
  wood 215
Rail aluminum pipe 199
  crane 254
  galvanized pipe 199
  guide 117
  guide/guard 117
  industrial 199
  pipe 199
  railroad 204
  stainless pipe 199
  steel pipe 199
  wall pipe 199
Railing bridge 119
  demolition 209
Railroad ballast 203
  bumper 203
  concrete tie 203
  derail 203
  joint bar 204
  rail 204
  realign 203
  relay rail 203, 204
  siding 203, 400
  spike 204
  switch tie 203
  tie plate 204
  tie step 108
  track 204
  track bolt 204
  track removal 36
  trackwork 203
  turnout 204
  wheel stop 204
  wood tie 203
Raise manhole 96
  manhole frame 96
Raising concrete 45
Rake topsoil 123
Rammer tamper 55
Ramp handicap 159
  temporary 16
Range shooting 244
Ratio water cement 10
Razor wire 114
Reactor substation 306
Ready mix concrete 157, 398
Realign railroad 203
Recirculating chemical toilet 244
Reclamation pavement 134
Recycled asphalt 30
  pavement 30
Recycling cold 134
  pavement 133
Redwood bark mulch 122
  tank 252
Reflective insulation 225
  painting 105
  sign 120
Regulator voltage 305
Rehabilitation concrete 166
Reinforced beam 338
  concrete pipe 100, 101
  culvert 100
Reinforcement concrete 151
Reinforcing 154, 155, 396
  accessory 151
  beam 154
  coated 154
  column 154
  dowels 155
  fiber 157
  footing 154
  girder 154
  joint 170
  ladder 170
  masonry 170
  metric 395
  prices 396
  slab 154
  splice 155, 156
  steel 151
  steel accessory 152, 153
  steel cost 396
  testing 10
  truss 170
  wall 155
Relay pipe 43
  rail railroad 203, 204
Release form 138
Removal air conditioner 256
  block wall 35
  boiler 256
  catch basin 36
  chain link fence 36
  concrete 37
  concrete pipe 36
  curb 37
  driveway 37
  fence 35
  floor 35
  guide rail 36
  hydrant 36
  lavatory 256
  masonry 36
  pipe 36
  plumbing fixtures 256
  railroad track 36
  runway skid marks 36
  shingle 222
  sidewalk 36
  sign 121
  sod 124
  steel pipe 36
  stone 36
  stump 42
  tank 32
  tree 41-43, 122
  urinal 256
  water closet 256
  water fountain 256
  window 232
Remove rock 123
  topsoil 123
Rental equipment 18, 26
  equipment rate 19
  generator 22
Repair crack 166
  pipe 132, 133
Replacement pavement 104
Resaturant roof 228
Reshoring 149
Residential bulkhead 65
  fence 111
  gutting 38
Resilient pavement 109
Resin epoxy 138
Resistor substation 307
Restoration concrete 166
  masonry 179
  mortar 169
Retaining wall 116, 160
  wall formwork 145
  wall segmental 115
  walls interlocking 115
  walls stone 116
Retarder vapor 227
Revolving door 247
Ribbed waterstop 150
Ridge board 215
  flashing 250
  vent 228
Rig drill 33
Rigging forklift 22
Rigid anchor 171
  conduit in-trench 299
  in slab conduit 299
  insulation 225
Ring split 211
  toothed 211
Rip rap 64
Ripping 56
  dozer 56
  grader 56
River stone 107
Rivet 188
  aluminum 188
  copper 188
  monel 188
  stainless 188
  steel 188
  tool 188
Road base 103
  berm 105
  mixer 25
  sign 121
  sweeper 25
  temporary 16
Roadway lighting 308
  plate 25
Rock bolt drilling 55
  bolting 46
  borrow 53
  drill 33, 55
  excavation 55
  remove 123
  salt 134
Rod backer 148, 229
  tie 45, 192
  weld 189
Roller compacted concrete 158, 163
  compaction 52
  earth 20
  vibrating 54
  vibratory 20
Rolling ladder 15
  topsoil 123
  tower 15
Roman brick 172
Roof accessories 228
  aggregate 139
  beam 219
  cant 215
  coating 228
  concrete 164
  curb 215
  deck 217
  deck insulation 226
  expansion joint 204
  frame opening 192
  framing removal 38
  gravel 139
  insulation 226, 250
  jack 215
  nail 210
  patch 228
  precast 166
  purlin 213
  rafters framing 215
  resaturant 228
  sheathing 217
  stressed skin 192
  truss 213, 219
  vent 251
Roofing demolition 222

INDEX
477

membrane 228
Root raking & loading 123
Rope steel wire 195
Rosin paper 228
Rotary drill 22
  hammer drill 22
Rough carpentry 212
  hardware 211
  stone wall 179
Rounds wood 107
Rubber coating 224
  control joint 172
  expansion joint 148
  pavement 109
  tired roller 20
  waterproofing 223
  waterstop 150
Rubberized asphalt 109
  asphalt sealcoat 109
Rubbing wall 162
Rubbish chute 38
  handling 38
Rumble strip 118
Runway centerline light 311
  skid marks removal 36
Rupture testing 10
Rustication strip 145

S

Safety net 13
  switch 307
Sagging conductor 303
Salamander 23
Sales tax 7, 368
Salt rock 134
  treatment lumber 209
Sand 169
  backfill 351
  screened 169
  seal 109
Sandblast masonry 181
Sandblasting concrete 162
  equipment 23
  metal 205
Sander floor 24
Sandl concrete 139
Sandstone flagging 108
Sash 251
Saw 24
  blade 39
  brick 180
  chain 24
  circular 24
  concrete 18, 39
  cutting 39
  joint 166
  masonry 24
Sawing 180
Scaffold specialties 15
Scaffolding 373
  masons 15
  plank 14, 15
  putlog 15
  tubular 13
Scarify 123
  concrete 236
  subsoil 123
Schedule 9
  critical path 9
  progress 9
School crosswalk 121
Scissor lift 254
SCR brick 172
Scraper 20

elevating 58
excavation 58, 59
excavation bulk 58
mobilization 52
self propelled 58
Screed base 153
Screed, gas engine, 8HP vibrating . 18
Screed holder 153
Screened loam 123
  sand 169
  topsoil 123
Screening plant 20
Screw anchor 153
  anchor bolt 153
  brass 210
  eye bolt 153
  jack 45
  lag 188
  sheet metal 210
  steel 210
  wood 210
Seal curb 108
  fog 108
  pavement 108
  slurry 109
Sealant 222, 228
  caulking 229
  tape 229
Sealcoat 108
Sealer concrete 138
  joint 229
Sealing foundation 326
Secondary treatment plant 243
Sectionalizing switch 304
Security fence 114
Sediment bucket drain 289
Seeding 392
  athletic field 126
Seeding, athletic fields 126
Seeding birdsfoot trefoil 126
  bluegrass 126
  clover 126
  crown vetch 126
  fescue 126
  hydro 126
  rye 126
  shade mix 126
  slope mix 126
  turf mix 126
  utility mix 126
  wildflower 126
Segmental retaining wall 115
Select borrow 53
Selective clearing 42
  demo concrete 138
  demo steel 184
  demolition 168, 208, 232, 246
Self propelled crane 26
Self-closing relief valve 283
Self drilling anchor 187
Self-propelled scraper 58, 59
Self-supporting tower 92
Sentry dog 17
Separator oil/water 243
Septic galley 84
  system 83
  tank 83
  tank concrete 84
  tank precast 84
  tanks 83
Service boring 69
Set accelerator 138
Setting stone 117
Sewage aeration 243
  ejector pump 291
  holding tank 244

municipal waste water 243
pipe 83, 97, 100
pumping station 243
treatment plants 243
vent 82
Shaft construction 69
Shale paver 107
Shear connector welded 189
  plate 211
  test 10
  wall 217
Sheathed nonmetallic cable 297
Sheathing 217, 218
  asphalt 218
  gypsum 218
  paper 228
  roof & walls plywood 217
Sheepsfoot compactor 54
Sheet metal screw 210
  piling 44, 65, 379
Sheeting 44
  aluminum 65
  driver 22
  open 46
  tie back 46
  wale 44
  wood 43, 45, 378
Shelf brick 144
Shield expansion 187
  lag screw 187
  light 117
Shielded aluminum cable 298
  cable 297
  copper cable 297
Shift work 7
Shingle removal 222
Shooting range 244
Shore protect/mooring structures . 39
  service 66
Shoring 44, 45, 149
  heavy duty 13, 14
  horizontal 149
  installation 149
  slab 15
  temporary 45
  vertical 149
  wood 46
Shot blaster 24
Shotcrete 163
Shoulder paver 25
Shovel 57
  front 20
Shrinkage test 10
Shrub bed fertilizer 134
  maintenance 134
  moving 122
  planting 124
  prune 129
  weed 134
Side beam 140
Sidewalk 106, 107, 237, 354
  asphalt 106
  bituminous 353
  brick 107
  bridge 15
  concrete 106
  removal 36
  temporary 16
Siding demolition 222
  nail 210
  railroad 203, 400
  removal 222
Sidings railroad 203
Sieve analysis 9, 10
Sign 17, 120
  directional 121

posts 121
project 17
reflective 120
removal 121
road 121
traffic 120
Signal traffic 121
Silicon carbide abrasive 161
  carbide aggregate 138
Silicone caulking 229
  coating 224
  water repellent 224
Sill 215
  anchor 211
  block 177
  formwork 146
  masonry 179
  precast 165
Silo 251
  slipform 146
Silt fence 65
Single hub pipe 259
  pole disconnect switch 306
  tee beam 165
Sink removal 256
Sister joist 214
Site clearing 41
  demolition 35
  demolition minor 36
  drainage 97, 243
  furnishings 119
  improvement 110, 120
  maintenance 134
  preparation 33
  utility 351
  work maintenance general . . 134
Sitework manhole 358
Skidder log 20
Skylight 251
  removal 222
Slab blockout 144
  bolster 151
  concrete 158, 161
  cutout 37
  deck form 198
  edge form 198
  elevated 158
  lift 159
  lifting insert 153
  on grade 143, 159
  on grade removal 36
  pan 159
  reinforcing 154
  shoring 15
  textured 159
  waffle 159
Slabjacking 45
Slab-on-grade sidewalk 354
Slate 179
  flagging 108
  pavement 179
  removal 222
  sidewalk 108
  sill 179
  stair 179
Sleeper 215
  clip 153
Sleepers framing 215
Sleeve 75, 76
  and tap 77
  cut in 75
  formwork 149
Sleeves tapping 75
Slide gate 111, 242
Slipform 146
  barrier 25

Index

building . . . 146
paver . . . 25
silo . . . 146
Slope grading . . . 52
protection . . . 64, 65
Slot channel . . . 171
Slotted framing channel . . . 192
pipe . . . 101
Sluice gate . . . 242
Slurry seal . . . 109
trench . . . 102
Small tools . . . 16
Smokestack . . . 175
Snap tie formwork . . . 149, 150
Snow fence . . . 111
guards . . . 228
Soccer goalpost . . . 120
Socket meter . . . 300
Sod . . . 127
Sodding . . . 127
Sodium high pressure lamp . . . 312
low pressure fixture . . . 308
low pressure lamp . . . 312
Soffit steel . . . 250
Soft ground shield driven boring . 68
Soil cement . . . 64
compaction . . . 52
compactor . . . 19
decontamination . . . 33
nailing . . . 46
pipe . . . 259
poisoning . . . 64
sample . . . 34
stabilization . . . 63, 64
stabilization asphalt . . . 63
stabilization cement . . . 64
stabilization geotextile . . . 64
stabilization lime . . . 64
tamping . . . 52
test . . . 9, 10
treatment . . . 64
Soldier beam . . . 46
Space heater rental . . . 22
Spacer . . . 153
line . . . 303
transmission . . . 303
Spade air . . . 22
tree . . . 21
Spandrel beam . . . 140
Specialties & accessories, roof . . . 228
scaffold . . . 15
Specific gravity . . . 10
gravity testing . . . 9
Speed bump . . . 117
Spike grid . . . 211
railroad . . . 204
Splice pile . . . 71, 72
reinforcing . . . 155, 156
Split ring . . . 211
Spotter . . . 60
Spray coating . . . 222
Sprayer emulsion . . . 22
Spread footing 143, 159, 321, 322
footings . . . 384
loam . . . 123
park soil . . . 123
soil conditioner . . . 123
topsoil . . . 123, 124
Spreader aggregate . . . 25
chemical . . . 25
concrete . . . 23
Sprig stolens . . . 127
Sprinkler grass . . . 110
system . . . 110
Squash court backstop . . . 120
Sread stone . . . 107
Stabilization asphalt . . . 63
base . . . 64
cement . . . 64
fabric . . . 103
geotextile . . . 64
lime . . . 64
soil . . . 63, 64
Stabilizer base . . . 25
Stadium bleacher . . . 248
cover . . . 247
Staging swing . . . 15
Stain lumber . . . 219
truss . . . 238
Stainless bar grating . . . 201
plank grating . . . 200
rivet . . . 188
weld rod . . . 190
Stair basement . . . 165
brick . . . 177
concrete . . . 160
finish . . . 162
formwork . . . 144
landing . . . 198
metal . . . 198
nosing . . . 150
pan . . . 150
precast . . . 165
railroad tie . . . 108
removal . . . 208
slate . . . 179
steel . . . 198
stone . . . 121
temporary protection . . . 13
tread . . . 150, 178
Stake nail . . . 150
Stamping concrete . . . 162
texture . . . 162
Standpipe steel . . . 252
Starter board & switch . . . 305
wall . . . 143
Static capacitor . . . 305
Station capacitor . . . 305
weather . . . 13
Steam clean masonry . . . 181
cleaner . . . 24
cleaning metal . . . 205
Steel anchor . . . 171
bar grating . . . 200
beam W-shape . . . 193
bin wall . . . 116
blocking . . . 212
bolt . . . 211
bridge . . . 118, 119
bridging . . . 212
building pre-engineered . . . 249
conduit in slab . . . 299
conduit in trench . . . 299
cutting . . . 39, 190
deck . . . 197
dome . . . 248
drill . . . 22
drilling . . . 186
fence . . . 111
fitting . . . 272
flange . . . 275
form . . . 141
frame formwork . . . 141
H pile . . . 333
hex bolt . . . 185
high strength . . . 195
joist . . . 196
ladder . . . 198
lintel . . . 192
members structural . . . 192
mesh box . . . 64
mesh grating . . . 201
painting . . . 205
pile . . . 71
pipe . . . 80, 89, 95, 98, 271
pipe pile . . . 331, 332
pipe removal . . . 36
prestressing . . . 157
primer . . . 205
projects structural . . . 194
reinforcing . . . 151
rivet . . . 188
salvage . . . 257
screw . . . 210
sheet pile . . . 44
shell pile . . . 327, 328
shore . . . 149
silo . . . 251
stair . . . 198
standpipe . . . 252
step-tapered pile . . . 335
structural . . . 191
tank . . . 252
tower . . . 92
underground duct . . . 93
valve . . . 285
weld rod . . . 189
well casing . . . 81
well screen . . . 81
window . . . 251
window demolition . . . 232
wire rope . . . 195
Step . . . 121
bluestone . . . 108
brick . . . 108
manhole . . . 97
masonry . . . 108
railroad tie . . . 108
tapered pile . . . 71
tapered steel pile . . . 335
Stiffleg derrick . . . 27
Stockade fence . . . 114
Stolen broadcast . . . 127
Stolens plant . . . 127
sprig . . . 127
Stone aggregate . . . 139
anchor . . . 171
base . . . 102
curbs . . . 106
dust . . . 107
filter . . . 64
gabion retaining walls . . . 116
mulch . . . 122
paver . . . 108, 178
pea . . . 139
removal . . . 36
retaining walls . . . 116
setting . . . 117
stair . . . 121
threshold . . . 165
tread . . . 178
wall . . . 116, 178
Stool window . . . 179
Stop corporation . . . 80
curb . . . 80
Stops gas . . . 90
Storage box . . . 13
dome . . . 248
dome bulk . . . 248
tank . . . 251, 252
tank cover . . . 247
tank demolition . . . 246
trailer . . . 24
Storm drainage manholes, frames . 97
window . . . 251
Strand grouted . . . 157
ungrouted . . . 157
Strap tie . . . 211
Straw . . . 122
Streetlight . . . 308
Strength compressive . . . 10
Stressed skin roof . . . 192
Stressing tendons . . . 157
Strip chamfer . . . 147
footing . . . 159, 320
rumble . . . 118
soil . . . 43
Striper paint . . . 25
thermal . . . 25
Stripping topsoil . . . 43
Structural aluminum . . . 195
backfill . . . 52
columns . . . 191
concrete . . . 157
excavation . . . 59
fee . . . 6
formwork . . . 140
insulated panel . . . 217
panel . . . 217
shape column . . . 191
steel . . . 191
steel galvanized . . . 205
steel members . . . 192
steel painting . . . 205
steel projects . . . 194
welding . . . 190
Structure fabric . . . 247
hydraulic . . . 242
moving . . . 135
pre-engineered . . . 248
tension . . . 248
Stucco demolition . . . 236
Stud demolition . . . 208
driver pneumatic . . . 188
partition . . . 214
wall . . . 214
welded . . . 189
Stump chipper . . . 20
chipping . . . 41
removal . . . 42
Subcontractor O & P . . . 8
Subdrainage fitting . . . 96
pipe . . . 95
piping . . . 95
plastic pipe . . . 95
system . . . 94
Subfloor . . . 218
plywood . . . 218
wood . . . 218
Subgrade chair . . . 153
stake . . . 153
Submersible pump . . 23, 81, 289, 292
Substation battery . . . 307
bus . . . 306
cable-tray . . . 306
conductor . . . 306
conduit . . . 306
CT . . . 306
equipment . . . 305
fuse . . . 307
insulator . . . 306
manhole . . . 306
PT . . . 306
reactor . . . 306
resistor . . . 307
transformer . . . 305, 307
transmission . . . 305
Subsurface exploration . . . 34
investigation . . . 33
Suction hose . . . 23
Sump hole construction . . . 43
pit . . . 43
pump . . . 292
Super high-tensile anchor . . . 46

Supply wells, pumps water 82
Support dowel 143
framing pipe 192
Surface treatment 108, 109
Surfacing 108
Survey aerial 6
crew 6
monument 6
preblast 55
property line 6
stake 16
topographic 6
Sweeper road 25
Swell testing 9
Swing check valve 282
gate 111
staging 15
Switch disconnect 306
electric 307
general duty 307
grounding 306
safety 307
sectionalizing 304
single pole disconnect 306
tie railroad 203
transmission 304, 306
Switchgear 304
breaker low-vol 305
cable lug 304
low-voltage components 305
pothead 304
temperature alarm 305
transformer 304
Synchronous capacitor 305
Synthetic erosion control 65
System grinder pump 290
irrigation 110
pipe conduit 85, 87, 88
septic 83
sitework catch basin 358
sitework trenching 348
sprinkler 110
subdrainage 94

T

Tack coat 109
Tamper 22, 62
compactor 55
rammer 55
Tamping soil 52
Tandem roller 20
Tank clean 32
cover 247
disposal 33
fixed roof 252
gasoline 252
holding 244
pillow 252
removal 32
septic 83
steel 252
storage 251, 252
testing 11
water 252
Tanks septic 83
Tap and sleeve 78
Tape barricade 17
detection 32
sealant 229
underground 32
Tapping main 75
sleeves 75
valves 75
Tar kettle 25
kettle/pot 24
paper 123
pitch emulsion 108
roof 228
Target range 244
Tarpaulin 13, 16
duck 16
mylar 16
polyethylene 16
polyvinyl 16
Tax 7
sales 7
social security 7
unemployment 7
Taxes 368
Technician inspection 10
Tee beam 164
cleanout 288
pipe 275
precast 165
Telephone manhole 93
pole 94
underground electrical 93, 94
Temperature relief valve 283
Tempering valve 283
Temporary building 12
connection 12
construction 12, 13, 17
facility 16
fence 17
guardrail 16
heat 12, 139
light 12
oil heater 22
power 11
road painting 105
shoring 45
toilet 24
utility 11
Tendons stressing 157
Tennis court backstop 120
court fence 114, 120
Tensile test 10
Tension structure 248
Termite pretreatment 64
protection 64
Terra cotta demolition 37, 237
Terrazzo demolition 236
Test beam 10
load pile 72
moisture content 10
pile load 72
pits 34
pressure 293
soil 10
ultrasonic 11
well 81
Testing 9
pile 72
pipe 293
sulfate soundness 9
tank 11
Texture stamping 162
Textured slab 159
Thermal striper 25
Thermoplastic painting 105
Thinning tree 41
Thinset pavers 107
Threshold stone 165
THW wire 406
Tie adjustable wall 170
back sheeting 46
bar 151
clip 151
column 171
flange 171
formwork snap 150
plate railroad 204
rafter 215
rod 45, 192
strap 211
wall 170, 171
wire 153, 171
Tile demolition 236
flue 174
wall 92
Tilling topsoil 124
Tilt up construction 165
Tilt-up precast 165
Timber connector 211
fastener 209
framing 213
guide rail 117
laminated 219
pile 72
Time lapse photography 9
Toilet 244
chemical 24
partition removal 237
temporary 24
trailer 24
Tool air 21
powder actuated 188
rivet 188
van 24
Tools small 16
Toothed ring 211
Toothing masonry 37, 180
Top demolition counter 209
dressing 124
Topographic survey 6
Topping floor 162
Topsoil 43, 54, 124
remove 123
screened 123
spread 123
stripping 43
Torch cutting 24, 39, 190
Tower anchor 302
control 248
crane 26, 374
footing 302
foundation 302
hoist 27
light 23
line 302
radio 92
rolling 15
transmission 302
Track bolt railroad 204
drill 21
railroad 204
Trackwork railroad 203
Tractor 20, 62
loader 20
truck 24
Traffic barriers, hywy. sound bar . 118
channelizing 118
cone 16
line 105
lines and markings painted ... 105
sign 120
signal 121
Trailer office 12
platform 24
storage 24
toilet 24
truck 21, 24
Tram car 24
Transceiver 312
Transformer 301, 305
base 309
current 306
grounding 305
instrument 306
oil filled 301
potential 306
power 305
silicon filled 301
substation 305, 307
switchgear 304
transmission 305
weight 408
Transit mixer 18
Translucent pipe 257
Transmission capacitor 305
circuit-breaker 305
fuse 307
insulator 303, 304
jumper 303
line 304
line pole 302
pole 302
spacer 303
substation 305
switch 304, 306
tower 302
transformer 305
Transparent pipe 258
Trap rock surface 162
Trash pump 23
Tread stair 150, 178
Treated pile 68, 72
wood pile 336
Treatment lumber 209
plant secondary 243
plants sewage 243
plywood 209
potable water 292
surface 109
wood 209
Tree and shrub removal 122
cutting 42
deciduous 127
fertilizer 135
grate 107
guying 129
maintenance 135
moving 122
planting 124
prune 129
pruning 128
removal 41-43, 122
spade 21
thinning 41
water 129
Trench backfill 62, 351, 352
box 24, 46
cover 202
drain 289
excavation 61-63
excavation and pipe bedding .. 380
formwork 144
grating frame 202
slurry 102
utility 62
Trencher 21
chain 21
Trenching 43, 55, 349
Trial batch 10
Trim demolition 209
lawn 128
Triple brick 172
Trowel coating 222
concrete 18
Truck boom 26
concrete 18
dump 21

flatbed . . . 21
hauling . . . 60
loading . . . 59
mounted crane . . . 26
off highway . . . 21
pickup . . . 24
rental . . . 21
tractor . . . 24
trailer . . . 21, 24
vacuum . . . 24
Trucking . . . 60
Truss bowstring . . . 219
demolition . . . 208
flying . . . 149
joist . . . 197
metal roof . . . 194
painting . . . 238
plate . . . 211
reinforcing . . . 170
roof . . . 213, 219
stain . . . 238
varnish . . . 238
Tubing copper . . . 265
corrugated . . . 95
Tubular scaffolding . . . 13
steel joist . . . 219
Tugboat diesel . . . 27
Tunnel excavation . . . 68, 383
grouting . . . 68
liner grouting . . . 68
linings . . . 68
ventilator . . . 26
Tunneling . . . 68
Turbine water meter . . . 286
Turnout railroad . . . 204
Tyton joint . . . 77

U

Ultrasonic test . . . 10, 11
Underbuilt circuit . . . 303
conductor . . . 303
Underdrain . . . 97
Underground duct . . . 93
duct fitting . . . 93
piping . . . 351
storage tank removal . . . 376
tape . . . 32
Underlayment . . . 218
hardboard . . . 218
Underpin foundation . . . 45
Underwater lighting . . . 311
Undisturbed soil . . . 10
Unemployment tax . . . 7
Ungrouted bar . . . 157
strand . . . 157
Union PVC . . . 282
Unit masonry . . . 172
Upstanding beam formwork . . . 140
Urethane foam . . . 229
Urinal removal . . . 256
U.S. customary units u.s. . . . 395
Utility accessories . . . 32
boxes . . . 30
brick . . . 172
duct . . . 93
electric . . . 93
excavation . . . 348
materials demolition . . . 39
pole . . . 94, 308
pump . . . 289
sitework . . . 93
structure . . . 358
temporary . . . 11
trench . . . 62

V

Vacuum circuit-breaker . . . 306
truck . . . 24
Valley gutter . . . 250
rafter . . . 215
Valve . . . 282-285
backwater . . . 287
ball . . . 284
bronze . . . 282
butterfly . . . 31, 284
check . . . 31, 75, 285
check swing . . . 284
cut in . . . 75
gate . . . 31, 284, 285
globe . . . 31, 283-285
grooved joint butterfly . . . 276
grooved-joint gate . . . 276
iron body . . . 284
lubricated . . . 91
materials . . . 403
piping . . . 90
plastic . . . 284
plug . . . 91
pressure regulator . . . 90
relief pressure . . . 283
selection considerations . . . 403
soldered gate . . . 282
steel . . . 285
swing check . . . 282
tempering . . . 283
water pressure . . . 283
Valves . . . 31
forged . . . 285
tapping . . . 75
Vapor barrier . . . 227
barrier sheathing . . . 218
retarder . . . 227
Varnish truss . . . 238
VCT removal . . . 236
Veneer ashlar . . . 178
brick . . . 174, 175
granite . . . 178
members laminated . . . 219, 220
removal . . . 168
Vent ridge . . . 228
roof . . . 251
sewage . . . 82
Ventilated metal deck . . . 197
Ventilator tunnel . . . 26
Vermiculite insulation . . . 224
Vertical shore . . . 149
Vibrating plate . . . 54
roller . . . 54
screed, gas engine, 8HP . . . 18
Vibration monitoring . . . 11
pad . . . 188
Vibrator concrete . . . 18
earth . . . 20, 52, 62
plate . . . 52
plow . . . 21
Vibratory equipment . . . 19
roller . . . 20
Vibroflotation . . . 45
compaction . . . 380
Vinyl coated fence . . . 112
fence . . . 113
plastic waterproofing . . . 223
Void form corrugated . . . 144
formwork . . . 143
Voltage regulator . . . 305

W

Waffle slab . . . 159
Wale sheeting . . . 44
Wales . . . 44
Walk . . . 106
maintenance . . . 134
Wall area . . . 233
brick . . . 177
cast in place . . . 343
coating . . . 239
concrete . . . 159, 161, 342
cutout . . . 37
demolition . . . 236
expansion joint . . . 204
finish . . . 162
formwork . . . 144
foundation . . . 176, 343
framing . . . 216
framing removal . . . 38
furring . . . 216
grout . . . 169
hydrant . . . 290
insulation . . . 176, 224, 225, 250
masonry . . . 116, 175
panel . . . 159, 165
panel brick . . . 177
plug . . . 172
radial . . . 145
reinforcing . . . 155
removal . . . 35
retaining . . . 116, 160
rubbing . . . 162
shear . . . 217
sheathing . . . 217, 218
starter . . . 143
steel bin . . . 116
stone . . . 178
stud . . . 214
tie . . . 170, 171
tie masonry . . . 170
tile . . . 92
type cleanout . . . 288
Walls and partitions demolition . . . 236
Warehouse . . . 247
Wash brick . . . 181
Washer . . . 211
Wastewater treatment system . . . 244
Watchdog . . . 17
Watchman service . . . 17
Water blasting . . . 180
blasting metal . . . 205
cement ratio . . . 10
closet chemical . . . 244
closet removal . . . 256
copper tubing connector . . . 285
distribution chilled . . . 84
distribution hot . . . 85
distribution pipe . . . 76, 78, 80
effect testing . . . 9
fountain removal . . . 256
heater removal . . . 257
hose . . . 23
hydrant . . . 290
meter turbine . . . 286
pipe . . . 77
pressure relief valve . . . 283
pressure valve . . . 283
pump . . . 23, 27, 81
pumping . . . 43
repellent coating . . . 224
repellent silicone . . . 224
softener potable . . . 292
supply detector meter . . . 286
supply domestic meter . . . 286
supply meter . . . 286
supply wells, pumps . . . 82
tank . . . 252
tank elevated . . . 251
tank sprayer . . . 24
tempering valve . . . 283
trailer . . . 24
treatment . . . 244
tree . . . 129
well . . . 81
Watering . . . 129
Waterproof membrane . . . 166
Waterproofing . . . 139, 222, 223
butyl . . . 223
cementitious . . . 223
coating . . . 222
elastomeric . . . 222
foundation . . . 326
integral . . . 139
masonry . . . 169
membrane . . . 108, 222, 223
neoprene . . . 223
rubber . . . 223
Waterstop . . . 150
barrier . . . 148
center bulb . . . 150
dumbbell . . . 150
fitting . . . 150
PVC . . . 150
ribbed . . . 150
rubber . . . 150
Wearing course . . . 103
Weather data & design . . . 370
station . . . 13
Wedge anchor . . . 187
Weed barrier . . . 122
shrub . . . 134
Weeding . . . 129
Weld rod . . . 189
rod aluminum . . . 190
rod cast-iron . . . 190
rod stainless . . . 190
rod steel . . . 189
X-ray . . . 293
Welded shear connector . . . 189
structural steel . . . 399
stud . . . 189
wire fabric . . . 156
Welding certification . . . 11
fillet . . . 190
machine . . . 25
structural . . . 190
Well . . . 43, 81
casing steel . . . 81
gravel pack . . . 81
pump shallow . . . 82
screen steel . . . 81
water . . . 81
Wellpoint . . . 44, 377
equipment . . . 27
header pipe . . . 27
pump . . . 27
Wheel stop railroad . . . 204
Wheelbarrow . . . 25
Widener pavement . . . 25
Winch truck . . . 24
Wind cone . . . 311
Window demolition . . . 232
framing . . . 250
metal . . . 251
removal . . . 232
sill . . . 179
sills precast . . . 165
steel . . . 251
stool . . . 179
Windrow elevator . . . 26
mixing . . . 134
Winter concrete . . . 158
protection . . . 13, 139
Wire brushing . . . 205

fabric welded 156
fence 17, 113
fences, misc. metal 113
fencing 114
mesh 114
razor 114
rope steel 195
tie 171
Wiring device & box 300
methods 297
Wood beam 212, 213
bench 120
block floor demolition 236
blocking 212, 215
bridge 119
bridging 212
chamfer strip 148
chips 123
column 213
deck 217
demolition 236
dock 67
dome 248
fascia 215
fastener 210
fence 114
fiber insulation 224
fiber sheathing 218
fiber underlayment 218
floor demolition 236
foundation 345
framing 212, 216
furring 216
girder 212
joist 213, 219
laminated 219
partition 214
pile 389
pile treated 336
pole 94
rafter 215
roof deck 217
roof deck demolition 222
roof truss 219
rounds 107
screw 210
sheathing 218
sheet piling 378
sheeting 43, 45, 378
shoring 46
sidewalk 107
siding demolition 222
sill 215
subfloor 218
tank 252
tie railroad 203
treatment 209
wall 345
window demolition 232
Wool fiberglass 224
lead 148
Work extra 8
Workers' compensation ... 7, 364-366
Wrapping pipe 131
Wrecking ball 25
Wrench impact 22

X

X-ray concrete slabs 11
pipe 293
weld 293

Z

Zinc plating 210

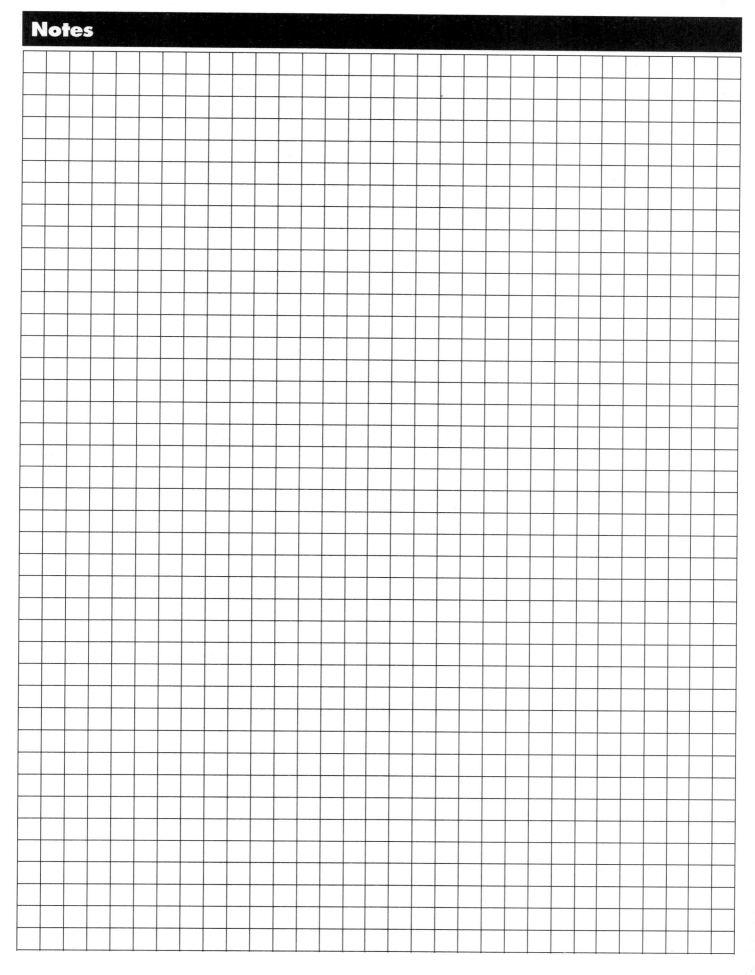

485

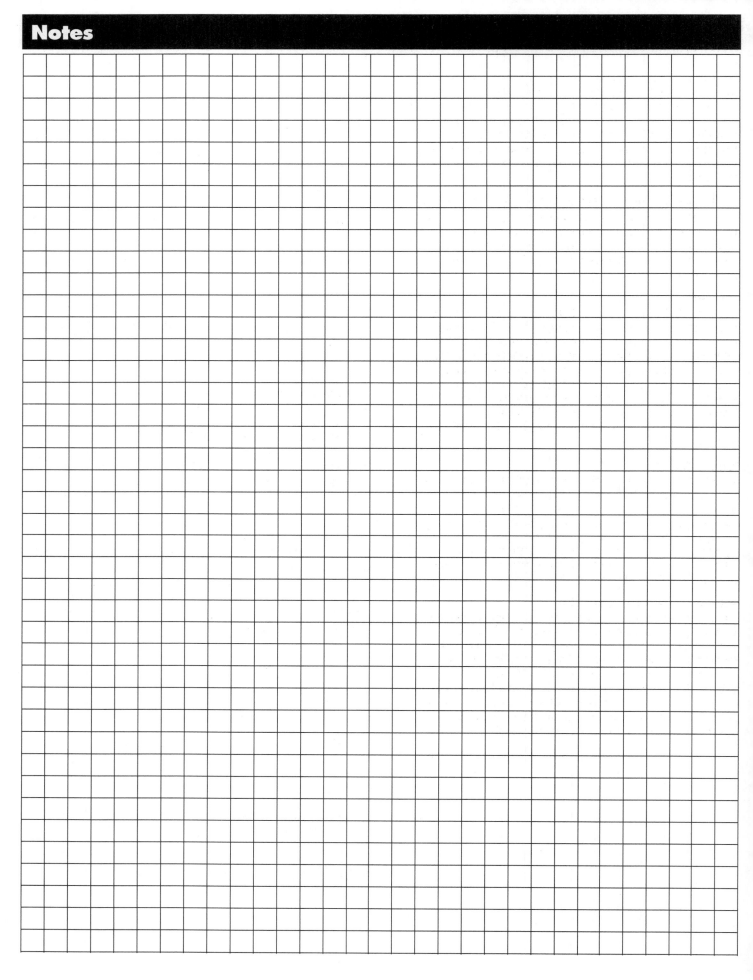

Reed Construction Data, Inc.

Reed Construction Data, Inc., a leading worldwide provider of total construction information solutions, is comprised of three main product groups designed specifically to help construction professionals advance their businesses with timely, accurate and actionable project, product, and cost data. Reed Construction Data is a division of Reed Business Information, a member of the Reed Elsevier plc group of companies.

The *Project, Product, and Cost & Estimating* divisions offer a variety of innovative products and services designed for the full spectrum of design, construction, and manufacturing professionals. Through it's *International* companies, Reed Construction Data's reputation for quality construction market data is growing worldwide.

Cost Information
RSMeans, the undisputed market leader and authority on construction costs, publishes current cost and estimating information in annual cost books and on the CostWorks CD-ROM. RSMeans furnishes the construction industry with a rich library of complementary reference books and a series of professional seminars that are designed to sharpen professional skills and maximize the effective use of cost estimating and management tools. RSMeans also provides construction cost consulting for Owners, Manufacturers, Designers, and Contractors.

Project Data
Reed Construction Data provides complete, accurate and relevant project information through all stages of construction. Customers are supplied industry data through leads, project reports, contact lists, plans and specifications surveys, market penetration analyses and sales evaluation reports. Any of these products can pinpoint a county, look at a state, or cover the country. Data is delivered via paper, e-mail, CD-ROM or the Internet.

Building Product Information
The First Source suite of products is the only integrated building product information system offered to the commercial construction industry for comparing and specifying building products. These print and online resources include *First Source,* CSI's SPEC-DATA™, CSI's MANU-SPEC™, First Source CAD, and Manufacturer Catalogs. Written by industry professionals and organized using CSI's MasterFormat™, construction professionals use this information to make better design decisions.

FirstSourceONL.com combines Reed Construction Data's project, product and cost data with news and information from Reed Business Information's *Building Design & Construction* and *Consulting-Specifying Engineer,* this industry-focused site offers easy and unlimited access to vital information for all construction professionals.

International
BIMSA/Mexico provides construction project news, product information, cost-data, seminars and consulting services to construction professionals in Mexico. Its subsidiary, PRISMA, provides job costing software.

Byggfakta Scandinavia AB, founded in 1936, is the parent company for the leaders of customized construction market data for Denmark, Estonia, Finland, Norway and Sweden. Each company fully covers the local construction market and provides information across several platforms including subscription, ad-hoc basis, electronically and on paper.

Reed Construction Data Canada serves the Canadian construction market with reliable and comprehensive project and product information services that cover all facets of construction. Core services include: *BuildSource, BuildSpec, BuildSelect,* product selection and specification tools available in print and on the Internet; Building Reports, a national construction project lead service; CanaData, statistical and forecasting information; *Daily Commercial News,* a construction newspaper reporting on news and projects in Ontario; and *Journal of Commerce,* reporting news in British Columbia and Alberta.

Cordell Building Information Services, with its complete range of project and cost and estimating services, is Australia's specialist in the construction information industry. Cordell provides in-depth and historical information on all aspects of construction projects and estimation, including several customized reports, construction and sales leads, and detailed cost information among others.

For more information, please visit our Web site at www.reedconstructiondata.com.

Reed Construction Data, Inc., Corporate Office
30 Technology Parkway South
Norcross, GA 30092-2912
(800) 322-6996
(800) 895-8661 (fax)
info@reedbusiness.com
www.reedconstructiondata.com

 Reed Construction Data

Means Project Cost Report

By filling out and returning the Project Description, you can receive a discount of $20.00 off any one of the Means products advertised in the following pages. The cost information required includes all items marked (✔) except those where no costs occurred. The sum of all major items should equal the Total Project Cost.

$20.00 Discount per product for each report you submit.

DISCOUNT PRODUCTS AVAILABLE—FOR U.S. CUSTOMERS ONLY—STRICTLY CONFIDENTIAL

Project Description (No remodeling projects, please.)

✔ Type Building _____

✔ Location _____

  Capacity _____

✔ Frame _____

✔ Exterior _____

✔ Basement: full ☐ partial ☐ none ☐ crawl ☐

✔ Height in Stories _____

✔ Total Floor Area _____

  Ground Floor Area _____

✔ Volume in C.F. _____

  % Air Conditioned _____ Tons _____

  Comments _____

Owner _____

Architect _____

General Contractor _____

✔ Bid Date _____

  Typical Bay Size _____

✔ Labor Force: _____ % Union _____ % Non-Union

✔ Project Description (Circle one number in each line)
  1. Economy 2. Average 3. Custom 4. Luxury
  1. Square 2. Rectangular 3. Irregular 4. Very Irregular

	✔ Total Project Cost			$
A	✔ General Conditions			$
B	✔ Site Work			$
BS	Site Clearing & Improvement			
BE	Excavation	(	C.Y.)	
BF	Caissons & Piling	(	L.F.)	
BU	Site Utilities			
BP	Roads & Walks Exterior Paving	(	S.Y.)	
C	✔ Concrete			$
C	Cast in Place	(	C.Y.)	
CP	Precast	(	S.F.)	
D	✔ Masonry			$
DB	Brick	(	M)	
DC	Block	(	M)	
DT	Tile	(	S.F.)	
DS	Stone	(	S.F.)	
E	✔ Metals			$
ES	Structural Steel	(	Tons)	
EM	Misc. & Ornamental Metals			
F	✔ Wood & Plastics			$
FR	Rough Carpentry	(	MBF)	
FF	Finish Carpentry			
FM	Architectural Millwork			
G	✔ Thermal & Moisture Protection			$
GW	Waterproofing-Dampproofing	(	S.F.)	
GN	Insulation	(	S.F.)	
GR	Roofing & Flashing	(	S.F.)	
GM	Metal Siding/Curtain Wall	(	S.F.)	
H	✔ Doors and Windows			$
HD	Doors	(	Ea.)	
HW	Windows	(	S.F.)	
HH	Finish Hardware			
HG	Glass & Glazing	(	S.F.)	
HS	Storefronts	(	S.F.)	

J	✔ Finishes			$
JL	Lath & Plaster	(	S.Y.)	
JD	Drywall	(	S.F.)	
JM	Tile & Marble	(	S.F.)	
JT	Terrazzo	(	S.F.)	
JA	Acoustical Treatment	(	S.F.)	
JC	Carpet	(	S.Y.)	
JF	Hard Surface Flooring	(	S.F.)	
JP	Painting & Wall Covering	(	S.F.)	
K	✔ Specialties			$
KB	Bathroom Partitions & Access.	(	S.F.)	
KF	Other Partitions	(	S.F.)	
KL	Lockers	(	Ea.)	
L	✔ Equipment			$
LK	Kitchen			
LS	School			
LO	Other			
M	✔ Furnishings			$
MW	Window Treatment			
MS	Seating	(	Ea.)	
N	✔ Special Construction			$
NA	Acoustical	(	S.F.)	
NB	Prefab. Bldgs.	(	S.F.)	
NO	Other			
P	✔ Conveying Systems			$
PE	Elevators	(	Ea.)	
PS	Escalators	(	Ea.)	
PM	Material Handling			
Q	✔ Mechanical			$
QP	Plumbing	(No. of fixtures	)	
QS	Fire Protection (Sprinklers)			
QF	Fire Protection (Hose Standpipes)			
QB	Heating, Ventilating & A.C.			
QH	Heating & Ventilating	(BTU Output	)	
QA	Air Conditioning	(	Tons)	
R	✔ Electrical			$
RL	Lighting	(	S.F.)	
RP	Power Service			
RD	Power Distribution			
RA	Alarms			
RG	Special Systems			
S	✔ Mech./Elec. Combined			$

Product Name _____

Product Number _____

Your Name _____

Title _____

Company _____
  ☐ Company
  ☐ Home Street Address _____

City, State, Zip _____

☐ Please send _____ forms.

Please specify the Means product you wish to receive. Complete the address information as requested and return this form with your check (product cost less $20.00) to address below.

RSMeans Company, Inc.,
Square Foot Costs Department
P.O. Box 800
Kingston, MA 02364-9988

For more information
visit Means Web Site
at www.rsmeans.com

Reed Construction Data/RSMeans . . . a tradition of excellence in Construction Cost Information and Services since 1942.

Table of Contents

Annual Cost Guides, Page 2
Reference Books, Page 6
Seminars, Page 12
Electronic Data, Page 14
New Titles, Page 15
Order Form, Page 16

Book Selection Guide

The following table provides definitive information on the content of each cost data publication. The number of lines of data provided in each unit price or assemblies division, as well as the number of reference tables and crews is listed for each book. The presence of other elements such as an historical cost index, city cost indexes, square foot models or cross-referenced index is also indicated. You can use the table to help select the Means' book that has the quantity and type of information you most need in your work.

Unit Cost Divisions	Building Construction Costs	Mechanical	Electrical	Repair & Remodel.	Square Foot	Site Work Landsc.	Assemblies	Interior	Concrete Masonry	Open Shop	Heavy Construc.	Light Commercial	Facil. Construc.	Plumbing	Western Construction Costs	Resi-dential
1	1104	822	907	1013		1045		847	1014	1102	1051	752	1528	874	1102	710
2	2505	1468	460	1450		7794		547	1387	2467	4978	692	4305	1584	2490	772
3	1401	113	100	726		1259		201	1782	1396	1404	229	1312	82	1399	248
4	835	18	0	647		657		574	1056	811	594	407	1058	0	821	334
5	1820	239	193	957		768		927	691	1788	1023	805	1802	316	1802	752
6	1489	82	78	1475		452		1406	318	1481	608	1626	1596	47	1826	1757
7	1264	158	74	1241		468		489	408	1264	349	956	1318	168	1264	752
8	1839	28	0	1907		313		1673	645	1821	51	1227	2034	0	1841	1193
9	1595	47	0	1424		233		1688	360	1547	159	1337	1809	47	1586	1216
10	861	47	25	498		193		700	170	863	0	394	906	233	861	217
11	1018	322	169	502		137		813	44	925	107	218	1173	291	925	104
12	307	0	0	47		210		1416	27	298	0	62	1435	0	298	64
13	1153	1005	375	494		388		884	75	1136	279	484	1783	928	1119	193
14	345	36	0	258		36		292	0	344	30	12	343	35	343	6
15	1987	13138	636	1779		1569		1174	59	1996	1766	1217	10865	9695	2017	826
16	1277	470	10063	1002		739		1108	55	1265	760	1081	9834	415	1226	552
17	427	354	427	0		0		0	0	427	0	0	427	356	427	0
Totals	21227	18347	13507	15420		16261		14739	8091	20961	13159	11499	43528	15071	21347	9696

Assembly Divisions	Building Construction Costs	Mechanical	Elec-trical	Repair & Remodel.	Square Foot	Site Work Landsc.	Assemblies	Interior	Concrete Masonry	Open Shop	Heavy Construc.	Light Commercial	Facil. Construc.	Plumbing	Western Construction Costs	Asm Div	Resi-dential
A		19	0	192	150	540	612	0	550		542	149	24	0		1	374
B		0	0	809	2480	0	5590	333	1914		0	2024	144	0		2	217
C		0	0	635	862	0	1220	1568	145		0	767	238	0		3	588
D		1031	780	693	1823	0	2439	753	0		0	1310	1027	896		4	871
E		0	0	85	255	0	292	5	0		0	255	5	0		5	393
F		0	0	0	123	0	126	0	0		0	123	3	0		6	358
G		465	172	332	111	1856	584	0	482		432	110	113	560		7	299
																8	760
																9	80
																10	0
																11	0
																12	0
Totals		1515	952	2746	5804	2396	10863	2659	3091		974	4738	1554	1456			3940

Reference Section	Building Construction Costs	Mechanical	Electrical	Repair & Remodel.	Square Foot	Site Work Landsc.	Assemblies	Interior	Concrete Masonry	Open Shop	Heavy Construc.	Light Commercial	Facil. Construc.	Plumbing	Western Construction Costs	Resi-dential
Tables	130	45	85	69	4	80	219	46	71	130	61	57	104	51	131	42
Models					102								43			32
Crews	410	410	410	391		410		410	410	393	410	393	391	410	410	393
City Cost Indexes	yes	yes	yes	yes	yes	yes	yes	yes	yes	yes	yes	yes	yes	yes	yes	yes
Historical Cost Indexes	yes	yes	yes	yes	yes	yes	yes	yes	yes	yes	yes	yes	yes	yes	yes	no

1

Annual Cost Guides

For more information
visit Means Web Site
at www.rsmeans.com

Means Building Construction Cost Data 2005

Available in Both Softbound and Looseleaf Editions

The "Bible" of the industry comes in the standard softcover edition or the looseleaf edition.

Many customers enjoy the convenience and flexibility of the looseleaf binder, which increases the usefulness of *Means Building Construction Cost Data 2005* by making it easy to add and remove pages. You can insert your own cost information pages, so everything is in one place. Copying pages for faxing is easier also. Whichever edition you prefer, softbound or the convenient looseleaf edition, you'll be eligible to receive *The Change Notice* FREE. Current subscribers receive *The Change Notice* via e-mail.

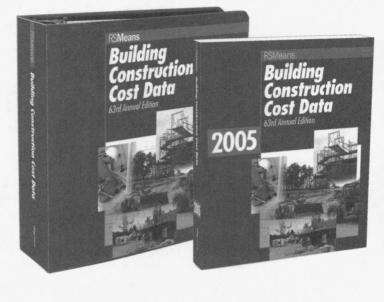

$115.95 per copy, Softbound
Catalog No. 60015

$144.95 per copy, Looseleaf
Catalog No. 61015

Means Building Construction Cost Data 2005

Offers you unchallenged unit price reliability in an easy-to-use arrangement. Whether used for complete, finished estimates or for periodic checks, it supplies more cost facts better and faster than any comparable source. Over 23,000 unit prices for 2005. The City Cost Indexes cover over 930 areas, for indexing to any project location in North America. Order and get *The Change Notice* FREE. You'll have year-long access to the Means Estimating **HOTLINE** FREE with your subscription. Expert assistance when using Means data is just a phone call away.

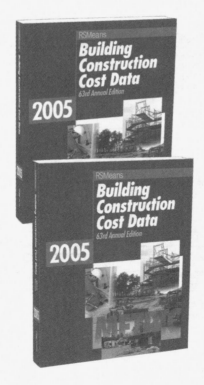

$115.95 per copy
Over 700 pages, illustrated, available Oct. 2004
Catalog No. 60015

Means Building Construction Cost Data 2005

Metric Version

The Federal Government has stated that all federal construction projects must now use metric documentation. The *Metric Version* of *Means Building Construction Cost Data 2005* is presented in metric measurements covering all construction areas. Don't miss out on these billion dollar opportunities. Make the switch to metric today.

$115.95 per copy
Over 700 pages, illus., available Dec. 2004
Catalog No. 63015

For more information
visit Means Web Site
at www.rsmeans.com

Annual Cost Guides

Means Mechanical Cost Data 2005

• HVAC • Controls

Total unit and systems price guidance for mechanical construction...materials, parts, fittings, and complete labor cost information. Includes prices for piping, heating, air conditioning, ventilation, and all related construction.

Plus new 2005 unit costs for:
• Over 2500 installed HVAC/controls assemblies
• "On Site" Location Factors for over 930 cities and towns in the U.S. and Canada
• Crews, labor and equipment

$115.95 per copy
Over 600 pages, illustrated, available Oct. 2004
Catalog No. 60025

Means Plumbing Cost Data 2005

Comprehensive unit prices and assemblies for plumbing, irrigation systems, commercial and residential fire protection, point-of-use water heaters, and the latest approved materials. This publication and its companion, *Means Mechanical Cost Data*, provide full-range cost estimating coverage for all the mechanical trades.

$115.95 per copy
Over 550 pages, illustrated, available Oct. 2004
Catalog No. 60215

Means Electrical Cost Data 2005

Pricing information for every part of electrical cost planning: More than 15,000 unit and systems costs with design tables; clear specifications and drawings; engineering guides and illustrated estimating procedures; complete labor-hour and materials costs for better scheduling and procurement; the latest electrical products and construction methods.
• A Variety of Special Electrical Systems including Cathodic Protection
• Costs for maintenance, demolition, HVAC/ mechanical, specialties, equipment, and more

$115.95 per copy
Over 450 pages, illustrated, available Oct. 2004
Catalog No. 60035

Means Electrical Change Order Cost Data 2005

You are provided with electrical unit prices exclusively for pricing change orders based on the recent, direct experience of contractors and suppliers. Analyze and check your own change order estimates against the experience others have had doing the same work. It also covers productivity analysis and change order cost justifications. With useful information for calculating the effects of change orders and dealing with their administration.

$115.95 per copy
Over 450 pages, available Oct. 2004
Catalog No. 60235

Means Facilities Maintenance & Repair Cost Data 2005

Published in a looseleaf format, *Means Facilities Maintenance & Repair Cost Data* gives you a complete system to manage and plan your facility repair and maintenance costs and budget efficiently. Guidelines for auditing a facility and developing an annual maintenance plan. Budgeting is included, along with reference tables on cost and management and information on frequency and productivity of maintenance operations.

The only nationally recognized source of maintenance and repair costs. Developed in cooperation with the Army Corps of Engineers.

$252.95 per copy
Over 600 pages, illustrated, available Dec. 2004
Catalog No. 60305

Means Square Foot Costs 2005

It's Accurate and Easy To Use!

• **Updated 2005 price information,** based on nationwide figures from suppliers, estimators, labor experts and contractors.

• "How-to-Use" Sections, with **clear examples** of commercial, residential, industrial, and institutional structures.

• Realistic graphics, offering true-to-life illustrations of building projects.

• Extensive information on using square foot cost data, including **sample estimates** and **alternate pricing methods.**

$126.95 per copy
Over 450 pages, illustrated, available Nov. 2004
Catalog No. 60055

Annual Cost Guides

For more information
visit Means Web Site
at www.rsmeans.com

Means Repair & Remodeling Cost Data 2005

Commercial/Residential

You can use this valuable tool to estimate commercial and residential renovation and remodeling.

Includes: New costs for hundreds of unique methods, materials and conditions that only come up in repair and remodeling. PLUS:

- Unit costs for over 16,000 construction components
- Installed costs for over 90 assemblies
- Costs for 300+ construction crews
- Over 930 "On Site" localization factors for the U.S. and Canada.

$99.95 per copy
Over 650 pages, illustrated, available Nov. 2004
Catalog No. 60045

Means Facilities Construction Cost Data 2005

For the maintenance and construction of commercial, industrial, municipal, and institutional properties. Costs are shown for new and remodeling construction and are broken down into materials, labor, equipment, overhead, and profit. Special emphasis is given to sections on mechanical, electrical, furnishings, site work, building maintenance, finish work, and demolition. More than 45,000 unit costs, plus assemblies and reference sections are included.

$279.95 per copy
Over 1200 pages, illustrated, available Nov. 2004
Catalog No. 60205

Means Residential Cost Data 2005

Contains square foot costs for 30 basic home models with the look of today, plus hundreds of custom additions and modifications you can quote right off the page. With costs for the 100 residential systems you're most likely to use in the year ahead. Complete with blank estimating forms, sample estimates and step-by-step instructions.

$99.95 per copy
Over 600 pages, illustrated, available Oct. 2004
Catalog No. 60175

Means Light Commercial Cost Data 2005

Specifically addresses the light commercial market, which is an increasingly specialized niche in the industry. Aids you, the owner/designer/contractor, in preparing all types of estimates, from budgets to detailed bids. Includes new advances in methods and materials. Assemblies section allows you to evaluate alternatives in the early stages of design/planning.

Over 13,000 unit costs for 2005 ensure you have the prices you need...when you need them.

$99.95 per copy
Over 650 pages, illustrated, available Nov. 2004
Catalog No. 60185

Means Assemblies Cost Data 2005

Means Assemblies Cost Data 2005 takes the guesswork out of preliminary or conceptual estimates. Now you don't have to try to calculate the assembled cost by working up individual components costs. We've done all the work for you.

Presents detailed illustrations, descriptions, specifications and costs for every conceivable building assembly—240 types in all—arranged in the easy-to-use UNIFORMAT II system. Each illustrated "assembled" cost includes a complete grouping of materials and associated installation costs including the installing contractor's overhead and profit.

$189.95 per copy
Over 600 pages, illustrated, available Oct. 2004
Catalog No. 60065

Means Site Work & Landscape Cost Data 2005

Means Site Work & Landscape Cost Data 2005 is organized to assist you in all your estimating needs. Hundreds of fact-filled pages help you make accurate cost estimates efficiently.

Updated for 2005!

- Demolition features—including ceilings, doors, electrical, flooring, HVAC, millwork, plumbing, roofing, walls and windows
- State-of-the-art segmental retaining walls
- Flywheel trenching costs and details
- Updated Wells section
- Landscape materials, flowers, shrubs and trees

$115.95 per copy
Over 600 pages, illustrated, available Nov. 2004
Catalog No. 60285

**For more information
visit Means Web Site
at www.rsmeans.com**

Annual Cost Guides

Means Open Shop Building Construction Cost Data 2005

The latest costs for accurate budgeting and estimating of new commercial and residential construction... renovation work... change orders... cost engineering. *Means Open Shop BCCD* will assist you to...
• Develop benchmark prices for change orders
• Plug gaps in preliminary estimates, budgets
• Estimate complex projects
• Substantiate invoices on contracts
• Price ADA-related renovations

$115.95 per copy
Over 700 pages, illustrated, available Dec. 2004
Catalog No. 60155

Means Heavy Construction Cost Data 2005

A comprehensive guide to heavy construction costs. Includes costs for highly specialized projects such as tunnels, dams, highways, airports, and waterways. Information on different labor rates, equipment, and material costs is included. Has unit price costs, systems costs, and numerous reference tables for costs and design. Valuable not only to contractors and civil engineers, but also to government agencies and city/ town engineers.

$115.95 per copy
Over 450 pages, illustrated, available Nov. 2004
Catalog No. 60165

Means Building Construction Cost Data 2005
Western Edition

This regional edition provides more precise cost information for western North America. Labor rates are based on union rates from 13 western states and western Canada. Included are western practices and materials not found in our national edition: tilt-up concrete walls, glu-lam structural systems, specialized timber construction, seismic restraints, landscape and irrigation systems.

$115.95 per copy
Over 650 pages, illustrated, available Dec. 2004
Catalog No. 60225

Means Heavy Construction Cost Data 2005
Metric Version
Make sure you have the Means industry standard metric costs for the federal, state, municipal and private marketplace. With thousands of up-to-date metric unit prices in tables by CSI standard divisions. Supplies you with assemblies costs using the metric standard for reliable cost projections in the design stage of your project. Helps you determine sizes, material amounts, and has tips for handling metric estimates.

$115.95 per copy
Over 450 pages, illustrated, available Dec. 2004
Catalog No. 63165

Means Construction Cost Indexes 2005

Who knows what 2005 holds? What materials and labor costs will change unexpectedly? By how much?
• Breakdowns for 316 major cities.
• National averages for 30 key cities.
• Expanded five major city indexes.
• Historical construction cost indexes.

$252.95 per year
$63.00 individual quarters
Catalog No. 60145 A,B,C,D

Means Interior Cost Data 2005

Provides you with prices and guidance needed to make accurate interior work estimates. Contains costs on materials, equipment, hardware, custom installations, furnishings, labor costs . . . every cost factor for new and remodel commercial and industrial interior construction, including updated information on office furnishings, plus more than 50 reference tables. For contractors, facility managers, owners.

$115.95 per copy
Over 600 pages, illustrated, available Oct. 2004
Catalog No. 60095

Means Concrete & Masonry Cost Data 2005

Provides you with cost facts for virtually all concrete/masonry estimating needs, from complicated formwork to various sizes and face finishes of brick and block, all in great detail. The comprehensive unit cost section contains more than 8,500 selected entries. Also contains an assemblies cost section, and a detailed reference section which supplements the cost data.

$105.95 per copy
Over 450 pages, illustrated, available Dec. 2004
Catalog No. 60115

Means Labor Rates for the Construction Industry 2005

Complete information for estimating labor costs, making comparisons and negotiating wage rates by trade for over 300 cities (United States and Canada). With 46 construction trades listed by local union number in each city, and historical wage rates included for comparison. No similar book is available through the trade.

Each city chart lists the county and is alphabetically arranged with handy visual flip tabs for quick reference.

$253.95 per copy
Over 300 pages, available Dec. 2004
Catalog No. 60125

Reference Books

Building Security: Strategies & Costs
By David Owen

Unauthorized systems access, terrorism, hurricanes and tornadoes, sabotage, vandalism, fire, explosions, and other threats... All are considerations as building owners, facility managers, and design and construction professionals seek ways to meet security needs.

This comprehensive resource will help you evaluate your facility's security needs — and design and budget for the materials and devices needed to fulfill them. The text and cost data will help you to:

- Identify threats, probability of occurrence, and the potential losses— to determine and address your real vulnerabilities.
- Perform a detailed risk assessment of an existing facility, and prioritize and budget for security enhancement.
- Evaluate and price security systems and construction solutions, so you can make cost-effective choices.

Includes over 130 pages of Means Cost Data for installation of security systems and materials, plus a review of more than 50 security devices and construction solutions—how they work, and how they compare.

$89.95 per copy
Over 400 pages, Hardcover
Catalog No. 37339

Green Building: Project Planning & Cost Estimating
By RSMeans and Contributing Authors

Written by a team of leading experts in sustainable design, this new book is a complete guide to planning and estimating green building projects, a growing trend in building design and construction – commercial, industrial, institutional and residential. It explains:

- All the different criteria for "green-ness"
- What criteria your building needs to meet to get a LEED, Energy Star, or other recognized rating for green buildings
- How the project team works differently on a green versus a traditional building project
- How to select and specify green products
- How to evaluate the cost and value of green products versus conventional ones — not only for their first (installation) cost, but their cost over time (in maintenance and operation).

Features an extensive Green Building Cost Data section, which details the available products, how they are specified, and how much they cost.

$89.95 per copy
350 pages, illustrated, Hardcover
Catalog No. 67338

Life Cycle Costing for Facilities
By Alphonse Dell'Isola and Dr. Steven Kirk

Guidance for achieving higher quality design and construction projects at lower costs!

Facility designers and owners are frustrated with cost-cutting efforts that yield the cheapest product, but sacrifice quality. Life Cycle Costing, properly done, enables them to achieve both—high quality, incorporating innovative design, and costs that meet their budgets.

The authors show how LCC can work for a broad variety of projects – from several types of buildings, to roads and bridges, to HVAC and electrical upgrades, to materials and equipment procurement. Case studies include:

- Health care and nursing facilities
- College campus and high schools
- Office buildings, courthouses, and banks
- Exterior walls, elevators, lighting, HVAC, and more

The book's extensive cost section provides maintenance and replacement costs for facility elements.

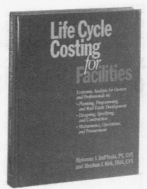

$99.95 per copy
450 pages, Hardcover
Catalog No. 67341

For more information
visit Means Web Site
at www.rsmeans.com

Reference Books

Value Engineering: Practical Applications

. . . For Design, Construction, Maintenance & Operations
By Alphonse Dell'Isola, PE

A tool for immediate application—for engineers, architects, facility managers, owners, and contractors. Includes: making the case for VE—the management briefing, integrating VE into planning and budgeting, conducting life cycle costing, integrating VE into the design process, using VE methodology in design review and consultant selection, case studies, VE workbook, and a life cycle costing program on disk.

$79.95 per copy
Over 450 pages, illustrated, Hardcover
Catalog No. 67319

Facilities Operations & Engineering Reference

By the Association for Facilities Engineering and RSMeans

An all-in-one technical referance for planning and managing facility projects and solving day-to-day operations problems. Selected as the official Certified Plant Engineer reference, this handbook covers financial analysis, maintenance, HVAC and energy efficiency, and more.

$109.95 per copy
Over 700 pages, illustrated, Hardcover
Catalog No. 67318

The Building Professional's Guide to Contract Documents

3rd Edition
By Waller S. Poage, AIA, CSI, CVS

This comprehensive treatment of Contract Documents is an important reference for owners, design professionals, contractors, and students.
• Structure your Documents for Maximum Efficiency
• Effectively communicate construction requirements to all concerned
• Understand the Roles and Responsibilities of Construction Professionals
• Improve Methods of Project Delivery
$64.95 per copy, 400 pages
Diagrams and construction forms, Hardcover
Catalog No. 67261A

HVAC: Design Criteria, Options, Selection

Expanded 2nd Edition

By William H. Rowe III, AIA, PE
Includes Indoor Air Quality, CFC Removal, Energy Efficient Systems, and Special Systems by Building Type. Helps you solve a wide range of HVAC system design and selection problems effectively and economically. Gives you clear explanations of the latest ASHRAE standards.

$84.95 per copy
Over 600 pages, illustrated, Hardcover
Catalog No. 67306

Cost Planning & Estimating for Facilities Maintenance

In this unique book, a team of facilities management authorities shares their expertise at:
• Evaluating and budgeting maintenance operations
• Maintaining & repairing key building components
• Applying *Means Facilities Maintenance & Repair Cost Data* to your estimating

Covers special maintenance requirements of the 10 major building types.

$89.95 per copy
Over 475 pages, Hardcover
Catalog No. 67314

Facilities Maintenance Management
By Gregory H. Magee, PE
Now you can get successful management methods and techniques for all aspects of facilities maintenance. This comprehensive reference explains and demonstrates successful management techniques for all aspects of maintenance, repair, and improvements for buildings, machinery, equipment, and grounds. Plus, guidance for outsourcing and managing internal staffs.

$86.95 per copy
Over 280 pages with illustrations, Hardcover
Catalog No. 67249

Builder's Essentials: Advanced Framing Methods

By Scot Simpson

A highly illustrated, "framer-friendly" approach to advanced framing elements. Provides expert, but easy-to-interpret, instruction for laying out and framing complex walls, roofs, and stairs, and special requirements for earthquake and hurricane protection. Also helps bring framers up to date on the latest building code changes, and provides tips on the lead framer's role and responsibilities, how to prepare for a job, and how to get the crew started.

$24.95 per copy
250 pages, illustrated, Softcover
Catalog No. 67330

Reference Books

For more information
visit Means Web Site
at www.rsmeans.com

Interior Home Improvement Costs,

New 9th Edition

Estimates for the most popular remodeling and repair projects—
from small, do-it-yourself jobs—to major renovations and new
construction. Includes: Kitchens & Baths; New Living Space from
Your Attic, Basement or Garage; New Floors, Paint & Wallpaper;
Tearing Out or Building New Walls; Closets, Stairs & Fireplaces;
New Energy-Saving Improvements, Home Theatres, and More!

$24.95 per copy
250 pages, illustrated, Softcover
Catalog No. 67308E

Exterior Home Improvement Costs

New 9th Edition

Estimates for the most popular remodeling and repair
projects—from small, do-it-yourself jobs, to major renovations
and new construction. Includes: Curb Appeal Projects—
Landscaping, Patios, Porches, Driveways and Walkways; New
Windows and Doors; Decks, Greenhouses, and Sunrooms; Room
Additions and Garages; Roofing, Siding, and Painting; "Green"
Improvements to Save Energy & Water

$24.95 per copy
Over 275 pages, illustrated, softcover
Catalog No. 67309E

Builder's Essentials: Plan Reading & Material Takeoff

By Wayne J. DelPico

For Residential and Light Commercial Construction

A valuable tool for understanding plans and specs, and
accurately calculating material quantities.
Step-by-step instructions and takeoff procedures based on a
full set of working drawings.

$35.95 per copy
Over 420 pages, Softcover
Catalog No. 67307

Planning & Managing Interior Projects

2nd Edition

By Carol E. Farren, CFM

Addresses changes in technology and business, guiding you
through commercial design and construction from initial client
meetings to post-project administration. Includes: evaluating
space requirements, alternative work models,
telecommunications and data management, and environmental
issues.

$69.95 per copy
Over 400 pages, illustrated, Hardcover
Catalog No. 67245A

Builder's Essentials: Best Business Practices for Builders & Remodelers:

An Easy-to-Use Checklist System
By Thomas N. Frisby

A comprehensive guide covering all aspects of running a
construction business, with more than 40 user-friendly checklists.
This book provides expert guidance on: increasing your revenue
and keeping more of your profit; planning for long-term growth;
keeping good employees and managing subcontractors.

$29.95 per copy
Over 220 pages, Softcover
Catalog No. 67329

Builder's Essentials: Framing & Rough Carpentry 2nd Edition

By Scot Simpson

A complete training manual for apprentice and experienced
carpenters. Develop and improve your skills with "framer-
friendly," easy-to-follow instructions, and step-by-step
illustrations. Learn proven techniques for framing walls, floors,
roofs, stairs, doors, and windows. Updated guidance on
standards, building codes, safety requirements, and more.
Also available in Spanish!

$24.95 per copy
Over 150 pages, Softcover
Catalog No. 67298A   Spanish Catalog No. 67298AS

How to Estimate with Means Data & CostWorks

By RSMeans and Saleh Mubarak
New 2nd Edition!

Learn estimating techniques using Means cost data.
Includes an instructional version of Means CostWorks
CD–ROM with Sample Building Plans.
The step-by-step guide takes you through all the major
construction items. Over 300 sample estimating problems
are included.

$59.95 per copy
Over 190 pages, Softcover
Catalog No. 67324A

Unit Price Estimating Methods

New 3rd Edition

This new edition includes up-to-date cost data and estimating
examples, updated to reflect changes to the CSI numbering
system and new features of Means cost data. It describes the
most productive, universally accepted ways to estimate, and
uses checklists and forms to illustrate shortcuts and
timesavers. A model estimate demonstrates procedures. A
new chapter explores computer estimating alternatives.

$59.95 per copy
Over 350 pages, illustrated, Hardcover
Catalog No. 67303A

Means Landscape Estimating Methods

4th Edition
By Sylvia H. Fee

This revised edition offers expert guidance for preparing accurate
estimates for new landscape construction and grounds maintenance.
Includes a complete project estimate featuring the latest equipment
and methods, and **two chapters on Life Cycle Costing, and Landscape
Maintenance Estimating.**

$62.95 per copy
Over 300 pages, illustrated, Hardcover
Catalog No. 67295B

Means Environmental Remediation Estimating Methods, 2nd Edition

By Richard R. Rast

Guidelines for estimating 50 standard remediation technologies.
Use it to prepare preliminary budgets, develop estimates,
compare costs and solutions, estimate liability, review quotes,
negotiate settlements.

A valuable support tool for *Means Environmental Remediation
Unit Price* and *Assemblies* books.

$99.95 per copy
Over 750 pages, illustrated, Hardcover
Catalog No. 64777A

For more information
visit Means Web Site
at www.rsmeans.com

Reference Books

Means Illustrated Construction Dictionary, Condensed Edition, 2ⁿᵈ Edition

By RSMeans

Recognized in the industry as the best resource of its kind, this has been further enhanced with updates to existing terms and the addition of hundreds of new terms and illustrations . . . in keeping with the most recent developments in the industry.

The best portable dictionary for office or field use—an essential tool for contractors, architects, insurance and real estate personnel, homeowners, and anyone who needs quick, clear definitions for construction terms.

$59.95 per copy
Over 500 pages
Catalog No. 67282A

Means Repair and Remodeling Estimating

New 4th Edition
By Edward B. Wetherill & RSMeans

Focuses on the unique problems of estimating renovations of existing structures. It helps you determine the true costs of remodeling through careful evaluation of architectural details and a site visit.

New section on disaster restoration costs.

$69.95 per copy
Over 450 pages, illustrated, Hardcover
Catalog No. 67265B

Facilities Planning & Relocation

New, lower price and user-friendly format.
By David D. Owen

A complete system for planning space needs and managing relocations. Includes step-by-step manual, over 50 forms, and extensive reference section on materials and furnishings.

$89.95 per copy
Over 450 pages, Softcover
Catalog No. 67301

Means Square Foot & Assemblies Estimating Methods

3rd Edition!

Develop realistic Square Foot and Assemblies Costs for budgeting and construction funding. The new edition features updated guidance on square foot and assemblies estimating using UNIFORMAT II. An essential reference for anyone who performs conceptual estimates.

$69.95 per copy
Over 300 pages, illustrated, Hardcover
Catalog No. 67145B

Means Electrical Estimating Methods 3rd Edition

Expanded new edition includes sample estimates and cost information in keeping with the latest version of the CSI MasterFormat and UNIFORMAT II. Complete coverage of Fiber Optic and Uninterruptible Power Supply electrical systems, broken down by components and explained in detail. Includes a new chapter on computerized estimating methods. A practical companion to *Means Electrical Cost Data.*

$64.95 per copy
Over 325 pages, Hardcover
Catalog No. 67230A

Means Mechanical Estimating Methods 3rd Edition

This guide assists you in making a review of plans, specs, and bid packages with suggestions for takeoff procedures, listings, substitutions and pre-bid scheduling. Includes suggestions for budgeting labor and equipment usage. Compares materials and construction methods to allow you to select the best option.

$64.95 per copy
Over 350 pages, illustrated, Hardcover
Catalog No. 67294A

Means Spanish/English Construction Dictionary

By RSMeans, The International Conference of Building Officials (ICBO), and Rolf Jensen & Associates (RJA)

Designed to facilitate communication among Spanish- and English-speaking construction personnel—improving performance and job-site safety. Features the most common words and phrases used in the construction industry, with easy-to-follow pronunciations. Includes extensive building systems and tools illustrations.

$22.95 per copy
250 pages, illustrated, Softcover
Catalog No. 67327

Project Scheduling & Management for Construction

New 3rd Edition
By David R. Pierce, Jr.

A comprehensive yet easy-to-follow guide to construction project scheduling and control—from vital project management principles through the latest scheduling, tracking, and controlling techniques. The author is a leading authority on scheduling with years of field and teaching experience at leading academic institutions. Spend a few hours with this book and come away with a solid understanding of this essential management topic.

$64.95 per copy
Over 300 pages, illustrated, Hardcover
Catalog No. 67247B

Reference Books

For more information
visit Means Web Site
at www.rsmeans.com

Concrete Repair and Maintenance Illustrated
By Peter H. Emmons
$69.95 per copy
Catalog No. 67146

Superintending for Contractors:
How to Bring Jobs in On-Time, On-Budget
By Paul J. Cook
$35.95 per copy
Catalog No. 67233

HVAC Systems Evaluation
By Harold R. Colen, PE
$84.95 per copy
Catalog No. 67281

Basics for Builders: How to Survive and Prosper in Construction
By Thomas N. Frisby
$34.95 per copy
Catalog No. 67273

Successful Interior Projects Through Effective Contract Documents
By Joel Downey & Patricia K. Gilbert
Now $24.98 per copy, limited quantity
Catalog No. 67313

Building Spec Homes Profitably
By Kenneth V. Johnson
$29.95 per copy
Catalog No. 67312

Estimating for Contractors
How to Make Estimates that Win Jobs
By Paul J. Cook
$35.95 per copy
Catalog No. 67160

Successful Estimating Methods:
From Concept to Bid
By John D. Bledsoe, PhD, PE
$32.48 per copy
Catalog No. 67287

Total Productive Facilities Management
By Richard W. Sievert, Jr.
$29.98 per copy
Over 270 pages, illustrated, Hardcover
Catalog No. 67321

Means Productivity Standards for Construction
Expanded Edition (Formerly Man-Hour Standards)
$49.98 per copy
Over 800 pages, Hardcover
Catalog No. 67236A

Means Plumbing Estimating Methods, 3rd Edition
By Joseph Galeno and Sheldon Greene
Now $29.98 per copy
Catalog No. 67283B

For more information
visit Means Web Site
at www.rsmeans.com

Reference Books

Preventive Maintenance Guidelines for School Facilities

By John C. Maciha

A complete PM program for K-12 schools that ensures sustained security, safety, property integrity, user satisfaction, and reasonable ongoing expenditures.

Includes schedules for weekly, monthly, semiannual, and annual maintenance with hard copy and electronic forms.

$149.95 per copy
Over 225 pages, Hardcover
Catalog No. 67326

Preventive Maintenance for Higher Education Facilities

By Applied Management Engineering, Inc.

An easy-to-use system to help facilities professionals establish the value of PM, and to develop and budget for an appropriate PM program for their college or university. Features interactive campus building models typical of those found in different-sized higher education facilities, and PM checklists linked to each piece of equipment or system in hard copy and electronic format.

$149.95 per copy
150 pages, Hardcover
Catalog No. 67337

Historic Preservation: Project Planning & Estimating

By Swanke Hayden Connell Architects

Expert guidance on managing historic restoration, rehabilitation, and preservation building projects and determining and controlling their costs. Includes:

• How to determine whether a structure qualifies as historic
• Where to obtain funding and other assistance
• How to evaluate and repair more than 75 historic building materials

$99.95 per copy
Over 675 pages, Hardcover
Catalog No. 67323

Means Illustrated Construction Dictionary, 3rd Edition

Long regarded as the Industry's finest, the Means *Illustrated Construction Dictionary* is now even better.
With the addition of over 1,000 new terms and hundreds of new illustrations, it is the clear choice for the most comprehensive and current information. **The companion CD-ROM that comes with this new edition adds many extra features: larger graphics, expanded definitions, and links to both CSI MasterFormat numbers and product information.**

$99.95 per copy
Over 790 pages, Illustrated, Hardcover
Catalog No. 67292A

Designing & Building with the IBC, 2nd Edition

By Rolf Jensen & Associates, Inc.

This updated comprehensive guide helps building professionals make the transition to the 2003 *International Building Code®*. Includes a side-by-side code comparison of the IBC 2003 to the IBC 2000 and the three primary model codes, a quick-find index, and professional code commentary. With illustrations, abbreviations key, and an extensive Resource section.

$99.95 per copy
Over 875 pages
Catalog No. 67328A

Residential & Light Commercial Construction Standards, 2nd Edition

By RSMeans and Contributing Authors

New, updated second edition of this unique collection of industry standards that define quality construction. For contractors, subcontractors, owners, developers, architects, engineers, attorneys, and insurance personnel, this book provides authoritative requirements and recommendations compiled from the nation's leading professional associations, industry publications, and building code organizations.

$59.95 per copy
600 pages, illustrated, Softcover
Catalog No. 67322A

Means Estimating Handbook, 2nd Edition

By RSMeans

Updated new Second Edition answers virtually any estimating technical question - all organized by CSI MasterFormat. This comprehensive reference covers the full spectrum of technical data required to estimate construction costs. The book includes information on sizing, productivity, equipment requirements, code-mandated specifications, design standards and engineering factors.

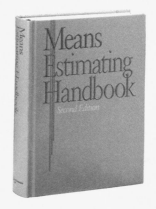

$99.95 per copy
Over 900 pages, Hardcover
Catalog No. 67276A

**For more information
visit Means Web Site
at www.rsmeans.com**

Seminars

Means CostWorks Training

This one-day seminar course has been designed with the intention of assisting both new and existing users to become more familiar with CostWorks program. The class is broken into two unique sections: (1) A one-half day presentation on the function of each icon; and each student will be shown how to use the software to develop a cost estimate. (2) Hands-on estimating exercises that will ensure that each student thoroughly understands how to use CostWorks. You must bring your own laptop computer to this course.

CostWorks Benefits/Features:
- Estimate in your own spreadsheet format
- Power of Means National Database
- Database automatically regionalized
- Save time with keyword searches
- Save time by establishing common estimate items in "Bookmark" files
- Customize your spreadsheet template
- Hot key to Product Manufacturers' listings and specs
- Merge capability for networking environments
- View crews and assembly components
- AutoSave capability
- Enhanced sorting capability

Unit Price Estimating

This interactive two-day seminar teaches attendees how to interpret project information and process it into final, detailed estimates with the greatest accuracy level.

The single most important credential an estimator can take to the job is the ability to visualize construction in the mind's eye, and thereby estimate accurately.

Some Of What You'll Learn:
- Interpreting the design in terms of cost
- The most detailed, time tested methodology for accurate "pricing"
- Key cost drivers—material, labor, equipment, staging and subcontracts
- Understanding direct and indirect costs for accurate job cost accounting and change order management

Who Should Attend: Corporate and government estimators and purchasers, architects, engineers…and others needing to produce accurate project estimates.

Square Foot and Assemblies Cost Estimating

This two-day course teaches attendees how to quickly deliver accurate square foot estimates using limited budget and design information.

Some Of What You'll Learn:
- How square foot costing gets the estimate done faster
- Taking advantage of a "systems" or "assemblies" format
- The Means "building assemblies/square foot cost approach"
- How to create a very reliable preliminary and systems estimate using bare-bones design information

Who Should Attend: Facilities managers, facilities engineers, estimators, planners, developers, construction finance professionals…and others needing to make quick, accurate construction cost estimates at commercial, government, educational and medical facilities.

Repair and Remodeling Estimating

This two-day seminar emphasizes all the underlying considerations unique to repair/remodeling estimating and presents the correct methods for generating accurate, reliable R&R project costs using the unit price and assemblies methods.

Some Of What You'll Learn:
- Estimating considerations—like labor-hours, building code compliance, working within existing structures, purchasing materials in smaller quantities, unforeseen deficiencies
- Identifies problems and provides solutions to estimating building alterations
- Rules for factoring in minimum labor costs, accurate productivity estimates and allowances for project contingencies
- R&R estimating examples are calculated using unite prices and assemblies data

Who Should Attend: Facilities managers, plant engineers, architects, contractors, estimators, builders…and others who are concerned with the proper preparation and/or evaluation of repair and remodeling estimates.

Mechanical and Electrical Estimating

This two-day course teaches attendees how to prepare more accurate and complete mechanical/electrical estimates, avoiding the pitfalls of omission and double-counting, while understanding the composition and rationale within the Means Mechanical/Electrical database.

Some Of What You'll Learn:
- The unique way mechanical and electrical systems are interrelated
- M&E estimates, conceptual, planning, budgeting and bidding stages
- Order of magnitude, square foot, assemblies and unit price estimating
- Comparative cost analysis of equipment and design alternatives

Who Should Attend: Architects, engineers, facilities managers, mechanical and electrical contractors…and others needing a highly reliable method for developing, understanding and evaluating mechanical and electrical contracts.

Plan Reading and Material Takeoff

This two-day program teaches attendees to read and understand construction documents and to use them in the preparation of material takeoffs.

Some of What You'll Learn:
- Skills necessary to read and understand typical contract documents—blueprints and specifications
- Details and symbols used by architects and engineers
- Construction specifications' importance in conjunction with blueprints
- Accurate takeoff of construction materials and industry-accepted takeoff methods

Who Should Attend: Facilities managers, construction supervisors, office managers…and other responsible for the execution and administration of a construction project including government, medical, commercial, educational or retail facilities.

Facilities Maintenance and Repair Estimating

This two-day course teaches attendees how to plan, budget, and estimate the cost of ongoing and preventive maintenance and repair for existing buildings and grounds.

Some Of What You'll Learn:
- The most financially favorable maintenance, repair and replacement scheduling and estimating
- Auditing and value engineering facilities
- Preventive planning and facilities upgrading
- Determining both in-house and contract-out service costs; annual, asset-protecting M&R plan

Who Should Attend: Facility managers, maintenance supervisors, buildings and grounds superintendents, plant managers, planners, estimators…and others involved in facilities planning and budgeting.

Scheduling and Project Management

This two-day course teaches attendees the most current and proven scheduling and management techniques needed to bring projects in on time and on budget.

Some Of What You'll Learn:
- Crucial phases of planning and scheduling
- How to establish project priorities, develop realistic schedules and management techniques
- Critical Path and Precedence Methods
- Special emphasis on cost control

Who Should Attend: Construction project managers, supervisors, engineers, estimators, contractors…and others who want to improve their project planning, scheduling and management skills.

Advanced Project Management

This two-day seminar will teach you how to effectively manage and control the entire design-build process and allow you to take home tangible skills that will be immediately applicable on existing projects.

Some Of What You'll Learn:
- Value engineering, bonding, fast-tracking and bid package creation
- How estimates and schedules can be integrated to provide advanced project management tools
- Cost engineering, quality control, productivity measurement and improvement
- Front loading a project and predicting its cash flow

Who Should Attend: Owners, project managers, architectural and engineering managers, construction managers, contractors…and anyone else who is responsible for the timely design and completion of construction projects.

Seminars

For more information
visit Means Web Site
at www.rsmeans.com

2005 Means Seminar Schedule

Note: Call for exact dates and details.

Location	Dates
Las Vegas, NV	March
Washington, DC	April
Phoenix, AZ	April
Denver, CO	May
San Francisco, CA	June
Philadelphia, PA	June
Washington, DC	September
Dallas, TX	September
Las Vegas, NV	October
Orlando, FL	November
Atlantic City, NJ	November
San Diego, CA	December

Registration Information

Register Early... Save up to $100! Register 30 days before the start date of a seminar and save $100 off your total fee. *Note: This discount can be applied only once per order. It cannot be applied to team discount registrations or any other special offer.*

How to Register Register by phone today! Means toll-free number for making reservations is: **1-800-334-3509.**

Individual Seminar Registration Fee $895. Individual CostWorks Training Registration Fee $349. To register by mail, complete the registration form and return with your full fee to: Seminar Division, Reed Construction Data, RSMeans Seminars, 63 Smiths Lane, Kingston, MA 02364.

Federal Government Pricing All Federal Government employees save 25% off regular seminar price. Other promotional discounts cannot be combined with Federal Government discount.

Team Discount Program Two to four seminar registrations: Call for pricing.

Multiple Course Discounts When signing up for two or more courses, call for pricing.

Refund Policy Cancellations will be accepted up to ten days prior to the seminar start. There are no refunds for cancellations received later than ten working days prior to the first day of the seminar. A $150 processing fee will be applied for all cancellations. Written notice of cancellation is required . Substitutions can be made at anytime before the session starts. **No-shows are subject to the full seminar fee.**

AACE Approved Courses The RSMeans Construction Estimating and Management Seminars described and offered to you here have each been approved for 14 hours (1.4 recertification credits) of credit by the AACE International Certification Board toward meeting the continuing education requirements for re-certification as a Certified Cost Engineer/Certified Cost Consultant.

AIA Continuing Education We are registered with the AIA Continuing Education System (AIA/CES) and are committed to developing quality learning activities in accordance with the CES criteria. RSMeans seminars meet the AIA/CES criteria for Quality Level 2. AIA members will receive (14) learning units (LUs) for each two-day RSMeans Course.

NASBA CPE Sponsor Credits We are part of the National Registry of CPE Sponsors. Attendees may be eligible for (16) CPE credits.

Daily Course Schedule The first day of each seminar session begins at 8:30 A.M. and ends at 4:30 P.M. The second day is 8:00 A.M.–4:00 P.M. Participants are urged to bring a hand-held calculator since many actual problems will be worked out in each session.

Continental Breakfast Your registration includes the cost of a continental breakfast, a morning coffee break, and an afternoon break. These informal segments will allow you to discuss topics of mutual interest with other members of the seminar. (You are free to make your own lunch and dinner arrangements.)

Hotel/Transportation Arrangements RSMeans has arranged to hold a block of rooms at each hotel hosting a seminar. To take advantage of special group rates when making your reservation, be sure to mention that you are attending the Means Seminar. You are, of course, free to stay at the lodging place of your choice. **(Hotel reservations and transportation arrangements should be made directly by seminar attendees.)**

Important Class sizes are limited, so please register as soon as possible.

Note: Pricing subject to change.

Registration Form   Call 1-800-334-3509 to register or FAX 1-800-632-6732. Visit our Web site www.rsmeans.com

Please register the following people for the Means Construction Seminars as shown here. Full payment or deposit is enclosed, and we understand that we must make our own hotel reservations if overnight stays are necessary.

☐ Full payment of $ _____ enclosed.

☐ Bill me

Name of Registrant(s)
(To appear on certificate of completion)

_____

_____

_____

P.O. #: _____
    GOVERNMENT AGENCIES MUST SUPPLY PURCHASE ORDER NUMBER

Firm Name _____

Address _____

City/State/Zip _____

Telephone No. fax No. _____

E-mail Address _____

Charge our registration(s) to: ☐ MasterCard ☐ VISA ☐ American Express ☐ Discover

Account No. _____ Exp. Date _____

Cardholder's Signature _____

Seminar Name	City	Dates

Please mail check to: Seminar Division, Reed Construction Data, RSMeans Seminars, 63 Smiths Lane, P.O.Box 800, Kingston, MA 02364 USA

MeansData™

CONSTRUCTION COSTS FOR SOFTWARE APPLICATIONS
Your construction estimating software is only as good as your cost data.

A proven construction cost database is a mandatory part of any estimating package. The following list of softwa providers can offer you MeansData™ as an added feature for their estimating systems. See the table below for w types of products and services they offer (match their numbers). Visit online at **www.rsmeans.com/demosource/** for more information and free demos. Or call their numbers listed below.

1. **3D International**
713-871-7000
venegas@3di.com

2. **4Clicks-Solutions, LLC**
719-574-7721
mbrown@4clicks-solutions.com

3. **Aepco, Inc.**
301-670-4642
blueworks@aepco.com

4. **Applied Flow Technology**
800-589-4943
info@aft.com

5. **ArenaSoft Estimating**
888-370-8806
info@arenasoft.com

6. **ARES Corporation**
925-299-6700
sales@arescorporation.com

7. **BSD - Building Systems Design, Inc.**
888-273-7638
bsd@bsdsoftlink.com

8. **CMS - Computerized Micro Solutions**
800-255-7407
cms@proest.com

9. **Corecon Technologies, Inc.**
714-895-7222
sales@corecon.com

10. **CorVet Systems**
301-622-9069
sales@corvetsys.com

11. **Discover Software**
727-559-0161
sales@discoversoftware.com

12. **Estimating Systems, Inc.**
800-967-8572
esipulsar@adelphia.net

13. **MAESTRO Estimator Schwaab Technology Solutions, Inc.**
281-578-3039
Stefan@schwaabtech.com

14. **Magellan K-12**
936-447-1744
sam.wilson@magellan-K12.com

15. **MC² - Management Computer**
800-225-5622
vkeys@mc2-ice.com

16. **Maximus Asset Solutions**
800-659-9001
assetsolutions@maximus.com

17. **Prime Time**
610-964-8200
dick@primetime.com

18. **Prism Computer Corp.**
800-774-7622
famis@prismcc.com

19. **Quest Solutions, Inc.**
800-452-2342
info@questsolutions.com

20. **RIB Software (Americas), Inc.**
800-945-7093
san@rib-software.com

21. **Shaw Beneco Enterprises, Inc.**
877-719-4748
inquire@beneco.com

22. **Timberline Software Corp.**
800-628-6583
product.info@timberline.com

23. **TMA SYSTEMS, Inc.**
800-862-1130
sales@tmasys.com

24. **US Cost, Inc.**
800-372-4003
sales@uscost.com

25. **Vanderweil Facility Advisors**
617-451-5100
info@VFA.com

26. **Vertigraph, Inc.**
800-989-4243
info-request@vertigraph.com

27. **WinEstimator, Inc.**
800-950-2374
sales@winest.com

TYPE	1	2	3	4	5	6	7	8	9	10	11	12	13	14	15	16	17	18	19	20	21	22	23	24	25	26	27
BID					●		●	●		●	●				●		●		●	●		●				●	●
Estimating		●			●	●	●	●	●	●	●	●	●		●		●		●	●	●	●		●		●	●
DOC/JOC/SABER		●			●		●					●				●					●	●					●
IDIQ		●									●	●				●					●						●
Asset Mgmt.											●			●		●		●						●			●
Facility Mgmt.	●		●								●			●	●	●	●			●			●				●
Project Mgmt.	●	●					●		●	●						●	●			●	●	●					
TAKE-OFF					●			●	●		●				●				●	●		●		●		●	●
EARTHWORK									●		●				●	●	●		●							●	
Pipe Flow				●							●						●										
HVAC/Plumbing					●			●			●						●										
Roofing					●			●			●						●										
Design	●										●						●								●	●	●
Other Offers/Links:																											
Accounting/HR		●			●						●								●	●		●					
Scheduling					●						●		●						●			●		●			●
CAD											●		●									●					●
PDA																						●		●			●
Lt. Versions		●						●									●					●					●
Consulting	●	●			●		●	●		●					●	●			●			●		●			●
Training		●			●		●	●			●				●	●	●	●		●	●	●	●	●	●	●	●

Reseller applications now being accepted. Call Carol Polio Ext. 5107.

FOR MORE INFORMATION
CALL 1-800-448-8182, EXT. 5107 OR FAX 1-800-632-6732

For more information
visit Means Web Site
at www.rsmeans.com

New Titles

Builder's Essentials:
Estimating Building Costs for the Residential & Light Commercial Contractor

By Wayne J. DelPico

Step-by-step estimating methods for residential and light commercial contractors. Includes a detailed look at every construction specialty—explaining all the components, takeoff units, and labor needed for well-organized, complete estimates covers:

Correctly interpreting plans and specifications.

Developing accurate and complete labor and material costs.

Understanding direct and indirect overhead costs... and accounting for time-sensitive costs.

Using historical cost data to generate new project budgets.

Plus hard-to-find, professional guidance on what to consider so you can allocate the right amount for profit and contingencies.

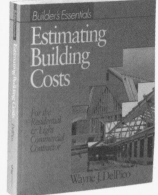

$29.95 per copy
Over 400 pages, illustrated, softcover
Catalog No. 67343

Building & Renovating Schools

This all-inclusive guide covers every step of the school construction process—from initial planning, needs assessment, and design, right through moving into the new facility. A must-have resource for anyone concerned with new school construction or renovation, including architects and engineers, contractors and project managers, facility managers, school administrators and school board members, building committees, community leaders, and anyone else who wants to ensure that the project meets the schools' needs in a cost-effective, timely manner. With square foot cost models for elementary, middle, and high school facilities and real-life case studies of recently completed school projects.

The contributors to this book – architects, construction project managers, contractors, and estimators who specialize in school construction – provide start-to-finish, expert guidance on the process.

$99.95 per copy
Over 425 pages
Catalog No. 67342

Means ADA Compliance Pricing Guide, New Second Edition
By Adaptive Environments and RSMeans

Completely updated and revised to the new 2004 Americans with Disabilities Act Accessible Guidelines, this book features more than 70 of the most commonly needed modifications for ADA compliance—their design requirements, suggestions, and final cost. Projects range from installing ramps and walkways, widening doorways and entryways, and installing and refitting elevators, to relocating light switches and signage, and remodeling bathrooms and kitchens. Also provided are:

Detailed cost estimates for budgeting modification projects, including estimates for each of 260 alternates.

An assembly estimate for every project, with detailed cost breakdown including materials, labor hours, and contractor's overhead.

3,000 Additional ADA compliance-related unit cost line items.

Costs that are easily adjusted to over 900 cities and towns.

Over 350 pages
$79.99 per copy
Catalog No. 67310A

2005 Order Form

ORDER TOLL FREE 1-800-334-3509
OR FAX 1-800-632-6732.

Qty.	Book No.	COST ESTIMATING BOOKS	Unit Price	Total
	60065	Assemblies Cost Data 2005	$189.95	
	60015	Building Construction Cost Data 2005	115.95	
	61015	Building Const. Cost Data–Looseleaf Ed. 2005	144.95	
	63015	Building Const. Cost Data–Metric Version 2005	115.95	
	60225	Building Const. Cost Data–Western Ed. 2005	115.95	
	60115	Concrete & Masonry Cost Data 2005	105.95	
	60145	Construction Cost Indexes 2005	251.95	
	60145A	Construction Cost Index–January 2005	63.00	
	60145B	Construction Cost Index–April 2005	63.00	
	60145C	Construction Cost Index–July 2005	63.00	
	60145D	Construction Cost Index–October 2005	63.00	
	60345	Contr. Pricing Guide: Resid. R & R Costs 2005	39.95	
	60335	Contr. Pricing Guide: Resid. Detailed 2005	39.95	
	60325	Contr. Pricing Guide: Resid. Sq. Ft. 2005	39.95	
	64025	ECHOS Assemblies Cost Book 2005	189.95	
	64015	ECHOS Unit Cost Book 2005	126.95	
	54005	ECHOS (Combo set of both books)	251.95	
	60235	Electrical Change Order Cost Data 2005	115.95	
	60035	Electrical Cost Data 2005	115.95	
	60205	Facilities Construction Cost Data 2005	279.95	
	60305	Facilities Maintenance & Repair Cost Data 2005	252.95	
	60165	Heavy Construction Cost Data 2005	115.95	
	63165	Heavy Const. Cost Data–Metric Version 2005	115.95	
	60095	Interior Cost Data 2005	115.95	
	60125	Labor Rates for the Const. Industry 2005	253.95	
	60185	Light Commercial Cost Data 2005	99.95	
	60025	Mechanical Cost Data 2005	115.95	
	60155	Open Shop Building Const. Cost Data 2005	115.95	
	60215	Plumbing Cost Data 2005	115.95	
	60045	Repair and Remodeling Cost Data 2005	99.95	
	60175	Residential Cost Data 2005	99.95	
	60285	Site Work & Landscape Cost Data 2005	115.95	
	60055	Square Foot Costs 2005	126.95	
		REFERENCE BOOKS		
	67147A	ADA in Practice	59.98	
	67310A	ADA Compliance Pricing Guide, 2nd Ed.	79.95	
	67273	Basics for Builders: How to Survive and Prosper	34.95	
	67330	Bldrs Essentials: Adv. Framing Methods	24.95	
	67329	Bldrs Essentials: Best Bus. Practices for Bldrs	29.95	
	67298A	Bldrs Essentials: Framing/Carpentry 2nd Ed.	24.95	
	67298AS	Bldrs Essentials: Framing/Carpentry Spanish	24.95	
	67307	Bldrs Essentials: Plan Reading & Takeoff	35.95	
	67261A	Bldg. Prof. Guide to Contract Documents 3rd Ed.	64.95	
	67342	Building & Renovating Schools	99.95	
	67339	Building Security: Strategies & Costs	89.95	
	67312	Building Spec Homes Profitably	29.95	
	67146	Concrete Repair & Maintenance Illustrated	69.95	
	67314	Cost Planning & Est. for Facil. Maint.	89.95	
	67317A	Cyberplaces: The Internet Guide 2nd Ed.	59.95	
	67328A	Designing & Building with the IBC, 2nd Ed.	99.95	
	67230B	Electrical Estimating Methods 3rd Ed.	64.95	

Qty.	Book No.	REFERENCE BOOKS (Cont.)	Unit Price	Total
	64777A	Environmental Remediation Est. Methods 2nd Ed.	$ 99.95	
	67160	Estimating for Contractors	35.95	
	67276A	Estimating Handbook 2nd Ed.	99.95	
	67249	Facilities Maintenance Management	86.95	
	67246	Facilities Maintenance Standards	79.95	
	67318	Facilities Operations & Engineering Reference	109.95	
	67301	Facilities Planning & Relocation	89.95	
	67231	Forms for Building Const. Professional	47.48	
	67260	Fundamentals of the Construction Process	34.98	
	67323	Historic Preservation: Proj. Planning & Est.	99.95	
	67308E	Home Improvement Costs–Int. Projects 9th Ed.	29.95	
	67309E	Home Improvement Costs–Ext. Projects 9th Ed.	29.95	
	67324A	How to Est. w/Means Data & CostWorks 2nd Ed.	59.95	
	67304	How to Estimate with Metric Units	9.98	
	67306	HVAC: Design Criteria, Options, Select. 2nd Ed.	84.95	
	67281	HVAC Systems Evaluation	84.95	
	67282A	Illustrated Construction Dictionary, Condensed	59.95	
	67292A	Illustrated Construction Dictionary, w/CD-ROM	99.95	
	67295B	Landscape Estimating 4th Ed.	62.95	
	67341	Life Cycle Costing	99.95	
	67302	Managing Construction Purchasing	19.98	
	67294A	Mechanical Estimating 3rd Ed.	64.95	
	67245A	Planning and Managing Interior Projects 2nd Ed.	69.95	
	67283A	Plumbing Estimating Methods 2nd Ed.	59.95	
	67326	Preventive Maint. Guidelines for School Facil.	149.95	
	67337	Preventive Maint. for Higher Education Facilities	149.95	
	67236A	Productivity Standards for Constr.–3rd Ed.	49.98	
	67247B	Project Scheduling & Management for Constr. 3rd Ed.	64.95	
	67265B	Repair & Remodeling Estimating 4th Ed.	69.95	
	67322A	Resi. & Light Commercial Const. Stds. 2nd Ed.	59.95	
	67327	Spanish/English Construction Dictionary	22.95	
	67145A	Sq. Ft. & Assem. Estimating Methods 3rd Ed.	69.95	
	67287	Successful Estimating Methods	32.48	
	67313	Successful Interior Projects	24.98	
	67233	Superintending for Contractors	35.95	
	67321	Total Productive Facilities Management	29.98	
	67284	Understanding Building Automation Systems	29.98	
	67303A	Unit Price Estimating Methods 3rd Ed.	59.95	
	67319	Value Engineering: Practical Applications	79.95	

MA residents add 5% state sales tax		
Shipping & Handling**		
Total (U.S. Funds)*		

Prices are subject to change and are for U.S. delivery only. *Canadian customers may call for current prices. **Shipping & handling charges: Add 7% of total order for check and credit card payments. Add 9% of total order for invoiced orders.

Send Order To: ADDV-1001

Name (Please Print) _____

Company _____

☐ **Company**
☐ **Home**   Address _____

_____

City/State/Zip _____

Phone # _____ P.O. # _____

(Must accompany all orders being billed)

Mail To: **RSMeans** P.O. Box 800, Kingston, MA 02364-0800